−110001

INFORMATION CENTRE

This item is due back on or before the date stamped below

2 4 DEC 2002		
− 5 FEB 2003		

TO RENEW THIS ITEM PLEASE RING THE INFORMATION CENTRE ENQUIRY DESK ON ext. 2272, QUOTING THIS NUMBER: −110001.

Information Centre, Central Science Laboratory, Sand Hutton, York, YO4 1LZ tel 01904 462272

5263

CALL No:	
DYNIX BARCODE:	− 110001
DYNIX BiB No:	
DATE:	

PRACTICAL POLYMER ANALYSIS

PRACTICAL POLYMER ANALYSIS

T. R. Crompton

North West Water Authority
Warrington, England

PLENUM PRESS • NEW YORK AND LONDON

Library of Congress Cataloging-in-Publication Data

Crompton, T. R. (Thomas Roy)
    Practical polymer analysis / T.R. Crompton.
        p.   cm.
    Includes bibliographical references and index.
    ISBN 0-306-44524-7
    1. Polymers--Analysis.   I. Title.
QD139.P6C77   1993
547.7'046--dc20   93-25290
                                                             CIP

ISBN 0-306-44524-7

©1993 Plenum Press, New York
A Division of Plenum Publishing Corporation
233 Spring Street, New York, N.Y. 10013

Printed in the United States of America

The aim of this book is to familiarize the reader with the practical aspects of polymer analysis. A wealth of practical detail, including some detailed methods is included. The book covers not only the analysis of the main types of polymers and copolymers now in use commercially, but also the analysis of minor non-polymeric components of the polymer formulation, whether they be deliberately added, such as processing additives, or whether they occur adventitiously, such as moisture and residual monomers and solvent. A broad scheme for the examination of polymers is discussed in Chapter 2.

Practically all of the major newer analytical techniques and many of the older classical techniques, have been used to examine polymers and their additive systems. As so many different polymers are now used commercially it is also advisable when attempting to identify a polymer to classify it by first separating it into pure polymeric and gross non-polymeric fractions (Chapter 2) and then carrying out at least a qualitative elemental analysis and possible a quantitative analysis (Chapters 3 and 4) and then in some cases, depending on the elements found, to carry out functional group analysis (Chapters 6 and 9). If a simple qualitative identification of the plastic is all that is required then it is examined by finger printing techniques, as discussed in Chapter 4, in order ascertain whether a quick identification can be made by comparing its infrared or NMR spectrum, its pyrolysis-gas chromatography pattern or its thermal behaviour with those of authentic specimens of known polymers.

Frequently, however, the identification of a polymer, especially copolymers or terpolymers, is not as simple as this, and it is necessary to obtain a detailed picture of the microstructure of the polymer before an identification can be made.

Examination of polymer microstructure can be an additional means of identifying the polymer or it can be a means of helping to explain particular chemical or physical features of the polymer. In the latter case the type of polymer might already be known but we are looking for structural detail that will explain these properties such as tacticity, geometrical isomerism, regeoisomerism or the types of unsaturated structures, end-groups and branching present; (Chapter 10). In copolymers and terpolymers measurements can be made of comonomer ratio and monomer unit sequences in addition to the measurements discussed above (Chapter 9). Techniques that might be used in addition to elemental and functional group analysis, include spectroscopic techniques such as infrared, NMR, PMR, and systematic investigations by pyrolysis-gas chromatography. Examination for the type of unsaturation present, the nature of side-chain groups and end-groups, the presence of oxygenated groups such as carbonyl and whether they are macro or micro constituents will all assist in building up a picture of the polymer structure. In many cases, considerable experience and innovative skills are required by the analyst in order to successfully identify polymers by these techniques, and it is hoped that this book will assist the analyst in developing such skills.

Examination for non-polymeric processing additives, monomers, oligomers, residual volatiles, water and catalyst remnants are discussed in Chapter 5.

In Chapter 6 is discussed the fractionation of polymers using classical techniques and more recently developed techniques such as hplc, size exclusion chromatography and supercritical fluid chromatography. Some of these techniques also provide molecular weight data which is discussed further in Chapter 7.

Other polymer physical properties which are discussed include thermal, oxidative or photo-oxidative stability, its combustion properties, measurement of glass transition temperature and other transitions, measurements of crystallinity and melting temperature and of polymer lifetimes under service conditions, (Chapters 11-13).

The book gives an up-to-date and thorough exposition of the present state of the art of polymer analysis and, as such should be of great interest to all those engaged in this subject in industry, university, research establishments and general education. It is also intended for undergraduate and graduate chemistry students and those taking courses in plastics technology, engineering chemistry, materials science and industrial chemistry. It will be useful reference work for manufacturers and users of plastics, the food and beverage packing industry, the engineering plastics industry, plastic components manufacturers, and those concerned with pharmaceuticals and cosmetics.

Before proceeding to the first two chapters, which deal respectively with the determination of elements and functional groups, it would be interest in Chapter 1 to discuss briefly the various types of polymers used commercially, and their properties and applications.

T. R. Crompton

ACKNOWLEDGEMENTS

        The author wishes to express his gratitude to the publishers of
various journals for permission to reproduce the following illustrations.

Figs 1-3   Squirrel, D.C.M., Analyst (London), 106, 1042 (1981).

Figs 5,6   Groten, B., Anal. Chem., 36, 1206 (1964).

Fig 7   Lehman, F.F., Branch, G.M, Anal. Chem., 33, 673 (1961).

Fig 10   Folmer, O.F. Anal. Chem, 43, 1057 (1971).

Table 15   Juvet R.S., Smith, L.J.S., Pang, K.K., Anal. Chem., 44,
                    49 (1972).

Fig 11,12   Jackson, M.T., Walker, J.Q., Anal. Chem., 43, 74 (1971).

Table 18   Osland, R., Laboratory Practice, 37, 73 (1971).

Fig 13   Brako, F.D., Wexler, A., Anal. Chem., 35, 1944 (1963).

Tables 19,20   Dobies, R.S.J., Chromatography, 40, 110 (1969).

Figs 14,15 &
Table 15   Schabon, J.T., Fenska, L.E., Anal. Chem., 52, 1411
                    (1980).

Figs 16,17 &
Table 23   Albarino, R.V., Applied Spectroscopy, 27, 47 (1973)

Fig 18   Soncek, J., Jelinova, E., Analyst (London), 107, 623
                    (1982).

Fig 19   Drushel, H.V., Sommers, A.L., Anal. Chem., 36, 836
                    (1964).

Fig 21   Krisher, A., Anal. Chem., 43, 1130 (1971).

Table 27   Simpson, D., Currell, B.R., Analyst (London), 96, 515
                    (1971).

Tables 29 & 43   Millingen, M.B., Anal. Chem., 46, 746 (1974).

Table 32   Robertson, M.M., Rowley, R.M., British Plastics, 26
                January (1960).

Fig 22   Majors, R.E.J., Chromatographic Science, 8, 338 (1970).

Table 36   Denning, J.A., Marshall, J.A., Analyst (London), 97 710
                (1972).

Table 37   Lappin, G.R., Zannucci, J., Anal. Chem., 41, 2076
                (1976).

Fig 23   Tittareli, P., Zerlia, T., Colli, A. and Ferrari, G.,
                Anal. Chem., 55, 220 (1983).

Table 41   Protivova, J., Posposil, S.J., Chromatography, 88, 99
                (1974).

Fig 24   Anderson, D.M.W., Analyst (London), 84, 50 (1959).

Fig 25(a) &
Table 45   Cambell, R.H., Wize, R.W., J. Chromatography, 12, 178
                (1963).

Fig 25(b)   Schabaum, J.F., Fenska, L.E., Anal. Chem., 52, 1411
                (1980).

Fig 26 &
Table 46   Vargo, J.D., Olsen, K.L., Anal. Chem., 57, 672 (1985).

Figs 27,28   McCoy, R.N., Fiebig, E.C., Anal. Chem., 37, 593 (1965).

Figs 29-31 &
Table 49   Juvet, R.S., Smith, L.S., and Li, K.P., Anal. Chem., 44,
                49 (1972).

Tables 56,57   Striechen, R.J., Anal. Chem., 48, 1398 (1976).

Fig 36 &
Table 58   Betso, S.R., McLean, J.D., Anal. Chem., 48, 766 (1976).

Table 59   Skelly, N.E., Husser, E.R., Anal. Chem., 50, 1959
                (1978).

Fig 40   Roper, J.N., Anal. Chem., 42, 688 (1970).

Figs 43,44   Jeffs, A.R., Analyst (London), 94, 249 (1969).

Table 66   Korenaga, T., Analyst (London), 106, 40 (1981.

Table 73   Baltiste, D.R., Butler, J.P., Cross, J.B., McDaniel,
                M.P., Anal. Chem., 53, 2232 (1981).

Table 74   Hyden, S., Anal. Chem., 35, 133 (1963).

Fig 45   Brako, F.D., Wexler, A.S., Anal. Chem., 35, 1944 (1963).

Table 75   Stelzer. R.S., Smullin, C.F., Anal. Chem., 34, 194
                (1962).

Table 76	Smith, R.M., Dauson, M., Analyst (London), <u>105</u>, 85 (1980).
Fig 46 & Table 77	Ho, F.T., Anal. Chem., <u>45</u>, 603 (1973).
Table 78	Kim, C.S.Y., Dodge, A.I., Hay, S., Kawasaki, A., Anal. Chem., <u>54</u>, 232 (1982).
Table 79	Schleuter, D.D., Siggia, S.S., <u>49</u>, 2343 (1977).
Fig 47	Fletcher, J.P., Persinger, H.E., J. Polymer Science, A-1, <u>6</u>, 1025 (1968).
Fig 48 Tables 81-83	Mori, S., Uno, Y., Anal. Chem., <u>59</u>, 90 (1987).
Table 84	Atkinson, E.R., Calouche A., Anal. Chem., <u>43</u>, 460 (1971).
Figs 49-51	Mourney, T.H., Smith, G.A., Snyder, L.R., Anal. Chem., <u>56</u> 1774 (1984).
Figs 52-53	Mourney, T.H., Anal. Chem., <u>56</u>, 1777 (1984).
Fig 54 & Table 85	Bode, R.K., Benningfidd, L.V., Mowrey, R.A., Fuller, E.N., International Laboratory, <u>40</u>, Nov/Dec (1981).
Fig 55 & Tables 87-91	Mori, S., Anal. Chem., <u>53</u>, 1813 (1981).
Fig 56 & Table 92	Mori, S., Anal. Chem., <u>55</u>, 2414 (1983).
Fig 57	Tally, C.P., Bowman, L.M., Anal. Chem., <u>51</u>, 2239 (1979).
Fig 58-61 & Table 93	Mori, S., Uno, Y., Suzuki, M., Anal. Chem., <u>58</u>, 303 (1986).
Fig 62	Williamson, G.R., Cervenka, A., European Polymer Journal, <u>8</u>, 1009 (1972).
Fig 66	Taylor, W.C., Tung, L.H., Paper presented at 140th Meeting of American Chemical Society, Chigaco, Illinois, September (1961).
Fig 67, 68	Machtle, W., Klodwig, H., Mackromal Chem., <u>180</u>, 2507 (1979).
Table 94	Williamson, G.R., Cervenka, A., European Polymer Journal, <u>10</u>, 295 (1974).
Table 95	Lattimer, R.P., Harmon, D.J., Houson, G.E., Anal. Chem., <u>52</u>, 1808 (1980).

x   AKNOWLEDGEMENTS

Fig 7 & Tables 105,106	Van Houwlinger, G.D.B., Analyst (London), $\underline{106}$, 1057 (1981).
Table 100	Altenau, H.G., Headley, L.M., Jones, S.O., Rensaw, H.C., Anal. Chem., $\underline{42}$, 1280 (1970).
Table 101	Anderson, D.G., Isakson, K.E., Snow, D.L., Tessari, D.J., Vandenberg, J.T., Anal. Chem., $\underline{43}$, 894 (1971).
Table 102	Barrall, E.M., Porter, R.S, Johnson, D.E., Anal. Chem., $\underline{35}$, 73 (1963).
Fig 72	Porter, R.S., Nickosic, S.W. and Johnson, J.F., Anal. Chem., $\underline{35}$ 1948 (1963).
Table 103	Van Lingen, R.L.M., Fresenius, Z., Anal. Chem., $\underline{169}$, 410 (1969).
Figs 73,74	Douglas, J., Timnick, A. & Guile, R.L., J. Polymer Science, A-1, 1609 (1963).
Fig 75 & Table 104	Johnson, D.E., Lyeria, J.R., Horikawa, T.T., & Pederson, L.A., Anal. Chem., $\underline{77}$, 49 (1977).
Fig 76	Sharp, J.L. & Patenson, G., Anal. Chem., $\underline{105}$, 517 (1980).
Fig 77 & Table 108	Anderson, D.G., Isakson, K.E., Vandenberg, J.T., Jao, M.Y.T., Tessari, D.J. & Afremow, L.C., Anal. Chem., $\underline{47}$ 1008 (1975).
Fig 78	Tsuji, K. & Kouishi, K., Analyst (London), $\underline{99}$ 54 (1974).
Fig 79 & Table 109	Swaraj, S., Ranby, B., Anal. Chem., $\underline{47}$, 1428 (1975).
Fig 80	Peltonen, K., Pfaffi, P. & Itkonen, A., Analyst (London), $\underline{110}$, 1173 (1985).
Fig 81 Tables 110-112	Hammerich, A.D. & Willeboodse, F.G., Anal. Chem., $\underline{45}$, 1696 (1973).
Fig 82 & Tables 113-115	Schleuter, D.D., Siggia, S., Anal. Chem., $\underline{49}$ 2349 (1977).
Fig 83	Tukuchi, T., Tauge, S. & Sigimura, Y., J. Polymer Science, A-1, $\underline{6}$, 3415 (1968).
Table 116	Krishen, A., Anal. Chem., $\underline{44}$, 494 (1972).
Fig 84 & Table 117	Tosi, G., Simonazzi, T., Die Augewandte Makromolecular Chemie, $\underline{32}$, 153 (1973).

Fig 85	Drushell, H.V., Iddings, F.A., Division of Polymer Chemistry, 142nd Meeting American Chemical Society, Atlantic City, September (1962).
Tables 118,119	Lomonte, J.N, Tirpak, G.A., J. Polymer Science, A2, 705 (1964).
Fig 86 & Table 120	Brown, J.E., Tyron, M. & Mandel, J., Anal. Chem., 35, 2173 (1963).
Table 121	Paxton, J.R., Randall, J.C., Anal. Chem., 50, 1777 (1978).
Figure 87	Neumann, E.W., Nadeau, H.G., Anal. Chem., 10, 1454 (1963).
Table 122,123	Anderson, D.G., Isakson, K.E., Snow, D.L., Tessari, D.J., Candleberg, J.T., Anal. Chem., 43, 894 (1971).
Fig 88 & Table 125	Allen, B.J., Elsea, G.M., Keller, K.P., Kinder, H.D., Anal. Chem., 49, 471 (1977).
Fig 89 & Table 126	Brame, E.G., Yeager, F.W., Anal. Chem., 48, 709 (1976).
Fig 90 & Table 127	Blackwell, J.T., Anal. Chem., 48, 1883 (1976).
Fig 91 & Table 129	Voigt, J., Kunststoffe, 54, 2 (1964).
Fig 92	Van Schooten, J., Evenhuis, J.K., Polymer (London), 6, 651 (1965).
Fig 93-95	Lehmann, F.A., Brauer, G.M., Anal. Chem., 33, 676 (1961).
Fig 96	Van Schooten, J., Evenhuis, J.K., Shell Chemical Co., Holland, private communication.
Tables 131,132	Bucci, G., Simmonazzi, T., J. Polymer Science, C-7, 203 (1964).
Fig 97 & Table 133	Van Schooten, J., Duck, E.W., Berkenbosch, R., Polymer (London), 2, 357 (1961).
Fig 98	Veerkamp, M.A., Veermans, A., Makromol. Chemie, 50, 147 (1961).
Fig 99	Porter, R.S., J. Polymer Science, A1, 4, 189 (1966).
Table 134	Paxton, J.R., Randall, J.C., Anal. Chem., 50, 1777 (1978).
Figs 100-102 & Tables 137,138	Sugimura, Y., Takenchi, T., Anal. Chem., 50, 1173 (1978).

Table 141   Porter, R.S., J. Polymer Science, A-1, 4, 189 (1966).

Fig 106   Randall, J.C., J. Polymer Science, 13, 889 (1975).

Fig 107   Binder, J.L., J. Polymer Science, A1, 42 (1963).

Table 143,144   Carlson, D.W., Ransaw, H.C., Altenau, A.G., Anal. Chem.,
                     42 1278 (1970).

Tables 147-149   Vodchnal, J., Kossler, I., Coll. of Czech. Communs. 29,
                     2428, 2859 (1964).

Fig 110   Luougo, J.P., Salovay, R., J. Applied Polymer Science,
                     7, 2307 (1963).

Fig 114   O'Mara, H.M., J. Applied Polymer Science, A-1, 8, 1887
                     (1970).

Tables 157-159   Van Schooten, J., Evenhuis, J., Polymer (London), 6, 651
                     (1965).

Fig 115   Seger, M., Barrall, E., J. Polymer Science, A-1, 13,
                     1515 (1975).

Table 160   Ahlstrom, D.H., Liebman, S.A., Abbass. J. Polymer
                     Science, Polymer Chemistry Edition, 14, 2479 (1976).

Figs 124-127   Kullik, E., Kaljurand, M. & Lainberg, M., Laboratory
                     Practice, p73, Jan-Feb (1987).

Fig 128 &
Tables 165,166   Shulman, G.P., Polymer Letters, 3, 911 (1965).

Figs 129,130,
133,144 &
Tables 167,168   Risby, T.H., Yergey, J.A., Anal. Chem., 54, 2228 (1982).

Figs 131-132 &
Tables 169-171   Undseth, H.R., Friedman, L., Anal. Chem., 53, 29 (1981).

Table 174   O'Mara, M.M., J. Polymer Science, A-1, 8, 1887 (1970).

Figs 135-138 &
Tables 175,176   Chang, T.L., Head, T.E., Anal. Chem., 43, 534 (1971).

Fig 139   Peltonen, K., Analyst (London), 111, 819 (1986).

Fig 144   Schole, R.G., Bednarczyk, J., Yamanchi, T., Anal. Chem.,
                     38, 331 (1966).

Fig 146   Joseph, K.T., Browner, R.E., Anal. Chem., 52, 1083
                     (1980).

Figs 151,152   Wohitjen, H., Dessy, K., Anal. Chem., 51, 1470 (1979).

Fig 153   Wohitjen, H., Dessy, K., Anal. Chem., 51, 1463 (1979).

Figs 173,174 &
Table 188   Bergmann, K., Polymer Bulletin, $\underline{5}$, 355 (1981).

Figs 176,177   Young, P.H., Spectroscopy International, $\underline{1}$, 50 (1989).

Fig 178   Clampett, B.H., Anal. Chem., $\underline{35}$, 1834 (1963).

Fig 181   Illers, K.H., Heckmann, W., Hambrecht, J., Colloid and
                     Polymer Science, $\underline{262}$, 557 (1984).

Fig 182   Simack, P., Makromal Chemie, $\underline{178}$, 2927 (1977).

CHAPTER 10 POLYMER MICROSTRUCTURE   437

CHAPTER 1

TYPES OF POLYMERS AND THEIR USES

Synthetic resins, in which plastics are also included, vary widely in their chemical composition and in their physical properties. The number of synthetic resins which can be made is vast; relatively few, however, have achieved commercial importance. Some of the polymers that have achieved commercial importance and their uses are tabulated in Table 1 and some of their important physical properties are listed in Appendix 2.

Well over 90% of all synthetic resins made today comprise no more than 20 different types, although there are certain variations to be found within each type. Synthetic resins are familiar to most people as plastics, but they have other uses, such as in the manufacture of surface coatings, glues, synthetic textile fibres, etc. The rapid growth of the synthetic resin industry has to a large extent, been made possible by the fact that ample supplies of necessary raw materials have become available from petroleum.

The synthetic resins may be divided into two classes, known respectively as 'thermosetting' and 'thermoplastic' resins, each class differing in its behaviour on being heated. The former do not soften; the latter soften but regain their rigidity on cooling. Both types are composed of large molecules, known as macromolecules, but the difference in thermal behaviour is due to differences in internal structure.

The larger molecules of the thermoplastics have a long-chain structure, with little branching. They do not link with each other chemically, although they may intertwine and form a cohesive mass with properties ranging from those of hard solids to those of soft pliable materials, in certain cases resembling rubber. On being heated, the chain molecules can move more freely relative to each other, so that, without melting, the material softens and can flow under pressure and be moulded to any shape. On cooling, the moulded articles regain rigidity. Some resins require the addition of liquid plasticizers to improve the flow of the plastic material in the mould. In such cases the moulded articles are usually softer and more flexible than the products made from the unplasticized resins.

The macromolecules of the thermosetting resins are often strongly-branched chains and are chemically joined by crosslinks, thus forming a complex network. On heating, there is less possibility of free movement, so that the material remains rigid.

Production of these resins also falls into two groups since there are, generally, two main types of chemical reaction by which they are made. These are polycondensation reactions and polymerization reactions.

TABLE 1

Polycondensation types		
Phenolformaldehyde	(a) Unfilled	(a) Adhesives, laminates, pulp mouldings, particle board.
	(b) Woodflour/cotton flock filled	(b) Bottle tops, electrical parts, fuse boxes, meter cases, heat-resistant close-tolerance mouldings, toilet seats, restricted in colours obtainable, coloured ashtrays.
Urea formaldehyde	Cellulose filled	As for, cellulose filled melamine formaldehyde but suitable for dinner ware. U/F resins are used for similar applications to those shown under unfilled phenol formaldehyde, white electric plugs.
Melamine formaldehyde	(a) Unfilled	(a) Usually occurs in laminate form as surfacing for tables etc.
	(b) Alpha-cellulose	(b) Noted for durability, hardness and good electrical properties, suitable for appliance housings, dinnerware, closures, writing equipment, clock housings, knobs, handles, lighting fixtures, appliances, instruction panels.
Polyesters	(a) Resin	(a) Unreinforced resin for buttons, surface coatings, embedding and potting and nut locking. Filled resin for imitation marble, flooring, pipe joints, mortars and body stoppers.
	(b) Dough moulding compound	(b) Protective housings, connectors, cowls, guards and ducts. Components often replace metal, offering non-corrosion, durability, good electrical performance and high strength.
	(c) Sheet moulding compound	(c) Outlets in the electrical, building, motor engineering and furniture industries that compete on a cost basis with die castings and sheet metal fabrications owing to ease of moulding complicated shapes and short moulding cycles.
Expoxides		Chemically resistant paints, adhesives, tools, PVC stabilizers, electrical insulation, chemical- and wear-resistant jointless flooring, road coatings, cements, laminates, powder coatings, stopping compounds, repair kits, printed circuits, filament wound pipes, tanks and pressure vessels.
Nylon	(a) Type 6	(a) Moulded mechanical parts, gear wheels, bushings, sliding parts for storm windows, automobile and refrigerator door closures, mixer valves, switch housings, grommets, cable clamps, pipes, tubing, filaments, aerosol bottles, stockings, clothing,

	(b)	Type 6/6	zips, curtain fittings. As for Type 6.
	(c)	Type 11	(c) Electro/mechanical components such as cams, housings, guides, terminal blocks, transformers and bobbins. Tubing and hose for automotive use, reinforced hose, technical components in copiers, calculators, machinery and tools.
	(d)	Type 12	(d) Injection moulded parts for automotive, electrical and electronic and precision machine industries. Semi-rigid or flexible tubing for fuel lines, air brakes, pressured air lines. Cable and wire sheathing, chill roll and blown film. Powder coatings.
Polymerization types Polyethylenes	(a)	Low density	(a) Housewares (bottles, bowls, buckets, containers), closures for squeeze tubes, spouts for detergent cans, carboys, shoe parts, toys, packaging film, garment bags, sheet, piping for domestic industrial and agricultural use, cable and wire insulation and sheathing, paper, cellulose and foil coating, carpet backing, monofilament, cold water storage tanks.
	(b)	High density	(b) As for low-density polyethylene. Material have greater rigidity, specially suitable for large carrying cases, housings, closures, appliance parts, packaging film, sheet, piping, carboys, containers, bottles.
Polypropylene			Domestic, hospital and laboratory ware, textile, automotive, electrical and industrial usages. Containers and closures, crates, toys, blown containers. Film fibre for baler twine, ropes, sacks and carpet backing. Fibre for carpet face yarns. Extruded sheet, pipe, film and filament.
Polystyrenes	(a)	Conventional	(a) Packaging (disposable and others), dishes and utensils, refrigerator parts, emblems, signs, displays, toys, novelties, combs, brush-backs, radio and TV cabinets, lighting fixtures, rigid containers, housewares, reels/spools for film/tape, appliance panels, handles and switches.
	(b)	Toughened	(a) Wheels, helmets, valve parts, refrigerator parts, bobbins, radio cabinets, electric fan blades, toys, housewares, containers, battery cases, refrigerator linings and trays, yoghurt pots, footwear.

TABLE 1 continued

Styrene acrylonitrile Acrylonitrile/butadiene/ styrene			Cups, tumblers, trays and general table, kitchen ware, toothbrush handles, refrigerator components, radio knobs and scales, lenses, cosmetic items, hi-fi covers and cases, packaging. Shoe heels, telephone handsets, housing for consumer durables, food containers, luggage, refrigerator liners, safety helmets, radio cabinets, tote boxes, car facia panels, instrument clusters, boat hulls, furniture.
Vinyl polymers	(a)	Rigid polyvinyl chloride	(a) Extrusion of piping, profiles and sheet in applications requiring chemical inertness and scuff-resistance combined with light weight. Plastisols for toys, leathercloth etc. Plastic guttering, high density bottles, formed packaging trays, moulded containers.
	(b)	Rigid vinyl chloride/acetate	(b) Similar applications to those for PVC but widely used in the manufacture of calendared sheet used for toys, novelties, wall coverings, displays, templates, etc. Gramophone records.
	(c)	Rubber modified PVC.	(c) Same as for PVC.
Polyacetals			Load-bearing mechanical parts, small pressure vessels, aerosol containers business machine parts, appliances, automobile, engineering and industrial products.
Polycarbonates			Business machine parts, camera components, electrical apparatus, sterilisable ware, draughtsman's instruments, lamp covers, safety helmets, tail-lights, de luxe housewares, engineering and industrial components, sterilisable transparent feeding bottles for babies.
Ionomers			Shoe heel tips, tool handles, hammer and mallet heads, bottles, skin packaging, coating, toys, shoe soles, shoe stiffeners, meat packaging, flexible packaging, packaging of wine and fruit juices.
Polyphenylene oxide			TV set components, valve bases, switches, housings, parts for domestic appliances, meter cases, machine housings, computer and camera parts, automotive grilles, ducts, light housing, instrumentation parts.
Polyphenylene sulphide			Engineering plastic, high heat and chemical resistance

Acrylics			applications. Automatic parts, control knobs, dials and handles, meter cases, lenses, pens and pencils, brush-backs, hospital equipment, display material, signs, light fittings, inspection panel coves, windscreens, machine guards, skylights, some telephones, sanitary ware, TV tube implosion guards.
Polyvinylacetate			Foodstuff packaging film.
Ethylene vinyl acetate			Flexible extrusions, tubing and hose, sachets, sheathing, cable coverings. Closures, gaskets, handle grips, shoe soles, teats, disposable gloves, box liners, packaging film, greenhouse film, inflatable toys.
Polyoxyalkylene glycols			Waterproofing of paper wrappings, wood preservation.
Polyacrylonitriles			Synthetic film (eg. Acrilan).
Polyurethanes			Thermal insulation panels, sealant material, eg. battery containers.
Polysulphones			Engineering plastic, replacements for stainless steel.
Fluorinated polymers	(a)	Fluorinated ethylene propylene	(a) Coil formers, terminal blocks, valve holders, wire insulation electronic components, terminal encapsulations and fluidised bed coatings, non stick valves.
	(b)	Polytetrafluro-chloroethylene	(b) Gaskets, packing, valves, sintered metal bearings, rigid and flexible pipes, membranes, wire insulations, electronic engineer--ing applications, non-stick coatings for kitchen utensils, heat sealing equipment and confectionery machinery tanks.
	(c)	Polytrifluoro-chloroethylene	(c) Extruded sheet, profile and film, electronic parts, gaskets, pump sealants, dispersion coatings, liquid level indicators of particular use where resistance to aggressive chemicals is needed.
Cellulose plastics	(a)	Cellulose acetate	(a) Toys, beads, cutlery handles, electrical parts, knobs, steering wheels, shoe heels, packaging sheeting, toothbrushes, cosmetics, windows in window cartons.
	(b)	Cellulose acetate butyrate	(b) Moulded or extruded parts for metallization (reflectors etc.) out-door signs, automobile tail-light covers, tool handles, toothbrushes, pipe inspection traps, piping.

In polycondensation reactions, two or more chemicals are brought together and a reaction between them is initiated by using heat or a catalyst or both. The reaction proceeds with the elimination of water and the molecules are joined by chemical bonds to form macromolecules, either long-chain or crosslinked structures of the thermoplastic or thermosetting types, respectively. Many resins obtained by polycondensation are the thermosetting type.

In the manufacture of these resins the chemical reactions are arrested at an intermediate stage in which the resins are temporarily thermoplastic; they are set in their final shape by the application of heat and pressure. At this stage the interlinking of the molecules takes place.

Important thermosetting synthetic resins made by polycondensation, using petroleum chemicals as raw materials, include the phenolformaldehyde ('Bakelite'), urea-formaldehyde, alkyd- and epoxy- types.

Resins with long-chain macromolecules obtained by polycondensation, have thermoplastic properties. Polyesters ('Terylene') and polyamides (nylon) are examples of polycondensations. The synthetic fibre 'Terylene' (known as 'Dacron' in the U.S.A.) is a polyester formed by the reaction of ethylene glycol with terephthalic acid; the terephthalic acid is obtained from para-xylene by oxidation.

$$CH_2OH\ |\ CH_2OH \quad + \quad \overset{COOH}{\underset{COOH}{\bigcirc}} \quad \xrightarrow{polycondensation} \quad \left[-CH_2-CH_2OOC- \bigcirc -COO \right]_n \ +\ 2nH_2O$$

Nylon type fibres (polyamides) are manufactured from adipic acid, which can be made from either cyclohexane or phenol the adipic acid is condensed with hexa-methylene diamine, which is a derivative of adipic acid.

$$\overset{OH}{\bigcirc} \xrightarrow{hydrogeneration} \overset{CHOH}{\bigcirc} \xrightarrow{oxidation} \overset{COOH}{\underset{COOH}{(CH_2)_4}}$$

phenol   cyclohexanol   adipic acid

$$\overset{COOH}{\underset{COOH}{(CH_2)_4}} \quad + \quad \overset{NH_2}{\underset{NH_2}{(CH_2)_6}} \quad \xrightarrow{polycondensation} \quad \left[NH(CH_2)_6 NH - CO(CH_2)_4 CO \right]_n \ +\ 2nH_2O$$

adipic   hexamethylene
acid   diamine

Resins produced by polymerization reactions, known technically as high polymers, are rapidly increasing in number and in importance as compared with the polycondensation resins. High polymers are usually made by joining together into long chains a number of molecules which have the same kind of reactive points or groupings in their structure. These individual molecules are usually olefins or other compounds with double bonds, and are called 'monomers'. The molecule of the polymer often contains hundreds of monomer units.

The manufacture of high polymers therefore takes place in two stages; first, the production of the monomer, or repeating chemical unit;

and second, the polymerization to a resin.

Thus, if we take the preparation of polyvinyl chloride as an example we have:

        1st stage
        $CH_2 = CH_2 + Cl_2 \longrightarrow CH_2Cl - CH_2Cl \longrightarrow CH_2 = CHCl + HCl$
          ethylene   dichloroethane vinyl chloride
                                                      monomer

        2nd stage
                                                        polymerization
        $CH_2 = CH_2 + CH_2 = CHCl + CH_2 = CHCl$ ------ $\longrightarrow$
        $- CH_2CHCl - CH_2 - CHCl - CH_2 - CHCl -$
              polyvinyl chloride

In some cases it is possible to form polymers from two or even three monomers which may differ from one another in chemical form and yet be capable of linking end-to-end to form mixed monomer chains. These are known as 'copolymers', and such polymers form the basis of the most important types of synthetic rubber.

Further examples of polymerizations:

Styrene butadiene copolymer
$CH = CH_2 + CH_2 = CH - CH = CH_2 \longrightarrow - CH - CH_2 - CH_2 - CH = CH - CH_2-$
$\quad |$ $\quad |$
$\quad Ph$ $\quad Ph$

Styrene   butadiene   Styrene - butadiene copolymer

        Polymethylmethacrylate
                                        $\begin{bmatrix} & COOMe \\ & | \\ - CH_2 - & C - \\ & | \\ & Me \end{bmatrix}_n$

        $nCH_2 = C + COOMe \longrightarrow$
               $|$
               Me

Buna N rubber
$CH_2 = CHCN + CH_2 = CH - CH = CH_2 \rightarrow CH_2 - CH - CH_2 - CH = CH - CH_2 -$
                                                    $|$
                                                   CN

Vinyl chloride - vinylidine chloride copolymer
$CH_2 = CH - Cl + CH_2 = CCl_2 \longrightarrow CH_2 - CCl_2 - CH_2 - CHCl -$
vinyl chloride vinylidine chloride

        Polyacrylonitrile
        $n\begin{bmatrix} CH_2 = CH \\ | \\ CN \end{bmatrix} \longrightarrow \begin{bmatrix} - CH_2 - CH - \\ | \\ CN \end{bmatrix}_n$

        Polyvinyl acetate

        $nCH_2 = CH \longrightarrow - CH_2 - CH -$
              $|$ $|$
            $OOC\ CH_3$ $OOCCH_3$ n

Polyurethanes

$$OCN - \bigotimes_{Me} - NCO \quad + \quad \underset{CH_2 - CH_2}{HO \quad OH}$$

eg toluene   eg ethylene
dissocyanate   glycol

$$- \bigotimes_{Me} - NH(COOCH_2 - CH_2OOCNH - \bigotimes_{Me} -$$

polyurethane

   The number of forms in which polymers are encountered in practice is
immense. The polymer might exist as a solid, or, a liquid, either in the
neat form or as an aqueous or solvent emulsion such a polyvinyl acetate or
as a gum, or adhesive.

   Solid polymers might be rigid or flexible. Examples of rigid
polymers are polyolefins, polystyrene and unplasticized PVC. Examples of
flexible polymers are rubbers, elastomers and plasticized PVC. The
polymer might be virtually 100% polymeric or might contain inert fillers
to impart desired properties. Examples of these range from additions of a
few percent of substances such as talc, zinc oxide or titanium dioxide to
materials where a major proportion of the sample is a filler, eg. the old
style automotive battery cases which contain only 10-20% of a polymeric
constituent the remainder being coal dust. Composite polymeric materials
containing high percentages of non-polymeric materials are used
extensively in the fabrication and engineering industries. The sample
might be in the form of a polymer impregnated paper or textile such as is
used in the packaging or clothing industries. Polymers may occur as a
constituent with other materials in, for examples, paints, varnish and
glass fibre reinforced plastics.

   Polymers can be arbitrarily divided into three types, i) those
completely soluble in organic solvents, (eg. conventional polystyrene),
ii) those which do not completely dissolve in organic solvents but which
do swell sufficiently to enable soluble additives to be removed prior to
further analysis (Chapter 3), (eg. low density polyethylene and
polypropylene or cross linked styrene-butadiene copolymer or
acrylonitrile-butadiene-styrene terpolymer in which the crosslinked gel
component is completely insoluble and the uncrosslinked material is
soluble) and iii) polymers which are completely or almost completely
insoluble in organic solvents, (eg. polytetrafluoroethylene, some
silicones and some epoxy resins).

   The simplest type of polymer sample to handle is, of course, the
type of polymer for which a good organic solvent exists and which contains
no fillers and little or no additives. In these instances a fractional
precipitation or a fractional extraction procedure using organic solvents
and non-solvents such as described in Chapter 3 will suffice to separate
and purify the gross polymer and gross additive fractions preparatory to
further analysis of each fraction. Partially soluble polymers of type ii
mentioned above can also often be handled by these procedures. Thus if
low density polyethylene is refluxed with toluene or xylene, although the
majority of the polymer does not dissolve it does soften and disperse in
the organic solvent leaving the additives in solution. Filtration or
centrifuging of the mixture gives a pure polymer extract and a solvent
extract containing additives. In the case of high impact polystyrene,

i.e. styrene-butadiene latices contact with chloroform or toluene dissolves the uncrosslinked copolymer and additives, leaving the undissolved gel which can be separated by filtration or centrifuging. Treatment of the chloroform or toluene phase with absolute ethanol precipitates the uncrosslinked copolymer leaving the additives in solution. Thus three cross fractions are obtained, crosslinked copolymer, uncrosslinked copolymer and gross additives each of which is amenable to further examination by various techniques. In some cases, eg. acrylonitrile-butadiene-styrene terpolymers, extremely powerful solvent systems such as boiling chlorobenzene are required to soften the polymer.

The separation of gross polymer from additives and fillers is important both from the points of view of obtaining fractions of these for further examination and also from the point of view of obtaining uncrosscontaminated fractions of each to facilitate these identifications. This applies particularly in the case of polymeric materials containing high proportions of non-polymeric constituents. Thus, heavily plasticized PVC might contain up to 60% of a plasticizer which it is essential to separate from the PVC, usually by solvent extraction, prior to attempted identification of the polymer or the plasticizer. In the case of a polymer containing percentage levels of inorganic metal oxides, solution of the polymer in an organic solvent followed by extraction with an aqueous acid might suffice to separate the polymer from the metallic constituent.

Polymers for which no solvent can be found are a different problem and present difficulties especially if appreciable amounts of fillers or additives are present. Direct spectroscopic examination by infra-red or NMR spectroscopy of thin cast films of the polymer will often yield some information, as will pyrolysis or photolytic techniques, especially if the latter is coupled to gas chromatography and/or mass spectrometry. Pyrolytic or photolytic techniques (Chapter 4.4.2, 4.4.3) operate by providing information regarding the polymer degradation products formed under these conditions, the nature of which will often yield information on the structure of the original polymer.

Much information is provide throughout this book on preparing samples for analysis especially in the case of the more commonly used polymers. In the case of the less commonly used polymers, especially the wide range of copolymers now being produced and particularly in the case of highly filled materials the analyst may have to exhibit considerable ingenuity in devising methods of preliminary work-up of the sample prior to the commencement of analysis. The general principals illustrated in this chapter and throughout the book will help in this respect.

CHAPTER 2

POLYMER CHARACTERISTICS REQUIRING ELUCIDATION

It is emphasized at the start that the examination of a polymer can fall into one of several broad categories;

a)   the identification of an unknown polymer also the identification and determination of other materials present in the formulation such as additives, fillers, monomers, volatiles, water, catalyst remnants and other impurities

b)   the examination of the micro structure of a polymer for information that would help to explain particular chemical or physical features of the polymer. In this case the type of polymer might be known at the start but we are looking for structural detail that will explain its physical properties (e.g. tacticity in polypropylene or the molecular weight distribution of a polymer), and would explain the difference in measurable physical properties obtained between different grades or sources of the same polymer

c)   the measurement of physical properties of a known polymer such as thermal oxidative or photo-oxidative stability, its combustion properties, measurement of Tg and other transitions and of crystallinity and Tm and finally estimation of the lifetime of a polymer to be expected under particular service conditions.

A scheme for the examination of polymer systems covering all these approaches is given in Table 2. In this table the different types of testing that might be required are identified with the appropriate Chapter numbers.

2.1   IDENTIFICATION OF UNKNOWN POLYMERS

To fully characterise an unknown polymer the following types of information are required.

i)   type of polymer
ii)   types and amounts of processing additives present
iii)   types and amounts of inert fillers present
iv)   types and amounts of monomers present
v)   types and amounts of organic volatiles present
vi)   water content
vii)   types and amounts of catalyst residues present
viii)   types and amounts of other impurities (e.g. metal and peroxide residues) present

i) Type of Polymer

To successfully identify the polymer, it should, if possible be separated from processing additives and fillers present. This will avoid the complication of interference by additives during the polymer identification and vice versa. In some instances, e.g. a suitable solvent for the polymer is not available this separation cannot be achieved and direct examination of the original polymer sample may have to be carried out.

Methods of separation of polymer and additives are discussed in Chapter 3. Frequently, one of two methods are adopted; either i) the polymer is dissolved in a good solvent for the polymer and the additives, then a non-solvent for the polymer added which precipitates the polymer free from additives and leaves the additives dissolved in the solvent-non solvent mixture (fractional precipitation), or, ii) the polymer is refluxed with a solvent which disperses it to provide a solvent phase containing additives (fractional extraction). Other methods that have been employed for separating additives from polymers are based on diffusion, dialysis or electro-dialysis. In Chapter 3 are discussed methods for separating the following polymers from additives, polyolefins, polystyrene, acrylics, PVC, rubbers, polacrylamide, polyurethane and some copolymers.

Having carried out this separation, some preliminary tests can be carried out on the gross additive-free polymer or copolymer fraction or the original polymer or copolymer as discussed in Chapter 4. Simple solvent solubility, ignition and colour reaction tests often do not provide the answer especially in the case of copolymers, but in many instances do provide useful clues in the identification e.g. an acrid smell of hydrogen chloride will suggest a chlorinated polymer. Analysis for elements is mandatory even if only carried out qualitatively and will often point to a particular type of polymer, e.g. nitrogen or sulphur containing.

Very useful information is often provided by running an infra-red or NMR spectrum or carrying out a pyrolysis (or photolysis) - gas chromatography scan on the sample and comparing the output with outputs obtained for a library of standard polymers obtained under the same or similar experimental conditions. Similarity of the outputs, especially if obtained for more than one of the techniques will often provide an identification.

Quantitative determination of elements Chapters 4.3. and 5.5 and functional groups (Chapter 6 polymers, Chapter 9.2 copolymers) will often provide confirmatory evidence regarding the identity of a polymer or copolymer as will the determination of conomomer ratios in copolymers, (Chapter 9.3)

ii) Type of Additive

The polymer-free gross additive extract prepared as described in Chapter 3 is now ready for the identification and determination of additives using methods discussed in Chapter 5. Usually, a polymer contains more than one type (up to six) of additive from the following list, antioxidants (phenolic, amine, dialkyldithiopropionate and tertiary phosphite types), plasticizers, ultraviolet absorbers, stabilizers (organotin, metal stearate and others), pigments, flame retardents, organic peroxides, antiozonants and accelerators. If the type of additive present is known in advance, then an appropriate method for the particular

Table 2 - Scheme for the Examination of Polymer Systems

			Chapter
A.	Preliminary tests	a) solubility in solvent b) elements qualitative and quantitative and ash content c) ignition tests d) colour reactions finger printing techniques comprising e) infrared spectroscopy f) Raman spectroscopy g) NMR h) pyrolysis and gas chromatography - mass spectrometry i) photolysis - gas chromatography - mass spectrometry	4
B.	Separation and analysis of gross polymer or copolymer and gross additive (if sample dissolves or swells in solvents)		1 & 3
B1. and	Gross crosslinked polymer fraction		
B2.	Gross uncrosslinked polymer fraction	Repeat preliminary A a) to i) tests (see above) Determine functional groups (polymers) Determine functional groups (copolymers) Determine comonomer contents (if copolymer)	4 6 9.2 9.3
B3.	Gross additives	Identify and determine individual additives	5.1
C.	Analysis of original polymer or copolymer	Identification of polymer (if not soluble or swollen by solvents). Carry out preliminary tests A a) to i) (see above) Determine functional groups (polymers) Determine functional groups (copolymers) Determine comonomer contents (copolymers)	4 6 9.2 9.3

polymer can be chosen. If, as is the normal case, the additive(s) present
are not known, or, if more than one type of additive is present, then a
chromatographic step has to be incorporated to separate the additives from
each other prior to their determination by various methods.

Complimentary techniques are frequently employed for the purpose of
identifying and analysing additive mixtures such as; gas chromatography -
infra-red spectroscopy, gas chromatography - mass spectrometry, pyrolysis
gas chromatography - mass spectrometry, hplc - infra-red spectroscopy,
hplc - mass spectrometry, thin-layer chromatography - infra-red spectroscopy and photolysis - gas chromatography.

iii) Inert Fillers

Usually these are inorganic materials such as zinc oxide, silica,
titanium dioxide, barium carbonate, calcium carbonate, alkali metal
stearates and lead stearate. The determination of metals is discussed in
Chapter 5. A commonly used method to separate fillers from polymer
involves dissolving or dispersing the polymer in an organic solvent and
then extracting the solvent phase with an aqueous solution of a reagent,
usually an acid, which dissolves the metal into the aqueous phase.
Alternatively, if the polymer is completely soluble in a solvent and the
filler is not, then simple filtration will suffice. Careful evaporation
of the polymer extract to dryness provides material for identification.

iv), v) Monomers and Organic Volatiles

In addition to monomers and oligomers, polymers might contain residual amounts of other organic volatiles. These might be adventitious such
as residual C_8-C_{10} polymerization solvent paraffins, in high density
polyethylene or deliberately added such as residual C_5 paraffin expanding
agents in expanded polystyrene granules.

Traces of such solvents and monomers can be of profound importance
in commercial grades of polymers as they can effect physical properties
and cause teinting in grades of polymer intended for packaging foodstuffs
and beverages. They also have implications in the packaging of cosmetics
and pharmaceuticals. Various methods, largely based on head space
analysis have been developed for identifying and determining volatiles in
plastics, (Chapter 5).

vi) Water Content

The determination of water in polymers is also of importance
(Chapter 5.4) as its presence above certain limits can cause moulding
problems.

vii) and viii Catalyst Residues and Impurities

Catalysts are used in many polymer manufacturing processes.
Examples include the use of aluminium alkyls and titanium halides as
catalysts in the manufacture of high density polyethylene and polypropylene and the use of various organic peroxides in the manufacture of
polystyrene. The finished polymer contains residual traces of breakdown
products of these catalysts. Thus, polyethylene contains 100 ppm amounts
of residual aluminium, titanium and chloride ion. It is usually necessary
to control the level of these in the final polymer in order that its
properties are not adversely affected. Thus, when high density polyethylene containing more than 50-100 ppm of chloride ion present as sodium

chloride is exposed to a damp atmosphere during storage it picks up water from the atmosphere to the point where water content exceeds 0.03% which is the maximum for trouble-free moulding of this polymer. The detection of traces of particular trace elements, aluminium, chromium, vanadium can provide information on the type of catalyst used in its manufacture. Traces of impurity elements in polymers are often of importance e.g. 0.1-0.2 ppm copper in polyolefins profoundly effect its oxidation resistance. In Chapter 5.5 is discussed the determination of traces of metals, halogens, sulphur, phosphorus and nitrogen in polymers.

The molecular weight distribution of a polymer, its number average molecular weight (Mn) and weight average molecular weight (Mw) all have profound effects on the physical properties and solubility of polymers and are influenced by the conditions under which is is manufactured, i.e. are controllable. The polymer chemist must, therefore, have means available for determining these parameters as discussed in Chapters 7 and 8. Fractionation, generally using high performance liquid chromatography nowadays, is used to separate higher molecular weight polymers into narrower molecular weight bands, to separate lower molecular weight polymers or oligomers into individual carbon number compounds and to separate mixtures of different polymers, (Chapter 7).

2.2 MICROSTRUCTURE

One aspect of microstructure in polymers, namely sequences in co-polymer chains, has already been discussed, (Chapter 9.4). There are other important aspects of microstructure related to various forms of isomerism which can have very important effects on the properties of polymers. These include stereoisomerism which governs the tacticity of polymers (Chapter 10.2) geometrical isomerism related to cis and trans configurations in unsaturated polymers (Chapter 10.3) and regioisomerism governed by head to head, tail to tail and head to tail configurations of monomer units in polymers (Chapter 10.4). A particular polymer can, commercially, be useless or highly suitable for its application depending on microstructural detail in any of the above respects. It is therefore extremely important to be able to measure such microstructural detail. Short chain branching is another aspect of microstructure which has received considerable attention (Chapter 10.6).

2.3 OTHER PHYSICAL PROPERTIES

Other polymer characteristics discussed include thermal stability (Chapter 11.1), oxidative stability (Chapter 11.2), photoxidative stability (Chapter 11.3) and the measurement of polymer shelf life under particular service conditions (Chapter 12.2.4), the combustion properties of polymers (Chapter 11.4, 11.5), the measurement of transition temperature (Tg) and other transition temperatures (Chapter 12) and melting temperatures (Tm) and the degree of crystallinity (Chapter 13).

CHAPTER 3

SEPARATION OF POLYMER AND ADDITIVES

3.1   INTRODUCTORY

Most polymers contain quite complex additive systems which are incorporated during manufacture to impart beneficial properties during manufacturing operations, e.g. protective antioxidants, slip agents etc, and during end-use, e.g. antioxidants, u.v. stabilizers, plasticisers and anti-static agents, flame retardants, antioxidants and thermal stabilizers. The first stage in the examination of a polymer, either from the point of view of identifying the polymer or identifying and determining additives present must be to separate gross polymer from gross additives. It may then be necessary as discussed in Chapter 5.1.16 to separate the gross additive fraction into individual additives by a chromatographic procedure in order to facilitate the identification of individual additives.

Separation of gross polymer from gross additives is necessary so that examination of the polymer is not interfered with by additives and conversely. Many different procedures have been studied for the removal of gross additives from the polymer and some of these are discussed below.

3.2   FRACTIONAL PRECIPITATION

(a)   A solution or dispersion of a plastic is stirred into at least 10 times its amount of a solvent which acts as a precipitant for the dissolved plastic but which dissolves completely in the first solvent. The precipitated plastic is pulverized and repeatedly extracted with the precipitant/solvent or re-dissolved and precipitated.

(b)   A non-solvent is added to a solution of the plastic until the first faint turbidities appear due to the high polymer plastic components; the solvent is now reduced on a water bath or by vacuum distillation. The solvent and the non-solvent must be completely miscible and the solvent must have a lower boiling point than the non-solvent.

(c)   The plastic is dissolved in a solvent which dissolves the plastic at higher temperatures; the plastic separates out on cooling while the additives remain in solution.

(d)   Plastic dispersions can often be separated by stopping the action of the emulsifier and/or dispersing agent, i.e. breaking the emulsion. The following methods can be used according to the type of emulsifier:

1.   Alteration of the solvent phase.

2.   Precipitation of the emulsifier by the addition of acids or a counter interfacially active ion.

3.   Freezing the dispersion out with solid carbon dioxide.

2   Subjection of the dispersion to dialysis, whereby the water-soluble inorganic salts and relatively low molecular emulsifiers are diffused through cellophane membranes so that the protective colloid and the polymer can now be easily separated quantitatively by centrifugal action.

3.3 FRACTIONAL EXTRACTION

Fractional extraction can be used in three ways, depending on the substance under examination:

(a)   Finely shredded solid plastics are extracted one by one in a Soxhlet apparatus with solvents of increasing solvent power.

(b)   To ensure that the plastic particles are fine enough to allow the components to be extracted to diffuse sufficiently quickly, a solution or dispersion of the plastic is poured over an inert carrier, such as silica gel or sterchamol, the solvent is allowed to evaporate and the carrier is then extracted in a heatable column with solvents of increasing solvent power.

(c)   The components of a plastic are distributed between two non-miscible liquids in separating funnels or distribution apparatus.

3.4 SEPARATION BY DIFFUSION METHODS

The plastic is precipitated from a solution or dispersion in a glass vessel as a thin film on the inner wall; meanwhile a solvent which is a non-solvent towards the plastic, is added in several stages. The solvent must be able to dissolve the additives. Additives or a second plastic can be isolated in a short time by diffusion alone if the mixture is left to stand in the solvent, occasionally shaken and the solvent is renewed four to six times.

3.5 DIALYSIS OR ELECTRODIALYSIS

This method has only been used for separating low-molecular weight salts and interfacially active substances from protective colloids and high polymer plastics and for fractionating pure plastics according to their molecular weight.

Some examples of these four separation procedures are tabulated in Table 3.

3.6 PRACTICAL EXAMPLES OF POLYMER/ADDITIVE SEPARATIONS

In practice, the most commonly used procedures for the separation of gross polymer and gross additives are based on fractional precipitation and fractional extraction. In the fractional precipitation procedure a good solvent for the polymer is required also a non-solvent for the polymer in which the additives remain soluble. Information on solvents and non-solvents for various polymers is tabulated in Table 4. It must be realized that the solubility of a polymer depends not only on the type of polymer but also the degree of polymerization, i.e. molecular weight and

Table 3 - Examples of separating procedures

Plastic	Operation	Aim or separating procedure

3.2 (a) Fractional precipitation procedures (precipitation by addition of non-solvent for polymer to solvent solution of polymer).

Plastic	Operation	Aim
Polyvinyl chloride	stirring of a concentrated solution of tetrahydrofuran into methanol	plasticizers, emulsifiers remain in solution, PVC and its co-polymers are precipitated.
Polyacrylate	stirring the solution in acetone into 20 times the amount of water	emulsifiers, watersoluble resins and salts remain in solution, polymerizate is precipitated

3.2 (b) Fractional precipitation (titration of solvent solution of polymer with non-solvent for polymer).

Plastic	Operation	Aim
Polyacrylate	water is added to the solution in acetone till faint turbidity occurs, acetone then distilled off under vacuum	emulsifiers, watersoluble resins and salts remain in solution, polymerizate is precipitated

3.2 (c) Fractional precipitation (thermal precipitation of polymer from solution at low temperature).

Plastic	Operation	Aim
Polyethylene	dissolve in enough benzene to make a clear solution appear when warmed then cool	polymerizate is precipitated, additives soluble in cold benzene, paraffin, waxes, resins remain in solution

3.2 (d) Emulsion breaking methods.

3.2 (d)1 Addition of solvent.

Plastic	Operation	Aim
Polymethacrylate-dispersions	dispersion is stirred into 20 times its amount of methanol or isopropanol	polymerizate is precipitated, emulsifier and any protective colloid remain in solution

3.2 (d)2 Emulsifier precipitation by acid addition.

Plastic	Operation	Aim
Rubber or butadiene co-polymers	dispersion is broken by acidifying as long as soap is used as the emulsifier	after acidifying, fatty or resin acid is extractable with ether; latex is precipitated

3.2 (d)3 Freezing out dispersion by cardice addition.

Polyvinyl acetate or silicone oil dispersions	action of emulsifier destroyed by cold	polymerizate is precipitated, can now be filtered off or centrifuged off

3.2 (d)4 Application of dialysis.

Polyvinyl acetate ester or polyacrylate dispersions	salts and low molecular emulsifiers are dialyzed	in the dialyzate, salts and emulsifiers, in the dialyzate residue, protective colloid polymerizate

3.3 Fractional extraction procudures.

3.3 (a) Solvent extraction of shredded plastic.

Polyvinyl chloride	consecutive extractions with CCl_4, ether, benzene, $CH_2 CCl_2$, tetrahydrofuran	isolation of stabilizers and plasticizers, recognition of copolymers or poly blends

3.3 (b) Solution or dispersion of plastic passed over inert carrier on which plastic adsorbs, solvent evaporation from carrier, then desorbtion from carrier with strong solvents.

Polyethylene	consecutive extraction of silica with gasoline, methanol, $CHCl_3$, water, methanol or acetone, then benzene or toluene (perhaps warm)	in the gasoline: waxes; in CH_3OH: emulsifiers; in $CHCl_3$: montan waxes in water: cellulose derivatives and polyvinyl alcohol; in benzene/ toluene: polyethylene

3.3 (c) Plastics components distributed between two non-miscible liquids.

Polyvinyl acetate	extracted from aqueous methanol solutions or dispersions with pentane/ether mixtures	isolation of plasticizers

3.4 Separation by diffusion.

Polyvinyl acetate	polymer precipitated with ligroin or ligroin/ether mixtures	isolation of the plasticizers, can also be used for traces

3.5 Separation by dialysis or electrodialysis.

Polyacrylate	see 3.2 (d)4

Table 4 - Summary of the solubility of plastics

Plastics	Typical solvents	Typical non-solvents
Polyethylene	benzene, xylene tetrahydronaphthalene	gasoline, alcohols, esters, hydrocarbon halides
Polypropylene	xylene, trichloro-ethylene, chloroform	alcohols, gasoline, esters
Polyisobutylene	gasoline, ether	alcohols, esters
Polybutadiene	benzene	gasoline, alcohols, esters, ketones.
Polyisoprene	benzene	gasoline, alcohols, esters, ketones
Natural rubber	hydrocarbons and hydrocarbon halides	gasoline, alcohols, esters, ketones
Polystyrene	dioxane, benzene, chlorohydrocarbons	gasoline, alcohols
Polyindene		
Regenerated cellulose	no solvent	
Polyvinyl alcohol	water, formamide	gasoline, aromatic hydrocarbons, alcohols
Phenolic plastics uncured	alcohol, ketones	gasoline, hydrocarbon halides
cured	insoluble and often infusible	
Cellulose methyl ether	ethylene chlorohydrin, water, dilute solution of caustic soda	aliphatic and aromatic hydrocarbons
Cellulose ethyl ether	methylene chloride/ methanol	water, aliphatic and aromatic hydrocarbons
Celulose benzyl ether	acetone, butanol	water, aliphatic and aromatic hydrocarbons
Polyvinyl methyl ether	water, methanol	gasoline

Polyvinyl ethyl ether	gasoline, aromatic hydrocarbons, hydrocarbon halides, alcohols, esters, ketones	water
Polyvinyl butyl ether	aromatic and hydrocarbon halides, ketones	methanol, ethanol
Polyglycols	water, hydrocarbon halides, alcohols	gasoline
Polyvinyl formal	tetrahydrofuran, ketones, esters	methanol, aliphatic hydrocarbons
Polyvinyl acetal	tetrahydrofuran, ketones, esters	methanol, aliphatic hydrocarbons
Polyvinyl butyral	tetrahydrofuran, ketones, esters, chloroform-methanol (9+1)	methanol, aliphatic hydrocarbons
Polymer aldehydes	ketones, esters	gasoline, alcohols
Polymer ketones	ketones, esters	gasoline, alcohols
Epoxy resins, uncured	alcohol, esters, ketones, dioxane	water, hydrocarbons
Natural resins	alcohol, benzene, hydrocarbon halides, esters, ether	gasoline
Modified phenoplasts, accord. to cure	ethanol	aliphatic and hydrocarbon halides
Polymethacrylates	esters, aromatic and chlorinated hydrocarbons, ketones, dioxane	aliphatic hydrocarbons, alcohols, ether
Polyacrylates	aromatic hydrocarbons, hydrocarbon halides, esters, ketones, tetrahydrofuran	gasoline
Polyvinyl acetate	aromatic and chlorinated hydrocarbons, methanol, ketones	gasoline
Polyesters	phenols, halogen phenols, nitrohydrocarbons, benzyl alcohol	hydrocarbons, alcohols, esters

Table 4 - Continued

Plastics	Typical solvents	Typical non-solvents
Alkyd resins	aromatic hydrocarbons, hydrocarbon halides, hydrocarbons, lower alcohols, ketones, esters	hydrocarbons
Cellulose esters	ketones, esters	aliphatic hydrocarbons
Polyvinyl chloride	tetrahydrofuran	hydrocarbons, butyl acetate, alcohols
Polyvinylidene chloride	tetrahydrofuran, dioxane, ketones, butyl acetate	alcohols, hydrocarbons
Polyvinyl chloride co-polymer with vinyl acetate	tetrahydrofuran, methylene chloride	hydrocarbons, alcohols
Acrylic esters	tetrahydrofuran, methylene chloride	hydrocarbons, alcohols
Maleic esters	tetrahydrofuran, methylene chloride	hydrocarbons alcohols
Chlorinated rubber	carbon tetrachloride, ketones, esters, tetrahydrofuran	aliphatic hydrocarbons
Rubber hydrochloride	ketones	aliphatic hydrocarbons, carbon tetrachloride
Neoprene	toluene, hydrocarbon halides	alcohols, esters
Trifluorochloroethylene	none	
Tetrafluoroethylene	none	
Polyamides	formic acid, phenols trifluoroethanol	alcohols, esters, aliphatic and aromatic hydrocarbons
Cellulose nitrate	lower alcohols, acetic esters, ketones, ether/alchohol (3:1)	ether, benzene, hydrocarbon halides
Polyacrylonitrile	dimethyl formamide, nitrophenol, ethylene carbonate	alcohols, esters, ketones, hydrocarbons
Polyacrylamide	water	alcohols, esters, hydrocarbons

degree of branching and cross linking and on its steric configuration. Thus a relatively low molecular weight unbranched polyethylene may have a high solubility in toluene whilst a high molecular weight highly branched polyethylene would have a low solubility even in hot toluene. In the case of styrene-butadiene copolymer, the uncross-linked polymer is soluble in aromatic solvents, whilst the highly cross-linked (gel) fraction is completely insoluble and, indeed, this can be used as the basis of a method for separating gel from uncross-linked polymer. Copolymers usually dissolve in a greater number of solvents than homopolymers. Thus whilst PVC is only slightly soluble in acetone or methylene chloride, its copolymers with vinyl acetate or acrylates easily dissolve.

<u>Fractional precipitation methods.</u> This method of extraction involves dissolution of the organic phase of the plastic composition in a suitable (sometimes hot) solvent, followed by precipitation of the polymeric constituents, often in a finely divided form in suspension with the inorganic fillers, by cooling or by means of another solvent in which the analyte compound is also soluble. This method is labour intensive but very effective and if it does not completely release any chemisorbed constituents from the polymer - filler matrix it will often leave them in a form very vulnerable to attack by the analytical reagent(s) finally used in the determination.

<u>Fractional extraction methods.</u> Most of these procedures can be carried out on material cut to a particle size of less than 2 mm diameter, it is often advantageous to produce material of a smaller size and with a large surface area to mass ratio. This is conveniently done by grinding at the temperature of liquid nitrogen using an efficient and easily cleaned cutter mill.

The extraction of additives strongly adsorbed or chemisorbed on the polymer, filler matrix must be carefully observed by the analyst, as a change of the method of manufacture of, for example, the filler or in the method of compounding the plastic formulation, can also markedly alter the degree of adsorption being produced and hence invalidate an established quantitative extraction procedure. The use of "stronger" extraction reagents can cause complications at the measurement stage and hence each system must be carefully screened and frequently checked.

In a modification of this procedure the polymer is refluxed with a reagent which decomposes additives present to put them in a form which is soluble in the solvent. Thus when polypropylene on which is chemisorbed fatty acid amides or amines is refluxed with a solution of ethylene dichloride and trichloro acetic acid then the amides are decomposed as follows:

$$RCONH_2R^1 + H_2O = RCOOH + R^1NH_2$$

The separation of particular polymers is now discussed in further detail and illustrated by practical examples.

3.6.1 Polyolefins

The scheme outlined in figure 1 has been proposed for the solvent extraction and examination of polythene[1]. Some useful solvents are listed in Table 5.

Details of particular solvent extraction schemes for polyethylene are described in more detail below:

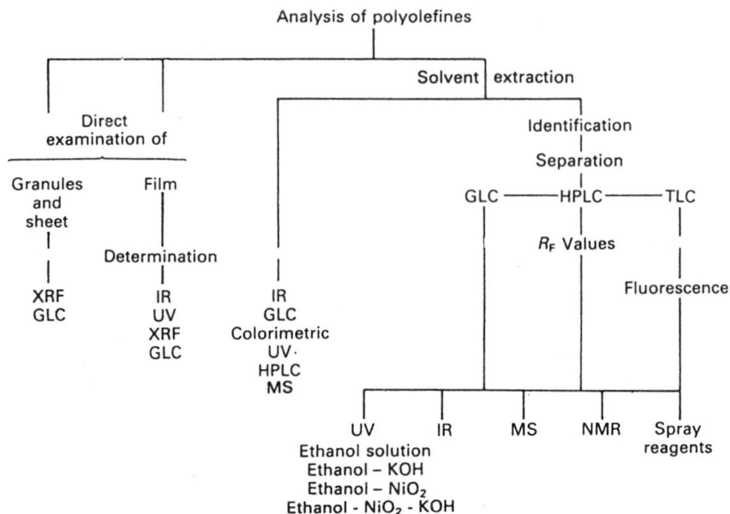

Figure 1 Analysis of polyolefins

Toluene or xylene extraction system.   In this scheme the polyethylene is refluxed with toluene which partly dissolves the polymer. Upon the addition of absolute ethanol the polymer is reprecipitated leaving the additives in solution in the toluene-ethanol solvent mixtures which is then filtered off and carefully evaporated to dryness to provide the additive extract which is essentially free from polymer.   Such extracts when prepared quantitively, are extremely useful either for direct analysis for known additives or for subsequent separation of their components by one of a number of chromatographic techniques prior to examination of the individual separated compounds by spectroscopy for identification and for quantitation.

Weigh a suitable amount of polyethylene powder, shavings or nibs, between 0.05 and 1 g, into a 250 ml conical flask.   Pipette in 50 ml toluene or xylene and add 2 g of broken porous pot.   Purge the interior of the flask with nitrogen and connect immediately to a condenser which has been previously purged with nitrogen.   Connect a T-piece to the open end of the condenser and pass a gentle purge of nitrogen through the T-piece during the reflux period.   Reflux gently for one hour in a boiling water bath.   Allow the contents of the flask to cool to approximately 40°C and add 20-30 ml of ethanol down the condenser to reprecipitate any dissolved polymer.

Support a Whatman No. 54 filter paper in a filter funnel standing in a 100 ml volumetric flask.   Carefully transfer the contents of the conical flask into the filter paper using a thin glass rod to assist the transfer. Loosen any polymer adhering to the bottom of the conical flask by means of a glass rod and rinse down the sides of the flask with 10 ml of ethanol. Heat to about 60°C and transfer the contents of the flask to the filter paper.   Rinse the polymer thoroughly with a jet of ethanol delivered from a glass wash bottle until the volumetric flask contains approximately 95 ml of solvent.   Allow the solvent to acquire room temperature, make up to 100 ml with ethanol and mix well.   Low density (i.e. high pressure) polyethylenes usually completely dissolve during the reflux.   High density

Table 5 - Solvent extraction methods of additive extraction from polymers

Polymer type	Substances extracted	Extracting solvent	Comments	References	Type of extraction
Polyethylene	cresolic and phenolic antioxidants	chloroform	heat at 50°C for 3 h in a closed container	2	fractional extraction
Polyethylene	cresolic anti-oxidants	hexane	heat at 50°C for 24 h	3	fractional extraction
Polyethylene	cresolic anti-oxidants	ether	in the dark at 20°C	4	fractional extraction
Polyethylene	phenolic anti-oxidants	chloroform		5	fractional extraction
Polyethylene	antioxidants	toluene	reflux to dissolve polymer in precipitate with methanol	6	fractional precipitation
Polyethylene	antioxidants	water	at 70°C under nitrogen	7	fractional extraction

polyethylenes (i.e. Zeigler low pressure) either completely dissolve or disperse sufficiently to allow full extraction of additives into the solvent phase.

<u>Ethanol-methanol extraction system.</u> The sample to be analysed must be very thinly sheeted or powdered. Weigh a 1.000 $\pm$ 0.0005 g sample. Wrap with extraction cloth which has been previously extracted to remove sizing, etc. Place in an Underwriter's extraction cup and extract for 16 hrs with 95% ethanol or methanol. Transfer the alcohol extract to a 100 ml volumetric flask. Cool to room temperatures and bring to the mark with the extraction solvent.

<u>Ethanol extraction.</u> Prepare the polymer compound containing the unknown stabiliser by cutting or grinding to small pieces, and extract about 5 g by heating under reflux for 24 hours with 50 ml of ethanol. Cool the extract to room temperature and filter off the polymer.

<u>Cyclohexane extraction</u>[8]. Grind a representative sample of polyethylene in the Apex Mill. Weigh about 1 g of the sample into a 100 ml round-bottomed flask and add 20 ml cyclohexane. Fit a condenser to the flask and allow the cyclohexane to reflux on a water bath for 1 hr. Wash down the condenser with 20 ml cyclohexane, remove the flask from the bath, cool to room temperature and shake well. Filter through a No. 802 filter paper into a 100 ml volumetric flask and make up to the mark.

Details of particular extraction systems recommended for poly-propylene and polyethylene are given below.

<u>Diethyl ether extraction.</u> Weigh 2-5 g of polypropylene plastics and extract with diethyl ether for 8 hours in a Soxhlet continuous extraction apparatus. The weight to be taken will vary with the types and amounts of additives expected. If these are not known take the maximum amount, if possible. Remove the diethyl ether by distillation on a water bath and add 5 ml of ethanol to each flask. Allow the flasks to stand for 10 minutes. If necessary, at the end of this time, filter the resulting solutions quantitatively to remove any precipitated polymer. If required, evaporate the contents of the flask to dryness and dissolve in a suitable solvent for further analysis of additives . Allow the extracted polymer to air dry. Alternatively, weigh 5 g of polymer into an extraction thimble, add 130 ml of the diethylether, extract for 3 h, allow the system to cool to room temperature and drain the solvent into the 250 ml flat bottom flask. Rinse the extraction thimble and extraction apparatus with 25-30 ml of extraction solvent and drain into a 250 ml flask (a turbid solution due to the precipitation of extracted polyolefin may be seen upon cooling; however, this is not uncommon nor should any special precautions be taken), Remove the extraction solvent by evaporation to leave a dry residue.

<u>Heptane extraction.</u> Extraction of polypropylene sheet. Weigh samples of 0.5 mm thick 50 x 50 mm polypropylene sheets and reflux in 50 ml of boiling heptane. The efficiency of extraction is better than 98% after 3 h of extraction.

<u>Acetonitrile extraction.</u> Cut the sample into small shavings with a drill bit prior to extraction. Extract approximately 1 g of the plastic shavings overnight with 5 ml of acetonitrile. Perform the extraction at ambient temperature in a sealed amber vial with constant stirring. The additives are sufficiently soluble in the acetonitrile extracting solvent due to the small amounts of analyte present. Filter the extract solutions prior to analysis.

<u>Chloroform or 1:1:1 trichloroethane extraction.</u> Weigh accurately into a 250 ml round-bottomed flask a portion of the polypropylene sample, (2-20 g). Add 50 ml of chloroform (or 1:1:1 trichloroethane), place the flask under a water filled reflux condenser, bring the mixture gently to the boil on a water bath and heat for 30 min. Allow the solution to cool and decant into a 100 ml volumetric flask after filtering through a No. 42 Whatman filter paper. Transfer any polymer collected in the filter paper to the round-bottomed flask and add 40 ml of chloroform. Repeat the extraction procedure. Filter the cooled solution into the volumetric flask containing the first extract. Wash the residue with a little more chloroform, filter into the volumetric flask and make up to the 100 ml mark with chloroform. Shake the solution well. Alternatively grind the sample in a Wiley Mill to pass through a 60 mesh screen. Extract a 10-20 g portion of sample in a Soxhlet apparatus with heptane for a minimum of 6 hours. Transfer the extract to a 300 ml tall-form beaker, evaporate to approximately 20 ml, and then dilute to 40 ml with chloroform.

3.6.2 Polystyrene

<u>Toluene extraction.</u> Weigh our 5.0 $\pm$ 0.01 g of sample and transfer to a 250 ml Pyrex glass centrifuge bottle. Measure 50 ml of toluene into the bottle. Drop a polythene coated magnetic stirrer rotor into the bottle and stopper with a cork (avoid rubber bungs). Stand the bottle on a magnetic stirrer and leave for several hours until the sample has completely dissolved or dispersed in the solvent. Accurately pipette into the gently stirred contents of the bottle 50 ml of absolute alcohol to precipitate the polymer. Centrifuge to isolate the solvent phase from the polymer phase.

<u>Diethyl ether extraction.</u> Weigh our 5 $\pm$ 0.1 g of sample and transfer to a 50 ml beaker containing 25 ml diethyl ether. Cover with a watch glass and leave overnight. Carefully decant the ether to a large beaker and wash the beads with two further 10 ml portions of ether, decanting each washing into the main extract. By means of an air line and warm water bath carefully evaporate off the ether until about 1 ml of solution remains. Transfer this by means of a 1 ml pipette to a 2 ml volumetric flask. Rinse round the walls of the beaker with 5-10 ml of ether and concentrate to about 0.5 ml. Again transfer this solution to the 2 ml volumetric flask and make the flask contents up to 2 ml with diethylether.

<u>Ethanol or methanol extraction.</u> The sample to be analysed must be very thinly sheeted (or powdered by passing through a Wiley Mill). A 2.00 $\pm$ 0.02 gram sample is accurately weighed and wrapped with extraction cloth which has been previously extracted to remove sizing, etc. The sample is placed in an Underwriter's extraction cup and extracted for 16 hours with 95% ethanol or methanol. The alcohol extract is transferred to a 100 ml volumetric flask, cooled to room temperature and brought to the mark with the extraction solvent.

3.6.3 Acrylic Polymers

A fractional reprecipitation procedure involving solution of the polymer in acetone, followed by reprecipitation with light petroleum can be used to separate additives from acrylics. The acetone-light petroleum extract can be used for the determination of plasticizers and lubricants, (Figure 2).

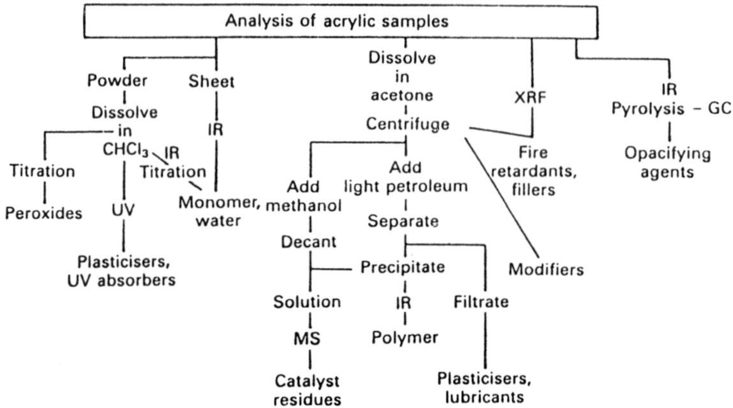

Figure 2 Analysis of acrylic samples

3.6.4 PVC

Diethyl ether is a favoured solvent for removing additives from this polymer. It has been used to extract stabilizers, lubricants and plasticizers from PVC[8]. In Figure 3 is shown a scheme involving ether extraction for the separation of additives from PVC[5-10].

Table 6 gives the results of following 4 h extractions with ether by extractions with other solvents. It can be seen that extraction for 4 h with carbon tetrachloride/methanol azeotrope gives better results than extraction for 15 h with methanol.

Haslam and Soppet[11] found that extraction with acetone, followed by precipitation of dissolved polymer with light petroleum, gave poor results: only 28.5% was recovered from a composition containing 31.8% tritolyl phosphate. Substitution of 1,2-dichloroethane for acetone gave no improvement, but diethyl ether extracted 31.8% of the sample, and the extract contained a negligible amount of polyvinylchloride. For routine extraction it was recommended by Haslam and Soppet that the sample be stood overnight in cold ether, then extracted for 6-7 h in the Soxhlet apparatus. This procedure has also been described by Doehring,[12] who specified that anhydrous ether should be used. Thinius[13] compared the rates of extraction of dioctyl phthalate by ether, carbon tetrachloride, and petroleum, at room temperature. In 70 min the ether extract from a composition containing 40% plasticizer amounted to 39.7%; in the same time the carbon tetrachloride extract was 33.1% and the petroleum extract 29.9%. Extraction with carbon tetrachloride for 64 h gave only 35.6% extract. Thinius also found that mixtures of phthalate and phosphate esters could be completely removed by ether, but light petroleum gave only 65% of the expected yield. Toluene dissolved some polyvinylchloride at room temperature, and very much more in a Soxhlet extraction.

For compositions containing polypropylene adipate, which is only partly extracted by ether, Haslam and Squirrel[14] used a 6-h ether extraction, followed by an 18-h methanol extraction. The ether extract was 34.5% from a composition containing 36.1% of a mixture of equal parts of dioctyl phthalate and tritolyl phosphate, but only 33.1% from a

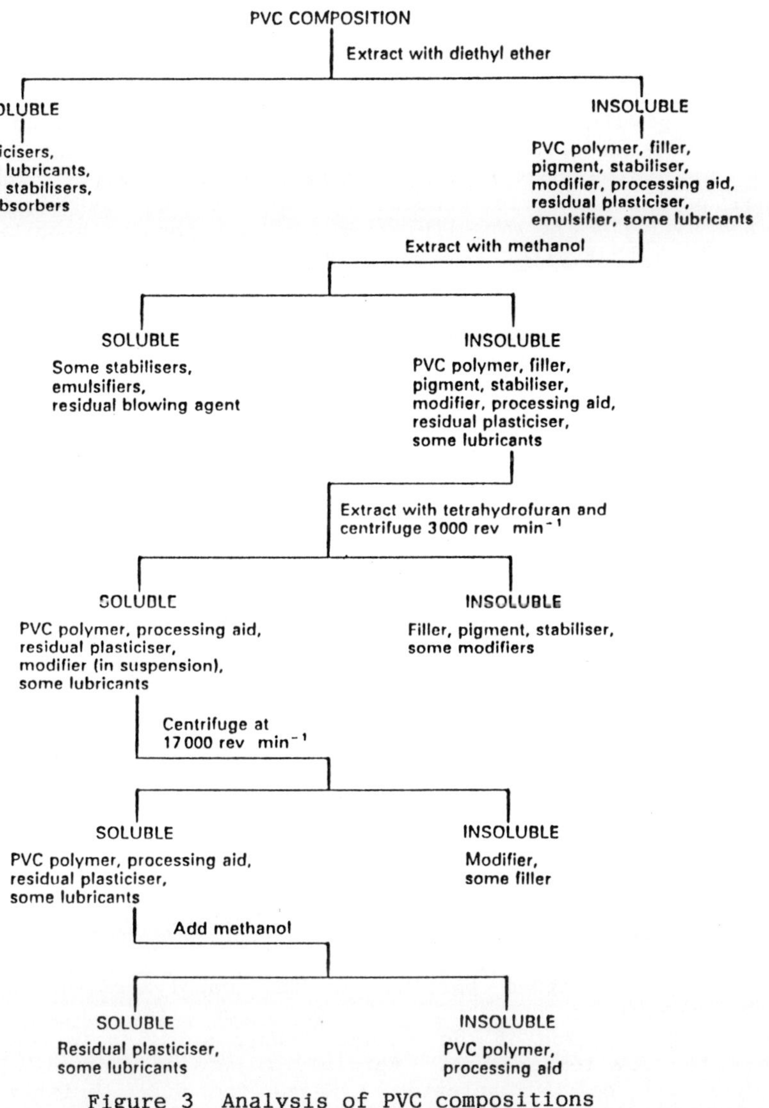

Figure 3   Analysis of PVC compositions

Table 6 – Multiple extraction

Plastisizer	Concentration %	Extracts (%)	
		4 h ether 5 h MeOH	4 h ether 4h CCl$_4$, MeOH
Bisoflex 791	28.5	27.7	28.7
Tritolyl phosphate	28.5	25.8	28.9
Mesamoll	28.5	25.5	28.1
Bisoflex 795	28.5	28.1	29.0
Reoplex 220	28.5	20.4	29.3
Hexaplas PPA	23.5	11.0	17.4

composition containing 45.7% of a mixture of equal parts of dioctyl phthalate, tritolyl phosphate and polypropylene adipate. The combined ether and methanol extracts amounted to 34.7% and 44.3% respectively. Wake[15] quotes the results of some unpublished work carried out at the laboratories of the Rubber and Plastics Research Association: methanol extracted 40.5% and 36.8% from compositions containing 42.3% polypropylene sebacate and 36.4% polypropylene adipate respectively. The extracts contained polyvinylchloride equivalent to 0.6% and 0.9% respectively. Clarke and Bazill[16] extract with ether for 15 h then with methanol for 8-h. They state that ether removes plasticizers whose molecular weight is less than 1000.

Korn and Waggon[17] used methanol or diethyl ether to extract diphenyl thiourea and 2-phenylindole dicyanamide from PVC.

3.6.5 Rubbers

Various extraction solvents for rubbers are listed in Table 7.

3.6.6 Polyacrylamide

Add a weighed quantity (1-10 g) of polymer to 50 ml of an 80/20 volume: volume methanol/water solution in a glass bottle. Add a magnetic stirring bar coated with Teflon, stir the solution and allow acrylamide monomer to extract for at least 3 h, preferably overnight.

3.6.7 Polyurethane

Cover the weighed sample of polyurethane foam with 75 ml of methanol in a 250 ml beaker and soak for 5 min with occasional compression. Decant the methanol, squeezing the foam as completely as possible to express the methanol. Repeat this twice with fresh methanol and concentrate the combined extracts to 25 ml.

3.6.8 Vinyl chloride, Butadiene, Acyrylonitrile, Styrene, 2 Ethylhexyl Acrylate Copolymers

Dissolve a weighed portion of the copolymer in septum vials containing measured aliquots of N, N-dimethylacetamide. Heat the vials to 90°C to aid dissolution of the polymer. When solution is complete, cool the vials to room temperature. Forcibly inject an aliquot of distilled water into each polymer solution. Shake the vials briefly to assure complete mixing of the water with the organic phase and to prevent the precipitated polymer from forming a film on top of the solution.

Special care and specialised techniques are required when dealing with laminates and surface-coated films. For major components the separation is made quantitatively and the analysis is completed by various techniques. For volatile components, separation, identification and quantification can often be carried out in one analytical process.

Steam-solvent distillation using diethyl ether has been used to remove and analyse for odour and teint from additives in food packaging films. Another technique that has been used is vacuum/thermal extraction. This procedure has been applied to nylons and fluorocarbon polymers. The procedure is used for the direct isolation or release of volatile components from a polymeric matrix and may involve the combined use of vacuum and heat, as for example in the mass spectrometer direct insertion probe or during dry vacuum distillation. Alternatively, the volatiles may

Table 7 - Extraction solvents from rubbers

	Type of additive	Extraction solvent	Extraction procedure	Reference
Rubbers	amine and phenolic antioxidants	ethanol/HCl	reflux, then steam-distill amines from extract	18
Rubbers	phenyl salicylate, resorcinol benzoate	ether		19
Rubbers	antioxidants	acetone		20
Rubbers	ketone-amine condensates, phenols, 2-mecaptobenzimidazole	acetone		22
General	p-phenylenediamine derivatives	95% methanol or ethanol	reflux 16 h in an extraction cup	23

be swept from the heated sample by a flow of inert gas for concentration by freeze trapping and/or collection on to a solid adsorbent prior to thermal or solvent desorption for gas-chromatographic or mass spectrometric examination.

IDENTIFICATION OF POLYMERS

4.1 INTRODUCTORY

The polymer analyst working in industry is frequently asked to identify a polymer or at least to give an opinion as to the type of a polymer. This might arise as a result of an interest in the materials being used by competitors or manufacturers in components or packaging materials. Obviously, if a detailed examination of a polymer or copolymer and its additive system is required then other chapters of this book should be referred to. However, if a quick opinion is all that is required, then some simple physical and chemical tests and the fingerprinting approach discussed in this chapter, might suffice. If on the other hand, the polymer does not yield to such tests, or , if a more detailed structural elucidation is required, then the full range of elemental and functional group analysis (Chapters 4.3 and 9.2 and microstructural and spectroscopic studies (Chapters 9 and 10) will be required.

4.2 SIMPLE PHYSICAL TESTS

Simple physical and chemical tests considered here are elemental analysis, saponification number or solvent solubility, ignition tests, colour reactions and burning/heating tests. These are of limited value nowadays because of the wide range of polymers now being manufactured and certainly are probably of no value in the case of copolymers. Conventional low impact polystyrene is soluble in hot toluene, whereas high density polyethylene or propylene have little or no solubility in this solvent. However, if the polystyrene contains some copolymerized butadiene, as occurs in the case of high impact polystyrene, then due to the presence of crosslinked gel, the polymer would not completely dissolve in hot toluene. So even in the case of simple polymers solubility tests are of limited value. Polystyrene, on the other hand, unlike the polyolefins, when it is held in a flame, due to its aromatic nature, will burn with a smoky flame. However, if it is a non-flame grade it will not burn. PVC will, when burnt, produce an acrid odour of hydrogen chloride, so will polyvinylidine chloride and many copolymers containing vinyl chloride or vinylidene chloride. If the polymer contained fluorine, swelling the odours produced on combustion might be a hazardous operation as it would be in the case of acrylonitrile and polyurethane polymers which, in these circumstances, are likely to produce hydrogen cyanide. Density measurement, i.e. whether the polymer sinks or floats in water is another simple parameter that can be observed. It will distinguish polyethylene with a density of less than one, from a highly chlorinated polymer with a density of greater than one. However, if the polyethylene contains lead stearate or iron oxide filler, its density may well exceed unity.

Enough has been said to indicate that in many instances simple physical tests just do not provide reliable information. Qualitative or quantitative examination for the presence of elements, glass transition or melting temperature measurements, and the pyrolysis gas chromatograph or infrared fingerprinting approach, as discussed below, do however provide more useful information and indeed, in many cases will lead to a successful resolution of the problem.

4.2.1 Classification by Solvent Solubility

As mentioned earlier, this technique is of limited value, especially when applied to copolymers. Solubility does not only depend on the type of the monomer base products but also to a large extent on the degree of polymerization (i.e. on the molecular weight) on the branching and cross-linking of the molecules and on the steric configuration of the polymerizate. Very efficient solvents can however be distinguished from the less efficient ones for different plastics. The solubility of polymers is discussed further in Chapter 3.

4.2.2 Classification by Ignition Tests

These tests are of limited value although Bird[24] has described the following simple thermal tests for the identification of plastic films.

Identification of plastic films - thermal methods[24].

__Burning test.__ Materials - bunsen burner or ordinary gas range, 1" x 3/8" specimen of material to be tested, pair of tweezers.

__Procedure.__ Hold material in tweezers over or at outer edge of a bunsen burner flame.

__What to look for.__ Colour of flame, visible changes in film sample, smoke formation.

__Interpreting observations.__ If film shrinks a great deal, this usually means material was stretched or oriented in the production process.

__Copper wire test for halogens.__ (Beilstein test). Materials - piece of strong copper wire, cork, bunsen burner.

__Procedure.__ Insert one end of copper wire into the cork. Holding the cork, heat wire in flame until you see only a very slight yellow colour in flame. Now melt the film sample onto the wire while it is still hot. Hold wire in outer part of a fully burning flame.

__What to look for.__ Green colour in flame indicates presence of halogens, e.g. chlorine or fluorine.

Interpreting observations - caution should be used in interpreting results because test sample may have been varnished or laminated with halogen containing products. In such cases, thorough laboratory analysis may be necessary to clarify inconclusive results (see footnote in table 8 for a simple method of determining whether the film is varnished or not).

__Heating tests.__ Materials - bunsen burner, test tube, small piece of material to be tested.

<u>Procedure.</u> Place sample piece at bottom of test tube. Heat very slowly over a bunsen burner flame.

<u>What to look for.</u> Observe process closely from melting to carbonization and check odour continuously.

<u>Interpreting observations.</u> To analyse chemical nature of fumes, insert strip of wet litmus paper in upper part of test tube. Alkaline reaction will turn paper blue, acid reaction will turn paper red. If acid reaction is very strong, insert a strip of Congo paper. If it turns red, a mineral acid has been produced. Check table 8 for identification.

Characteristic results obtained by these test are shown in table 8.

4.2.3 Classification by Colour Reactions

Colour reactions for particular homopolymers are listed in table 9. These tests are rather laborious and are perhaps best regarded as confirmatory tests for polymer identifications achieved by other means.

4.2.4 Classification by Qualitative Elemental Analysis

It is advisable when commencing the analysis of a polymer for unknown additives to determine first its content of various non-metallic and metallic elements In the first place, these tests could be qualitative, simply to indicate the presence or absence of the element. All that is required here is that the test is of sufficient sensitivity such that elements of importance are not missed. If in these tests, an element is found it may then be necessary to determine it quantitatively as discussed below. Any element found to be present must be accounted for in the subsequent examination for, and identification of, additives. Hence, elemental analysis reduces the possibility of overlooking any additive which contains elements other than carbon, hydrogen and oxygen. The analytical methods used to determine elements should be sufficiently sensitive to determine about 10 ppm of an element in the polymer, i.e. should be able to detect in a polymer a substance present at 0.01% and containing down to 10% of the element in question.

This requirement is met for almost all the important elements by use of optical emission spectroscopy and x-ray fluorescence spectrometry. X-ray fluorescence spectrometry is applicable to all elements with an atomic number greater than 12. Using these two techniques, all metals and non-metals down to an atomic number of 15 (phosphorus) can be determined in polymers at the required concentrations. (Cook et al[50]; Houk and Silverman[51]; Mitchell and O'Hear[52] and Bergmann et al[53].) Nitrogen is determinable at these levels by micro Kjeldahl digestion techniques. A cautionary note is that, in addition to the polymer itself, the polymer additive system may contain elements other than carbon, hydrogen and oxygen. The detection of an element such as nitrogen, sulphur, halogens, phosphorus, silicon or boron in a polymer is indicative that the elements originate in the polymer and not the additive system if the element is present at relatively high concentrations such as several per cent. This is highlighted by the example of a high density polyethylene which might contain 0.2 - 1% chlorine originating from polymerization residues and PVC homopolymer which contains more than 50% chlorine.

When looking for elements which would indicate the presence of particular polymers or additives, it is also worth bearing in mind that low levels of elements can be present which originate not in the polymer or its additives but in adventitious impurities or catalyst remnants as discussed in Chapter 5.5.

Table 8 - Identification of Polymer Films

Film Type	Physical Characteristics	1.Burning Test: ...Watch both flame and film	2.Copper-Wire Test ...Look for flame and film	3.Heating Tests ...Check odour and acid-alkalinity	4.Solubility Test: Dissolves and may swell in
EXTENSIBLE					
Fluorocarbon:	Waxy feel	No flame produced; sample deforms slowly	Flame turns green	Fumes may be toxic – do not inhale	Virtually inert
Polyolefins: (low and high density PE,PP)[a]	Wax-like feel; good extensibility	Burns freely; smokeless, bluish flame; melts, drips like wax[c]	None	Burnt paraffin wax aroma	Tetralin at 275 F
Polyvinyl Chloride: (flexible)	Softness range is that of rubber; extensibility proportionate to plasticizer used	Burns with yellow-ish sooty flame; melts freely to form pearl like drops	Flame turns green	Sharp acrid-fumes turn Congo paper blue	Cyclohexane, tetrahydro-furane and pyridine; solu-tion turns red in presence of alcoholic potassium hydroxide solution

Table 8 - (Continued)

Rubber Hydro-chloride	Clear and tough; relatively good extensibility[c]	Won't burn; edges tend to singe; dark smoke given off; green-blue seam in flame	Flame turns green	Rubber odour; fumes turn Congo paper blue	Chloroform; solution shows no change in presence of pyridine and alcoholic potassium hydroxide solution
SEMIEXTENSIBLE					
Cellulose Acetate:	Poor extensibility	Melts and drips	None	Acetic odour	Acetone
Cellophane:[b]	Transparent; paper-like feel when dry; crumples easily; poor extensibility	Burns like paper, with light flame	None	Burnt paper aroma	Schweitzer's reagent (cupro-ammonia)
Cellulose Acetate Buty-rate:	Horn-like feel	Melts and drips; drippings may burn	None	Rancid butter	Ketones, lactates, Me, Et propyl chlorides
Cellulose Nitrate	Similar to cellulose	Burns rapidly with an intense, white flame	None	Odour of camphor	Acetone at high nitrogen content

Polyamide: (Nylon)	Opaque to clear; hard feel; poor extensibility	Burns with yellowish, sooty flame; melts to form pearl-like drops	None	Sour aroma, like burnt horn or hair; fumes turn litmus paper red	Formic acid
Polyester Film: (Polyethylene terephthalate type)	Clear, very tough; non-tearable; poor extensibility	Ignites, but extinguishes on removal of flame; shrinks without singeing of edges; melts to form pear-shaped drops	None	Faintly sweet	Chlorophenol
Polyvinyl Chloride: (rigid)	Hard, greasy feel; white fracturing will show up in creasing line	Self-extinguishing	Flame turns green	Sharp, acrid odour; fumes turn Congo paper blue	Cyclohexane, tetrahydrofurane and pyridine; solution turns red in presence of alcoholic potassium hydroxide solution
Polyvinylidene Chloride:	Either milky or clear film; softer feel than PVC;	Self extinguishing	Flame turns green	Sharp, acrid odour; fumes turn Congo paper blue	Cyclohexane; solution black in presence of pyridine and alcoholic potassium hydroxide solution

Table 8 – (Continued)

NONEXTENSIBLE

Oriented Polystyrene:	Clear; nonextensible; when crumpled, gives metallic sound; sharp folds displays mother-of-pearl effect	Burns with luminous, soot-producing flame; smoke is dense	None	Repugnantly sweet, like artificial illuminating gas	Benzene, hydrocarbons

[a] Oriented polyolefin films will shrink substantially on initial heating.

[b] Absence of condensation when breathing on film indicates film is unvarnished. To test for commonly applied nitrocellulose varnish, apply several drops of a mixture of one part diphenyl and 100 parts concentrated sulfuric acid to film surface. Appearance of blue colour indicates presence of varnish. Polymer varnishes are readily identified by the Beilstein test for halogens.

[c] Extensible > 200% elongation
Semi-extestensible 25-200% elongation
Non-extensible < 25% elongation

Table 9 - Colour Reactions to Show Structures and Functional Groups

Plastics	Reagent	Reaction	Reference
Polyvinyl alcohol	Boric acid/iodine	greenish blue	25,26
Cellulose derivatives	alpha-naphthol/H_2SO_4	reddish violet	27
	Aniline acetate	turns filter paper red	28
	Anthrone	violet	29
Polyvinyl ether	Solution in acetic anhydride + conc. H_2SO_4	blue-violet	30-32
	Dichloroacetic acid	blue	26
Coumarone resins	Bromo-glacial acetic acid with solution of $CHCL_3$	turns permanently red	
Phenolic resins	Diazotized 2-nitro-4-chloroaniline	red	33,34
	2,6 dibromoquinone chlorimide	blue	35
	$FeCl_3$	green	36
	Millon's reagent	violet	37
Furfural resins	Aniline acetate	turns filter paper red	28
Epoxy resins	Nitration reactions		38,39
	Mercuric sulphate	precipitate	38,39
	Paraformaldehyde/ conc. H_2SO_4	bright blue	40
	Reaction with pyridine or quinoline bases	blue	41,42
Natural resins	Acetic anhydride/H_2SO_4	reddish violet	43,44
Polymethacry-lates	Reaction of depolymeriz-ates with conc. HNO_3 and Zn powder	blue colouring which can be removed with $CHCl_3$	45
HCHO amino plastics	Carbazole/H_2SO_4	dark blue	46
Hydrocarbon halides	Solution in pyridine + methanolic KOH	red-brown	47-49

An undoubtedly incomplete classification of polymers based on elemental classification and saponification number is provided in table 10. Note that copolymers are not included in this list.

<u>Qualitative Determination of Sulphur - Barium Chloride Method</u>

<u>Reagents.</u> Hydrochloric acid, N.

Precipitating reagent solution A - dissolve 0.2 g of peptone in 50 ml of 1 percent barium chloride ($BaCl_2 2H_2O$) solution. Buffer to a pH of 5.0 with 0.02 N hydrochloric acid, add 20 g of sodium chloride (Analar) and dilute to 100 ml. Heat on a water bath for 10 minutes, and add a few drops of chloroform. Filter if necessary.

Precipitating reagent solution B - dissolve 0.04 g of gum ghatti in 200 ml of distilled water by warming slightly. When solution is complete, ad 2.0 g of barium chloride ($BaCl_2 2H_2O$). Filter if necessary.

Store solutions A and B separately and prepare the final reagent just before use by diluting 10 ml of solution A to 100 ml with solution B.

Hydrogen peroxide, 100 volume - AnalaR

<u>Method.</u> Transfer 5 ml of the test solution to a 6" x 1" test tube and add 2 drops of 100 volume hydrogen peroxide, add then 1.2 ml of N hydrochloric acid. Mix well and add 2.0 ml of precipitating reagent with continued shaking. A distinct turbidity will be produced in the mixed solution if hydrogen peroxide decomposable sulphur is present in the sample; the blank test under the same conditions will be perfectly clear. If a semi-quantitative estimation of the sulphur content is required, add 5 ml of distilled water to both blank and test solutions, mix, and set aside for 30 minutes. Mix the solutions and measure the optical density of the test solution in a 4 cm cell at 700 nm with the blank solution in the comparison cell.

<u>Qualitative Determination of Sulphur - 4-Amino -4' Chlorodiphenyl Hydro-
chloride Method</u>

<u>Reagents.</u> 4-amino-4'chlorodiphenyl hydrochloride - peptone solution (solution A). Dissolve 0.025 g of peptone (Hopkin and Williams Ltd.) and 0.125 g of 4-amino-4'chlorodiphenyl hydrochloride (L. Light and Co. Ltd.) as completely as possible by warming with 50 ml of 0.05 N hydrochloric acid. Cool, and filter through a double Whatman No. 40 filter paper.

4-amino-4'chlorodiphenyl hydrochloride - gum ghatti solution (solution B). Dissolve 0.20 g of gum ghatti (Hopkin and williams Ltd. finely ground) and 1.00 g of 4-amino-4'chlorodiphenyl hydrochloride as completely as possible by warming to about 70°C with 400 ml of 0.05 N hydrochloric acid. Cool and filter through a double Whatman No. 40 filter-paper.

Final precipitation reagent - dilute 10 ml of solution A to 100 ml with solution B. This solution should be freshly prepared.

Hydrochloric acid, N.

<u>Method.</u> Transfer 5 ml of the test solution to a 6" x 1" test tube and add 1-2 ml of N hydrochloric acid. Mix well and add 10.0 ml of the prepared final precipitation reagent. A distinct turbidity will be produced in the mixed solution if hydrolysable sulphur is present in the

Table 10 - Plastics Groups

I   Principal elements C,H

Aliphatic	Aromatic
Polyethylene	Polystyrene
Polypropylene	Polyindene
Polyisobutylene	Polyxyleneyls
Polybutadiene	Polymer petroleum fractions
Polyisoprene	
Natural rubber	
Butyl rubber	

II   Principal elements C,H,O

Saponification number = 0	Saponification number = < 200	Saponification number = > 200
Regenerated cellulose	Natural resin	Cellulose acetate
Polyvinyl alcohol	Modified phenoplasts	Cellulose butyrate
Phenoplasts		Cellulose acetate butyrate
Phenol-furfural resins		Polyvinyl acetate and its co-polymers
Xylene-formaldehyde resins		Polyvinyl propionate
Cellulose ether		Polyacrylate
		Polymethacrylate
Polyvinyl ether		Alkyd resins
Coumarone resins		Polyester
Polyglycols		Polycarboxylic acid anhydride
Polyvinyl acetals		Polycarbonic ester

Table 10 - Continued

| Polymer aldehydes |
| Polyketones |
| Epoxy resins |

III Principal element, halogen

Polymer halogen olefins	Rubber derivatives	Others
Polyvinyl chloride (PVC)	Chlorinated rubber	Clophen resins
PVC co-polymers	Rubber hydrochloride	Chloronaphthalenes
Polyvinylidene chloride	Chlorinated synthetic rubber	Chlorinated paraffins
Poly-2-chlorobutadiene		
Polychlorostyrene		
Polytetrafluoroethylene		
Polytrifluorochloro-ethylene		
Polyvinyl fluoride		

IV Principal elements N or N and O

Polyacrylic and poly-vinyl compounds	Basic components of formaldehyde aminoplasts	Others
Polyacrylonitrile	Urea	Nitrocellulose
Polyacrylamide	Ethylene urea	Polyamides
Polymethacrylamide	Propylene urea	Polyurethanes
Polyvinylidene cyanide	Dicyandiamide	Polyureas
Polyvinyl pyridine	Melamine	Phenoplasts and epoxy resins cured with amines
Polyvinyl pyrrolidone	Acetylene diurea	
	Glyoxylureides	
	Anilines	

V Principal elements, S in addition to O

Products of synthesis	Modified natural products
Polyethylene polysulphide	Vulcanized rubber
Polydiethylether-polysulphides	Sulphurized stand oils
Polythioether	

VI Principal element, silicon

Silicon oils and rubbers

Silica acid esters

VII Principal elements, nitrogen and sulphur

Thiourea formaldehyde resins

Sulphonamide resins

Polyacrylamides, polyacrylimides

VIII Principal elements, halogen and sulphur

Sulphochlorinated polyethylene and its vulcanizates

Polychlorobutadiene, vulcanized with compounds containing sulphur

IX Principal elements, nitrogen, sulphur and phosphorus

Casein condensates

X Principal elements, phosphorous, nitrogen and halogen

Polyphosphoric nitrile chloride

---

sample. A blank test under the same conditions will be perfectly clear. If a semi-quantitative estimation of the sulphur content is required, set the solutions aside for 30 minutes, then mix and measure the optical density of the test solution against the blank in 2 cm cells at 700 nm.

Qualitative Determination of Hydrolysable Phosphorus

Reagents. Ammonium molybdate solution - dissolve 10 g of Analar ammonium molybdate $(NH_4MO_7O_{24}4H_2O)$ in about 70 ml of water and dilute to 100 ml. Add this solution with stirring to a cooled mixture of 150 ml of sulphuric acid and 150 ml of water.

Ascorbic acid - B.D.H. laboratory reagent grade.

Method. Transfer 2 ml of the test solution to a 100 ml beaker. Add 40 ml of distilled water and 4 ml of ammonium molybdate solution. Mix thoroughly, then add 0.1 g of ascorbic acid and boil the solution for 1 minute. Cool in running water for 10 minutes and dilute to 50 ml with distilled water. Treat the blank solution in a similar manner. When hydrolysable phosphorus is present in the sample, a blue colour will be developed in the test solution as compared with a pale yellow in the blank. If a semi quantitative estimation of the phosphorus is required, measure the optical density of the test solution against the blank solution at 820 nm in 2 cm cells.

Haslam et al[7] have adopted the oxygen flask combustion technique to the qualitative detection in polymers of nitrogen, chlorine, fluorine, phosphorus and sulphur in amounts down to 0.25%. Because of the importance of these tests they are reproduced in full below.

Qualitative Detection of Elements in Polymers - Oxygen Flask Combustion[54]

Apparatus and reagents.

Combustion unit - the electrically fired combustion unit was used in this work, although there is no reason why other forms of oxygen flask should not be used, with appropriate calibration.

Filter paper - Whatman No. 541 filter paper. When not in use, store filter paper in a sealed container out of contact with the laboratory atmosphere.

Cotton wool - B.P.C. Super Quality.

Sodium hydroxide, M

Procedure. Weigh out approximately 20 mg of sample and transfer it to the centre of a small piece of the Whatman No. 541 filter paper weighing approximately 0.1 g. Fold the filter paper so that the sample is completely enclosed and before making the final fold, insert a small wick of cotton wool weighing about 6 mg.

Carry out the combustion using 5 ml of N sodium hydroxide in the bottom of the flask as absorption solution. Set the flask aside for 15 minutes to allow the gases formed in the combustion to be absorbed and then wash the contents of the flask quantitatively into a 25 ml measuring cylinder with distilled water. Dilute the solution to 25 ml and mix well. This constitutes the test solution, aliquots of which are taken for the detection of the individual elements by the colorimetric methods detailed below. For comparison purposes in the tests, prepare a blank test solution by carrying out the combustion procedure on the filter paper and cotton wool only.

Nitrogen

Reagents. Resorcinol - AnalR

Acetic acid, glacial

Ammonium ferrous sulphate - AnalR

Procedure. Weigh 0.1 g of resorcinol into a clean dry 50 ml beaker and dissolve in 0.5 ml of glacial acetic acid. Add 5 ml of the test solution and, after mixing, add 0.1 g of ammonium ferrous sulphate. Carry out the same test on the blank test solution. The development of a green colour in the sample test solution, compared with a pale yellow in the blank, indicates the presence of nitrogen in the sample.

If a semi-quantitative estimation of the nitrogen content of the sample is required, set both sample and blank solutions aside for 20 minutes. Add 10 ml of distilled water to each, mix and measure the optical density of the sample solution against the blank at 690 nm in a 4 cm cell. The blank in this test is low, giving an optical density of 0.07 measured against water at 690 nm in the 4 cm cells.

<u>Fluorine</u>

<u>Reagents.</u> Buffered alizarin and complexan solution - Weigh 40 mg of 3-amino-ethylalizarin-NN-diacetic acid (Hopkins and Williams Ltd.) into a beaker, and add 2 drops of N sodium hydroxide and approximately 20 ml of distilled water. Warm the solution to dissolve the reagent, cool and dilute to 208 ml. Weigh into another beaker 4.4 g of sodium acetate ($CH_3COONa3H_2O$) and dissolve in water. Add 4.2 ml of glacial acetic acid and dilute to 42 ml. Pour this sodium acetate solution into the alizarin complexan solution and mix to give the final buffered alizarin complexan solution.

Cerous nitrate, 0.0005 M - dissolve 54.3 mg of cerous nitrate ($Ce(NO_3)_3 6H_2O$) in water, and dilute to 250 ml.

<u>Method.</u> Transfer 20 ml of distilled water and 2.4 ml of buffered alizarin complexan solution to a 50 ml beaker. Add 1 ml of test solution and mix swirling the solution. Finally, add 2 ml of cerous nitrate solution and mix again. Treat the blank solution in a similar manner. When fluorine is present in the sample mauve colour will be developed in the test solution compared with the pink coloured blank solution. If a semi-quantitative estimation of fluorine is required, set the solutions aside for 10 minutes and measure the optical density of the test solution against the blank solution at 600 nm in 1 cm cells. Sulphur, chlorine, phosphorus and nitrogen do not interfere in this procedure.

<u>Chlorine</u>

<u>Reagents.</u> Ammonium ferric sulphate solution - dissolve 12 g of AnalaR ammonium ferric sulphate in water and add 40 ml of AnalaR nitric acid. Dilute to 100 ml and filter.

Mercuric thiocyanate solution - dissolve 0.4 g of recrystalized mercuric thiocyanate in 100 ml of absolute ethanol.

<u>Method,</u> Transfer 5 ml of the test solution to a 50 ml beaker and add 1 ml of ammonium ferric sulphate solution. Mix the solution and add 1.5 ml of mercuric thiocyanate solution. Mix and dilute to 10 ml. Treat the blank solution in a similar manner. When chlorine is present in the sample an orange colour will be developed in the test solution compared with yellow coloured blank solution. If a semi-quantitative estimation of chlorine is required, set the solutions aside for 10 minutes and measure the optical density of the test solution against the blank solution at 460 nm in 2 cm cells.

<u>Bromine</u>

<u>Reagents.</u> Fluorescein solution - dissolve 0.1 g of fluorescein in 25 ml of 0.1 N sodium hydroxide and dilute to 100 ml with water.

Sodium acetate buffer solution - mix 100 ml of N sodium acetate with 15 ml of N acetic acid.

Chloramine-T solution - dissolve 12 g of chloramine-T in 100 ml of distilled water.

Sodium thiosulphate solution - prepare a 0.5% w/v solution of sodium thiosulphate in 5% w/v sodium hydroxide solution.

Hydrochloric acid, N.

Method. Transfer 5 ml of the test solution to a 50 ml beaker. Add 1 ml of N hydrochloric acid and then 0.5 ml of sodium acetate buffer solution and 1 drop of fluorescein solution. Mix thoroughly and then add 1 drop of chloramine T solution. Mix by swirling and set aside for 30 seconds, then stop the reaction by adding 2 drops of alkaline thiosulphate reducing agent. Treat the blank solution in a similar manner. When bromine is present in the sample, a rose-pink colour will be developed in the test solution compared with the yellow-green blank solution.

Note: This test is also given by iodine.

Iodine

Reagents. Starch solution - dissolve 0.2 g of soluble starch in 100 ml of distilled water.

Sulphuric acid, dilute - prepare an approximately 10% w/v solution of concentrated sulphuric acid in distilled water.

Method. Transfer 5 ml of test solution to a 50 ml beaker and add a few drops of starch solution. Mix the solution and then acidify with dilute sulphuric acid. Treat the blank in a similar manner. When iodine is present in the sample, the characteristic blue colour of starch iodide will be developed in the test solution compared with the colourless blank solution.

4.3 QUANTITATIVE ELEMENTAL ANALYSIS

Some methods for the quantitative determination of elements in polymers are listed in Table 11. Individual methods are described in detail below.

4.3.1 Determination of Chlorine, Bromine and Iodine

Determination of Chlorine, Bromine and Iodine in Chlorobutyl and other Chlorine Containing Polymers. Oxygen Flask Combustion - Turbidimetry[55]

Apparatus. The conventional 1 and 2 litre Schoniger type oxygen combustion apparatus. The absorbance measurements were obtained using a Beckman Model DB-G grating spectrophotometer with a 1 cm cell and a tungsten lamp as the energy source, or equivalent.

Reagents. Standard aqueous potassium chloride solution in demineralized water to contain 1000 ppm chloride. Dilutions made from this standard stock solution to those containing very low chloride levels.

Silver nitrate, 0.01M, aqueous.

Nitric acid, 0.01M - potassium nitrate 0.01M, aqueous.

Procedure. The preparation of the calibration curve involves the preparation of six standard chloride solutions to contain 0 to 4 ppm chloride. Acidified 0.01M silver nitrate is added to each solution to form a silver chloride suspension. All the solutions are swirled briefly and allowed to stand at least 35 minutes with occasional shaking. The absorbance of each of these solutions is then measured at 420 nm in a 1 cm cell. A straight line curve passing through the origin is obtained covering chloride concentrations from 0 to 4 ppm. The ideal absorbance is approximately in the range of 0.03 to 0.2 absorbance unit.

During the preparation of standards swirling each solution for a few seconds is sufficient. Prolonged agitation tends to agglomerate the silver chloride particles.

Generally, about 100 mg of a weighed sample are combusted in a 2 litre Schoniger flask containing 10 ml of 0.01 M nitric acid. After the combustion, the flask is allowed to stand at least 15 minutes with occasional shaking before 5 ml of 0.01 M silver nitrate solution are added.

If the polymeric sample is difficult to combust, the normal platinum sample basket may be wrapped with 52 mesh platinum gauze to prevent hot and partially burned sample from dripping out of the sample basket. For subsequent combustions of this sample, a smaller sample size and the use of twice the normal volume of nitric acid absorbent are recommended.

The measurement of the absorbance of the silver chloride suspensions should be done within 35-60 minutes of turbidity development of both the sample and standard chloride solutions. The absorbances of the standard solutions are measured first, followed by the sample solutions, then the standard solutions are measured again.

The calculation of results is made in two different ways. For samples of low halogen level and if the combustion results in a clear solution, no removal of a solution aliquot is necessary and the calculation is as follows:

$$\% \; Cl = \frac{ppm \; chloride \times 15}{sample \; weight \; (mg) \times 1000} \times 100$$

The value for ppm chloride is obtained from the calibration curve.

If combustion of the sample is such that a clear solution is not obtained and an organic film results in the combustion flask, an 8 ml aliquot is usually removed and only 4 ml silver nitrate is added to the clear aliquot. Here again, the ratio of sample solution aliquot volume to that of silver nitrate is still 1:1 and the calculation for chloride is the same as above.

Chlorine, bromine and iodine. An alternate oxygen flask method is given below for the determination of between 2 and 80% of total chlorine, bromine and iodine in polymers based on titration with standard silver nitrate.

Determination of up to 80% Chlorine, Bromine and Iodine in Polymers. Oxygen Flask Combustion - Titration Procedure

Scope. The method describes a procedure for the determination of halogens, either singly or in a mixture, in concentrations of 2 - 80%.

Method summary. The sample is burnt in oxygen in a sealed Schoniger flask and the resultant gases are absorbed in a suitable absorption

Table 11 - Methods of determination of traces of various elements in polymers

Element	Procedure	Reference	Interferences
Sulphur up to 30 mg	Combusted in oxygen-filled flask over dilute hydrogen peroxide solution	56,57	Chloride, fluoride, phosphate, nitrogen, boron and metals all interfere
	Potentiometric titration of sulphuric acid with N/100 sodium hydroxide or photometric titration of sulphate with N/100 barium perchlorate		Up to 2 mg chlorine, fluorine, nitrogen, boron and metals do not interfere in the determination of 1 mg sulphur. Phosphorus (up to 2 mg) interferes in the determination of sulphur (1 mg) but this interference can be overcome using the procedure of Colson (92-94)
Chlorine or bromine	Combustion as above		
	Chloride titrated potentiometrically with N/100 silver nitrate in presence of nitric acid in acetone		
Chlorine or bromine or iodine	Combustion as above	58	Up to 8 mg phosphorus, fluorine, sulphur, do not interfere in determination of 2 mg chlorine
	Halide titrated with mercuric nitrate		
Phosphorus 0.1 g polymer	Digested with sulphuric acid/perchloric acid	59	No interference by sulphur, chlorine, fluorine, nitrogen
	Digested diluted and ammonium vanadate/molybdate added. Yellow phosphovand-molybdate complex evaluated at 430 nm		

Nitrogen	Digestion carried out on 0.5-1.0 g polymer a) Kjeldahl digest made alkaline and distilled into 4% boric acid. Ammonia estimated by acid titration b) Spectrophotometric estimation at 630 nm of phenol indophenol derivative	61	No interference by chlorine and nitrogen
Fluorine	30 mg polymer combusted in an oxygen filled flask over distilled water Reacted with buffered alizarin complex/cerous nitrate, blue colour produced evaluated at 610 nm	60	No interference by large excesses of sulphur, chlorine, phosphorus and nitrogen
Silicon	30 mg polymer in gelatin capsule and combusted with sodium peroxide, sucrose and benzoic acid in 22 ml capacity Parr Bomb		No interference by sulphur, halogens, phosphorus, nitrogen and boron
Boron	0.1 g polymer digested with concentrated nitric acid in a sealed ampoule to convert organo boron compounds to boric acid		
	Digest dissolved in methyl alcohol and boron estimated flame-photometrically at 5.9.5nm	61	No interference by chlorine and nitrogen

solution. This solution is acidified and titrated potentiometrically with standard silver nitrate solution.

<u>Apparatus.</u> Combustion flask - Pyrex, 500 ml capacity conical flask with B24 conical ground stopper.

Stopper B24 - with a fixed in platinum wire (30 mm long, 0.8 mm diameter) carrying a 15 x 20 mm piece of 40 mesh platinum gauze or carrying a 5/8" long x 1/4 " diameter bucket fabricated in 40 mesh platinum gauze.

Safety jacket - for combustion flask to serve as a protection during the combustion. Conical shaped, detachable metal wire gauze fitting around the conical flask.

Automatic titrimeter - with silver/glass electrode system. Wick lighter, small burner fed with sulphuric and halogen free fuel.

Cellulose capsules - halogen free.

<u>Reagents.</u> Silver nitrate solution standard, 0.1 N.

Sodium hydroxide solution, 0.01 N approximately, dissolve 0.4 g of sodium hydroxide in distilled water and dilute to 1 litre.

Sodium metabisulphite solution, 0.05% w/v, (Micranalytical Reagent Grade, available from British Drug Houses Ltd., Poole, Dorset) dissolve 0.05 g of sodium metabisulphite in distilled water and dilute to 100 ml.

Barium nitrate, Analar, solid.

Oxygen, cylinder.

Nitric acid, Analar, concentrated.

Methyl alcohol, Analar.

Acetone, pure grade, redistilled.

Water, deionized halogen content less than 0.05 ppm

Filter paper, any grade with a halogen content less than 50 ppm.

<u>Procedure.</u> The method requires different absorption solutions for the various halogens as follows:

Chlorine	10 ml of 0.01 N sodium hydroxide solution.
Bromine and or/iodine	10 ml 0.05% w/v sodium metabisulphite solution.
Chlorine and bromine and/or iodine	5 ml 0.01 N sodium hydroxide solution and 5 ml 0.05% w/v sodium metabisulphite solution.

Pipette the required absorption solution into the flask and fill with oxygen, stopper securely. Accurately weigh out 10-30 mg of sample into a cellulose capsule and place the capsule in the Schoniger basket with a strip of filter paper, to act as a fuse. Light the fuse and quickly insert the basket into the oxygen filled flask. During the combustion hold the basket in the flask firmly and at the same time lift the flask off the bench. Allow the mist formed in the flask to subside with periodic shaking over 15 to 30 minutes. Transfer the solution to a 250 ml beaker using 50 ml of distilled water and 120 mg of methanol. (If the halogen content of the sample is less than 5% then 120 ml acetone should be used instead of methanol).

If any carbonaceous matter is evident in the solution then the determination should be abandoned and a further combustion carried out using a smaller sample weight.

Carry out a reagent blank combustion including the paper and all reagents, but omitting the sample.

Add 1 drop of concentrated nitric acid and, if a mixture of halogens is present, a few crystals of barium nitrate. Titrate the solution potentiometrically using 0.1 N silver nitrate solution.

<u>Calculations.</u> The halogens are titrated in the following order: Chlorine, bromine, iodine.

Calculate the halogen content as follows:

$$\text{Chlorine \% w/w} = \frac{(Ts - Tb) \times N \times 35.46 \times 100}{W \times 1000}$$

$$\text{Bromine \% w/w} = \frac{(Ts - Tb) \times N \times 79.92 \times 100}{W \times 1000}$$

$$\text{Iodine \% w/w} = \frac{(Ts - Tb) \times N \times 126.91 \times 100}{W \times 1000}$$

Where:

Ts = sodium nitrate titration (mls) obtained in sample combustion.
Tb = silver nitrate titration (mls) obtained in blank combustion.
N = normality of silver nitrate.
W = weight (g) of sample taken for analysis.

<u>Determination of Chlorine in Polymers containing Chloride and Sulphur and/or Phosphorus and/or Fluorine.</u> <u>Oxygen Flask Combustion - Mercurimetric Titration</u>

<u>Scope.</u> This method for the determination of chloride in polymers involves oxygen flask combustion over water, addition of ethanol and titration to the diphenylcarbazide indicator end point with 0.005 mercuric nitrate.

$$M\ Cl' + Hg\ (NO_3)_2 = HgCl_2 + 2HNO_3$$

This method is without interference from phosphorus, sulphur or fluorine.

<u>Apparatus.</u> Oxygen flask - 500 ml capacity.

A conical flask provided with B24 ground glass stopper, sealed at the lower end to a glass tube 5 mm in external diameter. The lower end of this tube is sealed and flattened to form a slight flange. The length from the lower edge of the ground portion of the stopper to the flanged end of the tube is 70 mm. The sample holder is constructed from 80 mesh platinum gauze in the form of a cylinder 6 mm in diameter and about 10 mm long. It is closed at one end and fastened to the flanged tube by a length of 0.5 mm diameter wire, one end of which is bent over the edge of the open end of the cylinder and pinched so that the gauze is firmly gripped. The arrangement has significant advantages over the usual type of sample holder. The platinum gauze and wire can be renewed with ease and the thermal capacity of the assembly is relatively small.

Magnetic stirrer.

Content pipette, 2.0 ml capacity - conforming to B.S. 1428.

Microburette, 10 ml capacity - conforming to B.S. 1428.

<u>Reagents.</u> Sodium hydroxide, 0.1 N.

Nitric acid, 0.1 N.

Bromophenol blue indicator solution, dissolve 50 mg of bromophenol blue in 50-0 ml of ethanol.

Diphenylcarbazone indicator solution, dissolve 20 mg of diphenylcarbazone in 20 ml of ethanol. Keep the solution in the dark and renew it after 2 weeks.

Sulphur dioxide solution, saturate 50 ml of water with sulphur dioxide. Renew it after 2 to 3 days.

Hydrogen peroxide, 100 volume, micro-analytical reagent grade.

Barium nitrate solution, a saturated aqueous solution of analytical reagent grade barium nitrate.

Thorium nitrate solution, dissolve 15.0 g of analytical reagent grade thorium nitrate tetrahydrate in water and dilute the solution to 1000 ml.

Ethanol, absolute.

Mercuric nitrate, approximately 0.005 M, dissolve 3.5 g of mercuric nitrate in 570 ml of 0.01 N nitric acid and set the solution aside for at least 2 days. Filter off any precipitate and dilute the filtrate to 2 litres.

Sodium chloride, 0.025 M.

<u>Standardisation of the mercuric nitrate solution.</u> Measure out 2.0 ml of 0.025 M sodium chloride solution by means of a content pipette. Dilute to 15.0 ml with water and add 0.05 ml of bromophenol blue indicator solution. Add 0.1 N nitric acid until the yellow colour of the indicator appears and then add a further 0.5 ml of acid. After the addition of 100 ml of ethanol and 0.5 ml of diphenylcarbazone indicator solution, titrate the solution with the mercuric nitrate solution to the first appearance of

a permanent violet colour. The solution should be stirred magnetically throughout the titration period. The use of a content pipette for measuring the sodium chloride solution is recommended, since it has been found that the reproducibility of repeated titrations is better than that obtained with, for example, a 5.0 ml delivery pipette, in conjunction with 0.01 M sodium chloride solution. The titration carried out as described must be corrected for the indicator blank value, obtained by the titration of 15.0 ml of water. This blank value is usually less than 0.05 ml of mercuric nitrate solution but much higher values have been obtained occasionally and these have been found to be due to the presence of chloride in the ethanol. In such instances, a suitable stock of ethanol may be mixed with a predetermined volume of the mercuric nitrate solution, sufficient to reduce the blank value to less than 0.05 ml.

<u>Procedure.</u> The procedures described below are restricted to the determination of chlorine in organic compounds containing phosphorus or fluorine, or both, in the presence or absence of sulphur. For compounds containing little if any hydrogen, a preferred alternative procedure is given. The determination of iodine or bromine in the presence of phosphorus or fluorine is not included since the determination of these halogens is in this instance best completed by other well known volumetric methods, such as Leipert's method for iodine and oxidation to bromate for bromine.

Compounds containing phosphorus or fluorine, or both: Weigh out an amount of sample corresponding to 1.0 to 2.0 mg of chlorine and decompose it in an oxygen flask containing 5.0 ml of water. Wash the stopper and sample holder with 10.0 ml of water, add 0.1 ml of bromophenol blue indicator solution and neutralise the solution with 0.1 N sodium hydroxide to the blue colour of the indicator. Add 0.75 ml of thorium nitrate solution, 0.5 ml of diphenylcarbazone indicator solution and 100 ml of ethanol and titrate the stirred solution with standard mercuric nitrate solution to the appearance of a permanent violet colour.

Compounds containing sulphur and either phosphorus or fluorine, or all three together: Proceed as before up to the washing of the stopper, etc. Boil the solution for 60 seconds, holding the flask directly over a small bunsen burner flame and swirling the solution continuously. Add 0.5 ml of barium nitrate solution, 0.1 ml of bromophenol blue indicator solution and sufficient 0.1 N sodium hydroxide to produce the blue colour of the indicator. After the addition of thorium nitrate, complete the determination as before.

Compounds containing phosphorus or fluorine, or both, but little if any hydrogen: Decompose the weighed sample in an oxygen flask containing 5.0 ml of water and 0.25 ml of saturated solution of sulphur dioxide. Wash the stopper etc. with 10.0 ml of water and complete the determination as described for the analysis of compounds containing sulphur. In this procedure the volume of barium nitrate should be increased if 0.5 ml is found to be insufficient for complete precipitation of the sulphate.

Blank determinations on the reagents, etc. should be carried out in conjunction with each of the procedures described above. These blank values should not exceed about 0.2 ml of 0.005 M mercuric nitrate.

The method has a precision of approximately 0.1 at the 22% chlorine level.

Mittenberger and Gross[63] determined chlorine in PVC by fusion with sodium peroxide and silver nitrate titration for the chloride produced.

Tanaka and Morzkawa[64] described a semi-micro technique for the determination of total chlorine in PVC using a semi-micro method based on the Schoniger flask and Fajans method.

4.3.2 Determination of Fluorine

Determination of Fluorine in Fluorinated Polymers - Oxygen Flask Combustion - Spectrophotometric Procedure

Summary. The method for the determination of fluorine is based on decomposition of the sample by oxygen flask combustion followed by spectrophotometric determination of the fluoride produced by a spectrophotometric procedure involving the reaction with the cerium III complex of alizarin complexan (1,2-dihydroxy-anthraquinone 3-ylmethylamine N, N-diacetic acid). The blue colour of the fluoride containing complex(absorption maximum, 565 nm) is completely distinguishable from either the yellow of the free dye (maximum absorption, 423 nm) or the red of its cerium III chelate (maximum absorption, 495 nm).

Apparatus. The combustion apparatus consists of silica or boron-free glass 500 ml Erlenmeyer flask constructed of suitable glass. Into the stopper is fused one end of a length of platinum wire, 1 mm in diameter, to the free end of which is attached a piece of 36 mesh platinum gauze, 1.5 cm x 2 cm to act a sample holder. The flask shall be essentially free from boron and aluminium. Low results are obtained using borosilicate glass flasks.

Optical densities were measured in 4 cm cells with a Unicam SP600 visual range spectrophotometer.

Reagents. Alizarin complexan, 0.0005 M, transfer 0.385 g of alizarin complexan to a 2 litre calibrated flask by means of 20 ml of recently prepared 0.5 N sodium hydroxide and set aside for 5 minutes with occasional swirling, to ensure complete solution. Dilute to about 1500 ml with water, add 0.2 g of hydrated sodium acetate and adjust the pH to about 5 by careful addition of 1 N hydrochloric acid. Dilute to the mark and filter into a brown glass bottle. This solution is stable for at least 4 months.

Cerous nitrate, 0.0005 M, standardise approximately 0.02 M cerous nitrate by titration against standard ethylenediaminetetra-acetic acid solution at pH 6 with xylenol orange as indicator. To a suitable volume of this solution (about 50 ml) add 0.2 ml of concentrated nitric acid, 0.1 g of hydroxylamine hydrochloride and sufficient water to produce 2 litres and filter.

Acetate buffer solution, pH 4.6, dissolve 150 g of hydrated sodium acetate in about 600 ml of water, add 75 ml of glacial acetic acid, dilute to 1 litre with water and filter.

Standard fluoride solution, 5 ug per ml, dissolve about 22 mg (accurately weighed) of dried analytical reagent grade sodium fluoride in water and adjust the volume to 2 litres. Store in a polythene container.

Preparation of calibration graph. In each of a series of 100 ml calibrated flasks place 50 ml of distilled water an accurately measured volume between 2 and 8 ml of standard fluoride solution, 10 ml of alizarin complexan solution and 3 ml of acetate buffer solution. Mix each solution thoroughly, add 10 ml of 0.0005 M cerous nitrate, dilute to the mark with distilled water and set aside, protected from direct light for 1 hour. At the same time, prepare a blank solution in a similar fashion by omitting the standard fluoride solution. Measure the optical densities of the test solutions against the blank in 4 cm cells at 610 nm and plot a graph of optical density against amount of fluoride present.

Procedure. Accurately weigh an appropriate amount of the sample (5 to 25 mg) on a strip of filter paper approximately 3 cm x 4 cm (Whatman No.1 grade is suitable) that has been folded into three along its length. Enclose the sample in the paper, insert a narrow strip of filter paper to act as a fuse and fix it in the platinum gauze sample holder. Place 20 ml of water in the combustion flask, fill the flask with oxygen, ignite the fuse and immediately insert the stopper. Carefully tilt the flask and when combustion is complete, set it aside the 10 minutes, with intermittent shaking. Quantitatively transfer the liquid to a 250 ml calibrated flask, dilute to the mark and treat an aliquot expected to contain about 25 mg of fluoride by the procedure described for colour development under Preparation of Calibration Graph. At the same time prepare a standard colour from 5 ml of standard fluoride solution to serve as a check on the calibration graph.

Liquid samples can be satisfactorily decomposed by burning in a small gelatin or, preferably, methylcellulose capsule containing approximately 30 mg of powdered cellulose.

For solutions derived from the combustion of sulphur containing compounds, boil gently for about 10 seconds with 1 ml of 100 volume hydrogen peroxide, neutralise to phenolphthalein with 1 N sodium hydroxide and then add 1 ml in excess; boil to destroy excess of peroxide, cool and adjust the pH to about 4 with 1 N hydrochloric acid.

Using this method Johnson and Leonard[65] obtained from polytetra-fluoroethylene a content of 75.8% fluorine using a silica or boron-free glass combustion flask against a theoretical value of 76%. Using a borosilicate glass combustion flask they obtained a low fluorine recovery of 72.1%.

4.3.3 Determination of Sulphur

Determination of Macro-amounts of Sulphur in Polymers . Sodium Peroxide Fusion - Titration Procedure[66]

Summary. The polymer is fused with sodium peroxide in a steel bomb. Sodium is removed from the fusion by means of a cation exchange resin and the sulphate is determined by titration with standardized 0.01 N barium perchlorate. Chlorine, fluorine and nitrogen in amounts up to 2 mg, in the sample, are without serious effect on the determination of sulphur. The effect of larger amounts of fluorine can be suppressed by the addition of boric acid.

Apparatus. Nickel "fluorine bomb" capacity 2 ml - obtainable from Charles W. Cook and Sons Ltd., 97 Walsall Road, Perry Barr, Birmingham.

Conical flask with ground glass stopper, capacity 150 or 200 ml.

Magnetic stirrer.

Microburette, capacity 10 ml - conforming to BS.1428.

<u>Reagents.</u> Sodium peroxide, MAR grade.

Ethanol, absolute.

Barium perchlorate, 0.01, N aqueous , adjust to about pH 3.0 by adding perchloric acid.

Thorin indicator solution, dissolve 25 mg of thorin in 5 ml of distilled water.

Methylene blue indicator solution, dissolve 15 mg of methylene blue in 50 ml of water.

Cation exchange resin, Amberlite IR-120 (H).

<u>Pre-treatment of the resin.</u> By the column method wash about 400 g of the analytical grade resin with about 700 ml of 3.0 N hydrochloric acid and then with 4 or 5 litres of water. Transfer the resin to a 1 litre flask, remove most of the water by decantation and shake the resin thoroughly for several minutes with three successive 100 ml portions of ethanol. Remove the residual ethanol by repeated washing with water and shake the resin vigorously with 400 ml of 0.5 N sodium hydroxide for about 10 minutes. Decant the slightly turbid solution, wash the resin with water until all suspended fine particles have been removed, and filter on a Buchner funnel. Press the resin between filter papers and store it in a glass stoppered bottle sealed with adhesive tape.

<u>Procedure.</u> Fusion of the sample with sodium peroxide: Place in the dry bomb 0.5 g of powdered sodium peroxide, a suitable weighed amount of the sample (5 to 15 mg) and a further 0.5 g of sodium peroxide. Close the bomb and mix the contents by rotation. Heat the bomb in a muffle furnace for 3 minutes at 650°C. Cool the bomb, remove the lid and extract the fusion product by placing the bomb in a small beaker containing 10 to 15 ml of water and warming until effervescence ceases. Remove the bomb, rinse it with water and then rinse the lid. Transfer the combined solutions quantitatively to a 50 ml calibrated flask, and dilute to the mark.

Removal of sodium from the fusion product extract and titration of the sulphate ion: Place about 30 g of the cation exchange resin in a conical flask. Add about 25 ml of ethanol, shake the stoppered flask for about 1 minute, remove the ethanol by decantation and repeat the operation with a further 20 ml of ethanol. Transfer 25 ml of the fusion product extract to the resin, shake for about 5 minutes and decant the solution into a suitable conical titration flask containing a magnetic stirrer bar. Rinse the resin with four successive 25 ml portions of ethanol, add all washings to the contents of the titration flask. Add 0.1 ml each of thorin and methylene blue indicator solutions and titrate with 0.01 N barium perchlorate to a pink end-point colour persisting for about 20 seconds. Carry out a blank determination in the same manner, omitting only the sample. The blank value should lie between 0.1 and 0.2 ml.

Precautions:

1. When the sodium peroxide fusion is carried out as described, only a small amount of insoluble matter should be found in the aqueous extract of the fusion product, but his material can interfere with the detection of the end-point of the titration. For the best results the extract should be set aside overnight, so that a clear portion can be withdrawn for titration.

2.   End-points in the barium perchlorate titration may be unsatisfactory
if the ethanol washed resin is set aside for more than about 2 hours
before use.   If such a delay is unavoidable, the resin should be washed
once more with about 20 ml of ethanol immediately before use.

3.   With some batches of resin it has been observed that, after pre-
treatment in the prescribed manner, unsatisfactory end-points are obtained
in the titration of the resin treated sample solution.   If this defect is
found with any portion of a given batch of resin, it is recommended that
all subsequent portions should be vigorously shaken with about 25 ml of
0.5 N sodium hydroxide for about 10 minutes and then washed with water and
ethanol in succession before use.

Determination of Sulphur in Polymers - Oxygen Flask Combustion

Scope.   This method describes a rapid procedure for the
determination of sulphur in polymeric materials.   The precision of the
method is such that a three-decimal place repeatability is obtained for
the sulphur range below 0.5%.   Chlorine and nitrogen concentrations in the
sample may exceed the sulphur concentrations several times without causing
interference.   Fluorine does not interfere unless present in concentrations exceeding 30 per cent of the sulphur content.   Phosphorus and
metallic constituents interfere even when present in moderate amounts.

Method summary.   The sample, contained in a piece of filter paper
and suitably suspended, is rapidly and completely burnt in a closed
conical flask filled with oxygen at atmospheric pressure.   The products of
combustion are allowed to enter hydrogen peroxide solution.   The amount
of sulphuric acid formed is determined by titration with barium
perchlorate, using thorin as the indicator.   The equivalence point is
obtained photoelectrically by comparing the optical density of the
solution with that of a standard reference solution.   Depending on the
sulphur content of the sample, two different strengths of titrant are
applied.

Apparatus.

a)   Conical flask - Pyrex or Jala glass, 500 ml capacity, provided with
conical ground joint BS-B 24.

b)   Stopper, BS-B 24, with fused in platinum wire (30 mm long, 0.8 mm
diameter) carrying a 15 x 20 mm piece of 40 mesh platinum gauze.

c)   Safety jacket, to serve as a protection during the combustion.
Detachable metal wire gauze jacket fitting around the conical flask.

d)   Syringe - 0.05 ml capacity, with 0.001 ml divisions.

e)   Lighter - any suitable small flame fed with sulphur-free fuel, e.g.,
alcohol.

f)   Micro burette assembly, with 10 ml burette and Pyrex supply bottle,
two required.

g)   Ultra micro burette, with 0.001 ml divisions and provided with a
glass capillary delivery tube.

h)   Photoelectric colorimeter according to Lange, model J, equipped with
two 100 ml optical cells and two 520 nm interference filters in rim, or
equivalent.

i)   Electric stirrer, of suitable dimensions to fit on and close the colorimeter compartment, provided with small glass propeller-shaped stirrer.

<u>Reagents and materials.</u>

a)   Barium perchlorate solution, standard 0.01 N, prepare as follows: Dissolve 1.7 g of barium perchlorate, cp, anhydrous, in 200 ml of deionized water and fill up to one litre with isopropanol. Adjust to pH 3.5 by adding a 10% perchloric acid solution. Standardize in an optical cell against sulphuric acid using 10 ml of the standard solution described in para j) of this section. Dilute with 10 $\pm$ 0.5 ml of deionised water and add 80 $\pm$ 2 ml of isopropanol. Add 1 drop (approx 0.05 ml) of perchloric acid solution 3% and 200 $\pm$ 1 ul (1 ul = 0.001 ml) of thorin indicator solution (measured from the ultra micro burette) and proceed according to section h). Run a blank titration in the same way using exactly the same amounts of reagents but omitting the sulphuric acid and applying 20 $\pm$ 0.5 ml of deionized water instead of 10 ml. Obtain the net normality of the barium perchlorate solution.

b)   Barium perchlorate solution, standard 0.001 N, prepare as follows: Dissolve 0.17 g of barium perchlorate, cp, anhydrous, in 200 ml of deionized water and fill up to one litre with isopropanol. Adjust to pH 3.5 by adding a 10% perchloric acid solution. Standardize in an optical cell against sulphuric acid using 1 ml of the standard solution described in para j) of this section. Dilute with 19 $\pm$ 0.5 ml of deionised water and add 80 $\pm$ 2 ml of isopropanol. Add 1 drop (approx 0.05 ml) of perchloric acid solution 3% and 200 $\pm$ 1 ul (1 ul = 0.001 ml) of thorin indicator solution (measured from the ultra micro burette) and proceed according to section h). Run a blank titration in the same way using exactly the same amounts of reagents but omitting the sulphuric acid and applying 20 $\pm$ 0.5 ml of deionized water instead of 19 ml. Obtain the net normality of the barium perchlorate solution.

Note, caution - the dry barium perchlorate should not be heated in the presence of organic substances or be contacted with acids, because this may give rise to explosion.

c)   Filter paper, ash-free. Schleicher and Schull No. 5892 is recommended. Store in a closed bottle.

d)   Hydrogen peroxide, 30%, AR.

e)   Isopropanol, cp, denatured.

f)   Oxygen, compressed, free of sulphur compounds.

g)   Perchloric acid, 10%, AR.

h)   Perchloric acid, 3%, AR.

j)   Sulphuric acid solution, standard, 0.01 N, prepare by accurate dilution of standardized 0.1 N sulphuric acid solution.

k)   Sulphuric acid solution, AR, 0.0005 N, prepare by accurate dilution of 0.01 N sulphuric acid solution.

l)   Thorin indicator solution, 0.2% aqueous solution of thorin (sodium salt of 2(2-hydroxy-3, 6-disulfo-1-napthylazo) benzene-arsonic acid). The thorin quality marketed by Merck, Darmstadt, West Germany, is recommended.

m)   Water, deionized.

<u>Solids.</u>   Weigh a filter strip to the nearest 0.1 mg.   Place approximately 30 mg of sample onto the middle of the filter strip and weigh again.   Wrap the sample up in the following way: First cover the sample with the raised edges of the strip and then roll up the body of the strip towards the end, which will serve as a fuse.   Now, clamp the packet in the platinum gauze of the stopper, keeping the fuse free and in-line with the platinum suspension wire.

<u>Procedure.</u>

a)   Introduce 4 $\pm$ 0.5 ml of deionized water and 3 drops of hydrogen peroxide solution 30% into the conical flask.

Install the safety jacket.

b)   Replace the air in the conical flask by oxygen by introducing a rapid stream of oxygen at a point near the bottom for 30 seconds. (Place a G 3 porosity glass filter in the outlet end of the oxygen line to prevent contamination of the absorption liquid).

c)   Place the flame of the lighter close to the top of the conical flask, ignite the end of the fuse of the sample packet and immediately insert the stopper.

d)   Keep the flask firmly closed and keep it upside down to prevent the flame from touching the walls or bottom of the flask.   When the combustion slows down bring the flask in an inclined position to promote complete combustion also of any dropping particles.   Remove the safety jacket, shake the bottle for one minute and allow it to stand for 15 minutes.

e)   In an optical cell prepare a colour reference solution by mixing 100 ml of deionized water and 200 $\pm$ 1 ul of thorin indicator solution (measured from the ultra-micro burette).   Measure the optical density using the deflection method.   This value should be 0.17 $\pm$ 0.01, adjust if necessary by adding a known amount of either 0.0005 N sodium hydrogen carbonate solution or 0.0005 N sulphuric acid solution.

f)   Prepare a double amount of the colour reference solution described in para. e) in a beaker.   Introduce half of it into each optical cell and check the optical density of the solution.   With both cells in the compartments and the photometer adjusted for highest sensitivity bring the pointer at position 50 of the linear scale marked "abs".   Check whether the cells are matched by interchanging them.   During the actual determination keep the right-hand cell in its place and titrate in the left-hand cell.

Note: a)   Prepare and measure a fresh colour reference solution at least once a day and whenever a new batch of deionized water or thorin indication solution is used.   b)   Switch on the photometer at least two hours before use.

g)   Introduce 1 ml of 0.0005 N sulphuric acid solution by means of a pipette and 15 $\pm$ 0.5 ml of deionized water into the optical cell.   Wet the rim between flask and stopper with some isopropanol and open the flask which the liquid is drawn in.   Use 80 $\pm$ 2 ml of isopropanol to wash the stopper and the platinum gauze and to transfer the solution under test quantitatively to the cell.   Add 1 drop (approx 0.05 ml) of perchloric acid solution 3% and 200 $\pm$ 1 ul of thorin indicator solution.

h)   Place the optical cell in the photometer and install the electric stirrer.   Immerse the burette tip into the cell solution and stir. Titrate at a slow rate with barium perchlorate solution, till the pointer reaches 50 again.   Use the 0.01 N barium perchlorate solution for sulphur contents exceeding 0.15% and the 0.001 N barium perchlorate solution for the lower sulphur concentrations.

Note: Especially in the lower sulphur range, a minimum titration time of two minutes is recommended.

<u>Blank.</u> Run a blank determination including all reagents used (also filter paper) but omitting the sample.

<u>Calculation.</u>   Calculate the sulphur content by means of the following equation:

$$\text{Sulphur, \% w} = \frac{(V - B) \times N \times 1603}{W}$$

where:

V =   volume of barium perchlorate solution consumed in the actual determination, millilitres.

B =   volume of barium perchlorate solution consumed in the blank determination, millilitres.

N =   normality of the barium perchlorate solution.

W =   weight of sample milligrams

<u>Precision.</u>   The following data should be used for judging the acceptability of results (95% probability).   Duplicate results by the same operator should not differ by more than the following amounts:

<u>Sulphur content</u>	<u>Repeatability</u>
Above 0.5%	0.05
Below 0.5%	2% of amount present but not better than 0.005

4.3.4 Determination of Nitrogen

Determination of 1 to 90% Organic Nitrogen in Polymers, Kjeldahl Digestion - Boric Acid Titration Method

<u>Scope.</u>   A method is described for the determination of nitrogen in polymers which contain between 1% and 90% of this element.   The accuracy of results obtained is of the order of $\pm$ 1.0% of the true nitrogen content.

The Kjeldahl digestion procedures described quantitatively decomposed amines, amino compounds, amino acids, amides, nitriles and their simple derivatives and also many refractory nitrogen compounds. Quantitative decomposition of nitrogen containing polymers, e.g. styrene-acrylonitrile copolymers and polyacrylonitrile, is achieved.

Modified digestion procedures are described for the conversion, to inorganic nitrogen, of nitro, nitroso and azo compounds, also hydrazones and oximes. The method is not reliable, however, for many diazo compounds or volatile non-basic compounds, e.g., nitroparaffins.

Summary. The procedure involving Kjeldahl digestion followed by collection of generated ammonia in boric acid.

A suitable weight of sample is digested with concentrated sulphuric acid, sodium or potassium sulphate and digestion catalysts using an electrical heater (note 1). A reagent blank determination is also carried out. The digested sample, after dilution, is transferred to a steam distillation apparatus. The addition of sodium hydroxide liberates ammonia equivalent to the nitrogen content of the sample. Liberated ammonia is collected in boric acid solution and determined by titration with standard hydrochloric acid.

Apparatus.

a)   Kjeldahl digestion apparatus consisting:

Kjeldahl digestion flasks - 100 ml "Pyrex" glass.

Electric heating "Bun-ray" electric bunsens suitable (minimum loading 375 watts) or Kjeldahl digestion flask heaters with flask supports (100 ml flask size).

b)   Ammonia recovery apparatus consisting of (see fig 4):

Condenser (B24 cone and socket), Leibig 12". Round bottomed steam generation flask, 1000 ml central B24 and 2 x B19 side. Three way stopcock, vertical arm connected to B24 cone. Separatory funnel 250 ml with B19 cone.

Recovery apparatus proper:

Consisting of steam stripping vessel (interior vessel large enough to hold 150 ml solution). Separatory funnel 50 ml with B14 cone and short delivery stem. Steam trap leading to vertical condenser and guard tube containing activated silica gel which is connected by a suitable adaptor (note 2.). Activated silica gel. Self indicating, mesh 6-20, available from British Drug Houses Ltd., Poole. To regenerate gel, heat for 6 hours at 150°C in a vacuum oven.

c)   Apparatus for determination of ammonia in distillate, miscellaneous glassware:

Burette, 50 ml pipettes.

Reagents.

a)   Sulphuric acid, preferably nitrogen-free quality supplied by British Drug Houses Ltd., Poole.

b)   Glucose, micro analytical reagent (MAR) grade, (British Drug Houses Ltd., Poole.

c)   Kjeldahl digestion catalysts, the following supplied by British Drug Houses Ltd., Poole.

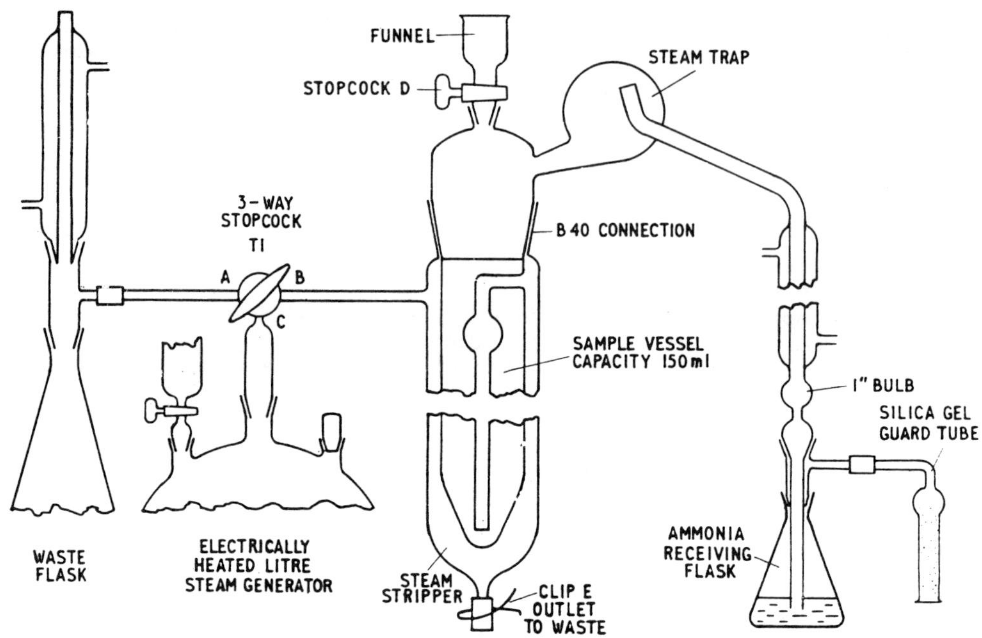

Figure 4   Ammonia recovery steam stripper assembly

Tablets each consisting of 1 g sodium sulphate and the equivalent of 0.05 g selenium (as selenium dioxide).

Tablets each consisting of 1 g sodium sulphate and equivalent of 0.05 g of mercury (as mercuric oxide).

d)   Cupric sulphate, 25% w, dissolve 25 g $CuSo_4 4H_2O$ Analar in water and make up to 100 ml (0.2 ml of this solution contains 0.05 g $CuSo_4 5H_2O$.

e)   Mercuric oxide, pure grade.

f)   Potassium sulphate, Analar.

g)   Sodium hydroxide, (use when mercury - free catalysts are used for Kjeldahl digestion). Pour 1600 ml distilled water into a 4 litre beaker and add 1500 g of sodium hydroxide Analar. When dissolved allow to cool and transfer to a winchester.

Sodium hydroxide - sodium thiosulphate (use when mercury containing catalysts are used for Kjeldahl digestion). Pour 1400 ml distilled water into a 4 litre beaker and 1500 g sodium hydroxide Analar. When dissolved allow to cool. In a further beaker dissolve 150 g sodium thisulphate $5H_2O$ Analar in 200 ml distilled water. Mix the two solutions and transfer to a winchester. Remove nitrogenous impurities from either reagent as follows: Add 5 g zinc powder, and pass a slow stream of nitrogen for 48 hours to sweep our ammonia. Filter the cold solution through a 40 cm Whatman No. 1 filter paper to remove sodium carbonate. Store the reagent in polyethylene bottles. Carry out these operations in a fume cupboard to avoid contamination of the solutions by any ammonia impurity present in the laboratory atmosphere.

h)   Boric acid, 4% w, dissolve 40 g boric acid (H_3BO_3) Analar in 1000 ml distilled water contained in a 2 litre round bottomed flask. Boil for 20 minutes to expel carbon dioxide. Connect a carbon dioxide absorbing guard tube and cool. Transfer to a 1 litre volumetric flask. Make up to 1 litre with cold boiled out distilled water and mix.

i)   Methyl red, bromo-cresol green indicator: Mix 5 volumes of 0.2% bromo-cresol green with 1 volume of 0.2% methyl red, both in 95% ethanol.

j)   Hydrochloric acid, 0.05 N

k)   Hydrochloric acid, 0.02 N, prepared daily by dilution of the 0.05 N hydrochloric acid stock.

Sampling.

a)   The weights of sample required for the determination of nitrogen in compounds containing between 1% and 90% of nitrogen are shown below. Also shown is the normality of the hydrochloric acid required for the titration of ammonia at the end of the determination.

Recommended sample size and normality of hydrochloric acid
used in final ammonia titration

Weighing with 4-place balance

Nitrogen content of sample %	Weight of sample for digestion (g)	Strength of hydrochloric acid to be used in titration of ammonia, normality
1-3	0.2	0.02
5	0.2 - 0.1	0.02
10	0.1	0.02
20	0.1 - 0.05	0.02
40	0.03 or 0.05	0.02 or 0.05
60	0.05	0.05
80-90	0.03	0.05

b)   Methods of sample transfer into Kjeldahl digestion - Solids: These may be transferred directly to the Kjeldahl flask or wrapped in a cigarette paper, (use potassium nitrate free brand of paper) which is dropped into the flask. Include a cigarette paper in order to avoid loss of sample on the walls of the flask.

    Aqueous solutions of polymers: Pipette a suitable aliquot of test solution into the flask and add 6 drops of 0.5 N sulphuric acid (include in blank). Heat at 50 °C under vacuum to reduce the volume to approximately 1 ml. The sample is now ready for analysis.

<u>Procedure.</u> Sample digestion:

a)   In the kjeldahl flask containing the sample introduce a small piece of clean porous pot. Depending on the weight of sample to be digested, add the recommended quantities of a suitable catalyst mixture and of concentrated sulphuric acid as indicated in Table 12 also 0.05 g glucose (note 3).

To the flask used for the reagent blank determination introduce the same amount of catalyst mixture, sulphuric acid and glucose. Suitable catalysts and length of sulphuric acid digestion time for the digestion of various types of nitrogen compounds and polymers are shown in Table 13.

b)   Mount the digestion flask containing sample, sulphuric acid, catalysts and glucose on an electrical heater (note 2) and insert a small funnel in the neck of the flasks. Digest gently at first and then increase the digestion temperature steadily, at five or ten minute intervals over the recommended digestion period (Table 13). After the first 45 minutes digestion the heating rate should be such that the acid is condensing about half way up the neck of the digestion flask (maximum digestion temperature 380 °C). Ensure that the carbon char becomes completely oxidized during digestion. Continue the digestion 20 minutes at least, at maximum temperature after the solution has clarified. When digestion is complete, switch off the heaters. Leave to cool at room temperature. At least half of the original sulphuric acid addition should remain in the digestion flasks at this stage (note 5). Add distilled water to the contents of each digestion flask.

c)   Ammonia recovery by steam distillation. Set up the distillation apparatus shown in fig 4. To the 1 litre steam generation flask add several pellets of sodium hydroxide and some boiling chips and three quarters fill with distilled water. Turn stopcock T1 to connect A to C. Boil the contents of the generator for 5 minutes (note 1) to sweep out dissolved ammonia into the waste flask. Then turn stopcock T1 to connect B to C thereby passing steam through the steam stripper (stopcock D and clip at E closed). Again divert the steam supply to the waste flask.

d)   Transfer the contents of a digestion flask with two or three 5 ml distilled water washings together with 0.2 ml 25% cupric sulphate solution via the funnel on the steam stripper to the interior sample vessel and close the stopcock D. Open the outlet E to waste and turn stopcock T1 to re-connect B and C. Pass steam until it exits at E. Meantime pour an excess of 40% sodium hydroxide into the (closed) funnel at D. Use 10 ml, 20 ml, or 40 ml sodium hydroxide respectively, depending on whether 2 ml, 5 ml or 10 ml of concentrated sulphuric acid was used for acid digestion. If the digestion catalyst used contains mercury or mercuric oxide use instead, the same volumes of sodium hydroxide - sodium thiosulphate solution (note 6). Use the same amount of alkali in both sample and blank determinations.

e)   Connect to the vertical condenser of the steam stripper a 250 ml flask. Connect a guard tube containing freshly activated silica gel in position above this receiving flask (note 2). When steam is emitting from the steam stripper outlet E close the clip thereby allowing steam to pass through the sample solution. Steam distil until about 50 ml distillate has been collected (note 7) and disconnect the flask from the apparatus.

Into a clean 250 ml ammonia receiving flask pipette 25 ml 4% boric acid solution and add 4 drops of bromo-cresol green - methyl red mixed

Table 12 - Catalyst and sulphuric acid additions

Weight of sample digested (g)	Catalyst and alkali sulphate mixture		Volume of concentrated sulphuric acid used for digestion, ml
0.1-0.3*	A**	0.4 g sodium sulphate 0.02 g selenium (as dioxide) 0.02 g $CuSo_4 5H_2O$	2
	or		
	B	0.4 g sodium sulphate 0.02 g selenium (as dioxide) 0.02 g mercury (as oxide)	2
	or		
	C	1.3 g potassium sulphate 0.04 g mercuric oxide	2
0.03-0.08*	A**	1g sodium sulphate 0.05 g selenium (as dioxide) 0.05 g $CuSO_4 5H_2O$	5
	or		
	B	1 g sodium sulphate 0.05 g selenium (as dioxide) 0.05 g mercury (as oxide)	5
	or		
	C	3 g potassium sulphate 0.1 g mercuric oxide	5
0.08-0.2	A**	2g sodium sulphate 0.1 g selenium (as dioxide) 0.1 g $CuSO_4 5H_3O$	10
	or		
	B	2 g sodium sulphate 0.01 g selenium (as dioxide) 0.01 g mercury (as oxide)	10

* Use a micro or semi-micro balance when weighing out 0.01-0.03 g samples.

** Recommended as a good general digestion catalyst. The sodium sulphate-selenium catalyst is available in tablet form from B.D.H. Ltd.

indicator. If this solution is blue-green in colour (due to the presence of a trace of dissolved ammonia), then add single drops of 0.01 N hydrochloric acid from a burette until the solution becomes neutral grey in colour. Connect this flask to the vertical condenser of the steam stripper, (silica gel guard tube still in position).

f)'   When steam is emitting from the steam stripper outlet E, close the clip thereby allowing steam to pass through the sample solution. Open stopcock D and slowly admit the alkaline reagent in several portions. Leave 1 ml of alkali in the funnel to act as a seal. It is necessary to ensure that an excess of alkali is added at this stage. Alkalinity is indicated by the formation of, first, a soluble deep blue cupramine salt than precipitate of brown copper oxide.

Table 13 - Selection of Suitable Catalysts and Digestion Times

Type of compound digested	Catalyst and alkali metal sulphate mixture recommended* (see Table 12)	Total digestion time with sulphuric acid (hrs)
Amines, amino compounds amino acids, amides and their simple derivatives	A, B or C	1½ - 2
Nitriles and their simple derivatives also; styrene-acrylonitrile copolymers and polyacrylonitrile	A	3
Refractory nitrogen compounds,	C	4
Nitro, nitroso and azo compounds hydrozones, oximes and some heterocyclic nitrogen compounds	See Note 4	
Many azo compounds, volatile nitro compounds, diazo ketones and certain semicarbazones	No successful Kjeldahl digestion procedure known	-

g)   Steam distil until the volume of liquid in the receiver is approximately 100 ml, i.e. for about 20 minutes. Remove the receiver flask from the apparatus. Clean the steam stripper ready for the next analysis (note 8).

h)   Titrate the contents of the receiver flask with either 0.05 N or 0.02 N hydrochloric depending on the amount of sample digested and its nitrogen content.

The colour change at the end-point is from green to neutral grey. The addition of a further drop of hydrochloric acid beyond this end-point should produce a pale pink colour.

Calculations.

The nitrogen content of the sample is given by:

$$\% \text{ N } (\%) = \frac{T \times f \times 14}{W \times 10}$$

Where:

T = Titration (ml) of hydrochloric acid

f = Normality of hydrochloric acid

W = Weight (g) of sample taken for analysis

Note 1 - Electrical heating.

Electrical heating during the Kjeldahl digestion and ammonia recovery is preferable to gas heating. The chance of contamination of the sample by nitrogenous impurities in coal gas is thereby avoided.

Note 2 - Pick up of ammonia impurity from laboratory atmosphere.

The apparatus used for steam distillation from alkali to recover ammonia from the acid digest is fitted with an activated silica gel guard tube. This prevents contamination of the distillate with any ammonia impurity that is present in the laboratory atmosphere.

Note 3 - Addition of glucose in Kjeldahl digestion.

Normally no organic matter would be present in the blank digestion flask. If the nitrogen impurity in the reagents is in both oxidized and reduced forms, therefore, it is necessary to add some organic material to the contents of the blank flask in order to ensure that nitrogen impurity is completely reduced. The addition of 0.05 g pure glucose to the blank (and sample) is sufficient for this purpose.

Note 4 - Reduction of nitrogen compounds prior to Kjeldahl digestion.

Nitro, nitroso and azo compounds also, hydrazones, oximes and some heterocyclic compounds are not quantitatively digested by the catalysts A, B or C shown in Table 12. If compounds of these types are being analysed either add 0.5 g pure sucrose (or glucose) to the mixture of catalyst A or B, sulphuric acid and a suitable weight of sample or carry out a preliminary reduction of the sample with hydriodic acid and red phosphorus as described below.

Transfer a suitable weight of sample (catalysts and sulphuric acid absent) and 5 ml of pure hydriodic acid (Analar about 55%) to a clean Kjeldahl flask with B19 quickfit socket. Warm gently and introduce 50 mg pure red phosphorous and some porous pot. Reflux for 45 minutes and then dilute with 5 ml water. Add 5 ml concentrated sulphuric acid and mix. Boil the mixture rapidly (without condenser) to remove hydriodic acid and iodine vapour. If iodine is not completely removed add a further 5 ml of water and boil again until the mixture fumes. Carry out an identical reagent blank omitting only the sample additions. Add a suitable quantity of Kjeldahl digestion catalysts A or B (Table 12), concentrated sulphuric acid and 0.05 g glucose (Note 3) to the reduced sample and blank solutions.

Suitable quantities of catalyst and acid for the digestion of various sample weights are shown in Table 12. Continue by the normal Kjeldahl digestion as described in section b) under procedure.

Various other methods have been proposed for the reduction of nitro groups prior to digestion. These include the use of thiosalicylic acid, salicylic acid - sodium thiosulphate, sodium hydrosulphite-ethanolic hydrochloric acid and zinc dust - pyrogallic acid.

To obtain satisfactory results in the digestion of pyridine it is necessary to leave the sample in contact with catalysts and sulphuric acid for 8 hours at room temperature. Then digest at a low temperature for 1 hour and at a higher temperature for 4 hours. Alternatively add a crystal of iodine to the sample, sulphuric acid and catalysts and digest in the normal manner.

No satisfactory general method for the digestion of nitrogen compounds with an N-N linkage is known.

Note 5 - Loss of nitrogen during digestion.

Some sulphuric acid is usually lost during digestion. The boiling temperature of the digestion mixture than increases due to the increase in the alkali metal sulphate concentration in the digestion mixture. If the temperature of the mixture becomes too high, due to loss of acid, then a partial loss of nitrogen also occurs. No loss of nitrogen occurs, however, provided the volume of sulphuric acid left at the end of the digestion is at least half of the amount originally added.

Note 6 - Modification of ammonia recovery stage when mercury containing catalysts are used for digestion.

Low ammonia recoveries result during steam distillation from sodium hydroxide when mercury containing catalysts are used in the Kjeldahl digestion. This is due to the partial formation of a mercuric-amine which is not completely decomposed by sodium hydroxide. The addition of a mixture of sodium hydroxide and sodium thiosulphate completely degrades this complex and quantitative ammonia recoveries result.

Note 7 - Volatile acids in digestion mixture.

It has been found that traces of steam volatile acids remain in the acid digestion mixture after digestion of certain organic materials. These are removed by passing steam through the diluted digestion mixture prior to sodium hydroxide addition. This acidic distillate is rejected.

Note 8 - Cleaning of steam distillation apparatus

Clean the steam stripper ready for the next determination as follows. With clip E closed, turn stopcock T1 to divert the steam supply to the waste flask (i.e. connect A to C). The sample solution now syphons from the sample vessel into the outer vessel and is disposed of by opening clip E. Now fill the sample vessel with water via the funnel (clip E again close) and pass steam through this water until it boils. Syphon the water from the sample vessel as before. Repeat this cleaning operation until the interior of the vessel is perfectly clean.

4.3.5 Determination of Phosphorus

Determination of 0.01 - 2% Phosphorus in Polymers - Oxygen Flask Combustion - Spectrophotometric Method

Scope. Two oxygen flask methods are given below for the determination of phosphorus in polymers. The first is applicable in the range 0.01 to 2% phosphorus and the second in the range 2 to 13% phosphorus.

Summary. The ground polymers mixed with sodium carbonate, is decomposed by igniting in an oxygen filled flask containing dilute sulphuric acid. The solution is treated with molybdate-hydrazine reagent and the phosphate measured colorimetrically as its heteropoly-blue equivalent.

Apparatus.

a)   500 ml conical combustion flask fitted with a reducing adaptor and joint carrying approximately 1" square platinum gauze.

b)   Unicam SP 500 spectrophotometric with 1 and 4 cm glass cells and tungsten lamp.

c)   Methyl cellulose capsules - No. 4 small, Arthur H. Thomas Co. Catalogue No. 6471-0.

d)   Sodium carbonate scoop - fabricated from glass tubing - 4 mm in outside diameter and marked to contain 50 $\pm$ 10 mg of anhydrous sodium carbonate powder.

e)   Filter paper fuses approximately $1\frac{1}{2}$ x 1/8" in size made from ashless filter paper.

f)   100 ml volumetric flasks; 150 ml conical flasks; pipettes, etc.

Reagents.

a)   Demineralised water used throughout.

b)   Ammonium molybdate solution, dissolve 40 g reagent grade ammonium molybdate $(NH_4)_6Mo_7O_{24}$. $4H_2O)$ in a cooled mixture of 450 ml concentrated sulphuric acid and 1 litre water. Dilute to 2 litres with water.

c)   Hydrazine sulphate solution, 1.5 g/litre.

d)   Molybdate-hydrazine reagent, dilute 50 ml of ammonium molybdate solution with 130 ml of water, add 20 ml of hydrazine sulphate solution and mix well. Use 50 ml for each determination and prepare no earlier than 1 hour before use as this mixture is unstable.

e)   Stock standard phosphorus solution, dissolve 4.393 g of dried Analar potassium dihydrogen phosphate (KH_2PO_4) in 150 ml of 1:10 sulphuric acid and dilute to 1 litre with water (1.0 mg/ml phosphorus). From this prepare solutions containing 0.01 mg P/ml and 0.001 mg P/ml.

f)   Oxygen

g)   Sodium carbonate, anhydrous, Analar

Procedure

a)   Place 1 scoop of anhydrous sodium carbonate in the bottom half of a methyl cellulose capsule. Weigh a sample of not more than 32 mg estimated to contain between 0.001 and 0.06 mg phosphorus, directly onto the sodium carbonate bed. Cover the sample with another scoop of sodium carbonate. Insert a filter paper fuse between the top and bottom halves of the capsule after making a small slit in the top half to give the clearance required. Secure the capsule in the platinum gauze with the wick towards the stopper.

b)   Pass oxygen into the flask containing 10 ml of 1:10 sulphuric acid for 30 seconds.

c)   Holding the flask horizontally remove the oxygen lead, light the

sample fuse and quickly insert the stopper in the flask. See Note 1.
Tilt the flask immediately at an angle of 135° from the vertical so that
the absorbing solution forms a liquid seal at the neck of flask. During
the period of maximum flame height completely invert the flask to prevent
impingement of the flame on the flask walls. Hold the stopper firmly in
place during the combustion (Note 1). After combustion allow the flask to
stand for 10 to 15 minutes.

d)   Open the flask and wash the platinum gauze and surfaces of the
reducing adaptor with a maximum of 35 ml of water. Remove the adaptor and
lower the gauze into the solution.

e)   Add 50 ml molybdate hydrazine reagent, (Note 2). Heat the solution
rapidly to boiling and boil for 2-3 minutes. Cool to room temperature in
an ice-water bath, transfer to 1000 ml volumetric flask and dilute to
volume.

f)   Measure the absorbance against water in appropriate cells at 830 nm.

g)   Carry our blank determination.

      Preparation of calibration curves. Prepare separate calibrations
from standard phosphorus solutions for 1 cm and 4 cm cells containing 0-
0.06 mg P/100 ml and 0-0.015 mg P/100 ml respectively as follows:

a)   Measure appropriate amounts of the phosphorus standards into 150 ml
conical flasks and dilute to approximately 30 ml with water. Add 10 ml of
1:10 sulphuric acid and 50 ml of molybdate/hydrazine reagent and develop
the colour as before.

b)   Measure the absorbance against water in appropriate cells at 830 mu.

c)   Carry out a blank determination.

d)   Prepare graphs of corrected absorbance against weight of phosphorus
(mg per 100 ml). Linear calibrations are obtained.

Note 1

      Gloves and goggles should be worn and the ignition carried out
behind a protective glass screen.

Note 2

      It is essential that the solution is diluted before the molybdate
hydrazine reagent is added.

Determination of 2 - 13% Phosphorus in Polymers - Oxygen Flask Combustion
- Spectrophotometric method

      Method summary. The ground polymer, mixed with sodium carbonate, is
decomposed by igniting in an oxygen-filled flask containing saturated
bromine water and sodium hydroxide. The solution is acidified, boiled and
made up to 100 ml. The phosphorus is determined colorimetrically as
molydovanadophosphoric acid.

      Apparatus.

a)   500 ml conical flask fitted with a reducing adaptor and joint carrying approximately 1" square of platinum gauze.

b)   Methyl cellulose capsules - No. 4 small, Arthur H Thomas Co.
Catalogue No 6471-0.

c)   Sodium carbonate scoop, fabricated from glass tubing 4 mm in outside
diameter and marked to contain 50 $\pm$ 10 mg of anhydrous sodium carbonate
powder.

d)   Filter paper fuses, approximately 1½" x 1/8" in size made from
ashless filter paper.

e)   Unicam SP 500 spectrophotometer with 1 and 4 cm cells and tungsten
lamp.

f)   100 volumetric flasks, 150 ml conical flasks, pipettes, etc.

   Reagents.

a)   Oxygen.

b)   Sodium hydroxide, 0.5 N.

c)   Saturated bromine water.

d)   Percolated water, prepared by percolating distilled water through a
mixed resin bed containing Amberlite IR 120 (H) and Amberlite IRA 400
(OH).

e)   Ammonium molybdate, 5%v.   Dissolve 50 g ammonium molybdate in a
litre of warm distilled water.

f)   Ammonium vanadate, 0.25% w. Dissolve 2.5 g of ammonium vanadate in
500 ml hot water. Cool and add 20 ml of concentrated nitric acid. Cool
and dilute to 1 litre.

g)   Aqueous sulphuric acid, 25% w.

   Procedure.

a)   Pass oxygen into a flask containing 5 ml of 0.5 N sodium hydroxide
and 4 ml saturated bromine water for 30 seconds.

b)   Place 1 scoop of anhydrous sodium carbonate in the bottom half of a
methyl cellulose capsule supported in a bored cork. Weigh a sample of not
more than 32 mg, estimated to contain between 0.05 and 2.5 mg phosphorus,
directly onto the sodium carbonate bed. Cover the sample with another
scoop of sodium carbonate. Insert a filter paper fuse between the top and
bottom halves of the capsule after making a small slit in the top half to
give the clearance required. Secure the capsule in the platinum gauze
with the wick towards the stopper.

c)   Holding the flask horizontally remove the oxygen lead, light the
sample fuse and quickly insert the stopper in the flask. Tilt the flask
immediately at an angle of 135° from the vertical, so that the absorbing
solution forms a liquid seal at the neck of the flask. During this period
of maximum flame height, completely invert the flask to prevent
impingement of the flame on the flask walls. Set aside until the mist has
cleared (10 to 15 minutes is usually sufficient).

d)   Open the flask and wash down the platinum gauze and surfaces of the reducing adaptor with demineralised water.   Remove the adaptor and lower the gauze into the solution.

e)   Add to the solution 6 ml of 25% v. sulphuric acid and boil the solution for 15 minutes.   Cool and transfer the solution to a 100 ml graduated flask.   Add 10 ml of ammonium vanadate, followed by 10 ml of ammonium molybdate and make up to 100 ml.   Allow 30 minutes for colour development.

f)   Using either 1 or 4 cells, read off the optical density at 460 nm.

g)   Carry out a blank determination.

4.4 FINGERPRINTING TECHNIQUES

4.4.1 Introductory

A very successful method of identifying an unknown polymer is to compare some of its physical characteristics with those of libraries and such properties which have been put together from information obtained on known polymers.   Various techniques have been employed for this purpose including:

i   Measurement of glass transition temperature (Tg) and melting temperature (Tm), (see Chapter 6).

ii   Preparation of pyrolysis (or photolysis) - gas chromatogram or pyrolysis (or photolysis) - gas chromatogram - mass spectrogram.

iii   Preparation of infrared spectrum.

iv   Preparation of NMR spectrogram.

4.4.2 Thermal Analysis Techniques

Glass transition temperature (Tg) and melting temperature Tm.   The identification of polymers is frequently based on a comparison of glass transition or melting temperature with literature values (Table 14), derived from differential scanning calorimetry or thermomechanical analysis (Chapter 6).

Very often thermogravimetric curves are characteristic for particular polymers and can therefore be used for their identification. Because of the poor heat conductivity, temperature gradients occur within the sample at high heating rates.   To obtain reproducible results, a standardization heating rate should be used, for example, 10°C/min when comparing an unknown polymer with a set of reference polymers.

Other thermal analysis techniques have been used to identify polymers by fingerprint approach.   These include differential thermal analysis (Chapter 6) which has been used to identify polymers including polyethylene, polypropylene, polymethyl methacrylate, polyesters, PVC, and polycarbonates[67], and pyrolysis or photolysis techniques as discussed below.

4.4.3 Pyrolysis Techniques

Pyrolysis is an analytical technique whereby complex involatile materials are broken down into smaller volatile constituent molecules by the use of very high temperatures. Polymeric materials lend themselves very readily to analysis by this technique. Providing that the pyrolysis conditions are kept constant, a sample should always degrade into the same constituent molecules. Therefore, if the degradation products are introduced into a gas chromatograph, the resulting chromatogram should always be the same and a fingerprint uniquely characteristic of the original sample should be obtained.

In one technique a pyrolyser probe is inserted into a purpose-designed adaptor which is installed in the injector unit of the gas chromatograph. Different types of adaptor are available for packed or capillary column work. Correct selection of the appropriate adaptor is a prerequisite for optimum system performance.

The choice of packed or capillary pyrolysis - gas chromatography is generally a matter of personal requirements. The type and range of samples to be analysed, the complexity of pyrogram required and the length of the analysis - all will play a significant role in decision-making. For the general fingerprinting or analysis of a wide range of routine rubbers or plastics, packed column pyrolysis is more than adequate. However, for comparison of one batch of rubber with another batch of the same rubber, the detail obtained from capillary column pyrolysis - gas chromatography will probably give the best results because minor details can be compared.

Applications for pyrolysis are vast and include all types of man-made polymers, rubbers and plastics, as well as latexes, paints and varnishes; in fact, almost any sample that contains involatile organic material that can be contained in a tube or coated onto a platinum ribbon so that it may by pyrolysed.

Pyrolysis chromatograms of several common plastics are shown in Appendix 1 which illustrate the value of pyrolysis - gas chromatography in the identification of polymers by fingerprinting.

Pyrolysis gas chromatographic procedures have generally followed one of two basic patterns. In the flash technique, a sample is deposited on an electrically heatable filament, or placed in a boat in a furnace, and the gas chromatographic carrier stream used to transport the pyrolysate directly to the column. The alternative procedure involves heating of the polymer in a separate enclosure, trapping the off-gases and admitting the pyrolysate to the chromatograph after a given collection interval. Each approach has certain advantages and the type of information desired from pyrolysis - gas chromatograms should therefore, dictate the sampling method to be employed.

Filament pyrolysis. For quantitative identification a thin film of sample is usually coated onto Nichrome or platinum spiral or is placed in a small boat so that the weight of the residue remaining after heating can be determined. Although the filament may catalyse the degradation, Jones and Moyles[68] showed that with 20-30 ug samples the pyrograms obtained with Nichrome, platinum or gold plated platinum filaments are identical. The power supply to the filament is generally controlled by a variable transformer and timed by a stopwatch, but exact measurement of the

Table 14 - Tg and Tm values of polymers

Polymer	Abbr.	Tg	Tm
Polybutadiene		-86	(-20)
Polyisobutylene	PIB	-73	(44)
Poly(ethylene vinyl acetate) copolymer	EVA	-20..20	40.100
Polyethylene, low d.	LDPE	(-100)	120
Polybutene	PB		130
Polyethylene, high d.	HDPE	(-70)	135
Poly(oxymethylene) copolymer	POM		164..168
Polypropylene	PP	(-30)	165
Poly(vinyl chloride) soft	PVC w	-40..10	
Poly(vinilidene chloride)	PVDC	-17	
Poly(oxymethylene)	POM		175..180
Poly(vinylidene fluoride)	PDF_2		178
Polyamide 11	PA11		186
Poly(vinyl acetate)	PVAC	30	
Poly(vinyl chloride)	PVC	85	(190)
Poly(butylene terephthalate)	PBTB	65	220
Polyamide 6	PA6	(40)	220
Polyamide 6,10	PA6,10	(46)	226
Poly(vinyl alcohol)	PVA	85	
Polystyrene	PS	90..100	
Poly(methyl methacrylate)	PMMA	105	
Epoxy resin	EP	50..150	
Poly(phenylene oxide)	PPO		230
Polycarbonate	PC	155	(235)

Polyamide 6,6	PA6,6	(50)	255
Poly(ethylene terephthalate)	PETP	(69)	256
Poly(ethylene tetrafluoro-ethylene) copolymer	ETFE		270
Poly(fluorethylenepropylene)	FEP		280
Poly(phenylene sulfide)	PPS	80	280
Polyacrylonitrile	PAN	100	(320)
Polytetrafluoroethylene	PTFE	(-20)	327

filament temperature is difficult. More elaborate automatic time and voltage controls have been suggested[69,70]. If desired, the pyrolyser temperature can also be manually programmed to obtain more equilibrium and to remove the pyrolysis products from the heated zone immediately after they are formed[71]. For quantitative studies of the mechanism and the kinetics of polymer degradation where accurate analysis of the volatile and non-volatile reaction products obtained at a certain temperature and under closely controlled conditions is required, it is preferable to employ preheated tube furnaces than to refine the design of the filament-type pyrolyser. Despite the fact that the filament-type pyrolyser does not allow optimum control of degradation conditions the pyrograms are entirely satisfactory for identification purposes when polymers of known composition are available for comparison.

Giacobbo and Simon[72] have described a useful pyrolysis unit for microgram samples. The material coated on a small ferromagnetic wire is pushed by means of a magnet into the pyrolysis capillary which is surrounded by an induction coil. Using a frequency of 450 KHz the high frequency induction oven will heat an iron wire of 0.6 mm diameter to the Curie temperature in 2 x 10-3 seconds. During the heating time, which can be controlled from 0.06 seconds to several seconds, the temperature of the wire remains constant. The pyrolysis temperature can be varied by choosing a ferromagnetic conductor with a suitable Curie point temperature. With a reactor capillary of 0.6 mm and wire of 0.5 mm diameter and 1 cm length the carrier gas will pass through the reactor in 5 x 10-3 seconds, assuming a flow rate of 10 ml/min.

Advantages of this pyrolysis technique include the use of microgram samples and the rapid establishment of a reproducible pyrolysis temperature that is essentially dependent on the composition of the ferromagnetic conductor. However, the possibility of catalytic side reactions due to the heating element should not be overlooked. With 0.1 ug samples it may be possible to coat the wire with a monomolecular layer of polymer film which would greatly reduce the possibility of secondary reactions between components of a mixture.

May et al[73] have used the Curie point filament pyrolyser to produce pyrolysis - gas chromatograms for various polymers used in paint production. Good interlaboratory reproducibility is claimed for this procedure. This Curie point unit gives rapid and reproducible heating to the Curie point of the pyrolysis wire. If pyrolysis wires of a fixed composition are used in different pyrolysers, identical pyrolysis temperatures are obtained that give highly reproducible pyrograms. A temperature of 610°C was chosen, as it is in the region in which most polymers give characteristic fragmentation patterns. May et al[74] have published a full reference collection of Curie point pyrograms.

<u>Method - pyrolysis - gas chromatography of polymers - Curie point filament pyrolyser technique.</u>

Apparatus and conditions:

Pyrolyser - Pye Unicam Curie point, attached directly to the column, pyrolysis temperature 610°C maintained for 10 seconds.

Gas chromatograph - Pye 104, Model 64 (dual columns, twin flame-ionisation detectors).

Columns - standard Pye 5 feet long x 4 mm i.d. glass (silanised).

Carrier gas - nitrogen with a flow rate of approximately 60 ml min-1.

Column packing - Porapak Q, 50 to 80 mesh or 80 to 100 mesh.

Temperature programme - from 100 to 200°C at 8°C min-1 (the temperature was maintained at 200°C for up to 25 minutes and the programmer started on completion of 10 seconds pyrolysis).

The columns are silanised before being packed by passing a 5 per cent solution of di-chlorodimethylsilane in toluene through the columns and then drying at 100°C. Packing is facilitated by applying suction from the detector end of the column, accompanied by gentle tapping. The columns are aged by heating at 250°C for 48 hours, with a nitrogen flow of 60 ml min-1. The hydrogen and air flows are adjusted to give approximately the maximum sensitivity.

The flow rate required for pyrolysis is determined by injecting through the pyrolyser head 1 ul of headspace gas from a retention time standard comprising methanol-n-propanol (50 + 50) with the column temperature at 100°C and immediately programming at 8°C min-1. The flow-rate of nitrogen is adjusted to give retention times of 2.6 and 9.1 minutes for methanol and n-propanol respectively.

After fixing the flow rate of carrier gas, the upper temperature limit is adjusted to give a retention time of 4.2 minutes for benzene or 4.4 minutes for cyclohexane when these compounds are injected at the upper temperature limit. These two compounds have retention times of 13.4 minutes (benzene) and 14.0 minutes (cyclohexane) when injected at 100°C and the oven is programmed as described.

The pyrolysis wire is prepared in the following way. A 5 to 10 mm length at one end of the wire is flattened until the flattened portion is approximately 1 mm wide. A "hook" is then made by doubling over about 2 mm of the flattened end. At this stage, the hook end plus a few centimetres of the wire adjacent to it are heated to red heat in a bunsen flame so as to remove contaminants. The sample is placed in the elbow of the hook, care being taken not to touch or otherwise contaminate that end of the wire. The sample is then pressed into close contact with the wire surface by crimping the hook.

In developing a systematic scheme of polymer identification, pyrolysis conditions must be employed so that all polymers degrade rapidly. However, at temperatures above 1000°C the pyrograms will also be less suitable for identification, since secondary reactions become predominant, leading to increasing amounts of simple molecules such as carbon dioxide, acetylene, ethylene or ethane which give less characteristic patterns than those observed for monomers or primary degradation products. Pyrolysis temperatures kept between 500 and 800°C for 10 seconds are usually recommended. Although pyrograms containing from 30 to 50 peaks (for polyolefins) have been reported (Voigt[75-77]) the relative retention times and peak height ratios of three to five major peaks are usually sufficient for identification. Thus, a single set of column operating conditions is desirable that will give characteristic pyrograms for degradation products of all types of polymers.

Voigt[75-77] employed a platinum pyrolyser which is attached directly to the gas inlet of the gas chromatograph for the examination of ethylene - propylene copolymers. The Pyrex glass pyrolysis cell is located in an oven, the carrier gas entering the cell from above and bringing with it the pyrolysate formed on the red hot platinum wire on leaving the cell and passing through a gas inlet and a heated feed line to the gas chromatograph.

Specimens weighing about 2 mg are heated for 18 seconds up to a maximum temperature of 550°C. During the pyrolysis, carrier gas flushes the products of pyrolysis into the separating column. The pyrolysis cell is cut out of the carrier gas flow 1 minute after starting the pyrolysis.

Pyrograms will differentiate between random copolymers and block polymers or polymer mixtures[78-81]. Presence of foreign monomer may interrupt chain transfer processes involved in the degradation. Similar products result but their quantities as determined from peak heights are different. Voigt demonstrated that even for closely related polyolefins the pyrograms will distinguish between poly(ethylene-propylene) block and random copolymers of the same composition[79].

Systematic identification for plastics was attempted by Nelson et al[82] by pyrolysing 0.2-0.5 mg samples for 10 seconds at 650-750°C in an argon gas chromatograph using a 4 ft column of 5% silicone oil on Chromosorb W. Similar plastics were distinguished from pyrograms obtained at two different temperatures.

Groten[83] used a platinum filament type pyrolyser and six-way gas sampling valve pyrolyser produced by Aminco in conjunction with an isothermal gas chromatograph. The pyrolysis was conducted by heating of a platinum coil at 950°C for 26 seconds and a 1/4 inch by 12 ft column containing 20% Carbowax on Diatoport P was used. Identifications were based mainly upon the moderately volatile products. For the 150 different polymers investigated, the individual members of a group generally gave pyrograms that allowed unambiguous identification of the original material. In the few cases where the standard conditions resulted in ambiguous pyrograms only slight variations in pyrolysis times and/or temperatures were necessary to establish an identity. The arrangement is shown in Fig. 5. The valve and pyrolyser assembly are mounted in ports cut through the oven insulation in the chromatograph. This placement exposes the block to oven temperatures and, thus heated, prevents condensation of the less volatile portions of the pyrolysate. The functions of the valve are to reduce pressure fluctuations, and to prevent large quantities of air from entering the hot systems when the pyrolyser is opened for a sample change. In addition, this arrangement also allows the instrument to be employed as either a pyrolysis apparatus or as a conventional gas chromatograph by simply turning the gas valve. Thus,

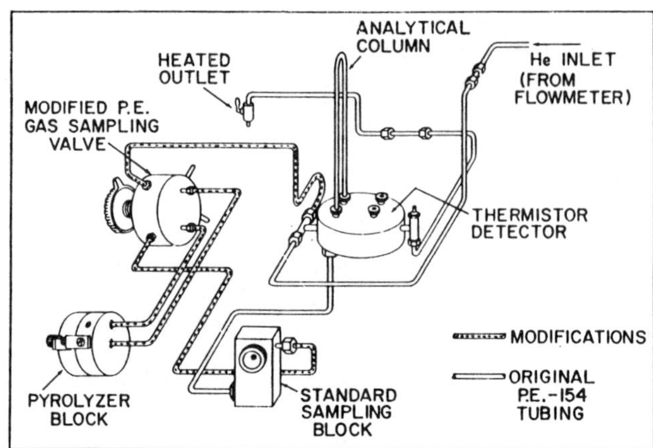

Figure 5   Perkin Elmer 154 gas chromatograph modified for
pyrolysis

when peak identification is required, known materials can be injected into
the gas chromatograph in the usual manner and retention times matched with
those found in the pyrolysate pattern (vide infra).

The pyrolyser chamber consists of a two-piece stainless steel block
containing carrier gas connections and the platinum heater coil. Samples,
of approximately 1 milligram, are weighed into 2 mm o.d. quartz capillary
tubes and placed in the centre of the coil. After a run is completed, the
tube is removed and any residue may then be recovered. Pyrolysis
temperatures are controlled by a variable transformer in the heater
circuit, and the duration of pyrolysis is regulated by an adjustable cam
operating a microswitch. The transformer is calibrated by measuring the
maximum temperature actually achieved within the quartz sample capillary
rather than the platinum coil temperature. Since the total heating-

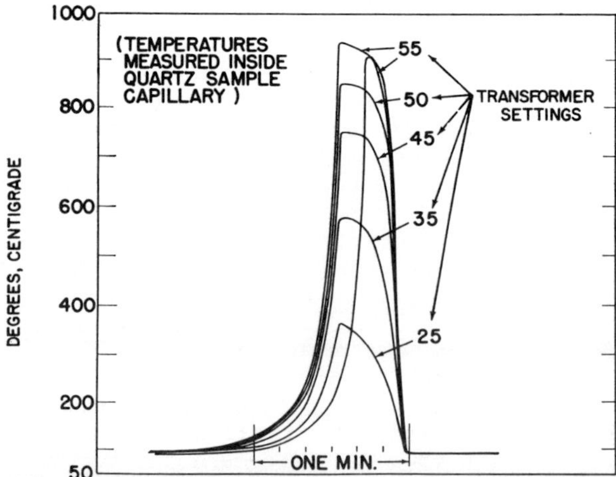

Figure 6   Temperature profiles achieved within quartz sample
capillary. Elapsed time (x-axis) reads from right to
left

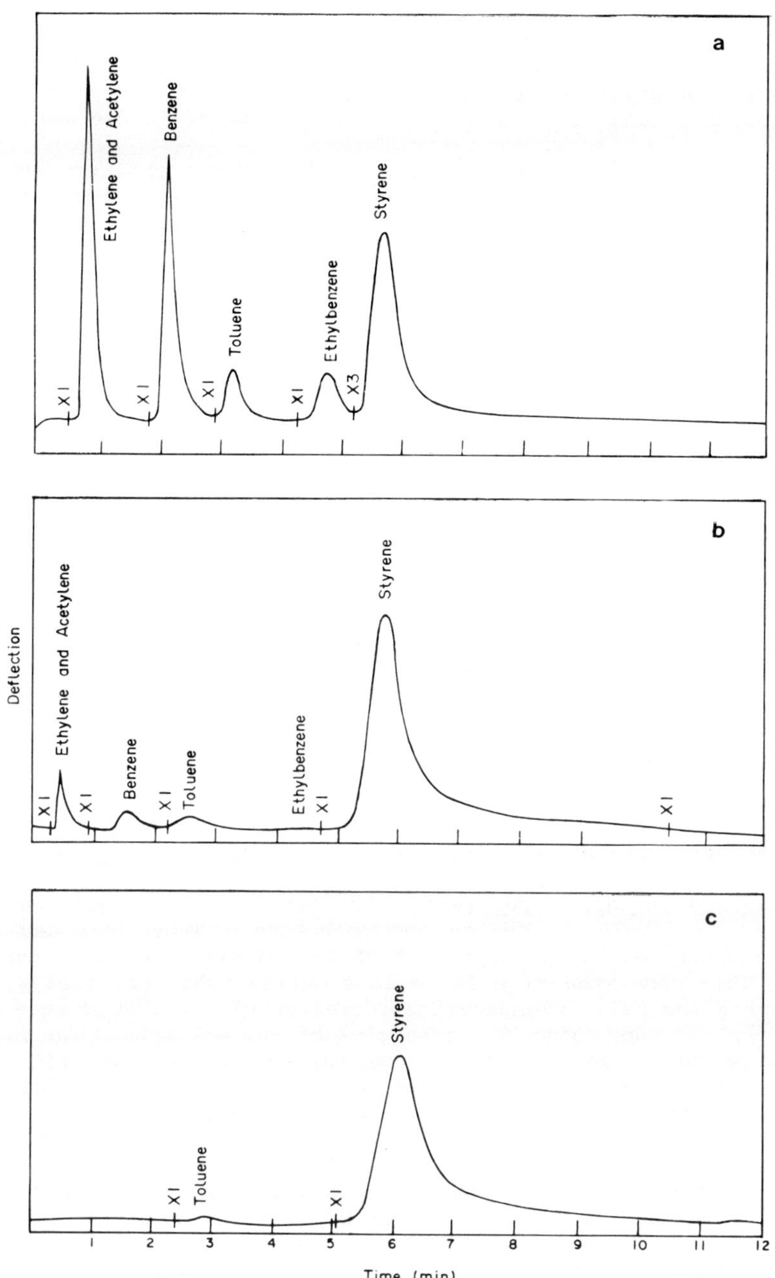

Figure 7 Pyrograms of polystyrene at various pyrolysis
temperature (a) 425°C, (b) 825°C, (c) 1025 °C, Column
Apiezon L, column temperature: 140°C, flow rate, 60 ml
min^{-1}, attenuation indicated by number in figure

peaking-cooling cycle is important in pyrolysis studies, a complete profile was obtained for each transformer setting and at various total heating times. Typical cycles are shown in Fig. 6. These curves show that the attainment of maximum temperature is quite rapid under all conditions. The cycle was fairly insensitive to minor changes in carrier flow rate. The column used in most of this work was 1/4 inch by 12 foot containing 20% Carbowax on Diatoport P. Water insoluble Ucon (5%) on an 8 foot Haloport F column was used for analysis of the cellulose esters.

Chromatographic conditions were: bridge, 8 volts; attenuation, 2 x unless otherwise specified; helium flow rate, 45 cc per minute; pressure, 10 psi; column temperature, 150°C; (100 for Ucon column).

Slight differences were found in the pyrograms obtained on degrading polyethylenes at various temperatures. The polymer was pyrolysed at 900°C and the volatile degradation products were chromatographed on 30-60 mesh silica gel at 150°C. The pyrograms consisted of 8-11 peaks. The elution profiles of these pyrolysis products; that is the plot of the normalized peak heights or areas versus pyrolysis temperature, allowed a distinction to be made between some commercial polyethylenes. Recent advances in instrumentation coupled with programmed temperature operation yield a much larger number of distinctive peaks due to the much greater efficiency of suitable columns to separate the complex mixture of the pyrolysis products.

Fig. 7 shows the pyrograms of polystyrene degraded at 1025, 825 and 425°C. Samples can be pyrolysed at a series of temperatures and the change in degradation behaviour as evidenced by the (1) "appearance temperatures" of various peaks and (2) the relative abundance of products as a function of temperature are noted. Changes of the most character- istic breakdown products versus temperature can also be plotted. For homopolymers the temperatures at which degradation products are first obtained may be more distinctive than the retention times of the products.

Wampler and Levy[85] have investigated the effect of slow heating rates on the pyrolysis of polyethylene.

<u>Combustion furnace pyrolysers.</u> Earlier pyrolysis equipment tended to be somewhat complex in design and operation. Thus, Cox and Ellis[86] described a micro-reactor pyrolyser which they applied to large numbers of polymers. This consisted of a 20 cm long quartz tube contained within an electrical heating coil. Temperature increases of 700-1000°C were used in order to completely pyrolyse 0.1 g samples of the polymers. The pyrolysis products were collected for 15 minutes then swept onto a gas chromato- graphic column equipped with a flame ionization detector. The elution times of the peaks were then expressed as percentage fractions of the elution time of the reference peak which is arbitrarily selected as the peak of the last product eluted during a 15 minute period after pyrolysis. Pyrograms may thus be related to those obtained on other instruments by measuring retention times of the reference peaks, with respect to the retention time for the reference peak, with the positions of the distinctive peaks represented by rectangles of width equal to twice the standard deviation in the relative retention times, averaged over at least five determinations.

This type of equipment has now been displaced by the more recently described equipment as discussed later.

Kolb and Kaiser[87] pyrolysed 4 mg of polyethylene in an oven furnace at 1000°C for 10 seconds. A packed column containing 15% by weight of OS-138 poly(phenyl ether) on Celite 545 was programmed linearly from 50°C up to 240°C, at a rate of 5°C/min; hydrogen was used as carrier gas. The first peak of the repeated groups of triplet peaks of the pyrogram was identified as the normal paraffin. When the degradation products were hydrogenated before chromatographic separation in a precolumn (4 parts of Celite 545 to one part of 5% Pd on Kieselguhr; the mixture was then coated with 5% silicone gum rubber SE 30), a homologous series of single peaks appeared. Thus, two peaks of the triplet groups on the original pyrograms must correspond to unsaturated normal hydrocarbons since the column employed separates iso from n-paraffins.

Cieplinski et al[88] have investigated extensively the thermal degradation products of polyolefins in an oven furnace using a Perkin Elmer chromatograph equipped with a differential flame ionization detector. Samples (50 - 79 mg) were pyrolysed in a reaction chamber and the volatile reaction products were passed through a linearly programmed temperature, open tubular (Golay) column (8 wt% pretreated Apiezon L on hexamethyldisilazane-treated Chromosorb W, 80-100 mesh). On pyrolysis at 690°C a large number of triplet peaks were obtained but with this column in the last peak in each group was identified as the normal paraffin. The peaks are evenly spaced up to C22, where the temperature programming ended. From here on the consecutive homolog peaks start to emerge in the usual logarithmic scale of an isothermal run. Comparison of pyrograms of high and low density polyethylenes indicates significant differences for peaks above C21, with the relative amounts of the two major peaks being reversed. Thus for the low density polyethylene the last peak corresponding to the n-paraffin is always larger, whereas for Ziegler types of high density polyethylene the first peak predominates. For polymethylene pyrolysed at 640°C the triplets are even better separated and many smaller peaks occur. Apparently due to the low molecular weight of the polymer the yield of hydrocarbons above C 17 is very small. On plotting the adjusted retention time versus carbon number or boiling point for the various peaks of each group, a linear relationship is found indicating that each peak in a particular group is representative of a member of a particular homologous series.

The pyrograms of polymethylene obtained by Ciepielsky[88] are greatly dependent on the pyrolysis temperature. When the pyrolysis is conducted at 620°C (instead of 640°C) the large composite peak for the low molecular weight hydrocarbons disappears and the number of isomeric degradation products is reduced. The $n-C_{14}H_{30}$ is the major component, whereas on pyrolysis at 640°C, the n-dodecane gives the largest peak. On pyrolysis at 600°C the composite peak corresponding to the light hydrocarbons becomes dominant and overlaps the C 8 to C 9 peaks. The height of the C 10 - C 13 peaks decreases, especially that representing the C 13 hydrocarbons. The pyrogram of isotactic polypropylene differs greatly from that of polymethylene with a large increase in the yield of light hydrocarbon isomers that are formed on scission at the more reactive tertiary carbon atoms. Another characteristic feature of this pyrogram is the presence of large concentrations of certain hydrocarbons, whereas others with an intermediate number of carbon atoms are nearly completely absent.

Two procedures have been described, thus far, for pyrolysing polymers. One approach is flash pyrolysis in which a small amount of polymer is coated onto a platinum filament which is then heated electrically to produce a pulse of volatiles which pass directly into the gas chromatograph. The second approach involves heating the polymer in a

micro-reactor and collecting the total volatiles produced. When volatilization is complete the mixed volatiles are passed into the chromatograph.

In yet another approach, the pyrolyser probe method, discussed below
the polymer is placed in a quartz tube which is then inserted into a
platinum coil heater element. The coil heater is switched on to release
the pyrolysis products which are then directly swept into the gas
chromatograph.

Chemical Data Systems UK supply a range of pyrolyser probes which
are discussed below. These probes can be used in conjunction with any
suitable gas chromatograph.

The Chemical Data System, Pyroprobe 190 platinum ribbon coil
pyrolyser. The two types of pyrolyser probe available are based on a
platinum ribbon or coil. If a sample can be dissolved in a suitable
solvent, it is best applied to the ribbon as a solution. This should be
prepared at a concentration of about 10 mg/1 ml (approximately 1%). About
1-5 ul (representing 10-50 ug of sample) is applied to the centre of the
ribbon using a micro syringe (this may take several applications). The
solution should be spread as evenly as possible so that the entire sample
will experience the same heating profile during pyrolysis. It is
essential to avoid the extreme ends of the ribbon as the points of
attachment to the probe mounting provide a large heat-sink and do not
reach the set pyrolysis temperature.

Before inserting the probe into the injection adaptor, the final
temperature should be set to a low value (such as 100°C) with an interval
of 5 to 10 seconds, so that the residual solvent is flashed off without
pyrolysing the sample. For most samples the solvent can be flashed off in
the atmosphere - but for extremely sensitive materials, the probe should
be located in the injector adaptor prior to solvent flash. After the
solvent has been removed and the detector response has returned to baseline, the sample can be pyrolysed.

Insoluble but meltable samples should be applied by placing fine
particles of material onto the ribbon and gradually raising the
temperature with the interval set at 1 to 5 seconds until the particles
adhere to the ribbon.

Samples that are insoluble and that will not melt should be
pyrolysed on a quartz tube located inside the coil probe. The quartz tube
must be inserted into the platinum coil probe very carefully to avoid
distortion and damage.

Pyrolysis - Gas Chromatography of Polymers, Platinum Ribbon or Coil
Procedures (Chemical Data Systems Pyroprobe 190)

Application of samples to the probe. Quartz tubes must be inserted
into the platinum coil probe very carefully to avoid distortion and
damage. This is best achieved by passing the coil probe in the probe
stand and holding the tube by one end with thumb and index finger, gently
inserting it into the coil, rotating it as it enters. When the tube is
through the coil part it can then be gently pushed all the way in so that
the end rests on the base of the coil. After insertion, a careful check
of the element is advisable to ensure that there is no shorting to any
parts of the probe assembly and that the coil is even and not distorted.

Samples can then be inserted directly into the quartz tube and, for
best results, should be between 10 and 50 ug and placed in the centre of

the tube. As far as possible, the sample size should be kept constant from analysis to analysis, but for optimum repeatability of pyrolysis patterns, better results will be achieved by placing the sample onto a small amount of quartz wool which is itself located in the centre of the tube. Sample solutions can also be analysed in this way by injecting 1-3 ul from a microsyringe of a 1% solution of sample onto the quartz wool. The solvent should be evaporated to dryness by flashing the coil at 100°C with interval set at 1 second. For most samples, the solvent can be flashed off into the atmosphere but for extremely sensitive material the probe should be located in the injector adaptor prior to solvent flash. After the solvent has been removed and the signal from the gas chromatograph has returned to base-line, the sample can be pyrolysed.

The coil probe generally requires a final temperature of about 150°C higher than that required by the ribbon probe.

<u>Cleaning the probes.</u> Between each sample the probes should be cleaned in the atmosphere by setting an interval of 2-5 seconds and a final temperature of 1000°C and pressing the RUN button.

<u>Selection of pyrolysis temperature.</u> An investigation to determine the lowest temperature at which pyrolysis will occur for a particular sample can be made with repeated runs on the same sample at increasing pyrolysis temperatures without removing the probe from the adaptor.

Alternatively, arbitrary pyrolysis conditions may be selected and the sample pyrolysed. The amount of residual organic material remaining can be determined by repyrolysing the sample at 980°C for 1 second for the ribbon probe and 10 seconds for the coil probe. A large residual peak under these conditions indicates that only a small part of the sample was initially pyrolysed. A higher final pyrolysis temperature should then be used.

<u>Packed column pyrolysis gas chromatography.</u> A large number of materials can be analysed by this technique. The chromatograms obtained are relatively simple and the analysis time comparatively short. These chromatograms are therefore especially suitable for fingerprinting. Relatively large amounts of sample can be introduced onto the column and therefore minor differences between batches of similar sample may be observed. Figs. 8a) and b) are examples of two types of acrylic material, both analysed using the same conditions. Note the similarity of the peak pattern produced from the identical thermal degradation of the two samples, but also note the relative difference in the size of the peak at 12.8 minutes and also the presence of the peak at 5.04 minutes in the modacrylic chromatogram. The results obtained from this type of analysis are qualitative. Quantitative analysis is difficult as the identities of the separated degradation products are usually unknown. Packed column analysis of pyrolysis products generally allows observation of the more volatile constituents produced by the thermal degradation.

<u>Capillary column pyrolysis - gas chromatography.</u> As mentioned earlier, packed columns, as opposed to capillary columns, are often adequate for polymer fingerprinting. The chromatograms are relatively simple and the analysis time relatively short, and the technique is particularly amendable to the examination the more volatile degradation products, such as monomers produced by thermal degradation. However, when wishing to distinguish between similar polymers such as polyethylene and an ethylene-polypropylene copolymer it may be necessary to use capillary columns so as to produce a thermogram of more complexity, so that differences can be seen between different polymers.

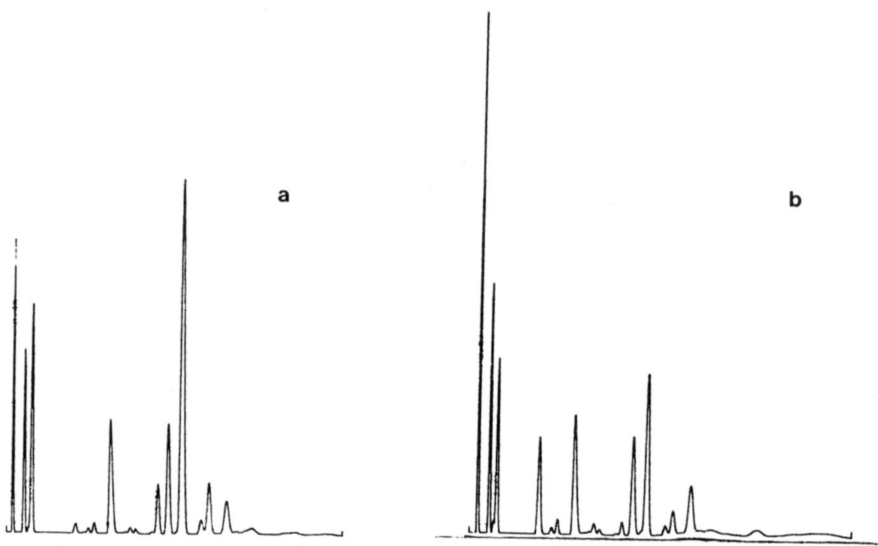

Figure 8 Pyrolysis - gas chromatography of modacrylic resin.
          Instrument: Perkin Elmer 8310 series; column 1 m
          Poropak OS. Conditions: 50°C (2 min), 15°C min^{-1} to
          230°C in 15 minutes Pyrolysis: 950°C for 10 sec
          (a) modacrylic resin, (b) acrylic resin

Separation of all the more volatile fragments is not possible
without cryogenic cooling of the oven. Under standard chromatographic
conditions it is the high boiling degradation products that are readily
separated. Therefore it is the latter part of the chromatogram that is
used for identification and fingerprinting.

Almost no peak separation is achieved in the first 2.5 minutes even
with an initial column temperature of 50°C. For the same samples, such as
polyethylene, the parent polymer thermally degrades into similar fragments
of varying chain length and thus gives a reiterant pattern. Other
samples, such as the ethylene-propylene copolymer, thermally fragment less
uniformly and such patterns are not observed.

Because it is the higher boiling degradation products that are being
observed, a number of polymers have very similar fingerprints. In the
case of polyethylacrylate and polyvinyl acetate, Figs. 9a) and b), the
pyrograms produced are almost identical. The only observable differences
are the presence of two additional peaks in the vinyl acetate pyrogram
(Fig. 9b) at 8.74 minutes and 12.79 minutes (marked with an asterix).

Therefore, capillary column pyrolysis - gas chromatography is
slightly less useful than packed column pyrolysis as a strict
identification technique. However, the complexity of the fingerprints
produced allows minute details to be observed. Analysis times are two to
three times longer than those for packed column pyrolysis - gas
chromatography; runs are generally in the order of 60 minutes in length.
The high split ratio and very small amounts of sample required to obtain
good results with capillary column pyrolysis - gas chromatography mean
that there is an obvious advantage in using this technique when only very
minute amounts of sample are available for analysis.

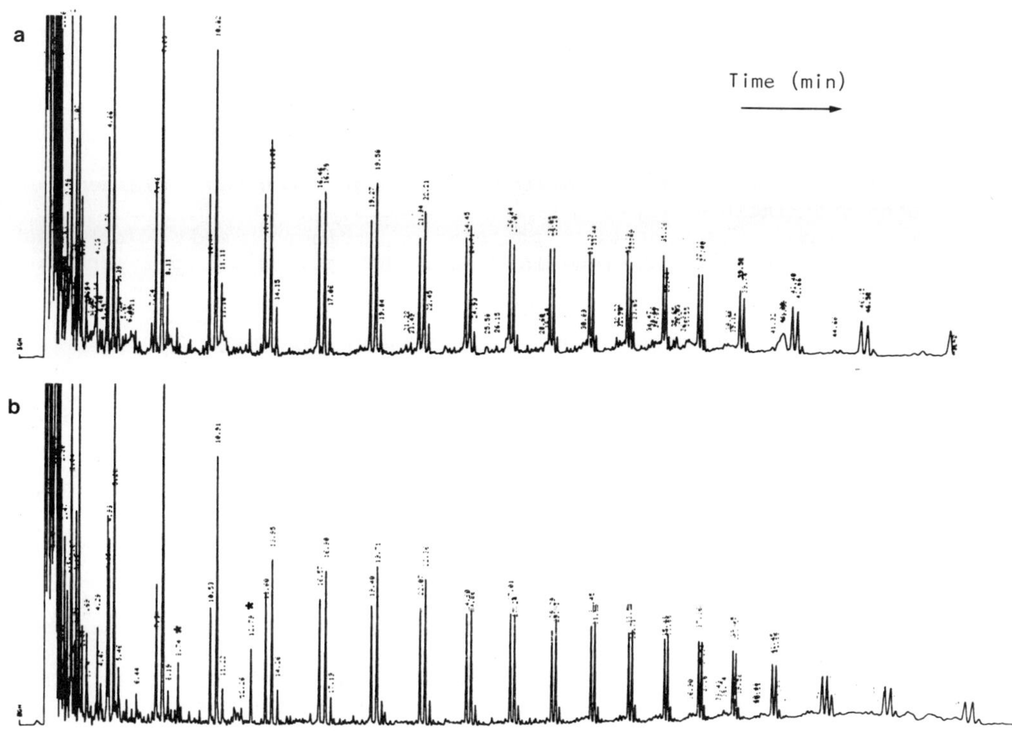

Figure 9   Pyrolysis - gas chromatography of (a) polyethyl
acrylate, (b) polyvinylacetate

The Chemical Data Systems - filament coil type pyrolyser, Pyro-probe
122.   The polymer samples are held in a quartz tube that was inserted into
the platinum coil.   If pyrolysis is to be conducted at slow rates the
pyrolyser is interfaced to a sample concentration (Model 320, Chemical
Data Systems) which collects the pyrolysates on a Tenax filled trap.   In
this way polymer samples can be processed for minutes or hours at slow
heating rates and the pyrolysates collected for a single gas chromatographic analysis.   When the pyrolysis is complete, the trap is pulse
heated and backflushed with carrier gas.   The desorbed pyrolysates are
transferred to a gas chromatograph (Model 3700 Varian) equipped with a 50
m x 0.25 mm i.d. SE-54 capillary column (Quadrex).   A 60:1 split was
established at the injection port of the gas chromatograph and the column
was programmed (from 50 to 280°C at a rate of 40°C min-1).   A flame
ionization detector is used and the area data obtained from a recording
integrator (3390 A Hewlett Packard).

Wampler and Levy[89] used this apparatus to study the effects of slow
heating rates on the products formed during the pyrolysis of polyethylene.
They showed that both the pyrolysis temperature and the rate at which that
temperature is achieved have significant effects on the formation of
pyrolysates from a solid polymer and this will have a bearing on the use
of pyrolysis - gas chromatography technique for identification of polymers
by the fingerprinting approach.   Pyrolysis at very slow rates (0°C min-1
rather and °C millisec-1) produce pyrograms that resemble those of pulse
pyrolysis at lower end-point temperatures.   It is possible that the slower
rates permit one to see more of the primary pyrolysis products, which may
be formed at lower temperatures then volatilized and removed from the
pyrolysis chamber before secondary pyrolysis reactions ensue.   Pyrolysis

at elevated temperatures or at a faster rate to the same final temperature, may enhance secondary pyrolysis reactions. For polyethylene, the general trend is for the production of more terminally unsaturated dialkenes at higher temperatures or faster rates of heating, whereas the corresponding alkane is enhanced at lower temperatures and slower rates.

<u>Laser pyrolysis - gas chromatography.</u> The advantages claimed for this technique include rapid heating and cooling of the sample and relatively simple fragmentation patterns. Folmer[90] and Folmer and Azarrage[91] studied this technique in detail and applied it to a range of polymers.

<u>Pyrolysis - Gas Chromatograpy of Polymers - Laser Pyrolysis Technique</u>

<u>Equipment.</u> A Gen-a-lite Model 3R laser (General) (Laser Corporation, Natick, Mass 01760) and a Focuscope (General) Laser Corporation, focussing device. The laser uses a $3\frac{1}{4}$" ruby rod with a maximum energy output of 2 joules with a pulse length of 600 micro-seconds. The instruments are mounted on an optical rail so that the focused beam entered a box containing a sample holder and a peak-diverting prism. The prism is mounted so that the beam can be moved in two directions and can thus be aimed at any desired portion of the sample. The sample holder is a borosilicate glass tube 6 mm o.d. by 25 mm in length. It is clamped between two Teflon (Du Pont) gaskets so that carrier gas can sweep through the tube in to the chromatograph. The carrier gas enters the chromatograph through a 1/16 inch o.d. tube which pierces the injection port septum and extends into the injection port.

Fig. 10 is a schematic drawing of the apparatus showing the light paths.

The chromatograph used was an F and M model 810 (F & M Scientific Corp., now Hewlett-Packard Corp., Avondale, Pa.) modified by the substitution of a pair of flame ionization detectors for the original thermal conductivity detectors. The flame ionization detectors and the associated duel electrometer were from a Varian Aerograph (Varian Aerograph, Walnut Creek, Calf. 94598) instrument. The output signal from the electrometer is connected to an Infotronics Digital Readout Systems Model CRs-104 (Infotronics, Inc., Houston, Texas 77042) and the output from this system went into a 10 mV Honeywell (Honeywell, Inc., Philadelphia, Pa. 19144) strip chart recorder.

<u>Column A.</u> A 223 cm (7 ft) 0.25 cm i.d. (3/16 inch o.d.) stainless steel column of 70-80 mesh Anakrom ABS coated with 10% UCW-98. This column is operated at room temperature for one minute, then heated to 70°C at the end of the second minute; its temperature is increased 20°C/min after a maximum temperature of 210°C is reached. This temperature is maintained until all of the components apparently eluted. Helium is used as a carrier gas at a flow rate of 40-50 ml/min. Hydrogen and air flows are optimized and maintained at those values.

<u>Column B.</u> 476 cm (15 ft) 0.25 cm i.d. (1/8 inch o.d.) stainless steel column of 70-80 mesh Anakrom ABS coated with 10% UCW-98. It is operated at room temperature for 2 minutes, at 70°C for another minute, then programmed at 10°C/min to the end of the analysis or to a temperature of 260°C which ever came first. Column flow rate is 22 ml/minute.

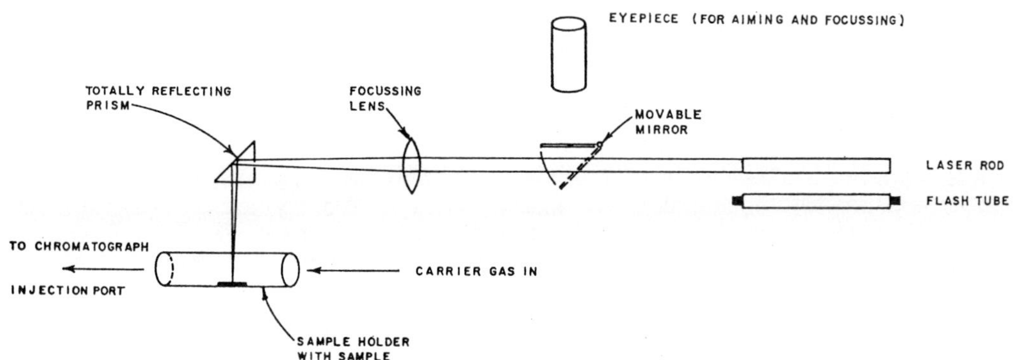

Figure 10 Schematic drawing of laser pyrolysis apparatus

Detector flows are the same as with the other column. Energy output of the laser rod is measured with a Quantromix Model 504 energy meter equipped with a Model 500 energy receiver.

Sample preparation. Three different types of sample preparation were used. All samples were run as solids.

1.   None. Small pieces of the sample were run as-is.

2.   Coated. Sample pieces were coated with graphite by rubbing them with a 6B drawing pencil. Gold coating was accomplished in a vacuum evaporator.

3.   Added carbon. Weighed amounts of sample and finely ground premium coke are melted and intimately mixed in a porcelain dish. In some cases the sample was ground and mixed with coke before melting and further mixing.

Folmer[90] studied the effects of different operating conditions and methods of sample preparation on fragmentation patterns. Clear or translucent samples give reproducible results if mixed with carbon. The concentration of carbon is critical; it has a great effect on the fragmentation pattern. Some polymers were run whose pyrolysis chromatograms show great difference in pattern. Others had patterns which were quite similar. To better compare these similar patterns and the patterns arising from different operating conditions, a statistical method was devised. An attempt was made to correlate these comparisons of patterns with some of the known characteristics of the polymers.

4.4.4 Photolysis - Gas Chromatography

Juvet et al[92] have used this technique for the preparation of finger print pyrograms of polymers. They claim that this technique yields considerably more simple and reproducible decomposition patterns than filament, furnace and Curie point pyrolysers and claim that this is due to greater control of energy input and a more predictable manner in which the polymers decompose photolytically.

Table 15 - Photolysis Products of Common Polymers

Polymer	Retention Index, I on SE-30	Rel. peak area[a]
Polyethylene	<500	1.0
	604 ± 12	0.1
	680 ± 8	0.02
	702 ± 6	0.1
	765 ± 5	0.02
	800 ± 5	0.15
	876 ± 5	0.01
	900 ± 5	0.07
Polystyrene	<500	0.02
	660 ± 10	1.0
	784 ± 8)	
	830 ± 8)	ca. 0.5[b]
	886 ± 6)	
Poly(methylmethacrylate)	<500	1.0
	720 ± 8	0.5
	1525 ± 5	0.2
	1550 ± 5	0.2
Poly(tetrafluoroethylene)	<500	...

[a] Irradiation time, 30 minutes.

[b] Total of three incompletely resolved peaks.

Photolysis is carried out on a pure thin film of the polymer which is then irradiated using a medium pressure mercury source. The photolysis products are swept onto a gas chromatogram to produce a pattern characteristic of the polymer.

A tabulation of the photolysis products of some common polymers is given in Table 15.

Polyethylene yields primarily a series of products eluting at integral multiples of I = 100; these are undoubtedly n-alkanes and the corresponding alkenes. Products at intermediate I values are probably due to branched chain hydrocarbon fragments.

Polystyrene yields benzene (I = 660) as the major photolysis product with apparently smaller amounts of toluene, ethylbenzene and styrene formed. Quantitation of the last three materials is complicated by variable amounts of residual monomer in the samples available and a small amount of thermal decomposition occurring in the injection port.

Ultraviolet absorbance of the polymer influences the response to photolytic degradation. Polyolefins, which are relatively transparent in the ultra-violet, receive essentially constant radiation throughout the sample and thus yield photolysis products as a function of total sample weight. Highly ultra-violet absorbing polymers, such as polystyrene, strongly attenuate the incident radiation. Measurements of the photolysis yield of benzene from polystyrene films over a ten-fold range of sample weight and film thickness show that in the case of polystyrene, photolysis is surface area controlled and products are formed within a thin surface layer. If the injection port temperature of the gas chromatograph is maintained below the glass transition point of polystyrene, less than 10% of the photolysis products formed are volatilized. Raising the temperature above the glass transition temperature lowers the viscosity of the polymer and allows photolysis products to diffuse out of the polymer and be swept into the column.

Photolysis of poly(methyl methacrylate) proceeds primarily by formation of the monomer, corresponding to the product observed at I = 720. As in the case of polystyrene, small amounts of residual monomer and monomer formed from thermal decomposition were observed in non-irradiated samples. Methanol and methyl formate, which have also been reported as photolysis products are not separated on SE-30 but analysis on Carbowax 20 M give two peaks corresponding to these substances.

4.4.5 Pyrolysis - Mass Spectrometry

Hughes et al[93] and Menzelaar[94] have shown that pyrolysis - mass spectrometry has considerable potential for the characterisation and discrimination of natural and synthetic polymers. Their equipment was specially built for the technique and combined a Curie point pyrolyser and quadruple mass spectrometer operated in an optimum geometric arrangement. They describe the linking of a Curie point pyrolyser to a magnetic sector mass spectrometer to provide a method of generating pyrolysis mass spectra. The pyrolysate is passed through an empty glass chromatographic column and jet separator before entering the mass spectrometer. Forty or more sequential mass spectral scans are integrated by computer processing to give a composite mass pyrogram.

A Pye Curie-point pyrolyser, a Varian 2700 gas chromatograph and a V.G. Micromass 12 F mass spectrometer were used. The pyrolyser was mounted on the front of the gas chromatograph and was purged with helium at a flow rate of 5 ml min-1. Samples were pyrolysed at 610°C (heating time 15 seconds) and the pyrolysate was swept into an empty 45 cm x 6.35 mm o.d. x 2 mm i.d. glass column maintained at 200°C in the gas chromatograph. A make-up flow of helium (10ml min-1) was introduced at the end of the column by using a Swagelok stainless-steel 1/4-1/6 inch reducing union with a length of 1/16 inch o.d. stainless steel tubing brazed into it. The mixed outlet flow as passed through a length of glass-lined stainless-steel microbore tubing to the jet separator of the mass spectrometer.

The mass spectrometer was operated under standard electron impact conditions, electron energy 70 eV, emission current 100 uA, accelerating voltage 45kV and source temperature of 240°C. About 50 scans of the pyrolysate were made with acquisition, storage and processing of spectra carried out with a V.G. 2040 data system, supplemented with an X - Y plotter to produce composite X - Y plots (mass pyrograms) of the integrated spectra after background subtraction.

The system was used to study a wide range of polymeric materials including bitumens, adhesives, putties and greases. The sensitivity is sufficiently high to allow samples of 5 ug or less to give adequate electron-impact spectra, but in the chemical ionisation mode larger samples are necessary. Mass pyrograms are usually characteristic of the sample type and frequently allow discrimination between samples of similar composition provided that the analysis is reproducible. In a few instances where this is not so, the irreproducibility appears to be associated with the partial segregation of the pyrolysate during passage through the glass column.

The advantages of pyrolysis - mass spectrometry over pyrolysis - gas chromatography for generating information about polymeric materials are its speed, sensitivity, ease of producing data that can be computer processed and the elimination of the variables associated with gas chromatography. A major disadvantage of pyrolysis - mass spectrometry is that a complex mixture is produced by a combination of pyrolysis and electron-impact fragmentation, which makes a mass pyrogram more difficult to interpret than the chromatograms produced in pyrolysis - gas chromatography, in which only a pyrolytic breakdown is involved.

Jackson and Walker[95] studied the applicability of pyrolysis combined with capillary column gas chromatography to the examination of phenyl polymers (e.g. styrene-isoprene copolymer) and phenyl ethers (e.g. bis[m-(m-phenoxy phenoxy)phenyl]ether). They examined the effect of varying parameters affecting the nature of products formed and relative product distribution in routine pyrolysis. These parameters include the effects of pyrolysis temperature rise times, pyrolysis temperatures up to 985°C and pyrolysis duration. Temperature rise time (0.1 to 1.5 seconds) is not a critical factor in the Curie point pyrolysis of styrene-isoprene copolymer, either with regard to the products formed or the relative distributions. Additionally, the variation of pyrolysis duration or hold time (2.0 to 12.5 seconds) at a fixed Curie temperature reflected no change in the nature of components formed; however, changes in product distributions were observed. Variations in Curie temperature at a fixed pyrolysis duration produced drastic changes in product distributions such as a three-fold change in isoprene dimer formation; however, temperature variance did not change the nature of the products formed. A "polymer like" compound, bis(M-[m-phenoxy phenoxy]phenyl) ether, produced two primary pyrolysis products, diphenylether and dibenzofuran.

Three pyrolysers were used for this work. The three systems showed that Curie point thermal fragmentation patterns were not reproducible, at least in the two laboratories. The three units consisted of the following: A pyrolyser (Phillips Electronics) equipped with a 0.05 to 10.0 second duration timer and a 30 W power supply; a pre-column pyrolyser (Pye-Unicam) equipped with a 0.2 to 15.0 second duration timer and a 30 W power supply, and a lab-built pyrolysis chamber equipped with a manual timer and a 2.5 kW rf generator (Lepel High Frequency Laboratories, Model T-1.5-1-MC-AP/13).

The data from the quantitative experiments were obtained using a dual flame detector gas chromatograph (Hewlett Packard Model 5750) and two digital integrators (Infotronics Models CRS-100 and CRS-108). The quantitative data were obtained with a dual flame detector gas chromatograph (Varian Aerograph Model 1700) and a zero dead-volume column effluent splitter was connected in tandem for gas chromatography/mass spectrometry (GC/MS) operation. A Biemann-Watson type molecular separator provided the interface between a mass spectrometer (Atlas CH-7) and the gas chromatograph. A block diagram of the pyrolysis GC/MS system is shown in Fig. 11.

The separation column(s) used for the polymer products were 50 ft x 0.02 inches (i.d) support coated open tubular (SCOT) column (Perkin-Elmer Corp.) containing 5 ring polyphenyl ether liquid phase with Apiezon L + Igepal Co 880 added; and 200 ft x 1.02 inches (i.d.) di-isodecylphthalate capillary column. The column used for the separation of the pyrolysis

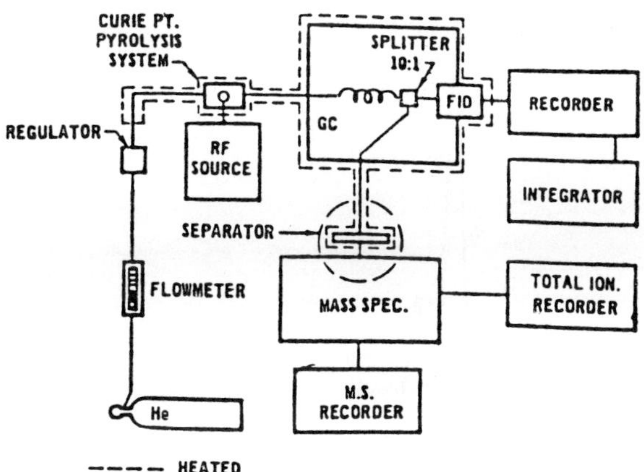

Figure 11 Pyrolysis GC-MS system. Note that the dashed line
         indicates the heated carrier gas transfer lines and the
         gas chromatograph heated areas

products from the bis(M-[m-phenoxy phenoxy]phenyl)ether pyrolysis was a
100 ft x 0.02 inches (i.d) SCOT column (Perkin Elmer Corp.) containing
free fatty acid (FFAP) liquid phase.

In the procedure the polymer sample is dissolved in a volatile
solvent (benzene) to form a 10 weight percent solution. The pyrolysis
Curie point temperature wire is dipped 0.25 inches into the polymer
solution. The polymer-coated wires are then placed in a vacuum oven at 75
to 80°C for 30 minutes to remove the solvent. Multiple coatings of
polymer on the wire were necessary for tandem GC/MS identification of
trace products; however, multiple coatings are not recommended for
quantitative data. After the sample is positioned in the pyrolyser, a
period of 5 minutes is allowed for flow rate equilibration.

An auxiliary heating device (heat gun or heating tape) was used to
pre-heat the carrier gas line before entry into the pyrolyser. The
pyrolysis chamber was maintained at 174°C. Asbestos tape and a heating
tape were used to eliminate cold spots between the pyrolysis chamber and
the gas chromatograph injection port so that sample and products would not
condense prior to entering the column.

Fig. 12 shows a pyrogram of the copolymer (isoprene-styrene)
resulting from a 10 second pyrolysis at 601°C. One might expect that such
a copolymer could yield a product distribution similar to the sum of the
two constituent product distributions. For example, when the polymer
polyisoprene is pyrolysed, C 2, C 3, C 4, isoprene and $C_{10}H_{16}$ dimers are
produced. When polystyrene is pyrolysed, styrene and aromatic hydrocarbons are the products. Fig. 12 shows that the copolymer product
distribution and relative area basis resemble the two individual polymer
product distributions.

Mattern et al[96] carried out laser mass spectrometry on polytetra
fluoro-ethylenes. They found a fragmentation mechanism common to each
fluoropolymer yields structurally relevant ions indicative of the
orientation of monomer units within the polymer chain. A unique set of
structural fragments distinguished the positive ion spectra of each
homopolymer, allowing identification.

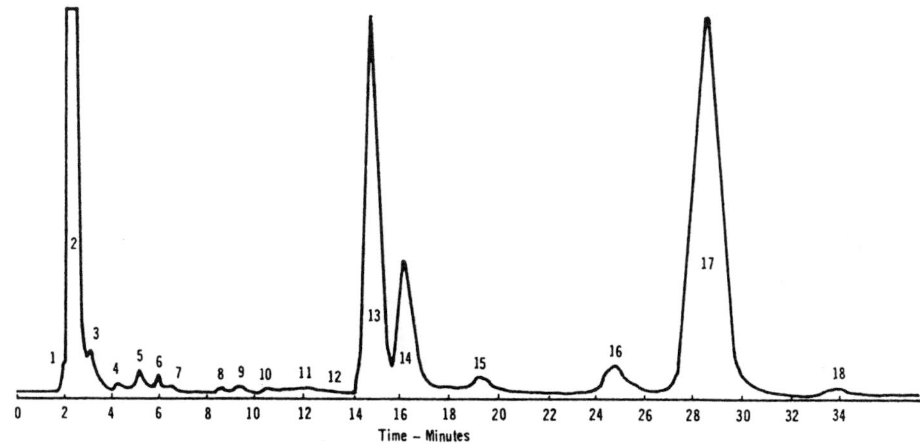

Figure 12 Pyrograms of isoprene-styrene copolymer at 600°C, 10 sec
         duration

Chin-An-Hu[97] has described a pyrolysis mass spectrometric method for
polymer characterisation.

<u>4.4.6 Infrared Spectroscopy</u>

Infrared spectra of thin films of polymer in the region up to 4000
cm^{-1} are characteristic of the polymer. Some infrared spectra are shown
in Appendix 1. Computerized retrieval from data in a library of standard
polymers have been used in the pyrolysis - gas chromatography and the
infrared fingerprinting techniques to facilitate polymer identification[98]. Where doubts exist it would be advisable to attempt
identification by both pyrolysis - gas chromatographic and infrared
spectroscopic techniques so as to obtain confirmation.

The spectra were run either as potassium bromide discs (1.5 - 2 ug
polymer per 400 mg potassium bromide) or as polymer film of varying
thickness up to 12 micron using a sodium chloride prism under the
following conditions[99]:

Adjustment 0 to 100 cent T at 2.00 u
Gain: 3.9 - 4.0 (on the brink of aperiodicity)
Time constant of the amplifier: 2
Speed of the optical attenuator: 1100
Resolution (slit programme): 927
Scanning speed: 1 x 10 + scan program (about 11 minutes from 1.25 to
15.25 micron
Speed suppression: 8
Nernst current: 0.30 to 0.35 A
Wavelength region scanned: 1.25 to 15.25 micron

Tables 16 and 17 tabulate further data on the infrared absorptions
occurring in polymers.

Table 17 in particular is useful when attempting to identify a
polymer or a mixture of polymers by the infrared fingerprinting technique.

Deschant[100] has discussed in detail the infrared spectra of 35 polymers and Hippe and Kerste[101] developed an algorithm for the infrared identification of vinyl polymers. Pyrolysis followed by infrared spectroscopy is a particularly useful technique for application to polymers which are rendered opaque or completely non-transparent by the presence of pigments or fillers.

Leukoth[102] used this technique to determine the chemical composition of plastics and rubbers containing high proportions of pigments or fillers.

Alexander[103] has described a method for obtaining spectra of thin films of polymer which are free of interference fringes. The method is based on measuring a transmission/reflection spectrum of the polymer using a specular reflectance accessory on the infrared instrument. The film is placed on a reflectance accessory with a mirror above it. The radiation transmitted through the film is returned by the mirror and so is measured together with the radiation reflected by the film. This arrangement provides a general method for eliminating interference and so is useful for the examination of polymer films. The most important practical consideration is that the film is flat and parallel to the mirror and this can be ensured by placing the film between windows or between a window and the surface of the mirror. Spectra of films may be subject to considerable errors when fringes are present. An improvement in the spectra is obtained by using transmission/reflection rather than simple transmission measurements.

Osland[104] has described a heated press for the preparation of plastics films for analysis by infrared spectroscopy. The press can produce films as thick as 500 um and of reproducible thickness. Earlier workers have used the hot press film method, here the sample is heated until molten, pressed to a thin film and allowed to solidify. Alternatively, if the identity of the plastic is known, the sample material can be dissolved in a suitable organic solvent and a film cast onto glass or a cell window.

These techniques are adequate for identifying the bulk polymer but polymer blends can contain many other components which must be identified or quantified and for this purpose they are generally unsuitable, the reason being that they cannot produce thick films or reproducible film thicknesses.

A new sample handling accessory with which films of constant thicknesses can be prepared has been introduced by Phillips Analytical. The plastics film press contains a thermostatically controlled oven unit which is calibrated up to 300°C (Table 18). It also contains a cooling facility which may be connected to a low pressure compressed air supply to cool the prepared films rapidly. Reproducible thickness is ensured by using a set of brass dies which can be heated and cooled quickly. The dies can produce films of 20 um, 50 um, 100 um, 200 um and 500 um thickness.

To prepare a sample film using the Precision Plastics Film Press, the required die is positioned in the base of the screw press and the appropriate mass of material placed on the centre of the die. The amount of material required depends on the thickness of the film being prepared, a 20 um film requiring 6 mm^3 while a 500 um film requires around 75 mm^3.

The probe of the thermometer passes through the handle of the screw

Table 16 - Some Infrared Spectral Lines of Polymers

Figure No.	Characteristic Absorptions (microns)
1. Polyacrylonitrile	C=N, 4.47 Aliphatic C - H, 3.41, 3.49 C = O stretching, 5.78 Others 6.01, 6.90, 3.34, 3.41
2. Polyformaldehyde	C - H stretching, 4,90, 6,80, 7.0, 7.23, 8.08, 9.15 Others, 3.34, 3.41, 11.15 (strong)
3. Polyisobutylene	Aliphatic CH) 3.35 - 3.50) 6.70 - 6.90 Others, 7.20, 7.32, 8.15 Doublet at 10.56, 10,87
4. Polyester resin	Aliphatic CH, 3.30, 3.40 C = O stretching, 5.80 (strong) C = O stretching ester, 7.65 - 8.03 Others, 6.09, 6.15, 6.50 6.61, 6.70, 7.65 and 12.90 and 14.35 (characteristics)
5. Buna Rubber	Aliphatic CH, 3.40 Double bond, 5.90 Others, 6.10, 6.25, 6.69, 6.90-6.96 Strong, 10.36, 11.0, 13.22, 14.33
6. Vinylite polymers (Union Carbide) copolymerized vinyl chloride - vinyl acetate	Vinylite VMCH aliphatic CH, 3.40 C = O stretching, 3.65, 5,76 H O H binding adsorbed water, 6.16

		C - O stretching, 7.00, 7.32, 7.56, 8.1
		Others, 9.17, 9.28
7.	Vinylite VAGH	Aliphatic CH, 3,45, 3.56
		C = O stretching, 5.79
		Adsorbed water, 6.16
		C - O stretching, 7.02, 7.55, 8.05
		Others, 9.16, 10.42
		On comparing with above spectra for Vinylite VMCH it is seen that there are sufficient differences to make positive identification possible in the region 7.0-7.5 and 9.5-10
8.	Vinylite XYHL	The difference between this and the spectra of Vinylite VMCH and Vinylite VAGH is the strong absorption at 8.73
9.	Silicone oil	Methyl in Si-CH3, 7.95, 9-10,12.6
		Si-O-Si, 12,6
		Aliphatic CH, 3.36, 3.43 - vibrations
10.	Methyl cellulose	OH stretching, 2.90
		Aliphatic CH stretching, 3.42, 3.51
		Adsorbed water, 6.15
		(similar spectrum to cellulose)
11.	Ethyl cellulose	Splitting of aliphatic CH stretching vibration at 3.37-3.50
		Others, 6.94, 7.27 (strong), 10.90, 11.37 (weak)
12.	Carboxyl methylcellulose sodium salt	Resembles spectum of cellulose

Table 16 - Continued

13.	Carboxy methyl cellulose (free)	C = O stretching, 5.57 (strong)
		Carboxy, 3-4
14.	Nitrocellulose	R-O - NO2, 6.05, 7.80 (characteristic)
15.	Phenol formaldehyde (asbestos filled)	C-H groups, 3.3 (weak)
		Double bond plus C-H absorption, 6.25 double peak
		-Si O (from asbestos), 9.5-10.0
16.	Melamine formaldehyde (asbestos filled)	N-H groups, 2.75
		Si O (from asbestos), 9.5-10.0
17.	Polycarbonate	OH, 2.90, 6.12 (adsorbed water) CH adjacent to double bond, 3.28 (aromatic CH)
		Aliphatic C-H, 3.37 3.48
		C = O stretching, 5.88 (strong)
		CH, 6.9, 7.10-7.37
		Carbonylester, 8.0
		1.4 disubstitution, 12.1, 13.8
18.	Polyethylene glycol	OH, 2.90 weak
		Aliphatic C - H, 3.45 (strong), 6.75-7.50
		- CH2 - O-CH2 band, 9
19.	Polypropylene glycol	OH adsorption, 2.90
		Triple C-H, 3.35 - 3.50
		CH, 6.9
		CH3, 7.3 (strong)
		C-O-C, 9 (strong)

		Others, 10.8, 11.6, 12.0
20.	Polyvinylidene chloride	Aliphatic C-H, 3.4-3.5
		C = O absorption, 5.75 (impurity)
		CH, 7.10, 7.31
		C-Cl, 13.35 (usually 13.3-14.3)
21.	Chlorinated polypropylene	Absorbed water, 2.93, 6.16
		Aliphatic C-H, 3.40, 6.96, 7.25
		Methyl C-Cl 13.15, 13.65, 12.75 (probably)
22.	Polyvinylacetal	OH, 2.91
		Aliphatic CH, 3.4
		C = O ester, 5.76
		Methyl groups, 7.28
		- CH_2 - O - CR, 8.8
23.	Polyacrylamide	C = O, 6.00
		C = N (probably), 4.5
		OH (moisture), 2.95
24.	Cellulose propionate	Moisture, 2.9, 6.13
		Triple C - H adsorption, 3.4
		Ester carbonyl, 5.7, 8.6
		Others (characteristic), 7.25-8.5, 11.4, 12.45, 13.45
25.	Cellulose acetate butyrate	Characteristic absorptions, 10.5, 11.1, 12.0, 12.6, 13.4, 14.5

Table 16 - Continued

26.	Butyl rubber	Aliphatic CH, 3.4 (strong), 6.8
		Two methyls attached to carbon, 7.19, 7.31
		CH adjacent to aliphatic double band, 10.55, 10.85
27.	Nitrile rubber	CH adjacent to double band, 10.3 (strong), 3.25, 6.00, 6.10, 6.25 (strong)
		Aliphatic C - H, 3.40, 3.49, 6.9
		C = N, 4.46
28.	Styrene-butadiene rubber	Polystyrene bands, 3.25, 3.30, 3.35
		(C - H adjacent to double bond)
		Monosubstituted benzene ring, 5.6, 13.2, 14.33
		C - H adsorption, 10.35
29.	Styrene-butadiene acrylonitrile	Moisture, 2.92, 6.12
		Aromatic C - H, 3.28, 3.31, 6.28 and C -H adjacent to double bond
		Aliphatic C - H, 3.41, 3.50, 6.88. C = N, 4.47
		Monosustituted benzene ring, weak 13.2, 14.35
30.	Ethylene-propylene copolymer	Closely resembles polyethylene spectrum plus double bonds, 10.3
		Aliphatic C - H, 3.4
		CH adjacent to double bond, 3.3
		Methyl groups, 7.25

31.	Ethylene-propylene-diene terpolymer	Closely resembles spectrum of ethylene propylene copolymer. Addition spectral lines
		Aliphatic double bonds, 3.3 (weak) 10.65
32.	Epoxy resin (araldite)	Hydroxy, 2.9, 9.04
		C - H, 3.27, 3.34, 6.20, 6.30
		Split methyl group, 7.20, 7.32
		1,4-di-substituted benzene ring, 5.6 (weak) 12.3
		Aliphatic C - H, 3.39, 3.43, 3.45, 6.84
33.	Coumarone-indene resin	Adsorbed water, 2.92
		Aromatic C - H, 3.31, 6.49
		Substituted benzene ring, 13.4, 14.3
		Aliphatic C - H, 3.42, 3.50, 6.90
34.	Polyvinylacetate	C = 0, 5.75
		Ester, 8.0, 9.8
35.	Polyvinyl butyral	Characteristic absorptions, 7.25, 8.05, 8.9, 10.0
36.	Polytetrafluorethylene	CF2, 8.4
		Others, 8.4, 4-5
37.	Polyethylene terephthate (Terylene)	OH stretching, 2.82
		C = 0, 2.90, 5.80
		H stretching, 3.25
		CH2 stretching, 3.35, 3.43
		CH2 deformation, 7.08, 7.20
		CH deformation, 7.90

Table 16 - Continued

C-O-C stretching, 8.51, 9.59

C-C, 6.80

Others, 5.80, 6.19, 11.45, 13.77,
6.65, 7.30, 7.45, 8.86, 9.10, 9.80,
10.3, 11.15, 11.45, 11.80, 12.62,
13.75

38. Polymethlmethacrylate

Ester carbonyl, 5.75

C-O-C stretching, 6.70-6.95, 8.30-8.80

Other, 10.00, 10.40

39. Cellulose (cellophane)

Water absorption, 2.8 - 3.4

OH stretching, 3.4

CH stretching, 3.4

C = O stretching, 5.8

COO - ion stretching, 6.28

CH2 deformation, 7.0

CH deformation, 7.3, 7.6, 11.1

OH deformation, 7.5, 8.9, 15.4

C - O stretching, 8.6

OH bending, 9.4

C - OH stretching, 9.8

C - OH, 8.9

C - O, C - C stretching, 10.0

Other absorption, 9.0, 6.1

40. Cellulose acetate

Adsorbed water, 2.9

Ester carbonyl, 5.75

		Other absorption, 3.5-5.35, 8.0, 7.5-10.0
41.	Melamine	Characteristic absorptions, 3.0, 6.0-6.5, 9.7, 12.3
42.	Urea formaldehyde	Characteristic absorptions, 6.0-6.2, 6.9, 8.0, 10.0
43.	Natural rubber	C = C, 6.08
		CH2 and CH3 deformation, 7.0, 7.25, 12.00
		Other absorption, 7.5 - 11

press to the die centre. The exact pressing temperature will depend not only on the type of polymer but also to some extent on the nature and concentration of any additives present and the actual temperature could be vastly different to those given in Table 18. After full pressure has been applied for about 30 seconds, the liquid plastic will have spread evenly over the die face and the press can then be transferred to the cooling section and the air flow valve opened.

The films produced have a diameter of approximately 16 mm which can be mounted on the standard two piece disc holder and mount. Alternatively, the films can be mounted over an aperture cut in a card frame, with adhesive tape.

Another approach to obtaining the infrared spectrum of polymers is the surface reflectance technique[105, 106].

Gardella and Grobe[106] compared attenuated total reflectance and photoacoustic sampling for surface analysis of polymer mixtures by Fourier transform infrared spectroscopy. They show that analysis by attenuated total reflectance is more suitable for smooth surfaces and is faster. Photoacoustic methods have shallower sampling depths than attenuated total reflectance but the latter technique is applicable over a range that is more controllable.

Particular studies on the infrared spectra of polymers include isotactic poly(1-pentene), poly(-4-methyl-1-pentene) and atactic poly(-4-methyl-pentene)[107], chlorinated polyethylene[108], aromatic polymers including styrene, terepthalic acid, isophthalic acid[109], polystyrene[110,111], styrene-glycidyl-p-isopropenylphenyl ether copolymers[112], styrene-isobutylene copolymers[113], vinyl chloride-vinyl acetate-vinyl fluoride terpolymers[114], vinyl chloride-vinyl acetate copolymers[115], styrene copolymers[116,117], ethylene-vinyl acetate copolymers, graft copolymers and butadiene-styrene and acrylonitrile-styrene copolymers[118].

Brako and Wexler[119] have described a useful technique for differentiating the presence or absence of functional groups such as hydroxyl, carboxylic acid or ester in polymers containing small percentage

Table 17

Identification of polymers by infrared absorption

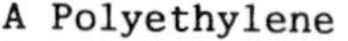

A Polyethylene
B Low pressure polyethylene
C Polypropylene
D Polystyrene
E Rigid PVC
F Plasticized PVC
G Polyvinyl alcohol
H Polyvinyl acetate
I Polyvinyl butyral
J Polyvinyl formal
K Polytetrafluoroethylene
L Polyamide
M Polyterephthalic acid
N Polymethacrylate
O Cellulose
P Cellulose acetate
Q Melamine
R Urea formaldehyde
S Natural Rubber
T Neoprene
a Polyacrylonitrile
b Polyformaldehyde
c Polyisobutylene
d Polyester resin
e Buna rubber
f Vinylite VMCH
g Vinylite VAGH
h Vinylite XYHL
i Silicone oil
j Methyl cellulose
k Ethyl cellulose
l Na-carboxymethyl cellulose
m Carboxymethyl cellulose
n Nitrocellulose
o Soluble starch
p Asbestos phenol
q Asbestos melamine

Table 18 - Approximate pressing temperatures for some polymers in °C

High Density Polyethylene	150
Low Density Polyethylene	110
Linear Low Density Polyethylene	130
Polypropylene	175
Polystyrene	140
Polymethylmethacrylate	150
Polycarbonate	210
Polyvinylchloride	160
Polyacetal	100
Polyamide 6	200
Polyamide 6,6	240
Polyamide 6,10	200
Polyamide 11	160
Polyamide 12	170

components of such groups. Films of latexes or polymers are subjected to chemical treatment which results in marked changes in the infrared spectrum and which can be associated with the disappearance of a functional group. Infrared data may be readily interpreted negatively so that one may definitely preclude the presence of hydroxyl, carbonyl, amine, amide, nitrile, ester, carboxylic, aromatic, methylene, tertiary butyl, and terminal vinyl groups if the corresponding group vibrations are absent in the infrared spectrogram. More difficult is the assignment of functional groups where multiple or several alternative possibilities exist, as in the mixture of a carboxylic and keto group or in the assignment of a band to an olefinic group.

Fig. 13a shows the infrared spectra of a sodium polyacrylate film before and after exposure to hydrochloric acid vapour. Exposure to acid results in the disappearance of the broad, intense band associated with the carboxylate group in the polyacrylate ion at around 6.25 micron. Appearance of a broad intense absorption at about 5.8 micron is associated with carbonyl of the carboxylic acid group in the polymer. Significant changes are also observed in the 9.09-8.33 micron region. Heating the acidified film resulted in minor changes in the spectrum. Figure 13b shows the changes in the infrared spectrum resulting from the exposure of acrylic acid-vinylidene chloride copolymer film to ammonia vapour. Bands associated with the carboxylic acid carbonyl stretching frequencies at 5.83-5.75 micron disappear on exposure to ammonia vapour. A well-defined carboxylate band appears at 6.37 micron. This change is sufficient to confirm that the copolymer contains carboxylic acid groups.

4.4.7 Raman Spectroscopy

Yashino and Shinomiya[120] have published Raman spectra of solutions of various polymers. The technique has also found a limited application in structural studies on polymers[122-133], for example the three C = C stretching bands in polybutadiene corresponding to the three possible configurations can be observed by Raman spectroscopy[121]. Raman spectroscopy has applications in the identification of polymers in which additives obscure the polymer peaks obtained in its infrared spectrum.

An example of the application of Raman spectroscopy is the identification of additives in fire retardant polypropylene. When a sample of polypropylene was examined by infrared spectroscopy the strongest bands (9.8 and 14.9 um) were due to a talc-type material and

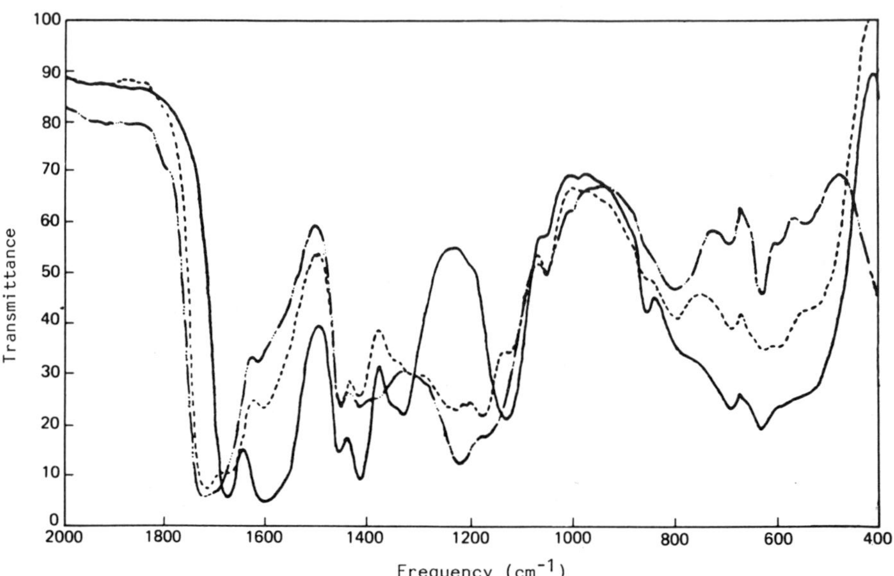

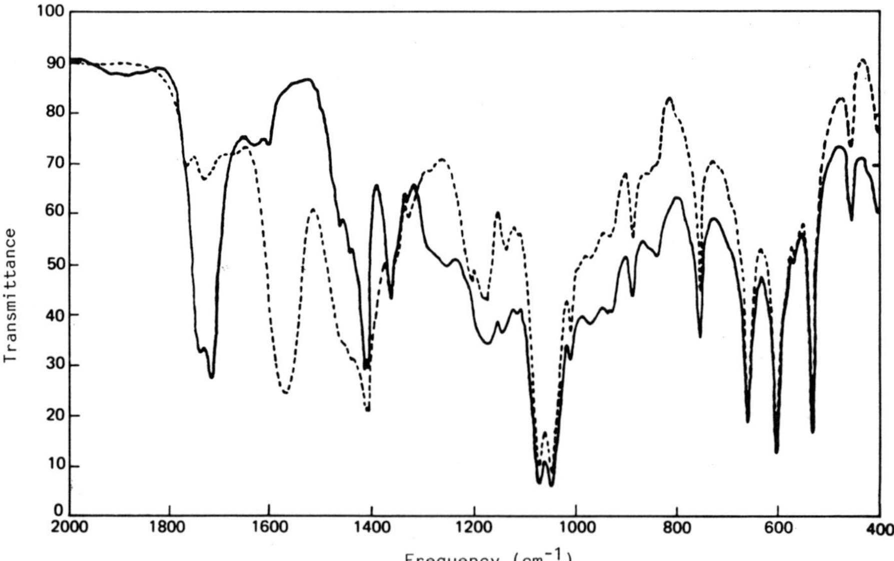

Figure 13   Infrared spectra of chemically treated polymers.
            (a) spectrum of sodium polyacrylate before and after
            exposure to hydrogen chloride vapour and the heated film
            of the hydrogen chloride treated acrylate, sodium
            polyacrylate film, sodium polyacrylate film exposed to
            HCL vapour, sodium polyacrylate film exposed to HCL
            vapour and then heated. (b) spectra of copolymer of
            acrylic acid and vinylidene chloride and the film after
            exposure to ammonia vapour, acrylic acid-vinylidene
            chloride copolymer film, acrylic acid-vinylidene
            chloride copolymer film exposed to ammonia vapour

bands of medium intensity were assigned to polypropylene and possibly antimony trioxide (13.4 um). Additional weak bands in the 7.3-7.7 um region were possibly due to decabromodiphenyl ether. In the Raman spectrum, however, the strongest bands (250 and 185 cm-1 shift) confirmed the presence of antimony trioxide and some bands of medium intensity confirmed the presence of decabromodiphenyl ether (doublet at 140, triplet at 220 cm-1 shift) and polypropylene (800, 835, 1150, 1325, 1450, and 2900 cm-1 shift). The silicate bands that obscured the regions of the infrared spectrum were not observed in the Raman spectrum.

Although both of these spectroscopic methods have a wide use in their own right, this example demonstrates well the complementary value of the two methods, taking advantage of the fact that elements of high atomic number, e.g. antimony and bromine, have relatively more intense Raman spectra but the lighter elements show up clearly in the infrared spectra.

4.4.8 NMR and PMR Spectroscopy

Some standard NMR and PMR spectra for polymers are illustrated in Appendix 1.

ADDITIVES, VOLATILES AND CATALYST REMNANTS

5.1 ADDITIVES

5.1.1 Introductory

In order to appreciate fully the techniques which have been developed for the analysis of additives in polymers, it is necessary to be familiar with the difficulties involved in such an undertaking, and also with the chemical and physical properties of the additives themselves.

Most of the analytical problems arise from three factors; the situation of the additive in a more or less insoluble polymer matrix, the high reactivity and low stability of many types of additives, especially anti-oxidants, and the low concentrations of additives present in many instances in the polymer matrix. The first factor severely limits the choice of analytical techniques that can be applied to the sample without prior separation of the additive from polymer, a procedure which is itself hindered by the nature of the polymer matrix. In addition, any extract of the polymer is liable to contamination by low molecular weight polymer 'wax' which may interfere with subsequent analysis and is difficult to remove. This aspect is discussed further in Chapter 3.

The second and third factors mentioned above combine to make the handling of extracts an exacting job if quantitative information is required. Antioxidants, particularly, are labile unstable compounds, forming complex decomposition products; this considerably complicates interpretation of analytical data, and any loss of material by decomposition is liable to be significant since the quantities present are initially so low. Crompton[134] and others, for example, have recommended that polymer extracts are kept in actinic glassware and used for subsequent analysis without delay. If any storage of solutions is necessary, this should be done under nitrogen, in the dark and in a refrigerator to minimize the effects of oxygen, light and heat on any labile compounds present. Lorenz et al[135] have published data on sample changes during handling of antioxidant extracts, including losses during concentration by evaporation. Generally, however, this aspect of additive analysis does not seem to have received the consideration it deserves.

Apart from these factors which complicate the processing of the sample, there are others which complicate the interpretation of the data obtained, the principal ones being the wide range of additives used nowadays in polymer technology, which makes positive identification difficult by all but the most sophisticated analytical techniques, the presence of several types of additives in a single polymer formulation, e.g., plasticizers, ultraviolet stabilizers, slip agents and possibly two antioxidants, one for processing and one for service, may all be present in a single formulation, and finally, depending on processing history and age of the polymer, the possibility that additive decomposition products may also be present to complicate the analytical problem in hand. The latter type of additive decomposition should be distinguished from that occurring during analytical processing. For example, a particular type of polymer additive may undergo partial thermal degradation during extrusion operations involved in its manufacture, and then during analysis may degrade by another route under the influence of light. The analysis of single additive types is discussed in Chapters 5.1.1 to 5.1.15 and the analysis of mixtures of additives and/or additive degradation products is discussed in Chapter 5.1.16.

To summarize then the determination of additives in polymers presents the analyst with some difficult problems. Only small concentrations are present, complex mixtures may be involved and, moreover, frequently the mixture is of compounds of completely unknown type. Most methods for the determination of additives in plastics come under one of three categories;

i)   direct examination of polymer, i.e. non destructive testing,
ii)   examination of solvent extracts of polymer,
iii)   examination of volatiles released on heating polymer.

An example of direct examination is the examination of the polymer film by infrared or ultraviolet spectroscopy or of thicker sections of polymer by attenuated total reflectance infrared spectroscopy. Such techniques have severe limitations in that, because the additive is in effect heavily diluted with polymer, detection limits are usually well above the low concentration of additive present, and this method is only applicable if the additive has distinct sharp absorption bands in regions where the polymer itself shows little or no absorption. In-situ spectroscopic techniques are not likely to be of value, then, in the analysis of samples of unknown composition. If known amounts of additive can be incorporated into additive-free polymer, however, these techniques are likely to be extremely useful in the study of solvent extraction procedures, and the study of additive ageing processes (i.e. the effects of heat, light, sterilization, radiation, etc.), since the rate of disappearance or of decay can be measured directly by the decrease in absorbance of the sample at a suitable wave-length.

The analysis of polyethylene additives by means of direct ultraviolet spectroscopy is limited by excessive beam dispersion due to light-scattering from the polymer crystalline regions. Additives at low concentrations (0.1%) require sample thicknesses such that analysis must be performed in the presence of a high level of light-scattering which may change unpredictably with wavelength. At lower levels of concentration and corresponding greater sample thickness, unacceptable signal-to-noise ratios exist. Nevertheless, ultraviolet spectroscopy remains an attractive method for analysis for many additives. Principal advantages over infrared spectroscopy include greater sensitivity arising from higher extinction co-efficients and a lack of interfering absorptions from the polymer matrix. These advantages can be realized, however, only if background scattering from the polymer can be reduced.

Some examples of this technique are included in the following section.

Methods for preparing solvent extracts of polymers for additive analysis are discussed in Chapter3 and in the examples which follow.

The third method, involving examination of volatiles released from the polymer upon heating, is obviously only applicable to additives or other non-polymer components such as monomers and residual solvents which are volatile. In this technique the polymer is heated under controlled conditions in a sealed container. A gas chromatographic carrier gas stream is then connected to the container to sweep the volatiles into a gas chromatograph for identification and for quantitation. A classical example of the application of this technique is the determination of styrene monomer and aromatic impurities in polystyrene. Up to 30 different volatiles have been identified in this way in polystyrene in amounts down to 5 ppm.

These techniques are discussed further in sections 5.2 (Monomers and Oligomers) and 5.3 (Other Volatiles).

Methods for the determination of various classes of additives in polymers are reviewed below. These methods are only applicable when dealing with a known additive.

<u>5.1.2   Phenolic Antioxidants, (see also Appendix 3) in Polyethylene, Solvent Extraction Procedures</u>

<u>Ultraviolet spectroscopy.</u> Straightforward ultraviolet spectroscopy is liable to be in error owing to interference by other highly absorbing impurities that may be present in the sample[136-139]. Interference by such impurities in direct ultraviolet spectroscopy has been overcome or minimized by selective solvent extraction or by chromatography[137]. However, within prescribed limits ultraviolet spectroscopy is of use and, as an example,[140-143] procedures are discussed below for the determination of Ionol (2,6-di-tert-butyl-p-cresol) and of Santonox R (4,4'-thio-bis-6-tert-butyl-m-cresol in polyolefins.

Certain additives, e.g. calcium stearate and lauryl thiodi-propionate, do not interfere in the determination. Other phenolic antioxidants, e.g. Ionox 330, Topanol CA and Santonox R, do interfere.

<u>Infrared spectroscopy.</u> Spell and Eddy[144] have described infrared spectroscopic procedures for the determination of up to 500 ppm of various additives in polyethylene pellets following solvent extraction of additives at room temperature. They showed Ionol (2,6-di-t-butyl-p-cresol) and Santonox R (4,4'-thio-bis-(6-t-butyl-m-cresol) are extracted quantitatively from polyethylene pellets by carbon disulphide in 2-3 hrs and by iso-octane in 50-75 hrs. The carbon disulphide extract is suitable for scanning in the infrared region between 7.8 and 9.3 micron, whilst the iso-octane extract is suitable for scanning in the ultraviolet between 250 and 350 nm.

Miller and Willis[145] obtained infrared spectra of antioxidants from polymer films. They compensated with additive free polymer in the reference beam. Infrared spectroscopy is more specific than ultraviolet spectroscopy, but some workers[146] find that the antioxidant level in polymers is too low to give suitable spectra.

In-situ spectroscopic techniques are not likely to be of value then, in the analysis of samples of unknown composition. Luongo[147] and Drushel and Sommers[148] described methods for putting known amounts of additives into polymer. Luongo prepares a master batch by milling a known amount of additive into the polymer and obtains standards by further milling known weights of master batch and additive-free polymer. He then hot-moulds his samples into approximately 0.25mm thick films, either in a standard laboratory metallographic mounting press, or in a larger press between water-cooled, polished aluminium platens.   Drushel and Sommers [148] prepared their standards by adding the inhibitor in hexane solution to the polymer, evaporating the resulting slurry to dryness and hot-pressing between aluminium foil.   In both cases, the films are mounted in frames before spectroscopic examination.

<u>Thin-layer chromatography.</u>   Thin-layer chromatography, like gas chromatography, comes into its own when dealing with mixtures of substances.   However, both techniques have been employed for the determination of phenolic antioxidants.   In determining Santonox R antioxidant in polyethylene, the polymer was refluxed with toluene for 2 hrs, then absolute ethanol added to reprecipitate any dissolved polymer. The filtrate was carefully evaporated to dryness and dissolved in a known volume of chloroform.   Portions of this solution together with standards, were applied to a 250 thick layer of Merck GF 254 silica gel and the plate eluted with 5:1 petroleum ether (40:60) ethyl acetate.   After development the plate was sprayed with a 2% ethanol solution of 2,6-dibromo-p-benzo-quinone-4-chorimine followed by 2% aqueous sodium tetraborate.   The Santonox R content is obtained by comparing the intensity of the purple spot produced bythe sample with those obtained with the standards.   Van der Heide and Wouters[149] examined the thin-layer chromatography of some anti-oxidants in polyethylene.   Slonaker and Sievers[150] and Waggon and Jehle[151] did similar work, and were able to detect between 300 and 900 ppm of antioxidants in polyethylene.   Schroeder reviewed work on the application of thin-layer chromatography to phenolic antioxidants[152].

Dobies[153] has described detailed thin-layer procedures for determining 0.02-0.2% of various antioxidants in polyethylene and polypropylene films.   A feature of this method is that it requires only a small polymer sample (5 g) for the determination of down to 0.02% antioxidant, 3% ethanolic phosphomolybdic acid was used as a spray reagent.   The six antioxidants he studed were: 4,4'-butylidene (2-tert-butyl-5-methyl)phenol; 4,4-thiobis (6-tert-butyl-m-cresol); pentaeryth-ritol tetrakis (3,5-di-tert-butyl-4-hydroxyhydrocinnamate); 2,2'-methylenebis (4-methyl-6-tert-butylphenol); octadecyl (3,5-di-tert-butyl-4-hydroxyphenol) acetate; 2,6-di-tert-butyl-p-cresol.

In Table 19 are reproduced Rf values obtained by Dobies[153] when 20 ul of standard solutions (100 mg/200 ml) of the above mentioned antioxidants were applied to thin-layer plates and eluted with the development solvent.   System A was used for identification purposes while System B was used for quantitative analysis.

In Table 20 are shown the results obtained by Dobies[153] in applying his method to samples of polyethylene and polypropylene containing approximately 0.1%.   2,2'-methylenebis (4-methyl-6-tert-butylphenol), compound with results obtained by an ultra violet spectrophometric method. Although there are minor differences in the results of the methods, the thin-layer procedure shows the presence of an additional zone, attributed to the presence of 2,6-di-tert-butyl-p-cresol.   The presence of this antioxidant, although not determined by the thin-layer chromatographic procedure, could interfere with the ultraviolet procedure thus giving higher results.

Below is described a typical quantitative thin-layer procedure for determining a single additive, namely Santonox R (4,4'-thio-bis-6-tert-butyl-m-cresol) in polyethylene in amounts down to 20 ppm in polymer with an accuracy of ± 20% of the determined amount.

Table 19 - R$_f$ Values of Antioxidants

System A: Silica Gel G, reverse phase 5% Dow Silicone; developing solvent, ethanol-water.

System B: Silica Gel G Uniplates; developing solvent, cyclohexane-methanol.

	R$_f$ Values	
Antioxidant	System A	System B
4,4'-Butylidenebis (2-tert-butyl-5-methyl) phenol	0.80	0.0
4,4'-Thiobis(6-tert-butyl-m-cresol)	0.84	0.0
Pentaerythritol tetrakis(3,5-di-tert-butyl-4-hydroxyhydrocinnamate)	0.60	0.29
2,2'-Methylenebis(4-methyl-6-tert-butyl-phenol)	0.76	0.34
Octadecyl (3,5-di-tert-butyl-4-hydroxy-phenyl) acetate	0.33	0.70
2,6-Di-tert-butyl-p-cresol	0.71	0.72

Determination of Santonox R Phenolic Antioxidant in Polyolefins - Thin Layer Chromatography

Polyethylene extraction procedure. Accurately weigh 5 g of polymer into a 500 ml round-bottomed flask. Add 65 ml of toluene and heat on a boiling water bath for 90 min using a reflux condenser.

Remove the flask from the water bath and immediately pour 85 ml of absolute ethanol down the condenser to precipitate the dissolved polyethylene from the hot toluene solution. Allow the flask to cool to room temperature, remove the condenser, stopper the flask and shake well. Filter the solution through a No. 42 Whatman filter paper into a 250 ml beaker. Wash the flask and residue with a further 100 ml of ethanol.

Evaporate the toluene/ethanol solution almost to dryness on a boiling water bath with the aid of a stream of nitrogen and finally to dryness using only the nitrogen stream. If more polymer is precipitated during the evaporation, refilter the solution into a smaller beaker. Wash the residue into a 2 ml volumetric flask with chloroform and make up to the mark.

Santonox R is extracted from polyethylene by precipitation of the polymer with ethanol from hot toluene solution as described above. The extract is evaporated to dryness, dissolved in chloroform and an aliquot applied to a thin-layer plate and the chromatogram developed. The concentration of Santonox R in the polyethylene is estimated by visually comparing the intensity of the spot obtained with corresponding spots from known quantities of Santonox R.

Apparatus.

Thin layer chromatography tank.
Thin layer plate 20 cm x 20 cm, pre-coated with a 250 u layer of Merck

Table 20

Replicate determination of approx. 0.1% 2,2'-methylenebis (4-methyl-6-tert-butylphenol) in polyethylene and polypropylene

Sample	% Antioxidant				
	1st day	2nd day	3rd day	Average	By U.V. analysis
Polyethylene	0.095, 0.094 0.096, 0.092	0.103, 0.100	0.087, 0.089 0.087	0.094	0.102, 0.105
Polypropylene	0.080, 0.084 0.073, 0.086	0.079, 0.078	0.078, 0.078 0.079	0.079	0.083, 0.083

GF254 silica gel, U.K. distributors, Anderman & Co. Ltd., Battlebridge House, Tooley Street, London SE1.
Hypodermic syring - 25 ul capacity.
Ultraviolet lamp - wavelength 254 nm.

Reagents.

Chloroform, analar, British Drug Houses.
Petroleum ether, 40/60 analar, British Drug Houses.
Santonox R - (4,4'-thiobis-3-methyl 6-6-butylphenol).
Ethyl acetate, Analar, British Drug Houses.
2,5-dibromo-p-benzoquinone-4-chlorimine, British Drug Houses.
Sodium tetraborate (borax), Analar, British Drug Houses.

Place 20 ul of the chloroform extract of the polyethylene sample as a spot on a thin layer plate. Also apply 20 ul aliquots of standard solutions of Santonox R in chloroform, 0.05%, 0.04%, 0.03%, w/v. Develop the chromatogram to a distance of 10 cm in a chromatography tank containing petroleum ether (40:60)/ethyl acetate (5:1 v/v mixture) as the eluent.

Inspect the plate under ultraviolet light (254 nm) and compare the intensity of the Santonox R spot from the polyethylene extract with the intensities of the standard spots. If the spot is of lower intensity than that of the 0.03% w/v standard than a new chromatogram should be developed using standards, 0.025%, 0.015% and 0.005% w/v of Santonox R in chloroform.

Visually compare under an ultraviolet lamp the intensity of the spot from the polyethylene extract with the standard spots and thus estimate the percentage level of Santonox R in the chloroform solution. Spray the thin-layer plate with a 2% w/v ethanol solution of 2,6-dibromo-p-benzoquinone-4-chlorimine. Allow the thin-layer plate to dry and re-spray using a 2% w/v aqueous solution of borax. Re-estimate the amount of Santonox R present in the chloroform solution by visually comparing the purple spots produced.

Calculation. Santonox R from 5 g of polyethylene is concentrated into a 2 ml chloroform solution.

Thus the weight % of Santonox R in high density polyethylene = 2/5 x the level of Santonox R present in the 2 ml chloroform solution (as determined by thin-layer chromatography in % w/v).

Gas liquid chromatography. This technique has been used extensively to determine individual phenolic antioxidants in polymers, particularly the polyolefins and polystyrene (Table 21).

Other phenolic antioxidants that can be determined by gas chromatography of suitable extracts of the polymer include: Ionol (2-di-t-butyl-p-cresol) in polystyrene[164], 2-t-butyl-4-methyl phenol, 2,6-di-t-butyl-4-methyl phenol and p-t-butyl phenol in polyethylene[163,165], butylated hydroxy anisole and butylated hydroxy toluene[166] and Ionox 330[167].

High performance liquid chromatography. Schabon and Fenska[168] have described a rapid extraction procedure followed by high performance liquid chromatography for the determination in polyethylene of down to 10 ppm of Irganox 1076 (octadecyl 3,5-di-tert-4-hydroxy-hydrocinnamate), Irganox 1010 (tetrakis(methylene(3,5-di-tert-butyl-4-hydroxhydrocinnamate)) methane), and butylated hydroxytoluene (2,6-di-tert-butyl-4-methyl-phenol).

Table 21 – Separation of Phenolic Antioxidants – Gas Liquid Chromatographic Techniques

Substances separated	Stationary phase	Column °C	Other details	Reference
2,5-Di-t-butyl-p-cresol,2-(2-hydroxy-5-methylphenyl)benzo-triazole	25% LAC-2R/446 (adipate ester) + 2% H_3PO_4 on Chromosorb	135	H_2 carrier gas, F.I.D. error $\pm$ 1%	154
2,6-Di-t-butyl-p-cresol,(I) 2,6-di-t-butyl-phenyl	10% Apiezon N on Celite 545	154	H_2 carrier, F.I.D. 10-3 M in presence of others can be detected	155
2,4,6-Tri-t-butyl phenol Diphenylamine 2,6-Di-t-butyl-p-cresol	Apiezon		F.I.D.	156
Halogenated bis-phenols	10% DC-710 Silicone oil on Chromoport 80-100 mesh	225-250	12 in glass column ¼ in o.d. Carrier: 130 ml He/min	157
Low b.p. phenols	Capillary column coated with 10% xylenol phosphate	125	F.I.D.	158
Phenols and 5-t-butyl derivatives	Silicone oil 550-Carbowax 400 (3:2)	200	Mean deviation 0.4%	159
Phenols and cresols	5% W/W of various phosphate esters of phenols	110	120 cm x 4.5 mm column, Pye-Argon Chromatography	160
Ionox 330	a) 20% DC-710 Silicone oil on Chromosorb b) 2% SE30 Silicone gum on Chromosorb mesh	200-300 in. 10 min	a) 12 x 3/16 in. column b) 12 x 1/16 in. stainless steel column	161
Low molecular weight phenol	Silione-coated capillary column		Converted to trimethyl silyl esters before chromatography	162
2,6-Di-4-methylphenol	20% SE30 on HMDS-treated 60 mesh Chromosorb W	200	E.C. detector	163

<u>Determination of Irganox 1076, Irganox 1010 and Butylated Hydroxytoluene</u>
<u>Phenolic   Antioxidants   in   Polyethylene   -   High   Performance   Liquid</u>
<u>Chromatography</u>[168]

<u>Apparatus.</u>

Waters Model 204 liquid chromatograph equipped with two Model 6000 A
pumps, a Model 660 solvent programmer, and a U6K injector. Elution was
monitored with a Waters Model 450 variable wavelength detector set at 280
nm and a 10 mv strip chart recorder.
Column - 3.9 mm x 30 cm u-Porasil column packed with 10 um porous silica
obtained from Waters Associates, Milford, Mass.
Thermolyne Type 1000 stir plates (Sargent Welch).
Stir bars 3/8 inch o.d. x 1½ inch Teflon coated magnetic stir bars.
Sample filtering apparatus - a Waters 20-30 nm stainless steel solvent
reservoir filter was connected to about a 5" length of 3 mm i.d. Teflon
tubing. The other end of the Teflon tubing was connected to a 1½ inch
long blunt 16 gauge Luer lock needle with a 1/16 inch stainless steel nut
and ferrule at the end of the needle. The needle was connected to a
Hamilton No. 1010 W gastight 10 ml syringe with Teflon plunger.

<u>Reagents.</u>

Heptane, spectro grade.
Chloroform, Mallinckrodt AR grade, Scientific Products.
Methylene chloride, distilled in glass.
The above mobile phase solvents were all filtered through Millipore Type
F-H 0.5 um filters prior to use.

<u>Procedure.</u> Heated standard solution: A 50 ml portion of a standard
solution containing about 0.02 mg/ml each of BHT, Irganox 1076 and Irganox
1010 is pipetted into a 100 ml beaker. A stirring bar is added and the
solution is heated to 110°C with a gentle stirring for 30 min. The
solution is transferred to a cool stirrer and cooled to room temperature.
This heated and cooled standard solution is used to obtain quantitative
data on the sample extract solutions.

About 2 g of polyethylene pellets are weighed into a 100 ml beaker.
A stirring bar is added and 50 ml of decalin is pipetted into the beaker.
The mixture is heated at 110°C on a hot plate with gentle stirring for
about 30 minutes or until dissolution is complete. The beaker is than
transferred to a cool stirrer and cooled to room temperature with stirring
to precipitate the polyethylene.

The precipitated polyethylene from the above extraction is pushed
aside with a microspatula. The porous metal filter portion of the filter
apparatus is inserted into the solution and about 5-10 ml of solution is
drawn into the syringe. The Teflon tube is removed from the ferrule on
the needle and the filtered solution is dispensed into a small vial. The
filter apparatus is rinsed with acetone and dried between samples. After
extensive use, the metal filter became partially clogged and is
regenerated by placing it in hot decalin and stirring.

The Model 660 solvent programmer is set at Program 6 (linear) going
from 100% heptane to 100% methylene chloride in 5 min. The total flow
rate is 3 ml/min. The Model 450 UV detector is set at 0.2 or 0.4
absorbance unit sensitivity and the recorder chart speed is 1 cm/min.
Duplicate injections of 100 ul of each of the standard and sample
solutions are made. The mobile phase gradient is started at the point of
injection.

The retention volumes (VR) for BHT, Irganox 1076 and Irganox 1010 are 10.3, 14.8 and 22.2 ml respectively. The amount of each additive is determined from each sample injection by comparing peak heights for samples and standards. A blank decalin injection is made to determine from what points on the base line, peak heights should be measured. Gradient reset is instantaneous, from 100% methylene chloride to 100% heptane. Sample injection could be made any time after the appearance of a refractive index peak from the UV detector, signifying the emergence of heptane from the column.

A chromatogram of a polyethylene extract is shown in Figure 14.

Minor peaks in this chromatogram are due to impurities in the decalin extraction solvent. Excellent antioxidant recoveries are obtained by this procedure, (Table 22).

Figure 15 shows high-performance chromatograms obtained by Majors[169], using zipax column packing with mixtures of N,N, diethylaniline, N-ethyl-aniline, diphenylamine and N-phenyl-2-napthylamine amine antioxidants used in rubber manufacture.

A number of comparisons can be made for the supports. Although the relative elution order is the same on all three columns, the selectivity for each peak relative to N,N-diethylaniline appears to be affected. Selectivity for each solute (elution volume of solute divided by that of N,N-diethylaniline) was the greatest on Durapak and the least on beta,beta'-oxydipropionitrile/Zipax.

<u>In Polyethylene, Direct Spectroscopy of Polymer Film</u>

The difficulties involved in extraction of additives from polymers have led to a search for analytical techniques not involving a prior solvent separation of an additive extract. Of all the techniques tried, only those based on spectroscopy can claim any measure of success. Luongo[170] has tried ultraviolet examination of thin, hot-pressed polymer films. Using a double-beam spectrophotometer with air in the reference beam, he was able to estimate antioxidant levels ranging from 0.002 to 1.00% in polyethylene. Such a procedure is limited in that the polymer must exhibit a relatively flat absorption curve in the wavelength range used; also many antioxidants exhibit similar or identical spectra.

Miller and Willis[171] obtained infrared spectra of antioxidants from polymer films in a similar way, except that they compensated with additive-free polymer in the reference beam. Infrared spectroscopy is more specific than ultraviolet spectroscopy, but some workers[171] find that the antioxidant level in polymers is too low to give suitable spectra. Drushel and Sommers[173] combined specificity with simplicity by using spectrofluorimetric and phosphorescence techniques. Again, they used a double-beam spectro-photometer; this time with a wedge of additive-free polymer in the reference beam. They admit that the method is only applicable if the antioxidant has distinct sharp bands, and if no other components exhibit intense absorption in the same region.

In-situ spectroscopic techniques are not likely to be of value, then in the analysis of samples of unknown composition. If known amounts of additive can be incorporated into additive-free polymer, however, these techniques are likely to be extremely useful in the study of solvent extraction procedures, and the study of additive ageing processes (i.e. the effects of heat, light, sterilization, radiation, etc.), since the

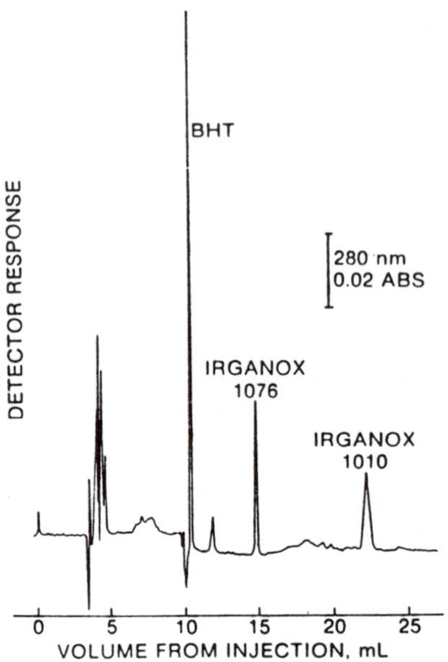

Figure 14 Chromatogram of 100 ul extract from 2.07g polyethylene

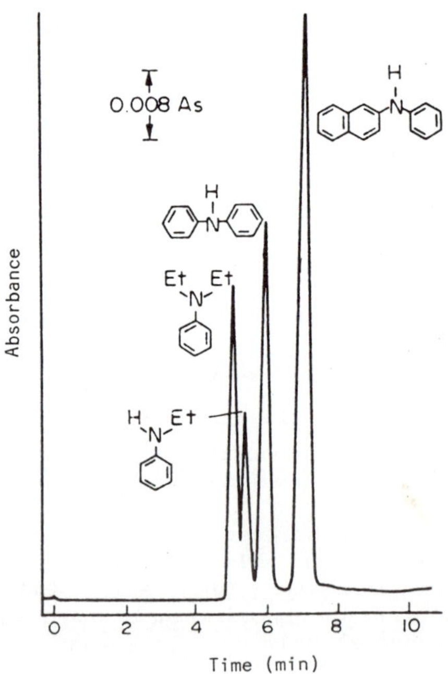

Figure 15 Separation of aromatic amine antioxidants using Zipax.
Column 1000 mm x 2.1 mm i.d.; packing, 0.5%, Beta beta-
oxydipropionitrile on 20 - 37 u Zipax support; carrier-
iso-octane; flow rate 0.31 ml/min; sample 10.6 ul of a
mixture of 9.5 ug/ml each of N, N-diethylaniline and N-
ethylaniline, 29 ug/ml of diphenylamine and 52 ug/ml of
N-phenyl-2-napthylamine in iso-octane.

Table 22 – Recoveries of BHT, Irganox 1010 from Three Polyethylene Samples

Sample[a]	amount (g)	Amount added (mg)			Amount found (mg)			Percent recovered		
		BHT	Irganox 1076	Irganox 1010	BHT	Irganox 1076	Irganox 1010	BHT	Irganox 1076	Irganox 1010
A	1.96	1.02	1.00	1.01	1.04	1.08	1.03	103	108	102
A	2.01	1.02	1.00	1.01	1.06	1.10	1.13	104	110	112
B	2.02	1.02	1.00	1.01	1.02	1.07	0.99	100	107	98
B	1.97	1.02	1.00	1.01	0.99	1.09	0.94	98	109	93
C	1.97	1.02	1.00	1.01	1.04	1.05	0.98	102	105	97
C	2.04	1.02	1.00	1.01	1.07	1.06	1.05	105	106	104

rate of disappearance or of decay can be measured directly by the decrease in absorbance of the sample at a suitable wavelength.

Albarino[174] has stated that analysis of polyethylene additives by means of ultraviolet spectroscopy is limited by excessive beam dispersion due to light-scattering from the polymer crystalline regions. Additives at low concentrations (0.1%) require sample thickness such that analysis must be performed in the presence of a high level scattering, which may change unpredictably with wavelength. At lower levels of concentration and correspondingly greater sample thicknesses, unacceptable signal-to-noise ratios exist. Nevertheless, ultraviolet spectroscopy remains an attractive method for analysis of many additives. Principal advantages over infrared analysis include greater sensitivity arising from higher extinction coefficients and a lack of interfering absorptions from the polyethylene matrix. These advantages can be realized, however, only if background scattering from the polymer can be reduced.

Albarino[174] demonstrated the feasibility of quantitative ultraviolet analysis of Irganox 1010 antioxidant in polyethylene at temperatures above the polymer melting point where the crystallites, which account for much of the scattering, are eliminated. Greater sample thickness and analytical sensitivity are possible compared to analysis of solid samples at room temperature. In this work, sample thickness was controlled by brass shims held between suprasil-grade silica windows (Amersil, Inc.) by a faceplate bolted to the cell body.

Polyethylene samples were prepared for analysis by calculating the weight required to fill the shim opening in the melt. Samples were inserted into the shim opening as pressed films cut to size; several layers were required for greater thicknesses. After gently tightening the faceplate, the cell was rapidly heated to 120-125°C by supplying about 65 W to the heater. By proper tightening of the faceplate the shim space was uniformly filled with polyethylene, after which the cell was transferred to the sample compartment of a spectrometer. Upon warm-up to the melt, an input power of 29 W maintained cell temperature within the limits given in Table 23 during scanning. Cell temperature was regulated only to the extent of maintaining the melt between 121 and 135°C. A small temperature increase, given by the intervals of Table 23, was generally allowed. Spectra were found to be insensitive to temperature in the intervals $128 \pm$ 4°C to 145 $\pm$ 4°C; a thermometer in contact with woods metal was used to indicate initial cell temperature and temperature upon completion of spectra.

Albarino[174] used standards consisting of polyethylene and Irganox 1010. These were made by milling at temperatures of about 127°C. Samples containing 0.051 and 0.010% Irganox 1010 were made from a master batch containing 0.101% Irganox 1010. These standards and an unstabilized control were moulded into sheets 0.0624-0.076 cm thick for use in the analysis.

The effect of sample melting on scattering is illustrated in Fig. 16(a) which is the spectrum of a 0.045 cm polyethylene specimen with 0.101% Irganox 1010 at room temperature; Fig. 16(b) was recorded at 122-126°C with a 0.058 cm specimen. A very substantial decrease in scattering, i.e. temperature increase, has resulted with little change in the antioxidant absorption at 280 nm. The effect of sample thickness on absorption at 122-127°C is shown in Figure 15(b). The extent to which scattering may be reduced in the melt is indicated by Fig. 16(c), where sample thickness was 0.780 cm and antioxidant concentration 0.010%.

Table 23 - Analysis of Irganox 1010 Antioxidant in Molten Polyethylene

Composition Irganox 1010 in polyethylene %	Thickness (cm)	Temperature (°C)	Sample absorbance at 280 nm	Baseline absorbance at 280 nm	Antioxidant absorbance at 280 nm
1 0.101	0.030	121-129	0.212	0.032	0.180
2 0.101	0.058	122-126	0.347	0.044	0.303
3 0.101	0.081	122-125	0.511	0.053	0.458
4 0.101	0.112	123	0.678	0.065	0.613
5 0.051	0.218	124-125	0.692	0.107	0.585
6 0.051	0.056	122-127	0.217	0.043	0.174
7 0.051	0.109	124-127	0.378	0.064	0.314
8 0.051	0.165	124	0.548	0.086	0.462
9 0.010	0.274	123-128	0.278	0.129	0.149
10 0.010	0.508	122-124	0.486	0.222	0.264
11 0.010	0.612	125-128	0.585	0.264	0.321
12 0.010	0.780	127-135	0.753	0.330	0.423
13 0	0.058	123-124		0.058	
14 0	0.266	126-129		0.130	
15 0	0.508	124-132		0.218	
16 0	0.780	123-131		0.333	

Spectral analysis on this sample in the solid state would not be possible because of its thickness.

Application of the technique for the purpose of quantitative analysis of additives requires proof of the validity of Beer's law (log I_o/I = A = abc) over the concentration range of interest. In the case of polyethylene antioxidants it is particularly important to establish constant absorptivity with concentration, as a fraction of the material is likely to exist in solution.

Spectra of molten polyethylene containing 0.051% (3.42 x 10^{-4} M) nominal concentration of Irganox 1010 are given in Figs 16(b) and 16(c) as a function of thickness. A similar set of curves was obtained for 0.101% (6.78 x 10^{-4} M) antioxidant concentration. Four thicknesses were studied at each concentration in order to establish linearity of absorbance with sample thickness and molar absorptivity over the concentration range.

Control spectra of unstabilized polyethylene in the melt are given in Fig. 16(d) as a function of thickness.

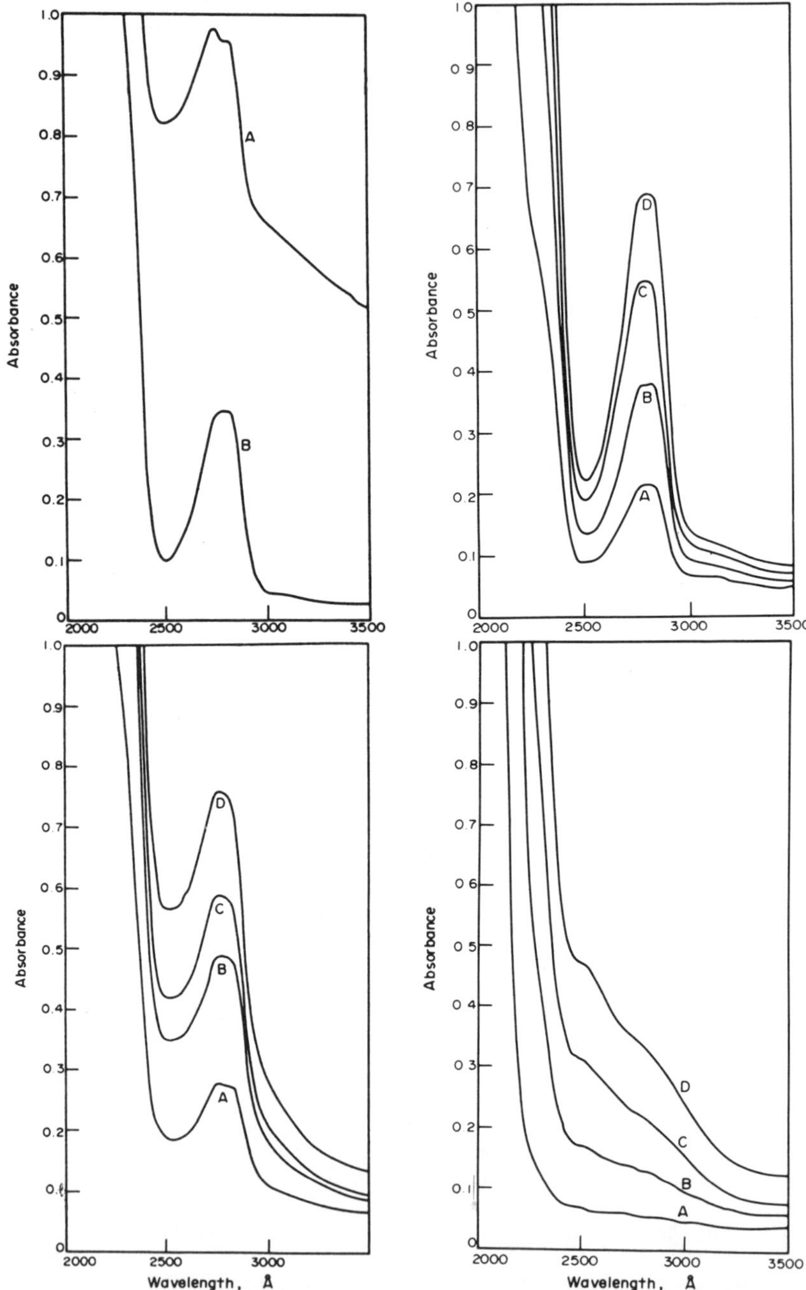

Figure 16 Ultraviolet spectroscopy of Irganox 1010 in polyethylene
(a) 0.01%, Irganox 1010 in polyethylene (A) 0.045 cm
20°C, (B) 0.058 cm, 122-126°C; (b) 0.51% Irganox 1010 in
polyethylene: (A) 0.056 cm, 122-127°C; (B) 0.109 cm 124-
127°C; (C) 0.165 cm, 124°C; (D) 0.218 cm 124-125°C, (c)
0.101 % Irganox 1010 in polyethylene (A) 0.274 cm 123-
128°C; (B) 0.508 cm, 122-124°C; (C) 0.612 cm, 125-128°C;
(D) 0.780cm, (d) Irganox-free polyethylene (A) 0.058 cm,
123-124°C; (C) 0.508 cm, 124-132°C; (D) 0.780 cm, 123-
131°C.

Figure 17(a) is a plot of absorbance at 280 nm against thickness for the unstabilized polyethylene samples of Fig. 16(d). From Fig. 17(a) the contributions of polyethylene and the quartz cell windows to total sample absorbance at 280 nm were determined. The finite intercept of Fig. 17 (a) represents scattering of the quartz windows at zero polyethylene thickness.

Graphs of total sample absorbance at 280 nm minus the baseline correction at 280 nm from Fig. 17(a) are given in Fig. 17(b) for the three stabilizer concentrations. Table 23 summarizes sample thickness, cell temperature, and absorbance data.

All graphs of Fig. 17(b) exhibit good linearity consistent with an intercept at the origin. In accord with Beer's law, the slopes of Fig. 17(b) divided by respective molar concentrations give values of 8150 (0.101%), 8100 (0.051%), and 7950 (0.010%), with an average of 8070 (1000 cm^2/mole) for the molar absorptivity. These values are based on the nominal concentra-tion of 0.101% and the dilutions made from it, and therefore subject to errors arising from stabilizer loss during initial compounding of the master batch and during subsequent dilutions to lower concentrations.

An estimate of ultimate sensitivity may be made by reference to Fig. 17(a), the baseline due to scattering, and Fig. 17(b) the absorbance at 0.01% antioxidant concentration. The slope for scattering in the melt predicts an absorbance of 1.0 for 2.54 cm of melted polyethylene. At this thickness, 0.01% Irganox 1010 would contribute 1.4 absorbance for a total sample absorbance of 2.4. Reducing concentration to 0.001% would result in 0.14 sample absorbance, which with the same baseline at 1.0 gives a total of 1.14. Absorbance error in this range is stated as 0.008 or about 10% of the sample signal.

Polypropylene

Ultraviolet spectroscopy. Determination of Santonox R antioxidant. In an attempt to overcome the difficulty of interference effects by other polymer additives in the ultraviolet spectroscopic determination of phenolic antioxidants Wexler[175] makes use of the bathochromic shift exhibited by phenols on changing from a neutral or acidic medium to an alkaline one. This shift is due to the change of absorbing species because of solute-solvent interaction. Using a double-beam recording spectrophotometer, he measured a difference spectrum by placing an alkaline solution of the polymer extract in the sample beam, and an identical concentration of sample in acid solution in the reference beam. The resulting difference spectrum is a characteristic and useful indication of the concentration and chemical identity of the phenolic substance. Possible interference due to non-ionizing, non-phenolic species is usually cancelled out in the difference spectrum, which should make the technique of interest to the polymer analyst. Close adherence to Beer's law is usually obeyed by the difference peak spectra.

Soncek and Jelinkova[176] have also used this differential principle to determine in polypropylene two antioxidants (2,6-di-tert-butyl-4-methylphenol and 4-substituted 2,6-xylenol) which have virtually identical ultraviolet absorption spectra in the absence of alkali. The antioxidants can be distinguished in alkaline medium, where 4 substituted 2,6-xylenol forms phenolate readily, thus allowing the utilisation of the bathochromic shift for its determination. The use of derivative spectroscopy reduces light scattering and matrix interferences when extracts from polypropylene samples are measured.

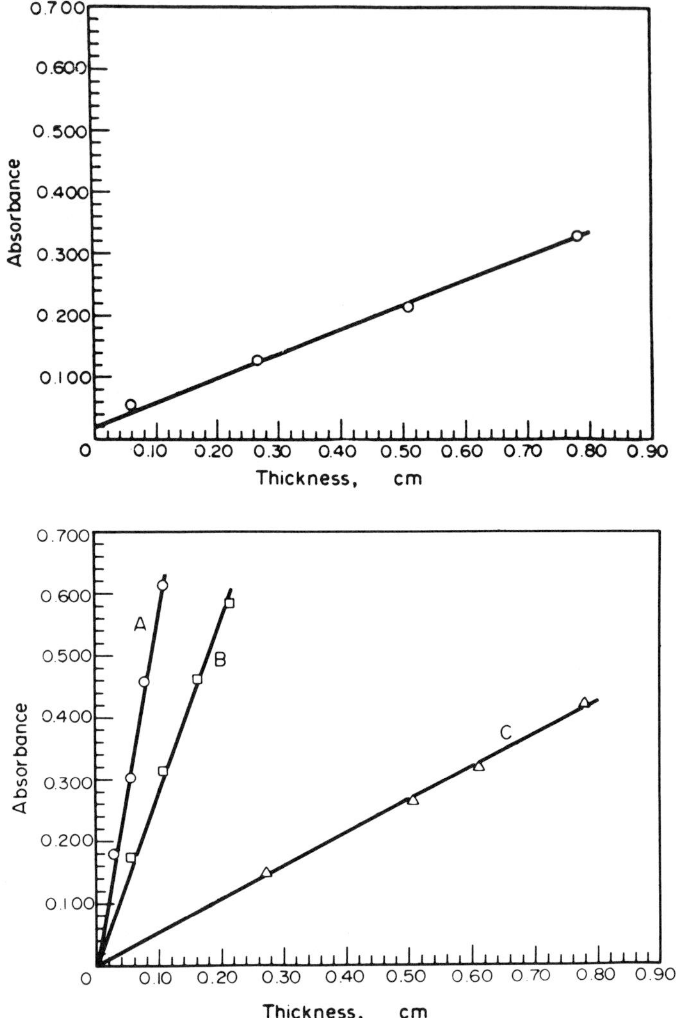

Figure 17   Direct ultraviolet spectroscopy of (a) polyethylene.
            Absorbence versus thickness.   (b) Irganox 1010 in
            polyethylene: (A) 0.101%; (B) 0.051%; (C) 0.01%.

        The principle of the 4-substituted 2,6-xylenol method is as follows.
First the phenolic component without steric hindrance is determined by
means of second derivative - difference spectroscopy of the
bathochromically shifted, basified extract using the non-basified extract
as a reference.   The second derivative spectral amplitudes of a non-
basified extract are additive for both antioxidants.   The contribution of
4-substituted 2,6-xylenol to the total second derivative amplitude is then
subtracted and the residual corrected amplitude corresponds to the 2,6-di-
tert-butyl-4-methyl phenol content.

        In Fig. 18 is shown the effect on the ultraviolet spectrum ofadding
alkali and nickel peroxide to an ethanolic solution of three phenolic
antioxidants.   Ruddle and Wilson[177] used the observed shifts in the
ultraviolet as the basis of a low interference method for determining
Topanol OC, Binox M and Ionox 330 in polyethylene.

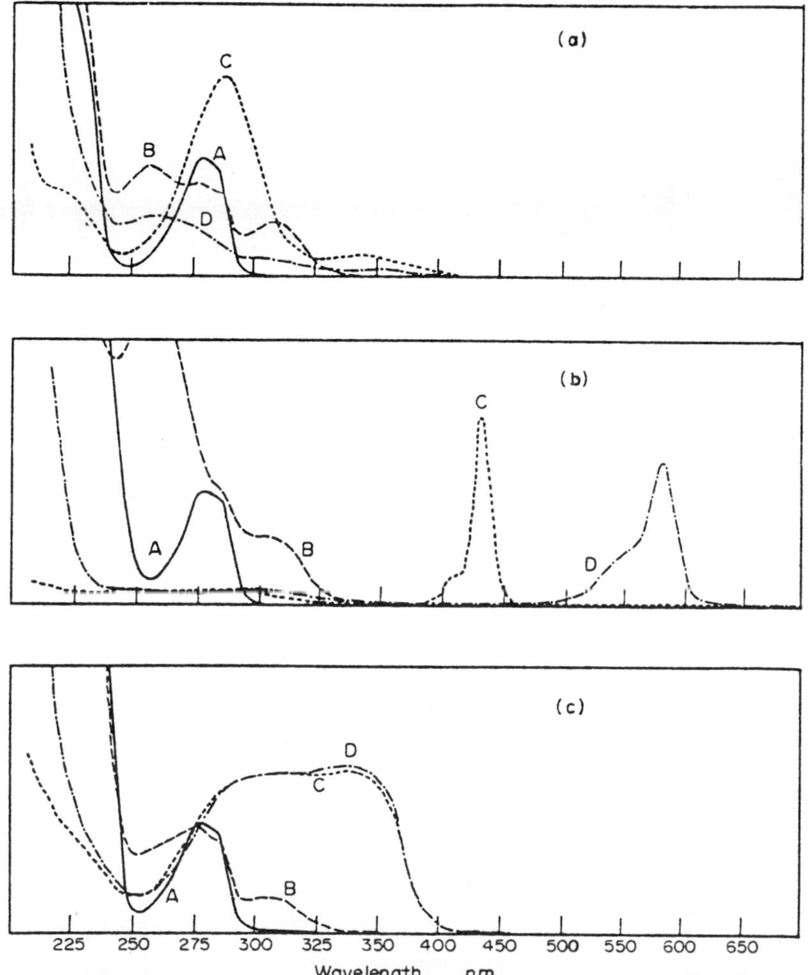

Figure 18   Ultraviolet spectra of (a) Topanol CA, (b) Binox M, (c)
Ionox 330. Curve A ethanolic solution; Curve B alkaline
ethanolic solution (10 ml ethanol solution plus 1 ml
water plus 2 ml ethanolic potassium dioxide solution);
Curve C, ethanolic solution after reaction with nickel
peroxide; Curve D nickel peroxide reaction product made
alkaline with 2 drops of ethanolic potassium hydroxide
solution.

Thin-layer chromatography.   Table 24 summarizes the results obtained
in some experiments carried out to determine the recovery from
polypropylene of compounds adsorbing at shorter and longer ultraviolet
wavelengths, respectively, di-n-butyl phthalate (222 mu) and Ionox 330
(277 mu).   These compounds were carried through the whole series of
operations involving application of sample to a silica gel plate, solvent
development, separation of adsorbent from plate and finally, solvent
extraction of the compound from the adsorbent.   Ionox 330 has a low Rf
value with 4:1 cyclohexane:benzene development solvent (Table 24) and
hence during solvent development, ultra violet-absorbing adsorbent
impurities are swept well away from this compound to the solvent front.

Table 24

Reproducibilty of recovery from polypropylene of Ionox 330 and di-n-butyl phthalate
from thin-layer plate (silica gel G254)

Compound	Absorbance maximum	No. of plates prepared	Sample size (ul)	Sample concentration (% w/v)	Development solvent	Plate drying time (h)	Section of adsorbent removed from plate R	Solvent used to desorb compound from adsorbent	Adsorbtivity of standard solution of compound (litres g cm)	Recovery of compound 2**	Standard deviation
Ionox 330	277u	10	1	1	4:1 cylco-hexane:benzene	18#	0.03–0.14	Methanol (extract made up to 1 ml)	8.0	100	1.2
Di-n-butyl-p-phthalate*	22u	5	1	5	9:1 iso-octane:ethyl acetate	18	–	Methanol (extract made up to 1 ml)	29.0	99.6	1.2

Applied by Hamilton PR600 repeating sample dispenser.
* Plate premigrated once with methyl alcohol then reconditioned for 1 hour at 120°C before application of sample.

To allow complete evaporation of benzene which would interfere in subsequent ultraviolet spectroscopy.

** Based on ratio of theoretical absorptivity and absorptivity of extract from thin-layer plate, methanol used in reference cell in all experiments.

Premigration of the plate with methyl alcohol was unnecessary and not used, therefore, prior to to application of Ionox 330 to the plate. Premigration of the plate with methanol before sample application was, however, carried out in the case of di-n-butyl phthalate. This was because this compound has a fairly high R_f value with the 9:1 iso-octane:ethyl acetate development solvent used, with consequent possible contamination of the di-n-butyl phthalate band with ultraviolet-absorbing adsorbent impurities near the solvent front. It is seen in Table 24 that satisfactory recoveries of both compounds were obtained in this procedure.

<u>Gas chromatography.</u>   An example of the application of gas chromatography to the determination of phenolic antioxidants in polymers is a method for the determination of a very high molecular weight compound Ionox 330 (1,3,5-tri-methyl-2,4,6 tri(3,5-di-t-butyl-4-hydroxybenzyl) benzene) in polypropylene[178].   The solvent extract of the polymer is analysed under the following conditions:

Chromatograph	Aerograph Model 600-C with flame ionization detector.
Column	1/8 inch x 18 inch stainless steel with 5% SE-30, Silicone Gum Rubber (Methyl, General Electric Co.), on 80/90 Anakrom ABS (Analabs, Inc.).
Temperatures: Injection port Column Detector	330°C 290°C 290°C
Carrier Gas	65 ml/min helium
Retention time for Ionox 330	8.5 min

Calibration standards representing 0.1g/l to 0.8g/l Ionox 330 were analysed with an average repeatability of $\pm$ 2.0%. The calibration curve obtained as a plot of peak area versus concentration was linear and passed through the origin, indicating that no Ionox 330 is lost through decomposition in the hot injection port.

Polyester Acrylates

Budyina et al[179] have described methods based on anodic voltametry for the determination of Ionol (2,6-di-t-butyl-p-cresol) and quinol in polyester acrylates. To determine Ionol the sample is dissolved in 25 ml of acetone and an aliquot (10 ml) is treated with 2.5 ml of acetone and 5 ml of methanol and diluted to 25 ml with a solution 0.1 M in lithium chloride and 0.02 M in sodium tetraborate. A polarogram is recorded with a graphite-rod indicator-electrode. To determine quinol, the sample (1 to 3 g) is dissolved in 80 ml of methanol or methanol:acetone (1:1) and the solution is diluted to 100 ml with the lithium chloride-sodium tetraborate solution. A polarogram is recorded under the same conditions. Concentrations are determined by the addition method. The $E_{\frac{1}{2}}$ values (vs. the S.C.E.) are 0.25 V for Ionol and 0.16 V for quinol.

Polystyrene

In a method described by Hilton[180] the polymer sample in the form of a powder or thin film is extracted with ethanol and the extracted phenolic antioxidant coupled with diazotized p-nitroaniline in strongly acidic medium. The solution is then made alkaline and the visible absorption spectrum determined. Many of the antioxidants studied have an absorption maximum at a characteristic wavelength. Hence, in some instances, it was possible to both identify and determine the antioxidant, provided a pure specimen of the compound in question is available for calibration purposes.

5.1.3 Amine Antioxidants (See also Appendix 3)

In Polyethylene

Spectrophotometric methods. In a procedure recommended by the British Standards Institution [181] for determining Nonox CI (N,N'-di-2 napthyl-p-phenylene-diamine) in low density and high density polyethylene, the polymer is refluxed with toluene, reprecipitated with ethanol and filtered to provide a clear extract. The antioxidant content of the filtrate is determined colormetrically by oxidation with hydrogen peroxide, in the presence of sulphuric acid. This reagent produces a green colour with Nonox CI which gradually reaches a maximum intensity. The colour is evaluated at 430 nm when the maximum depth of colour is reached.

Crompton[182] has described a modification to this procedure for the determination of Nonox CI in high density polyethylene which, compared to low density polymer, has only a small solubility in toluene. He has also extended the procedure to the determination of oxidized Nonox CI, and a further product which he calls degraded Nonox CI.

High performance liquid chromatography. Majors[183] used high performance liquid chromatography to separate three hindered phenolic antioxidants extracted from polyethylene on a Corasil II column. The additives were extracted from the polymer by grinding them in a freezer mill then extracting with diethyl ether in a Soxhlet extractor for two days. The ether was then evaporated off and the residues dissolved in 25 ml of 1% by volume of isopropanol in hexane. Using the chromatographic system of a 1% isopropanol in hexane solvent and Corasil II adsorbent the amount of hindered phenolic antioxidant in the ether extract of the polymer was determined. Calibration curves over the concentration range of interest were obtained for the antioxidants known to be present.

In Synthetic and Natural Rubbers

Gas chromatography. Gaeta et al[184] have described a gas chromatographic method for determining a number of antioxidants such as N-phenyl-2-naphthylamine and N,N'-sec-heptyl phenyl p-phenylene diamine in oil-extended synthetic polymers such as polybutadiene or styrene-butadiene rubber. This involves extracting the antioxidant from the polymer with ethanol, taking up the concentrated extract with the appropriate solvent and analysing the resulting solution by gas chromatography. They found - by use of standard solutions of the antioxidants in carbon tetrachloride, acetone or carbon disulphide and the careful choice of chromatographic conditions - that the elution of these materials had little or no interference from the extending oil present. In addition, the oil was completely soluble in the solvent and all but the heaviest fractions eluted prior to that of the antioxidant.

The instrument used in this work was a dual-column flame ionization detector chromatograph equipped with 5 ft columns of ½ in stainless-steel tubing packed with 60-80 mesh Chromosorb Z support containing 5% Apiezon N grease. The detector temperature was 205°C while that of the injection port was 275°C with a carrier gas flow of 30 ml of helium per minute. The solvents used (carbon tetrachloride, acetone and carbon disulphide) were of reagent grade. A different solvent was needed for each antioxidant to ensure complete solubility.

The small peaks which appear near to that of the antioxidant are those due to the extending oil present. No elution of the oil occurred during that of N-phenyl-2-napthylamine as a chromatogram of the oil in acetone was obtained previous to that of the antioxidant. At column temperatures above 170°C the different component fractions of the oil are manifested as separated chromatographic peaks as they pass through the column. Since these polymer extenders are wide petroleum cuts there are some heavy fractions which elute even after such high boiling materials as the substituted p-phenylenediamine antioxidants.

Wize and Sullivan[185-187] used high temperature gas chromatography for the analysis of mixtures of amine type antidegradents in rubber. These workers used a separation column constructed of aluminium packed with 20% Apiezon L on 30-60 mesh Chromasorb W. Analysis was carried out on an acetone extract of the rubber sample, employing diphenylamine as an internal standard. Using column temperatures up to 310°C they were able to separate a range of antidegradants including 1,2-dihydroxy-2-2,4-tri-methyl-6-ethoxy-quinoline, N-isopropyl-N'-phenyl-p-phenylene-diamine and N,N-diphenyl-p-phenylenediamine. Near quantitative determinations were obtained for all these substances. Apiezon L was found to be distinctly superior as a mobile phase to other substances tried. Dow-Corning 710 Silicone fluid and butanediol succinate were too volatile at operating temperatures up to 310°C whilst silicone rubber, although sufficiently non-volatile, did not give the high degree of resolution obtained with Apiezon L.

5.1.4 Phenolic and Amine Type Antioxidants (See also Appendix 3)

Many of the published procedures are not specific for phenolic type or amine type antioxidants but will determine both. These are discussed below.

Polyethylene

Spectrophotometric methods. Several chromogenic reagents have been used for the spectrophotometric determination of phenolic and amine antioxidants in polyethylene. These include p-diazobenzene sulphonic acid[188] alpha, alpha'diphenyl-beta-picrylhydrazyl[189], benzothiazolin-2-one-hydrazone[190] and 2,6 dichloro-p-benzoquinone -4-chlorimine[191].

Glavind[192] and Blois[193] have devised methods for estimating total antioxidants, irrespective of type, by coupling them with the free-radical alpha,alpha'-diphenyl-beta-picrylhydrazyl. The decrease in absorption of the reagent solution upon mixing with the polymer extract is related to the amount of antioxidant present.

A redox spectrophotometric procedure[191] involves oxidation of the antioxidant (A) under controlled conditions with an absolute ethanol solution of ferric chloride, followed by reaction of

$$A \text{ reduced} + Fe^{3+} = A \text{ oxidized} + Fe^{2+}$$

the ferrous iron produced with 2,2'-dipyridyl to form a coloured complex, the intensity of which is proportional to the concentration of antioxidant

present.   This spectrophotometric procedure has been applied to various
phenolic and amine-type antioxidants, viz. Succonox 18, butylated
hydroxytoluene, Ionol (2,6-di-teri-butyl-p-cresol), and Nonox CI (N-N'-di-
beta-naphthyl-p-phenylenediamine).   An advantage of this redox procedure
is that it determines only the antioxidant which is present in the polymer
in the reduced form, and does not include any oxidized material produced,
for example, by atmospheric oxidation during polymer processing at
elevated temperatures.   Total reduce plus oxidized antioxidant can be
determined by ultraviolet spectroscopic procedures and oxidized Santonox
obtained by difference from the two methods.

A British Standard method[198] for estimating the total phenolic
antioxidant content of polyethylene involves a preliminary extraction of
the polymer with hot toluene to extract the additive, followed by addition
of ethanol to precipitate any dissolved polymer, then coupling the extract
with diazotized sulphanilic acid to produce a colour which can be compared
with standards by visible spectrophotometry.   Decresylol propane and
Santonox R (4-4'-thiobis-(3-methyl-6-tert-butyl-phenol) and various amine
antioxidants in particular are mentioned as additives than can be
determined by this technique.

Hilton[194,195,199] published a coupling method for the determination
of phenolic and amine antioxidants in polymers based on the preparation of
a methanol or ethanol extract of the polymer, followed by coupling the
extracted phenol with diazotized p-nitroaniline in strongly acidic medium.
The solution is then made alkaline and the visible absorption spectrum
determined.   Many of the antioxidants studied have an absorption maximum
at a characteristic wavelength.   Hence, in some instances, it was possible
to both identify and determine the antioxidant, provided a pure specimen
of the compound in question is available for calibration purposes.   Table
25 shows absorptivity and wavelength maxima data for some phenolic
antioxidants.

<u>Fluorescence spectroscopy.</u>   Aromatic amines and phenols are among
the few classes of compounds in which a large proportion of their numbers
exhibit sensible fluorescence.   Parker and Barns[200] found that in solvent
extracts of rubbers the strong absorption by pine-tar and other
constituents masks the absorption spectra of phenyl naphthylamines,
whereas the fluorescence spectra of these amines are sufficiently
unaffected for them to be determined directly in the unmodified extract by
the fluorescence method.   In a later paper[201] Parker discussed the
possibility of using phosphorescence techniques for determining phenyl-
naphthylamines.   Drushel and Sommers[202] have discussed the determination
of Age Rite D (polymeric dihydroxy quinone) and phenyl 2-naphthylamine in
polymer films by fluorescence methods and Santonox R (4,4'-thio-bis-(6-
tert-butyl-m-cresol) and phenyl-2-naphthylamine by phosphorescence
methods.   They emphasize the freedom that such techniques have from
interference by other polymer additives and polymerization catalyst
residues.   With practical samples of polyethylene film, difficulty was
found in obtaining a reliable correlation between the concentration of
stabilizer present in the film and its phosphorescence intensity at 77K by
this technique.   This may be attributable to variations in the degree of
crystallinity which affect the optical properties of the polyethylene film
matrix in these samples.

Table 25
Composition and Absorptivity Data for Phenolic Antioxidants
after Hilton[194,195]

Antioxidant	Composition	Absorptivity A max-A 700	Wavelength max, mu
AgeRite Alba	Hydroquinone monoben-zyl ether	31.48	565
AgeRite Spar	Styrenated phenol	44.06	548
AgeRite Superlite	A polyalkyl polyphenol	23.40	560
Antioxidant 5	Not disclosed	18.81	585
Antioxidant 425	2,2'-Methylene-bis (6-tert-butyl-4-methylphenol)	22.30	585
Antioxidant 2246	2,2'-Methylene-bis (6-tert-butyl-4-methylphenol)	20.60	578
Deenax	2,6,Di-tert-butyl-p-cresol	Does not couple	
Ionol	2,6,Di-tert-butyl-p-cresol	Does not couple	
1-Napthol	1-Napthol	120.2	598
2-Napthol	2-Napthol	115.1	540
Naugawhite	Alkylated phenol	8.20	580
Nevastain A	Not disclosed	12.44	550
Nevastain B	Not disclosed	6.62	550
Nonyl phenol	Nonyl phenol	36.25	538
p-Phenyl phenol	p-Phenyl phenol	80.80	548
Polygard	Tris (nonylated phenyl) phosphite	Must be hydrolysed before it will couple	
Santovar A	2,5-Di-tert-amyl-hydroquinone	Colour too weak	
Santovar O	2,5-Di-tert-amyl-hydroquinone	Colour too weak	
Santowhite crystals	4,4'-Thio-bis (6-tert-butyl-2-methyl phenol)	78.84	565
Santowhite MK	Reaction product of 6-tert-butyl-m-cresol and SCl_2	66.94	560
Sanotwhite powder	4,4'-Butylidene-bis (3-methyl-6-tert-butyl phenol)	Colour too weak	
Solux	N-p-Hydroxyphenyl-morpholine	Colour too weak	
Stabilite white powder	Not disclosed	Colour too weak	
Styphen 1	Styrenated phenol	22.61	558
Wingstay S	Styrenated phenol	50.82	545
Wingstay T	A hindered phenol	10.27	590

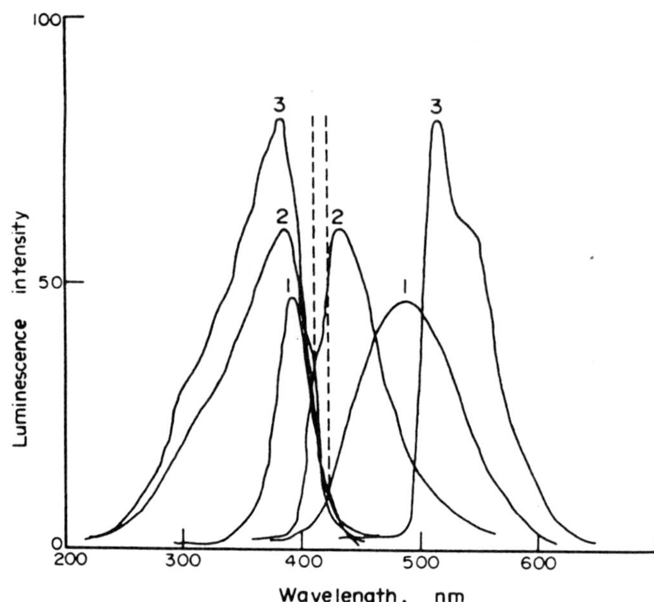

Figure 19 Spectral characteristics of luminescence observed for
Nonox CI: (1) excitation and emission spectra at room
temperature, sensitivity scale 0.01; (2) excitation and
emission spectra at 77K for total luminescence,
sensitivity scale 1.0; (3) excitation and emission
spectra at 77K for phosphorescence.

Figure 19 shows the spectral characteristics of a 1000 ppm solution
of Nonox CI antioxidant dissolved in diethylamine at 77K and at 293K.
Calibration curves for both low-temperature luminescence intensity
(phosphorescence plus fluorescence) and phosphorescence intensity versus
concentration of antioxidant in ppm of the compound in the solvent can be
obtained[203].

The wavelengths of maximum excitation and emission are set on the
spectrophosphorimeter, and the relative intensity readings from the photo-
multiplier microphotometer recorded for a series of standard solutions
containing 0.001-1000 ppm of each compound.

In conclusion, measurement of the phosphorescence characteristics of
samples obtained after extraction of polymers with organic solvents yields
useful information regarding the nature and concentration of the
stabilizer compounds present. It should be possible to obtain good
selectivity, with a sensitivity which compares favourable with that of
ultraviolet absorption spectrophotometry, in the determination of two or
more stabilizer compounds simultaneously by correct choice of excitation
and emission wavelengths and phosphorescence speeds.

Thin-layer chromatography. For additives extracted from
polyolefins, usually with diethyl ether, the extract is refluxed with
ethanol and the solution is decanted off from the insoluble residual
polymer. On cooling, additives such as dilauryl and distearyl
thiodipropionate separate out and are identified by infrared examination.

A 30 ul volume of the ethanolic solution is then spotted on to a thin Kieselgel 60 TLC plate and eluted with a suitable solvent, usually 98.5 + 1.5 toluene-ethyl acetate.   The eluted plate is dried and sprayed with colour-developing reagents and the spots are examined.   Table 26 illustrates R_F values (based on Topanol OC = 1.00), colours and suitable solvent systems.   If the spots are to be submitted to mass spectrometric examination, methanolic iodine is used as the colour-developing reagent as this does not over-complicate the mass spectrometry.

The normal spray procedure uses a 0.5% solution of 2,6-dibromo-p-benzoquinone-4-chlorimine in ethanol followed, after drying, by a 0.5% borax solution.   After this spray, the plate is dried at 120°C for 5 min to develop the colours (Table 26).

Simpson and Currell[204] use thin-layer chromatography in the determination of additives such as phenolic and amine antioxidants, ultraviolet absorbers and organotin stabilizers in polyolefins and other polymers.   Comparatively small samples of the plastics materials are required, and, by means of the techniques described, it is possible to identify additives in extracts containing several different components (Table 27).   The method described below can be used to detect additives down to between 10 and 1 ug per sample and both qualitative and more accurate quantitative determinations are made possible.   Some typical applications of thin-layer chromatography to the determination of antioxidants in polymers, are tabulated in Table 28.

<u>Electrochemical methods.</u>   Electrochemical methods have been investigated but in general are only to be recommended where a simpler method is not available.   Mocker[218-220] and Mocker and Old[221] have explored the use of polarography and find the technique to be more applicable to rubber accelerators than to antioxidants.   They included both phenolic and amine types of antioxidants in their study. Difficulties arise because the dropping-mercury electrode cannot be used at potentials more positive than +0.4V with respect to the saturated calomel electrode, and since many aromatic phenols (and amines) can only be oxidized at electrodes, positive voltages have to be applied in their analysis.   Nevertheless, the polarography of some amines and phenols has been studied[222-225] , and whilst no electrode is as suitable for polarography as the dropping-mercury electrode, antioxidants have also been studied with other types of electrodes, notably the graphite[226-229] and the platinum[230,231] electrode.   These methods include the determination by voltametric methods of N,N'-di-sec-butyl-p-phenylene-diamine and N-butyl-p-amino phenol[228] and of Ionol (2,6-di-tert-butyl-p-cresol)[226].   In addition, at least two commercially available antioxidants have been shown by differential cathode-ray polarography to exhibit reduction waves: Santonox R (4,4'-thiobis-(3-methyl-6-t-butyl phenol) gives a poorly shaped wave at -0.6 V in a electrolyte consisting of ammonia and ammonium chloride in methanol-water[232] and 3,5-di-t-butyl-4-hydroxytoluene gives wave at -0.65 V in aqueous was adequate for quantitative analysis.

<u>Electrophoresis.</u>   Electrophoresis is a technique worthy of further consideration for the analysis of antioxidants[234].   Sawada et al[235] report successful separations by coupling the antioxidants with p-diazobenzene sulphonic acid before electrophoresis.   Amine antioxidants are coupled in acetic acid and phenolic antioxidants in sodium hydroxide-ethanol. Electrophoresis was carried out in 1% w/v methanolic sodium borate.

Table 26 – Thin-layer Chromatography of Polymer Additives

| Additive | \multicolumn{4}{c}{R_f value*} | Colour of spot |

Additive	Solvent 1	Solvent 2	Solvent 3	Solvent 4	Colour of spot
Topanol OC	1.00	1.00	1.00	1.00	Pale yellow centre with pink outer
DLTDP	0.40	1.00	0.05	0.30	Yellow-brown
DSTDP	0.45	1.00	0.10	0.35	Yellow-brown
Distearyl disulphide	1.10	1.05	1.20	1.20	Bright yellow
DTB glycol ester	0.00	0.75	0.00	0.00	Brown centre with mauve outer
Ionox 330	1.05	1.05	0.80	0.80	Pinkish brown
Irganox 259	0.30	1.05	0.00	0.65	Brown
Irganox 288	0.25	1.05	0.00	0.60	Brown
Irganox 1010	0.30	1.10	0.00	0.60	Brown
Irganox 1076	0.85	1.05	0.60	0.95	Brown
Nonox WSP	0.85	0.95	0.45	0.35	Bright yellow centre with pink outer
Polygard	0.30, 0.15	0.80, 0.70	0.10	0.05	Blue and red

Santonox R	0.35	0.80	0.10	0.00	Purple
Tinuvin 326	1.05	1.05	0.95	1.05	Yellow
Tinuvin 327	1.05	1.05	1.05	1.10	Yellow
Tinuvin 328	1.00	1.00	1.00	1.05	Yellow
Topanol CA	0.10	0.60	0.00	0.00	Brown
UV 531	0.75	1.00	0.35	0.80	Blue
Hoechst D55	0.85, 0.75, 0.55, 0.35 0.30, 0.20	0.90, 0.85	0.55, 0.45 0.25, 0.15 0.10, 0.00	1.10, 0.45 0.25, 0.15	Blue and red
Oleamide	0.00	0.00	0.00	0.00	These additives give
Ecrucamide	0.00	0.00	0.00	0.00	only a very faint
Ethomeen T12	0.00	0.00	0.00	0.00	brown coloration
Stearic acid	0.00	0.00	0.00	0.00	

*R_f values quoted to the nearest 0.05. Solvent 1 = toluene - ethyl acetate (98.5 + 1.5); solvent 2 = toluene - isopropanol (88 + 12); solvent 3 = toluene - light petroleum (b.p. 60-80 °C) (1 + 1); and solvent 4 = cyclohexane - toluene - methanol (88 + 10 + 2).

Table 27 - Results Obtained for Each Additive in each Mobile Phase

Substance	R_f value in benzene-ethyl acetate-acetone hexane	R_f value in chloroform-hexane
Antioxidants:-		
Nonox SP	0.61, 0.93	0.50
Nonox TBC	0.98	0.62
Nonox WSP	0.85	0.52
Nonox EX	0.56	0.38
Nonox WSL	0.82	0.54
BHT	0.98	0.91
2246	0.75	0.56
Nonox Cl	0.65, 0.84	0.49
Nonox DPPD	0.66	0.48
Nonox OD	0.56	0.40
Nonox ZA	0.43, 0.38	0.38
Santoflex	0.15, 0.61, 0.74	0.47, 0.54
Santoflex AW	0.80, 0.51, 0.55	0.54, 0.45
Santoflex R	0.26, 0.53	0.39
DLTP	0.61	0.41
Nonox NS	0.52	0.4
Superlite	0.64, 0.93	0.58
Polygard	0.52, 0.61	0.53
Nonox HO	0.46, 0.64, 0.60, 0.99	0.52
Nonox WSO	0.58	0.39
Irganox 1076	0.57, 0.75	0.56
Ultraviolet absorbers:-		
Eastman DOBP	0.76	0.39
Uvinul 400	0.22	0.08
Permyl B100	0.24	0.08
Salol	0.73	0.45
Tinuvin HE	-	0.71
Tinuvin P	0.76	0.59
Eastman OPS	0.79	0.65
UV 318	0.26	0.14
Cyasorb 1988	0.26	0.12
Cyasorb 1084	0.24	0.08
Uvinul N 35	0.28	0.16

<u>Polyvinyl Chloride</u>

<u>Polarography.</u> Vasil'eva et al [236] described a method for the determination of Ionol and 4,4-isopropylidenediphenol in PVC by anodic voltametry. The sample (1 g) is dissolved in 10 ml of dimethylformamide and the solution is treated with 5 ml of methanol to precipitate PVC and diluted to 25 ml with 1.5 M aqueous sodium acetate. A polarogram of an aliquot (10 ml) of the clear solution is recorded (graphite rod indicator-electrode and 0.5 M cadmium sulphate-cadmium reference electrode) to include the steps for Ionol (2,6-di-t-butyl-p-cresol) at 0.21 V versus the SCE, and for 4,4'-isopropylidenediphenol at 0.53 V versus the SCE. The application of voltametry to the determination of antioxidants has also been discussed by Barendrecht[226].

Table 28 - Separation of Antioxidants - Thin-layer Chromatographic Methods

Substances separated	Stationary phase	Mobile phase	Detection	Refs
Phenolic antioxidants	Silica Gel G	Methanol-cyclohexane (1:24)	30% Molybdophosphoric acid + ammonia vapour	205
Organo-tin stabilizers	Not stated	Acetic acid-isopropyl ether (1.5:98.5)	20% Molybdophosphoric acid + ammonia vapour	206
Antioxidants	Not stated	Light petroleum-ethyl acetate (9:1)	a) Ethanolic 2,6-dichloro-p-benzo-quinone-4 chlorimine + 2% aq. $Na_2B_4O_7$ b) Diatzotized p-nitroaniline	207
Organic stabilizers	Kieselgel G	Ethanol-free chloroform		208
Phenolic antioxidants	Polyamide powder	Methanol-water (3:2) or Methanol-carbon tetrachloride (1:9)	Diazotized sulphanillic acid	209
Phenyl salicylate Resorcinol benzoate	Kieselgel G	Dichloromethane or isopropyl ether -light petroleum (40-60°) (7:3)	Ultraviolet	210
BHA,2,6-di-t-butyl-p-cresol	Silica gel	Chloroform	20% Molybdophosphoric acid + ammomia vapour	211
Antioxidants	Polyamide powder	Methanol-acetone-water (6:1:3)	Diazotized sulpanilic acid or molybdophosphoric acid	212
Antioxidants	Kieselgel G		alpha, alpha'-Diphenyl-beta-picryl hydrazyl (free radical)	213
Antioxidants	Alumina + 5% plaster of paris on microscope slide	Petrol-dioxan (10:1)	5% Ethanol, phosphomolybdic acid	214

Antioxidants	Silica gel	Acetone, chloro-form, benzene, carbon tetrachloride or binary mixtures		215
Antioxidants	a) 10% starch in polyamide powder b) 10% PVC in polyamide	Methanol-acetone-water (3:1:1) light petroleum-benzene acetic acid-DMF (40:40:20:1)		216
Antioxidants	Silica Gel G	Benzene	0.5% $Fe_2(SO_4)_3$ in sulphuric acid + 0.2% $K_4Fe(CN)_6$ (1:1)	217

Other procedures which have been described include the conversion of an antioxidant into a polarographically reducible form,[237] and a general method for antioxidants which involved measuring a decrease in the height of the wave, due to the reduction of dissolved oxygen by antioxidants[231]. Ward[238] has discussed in some detail the determination of phenolic and amine types of antioxidants and antiozonants in PVC by the chronopotentio-metric technique, using a paraffin wax-impregnated graphite indicating electrode[239,240] and solutions of lithium chloride and lithium perchlorate and acrylonitrile in 95% ethanol as supporting electrolytes. Precision obtainable for repeated chronopotentiometric runs in acetonitrile was found to be better than $\pm$ 1.0% in cases in which electrode fouling did not occur, and $\pm$ 1.7% when the electrode was fouled by electrolysis products.

Phenolic (and amine) antioxidants in extracts of PVC have been titrated electrometrically with lithium aluminium hydride, with platinum or silver electrodes[241]. Small amounts of water in the sample or analysis solvent have an influence on the results obtained by these procedures.

<u>Thin-layer chromatography.</u> Procedures involving solvent extraction followed by thin-layer chromatography have been described for the determination of phenolic and amine types of antioxidants[242] and optical whiteners in PVC.

<u>Synthetic and Natural Rubbers</u>

Millingen[243] has described a thin-layer chromatographic method for the identification of phenolic and amine antioxidants in unvulcanized natural rubber. Overspraying the developed plate with 1% dibromo benzoquinone chlorimide in 4% sodium hypo-chlorite and 1% sodium bicarbonate followed by drying at 65°C reveals amine and phenolic antioxidants (Table 29) and rubber accelerators. On overspraying with a reagent comprising 0.15% platinum chloride in 3% potassium iodide, the amine antioxidant spots retain their colour, while the accelerator spots fade or became a darkened centre with an intense white halo. Only those antioxidants based on phenylene diamine gave a colour reaction with iodoplatinate reagent.

Table 29 - Reactions[a] of Antioxidants

Chemical	Trade name or abbreviation	Spray 1 iodoplatinate	Spray 2 chlorimide	Spray 2 plus spray 1	Primary solvent system R_f
Acetone diphenylamine condensation product	Nonox BL		Pale green	Green blue	0.53
Butraldehyde aniline condensation product	Antox		Brown yellow edge	Grey purple	0.40
4,4'-Butylidene-bis(6-tert-butyl-3-methylphenol)	Santowhite powder		Purple	Grey	0.37
4-Phenyl-N'cyclohexyl-p-phenylene diamine	Antioxidant 4010	Grey blue	Purple	Purple	
2,2'-Methylene-bis(4-ethyl-6-tert-butylphenol)	Antioxidant 425		Green white centre	Rust yellow	0.58
Diphenylamine acetone reaction product	Ble	Green blue (weak)	Yellow	Rust	0.48
N-Phenyl-N'-1,3-dimethyl butyl p-phenylenediamine	Santoflex 13	Blue brown	Brown yellow edge	Purple	0.48
6-Dodecyl-1,2-dihydro-2,2,4-trimethylquinoline	Santoflex DD		Yellow	Rust yellow centre	0.38
Diphenyl-p-phenylene-diamine	DPPD	Green	Purple	Purple	0.47
Mixture of diaryl p-phenylenediamines	Wingstay 100	Brown green	Brown yellow edge	Purple	0.50
6-Ethoxy-1,2-dihydro-2,2,4-trimethylquinoline	Santoflex AW		Cyclamen	Purple	0.40
N,N'-Bis(1-ethyl-3-methyl-pentyl)-p-phenylenediamine	UOP88	Cyclamen	Cyclamen	Cyclamen	
N,N'-Bis(1-methylheptyl)-p-phenylenediamine	Santoflex 217	Purple	Pink, grey edge	Purple	0.25
N,N'-Bis(1,4-dimethyl pentyl)-p-phenylenediamine	Santoflex 77	Purple	Brown	Purple	0.23
2,2'-Methylene-bis(6 alpha-methylcyclohexyl-4-methyl-phenol)	Nonox WSP		Yellow	Yellow orange edge	

Table 29 - Continued

4-Isopropylaminodiphenylamine	Nonox ZA	Grey green	Brown yellow edge	Purple	0.33
A substituted phenol	Nonox WSL		Rust	Yellow purple edge	0.55
beta-Napthylamine			Brown	Purple	0.22
Tri-(nonylated phenyl)-phosphite	Polygard	White	Purple,orange	White purple edge	0.57, 0.33
Substituted styrenated phenol	Wingstay S		Purple, blue	Purple	0.60, 0.37
2,6-di-tert-butyl-4-methylphenol	Ionol		Rust	Yellow brown edge	0.90
2,2'-Methylene-bis(4-methyl 6-tert-butylphenol)	Antox 2246		Yellow	Orange brown edge	0.57
Modified phenyl beta-naphthylamine	Nonox DN		Pink	Purple	0.53
A phenol condensation product	Nonox EX		Yellow	Yellow purple streak	0.38
A blend of arylamines	Nonox HFN		Pink	Purple	0.53
Sym. di-beta-naphthyl-p phenylenediamine	Nonox CI		Brown streak	Lavender streak	0.00
Phenyl-beta-naphthylamine + diphenyl-p-phenylenediamine	Nonox HP	Weak pale green	Yellow,purple	Purple	0.47, 0.53
A phenol condensation product	Nonox WSP		Yellow	Rust	0.37
Octylated diphenylamines	Octamine		Pale pink	Brown green	0.99
2,2,4-Trimethyl-1,2-dihydroquinoline	Flectol Flakes	(Weak response)	Turquoise	Green	0.45, 0.35 0.20, 0.13, 0.05
A phenol condensation product	A.F.D.		Grey		0.60
N,N'-Bis(1-methyl heptyl)-p phenylenediamine	UOP288	Cyclamen	Pink, blue	Purple	0.50, 0.08

[a] Colour reactions given in the tables refer to pure compounds. At the dilution level in a rubber stock, many of the positive responses given by antioxidants to the idodoplatinate reagent are weak and in many instances, could be regarded as negative. Those antioxidants which remain clearly evident at such dilution are Santoflex 13, Santoflex 77, Santoflex 217, and Nonox ZA.

Thin-layer chromatography. Solvent extraction - gas chromatographic methods have been described[244] for determining alkylated cresols (2,6-di-t-butyl-p-cresol) and amine antioxidants (N-phenyl-2-naphthylamine, p-phenylene diamine type) and Santoflex (NN'sec-heptyl phenyl-p-phenylene diamine) in polybutadiene.

Polyester Acrylates

Anodic voltametry. Budyina et al[245] have described methods based on anodic voltametry for the determination of Ionol (2,6-di-t-butyl-p-cresol) and quinol in polyester acrylates. To determine Ionol a sample is dissolved in 25 ml of acetone and an aliquot (10 ml) is treated with 2.5 ml of acetone and 5 ml of methanol and diluted to 35 ml with a solution 0.1 M in lithium chloride and 0.02 M in sodium tetraborate. A polarogram is recorded with a graphite-rod indicator-electrode and a 0.5 cadmium sulphate-cadmium reference electrode (Vasil'eva et al[236]). To determine quinol, the sample (1-3 g) is dissolved in 80 ml of methanol or methanol:acetone (1:1) and the solution is diluted to 100 ml with the lithium chloride-sodium tetraborate solution. A polarogram is recorded under the same conditions. Concentrations are determined by the addition method. The $E_\frac{1}{2}$ values (versus the SCE) are 0.25 V for Ionol and 0.16 V for quinol.

5.1.5   Dialkyl Thiodipropionate Secondary Antioxidants (See also Appendix 3)

Thin-layer chromatography. A procedure has been described[246] for the determination of dilauryl:beta, beta'-thiodipropionate antioxidant in polyolefins, ethylene-vinyl acetate copolymer, acrylonitrile-styrene-butadiene terpolymer and polystyrene. In this procedure the total additives are first extracted from the polymer by extraction under nitrogen with a 1:1:4 mixture of chloroform, ethanol and n-hexane. To separate dilauryl-beta, beta'-thiodipropionate from its own oxidation products and from other additives, the extract was subject to thin-layer chromatography on a silica gel-coated plate. The spot containing dilauryl-beta,beta'-thiodipropionate is refluxed for 30 min at 80°C with methanolic 5N potassium hydroxide to hydrolyse the ester to lauric acid, which is then extracted from the aqueous alcoholic phase with chloroform containing n-octadecene internal standard. This extract is gas chromatographed on an isothermal column packed with 1.5% fluorosilicone oil FS 1265 on 60 to 100 mesh Chromasorb W operated at 300°C and utilizing a flame ionization detector. Suitable reference solutions of dilauryl-beta,beta'-thiodipropionate are run in parallel through the whole procedures for calibration purposes.

In Figure 20 is shown a thin layer chromatogram obtained for extracts of polypropylene suspected to contain the following additives; Topanol CA (4,4',4''-butylene -1,1 tris (methyl-6-tert butyl phenol), DLTDP (dilauryl thiodipropionate) and Ionox 330 (1,3,5 trimethyl -2,4,6-tri (3,5-di-tert-butyl)-4-hydroxy benzyl) benzene.

Additives were removed from 30 g of each of the polymers by Soxhlet extraction overnight with diethyl ether. The extract residue was made up to 10 ml with chloroform and portions of these solutions and of solutions of the trace additives mentioned above were spotted on to silica gel plates. Viewing the developed plates under 254 nm ultra-violet light located Ionox 330 and Topanol CA but not DLTDP. All three additives were however, located with aggressive spray reagents and with a reagent for

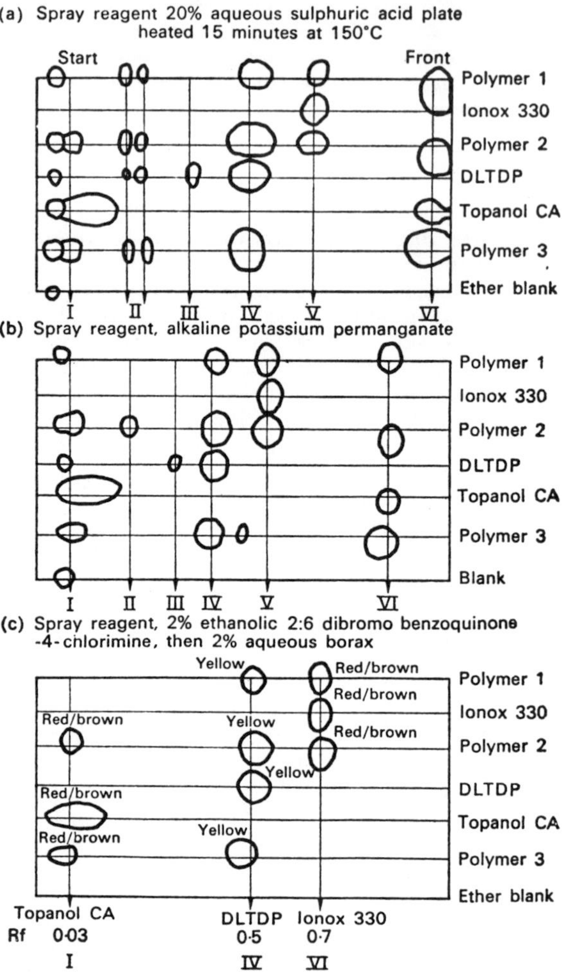

Figure 20   Thin layer chromatograms of extracts of three
polyolefins.
Plate: Merck silica gel GF 254. development solvent;
petroleum ether (40:60); ethylacetate 0:1 v/v.

phenols (2:6 dibromo benzoquinone-4-chlorimine) which gives a grey colour
with hindered phenols and also a yellow colour with DLTDP.   The
chromatograms in Fig 20 show that polymers 1 and 2 both contained DLTDP
and Ionox 330, polymer 2 also contained Topanol CA; and polymer 3
contained DLTDP and Topanol CA.   Both of the aggressive spray reagents
revealed the presence of low molecular weight polymer at the solvent front
(spot vi).   These reagents also revealed the presence in Topanol CA of
impurity which appears at the solvent front.   Obviously, this impurity is
of low polarity and is not phenolic as evidenced by the fact that it does
not produce a colour with the phenol reagent.   It is probably a
hydrocarbon.

Gas chromatography.   Gas chromatography has been used to determine
dilauryl beta, beta' thiodipropionate, lauryl sulphonyl beta, beta'

dipropionate and dilauryl sulphonyl beta, beta' dipropionate in polyolefins. The polymer extract is evaporated to dryness, then refluxed with 5N potassium hydroxide in 2:10 aqueous methanol. The free acids are extracted with chloroform. Gas chromatography of the chloroform extract is carried out on a 6' x 1" column packed with 1.5% fluorosilicone oil FS-1265 on 80/100 mesh Chromasorb W AW-SMCS operated isothermally at 165°C, Argon is used as carrier gas and n-octadecane as a chromatographic marker. Calibration is carried out against standard solutions of lauryl alcohol in chloroform.

<u>Selective oxidation technique.</u> Kellum[247] has described a method, based on selective oxidation, for the determination of diorganosulphide (and tertiary phosphite) types of secondary antioxidants in polyolefins. Some of these types of antioxidants are listed below:

       distearyl thiodipropionate
       dilauryl thiodipropionate
       4,4'-thiobis(6-tert-butyl-m-cresol)
       1,1'-thiobis(2-naphthol)
       triphenyl phosphite
       triethyl phosphite
       tri-p-tolyl phosphite
       tris(dinonyl phenyl)-phosphite
       2,2'-thiobis(6-tert-butyl-p-cresol)
       tri-isopropyl phosphite

These two classes of compounds were selectively determined in the presence of each other by oxidation using m-chloro-peroxy-benzoic acid to sulphones and phosphates. In this method, a heptane extract of the polyolefins containing the antioxidants is treated with a two-fold excess of the oxidant and allowed to react for 45 minutes before the unreacted oxidant is decomposed with sodium iodide to produce iodine which is estimated by sodium thiosulphate titration.

$$\underset{\text{Cl}}{\bigcirc} - \text{COOOH} + 2\text{NaI} \longrightarrow \underset{\text{Cl}}{\bigcirc} - \text{COOH} + \text{H}_2\text{O} + \text{I}_2$$

This method has the advantage of being free from interference by hindered phenols, benzophenones, triazoles, fatty acid amides and stearate salts, all of which might be present in the polymer extract. Recoveries of the following antioxidants were usually in excess of 99%. Stabilizers and additives, other than secondary antioxidants did not interfere in the procedure, these materials included phenolic antioxidants (2,6-di-tert-butyl-p-cresol, 2,2'-methylene-bis(6-tert-butyl-p-cresol), 4,4'-butylidenebis(6-tert-butyl-m-cresol), benzophenone light stabilizers (2-hydroxy-4-octoxy benzophenone), substituted hydroxyphenyl benzotriazoles, erucamide (a slip agent), and calcium stearate (an anti-block agent).

Table 30 shows the results obtained by applying the oxidation method to blends of antioxidants in unstabilized polyethylene. The results are in excellent agreement with theoretical antioxidant content of these polymers.

Table 30 - Analysis of typical polypropylene samples with
various stabilizers present

Additive	Percentage added	Percentage found
Distearyl thiodipropionate	0.20	0.19
Distearyl thiodipropionate	0.40	0.36
Distearyl thiodipropionate	0.10	0.095
4,4'-Thiobis (6-tert-butyl-m-cresol)	0.10	0.096
4,5'-Thiobis (6-tert-butyl-m-cresol)	0.20	0.19
Distearyl thiodipropionate	0.20	0.39 total
4,4'-Thiobis (6-tert-butyl-m-cresol)	0.20	0.39 total
Distearyl thiodipropionate	0.20	0.39 total

5.1.6 Tertiary Phosphite Antioxidants, (See also Appendix 3)

Styrene Butadiene Latices

Spectrophotometric procedures. Nawakowski[248] has described a
colorimetric method for determining Polygard (trisnonylphenyl phosphite)
based on hydrolysis to nonyl phenol, followed by coupling with p-
nitrobenzene-diazonium fluoroborate and colorimetric estimation at 550 nm.

Various other phenolic antioxidants produced dyes under these
conditions, viz. Wingstay S, Agerite Superlite and Nevastain A.

The procedure was applied with good precision to the determination
of Polygard in styrene-butadiene latexes. Good agreement was obtained
between this procedure and direct determinations of phosphorus by
elemental analysis.

A very popular method of estimating antioxidants in polymer extracts is by coupling or oxidizing them to form coloured products and measuring the resulting absorbance in the visible region of the spectrum. This technique is not particularly specific for individual antioxidants, but is specific for phenolic antioxidants and amine antioxidants, and hence can often be applied without interference from other types of polymer additives.

In one procedure[249] the polymer sample in the form of a powder or thin film is extracted with ethanol and the extracted phenolic antioxidant coupled with diazotized p-nitroaniline in strongly acidic medium.

$$\text{(phenyl)} - OH \; + \; \text{(nitrophenyl)} \; N - N^{+} \;\longrightarrow\; \text{(phenyl)} - OH \;\; N = N - \text{(phenyl)} - NO_2$$

The solution is then made alkaline and the visible absorption spectrum determined. Many of the antioxidants studied have an absorption maximum at a characteristic wavelength. Hence, in some instances it is possible to both identify and determine the antioxidant, provided a pure specimen of the compound in question is available for calibration purposes.

Ultraviolet spectroscopy. Brandt[250] has described an alternative method for the determination of Polygard (trisnonylyphenyl phosphite) in styrene-butadiene lattices, which utilizes the bathochromic shift in the spectrum of phenols resulting from the formation of phenolate ions in alkaline solution.

The latex is flocculated by the addition of acid and the antioxidant extracted with iso-octane.

Polygard in iso-octane has an ultraviolet spectrum with a peak at 273 mn in neutral solution. By adding a strong base (tetrabutylammonium hydroxide) the Polygard is hydrolysed and the peak is shifted to 296 nm. The difference in absorbance at 299 nm between the neutral and alkaline solutions is directly proportional to the amount of Polygard present. By use of this bathochromic shift, interference of non-phenolic impurities is eliminated, and a background correction factor is not required. Results obtained by this procedure agree well with these based on direct determination of elemental phosphorus, (Table 31).

5.1.7 Plasticizers

Polyvinyl Chloride

Gas chromatography. Due to their volatility the esters used as plasticizers in polymers such as flexible PVC can be determined by gas chromatographic analysis of a solvent extract of the polymer. Most published methods of analysis are based on this technique.

Robertson and Rowley[251] have published an excellent detailed description of methods for the solvent extraction of plasticizers from polyvinylchloride and other polymers prior to their determination by weighing or gas chromatography of the extract. They state that the quantitative separation of plasticizers from the other ingredients is the first and most important step in the analysis of plasticized polyvinylchloride compositions. The most effective and convenient method of separation is by extraction with a suitable solvent, using Soxhlet apparatus. The efficacy of this procedure depends mainly on the choice of solvent. The ideal solvent would not dissolve any of the polyvinylchloride, but would remove all the plasticizer, and all, or none, of the other ingredients of the composition.

Robertson and Rowley[251] compared the efficacies of ether, carbon tetrachloride and methanol, in the Soxhlet extraction of a number of polyvinylchloride compositions. Binary azeotropes of methanol with carbon tetrachloride, chloroform, 1,2-dichlorethane, and acetone, and of diethyl ether with 1,2-epoxypropane, were also compared.

Table 32 gives the results of following 4 hour extractions with ether by extractions with other solvents. It can be seen that extraction for 4 hours with carbon tetrachloride/methanol azeotrope gives better results than extraction for 15 hours with methanol.

Haslam and Soppet[252] found that extraction with acetone, followed by precipitation of dissolved polymer with light petroleum, gave poor results: Only 28.5% was recovered from a composition containing 31.8% tritolyl phosphate. Substitution of 1,2-dichlorethane for acetone gave no improvement, but diethyl ether extracted 31.8% of the sample, and the extract contained a negligible amount of polyvinylchloride. For routine extraction it was recommended by Haslam and Soppet that the sample be stood overnight in cold ether, then extracted for 6-7 hours in the Soxhlet apparatus. This procedure has also been described by Dohring[253], who specified that anhydrous ether should be used. Thinius[254] compared the rates of extraction of dioctyl phthalate by ether, carbon tetrachloride, and petroleum, at room temperature. In 70 minutes the ether extract from a composition containing 40.0% plasticizer amounted to 39.7%; in the same time the carbon tetrachloride extract was 33.1% and the petroleum extract 29.0%. Extraction with carbon tetrachloride for 64 hours gave only 35.6% extract. Thinius also found that mixtures of phthalate and phosphate esters could be completely removed by ether, but light petroleum gave only 65% of the expected yield. Toluene dissolved some polyvinylchloride at room temperature, and very much more in a Soxhlet extraction.

For compositions containing polypropylene adipate, which is only partly extracted by ether, Haslam and Squirrel[255] used a 6 hour ether extraction, followed by an 18 hour methanol extraction. the ether extract was 34.5% from a composition containing 36.1% of a mixture of equal parts of dioctyl phthalate and tritolyl phosphate, but only 33.1% from a composition containing 45.7% of a mixture of equal parts of dioctyl phthalate, tritolyl phosphate and polypropylene adipate. The combined ether and methanol extracts amounted to 34.7% and 44.3% respectively. Wake[256] quotes the results of some unpublished work carried out at the laboratories of the Rubber and Plastics Research Association: Methanol extracted 40.5% and 36.8% from compositions containing 42.3% polypropylene sebacate and 36.4% respectively. The extracts contained polyvinylchloride equivalent to 0.6% and 0.9% respectively. Clarke and Bazill[257] extract with ether for 15 hours then with methanol for 8 hours. They state that ether removes plasticizers whose molecular weight is less than 1000.

Figure 21(a) shows a gas chromatogram of a mixture of plasticizers of the type used in PVC formulations. The polymer extract was obtained[258] by dissolving the PVC in tetrahydrofuran, then adding 4 volumes of methanol to precipitate the polymer and leaving the plasticizers in

Table 31 - Determination of Polygard: comparison of ultraviolet
with perchloric acid methods

SBR latex type 6101	Perchloric acid (phospho- rus) method		Ultraviolet method	
Sample 1	1.31	1.30	1.18	1.30
Sample 2	1.26	1.28	1.24	1.36
Sample 3	1.23	1.24	1.36	1.32
Sheet rubber				
Type 1019	1.25	1.28	1.04	1.06
Type 1503	1.62	1.63	1.53	1.60
Type 1018	1.59	1.56	1.57	1.58
Type 1022	1.27	1.31	1.16	1.14

Table 32 - Multiple Extractions

Plasticizer	Concentration (%)	Extracts(%)	
		4h ether, 5h MeOH	4h ether, 4h CCl$_4$ MeOH
Bisoflex 791	28.5	27.7	28.7
Tritolyl phosphate	28.5	25.8	28.9
Mesamoll	28.5	25.5	28.1
Reoplex 220	28.5	20.4	29.3
Hexaplas PPA	23.5	11.0	17.4

solution. Many of the commonly used plasticizers are well separated and
can be identified by the retention times.

Although the relative retentions of the plasticizers are helpful for
the identification, the complexity of mixtures, normally encountered,
necessitates hydrolysis and esterification to obtain information on the
components of plasticizers. A definite identification of the original
plasticizer can be obtained only after the identification of the
constituent alcohols and acids has been made. To hydrolyse the esters a
freshly cut 0.1 g piece of lithium metal was added to the methanol
solution of the plasticizers: and the mixture was refluxed for 2 hours.
At the end of this time the solution was allowed to cool, and then it was
carefully acidified by dropwise additions of concentrated sulphuric acid.
When the solution became acidic, as noted by pH indicating paper, it was
refluxed again for 1 hour in order to convert the acids to their methyl
esters. After cooling, the solution was neutralized by addition of dry
sodium carbonate, and the pH was checked. A suitable sample of this
solution was injected into the gas chromatograph for detection of the

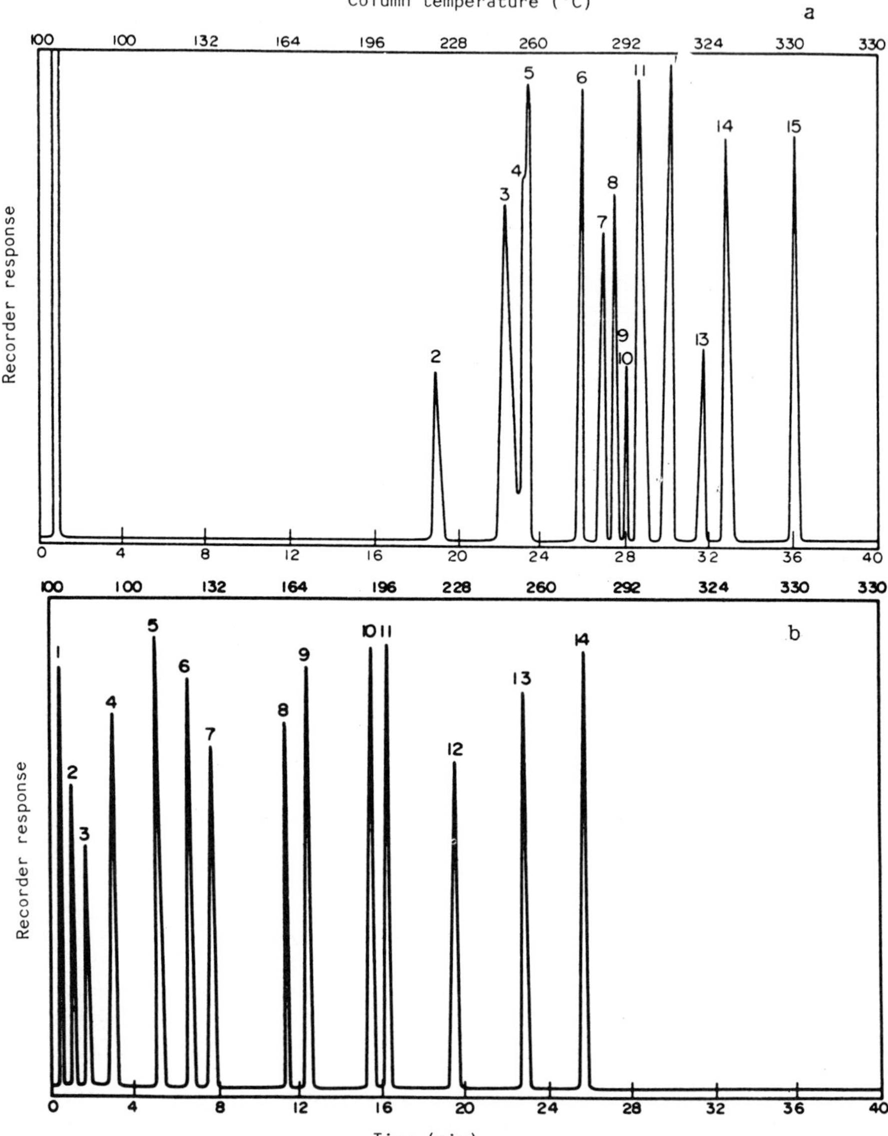

Figure 21 Gas chromatograpy of (a) ester plasticizers: (1) tetrahydrofuran; (2) triethyl citrate; (3) methyl-
phthalylethyl-glycolate; (4) ethylphthalyethyl
glycolate; (5) dibutyl-phthalate; (9) butylbenzyl
phthalate; (10) triocytl-phosphate; (11) di (2-
ethylhexyl)adipate; (12) di(2-ethyl-hexyl)phthalate;
(13) di(2-ethylhexyl); (14) di (2-ethyl-hexyl)sebacate;
(15) di-n-decyl phthalate. (b) gas chromatography of
alcohols and methyl esters of acids resulting from
hydrolysis of plasticsers, (1) methanol; (2) butanol;
(3) pentanol; (4) hexanol; (5) heptanol; (6) 2-
ethylhexanol; (7) octanol; (8) dimethyladipates; (9)
decanol; (10) di-methyl-o-phthalate; (11) dodecanol;
(12) tetradecanol; (13) methyl palmitate; (14) methyl
stearate.

alcohols and methyl esters of the acids.   The gas chromatographic
conditions used were identical to those for chromatographing the original
plasticizers (Figure 21(b)).

This scheme is helpful in identifying these products.   Since the gas
chromatographic conditions used for the alcohols and the methyl esters of
the acids are identical to those used for the original plasticizers, it is
very easy to establish whether the plasticizers have been completely
reacted.   The presence of non-hydrolysable components in a mixture can
also be detected by examining the gas chromatograms of plasticizers before
and after hydrolysis, as the original gas chromatographic peak will still
be observed.

High performance liquid chromatography.   Majors[259] describes some
separations obtained on several commercially available support materials
using a liquid chromatographic system capable of operation up to 5000 psi.
Because of the nature of these solid-core supports, unusually large
pressure drops across the chromatographic column do not occur unless one
uses several columns in series, or extremely fast flow rates.

A schematic diagram of a typical high-pressure liquid
chromatographic system is shown in Figure 22.   The system employs a
microregulating high-pressure feed pump with an inlet pressure 5000 psi.
The solvent reservoir is placed a few feet above the pump since a slightly
positive inlet pressure is required for operation.   The degassed solvent
is slowly stirred by means of a magnetic stirrer and is heated externally
slightly above room temperature to keep the solvent degassed.   A valve is
placed on the high-pressure side of the pump to aid in priming the pump or
to shut off the solvent flow when required.   Downstream from the pressure
gauge, the entire system is connected with 0.04 inch i.d., 0.063 inch o.d.
stainless-steel tubing to ensure a minimal dead-volume in the system,
which is particularly desirable when changing solvents or solvent
programming.

When necessary, the liquid flow can be split between the analytical
column and the reference column.   A valve placed before the reference
column permits the flow rate through that column to be completely shut off
or varied, depending on the requirements of the system.   This is
particularly useful when using the refractive index monitor at higher flow
rates.   For column flow rates up to 2 ml/min good stability could be
obtained by careful balance of flow.   The reference column was either
filled with the same liquid-liquid support employed in the analytical
column or merely filled with uncoated glass beads.   With the ultraviolet
detector the reference column was normally not used.

The separation of three plasticizers isolated from PVC is summarized
in Table 33.   At the same flow rate the separation times on Corasil are
considerably longer and the plate heights greater although not excessive.
The elution bands are very symmetrical.

Nuclear magnetic resonance spectroscopy.   Wide-line nuclear magnetic
resonance spectroscopy has been used (Mansfield[260]) for the determination
of the di-iso-octyl phthalate content of polyvinylchloride.   The principle
of the method is that the narrowline liquid-type NMR signal of the
plasticizer is easily separated from the very broad signal due to the
polymer; integration of the narrow-line signal permits determination of
the plasticizer.   A Newport Quantity Analyser Mk I low-resolution
instrument, equipped with a 40 ml sample assembly and digital read-out,

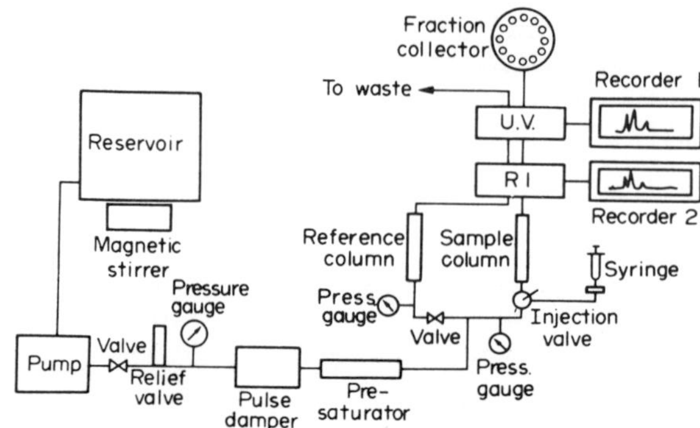

Figure 22 Schematic of high pressure liquid chromatographic
system.

Table 33 - Evaluation of Solid Core Supports for Phthalate Plasticizers

	Zipax		Corasil 1		Deactivated Corasil	
Solute	V_R,ml	HETP,mm	V_R,ml	HETP,mm	V_R,ml	HETP,mm
Didecyl Phthalate	2.0	0.38	3.1	1.9	3.1	1.8
Decyl Benzyl Phthalate	2.4	0.93	5.1	3.3	4.7	2.7
Dibenzyl Phthalate	3.8	0.51	11.6	1.9	8.4	2.5

has been used to determine 20-50% of plasticizer in polyvinylchloride.
The sample may be in any physical state without significantly affecting
the results; e.g. sheet samples are cut into strips 50 mm wide, which are
rolled up and placed in the sample holder. A curvilinear relationship
exists between the signal per g and the percentage by weight of the
plasticizer. For highest precision it is necessary to know the type of
plasticizer present; use of the appropriate calibration graph gives a
precision of $\pm$ 0.5%. However, one general calibration graph can be used;
the precision is then approximately $\pm$ 3%. As the NMR signal is
temperature-dependent the temperature of calibration and of analysis
should not differ by more than 4°C.

Miscellaneous. Other work on the determination of plasticizers in
PVC is reviewed in Table 34.

5.1.8 Ultraviolet Absorbers

Among the numerous additives commonly used in plastic materials, the
ultraviolet absorbers are increasing in importance because they are often
used in food packaging materials to protect the plastic material, usually
polyolefins or polystyrene, as well as the foodstuff packaged from the
actinic action of ultraviolet radiation. Actinic effects may cause
discoloration of both the plastic material and the foodstuff and on
occasion, also changes in taste and loss of vitamins in the food.

Table 34 - Plasticizers in PVC

		Reference
Solvent extraction methods		
e.g.		
diocytylphthalate tritolylphthalate	diethyl ether extraction, weighing	261, 262
plasticizers	solvent extraction, weighing	263
polypropylene adipate diocylphthalate tritolyphthalate	diethyl ether then methanol extraction, weighing	264
polypropylene sebacate polypropylene adipate	methanol extraction	265
Gas chromatography		
plasticizers	solvent extraction, glc	266, 267
alkyl adipate alkyl phthalate	solvent extraction, glc	268, 269
benzylbutyl phthalate	solvent extraction, glc	270
alkyl phthalates fatty acid esters	solvent extraction, glc	271,272
dimethylphthalate dimethyl sebacate triacetin diacetin diethylphthalate	dichloromethane extraction, glc	273
adipate and phthalate type	mild pyrolysis - gas chromatography	274
plasticizers	solvent extraction, gas chromatography	275
plasticizers	hydrolysis to alcohols, gas chromatography	276
dibutylphthalate diisobutylphthalate bis(2-ethylbexyl)phthalate di-n-octyl phthalate	pyrolysis - gas chromatography	277, 278
alkyl phthalates alkyl adipates alkyl azealates alkyl citrates alkyl sebacates alkylglycollates alkyl phosphates	tetrahydrofuran extraction, glc (includes determination of alcohol degradation products)	279
plasticizers	pyrolysis - gas chromatography of polymer	280

Table 34 - Continued

<u>Thin layer chromatography</u>

plasticizers	dimethoxymethane extraction, thin-layer chromatography	281
phthalic acid esters sebacic acid esters phosphoric acid esters epoxy esters amino crotonic esters diphenylthio adipic acid type	solvent extraction, thin-layer chromatography for identification and determination	282-301
chlorinated paraffin waxes tricresyl phosphate dibutuyl succinate di-n-octylphthalate bis(2-ethylhexyl)phthalate tributyl phosphate esters and carboxylic acids and dicarboxylic acids dicarboxylic acids diethyl phthalate dimethylsebacate	solvent extraction, glc	302
phthalate type	solvent extraction, thin-layer chromatography and infrared identification	303

<u>Infrared and NMR spectroscopy</u>

polypropylene adipate polypropylene sebacate tritolyphosphate bisoflex 791 bisoflex 220 bisoflex 795	extraction diethyl ether then carbon tetrachloride then with methanol followed by infrared spectroscopy	304
tricresyl phosphate dibutyl succinate di-n-octylphthalate bis(2-ethylhexyl)phthalate tributylphosphate esters of carboxylic acids dicarboxylic acids diethylphthalate dimethylsebacate	solvent extraction, infrared spectroscopy	302
diiso-octylphthalate	solvent extraction, NMR spectroscopy	305

<u>Miscellaneous methods</u>

plasticizers	benzene extraction, measurement of saponification value	306
plasticizers	diethyl ether extraction, various instrumental finishes	307
phthalate plasticizers	extraction with boiling hydrochloric acid, polarography	308, 309

The ultraviolet absorbed can be divided in different groups (Table 35).

a)   Benzophenone derivatives

b)   Salicylic acid esters

c)   Resorcinol esters

d)   Benzotriazole compounds

e)   Coumarin derivatives

Methods for the determination of ultraviolet absorbers and optical brighteners are based principally on fluorimetry.

In many instances visible fluorescence techniques are less subject to interference by other polymer additives present in a polymer extract than are ultraviolet methods of analysis. Therefore, in some instances visible fluorimetry offers a method of determining a polymer constituent without interference from other constituents, when this would not be possible by ultraviolet spectroscopy. Apart from specificity, fluorescence techniques are more sensitive than absorption spectroscopic techniques.

<u>Polyolefins</u>

<u>Direct film infrared spectroscopy.</u> An example is given below of a method based on direct polymer film infrared spectroscopy for the determination of the ultraviolet absorber Cyasorb UV531 (2-hydroxy-4-n-octoxybenzophenone) at concentrations of 0.1 to 1% in unpigmented high-density polyethylene. Antioxidants such as Polygard and Santonox R do not interfere in this procedure.

<u>Apparatus.</u> Double beam spectrometer covering the 15-17 micron region (e.g. Grubb Parsons GS2A). Hydraulic press with heated and water-cooled platens. Stainless-steel moulding plates (6" x 6" x 1/6"). Shims 0.06 cm thick (circular 1" diameter or rectangular 1" long). Aluminium foil. Clear plastic rule calibrated in millimetres. Dial gauge calibrated in 0.01 mm divisions.

<u>Preparation of sample film.</u> Cover two stainless steel moulding plates with aluminium foil and place up to six 0.06 cm thick shims on one of these. Place approximately 0.2 g of polymer sample in the centre of each shim and carefully place the second moulding plate on top. Position the two plates in the press and apply contact pressure. Switch on the heating supply and set the thermostat to 120°C. When the temperature reaches 120°C increase the pressure to 3000lb/in^2, switch off the heating supply and water cool to room temperature. Carefully strip off the aluminium foil from the polymer films and push out the films from the shims. Carefully strip off the aluminium foil from the polymer films and push out the films from the shims. Check the thickness of each film by means of the dial gauge. Six readings on each film should not vary by more than 0.03 mm. Reject any which has air bubbles, is uneven or is wedge-shaped. Shape the film to fit the spectroscopic sample holder and gently scrape one of the surfaces with a fine emery board to produce a series of fine parallel lines. This reduces the incidence of interference fringes.

Table 35 - Some Ultraviolet Absorbers Used in Plastic Materials

Chemical Formula	Trade Name	Manufacturer
1. 2-hydrocy-4-methoxy-benzophenone	Uvinal M 40 Uvistat 24 Cyasorb UV 9	General Aniline Co Ward & Blenkinsop Cyanamid
2. 2,4-dihydroxy-benzophenone	Uvistat 12 Uninul 400	Ward & Blenkinsop General Aniline Co
3. 2-hydroxy-4-methoxy-4-methyl-benzophenone	Uvistat 2211	Ward & Blenkinsop
4. 2,4,5-trihydroxy butyrophenone	Inhibitor THBP	Eastman
5. 4-dodecyloxy-2-hydroxy benzophenone	Inhibitor DOBP	Eastman
6. 2-hydroxy-5-n-octoxy-benzophenone	Cyasorb UV 531	Cyanamid
7. 2,2'dihydroxy-4-methoxy benzophenone	Cyasorb UV 24	Cyanamid
8. 2,2'dihydroxy-4,4' dimethoxy benzophenone	Uvinul D49	General Aniline Co
9. p-tert-butylphenylsali-cilate		
10. resorcinol mono benzoate	Inhibitor RMB	Eastman
11. hydroxyphenylbenzo-triazole	Tinuvin P	Geigy
12. 7-diethylamino-4-methyl coumarin		Ward

Recording the infrared spectrum. Place the film in the sample holder and position in the infrared instrument so that the beam passes through the film at right angles to the scratch marks.

Record the infrared spectrum from 15.5 to 16.5 micron using a scanning speed of ½ micron per minute.

Before removing the film from the instrument, mark the position of the infrared beam. Remove the sample from the holder and measure the thickness to the nearest 0.01 mm by means of the dial gauge at six points within the marked area. Calculate the mean of these six measurements.

Measurement of absorbance. Remove the chart from the spectrometer and with a sharp pencil rule a base-line from approximately 15.8 micron to approximately 16.2 micron. With a ruler measure I and I_o to the nearest 0.1 mm at the wavelength of the peak maximum. Calculate the absorbance at 15.94 micron and hence the absorbance per unit thickness by means of the following expression:

$$\text{Absorbance per unit thickness} = \frac{\log_{10} I_0}{\text{Film of thickness (in cm)}}$$

Prepare duplicate films from the standard sheets containing 0.1, 0.3, 0.5 and 1.0 wt%. UV 531 as above. Record the infrared spectrum as described above, and calculate the absorbance per unit thickness as described under measurement of absorbance. Construct a calibration curve by plotting absorbance per unit thickness against percentage weight UV 531 for each standard film. Use this calibration curve to obtain the UV 531 content of the polyethylene sample.

<u>Thin-layer chromatography.</u> Thin-layer chromatography is useful for the determination of light stabilizers in polymers. In one procedure for determining Cyasorb UV 531 light stabilizer (2-hydrocy-4-n-octoxy-benzophenone) in high-density polyethylene the stabilizer is extracted from the polymer by precipitation of the polymer with ethanol from a hot toluene solution. An aliquot of the extract is applied to a GF245 silica-gel-coated thin-layer plate and chromatographed using methylene dichloride as an eluant. The zone corresponding to UV531 is identified as a dark blue spot under ultraviolet light and then the silica gel is removed from the plate and extracted with ethanol. The concentration of UV 531 is determined by measuring the ultraviolet absorption peak near 295 nm in ethanol solution and referring to a prepared calibration graph.

No interference is encountered from a number of additives in the polymer viz. dilauryl thiodipropionate, Ionox 330, Ionol CP, Topanol CA, Santonox R, polygard.

<u>Gas chromatography.</u> Denning and Marshall[310] devised a method in which toluene extracts of polyethylene are examined for ultraviolet absorbers (and antioxidants) at two column temperatures in order to overcome the problem that, whereas some compounds have a relatively low retention time, others of higher molecular weight have very long retention time at 250°C. Santonox R was used as an internal standard in this method. Denning and Marshall[972] used a Pye Series 104, Model 64, dual-column chromatograph, equipped with a heated flame-ionization detector and an isothermal column oven. The oven was operated isothermally at 250 or 300°C. Relative retention data obtained for ultraviolet absorbers are given in Table 36.

Lappin and Zannucci[311] showed that a number of ultraviolet light absorbers could be quantitatively determined on an SE-30 column. SE-30 has a maximum operating temperature of about 350°C. This high temperature permits elution of some high molecular weight additives.

They examined two methods of separating the additives from the polymer. Gas chromatography of a hexane extract of the polymer produced numerous extraneous peaks, probably owing to a decomposition of dissolved amorphous polymer in the injection port, rendering the chromatogram useless. When the polymer was dissolved in p-xylene and reprecipitated with an equal volume of p-dioxane, a relatively clean chromatogram was obtained from the filtrate. The position of decomposition peaks from unstable compounds, such as dilauryl 3,3'-thiodipropionate and distearyl pentaerythritol diphosphite, are predictable and do not interfere with the determination of those additives studied by Lappin and Zannucci[311]. They found that the optimum polymer sample size was 3 g, although runs were possible on the samples as small as 200 mg. Benzophenone is another possible internal standard. Lappin and Zannucci determined 4-(dode-cyloxy)2-hydroxybenzophenone and 2,6-di-tert-butyl-p-cresol in polypropylene by this technique (Table 37). The precision of the 2,6 di-

Table 36 - Separation of Ultraviolet Absorbers

Ultra violet absorber	Class of compound	Relative retention time (relative to Santonox R)	Column Temperature (°C)
Tinuvin P	Benzotriazole	0.20	250
Tinuvin 326	Benzotriazole	0.64	250
Tinuvin 327	Benzotriazole	0.85	250
Cyasorb UV 531	Aromatic ketone	1.0	250
Tinuvin P	2-(2'-Hydroxy-5'-methylphenyl benzotriazole)		
Tinuvin 326	2-(2'-Hydroxy-3'-t-butyl-5'methyl ethyl phenyl-3-) chlorobenzo triazole		
Tinuvin 327	2-(2'-Hydroxy-3',5'-di-t-butylphenyl)-3-chlorobenzo triazole		
Cyasorb UV 531	2-Hydroxy-4-n-octoxybenzophenone		

Table 37 - Analyses of Polypropylene for BHT and DOBP

Sample composition	Additive found (%)			
	Run 1	Run 2	Run 3	Run 4
0.05% 2,6-di-tert-butyl-p-cresol	0.02	0.02	0.03	0.02
0.30% 4(dodecyloxy) 2-hydroxy-benzophenone	0.25	0.32	0.20	0.35

tert butyl-p-cresol determination is good but the quantity found (0.02%) was less than the amount added (0.05%), possibly due to some losses of volatile antioxidant during polymer compounding.

4-(dodecyloxy)2-hydroxybenzophenone determinations ranged from 0.20 to 0.35% for a sample originally containing 0.30% of the additive.

Polystyrene

Fluorescence techniques. Uvitex OB has an intense ultraviolet absorption at a wavelength of 378 nm, which is high enough to be outside the region many potentially interfering substances present in the polymer extract would be excited to fluorescence. This is illustrated by a fluorimetric procedure for the determination of down to 10 ppm Uvitex OB in polystyrene. Antioxidants such as Ionol CP (2,6-di-tert-butyl-p-cresol), Ionox 330 (1,3,5-tri-methyl-2,4,6-tri(3,5-di-t-butyl-4-hydroxy-benzyl)benzene), Polygard (tris(nonylated phenyl) phosphite), Wingstay T (a butylated cresol), and Wingstay W and many others, do not interfere in this procedure. In this procedure the polystyrene is dissolved in chloroform and the solution excited by ultraviolet radiation of wavelength 370 nm from a mercury vapour lamp and the fluorescence spectrum of the sample recorded over the range 400-440 nm. The reading from the fluorimeter is noted, and the Uvitex OB concentration in the polystyrene determined by reference to a prepared calibration graph.

Ultraviolet spectroscopic procedure. By their nature, many of these types of compounds are amenable to analysis by fluorimetric analysis, thus Uvitex OB has an intense ultraviolet absorption at a wavelength of 435 um, which is high enough to be outside the region where many potentially interfering substances present in the polymer extract would be expected to fluoresce. Thus, using a 0.5% solution of polystyrene in chloroform a limit of detection of 10 ppm Uvitex OB in polystyrene is achieved. Antioxidants such as Ionol CP, Ionox 330, Polygard, Wingstay T and Wingstay W and many others do not interfere in this procedure.

Thin-layer chromatography. To estimate 7(-6 butoxy-5-methyl-benzotriazol-2-yl) 3-phenyl coumarin in polymer granules, the sample was extracted from the ground or chopped sample by heating under reflux with chloroform[312]. The extract, together with a chloroform solution of an authentic sample of the brightener was applied to two Kieselgel G plates, and chromatograms developed with benzene-chloroform (2:3) and benzene, respectively; the spots were detected by their fluorescence in ultraviolet radiation.

Miscellaneous. Other methods for the determination of UV absorbers in polymers are reviewed in Table 38.

5.1.9 Organotin Stabilizers

Polyvinylchloride

Titration methods. Organotin compounds in solvent extracts of PVC can be determined by potentiometric and manual titration procedures[329,330]. Potentiometric titration of non-sulphur-containing organotin compounds is achieved by titrating the carboxylate groups with sodium methoxide in pyridine by using antimony and calomel electrodes. The end-point roughly coincides with the change of colour when thymolphthalein is used as indicator. The titrant, i.e. sodium methoxide, is standardized against benzoic acid and the titration carried out under nitrogen as a precautionary measure. Dialkyl tin thio compounds can be titrated[329] with a solution of silver nitrate in 1+1 isopropanol-water. The sample is dissolved in a solvent consisting of a 1+1 mixture of benzene-methanol, containing sodium acetate trihydrate (13.7 g l^{-1}). The indicator electrode is a length of silver wire coated with sulphide. It is prepared by immersion of a silver wire into an alkaline solution of sodium sulphide followed by addition of a silver nitrate solution (0.1 N) and stirring. The electrode is wiped clean and polished with a clean cloth and is then ready for use. The reference electrode consists of a copper wire immersed in a pool of mercury covered with a solution of 0.1 N sodium acetate. Contact with the test solution is made through an agar-gel bridge. More than one end-point is sometimes produced. However, titration of reference samples under identical conditions helps in the interpretation of the results.

Thin-layer chromatography. Quantitative thin-layer chromatography has been used [331] for the determination of organotin stabilizers.

Hexane-glacial acetic acid (12+1) as eluting agent is satisfactory for thioacids and thiols. A small amount of the polymer extract dissolved

Table 38 - Ultraviolet Absorbers in Polyethylene

		Reference
Thin-layer chromatography		
benzophenone and salicylic acid types	solvent extraction, thin-layer chromatography	313
benzophenone type salicylate type	solvent extraction, thin-layer chromatography	314
benzotriazole type	for quantitative estimation	
salicylate type substituted acrylonitriles organonickel type ultraviolet absorbers and optical brighteners	solvent extraction, thin-layer chromatography	315-324
optical whiteners	solvent extraction, thin-layer chromatography	325
Miscellaneous methods		
Tinuvin P (2'hydroxy 5'-methylphenylbenzotriazole) Tinuvin 326 (2-(2'hydroxy -3'-t-butyl 5' methylethyl-phenyl-3 chlorobenzotriazole)) Tinuvin 327 (2-(2'hydroxy-3' -5'-di-6-butyl(phenyl)-3-chlorobenzotriazole)) Cyasorb UV531 (2 hydroxy-4-n octoxy benzophenone)	toluene extraction, glc	326
benzophenone type	solvent extraction, potentio-metric titration with sodium methoxide in dimethyl-formamide	327
antioxidants	solvent extraction, column chromatography	328

in the elution solvent is applied to a thin-layer chromatographic plate coated with a 1.25 mm thick layer of Kieselgel G as the stationary phase, and eluted with the hexane-glacial acetic acid mixture. After drying, the stationary phase is sprayed with a 0.1% solution of catechol violet in 95% ethanol, blue spots appearing where tin compounds are present.

Separations have been achieved[332] of various organotin stabilizers on 0.25 mm layers of Merck Kieselgel GF-254 using butanol-glacial acetic acid (97+3) as the mobile phase and spraying the plate with catechol violet solution followed by irradiation of the plate for 10 min with ultraviolet radiation and a respray with catechol violet solution to detect the separated compounds. This procedure can be applied to the organotin compounds in PVC using a preliminary extraction of the additives from 5 g of polymer with diethyl ether for 8 hours. Ether is removed by evaporation and ethanol added to precipitate polymer, which is then filtered off. Down to a few parts per million of organotin can be determined in polymers.

Determination of tin and sulphur. Organotin compounds are widely used in the plastics industry as stabilizers for poly(vinyl chloride) compositions. The most important compounds used for this purpose are based on dialkyltin groups

$$\begin{matrix} R \\ >Sn< , \\ R \end{matrix}$$

especially where R = butyl or octyl.

It is convenient to divide tin stabilizers into those containing sulphur and those without sulphur. To the first group belong compounds, like dialkyltin mercaptides, mercapto-esters and mercapto carboxylates, to the second, dialkyltin carboxylates and their esters.

A very useful preliminary to the identification of organotin compounds is the determination of the tin and sulphur content of an extract of the polymer which has been purified so that only the organotin compound is present. Tin can be determined[333] by wet oxidation of the sample followed by precipitation of tin with cupferron. When an accuracy of $\pm$ 0.2% absolute is sufficient the sulphated ash procedure may be combined with spectrographic examination to check that other metals are absent.

Tin can be determined in PVC by polarography and atomic absorption spectroscopy and in PVC foils by X-ray fluorescence spectroscopy[335]. In the polarographic method[334] the sample is decomposed with sulphuric acid-nitric acid (1:1), then hydrochloric acid is added and ammonia added until the solution is made 4 M to ammonium chloride. Tin was determined polarographically at $E_{\frac{1}{2}}$ = -0.52 V. In the atomic absorption method, the sample is decomposed, and tin determined as the acetate using an air-acetylene flame at 224.6 nm. Both methods can determine tin in PVC in amounts down to 300 ppm.

In the X-ray fluorescence method the polymer extract is diluted with 2-ethyl-hexanol to give a tin concentration of 0.9-1.7%, and about 40 ml of the solution is taken for the determination. The Sn K radiation (25.2 keV) excited by low-energy gamma-radiation from a Am source is measured by means of a NaI(Tl) crystal, a single beam gamma-spectrometer being used for evaluation. Sulphur can also be determined by X-ray fluorescence spectroscopy[333].

Infrared spectroscopy. The identification of acids and alcohols present in tin stabilisers containing no sulphur by infrared spectroscopy is usually straightforward and the identification of alcohols from stabilisers containing mercapto esters also presents little or no difficulty. However, the identification of the thioacid and of the alkyl groups attached directly to tin can prove more difficult, especially if long chain acids form part of the compound or if the stabiliser is not pure, e.g. if it contains plasticisers. The thioacid may also tend to decompose during hydrolysis procedures.

Udris[336] has described two schemes, based on chromatography and infrared spectroscopy, for the identification, respectively in PVC extracts of;

i)   dialkyltin dialkylthioglycollates,
R   SCH$_2$ COOR'
  Sn
R   SCH$_2$ COOR''

dialkyltin dilauryl mercaptides
R   SCH$_2$ (CH$_2$)$_{10}$ CH$_3$
 Sn
R   SCH$_2$ (CH$_2$)$_{10}$ CH$_3$

(R, R', R'' = alkyl) stabilizers in which the dialkyltin group is combined with both thiol and carboxyl groups.

ii)   dialkyltin maleates
R   OO CCH
  Sn
R   OO CCH

dialkyltindialkyl maleates
R   OO CCH = CH COOR'
  Sn
R   OO CCH = CH COOR''

dialkyltindilaurates
R   OOC (CH$_2$)$_{10}$ CH$_3$
  Sn
R   OOC (CH$_2$)$_{10}$ CH$_3$

It is first of all necessary to prepare an extract of the PVC in which the organotin stabiliser is quantitatively recovered.

For the extraction of the first set of compounds, i) the polymer is refluxed with acetone and the silver salts precipitated by the addition of aqueous silver nitrate. After drying the residue is examined by infrared spectroscopy to identify alcohols, thioacids, thiols, and alkyl groups attached to tin. For the extraction of the second set of compounds ii) the polymer is refluxed with 10% aqueous sodium hydroxide prior to the identification of alcohols by gas chromatography then the identification of alkyl groups attached to tin and carboxylic acids.

Udris[336] examined methods for the recovery and identification of tin stabilisers in excess plasticiser (90 to 95%). The plasticisers included diisooctyl phthalate and mixtures of diisooctyl phthalate and tritolyl phosphate in various proportions.

Good samples of dialkyltin dichlorides were isolated by this method from mixtures of tin stabilisers with diisooctyl phthalate and tritolyl phosphate; the dialkyltin groups have been identified in dibutyl tin dinonylmaleate, dioctyltin dilaurate, dibutyltin dinonylthioglycollate and dioctyltin thioglycollate.

Miscellaneous.   Other work on the determination of organotin stabilizers in PVC is reviewed in Table 39.

Table 39 - Organotin Type Stabilizers in Poly(vinylchloride)

Thin-layer chromatography		Reference
dibutyltin-dilaurate dioctyltin-dilaurate butyltintrichloride dimethyltinchloride diphenyltin dichloride hexabutylditin tributyltin laurate dibutyltin bis-(2-ethyl- hexylthioglycollate)	solvent extraction, thin-layer chromatography for identification of organotin compounds	337
dibutyltin bis(2-ethyl hexyl) thioglycollate	solvent extraction, thin-layer chromatography	337
di-n-octyltin maleate	solvent extraction, thin-layer chromatography	338
stabilizers	various methods	339-344
dibenzyltin bis (isooctyl- mercaptoacetate) dibenzyltin bis (butyl- mercaptoacetate) dibenzyltin bis(cyclohexl- mercaptoacetate	solvent extraction, hydrolysis to glycol, thin-layer chromatography of 3,5 dinitrobenzoates	345
Titration methods		
stabilizers	titration of carboxylate end- groups with sodium methoxide in pyridine	346, 347
stabilizers	determination of sulphur by potentiometric titration with potassium periodate in glacial acetic acid	348
dialkyltin thio compounds	potentiometric titration with silver nitrate in aqueous iso- propanol. Also, identification by infrared spectroscopy	346
Miscellaneous methods		
stabilizers	heptane-acetic acid extraction, various instrumental finishes	349
dialkyltin types dialkylthioglycollates dialkyltin laurylmercaptides e.g. dibutyltin oxide, dioctyltin oxide, dibutyl- tin maleate, dioctyltin laurate, dioctyltin thio- glycollate	identification by infrared spectroscopy and chemical methods	346

5.1.10 Metal Stearate and Other Types of Stabilizers

Polyvinylchloride

Other types of non-tin organometallic stabilizers have been used in the formulation of PVC. Mal'kova et al[350] described an alternating current polarographic method for the determination of cadmium, zinc and barium stearates or laurates in PVC. The samples are prepared for analysis by being ashed in a muffle-furnace at 500°C, a solution of the ash in hydrochloric acid being made molar in lithium chloride and adjusted to pH 4.0 $\pm$ 0.2. Alternatively, the sample solution can be prepared by boiling with 2 M hydrochloric acid for 3 minutes, cooling and adjusting the pH of a portion of the solution to 4 with 2 M lithium hydroxide. The solution obtained in either instance is de-aerated by passage of argon and the polarogram is recorded. Cadmium, zinc and barium give sharp peaks st -0.65, -1.01 and -1.90V respectively, vs the mercury-pool anode. The first digestion procedure is recommended if the sample contains esters of phosphorus acid.

Other techniques that have been used for the examination of organometallic stabilizers extracted from PVC include column chromatography (barium, cadmium and zinc salts of fatty acids[351]), paper chromatography (cadmium, lead and zinc salts or fatty acids[352] and polarography (cadmium, lead and zinc salts of fatty acids[353]).

Other procedures for the analysis of metal stearate and non-tin types of stabilizers in PVC are reviewed in Table 40.

5.1.11   Organic and Inorganic Pigments in Polymers

Organic and inorganic pigments are used for coloration of polymers, polymer films and polymer coatings on metal containers. Vapour phase ultraviolet absorption spectrometry at 200 nm has been used[376] to identify such pigments. In this method powdered samples are directly vaporized in the heated graphite atomizer. Thermal ultraviolet profiles of organic pigments show absorption bands between 300 and 900°C, while profiles of inorganic pigments are characterized by absorption bands at temperatures above 900°C. Temperature, relative intensity, and width of the bands allow the identification of the pigments. The technique shows fast acquisition of thermal ultraviolet profiles (2-3 min for each run) good repeatability and wide thermal range (from 150 to 2300°C). The method has been applied to a variety of polymers.

A practical example of the identification of pigments is given in Figure 23. A 1:1 mixture of organic pigment yellow (2-nitro-p-toluidine coupled with acetoacetanilide) and inorganic P.Y. 34 (lead chromate) was vaporized using the conditions quoted under Figure 23(a). The thermal ultraviolet profile shows clearly two absorption bands at about 500°C and 1250°C. The first band is attributable to the vapours which originate from the decomposition and pyrolysis of the organic pigment, the second band corresponds to the decomposition and vaporization of lead chromate at high temperature (mp 844°C). It is possible therefore to determine by a rapid run whether the pigment is a mixture or belongs to the organic or inorganic group.

Blue and green organic pigments are commonly derivatives of copper phthalocyanine, which is characterized by remarkable resistance to thermal treatments. In fact these pigments show ultraviolet absorption at temperatures (about 800 - 900°C) higher than those observed for yellow and red pigments, Figure 23(b). Alpha and beta forms of copper phthalocyanine

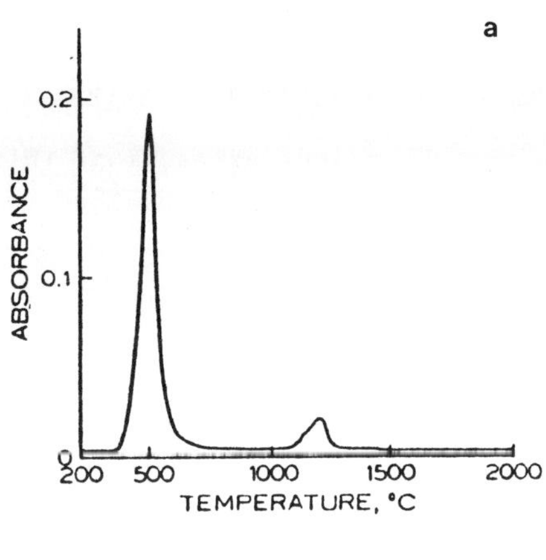

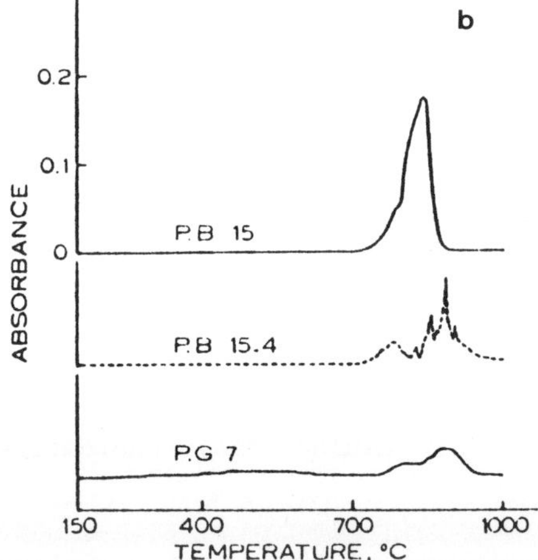

Figure 23   TUV profile of (a) 1.1 mixture of organic pigment yellow
(2-nitro p-toluidine coupled with acetoacetanilide and
inorganic pitment PY 34 lead chromate. (b) PB 15.4
copper phthalocyonine (beta form); PG 7 polychlorinated
copper phthalocyanine (14-16 chlorine atoms).

Thermal cycle	1	2	3
temp, °C	200	2000	2650
ramp times, S	10	8	2
hold temp, °C	25	10	5
time constant		0.5	

Table 40 - Metal Stearate Stabilizers in Poly(vinylchloride)

		Reference
Metal Stearate Stabilizers in Poly(vinyl chloride)		
Metal stearates	solvent extraction, thin-layer chromatography	354
Ba Cd stearates Zn	solvent extraction, column chromatography	355
Cd Pb stearates Zn	solvent extraction, paper chromatography	356
Cd Pb stearates Zn	solvent extraction or ashing, polarography	357
Cd Pb stearates Zn	solvent extraction, infrared spectroscopy	358
Na K stearates Ba	solvent extraction, chemiluminescence techniques	359
Cd Zn stearates and laurates Ba	ashing and elemental analysis by polarography	360
Na K Ba stearates Cd Pb	ashing, flame photometry	361
Non-tin Type Stabilizers in Poly(vinylchloride)		
Thin-layer chromatography		
3 amino crotonic ester of 2,2' thiodiethanol	mascerate with water, acetic acid-ethanol or heptane, then thin-layer chromatography	362
phenylamine 2 phenylindal dicyandimide aminocrotonic esters	solvent extraction, thin-layer chromatography	363-366
stabilizers	solvent extraction, thin-layer chromatography	367
epoxy types alkyl epoxy stearates epoxidized soyabean oil epoxidized linseed oil	solvent extraction, thin-layer chromatography	368

| epoxy types
epoxidized soyabean oil
epoxidized linseed oil
isooctylepoxystearate
2-ethylhexylepoxystallate
butyl epoxy tallate
butyl ester of epoxy -
linseed oil
di(isoamyl)4,4,epoxytet-
rahydrophthalate | methanol extracton, thin-layer
chromatogaphy | 369, 370 |

Miscellaneous methods

stabilizers	solvent extraction, glc	371-373
stabilizers	diethyl ether extraction, various instrumental finishes	374
diphenylthioureas 2-phenylindole	methanol or diethyl ether extraction, spectrophotometric finish	375

(P.B. 15 and P.B. 15:4) produce different thermal ultraviolet profiles. In any case, it is possible to distinguish organic pigments from their decomposition temperature and ultraviolet absorption. However, the unequivocal identification of pigments is strongly influenced by the repeatability of the thermal ultraviolet profile.

5.1.12 Flame Retardants

Surface tris (2,3-dibromopropyl) phosphate has been determined on the surface of retardant polyester fabrics[377]. The technique used to determine these involved extraction of the fabric with an organic solvent followed by analysis of the solvent by X-ray fluorescence for surface bromine and by high pressure liquid chromatography for molecular tris (2.3-dibromopropyl) phosphate.

Cope[378] has described a pyrolysis - gas chromatographic procedure for the determination of tetrakis(hydroxymethyl)phosphonium hydroxide and tris (2,3-dibromo-propyl)phosphate flame retardants on polyesters and cotton.

5.1.13 Organic Peroxides

Polystyrene

Certain types of peroxides used in polymer formulations are extremely stable and unreactive. This applied to substances such as dicumyl peroxide used as an ingredient of some self-extinguishing grades of polymers.

$$CH_3 - \underset{\underset{Ph}{|}}{\overset{\overset{CH_3}{|}}{C}} - O - O - \underset{\underset{Ph}{|}}{\overset{\overset{CH_3}{|}}{C}} - CH_3$$

This substance cannot be determined by polarography, and will not react with many of the reagents normally used for determining organic peroxides. Brammer et al[379] have describe a method for determining dicumyl peroxide in polystyrene, which is not subject to interference by other organic peroxides or additives that may be present in the polymer. The dicumyl peroxide is extracted from the polymer with acetone and then separated from any other additives present by thin-layer chromatography on silica gel. The gel in the area of the plate containing dicumyl peroxide is then isolated and digested with potassium iodide in glacial acetic acid followed by titration of the liberated iodine with very dilute sodium thiosulphate solution.

$$CH_3 - \underset{\underset{Ph}{|}}{\overset{\overset{CH_3}{|}}{C}} - OO - \underset{\underset{Ph}{|}}{\overset{\overset{CH_3}{|}}{C}} - CH_3 + 2HI = 2CH_3 - \underset{\underset{Ph}{|}}{\overset{\overset{CH_3}{|}}{C}} - OH + I_2$$

This procedure has a precision of $\pm$ 12% of the determined value with polymers containing 0.25-0.5% dicumyl peroxide. It is a rather time-consuming procedure but has the advantage of avoiding all risk of interference from other types of peroxides present in the sample.

5.1.14 Antiozonants

Rubbers and Rubber Vulcanizates

Protivova and Pospisil[380] have reported on the behaviour of some amine antioxidants and antiozonants during gel permeation chromatography using a 254 ml UV detector and have applied this technique to the analysis of rubber extracts.

Since the gel permeation method does not allow a direct determination of the molecular weights or molar volumes of the samples under investigation, calibrations are made by using standard compounds in the form of a graphic dependence of their molar volume on the elution volume[381-384]. Normal hydrocarbons (pentane, hexane, heptane, dodecane, hexadecane, octadecane) and aliphatic esters (octyl adipate and octyl sebacate) are used as standards. The molar volumes (ml/mole) were plotted against elution volumes, V_e (= ml) in the calibration curves. The molar volumes were calculated from the atomic volumes and structural coefficients[382].

The results of the gel permeation chromatography measurements by Protivova and Pospisil[380] of the elution volumes of aromatic amines, their molecular weights, calculated molar volumes and the effective molar volumes observed and read from the calibration curves are given in Table 41.

Lattimer et al[385] have applied mass spectrometry to the determination of organic additives (antioxidants and antiozonants) in rubber vulcanizates. Direct thermal desorption were used with three different ionization methods (electron impact, I; chemical ionization, CI; field ionization, FI). The vulcanizates were also examined by direct fast atom bombardment mass spectrometry (FAB-MS) as a means for surface desorption/ionization. Rubber extracts were examined directly by the four ionization methods. Of the various vaporization/ionization methods, it appears that field ionization is the most efficient for identifying typical organic additives in rubber vulcanizates. Other ionization methods may be required, however, for detection of specific types of additives. There was no clear advantage for direct analysis as compared to extract analysis. Antiozonants examined include aromatic amines (HPPD, DOPPD, DODPA and poly-TMDQ) and a hindered bisphenol (AO 425). These compounds could be identified quite readily by either extraction or direct analysis and by use of any vaporization/ionization method.

Further methods for the determination of antiozonants in rubber vulcanizates and polyethylene are reviewed in Table 42.

5.1.15 Rubber Accelerators

Millingen[398] has described a method (thin-layer chromatographic) for the identification of accelerators in unvulcanizes (natural) rubber compounds. This procedure highlights interference from other compounding ingredients and describes, in detail, means for overcoming such interferences.

The procedure is recommended of running two plates simultaneously, per analysis. One plate is sprayed with iodoplatinate reagent revealing visualizing accelerators as white, yellow, orange and brown spots on a brick red background. The second plate is sprayed with chlorimide reagent. Both accelerators and antioxidants produce coloured spots with this reagent. On overspraying with the iodoplatinate, the antioxidant spots retain their colour, while the accelerator spots fade or become a darkened centre with an intense white halo. The bonding effect of the antioxidant present in the extract is immediately evident. Accelerators studied were representative of a wide range of chemical types including guanidines, thiazoles, thiurams sulfenamides, diethiocarbamides and morpholine disulfides. All the accelerators tested gave a positive response to the iodoplatinate reagent. In Table 43 are shown the colour reactions of a wide range of accelerators tested involving both phenolic and amine types. Only those based on phenylene diamine gave a colour reaction with iodoplatinate reagent. Colours produced are distinctively different-cyclamens, purples and greens.

An overspray of sodium bicarbonate immediately following the chlorimide spray and then overspraying with iodoplatinate, can be useful for confirming the presence of the following accelerators: Tetramethyl thiuram monosulfide (TMTM), tetramethyl thiuram disulfide (TMTD), zinc diethyl dithiocarbamate(ZDC), 2-(morpholinothio)benzothiazole (Santocure MOR) and mercaptobenzothiazole (MBT).

Altenau et al[399,400] used mass spectrometry to qualitatively identify volatile antioxidants in 0.02-0.03 inch thick sheeted out samples of synthetic styrene-butadiene rubbers and rubber type vulcanizates. They extracted the polymer in acetone in a Soxhlet apparatus, removed excess solvent and dissolved the residue in benzene. Substances identified and determined by this procedure include N-phenyl-beta-napthylamine, 6-dodecyl-2,2,4-trimethyl 1,2-dihydroquinolines, trisnonylphenylphosphite, isobutylene-bisphenol a reaction product, 2-mercaptobenzothiazole sulpheramide (accelerator), N-cyclohexyl-2-benzothiazole sulphenamide, N-tert-butyl-2-benzothiazole sulphenamide, 2-(4-morpolinothio)benzothiazole, 2-(2,6-dimethyl-morphalinothio)benzothiazole, N,N-diisopropyl 2-benzothiazoles, 2-mercaptobenzothiazole and N,N'-dicylcohexyl-2-benzothiazole sulphamide.

Other methods for the determination of accelerators in rubbers are reviewed in Table 44.

Table 41 - The Behaviour of Amine Antioxidants, Antiozonants and
Model Compounds in Gel Permeation Chromatography

Chemical structure	Molecular weight	V_e (ml)	Molar volume (ml/mole) calculated	Effective	Deviation
Aniline	93.12	238	110.2	150.3	+ 40.1
4-Methylaniline	107.15	238	132.4	130.3	+ 17.9
2,3-Dimethylaniline	121.18	247	154.6	115.6	- 39.0
2,4,6-Trimethylaniline	135.20	232	168.8	183.2	+ 14.4
2,3,5,6-Tetramethylaniline	149.24	247	199.0	115.6	- 83.4
N-Methylaniline	107.15	253	133.9	95.9	- 38.0
N,N-Dimethylaniline	121.18	278	156.1	44.2	-111.9
1-Naphthylamine	143.18	242	161.8	134.9	- 26.9
2-Naphthylamine	143.18	241	161.8	139.0	- 22.8
Diphenylamine	169.22	229	200.3	201.0	+ 0.7
Phenyl-2-naphthylmine[a]	219.27	235	251.9	166.7	- 85.2
o-Phenylenediamine	108.14	238	124.4	150.3	+ 25.9
m-Phenylenediamine	108.14	220	124.4	266.1	+141.7
p-Phenylenediamine	108.14	248	121.4	111.4	- 10.0
4-Aminodiphenylamine	184.23	221	214.5	257.6	+ 43.1
4,4'-Bis-(dimethylamino) diphenylamine	255.41	228	320.5	212.3	-108.2
Benzidine	184.23	217	213.0	291.7	+ 78.7
o-Tolidine	212.28	224	257.4	234.4	- 23.0
N,N'-Dimethyl-p-phenylene-diamine	136.22	247	171.8	115.6	- 56.2
N,N'-Diethyl-p-phenylene-diamine	164.14	222	216.2	249.5	+ 33.3
N,N'-Di-sec-butyl p-phenylenedimine[b]	220.38	236	305.0	162.2	-142.8
N,N'-Di-iso-hepyl-p-phenylendiamine[c]	305.4	202	438.2	462.4	+ 24.2
N,N'-Di-iso-octyl-p-phenylenediamine[d]	332.58	200	482.6	495.5	+ 12.9
N,N,N'-Trimethyl-p-phenylenediamine	150.28	252	186.0	98.9	- 87.1
N,N'-Dimethyl-2-methyl-p phenylenediamine	150.28	250	186.0	105.2	- 80.8
N,N'-Diphenyl-p phenylendimine[e]	260.36	208	304.6	384.5	+ 79.9
N,N'-Dinaphthyl-p phenylenediamine[f]	360.46	205	415.2	421.7	+ 6.5
N-iso-Propyl-N'-phenyl-p-phenylenediamine[g]	226.34	222	282.6	249.5	- 33.1
N-iso-Butyl-N'-phenyl-p-phenylenediamine[h]	240.36	214	304.8	319.9	+ 15.1
N-Cyclohexyl-N'-phenyl-p-phenylenediamine[i]	266.41	206	326.8	410.2	+ 83.4
N-octyl-N'-phenyl-p-phenylenediamine[k]	296.47	206	393.6	410.2	+ 16.6
N,N'-Bis-4-(beta,N-dimethyl-amino)-phenyl-p-phenylene-diamine	346.55	218	424.8	283.1	-141.7

[a] Age Rite Powder (Anchor Chemical Company Ltd., Manchester, Great Britain)

[b] Topanol M (ICI, Macclesfield, Great Britain)

[c] Santoflex 77 (Monsanto, St. Louis, Mo., USA)

[d] UOP 88, UOP 288 (UOP chemical Company, East Rutherford, N.J. USA)

[e] DPPD (Monsanto), Altofane DIP (Etablissements Kuhlmann, Paris, France) JZF

[f] Santowhite CI (Monsanto), Antioxidant 123 (Anchor), DNPD (Chemical works of J. Dimitrov, Bratislava, Czechoslovakia), ASM DNP (Bayer, Leverkusen, Germany)

[g] ASM 4010 NA (Bayer), Nonox ZA (Arnold, Hoffman & Co., Providence, R.I., USA)

[h] Santoflex 13 (Monsanto)

[i] Flexone 6-H (U.S. Rubber Co., Naugatuck Division)

[k] UOP 688 (UOP)

Table 42 - Antiozonants in Rubber Vulcanizates and Polyethylene

		Reference
	Vulcanizates	
	Antiozanants	
	Thin-Layer Chromatography	
antiozonants	solvent extraction, TLC	386
	Paper Chromatography	
antiozonants	solvent extraction, paper chromatography for identification	387,388,390
	Polyethylene	
antiozonants	thin-layer chromatography	391-397

5.1.16 Mixtures of Additives, Complimentary Techniques

A necessary prerequisite to the methods discussed earlier in this chapter is that the analyst has a full knowledge of all the types of additive present in the polymer. This is necessary so that, in selecting a method for determining a particular constituent, due allowances can be made for other types of additive constituents present or of any decomposition products of additives present.   Whilst this

Table 43 – Reactions of Accelerators (Included are a few peptizers)

Chemical	Trade name or abbreviation	Spray 1 iodoplatinate	Spray 2 chlorimide	Spray 2 plus spray 1	Primary solvent system R_f
Diphenylguanidine	DPG	Brown, white edge	Blue	Green, white edge	0.00
Di-o-tolyguanidine	DOTG	Brown, white edge	Blue	Green, white edge	0.00
2-Mercaptobenzothiazole	MBT	Yellow	Orange	Rust, white edge	0.30
Benzothiazyl disulfide	MBTS	Cream streak	Pale yellow	Orange streak, white edge	0.50
Tetramethylthiuram disulfide	TMTD	Yellow,brown white edge	Dark red pink edge	Red, white edge	0.45
Tetramethylthiuram monosulfide	TMTM	Yellow,brown white edge	Dark red pink edge	Yellow, red edge	0.45
N,N-Dicylcylohexylbenzothiazole-2-sulfenamide	Vucazit DZ	Yellow, white edge	Yellow orange	Orange, white edge	0.70
Phosphorus-containing polysulfide	Vulcadone 3 SN	white, grey edge	Yellow	Rust, white edge	0.85
N-Cyclohexyl-2-benzothiazole sulfenamide	Santocure	Yellow, pale edge	Yellow orange	Orange, white edge	0.63
N-tert-butyl-2-benzothiazole sulfenamide	Santocure NS	Yellow, pale edge	Yellow orange	Orange, white edge	0.55
Zinc diethyl-dithiocarbamate	ZDC	Yellow	Brown, red edge	Rust, white edge	0.73
Zinc salt of 2 mercaptobenzo-thiazole	ZMBT	Yellow white	Red	Rust	0.25
2-Benzothiazyl-N,N-diethylthio-carbamyl sulfide	Ethylene	Yellow, white edge	Yellow orange	Rust, white edge	0.39

4-Morpholine disulfide	Morfax	White	Yellow	Brown, white edge	0.35
Copper dimethyl-dithiocarbamate	Cumate	Pale brown streak	Red streak	Brown streak	0.20-0.70
Diphenylmethylene thiuram tetrasulfide	Tetrone A	Brown streak cream centre	Orange yellow	Red streak	0.30-0.80
2-(Morpholinothio)-benzothiazole	Sanocure MOR	Yellow, cream	Yellow	Orange, white edge	0.30,0.63
Hexamethylene tetramine	Vulcacit H30	Red, white edge	Yellow	Red, white edge	0.00
Morpholine disulfide	Sulfasan R	Grey, white edge	Yellow	Rust, white edge	0.38,0.20
2-Mercapto imidazoline	Na 22	White	Brown	Brown orange	0.00
2,2'-Dibenzamido-diphenyl disulfide	Pepton22	Cream, brown edge	Pale orange	Red, white edge	0.35

Table 44 - Accelerators in Rubber Vulcanizates

Reference

Column chromatography

antioxidants and acceleratores, guanidine type, thiocarbanilide mercapto-benz-thiazole-2-disulphide 2-benzthiazolyl-N' cyclo-hexyl sulphememide phenyl napthylamines sym-di-beta-napthyl-p-phenylene diamine zinc dialkyldithiocarbamates formaldehyde-aniline conden-sation products aldol napthylamine conden-sation products polymerized trimethyl di-hydroquinone	solvent extraction, column chromatography. Column streaked with detection reagents UV and IR examination of extracts	401-403

Accelerators - thin-layer chromatography

accelerators	solvent extraction, TLC	404 405 406
guanidines, thiuram type carbamate type e.g tetra-methylthiuram monosulphide, dipentamethylene thiuram-tetrasulphide, cylclic thiuram piperidinium pentamethylene-dithiocarbamate, zinc (bismuth and cadmium)-dimethylthiocarbamate, 2-benzothiazyl N,N' diethyl-thiocarbonyl sulphide, 2-mercaptobenzthiazole, benzothiazyl disulphide	solvent extraction, TLC for identification	407

Paper chromatography

accelerators	solvent extraction, paper chromatography	408, 409

Mass spectrometry

2-mercapto-benzothiazole accelerator sulphenamide e.g. N-tertbutyl-2- benzoth-iazole sulphenamide, 2-(4 morpholinothio)benzoth-iazole, 2-(2,6-dimethyl morpholino-thio)benzothiazole,	volatile preconcentration/ mass spectrometry down to 2%	410

N,N-diisopropyl-2-sulphen-
amide,
N,N-dicyclohexyl-2-benzo-
thiazole-sulphenamide

rubber accelerators	in rubbers, polyisoprene solvent extraction, polarography	411-417
ketone-amine condensates 2-mercapto-benzinidazole	acetone, extraction, spectroscopy	418

---

information might be to hand if an analyst is examining materials of known origin, this would not always be so. In such cases it is mandatory that the first step must be to completely identify the additives present, before any consideration can be given to the problem of selecting or devising a method of quantitative analysis for any constituent present in the polymer.

The problem resolves itself into the preparation of a total solvent extract of the polymer in which all additives are completely recovered, (Chapter 3) followed by separation of the mixture into pure single components by a suitable form of chromatography and, finally, by identification of each separated pure component by suitable means, usually involving visible, infrared, ultraviolet, mass or nuclear magnetic resonance spectroscopy and perhaps microanalysis for elements present. Only after this stage can the details of the quantitative determination of particular polymer constituents be considered.

It is advisable, when commencing the analysis of a polymer for unknown additives, to determine first its content of various non-metallic and metallic elements. Any element found to be present must be accounted for in the subsequent examination for, and identification of, additives. Hence elemental analysis reduces the possibility of overlooking any additive which contains elements other than carbon, hydrogen and oxygen. The analytical methods used to determine elements should be sufficiently sensitive to determine about 10 ppm of an element in the polymer i.e. should be able to detect, in a polymer, a substance present at 0.01% and containing down to 10% of the element in question.

This requirement is met for almost all the important elements by use of optical emission spectroscopy and X-ray fluorescence spectrometry. Using these two techniques, all metals and non-metals down to an atomic number of 15 (phosphorus) can be determined in polymers at the required concentrations (Cook et al[419], Hank and Silverman[420], Mitchell and O'Hear[421] and Bergmann et al[422]) Nitrogen is determinable at these levels by micro-Kjeldahl digestion techniques. Various forms of chromatography are worthy of serious consideration for the separation of additive mixtures. These include gas chromatography, high performance liquid chromatography, and thin-layer chromatography. The application of these techniques where relevant is discussed earlier in this chapter. A difficulty frequently encountered in chromatographic techniques is that more than one compound may occur in a particular peak with the consequence that the presence of particular polymer additives may be missed or confused. In these circumstances distinctly more reliable results will be obtained if the chromatographic technique is coupled to a complimentary technique such as infrared spectroscopy or mass spectrometry. As each peak emerges from the column its spectrum is run and particularly in the case of mass spectrometry unequivocal identification is possible even when more than one substance is present.

Complimentary Gas Chromatographic Techniques

Gas chromatography - infrared spectroscopy. The direct application of gas chromatography to the identification and determination of unknown additives in polymers has its limitations. Amongst these are the fact that retention times are purely relative and not always exactly reproducible; even when the chemical class of the sample constituents is known, identification by comparison of the retention times of the unknown with those of standards requires exact reproduction of column operating conditions. Difficulty can also arise when non-symmetrical peaks are produced; the peak-maximum retention time of a component then depends on concentrations. Peaks can also overlap and certain combinations of substances are often difficult to separate.

Chromatographically, a single peak is no criterion of purity, since more than one substance may be present, the components present could often be resolved if their presence was suspected and alternative column operating conditions were selected.

Confirmation of homogeneity of fractions is therefore required together with unequivocal identification and accurate quantitative determination of the product. The analytical method used should involve some property of the molecule other than boiling point. Often, only fractional milligram amounts of material can be recovered from a column. Of the few techniques that are applicable to these small quantities of material, mass and infrared spectroscopy are particularly suited.

The use of fraction collecting techniques in conjunction with gas chromatography is now well established[423-429] and is and attractive proposition in additive identification problems, despite the fact that little published work has yet appeared on the application of this technique to polymer additives.

In this technique the separated compound as it emerges from the gas chromatographic column is swept by the carrier gas through a cold trap where it condenses. The material in the trap is then either transferred to an infrared gas cell for examination in the vapour phase or is transferred as a liquid to a suitable micro cell or may be condensed on a cold surface as a solid for examination by conventional spectroscopic techniques.

Anderson[430] has described in detail a technique for collecting individual products separated from mixtures by gas chromatography and the subsequently used technique for their identification by vapour phase infrared spectroscopy using Hilger H800 double beam spectrometer. The trap and infrared cell used by Anderson are shown in Figure 24(a). He claimed that the technique would identify and determine, to within $\pm$ 2%, amounts as low as 5 umole of all substances having a boiling point up to about 175°C. Heated gas cells enable liquids of higher boiling point to be examined. Naturally, as the infrared spectrum of a compound in the liquid and vapour form are different, it is necessary to compare spectra obtained by the above technique with spectra of standard vapours. Generally speaking, the additive constituents of polymers have a boiling point which is appreciably higher than 200°C and hence cannot be handled in infrared gas cells. For such substances, infrared examination as a liquid or a solid, as is discussed below, is more relevant. Obviously, in order to avoid volatilization during polymer processing and subsequent life, most types of additives used in polymer formulations are fairly high melting point solids. Volatile constituents are, however, sometimes

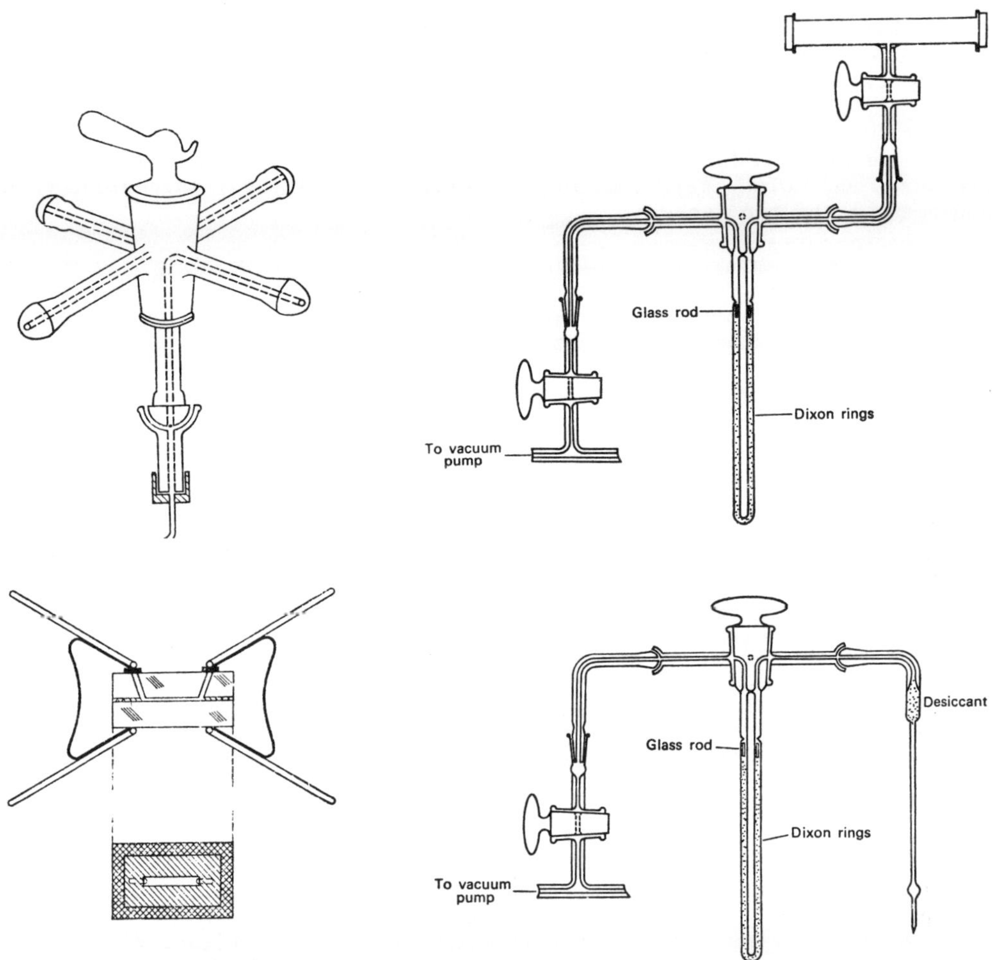

Figure 24   Apparatus used for trapping as chromatographic fractions
             and infrared spectroscopy; (a) trap for collecting
             products separated by gas chromatographs; (b) micro
             infrared cell for liquid samples; (c) apparatus for
             transfer of volatile components from trap to infrared
             gas cell; (d) apparatus for transfer of liquid component
             from trap to tip of side arm.

encountered viz. expanding agents, plasticizers, lubricants, adhesives,
solvents, monomers and degradation products of additives or the polymer
itself, and infrared gas cell techniques can be of value in the
examination of gas chromatographic fractions containing these types of
substances.

Haslam et al[431] have studied in detail the collection of volatile
and liquid fractions emerging from a gas chromatographic column and their
subsequent identification by infrared spectroscopy. They applied these
techniques to various polymeric materials. They incorporated several of
the techniques developed by Anderson[430] into their procedures but, unlike
him, found that his technique could not be applied effectively to
substances with a boiling point much in excess of 120°C. Moreover, they

found that many of the standard infrared spectra of substances encountered in the plastics industry were recorded only for liquid compounds and not for the corresponding vapours.

All the fraction-collecting apparatus described by Haslam[431] is used in conjunction with gas-chromatographic columns operated under reduced pressure and with katharometer detection. This gas-chromatographic equipment is of two types:

<u>Type 1.</u> Apparatus involving a column thermostatically controlled within the range 0°C to 130°C and a katharometer at room temperature. This apparatus is used for separating mixtures when the highest-boiling constituents boils below 160°C at atmospheric pressure.

<u>Type 2.</u> Apparatus involving a column and katharometer contained within the same thermostatically controlled chamber and maintained at some temperature in the range 100°C to 200°C. This apparatus is used for the separation of mixtures with constituents boiling up to 250°C which give trouble in apparatus of type 1 owing to condensation in the katharometer.

The trap shown in Fig. 24(b) has been successfully used with apparatus of type 1 for the condensation of substances boiling in the range - 100°C to + 160°C. The efficiency of this type of trap varies between 85 and 95% depending on the chemical class of the substance.

A preliminary chromatogram indicates the complexity of the mixture. In a second run, the traps are switched in by following the recorder trace and any overlap between peaks is allowed to go to waste. The delay time between the recorder signal and the component reaching the trap is negligible.

Having condensed the desired component in a trap it has to be decided whether to record the spectrum in the vapour state (method 1 below) or the liquid phase (method 2 below).

<u>Method 1 - liquids of boiling point up to 60°C.</u> Figure 24(c) shows how the low-boiling component is transferred to the infrared gas cell. the design of the apparatus is different, but the procedure for transference is identical to that described by Anderson[430] except that no heat is applied to the trap.

The 9 ml at SPT infrared gas cell is designed to obtain the strongest possible absorption from a small amount of sample. The intensity of absorption depends only on the number of molecules in the path of the radiation beam; this is increased by reducing the cross-sectional area of the cell, which however, must not be reduced to an extent such that the energy transmitted is insufficient to record the spectrum.

<u>Method 2 - liquids boiling above 60°C.</u> If the gas chromatographic evidence indicates that the substance boils above 60°C, the fraction condensed in the trap is treated as shown in Fig. 24(d). The trap and contents are held in liquid nitrogen whilst the apparatus is completely evacuated. After testing the system for leaks, the source or vacuum is cut off. The flask containing the liquid nitrogen is removed from the trap and placed over the side-arm so that the bulb is just immersed in the refrigerant. As the trap attains room temperature the small amount of liquid in it distils through the desiccant and is condensed round the sides of the bulb of the side-arm. The liquid is encouraged into the capillary tip by lowering the vacuum flask containing the liquid nitrogen

until only the last ¼" of the tip is immersed. The bulb is then warmed with the fingers. During this distillation the trap is not warmed, as this may cause some of the liquid to distil in the wrong direction. The apparatus may be left for 1 or 2 hours in order to achieve complete distillation of higher boiling liquids. Having obtained the liquid in the capillary tip, this tip is broken off at the construction just below the bulb.

The micro infrared cell design is shown in fig 24(b). The two rock-salt plates are about 20 mm x 10 mm x 3 mm in size. Two holes are drilled in one plate with a No. 86 drill; these holes taper inwards, as shown, being 7 mm apart on the inside and 12 mm apart on the outside of the plate. A gold-foil washer (25 micron thick) is cut, the inner space being 1.5 mm wide and approximately 7 mm in length (just sufficient to clear the outer edges of the filling holes); the width of this washer is about 2 mm. The cell is first stuck together with gold amalgam. The washer is immersed in mercury for a short period and withdrawn after a layer of amalgam has formed on its surface. The cell is then carefully assembled in a small clamp and left under pressure for 24 hours for the amalgam to set. The space between the plates outside the washer is then filled with Araldite 700 resin and a surplus of resin is allowed to bridge the edges of the plates on all four sides.

The cell is filled by inserting a capillary needle containing the liquid into one filling hole. Provided that the thickness of the cell is less than the diameter of the needle, the liquid will run in to fill the cell area. It is essential that no free-space be left in the cell behind the filling holes; such space will not immediately fill by capillary action and air bubbles trapped here may subsequently run into the useful part of the cell and are most difficult to remove. When the cell is filled, the outer ends of the filling tubes are covered by two small squares of nitrile-rubber sheet, held in position by "butterfly" spring clips.

The amount of liquid required to fill the cell by the method described above (as opposed to the theoretical capacity of the cell calculated from the dimensions) is determined experimentally by weighing the cell on a microbalance before and after filling with n-dodecane. With this cell, infrared spectra can be obtained from 0.5-1.0 micron portions of liquid samples.

<u>Gas chromatography - mass spectrometry.</u> This technique is finding increasing use in the plastics industry. It has been used to determine additives in polypropylene[432], cross linked rubbers[433] and other polymers[434]. The gas chromatography - mass spectrometry technique has been used for various identification problems including examination for the presence of isobutyronitrile and benzoylperoxide and lauryl peroxide and tetramethylsuccindinitrile, in polystyrene and the examination of headspace gases in plastic food packaging containers[435].

<u>Pyrolysis - gas chromatography - mass spectrometry.</u> Pyrolysis - gas chromatography has been used for many years for the characterisation of plastics materials, particularly when they are of an intractable nature owing to cross-linking or are very heavily filled. The technique has now been extended to include mass spectrometry and is of particular value when minor components need to be identified. An example has been described by Sharp and Paterson[436] for the identification of small amounts (1-10%) of decopolymerised unsaturated acids in acrylic polymers. The method can be summarised as follows:

$$\left[\begin{array}{c} CH_3 \\ | \\ -C-CH_2 \\ | \\ C=O \\ | \\ OCH_3 \end{array}\right]_n - \left[\begin{array}{c} R \\ | \\ -C-CH_2 \\ | \\ C=O \\ | \\ OH \end{array}\right]_m - \text{propylation} \left[\begin{array}{c} CH_3 \\ | \\ -C-CH_2 \\ | \\ C=O \\ | \\ OCH_3 \end{array}\right]_n - \left[\begin{array}{c} R \\ | \\ -C-CH_2 \\ | \\ C=O \\ | \\ OC_3H_7 \end{array}\right]_m -$$

$$\text{pyrolysis} \quad \begin{array}{c} CH_3 \\ | \\ C=CH_2 \\ | \\ C=O \\ | \\ OCH_3 \end{array} \quad \begin{array}{c} R \\ | \\ C=CH_2 \\ | \\ C=O \\ | \\ OC_3H_7 \end{array} \quad + \text{ alcohols, alkines, } CO_2\text{, etc.}$$

The copolymerised acid is propylated by treatment of the sample with Propyl 7 reagent, the resultant polymer is pyrolysed and the propyl ester of the acid, if present, is identified by gas chromatography - mass spectrometry. By this procedure copolymerised acrylic or methacrylic acid has been identified in terpolymers with (a) butyl acrylate and styrene, (b) methyl methacrylate and ethyl acrylate and (c) ethylene and propylene. A methyl methacrylate-alpha-methylstryrene-maleic acid terpolymer, when examined by this propylation - pyrolysis procedure, yielded dipropyl fumarate and a smaller amount of dipropylmalate. Attempts to detect copolymerised itaconic acid by this technique failed probably owing to the difficulty in esterifying the acid group.

Hlpc - infrared spectroscopy. One of the difficulties of column chromatography is the problem of identifying the fractions in which the separated compounds are concentrated. This can be achieved by the laborious process of examining all the fractions, for example by infrared or ultraviolet spectroscopy or by evaporating to dryness and weighing the residues: or by the less laborious process of monitoring the effluent as it leaves the chromatographic column so that solute containing fractions from the fraction collector can be picked out from the fractions which do not contain any substances. Several types of effluent monitors are available, based on the measurement of the ultraviolet absorption, conductivity, etc. These have the disadvantage of being too specific for dealing with mixtures of compounds of unknown type. For example, compounds which do not either absorb in the ultraviolet or do not ionize would be missed using these detectors. The most useful general-purpose monitors are those based on the measurement of refractive index and on thermal effects. The later operates on the principle that as each separated compound moves down the column it is accompanied by heat of absorption and desorption due to interaction between solute molecules and the stationary phases. These heat pockets (i.e. separated compounds) are detected by a thermistor at the column outlet and recorded on a strip chart which can be operated in conjunction with a fraction collector. Thus, separated fractions can be readily located and bulked if necessary for further examination.

The most recent development in liquid chromatography, namely high-pressure liquid chromatography, combines the advantages of built-in detectors with improved resolution in the separation of mixtures due to improved column packings and operation at an elevated pressure.

Recent advances have improved the speed and efficiency of high-pressure liquid chromatography to the point that they are now rapidly approaching the limits achievable by gas chromatography.

Analysis times in liquid chromatography can be shortened considerably without loss of peak resolution by optimizing the parameters of column length and diameter, flow rate, sample size, and support particle size. The theoretical groundwork for high-efficiency liquid chromatographic separations has been established by a number of investigations[437-442]. Because of their porous nature conventional liquid chromatographic absorbents of small particle diameter give rise to poor rates of mass transfer of solutes under rapid flow conditions giving poor column efficiency. Also porous absorbents tend to impact under the high pressures needed to achieve adequate flow. A number of high-efficiency liquid chromatographic supports have recently been introduced. These include Zipax (DuPont's CSP support[443,444]) Corasil I and II, and Durapak[445] (Waters Associates). With exception of Durapak, these materials, in the micron particle range consist of particles with a solid core and thin porous coating. This unique combination gives very high coefficients of mass transfer. Durapaks consist of conventional liquid phases, such a beta, beta'-oxy-dipropionitrile, chemically bonded to a rigid porous bead. Textured glass beads for liquid-liquid chromatography similar to those reported for gas chromatography[447] are being developed by Corning[448].

The Zipax material has been reported to have a relatively inert surface[446], whereas the Corasil support is also recommended for liquid-solid column chromatography[449]. Little has been reported on the use of these solid-core silcaceous backboned Corasil supports with a polar liquid coating in liquid-liquid chromatography. Kirkland[443,444,450] has studied the support in some detail for several model systems, and has reported much-improved performance for this material in liquid-liquid chromatography when compared to coated glass beads or diatomaceous earth. A number of separations have been described by Halasz[445] on Durapak type supports and by Majors[451] using various solid core supports operating at 5000 psi.

In a procedure for identifying the type of plasticizer present in PVC[452] an ether extract of the polymer is separated on a column of celite-silica gel and the fractions weighed after removal of the ether. Those fractions which contained any weighable amount of material were examined by infrared spectroscopy, which enabled an identification to be made.

Figures 25(a) and (b) contrast the type of chromatograms obtained by the two types of column chromatography. In the conventional labour-intensive manual method[253] (Fig.25(a)), the individual fractions are each made up to a standard volume and analysed. In the high-pressure method[454] the detector picks up each separated component to provide a chromatogram of the type shown in Fig.25(b).

The correct selection of elution solvents is very important when attempting to separate mixtures on a column. This is illustrated in Table 45, which shows how a system using three solvents can be used to separate nine plasticizers.

<u>Hplc - mass spectrometry.</u> Like other forms of molecular spectroscopy, mass spectrometry may be used as a "fingerprint" technique to identify the components of additive systems extracted from polymer compositions. The strengths of mass spectrometry are high sensitivity and the ability to distinguish between closely related compounds of differing

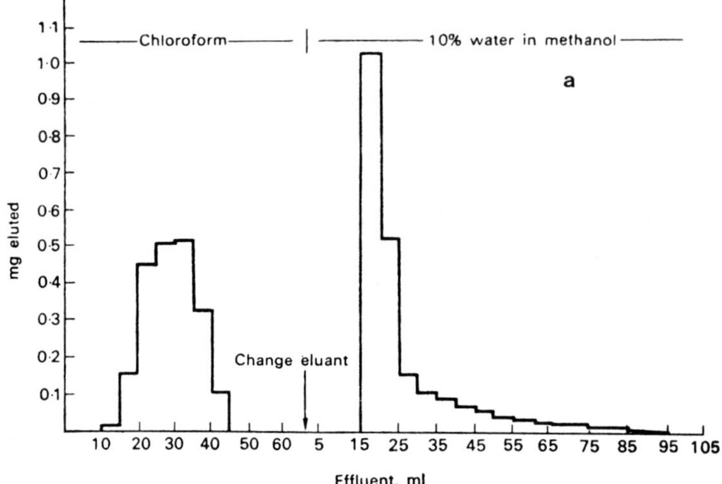

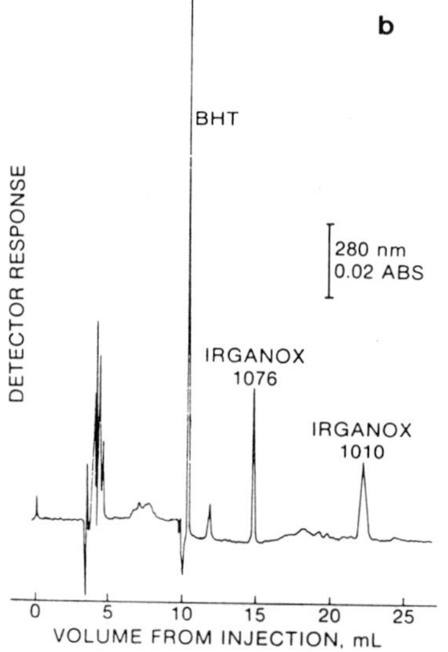

Figure 25   Column chromatography of PVC plasticisers: 9(a) elution
chromatogram of butylated hydroxy toluene and Santonox
R, polyethylene sample extracted with chloroform -
separation on alumina column. Antioxidant content of
each separated fraction measured by ultra violet
spectroscopy; (b) High performance liquid chromatogram
of 100 ul extract from 2.07 g polethylene - separation
on alumina.

Table 45 - Column Chromatography of PVC plasticizers. Elution
sequence for plasticizer mixture-column 1:1 by weight Celite
545 (100-200 mesh): silica gel (100-200) mesh

(Ether extract of PVC) eluent	Fraction	Plasticizer found
Carbon tetrachloride	1	Cereclor (chlorinated hydrocarbon)
	2	-
	3	-
	4	Mesamol (alkyl ester of a sulphonic acid)
Carbon tetrachloride/di-isopropyl ether (2%)	5	-
	6	Tritolyl phosphate
	7	Di-n-butyl phthalatc
	8	
Carbon tetrachloride/di-iso propyl ethr (5%)	9	Di-n-butyl sebacate
	10	-
	11	-
	12	-
Carbon tetrachloride/acetone (2%)	13-16	-
Carbon tetrachloride/acetone (5%)	17	Diethylene glycol
	18	Dibenzoate
	19	Abrac 'A' (epoxidized) vegetable oil
	20	-
Carbon tetrachloride/acetone (7.5%)	21	-
	22	Polypropylene secacate
	23	Polypropylene sebacate (trace)
	24	-

relative molecular mass, e.g. the various alkyl thiodipropionates used as
synergistic stabilisers in polyolefins and the ultraviolet absorbing
benzotriazole derivatives. Often is is not necessary to separate the
components before examination as some separation may be achieved by
careful variation of the sample probe temperature to produce in effect, a
fractional distillation of the components. The presence, however, of
large amounts of low relative molecular weight polymers such as
polyethylene and polypropylene can cause interference by producing a high
hydrocarbon background extending to several hundred relative atomic mass
units. In such instances thin-layer chromatographic separation can be
used as a clean-up procedure.

Rudewicz and Munson[455] have described selective mass spectrometric
(SIMS) procedure for the determination of additives in polypropylene
without prior separation by solvent extraction or precipitation
techniques. The additives are vaporized from polypropylene samples in a
heatable glass probe under chemical ionization conditions using a 1.1%
ammonia in methane reagent gas mixture. The dominant ion in this mixture,
NH_4 + is a low-energy reagent ion that reacts with the additives to give
very simple spectra of $(M + H)^{+1}$ or $(M + NH_4)^+$ ions and little
fragmentation. The detection of additives in a hydrocarbon matrix is very
selective.

Using this technique Rudewicz and Munson[455] determined Ionox 330 and

Irganox 168 antioxidants and UV 531 ultraviolet absorber in polypropylene. Both additives begin to vaporize rapidly at temperatures near the melting point of polypropylene, about 176°C. The two additives, however, can clearly be distinguished by their different masses and their very different heating profiles. The lower molecular weight additive gives a relatively sharp peak with a maximum at approximately 210°C, and the higher molecular weight additive gives a broader peak that maximizes near 300°C.

It is possible to identify these additives from a small set of likely compounds from the masses of their characteristic (M + H)+ or (M + NH₄)+ ions in the ammonia chemical ionization spectra. However, greater confidence in the identification would be provided by spectra obtained with a more reactive chemical ionization reagent gas, like methane, which gives significantly more fragmentation than does ammonia. The ammonia chemical ionization spectrum of Ionox 330 contains (M + H)+ as the most abundant ion and abundant fragment ions at m/z 569, 219, 102, 163. This compound may be readily identified in a simple mixture by noting that these ions have the same temperature profiles. Similarly, the methane chemical ionization spectrum of Irganox 168 contains a reasonably abundant (M + H)+ ion, the base peak at m/z 441 and another fragment ion at m/z 147. Similar temperature profiles for these ions indicate the presence of this additive. One can possibly identify the additives from a limited set of likely possibilities by comparing temperatures (or time) plots of characteristic ions, even when several additives are present.

The information obtained by absorbance detection when coupled with liquid chromatography is usually not specific enough to allow the qualitative indentification of compounds present in complex polymer samples with any degree of certainty. With the use of mass spectrometric detection Vargo and Olson[456] showed that characteristic mass spectral data can be obtained which greatly aid in elucidating the identity of unknown compounds. They used mass spectrometric detection in series with absorbance detection to identify or characterise antioxidants and ultraviolet light stabilizing additives in acetonitrile extracts of polypropylene, which were separated by liquid chromatography. The two detectors were operated in series without any significant loss of chromatographic resolution. Nonogram quantities of additives could be detected. The gradient elution scheme used by these workers is shown in Table 46.

Mass Spectrometric detection was achieved with a Finnegan MAT Model 4165 quadruple mass spectrometer which was equipped with a moving belt LC/MS interface (Finnegan-MAT, San Jose, C.A.). Methane chemical ionization is used. The ion source is pressurized to 0.3 torr with methane reagent gas, which is ionized with 70 eV electrons. Solutes are desorbed from the polyamide LC/MS interface belt at 230°C. The ion source temperature is 120°C. Chemical ionization spectra from m/z 200 to 1200 were recorded repetitively at 3 s/scan. Electron ionization spectra were recorded from m/z 50 to 800 at 3 s/scan.

In Fig. 26 are shown absorbance chromatograms at 280 nm and total ion current chromatograms for a mixture of nine antioxidants extracted from polypropylene beads. Detections limits of between 2 and 30 ng were obtained.

The procedure was used to characterize additives in polypropylene samples of unknown composition. Butylated hydroxy toluene and Irganox 1076 were identified in the extract of an automative component moulding.

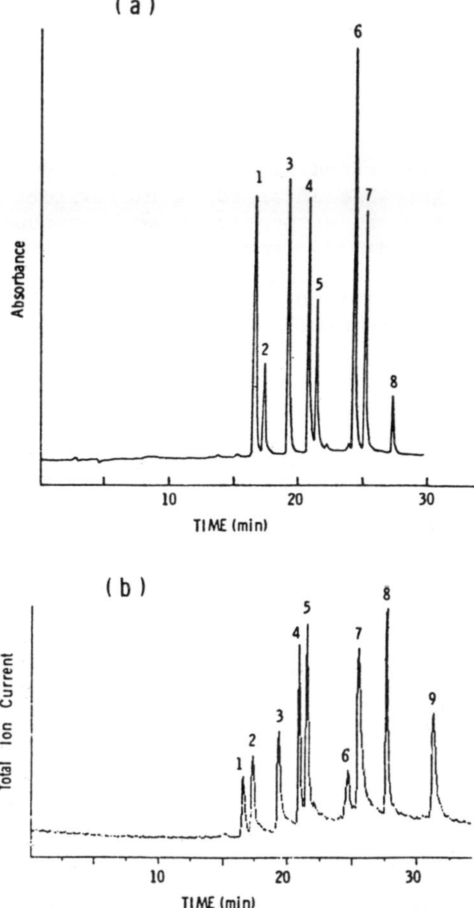

Figure 26 Absorbance chromatograms at 280 nm: (a) and total ion
current chromatography: (b) of antioxidancts and ultra
violet light stabilizers: (1) BHT 3.1 ug; (2)
Santowhite; 0.9 ug; (3) Topanol CA, 2.7 ug: (4) UV 531,
0.66 ug; (5) UVS 411, 0.68 ug; (6) Irganox 1010, 4.5 ug;
(7) Ionox 330, 214 ug; (8) Irganox 1076, 1.2 ug; (9) DST
DP, 1.7 ug. DSTDP (distearyl thiodipropionate) was not
detected with the absorbance detector at 280 mn.

Table 46 - Gradient Elution Scheme

Time, min	% solvent A[a]	% solvent B[b]
0	100	0
10	60	40
20	0	100
30	0	100
32	100	0

[a] Solvent A, 75% acetonitrile and 25% water
[b] Solvent B, 50% THF and 50% acetonitrile

In addition, several other additives were identified (palmitic acid, dioctylphthalate, stearic acid and octadecanol).

    <u>Thin-layer chromatography - infrared spectroscopy.</u> Thin-layer chromatography using plates coated with 250 micron absorbent is an excellent technique forseparating quantities of up to 20 milligrams of additive mixtures into their individual components and provides enough of each to prepare a recognisable infrared or ultraviolet spectrum which can be compared with spectra of authentic known compounds. However, this technique does not conveniently handle larger quantities and in these cases separation on the 50 to 500 milligram scale can be carried out with only a small loss of resolution by chromatography on a silica gel or aluminium packed column. The separated bands are marked under ultraviolet light, removed from the plate and extracted with diethyl ether. Separations on this scale provide sufficient of each fraction for full characterisation by nuclear magnetic resonance spectroscopy and mass spectroscopy, as well as infrared spectroscopic techniques. Column chromatography is more time-consuming than thin-layer chromatography and for this reason the latter technique should be used in preliminary scouting experiments aimed at finding suitable adsorbents and development solvents for achieving a satisfactory separation. These conditions can then usually be translated to a column without difficulty. Similarly, the thin-layer technique is useful as a rapid method of monitoring the purity of fractions obtained in column separations.

    Numerous works have been published on the experimental technique of thin-layer chromatography, which will not be discussed further except in so far as is relevant to its application to additive identification in polymers. Dohmann[457] carried out an excellent short review of current techniques. He discusses thin-layer chromatography in the normal sense of the word, i.e. with plate layers up to 250 micron thick and 20 cm x 20 cm or 20cm x 8 cm in area.

    Polymer extraction procedures using organic solvents do not extract all types of organic additives from polymers; many inorganic compounds and metalorganic compounds (e.g. calcium stearate) are also insoluble. The presence of metal will have been indicated in the preliminary examination of the polymer. Most types of organic polymer additives, however, can be readily extracted from polymers with organic solvents of various types. The first step is to solvent-extract the total additives from the polymer in high yield and with minimum contamination by low molecular weight polymer. Extracts should be used for analysis without delay, as they may contain light-or oxygen-sensitive compounds. When delay is unavoidable, storage in actinic glassware under nitrogen in a refrigerator minimizes the risk of decomposition.

    Total internal plus external additives can be extracted from low- and high-density polyethylene and polystyrene by procedures involving solution or dispersion of the polymer powder or granules (3 g) in cold redistilled sulphur-free toluene (50-100 ml), followed in the case of polyethylene by refluxing for several hours. Rubber-modified polystyrene does not completely dissolve in toluene if it contains gel. Methyl ethyl ketone or propylene oxide are alternative suitable solvents for polystyrene. Dissolved polymer is then reprecipitated by the addition of methyl alcohol or absolute ethanol (up to 300 ml), and polymer removed by filtration or centrifuging. The additive-containing extract can then be concentrated to dryness as described previously. Alternative procedures for the extraction of polyethylene and polypropylene involve refluxing with chloroform for 6 hours or contacting with cold diethyl ether for 24 hours or Soxhlet extraction with diethyl ether, methylene dichloride, chloroform or carbon tetrachloride for 6-24 hours followed by concentration of the extract. Methylene dichloride is a particularly good solvent for polypropylene extractions because of its high volatility and also its small extractions of atactic material from the polymer, compared with other solvents. In addition to additives, most solvents also extract some low molecular weight polymer with subsequent contamination of the extract. The overcome this, Slonaker and Sievers[458] have described a procedure for obtaining polymer-free additive extracts from polyethylene based on low-temperature extraction with n-hexane at 0°C. This procedure is also applicable to polypropylene and polystyrene.

Thin-layer chromatographic grades of silica gel usually contain traces of organic impurities. If, during development of chromatogram, these impurities migrate to regions of the plate which coincide with the R_f values of separated additives, then the impurities will interfere in the interpretation of the plate following spraying with aggressive detection reagents, such as concentrated sulphuric acid and antimony pent-achloride, as the organic impurities in the adsorbent also react with these reagents and show up on the sprayed chromatograms. Also the impurities absorb strongly in the ultraviolet region, especially below 250 nm. The adsorbent impurities do not have an appreciable adsorption in the infrared region. As these impurities are extracted from the silica gel with organic solvents such as diethyl ether, ethanol, acetone, benzene, chloroform and many others, they could occur as contaminants in some of the fractions of separated additives isolated from the plate by solvent extraction of the gel and consequently interfere in the interpretation of the spectra of the additives, particularly in the case of additives which absorb below 250 mu in the ultraviolet. For this reason an identical blank chromatogram (only sample absent) should always be run in parallel with the sample chromatogram in order to check whether such interference effects exist.

Ultraviolet absorbing impurities in adsorbents are influenced by the nature of the migration solvent. Depending on the polarity of the migration solvent used the impurities migrate to a greater or lesser extent up the plate towards the solvent front, with the result that the lower part of the chromatogram nearer the baseline is cleared of impurities, and the impurities become concentrated in the upper section of the plate nearer the solvent front. Subsequently, when a section of the absorbent is removed and eluted with a further solvent to recover a separated compound the amount of impurities contaminating the compound will depend on the location of the compound on the chromatogram (R_f value) and contamination may range from negligible to substantial.

In circumstances where slow-moving compounds are being separated the impurities may move away from the polymer additives towards the solvent front, and thus not interfere in the subsequent examination of the separated compounds. This behaviour leads to a convenient method for moving the impurities beyond the section of the chromatogram to be used for the separation of polymer additives by migrating the chromatogram with appropriate washing solvents before applying and migrating the sample mixture. If this premigration washing covers a longer distance on the plate than is to be used in the sample migration, then it is possible to move interfering impurities out of the way.

The polymer extract is then applied as a band along the base of the plate and the plate developed with suitable solvents.

Finding a chromatographic development solvent or a mixture of solvents for the separation of unknown mixture of additives is not always easy. In some cases a complete separation is not obtained with a single solvent or solvent combination, but necessitates the preparation of several chromatograms using different solvents.

An unknown mixture should first be chromatographed on 20 cm x 5 cm plates with solvent of different polarities to obtain an idea of the types of compounds present in the sample and to reduce the possibility of missing any of the sample components. Solvents of low polarity, such as n-hexane, tetrachlorethylene and carbon tetrachloride, cause polar sample constituents to migrate more readily. Solvents of intermediate polarity such as toluene, benzene, chloroform and methyl cellosolve have a greater effect on polar sample components, whereas highly polar solvent such as dioxan, methylene dichloride, ethyl acetate, nitromethane, acetone, lower alcohols and water elute polar sample constituents towards the solvent front, i.e. R_f values near unity. Mixtures of 40/60 petroleum spirit and up to 10% (v/v) ethyl acetate are useful general solvents for the separation of unknown mixtures.

Detection techniques should be carried out immediately after the chromatogram has been developed, in order to reduce to an absolute minimum any opportunity for volatile sample constituents to be lost by evaporation from the plate. Detection of the separated compounds on the plate is achieved by examination under ultraviolet light which locates some, but not all, types of compounds; and by spraying with a range of general or specific spray reagents (Table 47 and 48).

A further general test for organic compounds on the plate involves holding an electrically heated 25 cm long copper wire, set at red heat, at a few millimetres above the plate along the length in which the chromatogram has been developed. After a few seconds' exposure, many types of organic compounds reveal themselves by charring or otherwise discolouring. Having full information on the R_f values of the different sample components using the preferred development solvents, the next stage is to run a chromatogram on fresh 20 cm x 20 cm plates using a suitably sized sample (say 1 ml of a 1% solution) and mark off with a sharp stylus the bands corresponding to the known positions of the separated compounds. No detection reagents are applied to these plates although they may be examined under the ultraviolet light to precisely locate any components which show up under these conditions. The simplest method of removing the zones containing the separated compounds (after allowing solvent to evaporate from the plate) is to hold the plate vertically, its side resting on a sheet of paper and to scrape off the desired zone with a spatula. Each separated adsorbent band can be bottled off in 5-ml polythene stoppered tubes and retained for further examination.

The separated portions of adsorbent, each, hopefully, containing a pure constituent of the original polymer extract, are now extracted with suitable solvents to isolate the additive preparatory to identification by physical and chemical methods.

Each portion of adsorbent is transferred from the storage bottle to a separate 12 mm x 60 mm sintered glass extraction thimble and the organic compounds leached out with a suitable solvent such as anhydrous absolute ethanol, diethyl ether or methylene dichloride. This solvent must:

1   be good for the additive;

2   be sufficiently polar to desorb the additive from the absorbent (successive desorption with different solvents may be necessary at this stage);

Table 47 - General Spray Reagents for location of compounds
on 20 cm x 20 cm silica gel coated thin-layer
chromatography plates

A.   Reagents applied to GF 254 plate without subsequent heating*

1.	Potassium permanganate	(0.1 N) in aqueous sodium carbonate (5% w/v)
2.	Potassium permanganate	(2% w/v in aqueous sulphuric acid (6% v/v)
3.	Potassium permanganate	0.1% w/v) in sulpuric acid (96%)
4.	Antimony pentachloride	(2% w/v) in carbon tetrachloride
5.	Phosphomolybdic acid	(3% w/v) in ethanol, then expose plate to ammonia vapour

B.   Reagents applied to G254 plate with subsequent heating*

1.	Sulphuric acid aqueous (20% w/v)	5-15 min at 120°C, then 5 min at 150°C
2.	As (2) under A	5-15 min at 120°C, then 5 min at 150°C
3.	Phosphoric acid (10%) methanolic	5-15 min at 120°C, then 5 min at 150°C
4.	Perchloric acid (2%) methanolic	5-15 min at 120°C, then 5 min at 150°C
5.	As (4) under A	5-15 min at 120°C, then 5 min at 150°C
6.	Phosphomolybdic acid (20% w/v) in methanol or methyl cellosolve, then expose plate to ammonia vapour	5-15 min at 120°C

---

* Spray 20 cm x 2 cm wide sections of plate with each reagent using an
aluminium or glass mask with suitable aperture.

3   have a low boiling point to facilitate subsequent removal of solvent
     and reduce to a minimum evaporation losses of any volatile sample
     constituents; and/or

4   not interfere in the subsequent spectroscopic examination of
     extracts.

Provided the desorption solvent is sufficiently powerful and polar, it
should recover between 50 and 100% of the additive present in the silica
gel fraction and provide sufficient material for examination by
ultraviolet or infrared spectroscopy, or mass spectroscopy.

McCoy and Fiebig[471] described a technique using a cavity microcell
for obtaining infrared spectra of 50-100 mg of organic compounds dissolved
in a suitable spectroscopic solvent.

Alternatively, especially if the compound is insoluble in the usual

Table 48 - Chemical Analysis of Additives in Plastics;
specific spray reagents for location of compounds
by thin-layer chromatography

Additive type	Spray reagent	Reference
Phenolic antioxidants	1) 2,6-dichloro-benzoquinone chlorimine (1-2% in ethanol followed 15 min later by 2% borax in 40% aqueous ethanol)	459, 460, 461
	2) alpha, alpha'-diphenyl picryl hydrazyl (0.1% in 95% aqueous ethanol)	461
	3) palladium chloride (mix 150 ml palladium chloride with 100 ml of 2 N hydrochloric acid)	461
	4) diazotized p-nitroaniline (mix 5 ml 0.5% p-nitroaniline in 2 N hydrochloric acid with 0.5 ml 5% sodium nitrite until colourless and 15 min later add 15 ml 20% sodium acetate)	460
Amine antioxidants	Diazotized p-nitroaniline (mix 5 ml 0.5% p-nitroaniline in 2 N hydrochloric acid with 0.5 ml 5% sodium nitrite, until colourless and 15 min later add 15 ml 20% sodium acetate)	
Dialkyl thiodipropionates	Potassium platinoiodide (mix 5 ml 5% platinum tetrachloride in 1 N hydrochloric acid with 45 ml 10% potassium iodide and 100 ml water)	461
Phthalate ester plasticizer	Resorcinol. (Spray with 20% aqueous resorcinol in 2% aqueous zinc chloride and heat to 150°C. Then spray with 4 N sulphuric acid and heat for 20 min at 120°C. Spray with 40% potassium hydroxide to produce orange spots). Phthalic acid and phthalates also react	462,463
Acids and bases	Bromocresol green - (0.5% in 50% aqueous ethanol)	464,465
	Bromocresol purple -(0.5% in 50% aqueous ethanol)	
	Bromophenol blue, Methyl red (0.5% in 50% aqueous ethanol)	464,466
Carboxylic acids Aliphatic (primary, secondary, tertiary) amines, long chain quaternary salts and amine oxides	Sodium dichlorophenolindophenol (1% ethanolic). Cobalt thiocyanate 10 g $Co(NO_3)_26H_2O$ and 10 g ammonium thiocyanate made up to 100 ml. Produces blue colour	464,467
Alkanolamines	Ninhydrin (heat plate for 5 min at 110°C, spray with 0.2%	468

	ninhydrin in acetone and heat 5 min at 110°C to produce colours. Further colours then produced upon spraying plate with 0.2% alizarin in acetone)	
Alkyl phenols	Phenols coupled as p-nitrophenol azo dyes applied to plate of silca gel impregnated with alkali. Separated azo dyes located as yellow/red colours upon exposure of plate to ammonia vapour	
Carbonyl compounds	Carbonyl compounds in sample converted to 2,4-dinitrophenyl-hydrazones, applied to thin-layer plate and plate developed. Separated 2,4-DNPH compounds located as yellow or brown colours upon spraying plate with 2% sodium hydroxide in 90% ethanol	469
Organic peroxides	Hydriodic acid (spray plate with a reagent comprising 40 ml glacial acetic and 0.2 g zinc dust added to 10 ml of 4% aq. potassium iodide, then spray with fresh 1% starch solution). Peroxides (and certain other types of oxidizing agents) revealed by liberation of free iodine. Alternatively use 2,6-dibromo-benzoquinone chlorimine	470

*Spray 20 cm x 2 cm wide sections of the plate with various reagents using an aluminium or glass mask with suitable aperture.

spectroscopic solvents, the solid can be dispersed in well-ground solid dry potassium bromide using a dental mixing machine, and the mixture pressed into a 1 mm thick, 5 mm diameter disc. This disc can then be mounted in a cardboard or plastic holder and used to prepare a spectrum. To prepare ultraviolet spectra the portions of adsorbent from the thin-layer plate are separately dissolved in a suitable spectroscopic solvent and made up to volume in 1 ml flasks.

When quantitative determinations are desired using these techniques, it is necessary to know, or to determine, the absorptivities of the particular compounds. This is done by preparing and measuring solutions of known concentration of the pure compounds. It is also advisable to chromatograph known amounts of the pure compounds to verify the applicability of the technique to the particular compounds. This is recommended because unexpected errors can occur if compounds have enough volatility to escape from the adsorbent, or are unstable and change during chromatography and drying.

An example of the application of this technique to the identification of additives in polypropylene is discussed below.

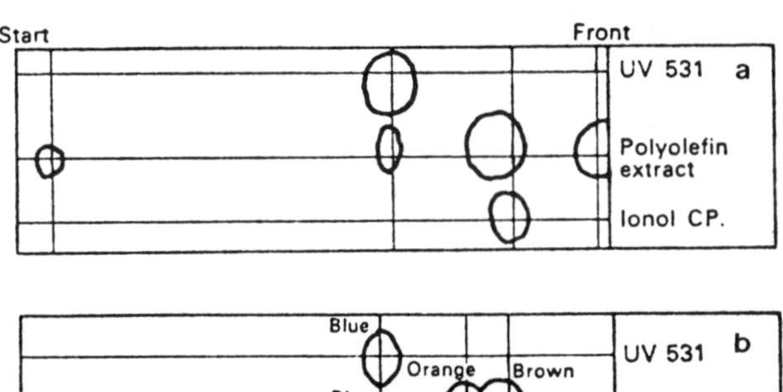

Figure 27 Thin layer chromatograms of solvent extracts of
polyolefins. Plate: Merck silica gel GF 254
development solvent 40:60 petroleum spirit:ethyl acetate
(9:1 v/v).
(a) Spray reagent; aqueous sulphuric acid, plate heated
for 15 minutes at 150°C   (b) Spray reagent; 2% ethanolic
2.6 dibromo benzoquinone-4-chlorimine; then 2% aqueous
borax.

The polymer was extracted with diethyl ether to isolate total
additives and a portion of a chloroform solution of the extract and of
various known light-stabilizers and antioxidants run in parallel on a
silical gel coated plate.

The chromatograms in Fig. 27 show what the polymer contained two
additives appearing at R_F 0.6 and 0.85, which coincided in R_F value and in
the colour obtained with 2,6-dibromo-benzoquinone-4-chlorimine with known
specimens of UV 531 (2-hydroxy-4-n-octoxy benzophenone) and Ionol CP (2,6-
di-tert-butyl-p-cresol).   Spraying the plate with 2,6-di-bromo-benzo-
quinone-5-chlorimine also revealed an additional orange-coloured spot at
R_F 0.8 which did not coincide with any of the known additives examined.
Next, an attempt was made to identify unequivocally these three polymer
components by comparing their infrared spectra with those of authentic
specimens of the suspected compounds.   Chloroform solutions (1 ml)
containing 15-30 mg of the polymer extract, and of authentic UV531 and
Ionol CP, were applied along the edge of three 20 cm x 20 cm plates and
the chromatograms developed using 40/60 petroleum spirit:ethyl acetate
(9:1 v/v).   The three ultraviolet adsorbing bands on each plate were
marked off and the silica gel corresponding to these zones removed from
the plate and the additive extracted from each portion of gel with
anhydrous diethyl ether.   After removing ether the residues were
intimately mixed with dry potassium bromide, and small discs prepared for
infrared spectroscopy.   As a control a further blank chromatogram was
developed, omitting the addition of chloroform solution of sample.   The
gel from this plate corresponding in R_F value and area to the R_F 0.6 and
0.8/0.85 bands observed in the polymer extract, were isolated and ether
extracted. Figures 28A (UV 531) and 28B (Ionol) show the infrared spectra
in the 2.5 to 15 micron region of the authentic additives (direct spectrum
a, spectrum after separation on the plate b, the blank run c and the
corresponding extract of the polymer d,e.   The spectra a and b of
authentic UV 631 are identical, as are spectra a and b in the case of
Ionol CP, i.e. it is valid to compare the spectra of these additives after

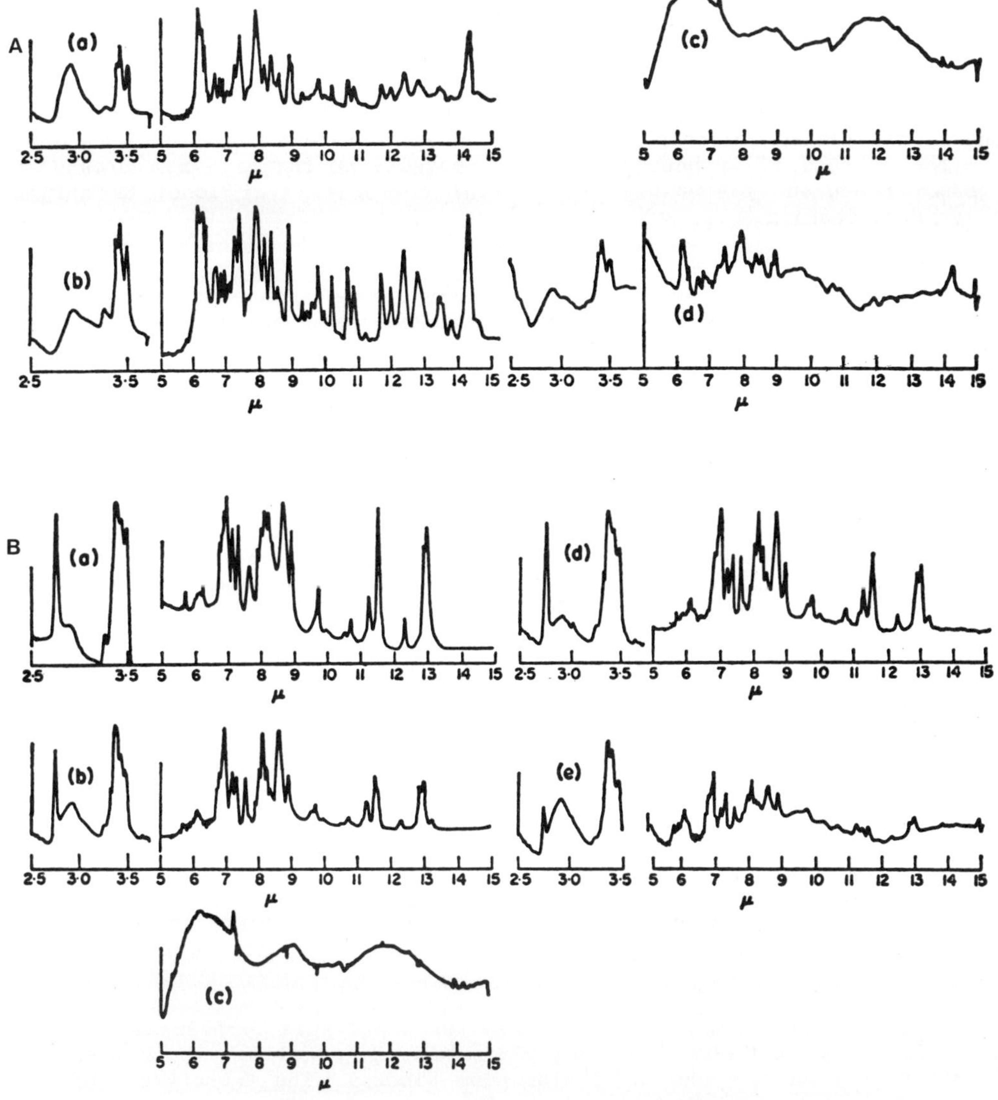

Figure 28 Infrared spectra of additives isolated from polyolefin
(KBr) discs). (A) UV 531 (2-hydroxy-4-n-octoxy
benzophenone, (B) Ionol and its degradation products.

chromatography with their direct infrared spectra, indicating that contact
with silica gel does not produce any structural alteration of these
substances. Also, the blank spectra in Figs. 28a and b show that only
minor infrared absorptions due to plate impurities occur at 6.1 micron
(water), and 9-10, 10.5, 13 micron (silica gel) and 7.2 micron (grease
from glassware). These absorptions would not interfere in the
interpretation of the additives spectra.

Comparison of Figs. 28A(b) and (d) reveals that the compound at R_f
0.6 is identical or very similar to UV 531. The light-stabilizer in the
polymer extract is certainly a substitute benzophenone, although it may
differ from UV 531 in the length of the alkoxy substituent which is known

to have little or no influence on the infrared spectrum of compounds of this class.

Comparison of Figs. 28B(b) and (d) confirms that the R_f 0.85 compound in the polymer extract is Ionol CP, and comparison of d and e shows that the R_f 0.8 component of the polymer extract has a spectrum very similar to that of authentic Ionol CP, suggesting that it is a breakdown product produced, presumably, by partial degradation of Ionol CP during polymer processing.

<u>Photolytic degradation - gas chromatography - mass spectrometry.</u> Juvet et al[472] have developed this technique in response to the accepted limitations of pyrolysis - gas chromatography as a means of identifying and determining additives in polymers. The latter technique, despite improvements in recent years, is still subject to several sources of error particularly when used in the quantitative mode. These include the effects of sample size, method of application of the polymer to the pyrolyser filament, the effects of trapped solvent in the polymer and the thermal conductivity of the polymer.

Photolytic degradation has been shown to yield considerably more simple and reproducible decomposition patterns due to greater control of input energy and the more predictable manner in which compounds decompose photolytically.

Juvet et al[472] showed that characteristic reaction products occur during photolytic degradation and polymer additives may be identified without separation from the polymer samples. The excellent reproducibility reported for photolytic degradation of liquid samples indicated the possibility of quantitative determination of polymer additives. They developed equations to predict the shape of calibration curves expected for both trace level additives and those present at higher concentration.

Since photolytic analysis does not require separative steps and can be performed with samples weighing less than a milligram, it provides a rapid, practical approach for the determination of additives in polymeric materials.

Pure polymer samples are prepared for photolysis by placing a sample of up to 50 mg between two plane glass microscope slides which are then inserted between the two large aluminium blocks. The temperature of the blocks is maintained slightly above the glass transition temperature of the polymer (usually ca 150°C). A 3/8" machine screw is tightened and the sample flattened into a thin film rapidly (about 5 seconds) to preclude thermal degradation of the polymer. The resulting film is places between two glass plates with one edge slightly protruding and sliced into strips ca. 0.8 mm wide with a scalpel. These strips are weighed to the nearest microgram and transferred to 1.0 mm i.d. quartz capillaries, previously sealed at one end. The capillary is evacuated for 5 minutes to remove air and is then sealed, care being taken to prevent thermal degradation in this operation.

In preparing samples of known composition for calibration purposes, weighed amounts of additives are mixed with pure polymer samples prior to casting into thin films. To provide maximum ultraviolet intensity for photolysis in the solid phase, an irradiation apparatus using an Engelhard Hanaovia Type 612C medium-pressure mercury source was employed. Photolysis periods of 5 minutes were used in quantitative determinations of polymer additives.

After irradiation, the samples are introduced onto the chromatographic column with a capillary crusher similar in design to the F and M (Hewlett-Packard) Model SI-4 solid sampler but constructed to a large scale to accommodate 1.0 mm i.d. quartz tubing.

A flame ionization detector was used to obtain chromatographic date. The column comprised 6 feet 1/8 inch o.d. stainless steel column packed with 10% w/w SE-30 on 80/200 mesh Chromport S, (injection port temperature; 150°C, temperature programming rate ca 6°/min, helium carrier gas flow rate of ca 30 ml/min).

To aid in the determination of the retention indices of the photolysis products, a device was used which automatically indicates column oven temperature on the recorder chart during temperature programming. Juvet et al[472] found that with low concentrations of additives, the photodegradation pattern of a polymer is not affected and new products appear from photolysis of the additives. As an example, Table 49 lists the retention indices on SE-30 for the additional photolysis products from two similar samples, 1.0% dilaurylthio-dipropionate in polyethylene and 1.0% distearylthiodipropionate in polyethylene. No difficulty is found in differentiating between these two additives by photolysis gas chromatography, although they are of closely related structure; indeed the similarity in structure is readily apparent from the constant difference of 600 I units in the homologous peaks of these two additives. The product from dilaurylthiodipropionate eluting at I = 1690 was identified from retention data and mass spectrometry as dodecyl propionate. The overlapping doublets which elute at I = 995-1000 and I = 1194-1200 on SE-30 are resolved on Carbowax 20M and the individual products elute at I values of 1000, 1045, 1200 and 1242. Considering the kinetic data presented in Figure 29, the decomposition of dilaurylthiodi-propionate is very likely to follow the mechanism:

$$C_{12}H_{25} - O - \overset{\overset{O}{\|}}{C} - CH_2 - CH_2 - S - CH_2 - CH_2 - \overset{\overset{O}{\|}}{C} - OC_{12}H_{25}$$

$$\swarrow \quad \text{LTDP} \quad \searrow$$

$$\downarrow$$

$$CH_3 - CH_2 - \overset{\overset{O}{\|}}{C} - OC_{12}H_{25} \longrightarrow C_{10}H_{20}$$

$$(I = 1690) \qquad (I = 1045)$$

$$C_{12}H_{24} \qquad + 2H^{\circ}$$

$$(I = 1242)$$

$$+ 2H^{\circ} \qquad C_{10}H_{22}$$

$$C_{12}H_{26} \qquad (I = 1000)$$

$$(I = 1200)$$

Addition of 1.0% dilaurylthiodipropionate to poly(methyl methacrylate) has no effect on the basic polymer photolysis pattern, but the formation of C_{10} and C_{12} hydrocarbons relative to the peak at I = 1690 is reduces to ca. 0.01 and 0.02 respectively. In this matrix the concentration of the product eluting at I = 1690 increases continuously throughout a 2 hour irradiation, further strengthening the argument that the hydrocarbon fragments obtained in the polyethylene matrix are secondary photolysis

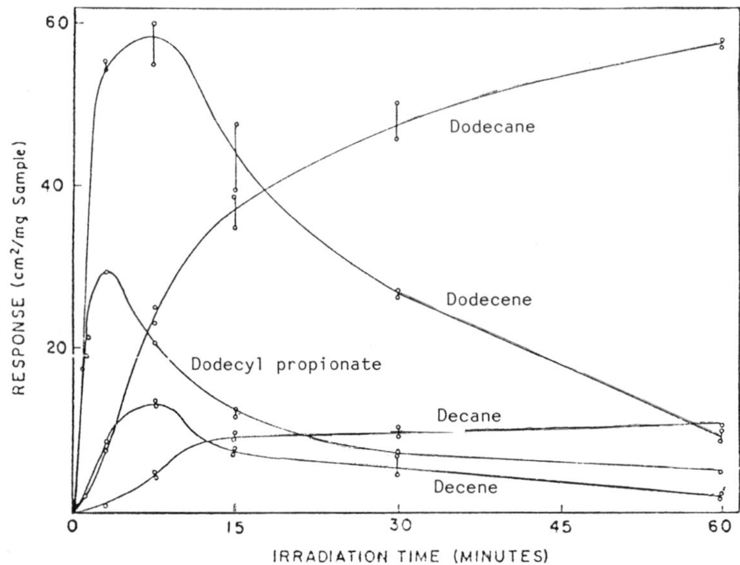

Figure 29 Radiation kinetics for products formed from 1% lauryl
thiodipropionate antioxidant in polyethylene.

Table 49 - Photolysis Products of Antioxidants in Polyethylene

Additive	Concentra-tion %	Additonal products, I values on SE-30	Relative peak area for 30 minutes irradiation
Dilaurylthio-dipropionate	1.0	995-1000 $\pm$ 4[a]	0.22
		1100 $\pm$ 6	0.09
		1194-1200 $\pm$ 4[a]	1.00
		1690-1200 $\pm$ 5	0.21
Distearlylthio-dipropionate	1.0	1592-1602 $\pm$ 4[a]	0.21
		1700 $\pm$ 6	0.34
		1792-1805 $\pm$ 6[a] 1.00	
		2298 $\pm$ 8	0.50

[a] Overlapping doublet peaks.

products which show a decrease in quantum yield in the methyl methacrylate
matrix toward production of the hydrocarbon fragments.

Quantitative measurements are carried out by irradiating the sample
for a carefully controlled time period using a light source of constant
intensity. In the analysis of the antioxidant, dilaurylthiodipropionate
in polyethylene an irradiation time of 5.00 $\pm$ 0.02 min was chosen and a
calibration curve constructed by plotting the peak area of the product
peak, dodecyl propionate at I = 1690 vs the per cent dilaurylthiodipropionate in the sample as illustrated in Fig. 30(a). The average
relative mean deviation of the 11 points used in constructing the graph is
2.7%. Since the product peak tails to a small extent under the

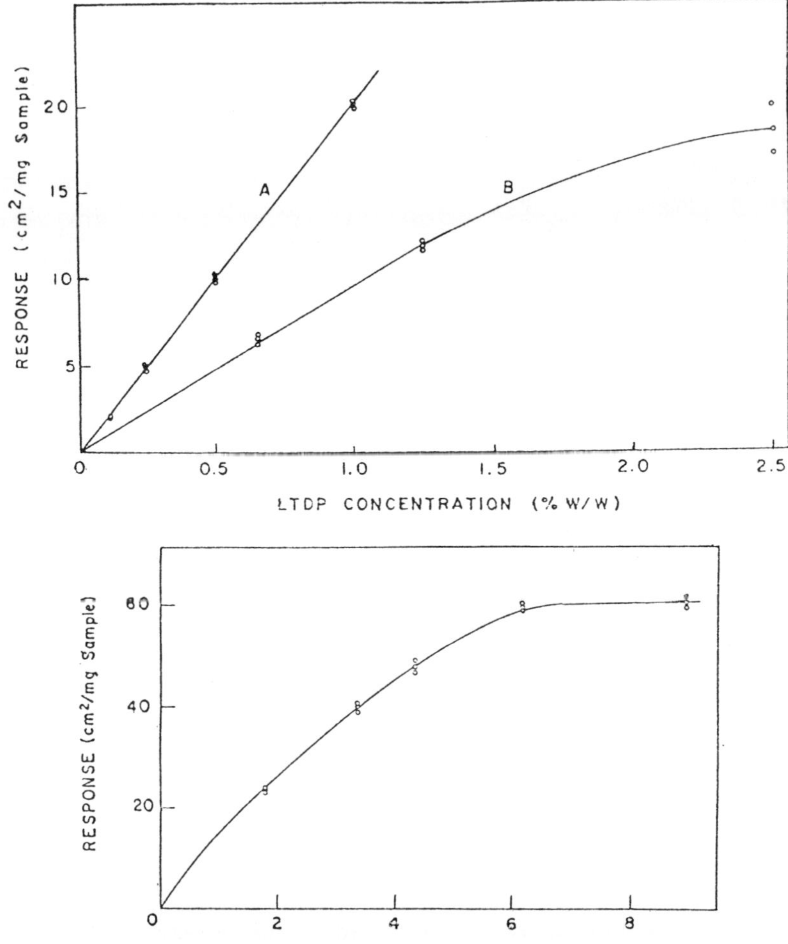

Figure 30 Photolytic detradation - gas chromatography of additives
in polymers: (a) determiation of lauryl thio-
diproprianate antioxidant in (A) polyethylene: (B)
polymethyl methacrylate using the dodecyl propionate
product peak at I = 1690 (measured on SE-30 under
programmed temperature conditions; irradiation time,
5.00 ± 0.02 min; (b) determination of "dinonyl"
phthalate plasticizer in polymethyl methacylate using
the I = 659 (Carbowax 20 M at 85°C) radiation product
peak. Irradiation time 15.0 ± 0.1 min.

chromatographic conditions used, the error in peak area measurement is
increased somewhat and is thought to be the major source of error in this
determination.

The quantitative determination of dilaurylthiodipropionate in
poly(methyl methacrylate) again using the dodecyl propionate peak at I =
1690 is shown in Fig. 30b. The resultant working curve is linear below
about 1% dilaurylthiodipropionate but shows marked non-linearity above
this value.

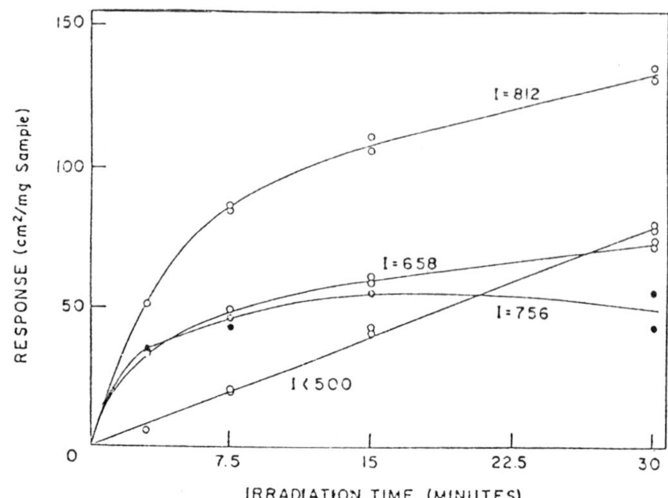

Figure 31   Reaction kinetics for 43% dinonyl phthalate plasticiser
in polymethyl methacrylate. Radiation product peak at I
= 658 from dinonyl phthalate; others from polymer
matrix. Retention indices measured on a Carbowax 20 M
column at 85°C.

As a typical example of the determination of plasticizers, dinonyl
phthalate at concentrations up to ca. 9% was studied in a poly (methyl
methacrylate) matrix.   The rate of formation of several photolysis
products from samples containing 4.30% dinonyl phthalate is shown in
Figure 31.   The peaks at I < 500, I = 756 and I = 812 are irradiation
products of the polymer and the product eluting at I = 658 is from
dinonylphthalate.   The nonlinear calibration curve produced using
irradiation times of 15.0 ± 0.1 min is shown in Fig. 30(b).   It is
interesting to note that dioctyl phthalate in poly(vinyl chloride)shows
linear dependence on concentration as high as 19 weight per cent for the
decomposition product peak appearing at I = 796 (probably n-octane) on SE-
30 liquid phase when an irradiation time of 5.00 ± 0.02 minutes is used.
In the case of dinonyl phthalate, variation in plasticizer concentration
affects the relative amounts of the polymer photolysis products at I = 756
and I = 812.   Since the relative product yield from the basic polymer is
altered it is apparent that the matrix in which photolysis of the additive
occurs is changing with plasticizer concentration.

5.2   MONOMERS AND OLIGOMERS (SEE ALSO APPENDIX 4)

5.2.1   Styrene and Alpha-Methyl Styrene in Polystyrene

The styrene monomer used in the manufacture of polystyrene generally
contains low concentrations of aromatic impurities.   These fall into two
categories (a) saturated compounds such as benzene, toluene, ethyl
benzene, xylenes, propyl benzene, butyl benzene, ethyl toluenes, diethyl
benzenes, traces of which remain in the final polymer and (b) unsaturated
impurities such as styrene and methyl styrenes which either wholly or
partially copolymerize with the styrene.   The presence of all these
substances in addition to residual styrene monomer in the final polymer
has implications from points of view of the mechanical properties of the
final polymer and toxicity in the case of foodstuff packing grades of
polystyrene.

<u>Direct - gas chromatography.</u> Crompton et al[473-474] have applied the gas chromatographic technique to the determination in polystyrene of styrene and a wide range of other aromatic volatiles in amounts down to the 10 ppm level. In this method a weighed portion of the sample is dissolved in propylene oxide containing a known concentration of pure n-undecane as an internal standard. After allowing any insolubles to settle an approximately measured volume of the solution is injected into the chromatographic column which contains 10% Chromasorb 15-20 M supported on 60-70 BS Celite. Helium is used as carrier gas and a hydrogen flame ionization detector is employed. Figure 32 shows a device[474] which is connected to the injection port of the gas chromatograph in order to prevent the deposition of polymeric material in the injection port of the chromatograph with consequent blockages. When a solution of polystyrene is injected into the liner, polymer is retained by the glass fibre, and volatile components are swept on to the chromatographic column by the carrier gas.

Figure 33(a) illustrates, by means of a synthetic mixture, the various aromatics that can be resolved, and Figure 33(b) illustrates a chromatogram obtained with a polystyrene sample, indicating the presence of benzene, toluene, ethyl benzene, xylene, cumene, propyl benzenes, ethyl toluenes, butyl benzenes, styrene and alpha-methyl styrene.

<u>Solution head-space gas chromatographic methods.</u> In this procedure a solution of the polymer in a suitable solvent is placed in a closed container and allowed to equilibrate at a controlled temperature so that volatile monomers or other impurities dissolved in the polymer solution partition between the solution and the gas phase. Subsequent analysis of the gas phase enables the concentration of monomers to be calculated. In a variant of this method the solid polymer is allowed to equilibrate with the head-space gas.

Although the solid head-space method, discussed later, provides about 10-fold more sensitivity than the solution head-space method (assuming a 10% sample solution), the solid method may be applied only to sample systems where equilibration with the head-space is rapid and complete. For example, residual styrene monomer in polystyrene does not reach equilibrium with the head-space after 20 hours at[475] and thus may not be determined by the solid head-space method. Furthermore, even if equilibration between the solid and head-space is obtained, the partition coefficient must also be determined for the component of interest in each type of sample matrix.

Various methods have been described for the determination of styrene monomer in polystyrene by solution-head space analysis[475,476]. One technique has been automated. In the automated procedure the polymer (0.2 g) is dissolved and the solution is maintained at 70°C. Standard solution of the compounds to be analysed, dissolved in dimethylformamide, are used for calibration. The detection limit of styrene is 10 mg per kg of polystyrene. The column (2m x 3mm) used 15% of Reoplex 400 on Embacel (60 to 100 mesh) and was operated at 100°C, using nitrogen as the carrier gas and flame ionization detection.

A further apparatus for carrying out solid polymer head-space analysis of polystyrene for styrene monomer and aromatic volatiles has been described by Crompton and Myers[473,474]. In this apparatus the polymer is heated up to 300°C in the absence of solvents, prior to their examination by gas chromatography. The technique avoids the disadvantages resulting from the use of extraction or solution procedures. The

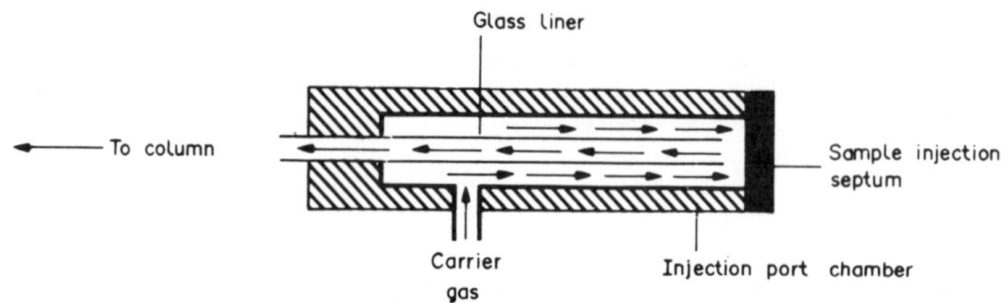

Figure 32   Injection port glass liner fitted to F & M Model 1609
            gas chromatograph, the glass liner measures 60 mm x 40
            mm o.d. and 2 mm i.d. and is very loosely packed with
            glass fibre.

apparatus is illustrated in Figure 34. A glass ignition tube is supported
as shown in a Wade ¼" diameter brass coupling unit, covered with a
silicone rubber septum and sealed with a Wade ¼" brass stop-end body. The
stop-end body has two 1 mm diameter holes drilled through the cap.   The
whole unit is placed in a slot in a cylindrical copper block (3" long x 2"
diameter) which is heated by two Bray (240 V, 85 W) cartridge heaters and
controlled at temperatures up to 300°C, from a variable transformer.   The
temperature is measured with a thermocouple capable of accurately
measuring temperatures in the 100°-300° temperature range with a maximum
error of ± 5%. The thermocouple is inserted in the slot adjacent to the
ignition tube, it has been shown that under these conditions the
thermocouple records the true temperature of the contents of the tube.
The provision of a slot in the copper block enables more than one ignition
tube to be heated simultaneously if required.

A sample of the polymer (0.25-0.50 g) is placed in an ignition tube
and sealed with Wade fittings and a septum, as described. If necessary,
the tube is then purged with a suitable gas by inserting two hypodermic
needles through the septum via the holes in the cap of the stop-end body
and passing the gas into the tube through one hole and allowing it to vent
through the other.   After purging the two needles are removed
simultaneously and the tube is then heated in the copper block under the
required conditions of time and temperature. A sample (1-2 ml) of the
head-space gas is withdrawn from the ignition tube into a Hamiliton gas-
tight hypodermic syringe via the septum and injected into a gas
chromatograph.   It is advisable the fill the syringe with the gas used
initially in the ignition tube and to inject this into the tube before
withdrawing the sample.   This facilitates sampling by preventing the
creation of a partial vacuum in the ignition tube or the syringe, or both.
It also minimizes any undesirable entry of air into the ignition tube.

With the apparatus, a polymer may be heated under any desired gas
and, while this may frequently be the carrier gas used with the gas
chromatograph, it is also possible to carry out studies in oxidising or
reducing atmospheres.   A polymer may also be heated to any known
temperature and samples of the head-space gas may be withdrawn at
intermediate temperatures and times to determine under what conditions any
particular volatile is liberated.

By using gas chromatographic detectors of suitable sensitivities and
selectivities, it is possible to examine polymers for the presence or
formation of monomers and other volatiles at both the percentage and
parts-per-million levels.

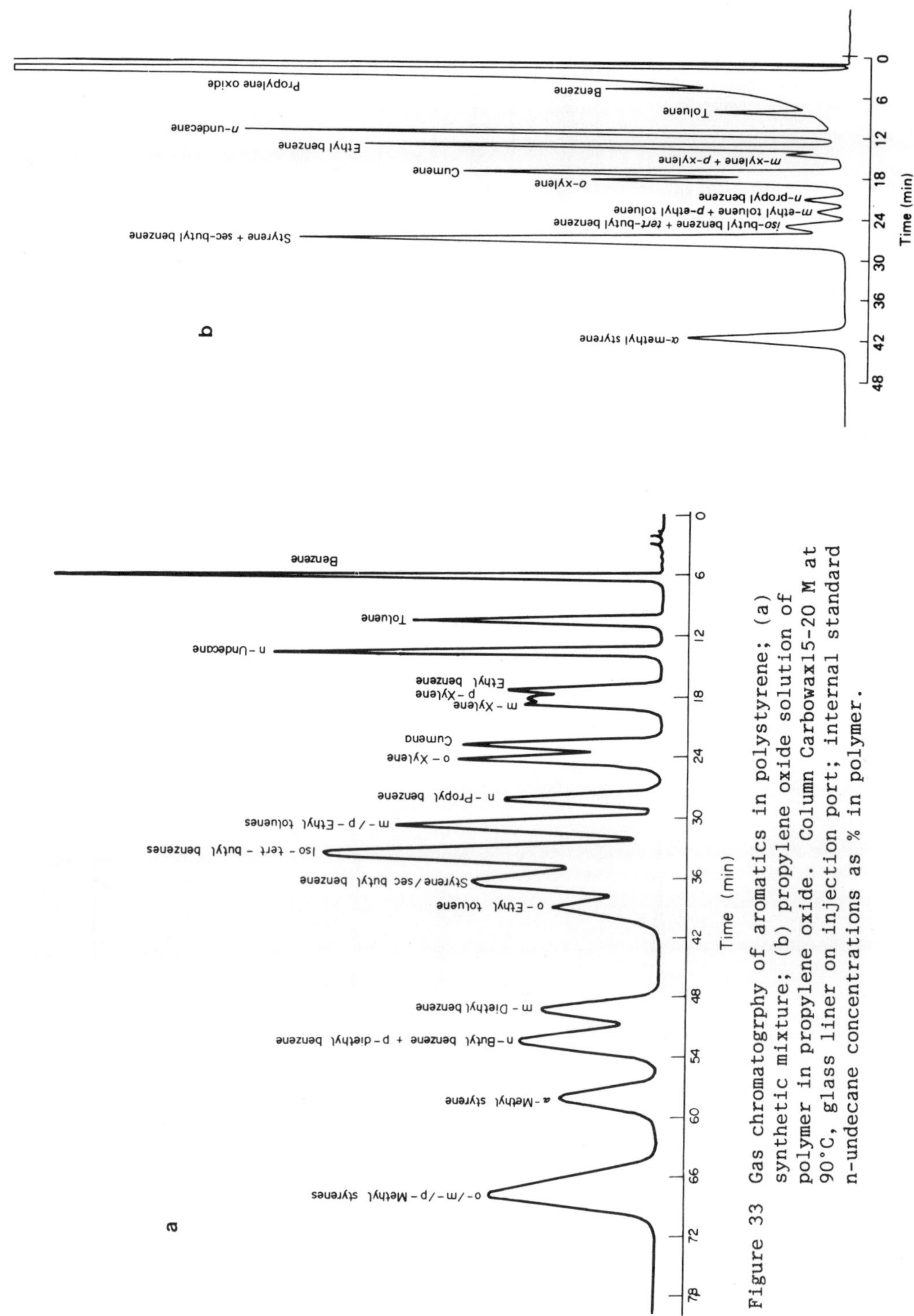

Figure 33 Gas chromatogrphy of aromatics in polystyrene; (a) synthetic mixture; (b) propylene oxide solution of polymer in propylene oxide. Column Carbowax15-20 M at 90°C, glass liner on injection port; internal standard n-undecane concentrations as % in polymer.

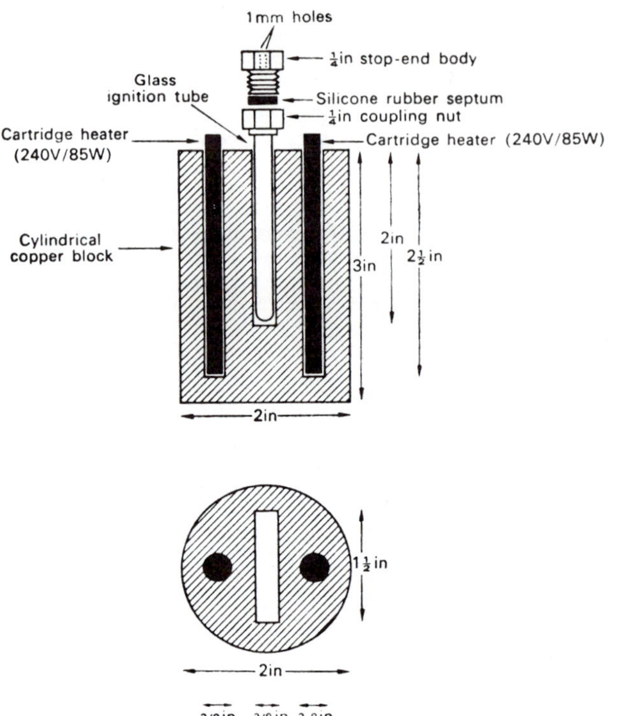

Figure 34 Apparatus for liberating volatiles from polymers.

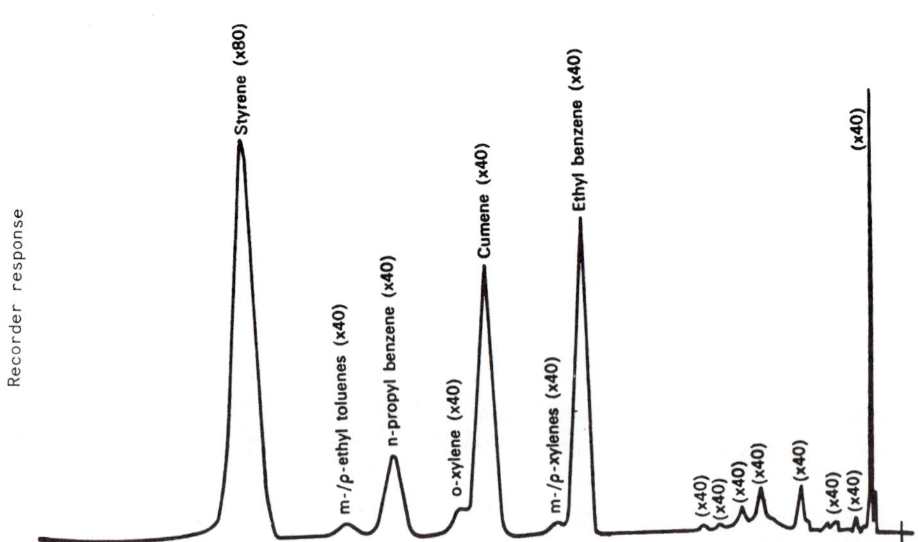

Figure 35 Gas chromatogram of volatiles liberated from commercial
polystyrene upon heating at 200°C. Chromatographed on
15 ft x 3/16 inch 10% Carbowax 15-20 M on 60-72 Celite
at 90°C and 100 ml min⁻¹ helium; flow rate, flame
ionization detector. Attenuations bracketted.

Using this method several polystyrenes from different manufacturers were heated at 200°C for 15 minutes under helium, and the liberated volatiles were examined by gas chromatography. A typical chromatogram is shown in Fig. 35. All the samples liberated the same range of aromatic hydrocarbons, these differing only in their relative concentrations.

The quantitative aspects of this technique itself were investigated by analysing the volatiles liberated by heating polystyrene at 200°C and then re-analysing the polymer as a solution in propylene oxide. Several samples of polystyrene were examined in this way and in each case component peak areas were normalized and compared. The results (Table 50) show that the values obtained by head-space technique differ by up to 20% from those obtained by the solution procedure which are known to be correct.

Table 50 – Comparison of normalized peak areas calculated
from chromatograms of polystyrene solutions and
polystyrene volatiles liberated by heating at 200°C

	Normalized peak area							
Component	Analysis of polymer as a solution in polypropylene oxide				Analysis of volatiles liberated by heating polymer at 200°C			
Manufacturer	A	B	C	D	A	B	C	D
Ethyl benzene	16.4	17.4	27.2	27.5	20.2	23.4	35.4	34.4
n-Propyl benzene	5.5	5.3	3.5	3.2	3.2	2.5	1.9	2.3
Cumene	17.7	18.1	10.4	10.8	15.4	12.7	10.2	10.7
m-/p-Xylenes	4.1	2.5	2.7	2.0	1.1	1.3	0.6	0.7
m-/p-Ethyl toluenes	1.7	1.7	2.0	1.5	0.8	0.7	0.5	0.6
Styrene	54.6	55.0	54.2	55.0	59.3	59.4	51.4	51.3

<u>Solid polymer head-space analysis techniques.</u> This technique has been applied to the determination of styrene monomer in polystyrene and its copolymers. The polymer (2 g) is placed in a 250 ml screw-top glass jar with a Teflon seal and left in an oven for 4-6 hours at 110°C prior to withdrawal of a portion of the head-space for gas chromatographic analysis using a flame ionization detector. A sample of monomer-free polystyrene was spiked with styrene monomer at concentrations of 0.5, 1.0, 1.5 and 2.0 ppm and 2 ppm of an internal standard of o-xylene was added. The ratio of the peak height of styrene to o-xylene was used to form a calibration curve. This curve has some scatter, but allows adequate accuracy for this concentration range.

In addition to gas chromatography, other techniques - particularly ultraviolet spectroscopy and polarography - have been applied to the determination of monomers in polymers.

<u>Ultraviolet spectroscopy.</u> Ultraviolet spectroscopy suffers from several disadvantages in the determination of monomers in polymers, particularly in the case of styrene monomer in polystyrene and its copolymers. In addition to lack of sensitivity, which limits the lower

detection limit to about 200 ppm styrene in polymer under the most favourable circumstances, ultraviolet spectroscopic methods are subject to interference by some of the types of antioxidants included in polystyrene formulations. Such interferences can only be overcome by applying a lengthy pre-treatment of the sample to remove antioxidants prior to spectroscopic analysis. In addition to residual styrene monomer, polystyrene may also contain traces of other aromatic hydrocarbons such as benzene, toluene, xylenes, ethyl benzene and cumene, which originate either as impurities in the styrene monomer employed to manufacture the polystyrene or they may have been used in small quantities as dilution solvents at some stage of the manufacturing process. Ultraviolet spectroscopic methods for determining styrene cannot differentiate between the various volatile substances present in polystyrene.

In the case of some types of polymer additives, interference effects can be overcome by the use of a baseline correction techniques. Thus polystyrene contains various non-polymer additives (e.g. lubricants) which result in widely different and unknown background absorptions at the wavelength maximum at which styrene monomer is evaluated (292 nm). The influence of the background absorptions on the evaluation of the optical density due to styrene monomer is overcome by the use of an appropriate baseline technique, claimed to make the method virtually independent of absorptions due to polymer additives. In this technique a straight line is drawn on the recorded spectrum across the absorption curve at a point close to the absorption minima occurring at 288 nm and 295-300 nm. A vertical line is drawn from the tip of the styrene absorption peak at 292 nm to intersect the baseline, and the height of this line is then a measure of the optical density due to the true styrene monomer content of the test solution.

This baseline correction technique can obviously be applied to the determination of styrene monomer in polystyrene only if any other ultraviolet absorbing constituents in the polymer extract (e.g. lubricant, antioxidants) absorb linearly in the wavelength range 288-300 nm. If the polymer extract contains polymer constituents other than styrene with non-linear absorptions in this region, then incorrect styrene monomer contents will be obtained. An obvious technique for removing such non-volatile ultraviolet-absorbing compounds is by distillation of the extract followed by ultraviolet spectroscopic analysis of the distillate for styrene monomer as discussed below.

In the distillation technique[473] the polystyrene is dissolved in chloroform or ethylene dichloride (20 ml) in a stoppered flask and the solution is poured into an excess of methyl alcohol (110 ml) to reprecipitate dissolved polymer. The polymer is filtered off and washed with methanol (120 ml) and the combined filtrate and washings gently distilled to provide 200 ml of distillate containing styrene monomer and any other distillable component of the original polystyrene sample. Non-volatile polymer components (viz. stabilizers, lubricants and low molecular weight polymer) remain in the distillation residue. The optical density of the distillate is measured at 292 nm or by the baseline method against the distillate obtained in a polymer-free blank distillation. Calibration is performed by applying the distillation procedure to solutions of known weights of pure styrene monomer in the appropriate quantities of methyl alcohol and the chlorinated solvent. Tables 51 and 52 show results obtained for styrene monomer determinations carried out on samples of polystyrene by the direct ultraviolet method, and by the distillation modification of this method. It is seen that the distillation method gives results that are consistently higher than those obtained by direct spectroscopy, indicating that additives present in the polystyrene are interfering in the latter method of analysis.

Table 51 - Comparison of direct ultraviolet and distillation/
ultraviolet methods for the determination of styrene monomer

Method	Solvent	Styrene monomer (% w/w) in polystyrene sample	
		No. 1	No. 2
Direct UV method	Chloroform/ carbon tetra-	<0.05	0.13
	chloride	<0.05	0.13 0.14
			0.16 0.18
	Ethyl acetate	<0.05	0.14 0.12
			0.14 0.14
			0.15
Distillation/UV	Ethylene di-	0.16 0.18	0.27 0.29
	chloride/	0.16 0.18	0.26 0.29
	methanol	0.20	0.29

Table 52 - Influence of phenolic antioxidant[a] on the determination
of styrene monomer by direct ultraviolet and by
distillation/ultraviolet methods

Styrene found (% w/w)

	Direct UV method[a]		Distillation/UV method	
Styrene added to polysty- rene (% w/w)	No phenolic antioxidant* addition	0.5% phenolic antioxidant* added on polymer	No phenolic antioxidant* addition	0.5% phenolic antioxidant* added on polymer
0.11	0.11 0.11	0.04 0.08	0.12	0.12
0.22	0.22	0.11	-	-
0.27	0.26	0.18	0.29	0.26
0.41	0.41 0.40	0.30 0.27	0.40	0.40

* Wingstay T.   [a] Chloroform used as a sample solvent.

Similar effects to these were also observed with polystyrene
containing other ultraviolet absorbing additives. Thus, the influence of
a mixture of 0.4% w/w tris-(nonylated phenyl phosphite (Polygard) and 0.2%
w/w 2,6-di-tert-butyl-p-cresol (Ionol CP) on the determination of styrene
monomer is shown in Table 53.

Direct ultraviolet spectroscopic methods are reliable only in the
case of polystyrene samples that do not contain ultraviolet absorbing
antioxidants or any other type of strongly ultraviolet absorbing additive.
The more lengthy distillation/ultraviolet spectroscopic method, however,
will give correct results in the presence of such additives unless the
additive is both sufficiently volatile to distil and absorbs in the same
region of the spectrum as styrene monomer.

Table 53 - Influence of Polygard/Ionol CP Mixture on Determination of
Styrene Monomer by Direct Ultraviolet and by
Distillation/Ultraviolet Methods

Styrene found (% w/w)

	Direct UV method[a]		Distillation/UV method[a]	
Styrene added to polystyrene (% w/w)	No additive	0.4% w/w Polygard and 0.2% w/w Ionol CP added on polymer	No additive	0.4% w/w Polygard and 0.2% w/w Ionol CP added on polymer
0.10	0.10 0.10	- [b]	0.11	0.11 0.09
0.26	0.25 0.26	0.16 0.16	0.26	0.27 0.25
0.41	0.42 0.41	0.29 0.29	0.41	0.40 0.42

[a] Chloroform used as a sample solvent.

[b] Not measurable due to strong interference by additives.

5.2.2 Styrene and Acrylonitrile Monomers

Polarography.   Residual amounts of styrene and acrylonitrile
monomers usually remain in manufactured batches of styrene-acrylonitrile
copolymers and acrylonitrile-butadiene-styrene terpolymers.   As these
copolymers have a potential use in the food-packaging field, it is
necessary to ensure that the content of both of these monomers in the
finished copolymers is below a stipulated level.

In a polarographic procedure[477,478] for determining acrylonitrile
(down to 2 ppm) and styrene (down to 20 ppm) monomers in styrene acrylonitrile copolymer, the sample is dissolved in 0.2 M tetramethylammonium-
iodide in dimethyl formamide base electrolyte and polarographed at start
potential of - 1.7 V and - 2.0 V respectively for the two monomers.
Excellent results are obtained by this procedure.

Table 54 shows the results obtained in determinations of
acrylonitrile monomer in some copolymers by the polarographic procedure.
Comparison of these results with those obtained by the dodecyl mercaptan
procedure show that acrylonitrile contents some 30% higher are obtained by
the latter method. The cause of the high results obtained by the dodecyl
mercaptan method was not discovered. However, it could be due to the
presence of interfering substances (e.g. polymerisation additives such as
residual traces of organic peroxide or of low polymers of acrylonitrile)
in the copolymers, which might also react with dodecyl mercaptan leading
to high results in this method of analysis.

The polarographic method for determining acrylonitrile was also
checked on some styrene-acrylonitrile copolymers containing less than 50
ppm of the monomer. Determination of acrylonitrile at these levels in
copolymers is necessary when they are being considered in connection with
food-packaging applications. Acrylonitrile was determined in styrene-
acrylonitrile copolymers that had been subjected to a treatment designed
to reduce their monomer content, namely re-precipitation with methanol
from chloroform solutions with subsequent removal of residual solvent from

Table 54 - Determination of Acrylonitrile: Comparison of
Polarographic and Chemical Methods

Acrylonitrile content, percent w/w, determined by:

polarograph	titration with dodecyl mercaptan
0.06 0.07	0.09 0.09
0.08	0.12
0.11	0.15
0.11	0.19
0.12	0.15
0.12	0.17
0.14	0.18
0.21	0.31 0.29

Table 55 - Determination of Acrylonitrile and Styrene
Monomers in Copolymers

Acrylonitrile determined by polarography per cent, w/w	Styrene, percent w/w determined by:	
	polarography	spectroscopy
0.015 0.014	0.20	0.21
0.015 0.015	0.15	0.13 0.15
0.035 0.036	1.60	1.53

the re-precipitated polymer by drying under vacuum. Polymers that
initially contained several hundred parts of acrylonitrile per million
were shown by the polarographic method to contain about 25 ppm of monomer
after a single re-precipitation step, which reduced to about 5 ppm of
monomer after a further re-precipitation from chloroform.

The acrylonitrile and styrene monomer contents of several copolymers
were determined by the polarographic procedure. The styrene monomer
content of these samples was also determined by a procedure involving
evaluation by ultraviolet spectroscopy at 292 nm of solutions of the
samples in carbon tetrachloride. The results obtained (see Table 55) show
the styrene contents determined by the two procedures are in good
agreement with each other.

<u>5.2.3 Vinyl Chloride, Butadiene, Acrylonitrile, Styrene and 2-Ethylhexyl
Acrylate in Copolymers</u>

<u>Head-space analysis.</u> Steichen[479] has discussed a modified solution
approach for the gas chromatographic determination of these monomers in
their associated polymers by head-space analysis. The more volatile
monomers (vinyl chloride, butadiene, and acrylonitrile) are determined by
dissolution of the polymer and analysis of the equilibrated head space
above the polymer solution. It is possible to determine vinyl chloride
and butadiene at the 0.05 ppm level and acrylonitrile down to 0.5 ppm.
The injection of water into polymer solutions containing styrene and 2-

ethyl hexyl acrylate monomers prior to head-space analysis greatly enhanced the detection capability for these monomers making it possible to determine styrene down to 1 ppm and 2-ethyl hexyl acrylate at 5 ppm. Incorporation of polymer into the calibration standards compensates for the effect which the polymer matrix has upon the equilibrium partitioning of the monomer between the solution and head-space. The relative precision and error in the determination of these monomers near the quantitation limit is less than 7%.

In this procedure a solution of the polymer in a suitable solvent is placed in a closed container and allowed to equilibrate at a controlled temperature so that volatile monomers or other impurities dissolved in the polymer solution partition between the solution and the gas phase. Subsequent analysis of the gas phase enables the concentration of monomers to be calculated. In a variant of this method, the solid polymer is allowed to equilibrate with the head-space gas.

Although the solid head-space method provides about 10-fold more sensitivity than the solution head-space method (assuming a 10% sample solution) the solid method may be applied only to sample systems where equilibration with the head-space is rapid and complete. For example, residual styrene monomer in polystyrene does not reach equilibrium with the head space after 20 hours[480] and thus may not be determined by the solid head-space method. Furthermore, even if equilibrium between the solid and head-space is obtained, the partition co-efficient must also be determined for the component of interest in each type of sample matrix.

The solution head-space approach is applicable to a much wider range of samples than the solid approach. When working with sample solutions, head-space equilibrium is more readily attained and the calibration procedure is simplified. The sensitivity of the solution method depends upon the vapour pressure of the constituent to be analysed and its solubility in the solvent phase. Vinyl chloride, butadiene and acrylonitrile are readily promoted from polymer solutions into the head-space by heating to 90°C. The head-space/solution partitioning for these constituents is not appreciably affected by changes in the solvent phase (vis. addition of water) since the more volatile materials favour the head-space at 90°C. Less volatile monomers such as styrene (b.p. = 145°C) and 2-ethylhexyl acrylate (b.p. = 214°C) may not be determined using head-space techniques with the same sensitivities realised for more volatile monomers. By altering the composition of the solvent phase to decrease the monomer solubility, the equilibrium monomer concentration in the head space can be increased. This resulted in a dramatic increase in the detection sensitivity for styrene and 2-ethylhexyl acrylate. Based on these principles, a procedure is described below, for the gas chromatographic analysis of residual vinyl chloride, butadiene, acrylonitrile, styrene, and 2-ethylhexyl acrylate in polymers by head-space analysis.

In this method, weighed portions of the polymer were dissolved in septum-sealed vials containing measured aliquots of N,N'dimethylacetamide. The vials were heated to 90°C to aid dissolution of the polymer. When solution was complete, the vials were cooled to room temperature. The solutions were swirled to mix and an aliquot of distilled water was forcibly injected into each polymer solution in order to decrease the solubility of monomers. The vials were shaken briefly to assure complete mixing of the water with the organic phase and to prevent the precipitated polymer from forming a film on top of the solution. The vials were equilibrated at 90°C for 60 minutes prior to head-space sampling and analysis by flame ionization gas chromatography. Standards comprising monomer-free polymer and known additions of standard monomer solutions were run in parallel. Greater sensitivities and shorter analysis times were obtained using the head space analysis methods than are possible by the direct injection of polymer solutions into a gas chromatograph (Table 56).

The response obtained for a given monomer using the head-space techniques was dependent upon several factors: the relative volatility of the monomer; the concentration of polymer in solution; and the solubility of the monomer in the solvent phase.

The greatest detection sensitivities using solution head-space analysis were obtained for monomers having relatively low boiling points. This is apparent from the relative sensitivities given in Table 57. These values were determined experimentally using polymer solutions with known amounts of monomer added.

The equilibrium head-space concentration for 2-ethylhexylacrylate at 90°C was not sufficient to allow the determination of residual 2-ethylhexylacrylate in the polymer, even when present at 1000 ppm level. The sensitivity achieved by the head-space method can be improved by decreasing the solubility of the 2-ethylhexylacrylate monomer in the solution of polymer in N,N'-dimethylacetamide through the introduction of a second solvent.

Water is the most effective solvent for this purpose (see Table 57). A greater than 200 fold increase in the 2-ethylhexylacrylate equilibrium head-space concentration resulted when water is injected into its polymer solutions. A significant increase in the styrene head-space concentration

Table 56 - Comparison of Quantitation Limited* for Residual
Monomers Using Conventional and Head-Space GC Methods

Monomer	Boiling point	Direct solution injection[a]	Solution head-space	Modified solution head-space
Vinylchloride	-13°C	1-2 ppm	0.05 ppm	-[b]
Butadiene	- 4°C	5 ppm	0.05 ppm	-[b]
Acrylontirile	76°C	10 ppm	0.50 ppm	-[b]
Styrene	145°C	10 ppm	20.00 ppm	1 ppm
2-ethylhexyl acrylate	214°C	200 ppm	1000.00 ppm	5 ppm

* The quantitation limit is defined as the monomer concentration necessary to produce a peak at least three times the baseline noise or 3% of full scale.

[a] Injection of a 10% polymer solution into a gas chromatograph.

[b] A 2- to 3-fold increase in monomer peak height resulted from the injection of water into the polymer solution. A baseline disturbance due to elution of water negated any real improvement in detection limit for these monomers.

Table 57 - Investigation of Solvents for Displacing 2-EHA
Monomer from DMA-Polymer Solutions into Headspace[a]

Displacing solvent	GC head-space response[b] for 2-EHA	Effect of displacing solvent upon polymer solution
None added	0	Clear solution
Water	652%	Polymer precipitates
Heptane	0	Turbid solution
Ethyl-Cellosolve	0	Clear solution
Ethylene glycol	54%	Polymer precipitates

[a] A 0.10 g portion of resin was dissolved in a septum-sealed vial with 2 ml of DMA. An aliquot of a standard solution containing 76 ug of 2-EHA was added. To this, 5.0 ml of the displacing solvent was injected.

[b] Percent of full scale response at lowest instrument attenuation.

is also obtained when polystyrene solutions are treated with water. In the region where styrene and 2-ethylhexylacrylate monomers elute, no increase in baseline noise results from the injection of water.

The response for styrene and 2-ethylhexylacrylate was affected by the relative amount of water injected into the polymer solution. Increasing the relative amounts of water caused a continuing increase in monomer peak height with no plateau.

Variations on the polymer weight had a much greater effect upon the monomer response for 2-ethylhexylacrylate and styrene than was observed for the other monomers. This greater dependence on polymer weight is apparently related to the precipitation of the polymer which occurs upon the injection of water. This may be attributed to either co-precipitation of the monomer with the polymer or entrapment of water in the polymer precipitate. Regardless of the mechanism, the precision was not affected.

The time required for the monomer in solution to equilibrate with the head-space was less then 60 minutes in all cases. When using the solution method (no water added), equilibrium was reached in less than 30 minutes.

5.2.4 Acrylic Acid and Acrylamide

Polarography. Betso and McLean[481] have described a differential pulse polarographic method for carrying out this determination. A measurement of the acrylamide electrochemical reduction peak current is used to quantitate the acrylamide concentration. The differential pulse polarographic technique also yields a well-defined acrylamide reduction peak at ca. - 2.0 V vs SCA, suitable for qualitatively detecting the presence of acrylamide. The procedure involves an extraction of the acrylamide monomer from the polyacrylamide, a treatment of the extraction solution on mixed resin to remove interfering cationic and anionic species and polarographic reduction in an 80/20 (v:v) methanol/water solvent with tert-n-butylammonium hydroxide as the supporting electrolyte. Acrylic acid is polarographically distinguishable from acrylamide in a neutral medium. Ethyl acrylate is an interference in the analysis. Acrylonitrile is removed, from interfering substances by treatment on mixed resin. The detection limit of acrylamide monomer by this technique is less than 1 ppm.

The recovery of acrylamide from the polymer is totally dependent on the extraction efficiency and the physical chemical interaction, e.g. absorption, between the monomer and the polymer. In evaluating the extraction procedure for an unknown polymer, it is imperative that the polymer extraction solution be spiked with a known quantity of acrylamide monomer and its recovery through extraction and resin treatment quantitated. Recovery of spiked acrylamide to polyacrylamides was greater then 90% with the extraction and resin treatment reported herein. The overall precision of the analysis was $\pm$ 5%.

The acrylamide reduction peak is well-defined and well-resolved from the background, no difficulty is encountered in either the detection of measurement of this peak. The differential pulse polarographic acrylamide reduction current is directly proportional to concentration as shown in Table 58. The polarographic detection limit for acrylamide in a clean system is less then 1 pg acrylamide/ml. Even at this low concentration, the acrylamide reduction peak is well-defined and resolved from the background.

The reduction of the acrylamide monomer occurs at ca. - 2.0 V vs. SCE. In this region of the polarogram, reduction of alkali cations occurs. The mixed resin treatment is used to remove polarographically interfering ionic species, such as sodium and potassium cations, which are major interferences. This is shown clearly in Fig. 36(a) which shows a polarogram of a methanolic solution containing sodium ions and acrylamide in a weight ratio of 4:1. After a mixed resin treatment for 20 minutes all the interfering sodium species are removed as seen in Fig 36(b), no acrylamide loss is detectable. Acrylic acid monomer does not interfere in the determination of acrylamide. Fig. 36(c) shows a polarogram of acrylic acid and acrylamide in an 80/20 methanol/water solution with tetra-n-butylammonium chloride as the supporting electrolyte. In this medium, the reduction of the associated acid occurs at ca. - 1.7 V vs SCE, 0.3 V more positive than the acrylamide reduction. The reduction peak of acrylic acid is easily resolved from acrylamide. The reduction current of this peak is directly proportional to the associated acid concentration. Acrolein, acetone, vinyl benzyl alcohol, vinyl benzyl chloride, also, do not interfere. Acrylonitrile, non-ionic species, nitrilotrispropionamide and some esters of acrylic acid do interfere.

<u>High performance liquid chromatography.</u> Brown[482] has described a high performance liquid chromatographic method for the determination of acrylic acid monomer in polyacrylates in which a known mass of polymer (500 mg) is added to 50 ml of a methanol:distilled water mixture (1 + 1) and allowed to stand overnight to complete the extraction of the monomer from the polymer. Various mixtures of methanol and distilled water (range 1 + 99 to 99 + 1, 50 ml) are added to the polymer for 1, 2, 8 and 24 hours.

An aliquot (12 ul) of the sample is injected into the liquid chromatographs Rheodyne valve and chromatographed under the following conditions: column Whatman PXS 10/25 um PAC (250 x 4 mm i.d.); mobile phase, 0.01% V/V orthophosphoric acid in distilled water; flow-rate, 4 ml min^{-1}; pressure, 2000 lb in^{-2}; detector wave-length 195 nm; chart speed, 0.5 cm min^{-1}; and absorbance scale 0.02. According to the sensitivity required, duplicate aliquots of the sample (1-100 ul) are injected. The concentration of acrylic acid is found by comparison with previously prepared calibration graph of total absorbance versus original acrylic acid concentration.

Table 58 - Differential Pulse Polarographic Response to
Acrylamide Monomer

Acrylamide concn[a],ppm	Peak current[b] uA	Peak potential,V vs SCE	$i$peak/concn, uA/ppm
0	0.00	...	...
1	0.35	-2.01	0.35
2	0.68	-2.01	0.34
3	1.07	-2.01	0.36
4	1.37	-2.02	0.34
5	1.67	-2.02	0.33
11	3.61	-2.02	0.33
15	5.51	-2.02	0.37
21	7.44	-2.02	0.35
31	11.30	-2.03	0.36
41	15.00	-2.03	0.37
51	19.20	-2.03	0.38
96	37.3	-2.04	0.39

[a] 9.5 ml 80/20 methanol/water + 0.5 ml 1 N tetra-n-butyl-ammonium hydroxide.

[b] Peak current is calculated as current above baseline.

High performance liquid chromatography, using the reverse phase mode, has been used[483] to determine acrylamide monomer and related compounds, including methacrylonitrile in polyacrylamide. Water soluble compounds such as acrylamide and methacrylamide have sufficient lipophilic character such that they can be retained and separated on HPLC reverse-phase columns using water as the eluent. By employing a low-wavelength ultraviolet detector, these compounds can be measured with high sensitivity. The relative precision of the 95% confidence level for acrylamide is $\pm$ 7.5%.

In this procedure the polymer is extracted for 4 hours with 80 + 20 methanol/water and the extract injected on a Partisil[-10] OD5[-2] 4 x 250 mm reverse phase column. These extracts can also be examined by ion exclusion liquid chromatography under the following conditions[484].

Column - Partial 10 PAC (250 x 4.6 mm), octadecyl loading = 5, acrylonitrile retention for 5 minutes.

Mobile phase - 15% methanol, 85% methylene chloride.

Flow rate - 1 ml/min

Pressure - 500 psi

Detector - 240 nm

The acrylamide response is linear from 1 - 500 ppm in solution using a 20 ul injection. This equates to 0.02 - 10 ug acrylamide injected. Area response obtained from a computing integrator is also found to be linear. Sensitivity of acrylamide detection is about 0.1 ppm in solution based on a 20 ul injection.

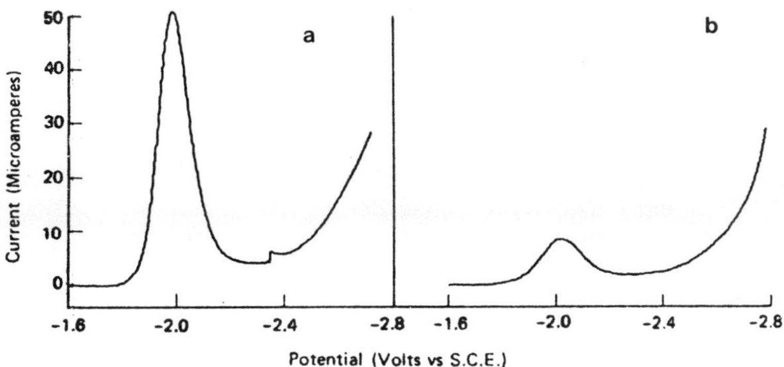

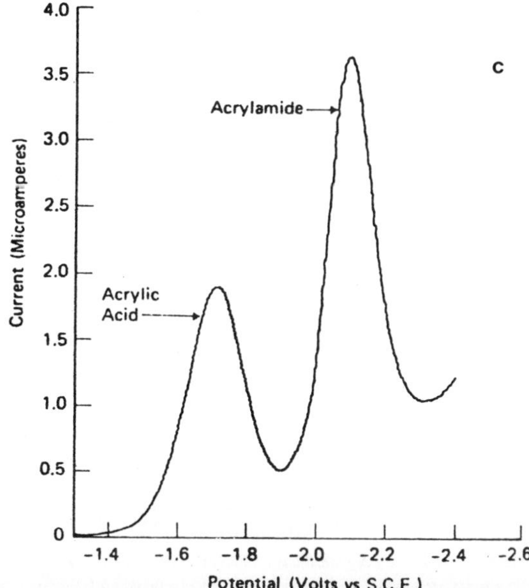

Figure 36 Removal of sodium ion by resin treatment; (a) 1 mH Na$^+$ +
10 ppm acrylamide; (b) After 20 min on mixed resin;
(c) differential pulse polarogram of a mixture of 10 ppm
acrylic acid + 10 ppm acrylamide in methanolic solvent
with tetra-n-butyl ammonium chloride as the supporting
electrolyte.

The retention times for acrylamide and related compounds are given
in Table 59. No known impurities are observed at the retention time of
acrylamide.

<u>5.2.5 Other Monomers (See Appendix 1)</u>

<u>5.2.6 Oligomers</u>

Oligomers are very low molecular weight polymers. This, polystyrene
usually contains low concentrations of monomer, dimer, trimer and tetramer
of the general formula $(C_6H_5 - CH = CH_2)_n$, where n = 1 to 4. Various
techniques have been used for the determination of oligomers including gel

Table 59 - HPLC Retention Times[a] for Acrylamide
and Related Compounds

Compound	Minutes
Acrylic acid	1.4
Beta-hydroxypropanamide	2.1
Acetamide	3.0
Acrylamide	5.4
Propanamide	7.3
Acrylonitrile	11.8
Methacylamide	18.0
Butanamide	20.8
Methacrylonitrile	46.0

[a] Partisil-10 ODS-2, water, 2.0 ml/min, 208 nm, 0.04 aufs.

permeation chromatography[485,486], (polyesters), thin-layer chromatography[487] (polystyrene and poly alpha-methyl styrene), liquid chromatography (p-alkylphenyl formaldehyde oligomers[490], polystyrene[491,492]), and gas chromatography[493].

High performance liquid chromatography has been used for the determination of oligomers in polyethylene terephthalate[488,489] and epoxy resins[488].

5.3 RESIDUAL VOLATILES

5.3.1 Introductory

Polymers, in addition to deliberately added non-polymeric additives such as described in Section 5.1 and monomers (Section 5.2), usually contain low concentrations of volatile constituents arising from their method of manufacture. The major types of substances in this category in addition to unreacted monomer and oligomers include non-polymerizable components of the original charge stock, residual polymerization solvents and water. In addition, residues of deliberately added volatiles may also be present, e.g. residual expanding agents in expanded polystyrene. The concentrations of these substances usually range from a few tens of parts per million to several hundred parts per million. Frequently, complex mixtures are present. Thus, the non-polymerizable fractions of styrene monomer (usually 1-2%) consist of several dozen aromatic hydrocarbons such as ethyl benzene. The polymerization solvent used in the manufacture of high-density polyethylene and polypropylene by the low-pressure catalysed route is usually a crude petroleum distillation cut with a complex composition.

It is important to be able to determine the concentrations of these substances for many reasons, two examples of which are the effects they have on the mechanical properties of polymers and the risk of tainting in the case of foodstuff- or beverage-packaging grades of polymers.

Polymers often contain substances of medium volatility such as residual monomers, residual polymerization solvents and expanding agents. In addition, when polymers are heated they may release volatiles as a result of the thermal degradation of either the polymers themselves or their additives or catalyst residues. These volatiles can have an important bearing on such properties as processability, the tendency to

form voids and, in the case of foodstuff-packaging grades, the possible
tendency to impart taste or odour to the packed commodity.

One way of identifying non-polymeric constituents of polymers is to
extract the polymer with a low-boiling-point solvent, remove the solvent
from the extract by evaporation or distillation and analyse the residue.
This procedure is, of course, inapplicable to the analysis of extracted
polymer constituents which are volatile enough to be lost during the
solvent-removal stage. Alternatively, an extract or a solution of the
polymer may be examined directly for volatile constituents by gas
chromatography, in which case losses of volatiles are less likely to
occur. In such a procedure, however, the large excess of solvent used for
extraction or solution might interfere with the interpretation of the
chromatogram, obscuring some of the peaks of interest. Trace impurities
in the solvent may also interfere with the chromatogram. Of course none
of these procedures is suitable for studying the nature of volatile
breakdown products which are produced only upon heating a polymer. For
the identification and/or determination of residual solvents in polymers
it is mandatory therefore to use solventless methods of analysis, i.e.
there must be no risk of confusing solvents in which the sample is
dissolved for analysis with residual solvents in the sample. Most methods
for the determination of residual solvents are based on the technique of
heating the solid polymer and examining the head-space over the polymer by
gas chromatography.

Due to their volatility and complex composition it is not surprising
that methods based on gas chromatography have emerged as being the most
suitable way of analysing for these parameters. Basically, three
different approaches have evolved in the application of gas
chromatography:

i)   solution of the polymer in a solvent and injection into a gas
       chromatograph,
ii)   heating the dry polymer and sweeping the volatiles released into a
       gas chromatograph using the carrier gas or heating the polymer in a
       closed system, then withdrawing the head-space with a syringe for
       direct injection into the gas chromatography, i.e. head-space
       analysis.

Some examples of these methods are discussed below:

<u>i)   Solution Analysis Methods</u>

<u>5.3.2   Expanding Agents in Polystyrene</u>

Expandable grades of polystyrene are manufactured by steeping
volatile polystyrene granules in a low boiling hydrocarbon solvent until
the polymer becomes saturated with the solvent. n- and iso pentane are
commonly used but, in addition, isobutane, isohexane, neohexane, isooctane
and cyclopentane have been used. When the granules are put in a mould and
are treated with steam, they undergo an expansion to many times their
original volume and the expanded granules coalesce. This process is used
for the manufacture of insulating polystyrene board and insulated hot
beverage vending vessels. It is important to be able to determine the
concentration of residual volatiles in the original and the expanded
polymer and methods for carrying out this analysis are discussed below.

In a typical method for n- and iso-pentane, the polystyrene (2 g) is
dissolved in 25 ml propylene oxide containing 1% internal standard. A
portion of the solution is chromatographed and the amounts of the pentanes

determined from previously calculated calibration factors, using the 2:2 dimethyl butane or cyclohexane as the internal standard. To calibrate, weigh in turn into a 100 ml volumetric flask 0.40 ml iso-pentane, 0.50 ml n-pentane and 1 ml internal standard. Seal the flask with a serum cap during each weighing. Dilute the mixture to 100 ml with propylene oxide, seal with a serum cap and mix thoroughly. Chromatograph 10 ul and measure either the peak heights or the integrated areas of the iso-pentane, n-pentane and internal standard.

Then

$$\% \text{ w/w isopentane} = \frac{25 \times W1 \times W4 \times P3 \times P4}{W3 \times W5 \times P1 \times P6}$$

$$\% \text{ w/w n-pentane} = \frac{25 \times W2 \times W4 \times P3 \times P5}{W3 \times W5 \times P2 \times P6}$$

where

W1, W2 - respectively are weights of iso and n-pentane in calibration blend.

W3, W4 - respectively are weights internal standard in calibration blend and sample solution.

W5 - weight of sample.

P1, P2, P3 - respectively, peak heights or area of iso-pentane, n-pentane and internal standard in calibration blend.

P4, P5, P6 - respectively, peak height or area of iso-pentane, n-pentane and internal standard in sample solution.

Fig 37 shows a gas chromatogram obtained in the analysis of an expandable polystyrene sample containing n- and iso-pentane. In this determination, cyclohexane ws used as the internal standard.

<u>5.3.3 Aromatics in Polystyrene</u>

Benzene, toluene, ethyl benzene, n-propyl benzene, cumene, isobutyl benzene, o-xylene, alpha-methylstyrene can all be quantitatively determined in amounts down to 10 ppm in polystyrene by a method based on solution of the polymer in propylene oxide containing n-undecane as internal standard and gas chromatography under the following conditions, (see Fig. 33)

Column	Copper tube (15ft x 2/16 in i.d.) packed with 10% wt/wt Carbowax 15-20 M on 60-72 BS mesh acid-washed Celite.
Gas flows	Helium, 30 psig, rotameter = 10.0 (100 ml/min). Hydrogen, 12 psig, rotameter = 10.0 (75 ml/min). Air, 7 psig, rotamter = 10.0 (650 ml/min).
Temperatures	Injection 155°C Column 80°C Detector 125°C Flame 200°C

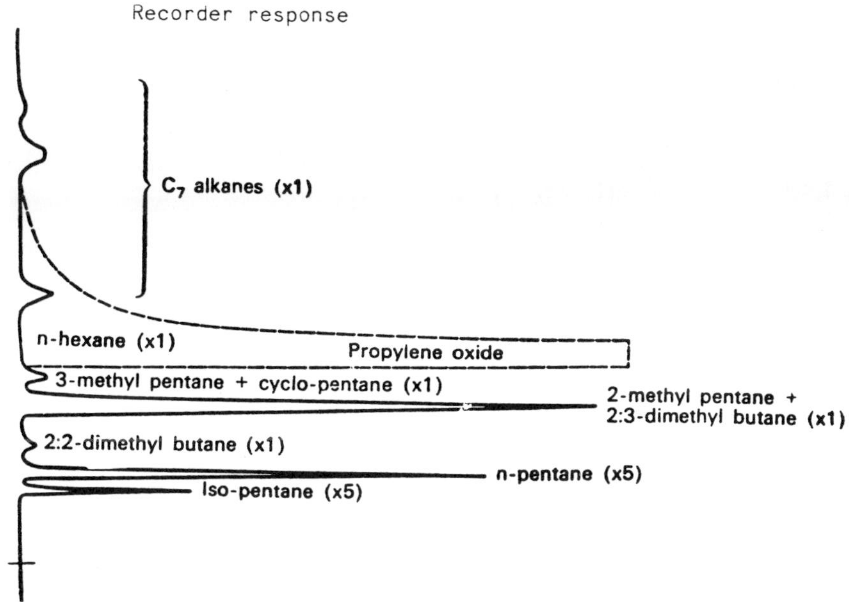

Figure 37 Gas chromatogram of pentanes in polystyrene.

ii) Head-space Analysis methods

The apparatus described by Crompton494-496 and discussed in Chapter 5.2 involving heating the polymer to 300°C in a small tube and sweeping the volatiles liberated into a gas chromatograph has been used to identify the volatiles released from polyolefins and polystyrene.

5.3.4 Aliphatic Hydrocarbons in Polyolefins

Food and drink containers extruded or moulded from polyethylene sometimes possess unpleasant odours which are likely to taint the packaged product and are unacceptable to the consumer. In one such case it was found that, by heating a sample of an odour-producing polyethylene for 15 min at 200°C under helium, the chromatogram of the liberated volatiles contained certain peaks which were absent from the corresponding chromatogram from a polyethylene which produced non-odorous food containers. The temperature of 200°C was chosen to stimulate extrusion temperature. The two chromatograms are shown in Fig 38, from which it may be seen that components A, B, D and I are present in the odorous sample but are absent from the non-odorous sample. These substances were always associated with the odorous polyethylene.

Polyethylene is not appreciably soluble in organic solvents and the determination of existing volatile impurities cannot be conveniently carried out by the analysis of solutions of the polymer. Many organic solvents will extract such volatiles from polyethylene, but these are likely to be lost during the solvent-removal stage prior to gas chromatographic analysis. A further problem is that extraction procedures

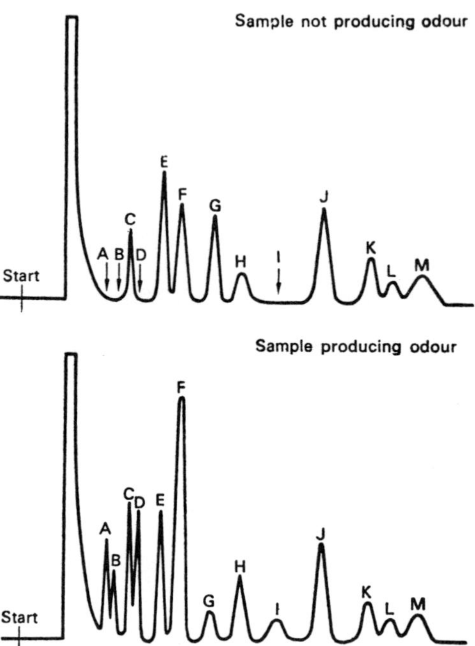

Figure 38 Gas chromatograms of volatiles liberated from odorous and non-odorous polyethylenes at 200°C for 15 min in helium. Chromatographed on 200 ft x 1/16 in i.d. dibutylphthalate coated copper column at 30°C and 100 ml per min helium flow with flame ionization detector.

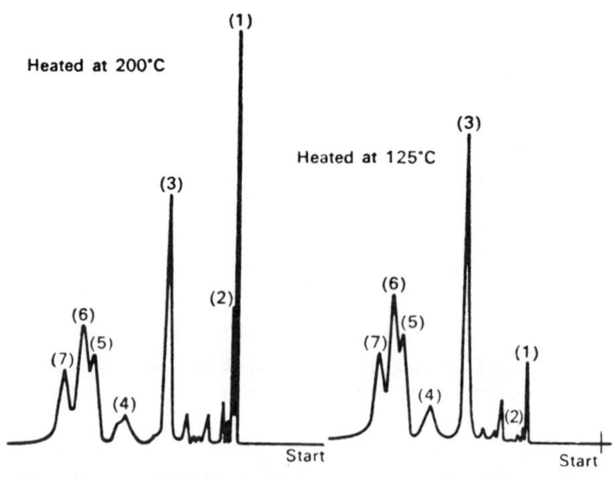

Figure 39 Gas chromatogram of volatiles liberated from polyethylene heated at different temperatures for 15 min in air. Chromatographed on 200 ft x 1/16 in i.d. dibutyl phthalate coated copper column at 30°C an 100 ml/min helium flow, with flame ionization detector.

cannot be used for investigating those volatiles that are not inherently present but are produced only when the polymer is heated. The following examples illustrate the use of this method for examining both types of volatiles released from polyethylene.

The technique was used to study the effect of temperature on the nature of the volatiles released and to investigate this polyethylene was heated for 15 min in air at 125°C, 150°C, 175°C and 200°C and the volatiles were examined by gas chromatography. Figure 39 shows the chromatograms of the low molecular weight paraffin hydrocarbons released at 125°C and 200°C and Table 60 gives the peak height ratios of the major components (1-7) release at all four temperatures. Replicate runs indicated good reproducibility.

The results in Table 60 show that the only significant variation of peak height ratios with temperature occurs when component 1 is compared

Table 60 - Peak Height Ratios of Volatiles Liberated by
Polyethylene at Various Temperatures

	Peak Height Ratio			
Component Number Ratio	125°C	150°C	175°C	200°C
1:2	4.4	5.4	3.1	3.1
1:3	0.3	0.3	0.3	1.7
1:4	2.1	3.2	2.4	14.6
1:5	0.7	1.0	0.8	4.7
1:6	0.5	0.8	0.6	3.5
1:7	0.9	1.2	1.0	5.9
2:3	0.06	0.06	0.08	0.06
2:4	0.5	0.6	0.8	4.6
2:5	0.2	0.2	0.3	1.5
2:6	0.1	0.1	0.2	1.1
2:7	0.2	0.2	0.3	1.9
3:4	8.1	9.6	9.1	8.3
3:5	2.9	3.1	3.1	2.7
3:6	2.1	2.3	2.3	2.0
3:7	3.4	3.7	3.8	3.4
4:5	0.4	0.3	0.3	0.3
4:6	0.3	0.2	0.3	0.2
4:7	0.4	0.4	0.4	0.4
5:6	0.7	0.7	0.7	0.8
5:7	1.2	1.2	1.2	1.3
6:7	1.7	1.6	1.6	1.7

with components 2-7 and component 2 is compared with components 4-7. As the temperature increases to 200°C components 1 and 2 increase while components 3-7 decrease. Components 1 and 2 were eluted in the C_2-C_4 hydrocarbon region and components 3-7 were eluted coincident with major components of the polymerization solvent known to be used in the polyethylene manufacturing process. These observations suggest that between 125°C and 200°C there is some thermal degradation of the polymerization solvent to C_2-C_4 hydrocarbons.

Thus the major volatile components of this polyethylene are residual polymerization solvent. When examining polyethylene for existing volatiles, it is necessary to use as low a temperature as possible for liberating the volatiles if thermal degradation is to be avoided. A temperature of 125°C appears to be suitable.

Roper[497] has discussed the problem of determining very low concentrations of volatiles in polymers. Methods for the determination of such volatiles frequently include application of heat to the sample and the sweeping action of an inert gas to separate the volatile components from the polymer. The volatiles are then analysed by gas chromatography. When there is a low concentration of volatile material, it is advantageous to concentrate it in order to improve the shapes of the chromatographic peaks.

To achieve this Roper[497] employed a trap-tube, shown in Fig 40(a), the capillary portion of the tube is packed with a gas chromatographic column packing consisting of 20-30% of a suitable liquid phase on a granular diatomaceous type support. The tube is fitted with a suitable hypodermic needle so that it can be connected to the gas chromatograph through the injection port. The apparatus is arranged as in Fig. 40(b). The gas chromatograph is equipped with a temperature programme and any suitable column which will separate the various volatile components.

The polymer sample to be analysed is weighed into the trap-tube, the sample size being chosen to give suitably sized peaks for measurement. The tube is connected to an inert-gas stream by means of butyl rubber tubing and heated to the proper temperature while the gas stream sweeps the volatile material into the packed section of the tube, which is usually cooled with dry ice. The temperature, sweeping rate, and sweeping time should be determined experimentally and will vary with the type of polymer being investigated. Melting the polymer is often necessary to rid it of volatiles. A micro combustion furnace is a suitable heater for the trap tube.

After the volatiles have been adsorbed in its packed section, the trap-tube is disconnected from the butyl rubber tube and moved to the chromatograph. With valve A open and valve B closed, the tube is connected to the butyl rubber tube with the heater positioned away from the capillary end. Then, with valve A closed too, the hypodermic needle is inserted in the injection port and valve B is opened. The heater, already at the proper temperature, is then moved to the capillary portion of the tube where the heat and carrier gas flow sweeps the volatiles from the packing through the hypodermic needle and into the gas chromatograph. In some cases it is necessary to programme the oven temperature of the chromatograph to get a suitable chromatogram.

It is advantageous to use an electronic integrator to determine the response in counts per microgram for the various components to be measured. With this information, the percentage of each volatile component of the sample can be calculated from the number of integrator counts in each peak.

In an alternative procedure the polymer is heated and the head-space atmosphere analysed by gas chromatography[498]. The film or sheet, together with internal standard, is placed in a 250 ml sealed container and heated at 100°C for 90 minutes.

The following results were obtained for the determination of toluene and ethyl acetate on adjacent pieces of a polythene adhesive-laminated to polypropylene double-coated with saran. The solvents originate from the adhesive.

									Mean	SD
Toluene/mg m^{-2}	127	135	119	127	140	122	130	119	127	7
Ethyl acetate/ mg m^{-2}	136	145	124	141	136	133	138	128	135	6

Results as follows were obtained for the determination of toluene on adjacent pieces of printed film (polypropylene with a single saran coating). The toluene originated from the printing ink.

Toluene/mg^{-2}	11.9	11.4	12.3	11.9	12.5	11.9	12.3

Mean 12.0 mg m^{-2}, standard deviation 0.34 mg m^{-2}.

This method has been used to determine many different solvents in several different substrates. The solvents include ethanol, ethyl acetate, ethyl methyl ketone, 2-ethoxyethanol, propan-1-ol and toluene. The substrates include polyethylene, polypropylene and cellophane, which occur individually, coated with saran or combined in laminates.

5.3.5 Aromatics in Polystyrene

In Figure 41 is a chromatogram illustrating the presence of aromatic volatiles in polystyrene obtained by the Crompton heated tube - gas chromatographic technique discussed in Section 5.2.

A variant on this method which has been applied to polystyrene is depicted in Figure 42. The gas chromatograph carrier gas is fed to the instrument via a gas sampling valve, coupled as shown. The sampling valve is attached to a glass tube, also as shown, and the latter is heated by means of a removable hot copper block.

The heater block is lowered and the sample valve turned to the "by-pass" position. 5-50 mg of the polymer is weighed into a rimless ignition tube which connects to the apparatus. The sample valve is turned to the "analysis" position for 1 minute to remove air, then the sample valve turned to the "by pass" position and the heater block (at 240°C) raised to surround the ignition tube which is then heated for 5 minutes. The sample valve (Fig. 42)is then turned to the "analysis" position to allow the chromatogram to develop.

Determine (F) for each aromatic hydrocarbon as follows:

$$F = \frac{\text{wt of hydrocarbon}}{\text{wt of n- undecane}} \times \frac{\text{pk. ht. of n-undecane}}{\text{pk. ht. of hydrocarbon}}$$

From the analysis chromatogram calculate the concentration of each aromatic hydrocarbon as follows:

$$\% \text{ wt/wt hydrocarbon} = \frac{\text{pk. ht. of hydrocarbon}}{\text{pk. ht. of n-undecane}} \times \frac{F \times P \times 10}{\text{wt of sample}}$$

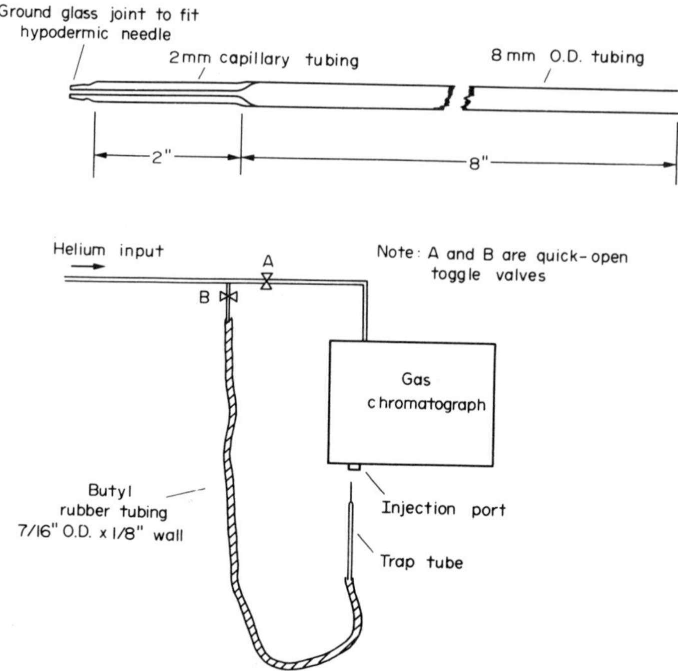

Figure 40   Head space method for volatiles in polymers: (a)
borosilicate glass trap tube; (b) arrangement of
apparatus.

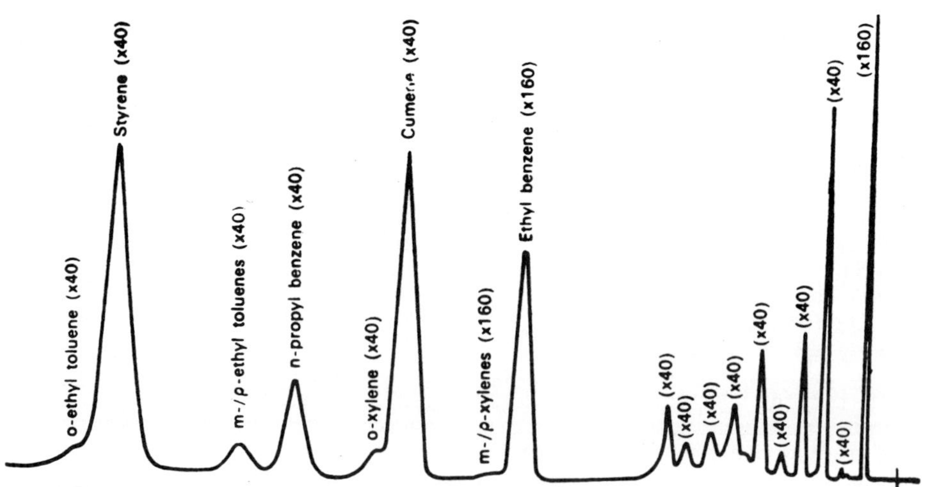

Figure 41   Gas chromatograms of volatiles liberated from different
polystyrenes at 200°C for 15 min in helium.
Chromatographed on 15 ft x 2/16 in 10% carbowax 15-20 M
on 60-72 Celite at 90°C and 100 ml/min helium flow with
flame ionization detector.

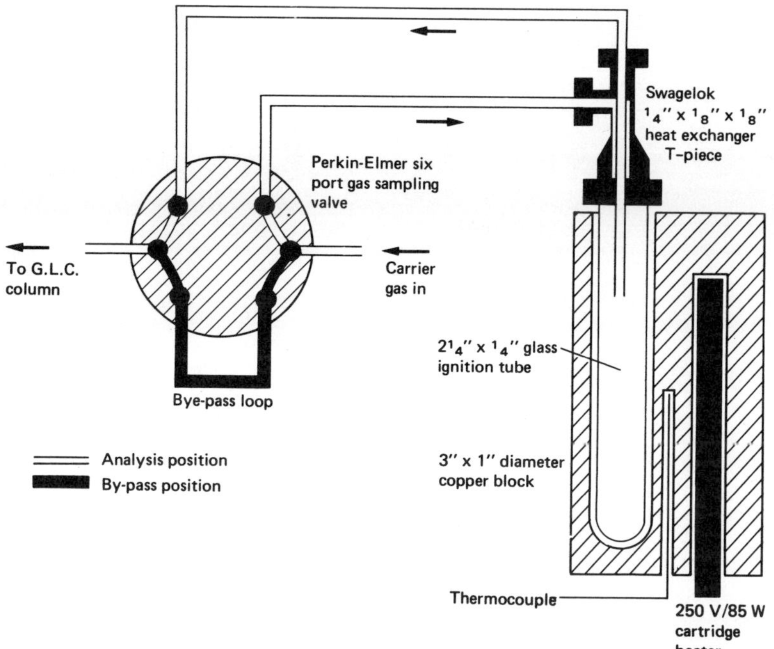

Figure 42   Apparatus for liberating expanding agents from
expandable and expanded polystyrene.

where P = % wt/vol concentration of n-undecane in the polymer solvent.

Column   Copper tube (15ft x 3/16 in i.d.) packed with 10%
                          wt/wt Carbowax 15-20 M on 60-62 BS mesh acid-
                          washed Celite.

Gas flows   Helium, 30 psig, rotameter = 10.0 (100 ml/min).
                          Hydrogen, 13 psig, rotameter = 10.0 (75 ml/min).
                          Air, 7 psig, rotamter = 10.0 (650 ml/min).

Temperatures   Injection 155°C Column 80°C
                          Detector   125°C Flame   200°C

Recorder   Honeywell-Brown, 1 mV full scale deflection, 1
                          second response, 10 in/hr chart speed.

5.3.6 Diaminotoluenes in Polyurethanes

     Guthrie and McKinney[399] have described a thin-layer chromatography –
fluorimetric method for the determination of 2,4 and 2,6 diaminotoluene in
methanol extracts of flexible polyurethane foams. The precision of this
method at the 20 ppm level is ± 30%. Diaminotoluenes can be detected in
amounts down to 1 ppm. The thin-layer plates are developed with
120:33:20:7 v/v chloroform: ethyl acetate, ethanol: glacial acetic and
after drying are sprayed with 0.015% fluram in acetone. The separated
aminotoluene spots can be visualized under a uv lamp spots and are
evaluated within 1 hour of development using an Aminco Scanner with
excitation and emission wavelengths of 390 and 500 nm respectively.

Table 61 - Residual Volatiles in Polymers

Volatile	Polymer	Method	Reference
Styrene o-xylene	polystyrene	head-space analysis, glc	500
Styrene	polystyrene	head-space analysis, glc	501-508
Dichloromethane	polycarbonate	head-space analysis	500
Styrene ethyl benzene	polystyrene	solution, glc	
Styrene	polystyrene	solution in o-dichlorobenzene or methylene dichloride, glc	509
Styrene	polystyrene	solution in DMF, glc	510
Styrene, o-methyl styrene	polystyrene	solution in propylene oxide, glc	511
Benzene, toluene, ethyl benzene, xylene, cumene, propyl benzenes, ethyl toluenes, butyl benzenes, styrene	polystyrene	solution in THF, glc	512 513-515
Alkanes	polystyrene	heating sample, glc	
Volatiles	styrene-buta diene	glc	516,517
Volatiles	PVC	glc	518,519
Volatiles	poly alpha-methyl styrene	glc	517,520
Volatiles	polycarbonates	glc	521
Volatiles	polyethylene		523
Volatiles	polypropylene	glc	524
Adhesives	rubber	glc	535-529
Methylene bis (aniline)	polyurethane	hplc	
Volatiles	polyether coated film	glc	531

| Propylene oxide, epichlorohydrin | polyesters | spectrophotmetric | 532 |
| Diols, adipic acid and cyclomonodiol adipates | aliphatic poly-esters | glc-ms | 533 |

5.3.7 Volatiles

Other methods for the determination of residual volatiles in polymers are reviewed in Table 61.

5.4 WATER

Conventional weight loss methods for determining water in polymers have several disadvantages, not the least of which is that any other volatile constituents of the polymer such as solvents, dissolved gases, volatile substances produced by decomposition of the polymer or additives therein are included in determination.

Specific methods for the determination of water in polymers can be based on the principle of heating a weighed amount of polymer in a boat in a tube furnace through which flows a gentle purge of nitrogen[534]. The wet nitrogen passes into an automated Karl Fisher titration unit which estimates the water as it is released from the polymer. In addition to being absolutely specific for water this method has an additional advantage in that it can provide information on the effects of temperature and time on the release of water from the polymers.

5.4.1 Water in Polyolefins, Vinyls and Acrylics

Jeffs[535] has described a rather complicated piece of equipment for the determination of water and other volatiles in vinyl, acrylic and polyolefin powder polymers. The instrument is shown diagrammatically in Fig. 43, and consists essentially of a sample tube forming an external loop, connected to a gas chromatograph. This loop can be isolated and the sample heated to the required temperature. After an initial heating period the volatile constituents liberated from the sample are "flushed" on to the chromatographic column by a flow of carrier gas through, or over, the sample, and the required components separated and determined quantitatively. A pneumatic switch valve located in the chromatographic oven to prevent the condensation of volatile constituents within the valve, and a split heater mounted on a horizontal travel in a plane at right angles, to the sample tube, are essential parts of the apparatus. The instrument is semi-automatic. The carrier gas flows through the copper sulphate in tube D, which imparts a constant amount of water (about 3 ppm w/v) to the helium. The "wet" carrier gas prevents the gas-flow lines from "drying out".

The pneumatic sample valve F, operates as a switch valve, directing the carrier-gas flow either around the internal loop, G, or the external loop, H. The pilot valve, N_1 operates sample valve, F. The sample split heater, J, consists of a cylindrical aluminium block 6 in. long with a 9 mm hole through the centre. The block is split axially and the two halves hinged. Each half of the block contains a thermocouple pocket to accommodate the thermocouple, P. The cartridge heaters are supplied by a

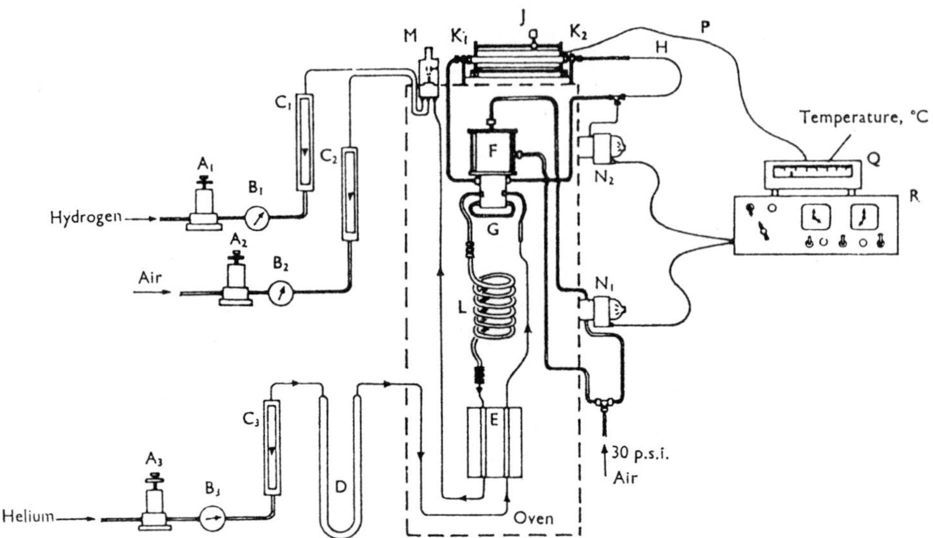

A₁, A₂ and A₃ = Edwards VPC I pressure controller

B₁, B₂ and B₃ = Pressure gauges 0 to 30 p.s.i.

C₁, C₂ and C₃ = Rotameter-type flow gauges

D = Clear plasticised PVC tubing (5 foot long × ¼ inch bore), packed with copper sulphate crystals, $CuSO_4.5H_2O$, >44 mesh

E = Katharometer

F = Pneumatic sample valve (Pye Cat. No. 12900); fitted with P 9904 change-over block

G = Internal loop

H = External loop made in part of 18-gauge stainless-steel capillary tubing

J = Sample split heater

K₁ and K₂ = Straight reducing couplings, captive seal type, for ¾ to ¼ inch o.d. tubing (Drallim, Cat. No. L/50/D/B)

L = Chromatographic column

M = Flame-ionisation detector

N₁ and N₂ = Electrically actuated 3-port pilot valves (Martinair, Type 557C/1Z)

P = Nickel - chromium/nickel - aluminium thermocouple embedded in the split heater

Q = Electronic temperature controller, proportional type

R = Sample-valve time-delay unit, containing three synchronous timers and a semi-conductor, proportional energy controller for the sample heater

Figure 43 Details of general purpose instrument for gas chromatographic determination of water and volatiles in polymers.

semiconductor energy controller contained in R, which is controlled by a galvanometer, two-position, temperature controller, Q. The heater is mounted on a horizontal travel in a plane at right angles to the sample tube. A jig for mounting the sample tubes is also part of the heater assembly. The 3/8" coupling K₁, is brazed to its mounting bracket, which is rigidly attached to the heater base. This coupling is accurately centred with the central hole through the aluminium block. The coupling, K₂, is brazed to the flexible carrier-gas inlet tube, and rests loosely in the second mounting bracket. The 3/8" couplings are supplied with neoprene, or butyl rubber captive seals.

The sample-heater assembly is placed on top of an oven, so that the L-bend of 1/8" o.d. S/S tubing disappears almost immediately into an opening on the top of the oven. In practice, both the inlet tube (6" long, 1/4" o.d and 1/8" i.d. copper) and the exit tube (1/8" o.d.) are wrapped with heating tape and lagging to maintain the temperature of the whole assembly at about 100°C. The inlet tube is wrapped to a length of 4-5" and the L-bend of the exit tube is wrapped to a point 3" inside the oven. Both tubes are wrapped up to, and including, the 1/4" thread of the Drallim coupling, leaving only the centre nut and the 3/8" coupling nut exposed so that the sample tubes can be readily changes. N₂ acts as a pressure release valve to the external loop H.

The apparatus shown in Fig. 43 is fitted with katharometer and flame-ionization detectors. Although only one detector is necessary for any one specific method, e.g. a katharometer for the determination of water in polymer powder, it is invaluable to have both available (with separate recorders) to establish the conditions, i.e. in the above case to ensure that no organic components are being eluted at the same time as water, and thus contributing to the peak measurement. Figure 44(a) shows chromatograms, obtained simultaneously from the katharometer and the flame-ionization detectors on a partially dried poly(vinylchloride) powder.

Figures 44(b) and 44(c), show some typical results obtained when this method is applied to vinyl polymers and polyolefins. The peaks due to water and monomers are clearly visible.

Jeffs[535] recommends that before carrying out any quantitative work on the volatile constituents obtained from a polymer powder, a preliminary gas-chromatographic investigation should be carried out, of their complexity. For example, Fig. 44(a) (flame-ionization detector trace) shows nine components other than air, water and the original monomer. These components are chlorinated hydrocarbons, such as 1,1- and 1,2-dichloroethane and cis- and trans-dichloroethylene, that are present as impurities in the original monomer.

Although these impurities are present only in ppm amounts in the original monomer they can be readily detected in the polymer. If water is to be determined, then a suitable column has to be chosen so that water is eluted free from organic constituents. At this stage the recording of simultaneous chromatograms with two different detectors is invaluable.

5.4.2 Water in Polyesters

Hoffman[536] has described a procedure for the determination of water in polyesters. In this method, water is allowed to react quickly with a mixture of hexamethyldisilazane and trimethylchlorosilane (2:1) in the presence of pyridine to form hexamethylidisiloxane.

$$2(Me - Si_2)_2NH + 3H_2O + 2Me_3SiCl = 3Me_3Si - O - SiMe_3 + 2NH_4Cl$$

The hexamethyldisiloxane is separated from other silanized active hydrogen compounds and determined by gas chromatography using an internal standard. The sensitivity and specificity is good. The limit of detection is 50 ppm and the relative standard deviation at the 200 ppm level is 13%.

Squirrel[537] has also reported on the fact that with some polymers water, in addition to the initial free water content of the polymer, is produced during heating. He observed this in the case of polyethylene terphthalate and for this reason, prefers to determine water by Karl Fischer titration.

Dry methanol is placed in the sample flask, excess of dilute Karl Fischer reagent (1 ml = 1 mg of water) is added and the contents are boiled under reflux for 20 mins to dry the apparatus. Slightly wet methanol is then added to the flask and the contents are immediately titrated to a visual end-point by means of the Karl Fischer reagent. The

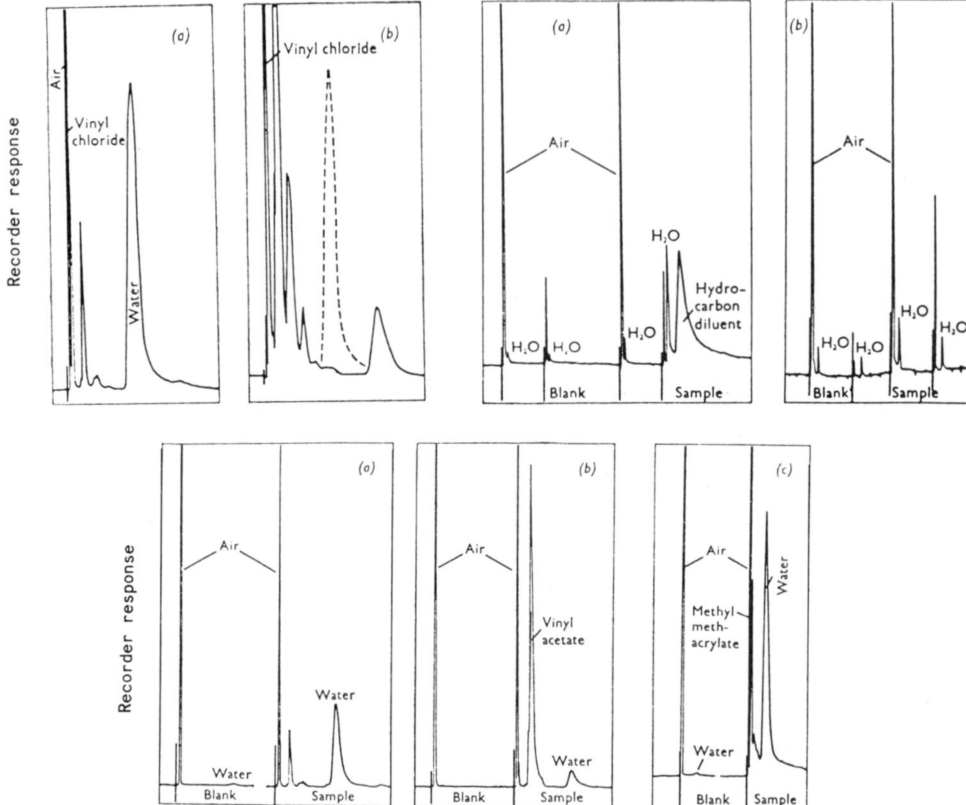

Figure 44   Determination of water in polymers: (A) gas
chromatograms obtained simultaneously with (a)
katharometer and (b) flame ionization detectors in a
partially dried PVC powder.   In (B) the dotted line
represents the portion of the water peak as transposed
from (a).   (B) gas chromatogram obtained for (a) PVC
powder, (b) PVC-PVA copolymer powder, (c) acrylic
moulding powder; (C) gas chromatogram obtained for (a)
polypropylene powder, (b) high pressure polyethylene
powder.

flask and contents are then refluxed for 6 hours and any water picked up
during this time is again titrated with the Karl Fischer reagent.   This
establishes that the equipment is giving a low blank value, which should
be less then the equivalent of 2 ml of Karl Fischer reagent.

For the analysis of the sample the apparatus is again pre-dried and
the sample introduced via the 70 ml top cup.   The cup is filled to a 50 g
mark with polymer granules and then the latter are allowed to drop into
the flask by releasing the clip on the widebore rubber tubing.   In this
way the exposure of the sample to the atmosphere is minimal.   The 6 hour
reflux period is then repeated and the water content of the sample is
calculated from the difference between the titre for sample and blank
experiments.

5.4.3 Water in Other Polymers

Other methods for the determination of water in polymers are reviewed in Table 62.

5.5 CATALYST REMNANTS AND OTHER IMPURITIES

5.5.1 Introductory

Two types of catalysts used in polymer manufacture are metallic compounds such as aluminium alkyls and titanium halides used in low pressure polyolefin manufacture and organic peroxides such as are used in the manufacture of polystyrenes. As the presence of residual catalysts can have important effects on polymer properties it is important to be able to determine trace elements which reflect the presence of these substances, (Sections 5.5.2 - 5.5.7).

Elemental analysis of polymers. Elements occurring in polymers and copolymers can be divided into 3 categories.

1. Elements which are a constituent part of the monomers used in polymer manufacture, such as nitrogen in acrylonitrile used in the manufacture of, for example, acrylonitrile-butadiene-styrene terpolymers, (i.e. major constituents, see Chapter 4.3).

2. Elements which occur in substances deliberately included in the polymer formulation such as, for example, zinc stearate which is incorporated in some polystyrene formulations, (Chapter 5.1.10).

3. Elements which occur as adventitious impurities in the polymer. For example, during the manufacture of polyethylene by the low-pressure process polymerization catalysts such as titanium halides and organo-aluminium compounds are used, and the final polymer would be expected to, and indeed does, contain traces of aluminium, titanium and chlorine residues, (Chapter 5.5.2 - 5.5.6).

Different types of techniques have been evolved for trace elemental analysis in polymers. Generally speaking the techniques used come under two broad headings.

1. Destructive techniques: these are techniques in which the sample is decomposed by a reagent such as those discussed below and then the concentration of the element in the extract is determined by a physical technique such as atomic absorption spectrometry, emission spectrography, flame photometry, polarography or any one of the increasing repertoire of physical methods of analysis.

2. Non-destructive techniques: these include techniques such as X-ray fluorescence and neutron activation analysis, in which the sample is not destroyed during the analysis.

These techniques are now briefly discussed with some examples of their application to particular problems.

Destructive Techniques

The classical techniques are generally based on one of three possible approaches to the analysis: i) dry ashing of the polymer with or without an ashing aid, followed by acid digestion of the residue, alternatively acid digestion of the polymer without prior ashing, ii)

Table 62 - Determination of Water in Polymers

Polymer	Procedure	Reference
PVC	Devolatilization - gas chromatography	538
Polypropylene	Heating under nitrogen - Karl Fischer titration	539
Anionic resins	Oven-drying - Karl Fischer titration	540
Polyamide and polyethane diol terephthalate		543
		545
Polypropylene	Removal of water in vacuo or a stream of nitrogen then Karl Fischer titration	544
Several polymers		541, 542

fusion of the polymer with an inorganic compound to effect solution of the elements, and finally, iii) bomb or oxygen flask digestion techniques.

Another method for avoiding losses of metals during ashing is the low temperature controlled decomposition technique using active oxygen. This method has been studied in connection with the determination of trace metals in PVC, polypropylene and polyethylene terephthalate[546].

5.5.2 Trace Metals in Polymers

One source of traces of metals in polymers are neutralizing chemicals added to the final stages of manufacture to eliminate the effects of acidic catalyst remnants on polymer processing properties (e.g. hydroscopicity due to residual chloride ion). A case in point is high density polyethylene and polypropylene produced by the aluminium alkyl-titanium halide route which is treated with sodium hydroxide in the final stages of manufacture. Sodium, potassium or lithium have been used in this context.

A technique, which involves combustion of the polymer under controlled conditions in a platinum crucible - followed by solution of the residual ash in a suitable aqueous reagent prior to final analysis by spectrophotometry, is of limited value. A quite complicated and lengthy ashing programme is necessary in this technique to avoid losses of alkali metal during ignition.

Time from start		
0-1 hour,	heat to 200°C
1-2 hours,	hold at 200°C
3-5 hours,	heat to 450°C
5-8 hours,	hold at 450°C

After ignition the residue is dissolved in 5 ml warm IN. nitric acid and made up to a standard volume prior to evaluation by flame photometry or atomic absorption spectrometry. Alternatively, the polymer is ashed overnight at 500°C with sulphur and a magnesium salt of a long chain fatty acid (Magnesium A.C. dope, Shell Chemical Co. Ltd.), and the ash mixed with twice its weight of carbon powder containing 0.1% palladium prior to emission spectrographic evaluation of the sodium:palladium 330.3/276.31 line pair.

The results in Table 63 show clearly that flame photometry following dope ashing at 500°C gives a quantitative recovery of sodium relative to results obtained by non-destructive method of analysis vis neutron activation analysis. Direct ashing without the magnesium ashing aid at 500°C causes losses of 10% or more of the sodium whilst direct ashing at 800°C causes even greater losses.

Dry ashing in platinum has been found to give reasonably good results in the determination of low concentrations of vanadium in ethylene propylene copolymer; 10 g of polymer is ashed in platinum by charring on a hot plate followed by heating over a Meker burner. Dilute nitric acid is added to the residue and any residue in the crucible dissolved by fusion with potassium persulphate. The vanadium is determined spectrophoto-metrically by the 3,3'-diaminobenzene method[547]. Table 64 compares results obtained by this method with those obtained by neutron activation analysis, which in this case can be considered to be an accurate reference method. Quite a good agreement was obtained between the two methods for samples containing 1 ppm vanadium.

Henn[548] has reported on a flameless atomic absorption technique with solid sampling for determining trace amount of iron, copper and chromium in polymers such as polyacrylamide with a detection limit of approximately 0.01 parts per million.

It has been shown[549,550] in studies using a radioactive copper isotope that, when organic materials containing copper are ashed, losses of up to some 10% of the copper will occur due to retention in the crucible, and this could not be removed by acid washing. Virtually no retention of copper in the silica crucible occurred, however, when copper was ashed under the same conditions in the absence of added organic matter. This was attributed to reduction of copper to the metal by organic matter present, followed by partial diffusion of the copper metal into the crucible wall. This effect also occurs with elements other than copper, and would be responsible for the return of low metal recoveries in polymers. One method of reducing or eliminating such losses is to ignite the polymer after it has been initially mixed with an inorganic salt solution such as magnesium nitrate which, upon ignition, provides a matrix of magnesium oxide in which the element being determined is retained and thereby prevented from being lost in the crucible material. The metallic oxide residue is then dissolved in a suitable acid and analysed by conventional techniques. To obtain intimate contact between polyethylene powder and the ashing acid during the preliminary mixing it was necessary to dissolve the magnesium nitrate in a 70:30 v/v mixture of distilled water and redistilled methyl alcohol. Following ignition at 600°C any residual carbon particles are dissolved by digestion with 1:3 concentrated sulphuric-nitric acid and the copper determined by the sodium diethyl dithiocarbonate spectrophotometric procedure. Distinctly higher copper determinations were obtained on polyolefins by the procedure involving the use on an ashing aid than are obtained without an ashing aid, or by the use of a molten potassium bisulphate fusion technique to take up the polymer ash.

Table 63 - The Effects of Modification of Ashing Procedure on the Flame
Photometric Determination of Sodium in Polyethylenes

Sodium ppm

By flame photometry

Sample	By neutron activation	By emis- sion spec- trography	Original (ashed between 650°C and 800°C)	Dope Ash at 500°C	Direct Ash at 500°C
1	99,96,99	95	60,76,55	100	75
2	256,247,259	258,259	160,178,271	225	208
3	343,321,339	339,287	250,312	282	265
4	213,210,212	218,212	140,196	210	191
5	194,189,192	209,198	80,158,229	196	169
6	186,191,198	191,191	96,173	193	173

Table 64 - Vanadium (ppm)

Sample	Dry ashing	Neutron activation
A	10.2	9.9 ± 0.2
B	14.0	14.1 ± 0.1
C	14.6	15.6 ± 0.3
D	0.5	0.14 ± 0.01
E	13.0	14.8 ± 0.2
F	0.9	0.27 ± 0.01
G	15.2	18.8 ± 0.3
H	18.2	17.9 ± 0.3

Table 65 - Analytical conditions.
Metals in polymers

Element	Wavelength nm	Band pass	Operating range (in polymer) ppm	Detection limit (in polymer) ppm	Concentra- tion of standard solution ug L^{-1}
Iron	248.3	0.3	5	0.57	500
Manganese	279.5	0.5	1.25	0.03	250
Chromium	357.9	0.5	2.5	0.03	500
Cadmium	228.8	1.0	0.5	0.015	50
Copper	324.7	1.0	1.25	0.045	250
Lead	217.0	1.0	5	0.15	1250
Nickel	232.0	0.15	2.5	0.07	500
Zinc	213.9	1.0	0.5	0.015	125

Atomic absorption spectrometry is a useful technique for the determination of traces of metals in polymers. Generally, the polymer is ashed at a maximum temperature of 450°C.

Time from start   0-1 hour,   heat to 200°C
                           1-3 hours,   hold at 200°C
                           3-5 hours,   heat to 450°C
                           5-8 hours,   hold at 450°C

Then digest the ash with warm I.M. nitric acid prior to spectrometric analysis. The detection limits for metals in polymers achievable by this procedure are given in Table 65.

Certain elements (such as arsenic, antimony, mercury, selenium and tin) can, after producing the soluble digest of the polymer, be converted to gaseous metallic hydrides by reaction of the digest with reagents such as stannous chloride or sodium borohydride, and these hydrides can be determined by atomic absorption spectrometry.

$$As_2O_3 + 3SnCl_2 + 6HCl = 2AsH_3 + 3H_2O + 3SnCl_4$$

$$NaBH_4 + As_2O_3 = 2AsH_3$$

To illustrate, let us consider a method developed for the determination of trace amounts of arsenic in acrylic fibres containing antimony oxide fire-retardant additive[551]. The arsenic occurs as an impurity in the antimony oxide additive and, as such, its concentration must be controlled at a low level.

In this method a weighed amount of sample (6 mg-1g) is digested with 3 ml each of concentrated nitric, perchloric and nitric acids and digested strongly until the sample is completely dissolved. Pentavalent arsenic in the sample is then reduced to trivalent arsenic by the addition of titanium trichloride dissolved in concentrated hydrochloric acid.

$$As^{5+} + 2Ti^{3+} = As^{3+} + 2Ti^{4+}$$

The trivalent arsenic is then separated from antimony by extraction with benzene, leaving antimony in the acid layer. The trivalent arsenic is then extracted with water from the benzene phase. This solution is then treated with a mixture of hydrochloric acid, potassium iodide and stannous chloride in a glass reaction vessel to convert trivalent arsenic to arsine (AsH_3), which is swept into the atomic absorption spectrophotometer. Arsenic is then determined at the 193.7 nm absorption line. Recoveries between 96 and 104% are obtained by this procedure in the 0.5-1.0 ug arsenic range. The detection limit was 0.04 ppm.

The results for the arsenic obtained with various acrylic fibre samples containing antimony oxide are given in Table 66. Antimony, present as the trioxide, has been accurately determined in a concentrated hydrochloric acid extract of polypropylene powder[552].

Table 66 - Results for the Determination of Arsenic in Acrylic
Fibres Containing Antimony Oxide

Sample no	Supplier	Content of Sb_2O_5*%	No of determinations	Arsenic Content, ppm
1	A	5.1	4	50 ± 1
2	A	5.1	3	84 ± 4
3	A	4.5	2	94 ± 7
4	A	4.6	5	10.3 ± 0.5
5	A	5.0	5	4.1 ± 0.2
6	B	3.0 (Sb_2O_3)	3	45 ± 0
7	C	-	2	180 ± 11
8	C	-	4	103 ± 6
9	D	2.4 (Sb_2O_3)	2	8.5 ± 0.5
10	E	1.0 (Sb_2O_3)	4	3.2 ± 0.1
11	-	Sb_2O_5 50 mg + acrylic fibre (no Sb) 1g	2	0.47 ± 0.03
12	-	Sb_2O_5 10 mg + acrylic fibre (no Sb) 1g	1	0.08

* Even when antimony (III) oxide was present in acrylic fibres (samples 6, 9 and 10), antimony (III) was easily and completely oxidised to antimony (V) by the wet digestion with a mixture of nitric, perchloric and sulphuric acids. Hence, arsenic in acrylic fibres containing antimony (III) oxide could be determined as well as that in acrylic fibre samples containing antimony (V) oxide by this method.

5.5.3 Trace Halogens in Polymers

Fusion Techniques

Fusion with sodium bisulphate. Sodium bisulphate is another useful reagent for the fusion of polymer samples prior to the determination of metals. It has the advantage of forming a very high-temperature melt in which the oxides of many metallic elements dissolve and are put in a chemical form suitable for subsequent analysis. The technique has been used, for example, to determine values down to one part per million of titanium, iron and aluminium in dried low-pressure polyolefins. The ignition takes place in clean silica crucibles over 4 hours at 800°C. All that is necessary to do is to dissolve the cooled residue in warm dilute sulphuric acid. The three metals can be determined by classical spectrophoto-metric techniques involving the use of Tiron, thioglycollic acid and aluminon, or by other suitable techniques.

Although, nowadays, these analyses would possibly be carried out by modern physical techniques such as X-ray fluorescence, there are no doubt circumstances (such as small one-off batches of samples) where the sodium bisulphate technique is still of value to the analyst. The technique would also be applicable to the solution of polymers preparatory to the application of techniques such as atomic absorption spectrometry.

Fusion with sodium carbonate. This method is very useful for the fusion of polymers which, upon ignition, release acidic vapours - for example polyethylene containing traces of chlorine, or polyvinylchloride, both of which, upon ignition, release anhydrous hydrogen chloride. To determine chlorine accurately in the polymer in amounts down to 5 ppm this acid must be trapped in a solid alkaline reagent such as sodium carbonate. In this method polyethylene is mixed with pure sodium carbonate and ashed in a muffle furnace at 500°C. The residual ash is dissolved in aqueous nitric acid, and then diluted with acetone. This solution is titrated potentiometrically with standard silver nitrate. Alternatively, chloride can be determined by another appropriate method.

Method - determination of chlorine in polyolefins. Accurately weigh 5 g of polymer into a metal crucible and cover the polymer with a layer of 2 g of sodium carbonate. Place the crucible in a cold muffle furnace and allow the temperature to increase gradually to 500°C, maintaining this temperature for 4 hours. Allow the crucibles to cool, then dissolve the residue in 10 ml of distilled water. Transfer this solution with crucible washings to a beaker (30 ml) and add 5 drops of screened methyl orange indicator solution. Neutralize the mixture by dropwise addition of 30% nitric acid to the purple-red coloured end-point. Add a further 10 drops of 30% nitric acid solution to the beaker and then add 30 ml of acetone. Titrate the resultant solution potentiometrically with silver nitrate solution (N/100), preferably using an automatic titrator equipped with a silver measuring electrode and a glass reference electrode. Carry out a blank determination exactly as described above, omitting only the sample addition.

Calculation ppm (w/w) chlorine in polymer

$$\frac{(T_s - T_B) \times N \times 35.46 \times 10^3}{W}$$

where T_B = titration of silver nitrate (ml) in blank determination
     T_s = titration of silver nitrate (ml) in sample determination
     N = normality of silver nitrate solution
     W = weight of polymer sample in grams.

Table 67 compares results for chlorine determinations in polyethylene obtained by this method with those obtained by an accurate independent referee method - in this case X-ray fluorescence spectroscopy. The averages of results obtained by the two methods agree satisfactorily within $\pm$ 15% of each other. The superior reproducibility of the chemical method of analysis is believed to be due to the fact that a considerably higher weight of sample is taken for analysis than is used in the X-ray method of analysis. Thus, whereas in the chemical method one obtains the average chlorine content of 5 g of polymer, in the X-ray method the amount analysed is only a few microns thickness of the disc of polymer taken for analysis. Any local variations in the chlorine content of the polymer will be picked up much more readily, therefore, by the X-ray method of analysis.

Table 67 - Comparison of Chlorine Contents by X-ray Method and chemical Method (averages in parentheses)

X-ray on discs	Chemical method on same disc as used for X-ray analysis*	Chemical method on powder*
865,814 (840)	700	786,761 (773)
535,570 (552)	606	636,651 (643)
785,675 (730)	598	650,654 (652)
625,675 (650)	600	637,684 (660)
895,870 (882)	733	828,816 (822)

* Analysis carried out on samples which had been treated with alcoholic potash to avoid losses of chlorine when preparing disc.

   <u>Fusion with sodium peroxide.</u>   Sodium peroxide is another useful reagent for the fusion of polymer samples preparatory to analysis for metal such as zinc, and non-metals such as chlorine[553] and bromine.   In this method the polymer is intimately mixed either with sodium peroxide in an open crucible or with a mixture of sodium peroxide and sucrose in a micro-Parr bomb.   After acidification with nitric acid chlorine can be determined complexometrically using EDTA[554].   In a method for the determination of traces of bromine in polystyrene in amounts down to 100 ppm bromine a known weight of polymer is mixed intimately with pure sodium peroxide and sucrose in a micro-Parr bomb which is then ignited.   The sodium bromate produced is converted to sodium bromide by the addition of hydrazine sulphate.

$$2NaBrO_3 + 3NH_2NH_2 = 2NaBr + 6H_2O + 3N_2$$

   The combustion mixture is dissolved in water and acidified with nitric acid.   The bromine content of this solution is determined by potentiometric titration with standard silver nitrate solution.   The organic bromo content of the polymer can then be calculated from the determined bromine content.

   <u>Oxygen flask combustion.</u>   This technique has been used to determine traces of chlorine in polyvinyl chloride[555] and in polyolefins and chlorobutyl rubber[556].

   Oxygen flask combustion methods have been used to determine traces of chlorine in polyolefins at levels between 0 and 500 ppm.   The Schoniger oxygen flask combustion technique requires a 0.1 g sample and the use of a 1 litre conical flask.   Chlorine-free polyethylene foil is employed to wrap the sample, which is then supported in a platinum wire attached to the flask stopper.   Water is used as the absorbent.   Combustion takes place at atmospheric pressure in oxygen.   The chloride formed is potentiometrically titrated in nitric acid/acetone medium in a small beaker using 0.01 N silver nitrate solution added from a syringe.

5.5.4 Trace Sulphur in Polymers

Oxygen flask combustion. To determine sulphur in amounts down to 500 ppm in polyolefins the sample is wrapped in a piece of filter paper and burnt in a closed conical flask filled with oxygen at atmospheric pressure. The sulphur dioxide produced in the reaction reacts with dilute hydrogen peroxide solution contained in the reaction flask to produce an equivalent amount of sulphuric acid. The sulphuric is estimated by visual titration with N/1000 or N/100 barium perchlorate using Thorin indicator.

$$H_2SO_3 + H_2O_2 = H_2SO_4 + H_2O$$

The repeatability of the method is $\pm$ 40% of the determined sulphur content at the 500 ppm sulphur level, improving to $\pm$ 2% at the 1% level. Chlorine and nitrogen concentrations in the sample may exceed the sulphur concentration several times over without causing interference. Fluorine does not interfere unless present in concentrations exceeding 30% of the sulphur content. Phosphorus interferes when present in moderate amounts. Metallic constituents also interfere when present in moderate amounts.

5.5.5 Trace Phosphorus in Polymers

Dry-ashing - digestion. Narasaki et al[557] have applied oxygen flask combustion to determination of traces of phosphorus in polymers. The sample is digested at 250-300°C with perchloric acid (0.5 ml 70%) and sulphuric acid (3 ml concentrated) for two minutes. The yellow phosphomolybdo complex is developed and this is evaluated at 430 nm.

Phosphorus has been determined[558] in thermally stable polymers by mineralizing with a nitric-perchloric acid mixture and subsequent titration with lanthanum nitrate or by photometric determination of the phosphomolybdic blue complex.

5.5.6 Trace Nitrogen in Polymers

Acid digestion. Apart from the classical Kjeldahl digestion procedure for the determination of organic nitrogen, acid digestion of polymers has found little application. One of the problems is connected with the form in which the polymer sample exists. If it is in the form of a fine powder, or a very thin film, then digestion with acid might be adequate to enable the relevant substance to be quantitatively extracted from the polymer. Thus reliable nitrogen contents can be obtained for fine polymer powders using Kjeldahl digestion procedure utilizing a mixture of concentrated sulphuric acid, potassium sulphate and an oxidation catalyst. However, low nitrogen results would be expected for polymers in larger granular form, and for the analysis of such samples classical microcombustion techniques are recommended.

Another reason why the Kjeldahl sulphuric acid digestion procedure is not usually applicable to the digestion of polymer powders, and certainly not to polymer granules prior to the determination of metals, is that the Kjeldahl digestion catalyst, usually a mercury, copper or selenium salt, would in all likelihood interfere in methods for the determination of traces of metals in digests.

A sample of polymer (up to 5 g) is digested with a mixture of sulphuric acid, mercuric oxide, copper sulphate and selenium catalysts to convert nitrogen containing organic compounds to ammonium sulphate. Digestion is completed by adding sufficient potassium sulphate to the digestion mixture to raise its boiling point to 380°C followed by heating to completely remove any carbonaceous matter present. The solution is then transferred to a distillation apparatus and excess sodium hydroxide added to convert ammonium sulphate to free ammonia. The ammonia is distilled off and collected in boric acid solution and the ammonia content of the steam distillate is determined by titration with standard 0.02 N hydrochloric acid solution to the methyl red - methylene blue (Tashiro) end-point.

Hernadez[559] has described an alternative procedure based on pyro-chemiluminescence which he applied to the determination of 250 to 1500 ppm nitrogen in polyethylene. In this technique the nitrogen in the sample is subjected to oxidative pyrolysis to produce nitric oxide. This when contacted which ozone produces a metastable nitrogen dioxide molecule which as it relaxes to a stable state, emits a photon of light. This emission is measured quantitatively at 700-900 nm.

5.5.7 Other Trace Elements

The determination of miscellaneous trace elements in polymers is reviewed in Table 68.

5.5.8 Non-destructive Techniques

X-ray spectrometry. X-ray fluorescence spectrometry has been used extensively for the determination of traces of metals and non-metals in polyolefins and other polymers. The technique has also been used in the determination of major metallic constituents in polymers, such as cadmium selenide pigment in polyolefins.

The X-ray fluorescence spectrometric method has several advantages over other methods. The analysis is non-destructive; specimen preparation is simple, involving compressing a disc of the polymer sample for insertion in the instrument; measurement time is usually less than for other methods; and X-rays interact with elements as such, i.e. the intensity measurement of a constituent element is independent of its state of chemical combination. However, the technique does have some drawbacks, and these are evident in the measurement of cadmium and selenium. For example, absorption effects of other elements present, e.g. the carbon and hydrogen of the polyethylene matrix, excitation of one element by X-rays from another e.g. cadmium and selenium mutually effect one another. The technique has been applied to the determination of metals in poly-butadiene, polyisoprene and polyester resins[566]. The metals determined were iron, cobalt, nickel, copper and zinc. The samples were ashed and the ash dissolved in nitric acid prior to X-ray analysis. No separation schemes are necessary and concentrations as low as 10 ppm can be determined without inter-element interference. A solution technique was used (rather than examination of the solid compressed polymer) to circumvent inter-element and matrix effects. Nitric acid was used because, unlike the other mineral acids, it does not have a strong absorbing effect on the X-ray fluorescence of these metals. In general the theoretical and found values agree within ± 10%. The determination of chromium tended to be the most erratic, relative to the other metals. Possibly the reason for the good recoveries after dry ashing found by these workers was due to the relatively low maximum temperature, 550°C, that was used.

Table 68 - Miscellaneous Trace Elements in Polymers

Element	Polymer	Method	Reference
Sulphur	misc	inductively coupled plasma atomic emission spectrometry	560,561
Zinc	polyesters	polarography	562
Aluminium	polyethylene	glc of aluminium trifluoro-acetonate or pivaloytri-fluoroacetonate	563
Zinc	chlorinated polymers	sodium peroxide fusion-titrimetric	564
Chorine	polystyrene	13 C NMR	565

Many investigators have found much higher recoveries using various aching aids such as sulphuric acid[570], elemental sulphur[567,568], magnesium nitrate[570,571] and benzene and xylene sulphonic acids[569] then by dry ashing. Possibly the reason for the good recoveries after dry ashing found by Cook et al[566] was due to the slowness and relatively low maximum temperature, 550°C, was used.

Layden[572] used X-ray fluorescence spectrometry to determine metals in acid digests of polymers. The aqueous solutions were applied to filter paper discs. He found that recoveries of metals by the X-ray technique was 101-110% compared to 89-94% by chemical methods of analysis.

X-ray fluorescence spectrography has been applied very successfully, industrially, to the routine determination in hot-pressed discs of polyethylene and polypropylene of down to a few parts per million of the following elements: chlorine, bromine, titanium, aluminium, sodium, potassium, calcium, magnesium, and vanadium.

An interesting phenomenon has been observed in applying the method to the determination of parts per million of chlorine in hot-pressed discs of low-pressure polyolefins. In these polymers the chlorine is present in two forms, organically bound and inorganic, titanium-chlorine compounds resulting as residues from the polymerization catalyst used in this process. The organic part of the chlorine is determined by X-ray fluorescence without complications. However, during hot processing of the discs there is a danger that some inorganic chlorine will be lost. This can be completely avoided by intimately mixing the powder with alcoholic potassium hydroxide, then drying at 105°C before hot pressing into discs. The results (Table 69) illustrate this effect. Considerably higher total chlorine contents were obtained for the alkali-treated polymers.

<u>Neutron activation analysis.</u> This is another non-destructive technique capable of determining a wide range of elements - e.g., chlorine in polyolefins, metals in polymethylmethacrylate[573], total oxygen in polyethylene-ethylacrylate and polyethylene-vinylacetate copolymers[574] and total oxygen in polyolefins. The advantages of the technique are extreme sensitivity and freedom from interference effects. A disadvantage is that, involving as it does the use of a nuclear reactor, the samples would usually have to be sent away for analysis. However, the technique is

Table 69 - Determination of Chlorine by X-ray Procedure

A: Polymer not treated with alcoholic potassium hydroxide before analysis, ppm chlorine X-ray fluorescence on polymer discs. Average of 2 discs (A)	B: Polymer treated with alcoholic potassium hydroxide before analysis, ppm chlorine, X-ray fluorescence on polymer discs. Average of 2 discs (B)	Difference between average chlorine contents obtained on potassium hydroxide-treated untreated samples (B) - (A)
510	840	330
422	552	130
440	730	290
497	650	153
460	882	422

extremely useful, and in many cases the results obtained can be considered as reference values for those materials, and these data are of great value when these samples are analysed by alternative methods in the originating laboratory.

To illustrate this, some work will be discussed on the determination of parts per million of sodium in polyolefins. During the manufacture of high-density polyethylene and polypropylene, sodium-containing alkalis are added to the polymer to neutralize residual acidity originating from the catalyst system. It was found that replicate sodium contents determined on the same sample by a flame photometric procedure were frequently widely divergent. Neutron activation analysis offers an independent non-destructive method of checking the sodium contents which does not involve ashing. Both methods are discussed below.

In the flame photometric procedure the sample is dry-ashed in a nickel crucible and the residue dissolved in hot water before determining sodium by evaluating the intensity of the line emission occurring at 589 nm.

Neutron activation analysis (flux 10^{12} neutrons cm^{-1} s-1) for sodium was carried out on polyethylene and polypropylene moulded discs containing up to about 550 ppm sodium which had been previously analysed by the flame photometric method. The results obtained in these experiments (Table 70) show that significantly higher sodium contents are usually obtained by neutron activation analysis, and this suggested that sodium is being lost during the ashing stage of the flame photometric method. Sodium can also be determined by a further independent method, namely emission spectrographic analysis, which involves ashing the sample at 500°C in the presence of an ashing aid consisting of sulphur and the magnesium salt of a long-chain fatty acid (compared to dry-ashing at 650-800°C without an aid as used in the flame photometric procedure[575-577]). In Table 71, results by neutron activation and emission spectrography are shown to agree well with each other, thereby confirming that low results can be obtained by flame photometry when a simple ashing process is employed. Unlike the neutron activation technique, emission spectrography involved a preliminary ashing of the sample at 500°C, thereby involving the risk of sodium losses, despite this, the technique gives correct results. The losses of sodium in the flame photometric ashing procedure were probably caused by the maximum ashing temperature used exceeding that used in the emission spectrographic method by some 150-300°C. This suggested that it

Table 70 - Interlaboratory Variation of Flame Photometric
Sodium Determinations in Polyolefins (Sodium Content, ppm)

| Neutron activation analysis | | Flame photometry |
Powder	Moulded discs	(moulded discs)
Polyethylene		
207	211,204	35,165,140
177	175,172	100,140,148
266	267,263	85,210,221
203	187,191	70,160,150
Polypropylene		
165	151,161	50,130,133
198	186,191	95,173
322	333,350	95,138

Table 71 - Comparison of Sodium Determination in Polyolefins by
Neutron Activation Analysis, Emission Spectrography and
Flame Photometry (sodium ppm)

Sample description	By neutron activation analysis	By emission spectrography	By flame photometry
Polyethylene	99,96,99	95	60,76,55
Polyethylene	256,247,256	258,259	160,178,271
Polyethylene	343,321,339	339,287	250,312
Polyethylene	213,210,212	218,212	140,196
Polypropylene	194,189,192	209,198	80,158,229
Polypropylene	186,191,198	191,191	95,173

might be possible to obtain more reliable sodium contents by flame
photometry if the magnesium AC dope ashing procedure could be used in
conjunction with flame photometry.

The results in Table 72 show clearly that flame photometry following
dope ashing at 500°C gives a quantitative recovery of sodium. Direct
ashing without an ashing aid at 500°C causes losses of 10% or more of
sodium, whilst direct ashing at 800°C causes even greater losses.

Commonly, nowadays, the active catalyst (based on chromium, titanium
or vanadium) used in high density polyethylene manufacture is adsorbed on
to a highly porous silica support. Determination of the silica catalyst
support content of the final polymer gives an assessment of the economic
productivity of the reactor, i.e. its output of polyethylene per g of
catalyst and also enables the very low concentration of active catalyst
metal in the polymer to be calculated.

Battiste et al[578] have described three methods based on neutron
activation analysis, infrared spectroscopy and ashing for the
determination of silica catalyst supports in polyethylene.

In the neutron activation technique approximately 2 g of polyethylene powder is irradiated with neutrons obtained with a 500 keV neutron activator according to the following reaction: $^3H(D,n)^4He$. Silicon is activated by the reaction $^{28}Si(N,p)^{28}Al$, and the concentration of silicon is then measured by counting the 1.78 MeV gamma-ray emission from the decay of ^{38}Al. A Conostan 5000 ppm Si standard is used for instrument calibration. Two 3 in sodium iodide detectors are used to measure the 1.78 MeV gamma-rays.

Silica can be estimated by direct measurement of the silica absorbance i.e. at the 21.27 micron absorbance band of silica. This is in the region of the infrared spectrum which is relatively free of polyethylene absorbance bands. The absorbance value of the 21.27 micron band is calculated for each standard by use of the peak height determined by means of the base line technique between minima near 28.57 and 17.24 micron.

Analysis results of polyethylene samples containing residual silica support of three different catalysts by infrared, neutron activation analysis, ashing and weight are shown in Table 73. Infrared results differ from neutron activation analysis results by 0-5% while ashing and weighing techniques differ from neutron activation analysis results by 5-21% and 5-28% respectively.

5.5.9 Trace Organic Peroxide Residues

Organic peroxides can occur in small amounts in some types of polymers such as polystyrene as a result of the fact that a peroxide has been used as a polymerization catalyst in polymer manufacture. Also, stable organic peroxides such as dicumyl peroxide have been used as synergists, in conjunction with bromine and or phosphorus-containing additives, to impart fire-resistance to cellular expanded polystyrene and other types of plastics.

In a solution of 0.6 M lithium chloride in 1:1 toluene:methanol benzoyl peroxide, p-tert butyl perbenzoate and lauryl peroxide have half-wave potentials respectively of 0.00 - 0.95 and -0.15 V in the presence of ethyl cellulose as a maximum suppressor. Based on this, a procedure has been described[579] for determining down to 30 ppm of these substances in polystyrene in which a suitable weight of polymer is dissolved in toluene, and then an equal volume of 0.6 M lithium chloride in methanol is added. Precipitated polymer is removed by centrifuging and peroxides determined in the clear extract by cathode-ray polarography. Polymerization additives, styrene monomer or antioxidants in the polymer do not interfere in the polarographic procedure.

Gas chromatography has also been used to determine certain types of organic peroxides. Bukata et al[580] for example, describe procedures involving chromatography of heptane solutions of peroxide (Table 74) on phthalate/diatomaceous earth or silicone/diatomaceous earth columns, and using helium as the carrier gas. No doubt this type of procedure could be easily adapted to the examination of solvent extracts of polymers.

Hyden[581] describes a gas chromatographic procedure for the determination of di-tert-butyl peroxide. This is based on the thermal decomposition of the peroxide in benzene solution into acetone and ethane when the solution is injected into the gas chromatographic column at 310°C. The technique is calibrated against standard solutions of pure di-tert-butyl peroxide of known concentrations.

Table 72 - The Effects of Modification of Ashing Procedure on the
Flame Photometric Determination of Sodium (sodium ppm)

By neutron activation	By emission spectrography	By flame photometry		
		original (ashed between 650°C and 800°C)	Dope ash at 500°C	Direct ash at 500°C
99,96,99	95	60,75,55	100	75
256,247,259	258,259	160,178,271	225	208
343,321,339	339,287	250,312	282	265
213,210,212	218,212	140,196	210	191
194,189,192	209,198	80,158,229	196	169
186,191,198	191,191	95,95,173	193	173

Table 73 - Catalyst Productivity for Some Test Samples

Sample	IR at 21.27 micron	NAA	Ashing at 650°C	Weight
1[a]	1893 (3.3)[d]	1923 (4.9)	1923 (4.9)	
2[a]	4165 (0)	4165	4545 (9.1)	3571 (14.3)
3[b]	1759 (2.9)	1709	2222 (30.0)	2007 (17.4)
4[c]	1727 (5.4)	1825	2000 (9.6)	2000 (9.6)
5[c]	2207 (3.5)	1288	2778 (21.4)	2941 (28.5)

[a] Catalyst A. [b] Catalyst B. [c] Cabosil S17.

[d] Values in parentheses are percent deviation from NAA.

Table 74 - Gas Chromatography, Retention Times for Organic Peroxides

Compound	Column*	Temperature (°C)	Helium pressure[1] (psi)	Retention time (min)
tert-Butyl hydroperoxide	2 m-A	80	20	22.7
tert-Pentyl hydroperoxide	1 m-A	80	15	19.9
tert-Butyl peracetate	1 m-A	100	20	6.5
tert-Butyl peroxyisobutrate	1 m-A	100	20	15.3
Di-tert-Butyl peroxide	2 m-A	80	20	8.1
Di-tert-Pentyl peroxide	1 m-A	80	15	15.4
2,5-Dimethyl-2,5di(tert-butylperoxy)-3-hexyne	1 m-O	138	15	4.9
n-Heptane	2 m-A	80	20	6.9
n-Dodecane	1 m-O	138	15	3.1
n-Pentane	1 m-A	100	20	0.4
n-Nonane	1 m-A	100	20	4.9

* Length (metres) and type of column used.

[a] Inlet pressure.

A, didecyl phthalate on diatomaceous earth; O, silicon grease on
diatomaceous earth.

Iodometric methods have been described for the determination of hydroperoxide residues in polypropylene[582] and organic peroxides in polyesters[583].

CHAPTER 6

FUNCTIONAL GROUPS IN POLYMERS

6.1 INTRODUCTORY

Functional groups can occur in polymers over a wide range of concentration, ranging from a few parts per million, as occurs for example in the case of end-groups, or micro-unsaturation functional groups to the percentage range. The occurrence of two or three double bonds per thousand carbon atoms in, for example, polyethylene can effect intrinsic polymer properties and can certainly help to distinguish between polyethylene manufactured by different manufacturers using different processes. As such, this type of determination falls within the province of micro-structure, which is discussed in Chapter 10. At the other end of the concentration scale a copolymer of, for example, ethylene and vinyl acetate will contain between 1 an 90% of either monomer. Analysis of functional groups in polymers in this concentration range is considered below.

A wide range of physical and chemical techniques have been employed in functional group analyses. The application of these techniques to the determination of particular functional groups is discussed under two main headings:

1. Chemical methods comprising techniques based on halogenation, titration, saponification values, procedures based on phthalation, acetylation, etc., hydrogenation and colorimetric procedures.

2. Physical methods comprising procedures based on infrared, Raman and nuclear magnetic resonance spectroscopy, saponification, and, finally, pyrolysis or alkali fission of the polymer followed by gas chromatography of the volatile products produced.

The determination of functional groups and conomoner ratios in co-polymers is discussed in Chapter 9. These methods are separated from Chapter 6 as in most cases the functional group method described in the literature discusses a particular copolymer or class of copolymers.

The determination of elements is discussed in Chapter 4.

6.2 UNSATURATION

Bromination procedure. In acidic medium potassium bromide and potassium bromate produce bromine stoichiometrically, and this is the basis of a titration method[584], which has been used to determine double bonds in polymethylacrylate. After bromination, excess bromine is estimated by the addition of potassium iodide and the iodine produced estimated by titration with standard sodium thiosulphate to the starch end-point.

$$KBrO_3 + 5KBr + 6HCl = 3Br_2 + 3H_2O + 6KCl$$

$$\left[-CH_2 = \underset{\underset{COOCH_3}{|}}{\overset{\overset{CH_3}{|}}{C}} - \right]_n + Br_2 \longrightarrow \left[-CH_2Br - \underset{\underset{COOCH_3}{|}}{\overset{\overset{CH_3}{|}}{C}} -Br \right]_n$$

Hydrogenation methods. Unsaturation in polymers is usually measured by physical techniques, as discussed later. This is especially so in the case of low levels of unsaturation, or in instances where a distinction has to be made between different types of unsaturation. Hydrogenation techniques have been used, however, to measure higher levels (0.5-5.0 mole %) of total unsaturation in polymers, a good example of which is the determination of terminal unsaturation in polystyrene oligomers[587-589] (low molecular weight polymers), e.g. polystyrene dimer.

$$2PhCH = CH_2 \longrightarrow PhCH_2 - CH = CH - CH_2Ph$$

Radiochemical methods. Radiochemical methods involving the use of ^{36}Cl have been used for the measurement of higher levels of unsaturation (0.5-2 mole %) in butyl rubber[585]. This method, with slight modification, is also capable of determining unsaturation at the 0.01 to 0.1 mole % level[586]. If the specific activity of the radiochlorine in the gas phase is known, the weight of chlorine in the polymer can be found by counting. The mole % unsaturation of the polymer was calculated from the weight per cent unsaturation by assuming that one atom of chlorine enters the polymer per double bond producing a monocholoro compound. With this assumption it was found that polyisobutene consumes a mean of 1.2 chlorine atoms per double bond producing a monocholoro compound. This is probably an indication that side-reactions such as addition of chlorine are occurring to a small extent. Since the true unsaturation is close to one double bond per chain it follows that the main chain-breaking reaction during the cationic polymerization of isobutene is proton transfer.

Infrared[590-594] and pyrolysis-infrared[595] methods have both been tried for determination of unsaturation in vulacanizates, but both methods have serious disadvantages. NHR spectroscopy has been used to determine unsaturation in 1,2 polybutadiene[596]. Brako and Wexler[597] have described a useful technique for testing for the presence of unsaturation in polymer films such as polybutadiene and styrene-butadiene. They expose the film to bromine vapour and record its spectrum before and after exposure (Figure 45). This results in marked changes in the infrared spectrum. Noteworthy is the almost complete disappearance of bands at 730, 910, 965 and 1640 cm^{-1} (13.691, 10.99, 10.36, and 6.10 microns) associated with unsaturation. A pronounced band possibly associated with a C-Br vibration appears at 550 cm^{-1} (12.73, 8.73 and 8.00 microns), which is due to exposure to bromine vapour. Exposure of butadiene-styrene copolymer (Figure 45) to bromine vapour results in the disappearance of bands at 910 and 965 cm^{-1} (10.99 and 10.36 microns) associated with unsaturation in the butene component of the copolymer. Some alteration of the phenyl bands at 700 and 765 cm^{-1} (14.28 and 10.37 microns) is evident. The loss of a band at 1550 cm^{-1} (6.45 micron) and the appearance of a band at 1700 cm^{-1} (5.88 micron) are probably due to the action of acidic vapours on the carboxylate purifactant of the latex.

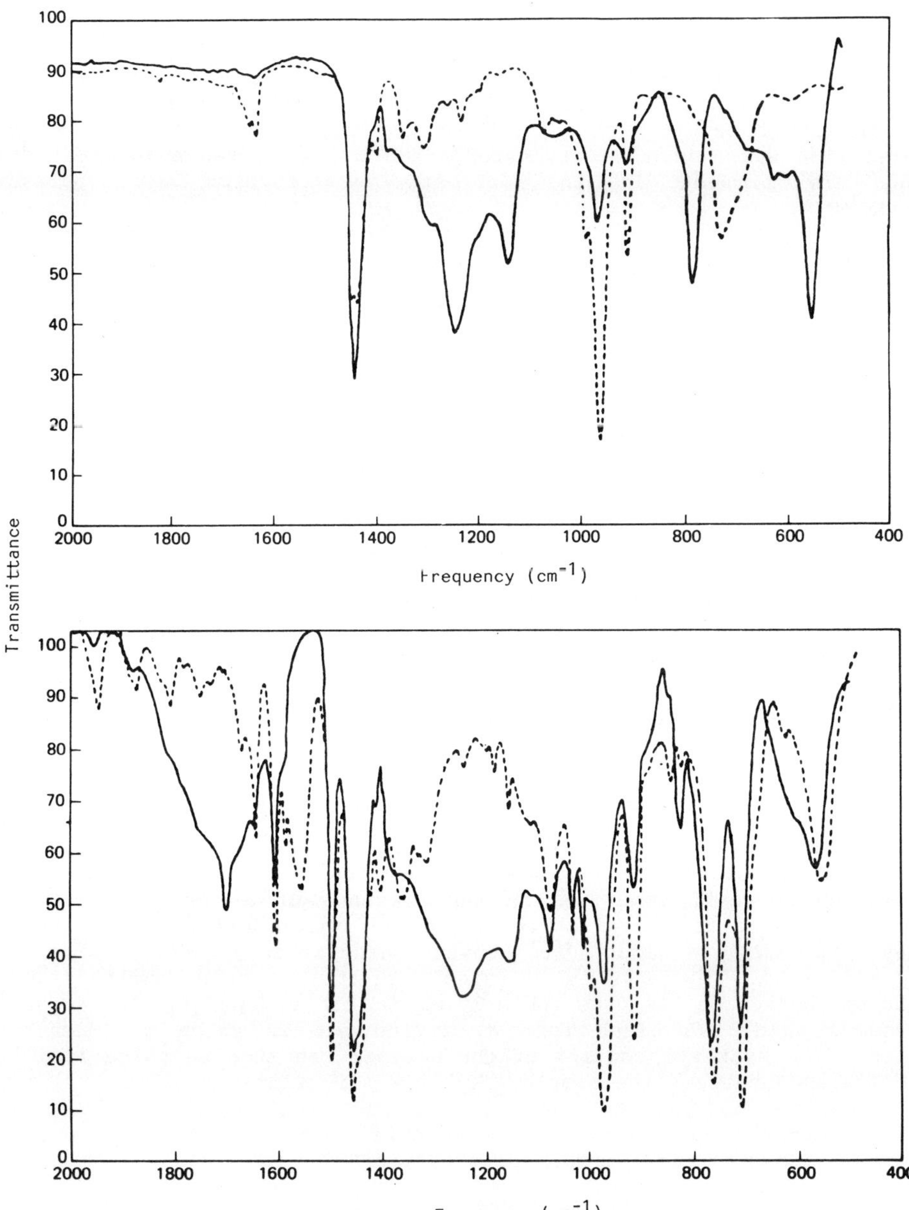

Figure 45   (a) Spectra of polybutadiene and of the film after
             exposure to bromine: ... polybutadiene; ___
             polybutadiene after exposure to bromine vapour;
             (b) Spectra of styrene-butadiene copolymer film and of
             copolymer film after exposure to bromine vapour: ...
             styrene-butadiene copolymer; ___ copolymer after
             exposure to bromine vapour.

6.3 HYDROXY GROUPS

Acetylation and phthalation procedures. Chemical methods for the determination of hydroxyl groups in polymers are based on acetylation[598-600] phthalation[599] perchloric acid catalysed acetylation[599,601,602] reaction with beta-nitro phthalic[603] anhydride, pyromellitic anhydride[604] and reaction with phenyl isocyanate[599,605] or, when two adjacent hydroxy groups are present in the polymers by reaction with potassium periodate[598].

The reactions on which these methods are based are as follows:

Acetylation
$$2R - OH + (CH_3CO)_2O \longrightarrow 2CH_3COOR + H_2O$$

Phthalation
$$2ROH + \text{(phthalic anhydride)} \longrightarrow \text{(phthalate diester)} + H_2O$$

Reaction with phenyl isocyanate
$$ROH + C_6H_5-NCO \longrightarrow C_6H_5-NHCOOR \text{ (urethane)}$$

Reaction with potassium periodate
$$R - \underset{OH}{CH} - \underset{OH}{CH} - R' + HIO_4 \longrightarrow RCHO + R'CHO + H_2O + HIO_3$$

In all procedures the hydroxyl-containing polymer is reacted with an excess of a standard non-aqueous solution of the acetylation or phthalation reagent, sometimes in the presence of a catalyst such a p-toluene sulphonic acid. After the formation of the ester is complete, an excess of water is added to convert excess anhydride to the free carboxylic acid. The acid is titrated with aqueous or alcoholic standard potassium hydroxide to the phenol phthalein end-point, to determine unconsumed acid. A blank run is carried out in which the sample is omitted. The hydroxyl content of the polymer can then be calculated from the difference between the sample and blank titrations.

Hydroxyl groups in epoxy resins have been determined by a method based on the use of lithium aluminium hydride[606].

Spectrophotometry using ceric ammonium nitrate has been used to determine hydroxyl groups in polycarbonates[607] and hydroxy-polybutadienes[608].

Acetylation and phthalation procedures. Stetzler and Smullin[599] found that for polypropylene glycols the classical phthalation procedure gave consistently low results, as did the perchloric acid catalysed acetylation procedure developed by Fritz and Schenk[601]. In addition to its greater intrinsic accuracy, the advantages claimed for the Stetzler and Smullin[599] toluene p-sulphonic acid catalysed procedure over phthalation include shorter reaction time and lower reaction temperature, also good reproducibility, indicating no reaction with the ether groups.

Table 75 - Hydroxyl values obtained on polypropylene glycol by
catalysed acetylation, phthalation and phenyl isocyanate reaction

Mol. wt.	p-Toluene sulphonic acid catalysed acetylation method	Phthalation method	Reaction with phenyl isocyanate
	Average hydroxyl value (mg KOH/g)	Average hydroxyl value (mg KOH/g)	Average hydroxyl value (mg KOH/g)
700	353	346	
1260	276	265	
5000	34.9	34.0	34.9
5000	34.3	33.8	35.1
2890	58.3	57.0	

Table 75 compares hydroxyl values obtained by three methods. It is
seen that in every case the acid-catalysed acetylation method gives a
higher result than that obtained by phthalation, the average difference
between the two methods being 2.5% of the determined value. It will be
seen from the table that the agreement between the p-toluene sulphonic
acid method and the phenyl isocyanate method is good.

The Stetzler and Smullin[599] method is applicable to polyoxyethylene,
polyoxypropylene and ethylene oxide tipped glycerol/propylene oxide
condensates. Compounds of this type in the molecular range 500 to 5000
can be analysed by this procedure.

Fritz et al[609] have described a method for the determination of
hydroxy groups in poly(ethylene glycols). The method has a sensitivity of
10^{-4} mol hydroxyl per kg polymer which is achieved by using silylation
with acrylsilyamines and subsequent photometric measurement of the
silylated polymer.

Kaduji and Rees[610] have described a direct injection enthalpimetric
method to measure the hydroxy value of glycerol alkylene oxide polythenes
and butane-1, 4-diol-adipic acid polyesters.

<u>Reaction with alkyl or aryldiisocyanates.</u> A kinetic method based on
the use of toluene diisocyanate for primary and secondary hydroxy groups
in polyols is discussed in Chapter 9.

<u>Gas chromatography.</u> Smith and Danson[611] determined traces of
polyethylene glycol by fission with hydrogen bromide to produce dibromo-
methane and dibromopropane followed by gas chromatographic determination
of these two substances.

The use of the electron capture detector can increase the
sensitivity of the determination of poly(ethylene glycol) to the 1-10
microgram range. If calibration standards are determined in parallel with
unknown samples the method could be expected to give good precision and
accuracy and would be suitable for the analysis of trace amounts of
poly(ethylene glycol) in extracts of environmental or other samples at the
1-10 microgram level.

A method for the determination of hydroxy groups in polyethylene
glycols has been described by Fritz et al[609]. The method consists of

substituting the hydroxy group with a chromophoric siloxy group, purification of the silyated polymer, and a photometric determination of the chromophor concentration. The silanization of the primary hydroxyls with dimethylaminosilanes proceeds quantitatively under very mild conditions and elimination of the excess reagent by precipitation is easy. The precision of the method is $\pm$ 4.5% (95% confidence level) down to 5 x 10^{-4} mol kg^{-1}. The method has about the same precision as acylation methods but yields a 1000 fold gain in sensitivity.

<u>Determination of Hydroxyl Groups in Polyethylene Glycol</u>

<u>Silation - Gas Chromatography Method[609]</u>

   <u>Apparatus.</u> Gas chromatographic analyses, a Packard Becker (Delft, Holland, Model 419) research gas chromatography by Fritz et al[609] equipped with flame ionization detectors. The carrier gas was helium. Spectrographs: IR Spectrometer (model 700, Perkin-Elmer), UV Spectrometer (model 635), Varian-Techtron or equivalent.

   <u>Reagents.</u>

1 napthyl dimethyl (dimethylamino) silane.

A solution of 1 equiv monochlorosilane (napthyl dimethyl chlorosilane) in hexane (in benzene if necessary for solubility reasons) is stirred in a flask equipped with a reflux condenser cooled at -20°C. From a communicating flask 2.2 equiv of dimethylamine vapour is slowly introduced over the solution and the mixture kept at room temperature overnight. After filtration and evaporation of the solvent, the residue is distilled.

$$\text{naphthyl} - \underset{\underset{CH_3}{|}}{\overset{\overset{CH_3}{|}}{Si}}Cl + HN(CH_3)_2 = \text{naphthyl} - \underset{\underset{CH_3}{|}}{\overset{\overset{CH_3}{|}}{Si}} - N(CH_3)_2$$

4.5 g dimethylamine, 11.1 g 1-napthylyldimethylchlorosilane, 100 ml hexane. Distillation of the raw product gave at 89°C/0.06 Torr; 7.8 g pure 1-napthyldimethyl aminosilane.

Poly(ethylene Glycols)

Purification of the poly(ethylene glycols) (Hydroxyl terminated) H-PEG.

A solution of 100 g of commercial practical grade polymer in 400 ml of 0.01 M sodium hydroxide is digested at 40°C for 24 h, then passed through two ion-exchange columns, the first containing 50 ml of Amberlite IR 120 (H+ form) and the second 50 ml of Lewatite MP 7080 (base form). After evaporation of the water, the residue is dissolved in 1000 ml of toluene and filtered on an aluminium oxide column (50 g). The agitated solution is then thermostated at the temperature indicated below and 5.0 L of petroleum ether added dropwise. The slurry is cooled to 0°C, allowed to stand for 15 min and the precipitate filtered. The temperature of precipitation is critical; it was 0°C for H-PEG of molecular weight 600 (H-PEG-600), 10°C for H-PEG-1000, 20°C for H-PEG-2000 and 30°C for polymers of higher molecular weight. The purified product is dried at room temperature/0.1 Torr.

    <u>Procedure.</u>   Determination of the hydroxyl content of polyethylene glycols.

For polymers with expected equivalent weights higher than 1000 a sample of 1 g (0.5 g for lower equivalent weights) is dissolved in 5 ml of dry 1,2-dimethyoxyethane. The dissolution is omitted if the melting point of the polymer is less than 60°C. The flask is purged with dry argon (or nitrogen) and three equivalents for each expected equivalent hydroxyl (but not less than 200 mg) of 1-napthyl dimethylaminosilane is added. The homogenized mixture is kept at 60°C for 30 min. The sample is then diluted at room temperature to 10 ml with 1,2-dimethyloxyethane. After cooling to an appropriate temperature, the silylated polymer is precipitated by adding dropwise 50 ml of low boiling petroleum ether to the stirred solution (the temperature of the precipitation is critical: see under section "Purification of H-PEG"). After precipitation, the mixture is stirred at 0°C for 15 min, then filtered on a sintered glass filter (G3) and washed with 5 ml portions of 20 ml of petroleum ether. The precipitate is redissolved on the filter in 5 ml of warm 1,2-dimethyoxyethane and precipitated a second time as above. If the expected hydroxyl content of the polymer is lower than 10^{-1} mol kg^{-1}, the precipitation should be repeated a third time. After the last precipitation the polymer is washed on the filter with 20 ml of Uvasol grade pentane, dried for 5 h at room temperature/0.1 Torr or in a stream of dry nitrogen. A portion of the dry sample is dissolved in Uvasol-grade ethanol so as to yield an absorbance of 0.2-0.8. The absorbance of the solution is measured vs. ethanol as reference at 282.5 mm. If the sample concentration is higher than 10 g L^{-1} an ethanolic solution of the pure poly(ethylene glycol) of about the same concentration must be used as a reference. The molarity of the silyl groups in the solution, M_{sol}, is given by:

$$M_{sol} = A \qquad = c/EW_{PSil} \qquad mol\ L^{-1}$$

where A is the absorbance of the solution, c is the concentration of the silylated polymer in g L^{-1}, A is the molar absorptivity of the silylating agent and EW_{PSil} g mol^{-1} is the equivalent weight of the hydroxyl terminated polymer. EW_{POH} the molecular weight of the silyl group MW_{Sil} and that of the substituted portion must be taken into account

$$EW_{POH} = \qquad A - (MW_{Sil} - 1) \qquad g\ mol^{-1}$$

The molarity of the hydroxyl groups in 1 kg of original polymer, m_{OH} is then given by:

$$m_{OH} = 1000/EW_{POH} \qquad mol\ kg^{-1}$$

For the 1-napthyldiemethylsilyl substituent, MW_{Sil} = 185.

The values of the molar absorptivities are given in Table 76.

1-napthyldimethyl (dimethylamino silane) silane quantitatively silated primary and secondary group at 60°C but little of tertiary hydroxyl groups.

The above procedure with 1-napthyldimethyl (dimethylamino) silane was applied under the above conditions to four polymers, two of them hydroxyl terminated, the other two with methoxy end groups of unknown, but certainly very low, hydroxyl content. Two nominal molecular weights were chosen for both types of polymers, 600 and 20,000 representing extreme

Table 76 - Physical Data of R', R" and R'" (2' methoxyethoxy) silanes

R',R",R'"	bp(°C)/Torr	mp(°C)	d20g cm^{-3}	20$_D$	$\epsilon(\lambda)$mol^{-1}(nm)
1-napthyl-dimethyl	111°/0.03	---	1.051	1.5582	7.33 x 10^3 (282.5)

The purity of the samples controlled by gas chromatography was better than 99.8%.

types regarding the difficulties involved in the purification of the silylated polymers by precipitation and filtration. For this operation, the best pair of solvents proved to be 1,2-dimethyloxyethane and low boiling petroleum ether. The samples were redissolved and precipitated repeatedly and each time a portion of the precipitate was subjected to photometric analysis. A plot of apparent hydroxyl content as a function of the number of precipitations shows that in the case of the hydroxyl terminated polymers a constant silyl concentration is attained after the first precipitation. Three precipitations were necessary however, for the methylated polymers in order to obtain consistent results.

Most of the chemical methods for determining hydroxy groups in polymers do not distinguish between primary, secondary or tertiary hydroxy groups. An exception to this is a kinetic approach to the determination of primary and secondary hydroxy groups as will be discussed in the Copolymer Section under ethylene oxide polyglycol condensates, (Chapter 9).

A further method for distinguishing between different types of hydroxy groups in polymers is the NMR method described by L'Ho[612]. This procedure is based on the reaction of the hydroxy compound with hexafluoroacetone to form an adduct which is amenable to F-19 NMR spectroscopy.

$$(CF_3)_2CO + ROH \rightleftharpoons CH_3 - \underset{\underset{OR}{|}}{\overset{\overset{OH}{|}}{C}} - CF_3$$

A fluorine resonance spectrum of a mixture of several hexafluoro-acetone adducts is shown in Figure 46. There are two interesting features in this spectrum. First, it illustrates clearly the high resolution and the information on structural aspects of the molecules one can obtain by this method. The hydroxyl adduct from each alcohol gives a sharp resonance, with the tertiary adducts at high field, followed by secondary and then primary. The chemical shift of the hexafluoroacetone alcohol adduct is determined by the structural environment of the hydroxyl. For example, the gradual upfield shift observed in the series methanol, ethanol, isopropanol and tert-butanol results from the increase in shielding caused by replacing a hydrogen atom with a methyl group. The high degree of resolution under Figure 46 was also observed for polymeric materials. Therefore, not only can the total hydroxyl concentration be determined but also the type or types present in the polymer.

Figure 46 shows that the presence of water and acid (trifluoracetic acid at 7406 Hz) does not interfere with the hydroxyl measurement. Amines

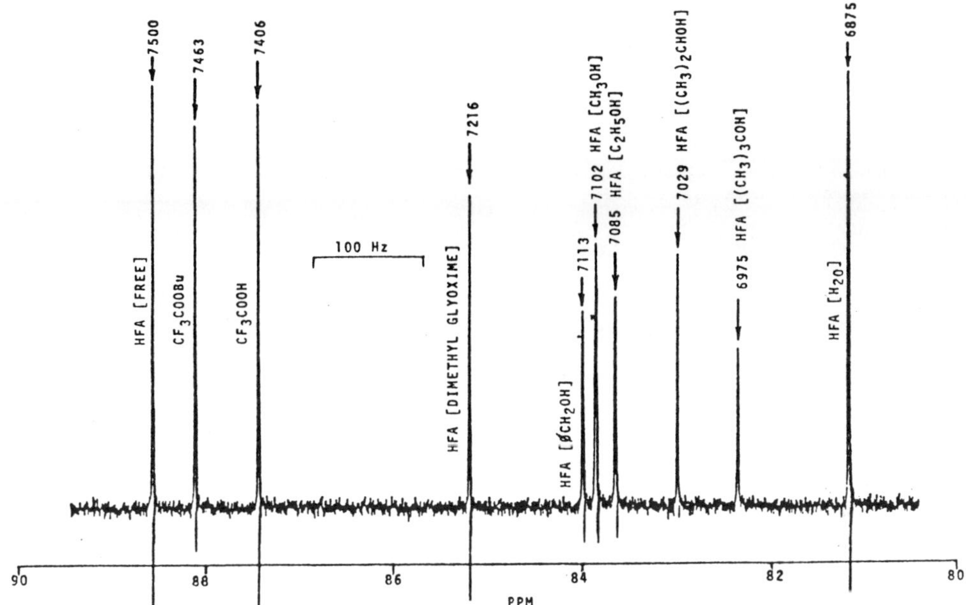

Figure 46   Fluorine-19 NMR spectra of several HFA adducts.   The
number on top are ᴴ3 downfield from C₆F₆ lock.

and amides also do not interfere.   This represents an important advantage
over the method based on chemical titration after acylation, in which
separate runs are needed to correct for the presence of water and acid.
Amines, aldehydes and epoxides would generally interact with the acylation
reagent to varying extents.

The quantative aspect of hydroxyl determination is illustrated in
Table 77 for the analysis of hydroxyl in some polymeric materials.   In
most cases, the fluorine resonance from n-butyl trifluoroacetate (at 7463
Hz in Figure 46) was used as internal standard to calibrate the spectral
integral.   From the date shown in Table 77, it seems that the adduct
formation is quantitative for primary and secondary hydroxyls.   The
reaction normally requires less than 20 minutes at room temperature.
However, as with the chemical acylation methods, tertiary hydroxyl was
found to react only partially, and the reaction for tert-butanol required
about 24 hours to reach equilibrium.   The method can be used for the
determination of hydroxyl groups in polymers of unknown structure.

<u>Infrared spectroscopy.</u>   Infrared spectroscopy is used to determine
hydroxyl groups in polymers[610-612] of the following structure:

Table 77 - Quantitative determination of hydroxyls in
several commerical polymers

Polymers	Hydroxyls	
	Found	Expected
Aliphatic polyethers		
Polyethylene glycol		
Carbowax 200	17.40	16.2-17.9[a]
Carbowax 1000	3.54	3.24-3.58[b]
Polypropylene glycol		
Voranol P1010	3.46	3.37[d]
Voranol P4000	0.85	0.85[e]
Polyepichlorohydrin[f]	0.84	0.85
Aromatic polyethers		
Bakelite phenoxy resin PKHC	5.8	6.0[g]
Monosanto R, J-100 resin	5.4	5.4-6.0
Polyesters		
Atlac 382E	0.92	0.93
Aromatic ester resin No. 1[h]	8.20	8.30
Aromatic ester resin No. 2[h]	12.20	12.35

[a] Based on literature molecular weight of 190 to 210.
[b] Based on literature molecular weight of 950 to 1050.
[c] Determined after addition of 4.7 molar equivalent of trifluroacetic acid and 7.6 molar equivalent of water.
[d] Based on an average molecular weight of 1010.
[e] Based on an average molecular weight of 4000.
[f] Polyepichlorihydrin 2000 from Shell Chemical Co.
[g] Calculated from the idealized molecular structure with a unit weight of 284.
[h] Experimental resin from Hercules Incorporated.

This method utilizes the strong infrared absorption band at 3450 cm^{-1} (2.90 micron). The hydroxyl concentration of approximately 30 mequiv/l is low enough that the hydroxyl groups are completely associated with the tetrahydrofuran spectroscopic solvents, and there are no apparent free or self-associated hydroxy peaks. It is essential in this method that the sample is dry, as water absorbs strongly in the 3450 cm^{-1} region of the spectrum. A further limitation of the method is that any other functional group in the sample, such as phenols, amides, amines and sulphonic acid groups which absorb in the 3400-3500 cm^{-1} (2.94-2.86 microns) region, are likely to interfere in the determination of hydroxyl groups.

Kim et al[613] have developed a method utilizing the strong infrared absorption band of tetrahydrofuran (THF) - associated hydroxyl groups at 3450 cm^{-1} in dinitropropyl acrylate prepolymer. The hydroxyl concentration level of approximately 30 mequiv/l is low enough that the O-H groups are completely associated with the solvent (THF), and there are no apparent free or self-associated with the solvent (THF), and there are no apparent free or self-associated OH peaks. They studied variables such as temperature, concentration, bulk dielectric properties, and the structure of the alcohols to determine their effects on the characteristics of the THF-associated OH absorption peak. These studies show that the infrared method has a general applicability. The hydroxy equivalent weight values by this method compare well with expected values on a variety of

Table 78 - OH equivalent weights of prepolymers: comparison
of IR method with other methods

| Prepolymers | Vendor's eq. wt. | IR method | chemical methods | |
			PA/PY *614	AA/NMIM 615 616*
Hydroxy terminated polybutadiene	1300	1280	1350	1440
Hydroxy terminated butadiene acrylonitrile copolymer	1820	1770		
Hydroxy terminated polycaprolactone	980	970	970	1000
Polyethylene glycol	1660	1630[a]	1670	1740
Hydroxy terminated polytetrahydrofuran	500	480	500	850

[a] Alkoxyethanols used for calibration.

* Reaction with excess acid anhydride in presence of base catalyst PA =
phthalic anhydride, AA = acetic anhydride. Excess anhydride hydrolysed at
end of reaction and carboxylate groups titrated with standard potassium
hydroxide.

commercial prepolymers. The molecules having hydroxy equivalent weight as
high as 3000 were analysed using 1.0 mm solution cells. Contrary to the
chemical methods, the infrared method requires a very dry sample to be
completely soluble in THF.

Kim et al[613] compared hydroxy equivalent weights obtained by this
method and chemical methods of analysis for a range of prepolymers. It is
shown in Table 78 that good agreement is obtained.

6.4 CARBOXYL GROUPS

A method for the determination of carboxyl groups in polymers and
copolymers is discussed in Chapter 9.

Carboxyl end-groups in poly-m-phenyleneisophthalamide have been
titrated potentiometrically in dimethylformamide using a non-aqueous
potassium hydroxide solution[617].

Titration with standard organic bases in non-aqueous medium has been
used to determine carboxyl groups in polyacrylic acid[618,619]. Precipitation as the copper or uranyl salt has been used to determine carboxyl
groups in carboxy methyl cellulose[620].

Hydroxy groups in carboxy terminated polybutadiene have been
determined by infrared spectroscopy[621,622].

6.5 CARBONYL GROUPS

Various physical techniques such as infrared and NMR spectroscopy
have been described for the determination of very low concentrations of
carbonyl groups in polymers. This is discussed in Chapter 9.

Germanenko[623] determined low concentrations for carbonyl groups in PVC and vinyl chloride-vinyl acetate copolymers. In this method the carbonyl groups in the polymer are reacted with 2,4-dinitrophenyl hydrazine to produce the corresponding phenyl-hydrazone. Excess reagent is washed away from the polymer, which is then digested with concentrated sulphuric acid to convert the bound hydrazone to ammonium sulphate, which is then estimated using Nessler's reagent.

Direct spectrophotometry in the visible and ultraviolet region has been used to determine low concentrations of functional groups in the surface of polymer films. Thus Kato[624] has followed the regeneration of carbonyl groups from 2,4-dinitrophenyl-hydrazones formed on the surface of irradiated polystyrene films by absorption measurements at 378 nm.

6.6 OTHER OXYGENATED GROUPS, ESTER, ALKOXYL, ETHER, EPOXY, OXIRANE

<u>Hydriodic acid reduction - gas chromatography.</u> Alkoxyl and ester groups have been determined in polymers and copolymers by the Ziesel procedure involving reaction with anhydrous hydrogen iodide at 100°C.

Alkoxy groups

e.g. $- CH_2 = CH - OR + HI \rightarrow CH_2 = CHI + ROH$
          | |
         Me   Me

Ester groups

e.g. $- CH_2 = CH - COOR + HI \rightarrow CH_2 - CHCOOH + RI$
          | |
         Me   Me

Hydrolysis using hydriodic acid has been used for the determination of the methyl, ethyl, propyl, and butyl esters of acrylates, methacrylates, or maleates[625,626], and the determination of polyethyl esters in methylmethacrylate copolymers[628,629]. First the total alcohol content is determined using a modified Ziesel hydriodic acid hydrolysis[627]. Secondly, the various alcohols, after being converted to the corresponding alkyl iodides, are collected in a cold trap and then separated by gas chromatography. Owing to the low volatility of the higher alkyl iodides the hydriodic acid hydrolysis technique is not suitable for the determination of alcohol groups higher then butyl alcohol. This technique has also been applied to the determination of alkoxyl groups in acrylate esters[626].

The Ziesel reaction has been extensively studied and has been used successfully for the determination of alkoxyl groups in cellulosic materials[630-632] and the determination of ether groups in cellulose and polyvinyl ethers[633]. Quantitative cleavage of the alkoxyl groups in polymers is routinely obtained. However, as discussed in the previous section on the determination of ester groups, hydriodic acid also cleaves any ester linkages on the polymer backbone, giving positive interference.

Epoxy groups in epoxy resins have been determined by spectrophotometric titration with halogen acid[634] and spectrophotometrically using 2,4-dinitrobenzene sulphonate as chromophore.

Saponification in alcoholic potassium hydroxide has been used to determine ester groups in alkyd resins and polyesters[640-642] and cellulose, polyvinyl ester, polyacrylates and polymethyacrylates[643-645].

NMR spectroscopy has been used to determine ethylene oxide units in polyethylene oxides[636].

Oxirane rings in epoxy resins have been determined by ring cleavage with pyridine-hydrochloric acid[639].

6.7 ALKYL AND ARYL GROUPS

The technique of alkali fusion reaction gas chromatography has been applied to the determination of alkyl and aryl groups in polysiloxanes[646,647]. The method involves the quantitative cleavage of all organic substituents bonded to silicon, producing the corresponding hydrocarbons.

$$
\begin{array}{c}
CH_3 \\
| \\
CH_3 - Si - \\
| \\
CH_3
\end{array}
\left[
\begin{array}{c}
CH_3 \\
| \\
O - Si - \\
| \\
CH_3
\end{array}
\right]_n
\begin{array}{c}
CH_3 \\
| \\
O - Si - CH_3 \\
| \\
CH_3
\end{array}
\xrightarrow{KOH}
$$

$$
\begin{array}{c}
OK \\
| \\
KO - Si - \\
| \\
OK
\end{array}
\left[
\begin{array}{c}
OK \\
| \\
O - Si \\
| \\
OK
\end{array}
\right]_n
\begin{array}{c}
OK \\
| \\
- O - Si - OK + 8CH_4 \\
| \\
OK
\end{array}
$$

Reactions are driven to completion, with no apparent decomposition, in less then 10 minutes by fusing the sample with potassium hydroxide in an inert atmosphere. After concentration of the volatile products, they are separated and determined by gas chromatography. Sample losses are minimised by performing the total analysis in a single piece of apparatus. Fluids, gum rubbers, and resins are handled with equal ease. The percent relative standard deviation of the method is 1.00%; the average deviation between experimental and theoretical or check method result is 0.5% absolute.

The technique for carrying out potassium hydroxide fusion procedure is the same as the analysis described in the previous section[646], for the analysis of polyamides and polyimides. The alkali fusion products of silicones were separated on matched 6 foot by 1/8 inch o.d. stainless steel columns packed with 50-80 mesh Poropak Q. A Perkin-Elmer Model 990 Gas Chromatograph, equipped with dual thermal conductivity detectors and a linear column temperature programmer, was used. The column temperature was maintained at 90°C during the reaction period. Upon injection of the products, time zero in the chromatograms the temperature was programmed from 90 to 260°C at 8°C per minute. Peak areas were measured with a Digital Integrator. The detector signal was displayed on a recorder.

Calibration curves are prepared for each compound produced. Products which are gases under normal conditions, are injected with a calibrated gas-tight syringe through the septum inlet of the reaction tube. Ambient temperature and pressure corrections are necessary to determine the actual amounts of gas injected. Standard solutions of liquid or solid reaction products are injected by syringe into an empty boat. These products are volatilized by pushing the boat into the heated reaction zone. Solid products having low vapour pressures at room temperature are weighed directly into platinum boats, covered with reagent and stored with the samples.

Table 79 - Reaction methods for the quantitative determination
of organic subtituents bonded to silicon

Group determined	Reagent	Reaction conditions	Product	Analysis	Ref.
Phenyl	60% aqueous KOH in DMSO	2 h at 120°C	Benzene	GC	648
Phenyl	Bromine in glacial acetic acid	Boiling solution	Bromoben-zene	Titration of excess bromine	649
Ethyl and phenyl	Phosphorus pentoxide and water	30-580°C over 45 min	Ethane and benzene	GC-FID	650
Methyl and ethyl	Powdered potassium hydroxide	2 h at 250-270°C	Methane and ethane	Gas buret	651
Methyl	Sulphuric acid	20 min at 280-300°C	Methane	Gas buret	652,653
Phenyl	Ethylbromide in the presence of aluminium chloride	...	Hexaeth-yl benzene	Gravimetric	654
Vinyl	Phosphorus pentoxide and water	80-600°C over 40 min	Ethylene	GC-FID	655
Vinyl	Phosphorus pentoxide	Ambient to 500°C	Ethylene	GC-FID	656
Vinyl	90% sulfuric acid	75-250°C at 10°C/min and 1 h at 250°C	Ethylene	GC-TC	657
Vinyl	Sodium hydroxide pellets	300°C for 15 min	Ethylene	Colorimetric	658
Vinyl	Potassium hydroxide pellets	Heat with Meker burner	Ethylene	GC-FID	659

6.8 Total Silanal, Silane Hydrogen Groups and Tetrapropoxy Silane and Diphenyl Silane Cross-linking Agents

The preferred approach is to use secondary standards, i.e. well characterised compounds or polymers that quantitatively react with the reagent to give the desired product. For example, when only methane is determined, GE Viscasil 60,000, a high molecular weight polydimethyl-siloxane, is used. DC 704 a polymethylphenylsiloxane, is used as the secondary standard when both methane and benzene are determined.

In all cases, the standards are trapped and chromatographed in exactly the same manner as the reaction products. The best straight line calibration curves are determined by a least-squares regression curve fitting computer program.

Various other reagents have been used for the determination of organic substituents bonded to silicon in organosilicon polymers (Table 79).

Total silanol (SiOH) silane hydrogen (SiH) groups and tetrapropyoxy-silane and diphenylmethyl silanol cross linking agents. Dubiel et al[660] described methods for determining these reactive components in room temperature vulcanized silicone foams. Total SiOH and SiOH are determined by Fourier transform infrared (FTIR) spectrometry, the SiOH peak at 3687 cm^{-1} and the SiH peak at 2168 cm^{-1} were used for quantitation. The tetrapropoxysilane content was determined by gas chromatography using a solid capillary open tubular (SCOT) column and linear programmed temperature control. The diphenylmethylsilanol content was determined by gel permeation chromatography using the tetrahydrofuran solvent.

6.9 AMINO GROUPS

Spectroscopy using p-dimethyl amino benzaldehyde[662] or ninhydrin[663] and fluorescence analysis[661] have been used to estimate primary amino groups in polymers.

6.10 NITRIC ESTER GROUPS

These have been determined in nitrocellulose by saponification, followed by reduction of the nitro group with devadas alloy and determination of the ammonia produced[664].

6.11 NITRILE GROUPS

These have been determined in polystyrene by a dye partition method[665].

FRACTIONATION OF POLYMERS

7.1 INTRODUCTORY

Polymers normally do not consist of a particular molecule with a unique molecular weight, but rather are a mixture of molecules with a molecular weight range which follows a distribution. With some types of polymers the picture is further complicated by the appearance of what are known as crosslinks. These are chemical bonds which link one polymer chain to another. Crosslinking will, therefore increase the molecular weight of a polymer and, incidentally, decrease its solubility in organic solvents. These are some of the features which make it possible to produce for a given polymer, say polypropylene, a range of grades of the polymer, each with different physical properties and end-uses and each characterized by a different molecular weight distribution curve and degree of crosslinking. The factors which control these parameters in a polymer are complex, and are linked with the details of the manufacturing process used. They will not be discussed further here. The measurement of the molecular weight is a task undertaken in its own right by polymer chemists, and is concerned with the development of new polymers and process control in the case of existing polymers. Additionally, however, it is necessary to separate a polymer not into unique molecules each with a particular molecular weight, but into a series of narrower molecular weight distribution fractions. This is required in order to obtain a more detailed picture of the polymer structure and these separated fractions may be required for further analysis by a wide range of techniques.

In the simplest case, discussed in section 7.2 below, it is required simply to separate, for example, the total gel fraction (i.e. crosslinked material) of a polymer from the total soluble fraction (non-crosslinked material). This is typified in the example discussed below on the separation of polystyrene into its gel and soluble fractions. In a more complicated case it may be required to carry out a separation of the original polymer into a series of fractions each with a narrower molecular weight distribution than the parent polymer. These methods are usually based on fractionation techniques, based on gradient elution, (Chapter 7.13), precipitation, (Chapter 7.14), or, more recently on one of the liquid chromatographic techniques, (high performance liquid chromatography), (Chapter 7.7), size exclusion chromatography, (Chapter 7.8) or supercritical fluid chromatography (Chapter 7.9). Thin layer chromatography (Chapter 7.5) and gas chromatography (Chapter 7.6) have been used to a limited extent for polymer fractionation. Finally, there is the case where it is required to determine the molecular weight distribution of the polymer. Depending on the type of polymer being examined many methods exist for carrying out these measurements which are discussed in Chapter 8.

Table 80 - Determination of monomer units in ABS terpolymers

Determined	Cycolac T 1000 natural nibs	Cycolac H 1000 natural nibs	Kralastic M H nibs	
Analysis of additive free polymer				
Butadiene (%)[1]	20.5	28.6	19.1	
Actylonitrile (%)[2]	23.8	20.9	20.9	
Styrene (%)[3]	54.0	48.5		
	(total = 98.3%)	(total = 98.0%)		
Analysis of insoluble fraction				
Insolubles in polymer (%)	56.0[1]	51.5[1]	63.5	11.0
Butadiene (%)	30.8	31.2	38.2	49.9
Acrylonitrile (%)	22.1	20.6	19.2	10.7
Sytrene (%)	48.0	48.0	-	-
	(total = 100.9%)	(total = 99.8%)		
Analysis of soluble fraction				
Solubles in polymer (%)	44.0	48.5	-	-
Butadiene (%)	10.3	7.0	-	13.9
Acrylonitrile (%)	27.6	28.7	-	21.5
Styrene	62.0	64.0	-	61.0
	(total = 99.9%)	(total = 99.7%)	(total = 96.4%)	

[1] Iodine monochloride

[2] Kjeldahl method

[3] Infrared method

7.2 MEASUREMENT OF CROSSLINKED GEL CONTENT

A good example of this is the separation for the two fractions from a styrene-butadiene-acrylonitrile (ABS) polymer and the determination of the monomer units in the separated fractions.

To remove non-polymer additives the polymer is dissolved (or dispersed) in chloroform. This solution is slowly poured into an excess of stirred methanol to reprecipitate the polymer and leave the soluble non-polymer additives in a clear chloroform-methanol phase which can be separated from the polymer by filtration. The polymer is then washed with methanol and vacuum-dried.

Contact with methyl ethyl ketone followed by high-speed centrifuging is an excellent method of separating the two fractions. Table 80 reports complete gel and soluble fraction analysis carried out on various ABS polymers. In both the whole ABS polymer and in its soluble and insoluble fractions the sum of the determined constituents usually adds up to 100 $\pm$ 2%.

Methods have been described for the determination of gel in PVC[666-668], vinyl chloride-propylene copolymer[668], polybutadiene-polyisoprene copolymer, styrene-butadiene copolymer, acrylonitrile-butadiene copolymer, polyacrylonitrile[667] and styrene-acrylonitrile copolymer[669].

7.3 GRADIENT ELUTION TECHNIQUES

Extraction solvents boiling point gradient. The time-consuming and laborious nature of the earlier fractionation procedures is illustrated well by the work of Nakajima[670] on the fractionation of polyethylene and its thermally degraded products, involving extraction with boiling hydrocarbons with increasing boiling points between 45°C and 95°C, and on the fractionation of polypropylene[671]. It was necessary to extract the polymer using a Soxhlet apparatus with 17 different hydrocarbon fractions based on normal paraffins with different boiling temperatures in the range from 35°C to 135°C. The extreme laboriousness of such procedures is self-evident.

Fractional extractions of polymers by the column technique is no less laborious. Two types of column extraction procedure are known. Gradient elution fractionation is achieved at a given temperature by making use of solvents with gradually increasing solvent power, (Chapter 7.3.2), or the increasing temperature fractionation, (Chapter 7.3.3) is performed with a given solvent at increasing temperatures. According to the findings on column techniques by Wijga et al[672] the gradient elution of the polymer separates fractions according to molecular weight, whereas the increasing temperature method fractionates the polymer mainly according to tacticity.

Gradient elution with solvents of increasing solvent power at constant temperature. The polymer is packed in a column and a solvent mixture passed down the column. Initially the solvent is a poor one for the polymer. Then increasing proportions of a better polymer solvent are incorporated in the solvent mixtures utilizing a gradient elution technique. A series of fractions is thus obtained, containing different molecular weight fractions of the original polymer. Methods have been described for the large-scale elution fractionation of ethylene-propylene copolymers[673], methyl methacrylate-styrene copolymers[674], polyethylene[675], and polypropylene[676].

Increasing (or decreasing) temperature gradient of extraction solvent fractionation. When a solid polymer packed in a column is contacted with a continuing flow of solvent in which it has a limited solubility at room temperature then, as the temperature of the column is raised in a controlled programme, fractions of the polymer with different tacticities will progressively dissolve. Collection of portions of the column eluate at various temperatures will provide a series of fractions.

Akutin et al[676] have shown that by a thermal precipitation method the molecular weight distribution of low-density polyethylene could be deduced from the precipitation curves using simple calculations. Ogawa and Hoshino[677] compared fractionations of isotactic polypropylene by using temperature and solvent gradient methods. Comparison of results agreed fairly well on fractionation of polyethylene-polypropylene blends between hypothetical calculations and experimental data by the solvent gradient method[678].

7.4   PRECIPITATION TECHNIQUES

   <u>Fractionation by polymer freezing.</u> Polystyrene dissolved in benzene
has been fractionated by slowly freezing the solutions with dry ice and
alcohol. A more detailed treatment of this method is given by Loconti and
Cahil[679,680]. The polymer in the first frozen-out portions was of higher
molecular weight than later. Ruskin and Parravano[681] were able to
fractionate polymer dissolved in cyclohexane by both the zone-melting and
freezing techniques.

   Bryson et al[682] investigated the fractionation of polystyrene by
slow freezing of dilute benzene solutions of the polymer using ice-water
mixtures instead of dry ice-alcohol. Bryson et al[682] concluded that no
fractionation of the polymer occurs according to molecular weight from
benzene solutions.

   <u>Solvent - non-solvent precipitation.</u> In this technique the polymer
is dissolved in a solvent and then gradual additions are made of a non-
solvent to the polymer. The polymers precipitated after each non-solvent
addition are separately collected. An example is the case of
vinylchloride-vinyl acetate copolymer using solvent - non-solvent systems
such as acetone-methanol, acetone-heptane and tetrahydrofuran-water[683].

   Precipitation with methanol from benzene solution has been used to
fractionate methacrylic acid-styrene copolymer[684], polychloroprene[685,686],
polybutadiene[685,686], and polyisoprene[685,686]. Other solvent - non-
solvent systems that have been used include nitrobenzene-tetrachloroethane
(polyethylene terephthalate[687]) and ethylene carbonate-ethylene
cyanohydrin or methyl ethyl ketone-cyclohexane (styrene-acrylonitrile
copolymers)[688].

7.5   THIN-LAYER AND PAPER CHROMATOGRAPHY

   Thin-layer chromatography is a useful technique for separating
polymers into molecular weight fractions on a fairly small scale. It has
been used to fractionate polyethylene terephthalate[688,691], styrene-
butadiene copolymers[692], styrene-acrylonitrile copolymers[695], polyoxy-
propylene glycols[696], Nylon-styrene graft copolymers, polymethylmethacrylate[697-699], styrene-methacrylate polymers[700], poly-alpha-methyl-
styrene[701], polyvinyl acetate-styrene copolymers and polyvinyl alcohol-
styrene copolymers[702]. Various solvents have been used for migration on
thin-layer plates including benzene-acetone,[694], toluene-acetone[693,694],
chloroform-diethyl ether,[691], chloroform-ethyl acetate and chloroform.

   Paper chromatography has been used to carry out molecular weight
distribution measurements on poly(ethylene glycols)[764].

7.6   GAS CHROMATOGRAPHIC METHODS

   Gas chromatography is, of course, limited to lower molecular weight
polymers which can be volatilized.

   Mikkelson[703] reported a gas chromatographic analysis of polyethylene
glycol 400; the original sample is injected directly into the gas
chromatograph, a technique also employed by Puschmann[704]. Celedes
Pacpuot[705] converted the polyethylene glycols into methyl ethers in a
reaction with dimethylsulphate before injection into the gas
chromatograph.

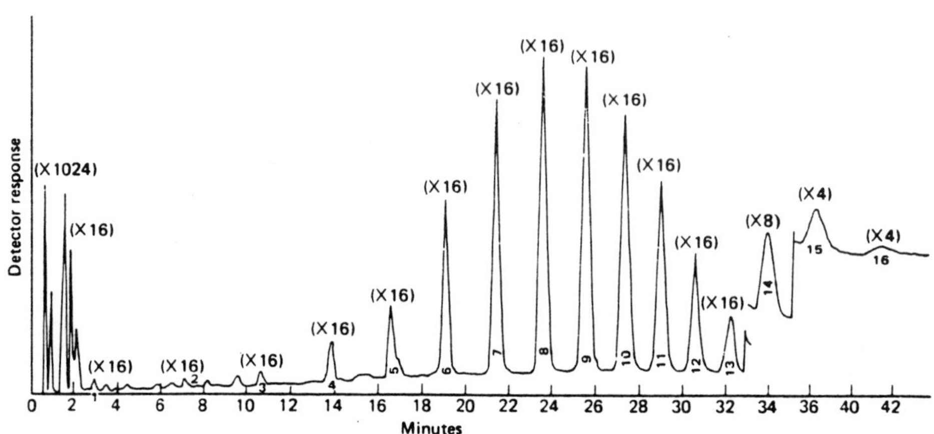

Figure 47   Gas chromatogram of trimethylsilyl derivative of PEG
            400. Dual columns 3 ft by 0.25 in o.d.: 5% SE-30 on
            Chromosorb G. Temperature programmed from 100 to 370°C
            at 10°C/min and then isothermal at 370°C.

The rapid formation of trimethylsilyl ether derivatives of
polyhydroxy compounds, followed by their separation and estimation by gas
chromatography, has been described by Sweeley et al[706]. This most useful
technique has been applied to the liquid polyethylene glycols 200, 300 and
400.   Fletcher and Persinger[707], studied this application using
poly(ethylene glycols) 200, 300 and 400 and reported the determination of
response factors relative to ethylene glycol for conversion of peak areas
to weight per cent when a thermal conductivity detector is used.   A
typical chromatogram of PEG 400 is shown in Fig 47, p-toluene sulphonic
acid catalysed acetylation has been used to produce volatile derivatives
of propylene oxide-glycerol condensates which are suitable for gas
chromatography[708]. Graphs relating initial retention distance on the gas
chromatogram with molecular weight give straight line relationships.

Uttenbach et al[709] characterized and determined formaldehyde
oligomers by capillary column gas chromatography. The homologous samples
of polyoxymethylene glycols and the corresponding mono methyl ethers were
derivativized with ammonia prior to gas chromatography.

7.7   LIQUID CHROMATOGRAPHIC METHODS

Three main forms of liquid chromatography have been used to
fractionate polymers and in some cases to determine average molecular
weights.   These techniques, discussed separately below, are high
performance liquid chromatography (Chapter 7.7.1), size exclusion chromatography (SEC) (alternatively known as gel permeation chromatography
(GPC)) (Chapter 7.7.2), and supercritical fluid chromatography, (SFC),
(Chapter 7.7.3).

7.7.1 High Performance Liquid Chromatography

Chromatography on activated silica gel columns.   This is another
technique for achieving fractionation of polymers.   It is based on
functionality distribution and has been applied to the fractionation of
polyoxyethylene[710], polyoxypropylene[710], polytetrahydrofuran[710], polyepi-

chlorohydrin, oligobutadienes[710], and carboxy- and hydroxy - terminated polybutadiene[711,712]. Isotactic polymethylmethacrylate has been separated from polymethylmethacrylate using competitive absorption on silica gel from chloroform solution. This method is unsatisfactory for unsaturated polymers due to poor recovery from the column, apparently caused by polymerization on the column. Law[713,714] circumvented this problem by using partially deactivated silica gel, and recycling the unfractionated portion through additional columns of increasing activity to obtain the desired resolution. Using this technique he separated carboxy-polybutadienes and hydroxy-polybutadienes according to functionality using stepwise elution from silica gel. Recoveries in the 95-199% range were achieved. Subjection of the fractions obtained from the silica gel separation to analysis via gel-permeation chromatography of chloroform solutions and infrared or near-infrared spectroscopy yielded not only functionality distribution data, but also provided the relationship between molecular weight distribution and functional type.

The accurate determination of the chemical composition distribution for copolymers is very important for the characterization of copolymers. Among several techniques to measure chemical composition distribution, high-performance liquid chromatography holds great promise because of its high efficiency. The following separations on a silica gel column have been reported; styrene methyl acrylate copolymers on a silica gel column[715], styrene acrylonitrile copolymers by precipitation liquid chromatography[716], styrene butadiene copolymers on a polyacrylonitrile gel column[717], styrene methyl methacrylate copolymers on a silica gel column[718,721,722], styrene methyl methacrylate block copolymers by column adsorption chromatography using a 50 mm i.d. cylindrical column[719] and styrene-n-butyl methacrylate copolymers by orthogonal chromatography[729]. High performance liquid chromatography on unmodified or activated silica has been used to separate prepolymer oligomers (defined as polymers of molecular weight less than 10.000) of polystyrene[723,724], p-alkylphenol-formaldehyde[725], and polyethylene terephthalate[726].

<u>Separation of styrene- methylmethyacrylate copolymers.</u> The separation has been reported[722] of styrene-methyl methacrylate random copolymers according to chemical composition by liquid adsorption chromatography. Silica gel was used as an adsorbent. The copolymers were separated by stepwise gradient elution using chloroform and 1,2-dichloroethane as mobile phases. When dichloroethane was used as the mobile phase, the copolymers adsorbed on the surface of silica gel, though polystyrene eluted from the column. When chloroform was used as the mobile phase instead of dichloroethane, the copolymers having a methyl methacrylate component less than 45% could be eluted as well as polystyrene. With increasing chloroform content in a mixture of chloroform and dichloroethane, the copolymers have been separated as a function of their compositions.

Mori and Uno[721], investigated in detail the effects of ethanol content in the mobile phase on the separation of the copolymers by linear gradient elution on a 30 A silica gel column and also investigated temperature effects. Chloroform without ethanol retained the copolymers in the column. By the addition of ethanol to chloroform, copolymers having less methyl methacrylate started to elute, and with increasing ethanol content in chloroform, those having more methyl methacrylate could be eluted. The copolymers tend to adsorb on the column at higher column temperature and those having more methyl methacrylate require a lower column temperature for elution. Ethanol content or column temperature did not affect peak retention volume for the copolymers. The effects of both ethanol concentration and column temperature were attributed to the change

Table 81 - Effect on column temperature on elution of
styrene-methylmethacrylate copolymer
(1% ethanol)(1:99 v/v ethanol:chloroform)

Copolymer composition % styrene	% recovery of copolymer in eluant. i.e. % of peak height obtained when copolymer is 100% eluted
85.5	100% between 10 and 80°C
73.4	100% between 10 and 55°C
66.6	100% between 10 and 25°C
57.4	100% between 10 and 15°C
48.7	35% at 10°C
42.1-15.2	nil at 10°C

of population of free silanol groups on the surface of silica gel, because
the hydrogen bonding of carbonyl groups in the copolymers to the silanol
groups was the main mechanism of this separation.

Measurements were performed on a Jasco Trirotar-VI high performance
liquid chromatograph (Japan Spectroscopic Co., Ltd., Hachioji, Tokyo 192,
Japan) with a variable-wavelength ultraviolet absorption detector Model
Uvidec-1001V at a wavelength of 354 nm. Attenuation of the detector was
0.64 AUFS. Silica gel with a pore size of 30 A and a mean particle
diameter of 5 um (Normura Chemical Co., Seto, Aichi 489, Japan) was packed
in 4.6 mm i.d. x 50 mm length stainless steel tubing. This column was
thermostated at a specified temperature to a precision of 0.1°C by using a
column jacket in which constant-temperature water was circulated.

Elution. The mobile phase was a mixture of ethanol-free chloroform
and ethanol or 1.2 dichloroethane and ethanol. In the isocratic elution
mode the flow rate of the mobile phase was 0.5 ml/min, linear gradient
elution was performed as follows: the initial mobile phase (A) was a
mixture of chloroform and ethanol (99.0:1.0 v/v); the composition of the
final mobile phase (B) was 95.5:4.5 (v/v) chloroform/ethanol; and the
composition of the mobile phase was changed from 100% A to 100% B in 15
minutes linearly after injecting a sample solution. Samples (0.05%) were
dissolved in the initial mobile phase.

The effect of column temperature on copolymer elution is interesting
and is illustrated in Table 81. Thus copolymers tend to adsorb on the
columns at higher column temperatures and copolymers with a higher methyl
methacrylate component require lower temperatures to fully elute from the
column

The influence of ethanol content of the eluant on the copolymer
recovery obtained is illustrated in Table 82. This demonstrates that at a
constant column temperature (10°C) copolymers with a higher methyl
methacrylate content can be eluted with increasing ethanol content in the
chloroform or 1,2 dichloroethane eluant. Thus, higher ethanol contents
and lower column temperature favour quantitative elution of copolymers.
However, when copolymer elution occurs, all the copolymers eluted at the
same retention volume. Linear gradient elution does separate the copolymers according to composition as seen in Figure 48.

Table 82 - Effect of ethanol content of eluant on elution of
styrene-methyl methacrylate copolymer

% ethanol in chloroform or 1.2 dichloroethane	Column temperature °C	% styrene in copolymer	
		100% copolymer eluted	copolymer does not elute or only partially elutes
0	10	-	85.5 - 15.2
<0.5	10	-	48.7
0.5	10	85.5	73.4 - 15.2
1.0	10	85.5 - 57.4	48.7
1.5	10	85.5 - 42.1	41.5 - 15.2
2.0	10	85.5 - 28.5	15.2
2.5	10	85.5 - 15.2	-

In adsorption chromatography with many eluants, an increase in column temperature results in a decrease of retention. Thermodynamics can predict the retention of a solute from the enthalpy change (ΔH) and the entropy change (ΔS) as

$$\ln K = -\Delta H/RT + \Delta S/R \tag{1}$$

where K is a partition coefficient and the sign of H is usually negative. In the case where the retention volume of a solute increases with an increase in column temperature, ΔH should be positive. Adsorption of several types of polymers (e.g., poly(ethylene glycol), poly(vinyl acetate), and poly(ethylene terephthalate)) on the surface of silica gel has been shown to increase with increasing column temperature[727]. In these cases, (ΔH) must be positive. The conformation of polymers in solution also changes with temperature, and the entropy change may contribute to the increase of K with increasing column temperature to some extent. However, in these cases, the change of K with temperature results in the change of retention volume. In the case of the styrene methyl methacrylate copolymers studied by Mori and Uno[721] the retention volumes of the sample copolymers were unchanged with temperature. The elution of the copolymers was very simple. The copolymers either eluted from the column at the exclusion limit, which is proportional to the interstitial volume in the column, or they did not; that is, the adsorption isotherm is on the Y axis when the copolymers eluted or is on the X axis when they were retained in the column. Therefore, in the case of these copolymers, we have to consider not the change of enthalpy or entropy, but other mechanisms.

Mori and Uno[721] found that in the case of a styrene (48.7%) - methyl

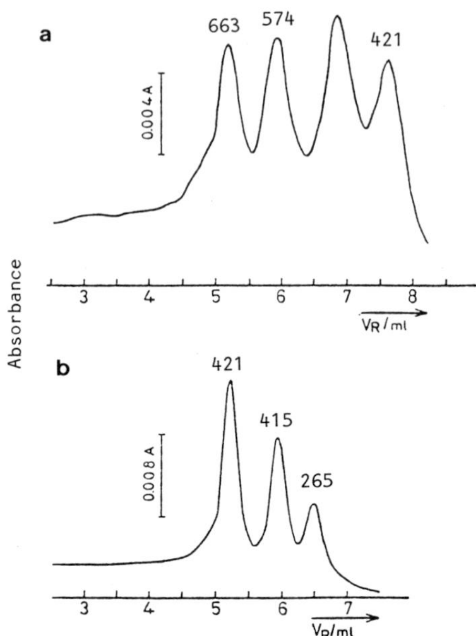

Figure 48 LAC chromatograms of P(S-MAA) copolymers obtained by the
          linear gradient elution method; column temperature (a)
          80°C (b) 30°C; sample (a) 0.013% each; (b) 0.017% each;
          injection volume 100 ul; for other conditions, see text.

methacrylate (51.3%) copolymer the fraction eluting (63%) in chloroform:
ethanol (99:1%) had different weight average molecular weights ($\overline{M}w$) and
number average molecular weights ($\overline{M}n$) to the fraction which remained
adsorbed on the column (37%), (Table 83). The fraction having the larger
methyl methacrylate content (fraction 2) had the higher average molecular
weight.

     The following mechanism is proposed for the separation of styrene-
methyl methacrylate copolymers eluted in chloroform-ethanol on silica gel.
A copolymer of a given composition elutes from the column when the
chloroform/ethanol mobile phase contains above a certain amount of ethanol
and is retained in the column below that value. The lower the methyl
methacrylate content of the copolymer, the lower is the ethanol content of
the mobile phase required to elute the copolymers from the column.
Adsorption will occur above a certain value of either the population of
free silanol groups on the surface of silica gel or the population of
carbonyl groups of the segment of the copolymers contacting the surface of
silica gel. A critical content of ethanol in the mobile phase will exist
for a copolymer of a certain content at the specified temperature, and a
critical composition of the copolymer will also exist for the mobile phase
of a certain composition.

     At elevated temperature, ethanol included as a moderator in
chloroform is desorbed from the silica gel stationary phase and removed
from the column[728]. Consequently, the ratio of ethanol in the stationary
phase to that in the mobile phase decreases with elevating column
temperature and free silanol groups on the surface of silica gel will
increase. As a result, the stationary phase is apt to adsorb more methyl

Table 83 - Composition and molecular weight averages of fractions
of copolymer V

Fraction no.	Obtained %	Styrene content mol%	mol wt average*	
			$\bar{M}_w$	$\bar{M}_n$
1	63	50.5	1.13×10^5	4.5×10^4
2	37	47.7	1.60×10^5	7.0×10^4
native	100	48.7	1.21×10^5	6.8×10^4

* determined by size exclusion chromatography using a calibration curve
obtained with polystyrene

methacrylate.   Therefore, an increase in adsorption of the copolymers with
increasing column temperature results from the change of ethanol content
in the stationary phase.

   <u>Separation of oligomers.</u>   High performance liquid chromatography has
been used to [729] functionate and determine oligomers of polyethylene
terephthalate containing from one to seven terephthalolyl repeat units as
shown below;

$$HOOC-C_6H_4-COOH$$
$$+$$
$$HOCH_2CH_2OH$$

$$HOOC-C_6H_4-COOCH_2CH_2OH \qquad HOCH_2CH_2-OCO-C_6H_4-COOCH_2CH_2OH$$

MHET   BHET
mono(2-hydroxyethyl)   bis(2-hydroxyethyl)terephthalate dimer
terephthalate   also trimer, tetramer, pentamer, hexamer
                              and heptamer

   Gas chromatographic methods[730,731] for determining unreacted
terephthalic acid, ethylene glycol and mono(2-hydroxy ethyl/terephthalate
are capable of determining only about 25% of the total prepolymer sample
as the higher molecular weight oligomers are not resolved.   High
performance liquid adsorptions on a Dupont Model 530 liquid chromatograph
with (100:1 v/v) chloroform:reagent alcohol eluent using a Dupont Zorbax
SIL Column and a 254 mm ultraviolet detector gives good separations of the
seven polyethylene terephthalate prepolymer oligomers.   The bis (2-
hydroxyethyl) terephthalate (BHET) content determined by gas
chromatography is used as the known standard for the high performance
liquid chromatographic separation.   All chromatograms were run at room
temperature.   The solvent system was chloroform (1250 parts by volume) and
reagent alcohol (12 parts by volume).   The reagent alcohol consisted of 90
parts ethanol and 5 parts each methanol and isopropanol.   Approximately 15
mg of sample, weighed to 0.1 mg accuracy, were placed in a 3-dram snap-cap

vial, and 5 ml chloroform/alcohol (9:1 by volume) were added. The sample was stirred on a magnetic stirrer until in solution. If complete solution was not attained, the solution was filtered through a 0.5 um Fibreglass filter using a Swinney filter holder. Valve injections were made. The higher oligomers are calculated from the following equation:

$$\text{Weight \% Oligomer} = \frac{\text{Area Oligomer x \% BHET}_{GC}}{\text{Area BHET}_{HPLC}}$$

where $\% \text{BHET}_{GC}$ = BHET by gas chromatography, and Area BHET_{HPLC} = BHET peak area by high performance liquid chromatography.

Chromophoric complexes of polyester monomer, dimer and trimer have approximately the same absorbance at a given wavelength per terephthalolyl equivalent. Since the oligomer equivalent weights per terephthaloyl group are also approximately the same, the equal UV weight response assumption appears to be valid.

A plot of log retention time vs. oligomer number illustrates that reagent alcohol consisting of 99% v/v chloroform, and 0.5% v/v each of methanol and isopropanol is more effective modifier for resolving the higher oligomers (n = 4-7) then is 99% v/v chloroform 1% methanol.

Table 84 lists the high performance liquid chromatographic results with precision data for two prepolymer samples. These results show the greatly increased assay of the prepolymer possible by using the high performance liquid chromatographic method. The prepolymer chromatogram shows the presence of oligomers higher than heptamer which are not well enough resolved for quantitation.

Mourney et al[732] separated vapour molecular weight anionically n-butyl lithium polymerized polystyrene standards (general structure CH_3 $(CH_2)_2$ CH $_2$ CH Ph)$_n$ CH_2 CH Ph) with $\bar{M}w$ values of 800, 2100 and 4800 on silica gel with 3:1 v/v n-hexane-tetrhydrofuran, n-hexane-ethyl acetate or n-hexane-dichloromethane gradients. Tetrahydrofuran and ethyl acetate elements gave separations according to the number of oligomer units and dichloromethane separated the stereoisomers of individual oligomers. The polystyrenes were dissolved in 3:1 hexane/B solvent where B was dichloromethane, tetrahydrofuran or ethyl acetate. Sample concentrations were 3.3-100 mg/ml. The samples (10 ml) were injected onto a 4.6 mm i.d. x 250 mm column packed with either LiChrosorb Si60 silica (E. Merck, 5 um particle diameter) or Hibar II LiChrospher Si500 (10 um particle diameter). Nominal pore diameters are 6 and 50 nm, respectively, for Si60 and Si500. The LiChrosorb Si60 column was packed by the stirred-slurry method, and the Hibar II column was obtained commercially. Both columns were thermostated at 30.00 $\pm$ 0.05°C. Polystyrene oligomers were gradient eluted by use of either two Waters Associates M6000A pumps with a 720 system controller or a Varian 5060 liquid chromatograph. Ultraviolet absorbance of the eluant was monitored at 265 nm with a Perkin-Elmer LC-55 variable-wavelength detector.

In Figure 49 are shown separations of $\bar{M}w$ 2100 polystyrene oligomers on 6 mm pore size silica achieved using the three solvent systems mentioned above. Oligomer retention times with ethyl acetate gradient are shorter than those obtained with tetrahydrofuran, but the distribution profiles obtained from both solvents are similar, and in both examples

Table 84 - Polyester prepolymer HPLC analytical results

Coponent		Lot no 1	Lot no 2
BHET	a	10.15	12.95
Dimer	$\bar{x}$	14.52	16.55
	std dev	±0.35	±0.71
	RSD	2.4%	4.3%
Trimer	$\bar{x}$	13.30	14.46
	std dev	±0.63	+0.79
	RSD	4.7%	5.5%
Tetramer	$\bar{x}$	10.42	9.14
	std dev	±0.76	±1.02
	RSD	7.3%	11.1%
Pentamer	$\bar{x}$	6.90	5.19
	std dev	±0.59	±0.39
	RSD	8.6%	7.5%
Hexamer	$\bar{x}$	5.00	3.47
	std dev	±0.25	±0.15
	RSD	5.0%	4.3%
Heptamer	$\bar{x}$	3.57	-
	std dev	±0.51	-
	RSD	14.3%	-
HPLC total (less BHET)		53.71	48.81
GC total		19.38	24.10
Total Assay		73.09	72.91
Number of determinations		5	4

a Results from GC analysis

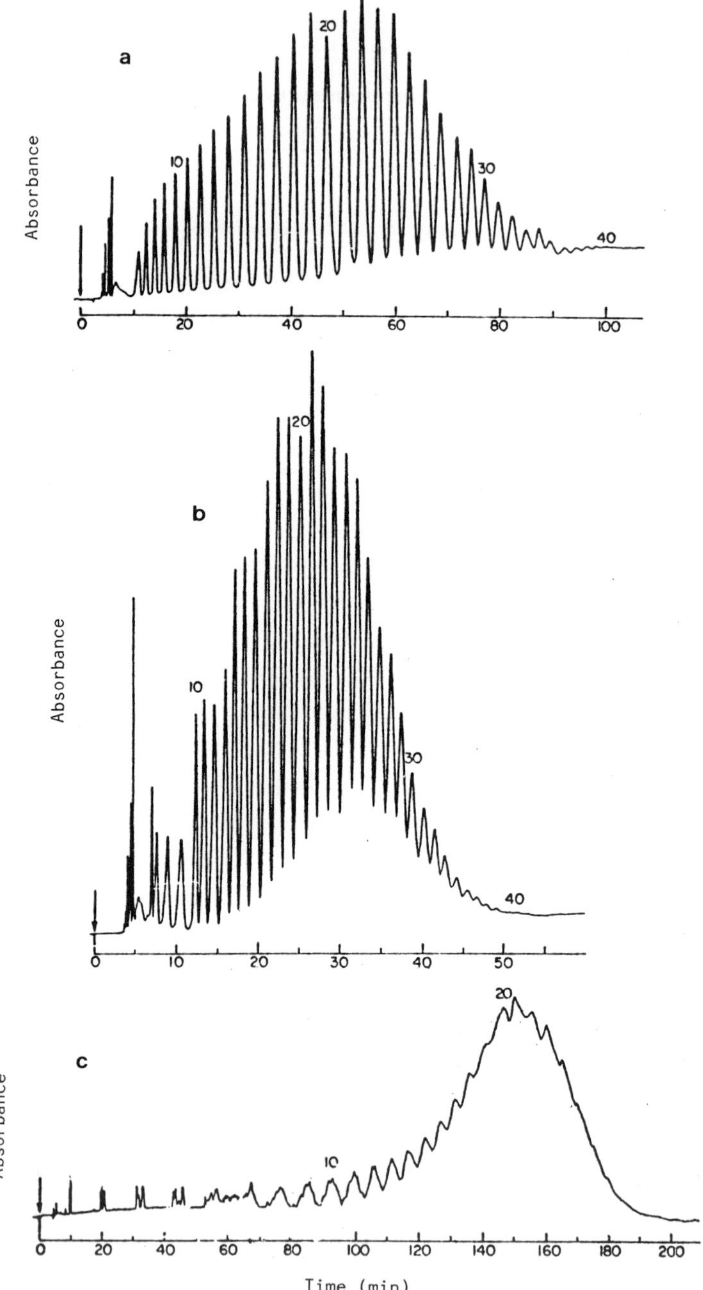

Figure 49 (a)   PS-2100, 1 mg injected in 10 ul separated on 4.6 x 250
mm LiChrosorb S160; absorbance units full scale (AUFS);
linear gradient increasing at 0.2%/ml from 97/3 (v/v) n-
hexane/THF at 1.0 ml/min; detection at 265 nm, 0.05
absorbance units full scale (AUFS).

Figure 49 (b)   PS-2100, same conditions as in Figure 49 (a) with linear
0.2% ml^{-1} gradient from 99/1 (v/v) n-hexane/ethyl
acetate: 0.1 AUFS.

Figure 49 (c)   PS-2100, same conditions as in Figure 49 (a) with linear
0.2% ml^{-1} gradient from 87/3 (v/v) n-
hexane/dichloromethane; 0.05 AUFS.

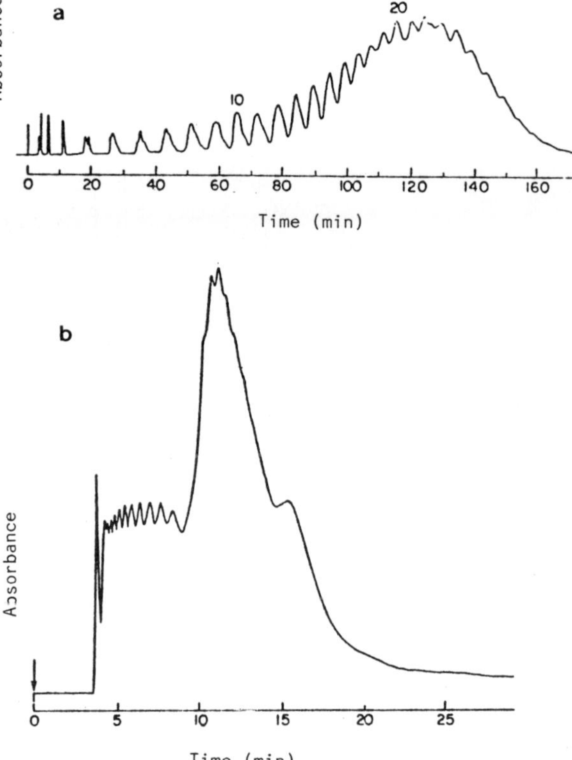

Figure 50 (a) PS 2100, 1 mg injected in 10 ul, separated on 4.6 x
           250 mm LiChrospher SI500 (50 nm pore size); n-
           hexane/ethyl acetate gradient, 0.04 AUFS.
           (b) PS2100, 1 mg injected to 10 ul separated on 4.6 x
           250 mm LiChrospher Si 500; n-hexane/dichloromethane
           gradient, 0.08 AUFS.

individual oligomer peaks are sharp and symmetrical. As many as 65
oligomers were separated in a 4800 weight average molecular weight
polystyrene using the tetrahydrofuran (Figure 49(a)) gradient. The
profile obtained from the n-hexane/dichloromethane gradient (Figure 49(b))
is noticeable different from profiles obtained from eluants containing
tetrahydrofuran and ethyl acetate. There is an apparent loss in
resolution of individual oligomer, and an additional splitting of oligomer
peaks gives the appearance of fine structure or oligomer packets,
particularly in oligomers 1-10. Oligomer retention times with ethyl
acetate (Figure 50(a)) and dichloromethane (Figure 50(b)) gradients on 50
nm pore diameter silica were less than those obtained from equivalent
gradients on 6 nm pore diameter silica (Figures 49(b) and 49(c)). The
reductions in retention times are proportional to the differences between
the phase ratios of these adsorbents.

     The gradients used to obtain chromatograms 49(a) to (c) and 50(a)
did not increase equally in solvent strength; they were volumetrically
equivalent gradients containing the second solvent (ethyl acetate,
tetrahydrofuran or dichloromethane) differing in strengths. Solvent
effects on selectivity and resolution are better compared by adjusting the
second solvent rates of increase to give equivalent solvent-strength

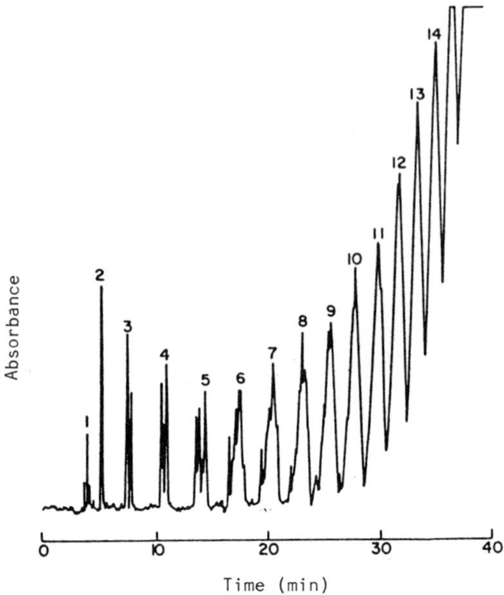

Figure 51 PS 2100 low molecular weight detail; 1 mg injected in 10
           ul; linear gradient increasing at 0.5%/ml, from 89/11
           (v/v) n-hexane/dichloromethane; detection at 265 nm,
           0.05 AUFS.

gradients. The resolution of individual oligomers decreased as expected
when the dichloromethane gradient was increased from 0.2%/ml to 0.5%/ml
(Figure 51) and produced retention times of oligomers 14-20 similar to
those obtained with 0.2%/ml tetrahydrofuran gradient (Figure 49(a)).

      Stereoisomers were semipreparatively separated in two steps.
Oligomers 1-6 of polystyrene-800 were isolated individually by separations
on a 10 mm i.d. x 500 column of 15-25 um LiChroprep Si60 (E.Merck). A
dichloromethane   solvent   gradient   beginning   at   92:8   (v/v)   n-
hexane/dichloromethane, programmed at a 0.3%/ml dichloromethane rate of
increase, gave base line separation of the first six oligomers. Three
injections of 500 mg in 1 ml were made, and each oligomer from 1 to 6 was
collected and the fractions of each of three separations were combined and
blown to dryness with nitrogen. Stereoisomers in each oligomer fraction
were then separated isocratically on a Partisil M-9 (Whatman) 10 mm i.d. x
250 mm column by using n-hexane/ dichloromethane binary eluants at a flow
rate of 4.2 ml/min. The splitting of peaks in the slower dichloromethane
gradient, which is not apparent in the THF separation (Figure 49(a)), is
still evident with the steeper dichloromethane gradient. This peak
splitting arises from the separation of stereochemical oligomer isomers.
It is not a result of end-group differences,since all oligomer chains have
equivalent n-butyl and proton terminal groups unable to separate
stereoisomers in oligomers larger than hexamer with any combination of
tetrahydrofuran or ethyl acetate gradient rates, but these isomer
separations were seen in oligomers containing as many as 20 repeat units
with various dichloromethane gradients. This behaviour is interpreted as
a difference between the selectivity of dichloromethane and tetrahydrofuran or ethyl acetate in the separation of stereochemical isomers.

Conditions used by Mourney et al[732] to obtain isocratic semi-preparative separations of stereoisomers are not optimal but do provide sufficient resolution and amounts of each isomer for identification by ^{13}C and ^{1}H NMR spectrometry. Assignments based on previous NMR studies of oligomers[733,734] helped in the identification of each stereoisomer and established an elution order for this separation to be, by polymer convention, syndiotactic (s), isotactic (i), followed by heterotactic (h) isomers of equivalent repeat unit number. The amounts of syndiotactic and isotactic isomers (first two peaks in each oligomer packet) decrease steadily with increasing oligomer length, until only heterotactic isomers are observed in oligomers longer than decamer.

In further work Mourney[735] investigated the adsorption chromatography of anionically and cationically prepared polystyrene oligomers on a 6 mm Lichrosorb Si60 silica (E. Merck) with n-hexane-dichlormethane eluants. End-group differences between the cationically prepared (Ph CH = ,CH (CH Ph CH$_2$)$_n$ CH Ph CH$_3$) and anionically (CH$_3$(CH$_2$)$_3$ CH$_2$ CH Ph)$_n$ CH$_2$ CH Ph) prepared polystyrenes produced significant differences in the retention of oligomers that are equivalent in length.

The solvent-displacement model[736-738] adequately explains the effects of solvents on the selectivity of oligostyrene stereoisomer separations[732] without considering in detail the effects of oligomer length and structural complexity on the adsorption. The more recent work of Mourney[735] defines the fundamentals of a potentially practical approach to the separation and analysis of homologues such as polystyrene oligomers and tests the adsorption theory for small-molecule separations when applied to large molecules. It focuses, in particular, on the adsorbed solute conformation and its influence on chromatographic retention.

A general expression for the chromatographic distribution coefficient K of polystyrene oligomer can be written as

$$\log K = \log V_a + \alpha[\overset{r}{\textstyle\sum} Q^\circ_r + Q^\circ_{e_1} + Q^\circ_{e_2} - \epsilon_{ab}(\overset{r}{\textstyle\sum} a_r + a_{e_1} + a_{e_2})] \tag{2}$$

where Q°_r, $Q^\circ_{e_1}$, and $Q^\circ_{e_2}$ are the standard adsorption potentials for adsorbed repeat units r and end groups e_1 and e_2, respectively, ϵ_{ab} is the mobile phase solvent strength, and a_r, a_{e_1} and a_{e_2} are the molecular areas of adsorption for adsorbed repeat units r and the oligomer end groups e_1 and e_2. Standard adsorption potentials are dimensionless quantities and molecular areas of occupation are in units of 8.5 A^2. The distribution coefficient is defined as $n_a V_1 / n_u W_s$, where n_a and n_u are the moles of adsorbed and unadsorbed solute, V_1 is the volume of the liquid phase, and W_s is the weight of the stationary phase. V_a is the adsorbent surface volume in ml/g, and alpha is the adsorbent activity[737].

In this equation, no provisions are made for stereochemical influences on oligomer conformation. This model simply assumes additivity of the individual oligomer-unit adsorption potentials and a planar adsorption conformation for all contributing groups. The additivity approach has accurately predicted the retention of a variety of small molecules and provides a starting point for the extension of adsorption chromatography to large, polyfunctional molecules. Equation (2) can be simplified to

$$\log K = \log V_a + \alpha(\overset{i}{\textstyle\sum} Q_i - \epsilon_{ab}\overset{i}{\textstyle\sum} a_i) \tag{3}$$

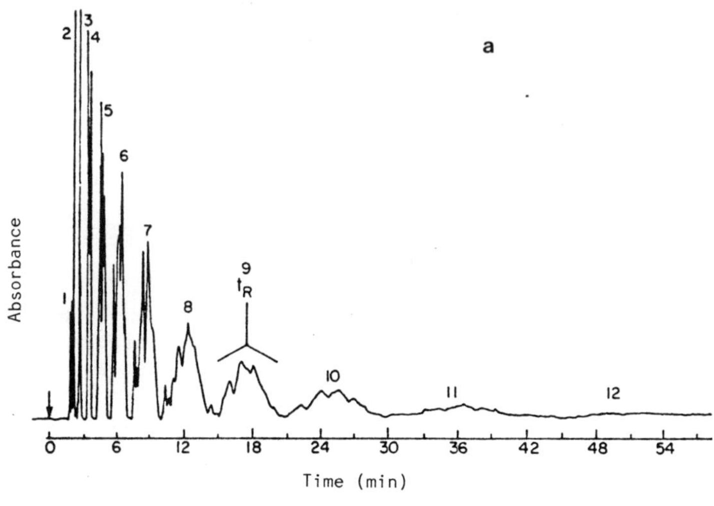

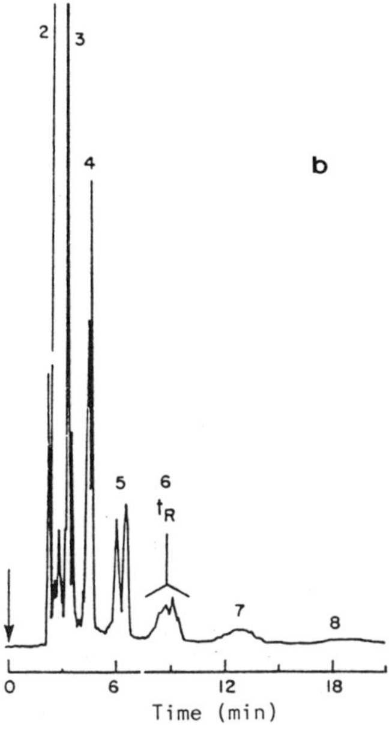

Figure 52   (a) Anionically polymerized polystyrene oligomers,
            isocratically eluted from 4.6 x 250 mm LiChrosorb Si60 5
            um silica column, mobile phase 78/22 (v/v) n-hexane-
            dichloromethane (E_ab = 0.18), 30 ug sample injected in
            10 ul. UV detection at 265 nm, 0.20 absorbance units
            full scale (AUFS). Oligomer Mw = 800, Mn = 650, 2.5
            mgl^{-1} in mobile phase. Column temperature 30± 0.05°C.
            (b) Cationically polymerized polystyrene oligomers, same
            conditions as in Figure 52 (a). 12.5 ug sample injected
            in 10 ul. 2.0 AUFS. Oligomer Mw 370, Mn = 200, 1.25
            mgl^{-1} in mobile phase; column temperature 30 ± 0.05°C.
            Column paramter alpha = 0.54, Log Va = -0.43 determined
            from elution of polyacrylic aromatic hydro carbons with
            n-hexane.

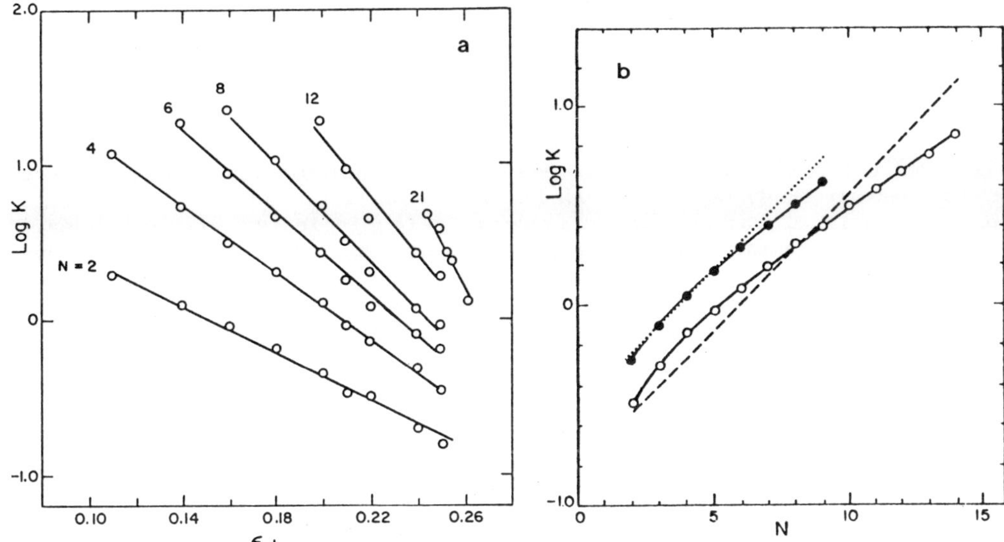

Figure 53   (a) Log K VS solvent strength ϵ_{ab} for anionically
prepared polystyrene oligomers. Values above lines
correspond to total number N of oligomer units.
(b) Log K vs N at ϵ_{ab} = 0.22 for ($\bullet$) cationically
prepared and ($\circ$) for anionically prepared PS-800 (...)
relationship predicted for planar adsorption.

where individual contributions from end groups and repeat units are
summed.   A plot of log K vs. solvent strength ϵ_{ab} is predicted to   give
a   straight   line   with slope $= -\alpha\sum^i a_i$ and intercept = log Va $+ \alpha\sum^i Q_i$.   If
the adsorbent parameters alpha and log Va are known, standard adsorption
potentials and molecular areas of occupation can be calculated for each
oligomer.   These values and the adsorption energy per unit of occupied
adsorbent surface area, $\sum^i Q_i/\sum^i a_i = S^\circ/A_s$,   can then be used to elucidate
the geometrical configuration of an oligomer on the adsorbent surface.

In Figure 52 are shown isocratic chromatograms obtained for an
ionically and cationically prepared polystyrene oligomers, obtained by
using an n-hexane/dichlormethane eluant with a solvent strength $_{ab}$ =
0.18.   The number of units in each oligomer (noted above each peak) was
confirmed by mass spectrometry on semipreparatively trapped fractions.
Substantial   fine   structure   within   each   oligomer   packet   results   from
partial separation of stereochemical isomers[732].   An average retention-
volume measurement was taken from the weighted mean of the individual
oligomer peak packets, as shown in the figures.   Distribution coefficients
calculated from the average retention volumes and void volume V_m, using K
= $(V_R - V_m)/W$, are plotted as a function of solvent strength, $_{ab}$, for all
retention   data   obtained   for   anionically   and   cationically   prepared
oligomers in Figure 53(a) and (b).

The separation of oligostyrene stereoisomers in the chromatograms
shown in Figures 52(a) and 52(b) may be qualitatively explained by the
influence of chain tacticity on the planarity of the adsorbed repeat
units.   Simple geometric models, without the aid of detailed rotational
isomeric state calculations, will show that planarity of the adsorbed

aromatic rings increases in the order syndiotactic, isotactic, to heterotactic isomers, which parallels the elution order for oligostyrene stereoisomers[732]. Chain tertiary structure also contributes to the planarity of adsorbed units in oligomers longer than pentamer. At large values of N, energy must be expended for both the uncoiling of the oligomer chain and the rotation of repeat units to a conformation that maximizes collectively the planar adsorption of all oligomer units. Based on this worker Mourney[735] concluded that the measured adsorption energies and occupational areas of oligomers 2-14 indicate tilted repeat-unit/adsorbent contact, which is a result of chain stiffness. Methylene backbone carbon atoms in oligomers longer than pentamer make only minor contributions to the adsorption process, as demonstrated by greater adsorption energy per unit of occupied adsorbent surface area for these oligomers than is calculated for planar solute conformation on the silica surface.

7.7.2 Size Exclusion Chromatography

This technique, formerly known as gel permeation chromatography was originated by Moore in 1964[739]. This is a technique where a solution of the polymer is passed down a column packed with a gel, which separates the polymer into fractions base on molecular weight by reason of the small pores present in the gel which, being of molecular dimensions, tend to pass through the smaller polymer molecules and slow down the larger molecules. The next problem is to detect the increasing higher molecular weight fractions of the polymer as they leave the column. Two principal types of detector are employed; one based on measurement of dielectric constant and the other on measurement of refractive index[740-743]. An important characteristic of detectors in this application is that they are universal and fairly non-selective as occurs, for example, in the case of ultraviolet or infrared detectors[744] which operate at a particular wavelength and rely on all individual constituents in the sample absorbing at a particular wavelength. The dielectric constant detector is a bulk property detector that complements, rather than competes with, the refractive index detector. The dielectric constant detector offers specific advantages for certain analyses, while the refractive index detector is preferable in other applications. Bode et al[745] used this technique for the examination of styrene-butadiene copolymers.

The dielectric constant detector has several advantages over the refractive index detector for some size-exclusion chromatography analyses. For example, the refractive index detector loses sensitivity faster as a function of temperature than does the dielectric constant detector. Therefore at elevated temperatures the dielectric constant detector may be more sensitive than the refractive index detector, even when the opposite is true at room temperature. This is caused by the reduced output of the light source and the reduced sensitivity of the photodiodes commonly used in refractive index detectors. As shown in Table 85, monomer ratios also greatly affect the response of the refractive index detector in some copolymers. In the case of styrene-butadiene copolymers, the refractive index response variation due to changes in the monomer ratio is approximately an order of magnitude higher than the variation observed with a dielectric constant detector. This effect can be significant for size-exclusion chromatographic analyses of copolymers with unknown or variable monomer ratio.

Figure 54(b) shows chromatographs of a polystyrene calibration standard. The composite chromatogram of Figure 54(b) shows the chromatographic separations of four different styrene-butadiene

Table 85 - Detector response

	RI	ΔRI*	ε	$\Delta\varepsilon$*
Polybutadiene	1.52	0.11	2.3	5.3
Polystyrene	1.60	0.19	2.6	5.0
Percentage		42%		5.6%

* Versus tetrahydrofuran; RI = 1.41, + 7.6.

copolymers. Figure 54(c) is a composite chromatogram showing the results obtained with a refractive index detector and dielectric constant detector in series. Some differences in the responses are evident. They both clearly show the bimodal molecular weight distribution of the copolymer, but the additive that elutes after the process solvent is barely seen by the refractive index detector. Note also the difference in response to the process solvent of the two detectors. The measurement of molecular weight averages and the distributions for polymers by gel-permeation chromatography requires the construction of a calibration curve using relatively monodisperse polymers such as a series of polystyrenes in the molecular weight range of 10^3-10^6 with $\bar{M}_w/\bar{M}_n$<1.10. However, to calculate the molecular weight averages and the distribution for any polymers other than polystyrene by this calibration curve, it is necessary to transform molecular weight units in the curve to those for the polymer specified. One of the techniques used to transform the molecular weight units curve to those for the polymer specified is so-called 'universal calibration'[746], and this method has been shown to have wide applicability. The procedure, however, requires accurate values of the Mark-Houwink parameters for both the sample and the standard used for calibration in the same solvent at the same temperature as the gel permeation chromatographic analysis, and therefore is not practical in most cases. The 'Q' factor method[747] also has many shortcomings, and should be used with caution. The 'Q' factor method can provide accurate results only when the Mark-Houwink exponents for both the sample and the standard are identical[748], and when the Q factor is determined experimentally in advance in the same solvent at the same temperature[749].

Many attempts have been undertaken to overcome the problem of the lack of suitable standards for each polymer type[750-754]. These attempts employ a known set of molecular weight averages of the broad molecular weight distribution polymers concerned. Some of the methods for calibration using polydisperse polymer samples were applied to linear calibration curves[755-757]. The calibration curve of a column given as a plot of log molecular weight versus the elution volume is generally a smooth curve with only slight curvature in the molecular weight range for which the column will be applied. Therefore, the calibration curve should be approximated by the third-order polynomial[758]. McCrackin presented a calibration method applied to the third-order polynomial[759], but it requires considerable calculation. The method described by Mahabadi and O'Driscoll[760] demands a calibration curve for polystyrene standards, the Mark-Houwink parameters for polystyrene and intrinsic viscosity and an average molecular weight for polymer samples.

Weiss and Cohn-Ginsberg[761], Belinskii and Nefedov[762], and Kalinsky and Janca[756] reported the method to calculate size-exclusion

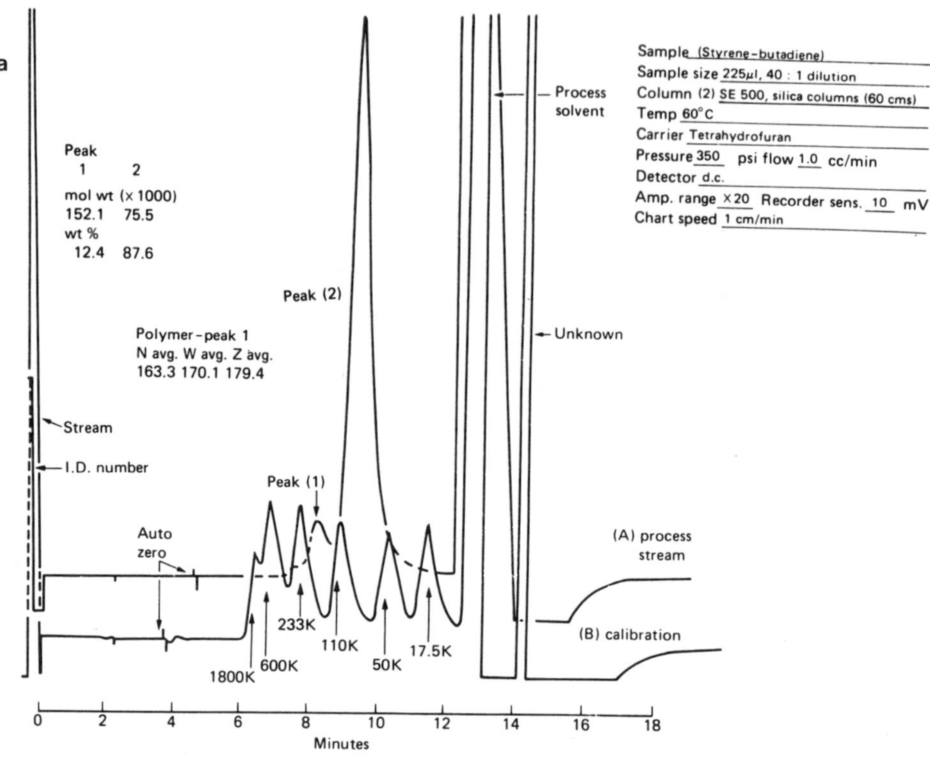
a
Sample (Styrene-butadiene)
Sample size 225µl, 40 : 1 dilution
Column (2) SE 500, silica columns (60 cms)
Temp 60°C
Carrier Tetrahydrofuran
Pressure 350 psi flow 1.0 cc/min
Detector d.c.
Amp. range ×20 Recorder sens. 10 mV
Chart speed 1 cm/min
Process solvent
Peak
1   2
mol wt (×1000)
152.1 75.5
wt %
12.4   87.6
Peak (2)
Polymer-peak 1
N avg. W avg. Z avg.
163.3 170.1 179.4
Unknown
Stream
I.D. number
Peak (1)
Auto zero
(A) process stream
(B) calibration
1800K
600K
233K
110K
50K
17.5K
0   2   4   6   8   10   12   14   16   18
Minutes

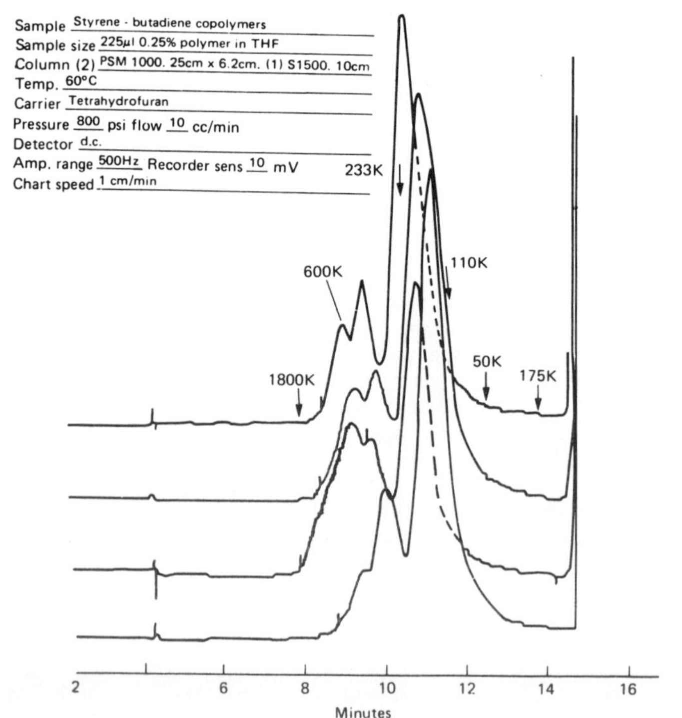
b
Sample Styrene - butadiene copolymers
Sample size 225µl 0.25% polymer in THF
Column (2) PSM 1000. 25cm x 6.2cm. (1) S1500. 10cm
Temp. 60°C
Carrier Tetrahydrofuran
Pressure 800 psi flow 10 cc/min
Detector d.c.
Amp. range 500Hz Recorder sens 10 mV   233K
Chart speed 1 cm/min
600K
110K
1800K
50K   175K
2   6   8   10   12   14   16
Minutes

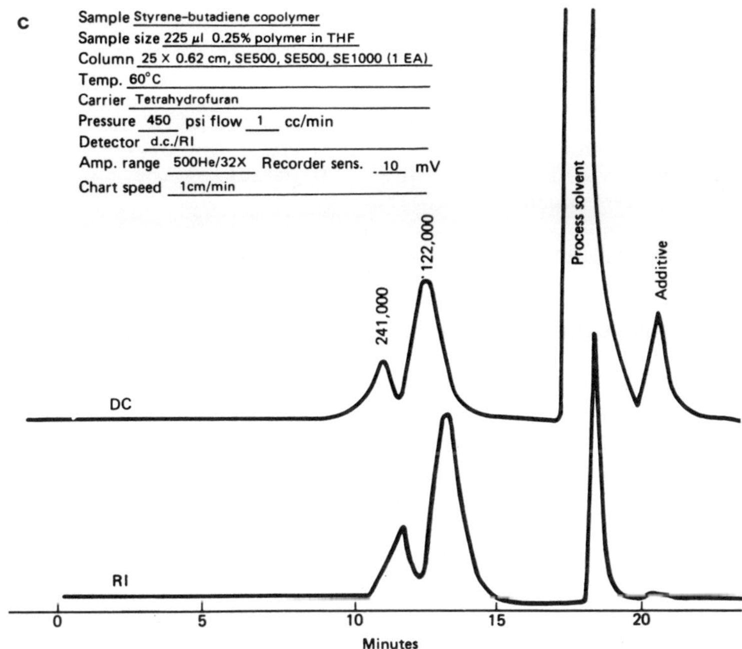

Figure 54   SEC separations: (a) styrene-butadiene copolymer and
polystyrene reference standards;   (b) four different
styrene-butadiene copolymers;   (c) RI and DC detectors
for the SEC separation of a styrene-butadiene copolymer.

chromatographic columns directly for the polymer under study[761].   The
universal calibration curve, log $[\eta]$ M versus elution volume, was first
established from narrower molecular weight distribution polystyrenes, and
its use was then extended to other systems to estimate the Mark-Houwink
parameters for the sample from data on broad molecular weight distribution
polymer (or polymers).   Two polymers have different intrinsic viscosities,
or one polymer for which the intrinsic viscosity and $\bar{M}_w$ (or $\bar{M}_n$) were known
was required to calculate the parameters with gel permeation
chromatography curves.   Once the parameters were obtained, one could
calculate molecular weight averages of the polymer under study.   The
validity of the method was estimated[764].   Recently, Hamielec and
Omoridion[765] presented an improved version which includes peak broadening
corrections, and they applied it to the non aqueous size exclusion
chromatography of polyvinylchloride and the aqueous chromatography of
polydextran[765].   A method for obtaining a calibration curve for polymer
solvent systems where polystyrene standards are not soluble was proposed.
The method requires the universal calibration curve based on polystyrene
in A solvent, the integral distribution curves of elution volumes for the
polymer under study in A and B solvents for the generation of a
calibration curve of the polymer in B solvent.

Mori[76] has discussed a modification of the methods of Weiss and
Cohn-Ginsberg[761] and Hamielec and Omoridion[765] for calibrating gel-
permeation chromatographic columns.

The modified method utilizes polydisperse calibrating samples for which any molecular weight averages are known, and a calibration curve of log molecular weight versus the elution volume constructed with polystyrene standards. Assumption of any formula for the calibration curve is not required. This method does not require prior knowledge of the specific Mark-Houwink parameters for both polystyrene standards and the polymers under study, nor does it require the determination of intrinsic viscosities of both polymers.

Basic theory of determination of molecular weight distribution. Let us first assume that the molecular weight $(M_b)_i$ of a species of a polymer B is related to the molecular weight $(M_A)_i$ of a species of a polymer A eluting at the same elution volume i by the expression.

$$(M_B)_i = s(M_A)^t_i \tag{4}$$

where s and t represent constants. Let polymer A be polystyrene standard experimentally used to establish a primary calibration curve and polymer B be a polymer requiring analysis. Suppose that the weight average molecular weight $\bar{M}_w$ and the number average molecular weight $\bar{M}_n$ of polymers B with a broad molecular weight distribution are given. These values of $\bar{M}_w$ and $\bar{M}_n$ should be measured by absolute methods, such as light-scattering and osmometry. Let the chromatogram of a polymer sample be given by its height h_i as a function of elution volume i. The weight- and number-average molecular weights of the polymer B (secondary standard) may be computed from the chromatogram h_i of the polymer and a calibration curve for polystyrene standards by the formulas

$$(\bar{M}_w)_c = \sum (h_i s(M_A)^t_i) / \sum h_i \tag{5}$$

$$(\bar{M}_n)_c = \sum h_i / \sum (h_i/s(M_A)^t_i) \tag{6}$$

where the subscript c indicates calculated molecular weight averages. Guess value t first, then the molecular weight averages of the polymer B may be computed by equations (5) and (6) from the chromatogram as follows:

$$(\bar{M}_w)_c = k_1 s \tag{7}$$
$$(\bar{M}_n)_c = k_2 s \tag{8}$$

where k_1 and k_2 are numerical values obtained from h_i and $(M_A)^t_i$. Let us assume

$$\bar{M}_w = (\bar{M}_w)_c; \quad \bar{M}_n = (\bar{M}_n)_c \tag{9}$$

then we obtain two values of s, which may be not equal in most cases. The goal of this calculation is to find a value t where a value of s from equation (7) equals that of s from equation (8). Trial value of t is assumed and values of s are calculated by equation (7) to (9). The process may be represented with other values of t until s of equation (7) - s of equation (8)] turns out to be sufficiently small. The numerical calculations were performed using a desk-top microcomputer. A hand calculation is also applicable. A calibration curve for polymer B can be constructed from equation (4), a final set of t and s, and a polystyrene calibration curve. Besides one polymer sample with broad molecular weight distribution and its $\bar{M}_w$ and $\bar{M}_n$ we can use two polymer samples with broad molecular weight distribution given either two $\bar{M}_w$, two $\bar{M}_n$, or one $\bar{M}_w$ and one $\bar{M}_n$. The experimental substantiation of equation (4) has been given by size-exclusion chromatography[767] of oligomers. For example, the following relation has been obtained between oligostyrene (A) and oligo (ethylene glycol) (B) in the range of molecular weight 200 and 3400

$$M_B = 0.967M_A^{0.945}$$ (10)

Similar relations have also been obtained for n-hydrocarbon epoxy resin, and p-cresol novolac resin as oligomer B.

The theoretical basis of this calibration method can be derived from the principle of 'universal calibration'. If it can be assumed that all polymers at a given elution volume have the same hydrodynamic volume and the same value of $[\eta]M$, then we can write finally, for any particular elution volume i:

$$(M_M)_i = (K_A/K_B)^{1/(1+aB)}(M_A)_i^{(1+aA)/(1+aB)}$$ (11)

where $[\eta]$ is the intrinsic viscosity in the solvent of the polymer having molecular weight M, the subscripts A and B refer to the polymer A and the polymer B, respectively, and K and a are the respective Mark-Houwink parameters (a coefficient an an exponent). As the values K and a are constants in the specified sample, solvent, and temperature, equation (11) can be regarded as identical with equation (4). Hence the values of the constants t and s are rearranged as

$$t = (1 + a_A)/(1 + a_B)$$ (12)

$$s = (K_A/K_B)^{1/(1+a_B)}$$ (13)

Thus, from equation (4) having evaluated the constants s and t, it is possible, knowing the molecular weight M_A of a standard polymer A to calculate the molecular weight (M_B) of a second polymer. It is then possible to calculate the weight average molecular weight $(\bar{M}_w)c$ of the polymer B and the number average molecular weight $(\bar{M}_n)c$ of the polymer B.

<u>Calculation of constants t and s (equation 4)</u>. In Table 56 are shown the known $\bar{M}_w$ and $\bar{M}_n$ values of seven standard polymers which were used to calculate values of the constants s and t in equation 4. The constants t and s in equation 4 calculated from $\bar{M}_w$ and $\bar{M}_n$ of one of four PVC samples, are shown in Table 87 with recalculated molecular weight averages and the ratios of them to the standard values in Table 86.

The calculated calibration curves of PVC which was generated using the constants t and s in Table 87 and equation 4 are shown in Figure 55(a) along with the experimental polystyrene calibration curve. A curve for PVC-3 falls between those for PVC-1 and PVC-4. The maximum discrepancy of molecular weight among three calibration curves for PVC-1 -3 and -4 is about 1-10% in the molecular weight range studied. The molecular weight range covered for the calibration curve should be limited to that of the species included in the secondary standard. To enable the calibration curve to be more fully covered and to make the discrepancy smaller, the average values of t and s were calculated. The average value of t from the three sets was obtained as the arithmetic mean and that of s as the antilogarithm of the arithmetic mean of logarithm s. The results are listed in Table 88. Agreement is excellent over the range of molecular weights studied.

A polymer having broader molecular weight distribution, that is, higher $\bar{M}_w$ and lower $\bar{M}_n$, will be more desirable as a secondary standard, but this requirement is not always readily achievable in practice. To overcome this difficulty, one possibility is to use two secondary standards, one

Table 86 - Molecular weight averages and instrinsic viscosity
data of sample polymers[a]

Sample	$\overline{M}_w$	$\overline{M}_n$	$[\eta]$	mol wt range
PVC-1	132,000	54,000	1.097	$1800 - 1.3 \times 10^6$
PVC-2	118,000	41,000	0.937	$1000 - 1.1 \times 10^6$
PVC-3	83,500	37,400	0.749	$1500 - 7.3 \times 10^5$
PVC-4	68,600	25,500	0.640	$700 - 6.8 \times 10^5$
PMMA	60,600	33,200	0.303	$2000 - 3 \times 10^5$
PVAc	331,000	83,000		$2500 - 3.2 \times 10^6$
PS(NBS 706)	257,800 (LS)	137,000	0.815	$3000 - 1.6 \times 10^6$
	288,000 (SD)			

[a](LS), measured by light scattering. (SD), measured by sedimentation
equilibrium ultracentrifugation. $\overline{M}_w$ and $\overline{M}_n$, manufacturer's data. $[\eta]$,
measured at 25°C in THF. Mol wt range, the lowest and highest molecular
weight of species included.

Table 87 - The values of t and s, the recalculated molecular
weight averages, and the ratios to the standard values,
from the combination of M_w and M_n of one polymer

Secondary standard	sample	$\overline{M}_w$	ratio	$\overline{M}_n$	ratio[a]
PVC-1	PVC-1	132,000	1.00	54,000	1.00
t = 1.151	PVC-2	107,300	0.91	44,930	1.10
s = 0.098	PVC-3	80,490	0.96	34,780	0.93
	PVC-4	64,250	0.94	22,950	0.90
PVC-2	PVC-1	140,000	1.14	51,900	0.96
t = 1.253	PVC-2	118,000	1.00	41,000	1.00
s = 0.031	PVC-3	85,000	1.02	31,360	0.84
	PVC-4	65,450	0.95	19,880	0.78
PVC-3	PVC-1	134,500	1.02	58,970	1.09
t = 1.127	PVC-2	110,210	0.93	47,880	1.17
s = 0.141	PVC-3	83,500	1.00	37,400	1.00
	PVC-4	67,090	0.98	25,130	0.99
PVC-4	PVC-1	140,860	1.07	57,730	1.07
t = 1.143	PVC-2	114,550	0.97	48,050	1.17
s = 0.120	PVC-3	85,930	1.03	38,240	1.02
	PVC-4	68,600	1.00	25,500	1.00

[a] Ratio is the value of the recalculated molecular weight average divided
by the standard value.

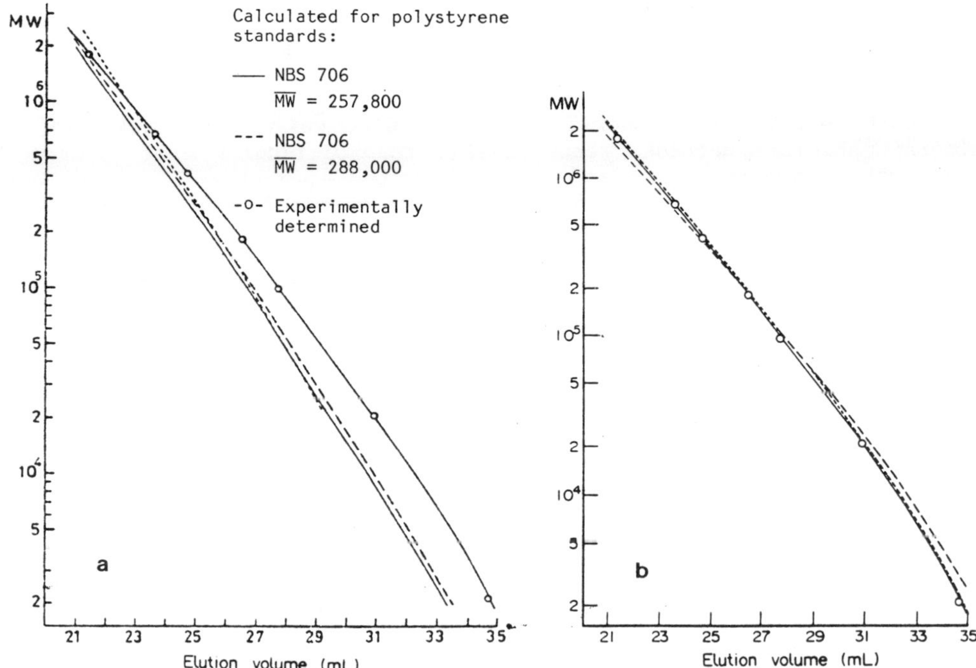

Figure 55 Calibration curves for polystyrene standards (O)
(experimentally determined) and (a) calculated for
polyvinyl chloride samples; (b) calculated for
polystyrene NBS 706 as MW = 257,800 (dashed line) and MW
288,000 (dotted line).

Table 88 - Recommended values of t and s, and the recalculated
molecular weight averages and the ratios to the standard values for PVC

Combination	sample	$\overline{M}_w$	ratio	$\overline{M}_n$	ratio[a]
mean from	PVC-1	134,650	1.02	56,160	1.04
PVC-1, PVC-3,	PVC-2	110,750	0.94	46,370	1.13
and PVC-4 in	PVC-3	83,300	0.99	36,650	0.98
Table 87	PVC-4	66,540	0.97	42,480	0.96
t = 1.140					
s = 0.118					
$\overline{M}_w$ of	PVC-1	132,000	1.00	56,560	1.05
PVC-1 and $\overline{M}_n$	PVC-2	107,900	0.91	47,270	1.15
of PVC-4	PVC-3	81,470	0.98	37,800	1.01
t = 1.099	PVC-4	65,320	0.95	25,500	1.00
s = 0.189					

has higher $\bar{M}_w$ and the lower $\bar{M}_n$. The values of t and s were calculated from $\bar{M}_w$ of PVC-1 and $\bar{M}_n$ of PVC-4 and the molecular weight averages of four PVC samples were recalculated. These results are shown in Table 88. Within the limits of the accuracy of the method, agreement between the recalculated and starting molecular weight has been found.

Both weight- and number-average molecular weights are measured by different absolute methods, respectively, thus difficulty will sometimes arise when both accurate $\bar{M}_w$ and $\bar{M}_n$ are required at a time. Usually, one of two molecular weight averages can be determined with considerable accuracy and reliability. Constants t and s, calculated from the combination of two $\bar{M}_w$ or $\bar{M}_n$ of two PVC samples and the ratios of the recalculated molecular weight averages to the standard values are compiled in Table 89. The results in the table suggest that if constants t and s are calculated by using only $\bar{M}_w$ (or $\bar{M}_n$) of two samples, then M_w (or M_n) of a sample recalculated from its chromatogram will be correct while its recalculated $\bar{M}_n$ (or $\bar{M}_w$) will be larger (or smaller) than its standard value. This can be ascribed mainly to the effect of peak broadening.

The molecular weight averages calculated from chromatograms uncorrected for peak broadening are obtained by

$$\bar{M}_w^{c} = \int_0^\infty F(v)\, M(v)\, dv \tag{14}$$

$$\bar{M}_n^{c} = 1 \, / \, \int_0^\infty (F(v)/M(v))\, dv \tag{15}$$

where F(v), which is normalized, is the measured chromatogram given by its height as a function of elution volume, and the superscript c indicates calculated molecular weight averages. Hamielec and Ray[768] have shown the molecular weight correction factor E under restricted conditions by

$$\bar{M}_w = E\bar{M}_w^{c} \quad \bar{M}_n = E^{-1}\bar{M}_n^{c} \tag{16}$$

where $\bar{M}_w$ and $\bar{M}_n$ are the true molecular weight averages measured by absolute methods, and $E = \exp[-(D_2 \sigma)^2/2]$ where D_2 is the slope of the calibration curve at the peak position of the chromatogram and σ^2 is the variance of the single species chromatogram at the peak position. The goal is to calculate t and s and equation 4 so as to obtain the true molecular weight averages by the expressions

$$\bar{M}_w = \int_0^\infty F(v)\, sM(v)^t\, dv \tag{17}$$

$$\bar{M}_n = 1/ \int_0^\infty (F(v)/sM(v)^t)\, dv \tag{18}$$

Substituting equation 14 and 17 in equation 16 gives

$$sM(v)^t = E^{-1}M(v) \tag{19}$$

Similarly, substituting equations 15 and 18 in equation 16 gives

$$sM(v)^t = E^{-1}M(v) \tag{20}$$

A solution of equations 17 and 18 is simultaneous values of t and s satisfying equations 19 and 20. A portion of higher molecular weight of the chromatogram is responsible for the value of $\bar{M}_w$ and the value of $\bar{M}_n$ is mostly influenced by the portion of lower molecular weight of the chromatogram. Thus the solution of equations 19 and 20 can exist experimentally.

Table 89 - Values of t and s and the ratios of the recalculated
molecular weight averages to the standard values obtained
from two $\bar{M}_w$ or $\bar{M}_n$ of two samples

Combination	sample	ratio for $\bar{M}_w$	ratio for $\bar{M}_n$
M_w of PVC-1 and PVC-3	PVC-1	1.00	1.21
t = 1.022	PVC-3	1.00	1.16
s = 0.498	PVC-4	0.98	1.19
M_w of PVC-1 and PVC-4	PVC-1	1.00	1.28
t = 0.928	PVC-3	1.02	1.24
s = 0.815	PVC-4	1.00	1.30
M_w of PVC-3 and PVC-4	PVC-1	0.94	1.36
t = 0.881	PVC-3	1.00	1.37
s = 2.684	PVC-4	1.00	1.50
M_n of PVC-1 and PVC-3	PVC-1	0.70	1.00
t = 0.899	PVC-3	0.75	1.00
s = 1.620	PVC-4	0.74	1.09
M_n of PVC-1 and PVC-4	PVC-1	0.80	1.00
t = 1.002	PVC-3	0.81	0.96
s = 0.512	PVC-4	0.79	1.00
M_n of PVC-3 and PVC-4	PVC-1	0.74	1.15
t = 1.094	PVC-3	0.80	1.00
s = 0.199	PVC-4	0.83	1.00

Equations 17 and 18 which include the effect of peak broadening, are
the formulae for calculating $\bar{M}_w$ and $\bar{M}_n$ neglecting peak broadening. When
the column is calibrated using secondary standards with known $\bar{M}_w$, the
correct value of $\bar{M}_w$ of any sample will be obtained from its chromatogram
neglecting peak broadening and using the constants t and s and equation 17
(if the sample has similar molecular weight averages and molecular weight
distribution). However, if $\bar{M}_n$ of the sample is calculated by using the
same constants t and s, and neglecting peak broadening, the result will be
rewritten by substituting equation 19 in equation 15 as

$$1/\int_0^\infty (F(v)/EM(v))\, dv = E\bar{M}_n^c = E^2\bar{M}_n \tag{21}$$

so that the calculated $\bar{M}_n$ will be E^2 times the true value. Similarly,
when the column is calibrated using samples with known $\bar{M}_n$, the correct
value of $\bar{M}_n$ of a sample calculated will be obtained, while its calculated
$\bar{M}_w$ will by $1/E^2$ times its true value. Mori[766] examined the problem by
associated with peak broadening effect by obtaining chromatogram of
polystyrene NBS 706 as a secondary standard and by calculating constants t
and s from equation 4. Molecular weight averages of PS NBS 706
experimentally calculated from a calibration curve of primary polystyrene
standards were $\bar{M}_w$ = 272 400 and $\bar{M}_n$ = 128 900. The molecular weight
correction factor E, which is defined as the ratio of the true $\bar{M}_w$ to the
calculated $\bar{M}_w$ or the ratio of the calculated $\bar{M}_n$ to the $\bar{M}_n$, was about
0.946, when the value $\bar{M}_w$ = 257 800 was applied as the standard value.

This result implies that the difference between the calculated molecular weight averages and the true ones was mostly due to the peak broadening effect. Constants t and s in equation 4 for PS NBS 706 were t = 0.922 and s = 2.57, when the value $\bar{M}_w$ = 257 800 was used as a standard value, and t = 0.006 and s = 1.106 when $\bar{M}_w$ = 288 000 was applied. Calibration curves for polystyrene experimentally obtained and calculated are shown in Figure 55(b). Though the curves, experimentally obtained and calculated by using the value of $\bar{M}_w$ = 257 800, coincide partly at the middle of the curves, which corresponds to the centre of the chromatogram of NBS 706, their extreme parts clearly diverge, implying this discrepancy may arise from the effect of peak broadening.

If the sample includes only smaller (or larger) molecular weight species as in the range of one extreme part and has narrower molecular weight distribution than NBS 706, then the molecular weight distribution will be higher (or smaller) then the true values. When the value $\bar{M}_w$ = 288,000 was used as a standard value, the results imply that the calibration curve might be shifted to some extent to correct the deviation of molecular weights for primary standards, rather than peak broadening.

As discussed above, the parameters t and s obtained from equations 17 and 18 are dependent on peak broadening. If the sample polymers have similar molecular weight averages and distributions to the secondary standard(s), the molecular weight averages of the samples calculated using the parameters t and s will be correct values neglecting peak broadening. To obtain the parameters correcting peak broadening effect, one should use the corrected SEC chromatogram W(v) instead of the observed SEC chromatogram F(v) in equation 17 and 18. The other method to correct peak broadening is to divide $\bar{M}_w$ in equation 17 by E or to multiply $\bar{M}_n$ by E in equation 18 before calculating the parameters[765]. If $\bar{M}_w$ in equation 17 is divided and $\bar{M}_n$ in equation 17 is multiplied by 0.946, then one will obtain t = 1 and s = 1 and the calibration curve (a dashed line) in Figure 55(b) will coincide with the curve experimentally determined from polystyrene standards (a full line). Since peak broadening is a function of molecular weight (greatest near the void and total permeation volumes), the parameters t and s obtained by the latter correction will still hold some approximation.

The above discussions may conclude that when an accurate $\bar{M}_w$ is required, constants t and s should be calculated by using two $\bar{M}_w$ of two samples. Similarly a combination of two $\bar{M}_n$ of two samples will give an accurate $\bar{M}_n$ of a sample. When both $\bar{M}_w$ and $\bar{M}_n$ of a sample are required, one should calculate t and s from $\bar{M}_w$ and $\bar{M}_n$ of one secondary standard sample or two. The useful range of molecular weights covered by a given set of s and t parameters will be estimated to be within the molecular weight distributions(s) of the secondary standard(s) as shown in Table 86.

It can similarly be derived that the molecular weight of polymethylmethacrylate is related to the molecular weight of polystyrene eluting at the same elution volume by the relationship

$$M_{PMMA} = 1.967 M_{PS}^{0.918} \tag{22}$$

Similarly, the molecular weight relationship between polyvinyl acetate and polystyrene is written as

$$M_{PVAc} = 1.009 M_{PS}^{1.058} \tag{23}$$

Table 90 - The values of t and s and molecular weight averages
of several oligomers

Sample	$\overline{M}_n$ from VPO	mol wt from from SEC		mol wt range
		$\overline{M}_w$	$\overline{M}_n$	
i) PEG (t = 0.916, s = 1.21)				$200\text{-}10^4$
PEG3400	3000	3060	3000[a]	
PEG5000	4055	4050	3940	
PEG6000	5920	6260	5920[a]	
ii) p-Cresol Novolac Resin				1100-8400
(t = 0.947, s = 1,41)				
sample 1	2910	3380	2910[a]	
sample 2	1960	1940	1890	
sample 3	1730	1780	1730[a]	
sample 4	2330	2410	2310	
iii) Epoxy Resin (t = 0.987, s = 0.84)				$900\text{-}10^5$
EPIKOTE-1004	1470	2900	1470[a]	
EPIKOTE-1007	3180	7220	3100	
EPIKOTE-1009	5130	15300	5130[a]	
iv) Phenol Novolac Resin (random)				-5000
(t = 0.955, s = 0.734)				
sample 1	163	230	163[a]	
sample 2	216	361	224	
sample 3	251	414	256	
sample 4	291	547	291[a]	

[a] The constants t and s for the oligomer concerned were calculated from
the chromatorgrams of these oligomers.

These constants t and s for polymethylmethacrylate and polyvinyl acetate
were obtained from $\overline{M}_w$ and $\overline{M}_n$ of one one sample.

Mori[766] calculated the constants t and s for several oligomers by
using two $\overline{M}_n$ of two samples. The results were listed in Table 90.
Molecular weight averages recalculated from their chromatograms and osmotic data are included in the table. The calibration curves constructed
from these constants are valid in the molecular weight range added in the
last column of Table 90. The calibration curves may be applied to
molecules having molecular weights above the molecular weight range.
However, for molecules having molecular weights below the molecular weight
range, a different calibration method, namely, the molecular weight
conversion equations derived from peak elution volumes and molecular
weights of the oligomer samples and oligostyrenes both of which can be
separated into individual species, should be applied[769].

Equations 12 and 13 permit the calculation of an exponent a and the
coefficient K of the Mark-Houwink equation of a polymer requiring analysis

providing the constants t and s in equation 4 for the polymer and the Mark-houwink parameters of polystyrene are known. The Mark-Houwink parameters of PVC were calculated by using constants t and s in Tables 87 and 88 and the Mark-Houwink parameters of polystyrene (a = 0.706, K = 1.60 x 10^{-4} at 25°C in tetrahydrofuran[770]). The results, which appear to be considerably less than literature values (a = 0.766, K = 1.63 x 10^{-4} [770]), are listed in Table 91. Parameters, a and K of polymethylmethacrylate, which were calculated by the same manner, are appreciably higher than the literature values (a = 0.73, K = 7.0 x 10^{-5} [771]). The parameters of polyvinyl acetate are in fairly good agreement with the literature values (a = 0.63, K = 3.5 x 10^{-4} [772]). This calculation technique to obtain the Mark-Houwink parameters involves the Mark-Houwink parameters of polystyrene, which seem to have considerable errors, resulting in inaccurate values.

The viscosity-average molecular weight of a polymer sample may be computed from the chromatogram h_i of the polymer by the formula

$$\bar{M}_v = \left(\sum h_i M_i^a / \sum h_i \right)^{1/a} \tag{24}$$

Substituting in the mark Houwink equation gives

$$[\eta] = K\left(\sum h_i M_i^a / \sum h_i \right) \tag{25}$$

where a and K are the Mark-Houwink parameters and $[\eta]$ is the intrinsic viscosity of the polymer. If there are two polymer samples whose intrinsic viscosity values are known and if a calibration curve for the polymer sample is already constructed, parameters K and a can be computed from equation 25 by the search technique. The results, which appear to be in fairly good agreement with the literature values, are shown in Table 91. These parameters are valid at 25°C in tetrahydrofuran. Polymer samples, polymethylmethacrylate and polystyrene had the following characteristics, as a pair polymer $\bar{M}_w$ = 1.20 x 10^5, $\bar{M}_n$ = 7.0 x 10^4 and $[\eta]$ = 0.505 (polymethyl methacrylate) and $\bar{M}_w$ = 7.5 x 10^4, $\bar{M}_n$ = 3.8 x 10^4 and $[\eta]$ = 0.319 (polystyrene. The Mark-Houwink parameters of any polymers calculated by this method are independent on those of the polystyrene standards.

Ogawa et al[773] have compared column fractionation and size exclusion chromatography methods for determining the molecular weight distribution of polypropylene. The calculated statistical parameters such as average molecular weight, standard deviation, skewness, and kurtosis for each distribution curve, and also the number-average and weight-average molecular weights were determined by osmometry and light scattering to compare with those from distribution curves. The results of their investigation on the comparison of the determination methods of molecular weight distribution for crystalline polypropylene are summarized as follows.

1. The molecular weight distribution curve obtained from column fractionation was narrower than that from size exclusion chromatography, and the D value from size exclusion chromatography was closer to that from absolute methods (osmometry and light-scattering).

2. It was confirmed that the distribution curve from column fractionation became similar to that from size exclusion chromatography on correcting the overlapping of the distribution of fractionated polymers.

Table 91 - The Mark-Houwink parameters calculated from equations
12, 13 and 25

equation no	table or equation no	sample or t and s	a	K x 10^4
12 and 13	Table 87	PVC-1	0.482	49.8
		PVC-2	0.361	182
		PVC-3	0.514	31.1
		PVC-4	0.492	37.6
12 and 13	Table 88	PVC t = 1.140 s = 0.118	0.497	39.2
12 and 13	Table 88	PVC t= 1.099 s = 0.189	0.552	21.2
12 and 13	Eqn 22	PMMA	0.858	0.455
12 and 13	Eqn 23	PVAc	0.613	1.58
25	Table 87	PVC-1 and PVC-3 t = 1.140 s = 0.118	0.753	1.59
25	Table 87	PVC-1 and PVC-4 t = 1.140 s = 0.118	0.706	2.79
25		PMMA	0.764	0.699
25		PS	0.705	1.23

3.   The broadening effect in size exclusion chromatography and thermal
      degradation during fractionations were both found to be of little
      importance.

4.   It was assumed by Crouzet et al[774] that a broader distribution curve
      had been obtained from size exclusion chromatography due to a
      broadening effect as compared with that from column fractionation
      and that the distribution curve from column fractionation was more
      accurate than that from size exclusion chromatography . However
      Ogawa et al[773] consider that the distribution curve obtained from
      size exclusion chromatography is more accurate and reliable than
      that from column fractionation.

To take the discussion further, consider a particular polymer,
polypropylene, in further detail. Crystalline polypropylene is mainly
characterized by three factors: tacticity, molecular weight and molecular
weight distribution. Molecular weight distribution is a very ambiguous
factor, in spite to the amount of attention it had received. This is
because the observed distribution curves depend on the determination
method, and no definite method has been established.

The molecular weight distribution of crystalline polypropylene is generally determined by column fractionation or size exclusion chromatography. To obtain the molecular weight distribution curve by column fractionation takes many hours, and the polymer must be protected from thermal degradation. The determination of the molecular weight distribution curve of polypropylene by size exclusion chromatography is, on the other hand, very easy.

Molecular weight of polyvinylpyrolidone and polyacrylonitrile. In size exclusion chromatography, it is first necessary to construct a calibration curve for the column system and a series of polystyrenes of relatively monodisperse character are mostly employed for this purpose. Because of the poor solubility of polystyrene in dimethyl formamide a retardation of retention volume (V_R) for polystyrene[775] has been observed. Consequently, there is a large divergence between universal calibration curves for polystyrene and polyacrylonitrile in dimethyl formamide-lithium bromide. This abnormality may be explained by partitioning or adsorption of polystyrene on the gel. Polar polymers such as poly(ethylene oxide) must be employed as standards.

To calculate the molecular weight averages for any polymers other than poly(ethylene oxide) by a poly(ethylene oxide) calibration curve, it is necessary to transform molecular weight units of poly(ethylene oxide) in the curve to those for the polymer specified. One of the techniques used to transform the molecular weight units is so-called "universal calibration"[777], and this method had been shown to have wide applicability. This procedure is only applicable in cases where the elution of both the sample and the standards; which depends on their hydrodynamic volumes ($[\eta]M$) in solution, no secondary effects occur. In addition, the Mark-Houwink parameters must be known for both polymers.

To overcome this difficulty Mori[778] applied his previously discussed procedure (see Reference 766) for preparing a calibration curve for a polymer where only broad molecular weight distribution samples were available. This method requires neither prior knowledge of the Mark-Houwink parameters nor the assumption of any formula for the calibration curve.

Let us first assume that the molecular weight $(M_B)_i$ of a species of polymer B is related to the molecular weight $(M_A)_i$ of a species of polymer A eluting at the same retention volume i by the expression

$$(M_B)_i = s(M_A)_i{}^t \qquad\qquad (26)$$

where s and t represent constants.

Let polymer A be a poly(ethylene oxide) standard experimentally used to construct a primary calibration curve and polymer B be polyacrylonitrile or polyvinylpyrolidone polymer requiring analysis. Suppose that two weight average molecular weights, $\bar{M}_{wB1}$ and $\bar{M}_{wB2}$, measured by an absolute method, such as light scattering, of two samples of polymer B with broad molecular weight distribution are known (secondary standards). Let the chromatograms of the polymer samples, B1 and B2, be denoted by height $h_{B1,i}$ and $h_{B2,i}$ as a function of retention volume i. The weight average molecular weights of two polymer, B1 and B2, may be computed from the chromatograms $h_{B1,i}$ and $h_{B2,i}$ of the polymers and a calibration curve for poly(ethylene oxide) using equation 26 and may be related to the true values measured by an absolute method by the formulas

$$\bar{M}_{w,B1} = \sum (h_{B1,i} s (M_A)_i^t) / \sum h_{B1,i} \tag{27}$$

$$\bar{M}_{w,B2} = \sum (h_{B2,i} s (M_A)_i^t) / \sum h_{B2,i} \tag{28}$$

Combining equations 27 and 28 and rearranging we obtain

$$\frac{\bar{M}_{w,B1} \sum h_{B1,i}}{\sum (h_{B1,i}(M_A)_i^t)} = \frac{\bar{M}_{w,B2} \sum h_{B2,i}}{\sum (h_{B2,i}(M_A)_i^t)} \tag{29}$$

The goal of this calculation is to find a value t where both sides of equation 29 are equal and to calculate the value s by substituting the value of t into equations 27 and 28.

Size exclusion chromatography was conducted on a high performance liquid chromatograph equipped with a differential refractometer detector. High performance size exclusion chromatography columns Shodex AD 80 M/S (Shako Co. Ltd. Minato-ku-Tokyo 105, Japan) were packed with a mixture of polystyrene gels of nominal exclusion limits of 10^3, 10^4, 10^5 and 10^6°A. Columns were thermostated at 60°C. The primary calibration curve was constructed with poly(ethylene oxide) standards and commercial poly(ethylene glycols). The ratio of $\bar{M}_w$ and $\bar{M}_n$ for these were between 1.02 and 1.10. The mobile phase was 0.01 M lithium bromide in N,N^1 dimethylformamide and the flow rate was adjusted to 1.0 ml/min. Sample concentrations were 0.1% w/v for calibration and 0.2% w/v for polyacrylonitrile and polyvinylpyrolidone. A 0.01 ml loop was used to inject these sample solutions.

Lithium bromide in the dimethyl formamide mobile phase has the beneficial effect of eliminating multiple peaks and increasing retention volume. No difference in retention volume and in elution behaviour for poly(ethylene oxide) was observed when lithium bromide was added to the dimethyl formamide.

Calculation of constants t and s in equation 26. The primary calibration curve constructed with poly(ethylene oxide) is shown in Figure 56(a). The value of M_A at each retention volume i where obtained from Figure 56(a) for poly(ethylene oxide) and those of h_{B1} and h_{B2} from chromatograms of polyacrylonitrile-1 (PAN-1, polymer B1) and polyacrylonitrile-2 (PAN-2, polymer B2) or polyvinylpyrrolidone-2 (PVP-2) and polyvinylpyrrolidone-3 (PVP-3).

The constants t and s in equation 26, calculated from equations 29 and 27, are shown in Table 92 with the recalculated weight average molecular weights for polyacrylonitrile and polyvinylpyrolidone in addition to number average molecular weights for reference. The calibration curve for polyacrylonitrile and polyvinylpyrrolidone, calculated by using the constants t and s in Table 92 and equation 26 are shown in Figure 56(a) along with the experimental poly(ethylene oxide) calibration curve.

The Mark-Houwink parameters for poly(ethylene oxide) in 0.01 M lithium bromide-dimethylformamide at 60° were obtained by measuring intrinsic viscosities of several poly(ethylene oxide) standards followed by relating them with molecular weights

$$[\eta]_{PEO} = (5.00 \times 10^{-4})M^{0.653} \tag{30}$$

According to the earlier work of Mori[758] if a calibration curve of molecular weight vs. retention volume for a polymer is known, the following equation can be derived.

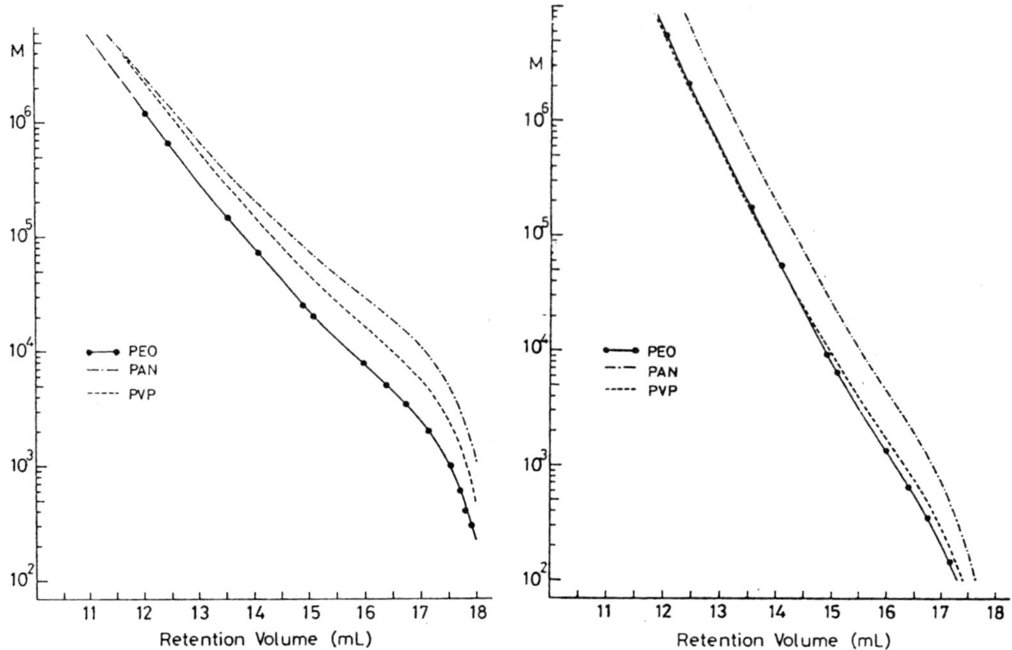

Figure 56 (a) Calibration curves of molecular weight (M) vs V_B experimentally determined for PEO and calculated for PAN and PVP samples.
(b) Universal calibration curves of ($\curlyvee$) M vs V_R for PEO, PAN, and PVP samples.

Table 92 – The values of t and s and the recalculated molecular weight averages for PAN and PVP

Sample	$\bar{M}_w$		$\bar{M}_n$	$\bar{M}_w/\bar{M}_n$
	nominal	found		
PAN=1	485,000	484,700	122,100	3.97
PAN-2	333,000	332,500	121,800	2.73
PVP-1	1,310,000	1,336,800	214,000	6.25
PVP-2	1,200,000	1,199,600	243,500	4.93
PVP-3	60,000	60,000	27,000	2.22
PVP-4	11,000	9,500	4,200	2.29

$$[\eta] = K(\textstyle\sum h_i M_i^a / \sum h_i) \tag{31}$$

where a and K are the Mark-Houwink parameters and $[\eta]$ is the intrinsic
viscosity of the polymer. The height of the chromatogram is expressed as
h_i and M_i is the molecular weight at the retention volume i. If there are
two polymer samples whose intrinsic viscosity values are known, one
parameter a can be calculated by

$$\frac{[\eta]_1}{[\eta]_2} = \frac{\sum h_{1,i} M_i^a / \sum h_{1,i}}{\sum h_{2,i} M_i^a / \sum h_{2,i}} \tag{32}$$

A trial value of a is assumed and the calculation of equation 32 is
repeated until the value of a minimizes the difference between both sides
of equation 31. Subsequently, a value of K can be derived from equation
31. When constructing molecular weight - retention volume curves for
polyacrylonitrile and polyvinylpyrolidone (Figure 56(a)) knowledge of the
universal calibration curve of [ ] M vs. retention volume of poly(ethylene
oxide) was not required, so that Mark-Houwink parameters calculated from
equations 31 and 32 are effective regardless of the secondary effects in
elution. The Mark-Houwink equations for polyacrylonitrile and polyvinylpyrolidone and PVP thus calculated are

$$\text{PAN} \quad [\eta] = (1.46 \times 10^{-5}) M^{0.897} \tag{33}$$

$$\text{PVP} \quad [\eta] = (1.85 \times 10^{-4}) M^{0.646} \tag{34}$$

The relationships between $[\eta]M$ and retention volumes obtained by
using these equations and Figure 56(a) are shown in Figure 56(b). The
curve for polyvinylpyrolidone is nearly on the same curve for polyethylene oxide, suggesting the effectiveness of hydrodynamic theory while
polyacrylonitrile still has secondary effects to polystyrene gels.
Therefore, the universal calibration method cannot be applied to the
calibration of selective size exclusion chromatographic systems for
polyacrylonitrile. The Mori method can be applied to such systems since
the effectiveness of universal calibration is not required.

There are several methods to obtain the Mark-Houwink parameters by
size exclusion chromatography. Most of them assume the effectiveness of
the hydrodynamic concept. One attempt is the calculation of the parameters from equation 30 and the respective s and t values for polyacrylonitrile and polyvinylpyrolidone and gives

$$\text{PAN, } K = 3.870 \times 10^{-6} \quad a = 0.9088$$

$$\text{PVP, } K = 6.910 \times 10^{-5} \quad a = 0.7236$$

These values are obviously different from those of equation 33 and 34.
However, the universal calibration curves of $[\eta]M$ and retention volumes
obtained by using these values and Figure 56 (a) piled up on the curve of
polyethylene oxide. This observation is a matter of course, because these
values of Mark-Houwink parameters are completely dependent on the
polyethylene oxide calibration curve and equation 30. The calculation of
the Mark-Houwink parameters independently on the parameters of
polyethylene oxide is useful in studies of the polymer-gel interaction as
well as for obtaining the precise Mark-Houwink parameters.

Tally and Bowman[779] have discussed the size exclusion chromatography
with refractive index detection of water soluble polymers such as poly(2-
vinyl-pyridine), dextrans and quarternized poly(4-vinylpyridine). They

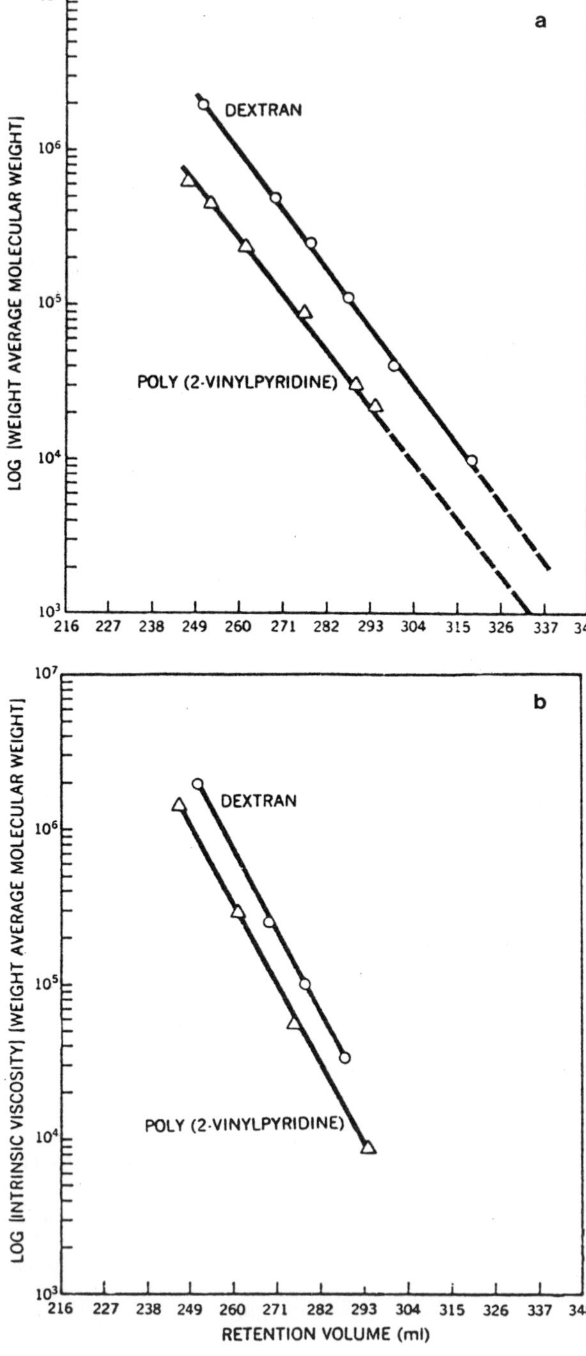

Figure 57   (a) Calibration curves for poly(2-vinylpyridine) and
dextran in 0.1 N HNO_3 + 0.1 N $NaNO_3$ on a 10 column set
of quaternized supports varying in porosity from 40 to
25,000 .   Flow rate = 2 ml/min.
(b) Log (hydrodynamic volume) vs. retention volume for
poly(2-vinylpyridine) and dextran in 0.1 N HNO_3 + 0.1 N
$NaNO_3$ on a 10 column set of quaternized supports varying
in porosity from 40 to 25,000 .   Flow rate = 2 ml/min.

used as a column packing a porous silica glass (40-25000A) to which quarternary ammonium groups had been bonded. Acidic salt solutions such as 0.1 N nitric acid-0.1 N sodium nitrate were used as eluants.

Figure 57(a) is the plot of log (weight average molecular weight) vs. retention volume for these polymers. Both series gave a linear relationship with retention volume which was an indication that separation by size was occurring. As expected, the two calibration curves are parallel rather than superpositioned. Poly(2-vinylpyridine) is charged, leading to a more extended structure in solution, whereas dextran would conform to a more compact structure. In other words, a particular molecular weight poly(2-vinylpyridine) eluted before the same molecular weight dextran.

Furthermore, a plot of the log (hydrodynamic volume) vs. retention volume for poly(2-vinylpyridine) and dextran does not superimpose as shown in Figure 57(b). If the universal calibration is to hold for these polymers in this chromatographic system, the two lines should coincide. Since the lines are parallel, it can be concluded that the universal calibration does not hold for comparison of neutral to charged polymers.

<u>Fractionation of copolymers.</u> It is well-known that copolymer properties are affected by composition in addition to molecular weight. Most copolymers have a chemical composition distribution and a molecular weight distribution. Although molecular weight distribution of homopolymers can be measured by size exclusion chromatography rapidly and precisely, accurate information on the molecular weight distribution of copolymers cannot be obtained by size exclusion chromatography alone. This is because separation in size exclusion chromatography is achieved according to the sizes of molecules in solution and the molecular weights of the copolymers are not proportional to molecular size unless the composition fluctuation is negligible and the chemical structure is the same across the whole range of molecular weights. Information on these distributions should be obtained by separating the copolymer by composition independently of molecular weight and then determining the molecular weight distribution of each fraction. Or, inversely, the molecular weight distribution is determined first, independently of composition and the chemical composition distribution of the same molecular weight species is measured. This is the principle of cross-fractionation, which can be performed by means of a combination of several chromatographic methods.

Some workers used size exclusion chromatography for the first fractionation, and as the second chromatographic method, thin-layer chromatography[780,781] and high-performance precipitation liquid chromatography[782,783] were applied. Balke and Patel[784,785] performed cross-fractionation by orthogonal chromatography (a combination of SEC-SEC) in which different mobile phases were used. Inagaki and his co-workers[786] used the combination of column adsorption chromatography (first fractionation) and size exclusion chromatography (second fractionation). They separated styrene-methyl methacrylate graft copolymer by using liquid adsorption chromatography using the mixture of ethyl acetate and benzene as the mobile phase. Styrene-methyl methacrylate random copolymer was also separated according to chemical composition by liquid adsorption chromatography on silica[787]. Mori et al[788] separated styrene-methyl methacrylate copolymers according to chemical composition and molecular size by liquid chromatography and size exclusion chromatography

They used a 1.2 dichloroethane-chloroform-silica gel system for liquid adsorption chromatography, i.e., chemical composition chromatography, and detection was achieved by ultraviolet at 254 nm. Next the molecular weight distribution of the liquid adsorption chromatography fractions was analysed by size exclusion chromatography using a refractive index detector. Silica gel having a pore size of 30 A was used as an adsorbent and mixtures of 1,2-dichlorethane and chloroform (including 1% ethanol as a stabilizer) were used as the mobile phase for liquid adsorption chromatography. The initial mobile phase for liquid adsorption chromatography was 1,2-dichloroethane and then the content of chloroform in the mobile phase was increased stepwise up to 100%. When 1,2-dichloroethane was used as the mobile phase, the copolymers adsorbed on the external surface of silica gel. With increasing a chloroform content in the mobile phase, the copolymer was desorbed as a function of their composition. The early-eluted fraction in liquid adsorption chromatography had lower molecular weigh averages and higher styrene content than those for the unfractionated copolymer. The late-eluted fraction had the opposite values.

Figure 58 is a gradient elution liquid adsorption chromatogram obtained in the chloroform:1,2-dichloroethane elution of methyl methacrylate copolymers containing respectively 85.5%, 73.4% and 66.3% styrene. From the elution behaviour of each copolymer, peak A is assigned to the copolymer I, B to the copolymer II, C to the copolymer III, and D to part of the copolymer I. In Figure 59 is shown a size exclusion chromatogram of a fraction containing the 85.5%:14.5% styrene-methyl methacrylate copolymer fractionated under various conditions. Peak elution volumes of two fractions were different from that of the native (unfractionated) copolymer, one was larger and the other was smaller, suggesting the difference in molecular weight averages. Polystyrene equivalent-molecular-weight averages were calculated by using a polystyrene calibration curve, and a styrene content in each fraction was determined by using a UV-RI dual-detector system. The results are listed in Table 93. The early-eluted fraction, which appeared at the 20/80 chloroform/1,2-dichloroethane mobile phase, has lower molecular weight averages and larger styrene content than those for the native copolymer. The late-eluted fraction appearing at 30/70 chloroform/1,2-dichloroethane has the opposite value. A copolymer containing 73.4% styrene remains unresolved by liquid adsorption chromatography, so the data of the fraction coincide with those of the native copolymer within experimental error.

In order to select the proper silica gel for liquid adsorption chromatography, polystyrene calibration curves were developed as shown in Figure 60. Polystyrene standards having narrow molecular weight distributions were used for the construction of calibration curves. Chloroform was used as the mobile phase. Two silica gels of different pore sizes, 100A and 30A, were packed into stainless steel tubes of 4.6 mm i.d. and 250 mm in length. The former has 5000 plates and the latter 8000 plates as measured by injecting 25 ul of a 0.5% benzene solution. Silica gel of pore size 30A has an exclusion limit of 3000 molecular weight as polystyrene and that of pore size 100A (Lichrosorb SI-100, 5 um) has a limit of 1×10^5. Therefore size exclusion cannot be ignored when silica gel of pore size 100A is used. Figure 61 shows the chromatograms of the copolymers II, III, and IV (each has $\bar{M}_w = (1.2-1.5 \times 5^5$ and $\bar{M}_n =)7-10 \times 10^4$) on these two silica gel columns. Mobile phase was chloroform. On 30A pore-size silica gel, all solutes eluted at the exclusion limit and peaks were sharp. However, peaks were broad, showing separation by size exclusion effect on the 100A pore-size silica gel. Therefore, silica gel at 30A pore size is a good choice for an adsorbent.

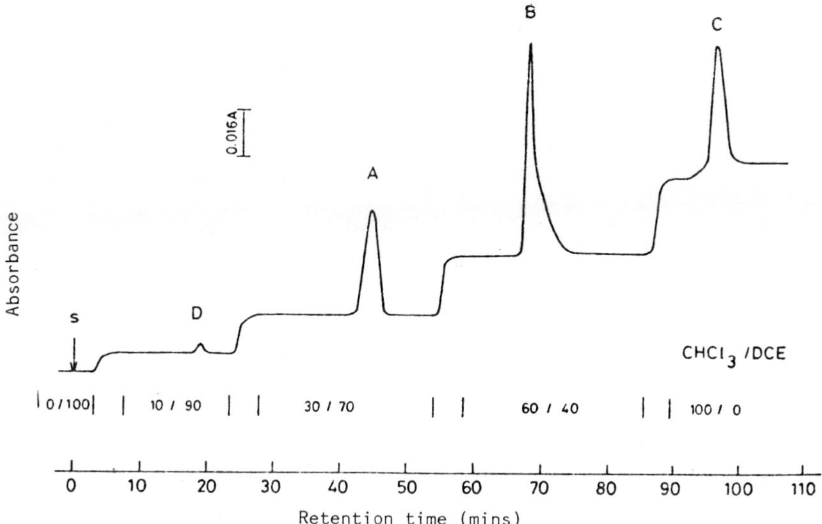

Figure 58   LAC chromatogram of a mixture of the copolymers I, II and III; 0.35% of each component, (A) copolymer I, (B) copolymer II, (C) copolymer II and (D) part of copolymer I.

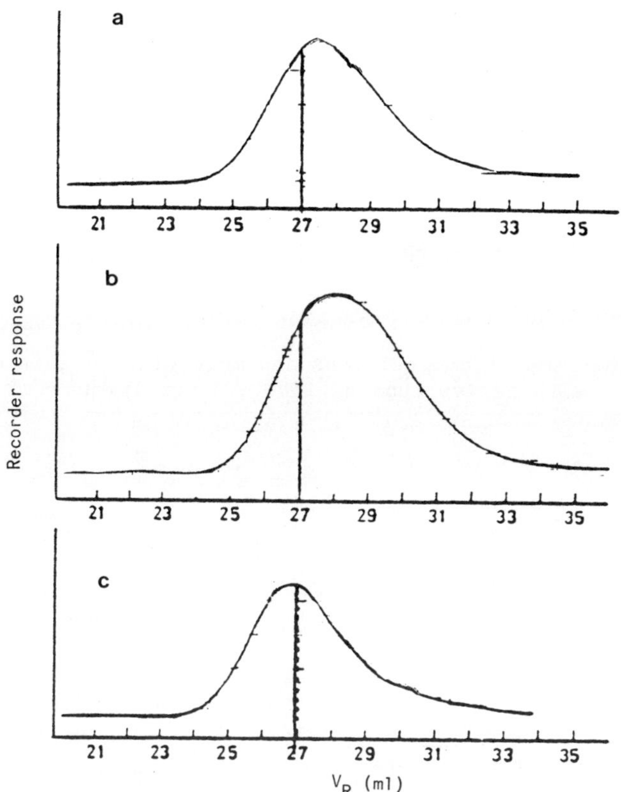

Figure 59   SEC chromatogram of the copolymer unfractionated and fractionated: (a) unfractionated; (b) fractionated at 20/80 CHCl₃/DCE mobile phase; (c) fractionated at 30/70 CHCl₃/DCE mobile phase.

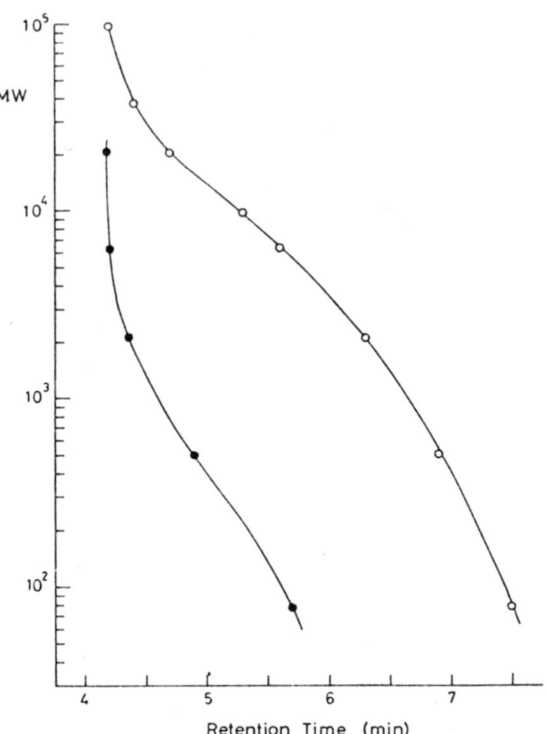

Figure 60   Calibration curves of silica gel columns obtained with
            polystyrene standards: (o) silica gel of pore size 100
            A; (●) silica gel of pore size 30 A; column, 4.6 mm i.d.
            x 250 mm; mobile phase, chloroform; flow rate, 0.5
            ml/min.

Table 93 - Comparison of compositions and molecular weights of fractions
           with native (unfractionated) copolymers

state	styrene content, mol %	molecular weight average	
		$\bar{M}_n$	$\bar{M}_w$
native copolymer I	85.9	77,000	128,000
fraction at $CHCl_3$/DCE (20/80)	92.0	60,000	97,000
fraction at $CHCl_3$/DCE (30/70)	82.4	92,000	153,000
native copolymer II	72.0	79,000	134,000
fraction at $CHCl_3$/DCE (70/30)	74.7	85,000	139,000

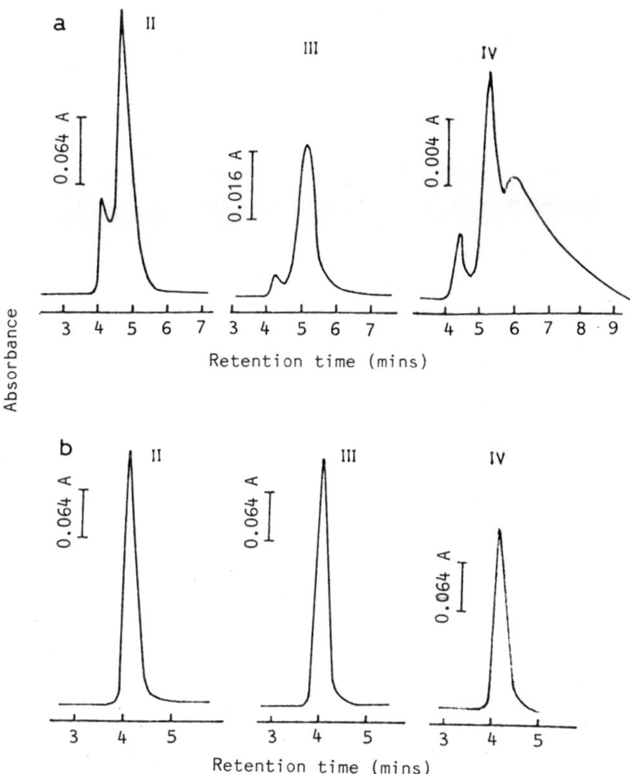

Figure 61   Chromatograms of P(S-MMA) copolymers on silica gels of
             different pores: (a) silica gel of pore size 100 A; (b)
             silica gel of pore size 30 A; column, 4.6 mm i.d. x 250
             mm; mobile phase, chloroform; flow rate, 0.5 ml/min;
             sample. 0.2%, 33 ul. (a), 17 ul; (b), II, III, IV,
             copolymers II, III and IV.

The elution behaviour of polystyrene poly(methyl methacrylate), and
their copolymers on 30A pore-size silica gel packed into a stainless steel
tube of 4.6 mm i.d. and 350 mm in length was investigated by using
chloroform, 1,2-dichloroethane, and their mixed solvents as mobile phases.
The 1,2-dichloroethane content in the mixed solvents increased in 10%
increment.   All samples were dissolved in the initial mobile phase.
Polystyrene eluted from the column with both chloroform and 1,2-
dichloroethane but polymethyl methacrylate and the copolymers did not
elute with 1,2-dichloroethane.   Furthermore, polymethyl methacrylate and
the copolymers including more than 50% methyl methacrylate did not elute
even with chloroform.   The copolymers having methyl methacrylate less than
50% eluted with chloroform and mixtures of chloroform and 1,2-
dichloroethane of certain ratios.   For example a copolymer containing
85.5% styrene re-eluted at the exclusion limit with chloroform and
mixtures of chloroform and 1,2-dichloroethane up to 60% 1,2-
dichloroethane.   However, the peak height of this copolymer decreased
gradually with increasing 1,2-dichloroethane content in the mobile phase
more than 60% and it disappeared from the chromatogram by using 1,2-
dichloroethane as a mobile phase.   The peak height of a copolymer
containing 73.4% styrene decreased with increasing 1,2-dichloroethane
content in the mobile phase to more than 40% and it disappeared when 1,2-

dichloroethane content in the mobile phase was 70%. The characteristics of these elution profiles are that all peaks have the same retention volume; e.g., they eluted at the exclusion limit of the column (about 2 ml) and the peak height of some copolymers started to decrease when the 1,2-dichloroethane content in the mobile phase increased to more than a certain ratio and finally they retained in the column. Whether the decrease in peak height depends on chemical composition or molecular weight of the copolymers was not resolved by Mori et al[788].

Williamson and Cervenka in a series of papers[789-794] have discussed the applications of size exclusion chromatography to a high and low density polyethylene and other polymers. These workers point out that, unlike polystyrene, for the vast majority of polymers a direct calibration procedure relating peak elution volume of a polymer and molecular weight require the availability of numerous polymer fractions of very narrow molecular weigh distributions and covering a broad range of molecular weights and frequently these are not available.

Indirect calibration is based upon establishing the so-called "universal calibration curve", log P= F(V), relating a universal parameter (P) controlling the molecular separation process to the gel permeation chromatography elution volume (V). Since the fractionation by gel permeation chromatography is believed to be governed by the molecular hydro-dynamic volume, the universal parameters recommended in the literature reflect dimensions of polymer molecules, which for flexible non-interacting chains, should be independent of the chemical structure of a polymer. The existence of a universal separation parameter is very important since it allows the polystyrene standards to be used for calibrating the gel permeation chromatograph for any polymer. However this is only true provided that:

1.   The chosen universal parameter (P) does indeed control both the molecular separation of the standard and the polymer to be studied.

2.   A relation between P an M is available so that the universal calibration curve log P = F(V) can be converted into the required calibration curve log M = f(V).

They demonstrated that the Benoit type universal calibration procedure[793] holds for widely diverse types of high density polyethylene (linear and also molecules containing short or long-chain branching).

They used the product of intrinsic viscosity $[\eta]$ and the weight average of molecular weight $\bar{M}_w$ as the universal parameter. They established the universal calibration curve log ($[\eta]$ $\bar{M}_w$) = F(V) and studied the effects of short and long-chain branching (SCB and LCB, respectively). They also established a Mark-Houwink equation $[\eta] = K \bar{M}_v^a$ permitting transformation of the experimental curve log ($[\eta]$ $\bar{M}_w$) into the desirable log M = f(V) in the solvent (1,2,4-trichlorobenzene) and under the operating conditions (135°C) used for gel permeation chromatography. The Mark-Houwink relationship obtained for high-density polyethylene in 1,2,4-trichlorobenzene at 135° and 145° was $[\eta] = 9.54 \times 10^{-4} \bar{M}_w^{0.64}$. In figure 62 is shown a universal calibration curve log $\bar{M}_w$ versus elution value established for polystyrene and polyethylenes of various structures. It is evident that a single calibration curve is obeyed with a few exceptions which occur mainly with the highly branched low-density polyethylene fractions. These deviations are believed to be caused mainly by inaccuracy in the light scattering measurements.

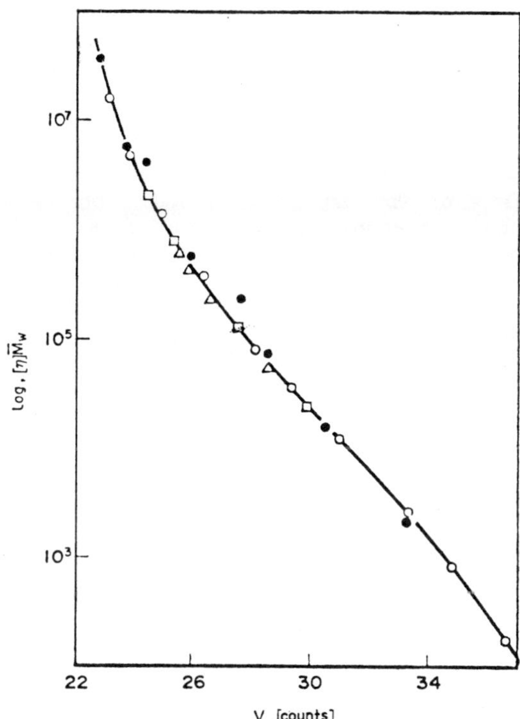

Figure 62   Universal calibration curve established on polystyrenes
(o) and polyethylenes of different structures (● -
fractions of low density polyethylene, ▲ high density
polyethylenes of narrow molecular weight distribution,
◻ fractions of ethylene-butene-1 copolymers) in TCB at
135°C under the gel permeation chromatography operation
conditions.

In further work[790] Williamson and Cervenka fractionated
polyethylenes of different structures and characterized the fractions by
light scattering gel permeation chromatography and viscometry. Intrinsic
viscosities were measured in solvents of different thermodynamical quality
including a theta-solvent (diphenyl at 118° for low-density polyethylene
and at 130° for high-density polyethylene and ethylene-butene-1
copolymer). The results were used for examining two aspects of the Drott
iterative procedure[795]: (a) the relationship between thermodynamical
quality of the solvent and depression in the intrinsic viscosity due to
branching; and (b) the analytical form of an expression relating the so-
called g-factor to the number of long-chain branches. The ratio of
intrinsic viscosities of branched and linear species at a given weight-
average molecular weight was clearly proved to be solvent independent, and
the equation relating the g-factor to the number of branches for polymer
monodisperse with respect to molecular weights appears to be a fair
representation of long-chain branching in low-density polyethylene. For
the polymers examined, the branching frequency was not independent of
molecular weight.

Cervenka and Williamson examined two features of the Drott iterative
procedure for the estimation of long chain branching in low density
polyethylenes, viz the proportionality between the number of long-chain
branches n per molecule and its molecular weight M, and the exponent ϵ

relating the ratio of intrinsic viscosities of branched and linear polymer of the same weight-average molecular weight to n. They replaced original Drott model describing the distribution of long-chain branches by a more general equation:

$$n = const\ M^B$$

and examined the effect of the exponents B and on the molecular weight distribution long chain branching characteristics. The analysis has revealed that, unless information on the long chain branching distribution is available, the method described by Drott for which B = 1 does not give meaningful information on molecular weight distribution and long chain branching for commercial low-density polyethylenes.

Size exclusion chromatography has been applied to determination of the molecular weight distribution of a wide range of other polymers including low- and high-density polyethylene, [796-807] polypropylene[809.81-] cis-1-4polybutadiene, ethylene-propylene copolymers[808], polystyrene,[810-120], isotactic polystyrene[821], styrene-butadiene copolymers[822-826], butadiene and methyl styrene copolymers[825.826], sulphated polystyrene[827], PVC[828-831], vinyl acetate-vinyl chloride copolymers[833], vinyl chloride-vinylidiene chloride copolymers[834], vinylidene chloride-methylmethyacrylate copolymers[834], acrylate-acrylonitrile copolymers[835], polyacrylonitrile[836.837], acrylonitrile-butadiene-styrene terpolymers[836], polyethylene terephthalate[838-841], polymethyl methacrylate[836.842-848], poly(2-methoxyethylmethacrylate)[848], polycarbonates[849.850], polyethylene glycol oligomers[832.851.852], polypropylene glycol[852.853] polyvinyl-pyrrolidone[837], polyvinyl acetate[836], epoxy resins[836], and phenol novalac resins.

7.7.3 Supercritical Fluid Chromatography

Supercritical fluid chromatography is a technique based on the observation that above the critical point a substance has density and solvating power approaching that of a liquid, but viscosity similar to the of a gas, and diffusivity intermediate between those of a gas and a liquid[857]. Supercritical fluids can dissolve a variety of solutes such as polymers with high molecular weight and low volatility, and the solubility may be easily varied by changing the density via the applied pressure. The low viscosity means that the pressure drop across the column is small, and the consequent permeability allows capillary columns with high efficiencies to be used. Solubility can be altered by increasing the temperature, and increased temperatures also increase solute diffusion coefficients in the mobile phase with a consequent increase in resolution.

The narrow columns necessary in supercritical fluid chromatography make small injection and detector volumes necessary, but supercritical chromatography is compatible with both gas chromatography and high performance liquid chromatography detectors. Coupling to spectrometric detectors such as mass, FTIR and even NMR spectrometers can also be carried out.

The technology for the preparation of capillary columns for supercritical fluid chromatography is now well advanced. Column diameters below 100 um are necessary because the solute diffusion coefficients are smaller than in gas chromatography. Practical efficiencies of up to -5000 plates per metre are possible for 50 um i.d. columns.

Retention of supercritical fluid chromatography is decreased by

increasing the mobile phase density. Density programming is made possible
via software for pump pressure control which incorporates the appropriate
pressure-density isotherms. The resolution of homologues decreases with
increasing density, and therefore asymptotic rather than linear density
programs may be required. Simultaneous density-temperature programming is
carried out to achieve variations in selectivity - indeed all the four
significant operating parameters in chromatography may be varied.

The high dissolving power of supercritical fluids for some polymers
and the high efficiencies of capillary columns make supercritical fluid
chromatography particularly suitable for analyses of oligomeric and
polymeric materials. The separation of 42 styrene oligomers has been
reported[854,855]. The separation of every oligomer in a mixture is
routinely possible, in contrast to size-exclusion chromatography in which
resolution is much lower. For example, Figure 63(a) shows the separation
of a polydimethysiloxane mixture, not only are the individual linear
oligomers resolved, but also a series of small peaks assigned to branched-
chain oligomers. The analysis of numerous other polymers and surface
active agents by supercritical fluid chromatography had been reported[856],
including methylphenylsioxanes, styrene and other vinyl aromatic polymers,
polyolefins and waxes, polyethers, polyglycols (underivatized, since
analysis temperatures are well below the decomposition temperature),
polyesters and more polar polymers such as epoxies (Figure 63(b), and
isocyanates.

Such analyses generally involve density programming to bring out
oligomers at fairly regular intervals; simultaneous temperature
programming also extends the range of compounds eluted and gives greater
chromatographic efficiencies by increasing solute diffusion coefficients
in the mobile phase. However, very highly polar polymers such as
carbohydrates may not be soluble enough in conventional mobile phases
unless derivatized. The silylation of maltrin (corn syrup solids)
enabled[858] supercritical fluid chromatography with CO_2 mobile phase to be
used to produce the elution of polyglucose oligomers with up to 18 units.
The resolution was sufficient for the separation of the alpha and beta
anomers.

For very high molecular weight polymers, hydrocarbon (e.g. pentane)
mobile phases modified with polar additives such as alcohols and ethers
are required. Many such separations have been reported by supercritical
fluid chromatography using packed columns, often with gradient elution,
but such an approach poses special problems in capillary supercritical
fluid chromatography because of the low flow rates. Capillary
supercritical fluid chromatography with solvent programming is a likely
future growth area.

Polymer additives - mould release agents, plasticizers, antioxidants and UV absorbers, with molecular weights extending beyond 1000 -
are generally unsuitable for gas chromatographic or liquid chromatographic
analysis because of their low volatility, lack of a chromophore or thermal
instability. Supercritical fluid chromatography is now the method of
choice[857,858,860] for the analysis of such compounds. Figure 64 shows the
chromatogram of a polymer containing - Tinuvin 1130.

A number of advantages accrue from combining the separating power of
capillary supercritical fluid chromatography with the explicit structural
information of the mass spectrometer. Most of the common ionization modes
have been shown to be compatible with supercritical fluid chromatography,
including electron impact, chemical ionization (CI) and charge

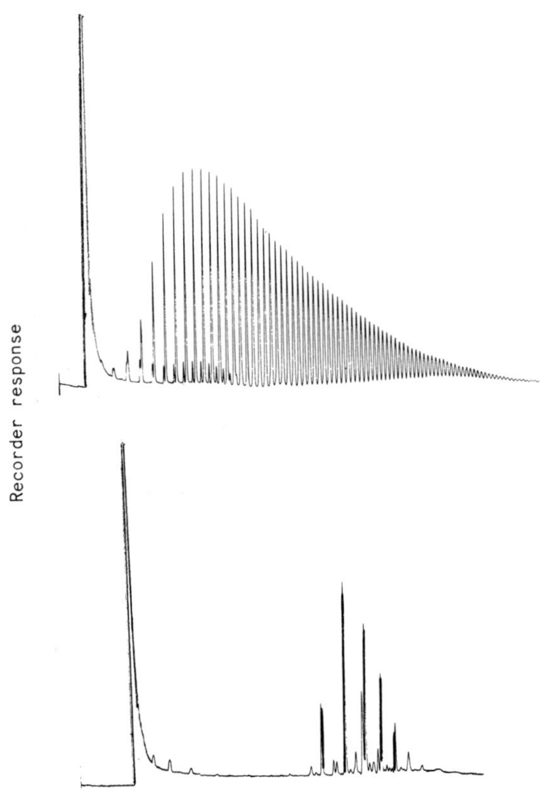

Figure 63   (a) SFC chromatogram of a polydimethylsiloxane.
             Conditions column, 20 m x 50 um i.d. SB-methyl; mobile
             phase, CO_2 at 120°C with asymptotic density programming,
             detector, FID.
             (b) SFC chromatogram of epoxy acrylate oligomers,
             Conditions: column, 20 m x 50 um i.d. SB-biphenyl-100;
             mobile phase, CO_2 at 70°C with linear density
             programming; detector FID.

exchange[861]. The variety of structural data available from supercritical
fluid chromatography with negative-ion CI detection has been demonstrated:
while methane reagent gas gave mainly the M + 1 ion (1112), both methane
with carbon dioxide and ammonia with carbon dioxide as reagent gases gave
many more fragment ions.

By changing the density of supercritical fluid, different fractions
may be selectively extracted from a complex mixture or sample matrix. On
decompression, the extracted solutes are precipitated and may be collected
for injection into a gas chromatograph or supercritical fluid
chromatography chromatograph and analysis. Figure 65 shows a simple
apparatus for on-line SFC/SFC: solutes extracted from the sample matrix
are deposited from the end of a restrictor into the internal loop of the
micro-injection valve of the capillary supercritical fluid chromatograph.
The valve loop contents are subsequently switched into the supercritical
fluid chromatography column by means of liquid or supercritical carbon
dioxide.

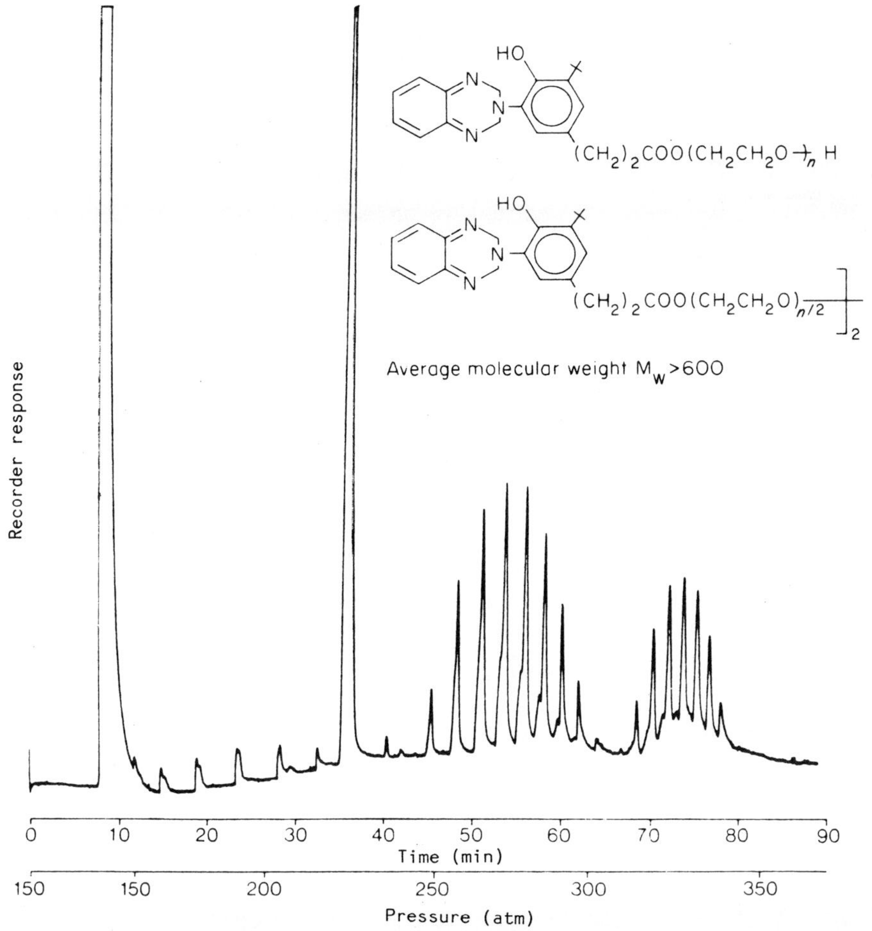

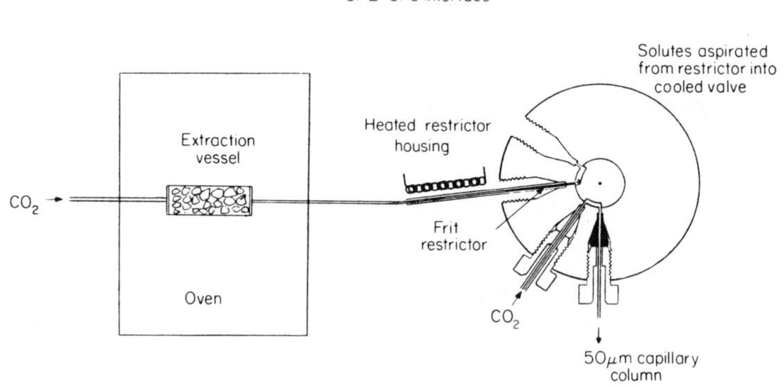

Figure 64   SFC chromatogram of polymer additive Tinuvin 1130. Conditions column 10 m x 50 um SB methyl, mobile phase, CO_2 at 150°C with pressure programming detector. FID.

Figure 65   Apparatus for coupled supercritical fluid extraction/SFC.

CHAPTER 8

MOLECULAR WEIGHT

8.1 INTRODUCTORY

Molecular weight obtained by size exclusion chromatography and other
chromatographic methods are not true determined volumes but are obtained
via a calibration process which involves the preparation of a universal
calibration curve relating log known molecular weight versus elution
volume obtained for standard polymer fractions of precisely known
molecular weight. Frequently standard polystyrenes are used for this
purpose as these are easily available. If the molecular weight $(MA)i$ of a
species of polymer A is related to the molecular weight $(M\ p/s)i$ of a
polystyrene standard eluting at the same retention volume i.

Then $$(MA)i = s(Mp/s)i^t$$

where $(MA)i$ is the molecular weight of the polymer requiring analysis.

$(M\ p/s)i$ = molecular weight of polystyrene standard used to establish
calibration curve,

s and t are constants.

If $(M\ p/s)i$, s and t are known $(MA)i$ can be calculated. To be able to
proceed with calculations of this type, it is necessary to have absolute
values for the weight average molecular weight $(\bar{M}_w$ and the number average
molecular weights $(\bar{M}_n)$ and these must be determined by independent
absolute methods such as light scattering $(\bar{M}_w)$ and osmometry $(\bar{M}_n)$ as
discussed in this Chapter.

The calculated weight average molecular weight $(\bar{M}_w)c$ of a polymer is
given by:

$$(\bar{M}_w)c = \frac{\Sigma h_i s(Mns)_i{}^t}{\Sigma h_i} \tag{1}$$

The calculated number average molecular weight $(\bar{M}_n)_c$ of a polymer is given
by:

$$(\bar{M}_n)_c = \frac{\Sigma h_i}{\sum \frac{h_i}{s(\bar{M}p/s)_i{}^t}} \tag{2}$$

where the subscript c indicates calculated molecular weight averages of
the weight average molecular weight $(\bar{M}_w)_c$ and the number average molecular
weight $(M_n)_c$.

h_i = peak height of polymer fraction on chromatogram.

i = elution volume.

s and t are constants.

True weight average molecular weights ($\overline{M}_w$) are given by:

$$(M_w) = \frac{E \sum h_i \; s(\overline{M}p/s)_i{}^t}{\sum h_i} \tag{3}$$

True number average molecular weights ($\overline{M}_n$) are given by:

$$(\overline{M}_n) = \frac{\sum h_i}{\sum \dfrac{h_i}{s(\overline{M}\,p/s)_i{}^t}} \tag{4}$$

where h_i, i and s have the meanings given above and

$$E = \exp - \frac{(D_2 \sigma)^2}{2}$$

where D_2 is the slope of the calibration curve at the peak position of the chromatogram and σ^2 is the variance of the single species chromatogram at the peak position.

8.2 ABSOLUTE WEIGHT AVERAGE MOLECULAR WEIGHT ($\overline{M}w$)

8.2.1 Light Scattering Methods

Two principal light scattering techniques are used to determine the molecular weight distribution of polymers. The first of these is the temperature gradient method, which is very similar to the thermal gradient method, mentioned earlier, for the preparation of polymer fractions. This method has been applied to the determination of the molecular weight distribution of polyethylene[862]. In this method a solution of alpha-chloro-naphthalene solvent and of a non-solvent containing a very low concentration of polyethylene is slowly cooled.

The high molecular weight species become insoluble and separate out, causing a small amount of turbidity. As the temperature continues to decrease, increasing amounts of polymer are precipitated out according to their molecular weight. Finally, a point is reached at which even the lowest molecular weight species become insoluble in the solution. At this point the turbidity is greatest, and ideally all of the polymer is precipitated but remains in suspension as very fine particles. If the increase in turbidity is plotted against the decreasing temperature, a cumulative plot is obtained which is similar to a cumulative wt% versus molecular weight. The increase in turbidity is related to the cumulative wt% and the molecular weight is related to the decrease in temperature. The ordinate in this is in terms of ΔE, the decrease in millivolts in phototube output from the starting value. Taylor and Tun[862] discuss the application to their results to procedures described by Morey and Tamblyn[863] and Claesson[864], to convert the experimental turbidity into molecular weight distribution data and the difficulties they encountered in this work.

Gamble et al[865] used a photoelectric turbidimeter for measuring molecular weight distribution of poly-alpha-olefins and ethylene-propylene copolymers. This instrument measures changes in turbidity as a function of temperature.

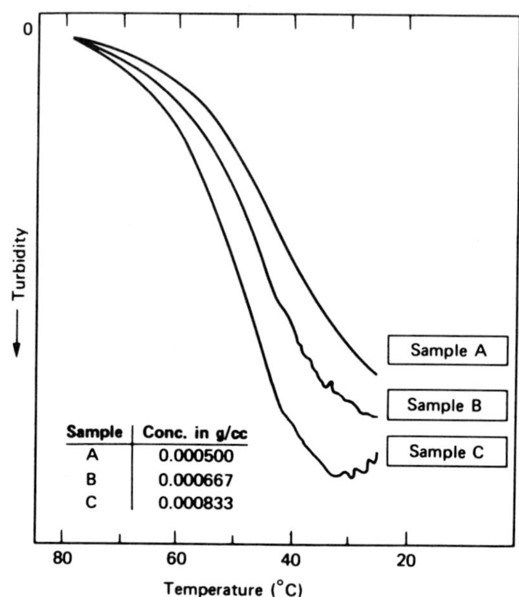

Figure 66   Effect of polymer concentration on turbidimetry.
Ethylene-propylene rubber.

A parameter designated as S was chosen by Gamble et al[865] to be the difference in temperature between points representing 20% of the maximum turbidity and 50% of the maximum turbidity as suggested by Taylor and Tung[862]. This portion of the curve is essentially linear for measuring polydispersity. The parameter S is correlated with a parameter determined from the Wesslau equation[866]. Because of the fairly narrow working temperature range, 80-25°C, of the solvent – non-solvent mixture used (heptane-n-propanol) it was necessary to study the effect of polymer concentration to provide a suitable, turbid system. Curves in Figure 66 demonstrate the effect of changing the concentration of ethylene-propylene rubber without changing the concentration of the non-solvent.

The second method of determining molecular weight distribution is turbidimetric titration. Beattie[867] developed an absolute turbidimetric titration method for determining solubility distribution (which is closely related to molecular weight distribution) of polystyrene. In this method polymer is precipitated from its solution in methylethyl ketone by addition of a non-solvent (isopropanol) of the same refractive index as that of the solvent. He showed, by the use of light-scattering theory, that under these conditions the concentration of polymer which is precipitated can be calculated from the maximum turbidity on the absolute basis. He also discusses the effect of particle size and particle size distribution. Beattie concluded that, with polystyrene, under the specified conditions, the reproducibility of turbidimetric precipitation curves is very good and that the method is accurate. Gooberman[868] also studied polystyrene but dissolved the polymer in butanone and titrated with isopropanol. He devised a correction procedure for the loss of precipitate during the titration and devised a method of location of the point of precipitation in relation to the weight-average molecular weight.

Tanaka et al.[869] have used the high-temperature turbidimetric titration procedure originally described by Morey and Tamblyn[870] for determining the molecular weight distribution of cellulose esters. This method has been applied to the measurement of the molecular weight distribution of polypropylene. They found that the type of molecular weight distribution of these polymers is a log-normal distribution function in a range of I(M), (cumulative wt%), between 5% and 90%. The effect of heterogeneity in the molecular weight distribution of polypropylene on the viscosity - molecular weight equation was examined experimentally; the results agreed with those calculated from theory. Strict temperature control ($\pm$ 0.15°C) is necessary in these determinations[871].

Taylor and Graham[872] have described a dual-beam turbidimetric photometer which they claim has distinct advantages over the earlier single-beam instruments[873,874], and have applied it to the determination of the molecular weight distribution of polyethylene, polypropylene, ethylene-propylene rubbers and other polymers[875].

Light-scattering measurements based on turbidimetric titration has been applied to a wide range of polymers including polyethylene[867], polypropylene[869], polystyrene[871,868,685-878], polymethylstyrene[879], PVC[880], chlorinated PVC-styrene[869], polyethylene oxide[882], polyphenylene-oxide[883], polyethylene terephthalate[884,885], and polyethylene glycol[871]. Other polymers studied by the light-scattering method include polystyrene[886-889], styrene-methyl methacrylate copolymer[889], water soluble acrylic acid and acrylamide polymers[890] and poly-alpha-methyl-styrene[891].

8.2.2 Sedimentation Methods

Some thirty years ago the analytical ultra centrifuge played an important part in characterising molecular weight distribution ($\bar{M}_w$), size distribution and density distribution in polymers. This technique was displaced by size exclusion chromatography. However, with improvements in instrumentation ultra centrifuge methods are to some extent making a comeback.

Machile and Klodwig[892] have reported on improvements in instrumentation in the analytical centrifuge technique, including an 8-cell interference optics multiplexer and an 8-cell Schlieren optics multiplexer, both based on the same light source, a modulable laser and a titanium rotor with a maximum speed of 60,000 rev min^{-1}. Figure 67 shows the optical arrangement of this combined interference and Schlieren pictures taken during a sedimentation run on polystyrene (M_w 111.000) dissolved in toluene whilst Figure 68(b) shows Schlieren pictures obtained for a range of polystyrenes in the M_w range 10^4 to 150×10^6.

Sedimentation methods have also been applied to polystyrene[893], polyglycols[894], polyesters[894].

Laser mass spectroscopy has been used[895] to determine the average molecular weight for polyethylene glycols and polypropylene glycols.

8.2.3 Size Exclusion Chromatography

Williamson and Cervenka[896] compared weight average molecular weight ($\bar{M}_w$) determinations on fractions of low density polyethylene and ethylene-butane-1 copolymers using size exclusion chromatography in conjunction with viscometry data and a light scattering method.

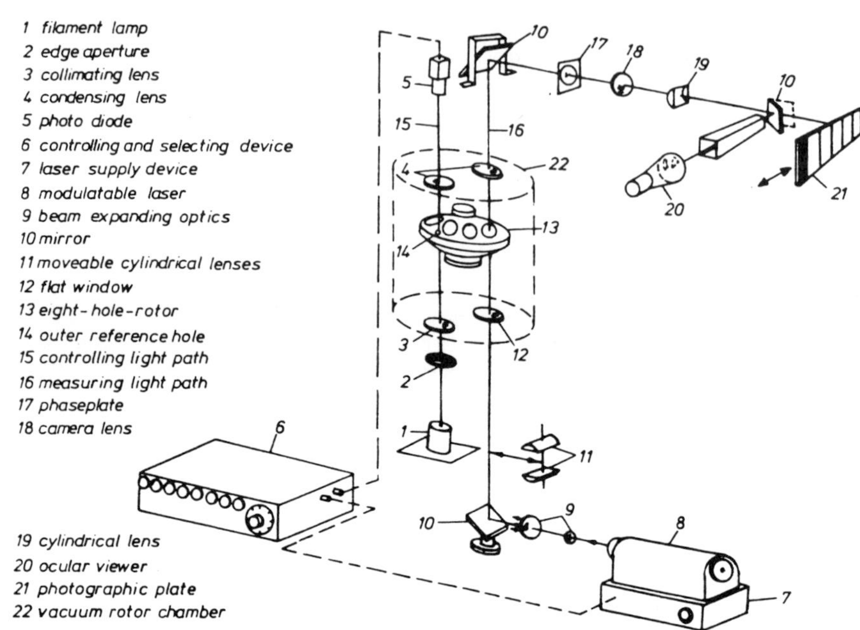

Figure 67   Optical arrangement of the combined interface and schlieren optics multiplexer for the analytical ultracentrifuge.

A comparison with $\bar{M}_w$ values using size exclusion chromatography for low-density polyethylene fractions, narrow distribution high-density polyethylenes and fractions from ethylene-butene-1 copolymer showed good agreement between the two methods (Table 94). Whereas high density polyethylene contains few long or short alkyl group branches, these do occur in high density (low pressure) polyethylene. The good agreement obtained for M_w between size exclusion and light scattering data is very satisfactory in that it infers that a chain branching has no serious effect on the determination of weight average molecular weights.

8.3   ABSOLUTE NUMBER AVERAGE MOLECULAR WEIGHT ($\bar{M}_n$)

8.3.1   Osmometry Methods

Osmotic pressure measurement of solutions of a polymer in a solvent can be employed to determine the molecular weight of the polymer. After approximately 1 hour the osmotic pressure of solutions of poly-alpha-methylstyrene in toluene become practically constant. From this osmotic pressure and the concentration of the polymer in solution it is possible to calculate the molecular weight of the polymer from the following relationship:

$$M = \frac{aRT}{PV}$$

where M = molecular weight
   R = gas constant
   T = temperature (K)
   P = osmotic pressure

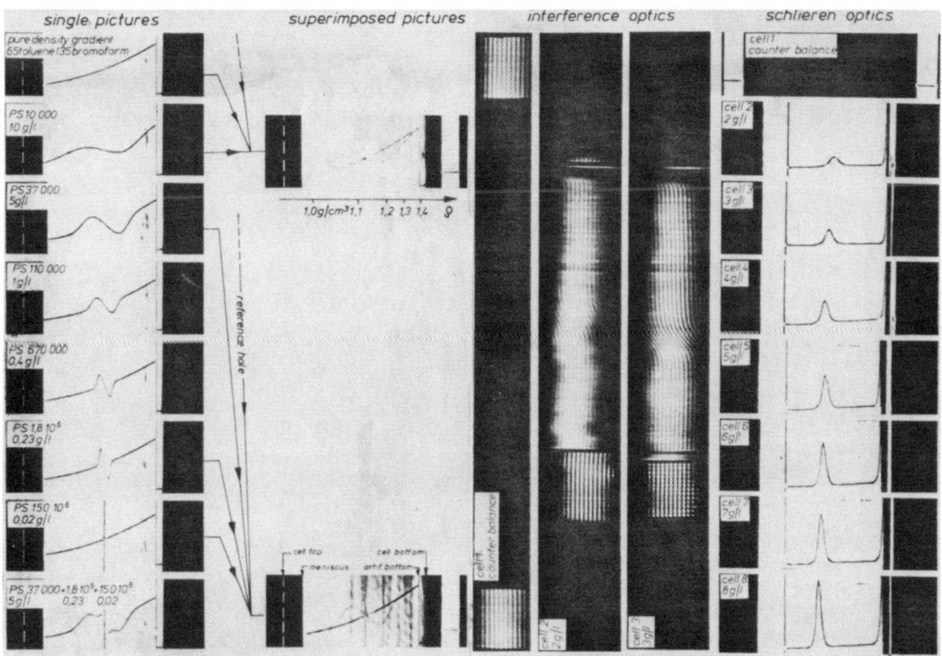

Figure 68   (a) Interference and schlieren pictures taken after a running time of about 93 min during an 8 cell sedimentation run of PS110000 dissolved in toluene, at the maximum rotor speed of 60000 rev/min.

(b) Schlieren pictures, taken by the laser schlieren optics multiplexer during an 8 cell density gradient run of PS samples with different molecular weights at the maximum rotor speed of 60,000 rev/min.

Table 94 - Molecular parameters and intrinsic viscosities for high-density
polyethylenes and ethylene-butene-1 copolymer fractions

Sample	$[\eta]$ (dl/g TCB at 135°)	$[\eta]$ (dl/g Decalin at 120°)	$[\eta]$ (dl/g Diphenyl at 130°)	Size exclusion chromatography			M_w x 10-4 (light scattering)
				M_w x 10^{-4}	M_n x 10^{-4}	M_w/M_n	
High-density polyethylene fractions							
B1	1.10	1.20	0.74	5.73	3.10	1.85	5.17
B5	1.47	1.78	0.97	9.70	4.97	1.95	-
B2	2.03	2.28	1.25	14.8	8.76	1.68	11.8
B3	2.65	3.17	1.44	24.2	12.2	1.98	16.8
B4	2.92	3.50	1.70	27.1	13.9	1.95	22.1
Fractions from ethylene-butene-1 copolymer							
C1	0.77	-	-	3.38	2.73	1.24	3.25
C2	1.60	-	-	8.9	5.86	1.52	8.44
C3	2.84	-	-	33.4	18.9	1.77	27.5
C4	4.70	-	-	54.3	22.6	2.40	45.0
Low density polyethylene fractions							
Sample D							
D1	0.23	-	-	-	-	-	0.94
D4	0.82	0.95	-	7.19	5.14	1.40	8.6
D5	1.06	1.34	0.64	11.8	8.87	1.33	22.3
D6	1.56	2.11	0.82	34.6	21.0	1.65	37.0
D7	1.91	2.50	0.89	118.0	28.4	4.15	216.0
D8	2.10	2.78	0.89	286.0	48.6	5.58	273.0

V = volume of sample solution (ml)
a = concentration of polymer in solution, gl^{-1}

Osmometry has been used to measure the molecular weight of a wide
range of polymers including PVC[897], polyethylene terephthalate[898-900],
Nylon[899], polystyrene[898], linear polyesters[900] including polytetramethylene terephthalate, polypentamethylene terephthalate, polyhexamethylene
terephthalate and polytetramethylene isophthalate.

8.3.2 End-Group Analysis

Examples of the occurrence of end-groups in polymers are quoted
below:

Polyethylene glycols
$$- CH_2 - CH_2 - OH + nCH_2CH_2O = CH_2CH_2O(CH_2CH_2O)_{n-1} CH_2CH_2OH$$

    ethylene glycol   polyethylene glycol hydroxy end-group

PVC
$$nCH_2 = 2CHCl = - (CH_2 - CHCl)_{n-1} - CH_2 = CHCl$$

    vinyl chloride   PVC   double bond end-group

End-group analysis is a useful method of measuring number average
molecular weight of certain types of polymers which are known to have one
or two terminal end-groups per molecule. Various techniques have been
used to determine end-groups. Obviously these methods must be highly
sensitive as, particularly with high molecular weight polymers , the
percentage of end-groups is very low. If 1 mole (62g) of diethylene
glycol is reacted with 20 moles (880g) of ethylene oxide according to the
following equation, then the molecular weight of the product is 62 + 880 =
942.

$$HO-CH_2CH_2OH + 20CH_2CH_2O =$$
$$HOCH_2CH_2(CH_2CH_2O)_9OCH_2CH_2O(CH_2CH_2O)_9CH_2-CH_2OH$$

Thus two hydroxy end-groups (34g) occur per 997g of polymer, i.e. 3.4%
hydroxy group. Similarly if 100 moles of ethylene oxide reacted with 1
mol of diethylene glycol then the final product would have a hydroxy end-
group content of 0.76%.

Kalinina and Motorina[901] developed a method for determination of the
molecular weight of compounds having terminal hydroxyl groups, e.g.
polyethylene glycols. This was estimated from treatment with ceric
ammonium nitrate reagent in nitric acid followed by spectrophotometric
measurement of the coloured complex.

If, for example, by this method 0.3% hydroxy end-groups was found in
an ethylene glycol-ethylene oxide condensate i.e.

$$HO - CH_2 - CH_2OH + nCH_2CH_2O = HO(CH_2CH_2O)_n CH_2CH_2OH$$

$$Percentage\ hydroxy = \frac{2 \times 17}{62 + n \times 44} = 0.3$$

$$n = \frac{3400 - 18.6}{13.2} = 255$$

Molecular weight of polymer is

$$HO(OCH_2CH_2)_{n/2 - 1}OCH_2CH_2O(CH_2CH_2O)_{n/2 - 1}OH =$$
$$HO(OCH_2CH_2)_{126.5}OCH_2CH_2O(CH_2CH_2O)_{126.5}OH =$$
$$11,226$$

Fritz et al[902] have described a method for the determination of
hydroxy groups in poly(ethylene glycols). The method has a sensitivity of
10^{-4} mol hydroxyl per kg polymer, which is achieved by using silylation
with acrylsilylamines and subsequent photometric measurement of the
silylated polymer.

Unsaturated end-groups in PVC have been estimated from NMR spectra,
using reference model compounds such as 1-chloropentene-2, 1,1-
dichloropentene-2, and 1,2-dichloropentene-2. Fourier transform proton
NMR studies of PVC indicated that the unsaturated end-groups contained
allylic chlorine atoms. The presence of the unsaturated end-groups would
partially explain the PVC degradation[903].

Terminal hydroxy groups were measured in polycarbonate following
complexation with ceric ammonium nitrate; absorbance has been measured at
500 nm and 530-540 nm[904].

Smirnova and co-workers[905], employed trichloroacetyl isocyanate and trifluoroacetic anhydride acetylations to determine hydroxyl end-groups in polyester polyols. The isocyanate reagent measured proton resonance and was best suited for samples having molecular weight of less than 4500; the anhydride method was more sensitive and applicable to higher molecular weights than a ^{19}F NMR method.

Law[906] has also described infrared methods for the determination of hydroxy equivalent weight and carboxyl equivalent weight of carboxyl and hydroxy polybutadienes.

Infrared spectroscopy has been used to study styrene oligomers containing acrylonitrile active ends[907]. The addition of acrylonitrile to oligostyrene alkylmetal salt solutions resulted in attachment of acrylonitrile across the double bond of the oligostyrene and by formation of the carbanion C-HCN.

Proton NMR signals for olefinic end-groups, $-CH_2C(CH_3) = CH_2$, $CH = C(Me)_2$, and $CH_2C - (CH_2)CH_2$ - were observed in high molecular weight polyisobutylenes by Manaff et al[908].

Valuev et al[909] separated oligomeric butadiene-isoprene block copolymers into fractions containing two, one, and no terminal hydroxyl groups, respectively, by thin-layer chromatography. Infrared spectroscopy was used for the determination of hydroxyl content.

Nissen and co-workers have described a method for carboxyl end-groups in poly(ethylene terephthalate). Hydrazinolysis led to formation of terephthalomono-hydrazide from carboxylated terephthalyl residues to provide a selective analysis for carboxyl groups via ultraviolet absorbance at 240 nm[910].

The application of dye-partition methods to the determination of end-groups is best illustrated by an example concerning the determination of end-groups in persulphate-initiated polymethylmethacrylate[911,912]. Sulphate and other anionic sulphoxy end-groups were determined by shaking a chloroform solution of the polymer with aqueous methylene blue reagent. The greater the anionic content, the more the methylene blue phase is decolorized. The blue colour is evaluated spectrophotometrically at 660 nm. The quantity of anionic sulphoxy end-group present in the polymer is obtained by comparing the experimental optical density values with a calibration curve of pure sodium lauryl sulphate obtained by following a similar procedure[913].

Results obtained by Ghosh et al[911,914] in end-group analyses of polymethylmethacrylate indicate that all the polymer samples exhibit a positive response to methylene blue reagent in the dye partition test, indicating the presence of at least some sulphate (OSO_3) end-groups.

Maiti and Saha[915] have described a dye partition technique[916-918] utilizing disulphine blue for the qualitative detection and, in some cases, the determination of amino end-groups in the free radical polymerization of polymethylmethacrylate. They found only 0.01-0.62 amino end-groups per chain in polymethylmethacrylate made by the amino-azo-bisbutyronitrile system, whereas in polymer made by the titanous chloride and acidic hydroxylamine systems they found 1.10-1.90 amino end-groups per chain.

Ghosh et al.[919,920] have also described a dye partition method for the determination of hydroxyl end-groups in poly(methylmethacrylate) samples prepared in aqueous media with the use of hydrogen peroxide as the photo-initiator. In this method the dried polymers were treated with chlorosulphonic acid under suitable conditions whereby the hydroxyl end-groups present in them were transformed to sulphate end-groups. Spectrophotometric analyses of sulphate end-groups in the treated polymers was carried out by the application of the dye-partition technique and thus a measure of hydroxyl end-groups in the original polymers was obtained (average 1 hydroxy end-group per polymer chain).

Saha et al.[921] and Palit[922,923] have developed a dye-partition method for the determination of halogen atoms in copolymers of styrene, methylmethacrylate, methylacrylate, or vinyl acetate with a chlorine-bearing monomer such as allyl chloride and tetrachloroethylene. The quarternized copolymers were quarternized with pyridine, then precipitated with petroleum ether or alcohol and further purified by repeated precipitation from their benzene solutions with a mixture of alcohol and petroleum ether as the non-solvent. The finally precipitated polymers were then washed with petroleum ether and dried in air. The test for quaternary halide groups in polymers was carried out with a reagent consisting of disulphine blue dissolved in 0.01 M hydrochloric acid and the colour evaluated spectrophotometrically at 630 nm. Saha et al.[921] and Palit[922,923] found that there may be uncertainty in the quantitative aspects of this method.

Stetnagel and Palit[924] applied dye partition end-group analysis procedures to the examination of sulphate end-groups in persulphate initiated polystyrene.

Ghosh et al.[925] have carried out end-group analysis of persulphate initiated polystyrene using a dye partition and dye interaction technique. Sulphate and hydroxyl end-groups are generally found to be incorporated in the polymer to an average total of 1.5 to 2.5 end-groups per polymer chain.

Banthia and co-workers[926] determined sulphate, sulphonate and isothironium salt end-groups in polystyrene by the dye-partition technique. Polymer polarity did not affect the results of the end-group determination. Nitrile groups incorporated in polystyrene by initiation or copolymerization have detected and estimated by dye-partition techniques after reduction to amino groups with lithium aluminium hydride in tetrahydrofuran[927].

Mukhopadhyay et al.[928] have reported a reverse dye-partition technique for the estimation of acid groups in water-soluble polymer such as copolymers of acrylamide and other carboxyl-bearing monomers (e.g. acrylic acid), and the monomer reactivity ratio, r, has been determined by measuring the carboxyl group content in those copolymers[929,930]. The carboxyl contents of the purified copolymers were determined by the reverse dye-partition method.

8.4 VISCOSITY AVERAGE MOLECULAR WEIGHT (Mv)

The viscosity-average molecular weight (M_v) of polymer sample may be computed from the chromatogram h_i of the polymer by the formula.

$$\bar{M}_v = \left(\sum h_i M_i^a / \Sigma h_i \right)^{1/a}$$

(5)

Substituting the Mark-Houwink equation gives

$$[\eta] = K(\sum h_i M_i^a / \sum h_i) \tag{6}$$

where a and K are Mark-Houwink parameters and $[\eta]$ is the intrinsic viscosity of the polymer. It is thus possible to calculate average molecular weights from viscosity data on the weighed fractions obtained by chromatography, or on the original unfractionated polymer provided appropriate calibration data are available for the calculation of constants in the above equation.

Intrinsic viscosity measurements have been used extensively for the determination of the molecular weight distribution of polyethylene[931-937], polypropylene[938-941], PVC[942], polyethylene terephthalate[943,944] and polyacrylanide[945,946].

Practical details for the determination of intrinsic viscosity have been given by Williamson and Cervenka[947] Trichlorobenzene, (at 135°C) decalin (at 120°C) and diphenyl (at 18 or 130°C) are commonly used viscometric solvents. Kinetic energy loss corrections are often required. These same workers[948] have studied the influence of the theta temperature of the polymer on viscometric results. They found, for example, that high density polyethylene had a theta temperature of almost 130°C which is different from that of branched low-density polyethylene, due to the different polymeric structures, most probably long chain branches in the low density material.

They investigated the intrinsic viscosity molecular weight relationship under theta conditions.

The $[\eta]$ - M_w relationship is shown in Figure 69. The Mark-Houwink equation for high-density polyethylene under Flory theta conditions (diphenyl at 130°C is:

$$[\eta]_\theta = K_\theta M_w^a = 3.02 \times 10^{-3} \overline{M}_w^{0.5}. \tag{7}$$

The constants in this equation are in good agreement with those found by other workers, thus "a" values of 0.50 to 0.507[949,950] and K_θ values of 3.05 x 10^{-3} [951] have been reported.

The agreement of "a" with the theoretical value, and K_θ [Eqn. 7] with values by other workers has considerable implications. This shows that not only are the intrinsic viscosity measurements and the size exclusion molecular weights substantially correct, but also that the different theta temperatures of 118°C and 130°C for the branched and linear polyethylenes in diphenyl are fully vindicated.

The effect of long chain branches in depressing the intrinsic viscosity below the Mark-Houwink relationship for linear polyethylene measured in the theta solvent is readily seen from Figure 69.

The $[\eta]$ - $\overline{M}_w$ relationships for fractions from low density polyethylenes in 1,2,4 trichlorobenzene at 135°C are shown in Figure 70(a) and for low density polyethylene in decalin at 120°C in Figure 70(b). The Mark-Houwink relationship from the narrow distribution branched high density polyethylene and high branched ethylene-butene copolymers is also included.

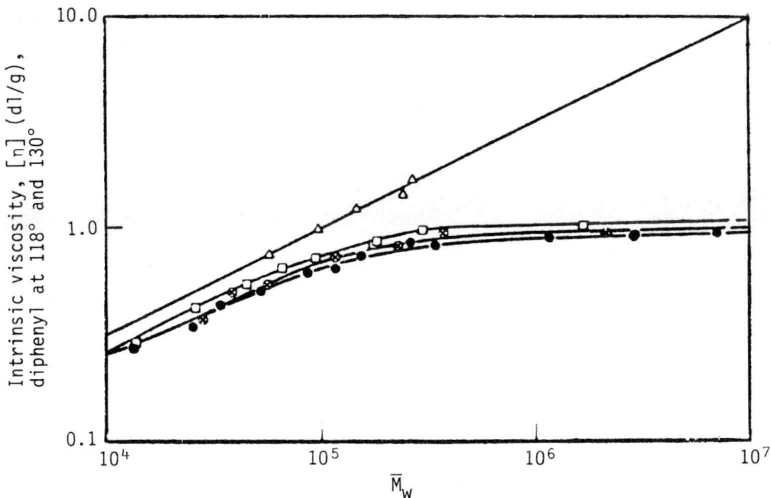

Figure 69   Relationship between intrinsic viscosity (theta
conditions) and weight average molecular weight (Mw) for
low and high density polyethylene. Δ high density
polyethylenes; low density polyethylene fractions o D
series; x E series; $\square$ F series.

     The effect of long-chain branching is readily seen. For molecular
weights above 100,000 long-chain branching becomes increasingly effective
in depressing the intrinsic viscosity and consequently the branching index
(g factor) decreases to the very low values. The level of long-chain
branching follows the polymer density order, i.e. 0.919, 0.925 and 0.930.
An interesting feature is the effect of long-chain branching on the
intrinsic viscosity at molecular weights below 100,000. The low-density
and high-density curves do not coincide at low molecular weights, showing
that the long-chain branching is still appreciable.

     Williamson and Cervenka[947] also used viscometric data to obtain
branching index data under theta (go) and non-theta (g solvent) conditions
and to obtain estimates of long-chain branching in low and high density
polyethylene.

8.5   FIELD DESORPTION MASS SPECTROMETRY

     Field desorption mass spectrometry has been shown to be a method of
choice for determining molecular weight of nonvolatile and higher
molecular weight chemicals[952-964]. Field desorption provides a gentle
means of ionization that imparts little excess energy to the molecule.
Consequently, molecular ions (or at times protonated or cation-adduct
molecular ions) are frequently the strongest ions observed. Fragment ions
and ions due to thermal decomposition products are often absent or else of
relatively low intensity. Field desorption mass spectrometry can be used
to obtain good qualitative distributions of oligomers for low molecular
weight polymers. Oligomeric mixtures studied in this regard include n-
paraffins[955], polypivalolactone[956,957], poly(2,2,4-trimethyl-1,2-
dihydroquinoline)[958,959], polystyrene[958-960], and poly(propylene
glycol)[958-960]. Components of several oligomeric polymer chemical
mixtures have also been qualitatively characterized by field desorption
mass spectrometry[961,962].

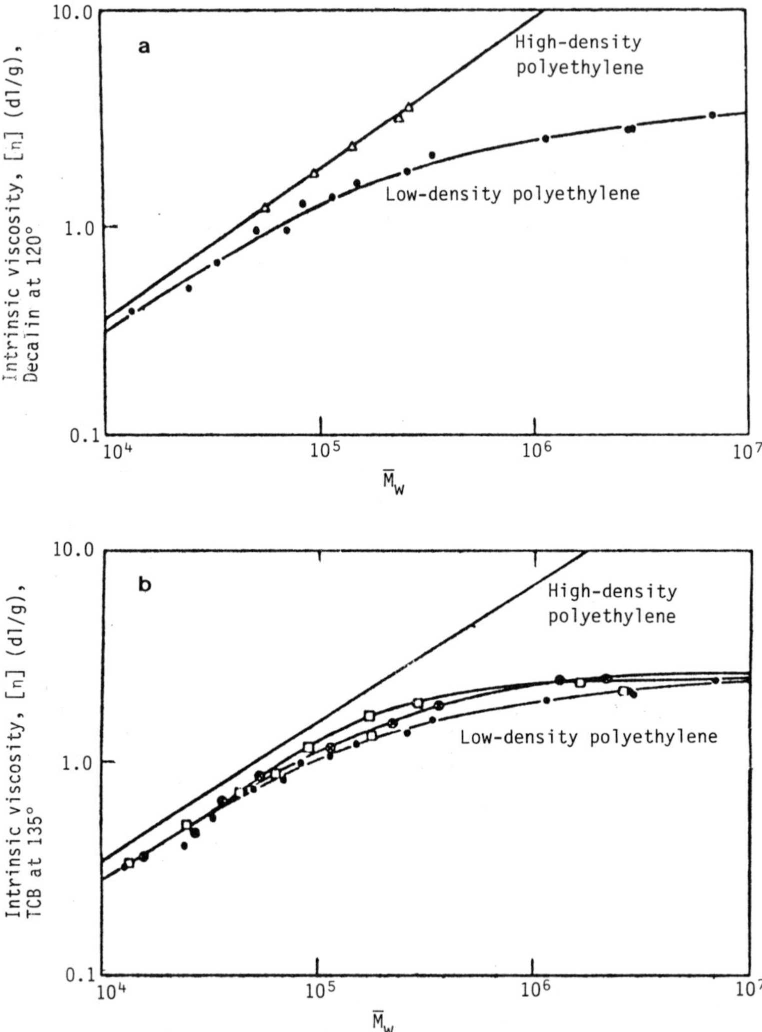

Figure 70 Relationship between intrinsic viscosity (n) and weight average molecular weight (Mw) for high and low density polyethylene and copolymers: (a) decalin at 120°C; (b) 1,2,4 trichlorobenzene at 135°C. Δ high density polyethylene; o low density polyethylene; □ ethylene-butylene copolymers.

In none of the above work was an attempt made to determine accurate molecular weight averages $\bar{M}_n$ and $\bar{M}_w$. Lattimer[963] applied the technique to the determination of molecular weight averages of narrow molecular weight distribution butyl lithium catalysed polystyrene oligomers with molecular weight up to 5300 atomic mass units. Molecular ion spectra were obtained by the latter method for polystyrene oligomers. Field desorption derived molecular weight averages ($\bar{M}_n$ and $\bar{M}_w$) compared favourably to values obtained by conventional techniques (vapour pressure osmometry, intrinsic viscosity, kinetic data and size exclusion chromatography). This agreement indicates that the relative intensities of the field desorption mass spectrometry oligomer molecular ions (with appropriate corrections for isotopic abundances) could be used directly to give reasonably accurate relative molar concentrations of the oligomer.

For mass spectrometric methods the number average molecular weight is defined as

$$\bar{M}_n = \sum N_i M_i / \sum N_i \tag{8}$$

The weight average is defined at

$$\bar{M}_w = \sum N_i M_i^2 / \sum N_i M_i \tag{9}$$

The viscosity average is defined as

$$\bar{M}_v = [\sum N_i M_i^{1+a} / \sum N_i M_i]^{1/a} \tag{10}$$

where "a" is the constant from the Mark-Houwink viscosity weight relationship and N_i and M_i refer, respectively to the number and masses of the ions detected.

$$[\eta] = kM^a \tag{11}$$

For a "most probable" distribution it can be shown[964] that

$$\bar{M}_n : \bar{M}_v : \bar{M}_w :: 1 : [(1 + a)\Gamma(1 + a)^{1/a}] : 2 \tag{12}$$

where $(1 + a)$ is the function of $1 + a$. When $a = 1$, the viscosity average (M_v) is identical with the weight average (M_w).

The desorption of oligomer molecular ions was dependent on the field desorption emitter heating current applied. That is, lower mass oligomers desorbed at lower heating currents (5-15 mA), while higher mass oligomers desorbed at higher heating currents (15-30 mA). Relative oligomer abundances were determined from the field desorption mass spectrometric data as follows. The height at peak maximum for each oligomer was recorded from each mass spectral trace. The "raw" peak intensities for each oligomer were then summed over the accumulated field desorption mass spectra (there were 10-15 scans for each polystyrene batch). The summed peak intensity for each oligomer was then multiplied by the appropriate correction factor to obtain a "corrected" intensity. Corrected intensities were then used to determine the percent relative intensities reported in Table 95. These percent intensities were used with the average oligomer molecular weights to obtain the molecular weight parameters (M_n and M_w) listed in Table 95.

It is clear that good molecular weight averages (M_n and M_w) have been obtained for field desorption mass spectrometry for the polystyrene standards. This indicates that the relative intensities of the field

Table 95 – Relative oligomer intensities and molecular weight averages for polystyrene

n	formula	nominal mass[a]	average mass[b]	(res. 400) correction factor[c]	(res. 500) correction factor[d]	batch 61004 liquid chromatography	batch 61004 311A	batch 61004 MS-50	batch 12a MS-50	batch 11b MS-50
								mass spectrometry		
1	$C_{12}H_{18}$	162	162.3	1.14	1.14	0.71		0.04		
2	$C_{20}H_{26}$	266	266.4	1.24	1.24	4.76	0.12	0.19		
3	$C_{28}H_{34}$	370	370.6	1.36	1.36	7.33	0.95	2.31		
4	$C_{36}H_{42}$	474	474.7	1.48	1.48	8.86	4.10	4.36		
5	$C_{44}H_{50}$	578	578.9	1.62	1.62	9.81	6.84	6.86	0.88	
6	$C_{52}H_{58}$	682	683.0	1.76	1.76	9.67	10.1	10.9	2.51	
7	$C_{60}H_{66}$	786	787.2	1.84	1.93	9.29	14.7	14.0	4.46	
8	$C_{68}H_{74}$	890	891.3	1.85	2.10	9.43	16.6	15.8	4.58	
9	$C_{76}H_{82}$	994	995.5	1.87	2.14	9.33	13.8	14.7	5.18	0.46
10	$C_{84}H_{90}$	1098	1099.6	1.79	2.15	8.48	12.2	10.9	6.03	1.29
11	$C_{92}H_{98}$	1202	1203.8	1.70	1.98	7.19	9.26	7.32	5.92	1.42
12	$C_{100}H_{106}$	1306	1307.9	1.65	1.88	5.57	5.24	5.45	6.52	2.35
13	$C_{108}H_{114}$	1410	1412.1	1.62	1.82	4.00	2.02	4.00	6.49	2.66
14	$C_{116}H_{122}$	1514	1516.2	1.60	1.78	2.62	1.94	1.34	6.11	2.90
15	$C_{124}H_{130}$	1618	1620.4	1.57	1.76	1.57	1.09	1.09	5.52	2.77
16	$C_{132}H_{138}$	1722	1724.5	1.56	1.75	0.86	0.50	0.48	4.74	2.36
17	$C_{140}H_{146}$	1826	1828.7	1.56	1.75	0.38	0.26	0.13	4.74	2.70
18	$C_{148}H_{154}$	1930	1932.8	1.55	1.74	0.14	0.10	0.07	4.89	3.02
19	$C_{156}H_{162}$	2034	2037.0	1.56	1.74		0.09	0.03	4.47	3.66
20	$C_{164}H_{170}$	2138	2141.1	1.54	1.74				4.23	3.80
21	$C_{172}H_{178}$	2242	2245.3	1.51	1.70				3.82	3.13
22	$C_{180}H_{186}$	2346	2349.4	1.48	1.66				3.19	3.39

23	$C_{188}H_{194}$	2450	2453.6	1.46	1.63				3.07	3.00
24	$C_{196}H_{202}$	2554	2557.7	1.45	1.61				2.56	3.32
25	$C_{204}H_{210}$	2658	2661.9	1.44	1.59				2.41	3.36
26	$C_{212}H_{218}$	2762	2766.0	1.43	1.58				1.68	2.80
27	$C_{220}H_{226}$	2866	2870.2	1.43	1.57				1.32	3.22
28	$C_{228}H_{234}$	2970	2974.3	1.42	1.56				1.12	2.56
29	$C_{236}H_{242}$	3074	3078.5	1.42	1.56				1.04	2.55
30	$C_{244}H_{250}$	3178	3182.6	1.41	1.56				0.90	2.86
31	$C_{252}H_{258}$	3282	3286.8	1.41	1.56				0.85	3.86
32	$C_{260}H_{266}$	3386	3390.9	1.39	1.54				0.52	3.75
33	$C_{268}H_{274}$	3490	3495.1	1.38	1.52				0.25	3.93
34	$C_{276}H_{282}$	3594	3599.2	1.37	1.50					3.09
35	$C_{284}H_{290}$	3698	3703.4	1.36	1.49					3.00
36	$C_{292}H_{298}$	3802	3807.5	1.35	1.48					3.16
37	$C_{300}H_{306}$	3906	3911.7	1.35	1.47					3.26
38	$C_{308}H_{314}$	4010	4015.8	1.34	1.46					2.85
39	$C_{316}H_{322}$	4114	4120.0	1.34	1.46					2.57
40	$C_{324}H_{333}$	4218	4224.1	1.34	1.45					2.35
41	$C_{322}H_{338}$	4322	4328.3	1.34	1.45					2.07
42	$C_{340}H_{346}$	4426	4432.4	1.34	1.45					1.37
43	$C_{348}H_{354}$	4530	4536.6	1.33	1.44					1.37
44	$C_{356}H_{362}$	4634	4640.7	1.32	1.43					0.82
45	$C_{364}H_{379}$	4738	4744.9	1.31	1.42					0.72
46	$C_{372}H_{378}$	4842	4845.0	1.30	1.41					0.53
47	$C_{380}H_{386}$	4946	4953.2	1.30	1.40					0.64
48	$C_{388}H_{394}$	5050	5057.3	1.30	1.39					0.51
49	$C_{396}H_{402}$	5154	5161.4	1.29	1.39					0.44
50	$C_{404}H_{410}$	5258	5265.6	1.29	1.39					0.11

Determined

$\overline{M}_n$						855	941	928	1690	2890
$\overline{M}_w$						1010	1018	1011	1940	3230
$\overline{M}_w/\overline{M}_n$						1.18	1.08	1.09	1.15	1.12

Table 95 – Continued

Nominal values

$\overline{M}_n$ (e)	811±5%	811±5%	811±5%	1710 ±7%	3100 ±5%
$\overline{M}_w$ (f)	1024 ±5%	1024 ±5%	1024 ±5%	2200 ±7%	3600 ±3%
$\overline{M}_w/\overline{M}_n$ (g)	$\leq$ 1.30	$\leq$ 1.30	$\leq$ 1.30	$\leq$ 1.10	$\leq$ 1.10

a) Nominal mass spectrometric molecular weight (C = 12, H = 1).
b) Number average accurate molecular weight for an oligomer isotopic cluster (determined by HIMAS computer program).
c) Correction factor determined by HIMAS computer program (M/ M = 400 for Varian MAT 311A).
d) Correction factor (M/ M = 500 for Kratos MS-50).
e) from data sheet.
f) from vapour pressure osmometry.
g) from frinsic viscosity in cyclohexane or benzene
h) determined by kinetic considerations and comparison with size exclusion chromatography data[965].

desorption mass spectrometry oligomer molecular ions (with appropriate corrections for isotopic abundances) can be used directly to obtain reasonably accurate relative molar concentrations of the oligomer. The field desorption mass spectrometry is appealing since it provides a means for direct determination of M_n and M_w. It is to be expected that other "good desorbing" polymer systems could also be characterized in this manner. Polymers containing highly polar groups would likely yield less accurate results (due to poor desorption characteristics).

8.6 NMR SPECTROSCOPY

High resolution magnetic resonance spectroscopy has been used for the determination of composition and molecular weight of polyester urethanes[966].

Slonim and co-workers[967] described NMR methods to determine chain structure, composition and molecular weight of unsaturated polyesters.

COPOLYMER COMPOSITION

9.1 INTRODUCTORY

The elucidation of the composition and structure of a copolymer or terpolymer is a challenging task for any polymer analyst. Firstly, it is necessary to be absolutely sure of the elemental composition of the sample, (Chapter 4.3), and preferably that the copolymer has been completely separated from any additives present, (Chapter 3). Accurate data on the type and concentration of elements present will often simplify the task ahead.

Spectroscopic or other data may have indicated the presence of various functional groups such as unsaturation; ester, carbonyl, hydroxy, alkoxy, oxyalkylene, epoxy, anhydride or amino. Various functional group analyses applicable to copolymers are discussed below in Chapter 9.2. Additionally various quantitative methods are available for the determination of comonomer constants and ratios as discussed in Chapter 9.3 including the following, styrene-butadiene, coolefin ratios e.g. ethylene-propylene, styrene-acrylate, carboxyl-glycol and hexa-fluoroproylene-vinylidene fluoride.

Having ascertained the composition of the copolymer it may now be required to obtain more detailed information regarding monomer unit sequences and branching as discussed, respectively in Chapters 9.4 and 9.5.

9.2 FUNCTIONAL GROUP ANALYSIS

9.2.1 Unsaturation

Most of the published work on the determination of functional groups, not unexpectedly, has been carried out on copolymers. This is because the determination of a functional group that is specific to one of the copolymer constituents is the key to the determination of the monomer ratios in the copolymer.

<u>Iodine monochloride procedure.</u> Styrene-butadiene copolymers contain residual double bonds which enable the butadiene content of the copolymer to be determined.

$$Ph - CH = CH_2 + CH_2 = CH - CH = CH_2 \rightarrow - \underset{\underset{Ph}{|}}{CH} - CH_2 - CH_2 - CH = CH - CH_2 -$$

In this procedure the polymer is reacted with an excess of standard iodine monochloride dissolved in glacial acetic acid (Wijs reagent).

$$- \underset{\underset{Ph}{|}}{CH} - CH_2 - CH_2 - CH = CH - CH_2 - + \; ICl \longrightarrow \; - \underset{\underset{Ph}{|}}{CH} - CH_2 - CH_2 - CHI - CH - CH_2 -$$

After completion of the reaction, excess iodine monochloride is reacted with potassium iodide and the liberated iodine estimated by titration with standard sodium thiosulphate.

$$ICl + KI = KCl + I_2$$

The double bond content of the original polymer can be then calculated from the measured consumption of iodine monochloride.

Crompton and Reid[968] have described procedures for the separation of high impact polystyrene, i.e. styrene-butadiene copolymer, into the free rubber plus rubber grafted polystyrene plus copolymerized rubber and a gel fraction and for estimating total unsaturation in the two separated fractions. To separate a sample into gel and soluble fractions it was first dissolved in toluene. Only gel remains undissolved. Methanol is then added, which precipitates the polystyrene-rubber graft, ungrafted rubber and polystyrene. Any styrene monomer, soap or lubricant remain in the liquid phase, which is separated from the solids and rejected.

The addition of toluene to the solids dissolved all the polymeric material with the exception of the gel. The toluene solubles are separated from the solid gel by centrifuging and made up to a standard volume with toluene. The gel is then dried in vacuum and weighed.

Both the gel and toluene soluble fractions are reserved for determination of unsaturation. The iodine monochloride method was used for measuring unsaturation. To determine unsaturation in styrene-butadiene rubbers with good accuracy using the iodine monochloride procedure it was found necessary to contact the sample with chloroform for 15 hours before reaction with iodine monochloride. Even with a 30 hour reaction period, a constant iodine value is obtained only when a five-fold excess of iodine monochloride reagent is used.

The solid gel, separated from a high impact polystyrene by solvent extraction procedure, is completely insoluble in chloroform and in the iodine monochloride reagent solution. A contact time with chloroform of 90 hrs with a 75 hr reaction period with reagent is required.

Crompton and Reid[968] used these procedures to study the distribution of rubber added in several laboratory preparations of high impact polystyrene containing 6 wt-% of a styrene-butadiene rubber and 94% styrene, i.e. theoretical 4.1% butadiene. The results in Table 96 show the way in which the added unsaturation of 4.1% butadiene distributes between the gel and soluble fractions; the butadiene content of the separated gel remains fairly constant, in the 20-25% region, regardless of the quantity of gel present in the sample. As the gel content increases, therefore, so more of the rubber becomes incorporated into the gel and less remains as free rubber or soluble graft. The recovered unsaturation lies mainly in the 90-95% region, indicating that loss of unsaturation due to grafting or cross-linking reactions occurs only to the extent of some 5-10%.

Table 96 - Distribution of butadiene between soluble and gel fractions
obtained from polystyrenes containing different amounts of gel

Gel content of sample (wt%)	Butadiene content isolated gel (wt%)	Soluble graft butadiene content A (calculated on original sample) (wt%)	Gel butadiene content B (calculated on original sample) (wt%)	Total butadiene content (A+B) (calculated on original sample) (wt%)	Amount of original rubber unsaturation in the sample C=(A+B) x 100%
	-	3.5	-	3.5	85
4.7	19.5	2.8	0.0	3.7	90
5.6	16.2	2.9	0.9	3.8	93
8.9	23.3	1.5	2.1	3.6	88
11.8	20.0	1.5	2.4	3.9	95

Table 97 - Rubber content of high-impact polystyrenes (based on PBD)

Sample	Polybutadiene %w (iodine monochloride method)	Polybutadiene %w (IR method)
Standard: 6.0%w diene 55	6.2	-
Standard: 12%w diene 55	12.2	-
Standard: 15%w diene 55	14.8	-
High-impact polystyrene 1	8.6	9.7
High-impact polystyrene 2	5.6	5.8
High-impact polystyrene 3	9.0	1.2
High-impact polystyrene 4	11.2	11.4
High-impact polystyrene 5	5.8	5.9

Albert[969] has compared determinations of butadiene in high-impact polystyrene by an infrared method and by the iodine monochloride method described by Crompton and Reid[968]. The infrared method is based on a characteristic absorbance in the infrared spectrum associated with the transconfiguration in polybutadiene:

Transpolybutadiene units

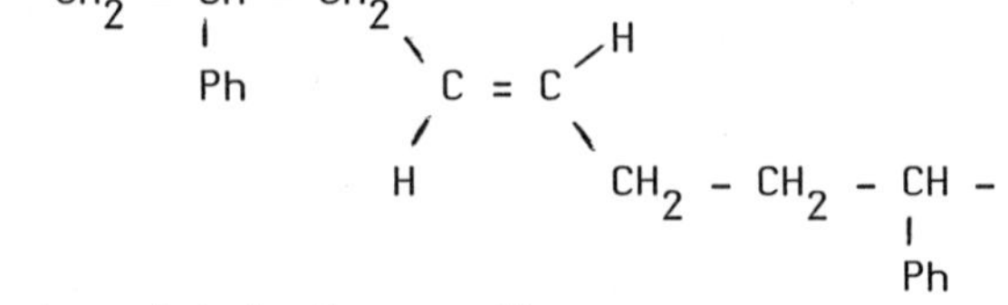

$$- CH_2 - CH - CH_2$$

cis polybutadiene units

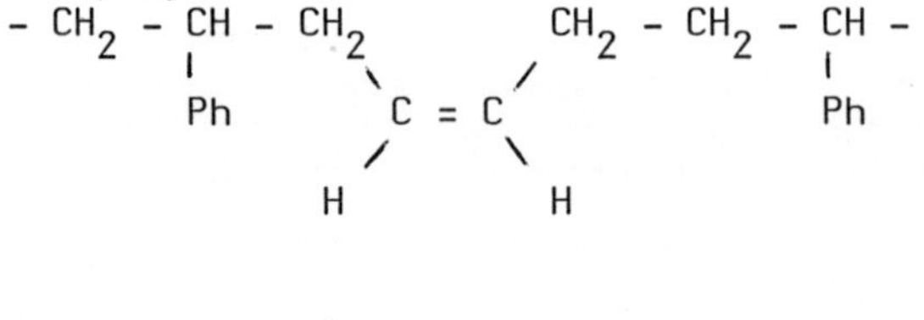

Since different grades of high-impact polystyrene may contain elastomers with different trans-butadiene contents, calibration curves based on the standard rubber are not always suitable for analysing these products. The results obtained by the two methods for several high-impact grades are compared in Table 97. The rubber content of high-impact polystyrene sample 1 determined by titration is lower than the value obtained by the infrared method. This is expected on interpolymerized polymers because of crosslinking, which reduces the unsaturation of the rubber. The other polymers (except sample 3), appear to contain diene-55 type rubber of high (trans-butadiene) content, since reasonable agreement was obtained between the iodine monochloride and infrared methods. High-impact polystyrene 3, however, must contain a polybutadiene of high cis-content to explain the low (1.2%w) amount of rubber found by the infrared method compared to the 9.0% found by the titration method.

The iodine monochloride method has been used for a variety of polymers. These polymers include those which are highly unsaturated, such as polybutadiene and polyisoprene[970-973] and polymers having low unsaturation such as butyl rubber[974] and ethylene propylene diene terpolymer. Considerable work has been done investigating the side reactions of iodine monochloride with different polymers[974]. These side reactions are substitution and splitting out rather than the desired addition reaction.

<u>Coulometric methods.</u>   Chemical methods for the determination of unsaturation are often based on halogenation or hydrogenation. An elegant method for determining unsaturated compounds is the in-situ generation of a halogenating agent (e.g. bromine) by constant-current coulometry. The amount of reagent consumed is proportional to the amount of electricity used to generate the reagent. The reaction between the double bond and bromine is catalysed by mercury (II) chloride. The advantages of a coulometric method are that a standard reagent is not needed, minimum side-reactions (substitution) occur because the bromine concentration is kept low and the method is precise, is suitable for low levels and can be readily automated.

A survey of applications has been given by Hirozawa et al[975].

Von Houwelingen[976] have used coulometric bromination to determine vinyl ester end-groups in polyethylene terephthalate formed by thermal chain scission with the subsequent liberation of acetaldehyde.

$$-\langle\!\bigcirc\!\rangle- \overset{O}{\overset{\|}{C}} - O - \overset{H}{\overset{|}{C}} = CH_2 + HO - CH_2 - CH_2 - O - \overset{O}{\overset{\|}{C}} -\langle\!\bigcirc\!\rangle- \longrightarrow$$

$$-\langle\!\bigcirc\!\rangle- \overset{O}{\overset{\|}{C}} - O - CH_2 - CH_2 - O - \overset{O}{\overset{\|}{C}} -\langle\!\bigcirc\!\rangle + CH_3 - CHO$$

The constant-current generation of bromine is carried out in a medium of dichloroacetic acid, hexafluoroisopropanol, water, potassium bromide and mercury II chloride. To this medium an amount of the polymer, previously dissolved in hexafluoroisopropanol and diluted with anhydrous dichloroacetic acid, is added and bromine is generated. The end of the reaction is detected biamperometrically. The suitability of this method was tested against methyl vinyl terephthalate.

$$CH_3 - O - \overset{\overset{\displaystyle O}{\|}}{C} - \langle\!\langle\bigcirc\rangle\!\rangle - \overset{\overset{\displaystyle O}{\|}}{C} - O - \overset{\overset{\displaystyle H}{|}}{C} = CH_2$$

Additions of 14.2 and 1.0 umol of methyl vinyl terephthalate (corresponding to 30 and 2 umol of vinyl ester end groups per kilogram of polymer) were recovered quantitatively (recoveries of 99.8 and 98.5% respectively). The coulometric analysis must be completed within 30 min because after longer times the hydrolysis of the vinyl ester end group is no longer negligible.

The vinyl ester end group is not the only reactive moiety in polyethylene terephthalate that consumes bromine, as impurities present also do so. To determine this background in a second sample the vinyl ester end group is previously hydrolysed at 80°C in a dichloro-acetic acid - water medium, Figure 71(a) shows the relationship between the bromine consumption and the hydrolysis time for methyl vinyl terphthalate and Figure 71(b) represents the same relationship for two polymers containing a high and negligible amount of vinyl ester end groups.

It can be seen that the vinyl ester end group in methyl vinyl terphthalate is hydrolysed completely after 4-6 hours at 80°C. For polymer A a sharp decrease occurs, which can mainly be attributed to the hydrolysis of the vinyl ester end group in the polymer. In addition, a much smaller effect is observed due to the hydrolysis of the bromine-consuming impurities (see relationship for polymer B, Figure 71(b). This small effect is corrected by extrapolating this relationship from time 6 hours to time zero (i.e. 0.7 mmol kg^{-1}).

The vinyl ester end-group content is calculated by subtracting the background (i.e. the value measured after 6 hours hydrolysis + 0.7 mmol kg^{-1}) from the content originally measured. The standard deviation of the method is 0.2 mmol kg^{-1}.

<u>Spectroscopic methods.</u> Spectroscopy is capable of distinguishing between the different types of unsaturation that can occur in a polymer.

Nuclear magnetic resonance spectroscopy has been used to determine unsaturation in acrylonitrile-butadiene-styrene terpolymers[980] and 1,2-polybutadiene[981]. Regarding acrylonitrile-butadiene-styrene terpolymers[980], NMR is capable of determining ungrafted butadiene rubber in solvent extracts of these polymers.

About 60-80% of the butadiene in the entire sample was present as ungrafted rubber. Using the compositional analysis of the grafted and ungrafted rubber, and the amount of ungrafted rubber extracted, one can calculate the composition of the graft. No aromatic protons of styrene or acrylonitrile protons are seen in the NMR spectra. The vinyl content of the polybutadiene is about 20%.

Chujo et al.[987,979] used PMR to estimate vinyl groups in vinyl chloride-vinylidene chloride copolymers.

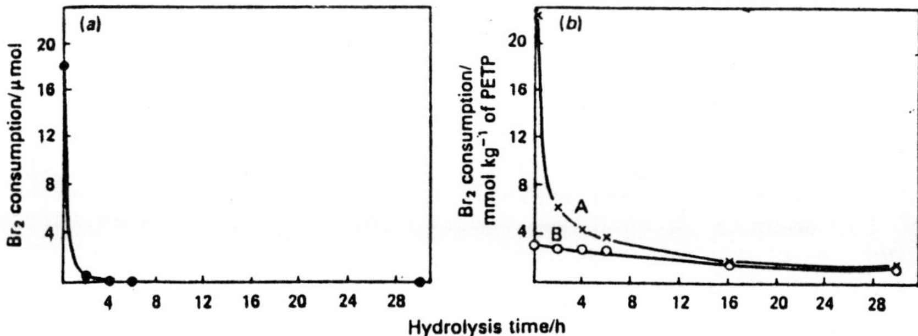

Figure 71   (a) Hydrolysis of MVT; (b) hydrolysis of VEE groups in
polyethylene terephthalate; (A) polyethylene
terphthalate with a high VEE group level; (B)
polyethylene terephthalate with a negligible VEE group
level.

Fraga[977] has also described an infrared thin film area method for
the analysis of styrene-butadiene copolymers. The integrated absorption
area between 6.6 and 7.2 microns has been found to be essentially
proportional to total bound butadiene, and is independent of the isomeric
type butadiene structure present. This method can be calibrated for bound
styrene contents ranging from 25 to 100%.

Other work on the determination of unsaturation in copolymers is
summarised in Table 98.

9.2.2 Diene Groupings in Ethylene-Propylene-Diene Terpolymers

Polymerization of ethylene and propylene results in a saturated
copolymer. In order to vulcanize this rubber, some unsaturation has to be
introduced. This is commonly done by adding a few per cent of nonconjugated diene (termonomer) such as dicyclopentadiene, 1,4-hexadiene, or
ethylidene norbornene, during the polymerization. Since only one of the
double bonds of the diene reacts during polymerization, the other is free
for vulcanization. The amount of unsaturation left in the ethylene

Table 98 - Unsaturation in copolymers

Unsaturation	styrene-butadiene methylmethacrylate terpolymer	Raman spectroscopy	981
Unsaturation	acrylonitrile-butadiene styrene terpolymers	NMR	982
Unsaturation	ethylene-propylene-diene terpolymer	NMR	983
Vinyl	vinylchloride-vinylidenechloride	NMR	984,985
Vinyl	styrene-divinyl benzene	Pyrolysis-mass	986

propylene diene terpolymer is of great interest, because the vulcanization properties will be affected.

Pyrolysis-gas chromatography has been used to determine the overall composition of ethylene-propylene-diene terpolymers[987]. In attempting to determine the third component in ethylene-propylene-diene terpolymers, difficulties might be anticipated, since this component is normally present in amounts around 5 wt%. However, dicyclopentadiene was identified in ethylene-propylene-diene terpolymers even when the amount incorporated was very low.

Van Schooten and Evenhuis[988,989] applied their pyrolysis (500°C)-hydrogenation-gas chromatographic technique to unsaturated ethylene-propylene copolymers, i.e. ethylene-propylene-dicyclopentadiene and ethylene-propylene-norbornene terpolymers. The pyrograms show that very large cyclic peaks are obtained from unsaturated ring: methyl cyclopentane is found when methyl norbornadiene is incorporated; cyclopentane when dicyclopentadiene is incorporated; methylcyclohexane and 1,2-methylcyclohexane when the addition compounds of norbornadiene with, respectively, isoprene and dimethylbutadiene are incorporated; and methylcyclopentane when the dimer of methylcyclopentadiene is incorporated. The saturated cyclopentane rings present in the same ring system in equal concentrations, however, give rise to peaks which are an order of magnitude smaller. Obviously, therefore, the peaks which stem from the termonomer could be used to determine its content if a suitable calibration procedure could be found.

Van Schooten and Evenhuis[988,989] subjected a number of terpolymers containing dicyclopentadiene, and having different amounts of unsaturation, to pyrolysis-gas chromatographic analysis, and plotted the height of the characteristic peaks (or ratio of the heights of these peaks to the height of the n-C peak) against unsaturation measured by ozone absorption[990]. A linear relationship was found between peak height ratio and ozone unsaturation up to about 16 double bonds per 1000 atoms. Similar curves were found for the methylcyclopentane or ethylcyclopentane peaks.

Sewell and Skidmore[991] used time averaged NMR spectroscopy at 60 MC/sec to identify low concentrations of nonconjugated dienes such as cyclopentadiene, 1,4-hexadiene or ethylidene norbornene introduced into ethylene-propylene copolymers to permit vulcanization. Although infrared spectroscopy[992] and iodine monochloride unsaturation methods[993] have been used to determine or detect such dienes, these two methods can present difficulties. Identification of the incorporated third monomer is not always practical in infrared spectroscopy at the low concentrations involved and in the high-resolution NMR spectra of these terpolymers the presence of unsaturation is not usually detected, since the signals from the olefinic protons are of such low intensity that they become lost in the background noise. The spectra obtained by time averaged NMR are usually sufficiently characteristic to allow identification of the particular third monomer incorporated in the terpolymer. Moreover, as the third monomer initially contains two double bonds, differing in structure and reactivity, the one used up in copolymerisation may be distinguished from the one remaining for subsequent use in vulcanization. Therefore information concerning the structure of the remaining unsaturated entity may be obtained. Table 99 shows the chemical shifts of olefinic protons of a number of different third monomers in the copolymers.

The cyclooctadiene and dicyclopentadiene terpolymers have alofinic protons with the same chemical shift, 4.55 ppm and so these cannot be

Table 99

Third Monomer	Chemical Shift ppm
Cyclooctadiene 1,5	4.55
Dicyclopentadiene	4.55
1:4 hexadiene	4.7
Methylene norbornene	5.25 and 5.5
Ethylidene norbornene	4.8 and 4.9

Table 100 - Determination of termonomer in ethylene-propylene diene terpolymers - comparison of methods (data shown as weight % terpolymer)

NMR	Lee[995] Koltoff and Johnson	Kemp and Peters[996]	Termonomer
7.3	3.0		dicyclopentadiene
1.1	1.6		1,4-hexadiene
5.7	5.9	9.0	ethylidene norbornene
2.8	3.6	4.5	ethylidene norbornene
1.7	2.3	4.8	ethylidene norbornene
4.6	5.4	6.0	ethylidene norbornene

differentiated by this technique but may be distinguished by the use of iodine monochloride. The hexadiene type of terpolymer may be identified by its olefinic resonance at 4.7 ppm. These three monomers have what appears as a single olefinic resonance in the terpolymer. On the other hand, the two norbornadiene types of monomer each show two characteristic resonances. In the methylene norbornene terpolymer the olefinic resonances arise from two protons, each giving a separate signal, whereas in the ethylidene norbornene terpolymer there is only one proton, the signal of which appears as a doublet. In view of these considerations it is more difficult to detect the olefinic resonance in the latter instance.

Altenau et al[994] applied time averaging NMR to the determination of low percentages of termonomer such as 1:4 hexadiene, dicyclopentadiene and ethylidene norbornene in ethylene-propylene termonomers. They compared results obtained by NMR and the iodine monochloride procedure of Lee et al[995]. The chemical shifts and splitting pattern of the olefinic response were used to identify the termonomer.

Table 100 compares the amount of termonomer found by the NMR method of Altenau et al[994] and by iodine monochloride procedures[995,996]. The termonomers were identified by NMR and infrared.

Table 100 shows that the data obtained by the NMR method agree more closely with the Lee, Kolthoff and Johnson iodine monochloride method[995] than with the iodine monochloride method of Kemp and Peters[996]. The difference between the latter two methods is best explained on the basis of side reactions occurring between the iodine monochloride and polymer because of branching near the double bond[995]. The reason for the difference between the NMR and Lee et al.[995] methods is not clear. The reproducibility of the NMR method was $\pm$ 10 to 15%.

Infrared spectroscopy has been used for the determination of unsaturation in ethylene-propylene-diene terpolymers[997]. Determination of extinction co-efficients for the various terpolymers is required if quantitative work is to be done.

9.2.3 Ester Groups

<u>Acrylate ester groups.</u> Most methods for the determination of ester groups in polymers are based on the following procedures:

a)   Saponification.

b)   Zeisel procedures, based on hydriodic acid.

c)   Pyrolysis - gas chromatography.

d)   Physical methods, e.g. infrared spectroscopy and nuclear magnetic resonance spectroscopy.

<u>Saponification methods.</u> Ester groups occur in a wide range of polymers, e.g. polyethylene terphthalate and in copolymers such as, for example, ethylene vinyl acetate. The classical chemical method for the determination of ester groups, namely, saponification, can be applied to some types of polymer. For example copolymers of vinyl esters and esters of vinyl esters and esters of acrylic acid, can be saponified in a sealed tube with 2 M sodium hydroxide. The free acids from the vinyl esters were determined by potentiometric titration or gas chromatography. The alcohols formed by the hydrolysis of the acrylate esters were determined by gas chromatography. Vinyl acetate ethylene copolymers can be determined by saponification with 1 N ethanolic potassium hydroxide at 80°C for 3 hours and back titration with standard acid[998,999] or by saponification with p-toluene sulphonic acid and back titration with standard acetic acid[1000,1001].

Polymethyl acrylate can be hydrolysed rapidly and completely under alkaline conditions, on the other hand, the monomer units in polymethyl methacrylate prepared and treated similarly are resistant to hydrolysis[1002] although benzoate end-groups react readily[1003]. Only about 9% of the ester groups in polymethyl methacrylate reacted even during prolonged hydrolysis; hydrolysis of polymethyl acrylate was complete in 0.5 hours. Although only about 9% of the ester groups in methyl methacrylate homopolymers are hydrolysed in several hours reflux with alcoholic sodium hydroxide, this proportion is increased by the introduction of comonomer units into the polymer chain. Thus, saponification techniques should be applied with caution to polymeric materials.

<u>Zeisel procedures.</u> Hydrolysis using hydriodic acid has been used for the determination of the methyl ethyl, propyl and butyl esters of acrylates, methacrylates or maleates[1004] and the determination of polyethyl esters in methyl methacrylate copolymers[1006,1007]. First the total alcohol content is determined using a modified Zeisel hydriodic acid hydrolysis[1005]. Secondly, the various alcohols, after being converted to the corresponding alkyl iodides, are collected in a cold trap and then separated by gas chromatography. Owing to the low volatility of the

higher alkyl iodides the hydriodic acid hydrolysis technique is not suitable for the determination of alcohol groups higher than butyl alcohol. This technique has also been applied to the determination of alkoxy groups in acrylate esters[1004].

Anderson et al[1008] have used combined Zeisel reaction – gas chromatography to analyse acrylic copolymers. Acrylic esters were cleaved with hydriodic acid and gas chromatography was used for analysing the alkyl iodides so formed.

$$
\left[\begin{array}{cc} H & H \\ | & | \\ -\ C\ -\ C\ - \\ | & | \\ H & Ph \end{array}\right]
\left[\begin{array}{cc} H & R' \\ | & | \\ -\ C\ -\ C\ - \\ | & | \\ H & C=O \\ & | \\ & OR \end{array}\right]
-\ +\ HI \longrightarrow
\left[\begin{array}{cc} H & H \\ | & | \\ -\ C\ -\ C\ - \\ | & | \\ H & Ph \end{array}\right]
-\left[\begin{array}{cc} H & R' \\ | & | \\ C\ -\ C \\ | & | \\ H & C=O \\ & | \\ & OH \end{array}\right]
$$

$$R = H, CH_3, C_2H_5, C_4H_9, R' = H, OH_3$$

Using this procedure, the recovery of alkyl iodides is greater than 95% for polymers containing between 10 and 90% of the methyl, ethyl and butyl esters of acrylic and methacrylic acid. In addition, the use of isopropylbenzene as the trapping solvent allows the determination of all C_1 to C_4 alkyl iodides. Quantitative cleavage by hydriodic acid of the acrylate and methacrylate esters is also observed in the presence of the modifying monomer units such as styrene, vinyl acetate, vinyl chloride and acrylamide.

<u>Method – Determination of Acrylic Ester Groups in Copolymers</u>

<u>Hydriodic Acid Reduction – Gas Chromatography</u>

<u>Apparatus.</u> Hewlett-Packard 5756 gas chromatograph or equivalent equipped with flame ionization detector. A stainless steel column, 10 ft x 0.125 in o.d. packed with 20% w/w DO-401 on 60-80 mesh Gas Pak WAB. helium flow rate 10 ml/min with an inlet pressure of 65 psi. Injection port and detector temperatures: 210 and 250°C respectively.

Column temperature 150°C

<u>Reagents.</u> Spectrograde acetone, methyl ethyl ketone-hexane, 1,1,2-trichloroethane and isopropylbenzene (Eastman Organic Chemicals). Tetrahydrofuran (E.I. du Pont de Nemoura) is distilled over triphenyl phosphite and nitrogen to remove any water and butylated hydroxy toluene inhibitor. After distillation, tetrahydrofuran is stored in amber bottles over thin strips of copper metal to prevent the formation of peroxides. Hydriodic acid (Fisher Scientific Company) is freshly distilled over hypophosphorus acid and nitrogen. Only the 57% hydriodic acid azeotrope boiling at 127°C is retained. The distillate is stabilised with hypophosphorus acid.

<u>Zeisel cleavage reaction.</u> Approximately 200 mg of the dried polymer is placed in the reaction flask. To that is added 10 ml of molten phenol, 5 ml of glacial acetic acid and 5 ml of acetic anhydride to swell and dissolve the polymer and scavenge any water that might be present. The reaction flask is then placed in a heating mantle held at 125°C until the

polymer is completely dissolved or sufficiently swollen to allow attack by
the hydriodic acid. While the sample mixture is heating, 5 ml of a 1% w/v
solution of 1,1,2-trichloroethane in isopropylbenzene is pipetted into the
receiver which is then placed in a dry ice-acetone bath. After the
solution is obtained, the contents of the reaction flask are cooled an 25
ml of hydriodic acid is added. The nitrogen flow rate is set at 10 ml/min
and the reaction flask placed in an oil bath at 132 $\pm$ 1°C. The position
of the reaction flask in the oil bath is extremely important. The
reaction mixture in the flask was level with the oil in the bath. Care is
taken to prevent the refluxing mixture from reaching the desiccant.

Analysis of alkyl iodides. Periodically, 2 ul samples of the
reaction products are removed from the receiver and analysed by gas
chromatography. Prior to taking a sample, the receiver is loosened and
the trapping solution shaken around the glass spiral to form a homogenous
solution of the alkyl iodides and the internal standard. Care is taken
not to raise the delivery tube above the liquid level in the receiver.
The reaction is allowed to proceed until the ratio of the longest chain
alkyl iodide to the internal standard became constant. Details of the
determination of relative response factors and calculation of copolymer
composition are given below.

Determination of response factors. An accurately known blend of the
alkyl iodides and the internal standard (1,1,2-trichloroethane) is
prepared and analysed using gas chromatography. The areas of the peaks in
the chromatograms are determined and the response factors of the alkyl
iodides calculated using the equation:

$$K_i = \frac{(W_i)\ (A_x)\ (K_x)}{(W_x)\ (A_i)}$$

where

K_i	=response factor of the respective alkyl iodide
K_x	=response factor of 1,1,2-trichloroethane (arbitrarily set equal to 1.00)
W_i	=weight of the respective alkyl iodide in the calibration blend
W_x	=weight of 1,1,2-trichloroethane in the calibration blend
A_i	=area of the respective alkyl iodide peak
A_x	=area of 1,1,2-trichloroethane peak

Response factors determined at two-week intervals were found to
deviate. Best results are obtained when response factors were determined
just prior to each analysis.

Determination of copolymer composition. Samples from the receiver
are periodically removed through the rubber septum and analysed using gas
chromatography. The areas of the alkyl iodide and internal standard peaks
are determined and the copolymer composition calculated using the
following equation:

$$\% \ acrylate = \frac{(K_i)\ (A_i)\ (W_x)\ (MA\ (100)}{(A_x)\ (W_{rs})\ (MW_i)}$$

where

K_1 =   response factor of the respective alkyl iodide
A_1 =   area of the respective alkyl iodide peak in the chromatogram
A_x =   area of 1,1,2-trichloroethane peak in the chromatogram
W_x =   weight of 1,1,2-trichloroethane added to the receiver
W_{rs} = weight of the sample used
MW_A =   weight of acrylate or methylacrylate being analysed for
MW_1 =   molecular weight of the respective alkyl iodide

Duplicate determinations should agree within ± 3% relative.

In Table 101 is shown some results obtained by applying this method to a range of acrylic polymers. The calculated recoveries are greater than 95%, for polymers containing between 10% and 100% acrylic monomer. The method has 99% confidence interval of 0.8. The presence of comonomers such as styrene, acrylonitrile, vinyl acetate, acrylamide or acrylic acid does not change the recovery of acrylate or methacrylate esters. Non-quantitative results are obtained, however, for polymers containing hydroxy-propyl methacrylate.

<u>Pyrolysis - gas chromatography.</u> Barrall et al[1009] have described a pyrolysis-gas chromatographic procedure for the analysis of polyethylene-ethyl acrylate and polyethylene-vinyl acetate copolymers and physical mixtures thereof. They used a specially constructed pyrolysis chamber as described by Porter et al[1010]. Less than 30 s is required for the sample chamber to assume block temperature. This system has the advantages of speed of sample introduction, controlled pyrolysis temperature and complete exclusion of air from the pyrolysis chamber. The pyrolysis chromatogram of poly(ethylene-vinyl acetate) contains two principal peaks. The first is methane and the second acetic acid.

$$- OCH_2 - CH_2 - CH_2 - \underset{\underset{\displaystyle OOCCH_3}{|}}{CH} \quad \xrightarrow[300 - 480^{\circ}C]{pyrolysis} \quad CH_3COOH + CH_4$$

Variations from 350°C to 490°C in pyrolysis temperature produced no change in the area of the acetic acid peak, but did cause an area variation in the methane peak. The pyrolysis chromatogram of poly(ethylene-ethyl acrylate) at 475°C shows one principal peak due to ethanol. No variation in peak areas was noted in the temperature range 300°C to 480°C. Table 102 shows the analysis of 0.05 g samples of poly(ethylene-ethyl acrylate) (FEEA) and poly(ethylene-vinyl acetate)(PEVA) obtained at a pyrolysis temperature of 475°C.

Pyrolysis - gas chromatography - mass spectrometry has been used to identify ester groups in acrylic polymers[1011].

Anderson et al[1012] have described an infrared procedure for distinguishing between copolymerised acrylic and methacrylic acids in acrylic polymers containing more than 10% of the acid. This method is based on precise measurement of the wavelength of the carboxylic acid absorption maximum at about 5.9 microns. The identification of the acid in compositions containing less than 10% of acid has hitherto not been possible, except when unpolymerised acid residues can be separated from the polymer[1013].

Haslam[1014] has discussed infrared methods for the determination of ester groups in acrylic copolymers.

Table 101 - Recovery of alkyl iodides from the
Zeisel cleavage of acrylic polymers[a]

Polymer	Methyl acrylate %	Methyl methacry- late %	Ethyl acrylate %	Ethyl methacry- late %	Butyl acrylate %	Butyl methacr- ylate %
1	33.7 (33.3)		33.0 (33.3)		32.7 (33.3)	
2		33.6 (33.3)		33.4 (33.3)		31.9 (33.3)
3			19.5 (20.0)			
4		50.1 (50.0)	30.5 (30.0)			
5		59.9 (60.0)			39.8 (40.0)	
6		29.7 (30.0)			29.8 (30.0)	
7		59.8 (60.0)				9.8 (10.0)
8	19.9 (20.0)		19.8 (20.0)			19.8 (20.0)
9		30.1 (30.0)		30.0 (30.0)	29.8 (30.0)	
10		29.9 (30.0)	29.6 (30.0)		39.7 (40.0)	
11		60.1 (60.0)	30.1 (30.0)			9.5 (10.0)

[a] All values are the average of at least three determinations and are
reported as: % monomer found (% monomer in polymer).

     NMR has been used to determine ethyl acrylate in ethyl acrylate
ethylene and vinyl acetate-ethylene copolymers[1015]. Measurements were
made on 10% solutions in diphenyl ester at an elevated temperature.
Resolution improved with increasing temperature and lower polymer
concentration in the solvent.

     Figure 72 shows NMR spectra for an ethylene-ethyl acrylate
copolymer. Spectra indicate clearly both copolymer identification and
monomer ratio. A distinct ethyl group pattern (quartet, triplet), with
methylene quartet shifted downfield by the adjacent oxygen, is observed.
The oxygen effect carries over to the methyl triplet which merges with the

Table 102 - Pyrolysis results on physical mixtures of poly
(ethylene-ethyl acrylate) and poly (ethylene-vinyl acetate)

Mixture	Acetic acid found	wt% calcd	Ethylene found	wt% calcd[a]	Oxygen found	wt% calcd
50% PEEA-1 and 50% PEVA-2	9.10	9.05	2.65	2.62	7.88	8.25
33.3% PEEA-2 and 66.6% PEEA-3	12.15	12.33	0.75	0.70	7.33	7.49

[a] Calculated from results for acetic acid and ethylene content for
individual samples on weight per cent basis.

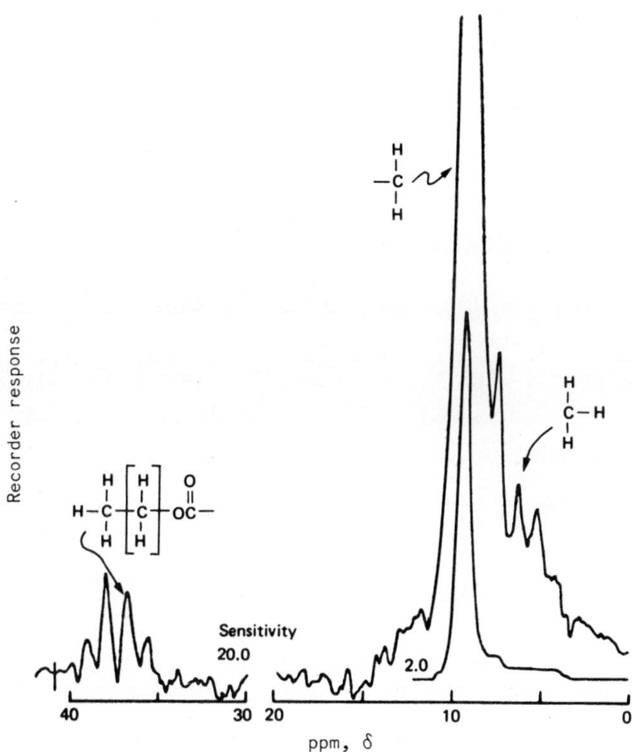

Figure 72   NMR spectra.   Polyethylene - ethyl acrylate.

aliphatic methylene peak. No other ester group would give this characteristic pattern. The area of the quartet is a direct and quantitative
measure of the ester content. All features in Figure 72 are consistent
with an identification of an ethylene-rich copolymer with ethyl acrylate.
Ethyl acrylate contents obtained by NMR (6.0%) agreed well with those
obtained by PMR (6.2%) and neutron activation analysis for oxygen (6.1%).

NMR spectroscopy has also been used to identify ester groups in
acrylic polymers[1014,1015] and copolymers.

Acetate ester groups. Saponification procedures can be applied to
the determination of ester groups in polymers. A copolymer of ethylene
and vinyl acetate has the following structure which, upon hydrolysis in
the presence of excess standard potassium hydroxide/p-toluene sulphonic
acid catalyst reacts as follows:

$$-(CH_2-CH_2-CH_2-CH-)_n + nKOH \rightarrow$$
$$\qquad\qquad\qquad\qquad |$$
$$\qquad\qquad\qquad OOCCH_3$$

$$-(CH_2-CH_2-CH-CH-)_n + nCH_3COOK$$
$$\qquad\qquad\qquad\quad |$$
$$\qquad\qquad\qquad\quad OH$$

Excess potassium hydroxide is then determined by titration with standard
acetic acid, and hence the vinyl acetate content of the polymer is
calculated from the amount of potassium hydroxide consumed. Vinyl acetate
ethylene copolymers can be determined by saponification with 1 N ethanolic
potassium hydroxide at 80°C for 3 hours[1018,1019].

Cambell[1020] has described an isotope dilution method for the
determination of vinyl acetate in vinyl acetate-vinyl chloride copolymers.
In this method a known amount of acetic 2-^{14}C acid is added to a solution
or suspension of the resin in methyl ethyl ketone and the ester is
hydrolysed with sodium hydroxide. The major portion of the sodium acetate
^{14}C is isolated and converted to 2-methyl benzimidazole-methyl^{14}C by means
of the Phillips reaction[1021] with O-phenylene diamine.

Methyl-labelled acetic acid is used to avoid the kinetic isotope effect
that would be expected with the reagent labelled in the carbonyl group.

Infrared spectroscopy has been applied to the determination of free
and combined vinyl acetate in vinylchloride-vinyl acetate copolymers[1022].
This method is based upon the quantitative measurement of the intensity of
absorption bands in the near infrared spectral region arising from vinyl
acetate. A band at 1.63 microns due to vinyl groups enables the free
vinyl acetate content of the sample to be determined. A band at 2.15
micron is characteristic for the acetate group and arises from both free
and combined vinyl acetate. Thus, the free vinyl acetate content may be
determined by difference at 2.15 micron. Polymerized vinyl chloride does
not influence either measurement.

Method – Determination of Combined Vinyl Acetate in Vinyl-Acetate-Vinylchloride Copolymers. Isotope Dilution – Derivative Method[1022]

Reagents. An aqueous solution of acetic $2^{14}C$ acid containing 0.5-1.0 millimole per gram is prepared from acetic anhydride. Its specific activity of acetic anhydride was determined by treating a portion with an excess of aniline to form acetanilide ^{14}C which was crystallized from water and assayed in dioxane-naphthalene scintillator solution.

Methyl ethyl ketone reagent grade, is stirred with calcium hydride and distilled.

Sodium hydroxide, 1 molar, is prepared in 9:1 methanol: water.

o-Phenylene diamine is crystallized twice from water with the aid of activated charcoal.

The scintillator solution is prepared by dissolving naphthalene (200 grams) PPO (2,5-diphenylaoxazole) (14.0 grams) and POPOD (1,4 bis-2-(5-phenyloxazolyl)benzene) (0.06 gram) in scintillation grade dioxane, diluting to 2 litres and adding 200 ml of methanol.

Apparatus. A Packard Model 3303 liquid scintillation spectrometer for radioassays.

Procedure. A specimen containing approximately one millimole of vinyl acetate, but not exceeding 4 grams, is weighed into a 125 ml Erlenmeyer flask and 40-50 ml of distilled methyl ethyl ketone is added. The solvent is added carefully so as to prevent agglomeration of the resin. A weighed aliquot of the solution of labelled acetic acid in water, containing 0.10-1.15 millimole of the acid, is then added, after which 5 ml of the sodium hydroxide solution is introduced. The stoppered flasks are swirled with periodic immersion in an oil bath at 60°C to redisperse any precipitated resin and then placed in the bath for 16 hours.

The contents of the cooled flasks are transferred to a 125 ml separatory funnel and the flasks are rinsed with two 15 ml portions of water. The funnel is shaken vigorously and the lower phase is removed to a 250 ml beaker. After evaporation of the contents of the beaker to dryness, the sodium acetate ^{14}C is extracted from the residue with 10-12 ml of hot methanol and the solution is decanted into a 50 ml beaker. The methanol is then carefully evaporated on a water bath.

The derivative is prepared in a thick-walled glass (polymerization) tube to which an excess of approximately 10% (130 mg) of purified o-phenylene diamine is added initially. The residue in the 50 ml beaker is dissolved in 3-5 ml of water and the solution is transferred to the tube, to which 1.5 ml of 10 N hydrochloric acid is added. The tube is then sealed, enclosed in a protective sleeve of glass cloth and heated one hour in an oven at 180°C. After cooling, the contents of the tube are transferred to a 30 ml beaker and a slight excess of 1:1 ammonium hydroxide is added with vigorous stirring to precipitate the 2-methylbenzimidazole. The beaker is then placed in an ice bath for 30-60 minutes.

The crude derivative is collected on qualitative filter paper cut to fit a micro (1 cm) Buchner funnel and washed with 2-3 ml of ice water. The compound is then transferred to a 30 ml beaker, dissolved in 10-12 ml of water by heating to incipient boiling, and the solution is decolorized with 30-35 mg of activated charcoal. The suspension is stirred briefly then filtered as before to remove the charcoal. The filtrate is transferred to a 30 ml beaker, which is placed in an ice bath and the solution is stirred vigorously with a glass rod to initiate crystallization. After 30-60 minutes, the purified derivative is collected by filtration, washed with 2-3 ml of ice water and dried in vacuum over phosphorus pentoxide. The product is assayed by weighing 4-6 mg into a counting vial and adding 15 ml of scintillator solution, or by weighing a larger amount into a 50 ml volumetric flask and removing aliquots.

$$\text{Vinyl acetate, \% by weight} = \left[W_1 \left(\frac{S_o}{S_t} - 1 \right) \right] \frac{8.609}{W_s}$$

where

W_1 = millimoles of acetic-2-^{14}C acid added
S_o = specific activity of the acetic-2-^{14}C acid, uCi/mmole
S_t = specific activity of the 2-methylbenzimidazole, uCi/mmol
W_s = weight of sample, grams

The vinyl acetate content of films of ethylene-vinyl acetate copolymers can be determined by methods based on the measurement of absorbances at 16.1 and 13.9 micron[1014] and at 5.73 micron[1025]. The acrylate salt in acrylate salt-ethylene ionomers has been determined from the ratio absorbances at 6.41 micron (asymmetric vibration of the carboxylate ion) and 7.24 micron[1023].

<u>Isophthalate ester groups.</u> NMR spectroscopy has been used for the determination of isophthalate in polyethylene terephthalate-isophthalate dissolved in 5% trichloroacetic acid. The NMR spectra of these polymers were measured on a high resolution NMR spectrometer at 80°C. A singlet at 7.74 ppm is due to the four equivalent protons attached to the nucleus of the terephthalate unit. The complicated signals which appear at 8.21, 7.90, 7.80, 7.35, 7.22 and 7.10 ppm are due to the four protons attached to the nucleus of the isophthalate unit. The content of the isophthalate unit can be calculated from the integrated intensities of these peaks. Aydin and co-workers[1026] hydrolysed polyesters and converted the product acids and glycols including 1,4-cyclohexane di-methanol and isophthalic acid to the corresponding trimethylsilyl esters and ethers which were then analysed by gas chromatography.

<u>Succinate ester groups.</u> Esposito and Swann[1027] published a technique involving methanolysis of a polyester resin with lithium methoxide as a catalyst: the methyl esters formed were separated from the polyols and identified by gas chromatography.

$$COOH - (CH_2)_2CO \left[O(CH_2)_2OCO(CH_2)_2CO \right] OH + 3CH_3OH \quad (\underset{\longrightarrow}{LiOCH_3})$$

succinic acid - diethylene glycol polyester

$$2CH_3OOC(CH_2)_2COOCH_3 + HO(CH_2)_2OH$$

This method was improved by Percival[1028] by using sodium methoxide as a catalyst and injecting the reaction mixture of the transesterification directly into the gas chromatograph (without any preliminary separation).

9.2.4 Carboxyl Groups

<u>Carboxyl groups in polyesters and polyamides.</u> The determination of these groups in high relative molecular weight polyesters and polyamides gives, in combination with other data, information about the degree of polymerisation, chain branching, degradation and thermal stability. The choice of the method depends on the solubility of the polymer. Most methods are based on titration in non-aqueous media with visual, potentiometric or photometric indication of the end-point.

A very convenient technique for the determination of carboxyl groups in polyethylene terephthalate and nylon-6 is photometric titration. The sample is dissolved in o-cresol at 125°C, cooled, diluted with chloroform and, after addition of bromophenol blue or bromocresol green, the titration is carried out in a spectrophotometer with tetrabutyl ammonium hydroxide. The change in absorbance is recorded continuously and the end-point is determined graphically. A detailed description has been given by Van Lingen[1-29].

Table 103 surveys applications of the photometric titration method together with a comparison with a potentiometric titration method. Generally the agreement is very satisfactory.

An attractive method for determining carboxyl groups is based on the Schmidt reaction, in which the carboxyl group is subjected to reaction with sodium azide insulphuric acid[1030]. The acylazide formed rearranges to an amine with the liberation of nitrogen and carbon dioxide.

$$\sim COOH \xrightarrow[H_2SO_4]{NaN_3} NH_2 + CO_2 + N_2$$

The carbon dioxide evolved is measured quantitatively by an automatic non-aqueous titration. This method is applicable to poly(m-phenylene isophthalmide) which is soluble in dimethylformamide, but is not applicable to poly (p-phenylene) terephthalamide for which no organic solvent could be found.

The suitability of this method has been investigated with the model compound NN'-bis-(p-carboxybenzol)-p-phenylenediamine.

$$HOOC - \bigcirc - \overset{\overset{\textstyle O}{\textstyle \|}}{C} - NH - \bigcirc - NH - \bigcirc - COOH$$

The conversion of this polymer in 100-101% sulphuric acid at 50°C. At reaction time of 2 hours gives a quantitative conversion, which also suffices for polymer solutions.

Table 103 - Application of the photometric titration method
to various samples

| Sample | | Content of carboxyl groups/mmol kg^{-1} | | | |
| | | Photometric titration | | Potentiometric titration | |
		$\bar{x}$ [a]	standard deviation	$\bar{x}$ [a]	standard deviation
PETP (M_N low	C	4.3	0.25	4.0[b]	0.10
	D	12.0	0.34	12.0[b]	0.30
	E	17.4	0.56	17.5[b]	0.49
PETP (M_N high	F	36.8	0.32	35.2	0.9
	G	68.3	0.29	68.2	1.2
	H	113.4	0.52	112.8	1.8
Nylon 6	I	73.3	0.31	72.4[c]	1.0
	J	53.2	0.22	52.5[c]	0.37
	K	35.3	0.45	36.2[c]	0.32

[a] Mean of four independent determinations.
[b] Titration in aniline at 40°C.
[c] Titration in benzyl alcohol-methanol-water.

Nissen and co-workers[1031] have described a method for carboxyl groups in poly(ethylene terephthalate). Hydrazinolysis led to formation of terephthalomonohydrazide from carboxylated terephthalyl residues to provide a selective analysis for carboxyl group via ultraviolet absorbance at 240 nm.

<u>Carboxyl groups in acrylate and methacrylate copolymers.</u> Most methods for the determination of carboxyl groups in polymers are based on titration techniques including, for example, the following copolymers, acrylic acid-itaconic acid[1032], acrylic acid-ethyl acrylate[1033] and maleic acid-styrene[1034]. High-frequency titration has been applied[1035] to the analysis of itaconic acid-styrene and maleic acid-styrene copolymers and ethyl esters of itaconic anhydride-styrene copolymers. The method can also be used to detect traces of acidic impurities in polymers and in the identification of mixtures of similar acidic copolymers. Titration indicates that the acid segments in the copolymers of itaconic acid-styrene and maleic acid-styrene, and the homopolymer polyitaconic acid, act as dibasic acids. The method has a sensitivity that permits identification and approximate resolution of two carboxylate species in the same polymer, e.g.:

$$\text{Polyitaconic acid} \quad \left[-CH_2 - \underset{\underset{COOH}{|}}{\overset{\overset{CH_2COOH}{|}}{C}} - \right]_n + nK^+ \; \rightarrow \; \left[-CH_2 - \underset{\underset{COOK}{|}}{\overset{\overset{CH_2OOK}{|}}{C}} - \right]_n + nH^+$$

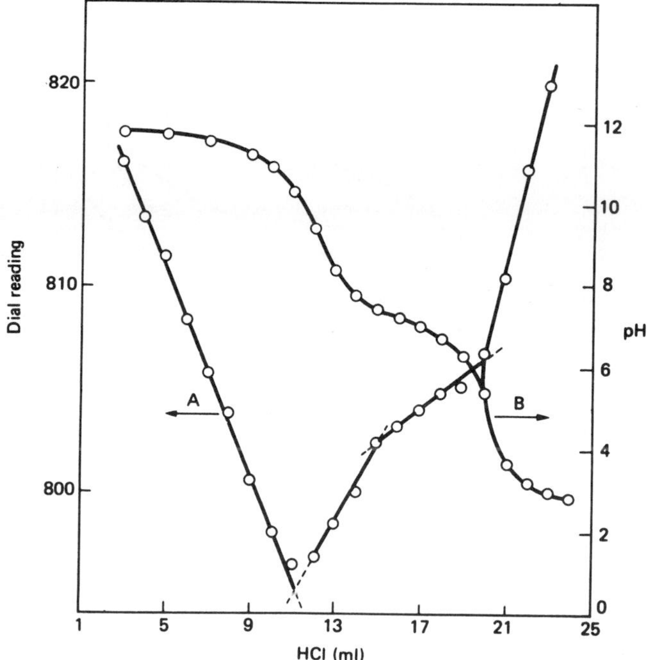

Figure 73   High frequency (A) and potentiometric (B) displacement
              titration of the monosodium salts of the monomethyl
              esters of poly(itaconic acid-costyrene). Titration of
              0.2345 g 57:43 anhydridestyrene copolymers + MeOH (heat)
              + excess NaOH with 0.1286 N HCI.

     High frequency titration gives a precise location of the inflection
points related to the polymer carboxyl groups and is a sensitive method
for the determination of the freedom of the copolymer samples from
monbasic acid impurities (comonomer acids), since mixtures of copolymer
acids with monobasic and dibasic acids show definite inflection points
that can be related to the individual carboxylate species present.

     A titration curve (Figure 73) is shown for a monomethyl ester of an
itaconic acid-styrene copolymer.

     Potentiometric titration provides a method of investigating changes
of conformation undergone by polyelectrolytes in solution, since the
environment of the dissociating groups is dependent on the conformation of
the polymer chain helix-coil transitions of polyacids. Thus, precise
potentiometric titration of solutions of high molecular weight polyacrylic
acid at constant ionic strength indicate the presence of such
conformational transition[1036-1038].

     Figure 74(a) shows the titration results for polyacrylic acid
plotted at   pH + $\log((1-\alpha)/\alpha)$ versus (degree of dissociation), as points
connected by full curves. The four curves at the different ionic
strengths all show the same features. The first short region, labelled A
in the figure, is probably due to some instability in the solution, such
as aggregation preceding precipitation. This region extends to higher
values of alpha at the higher ionic strengths. The second region, B,

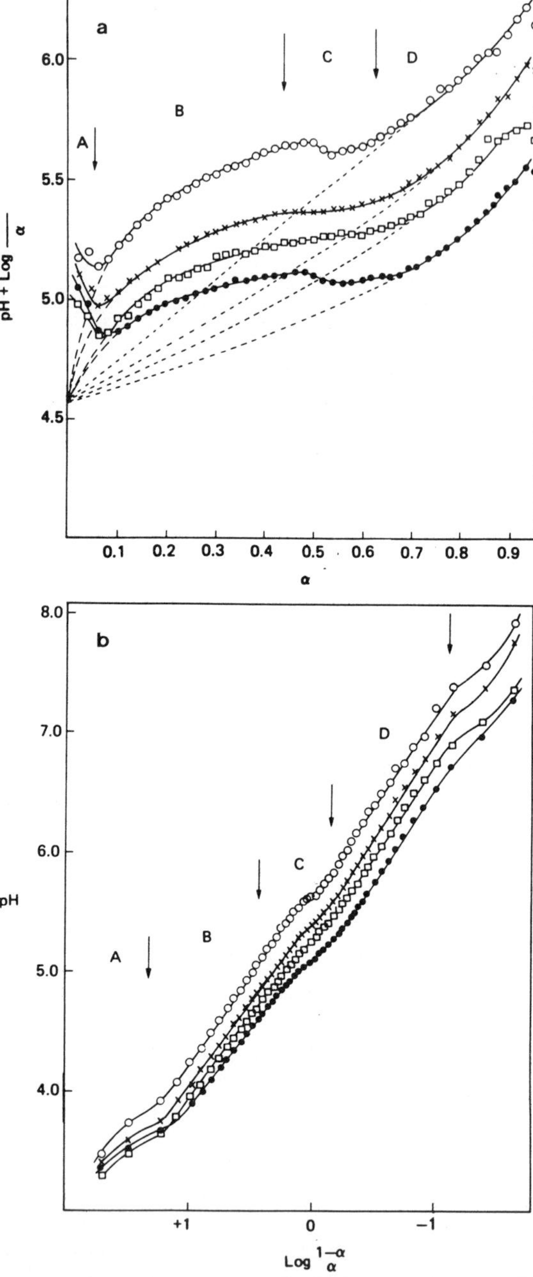

Figure 74   Titration of polyacrylic acid: (a) dependence of the
            function pH + log $\frac{(1-alpha)}{alpha}$ at various ionic strengths
            u: (o) U = 0.02; (x) u – 0.065; (□) u = 0.11; (●) u =
            0.20.
            (b) pH versus log $\frac{(1-alpha)}{alpha}$ at various ionic strengths
            u: (o) u = 0.02; (x) u = 0.065; (|□) u = 0.11; (●) u =
            0.20.

represents the ionization of the first conformation of the polymer; the third, C, the conformational transition; and the fourth, D, the ionization of the second conformation. The first conformation, which exists at the lower degree of dissociation, has presumably the more tightly coiled, is denoted PAA(b). The four curves of PAA(a) in Figure 74(a) have extrapolated (dashed curves) semi-empirically to zero alpha, and they meet there at a value of pH + $\log((1-\alpha)/\alpha)$ shown in Figure 74(b). These plots, though almost linear over the whole range, as previously reported by Mandel and Leyte[1039,1040], are not quite so, and regions A,B,C and D can be distinguished here also. Linear extrapolation of region B, which represents PAA(a), to higher values of $\log((1-\alpha)/\alpha)$ was used to obtain the extrapolations of the PAA(a) curves of Figure 74 to zero alpha.

In titrating copolymers of methyl methacrylate and methacrylic acid with standard base to determine composition, a number of deficiencies have been encountered. For example, there were two common sources of "contamination" that gave rise to underdetermination of the acid content; 1), the copolymers tended to be hygroscopic and hence, could on occasion contain 5% absorbed moisture and 2) they could, on occasion, retain solvents and/or monomers. Additionally, the method has been found to be inapplicable to copolymers of high molecular weight (MW>1,000,000) and/or high acid content (> 60% acid) because of the tendency of such systems to reprecipitate during the titration procedure. Because of these limitations, Johnson et al[1041] examined the applicability of Fourier transform 13 C NMR to the compositional analysis of methylmethacrylate acid copolymers.

By using the integral of the ester methoxy protons and combining this result with the total integral for CH_2 and CH_3 protons (the overlap between CH_2 and CH_3 resonances was enough at 100 MHz to prevent separate determination of these integrals), the copolymer composition could be ascertained. However, it was necessary to carry out the determinations at 100°C or higher to attain resolution sufficient for reliable integrals. An additional problem was that the reaction solvents (toluene and hexane) and comonomers had resonances that overlapped those of the CH_2 and CH_3 of the copolymers introducing considerable inaccuracy in the total CH_2 CH_3 integral. For this reason they investigated the applicability of ^{13}C NMR. Because of the greater spectral dispersion and narrower resonance lines obtained with ^{13}C NMR relative to proton NMR problems associated with resonance overlap can be resolved. Excellent agreement was obtained between carboxyl values obtained by this procedure and conventional titration in 1:1 ethanol water with standard potassium hydroxide to the phenol phthalein end-point even the acid content range 13% to 100%.

In pyridine solutions of these copolymers, the resonances arising from acid carboxyl and ester carbonyl carbons are sufficiently resolved to allow the determination of relative integrals.

<u>Method - Determination of Compositional Analysis of Methylmethacrylate-Methacrylic Acid Copolymers. Fourier Transform 13C NMR Spectroscopy[1041]</u>

Samples are prepared by dissolving or swelling ca. 0.3 g of copolymer in 2 g solvent - both components being weighed directly into a 10 mm NMR sample tube. A 50/50 mixture of pyridine and pyridine -d_5 is used as the solvent and provided the deuterium internal lock for the spectrometer. Spectra are obtained on a Varian CFR-20 pulse Fourier transform spectrometer operating at 18.7 kg (20.0 MHz 13-C frequency). The pulse power delivered to the single coil 10 mm o.d. probe is sufficient to rotate the 13-C magnetization by 90° in 15 us. Probe temperature under these conditions was 38°C. Spectrometer parameters for the determination of spectra are: ca 65 degree (10 us) pulse, 4 KHz spectral width, 1 second data acquisition time and 3 seconds delay between repetitive pulses (i.e. a total experiment recycle time of 4 seconds). Typically 12,000-15,000 free induction decays (FID) are accumulated (i.e. an "overnight" run mode was employed) for each quantitative determination. The accumulated FID is digitally-filtered with a time constant that produces a 1.7 Hz line broadening. The 8 K data table is then Fourier transformed to yield a 4 KHz spectrum defined by 4K real points (i.e. digital resolution ca 1 Hz).

Spin lattice relaxation times (T 1) for the carbonyl carbides are estimated from 180-5-90 pulse sequences and found to be in the range 0.8-1.1 seconds. It should be noted that the solutions were not degassed and that the presence of dissolved oxygen may influence the observed relaxation rate. If samples were degassed, the carbonyl relaxation could be longer in which case a longer delay between repetitive pulses would be required). The imposition of a 3 second pulse delay and employment of a 65° pulse ensured the recovery of the carbonyl magnetization during the recycle time, thus eliminating possible errors in quantitation resulting from differential relaxation times. Spectra are obtained under fully proton decoupled conditions using a gated-decoupling scheme in which the proton decoupler was off during the 3 seconds delay time and gated on at the start of the 1.0 seconds acquisition period. In this manner, problems associated with differential nuclear Overhauser enhancements (NOE) for different carbonyl carbons are avoided.

A method of checking that the spectrometer conditions are sufficient to yield quantitative data is to compare the total integral of the carbonyl region with those obtained for alpha-CH 3>C< CH2 since all integrals should be the same under quantitative conditions. In all cases examined, by Johnson et al.[1041], the integral values for all four resonance regions are within 2-3% with the deviations being random from sample to sample - i.e. in one case the carbonyl might be the greatest of the four in magnitude and in another case the lowest.

A 13 C NMR spectrum of a polymethylmethacrylate - methacrylic acid copolymer is shown in Figure 75(a). The spectrum demonstrates the acidic group and the ester group resonances.

In Figure 75(b) is shown the fully proton decoupled 13 C spectrum obtained in pyridine at 38°C of a (MMA/MAA) copolymer of ca. 33% acid composition. While the spectrum resolves into alpha CH3, CH2, OCH 3 >C<, and C = O regions, the structural similarity of the comonomers results in resonance overlap (at 20 MHz) between ester and acid carbons in all regions but the carbonyl. The complete structure in the carbonyl region arises from the sensitivity of the carbons to the microstructural features of the copolymer chain, i.e. sequence and tacticity effects. To assure these effects do not lead to overlap of ester and acid carbonyl resonance, chemical shifts of both homopolymers and homosteric copolymers were examined with the result that the carbonyl resonance region does separate into distinct acid and ester regions in pyridine.

In order to test the reliability of compositional results obtained by the [13]-C NMR, analyses of several copolymers were carried out by both [13] C NMR and titration. Titrations were carried out in 50% aqueous ethanol medium to the phenol phthalein end-point using 0.15 N aqueous potassium hydroxide as the titrant. As evidenced in Table 104 there is an

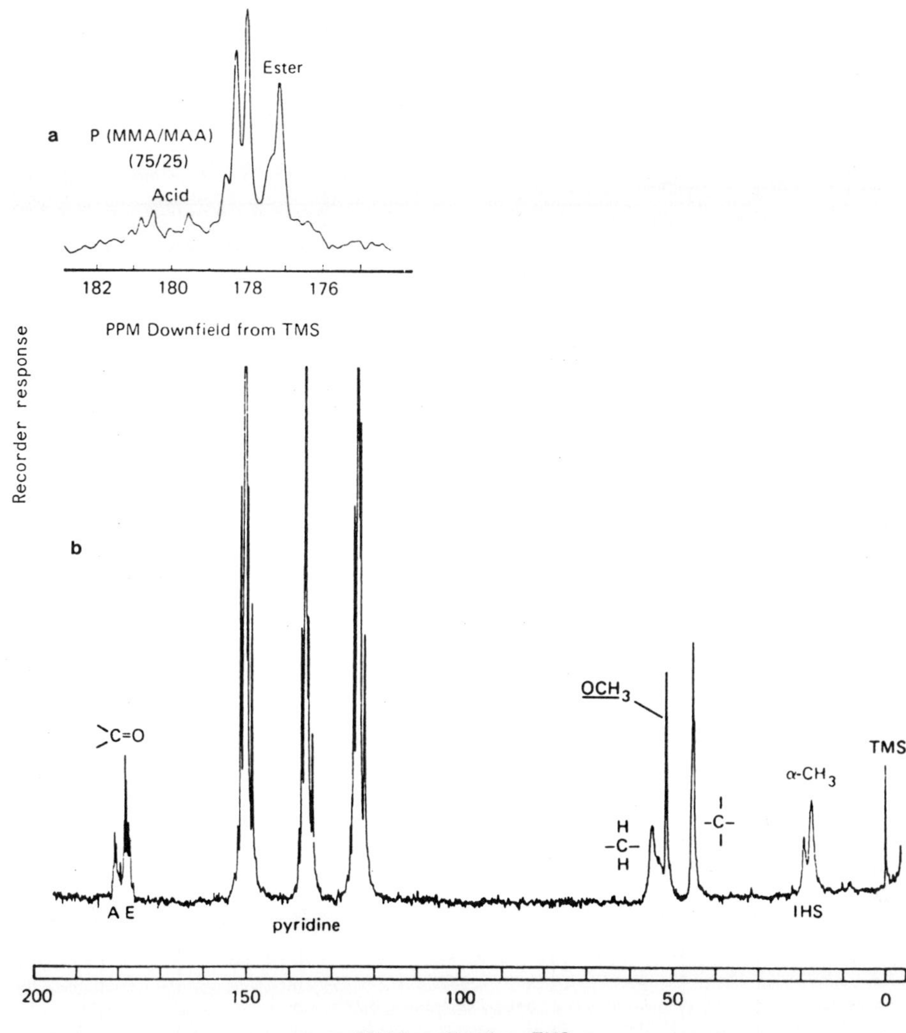

Figure 75   20 MHz ^{13}C NMR spectra of:
(a) P(MMA-COHAA) - 25% acid dissolved in pyridine.
(b) (MMA-COMAA) - 33% acid dissolved n 50:50 v/v
pyridine; pyridine ds. Assignment of the various
resonance regions is given in the figure; the I, H, S
rotation refers to isotactic, heterotactic and
syndiotactic stereochemical triads; A and E refer to
ester and acid.

excellent correlation between the compositional analysis by NMR and
titration (correlation coefficient = 0.998) with most of the comparative
analyses differing by 2% or less.   The results in Table 104 show the
excellent agreement obtained between titration and NMR methods.

Sharp and Paterson[1042] have described a pyrolysis - gas chromatographic - mass spectrometric procedure for the determination of 1-10% of

Table 104 - Compositional analyses of poly(methyl methacrylate-co-methacrylic acid) by titration and ^{13}C NMR

| | Acid content | |
Test sample number	Base titration %	^{13}C NMR %
1	14	13
2	15	15
3	21	22
4	23	24
5	23	26
6	24	27
7	24	24
8	25	27
9	24.5	23.5
10	25	24
11	25.5	25.5[b]
12	29	28
13	33	33
14	36	33
15	48	48
16	48	49
17	45	49
18	50	49
19	97	100

copolymerized acrylic acid and methacrylic acid in acrylic polymers. The acid groups are propylated and the polymer pyrolysed according to the following reaction scheme:

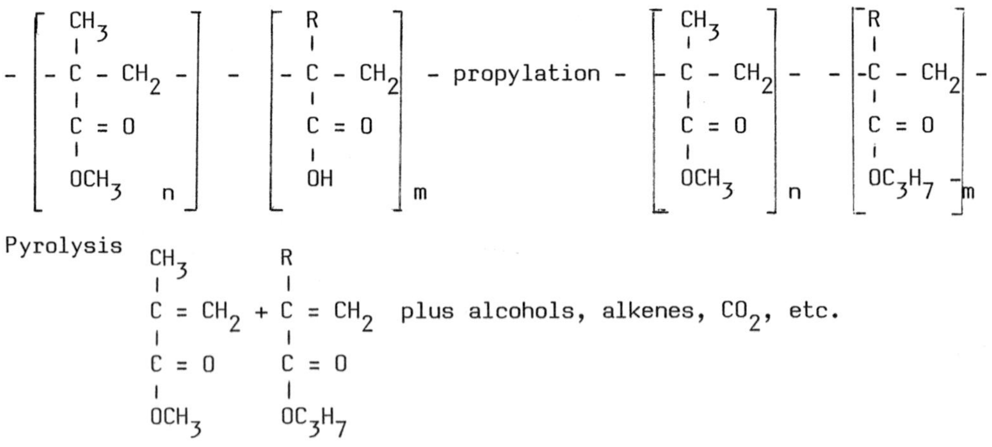

$$
-\left[\begin{array}{c} CH_3 \\ | \\ -C-CH_2- \\ | \\ C=O \\ | \\ OCH_3 \end{array}\right]_n - \left[\begin{array}{c} R \\ | \\ -C-CH_2 \\ | \\ C=O \\ | \\ OH \end{array}\right]_m - \text{propylation} - \left[\begin{array}{c} CH_3 \\ | \\ C-CH_2 \\ | \\ C=O \\ | \\ OCH_3 \end{array}\right]_n - \left[\begin{array}{c} R \\ | \\ -C-CH_2- \\ | \\ C=O \\ | \\ OC_3H_7 \end{array}\right]_m
$$

Pyrolysis

$$
\begin{array}{c} CH_3 \\ | \\ C=CH_2 \\ | \\ C=O \\ | \\ OCH_3 \end{array} + \begin{array}{c} R \\ | \\ C=CH_2 \\ | \\ C=O \\ | \\ OC_3H_7 \end{array} \quad \text{plus alcohols, alkenes, } CO_2, \text{ etc.}
$$

<u>Method - Identification of Acrylic Acid and Methacrylic Acid in Acrylic Copolymers. Propylation - Pyrolysis - Gas Cromatography</u>[1042]

<u>Apparatus.</u> Glass vials of capacity 15-30 ml with Teflon-lines septa (Pierce Hypo-vials) are suitable.

A thermostatically controlled oven at 60°C capable of being evacuated in less then 2kPA, was used.

Pyrolysis - gas chromatographic - mass spectrometric equipment, Sharp and Paterson[1042] used a Perkin Elmer filament pyrolysis unit fitted in a Perkin Elmer F11 gas chromatograph (pyrolysis temperature control 250-550°C). The gas chromatographic column used is a 1.7 m x 3 mm o.d. stainless steel column packed with 30% m/m silicone oil (Embaphase) on acid washed Celite, operated at 80°C with a helium flow rate of 30 ml/min^{-1}. The column effluent is split in the ratio 2:1 between a flame ionization detector and an AFI MS12 mass spectrometer equipped with a glass fit type of molecular separator at 150°C. Mass spectra are scanned from m/e 200 to 20 at 8 seconds per decade under standard electron bombardment conditions, electron energy 70 eV, emission current 500 uA, accelerating voltage 8 kV and source temperature 200°C.

<u>Reagents.</u> Dimethylformamide dipropyl acetal (2 mequiv ml^{-1} in pyridine) (Propyl-8, Pierce Chemical Co.).

<u>Procedure. Direct Pyrolysis.</u> Transfer 1 mg of polymer or an amount of latex or solution containing this mass of polymer on the pyrolysis filament. Remove any water or solvent by blowing with hot air from a hair-drier. Insert the filament into the pyrolysis chamber and connect the carrier gas and gas chromatographic apparatus. Heat the filament at 250°C for 15 seconds to ensure the removal of water, solvent or volatile additives. Pyrolyse the polymer by heating the filament at 550°C for 15 seconds.

<u>Propylation and pyrolysis.</u> Transfer about 0.25 g of polymer or an amount of the acrylic latex or solution containing this mass of polymer into a glass vial and dry by evacuation at 60°C for 15 hours. Seal the vial and inject by means of a hypodermic syringe 1 ml of the Propyl-8 reagent on the the dried polymer film. Heat at 60°C for 16 hours then let the vial and contents cool to ambient temperature. Transfer a smear of the gel or viscous solution produced on to the pyrolysis filament and remove the bulk of the propylation reagent by blowing with warm air from a hairdryer. Heat the filament for 15 second periods at 250°C in the stream of carrier gas until the chromatogram indicates the complete removal of reagent residues and finally pyrolyse at 550°C. Identify by mass spectrometry any eluted compounds that were not observed in the original direct pyrogram or that occur to a significantly greater extent in the pyrogram of the propylated polymer.

The presence of propyl acrylate or propyl methacrylate in the pyrogram of the propylated polymer indicates that the original polymer contained acrylic or methacrylic acid respectively.

The mass to charge (m/e) ratio and relative abundance of the molecular and fragment ions observed are listed below:

Propyl acrylate ($C_6H_{10}O_2$)M$_n$ = 114

m/e	55	73	29	42	41	43	27	85	31	39	59
relative abundance	100	46	15	11		9	7	6	55	5	4

m/e	15	99	114
relative abundance	1	1	0.2

Propyl methacrylate ($C_7H_{12}O_2$) M_n = 128

m/e	41	69	43	87	39	42	27	59	29	86	70
relative abundance	100	91	66	66	38	34	25	9	6	6	5

m/e	31	38	88	26	99	13	128
relative abundance	4	3	3	2	2	2	2

Figure 76 shows a pyrogram of poly(methylmethacrylate) copolymerized to contain 1 and 10% of acrylic or methacrylic acid. By this procedure, copolymerised acrylic or methacrylic acid has been identified in terpolymers with (a) butyl acrylate and styrene (b) methyl methacrylate and ethyl acrylate and (c) ethylene and propylene. A methyl methacrylate - methylstyrene - maleic acid terpolymer, when examined by this propylation - pyrolysis procedure, yielded dipropyl fumerate and a smaller amount of dipropyl maleate.

9.2.5 Carbonyl Groups

A spectrophotometric method using 2,4-dinitrophenyl hydrazine has been used to determine carbonyl groups in vinyl chloride vinyl acetate copolymers[1043].

9.2.6 Hydroxyl Groups

In polyesters and polyamides. Houwelingen[1044,1045] has described a procedure for the determination of hydroxyl groups in the following types of polyesters:

$$HO \left[-(CH_2)_n - O - \overset{O}{\overset{\|}{C}} - \langle \bigcirc \rangle - \overset{O}{\overset{\|}{C}} - O(CH_2)_n - O - \right]_n H$$

with n = 1-100 and x = 2 (polyethylene terephthalate) or 4 (polybutylene terephthalate) and ester-interchange elastomers of 4-polybutylene terephthalate and polypropylene glycol.

The hydroxyl groups in these products are determined by acetylation with an excess of dichloroacetic anhydride in dichloroacetic acid and measurement of the amount of acetylation by a chlorine determination. Owing to the low concentration of hydroxyl groups, especially in high relative molecular weight materials, the determination of the excess is inaccurate and the determination of the amount of reagent incorporated is much more attractive.

The derivatisation is carried out in a 10% m/m solution of dichloroacetic anhydride in dichloroacetic acid at 60°C. A reaction time of 1 hour suffices. After the reaction the solution is poured into water and the precipitated polymer is washed out. To remove the last traces of solvent and acetylation agent, re-precipitation of the derivatised polymer from a hexafluoroisopropanol solution into water is carried out. For polymers with a low hydroxyl content (below 100 mmol kg^{-1}) the reprecipitation is carried out from a solution in nitrobenzene into cold light petroleum, in order to obtain more reproducible results.

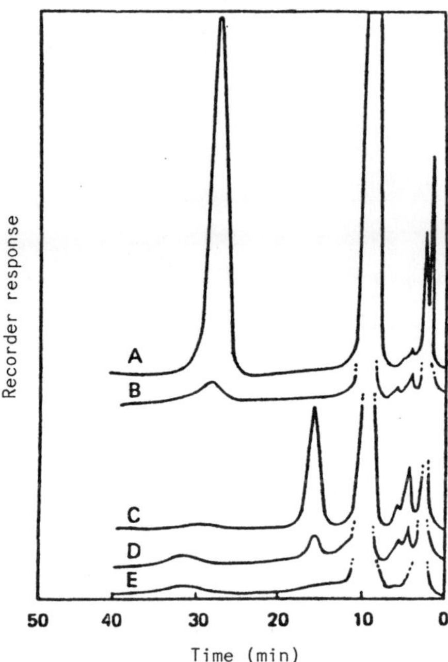

Time (min)

Figure 76   Pyrograms of polymethylmethacrylates and some acid
             copolymers after propylation. Polymethylmethacrylate
             containing the following amounts of copolymerized acid:
             A, 10% methacrylic; B, 1% methacrylic acid; C, 10%
             acrylic acid; D, 1% acrylic acid; E, no acid; elution
             times: methylmethacrylate, 9 mins, propyl acrylate, 15
             mins, propylmethacrylate, 28 min.

The content of hydroxyl groups is subsequently determined by
measurement of the chlorine content of the purified derivative. This is
done either by potentiometric titration with silver ions after combustion
or by x-ray fluorescence spectroscopy of a compressed disc of the polymer.

The suitability of this method for a number of samples is
demonstrated in Table 105.

The standard deviation of the method for high relative molecular
weight polymers is 0.7 mmol kg^{-1} whereas for the low relative molecular
weight materials, it is about 10 mmol kg^{-1}.

In an alternate method described by these workers, the polyester is
reacted with a dimethyl formamide -monochlorbenzene solution of vanadium
8-hydroxyquinolinate (V8HQ). According to Tanaka and Kojima[1046] a
coloured complex with the following structure is formed:

$$Q - V - Q + ROH \quad Q - U - V + H_2O$$

Table 105 - Hydroxyl end group content of several PETP,
PBTP and PBTP-PPG samples

Sample	Viscosity ratio[a]	Hydroxyl end groups/ mmol kg^{-1}	Remarks
PETP A	1.80	30.9-29.8 30.6-30.4	Total end groups[b] 76.6 mmol kg^{-1}
PETP B	1.82	14.4-14.8	Total end groups 79.1 mmol kg^{-1}
PBTP A	2.07	70.7-70.6	Total end groups 95.2 mmol kg^{-1}
PBTP B	2.08	50.6-49.3	Total end groups 92.6 mmol kg^{-1}
PBTP C	1.24	664-653	
PBTP-PPG A	1.25	742-729	M_N (calc)[c] 2700; M_N (measured) 2400[d]
PBTP-PPG B	1.34	329-345	
PBTP-PPG C	1.50	229-212	M_N (calc) 9000; M_N (measured) 10100
PBTP-PPG D	2.38	51-47	M_N (calc) 35000; M_N (measured) 44800

[a] Measured for a 1% m/m solution in m-cresol at 25°C (PETP and PBTP), or for a 1% m/V solution in o-chlorophenol at 25°C (PBTP-PPG).

[b] Total end groups in sum of OH + COOH + methyl ester end groups.

[c] M_N measured by gel-permeation chromatography in m-cresol as a solvent.

d Calculated from OH + COOH content.

Table 106 - Comparison of the V8HQ method with the acetylation method

	OH content/mmol kg^{-1}	
Sample	V8HQ	Acetylation
I	47.9-47.6 43.1-46.8	50-70
II	35.5-34.9	35-50
III	130.1-126.7	170-120
IV	73.8-74.9	50-70

After removal of the excess of reagent by extraction, the complex is acidified with dichloroacetic acid and the blue colour formed is measured at 620 nm. Calibration is carried out with an alcohol as internal standard.

The application of this method to some esters of new types of acids showed that precise determination is possible (Table 106). The standard deviation of this method is 1.5 mmol kg^{-1} which is about ten times more precise than that of the classical acetylation procedure.

Groom and co-workers[1047] employed trichloroacetyl isocyanate and trifluoro acetic anhydride acetylations to determine hydroxy end-groups in polyester polyols. The isocyanate reagent measured proton resonance and was best suited for samples having molecular weights of less than 4500; the anhydride method was more sensitive and applicable to higher molecular weights than is a ^{19}F NMR method.

Dickie et al[1048] derivatised surface hydroxy functional groups on acrylic copolymers prior to their characterization by x-ray photoelectron spectroscopy.

<u>Determination of primary and secondary hydroxyl groups in ethylene oxide tipped glycerol-propylene oxide condensates.</u> Crompton[1049] has developed a kinetic method for the evaluation of the reactivity of polyols in the range 3000-5000. This procedure may be used to determine the degree of ethylene oxide 'tipping' produced by the addition of ethylene oxide to glycerol/propylene oxide condensates. The method was developed to measure the relative reactivity with isocyanates of such condensates as a function of their ethylene oxide tipping content. The method is based on the observation that primary hydroxyl groups react with phenyl isocyanate to form a urethane faster than do secondary hydroxyl groups. Hence a 'tipped' polyol will react more completely in a given time with an equivalent amount of phenyl isocyanate than an 'untipped' polyol reacting under the same conditions. The greater the amount of ethylene oxide tipping the greater its rate of reaction with phenyl isocyanate.

The reaction is carried out under standard conditions in which a calculated weight of the polyol (depending on its hydroxyl number) is reacted under standard conditions with an excess of a standard toluene solution of phenyl isocyanate in the presence of a basic catalyst.

Unconsumed phenyl isocyanate is then reacted with excess standard potassium hydroxide.

$$PhCNO + 2KOH = K_2CO_3 + PhNH_2$$

Excess sodium hydroxide is estimated by titration with standard acetic acid to the phenol phthalein end-point. A blank run is performed in which the sample is omitted. From the difference between the sample and the blank titrations it is possible to calculate the hydroxyl content of the original polymer:

1 mole hydroxyl groups $\equiv$ 1 mole phenyl isocyanate $\equiv$ 2000 ml N KOH

Reactivity is calculated by a kinetic procedure in which the percentage of the original phenyl isocyanate addition, which reacts with the sample in a given time, it taken as an index of its reactivity. A calibration graph can be prepared in which the determined phenyl isocyanate reactivity is plotted against the ethylene oxide content, and this enables a determination to be made of the ethylene oxide content of unknown samples from the calibration graph by interpolation.

Crompton applied this method to a range of glycerol/propylene oxide adducts containing various accurately known amounts of ethylene oxide tipping, up to 5.3 moles (Table 107). As expected, 'untipped' polyols which are relatively free from primary hydroxyl groups, i.e. curves A and B, react comparatively slowly with phenyl isocyanate. Decinormal solutions of 'untipped' glycerol/propylene oxide condensates of molecular weight 3000 and 5000 had an identical rate of reaction with phenyl isocyanate. Thus the rate of reaction with phenyl isocyanate of the terminal isopropanol end-groups in polyols is independent of molecular weight in the molecular weight range 3000 to 5000 and depends only on proportions of primary and secondary hydroxyl end-groups present.

A calibration curve is prepared by plotting moles ethylene oxide per mole of glycerol for the range of standard tipped polyols of known ethylene oxide content against percentage of original phenyl isocyanate addition consumed after 60 and 100 min, i.e. P60% and P100%. This curve can be used to obtain from P60% and P100% data obtained for tipped glycerol propylene oxide polyols of unknown composition their tipped ethylene oxide contents (in moles ethylene oxide).

9.2.7 Alkoxy Groups

Anderson et al.[1050] identified and determined the etherifying alcohols present in thermosetting acrylamide interpolymers of the type shown below via alcohol exchange when interfering ester linkages are present in the polymer backbone. No interference is encountered from alkyl esters present in the sample.

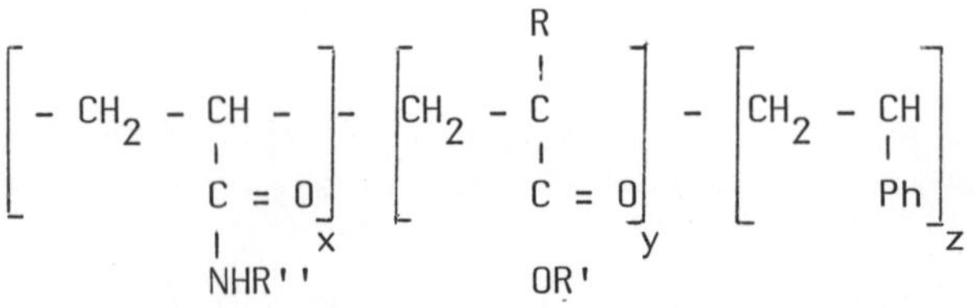

$$\left[\begin{array}{c} -CH_2-CH- \\ | \\ C=O \\ | \\ NHR'' \end{array}\right]_x - \left[\begin{array}{c} R \\ | \\ CH_2-C \\ | \\ C=O \\ | \\ OR' \end{array}\right]_y - \left[\begin{array}{c} CH_2-CH \\ | \\ Ph \end{array}\right]_z \qquad \begin{array}{l} \text{where } R = HOrCH_3 \\ R' = H, CH_3, C_2H_5 \\ C_4H_9 \text{ or } C_8H_{17} \\ R'' = H, CH_2OC_4H_9 \\ \text{or } CH_2OH \end{array}$$

Table 107 - Application of reactivity method to standard ethylene oxide
tipped polyols

Sample identification	Approximate molecular weight	Hydroxyl number mg KOH/g polyol	Ethylene oxide tipping, moles ethylene oxide/mole glycerol (by weight addition)	Reactivity, i.e. percentage of original phenyl isocyanate addition consumed in:	
				60 min reaction	100 min reaction
A	3000	59.0	0.0	25.2	35.2
B	5000	34.9	0.0	27.1	36.8
C	5000	34.3	3.0	41.5	46.6
D	5000	35.9	3.5	44.4	51.0
E	5000	36.0	4.3	47.5	54.1
F	5000	33.3	5.3	51.9	57.4

Method - Determination of Etherification Levels in Acrylamide Interpolymers. Alcohol Exchange - Gas Chromatography[1050]

Apparatus. Gas chromatographic separations are performed using a Hewlett-Packard 5760 gas chromatograph or equivalent equipped with flame ionization detectors. A 12 ft x 0.125 in o.d. stainless steel column is packed with 20% Carbowax 20 M on 60-80 mesh Gas Pak WAB. Helium flow rate: 20 ml/min with an inlet pressure of 65 psi. Injection port and detector temperatures: 220 and 240°C respectively. The column temperature: Held at 105°C until the elution of the pentanol peak, then programmed from 105 to 205°C at 15°C/min and held until the 2-ethylhexanol eluted.

The glass assembly used to carry out the alcohol exchange reaction consists of a 50 ml erlenmeyer flask equipped with a 14/20 joint topped with a Liebig condenser with a 14/20 joint.

Reagents. Reagent grade 2-ethylhexanol, n-butanol, n-pentanol, butyl cellosolve and acetone (Eastman Organic Chemicals). Paraformaldehyde, p-toluenesulfonic acid and propionamide (Eastman Organic Chemicals) used without further purification.

Drying of polymer samples. All acrylamide interpolymer samples are initially diluted to 10% nonvolatile resin with acetone. Thin films of the resin solutions are uniformly cast on clean glass plates using a smooth glass rod and immediately dried at 50°C under 5 cm Hg for 2½ hours to remove all solvents. The film thickness of the dried polymers is kept below 0.5 ml to ensure complete solvent removal. The films are removed from the glass plates using oil free razor blades and stored in vials.

Alcohol exchange reaction. The etherifying alcohols present in the polymer backbone are exchanged according to the reaction:

$$
\begin{array}{ccc}
\begin{array}{l}
- CH_2 - CH - \\
\qquad | \\
\qquad C = O \\
\qquad | \\
\qquad NH - CH_2OC_4H_9
\end{array}
& + C_8H_{17}OH \longrightarrow &
\begin{array}{l}
- CH_2 - CH \\
\qquad | \\
\qquad C = O \\
\qquad | \\
\qquad NHCH_2OOC_8H_{17}
\end{array}
& + C_4H_9OH
\end{array}
$$

Approximately 200 mg of dried polymer is placed in a reaction flask along with 20 ml of 2-ethylhexanol. To this solution 5 ml of a 0.4% w/v solution of p-toluenesulfonic acid in 2-ethylhexanol and 5 ml of a 1% w/v solution of n-pentanol in 2-ethylhexanol are added. A magnetic stirring bar is placed in the flask, the condenser is fitted to the flask and placed on a magnetic stirrer hot plate. Heating and stirring are employed until a vigorous reflux begins. The reaction is allowed to continue for four hours with the reaction flask being checked periodically to ensure constant refluxing. After the reaction period is complete, the reaction mixture is allowed to cool to room temperature. The condenser is thoroughly washed with fresh 2-ethylhexanol to wash all products into the reaction flask. After allowing total drainage from the condenser, the reaction flask is stoppered until gas chromatographic analysis.

<u>Analysis of exchanged alcohols.</u> Determination of response factors – an accurately known blend of the etherifying alcohols and internal standard (n-pentanol) is prepared and analysed by gas chromatography. The areas of the peaks in the chromatograms are determined by triangulation and the response factors of the etherifying alcohols calculated using the following equation:

$$ K_1 = \frac{(W_1) \, (A_x) \, (K_x)}{(W_x) \, (A_1)} $$

where K_1 = response factor of etherifying alcohol, K_x = response factor of n-pentanol (arbitrarily set equal to 1.00), W_1 = weight of respective etherifying alcohol in the calibration blend, W_x = weight of n-pentanol in the calibration blend, A_1 = area of respective etherifying alcohol peak and A_x = area of n-pentanol peak.

Response factors determined at two-week intervals were found to deviate. Best results were obtained when response factors were determined just prior to analysis.

<u>Calculation of etherifying alcohol concentration.</u> A 2 ul sample of exchanged alcohols is removed from the reaction flask and is analysed by gas chromatography. Prior to sampling, the reaction mixture is thoroughly shaken to ensure uniform composition. A typical reaction mixture yielded the gas chromatogram given in Figure 77. The areas of the etherifying alcohols and internal standard peaks are determined by triangulation and the moles of etherifying alcohol calculated using the following equation:

$$ M_A = \frac{(A_A) \, (K_1) \, (W_x)}{(A_x) \, MW_A) \, (W_A)} $$

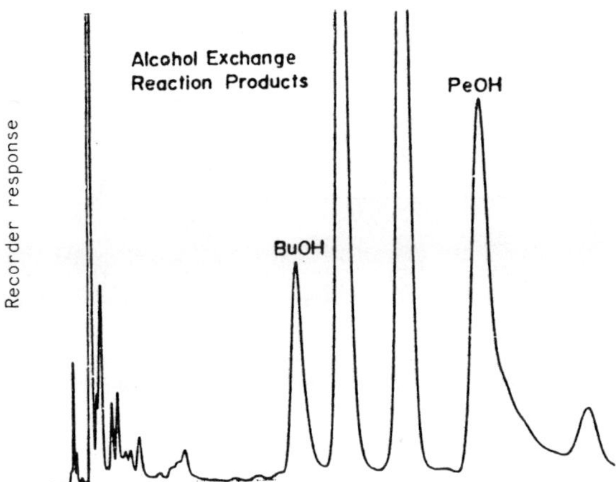

Figure 77   Gas chromatogram of exchanged butanol and pentanol
             internal standard.

where M_A = moles of etherifying alcohol per gram of dried polymer, A_A =
area of n-pentanol peak, W_x = weight of n-pentanol added to flask, W_A =
weight of dried polymer added to flask and MW_A = molecular weight of
etherifying alcohol.

Each reaction mixture is analysed in duplicate and the average value
used for further calculations.   Duplicate determinations do not vary by
more than $\pm$ 2% relative. 99.7% recovery was obtained with this procedure.

The moles of acrylamide present in the polymer are determined from
the nitrogen content of the dried polymer.   The acrylamide content is
calculated using the following equation:

$$M_a = \frac{(\%N)}{(14.01)\ (100)}$$

where Ma = moles of acrylamide per gram of dried polymer, 14.01 =
molecular weight of nitrogen and %N = percent nitrogen in dried polymer.

The percent etherification of the acrylamide in the polymer may be
calculated as follows:

$$\% \text{ etherification} = \frac{(M_A)\ (100)}{(M_a)}$$

where

M_A = moles of etherifying alcohol per gram of dried polymer
M_a = moles of acrylamide per gram of dried polymer

The total etherification is calculated by adding the % etherification due to all etherifying alcohols.

Table 108 - Correlation of alcohol exchange and Zeisel cleavage

Sample	Percent butylated acrylamide[a]	
	Alcohol exchange	Zeisel cleavage
14% Acrylamide	100	101
23% Acrylamide	101	100

[a] Based on experimentally determined butoxy and nitrogen content.

The level of alcohol obtained during the exchange reaction on acrylamide is correlated with data obtained using Zeisel cleavage of the alkoxy groups. These data are summarised in Table 108. Comparable results are obtained using both procedure, however, the Zeisel cleavage reaction will also cleave ester linkages as well as ether functionalities. This presents no problem with polymers which do not contain ester groups. In systems employing both butyl esters and butyl ethers, the Zeisel cleavage reaction gives a total $O-C_4H_9$ content in the sample. Alcohol exchange, on the other hand, will cleave only the butyl ether groups in the sample. By subtracting the alcohol exchange data from that obtained from Zeisel cleavage, one can assess the relative amounts of alkyl ester and alkyl ether in the sample.

Haslam et al.[1051] employed a procedure based on pyrolysis for the determination of polyethyl esters in methacrylate copolymers. The alkoxy groups in the polymer were reacted with hydrogen iodide and pyrolysed to their corresponding alkyl iodides which were then determined by chromatography on a dinonyl sebacate column at 75°C. Similarly, Miller[1052] determined acrylate ester impurities in polymers by converting the alkoxy groups to alkyl iodides which were gas chromatographed on a di-2-ethyl hexyl sebacate column at 70°C.

9.2.8 Oxyalkylene Groups

These polymers are manufactured by the reaction of alkylene oxides with a glycol.

$$CH_2OH + CH_2CH_2O + CH_3CHCH_2O \longrightarrow$$
$$|$$
$$CH_2OH$$
$$CH_2(CH_2CH_2O)n(CH_3CHCH_2O)mCH_2CH_2OH$$
$$|$$
$$CH_2(CH_2CH_2O)n(CH_3CHCH_2O)mCH_2CH_2OH$$

A method for determining alkoxy groups is based on thermal degradation of oxyalkylene groups to the corresponding olefin which is determined by gas chromatography[1055-1058].

Neumann and Nadeau[1053] and Swann and Dux[1054] studied the pyrolysis of ethylene oxide, propylene oxide condensates containing various proportions of ethylene oxide.

$$- CH_2 - CH_2O - CH - CH_2O \rightarrow CH_2 = CH_2 + CH_3CH = CH_2$$
$$\overset{|}{\underset{CH_3}{}}$$

There were no significant differences in the pyrolysis chromatograms of samples heated for 0.5 to 2 hours. The only effect of time of pyrolysis is the total amount of gas produced. The relative concentrations of components are not significantly changed. The temperature of pyrolysis, however, does play a large part in both the amount and type of components in the volatile gases. Between 390°C and 410°C there was no noticeable change in products. As the ethylene oxide content of the copolymer increases, so does the amount of ethylene produced. The curve is linear up to about 50% ethylene oxide, then turns sharply upward to an ethylene content of 38.6% for pure polyethylene glycol. The relative contents of ethylene oxide and propylene oxide in polyethylene-polypropylene glycols has been determined using combined pyrolysis-gas chromatography calibrated with polyethylene glycol and polypropylene glycol standards[1055].

<u>Ethylene oxide - propylene oxide - polyhydric alcohol and ethylene oxide - propylene oxide amine condensates.</u> Ethylene oxide and propylene oxide adducts of polyhydric alcohols and amines are widely used as polyethers in the production of polyurethane foams by reaction with diisocyanate. The physical properties of the foams depend to a certain extent on the chemical structure of these polyethers, so it is very important to establish a method for the identification of the base compounds and determination of the proportions of their oxyethylene and oxypropylene groups.

Mathias and Mellor[1059] split the polyethers with hydrobromic acid - acetic acid to give bromides, which were analysed by gas chromatography. In this way, the content of oxyethylene groups, and therefore, the original polyhydric alcohols, can be determined. Stead and Hindley[1060] modified this method and obtained good results for the determination of oxyethylene group contents of ethylene oxide - propylene oxide copolymers. The determination of the content of oxyethylene groups in these copolymers can easily be carried out by nuclear magnetic resonance spectrometry[1059,1061] without chemical splitting of the ether linkage. However, it is difficult to identify the base compounds by this method. A number of methods for the cleavage of ethers have been studied but few were applied to the identification of the base compounds of the polyurethane polyethers.

Several mixed anhydrides of carboxylic and sulphonic acids, as proposed by Karger and Mazur[1062], act as reagents for the cleavage of ether linkages, particularly that of acetic anhydride toluene-p-sulphonic acids, which is not only a powerful reagent for the cleavage of ether linkages but is also an active acetylating agent. For example, when the propylene oxide adduct of glycerol is treated with this reagent, the polyether is split, thus giving glycerol triacetate and propylene glycol diacetates, which are easily identified by gas chromatography. Tsuji and Kounishi[1063] extended this method to the identification base compounds and the determination of their oxyethylene and oxypropylene group contents.

<u>Method - Determination of Oxyalkylene Groups in Polyether Bases, Polyurethane Polymers. Mixed Anhydride Cleavage - Gas Chromatography</u>[1063]

<u>Cleavage reagent.</u> Acetic anhydride (80 g) is added dropwise to 120 g of toluene-p-sulphonic acid contained in a 300 ml round-bottomed flask

at room temperature and the mixture is refluxed at 120°C for 30 minutes. The product obtained is used as the reagent without removal of the acetic acid produced and the excess of acetic anhydride.

    <u>Apparatus.</u> Hitachi gas chromatograph, model 063, equipped with a thermal conductivity detector or equivalent. The column consists of a 1 m length of 1 mm i.d. stainless steel tubing packed with 15 % m/m FFAP (free fatty acid phase) or 10 % m/m SE-30 silicone) coated on 60-80 mesh Uniport B. Helium is used as a carrier gas.

    <u>Procedure.</u> A 100 mg amount of sample and 2 ml of reagent are mixed in a 20 ml round-bottomed flask and the mixture refluxed on an oil-bath at 120°C for 2 hours. The contents of the flask are cooled to room temperature and then neutralised with 50% aqueous sodium carbonate solution, followed by extraction with about 20 ml of diethyl ether. The ether layer is washed with de-ionised water and concentrated to a small volume on a steam-bath; the concentrate is then injected into the gas chromatograph. The FFAP column is operated isothermally at the appropriate temperature for the purpose of both identifying the base compounds and determining their oxyethylene group contents. For the analysis of the polyether based on 1,2-diaminoethane or 2,2'-diaminodiethylamine the SE-30 column is used.

    <u>Identification of polyol base compounds.</u> figure 78(a) shows gas chromatograms obtained with reaction products of several polyurethane polyethers.

    The acetate peaks are satisfactorily separated from each other and the peak of propylene glycol diacetate produced by the cleavage of the polyoxypropylene groups did not overlap those of the derivatives from polyol base compounds except that for the polyether based on propylene glycol. These base compounds could therefore be easily distinguished and identified. By decreasing the temperature of the gas chromatographic column, the peak for propylene glycol diacetate can be accurately identified.

    Polyethers based on sorbitol yielded complex products that consist of the acetates of sorbitans and sorbides produced by dehydration. However, the gas chromatograms always showed similar patterns, so that the base compound (sorbitol) can be identified from the chromatogram.

    In order to obtain the peaks of the reaction products of polyethers based on 1,2-diaminomethane and 2,2'-diaminodiethylamine, the SE-30 column is operated isothermally at 230°C and the flow-rate of the carrier gas was maintained a 60 ml/min^{-1}. A typical gas chromatogram is shown in Figure 78(b). These base compounds can also be distinguished and identified.

    <u>Determination of oxyethylene and oxypropylene groups.</u> The ethylene oxide propylene oxide copolymers, and polyols that were oxypropylated and then oxyethylated, were decomposed as described above for the determination of the polyol base compounds. The FFAP column is operated isothermally at 65°C and the flow-rate of the carrier gas is regulated at 60 ml/min^{-1}.

    A typical chromatogram for the reaction of the 1,2-diaminoethane ethylene oxide-propyleneoxide adduct is shown in Figure 78(b). The ethylene glycol diacetate and propylene glycol diacetate peaks, produced from polyoxyethylene and polyoxypropylene groups respectively are completely separated. The proportions of ethylene and propylene oxide can

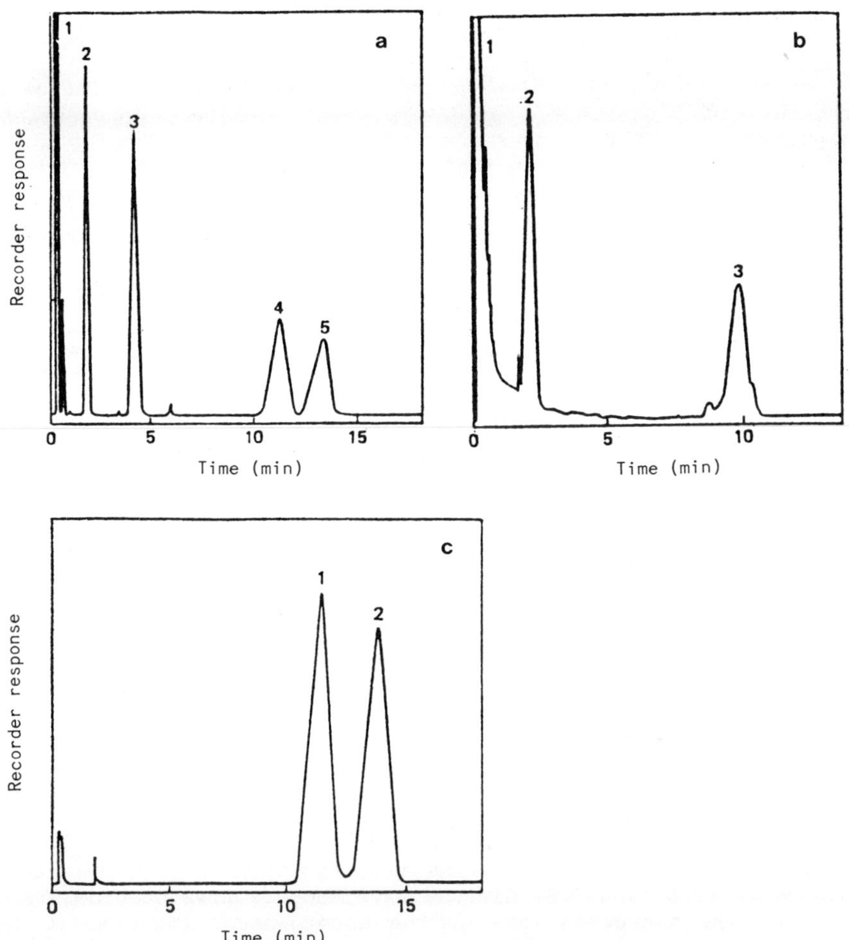

Figure 78   Reaction of polyethers with hydrobromic acid - gas
              chromatography: (a) Gas chromatogram of reaction
              products of polyurethane polyethers (1) propylene glycol
              diacetate; (2) glycerol triacetate; (3) trimethylpropane
              triacetate; (4) triethylanolamine triacetate; (5)
              pentaerythritol tetracetate; Gas chromatographic
              conditions, column 1 mm 3 mm, coated with 15% FFAQ on
              Uniport B, oven 170°C; chart speed 1 cm sec^{-1}.
              (b) Gas chromatogram of reaction products of
              polyurethane polyethers based on amines (1) pyropylene
              glycol diacetate; (2) derivative from 1,2 diaminoethane;
              (3) derivative from 2,2'diaminodiethylamine. Gas
              chromatographic conditions: Column 1 m x 3 m packed
              with 15% SE-30 on Uniport B; oven 230°C; chart speed, 1
              cm min^{-1}.

be determined by measuring the two peak areas (i.e. of the ethylene glycol and propylene glycol diacetates) and applying the appropriate calculations. The derivative from the base compound itself (1,2-diaminoethane) do not appear in the chromatogram and so do not interfere in the determination of the ratio of oxyethylene to oxypropylene groups.

Cervenka and Merrall[1064] have investigated the application of acidic dehydration of ethylene oxide-propylene oxide condensates in bromonaphthalene in the presence of p-toluene sulphonic acid and methyldioxane-1,4 to the elucidation of the molecular structure and monomer sequence of these polymers. Gas chromatography was used to determine dehydration products.

Studies on poly(ethylene glycol) and poly(propylene glycol) homopolymers showed that dioxane and its derivatives are not the only reaction products. Dehydration of poly(ethylene glycol) gave three products; that of poly(propylene glycol) six products. The majority of them were indentified.

Cervenka and Merrall[1064] conclude that results on homopolymers, their blends, and model copolymers of different chain architecture demonstrate that acidic dehydration is capable of distinguishing ethylene oxide-propylene oxide copolymers of different structures, giving correct absolute values of overall monomer contents and also ranking polyols according to their degrees of randomness.

9.2.9 Epoxy Groups

Swaraj and Ranby[1065] used infrared spectroscopy to determine epoxy groups in methylmethacrylate - glycidyl methacrylate copolymers.

The infrared analysis was performed on dried potassium bromide pellets containing 0.5 mg sample in 200 mg potassium bromide. An infrared spectrum of a methyl methacrylate - glycolymethyacrylate copolymer is shown in Figure 79. The peaks at the wave numbers 11.02 and 5.82 microns are the most suitable ones for analysis of epoxy and carbonyl groups respectively. Using the "base line density" method, the values of the absorbances at 11.02 and 5.82 microns wave numbers have been determined in triplicate. The average values of the absorbances, their ratio and the glycidylmethylacrylate mole fraction determined chemically are presented in Table 109.

The absorbance ratio at 11.02 vs 5.82 micron is linearly related to the the glycidyl methacrylate content in the copolymer and can be expressed by the following equation:

$$R = 0.250\ X_G = 0.033$$

where R is the absorbance ratio at 11.02 and 5.82 microns, X_G is the mole fraction of glycidylmethacrylate containing the epoxy group in the copolymer. The term 0.033 is considered to be a correction factor arising from the very weak absorption due to poly(methyl methacrylate) at wave numbers near 11.02 micron.

Hard epoxy resins of the diglycidylether-bisphenol A type (e.g. Epikote 1004 (Shell Chemicals) and Bakelite epoxy resin ERL 1774) are manufactured by the reaction of bisphenol and epichlorohydrin:

$$CH_2 - CH_2Cl + HO - C_6H_4 - C(CH_3)_2 - C_6H_4 - OH \longrightarrow$$
(epoxide)

$$CH_2 - CH - \left[O - C_6H_4 - C(CH_3)_2 - C_6H_4 - OCH_2 - CH_2 \right]_n - O - C_6H_4 - C(CH_3)_2 - C_6H_4 - OCH_2 - CH_2$$

Peltonen et al.[1066] have described an infrared method for determining epoxy resins and their thermal degradation products in workspace air samples which had been collected by filtration on glass-fibre filters. Epoxy residues were extracted from the filters with chloroform, the extract evaporated to dryness and the residue dissolved in 0.5 ml deuterochloroform ($CD Cl_3$). An infrared spectrum of an elute epoxy resin (Epicote 1004) is shown in Figure 80. If thermal degradation products of epoxy resins were present then the initial chloroform extract was washed with 1 N sodium hydroxide, then water to remove phenolic products, then evaporated and the residue dissolved in deuterochloroform prior to infrared spectroscopy. Deuterochloroform was used instead of chloroform for preparing the extract for infrared spectroscopy as chloroform has a strong absorption at 1510 cm^{-1} which precludes the use of this aromatic absorption for the determination of epoxy resin in the extract. The absorbance at 1510 cm^{-1} is linear with concentration in the concentration range 50-740 ug epoxy resin per 0.5 ml with a detection limit of 50 ug per 0.5 ml and recoveries ranging from 70% at the 72 ug per 0.5 ml level to 97% at the 288 ug per 0.5 ml level.

The high relative molecular mass fraction, derived from the cracked epoxy network, contains the same aromatic substances as the intact epoxy resin. This enabled Peltonen et al.[1066] to use pure epoxy resin as a

Table 109 - Analytically determined mole fraction of GMA in the MMA-GMA copolymers and the infrared absorbanced at 907 and 1717 cm^{-1}

Expt. no.	Mole fraction of GMA in the copolymer determined chemically	A, 5.82 micron	A, 11.02 micron	Absorbance ratio $\frac{A, 11.02\ micron}{A, 5.82\ micron}$
R50	0.218	1.366	0.128	0.093
R51	0.394	0.911	0.120	0.131
R52	0.584	0.629	0.115	0.182
R61	0.623	0.921	0.177	0.192
R53	0.706	0.955	0.204	0.213

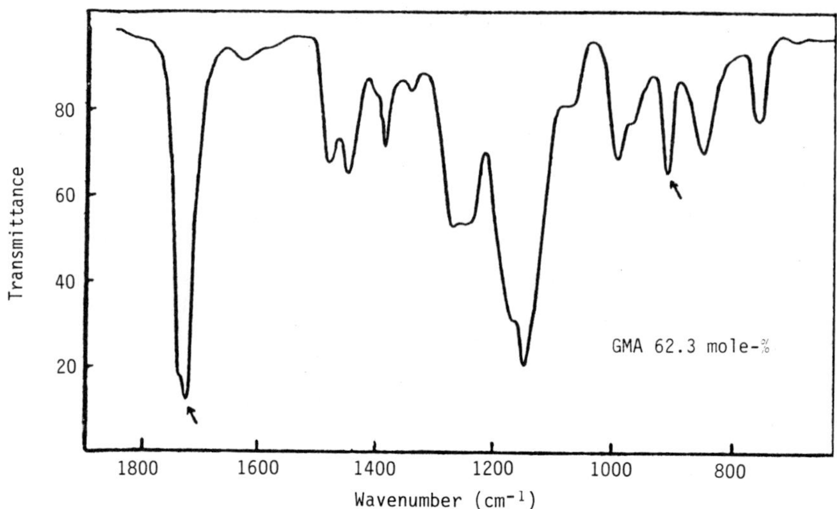

Figure 79   Infrared spectrum of a MMA-GMA copolymer containing 62.3 mole % GMA chemically determined.

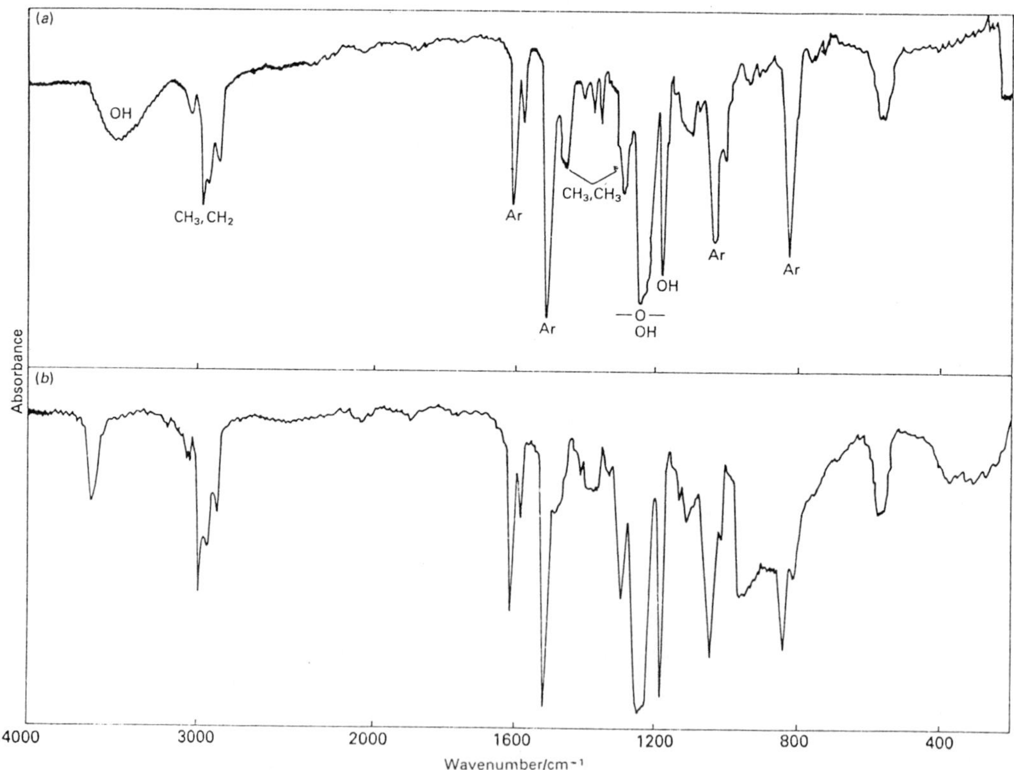

Figure 80   IR spectrum of a solid epoxy resin (Epicote 1004) in (a) potassium bromide and (b) chloroform-d.

standard for quantification of the high relative molecular mass fraction
of the thermal degradation products. High performance liquid chromatography revealed the presence of phenolic compounds. These were washed off
because they also absorb strongly at 6.62 microns. The removal efficiency
of phenols (100%) was confirmed by high performance liquid chromatography.
The removal of phenols was also observed in the infrared spectrum as
decreased absorption by hydroxy groups. The remaining absorption was
probably due to the secondary aliphatic alcohol groups in the chains.
Secondary aliphatic alcohols are not acidic and therefore they are not
removed by sodium hydroxide extraction.

Bis-phenol-A-epichlorohydrin condensates of the type mentioned above
have been determined by NMR spectroscopy[1067], also polycaprolactone polyols (see below), diethylene glycol started caprolactone ester diols:

$$\text{H } [OCH_2(CH_2)_3-CH_2-\overset{\overset{O}{\|}}{C}]_x OCH_2CH_2-OCH_2CH_2O[\overset{\overset{O}{\|}}{C}-CH_2(CH_2)_3CH_2O]$$

trimethylol propane started caprolactone, ester triols;

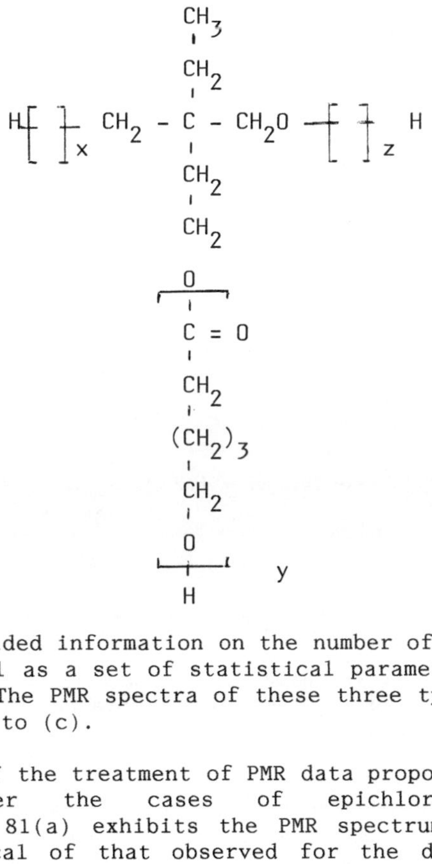

$$H\left[\;\right]_x CH_2 - \underset{\underset{\underset{(CH_2)_3}{|}}{\underset{\underset{CH_2}{|}}{\underset{\underset{C=O}{|}}{\underset{\underset{O}{|}}{\underset{\underset{CH_2}{|}}{\underset{\underset{CH_2}{|}}{\overset{\overset{CH_3}{|}}{\overset{\overset{CH_2}{|}}{C}}}}}}}} - CH_2O \left[\;\right]_z H$$

NMR spectroscopy provided information on the number of repeat units (n) in
these polymers as well as a set of statistical parameters inherent to the
choice of ratioing. The PMR spectra of these three types of polymers are
shown in Figure 81(a) to (c).

As an example of the treatment of PMR data proposed by Hammerich and
Willeboordse consider the cases of epichlorohydrin-bisophenol-A
condensates. Figure 81(a) exhibits the PMR spectrum of bakelite epoxy
resin ERL-2774, typical of that observed for the diglycidyl ethers of
bisphenol A where the multiplet at 2 ppm is attributable to
pentadeuteroacetone. If n represents the number of repeat units, then the
assignment and the number of protons contributing to each lettered area
are portrayed in Table 110.

Table 110 - Peak assignments for DGEBA Polymers

Area	Assignment	No. of protons
A	aromatic	$6n + 8$
B	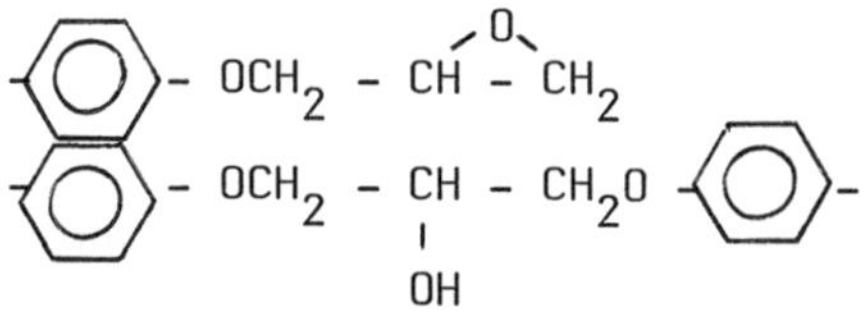 phenyl$-OCH_2 - CH \underset{\diagup O \diagdown}{} CH_2$	$6n + 4$
	phenyl$-OCH_2 - \underset{\underset{OH}{\vert}}{CH} - CH_2O-$phenyl$-$	
C + D	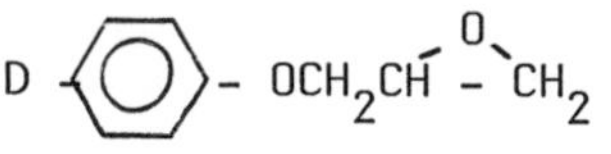 phenyl$- OCH_2CH \underset{\diagup O \diagdown}{} CH_2$	6
E	methyl	$6n + 6$

As the unreacted epoxy protons yield two separable areas, areas C and D in Figure 81(a) then a total of five distinct areas are observed for DGEBA polymers. Every conceivable ratio of a linearly independent combination of these areas yields a function, f(R), for n. (The ratio of aromatic to gem dimethyl protons is linearly dependent with a constant value of 4/3.) If we proceed under the assumption that all observed proton signals are attributable to some functional group portrayed in the idealized average molecular structure, then the entries in Table 111 may be computed. For these computations the areas corresponding to the unreacted epoxy group, areas C and D, were summed together. As an example, consider the first ratio for ERL-2774 portrayed in Table 111, comparing the area of the aromatic protons to that of the aliphatic. For this case,

$$R = \frac{x}{y} = \frac{A}{B + C + D + E} = \frac{2n + 2}{3n + 4}$$

$$\text{so that } n = f(R) = \frac{2 - 4R}{3R - 2}$$

$$f_1 = \frac{\partial f}{\partial R} = \frac{2}{(3R - 2)^2}$$

Hammerich and Willeboordse[1067] compared repeat unit (n) values obtained by PMR spectroscopy and those obtained by chemical methods of analysis. In the PMR method areas were excluded whose integrals were overlapped.

Though the NMR determinations are consistently lower than the corresponding titrimetric determinations, except for the first epoxy sample, if one assumes the titrimetric value to be representive of the correct result, the highest relative accuracy portrayed in Table 112 is 10% for EPON 1001. Maximum relative accuracies of 4 and 5% are observed for the diols and triols, respectively. While chemical methods prove to be superior for high molecular weight resins where the NMR ratios for calculating n are prone to greater inaccuracies, the methods are still

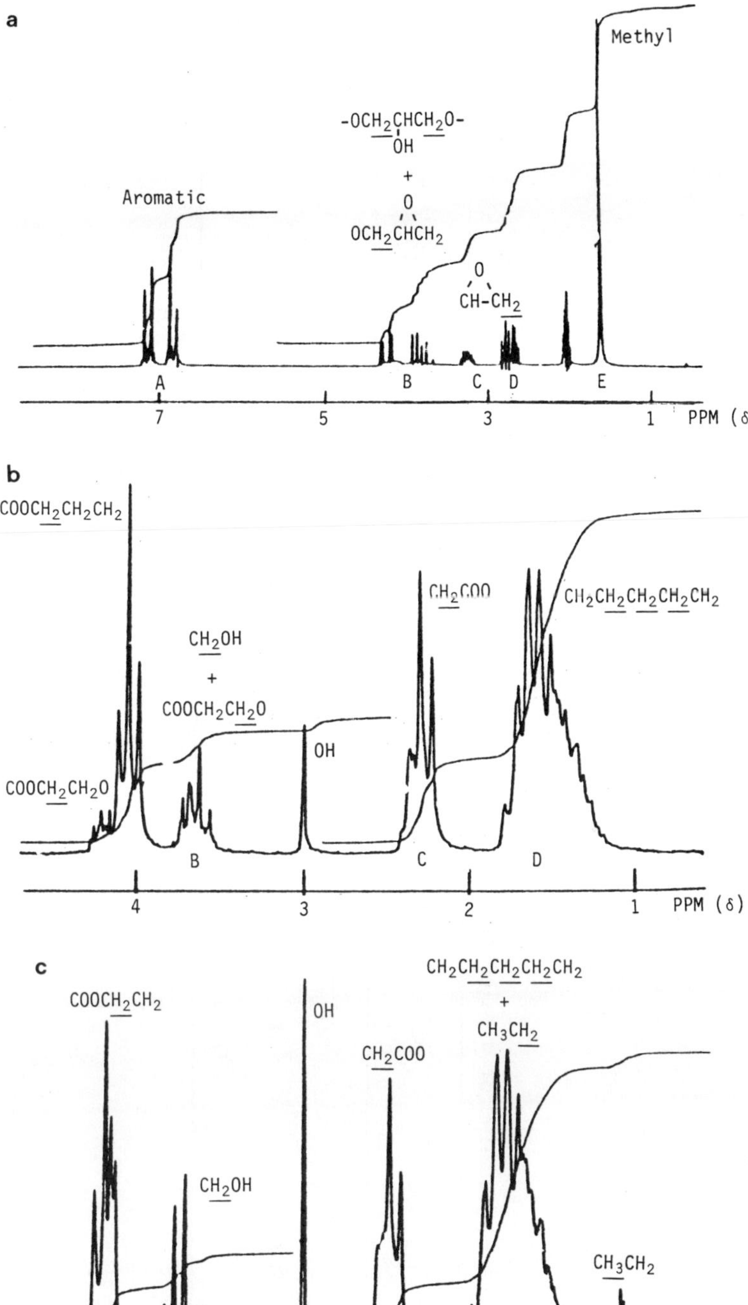

Figure 81   PMR spectra of: (a) bisophenol-epichlorohydrin condensate (Bakelite epoxy ERL 2774). (b) Ethylene glycol started caprolacone ester diol (Niox polyol, PCP-0230). (c) Trimethyl-ol propane started caprolactone ester triols (Niax polyol, PCP-0310).

Table 111 – Precision of n calculated for various NMR areas of bakelite epoxy resin ERL-2774

	$A/(B+C+D+E)$	$A/(B+C+D)$	A/B	$A/(C+D)$	$B/(C+D+E)$	$B/(C+D)$	B/E	$(B+C+D)/E$	$(C+D)/E$
n	−0.11	−0.10	0.62	0.06	0.25	0.17	0.59	−0.09	0.06
s_n	0.044	0.030	0.096	0.017	0.017	0.011	0.144	0.039	0.021
$100s_n/n$	42.4	31.3	15.5	30.7	6.93	6.60	19.5	45.5	33.7

Table 112 - Values determined by NMR minimization criteria
and from titration results

Sample	NMR	Titration[a]
ERL-2774[d]	0.26	0.25[b]
EPON 1001[f]	2.02	2.24
PCP-0200[d]	3.89	3.99[c]
PCP-0210[d]	6.34	6.52[c]
PCP-0230[d]	10.7	11.1[c]
PCP-0240[d]	16.1	16.5[c]
PCP-0300[e]	3.35	3.54[c]
PCP-0310[e]	6.29	6.47[c]

[a] obtained via epoxide equivalents and hydroxy numbers.

[b] pyridene-hydrochloric acid method[1068].

[c] acetylation method[1068].

[d] diethylene glycol started caprolactone ester diols.

[e] trimethylol propane started caprolactone ester triols.

[f] epichlorohydrin-bisphenol-A condensates.

subject to a multitude of interferences from chemically foreign materials
and an inaccurate presumed average molecular structure. Hence, the
presence of foreign species which offsets the accuracy and precision of
the NMR analysis also contributes to the error of the titrimetric
analysis. However, unlike the titrimetric method, one can selectively
choose the areas in the NMR approach to minimize this interference.
Furthermore, the NMR approach has the potential of being able to quantify
partially cured epoxies and polyols.

9.2.10 Anhydride Groups

Van Houwelingen[1069] discussed the determination of anhydride groups
in a resin derived from octadecene-1 and maleic anhydride.

$$
\left[\begin{array}{c}
\text{H} \qquad\qquad \text{H} \quad \text{H} \\
| \qquad\qquad\quad | \quad\ | \\
-\ \text{C}\ -\ \text{CH}_2\ -\ \text{C}\ -\ \text{C}\ - \\
| \qquad\qquad\quad | \quad\ | \\
\text{CH}_3 \qquad\quad\ \text{C}\quad\ \text{C} \\
\qquad\qquad\qquad // \,\backslash\,/\,\backslash \\
\text{O} \qquad \text{O} \qquad \text{O}
\end{array}\right]_n
$$

The acid groups are formed by hydrolysis of the anhydride group.

The common method for anhydride groups involving reaction with an excess of aniline and subsequent backtitration of the excess[1070] is unsuitable, as the reactivity of the anhydride group is low. Even after hydrolysis with aqueous pyridine (containing 40% V/V of water) in a Parr bomb at 150°C for 4 hours, anhydride groups are still seen in the infrared spectrum.

A suitable method for determining the anhydride group is titration with aqueous potassium hydroxide in pyridine after previous esterification of the carboxyl group with diazomethane. This esterification is carried out in diethyl ether methanol (9 + 1) after methylation, which takes about 10m for 0.5 g of sample. The solvents are removed by evaporation and a portion of the derivatised polymer is dissolved in pyridine and titrated.

In the infrared spectra of the resin before and after methylation it can be seen that the absorption band of the acid group at 1710 cm^{-1} disappears and a carbonyl band of the ester at 1740 cm^{-1} is formed. The acid content of the sample is found from the difference in titres of an unmethylated and a methylated product.

9.2.11 Amino Groups

Heterogeneous derivitivisation with 1-fluoro,2,4-dinitrobenzene has been used to determine residual groups in amino phenyl terephthalamide polymers[1069].

Polymerisation

$$NH_2\text{-}\langle\bigcirc\rangle\text{-}NH_2 \;+\; RCOO\text{-}\langle\bigcirc\rangle\text{-}OOCR \;+\; NH_2\text{-}\langle\bigcirc\rangle\text{-}NH_2 \;+\; RCOO\text{-}\langle\bigcirc\rangle\text{-}OOCR \;+\; \text{-----}$$

$$NH_2\text{-}\langle\bigcirc\rangle\text{-}\left[\langle\bigcirc\rangle\text{-}NHCO\right]_n\text{-}\langle\bigcirc\rangle\text{-}NH_2$$

Reaction with 1-fluoro, 2.4-dinitrobenzene

$$NH_2\text{-}\left[\langle\bigcirc\rangle\text{-}NHCO\right]_n\text{-}\langle\bigcirc\rangle\text{-}NH_2 \;+\; F\text{-}\langle\bigcirc\rangle\text{-}NO_2\;(NO_2) \;\longrightarrow$$

$$NO_2\text{-}\langle\bigcirc\rangle(NO_2)\text{-}NH\text{-}\langle\bigcirc\rangle\text{-}\left[\langle\bigcirc\rangle\text{-}NHCO\right]_n\text{-}\langle\bigcirc\rangle\text{-}NH\text{-}\langle\bigcirc\rangle(NO_2)NO_2 \;+\; 2HF$$

The polymer is treated for 4 hours at 80°C with 1-fluoro,2,4-dinitro-benzene in an ethanol-hydrogen chloride medium. After washing out the excess of reagent, the dinitrophenyl group introduced is measured spectrophotometrically after dissolution of the derivatised polymer in methane sulphonic acid. The measurement is carried out at 430 nm against a solution of the same amount of underivatised polymer. Fortunately, the molar absorptivity of unreacted 1-fluoro,2,4-dinitrobenzene at this wavelength is very low, which means that trace amounts of non-washed 1-fluoro,2-4-dinitrobenzene will contribute to only a small extent.

The spectrophotometric method was calibrated with the 1-fluor,2,4-dinitrobenzene derivative of the model compound NN'-bis(p-aminophenyl)terephthalamide:

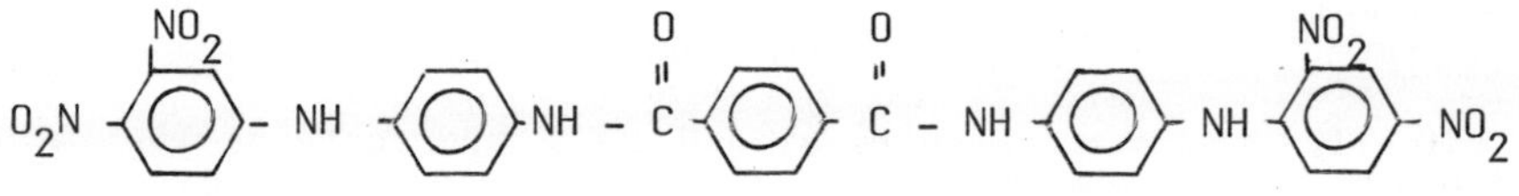

and the correctness of the whole procedure was confirmed by the use of carbon-14-labelled 1-fluoro,2,4-dinitrobenzene and measurement of the incorporated activity.

9.2.12   Amid and Imido Groups in Aromatic Polyamides, Polyimides and Poly(amides-imides)

Schleuter and Siggia[1071-1073] and Frankoski and Siggia[1074] used the technique of alkali fusion reaction gas chromatography for the analysis of imide monomers and aromatic polyimides, polyamides and poly(amide-imides). Samples are hydrolysed with a molten potassium hydroxide reagent at elevated temperatures in a flowing inert atmosphere. Volatile reaction products are concentrated in a cold trap before separation by gas chromatography. The identity of the amine and/or diamine products aids in the characterization of the monomer or polymer; the amount of each compound generated is used as the basis for quantitative analysis. The average per cent relative standard deviation of the method is $\pm$ 1.0%.

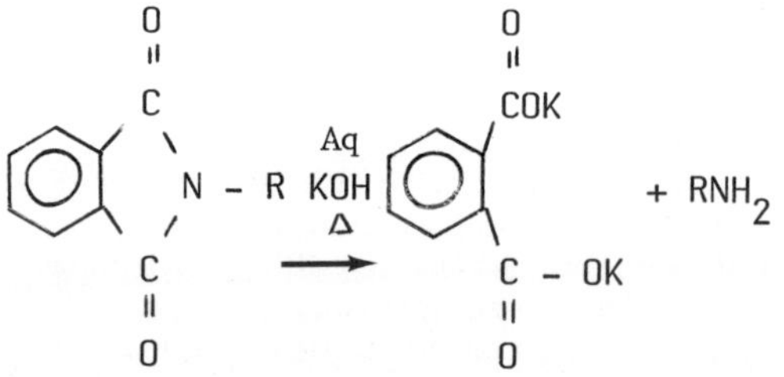

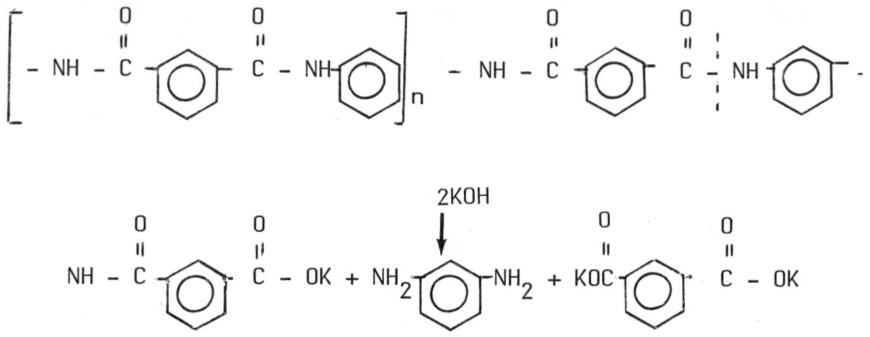

Method - Determination of Amino Groups in Aromatic Polyamides, Polyimides
and Poly(amides-imides).   Potassium Hydroxide Fusion Gas Chromatography[1071]

   Reagents. Fusion reagent: The alkali fusion is a prefused mixture
of potassium hydroxide (J.T. Baker, Analytical reagent grade) and 0.5%
sodium acetate.

   Typically, the mixture melts around 110°C and contains 13-14% water.
When preparing this reagent, it is important to avoid excessive heating.
If too much water is lost, the molten reaction mass may solidify before
the hydrolysis reaction is complete. On the other hand, too much water
will make the reagent sticky and difficult to handle. Additionally, it is
possible to plug the cold trap if large amounts of water are released. To
prevent adsorption of moisture from the atmosphere, the powdered reagent
is stored in a small desiccator within a nitrogen-filled glove bag.

   Apparatus. The total analysis (reaction, trapping, separation and
quantitation) is performed in a single piece of apparatus. The reaction
portion consists of a pyrolysis tube, combustion furnace assembly, and a
mounting frame taken from a commercial unit (Perkin Elmer Pyrolysis
Accessory 154-0825, Norwalk, Conn. 06856). Samples, standards and reagent
are contained in miniature platinum boats (10 mm x 4 mm x 4 mm, Fisher
Scientific Co., Pittsburgh, Pa. 15219) and are magnetically manipulated
within the confines of the glass tubing with metal cylinders and a hoe-
shaped retriever. The combustion furnace assembly monitors and controls
the temperature within the reaction zone.

   A loop-shaped piece of 316 grade stainless steel tubing (0.125 inch
o.d 0.093 inch i.d.) connected between the outlet of the reaction furnace
and a special low dead volume GC injector assembly (H) (Perkin Elmer Part
No. 009-1276) serves as a trap for efficiently concentrating most volatile
reaction products. All of the transfer tubing from the reaction zone to
the injection port, except the lower section of the loop, is heated to
300°C by a clam-shell type combustion furnace clamped around the tubing.
The exposed piece of tubing, which is loosely packed with quartz wool, is
cooled in a liquid nitrogen-filled Dewar flask when concentrating the
evolved products and heated with a 400°C nichrome wire heater when
"injecting" the products into the gas chromatography. Helium carrier gas
flows (60 ml/min) through the reaction unit and cold trap before entering
the chromatographic column.

   Procedure. One to five milligrams ($\pm$ 1 ug) of powdered sample or
standard is weighed into the tared platinum micro-boats. The boats are
then filled with the powdered caustic reagent in the nitrogen-purged glove
bag and quickly loaded into the storage arm of the reaction tube with a
metal cylinder behind each. When making quantitative analysis, the first
sample reacted in a series of runs is used to condition the apparatus.
Boats are usually loaded at the end of a work day, thus permitting
entrapped air to be purged from the system overnight. When samples are
loaded in the morning, the tube is purged for 1 hour before turning on the
thermal conductivity detector current.

   Once a stable baseline is obtained, the liquid nitrogen-filled
Dewar flask (- 196°C) is position around the unheated section of the trap.
The first metal cylinder is magnetically moved forward, pushing the boat
in front of it into the heated reaction zone. The cylinder is removed to
storage area. The optimum reaction temperature profile for a fusion
reaction will depend on the reactivity of the samples. The furnace

temperature was programmed from 100 to 300°C over a period of 10 min. As
the reaction occurs, the volatile reaction products and water, liberated
from the molten mixture, are carried by the flowing helium carrier gas and
concentrated in the trap. At the end of the reaction period, i.e. 10
minutes, the boat is withdrawn from the furnace with the magnetic
retriever and deposited with its cylinder. After adjusting the furnace
control to its initial temperature setting, the trapped compounds are
revolatilized and directed into the gas chromatograph by replacing the
Dewar flask with the heater (400°C). When the separation and integration
are completed and the initial conditions re-established, the procedure is
repeated for each sample and standard. Matched 6 foot by ¼ inch o.d.
stainless steel columns packed with 60/80 mesh Chromosorb 103 (Johns-
Manville) were used to separate the amine reaction products. Diamines
were chromatographed on 4 foot by 1/8 inch o.d. stainless steel columns
packed with 10% FFAP on 60/70 mesh Anakrom AB.

Calibration curves were prepared daily for each compound determined.
Anhydrous ammonia was injected with a calibrated 1000 ul gas tight syringe
through the septum inlet of the reaction tube. Ambient temperature and
pressure corrections were made to determine the actual amount of gas
injected. Benzylamine was generated from reaction of the hydrochloride
salt with potassium hydroxide. The diamine standards were weighed into
boats, covered with caustic reagent, and volatilized by moving the boat
into the heated furnace. In each case, the standard compounds were
trapped and chromatographed in the same manner as the volatile reaction
products. The best straight line calibration curves were determined by a
least-square regression curve fitting computer programme.

Table 113 summarizes the chemical structures and sample designations
of the polymers studied. The weight percent of adsorbed water and the
decomposition temperature, both determined by thermogravimetric analysis,
are also included. The water content was needed to calculate the dry
weight of polymer in each sample. It was not possible to simply dry and
weigh the polymers, because they tended to readsorb moisture from the
atmosphere too rapidly. Table 114 summarizes the diamine recoveries
obtained with isothermal (5 min at 380°C) and programmed (100 to 390°C
over 30 min) reaction temperatures. These numbers represent the mole per
cent of theoretical diamine based on the dry weight of sample and the
idealized linear polymer repeat units depicted in Table 113. Recoveries
were, in all cases between 90 and 101% of theory. The amount of diamine
recovered from a polymer reacted isothermally, was within ± 3% of the
amount produced by the programmed temperature approach. Average percent
relative standard deviations were similar in both cases (± 0.9% vs. 1.1%).

Figure 82(a) shows the reaction gas chromatogram obtained from
poly(amide-imide) PAI-2. Only one diamine was produced. The chromatogram
in Figure 82(b) obtained from the fusion reaction of the polyamide
terpolymer PI-3, contains peaks for each diamine. The quantitative
analysis of both products revealed that the weight ratio of 2,4 toluene
diamine to 4,4'-methuylenedianiline in the polymer was 2:1.

In an attempt to correlate the alkali fusion recoveries (expressed
as weight percent nitrogen) with the elemental nitrogen analyses of the
polymers, it was noted that the elemental nitrogen analyses of the
polymers were higher. Analysis of the fusion residue indicated that no
detectable amounts of nitrogen were present. This led to the discovery
that ammonia was also produced from the samples. When the nitrogen
contribution from ammonia was added to that of the diamine, the total was
in good agreement with the elemental value. Table 115 lists the results

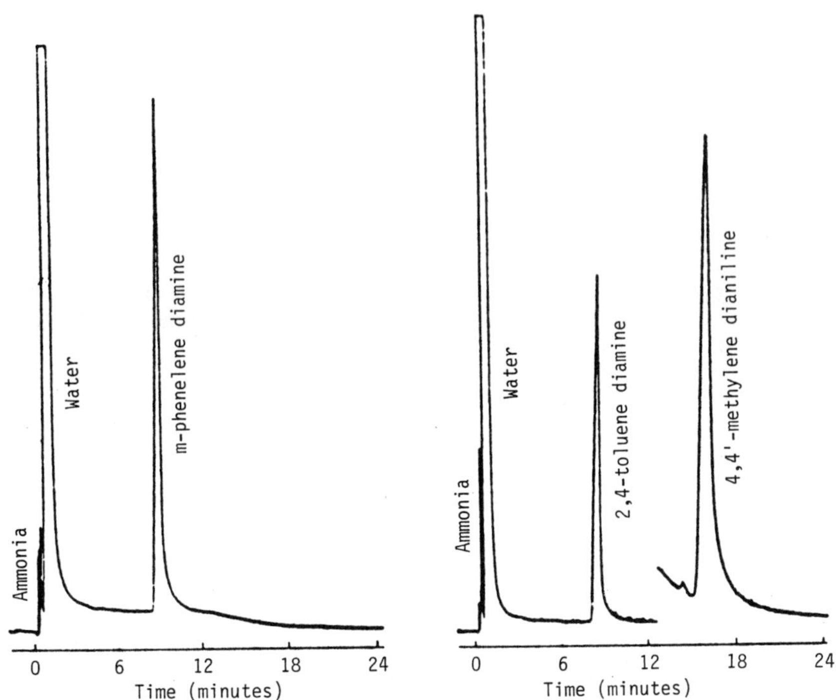

Figure 82   Alkali fusion reaction gas chromatogram produced from (a) poly(amide-imide) PAI-2). The column temperature was programmed from 100 to 250°C at 12°C min⁻¹. (b) polyamide terpolymer PI-3. The column temperature was programmed for 100 to 250°C at 12°C min⁻¹.

obtained from the analysis of mixtures of polymers. The amount of m-phenylenediamine produced was, in all cases, within 1.2% relative of the theoretical value.

Two indirect titration procedures have been described for the determination of amino end-groups in aromatic polyamides. One method involves the reaction of the amine with salicyladehyde to form the Schiff base. After precipitation of the polymer, the excess, unreacted aldehyde is titrated with potassium hydroxide[1075]. Amino groups have also been acetylated with acetic anhydride in dimethylacetamide. Diethylamine was added, and the excess amine was titrated potentiometrically[1076]. The sequence distribution of an aromatic polyamide terpolymer prepared under various reaction conditions was determined by nuclear magnetic resonance spectrometry[1077]. Infrared spectroscopy[1078,1079] and mass spectrometry[1080] have been used to estimate the degree of conversion of polyimides, i.e. the extent of polyamic acid ring closure.

Gomoryova[1081] hydrolysed amide and imide linkages with hydrochloric acid and identified the diamines using paper or thin-layer chromatography. The diamine portion or polyamides has also been determined by fusing the sample with an alkali reagent and separating the products by thin-layer chromatography[1082]. Acidification of the melt allowed the separation and identification of the diacid components. Polyimides have been decomposed with hydrazine hydrate and the diamine products identified by gas chromatography. Polyamides and poly(amide-imides) required a prehydrolysis step. The di, tri- or tetracarboxylic acid segments were determined

Table 113 – Structure, water content and decomposition temperature of the polymers studied

Designation	Structure of repeat unit	Wt% water	Decomposition temperature, °C
PI-1	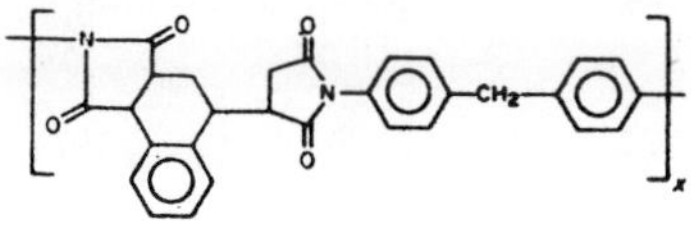	6.5	385
PI-2	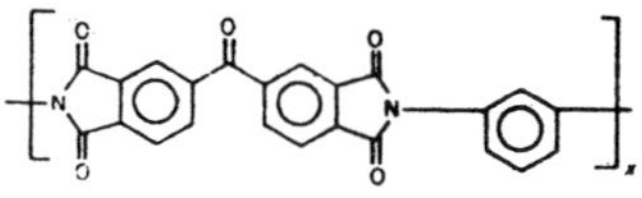	2.2	410
PI-3	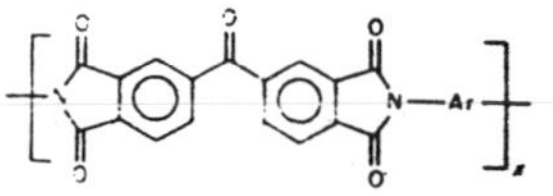 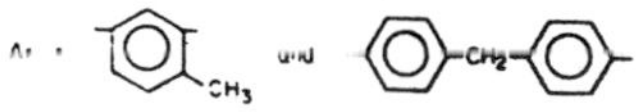	3.4	410
PI-4	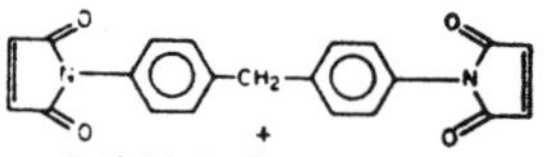 Ar = (with CH₃); and (with CH₂) · 4,4′-Methylenedianiline	0.6	310
PA-1	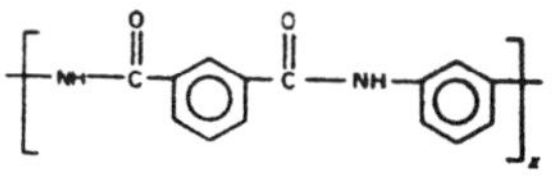	5.8	315
PA-2	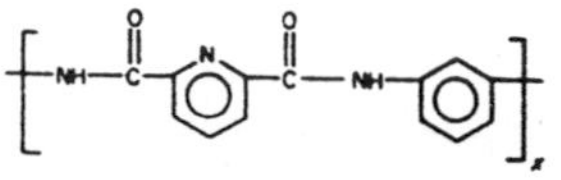	9.7	330
PAI-1	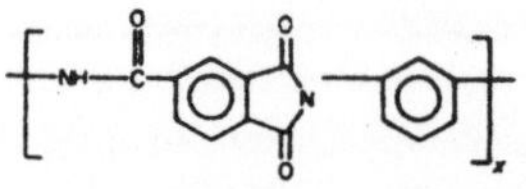	6.2	340
PAI-2	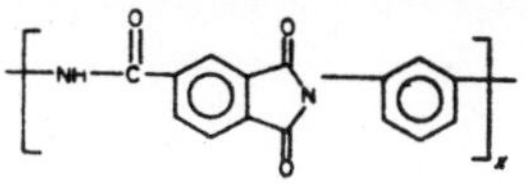	12.6	395
PAI-3	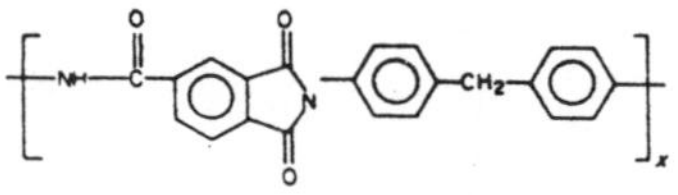	8.9	340

by reaction of the polymer with a 10% aqueous solution of tetramethylammonium hydroxide, pyrolysis of the resulting salt and identification of the volatile methyl ester by gas chromatography[1083]. Kalinina and Doroshina[1084] have reviewed the qualitative and quantitative methods for the analysis of polyamides and polyimides.

Table 114 - Analysis of polyimides, polyamides and poly(amide-imides by alkali fusion reaction gas chromatography

Sample	Diamine produced	Mol % of theoretical[a] $\pm$ RSD[b]	
		5 min at 380°C	30 min from 100 to 390°C
PI-1	4,4'-Methylenedianiline	98.3 $\pm$ 0.8	97.5 $\pm$ 0.9
PI-2	m-Phenylenediamine	89.1 $\pm$ 0.7 91.2 $\pm$ 0.9 91.6 $\pm$ 1.3	91.0 $\pm$ 0.8 91.0 $\pm$ 0.7
PI-3	2,4-Toluenediamine	73.1 $\pm$ 0.2 97.8 $\pm$ 0.5	73.2 $\pm$ 1.0 99.1 $\pm$ 1.2
	4,4'-Methylenedianiline	24.8 $\pm$ 0.3	25.9 $\pm$ 0.2
PA-1	m-Phenylenediamine	100.7 $\pm$ 0.6	97.0 $\pm$ 1.7 97.7 $\pm$ 1.7 97.9 $\pm$ 1.0 99.6 $\pm$ 2.6
PA-2	m-Phenylenediamine	95.9 $\pm$ 0.6 96.4 $\pm$ 1.5	92.9 $\pm$ 1.0 93.3 $\pm$ 0.6
PAI-1	m-Phenylenediamine	93.5 $\pm$ 0.9	93.1 $\pm$ 0.7 95.3 $\pm$ 0.5
PAI-2	m-Phenylenediamine	93.5 $\pm$ 1.1 94.1 $\pm$ 0.6 95.1 $\pm$ 1.2	97.4 $\pm$ 0.9
PAI-3	4,4'-Methylenedianiline	98.0 $\pm$ 1.1	98.8 $\pm$ 1.1

[a] These recovery values are based on the structures shown in Table 113 , which assume that one mole of diamine is produced for each mole of repeat unit. [b] The relative standard deviation is based on five or more determinations.

Table 115 - Analysis of polymer mixtures by alkali fusion
reaction gas chromatography

Polymer mixture	Micromoles of m-phenylene-diamine		Recovery, %
	Taken[a]	Found	
PI-1	2.24		
		11.50	11.36	98.8
PAI-2	9.26		
PA-1	7.23		
		14.66	14.71	100.3
PA-2	7.43		
PAI-1	4.99		
		12.66	12.58	99.4
PI-2	7.67		
PAI-1	5.28		
		12.37	12.48	100.9
PAI-2	7.09		
PA-1	5.51		
PAI-2	5.81	14.21	14.33	100.8
PI-2	2.89		
PA-2	8.79		
PAI-1	3.25	14.80	14.98	101.2
PA-1	2.76		

[a] These values were calculated from the weight of polymer taken and were
corrected for the weight of adsorbed water and the average
recovery (Table 114).

[b] For the key to type of polymer, see Table 113.

9.2.13 Nitrile Groups

An infrared method has been described[1088] for the compositional
analysis of styrene-acrylonitrile copolymers. In this method relative
absorbance between a nitrile $v(CN)$ mode at 4.4 microns and a phenyl $v(CC)$
modes at 6.2 microns is used.

A near-infrared method using combination and overtone bands has been
used[1086] for carrying out the same analyses. Near-infrared spectra of
random copolymers and homopolymers are shown in Figure 83. The bands
which occur in the region 1.6-2.2 microns result from overtones and
combination tones which occur, respectively, in the regions 1.6-1.8
microns and 1.9-2.2 microns.

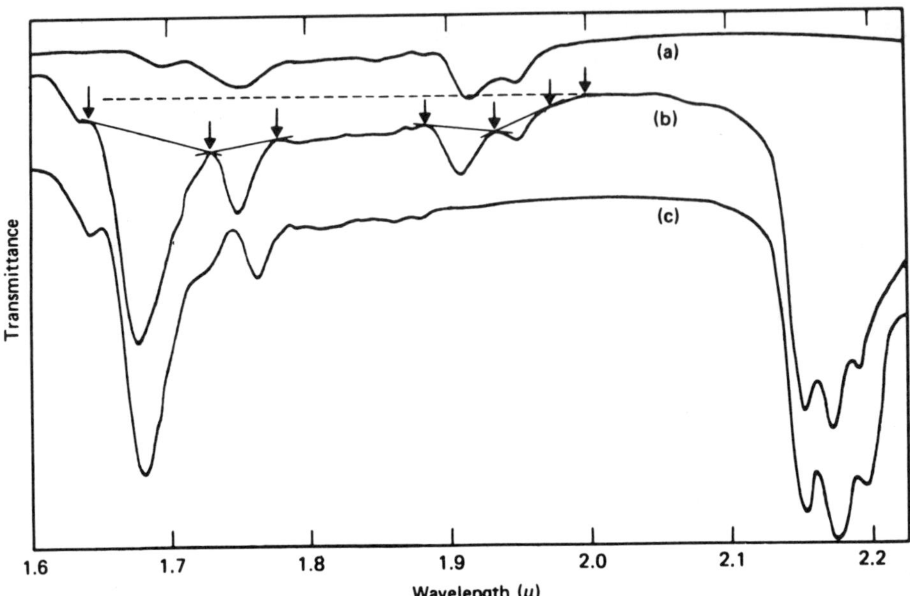

Figure 83   Near infrared spectra: (a) polyacrylonitrile; (b)
             styrene-acrylonitrile copolymer (acrylonitrile content
             25.7 wt.%); (c) polystyrene.

The band near 1.68 microns is assigned as an overtone of phenyl
$v(CH)$ mode near 3.33 microns and its absorbance is directly proportional
to the styrene content of the copolymer and the film thickness. The band
near 1.75 microns is assigned as an overtone of the aliphatic $v(CH)$ mode
near 3.45 microns and both acrylonitrile and styrene absorb at this
wavelength Bands at 1.910 and 1.952 microns and are combination tones of
acrylonitrile and are assigned as respectively, $v(CN) = v_{asym}(CH_3$(2237 +
2940 = 5177 cm^{-1} = 1.932 microns) and $v(CN) + v_{asym}(CH_3)$2237 + 2870 = 5107
cm^{-1} = 1.958 microns). For calculation of the absorbance of these four
characteristic bands, two different baseline methods were used, as shown
in Figure 83. The method using the extrapolated baseline from 2
introduces less deviation than the other baseline method. The absorbance
ratio $A_{1.675}/A_{1.910}$ proved to be the best one for analytical measurements.

9.3 COMONOMER CONTENTS AND RATIOS

Functional group analysis and the determination of monomer unit
concentrations can be used to determine the monomer composition and hence
the empirical formula of a copolymer or terpolymer. Thus, if a styrene-
butadiene copolymer contains 6% (6/56 moles per 100g) bound butadiene and
94% (94/104 moles/100g) bound styrene, then the empirical formula is:

$$Sty \ \frac{56 \times 94}{6 \times 104} \ Bu = Sty \ 8.44 \ Bu$$

i.e. there are 8.44 moles of styrene per mole of butadiene in the
copolymer.

Some examples are discussed below of the application of these
principles to copolymer and terpolymer analysis.

9.3.1 Acrylonitrile - Butadiene - Styrene Terpolymers

Acrylonitrile can be determined via a determination of organic nitrogen by the Kjeldahl procedure (Chapter 4.3.4). Styrene units can be determined by infrared spectroscopy. Butadiene units can be determined by the iodine monochloride procedure (Chapter 9.2.1). Some typical results obtained on the chlorobenzene soluble and insoluble fractions of an ABS terpolymer are presented below:

	Analysis of original polymer	Analysis of chlorobenzene soluble fraction (56%) of original polymer	Analysis of isolated-chlorobenzene insoluble gel (44%) of original polymer
Butadiene (B)	20.5	30.8	10.3
Acrylonitrile (A)	54.0	48.0	62.0
Styrene (S)	54.0	48.0	62.0
Total	98.3	100.9	99.9
Empirical formula	$AB_{0.813}S_{1.555}$	$AB_{1.320}S_{1.108}$	$AB_{0.353}S_{1.146}$

Oxidation with osmium tetroxide has been used as the basis for a method of determining bound butadiene in acrylonitrile - butadiene - styrene terpolymers[1087].

9.3.2 Natural Rubber. Styrene-Butadiene Rubber and Ethylene-Propylene-Diene Terpolymer in Cured Stocks

Krishen[1088] has described a procedure for the determination of these monomer units. He qualitatively analysed the gaseous pyrolysis products from natural rubber, styrene-butadiene rubber and ethylene-propylene diene terpolymer rubber by gas chromatography. He showed that the 2-methyl-2-butene peak was linear with the natural rubber content of the sample. Styrene-butadiene rubber was determined from the peak area of the 1,3-butadiene peak. The ethylene-propylene-terpolymer content was deduced from the 1-pentane peak area of the pyrolysis products.

Method - Determination of Natural Rubber, Styrene-Butadiene Rubber and Ethylene Propylene Diene Terpolymer in Compounded Rubber Stocks. Pyrolysis - Gas Chromatography

Pyrolysis Unit. Loenco Division Infotronics Corp., Altadena, Calif. this Unit, Model 260, uses a 3" x 1/8" quartz U-tube as the pyrolysis chamber. It is heated by sliding a thermostatically controlled heated furnace over it for a fixed time. A sliding gas valve arrangement in the unit allows uninterrupted carrier gas flow through the gas chromatographic column while the sample chamber is being reloaded or purged with helium.

<u>Stripping and backflushing valve.</u> A Carle Micro Volume Eight Port Gas Chromatography Valve No. 2012 (Carle Instruments, Fullerton, Calif.) is used to avoid the introduction of high boiling pyrolysis products onto the chromatographic column. It is connected to the columns and the gas chromatograph. The pyrolysis products are allowed to go onto the main column for the first 15 minutes only.

<u>Gas Chromatograph.</u> The pyrolysis products are analysed with an F & M 1609 gas chromatograph equipped with a flame ionization detector. The chromatographic peaks are recorded at electrometer range of 100 and a chart speed of $\frac{1}{2}$ inch per minute. The area measurements are made by the triangulation method (peak height x width at half peak height). The main column was aluminium tubing 40 ft x 3/16 inch o.d. 0.032 in wall thickness. It is packed with 10% tricresylphosphate on 60-80 mesh Chromosorb P, the stripper and restrictor columns are 10 ft lengths of the 2/16 inch aluminium tubing packed as the main column. The columns are operated at 35°C. The injection port and the detector block are maintained at 175°C. The helium, hydrogen and air pressure are 30 psi (flowrator at 4.0), 15 psi (flowrator at 7.50) and 20 psi (flowrator 9.0) respectively. Relative retention data for the pyrolysis products are shown in Table 116.

<u>Reagents.</u> Tricresylphosphate stationary phase. Chromosorb P, 60-80 mesh.

<u>Procedure.</u> The cured rubber samples are subdivided into small pieces. A 0.1 to 1 gram sample of these pieces is extracted with methanol in Underwriters' extraction apparatus for 8-12 hours. The solvent is completely removed from the rubber by placing the sample in a vacuum oven at 60°C for 30 minutes. A 500 to 800 ug sample of the dried rubber is weighed on the microbalance and carefully placed inside the quartz pyrolysis tube. Small borosilicate glass wool plugs are put on each end of the U-tube. This tube is then installed in the pyrolyser unit, purged with a helium flow of 30 ml per minute for 2 minutes, and then heated for 30 seconds at 700°C by positioning the furnace over the tube. At the end of 30 seconds, the furnace is moved away; the pyrolysis tube is connected in series with the column carrier flow using the appropriate valve on the pyrolysis unit; and the start of the chromatogram marked on the recorder chart. At the end of about 60 seconds, the carrier gas flow through the pyrolysis tube is terminated by sliding up the valve on the pyrolysis unit. The tube is removed from the pyrolysis unit, glass wool plugs are removed and the tube cleaned by heating it thoroughly in a burner flame to burn of all combustible material. Another weighed sample is then placed in the tube and it is installed in the pyrolysis unit as before ready for the next run. After 15 minutes, the flow of the carrier gas through the stripper column is terminated and the restrictor column placed in series with the main column by turning the Carle valve knob to the counter clockwise position. The stripper is backflushed while the chromatogram is being recorded. At the end of the chromatographic run, the Carle valve knob is turned back to the clockwise position and the apparatus is ready to accept the next run.

9.3.3 Ethylene and Propylene Units in Ethylenepropylene Copolymers

<u>Infrared spectroscopy.</u> Several methods have been reported for determining propylene in ethylene-propylene copolymers. The basis for calibration of many of these methods is the work published by Natta[1089] which involves measuring the infrared absorption of the polymer at 7.25 micron presumably due to methyl vibrations, is related to the propylene

Table 116 - Peak indentification and relative retention data.
(tricresylphosphate column at 35°C

Peak no. in figure	Compound	Relative retention nonane = 1.0000
1	Methane	0.0009
2	Ethane + Ethane	0.0082
3	Propane	0.0268
4	Propene	0.0347
5	2-Methylpropane	0.0849
6	Propadiene + Butane	0.0948
7	1-Butene + 2-Methylpropene	0.1048
8	trans-2-Butene	0.1409
9	cis-2-Butene	0.1649
10	1,3-Butadiene	0.1769
11	3-Methyl-1-butene	0.2074
12	1-Pentene	0.3038
13	2-Methyl-1-butene	0.2074
14	trans-2-Pentene	0.3848
15	cis-2-Pentene	0.3993
16	2-Methyl-2-butene	0.4476
17	Isoprene	0.5397

Table 117 - Calibration points for the composition analysis
of C_2-C_3 copolymers (spectra recorded at 160°C)

Sample	C_3 (wt-%)[a]	$^7.25/^6.85$
3137-25	40.5	0.652 ± 0.011
3629-48	32.9	0.561 ± 0.053
3629-44	32.0	0.548 ± 0.032
3137-38	30.7	0.606 ± 0.024[c]
3137-31	20.7	0.416 ± 0.018
3297-55	18.2	0.365 ± 0.037
3297-59	18.0	0.357 ± 0.015
3274-53	14.2	0.318 ± 0.015
Blend No. 1	11.0	0.315 ± 0.022
Blend No. 2	6.0	0.204 ± 0.013
Polyethylene	6.0	0.036 ± 0.007

[a] Radiochemical analysis

[b] The standard deviation has been multipled by the student's $t_{95\%}$ coefficient to the number of replications (on the average 5) for each calibration point.

concentration in the copolymer. In some cases the dissolution of copolymers with low propylene content or some particular structures is difficult[1090,1091]. Moreover, Natta's solution method was calibrated against his radiochemical method[1089] for which the precision of the method was not stated; a considerable amount of scatter is evident in the data presented. Typical methods that have used Natta's solution procedure[1089], for calibration are described in publications by Wei[1092] and Gosl[1093]. These infrared methods avoid the solution problems by employing intensity measurements made on pressed films. The ratio of the absorption at 13.95 micron to that at 8.70micron is related to the propylene content of the copolymer. Some objections[1094] to the use of solid films have been raised because of the effect of crystallinity on the absorption spectra in copolymers with low propylene content. These film methods are reliable only over the range of 30 to 500 mole % propylene.

Tosi and Simonazzi[1095] have described an infrared method for the evaluation of the propylene content of ethylene rich ethylene-propylene copolymers. This is based on the ratio between absorbances of the 7.25 micron band and the product of the absorbances by the half width 6.85 micron obtained on diecast polymer film at 160°C.

The calibration curve, based on a series of standard copolymers prepared with 14-C-labelled either ethylene or propylene is obtained by plotting the A 7.25:A 6.85 ratio against the C_3 weight fraction. To improve the precision, Tosi and Simonazzi[1095] ratioed A 7.25 to the product of the absorbance of the 6.85 micron band by its half-width. For the sake of simplicity, the latter quantity was expressed in wavelengths instead of wave numbers. $\Delta \frac{1}{2}A$ 6.85 is approximately 0.10 micron at room temperature and 0.14 at 160°C.

Table 117 lists the calibration data of Tosi and Simonazzi[1095]. Tosi and Simonazzi[1095] compared the upper curve of Figure 84 to calibration curves obtained by other workers for C_2 - C_3 copolymers based on the A 7.25/A 6.85 ratio. This comparison when the C_3 content is not too low should hold irrespective of the recording temperature: in fact they found only minor differences in the A 7.25/ A 6.85 ratio at room temperature and at 160°C.

Ciampelli et al[1096] have developed two methods based on infrared spectroscopy of carbon tetrachloride solutions of the polymer at 7.25 micron, 8.65 micron and 2.32 micron for the analysis of the ethylene-propylene copolymers containing greater than 30% propylene. One method can be applied to copolymers soluble in solvents for infrared analysis; the other can be applied to solvent insoluble polymer films. The absorption band at 7.25 micron due to methyl groups is used in the former case, whereas the ratio of the band at 8.6 micron to the band at 2.32 micron is used in the latter. Infrared spectra of polymers containing between 55.5 and 85.5% ethylene are shown in Figure 85.

This approach has also been discussed by Bucci and Simonazzi[1097]. This method is based on measurement of absorption of solutions of the polymer at 1.691 microns^{-1} (CH 3 groups) 1.764 micron^{-1} (CH 2 groups) and is applicable to polymers containing as high as 52 mole % propylene and as low as 15% mole % propylene.

Lamonte and Tirpak[1098] have developed a method for the determination of per cent ethylene incorporation in ethylene-propylene block copolymers. Standardisation is done from mixtures of the homopolymers. Both standards and samples are scanned at 180°C in a spring-loaded demountable cell. The

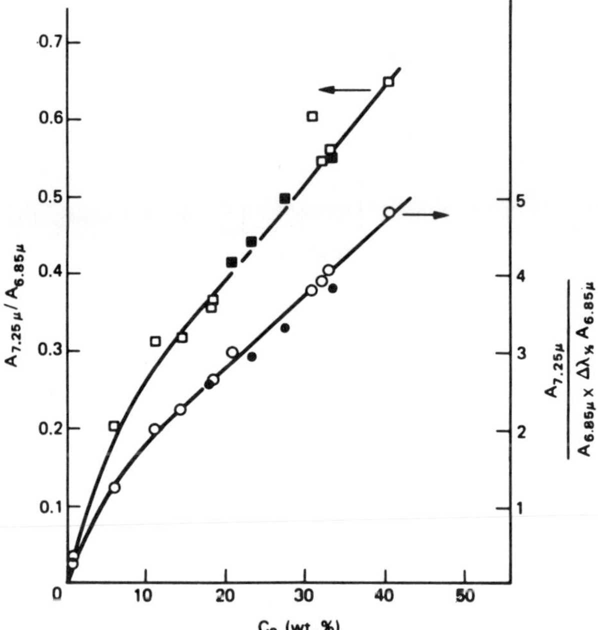

Figure 84   Calibration curves for the determination of C_3 in high
                C_2 copolymers. Full points refer to model hydrogenated
                poly-3 methyl alkenamers.

standardization was confirmed by the analysis of copolymers of know
ethylene content prepared with C 14-labelled ethylene. By comparison of
the infrared results from the analyses performed at 180°C and also at room
temperature, the presence of ethylene homopolymer can be detected. These
workers derived an equation for the quantitative estimation of per cent
ethylene present as copolymer blocks.

The method distinguishes between true copolymers and physical
mixtures of copolymers. This method makes use of a characteristic
infrared rocking vibration due to sequences of consecutive methylene
groups. Such sequences are found in polyethylene and in the segments of
ethylene blocks in ethylene-propylene copolymers. This makes it possible
to detect them at 13.70 micron and 13.89 micron. There are bands at both
these locations in the infrared spectrum of the crystalline phase but only
at 13.89 micron in the amorphous phase. The ratio of these two bands in
the infrared spectrum of a polymer film at room temperature is a rough
measure of crystallinity. As seen by this ratio, the infrared spectra of
the copolymers show varying degrees of polyethylene type crystallinity,
dependent on ethylene concentration and method of incorporation. It is
this varying degree of crystallinity which allows the qualitative
detection of ethylene homopolymer in these materials. A calibration curve
of absorbance at 13.89 micron versus ethylene was made from known mixtures
for both hot and cold runs. Both plots resulted in straight lines from
which the following equations were calculated.

$$\% \text{ Ethylene at } 180°C = A/(0.55\ b)$$

where A = absorbance measured at 13.89 micron an b = thickness of wire
spacer in centimetres and

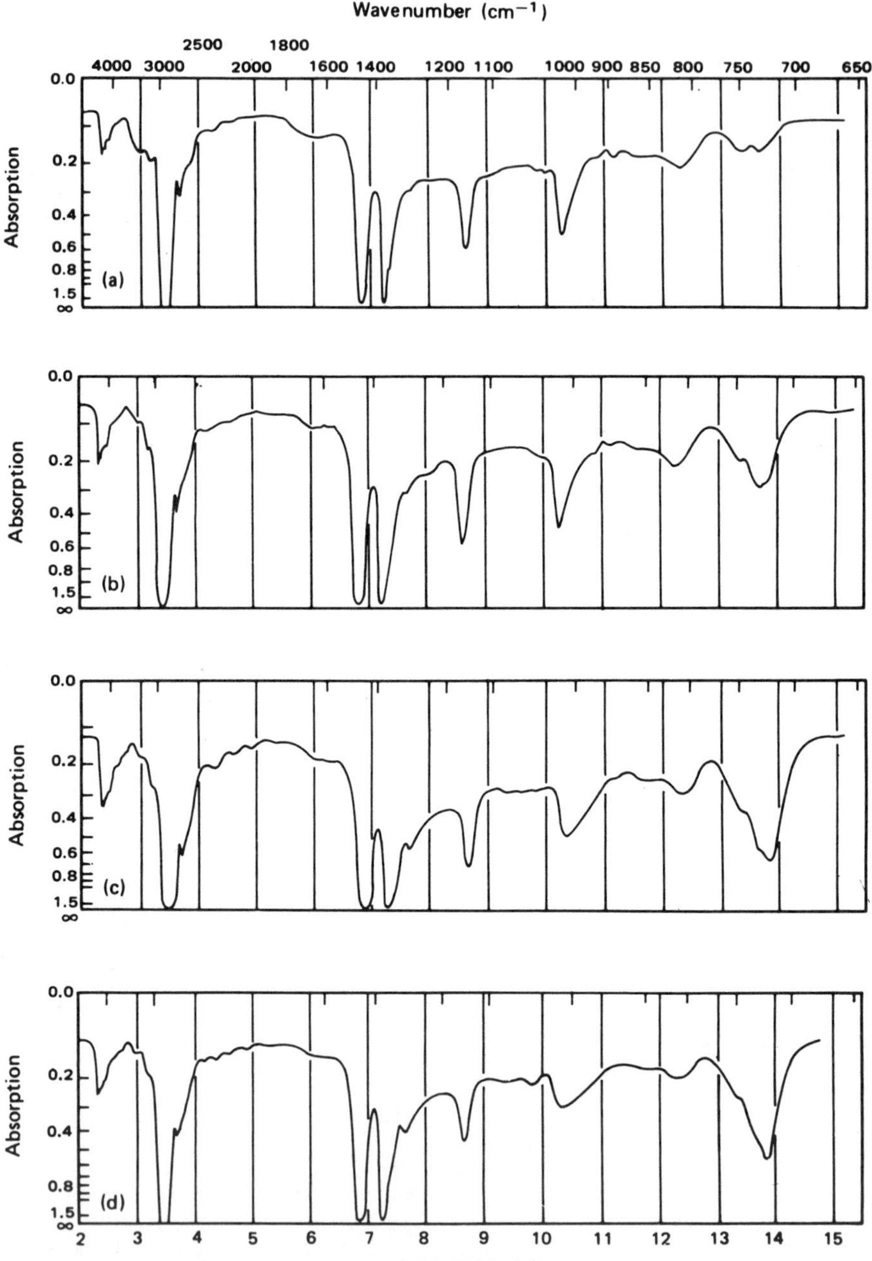

Figure 85   Infrared spectra of ethylene-propylene copolymers of
             various compositions: (A) 85.5% polyethylene; (B) 74.0%
             polyethylene; (C) 65.9% polyethylene; (D) 55.5%
             polyethylene.

% Ethylene at room temperature = A/(3.0 b)

A series of ethylene propylene block copolymers prepared with C 14 labelled ethylene was analysed for percent ethylene incorporation by radiochemical methods. These samples when scanned at 180°C gave results which checked with the radiochemical assay quite well. However, when the cooled samples were scanned, the results from the cold calibration were low in comparison with the known ethylene content. These data are shown in Table 118.

A pair of samples were prepared in which the active sites on the growing propylene polymer were eliminated by hydrogen before addition of ethylene. Practically identical values for percent ethylene incorporation were calculated for both hot and the cold scans. These data are shown in Table 119.

Drunstel[1099] and Tosi and Ciampelli[110] have discussed the application of infrared spectroscopy to the analysis of ethylene-propylene copolymers containing predominantly (⟩7.95%) propylene.

Brown et al.[101] showed that pyrolysis of ethylene propylene copolymers at 450°C produces derivatives that are rich in unsaturated vinyl and vinylide groups, similar to the pyrolysis of natural rubber and styrene-butadiene rubber mixtures[1102], which produces vinyl groups derived from the butadiene part of the molecule and the vinylidene groups from the methyl branch of the isoprene units. This unsaturation exhibits strong absorption of the infrared region. The ratio for the absorption of the vinyl groups to that of the vinylidene groups varies with the mole fraction of propylene in saturated ethylene-propylene copolymers. Making use of this ratio, they developed an analytical method for determining propylene in both raw and vulcanized ethylene-propylene copolymers. The vinyl group absorbs at about 11.00 micron and the vinylidene at about 12.25 micron.

Figure 86 shows some typical spectra obtained on the pyrolyzates in the region of $10.50-11.76^{-1}$ micron. The values of the ratio, R (X 100) range from 9.977 to 0.0290 respectively, for 0 to 100 mole% propylene for the raw samples and from 5.400 to 0.0431 respectively for 10 to 100 mole % propylene for the vulcanized samples. The common logarithm of the ratio, R, can be represented by a liner function of the mole % of propylene in the copolymer.

Table 120 lists the results expressed as common logarithms of 100R for unvulcanized samples.

Raman and infrared spectra of ethylene-propylene copolymers indicated that overall amounts of propylene altered crystallinity in blocks of ethylene[1103]. Tosi[1104] shows that experiment agreed with theory in ethylene-propylene copolymer composition and intensity of CH_2 rocking vibrations of methylene sequences. Popov and Duvanov[1105] used a comparable approach to analysis of ethylene-propylene block copolymers.

<u>Nuclear magnetic resonance spectroscopy.</u> Carbon 13 NMR has proved to be an excellent technique for analysis of sequence distributions and comonomer contents in ethylene-propylene copolymer[1106-1114,1116,1117]. These analyses are particularly straightforward if one of the monomer units is present at a level of 94% or greater because the other monomer will then occur primarily as an isolated unit.

Table 118 - C^{14} Labelled ethylene-propylene copolymers ethylene, %

Sample	Radiochemistry	Hot infrared scan	Cold infrared scan
3401	2.4	2.9	0.9
3402	4.0	3.65	1.3
3403A	22.4	20.7	14.7
3403B	24.5	22.2	15.3
3404	12.4	13.0	7.1
3405	14.0	14.1	7.5

Table 119 - Samples with hydrogen-reduced active sites

	Ethylene %	
Sample No	Hot infrared scan	Cold infrared scan
1487	5.0	5.3
1553	7.1	6.9

Table 120 - Log_{10} (100 R) for polymer pyrolyzates (raw samples)
(R = % mole propylene)

Log_{10} (100R) at various propylene concentrations, mole %

R	0	10	20	31	40	50	100
Sample No:							
1	2.827	2.584	2.309	2.041	1.931	1.620	0.695
2	2.840	2.525	2.309	2.048	1.896	1.614	0.743
3	2.946	2.604	2.318	2.060	1.940	1.592	0.596
4	2.999	2.587	2.376	2.047	1.908	1.596	0.580
5	2.996	2.552	2.238	2.097	1.886	1.589	0.542
6	2.954	2.568	2.327	2.055	1.896	1.588	0.432
7	2.989	2.562	2.315	2.063	1.902	1.589	0.542
8	2.951	2.578	2.301	2.048	1.916	1.582	0.461
9	2.897	2.567	2.340	2.053	1.933	1.620	0.591
10	2.964	2.561	2.362	2.068	1.904	1.588	0.658
Average	2.9365	2.5687	3.3193	2.0579	1.9114	1.5979	0.5840

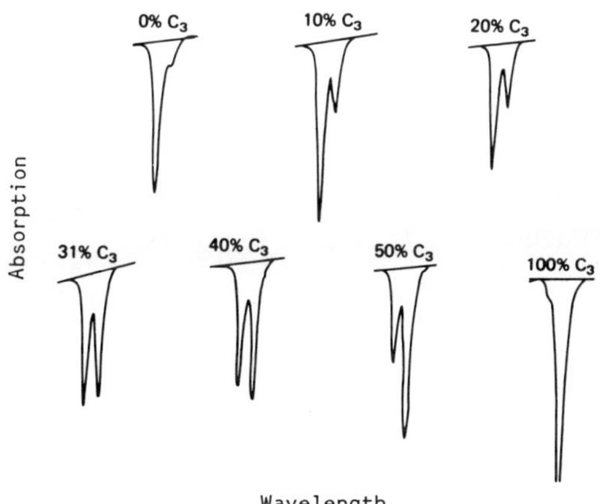

Figure 86   Typical infrared spectra of pyrolysates obtained from
             raw ethylene-propylene copolymers. The position of the
             absorption peaks from left to right are near 909 cm⁻¹
             and 889 cm⁻¹ respectively for each pair and the nominal
             compositions are indicated in mole % propylene.

Paxton and Randall[1115] used carbon-13 NMR and infrared spectroscopy
to measure the concentration of ethylene in ethylene propylene copolymers.
These polymers contained greater than 95% propylene, with the ethylene
units present as isolated entities between two head-to-tail propylene
units. These workers point out that most infrared bands used for
determining copolymer compositions are sensitive to sequences of both
monomers. This infrared method for compositional analysis can be
calibrated if: (a) known standards of similar constitution to the
copolymers being analysed are available add (b) assignments and behaviour
of the calibration bands are well established; preferably the
absorptivities of these bands should be relatively independent of the
position of monomer units in the chain. Thus, quantitative infrared
analysis of copolymers depends primarily on the standards employed whose
composition can be determined directly and reliably. Paxton and
Randall[1115] used 13 C-NMR to provide such reference standards for the less
time-consuming infrared measurements. Because it is relatively
inexpensive and easy to operate for copolymer analysis they showed that an
excellent correlation is obtained between 13 C NMR and infrared results on
a series of ethylene-propylene copolymers containing greater then 95% wt%
propylene.

<u>Method - Determination of Ethylene in Ethylene-Propylene Copolymers (≥95%
Propylene) - NMR Spectroscopy[1115]</u>

NMR spectra are measure from polymer samples dissolved in a mixture
of 1,2,4-trichlorobenzene and perdeuterobenzene at concentrations between
10-15 wt%. Sufficient perdeuterobenzene is added to maintain a lock
signal on a Varian XL0100 15 NMR spectrometer at a temperature of 125°C.

The spectrometer system is equipped with Varian's FT-100 pulsed NMR-
Fourier transform system and a disk accessory. The 13-C NMR spectra is
obtained utilizing the following conditions for the pulse sequences:

A pulse interval of 10 seconds is sufficient to satisfy the spin-lattice relaxation time, T_1 p of the methyl group, which is the slowest relaxing nucleus with a T_1 of 2 seconds at 125°C. The number of transients accumulated depended upon the ethylene content and in each case, accumulations are allowed to continue until a satisfactory signal-to-noise ratio was achieved. As shown by a typical example in Figure 188.10 (Appendix 1) each 13 C NMR spectrum is recorded with proton noise-decoupling to remove unwanted 13 C - 1 H scalar couplings. No corrections are made for differential nuclear Overhauser effects since these have been shown to be constant for the major resonances in low ethylene-propylene copolymers[118].

The best results are obtained using methane resonances 4 and 5 (Table 121) to determine copolymer composition. This is because the methane carbon resonance is the least sensitive towards configurational differences and also the least affected by overlap from neighbouring resonances. Similar results are obtained whether one uses peak heights or peak areas.

<u>Method - Determination of the Bound Ethylene in Ethylene-Propylene Copolymers ≥95% Propylene - Infrared Spectroscopy</u>

Infrared spectra are obtained on moulded films of a nominal 1 mm thickness. Approximately 0.52 g of sample is placed between 31.8 mm diameter aluminium foil disks on a 1 mm brass spacer of 19.0 mm i.d. and 31.8 mm o.d. resting in a Buehler 20-2112 specimen mould. The temperature of the assembled mould is maintained at 175°C by a standard thermocouple temperature controller. After the sample is pressed to 6000 psig in buehler 1315 AB specimen mould and cooled to 35°C under pressure, the film is removed and mounted in a holder. Five measurements with a micrometer are averaged to provide a film thickness for the ethylene calculations. Since small variations in thickness introduce substantial errors in final calculations, the film is discarded if the readings varied by more than 0.05 mm and another sample is moulded.

Infrared spectra are obtained on a Digilab Model 15 Fourier Transform system at four wavenumber resolution in double beam operation. An adequate signal-to-noise ratio resulted from 1000 scans. Standard double precision Digilab/Data General computer software are used to present data properly scale expanded in absorbance form. The absorbance of the 13.66 micron infrared band, attributed to γ_r $(CH_2)_3$ is recorded. This band is characteristic of an ethylene unit isolated between two head to tail propylene units. The equation relating the infrared absorbance to the ethylene content is:

$$Y = 2.465\ X + 0.451$$

where Y is the infrared absorbance at 732 wavenumbers divided by the film thickness in centimetres and X is the wt% ethylene. The standard errors are 0.110 and the intercept and 0.051 for the slope.

The infrared absorbance at 13.66 micron is sufficiently sensitive to the ethylene incorporation to determine the wt% ethylene within 0.1-0.2% at the 95% confidence level. From the 13-C NMR data, it can be concluded that the propylene units occur in predominantly isotactic, head to tail sequences and that the ethylene units are incorporated as isolated units only. Thus, this structural prequisite is a requirement for application of this method because it has not been tested on copolymers containing propylene configurational irregularities or ethylene sequences two units and longer.

Table 121 - Observed and reference ^{13}C NMR chemical shifts in ppm
for ethylene-propylene copolymers and reference polypropylenes as
measured with respect to an internal TMS standard

Resonance line	carbon	3/97 E/P[a]	E/P	97/3[a] E/P	sequence assignment	reference crystalline	PP amorphous
1	alpha alphaCH$_2$	46.4	46.3		PPPP	46.5	47.0-47.5 r 46.5 m
2	alpha alphaCH$_2$	46.0	45.8		PPE		
3	alpha gammaCH$_2$	37.8	37.0		PPEP		
4	CH	30.9	30.7		PPE		
5	CH	28.8	28.7		PPP	28.5	28.8 mmmm 28.6 mmmr 28.5 rmmr 28.4 mr+rr
6	beta betaCH$_2$	24.5	24.4		PPEPP		
7	CH$_3$	21.8	21.6		PPPPP	21.8	21.3-21.8 mm 20.6-21.0 mr 19.9-20.3 rr
8	CH$_3$	21.6	21.4		PPPE		
9	CH$_3$	20.9	20.7		PPPEP		
	CH$_3$		19.8	19.8	EPE		
	CH		33.1	33.1	EPE		
	alpha betaCH$_2$			37.4	EPE		
	alpha betaCH$_2$		EPE	27.3			
	alpha beta(CH$_2$)$_n$0		29.8	29.8	EEE		

[a] As measured by Paxton and Randall[1115]

pulse angle	90°	transients accumulated	20000-5000
pulse delay	8.5 s	spectral width	50000 Hz
aquisition time	1.5 s	no signal enhancement used	

Pyrolysis - gas - chromatography. The consensus of opinion is that, certainly in the cases of those polymers such as the polyolefins where complex pyrograms are produced, filament pyrolysis is the preferred method. For the purposes of fundamental studies, pyrolysis at a variety of temperatures and heating rates is preferred. Smaller sample weights of the order of 1-2 mg are preferred as these prevent or reduce the occurrence of secondary side reactions in the pyrolysis which might confuse the interpretation of the pyrogram when carrying out polymer structural studies. Certainly, sample weights above 3 mg should be avoided. The use of small sample sizes necessitates the use of the more sensitive types of gas chromatograph detectors such as flame ionization. In particular circumstances where the occurrence of a microstructural feature is being studied, failure to use a sensitive enough detector could result in the pyrolysis product being missed. In such studies, to improve sensitivity, a limited increase in sample weight, say from 1-2 mg to 4 might be permitted.

In-line hydrogenation is a useful innovation for simplifying the pyrogram obtained for those polymers which produce complicated mixtures, but should be used with caution in fundamental studies. More information might be obtained by carrying out studies with and without in-line hydrogenation. Finally, the type of gas chromatography separation column used should be the subject of close scrutiny. The information gained in pyrolysis studies is only as good as the degree and type of separation achieved on the column and, certainly in the early stages of investigation work, a variety of columns should be studied.

Quantitative measurements of the amounts of various pyrolysis products can in many instances be correlated with the percentage composition of a copolymer, or with the concentration or a particular microconstituent in the polymer. Thus, Van Schooten and Evenhuis[1119,1120] applied their pyrolysis-hydrogenation-gas chromatography technique to the quantitative determination of copolymer composition of ethylene-propylene copolymers, an analysis which presents difficulties in solvent solution-infrared methods, especially with samples that are only partly soluble in suitable solvents such as carbon tetrachloride. Since the hydrogenation pyrogram of polyethylene consists almost exclusively of normal alkanes and that of polypropylene isoalkanes, the ratio of the peak heights of a n-alkane to an iso-alkane is a good measure of the copolymer composition. The ratio $n-C_7(2-methyl\ C_7 + 4-methyl\ C_7)$ was good measure of ethylene-propylene ratio in copolymers.

9.3.4 Butene-1 Units in Ethylene-Butene Copolymers

In addition to polyethylene and polypropylene, a wide range of olefin comonomers are produced which consist of copolymers of C_2 to C_8 olefins. A case in point is a copolymer of ethylene and butene-1 containing up to 10% butene-1.

Numerous methods have been applied to the analysis of 1-butene/ - olefin copolymers, including pyrolysis gas chromatography[1121-1123,1132], differential thermal analysis[1124] x-ray crystallography[1125] melting-point fractionation[1126] and NMR spectroscopy[1132]. The most widely adopted technique is infrared spectrometry[1125-1129,1132] either using a heated cell to eliminate the effects of crystallinity, or more simply be scanning a film of known thickness and comparing the absorbance to those of standards. Infrared is the easiest method to run and the least demanding in equipment. It requires, however, a set of standards for calibration of the instrument used. These standards have in the past been frequently obtained by copolymerization of 14-C tagged monomer and radioassay[1130-1131]. This technique introduces additional manipulations and the possibility of isotope effects and is time-consuming.

The preparation of calibration standards presents a difficulty in infrared methods for analysing such copolymers. Physical blends of the two homopolymers, polyethylene and polybutene-1 will not suffice, as these have a different spectrum to a true copolymer with the same ethylene-butene ratio. An excellent method for preparing such standards is to copolymerize blends of ethylene and ^{14}C-labelled butene-1 of known activity. From the activity of the copolymer determined by scintillation counting its butene-1 content can be calculated. Standards prepared by this method are suitable for the calibration of the more rapid infrared method, which involves measurements of the characteristic absorption of the ethyl branches at 13 micron[1133]. Absorbance at 13 micron is directly proportional to the concentration of ethyl branches up to 10 per 1000°C.

Newmann and Nadeau[1134] applied pyrolysis-gas chromatography to the examination of ethylene-butene copolymers. Pyrolyses were carried out at 410°C in an evacuated gas vial and the products swept into the gas chromatograph. Under these pyrolysis conditions it is possible to analyse the pyrolysis gas components and obtain data within a range of about 10% relative. The peaks observed in the chromatogram were methane, ethylene, ethane, combined propylene and propane, isobutane, 1-butene, n-butene, trans-2-butene, cis-2-butene, 2-methylbutane, and n-pentane.

Figure 87 shows the relationship between the amount of ethylene produced on pyrolysis and the amount of butene in the ethylene-butene copolymers, which was determined by an infrared analysis for ethyl branches. The y intercept of 16.3 ethylene represents the amount of ethane which would result from a purely linear polyethylene. An essentially unbranched Phillips type polyethylene polymer yielded 14.5% ethylene which is in fairly good agreement with the extrapolated value.

As one would expect, ethyl branches will increase with the amount of butene copolymerized with the ethylene. This is in agreement with the work of Madorsky and Straus[1135], who found that the thermal stability and breakdown products obtained on pyrolysis can be related to the strength of the C-C bonds in the polymer chain, i.e. secondary>tertiary>quaternary.

9.3.5 Styrene and Acrylate Units in Styrene-Acrylate Copolymers

Anderson et al.[1136] determined acrylate units and styrene units in copolymers of acrylic acid or methacrylic acid with styrene, vinyl chloride and acrylamide by a combination of techniques. Acrylate groups were determined by carrying out a Zeisel reaction with hydrogen iodide to convert acrylate groups to alkyl iodides which were determined by gas chromatography. This method is discussed in further detail on the earlier section 9.2.3 on the determination of acrylate ester groups.

Styrene units were determined, as described below, by infrared spectroscopy at 14.28 micron which is the phenyl ring out of mode. This frequency provides specificity, freedom from interferences and an absorption that is directly proportional to the styrene content.

Method - Determination of Bound Styrene in Styrenated Alkyl Resins, Infrared Spectroscopy

To determine styrene units, solutions of the polymers are prepared in volumetric flasks by dissolving known weights of the dried polymers in

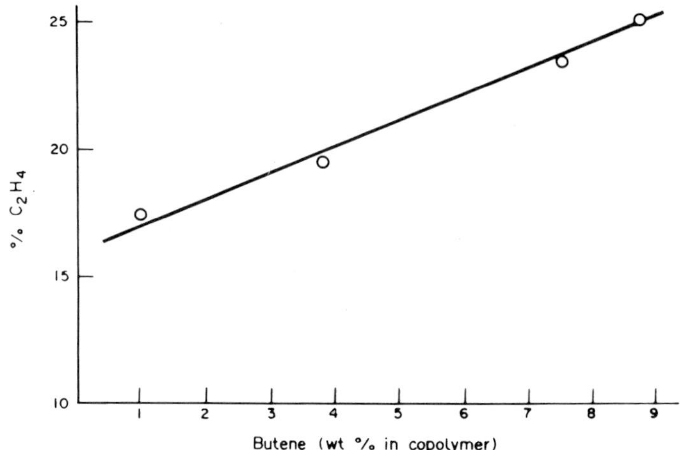

Figure 87   Percentage C₂H₆ as function of butene content of
                 ethylene-butene copolymers.

tetrahydrofuran or acetone. The infrared cell is rinsed twice with the
sample solution before use. A liquid cell, having potassium bromide
windows and a path length of 0.062 cm was used. The optimum region for
performing quantitative infrared analysis is between 0.3 and 0.6
absorbance unit. Typical polymer concentrations were 3 to 8 wt% within
the optimum region. Before scanning the spectrum, the filled infrared
cell is placed in the instrument for 15 minutes to allow the sample to
reach temperature equilibrium. The region of the spectrum between 15.38
and 13.33 micron is scanned at 17 cm -1/min and the absorbance of the
13.33 micron band is determined using the "basic line technique".

<u>Determination of absorptivity value.</u> Polystyrene solutions are
prepared between 10 and 60 g/1, and their infrared spectra determined
15.38 to 13.33 microns. The true absorbance of the 14.28 micron band is
then calculated using the following equation:

$$A = A_T - A_O$$

where A = true absorbance

   A_T = the total recorded absorbance of the absorption maxima

   A_O = the general background absorption at the absorption maxima

The true absorbance of the 14.28 micron styrene band is then plotted
against sample concentration. Using Beer's law an absorptivity value is
readily calculated. The absorptivity values determined in acetone and
tetrahydrofuran are given in Table 122.

<u>Determination of styrene content.</u> By knowing the absorbance, cell
path length and absorptivity value, the concentration (g/1) of styrene in
a solution is calculated using the equation:

$$c = A/ab$$

where c = styrene concentration (g/1) in the solution

Table 122 - Absolute absorptivity values obtained for styrene

Component	Absorptivity value 1/cm g	Solvent
Styrene	1.903	Acetone
Styrene	1.911	Tetrahydrofuran

A = true absorbance of 14.28 micron band

b = cell path length (cm)

a = absorptivity (1/g cm)

The concentration of styrene in the polymer is calculated using the equation:

$$\% \text{ styrene} = \frac{C}{W} (100)$$

where C = styrene concentration (g/l) in solution

W = sample concentration in solution (g/l)

The accuracy of this method on typical acid modified styrenated acrylic polymers is shown in Table 123. In the 5 to 70% styrene range, determined values are better than 95% relative.

9.3.6 Methacrylate and Styrene Units in Styrene-Methacrylate Copolymers

Evans et al.[1137] have described techniques employing pyrolyses - gas chromatography, proton NMR and carbon analysis for the determination of styrene and methacrylate units in styrene-methacrylate and styrene-n-butyl methacrylate copolymers. Agreement between the three independent methods is excellent. The comparison stresses the complementary nature of all three methods. The pyrolysis gas chromatography method possesses advantages such as simplicity and rapidity.

Method - Determination of Styrene and Methacrylate Units in Styrene Methyl-Methacrylate and Styrene-n-Butyl Methacrylate Copolymers - NMR Spectroscopy[1137]

The proton NMR spectra are obtained using a Jeol C-60H MHz spectrometer. Sweep width of 50 Hz and sweep time of 2.5 min are employed. For integration, sweep time is increased to 1.25 min. Deuterated chloroform is used as a solvent and trimethylsilane is used as an internal standard. The spectra are run at ambient temperature using 10-15 wt% solutions. Small variations in room temperature do not affect the precision of spectra as demonstrated by 1% standard deviation. Solutions of 10-15 wt% of copolymer are used since this gives the best spectra. If the number average molecular weight (M_n) of a copolymer is high, a lower wt% gives better results and vice versa. The sample size is optimized by varying the sample concentration to obtain the best possible spectra. Too high a concentration dampens the signals. One can easily

Table 123 - Infrared analysis for bound styrene in styrenated
acrylic polymers

| | Styrene | |
Sample	Calculated %	Observed %
1	5.0	5.0, 4.9
2	7.5	7.1, 6.9
3	10.0	19.6, 19.5
4	24.3	24.4, 24.3
5	24.3	25.6, 24.6
6	25.0	25.1, 25.8, 24.2[a]
7	35.0	36.3, 35.6
8	35.0	34.7, 35.2
9	40.0	38.9, 40.1
10	60.7	59.1, 60.3, 59.7[a]
11	70.0	69.3, 70.3, 69.5[a]

[a]Percent styrene values, determined in tetrahydrofuran. All the remaining
values were determined using acetone as the solvent.

differentiate the aromatic protons from the aliphatic protons since the
former absorb at 8 seconds whereas the latter absorb between 1-4 . Areas
under these peaks are integrated. The integrated areas are related to
either the aliphatic or aromatic protons of the copolymer under study.
These areas can then be converted to weight per cent of the monomers using
simultaneous equations.

<u>Method - Determination of Styrene and Methacrylate Units in Styrene Methyl
Methacrylate and Styrene n-Butyl Methacrylate Copolymers - Pyrolysis - Gas
Chromatography</u>

The <u>pyrolyser.</u> Chemical Data System Pyroprobe Model 100 are used as
the pyrolyser. Quartz sample tubes, 30 mm long and 1 mm i.d. are placed
in the platinum coil heating probe. Pyrolysis conditions are varied until
the most reproducible pyrograms were obtained. The optimum operation
conditions are interface temperature, 120°C; ramp, 20°C; interval, 20 s;
final temperature 600°C. The ramp indicates rise in temperature °C/m.
Interval indicates the period for which the final temperature is held.

The <u>gas chromatograph.</u> Hewlett Packard 5750 gas chromatograph with
12 ft x 1/8 inch stainless steel column packed with 80/100 mesh Chromosorb
P (Applied Science) coated with 10% W-982 are used. Operating conditions:
helium flow 25 ml/min; injection port temperature, 150°C; detector
temperature, 250°C; flame ionization detector range 10   with 4 x
attenuation; oven temperature programmed from 70-270°C at 8° C/min.

The copolymer material (100 mg) is weighed in a 10 ml volumetric
flask and dissolved in 10 ml methylene chloride. Two microlitres of this
solution are placed at the centre of a quartz sample tube and allowed to
dry. The sample tube is placed inside the platinum coil of the pyrolysis
probe which is then inserted into the heated interface. Before initiating
the run, 10 min are allowed for any entrapped air, residual solvents, or
monomers to escape.

For calculation of percent styrene and methacrylate in the sample, the ratio of integrated areas of acrylic monomer to that of the styrene monomer is obtained from the gas chromatographic scan. This ratio is then compared to the one obtained from a standard copolymer of known composition which is prepared by 100% conversion of the starting monomers.

The analysis of comonomer composition does not appear to be significantly affected by the extent of randomness observed for these materials. The difference between NMR and gas chromatography pyrolysis results are in the range of 0.4% and 0-4.8% for styrene/n-butyl methacrylate and styrene/methyl methacrylate, respectively. The difference in carbon analysis and pyrolysis gas chromatography results is also in the same range for both the copolymers. Standard deviation for pyrolysis gas chromatography ranges from 1.2-2.1%. Precision for NMR analyses is better than 1%.

Pyrolysis of methyl methacrylate-ethylene dimethacrylate copolymer gives only one major peak (methyl methacrylate) using a hot filament detector. The composition of methyl methacrylate-ethylene dimethacrylate copolymer can, however, be determined by pyrolysing a weighed sample and using the ratio of sample weight to area of the methyl methacrylate peak for obtaining a standard analysis curve. Under favourable conditions and with careful control of pyrolysis column, and detector variable, constituents can be determined within $\pm$ 0.5%.

The differences in the pyrograms of block and random copolymers allows estimation of the comonomer distribution. Random copolymers of ethylene with methyl acrylate or methyl methacrylate yield on pyrolysis a lower ratio of methanol/methyl acrylate or methyl methacrylate, respectively, than block polymers of the same composition. Differential thermal analysis measurements give a first-order transition for block polymers only, and by measuring the area under the transition an indication of the minimum chain length between acrylate units can be obtained.

Various other workers[1138-1143] have used NMR to determine methyl-methylmethacrylate in styrene - methmethacrylate copolymers.

In Table 124 is presented information concerning the measurement of comonomer ratios in other methyl acrylate copolymers.

9.3.7 Ethylene Glycol, 1,4-Butane Diol, Terephthalic Acid and Isophthalic Acid Repeat Units in Terylene

Terylene manufactured from terephthalic acid and ethylene glycol as follows:

$$\begin{array}{c} CH_2OH \\ | \\ CH_2OH \end{array} + \underset{COOH}{\overset{COOH}{\bigcirc}} \longrightarrow \left[- CH_2 - CH_2OOC - \bigcirc - COO \right]_n$$

Such polymer contain repeat units based on ethylene glycol (and indeed other glycols such a 1,4-butane diol) and terephthalic acid and

Table - 124

The determination	The technique	Reference
Glycidylmethacrylate in methylmeth-acrylate glycidyl methacrylate	IR	1144
Methacrylic acid-methylmethacrylate	^{13}C Fourier transform NMR	1145
Methylmethacrylate in styrene-methylmethacrylate copolymers	NMR	1146-1150

proportions of isophthalic acid. Allen et al.[1151] developed a precise quantitative method for determining such units in Terylene. The sample is subject to an alkaline hydrolysis and glycol and acidic products after conversion to the trimethylsilylderivatives are analysed by gas chromatography.

Method - Determination of Acid and Glycol Units in Terylene and Other Polyesters. Hydrolysis - Gas Chromatography[1151]

Apparatus. Hewlett-Packard Model 5711A gas chromatograph equipped with a flame ionization detector and a Hewlett-Packard Model 3352B laboratory data systems were used.

Reagents. Redistilled Eastman Kodak 2-ethoxyethanol (Practical Grade). Internal standard solutions:

1.   Dodecane, 1 g diluted to 100 ml with 2-ethoxyethanol.

2.   Nonyl alcohol, 5 g diluted to 100 ml with2-ethoxyethanol.

3.   Propyl alcohol, 0.4 g diluted to 100 ml with 2-ethoxyethanol.

4.   Heptadecane, 1 g diluted to 100 ml with 2-ethoxyethanol.

5.   N,O-bis (trimethylsilyl) trifluoroactamide.

Procedure. The amount of sample used for the alkaline hydrolysis is controlled by the type of determination. For example, a 1 g sample is sufficient for the determination of the glycols and acids comprising the major portion of the repeat units. A 4 g sample is used when determining trace components such as the methyl ester end-groups.

The sample is weighed into a 100 ml flask and 50 ml of 1 N potassium hydroxide in 2-ethoxyethanol is added to the flask. A condenser cooled by chilled water is attached to the flask and the reaction mixture is protected from carbon dioxide by means of a tube packed with Ascarite absorbent and Drierite desicant. The contents of the flask are heated and maintained at reflux temperature for 10 min with constant stirring. The flask is allowed to cool to room temperature and the hydrolysate is adjusted to a pH of 1 using concentrated hydrochloric acid (5 ml). An internal standard is added to the flask. Pyridine (25 ml) is added to dissolve the acids present and an aliquot sample is centrifuged to remove the potassium chloride. Approximately 50 ul of hydrolysed sample is allowed to react at room temperature for 5 to 10 min with 500 ul of N,O-bis(trimethylsilyl)trifluoracetamide to form silyl ethers and esters of the glycols and acids, respectively. The silyated hydrolysate is chromatographed by injecting 0.1 ul of sample into the gas chromatograph.

The silyl derivative of diethylene glycol is separated from the silyl derivative of ethylene glycol using a 10 ft x 1/8 inch stainless steel column packed with 100-200 mesh Chromosorb G-HP solid support containing 3% by weight of Versilube F-50 liquid phase, Versilube F-50 is the trademark for a mixture of trichlorophenyl silicone (10%) and methyl silicone manufactured by General Electric.

Figure 88(a) is a chromatogram obtained for the silyl derivatives of the hydrolysate of an experimental polyethylene terephthalate. Dodecane is used as an internal standard and the column is operated isothermally at 127°C with a nitrogen carrier gas flow rate of 10 ml/min.

Figure 88(b) is a chromatogram of the silyl derivatives of ethylene glycol, 1,4-butanediol and the cis and trans isomers of 1,4-cyclohexanedimethanol from the hydrolysate of an experimental polyester. A 6 ft x 1/4 inch glass column packed with 100-200 mesh Chromosorb W-HP solid support containing 10% by weight of Versilube F-50 liquid phase is used for this separation. Nonyl alcohol is used as an internal standard and the column is operated at 120°C for 8 minutes and programmed to 210°C at 4° C/min with a nitrogen carrier gas flow rate of 20 ml/min.

High-boiling acids, such as terephthalic and isophthalic acids, are separated using a 6 ft x 1/8 in stainless steel column packed with 100-200 mesh Chromosorb W-HP solid support containing 10% by weight Versilube F-50 liquid phase. Figure 88(c) is the chromatogram obtained for the separation of the silyl derivatives of isophthalic and terephthalic acid from an experimental polyester. The column is operated isothermally at 183°C with a nitrogen carrier gas flow rate of 33 ml/min. N-Heptadecane was added as an internal standard to permit the calculations of the weight per cent acids.

Another important use of the alkaline hydrolysis/gas chromatographic procedure is the determination of methyl ester end groups. If the polyester is terminated with methyl ester end groups, then methyl alcohol is produced when the polyester is hydrolysed. By determining the concentration of methyl alcohol, one can calculate the number of methyl ester end groups in polyethylene terephthalate. The hydrolysis procedure for determining methyl ester end-groups is the same as discussed for the glycols and acids. The gas chromatographic procedure is different in that no silylation reagent is used. The hydrolysed sample is separated using a 6 ft x ¼ inch glass column containing 60-80 mesh Chromosorb 102 column packing.

Propyl alcohol is used as the internal standard, and the column is operated isothermally at 145°C with a nitrogen carrier gas flow rate of 10 ml/min.

The data given in Table 125 show that the precision of the overall method is good. This method is relatively simple and fast. The method has been used to analyse experimental polyester for the amount of monomers present in the repeat units and for determining the composition of copolymers and polymer blends. The relative standard deviation varies from 0.8% for the determination of 1,4-butanediol to approximately 2% for 1,4-cyclohexanedimethanol. For the determination of different concentrations of isophthalic acid, the standard deviation varies from 0.7 to 2.4%.

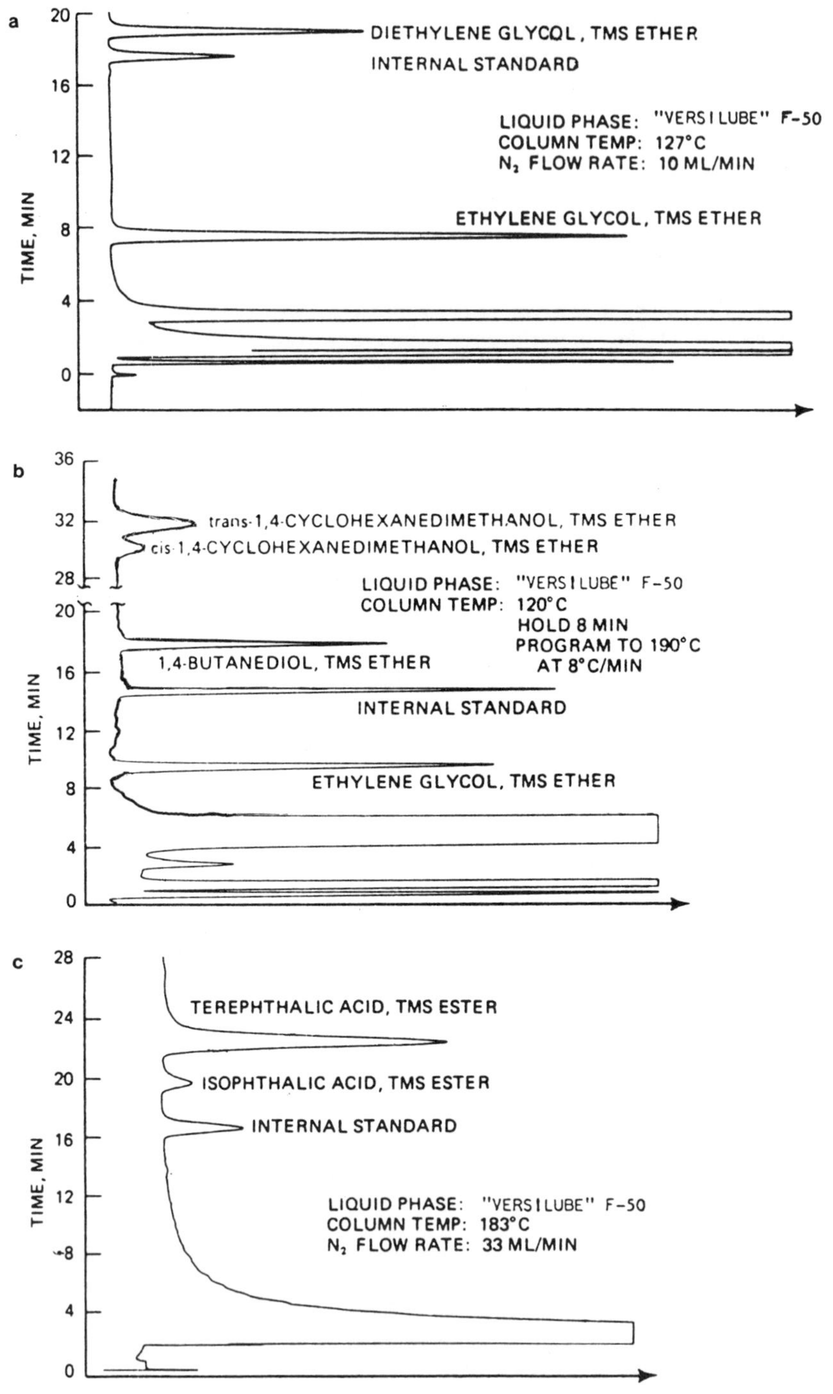

Figure 88   Chromatogram of (a) trimethylsilyl derivatives of the glycols from the hydrolysis of polyethylene terephthalate; (b) silylated glycols from experimental polyesters; (c) iso and terephthalaic acids, (n-heptadecane internal standard).

Table 125 - Precision of hydrolysis-gas chromatographic methods

Component	Concn, %	SD	No
Diethylene glycol, TMS ether	1.781	0.019	40
1,4-Butanediol, TMS ether	6.59	0.050	20
1,4-Cyclohexanedimethanol TMS ether	20.45	0.40	20
Isophthalic acid, TMS ester	5.00	0.12	20
Isophthalic acid, TMS ester	60.00	0.44	20
Methyl alcohol	0.088	0.002	20

9.3.8 Hexafluoropropylene and Vinylidene Fluoride Repeat Units in Hexafluoropropylene - Vinylidene Fluoride Copolymers

Two methods have been described for determining the compositional analysis of these copolymers, one based on high resolution continuous Fourier transform 19 F NMR[1152] and the other on pyrolysis - gas chromatography[1153].

19 F NMR Spectroscopy. When it is desired to measure the composition of a single component in a mixture, it is necessary to relate the component resonance to that of another compound, an internal standard, which is of known chemical composition and has been added in known weight to a known weight of unknown. Brame and Yeager[1152] used dichlorobenzotrifluoride as an internal standard in the continuous wave method for determining the compositional analysis of both repeat units, in hexafluoro-propylene-vinylidene fluoride copolymers. This work demonstrated the utility of the Fourier transform NMR method in quantitative analysis of the copolymer in relation to results obtained by continuous wave 19 F NMR and proton NMR. Brame and Yeager[1152] used a Varian XL-100-15NMR spectrometer and either a Varian S-124 XL pulse Fourier transform system or a Nicolet Technology Inc. TT-100 Pulse Fourier transform system. The usual spectral widths encountered covered the range to 15 kHz. Both 5 mm and 12 mm o.d. sample tubes were used in the analyses.

In the FT-19F NMR spectroscopy, the conditions generally used were:

Pulse width between 10 and 20 us, data acquisition time of 0.3 s or larger, pulse repetition rate for 0.3 s or larger, and frequency cut off of the audio filter must be in the range of 15 to 20 kHz.

A 19 F Fourier Transform NMR spectrum of polyhexafluoropropylene - vinylidene fluoride copolymer is shown in Figure 89. In this spectrum, the carrier was placed at low field and the number of transients taken was 128. the lines observed are attributed to the following:

CF 3 group (δ = -70 to -75), CF 2 groups (δ = -90 to -120) and CF group (δ = -180 to -185).

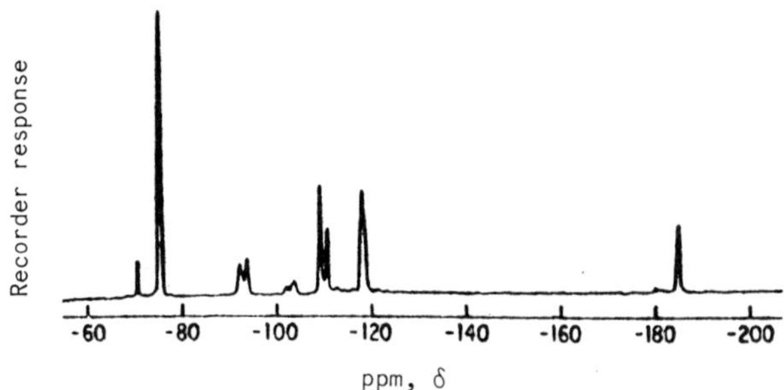

Figure 89   94.1 MHz ^{19}F - NMR spectrum of hexafluoropropylene-
              vinylidene fluoride copolymer containing - 40% w/w
              vinylidine fluoride.

The following equations were used to calculate the composition from
the integrated spectrum:

    kI = 3 (HFP)
    KII = 2 (HPP) + 2 (VF$_2$)

where k is a proportionality constant, area I (δ = -65 to -80) and area
II(δ = -80 to 130) are the integrated intensity measurements over the
indicated ranges and (HFP) and (VF) are the mole fractions of HFP and VF$_2$
respectively. The results of the determination on a copolymer sample are
given in Table 126. The value obtained is in excellent agreement with
those obtained by mass balance.

    Pyrolysis - gas chromatography. Blackwell[1153] used a Curie point
pyrolyser to carry out quantitative analysis of monomer units in
polyhexafluoropropylene - vinylidene fluoride. The polymer composition is
calculated from the relative amounts of monomer regenerated and the
trifluoromethane produced during pyrolysis. The exact mechanism by which
trifluoromethane is produced during pyrolysis is not known but it is
presumed that the free trifluoromethyl group is cleaved from the polymer
backbone. The trifluoromethyl group then extracts a proton from the
polymer chain to form trifluoromethane. A calibration curve was obtained
using samples whose compositions were measured by 19 F NMR as standards
and a least square fit calculated.

    The reproducibility of the pyrolysis step, achieved by the Curie
point pyrolyser permitted the monomer composition to be determined with a
reproducibility of ± 1%.

Method - Determination of Hexafluoropropylene and Vinylidene Fluoride
Units in Polyhexa-fluoropropylene - Vinylidene Fluoride. Pyrolysis - Gas
Chromatography[1153].

    Apparatus. The magnetic induction coil in the Curie point pyrolyser
is positioned as close to the injection port as possible. The solenoid
gas valve is operated by a switch on the control panel. The HP-5750 gas
chromatograph is equipped with a gas sampling valve and the solenoid valve
is connected in place of the sample loop. With the gas sample valve open,
the solenoid valve can be used to divert the carrier gas flow either
directly through the injection port or through the pyrolysis chamber. The
pyrolysis chambers supplied by the manufacturer were unsuitable and

Table 126- HFP/VF$_2$ copolymer compositions

Sample	Mass balance, wt % VF$_2$	FT-NMR	
		wt% VF$_2$	wt% HFP
A	48 $\pm$ 2	46.6 $\pm$ 1.1[a]	53.4 $\pm$ 1.2[a]

[a] Precision at 95% confidence level

replacements were fabricated. The needles used were 24 gauge Huber point hypodermic needles from which the syringe fitting was removed. These were joined to 3 mm o.d. quartz tubes and sealed with Toor Seal, an epoxy resin used in vacuum systems.

Sample preparation. A 10% solution of each hexafluoropropylene-vinylidene fluoride copolymer is prepared by dissolving 10 g of polymer in 100 ml of reagent grade acetone. A vial 58 mm high is filled with the sample solution. A pyrolysis wire, previously cleaned in a bunsen burner until red hot, then cooled, is dipped into the vial. The dipped wire is then dried in a vacuum oven at 90°C for 30 min. Use of lower drying temperatures or shorter times results in an acetone peak in the pyrogram. This is due to mechanically trapped acetone being liberated. This procedure will provide a pyrolysis wire coated with a thin film of uniform thickness and reproducible length.

Procedure. Pyrolysis conditions are varied until the most reproducible pyrogram is obtained. The coated wire is placed in the pyrolysis chamber and allowed to come to equilibrium for 5 min with the carrier gas flowing through the chamber. The samples are pyrolysed at 800°C for 4 seconds. The gaseous pyrolyzate is separated on a 10 foot by 1/8 inch o.d. stainless steel column packed with beta, beta' oxydipropionitrile Porasil C Durapak (Waters Associates), 80 to 100 mesh. The carrier gas is argon flowing at the rate of 15 cm 3/min. The temperature of the column oven is raised to 120°C and maintained there. The flame ionization detector is optimized for fluorocarbons. The complete elution of all fragments takes 30 min.

The components of the pyrolyzate are identified by mass spectrometry. The Curie point pyrolyser is coupled to the gas chromatography system (which utilizes a Varian 1800 GC) in the same way as it was coupled to the HP-5750 gas chromatograph. The first three peaks in Figure 90 are identified as vinylidene fluoride, trifluoromethane and hexafluoropropylene respectively. The other peaks are identified as oligomers or hexafluoropropylene and vinylidene fluoride. The output from the HP 5750 is integrated by a computer system. The relative peak areas for vinylidene fluoride, and hexafluoropropylene is calculated.

Two calibration plots are made, one of the sum of the relative peak areas for trifluoromethane and hexafluoropropylene produced during pyrolysis against the weight per cent of hexafluoropropylene calculated from 19 F NMR data and the other of the relative peak area for vinylidene fluoride produced during pyrolysis against the weight per cent vinylidene fluoride calculated from 19 F NMR data. A least square fit was calculated for each calibration plot and the slope and intercept were determined.

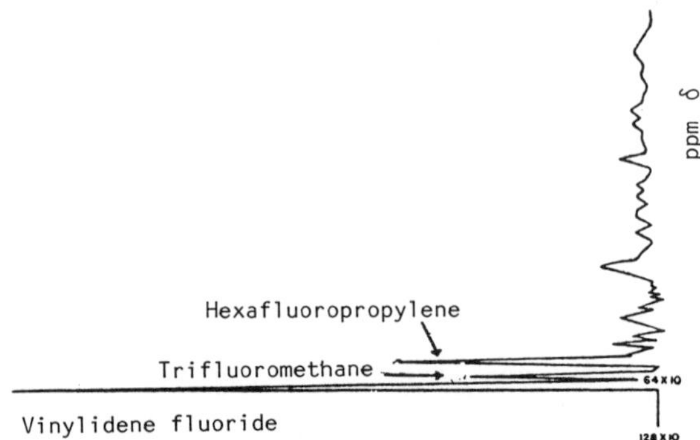

Figure 90   Typical chromatogram of the pyrolysis products for
              hexafluoroproylene-vinylidene fluoride copolymer.

The two calibration plots demonstrate a linear relationship between
the pyrolysis products and the NMR data. The slopes of the two
calibration plots are close to 1, the slope for weight percent
hexafluoropropylene is 0.99 and that for vinylidene fluoride is 0.97.

The monomer composition of the series of copolymers was calculated
from the pyrolysis data and compared too that calculated from the NMR
data. These data are summarized in Table 127. The difference in weight
percent calculated for the two techniques, averaged for
hexafluoropropylene $\pm$ 0.60% and for vinylidene fluoride $\pm$ 0.85%/.

9.3.9 Monomer Units in Other Olefin Copolymers. Fisch and Dannenberg[1154]
have reported that for the analysis of propylene butene-1 copolymers
containing up to 11% bound propylene, satisfactory calibration standards
can be obtained using 13-C NMR analysis.

Various other workers[1155-1157] have studied the 13 C NMR spectroscopy of propylene-butene copolymers.

In Table 128 is presented information concerning the measurement of
comonomer ratios in other olefin copolymers.

9.4 MONOMER UNIT SEQUENCE DETERMINATION IN COPOLYMERS

9.4.1 Introduction

A further very important aspect of microstructure is the sequence of
monomer units in a polymer. This applies whether the polymer is based on
a single monomer which is capable of polymerizing in different ways, e.g.
head-to-head or head-to-tail polymerization, or whether it is based on two
or more different monomers when, obviously, many variants of monomer
sequence are possible. Sequence distribution has an important bearing on
the tacticity (i.e. stereochemistry) and other properties of polymers, as
discussed in Chapter 10.2. Three major techniques have been used to study
sequence problems in polymers, they are pyrolysis-gas chromatography, NMR
and infrared spectroscopy.

9.4.2 Sequencing in Ethylene-Propylene Copolymers

A good example of a polymer in which sequences of different monomer
units occur is ethylene-propylene, and this will be discussed here in some
detail. Processes for the manufacture of this polymer can produce several
types of polymer which, although they may contain similar proportions of

Table 127 - Comparison of the monomer composition of HFP/VF$_2$ copolymer
as calculated by pyrolysis data and NMR data

Weight % HFP			Weight % VF$_2$			Std dev for Curie point pyrolyser
^{19}F NMR	Pyrolysis	Δ	^{19}F NMR	Pyrolysis	Δ	
28.6	28.8	0.2	71.4	70.9	0.5	1.04
29.0	28.8	0.2	71.0	71.0	0	0.34
34.79	36.04	1.25	65.21	62.63	2.58	0.52
36.46	36.73	0.27	63.96	62.91	1.05	0.82
37.24	37.33	0.11	62.76	62.61	0.15	0.39
39.04	39.07	0.03	60.96	60.88	0.08	0.67
39.04	38.35	0.69	60.96	61.55	0.59	0.63
39.82	38.46	1.36	60.18	61.89	1.71	0.30
40.10	39.45	0.65	59.90	61.49	0.59	0.10
42.4	43.2	0.8	57.6	56.7	0.9	0.39
44.8	46.2	0.2	55.2	53.8	1.4	0.48
44.8	45.4	1.4	55.2	54.3	0.9	0.59
47.3	47.0	0.3	52.7	52.8	0.1	0.92
49.95	49.45	0.50	50.05	50.55	0.50	0.30

Table 128

The determination	The technique	Reference
Butene-ethylene-propylene	NMR	1158
Alpha-olefins in ethylene-butene	NMR	1159
Methylinethylene-alpha-olefin	IR	1160
4-Methylpentene-1-1-pentene	NMR	1161
Ethylene-vincyclohexane	IR	1162
Vinylchloride-propylene	IR	1163
	^{1}H NMR	1164,1165
Vinyl acetate-polyethylene	IR	1166
4-Methyl-1-pentene		
4-Methyl-1-pentene-1-pentene copolymer	NMR	1167
Methyl methacrylate-vinylidene chloride	NMR	1168
High molecular weight polymers	NMR	1169

the two monomer units, differ appreciably in their physical properties. The differences in these properties lie not only in the ratio of the two monomers present but also, and very importantly, in the detailed microstructure of the two monomer units in the polymer molecule. Ethylene-propylene copolymers might consist of mixtures of the following types of polymer:

i)   Physical mixture of ethylene homopolymer and propylene copolymer:

       E-E-E-E-E- P-P-P-P

ii)   Copolymers in which the propylene is blocked, e.g.:

       E-E-P-P-P-P-P-E-E-E-E-E-P-P-P-

iii)   Copolymers in which the propylene is randomly distributed, e.g.:

       -E-P-E-P-E-P-E-P-(alternating e.g. pure cis-1,4 polyisoprene),
       (Chapter 10.3) or -E-E-E-P-E-E-E-E-E-P-E-E-E-E

iv)   Copolymers containing random (or alternating) segments together with blocks along the chains, i.e. mixtures of (iii) and (ii) (random and block) or (iii) and (ii) (alternating and block), e.g.;

       -E-P-E-P-E-P-

v)   Containing tail-to-tail propylene units in propylene blocks:

       i.e. head-to-head and tail-to-tail addition giving even-numbered sequences of methylene groups, (Chapter 10.4)

$$
\begin{array}{c}
\mathrm{Me} \qquad\qquad\quad \mathrm{Me} \qquad\qquad\qquad\qquad\quad \mathrm{Me}\ \ \mathrm{Me} \qquad\qquad\qquad\qquad\qquad\qquad\quad \mathrm{Me} \\
\ \ |\ \qquad\qquad\qquad |\ \qquad\qquad\qquad\qquad\qquad\quad |\ \ \ \ | \qquad\qquad\qquad\qquad\qquad\qquad\quad | \\
-\,CH - CH_2 - CH_2 - CH - CH_2 - CH_2 - CH_2 - CH - CH - CH_2 - CH_2 - CH_2 - CH_2 - CH
\end{array}
$$

vi)   Graft copolymers

```
          P
          |
       -E-E-E-E-E-E-E
          |         |
          P         P
          |         |
                    P
                    |
                    P
```

       Various methods are available for determining the percentage of ethylene and of propylene, the ethylene propylene ratio and the type of ethylene and propylene unit sequencing in these polymers, and these methods are discussed in further detail in this section.

       In fact the problem is somewhat more subtle than expressed above the types of measurements that are commonly required are listed below:

a)   determination of total percentage propylene in the above polymers, regardless of the manner in which the propylene is bound;

b)   determination of total percentage ethylene in the above polymers, regardless of the manner in which the ethylene is bound;

c)   determination of proportion of total propylene content of polymer which is blocked and that which is randomly distributed along the polymer chain.

Several possible techniques are available for carrying out these analyses, and these are now discussed:

Pyrolysis of sample followed by gas-liquid chromatography of pyrolysis products.

Direct infrared spectroscopy of polymer, either at room temperature or at elevated temperatures.

Pyrolysis of sample followed by infrared spectroscopy.

Nuclear magnetic resonance spectroscopy and proton magnetic resonance spectroscopy.

Pyrolysis-gas chromatography. Before discussing the applications of this technique to microstructural studies of polymers it is necessary to understand the principles of the technique, and the factors which affect the results obtained. These aspects are discussed first, followed by some examples of the application of pyrolysis-gas chromatography which show its usefulness in microstructural studies.

In this technique a small quantity of the polymer is mounted on an inert metal support and either an electrical current is passed through the support (filament method) or external heat is supplied to the support (furnace method) so as to rapidly heat up and break down (i.e. pyrolyse) the polymer into a mixture of smaller molecules which, under standard pyrolysis conditions, are characteristic of the polymer being examined. The products are swept from the pyrolysis chamber by a stream of carrier gas onto a gas chromatographic column and separated into their individual components prior to passing through the detector which records their retention time (time taken, under standard conditions, to travel from pyrolysis chamber to detector) and quantity (peak height under standard conditions).

This in principle is the essence of the pyrolysis-gas chromatography technique. It is, of course, possible to then pass then separated pyrolysis products one at a time into a mass spectrometer in order to obtain definitive information regarding their precise identity, i.e. pyrolysis-mass spectrometry.

It is emphasized that the technique of pyrolysis-gas chromatography can be used in two ways for the examination of polymers. It can be used as a fingerprinting technique (discussed in Chapter 4.4.2) in which the pyrogram of an unknown polymer prepared under standard conditions is simply visually compared with a library of pyrograms of various known polymers prepared under the same conditions. Also, as will be discussed below, the technique can be used to obtain fine detail regarding minor, but often very important, detail concerning polymer structure such as monomer sequences and, also, branching, crosslinking copolymer structure and the nature end-groups, and can also be used to elucidate polymer structure between adjacent monomer units in copolymers and to determine monomer ratios in copolymers. Some important general points concerning this technique below.

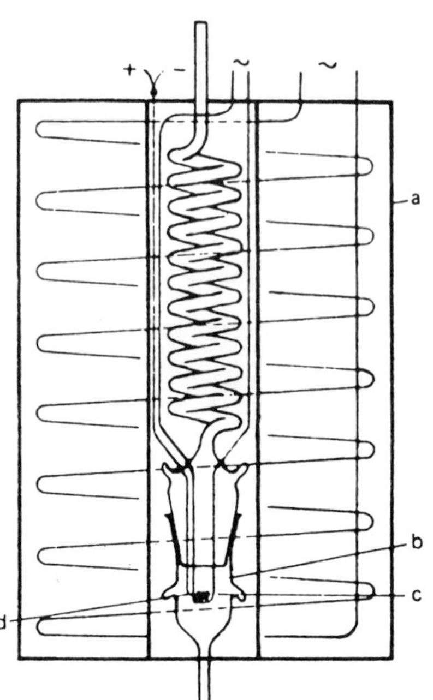

Figure 91   Pyrolysis apparatus: (a) jacket; (b) pyrolyser; (c)
platinum filament; (d) thermoelement.

Design of heated platinum filament pyrolyser unit.   Voigt[1170-1172]
employed the apparatus shown in Figure 91 attached directly to the gas
inlet of the gas chromatograph for the examination of polyethylene,
polypropylene and ethylene-propylene copolymers.   Figure 91 too shows the
Pyrex glass pyrolysis cell b located in the oven a, the carrier gas
entering the cell from above, and bringing with it the pyrolyzate formed
on the red-hot platinum wire c on leaving the cell and passing through the
gas inlet and a heated feed line to the gas chromatograph; d is the
junction of chromel-alumel thermocouple, the lead wires of which are
embedded in a Ni-Cr-Fe alloy tube filled with magnesium oxide.   Like the
platinum wire of the heating spiral, this tube has a diameter of only 0.5
mm.

Specimens weighing about 2 mg were heated for 18 s up to a maximum
temperature of 550°C.   During the pyrolysis the carrier gas flushes the
products of pyrolysis into the the separating column.   The pyrolysis cell
is cut out of the carrier gas flow 1 min after starting the pyrolysis.   In
this apparatus the pyrolysis gases were not hydrogenated, and hence a
mixture of unsaturated hydrocarbons was obtained.

In the above technique the polymer is pyrolysed with a certain
voltage applied to the platinum filament to produce a specified maximum
pyrolysis temperature.   It might be assumed that precise setting of the
voltage provides sufficient control to reproduce pyrolysis conditions.
However, it was found that it is the heating rate of the filament, rather
than its maximum temperature, that is the primary factor governing
reproducibility of the product distribution.   Poor control of the heating
rate may well explain the reported non-reproducibility of the filament,
compared with the furnace pyrolyser.   It is to be expected that variations
in heating rate of the filament can result certainly in the case of some
types of polymers, in different depths of cracking and different amounts
of secondary pyrolysis reactions.

Table 129 - Effect of filament heating rate on
pyrolysis pattern

Sample	Heating rate ($°C/s$)	Maximum temp ($°C$)	C_1 plus C_2	Area %* branched + cyclic C_4-C_{10}	Ratio ($1C_6$/n C_6)
Filament					
Linear polyethylene	8	550	7	9	3
(Marlex 6009)	20	550	7	7	2
	20	800	7	5	-
	70	550	9	6	3
	170	650	25	8	6
	280	800	47	9	12
Furnace					
	-	550	12	13	8
	-	650	28	21	19
	-	800	52	83	73

*Pyrogram-heated silicone column; peak areas normalized through C_{10}

The effect of heating rate on the pyrolysis patterns is illustrated by the data given in Table 129 which were obtained by pyrolysis-hydrogenation-gas-chromatography of polyethylene. Results are compared for a filament and a surface type of pyrolyser both run at maximum temperatures between 550 and 880°C. The patterns for the linear polyethylene are compared on the basis of C_1 plus C_2, and the non-normal hydrocarbons. The filament pyrolyser results are fairly consistent for heating rates of 70°C per second and lower. Thus, below a certain heating rate one does not need to control the rate precisely. At 170°C per second and above the amounts of C_1 plus C_2 and non-normals are increased markedly.

The filament temperature programming technique has the following advantages: (1) minimizes secondary pyrolysis reactions; (2) provides a means of duplicating product distributions with different filaments: (3) offers a standard method of thermal decomposition which is suitable for polymers of widely varying thermal stabilities.

The hydrogenation technique is used to convert the unsaturated olefinic pyrolysis products to saturated alkanes. As several different olefins, upon hydrogenation, will be converted to the same alkane, hydrogenation has the effect of reducing the number of components in the pyrolysis mixture and consequently of simplifying the task of interpreting the gas chromatogram. Thus, for example, the three olefins hexen-1, hexen-2 and hexene-3, upon hydrogenation, are converted to n-hexane. Of course, if the object is to deduce structural detail of the polymer, then hydrogenation and the subsequent alterations in pyrolysis product composition might mask some important feature, and in such instances it would be advisable not to use hydrogenation in the preliminary studies. The hydrogenation technique is used principally in the pyrolysis of polyolefins and PVC, and is not used in the case of many other types of polymers such as acrylates or polystyrene, where the pyrolysis products are either not olefinic, or are fewer in number then occurs in the case of polyolefins.

   The furnace pyrolyser. This type of pyrolyser, mentioned above, is different in principle from the filament type in which the filament and consequently the polymer, is heated to a predetermined temperature by the application of a known voltage to the filament. In the furnace pyrolyser a weighed quantity of polymer is placed in a small platinum dish. A small furnace is brought to a predetermined temperature by application of an electrical current, and the platinum dish then introduced into the furnace so that the polymer is very rapidly heated up to the required temperature. The pyrolysis products are then swept into the gas chromatogram by the carrier gas.

   Classic work on the pyrolysis of polyethylene, polypropylene and ethylene-propylene copolymers containing between 0 and 100% propylene, was reported by Van Schooten and others[1173-1177]. In their earlier work on the furnace method the sample (20 mg) in a platinum dish was submitted to controlled pyrolysis in a stream of hydrogen as carrier gas. The pyrolysis products were then hydrogenated at 200°C by passing through a small hydrogenation section containing 0.75% platinum on 30/50 mesh aluminium oxide. The hydrogenated pyrolysis products are then separated on a squalane on fireback column and the separated compounds detected by a katharometer. Under the experimental conditions used in this work only alkanes up to C_9 could be detected.

   Using this procedure, however, a large sample (20-30 mg) was required, which favoured the occurrence of consecutive and side-reactions of the primary pyrolysis products. In their improved method[1178,1179], discussed below, Van Schooten and Evenhuis pyrolysed a much smaller polymer weight (0.4 mg) in a stream of hydrogen on an electrically heated nichrome filament at 500°C and led the pyrolysis products onto a hydrogenation catalyst to convert all products to saturated hydrocarbons. Operation with such small samples was made possible by using a sensitive flame ionization GLC detector. This and the use of programmed heating of the Apiezon N/firebrick GLC column extended the range of detectable volatile products from C_9 to iso C_{13}, with considerable improvement in resolution of the pyrograms, especially of the more volatile components.

   These modifications of experimental techniques minimized the occurrence of secondary reactions, with the result that the relative amounts of lowest molecular weight fragments are considerably reduced. This is illustrated in Figures 92(a) and (b), where peak surface areas are given for the n-alkanes in the pyrograms of polyethylene and polypropylene as obtained by the old, (i.e. 20 mg polymer pyrolysed) and the new (i.e. 0.4 g polymer pyrolysed) pyrolysis apparatus. The modified technique therefore provides a much more reliable picture of the primary pyrolysis reactions, and enables some conclusions to be drawn regarding primary reaction mechanism.

   Lehmann and Brauer[1180] investigated the pyrolysis-gas chromatography of polystyrene under helium at temperatures ranging from 400 to 1100°C, utilizing for the pyrolysis a silica boat surrounded by a platinum heating coil[1182]. Separations were carried out on a column of Apiezon (30%) on Chlorowax 70 and dinonyl phthalate on firebrick operated at 100-140°C.

   Typical chromatograms obtained on pyrolysing polystyrene are shown in Figure 93. At 425°C only styrene monomer is eluted from the column; degradation at 825°C produces a number of additional products, presumably

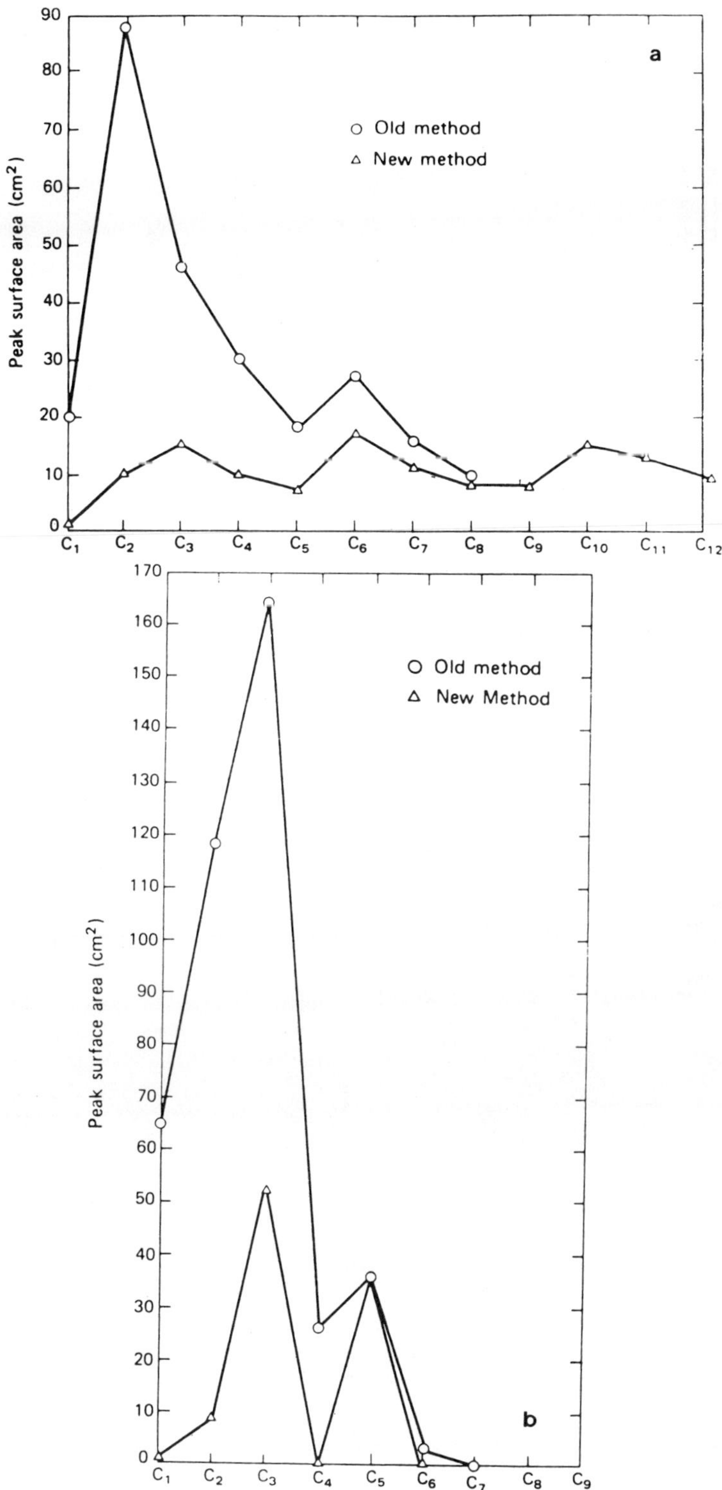

Figure 92   Peak surface area - n-alkane distributions of (a) polyethylene (b) polypropylene.

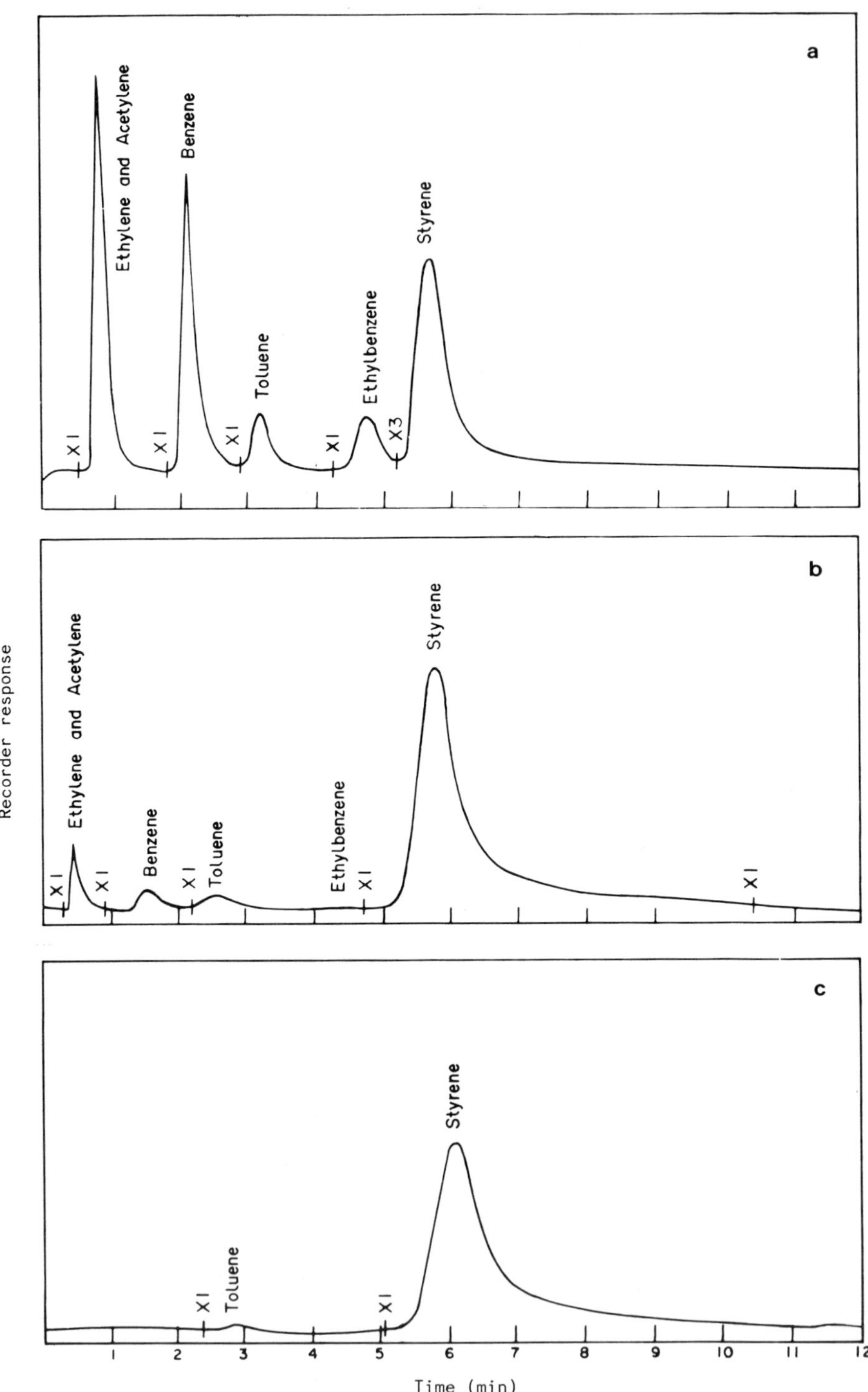

Figure 93   Chromatograms of pyrolysis products of polystyrene.
Column, Apiezon L; col. temp. 140°C; flow rate, 60
ml/min; pyrolysis temp; °C, top, 1025, middle 825,
bottom 425. (Attenuation scale indicated by numbers in
figure).

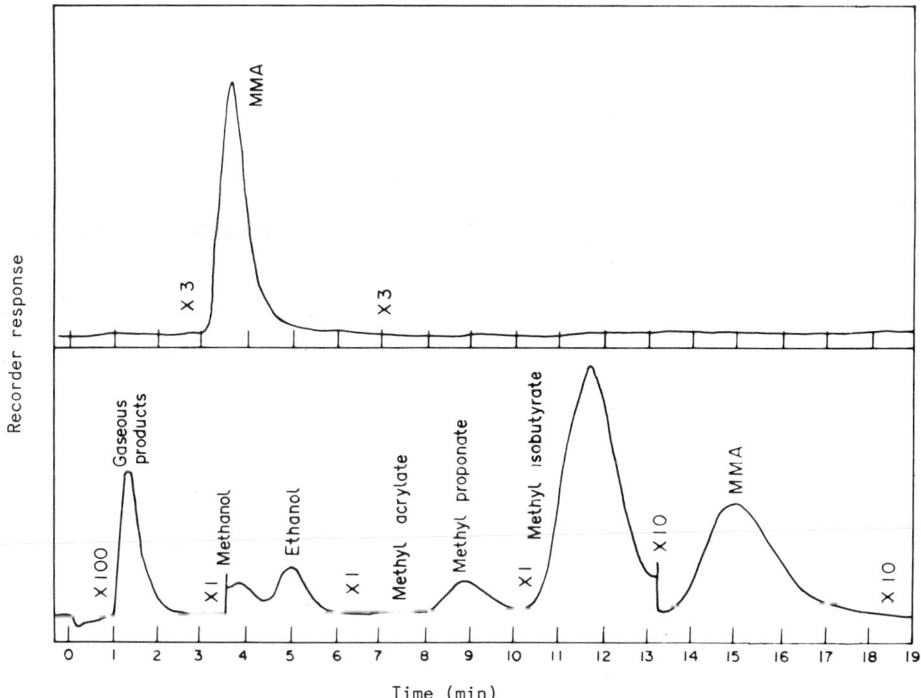

Figure 94   Chromatograms of pyrolysis products of poly(methyl-
             methacrylate); column, dinoyl phthalate: (a) top
             pyrolysis temp., 425°C; col. temp., 128°C; the flow
             rate, 60 ml/min. (b) bottom pyrolysis temp., 1025°C;
             col. temp. 100°C; the flow rate, 20 ml/min (MMA-methyl-
             methyacrylate monomer).

produced by secondary reactions undergone by the styrene monomer.
Reactions leading to these products are even more important at a
temperature of 1025°C.

Lehmann and Brauer[1186] and Brauer[1181] investigated the pyrolysis-gas
chromatography of polymethylmethacrylate at temperatures between 400°C and
1100°C utilizing for the pyrolysis a silica boat surrounded by a platinum
heating coil.

Chromatograms obtained from pyrolysing poly(methylmethacrylate) at
425°C and 1025°C using a dinonyl phthalate column are shown in Figure 94.
Monomer is formed nearly exclusively (99.4%) at 425°C reducing to 20% at
879°C, whereas a number of additional compounds are detected at 1025°C.
Non-volatile products are retained in the column. In view of these
findings regarding the effect of pyrolysis temperature on product
composition it is always desirable to investigate temperature effects at
an early stage and, indeed, studies performed at different pyrolysis
temperatures might provide additional information in polymer
microstructure studies.

In the case of polyethylene, it has been shown that increasing the
pyrolysis temperature to 650°C increased the yield of C_1 and C_2 and non-
normal hydrocarbons due to the occurrence of secondary reactions. In the
case of polystyrene and polyacrylates increase in pyrolysis temperature

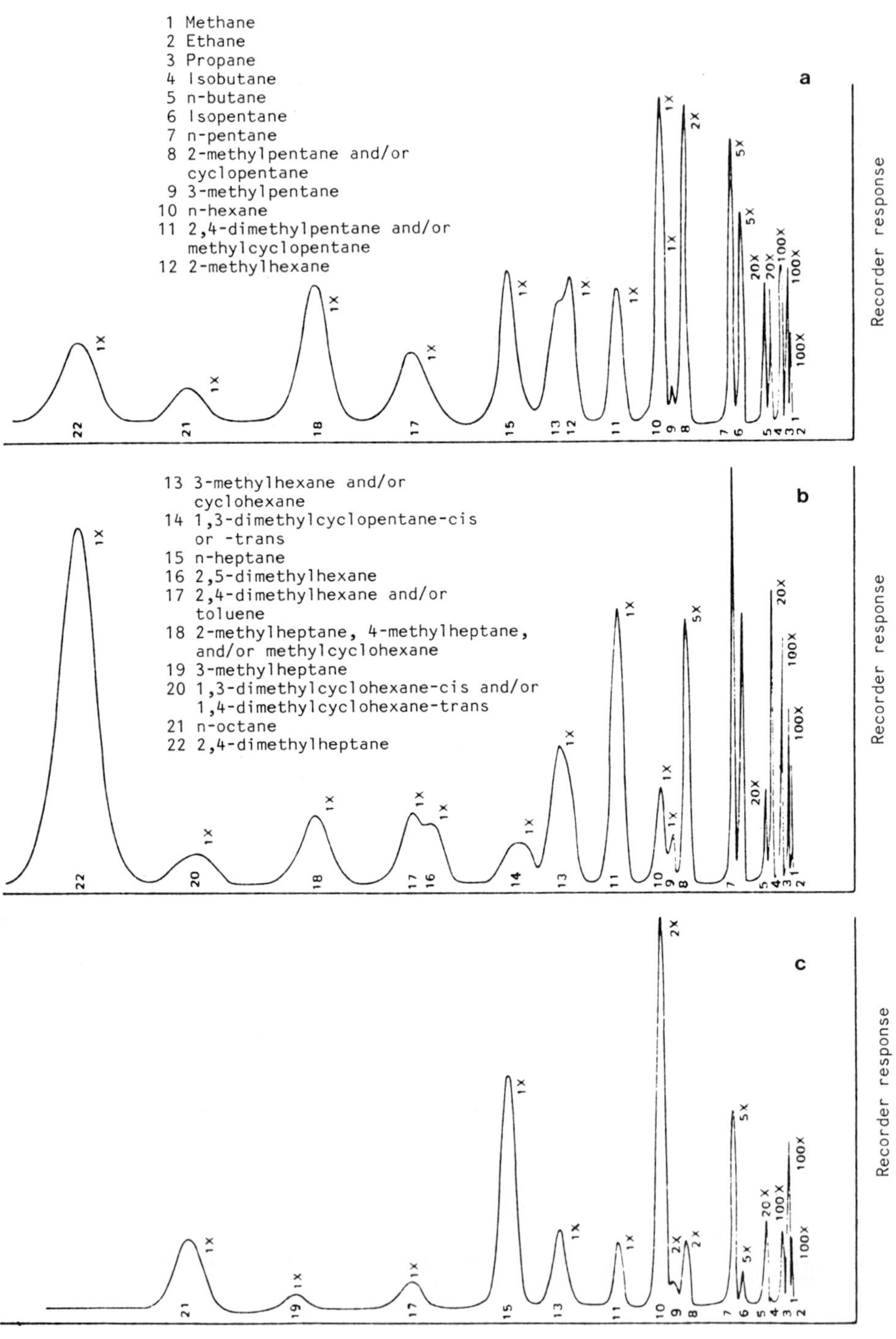

Figure 95   Gas chromatograms of pyrolysates of (a) polyethylene,
(b) polypropylene and (c) ethylene-propylene copolymer.

leads to a decrease in monomer yield and an increase of the amount of products formed in secondary reactions.

It is seen then that in order to successfully apply pyrolysis-gas chromatography to a study of sequence distribution in copolymers, it is necessary to carefully study and control experimental parameters such as sample weight, pyrolysis heating rate and temperature and the type of gas chromatic column and chromatographic conditions used.

Under pyrolysis conditions, different types of polymers break down in different ways. Two important different methods of decomposition are: (i) those involving the formation of free radicals with consequent hydrogen atom transfer followed by subsequent reactions and (ii) the unzipping mechanism. Of course, both mechanisms can occur simultaneously, and the representations below are oversimplified.

Free radical formation
polyethylene

$$- CH_2 - CH_2 - CH_2 - CH_2 \longrightarrow - CH_2 - CH_2^{\bullet} + {}^{\bullet}CH_2 - CH_2 -$$

depoloymerisation

unzipping mechanism

polyacrylic acid

$$\left[\begin{array}{c} - CH_2 - CH - \\ | \\ COOH \end{array} \right]_n \longrightarrow \quad n\ CH_2 = \underset{\underset{COOH}{|}}{C} - H$$

An example of the results obtainable by pyrolysis-gas chromatography is shown in Figure 95, which compares the pyrograms of polyethylene, polypropylene and an ethylene-propylene copolymer. To obtain these results the sample (20 mg), in a platinum dish, was submitted to controlled pyrolysis in a stream of hydrogen as carrier gas. The pyrolysis products were then hydrogenated at 200°C by passing through a small hydrogenation section containing 0.75% platinum on 30/50 mesh aluminium oxide. The hydrogenated pyrolysis products are then separated on a squalane on fireback column, and the separated compounds detected by a katharometer. Under the experimental conditions used in this work only alkanes up to C_9 could be detected.

It can be seen that major differences occur between the pyrograms of these three similar polymers. Polyethylene produces major amounts of normal C_2 to C_8 alkanes and minor amounts of 2-methyl and 3-methyl compounds such as isopentane and 3-methylpentane, indicative of short chain branching on the polymer backbone. In the case of polypropylene, branched alkanes predominate, these peaks occurring in regular patterns, e.g. 2-methyl, 3-ethyl and 2,4-dimethylpentane and 2,4-dimethylheptane, which are almost absent in the polyethylene pyrolysate. Minor components obtained from polypropylene are normal paraffins present in decreasing amounts up to normal hexane. This is to be contrasted with the pyrogram of polyethylene, where n-alkanes predominate. The ethylene-propylene copolymer, on the other hand as might be expected, produces both normal and branched alkanes. 2,4-dimethylpentane and 2,4-dimethylheptane concentrations are lower than occur in polypropylene.

Van Schooten and Evenhuis[1183] prepared gas chromatograms of hydrogenated pyrolysis products of polyethylene, polypropylene, ethylene-propylene copolymers and polyisoprene (hydrogenated natural rubber). These indicate a high degree of alternation in the ethylene-propylene copolymers. They identified most peaks in the chromatograms and ascribed them to a single component, some to two or three different iso-alkanes or cyclo-alkanes. A survey of the relative concentrations of the pyrolysis products is presented in Figures 96(a) and (b) where peak areas have been given for the pyrolysis products of polyethylene, polypropylene, polyisoprene and for a copolymer containing about 50% methylene. A cursory examination of these figures shows that, apart from a few minor differences, the pyrolysis spectra of ethylene-propylene rubber of 50% mC_2 and polyisoprene are very similar. This would indicate a high degree of alternation in the ethylene-propylene copolymer, because presumably the spectrum of a C_2/C_3 copolymer, having poor alternation, resembles that of a 1:1 mixture of polyethylene and polypropylene. The same conclusion was drawn from the size of the 2,4-dimethylheptane peak.

The main feature of the pyrolysis spectra of polyethylene, polypropylene and isoprene can be interpreted on the basis of simultaneous breakage of carbon-carbon bonds in the main polymer chain. For a more detailed interpretation of the pyrolysis spectrum it is necessary to assume a series of radical-chain reactions in which intramolecular hydrogen abstraction plays an important role.

High-vacuum pyrolysis at 40°C of ethylene-propylene copolymers carried out by Van Schooten et al[1184] followed by trapping of released volatiles in dry ice-acetone, produced a mixture of olefins and alpha-olefins. The most volatile fractions, collected in dry ice-acetone, were hydrogenated to saturated hydrocarbons, which were analysed by gas-liquid chromatography. For samples of copolymer prepared using either titanium trichloride or vanadium oxychloride catalysts the chromatogram of this fraction showed peaks of 2,4-dimethylheptane, 2-methylheptane, 4-methylheptane,2,4-dimethylhexane, 3-methylhexane and 2-methylhexane, but only in the chromatogram of the volatile fraction from the copolymer produced using vanadium was a peak of 2,5-dimethylhexane found. This is an indication that polymers prepared with a catalyst containing vanadium oxychloride contain methylene sequences of two units between branches. Van Schooten et al[1184] conclude that ethylene-propylene copolymers prepared with vanadium-containing catalysts, especially those with $VOCl_3$ or $VO(OR)_3$, have methylene sequences of two and four units.

<u>Infrared spectroscopy.</u> The infrared absorption of ethylene copolymers in the 14.28-11.76 micron region can provide information about their sequence distributions. Studies on model hydrocarbons by Van Schooten and others[1185-1188] have shown that the absorption of methylene groups in this region is dependent on the size of methylene sequences in the compounds. Methylene absorptions observed in this region, and their relation to structures occurring in ethylene copolymers, are shown in Table 130. The absorptions at 13.81 and 13.68 micron of several ethylene-propylene copolymers have been assigned[1189] to structures (6) and (4), respectively (Table 130). The relative intensities observed were qualitatively consistent with sequence distributions calculated for the copolymers, assuming reactivity ratios of 7.08 and 0.088 for ethylene and propylene, respectively. Veerkamp and Veermans[1190] developed a technique to measure the intensities of these absorptions accurately. By assuming similar extinction coefficients for the two bands, the ratio of methylene units in $(CH_2)_3$ and larger methylene sequences in a number of copolymers was determined. The results agreed reasonably well with theoretical

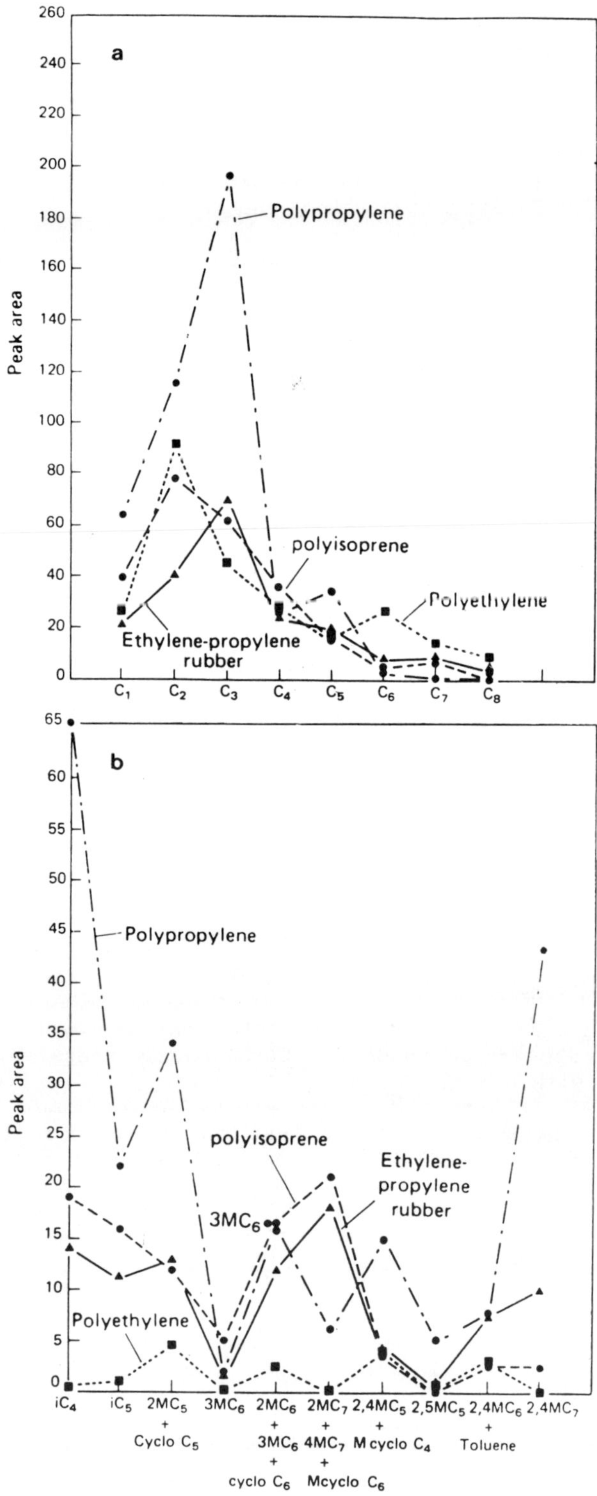

Figure 96   Peak areas obtained after pyrolysis of polyolefins; (a) normal alkanes; (b) iso-alkanes.

Table 130 - Methylene absorption and copolymer structure
responsible for absorption in ethylene copolymers

$-(CH_2)_n-$	cm^{-1}	microns	Sequence	
$-(CH_2)_1$	815	11.76	$\sim CH_2CHRCH_2CHRCH_2CHR\sim$	(2)
$(CH_2)_2$	751	13.31	$\sim CH_2CHRCH_2CH_2CHRCH_2\sim$	(3a)
			or	
			$\sim CHRCHRCH_2CH_2CHRCHR\sim$	(3b)
$(CH_2)_3$	733	13.64	$\sim CH_2CHRCH_2CH_2CH_2CHR\sim$	(4)
$(CH_2)_4$	726	13.77	$\sim CH_2CHR(CH_2CH_2)_2CHRCH_2\sim$	(5)
			or	
			$\sim CHRCHR(CH_2CH_2)_2CHRCHR\sim$	(5b)
$(CH_2)_4$	722	13.85	$\sim CH_2CHR(CH_2CH_2)_nCH_2CHR\sim$	(6)

values. The presence of $(CH_2)_2$ sequences in ethylene propylene copolymers has been considered in studies of their structure[1191]. Such sequences could result from tail-head incorporation of propylene units into the copolymers (structure (3a), Table 130).

Bucci and Simonazzi[1192] also investigated the assignment of infrared bands of ethylene-propylene copolymers in the spectral region 11.11-15.38 micron. They calculated absorbances at various frequencies and attempted a numerical evaluation of the distribution of monomeric units in the copolymer. The spectra of ethylene-propylene copolymers in this frequency range show peaks or shoulders at 12.27, 13.30, 13.64 and 13.85 microns, whose intensity changes with composition. For the assignment of these bands they examined spectra of several model compounds which contain (-CH_2-)$_n$ sequences with different values of n (Table 131). In the same table they compare their results with those given in the literature[1197,1195,1196,1195]. On the basis of this comparison they assign bands at 12.27, 13.30, 13.34 and 13.85 microns to the (-CH_2-) sequences with n = 1,2,3, and 5 or more, respectively (see Table 132).

This assignment does not agree with that proposed by Van Schooten and co-workers, who assign bands at approximately these wavelengths to (-CH_2-) sequences with n = 1,2,3 and 4.

Bucci and Simonazzi[1192] think that the difference lies in the fact that these authors assigned the band at 13.64 microns to the sequence (-CH_2-)$_4$. Bucci and Simonazzi were not able to detect in ethylene-propylene copolymer any absorption at 13.79 micron where the sequence (-CH_2)$_4$ should occur. They recorded the spectrum of squalane compensated with 2,6,10,14-tetramethylpentadecane, and did find a band due to (-CH_2)$_4$ at 13.77 micron.

Natta et al.[1203] determined the degree of alternation of ethylene and propylene units in ethylene-propylene copolymers from the infrared spectrum, using peaks at 13.35, 13.70 and 13.83 micron the one at 13.70 micron being attributed to a sequence of three methylene groups between branch points, presumably due to the insertion of one ethylene between two similarly oriented propylene molecules.

In order to check the correctness of this interpretation Van Schooten et al.[195] examined the infrared absorption spectra between 13 and

Table 131 - Assignment of infrared bands of ethylene-propylene
copolymers (frequency, cm^{-1})

Sequences and compounds	Bucci and Simonazzi [1190]	Mc Murray and Thornton [1193]	Van Schooten et al [1191]	Natta et al [1189, 1194]
$(-CH_2-)_1$	815 (12.27)	785-770 (12.74-12.99)		
Atactic polypropylene	815 (12.27)			
$(-CH_2-)_2$	752 (13.30)	743-734 (13.86-13.62)		
2,5-Diemthylhexane			753 (13.28)	
Ethylene-cis-butene-2 copolymer	752 (13.30)			752 (13.30)
Hydrogenated poly-2,3-dimethyl-butadiene			752 (13.30_	
$(-CH_2-)_3$	733-735 (13.64-13.60)	729-726 (13.72 13.77)		
2,6,10,14-Tetramethylpentadecane	735 (13.60)			
Squalane	735 (13.60)			
Hydrogenated natural	735 (13.60)		739 (13.53)	
$(-CH_2-)_4$	726 (13.77)	724-721 (13.77-13.86)	728 (13.73)	
Squalane compensated with 2,6,10,14-tetramethylpentadecane $(-CH_2-)_5$ or more	722-721 (13.85-13.86)	724-722 (13.81 13.85)	722 (13.85 13.85)	
n-Heptane	722 (13.85)			
n-Decane	721 (13.86)			
n-Nonadecane	721 (13.56)			

Table 132 - Assignment of infrared bands of ethylene-propylene copolymers

Sequences	Absorption frequency		Model compounds	Absorptivities for CH_2 (Kvi) $(cm^{-1}mol^{-1}cm^{-1}ml$
	cm^{-1}	Micron		
$(-CH_2-)_1$	815	12.27	Atactic polypropylene with head to tail linking	0.6 0.10 x 10^4
$(-CH_2-)_2$	752	13.30	Atactic polypropylene with a definite amount of head-to-head linking	1.0 0.15 x 10^4
$(-CH_2-)_3$	733	13.34	Hydrogenated natural rubber, squalane, 2,6, 10,14-tetra-methylpentadecane	1.2 0.20 x 10^4
$(-CH_2-)_5$	722	13.85	Linear C_{10}-C_{19} hydrocarbons	1.2 0.25 x 10^4

14 micron of ethylene-propylene copolymers prepared with various catalyst systems and compared them with the spectra of some model compounds, namely 2.5-dimethylhexane, 2,7-dimethyloctane, 4-methylpentadecane, 4-n-propyltridecane, polypropylene, polyethylene, polybutene-1 and hydrogenated polyisoprene, the last being considered as an ideally alternating ethylene-propylene copolymer.

The infrared absorption bands between 14.28 and 12.99 micron of various samples are given in Figure 97. These bands are due to rocking modes of the CH_2 groups, and their frequency depends on the length of the CH_2 sequences. In 2,5-dimethylhexane, hydrogenated polyisoprene and 2,7-dimethyloctane there are two, three and four CH_2 groups between branches, respectively, corresponding to peaks in the infrared spectra at 13.28 micron (ε spec = 0.0231/g.cm), 13.53 micron (spec. = 0.0581/g.cm) and 13.75 micron (ε spec. = 0.0421/g.cm). Amorphous and crystalline polypropylene do not show any absorption at all in this region, whereas crystalline polyethylene shows two peaks at 13.65 and 13.85 micron. The latter peak is also found in liquid low molecular weight hydrocarbons[1197] and in amorphous polymers containing long sequences of CH_2 groups[1197]; the 13.62 micron peak is a crystalline band, attributed to the CH_2 rocking in the polyethylene crystal[1197-1202]. From the spectra of an amorphous polybutene-1 it was concluded that ethyl side-groups give rise to a peak at around 13.1 micron and from the spectra of 4-methylpentadecane and 4-n-propyltridecane that n-propyl groups absorb at 13.53 micron, proving that the shoulder at 13.3 micron found in some polymers is not due to ethyl or n-propyl end-groups. A survey of the various peaks and shoulders present in the spectra is given in Table 133.

In the spectra of all copolymers, except sample 6, they found a sharp peak at 13.80-13.85 micron, which must be ascribed to sequences of CH_2 groups longer than four, i.e. sequences of more than two ethylene units.

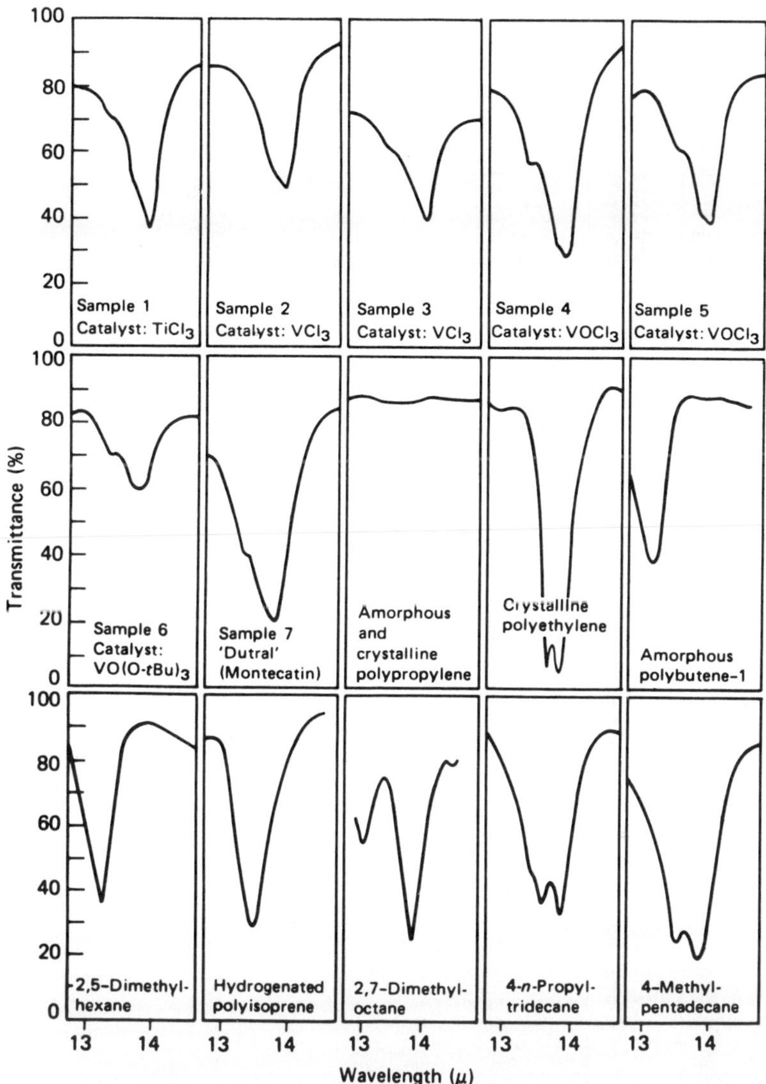

Figure 97   Infrared spectra between 700 and 770 cm⁻¹ (13-14 u) of
              C₂-C₂ copolymers and model compounds.

Significant differences are, however observed between the various
spectra in the wavelength region from 13.0 to 13.7 micron. None of the
copolymer spectra shows a clear shoulder at 13.53 micron where
hydrogenated polyisoprene shows maximum absorption.

This means that in all the samples the content of $(CH_2)_3$ sequences
is low. there are, however, several samples showing a pronounced shoulder
at around 13.3 micron viz. samples 3,4,5,6, and 7 (Table 133). This
shoulder should probably be assigned to sequences of two CH_2 groups
between branch points (cf. spectrum of 2.5-dimethylhexane). Of these
samples two have been prepared with a $VOCl_3$-containing catalyst and one
with a $VO(O-t-Bu)_3$-containing catalyst. Only a few of the samples
prepared with a catalyst containing VCl_3 (No.3), showed the shoulder at

Table 133 - Peaks and shoulders between 770 and 770 cm^{-1} (13-14) in infrared spectra of C_2-C_3 copolymers

Sample	13.8-13.9 ($725-719cm^{-1}$)	13.7-13.8 ($730-725cm^{-1}$)	13.6-13.7 ($735-730cm^{-1}$)	13.50-13.60 ($741-731cm^{-1}$)	13.0-13.35 ($769-749cm^{-1}$)
2,5-Dimethyl-hexane					peak at 13.28 ($753cm^{-1}$)
Hydrogenated polyisoprene				peak at 13.53 ($739cm^{-1}$)	
2,7-Dimethyloctane			peak at 13.73 ($727cm^{-1}$)		
Amorphous butene-1 polymer					peak at 13.12 ($762cm^{-1}$)
4-n-Propyltri-decane	peak at 13.87 ($721cm^{-1}$)			peak at 13.57 ($737cm^{-1}$)	
4-Methylpenta-	peak at 13.85 ($722cm^{-1}$)			peak at 13.53 ($739cm^{-1}$)	
Cryst. polyprop-ylene					
Cryst. polyeth-ylene ethylene propylene co-polymers	peak at 13.85 ($722cm^{-1}$)		peak at 13.65 ($733cm^{-1}$)		

Sample			
Sample 1 TiCl$_3$ catalyst	peak at 13.86 (721cm^{-1})		vague shoulder
Sample 2 VCl$_3$ catalyst	peak at 13.85 (722cm^{-1})	vague shoulder around 13.60 (735cm^{-1})	
Sample 3 VCl$_3$ catalyst	peak at 13.86 (721cm^{-1})		shoulder at 13.3 (752cm^{-1})
Sample 4 VOCl$_3$ catalyst	peak at 13.85 (722cm^{-1})	vague shoulder at 13.7 (730cm^{-1}) (752cm^{-1})	pronounced shoulder at 13.3
Sample 5 VOCl$_3$ catalyst	peak at 13.86 (721cm^{-1})	shoulder at 13.75 (727cm^{-1})	shoulder at 13.35 (749cm^{-1})
Sample 6 VO(O-t-Bu)$_3$ catalyst		broad peak at 13.75 (727cm^{-1})	peak at 13.35 (749cm^{-1})
Sample 7 dutral	peak at 13.83 (723cm^{-1})		pronounced shoulder at 13.3 (752cm^{-1})

13.3 micron. This shoulder was always observed in samples which had been
prepared with a catalyst obtained from $VOCl_3$ or $VO(OR)_3$. It might well be
that this band arises from the presence of $(CH_2)_4$ sequences. This holds
for samples 4,5 and 6, which were prepared with catalysts containing VO(O-
t-Bu)$_3$ or $VOCl_3$ and which, as we have seen, also display absorption bands
that indicate the presence of C_2 sequences. None of the copolymer spectra
obtained thus far indicates the presence of the structure.

$$
\begin{array}{cc}
H & H \\
| & | \\
-\,C\,-\,C\,- \\
| & | \\
Me & Me
\end{array}
$$

which would give rise to absorption at 8.9 micron. This means that head-
to-head orientation of propylene occurs only after addition of an ethylene
unit.

Van Schooten et al. [1195] have shown that ethylene-propylene
copolymers prepared with $VOCl_3$-or $VO(OR)_3$-containing catalysts not only
contain odd-numbered methylene sequences, as was expected, but also
sequences to two units. Some indications for the presence of methylene
sequences of four units were found. They[1196] also investigated copolymers
with much lower ethylene contents and pure polypropylenes prepared with
catalysts consisting of $VOCl_3$ or $VO(OR)_3$ and alkylaluminium
sesquichloride.

The infrared spectra of the ethylene-propylene copolymers containing
15 to 30 mole% of ethylene show little or no absorption above 13.70
micron. This means that very few methylene sequences of four or more
units are present. The absorption peaks at 13.30 and 13.60 micron reveal
the presence of sequences of two and three units respectively. Thus these
copolymers contained nearly all of the ethylene in isolated monomer units.

Veerkamp and Veermans[1190] have reported on the differential
measurement of $(CH_2)_2$ and $(CH_2)_3$ units at 13.42 and 15.87 microns
respectively, in ethylene-propylene copolymers. The band at 13.89 microns
was assigned to units of 5 or more CH_2 units. Thus, the absorbance ratio
A13.42/A13.89 microns provides a measure of the number of isolated
ethylene units, and when corrected for the effect of composition (by
multiplying by the ratio mole%C$_3$/mole5C$_2$) is related to the inverse of the
ratio A10.33 micron/A8.69 micron times the ratio mole/C$_2$/mole%C$_3$. the
more random the copolymer, as indicated by the absorbance ratio A13.42
micron/A13.89 micron for the relative number of isolated ethylene units,
the more diffuse the 10.33 micron band becomes.

In summary, the appearance of contiguous vs. isolated monomer units
may be studied by the spectral characteristics of the methyl and methylene
group rocking and wagging bands. The C-H stretching frequencies are
useful for measurement and copolymer composition when spectrometers of
sufficient resolution are available. A typical infrared spectrum of a
cast film of ethylene-propylene copolymer, showing the base line, is
reproduced in Figure 98. The ratio between the CH_3 and CH_2 asymmetrical
C-H stretching band was not used because of the large difference in
relative intensities. The ratios between the asymmetrical CH_3 and the
symmetrical CH_2 bands, which have nearly the same relative intensities,

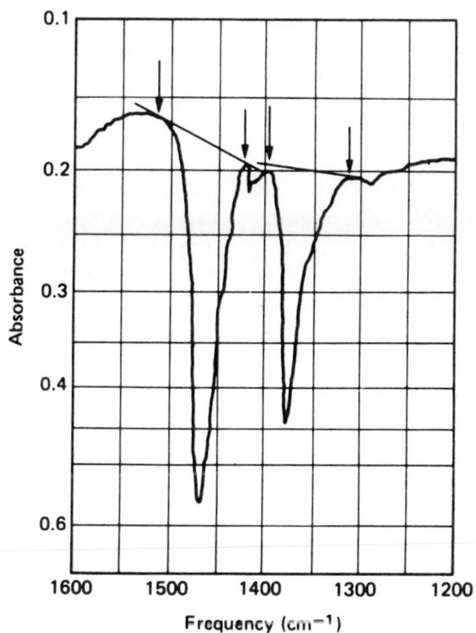

Figure 98   Typical infrared spectrum of an ethylene-propylene
              copolymer in the C-H bending region of the spectrum.
              Base line points shown by arrows. Grating to prism
              changeover at 1416 cm^{-1} (7.06 u).

produced more reliable results.   Resolution of the CH_3 from the CH_3
asymmetrical as well as the symmetrical bands must be satisfactory, as no
curvature of the calibration curve was seen.   If the CH_3 symmetrical band
was not resolved from the CH_2 band, a change in the slope of the curve at
high propylene contents would have been observed.

Liang et al[1204] on the basis of frequencies, relative intensities,
polarization properties, and effects on deuteration of polypropylene, have
tentatively assigned the 10.33 micron band in polypropylene to the methyl
rocking mode mixed with CH_2 and CH rocking vibrations.   From this
assignment, perhaps it is a change in the magnitude of the mixing with the
CH_2 and CH modes in the case of the isolated propylene units in the
copolymer which gives rise to a more diffuse absorption band at 10.33
microns.   In this connection, Liang and Watt[1205] prepared non-
stereospecific polypropylene in which 10.33 and 8.69 micron bands are
either missing or very weak.   Also, the occurrence of head-to-head
arrangements of contiguous propylene units would be expected to influence
the behaviour of the methyl rocking mode and any mixing which may be
involved.

For the purpose of polymer characterization the ratio of absorbances
at 10.33 and 8.69 microns should provide a measure of the degree of
randomness with respect to the introduction of propylene units into the
copolymer.   On the basis of the initial observations, the ratio A10.33
micron/A8.69 micron should decrease as the randomness increases.   In
addition, an increase in ethylene content is expected to increase the
probability of producing propylene units isolated by ethylene units.   For

comparison of the 10.33 to 8.69 absorbance ratio with other data, correction for the effect of composition is made by multiplying by the ratio mole%CH_2mole%CH_3, the molar ratio of ethylene to propylene in the copolymer. Relationships have been found between this modified absorbance ratio and other established physical or spectral properties indicative of randomness (or conversely, block polymer)

   <u>Nuclear magnetic resonance spectroscopy.</u>   Ethylene-propylene copolymers can contain up to four types of sequence distribution of monomeric units. These are propylene to propylene (head-to-tail and head-to-head), ethylene-propylene and ethylene to ethylene. These four types of sequences and the average sequence lengths of both monomer units, i.e. the value of n, below can be measured by the Tanaka and Hatada[1206] method.

$$\left(-\overset{\overset{\textstyle H}{|}}{\underset{\underset{\textstyle H}{|}}{C}}-\overset{\overset{\textstyle H}{|}}{\underset{\underset{\textstyle Me}{|}}{C}}-\overset{\overset{\textstyle H}{|}}{\underset{\underset{\textstyle H}{|}}{C}}-\overset{\overset{\textstyle H}{|}}{\underset{\underset{\textstyle Me}{|}}{C}}-\right)_n$$

$$\left(-\overset{\overset{\textstyle H}{|}}{\underset{\underset{\textstyle H}{|}}{C}}-\overset{\overset{\textstyle H}{|}}{\underset{\underset{\textstyle Me}{|}}{C}}-\overset{\overset{\textstyle H}{|}}{\underset{\underset{\textstyle Me}{|}}{C}}-\overset{\overset{\textstyle H}{|}}{\underset{\underset{\textstyle H}{|}}{C}}-\right)_n$$

$$\left(-\overset{\overset{\textstyle H}{|}}{\underset{\underset{\textstyle Me}{|}}{C}}-\overset{\overset{\textstyle H}{|}}{\underset{\underset{\textstyle H}{|}}{C}}-\overset{\overset{\textstyle H}{|}}{\underset{\underset{\textstyle H}{|}}{C}}-\overset{\overset{\textstyle H}{|}}{\underset{\underset{\textstyle H}{|}}{C}}-\right)_n$$

$$\left(-\overset{\overset{\textstyle H}{|}}{\underset{\underset{\textstyle H}{|}}{C}}-\overset{\overset{\textstyle H}{|}}{\underset{\underset{\textstyle H}{|}}{C}}-\overset{\overset{\textstyle H}{|}}{\underset{\underset{\textstyle H}{|}}{C}}-\overset{\overset{\textstyle H}{|}}{\underset{\underset{\textstyle H}{|}}{C}}-\right)_n \cdot$$

Measurements were made at 15.1 MHz. Assignments of the signals were carried out using the method of Grant and Paul[1207] and also by comparing the spectra with those of squalane, hydrogenated natural rubber (isoprene), polyethylene and atactic polypropylene. The accuracy and precision of intensity measurements were acceptable at 12% maximum.

   Tanaka and Hatada[1206] studied the accuracy and precision of such intensity measurements in ^{13}C NMR spectra and elucidated the distribution of monomeric units in ethylene-propylene copolymers. These analyses are particularly straightforward if one of the monomer units is present at a level of 95% or greater, because the other monomer will then occur primarily as an isolated unit.

   The investigation of the chain structure of copolymers by ^{13}C NMR has several advantages because of the large chemical shift involved. Crain et al[1208] and Cannon and Wilkes[1209] have demonstrated this for the determination of the sequence distribution in ethylene-propylene copolymers. In these studies the assignments of the signals were deduced by using model compounds and the empirical equation derived by Grant and Paul for low molecular weight linear and branched alkanes.

Ray et al.[210] determined comonomer content in isotactic ethylene-propylene copolymers, complete diad and triad content as well as partial tetrad and pentad distributions using ^{13}C NMR.

Proton magnetic block copolymers have been used to characterize ethylene-propylene block copolymers, containing from zero to 40% ethylene, either as homopolymer or copolymer blocks.[1211]. The test is independent of tacticity and provides qualitative information on copolymer sequencing and propylene chain structures. The analysis was developed using a series of standard reference polymers synthesized to contain various ratios of ^{14}C-tagged ethylene and propylene. All measurements were made on 10% polymer solutions in diphenyl ether. Analyses are accurate to about $\pm$ 10% at higher ethylene concentrations. The method is sensitive, with less precision, to below 1% ethylene either as blocks or homopolymer. Porter[1211] found that resolution improves with increasing temperature and with decreasing polymer concentration.

In addition to measuring the concentration of branching groups up to octyl in polyethylene ^{13}C NMR can be used to elucidate the composition and structure of olefin copolymers by measurement of the concentration of branched groups. Thus in the case of ethylene-propylene copolymers containing between 97 and 99% propylene it is possible to measure propylene groups and elucidate polymer microstructure.

As shown by a typical example in Figure 99, each ^{13}C NMR spectrum was recorded with proton noise-decoupling to remove unwanted [13]C-^{1}H scalar couplings. No corrections were made for differential Nuclear Overhauser effects (NOE) since constant NOEs were assumed[1212,1213] in agreement with previous workers[1214,1215,1211,1210]. Constant NOEs for all major resonances in low ethylene content ethylene-propylene copolymers have been reported[1216].

The ^{13}C NMR spectrum of an ethylene-propylene copolymer, containing approximately 97% propylene in primarily isotactic sequences, is shown in Figure 99. The major resonances are numbered consecutively from low to high field. Chemical shift data and assignments are listed in Table 134. Greek letters are used to distinguish the various methylene carbons and designate the location of the nearest methane carbons.

Paxton and Randall[1217] in their method use the reference chemical shift data obtained on a predominantly isotactic polypropylene and on an ethylene-propylene copolymer (97% ethylene). They concluded that the three ethylene-propylene copolymers used in their study (97-99% propylene) contained principally isolated ethylene-ethylene linkages. Knowing the structure of their three ethylene-propylene copolymers, they used the ^{13}C NMR relative intensities to determine the ethylene-propylene contents and thereby establish reference copolymers for the faster infrared method involving measurements at 13.66 microns. After a detailed analysis of resonances Paxton and Randall[1217] concluded that methine resonances 4 and 5 (Table 134) gave the best quantitative results to determine the comonomer composition. The composition of the ethylene-propylene copolymers was determined by peak heights using the methine resonances

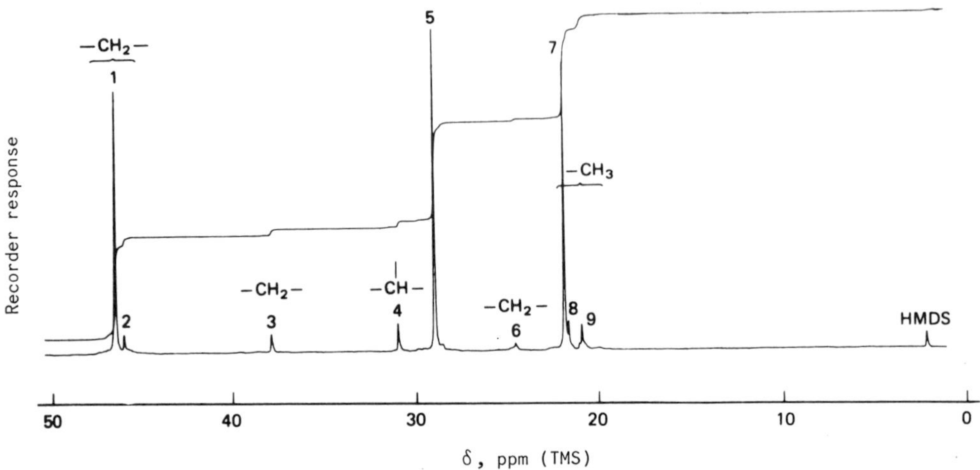

Figure 99   A ^{13}C NMR spectrum at 25.2 MHz and 125°C of a 97/3 propylene-ethylene copolymer in 1,2,4-trichlorobenzene and perdeuterobenzene. The internal standard copolymer is 1,2,4-trichlorobenzene and perdeuterobenzene.

only.   In no instance was there any evidence for an inclusion of consecutive ethylene units. Thus, composition data from ^{13}C NMR could now be used to establish an infrared method based on a correlation with the 13.66 micron band which is attributed to a rocking mode, r, of the methylene trimer, $-(CH_a)_3$.

Randall[1218] has developed a ^{13}C NMR quantitative method for measuring ethylene-propylene mole fractions and methylene number average sequence lengths in ethylene-propylene copolymers. He views the polymers as a succession of methylene and methyl branched methine carbons, as opposed to a succession of ethylene and propylene units. This avoids problems associated with propylene inversion and comonomer sequence assignment. He gives methylene sequence distributions from one to six and larger consecutive methylene carbons for a range of ethylene-propylene copolymers and uses this to distinguish copolymers which have either random, blocked or alternating comonomer sequences.

Various other workers have discussed the application of ^{13}C NMR to the analysis of sequence distribution in ethylene-propylene copolymers[1207,1208,1210,1211,1214,1215,1218-1222].

9.4.3 Sequences in Polyethylene

Voigt[1170] states that the formation of decomposition products of polyethylene is a result of a primary step in a statistical thermal breaking down of a chain:

$$- CH_2 - CH_2 - CH_2 - CH_2 - \longrightarrow - CH_2 - \overset{\bullet}{CH_2} + \overset{\bullet}{CH_2} - CH -$$

in which the free radicals formed can react further either by depolymerization (unzipping) or by the transfer of hydrogen atoms. The

Table 134 - Observed and reference ^{13}C NMR chemical shifts in ppm for ethylene-propylene copolymers and reference polypropylenes with respect to an internal trimethylsilane standard

Line	Carbon	3/97 E/P[a]	E/P	97/3[a] E/P	Sequence assignment	Reference cryst. (ref 1212)	PP amorphous (ref 1212)
1	alpha-alpha CH$_2$	46.4	46.3		PPPP	46.5	47.0-47.5r 46.5
2	alpha-alpha CH$_2$	46.0	45.8		PPPE		
3	alpha-alpha CH$_2$	37.8	37.8		PPEP		
4	CH	30.9	30.7		PPE		
5	CH	28.8	28.7		PPP	28.5	28.8mmmm 28.6mmmr 28.5rmmr 28.4mr+rr
6	beta-beta CH$_3$	24.5	24.4		PPEPP		
7	CH$_3$	21.8	21.6	P	PPPPP	21.8	21.3-21.8mm 20.6-21.0mr 19.9-20.3rr
8	CH$_3$	21.6	21.4		PPPE		
9	CH$_3$	20.9	20.7		PPPEP		
	CH$_3$	20.9	20.7		PPPEE		
	CH$_3$		19.8	19.8	EPE		
	alpha-CH$_2$			37.4	EPE		
	beta-CH$_2$			27.3	EPE		
	-(CH$_2$)n		29.8	28.8	EEE		

[a] As measured by Paxton and Randall[1217]

first type of reaction leads to the corresponding monomers, while the latter competes with the continuing decomposition of the chains to form a spectrum of hydrocarbons with different chain lengths. This spectrum contains, apart from saturated hydrocarbons and diolefins, olefins which, in their structure and quantitative distribution, are dependent in characteristic manner on the nature of the initial polymers, as might well be expected. The very marked depolymerization which takes place with other polymers - such as polymethacrylates, polystyrene, and the like - does not appear to play any major role in the cases of polyethylene or other polyolefins, since the corresponding monomers do not exist in any considerable excess as compared with the other decomposition products. As is to be expected, only unbranched fractions arising from statistical decomposition arise in the case of polyethylene, which breaks down with reactions of the following type:

$$- CH_2 - CH_2 - \overset{\bullet}{CH_2} + {}^{\bullet}CH_2 - CH_2 - \longrightarrow CH_2 - CH = CH_2 + CH_3 - CH_2 -$$

$$- CH_2 - CH_2 - CH_2 - + {}^{\bullet}CH_2 - CH_2 - \longrightarrow - CH_2 - CH - CH_2 - + CH_3 - CH_2 -$$

$$- CH_2 - CH_2 - CH_2 - CH_2 - CH_2 - \longrightarrow - CH_2 - CH = CH_2 + {}^{\bullet}CH_2 - CH_2 -$$

The very small amounts of branched molecules which are present even in low-pressure polyethylenes do not make themselves noticeable on the gas chromatogram under the conditions employed by Voight[1170]. Branched molecules are, however, found under other more searching gas chromatographic conditions (see later). In this connection, it should be pointed out that high-pressure polyethylenes produced completely identical gas chromatograms.

According to Wall et al.[1223] the pyrolysis of polyethylene proceeds by a radical chain mechanism. The products formed result from the process of random-chain cleavage, followed by intermolecular or intramolecular hydrogen abstraction. Hydrogen abstraction occurs preferentially at tertiary carbon atoms, and product formation results from homolysis of the carbon-carbon bond beta to the radical site. The major products formed are the n-alkanes and alpha-omega-diolefins. The peaks between the triplets result from chain branching.

From an examination of the products formed from the pyrolysis of high-density polyethylene and polymethylene, Tsuchiya and Sumi[1224,1225] proposed that the major hydrogen-abstraction reaction is due to an intramolecular cyclization. They proposed that, following initial radical formation at C_1, successive intramolecular hydrogen abstractions occur along the chain resulting in the formation of new radicals at C_5 C_9 and C_{13} as shown in equation 1-4.

$$R - R \longrightarrow 2 R \tag{1}$$

$$R' = (3)CH_2 \overset{CH_2}{\diagup \diagdown} \overset{\bullet}{CH} \rightsquigarrow \overset{5}{CH_3(CH_2)_3 - CH} - CH_2 - R \tag{2}$$

with $(2)CH_2$ and H substituents, and $(1)CH_2$

$$\overset{5}{CH_3 - (CH_2)_3 - CH} - CH_2 - R \longrightarrow \overset{9}{CH_3 - (CH_2)_6 - CH_2 - CH} - CH_2R \tag{3}$$

$$\overset{9}{CH_3 - (CH_2)_6 - CH_2 - CH} - CH_2 - R \longrightarrow \overset{13}{CH_3 - (CH_2)_{11} CH} - CH_2 - R \tag{4}$$

Cleavage of the carbon-carbon bonds beta to these macroradicals results in the formation of increased amounts of C_6, C_{10} and C_{14} alpha olefins an C_3, C_7 and C_{11} n-alkanes over that which could be predicted from statistical-chain cleavage.

9.4.4 Sequences in Polypropylene

The studies of Wall and Straus[1226] indicate that pyrolysis of polypropylene proceeds, principally by random cleavage of the polymer chain. Voigt[1170] observed no straight-chain decomposition products of polypropylene with a chain length greater than C_5, a fact that is in agreement with the prediction readily deduced from the head-to-tail structures of the polymer. The larger decomposition products all contain methyl branches, and the multiple-branched chains the expected 2,4-branching.

Van Schooten and Evenhuis[1178,1179] found identical pyrolysis patterns for various polypropylene samples. Depolymerization is much more important for polypropylene pyrogram can be interpreted as originating from intramolecular hydrogen transfer with the fifth carbon atom of the secondary radical:

$\overline{1}$,n-pentane + $\overline{1}$ $\overline{1}$,2,4 dimethyl heptanc

```
                     H
         |           |              |
    - C - ¦ - C - C - C - ¦ - C - C - C•     ⟶
      |   ¦           ¦           |       |
      C      11 C   1         C       C
```

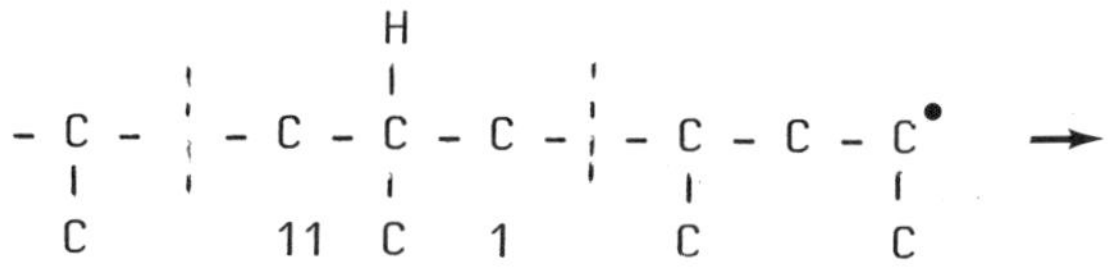

The primary radical gives it this way:

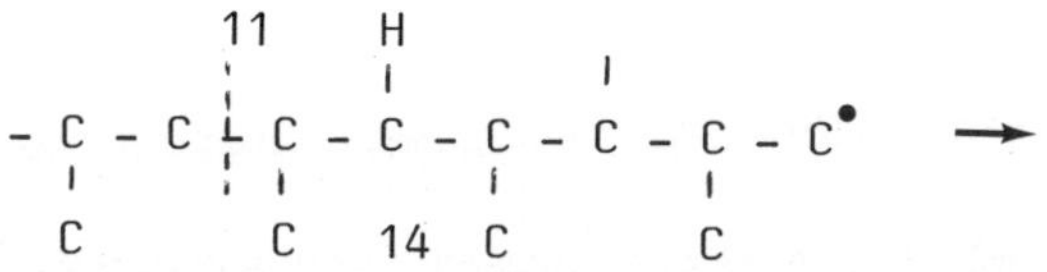

```
         11         H
         ¦          |          |
    - C - C ┴ C - C - C - C - C - C•     ⟶
      |     ¦ ¦           ¦       |
      C       C  14   C           C
```

The isobutane and especially the methane peaks are rather small, indicating that this reaction is not very frequent for the primary radical. For this radical, hydrogen transfer with the sixth carbon atom might be more important, as the 2-methylpentane peak could be explained in this way. Another possibility is that this peak has to be ascribed to intra-molecular hydrogen transfer reactions (i.e. formation of 2-methylpentadiene). Table 135 lists the products of expected from intramolecular hydrogen transfer during the pyrolysis of polypropylene.

9.4.5 Sequences in Polyisoprene

Polyisoprene (hydrogenated natural rubber) is a completely alternating ethylene propylene copolymer (i.e. has not ethylene or propylene blocking) and is therefore an interesting substance for pyrolysis gas-chromatography studies. The surface area of the main peaks up to C_{13} obtained by Van Schooten and Evenhuis[1178,1227] indicate that the

Table 135 – Products expected from intramolecular hydrogen transfer
during pyrolysis of polypropylene (main peaks found in pyrogram
in Figure 95 are underlined)

	Number of the hydrogen-donating carbon				
	5	6	7	8	9
Secondary radical C-C-C-C-C-C-C | | | | C C C C	$\underline{nC_5}$ $\underline{2.4MC_7}$	CH_4 $\underline{2MC_5}$ $4MC_7^*$ $4.6MC_9$	$4MC_7$ $2.4.6MC_9$	CH_4 $2.4MC_7$ $4.6MC_9^*$ $4.6.8MC_{11}$	$4.6MC_9$ $2.4.6.8MC_{11}$

Primary radical
C-C-C-C-C-C-C-
  | | |
  C   C   C
	$1C_4$
	$\underline{2MC_5}^*$
	$\underline{2.4MC_7}$
	CH_4
		$\underline{2MC_5}$
		$2.4.6MC_7$
			CH_4
			$2.4MC_5$
			$2.4MC_7^*$
			$2.4.6MC_9$
				$2.4MC_7$
				$2.4.6.8MC_9$
					CH_4
					$2.4.6MC_7$
					$2.4.6MC_9^*$
					$2.4.6.8MC_{11}$

* Formed by hydrogen exchange with a methyl group.

unzipping reaction which would yield equal amounts of ethylene and
propylene in the hydrogenated pyrolysate, evidently takes place to some
extent, but is less important than the hydrogen transfer reactions.

$$\text{Unzipping} \quad \sim CH_2 - \underset{\underset{Me}{|}}{C} = CH\text{-}CH_2\text{-}CH_2\text{-}\underset{\underset{|}{Me}}{C} = CH - CH_2\sim$$

$$Me - C + CH_2 = CH_2$$

The large numbered peaks produced upon pyrolysis-gas chromatography
reflect the many possible transfer reactions for this polymer some of
which are illustrated below. Hydrogen transfer reactions occur from the
fifth carbon atom predominate as indicated by the large butane, 3-methyl
hexane and 2-methyl heptane peaks:

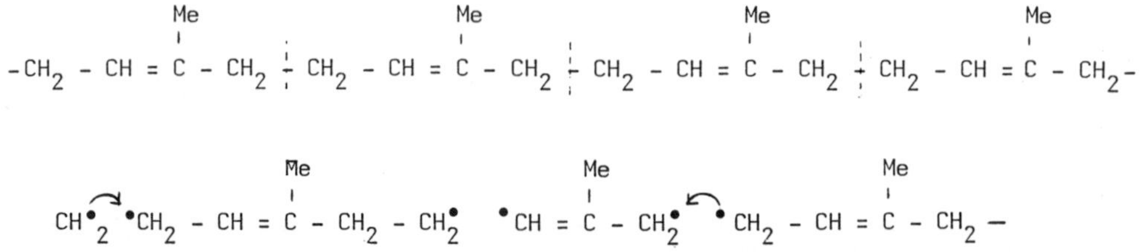

$$-CH_2 - CH = \underset{\underset{Me}{|}}{C} - CH_2 \,\vdots\, CH_2 - CH = \underset{\underset{Me}{|}}{C} - CH_2 \,\vdots\, CH_2 - CH = \underset{\underset{Me}{|}}{C} - CH_2 \,\vdots\, CH_2 - CH = \underset{\underset{Me}{|}}{C} - CH_2-$$

$$CH_2^{\bullet} \,\,{}^{\bullet}CH_2 - CH = \underset{\underset{Me}{|}}{C} - CH_2 - CH_2^{\bullet} \quad {}^{\bullet}CH = \underset{\underset{Me}{|}}{C} - CH_2^{\bullet} \,\,{}^{\bullet}CH_2 - CH = \underset{\underset{Me}{|}}{C} - CH_2 -$$

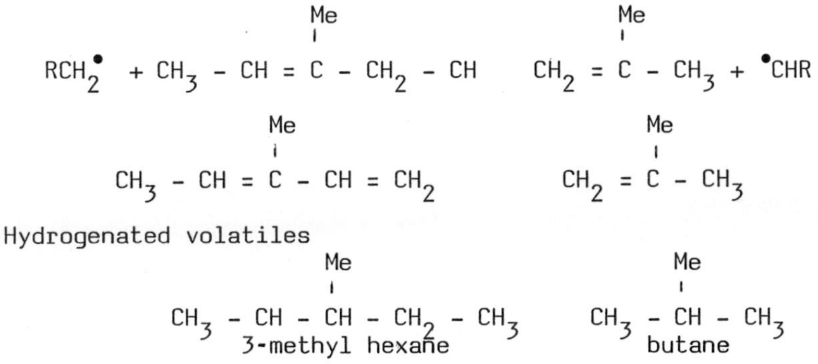

$$RCH_2^{\bullet} + CH_3 - CH = \overset{\overset{\displaystyle Me}{|}}{C} - CH_2 - CH \qquad CH_2 = \overset{\overset{\displaystyle Me}{|}}{C} - CH_3 + {}^{\bullet}CHR$$

$$CH_3 - CH = \overset{\overset{\displaystyle Me}{|}}{C} - CH = CH_2 \qquad CH_2 = \overset{\overset{\displaystyle Me}{|}}{C} - CH_3$$

Hydrogenated volatiles

$$CH_3 - \overset{\overset{\displaystyle Me}{|}}{CH} - CH - CH_2 - CH_3 \qquad CH_3 - \overset{\overset{\displaystyle Me}{|}}{CH} - CH_3$$
$$\text{3-methyl hexane} \qquad\qquad \text{butane}$$

Hydrogen transfer also from the ninth carbon atom, which is shown by the
size of the 3-methyl nonane, 3,7-dimethyldecane and the 2,6-
dimethylrendecane peaks:

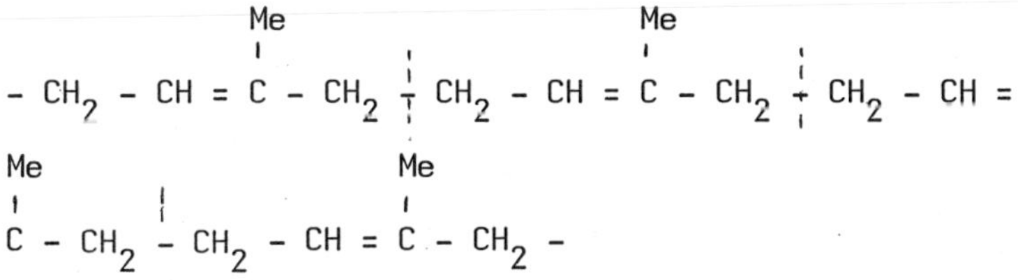

$$- CH_2 - CH = \overset{\overset{\displaystyle Me}{|}}{C} - CH_2 \overset{\vdots}{} CH_2 - CH = \overset{\overset{\displaystyle Me}{|}}{C} - CH_2 \overset{\vdots}{+} CH_2 - CH =$$

$$\overset{\overset{\displaystyle Me}{|}}{C} - CH_2 - CH_2 - CH = \overset{\overset{\displaystyle Me}{|}}{C} - CH_2 -$$

products after hydrogenation

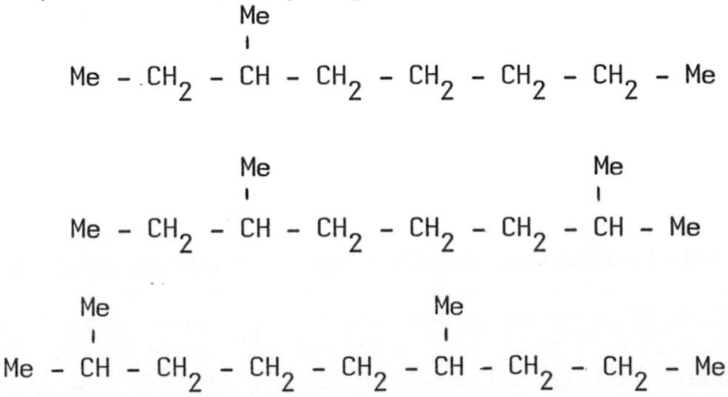

$$Me - CH_2 - \overset{\overset{\displaystyle Me}{|}}{CH} - CH_2 - CH_2 - CH_2 - CH_2 - Me \qquad \text{3-methyl nonane}$$

$$Me - CH_2 - \overset{\overset{\displaystyle Me}{|}}{CH} - CH_2 - CH_2 - CH_2 - \overset{\overset{\displaystyle Me}{|}}{CH} - Me \qquad \text{3,7-dimethyl decane}$$

$$Me - \overset{\overset{\displaystyle Me}{|}}{CH} - CH_2 - CH_2 - CH_2 - \overset{\overset{\displaystyle Me}{|}}{CH} - CH_2 - CH_2 - Me \qquad \text{2,6-dimethyl undecane}$$

9.4.6 Sequences in Butadiene - Propylene Copolymers

The sequence structure of this copolymer has been elucidated by
ozonolysis techniques[1228-1230]. Samples of highly alternating copolymers
of butadiene and propylene yielded large amounts of 3-methyl-1,6-hexane
dial when submitted to ozonolysis. The ozonolysis product from 4-methyl-
cyclohexane-1

$$(\text{i.e. 3-methyl-1,6-hexane dial,} \quad O = C - C - C - C - C - \overset{\overset{\displaystyle }{}}{C} = CO$$
$$\overset{|}{C}$$

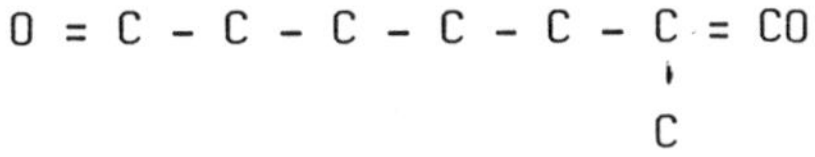

was used as a model compound for this structure. Ozonolysis of these polymers occurs as shown below:

$$
\begin{array}{ccccc}
\text{BD} & \text{prop} & \text{BD} & \text{prop} & \text{BD} \\
\end{array}
$$

- C - C = C - C - C - C - C - C = C - C - C - C - C - C = C - C -

$$\uparrow \qquad | \qquad\qquad \uparrow \qquad | \qquad\qquad \uparrow$$

$$O_3 \qquad \text{Me} \qquad O_3 \qquad \text{Me} \qquad O_3$$

triphenylphospine

O = C - C - C - C - C - C = O

|

C

The amount of alternation in these polymers can be determined if the amounts of 1,4 and 1,2 polybutadiene structure and total propylene have been determined by infrared or NMR spectroscopy (Chapter 10.3). Table 136 shows results obtained for several butadiene-propylene copolymers having more or less alternating structure. Similar polymers have been analysed by Kawaski[1229] by use of conventional ozonolysis methods with esters as the final products.

9.4.7 Sequences in Ethylene-Methyl Methacrylate Random Copolymers

Sigmura and Takenchi[1231] characterized subtle differences in two types of ethylene (92.4-83.1%) methyl methacrylate (7.6-16.9%) random copolymers of differing thermal stabilities by conventional and step wise pyrolysis -gas chromatography. These methods provided information on the degree of branching and the localized distribution of the methyl methacrylate unit in the copolymer chain. Figure 100 shows conventional pyrograms of (a) a copolymer B-2 containing 12.9% methyl methacrylate and 8.31% ethylene, (b) a copolymer A-3 containing 13.2% methyl methacrylate and 84.8% ethylene and (c) high density polyethylene. The general properties of these pyrograms are similar and this fact suggests the existence of fairly long ethylene sequences in both copolymers. The main degradation products on the pyrograms consist of mainly mono-olefins with a terminal double bond (mainly butadiene and butene-1) while the methyl methacrylate monomer is included in the first cluster of the peaks by this separation column. A small amount of n-paraffins mostly overlaps with the peaks of the olefins with the same carbon, number. On the other hand, branched alkenes and diene compounds usually appear between the main peaks. In the pyrogram of copolymer (a) (Figure 100 (a)), the intensity of those peaks is significantly stronger (shown by arrows) than those for copolymer (b) (Figure 100(b). This result suggests the higher degree of branching for the copolymer (a) since the formation of both isoalkanes and diene compounds are closely related to the branching in polyolefins.

Further work suggested that when the methyl methacrylate contents are in the 7.6 to 16.97% range both copolymers (a) and (b) have fairly randomly distributed short methyl methacrylate sequences surrounded by ethylene units in the copolymer chain.

Table 137 shows the observed relative yield of butadiene to 1-butene from the two copolymers pyrolysed at 690°C. These differences in the relative yield to butadiene from the two types of the copolymer might be

Table 136 - Microzonolysis of butadiene-propylene copolymers

	1,4 (%)	1,2 (%)	Propy-lene (mole%)	Succin-alde-hyde	3-meth-yl-1,6-hexane-dial	3-form-yl 1,6-hexane-dial	4-oct-ene-1, 8-dial	Alter-nating BD/Pr(%)
	Area (from gas chromatography) %							
A	45	5.7	49.3	5	92	1	2	77
B	47.8	2.2	50	11.5	85	0.5	3	71
C	53.1	3.2	43.7	25	61	6	8	48
D	-	-	30	49	38	1	12	33

Table 137 - Relative ycilds of butadiene to 1-butene from the copolymers at 690°C

MMA Content %w	relative peak area (butadiene/1-butene)
7.6	0.767
11.2	0.867
13.2	0.994
11.5	1.059
12.9	1.031
13.1	1.055
14.3	1.042
16.1	1.018

Table 138 Total MMA recovery from the copolymers during repeated stepwise pyrolysis

Pyrolysis temp. (°C)	Total MMA recovery (%)[a]	
	Methyl methacrylate % 13.2	B-2 12.9
400	15.7	21.9
420	16.7	17.3
440	14.1	13.8

[a] Observed total yield against the MMA content in the sample pyrolysed.

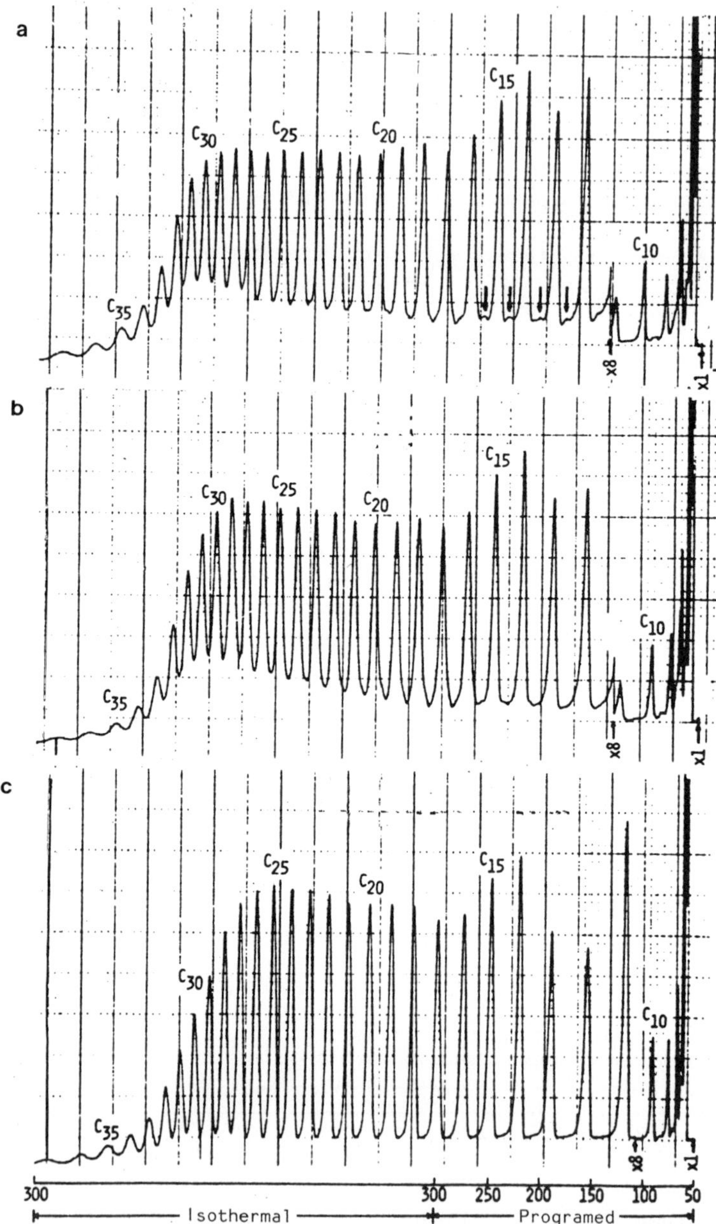

Figure 100   Pyrograms of E–MMA copolymers and polyethylene at 690°C
(fragments with a wide range of boiling points).
Separation column; 5% OV-17 on 80/100 mesh Chromosorb W,
DCHS treated, temperature programmed from 50 to 300°C:
(a) methylmethacrylate-ethylene (12.9:83.1% w/w)
copolymer B-2; (b) methyl methacrylate – ethylene 132:
(84.8% w/w) copolymer A3 (former polymer more highly
branched than latter) and (c) high density polyethylene.

Table 139 - Sequence measurement in polymers

Polythene	Sequence measured	Reference
Ethylene-propylene	$-(CH_2)-$	1233-1236
1-Hexene-4-methyl-1-pentene	block structure	1237
C_1-C_2O alkylethylene	methylene sequences	1238
Ethylene-2-butene	$-(CH_2-CH_2)$ units	1239
Acrylonitrile-2-substitued propene	$-CH_2-CCH_3-$	1240
Acrylonitrile-styrene	sequences	
Styrene-acrylonitrile	sequence distributions of styrene	1241,1242
Styrene-vinyl acetate	sequence distributions	1243
Alpha-methylstyrene-methylene block copolymers	alpha-methyl styrene sequences	1244
Styrene-methacrolein	sequence length of styrene units	1245
Styrene-maleic anhydride	styrene sequences	1246
Styrene-methylmethacrylate	styrene sequences	1247
Vinyl chloride-vinylidene-chloride	methylene sequences	1248,1249
Chlorinated polyethylene	methylene sequences	1250
Vinyl chloride-isobutylene	head-to-tail structure	1251
Polypropylene oxide	sequence distribution	1252

attributed to the differnce in the degree of branching, since the yield of butadiene from ethylene-propyrene copolymers increases as a function of the pyrogen content.

Stepwise pyrolysis was carried out for a single sample by a series of successive degradations in a fixed period of time either at a constant or standardized increasing temperatures. In this study the image furnace pyrolyser was used for the stepwise pyrolysis of the copolymer samples at a constant temperature of 400, 420 or 440°C. During each shot of the heat radiation, the resulting degradation products were continuously transferred to a trap capillary column (0.5 mm i.d. x 20 cm, stainless steel tubing) which was inserted in the column oven between the pyrolyser outlet and a separation column (comprising 30% Otoil-s on Celite S45 operated isothermally at 80°C). The trap capillary was maintained at

Table 140 - Application of NMR spectroscopy to the determination
of sequence distribution in polymers

Polymer	Reference
Butadiene-propylene	1253
1,4 poly(2,3 dimethyl-1,3-butadiene)	1254
poly 1,4-butadiene	1255,1256
2-chlorobutadiene-1,3,2,3-dichlorobutadiene	1257
butadiene-acrylonitrile	1258,1259
butadiene-vinyl chloride	1260
chlorinated PVC	1261-1263
ethylene glycol-sebacic acid-terephthalic acid	1264
chlorinated polyethylene	1265,1266
chlorosulphonated polyethylene	1267
styrene-1-chloro 1,3 butadiene	1268
Styrene-maleic anydride	1269
ethylene-vinyl acetate	1270-1273
methyl acrylate-methyl methacrylate	1274
styrene methacrylic acid	1275
ethylene-methylacrylate	1274
ethylene-2-chloroacrylate and ethylene glycidyl acrylate	1277
methacrylphenone-methyl methacrylate copolymers	1278
chloroprene-methyl methacrylate	1279
ethylene-carbon monoxide	1280

about 0°C by an ice-water bath during the pyrolysis of the sample, and
then exposed to the temperature of the column oven which was temperature
programmed from 25 to 80°C at a rate of 20°C/min.

Figure 101 shows a series of the pyrograms obtained by the stepwise
pyrolysis of copolymer (a) (copolymer B-2 more highly branched) using
repeated 20-s heat radiation of 400°C. In this case, only the fragments
around the methyl methacrylate monomer were separated using the octoil-s
on Celite separation column. In Figure 102 the cumulative weight loss of
the sample by the successive shot of the pulse is plotted vs. the number
of the shot. These data also indicate that copolymer (a) is more suscept-
ible to thermal degradation then copolymer (b).

On the other hand, as shown in Table 138, the totals of the methyl
methacrylate monomer from the two copolymer systems are fairly different
from each other. The difference is largest at 400°C and becomes almost
negligible small at 440°C. These results coincide with the fact that the
instantaneous pyrolysis between 590 and 764°C gave almost the same methyl
methacrylate recovery from the two copolymer systems. Therefore, in the
stepwise pyrolysis, the copolymer samples were pyrolysed at 400°C to
elucidate the subtle differences between the two types of the random
copolymers. These measurements confirmed that the relative thermal
instability of copolymers (a) compared to copolymers (b) could be
attributed to the more localized methyl methacrylate distributions which
pertain to the thermally weak link in the chain in copolymers (a) together
with the presence of grater amounts of low molecular weight material in
copolymers (a) compared to copolymers (b).

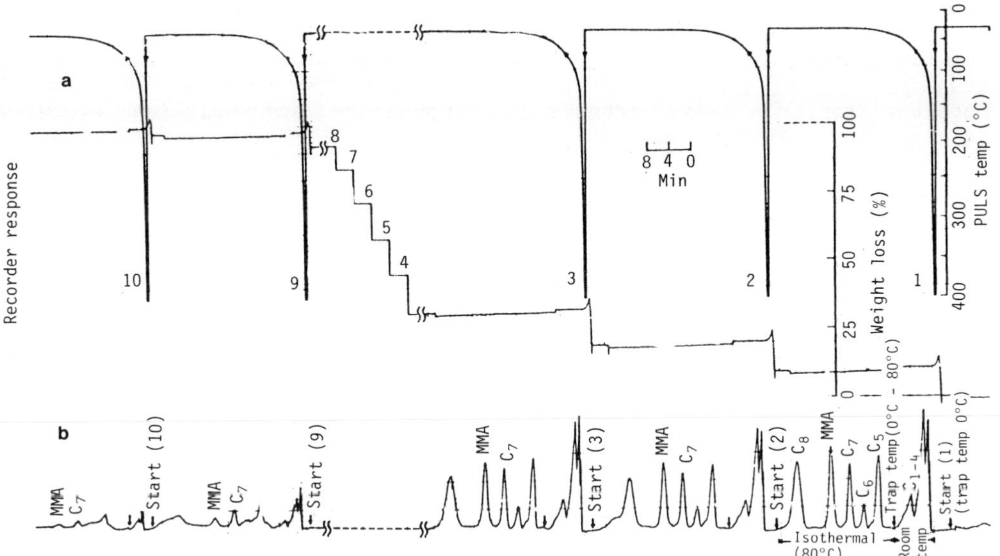

Figure 101   A series of pyrograms by stepwise pyrolysis of E-MMA
copolymer using repeated 20 s heat pulse of 400°C.
Separation column: (B) isothermal at 80°C. Sample, B-2
polymer (methylmethacrylate ethylene 12.9:83.1% w/w);
C_1-C_8: mono-olefins, 1, 2, 3...9, 10: shot number.

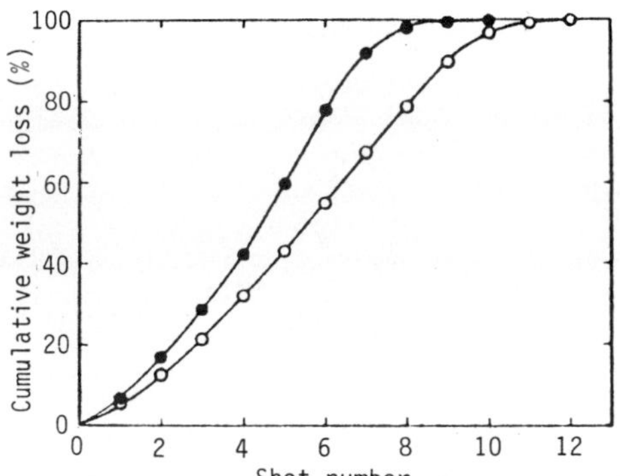

Figure 102   Cumulative weight loss of E-MMA copolymer sample by
successive shot of heat pulse (400°C) as a function of
shot number. O: A-3 copolymer methylmethacrylate-
ethylene (13.2:84.8% w/w) more highly branched than
copolymer B2; O: B-2 copolymer (methylmethacrylate
ethylene 12.9:83.1% w/w).

9.4.8 Sequences in other Polymers

Harwood[1232] has reported a detailed theoretical investigation of methods of ascertaining the sequence distribution of monomers in various types of copolymers. His discussion includes methods based on infrared and ultraviolet spectroscopy, NMR, electrical polarization, thermochemical methods, melting point and crystallinity measurements and entropies of copolymerization, glass transition temperature and solution property measurements. Further work on the application of infrared spectroscopy to sequence measurement in polymers is summarized in Table 139.

Nuclear magnetic resonance spectroscopy has been used in many other applications related to the measurement of sequences in polymers. Some examples are given in Table 140.

CHAPTER 10

POLYMER MICROSTRUCTURE

10.1 INTRODUCTORY

Microstructural features are often of tremendous importance in deciding the physical, and to some extent the chemical, properties of polymers. The term microstructure does not necessarily infer that the feature of concern occurs at low concentrations. It is more concerned with the particular detail of the structure of the polymer. Thus polybutadiene unsaturation exists in various cis and trans forms, all present as constituents of the polymer. Some forms of unsaturation might exist at high concentrations, and some at low. In the case of polyethylene, however, butyl hexyl and octyl side-groups, which are of greater interest from the microstructural point of view, all exist at low concentrations usually expressed as the number of such groups present per 1000 carbon atoms in the polymer chain.

In this chapter various aspects of microstructure are discussed in turn.

Changes in the branching of a polymer such as the methyl group in polypropylene lead to changes in its stereochemical configuration and this in turn is a fundamental polymer property affecting its physical and other mechanical characteristics. A specific polymer unit configuration can be converted to its opposite configuration by a simple end to end rotation and subsequent translation. Natta introduced the term isotactic to describe adjacent units with the same configuration, and syndiotacitc to describe adjacent units with opposite configurations. The measurement of isotacticity, syndiotacticity and the ratio between them, known as tacticity, is discussed in this chapter as is the measurement of diad, triad etc. concentrations.

As well as stereoisomerism, various other forms of isomerism exist in polymers. Thus, geometrical isomerism such as exists in simple organic compounds also exists in polymers, eg. cis and trans 1.4, 1,2 and 3.4 placements in polydienes. These measurements, which are initially tied up with the measurement of total insaturation (Chapters 6.2 and 9.2.1) are discussed here.

A further microstructural feature in polymers discussed is regio-isomerism. This is tied up with structural changes brought about by head to head, tail to tail and head to tail placements of adjacent pairs of monomer units in a polymer chain and is exhibited by a wide variety of polymers such as polypropylene, polyvinylidene halides, polydienes and polyglycols.

Various forms of unsaturation, external vinylidene, trans olefinic unsaturation (internal double bonds), terminal vinyl occur in polyethylene and polypropylene and these are of great interest from the micro-structural point of view as the type and amount of unsaturation can profoundly effect polymer properties. Electron-irradiation - infrared spectroscopic, pyrolysis - gas chromatographic and ESR methods all play their part in elucidating unsaturation in polyolefins and other polymers.

Branching is another aspect of polymer microstructure which is of great interest. Polyethylene, for example, can contain side-chain alkyl groups ranging from methyl to octyl or even higher, which can be identified and determined by techniques such as NMR and infrared spectroscopy. Such groups often have a profound effect on the physical properties of polyethylene and the presence of different types of alkyl side-groups accounts for the fact that many different grades of this polymer are available, each with its own particular physical properties. These techniques are also applicable to copolymers of ethylene with other olefins, e.g. ethylene-butene.

10.2 STEREOISOMERISM AND TACTICITY

During polymerization it is possible to direct the way in which monomers join on to a growing chain. This means that side-groups (X) may be placed randomly (atactic) or symmetrically along one side of the chain (isotactic) or in a regular alternating pattern along the chain (syndiotactic) as shown below.

Along with chemical composition, molecular weight, molecular weight distribution and type and amount of gel (see Chapter 8), branching is considered to be one of the fundamental parameters needed to characterize polymers fully, and this latter property, which is a microstructural feature of the polymer, has very important effects on polymer properties. Changes in branching of a given polymer such as polypropylene lead to changes in its stereochemical configuration and this, in turn, is a fundamental polymer property to formulating both polymer physical characteristics and mechanical behaviour. Technically each methane carbon in a poly(1-olefin) is asymmetric.

```
          R    H    H
          |    |    |
        - C -  C -  C -
          |    |    |
          H    H    H
                 \__/
        methine  methylene
        carbon   carbons
```

However, this symmetry cannot be observed because two of the attached groups are essentially equivalent for long chains. Thus a specific polymer unit configuration can be converted into its opposite configuration by simple end-to-end rotation and subsequent translation. It is possible, however, to specify relative configurational differences, and Natta introduced the terms 'isotactic' to describe adjacent units with the same configurations, and 'syndiotactic' to describe adjacent units with opposite configurations[28]. Tacticity is defined as the ratio of syndiotactic to isotactic structure. Although originally used to describe diad configurations, 'isotactic' now describes a polymer sequence of any number of like configurations and 'syndiotactic' describes any number of

alternating configurations. Diad configurations are called meso if they
are alike and racemic if they are unalike[1282].   Thus from a
configurational standpoint, a poly(1-olefin) can be viewed as a copolymer
of meso and reacemic diads.

10.2.1   Tacticity of Polypropylene

A good example of tacticity is exhibited by polypropylene, which in
the atactic form is an amorphous material of little commercial value but
in the isotactic form is an extremely versatile large tonnage plastic
material.

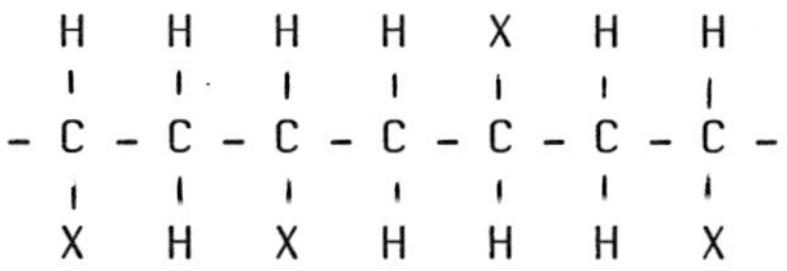

```
      H   H   H   H   X   H   H
      |   |   |   |   |   |   |
    - C - C - C - C - C - C - C -
      |   |   |   |   |   |   |
      X   H   X   H   H   H   X
          atactic


      H   H   X   H   H   H   X
      |   |   |   |   |   |   |
    - C - C - C - C - C - C - C -
      |   |   |   |   |   |   |
      X   H   H   H   X   H   H
          syndiotactic


      H   H   H   H   H   H   H
      |   |   |   |   |   |   |
    - C - C - C - C - C - C - C -
      |   |   |   |   |   |   |
      X   H   X   H   X   H   X
          isotactic
```

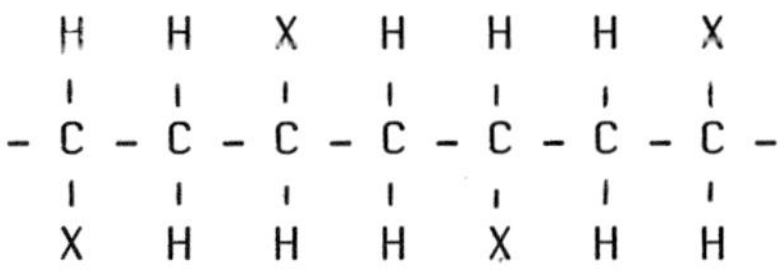

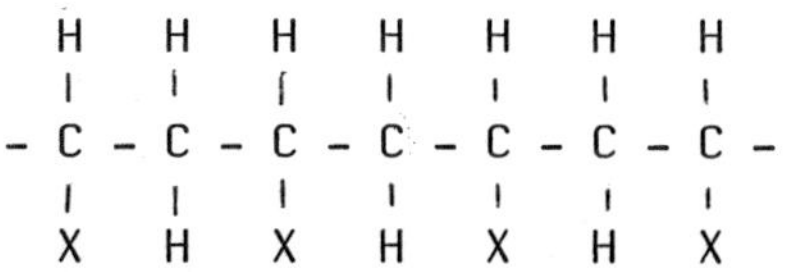

Three-dimensionally, atactic and isotactic polypropylene may be
represented as shown in Figure 103. Multiple sequences of syndiotactic or
isotactic units can exist in polypropylene.   Thus diad and triad of
isotactic polypropylene would have the structures:

```
                   H   H   H   H
                   |   |   |   |
         Diad    - C - C - C - C -
                   |   |   |   |
                   Me  H   Me  H


                   H   H   H   H   H   H
                   |   |   |   |   |   |
         Triad   - C - C - C - C - C - C -
                   |   |   |   |   |   |
                   Me  H   Me  H   Me  H
```

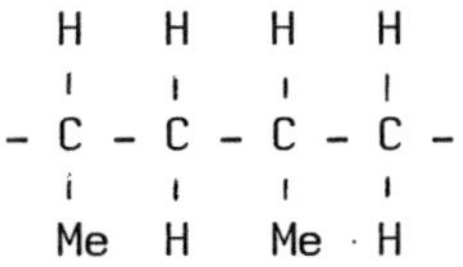

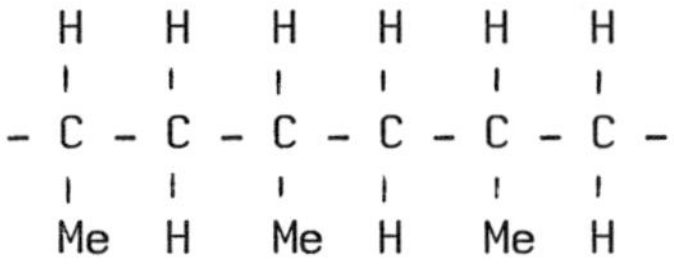

Similarly a pentad and hexad of syndiotactic polypropylene would have the structures:

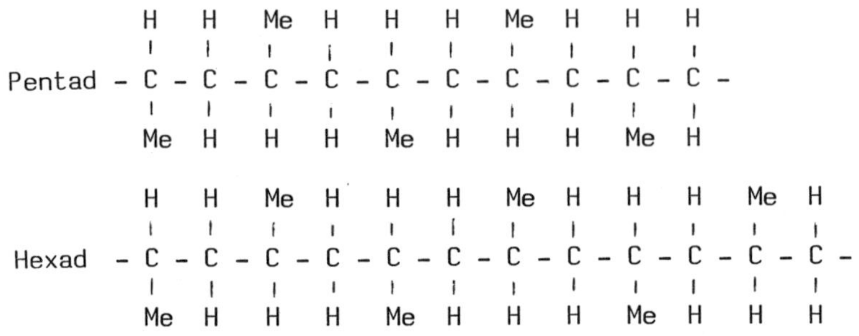

```
               H    H    Me   H    H    H    Me   H    H    H
               |    |    |    |    |    |    |    |    |    |
Pentad  - C -  C -  C -  C -  C -  C -  C -  C -  C -  C -
               |    |    |    |    |    |    |    |    |    |
              Me    H    H    H    Me   H    H    H    Me   H


               H    H    Me   H    H    H    Me   H    H    H    Me   H
               |    |    |    |    |    |    |    |    |    |    |    |
Hexad   - C -  C -  C -  C -  C -  C -  C -  C -  C -  C -  C -  C -
               |    |    |    |    |    |    |    |    |    |    |    |
              Me    H    H    H    Me   H    H    H    Me   H    H    H
```

As well as this the monomer units in a polymer can exist in head to head, head to tail and tail to tail configurations as illustrated below in the case of isotactic propylene (discussed further under regioisomerism, Chapter 10.4.

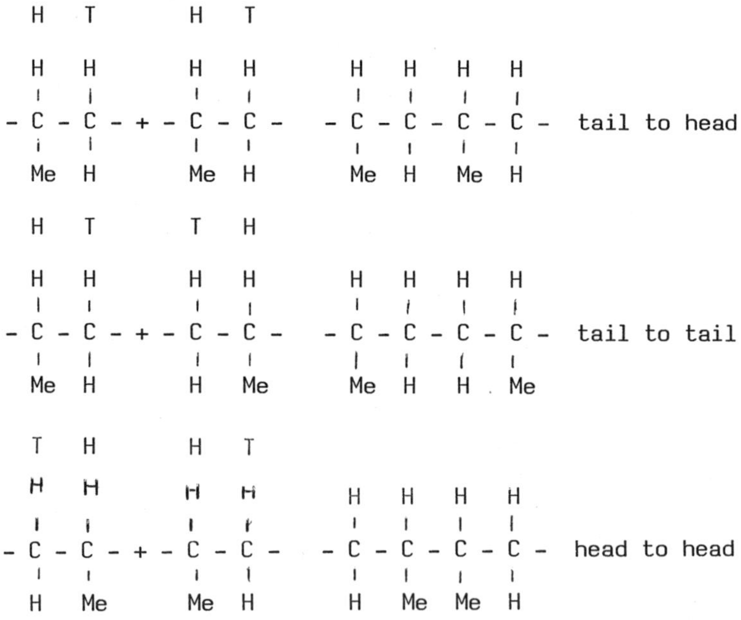

```
    H    T           H    T

    H    H           H    H          H    H    H    H
    |    |           |    |          |    |    |    |
  - C -  C -  +  -  C -  C -       - C -  C -  C -  C -    tail to head
    |    |           |    |          |    |    |    |
   Me    H          Me    H         Me    H   Me    H


    H    T           T    H

    H    H           H    H          H    H    H    H
    |    |           |    |          |    |    |    |
  - C -  C -  +  -  C -  C -       - C -  C -  C -  C -    tail to tail
    |    |           |    |          |    |    |    |
   Me    H           H   Me         Me    H    H   Me


    T    H           H    T

    H    H           H    H          H    H    H    H
    |    |           |    |          |    |    |    |
  - C -  C -  +  -  C -  C -       - C -  C -  C -  C -    head to head
    |    |           |    |          |    |    |    |
    H   Me          Me    H          H   Me   Me    H
```

Although molecular symmetry is well understood, until the development of proton NMR, and later ^{13}C NMR, a study of this aspect of polymer structure presented problems.

The advantages of ^{13}C NMR in measurements of polymer stereochemical configuration arise primarily from a useful chemical shift range which is approximately 20 times that obtained by proton NMR.   The structural

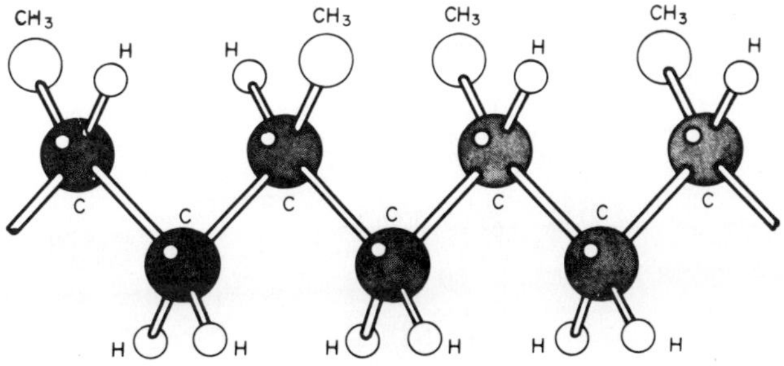

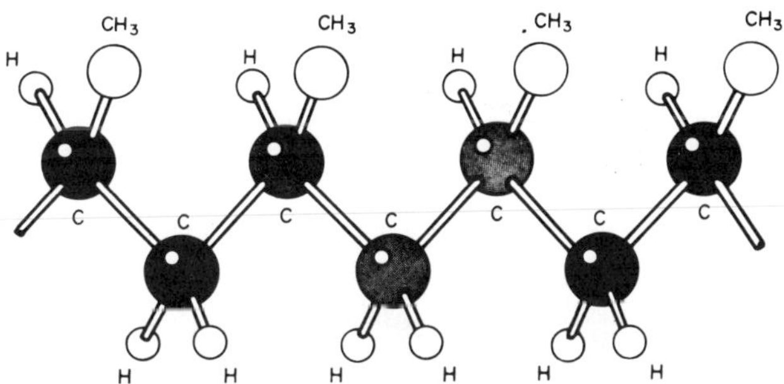

Figure 103   Polypropylene structures.

sensitivity is enhanced through an existence of well-separated resonances for different types of carbon atoms. Overlap is generally not a limiting problem.   The low natural abundance (~1%) of ^{13}C nuclei is another favourable contributing factor.   Spin-spin interactions among ^{13}C nuclei can be safely neglected, and proton interactions can be eliminated entirely through heteronuclear decoupling. Thus each resonance in a ^{13}C NMR spectrum represents the carbon chemical shift of a particular polymer moiety. In this respect, ^{13}C NMR resembles mass spectrometry because each signal represents some fragment of the whole polymer molecule. Finally, carbon chemical shifts are well behaved from an analytical viewpoint because each can be dissected, in a strictly additive manner, into contributions from neighbouring carbon atoms and constituents.   This additive behaviour led to the Grant and Paul rules[1283], which have been carefully applied in polymer analyses, for predicting alkane carbon chemical shifts.

The advantages so clearly evident when applying ^{13}C NMR to polymer configurational analyses are not devoid of difficulties. The sensitivity of ^{13}C NMR to subtle changes in molecular structure creates a wealth of chemical shift-structural information which must be 'sorted out'. Extensive assignments are required because the chemical shifts relate to sequences from three to seven units in length. Model compounds, which are often used in ^{13}C NMR analyses, must be very close structurally to the polymer moiety reproduced. For this reason, appropriate model compounds are difficult to obtain. A model compound found useful in polypropylene configurational assignments was a heptamethylheptadecane, where the relative configurations were known[1284]. To be completely accurate, the

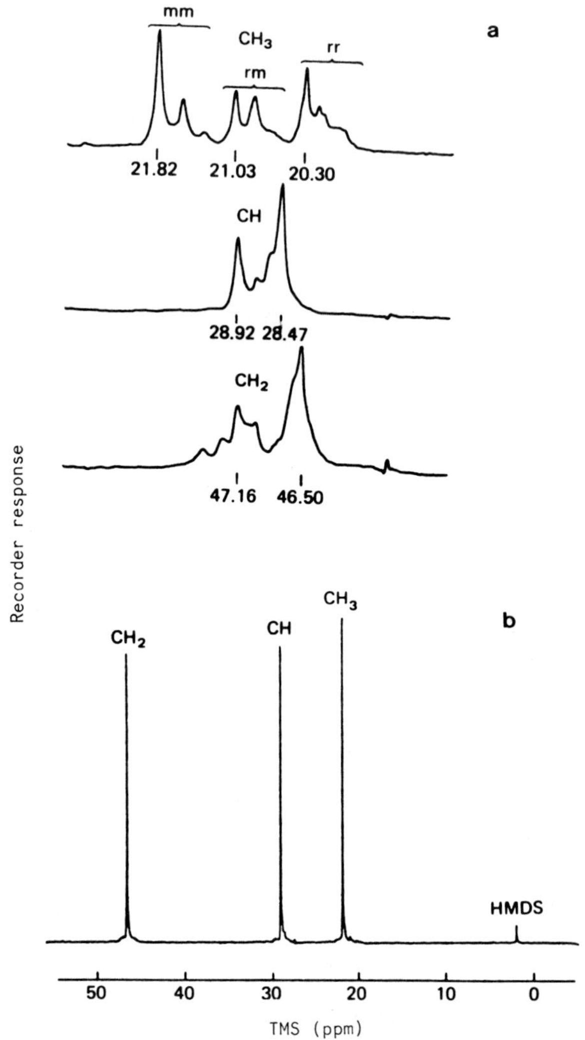

Figure 104 ^{13}C NMR spectroscopy of polypropylene (a) methyl, methine and methylene regions of the ^{13}C NMR spectrum of (a) non-crystalline polypropylene; (b) ^{13}C NMR spectrum at 25.2 MHz of crystallien polypropylene.

model compounds should reproduce the conformational as well as the configurational polymer structure. Thus reference polymers such as predominantly isotactic and syndiotactic polymers form the best model systems. Even when available, only two assignments are obtained from these particular polymers. Pure reference polymers can be used to generate other assignments[1285].

To obtain good quantitative ^{13}C NMR data one must understand the dynamic characteristics of the polymer under study. Fourier transform techniques, combined with signal averaging, are normally used to obtain ^{13}C NMR spectra. Equilibrium conditions must be established during signal averaging to ensure that the experimental conditions have not led to distorted spectral information. The nuclear overhauser effect (NOE), which arises from ^{1}H, ^{13}C heteronuclear decoupling during data acquisition, must also be considered.

Energy transfer, occurring between the ^{1}H and ^{13}C nuclear energy levels during spin decoupling can led to enhancements of the ^{13}C resonances by factors between 1 and 3. Thus the spectral relative intensities will only reflect the polymer's moiety concentrations if the differentiated Nuclear Overhauser effect (NOE) are equal or else taken into consideration. Experience has shown that polymer NOEs are generally maximal, and consequently equal, because of a polymer's restricted mobility[1286,1287]. To be sure, one should examine the polymers NOEs through gated decoupling or paramagnetic quenching, and thereby avoid any misinterpretation of the spectral intensity data.

The ^{13}C configurational sensitivity falls within a range from triad to pentad for most vinyl polymers. In non-crystalline polypropylenes, three distinct regions corresponding to methylene (-46 ppm), methine (-28 ppm) and methyl (-20 ppm) carbons are observed in the ^{13}C NMR spectrum. (The chemical shifts are reported with respect to an internal tetramethylsilane (TMS) standard.) The ^{13}C spectrum of a 1,2,4-trichlorbenzene solution at 125°C of a typical amorphous polypropylene is shown in Figure 104 (a). Although a configurational sensitivity is shown by all three spectral regions, the methyl region exhibits by far the greatest sensitivity and is consequently of the most value. At least ten resonances assigned to the unique pentad sequences, are observed in order, mmmm, mmmr, rmmr, mmrr, mmrm, rmrr, mrmr, rrr, rrrm and mrrm, from low to high field[1283,1284,1288].

The ^{13}C NMR spectrum of crystalline polypropylene shown in Figure 104(b) contains only three lines which can be identified as methylene, methine and methyl from low to high field by off-resonance decoupling. An amorphous polypropylene exhibits a ^{13}C spectrum which contains not only these three lines but additional resonances in each of the methyl, methine and methylene regions as shown in Figure 104(a). The crystalline polypropylene must therefore be characterized by a single type of configurational structure. In this case the crystalline polypropylene structure is predominantly isotactic, thus the three lines in Figure 104(b) must result from some particular length of meso sequences. This sequence length information is not available from the spectrum of the crystalline polymer, but can be determined from a corresponding spectrum of the amorphous polymer. To do so one must examine the structural symmetry of each carbon atom to the various possible monomer sequences. Randall[1289,1290] carried out a detailed study of the polypropylene methyl group in triad and pentad configurational environments. Figure 105 shows the PMR spectrum of hot n-heptane solution of atactic polypropylene containing up to 40% syndiotactic placement, and which by Natta's definition may be stereoblock[1211]. The spectrum is inherently complex, as a first-order theoretical calculation, and this indicated the possibility of at least 15 peaks with considerable overlap between peaks, because differences in chemical shifts are about the same magnitude as the splitting due to spin-spin coupling.

The largest peak, at high field in Figure 105(a) represents pendant methyls in propylene units. It is characteristically split by the tertiary hydrogen. By area integration, about 20% of the nominal methyl proton peak is due to overlap of absorption from chain methylenes. This overlap is consistent with a reported syndiotactic triplet, two peaks of which are close to A and B in Figure 105(a) and a third peak which falls with the low field branch of the methyl split[1212]. The absence of a

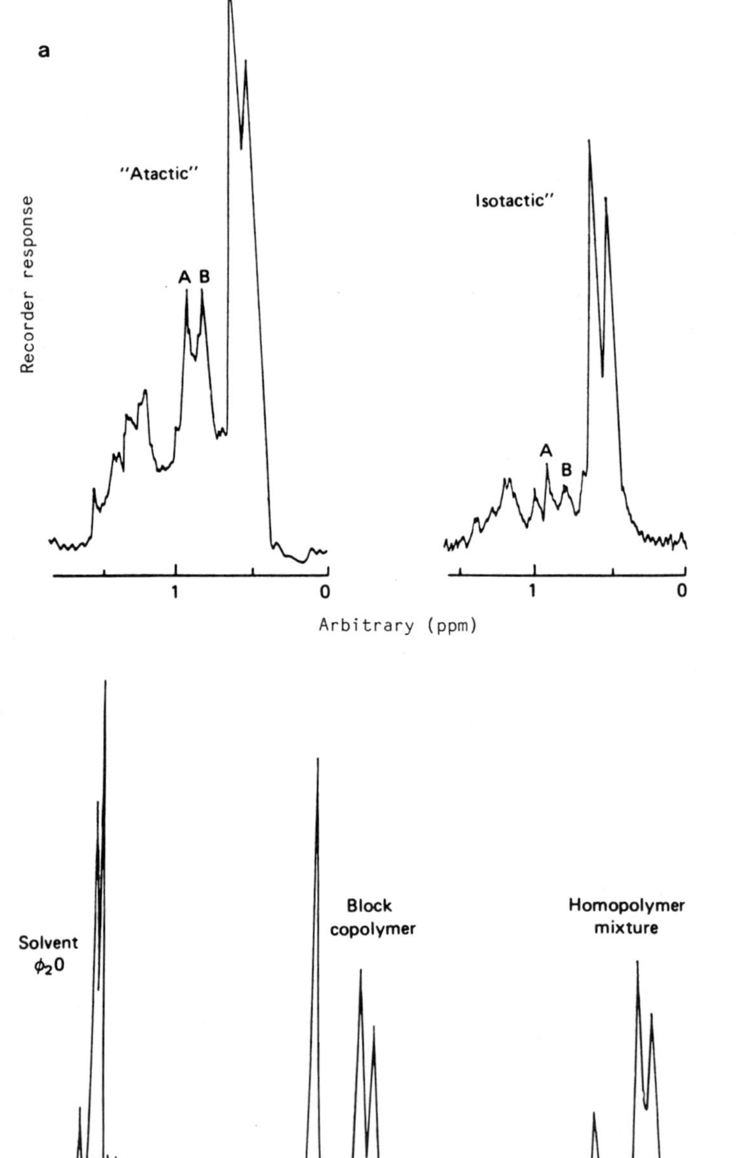

Figure 105   PMR spectra of: (a) polypropylenes; (b) ethylene-propylene copolymers.

strong singlet peak in the methylene range indicate the virtual absences of 'amorphous' polymer in the atactic polypropylene shown in Figure 105(a)[1211,1212], which could possibly be due to the head-to-head and tail-to-tail units. The low field peak represents the partial resolution of tertiary protons which are opposite the methyls on the hydrocarbon chain.

Figure 105(a) also shows a spectrum for an isotactic polypropylene, >95% isotactic by solubility (<5% soluble in boiling heptane). This spectrum has the same general character as the atactic polypropylene. The important difference is a marked decrease of peak intensity in the chain methylene region. This decrease is caused by extensive splitting, and the difference in chemical shifts for the non-equivalent methylene hydrogens in isotactic environments. This is in accord with the study of Stehling[1291] on deuterated polypropylenes, which indicated that much of isotactic methylene absorption is buried beneath the methyl and tertiary hydrogen peaks. The fractional area in the nominal methylene region of the spectrum is thus sensitive to the number of isotactic and syndiotactic diads, and therefore may be used as a measure of polypropylene and linear polyethylene. The low field absorption in Figure 105(b), characteristic of aromatic hydrogens, is due to the polymer solvent, diphenyl ether, which was used throughout. Polymer concentrations in solution can be readily calculated from the ratios of peak areas adjusted to the same sensitivity. The superpositions of spectra that were obtained separately on the homopolymers show the same pattern, with different intensities, as spectra on physical mixtures. The polyethylene absorption falls on the peak marked A of the chain methylene complex in polypropylene. Peak B, also due to chain methylenes in polypropylene, is resolved in both spectra in Figure 105(b) which also gives the spectrum of an ethylene-propylene block copolymer. The ethylene contribution again falls on peak A.

Various workers have developed analyses for physical mixtures and block copolymers based on the ratio of the incremental methylene area to the total polymer proton absorption. This concept has been tested by Barrall et al.[1292] using PMR analyses on a series of physical mixtures and block copolymers synthesized with ^{14}C-labelled propylene and others with ^{14}C=labelled ethylene. A most important feature of this analysis is that the methylene peaks A and B have virtually the same relative heights in polypropylenes with a variety of tacticities (note Figure 105(a)). This is also true for PMR spectra given by Satoh and others for a tactic series of polypropylenes[1293]. This suggests that PMR analyses for ethylene are independent of tacticity, since the area increment of peak A above peak B has been used for analysis.

Qualitative polypropylene tacticities can be estimated by PMR not only for homopolymers (Figure 105(a)) but also in the presence of polyethylene and ethylene copolymer blocks. The relative heights of the peaks for secondary and for tertiary hydrogen in Figure 105(b) indicate that the polypropylene in the copolymer and the physical mixture is dominantly isotactic.

Table 141 gives analyses carried out by Porter[1211] for a series of physical mixtures made up with tagged polypropylene standardized by radiocounting. The three sets of values are in good agreement.

Reilly[1294] has investigated an NMR method for determining the syndiotactic content of polypropylene. The results are not entirely consistent with the syndiotactic crystallinity as determined by alternative methods such as the density method. A possible explanation for the lack of consistency is that the NMR method does not require as

Table 141 - Tacticity of linear polyethylene-polypropylene
physical mixtures (polyethylene, wt%)

Sample	Made up	PMR	Tracer
6	5.0	4.6	4.2
7	9.0	8.6	7.6
5	17.0	16.7	16.3
8	34.0	31.1	31.0

long a block in the chain in order for the syndiotactic placement of
methyl groups to be detected. He attempted to determine the syndiotactic
content of some experimental polypropylenes. NMR spectra of samples
dissolved in o-dichlorobenzene were obtained at 170°C at 100Mc/s. Spectra
of syndiotactic and isotactic samples gave the following NMR parameters:

	Syndiotactic	Isotactic
δCH_3	0.835 ppm	0.895 ppm
CH_2	1.075 ppm	1.895 and 1.310 ppm
δCH	1.570 ppm	1.615 ppm
J_{CH_3-CH}	6.0 cps	6.0 cps
J_{CH-CH_2}	6.0 cps	6.0 cps
$J_{H-H \, (geminal)}$	Indeterminate	-14.0 cps

The reference for the chemical shift measurement was hexamethyl
disiloxane. The methylene hydrogens in the isotactic material are
nonequivalent - as expected on geometrical grounds. Spectra calculated
with the above parameters agreed reasonably well with the observed
spectra[1295]. High-resolution proton magnetic resonance spectroscopy has
been applied to an examination of very highly isotactic, very highly
syndiotactic, and stereoblock polypropylenes in o-dichlorobenzene
solutions. They discuss the effects of stereoregulation on proton
shieldings and some of the complexities of the methylene proton resonances
and determine tactic placement contents for several polymers by a method
based on the methylene proton resonances. Tactic pair contents were
determined for two stereoblock fractions by a method based on the methyl
proton resonances. Their results revealed much higher stereoblock
characters than those determined for the same fractions from melting data.
All the PMR results were in very good accord with the results obtained on
several polymers by infrared, x-ray diffraction, and differential thermal
analysis.

Mitani[1295] found that the diad and triad content of isotactic
polypropylene, syndiotactic polypropylene, and atactic polypropylene, as
determined from 100MHz NMR spectra, were in agreement with the values
determined from the 220-MHz NMR and ^{13}C NMR spectra.

Inoue et al.[1296], Zambelli at al.[1297] and Randall[1298] have shown that [13]NMR is an informative technique for measuring stereochemical sequence distributions in polypropylene. These workers reported chemical shift sensitivities to configurational tetrad, pentad and hexad placements for this polymer.

The sequence lengths of stereochemical additions in amorphous and semicrystalline polypropylene were accurately measured using ^{13}C NMR[1299]. The method has some limitations for addition polymers having predominantly isotactic sequences.

Randall[1300] has described work on the application of quantitative ^{13}C NMR to measurements of average sequence length of like stereochemical additions in polypropylene. He describes sequence lengths of stereochemical addition in vinyl polymers in terms of the number-average lengths of like configurational placements. Under these circumstances a pure syndiotactic polymer has a number-average sequence length of 1.0; a polymer with 50:50 meso:racemic additions has a number-average sequence length of 2.0 and polymers with more meso than racemic additions have number-average sequence lengths greater than 2. Amorphous and crystalline polypropylenes were examined using ^{13}C NMR as examples of the applicability of the average sequence length method. The results appear to be accurate for amorphous and semi-cryotalline polymers but limitations are present when this method is applied to highly stereoregular vinyl polymers containing predominantly isotactic sequences[1301]. Randall has measured the ^{13}C NMR spin lattice relaxation times of isotactic and syndiotactic sequences in amorphous polypropylene. Spin-lattice relaxation times for methyl, methylene, and methine carbons in an amorphous polypropylene were measured as a function of temperature from 46 to 138°C. The carbons from isotactic sequences characteristically exhibited the longest spin relaxation times of those observed. The spin relaxation time differences increased with temperature with the largest differences occurring for methine carbons, where a 32% difference was observed. Randall determined activation energies for the motional processes affecting spin relaxation times for isotactic and syndiotactic sequences. Essentially no dependence upon configuration was noted. High-resolution NMR spectra of isotactic and syndiotactic polypropylene have been used by Cavelli[1302] to provide conformational information. Brosio et al[1303] reported on ^{13}C NMR spectra for measurements of tacticity, terminal conformation and configuration of polypropylene.

Sternling and Knox[1304,1305] defined the stereochemical structure of polypropylene from the PMR spectrum of normal deuterated and epimerized polypropylene.

To determine the isotactic content of polypropylene Peraldo[1306] carried out a normal vibrational analysis. He considered the primary unit as an isolated three-fold helix. From this work and a number of subsequent publications[1307-1310] it was suggested that the absorptions at 8.57, 10.02 and 11.90 micron were indicative of the helical conformation of the isotactic form. Measurements of the isotactic contents of a series of polypropylene fractions based on these bands were made[1310] and compared with results from Flory's melting point theory[1311]. Melting points were determined as the point of disappearance of the birefringence on highly annealed samples. All three bands give qualitative agreement with the melting point data; however, only the method based on the 8.57 micron band gives quantitative agreement. Therefore, the method based on this band appears to give a good measure of the isotactic content in polypropylene at least in the 60-100% range.

10.2.2 Tacticity of Polyvinyl Chloride

The study of stereochemical configuration by ^{13}C NMR has not been limited to the polyolefins. Schneider et al[312] showed that the absorption around 14.49 microns was proportional to the number of isotactic diads, and in the region of 16.66-15.62 microns to syndiotactic diads.

```
      H   H   H   H              H   Cl  H   Cl
      |   |   |   |              |   |   |   |
    - C - C - C - C -          - C - C - C - C -
      |   |   |   |              |   |   |   |
      H   Cl  H   Cl             H   H   H   H
       isotactic diad           syndiotactic diad
          (h - t)                    (h - t)
```

Based on this finding they proposed a method for determining the tacticity of amorphous samples of PVC. As some samples cannot be easily transformed into an amorphous state Schneider et al[313] devised an infrared method of tacticity determination which is independent of the crystallinity of the samples. From the temperature dependence of infrared spectra of poly(vinylchloride) samples prepared by different methods, the intensity of the band at 14.40 microns (proportional to the number of isotactic diads in the sample), as well as that of the tacticity-independent C-H stretching band, was found to be independent of the crystallinity of the sample. These lines were applied for the tacticity determination in poly(vinylchloride), measured in the form of potassium bromide pellets. The numerical tacticity value was obtained from the known values of absorbance coefficients of S_{CH} and S_{HH} type C-Cl stretching bands in solution, and from the shape of the spectrum.

Abe et al[314] have investigated the NMR spectra of model compounds of poly(vinylchloride) in the hope that these investigations may offer useful information for the analysis of vinyl polymer spectra. They studied the NMR spectra of three stereoisomers of 2,4,6-trichloroheptane as model compounds of poly(vinylchloride).

```
            H   H   H   H   H
            |   |   |   |   |
     Me  -  C - C - C - C - C  -  Me
            |   |   |   |   |
            Cl  H   Cl  H   Cl


            H   H   H   H   Cl
            |   |   |   |   |
     Me  -  C - C - C - C - C  -  Me
            |   |   |   |   |
            Cl  H   Cl  H   H


            H   H   Cl  H   H
            |   |   |   |   |
     Me  -  C - C - C - C - C  -  Me
            |   |   |   |   |
            Cl  H   H   H   Cl
```

Spectra were observed at 60 and 100 Mc/sec., both at room temperature and at high temperatures and spin-decoupling experiments were performed. The difference of the chemical shifts of the two meso methylene protons at 60 Mc/sec. was found to be ca. 7 cps. for the isotactic three unit model while it was ca. 16 cps. for the isotactic two-unit model or heterotactic three-unit model. The spectra of poly(vinyl-chloride) can be interpreted reasonably on the basis of this result. Observed values of vicinal coupling constants of model compounds were interpreted as the weighted means of those for several conformations, and the stable conformations of the models were determined.

Chemical shifts of PVC and model compounds such as meso- and reacemic-2,4-dichloropentane have been measured from NMR spectra[1315].

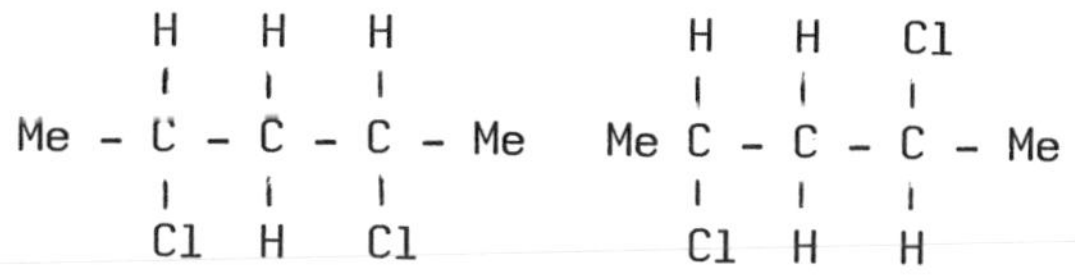

$$
\begin{array}{ccc}
\text{H} & \text{H} & \text{H} \\
| & | & | \\
\text{Me} - \text{C} - \text{C} - \text{C} - \text{Me} \\
| & | & | \\
\text{Cl} & \text{H} & \text{Cl}
\end{array}
\qquad
\begin{array}{ccc}
\text{H} & \text{H} & \text{Cl} \\
| & | & | \\
\text{Me} \ \text{C} - \text{C} - \text{C} - \text{Me} \\
| & | & | \\
\text{Cl} & \text{H} & \text{H}
\end{array}
$$

10.2.3 Other polymers

Randall[1316] and other workers[1317] studied amorphous polystyrene. Randall[1316] concluded that nine methylene resonances could be resolved, numbered A through I in Figure 106. A high-field strong methine resonance is also present, but shows no apparent configurational splitting. The observation of nine resonances is in itself interesting, since a ^{13}C NMR sensitivity to just tetrad sequences would have produced six resonances, while a complete hexad sensitivity would have produced twenty resonances.

The approximate degree of isotacticity of crystalline poly(3-methyl-1-butene) has been determined[1319] from the ratio of the absorbance at 12.85 to 7.47 microns.

Tacticity measurements in other polymers are reviewed in Table 142.

Strasilla and Klesper[1320] showed that in syndiotactic polymethyl methacrylate only the units once removed are responsible for resolving the methoxy resonance into 3 peaks.

Mass NMR infrared, and kinetic data were obtained by Richards and Wilhams[1321] to determine the structure of the tetramer of alpha-methylstyrene. The various proton resonances in the NMR spectra of poly(p-isopropyl-alpha-methylstyrene) have been assigned by Leonard and Malhotra[1322] to isotactic, heterotactic, and syndiotactic triads.

Tacticity measurements have been performed on polytrifluorochloro-ethylene[1323], acrylic acid copolymers[1324], polymethylvinyl ethers[1325], polyacrylonitrile[1326,1327] and polyvinyl trifluoroacetate[1328] and the following polymers have been subject of measurements of stereoregularity; polyvinyl alcohol and esters[1329,1330], polymethacrylonitrile poly-methylacrylic acid[1335], polypropylene oxide[1336] and isobutene-maleic anhydride, isobutene-dimethyl fumarate and isobutene-dimethyl maleate copolymers[1228]. Triad tacticities have been determined on 1:1 diphenyl ethylene-methyl acrylate copolymer[1337] and poly-acrylonitrile[1327].

Table 142 - Application of ^{13}C NMR spectroscopy to tacticity
sequence measurement in polymers

Polymer	Measurement	Reference
Isotactic-I-butene-polypropylene	dia, triad and tetrad	1339
Butadiene-propylene	measurement of monomer sequence distribution	1340
Polybutene	measurement of isotactic content	1341
Polychloroprene	measurement of diad and triad sequences	1342
PVC	measurement of distribution of stereochemical sequences tetrad, pentad and hexad placements	1344, 1345, 1346-1355
Ethylene-vinylchloride	measurement of diads	1356
Chlorinated polyethylene	measurement of methylene sequences	1357,1358
Poly alpha-methylstyrene	measurement of tacticity	1359
Polystyrene	measurement of stereochemical sequence distribution tetrad, pentad, and hexad placements.	1360-1362
	measurement of average sequence length of like stereochemical configurations	1363
Styrene-acrylic acid	measurement of isotacticity	1364
alpha-Methyl styrene-methacrylonitrile	measurement of syndiotacticity	1365
Styrene-acrylonitrile	measurement of diad and triad distribution	1366
Styrene-isobutene	measurement of randomness of structure	1367
Poly(p-fluoro-alpha-methyl styrene	measurement of syndiotactic and heterotactic triads	1368
Polyvinylacetate	measurement of diad tacticities, triad and pentad contents	1369,1370

Table 142 - Continued

Ethylene-vinyl acetate	measurement of tacticity-measurement of sequence distribution	1370-1373
Polymethylmethacrylate	measurement of isotactic triads,	1374
	tetrad, pentad and hexad resonances	1375
	conformation	1376
	tacticity	1377
Polymethylacrylate	measurement of tetrad configurations of triad, tetrad and pentad configurations	1378
Polyalkylmethacrylates	measurement of tacticity	1379
Polyallylmethacrylate	measurement of tacticity and sequence length	1380
Methylacrylate-butadiene	measurement of isotactic configuration	1383
Methylmethacrylate-butadiene	measurement of triad and pentad configuration	1384
Styrene-methylmethacrylate	measurement of diad, triad and pentad configurations	1385
Methyl alpha-chloro acrylate-methylmethacrylate	measurement of diad configurations and mean sequence length	1386
Polymethyl(vinylether)	measurement of diad, triad and pentad sequences	1387
Ethylene oxide-propylene oxide	measurement of diad, triad sequences	1388
Copolyesters	randomness	1389
Polypropylene oxide	stereochemistry	1390-1392
Polyalkylvinylesters	stereochemistry	1393-1394
Ethyl 2-chloroacetate	stereochemistry	1395
Ethylene-glycidyl acrylate	stereochemistry	1396
Poly-trans 1,3-pentadiene	stereochemistry	1397

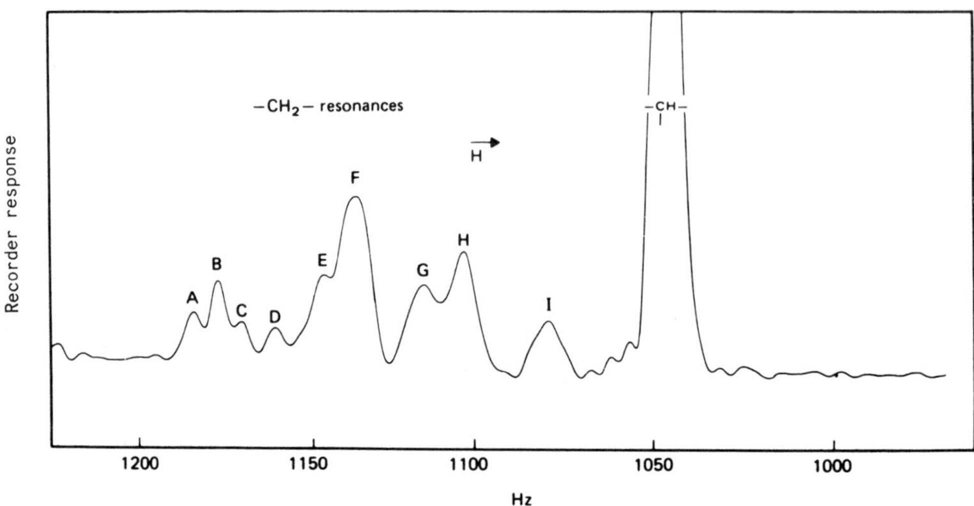

Figure 106   The methylene and methine region of a ^{13}C NMR spectrum
of an amorphous polystyrene at 25.2 MHz and 120°C. The
MHz values are relative to an internal tetramethylsilane
(TMS) standard.

10.3   GEOMETRICAL ISOMERISM

10.3.1   Polybutadiene

Infrared spectroscopy is a very useful technique for the measurement
of different types of unsaturation in polymers. Butadiene has the
following structure.

$$CH_2 = CH - CH = CH_2$$
$$\;\;1 \qquad 2 \qquad 3 \qquad 4$$

Its polymers can contain the following types of unsaturation:

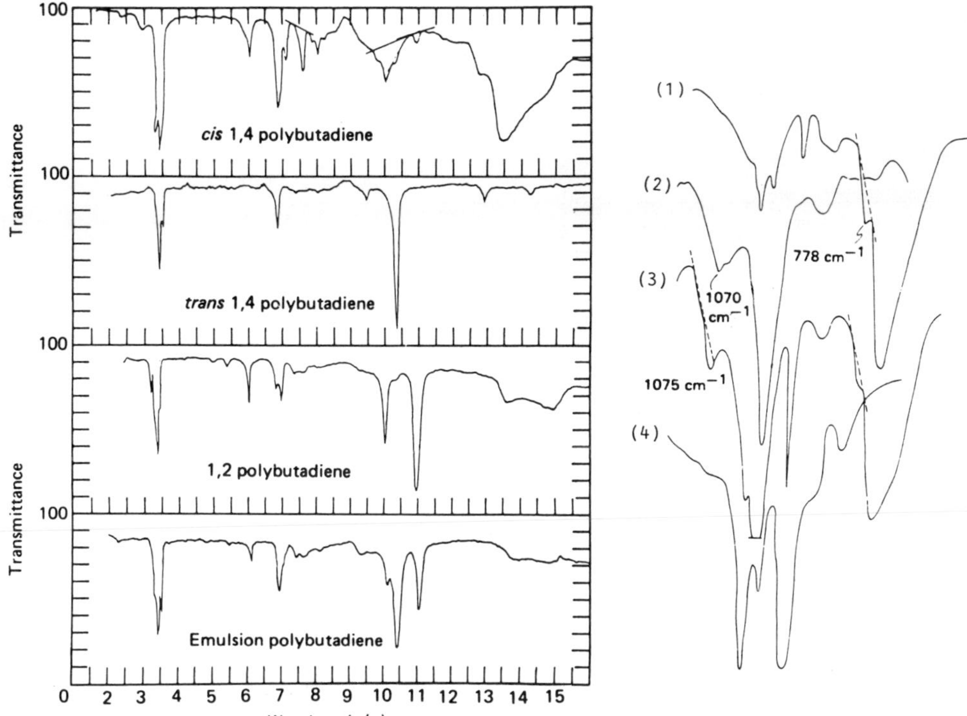

Figure 107   (a) Infrared spectra of (1) cis-1,4-PBD; (2) trans-1,4-
PBD; (3) mixed PBD structure: (4) atactic 1,2-PBD.
Figures displayed vertically for viewing. All samples
in carbon disulphide solutions; 2 mm light bath cell
with NaCl window.

Figure 107(a) shows infrared spectra of various kinds of polybutadienes[1398,1399] which illustrate the usefulness of infrared spectroscopy for distinguishing between different types of unsaturation.

Fraga[1400] has developed an infrared-near-infrared method of analysis of carbon tetrachloride solutions of polybutadienes suitable for the evaluation of cis-1,4, (12-14 microns) trans-1,4, (10.3 microns) and vinyl (11.0 microns) structures. Only polybutadiene is required for calibration purposes. The method is applicable to carbon tetrachloride soluble polybutadienes containing 0-97% cis-1,4 structure, 0-70% trans-1,4 structure, and 0-90% vinyl structure. Some typical spectra are shown in Figure 107(b) for high-cis-1,4 polybutadiene, and high-trans-1,4 polybutadiene, and a high-1,2 (atactic) polybutadiene, and other samples of different cis and trans compositions. Carlson and Atenau[1401] investigated NMR spectroscopy as a means of measuring different types of unsaturation in polybutadiene, butadiene styrene copolymers and natural rubber(polyisoprene) vulcanizates. The samples were analysed in hexachlorobenzene solution (trimethylsilane internal standard).

Carlson and Altenau[1401] used the equations to calculate polymer compositions by NMR which have been published by Mochel[1402]. The quality of the spectra was comparable to that of uncompounded polymers. Good agreement was obtained between the calculated amounts of butadiene,

Table 143 - Determination of polymer composition

Calculated compounded formula				Found compounded formula (infrared)			
Styrene	cis	Butadiene trans	1,2	Styrene	cis	Butadiene trans	1,2
	37	55	8		35	55	10
	14	30	56		16	27	57
15	20	53	12	16	21	51	12
21	13	52	14	23	15	49	13
26	8	53	14	28	12	48	12
17	17	43	23	22	17	39	22

Table 144 - Determination of polymer composition

	Butadiene					Butadiene			
Sty-rene	cis	trans	1,2	Natural rubber	Sty-rene	cis	trans	1,2	Natural rubber
	19	28	4	50		13	18	3	66
	37	55	8	–		37	54	10	–
19	4	22	20	35	14	4	20	12	50
29	6	35	30	–	31	11	39	19	–
14	5	32	9	40	10	5	30	6	49
24	8	54	15	–	30	10	49	11	–
13	4	27	7	50	8	6	22	6	59
24	7	53	14	–	32	9	49	10	–

natural rubber and styrene, and the found values; however, the method will not distinguish between butadiene from polybutadiene and butadiene-styrene copolymer. The relative amount of 1,2-butadiene and 1,4-butadiene were found in the case of polybutadiene and styrene butadiene rubber blends. The individual amounts of cis and trans-1,4-butadiene could not be calculated because of complete overlapping of the two signals, as is also the case in uncompounded polymers.

Determination of the different microstructures in a blend of polybutadiene and/or styrene butadiene rubber with natural rubber is very difficult because the signals from the olefinic protons severely overlap one another. Carlson and Altenau[1401] did not attempt to calculate the percentages of the different microstructures in these blends of rubber. Natural rubber is essentially 100% cis-1,4-polyisoprene.

In further work, Carlson et al.[1403] discuss a disadvantage of their NMR method connected with the fact that it provides only limited microstructural data. This was due to the lack of resolution of the 60-megacycle NMR. They developed alternative carbon disulphide extraction techniques to overcome these limitations and applied infrared methods as described by Binder[1414,1405].

Table 143 shows infrared analyses of vulcanizates containing either polybutadiene or styrene-butadiene copolymer, or a blend of these as reported by Carlson et al.[1403]. The relative amounts and types of polybutadiene and styrene-butadiene rubber varied in these blends. These differences are reflected in the calculated percentage for the samples.

Table 144 shows infrared analyses of vulcanizates containing natural rubber, polybutadiene and styrene-butadiene rubber. The infrared results indicated that the concentration of natural rubber in the carbon disulphide solution is considerably higher than that present in the vulcanizate.

Other workers who have applied spectroscopy to the study of unsaturated polybutadiene and styrene butadiene copolymers include Braun and Canji[1406,1407], Hast and Deur Siftar[1408], Silas et al.[1409], Cornell and Koenig[1410], Neto and Di Lauro[1411], Binder[1412], Clarke and Chen[1413] and Harwood and Richey[1414]. C^{13} NMR spectroscopy has also been used to study sequence distribution[1415] and cis-1,4, trans-1,4 and 1.2 units in polybutadiene and trans-1,4 moities in polyper fluorobutadiene[1416].

Figure 108 shows ^{13}C NMR spectra at 25.1 MHz of a hydrogenated polybutadiene that contained 73% 1,2 additions (or 183 branches per 1000 carbon atoms). The spectra were recorded at a temperature of 37 and 125°C in approximately 10 wt% polymer solutions of 1,2,4-trichlorobenzene and perdeuterobenzene. The temperature sensitivity of some of the resonances in Figure 108 is clearly indicated by the overlap that occurs at one temperature but is resolved at the other. In all, 19 resonances, could be identified even though only 18 were visible at 37°C and only 16 at 125°C.

The ozonolysis of double bonds in organic compounds and polymers in a non-aqueous solvent leads to the formation of ozonides which, when acted upon by water, are hydrolysed to carbonyl compounds

$$R - CH = CH - R' + O_3 = R - CH \underset{O}{\overset{O - O}{\diagup \diagdown}} CH - R'$$

$$R - CH \underset{O}{\overset{O - O}{\diagup \diagdown}} CHR' + H_2O = RCHO + R'CHO + H_2O_2$$

Triphenyl phosphine is frequently used to assist this reaction. Clearly, when applied to complex unsaturated organic molecules or polymers this reaction has greater potential for the elucidation of the microstructure of the unsaturation. Applications of the technique to various model compounds are illustrated in Table 145. Examination of the reaction products, for example by conversion of the carbonyl compounds to carboxylates then esters followed by gas chromatography, enables identifications of these products to be made.

An example of the value of the application of this technique to a polymer structural problem is the distinction between polybutadiene made up of consecutive 1,4-1,4 butadiene sequences I, and polybutadiene made up of alternating 1,4 and 1,2 butadiene sequences. II, i.e. 1,4-1,2-1,4.

456

$$
\begin{array}{l}
\text{I} \quad \overset{1.4}{-CH_2 - CH_2 - CH = CH} \quad -CH_2 - CH_2 - CH = CH \\[1em]
\text{II} \quad \overset{1.4}{-CH_2 - CH_2 - CH = CH_2} \quad \overset{1.2}{-CH_2 - \underset{\underset{\overset{\|}{CH}}{CH}}{CH} -} \quad \overset{1.4}{CH_2 - CH_2 - CH = CH_2}
\end{array}
$$

Upon ozonolysis, followed by hydrolysis, these in the case of 1,4-1,4 sequences produce succinaldehyde ($CHO-CH_2-CH_2CHO$) and in the case of 1,4-1,2-1,4 sequences produce formyl 1,6-hexane dial and formaldehyde.

$$
(CHOCH_2 - \underset{\underset{CHO}{|}}{CH} - CH_2 - CH_2 - CHO)
$$

Analysis of the reaction product for concentrations of succinaldehyde and 3-formyl-1,6-hexane-dial can show whether the polymer is 1,4-1,4 or 1.4-1,2-1,4, or whether it contains both types of sequence.

Various workers[1417-1422] have applied this technique to the elucidation of the microstructure of polybutadiene. They found that the 3 formyl-1,6-hexane-dial content was directly proportional to the 1,2(vinyl) content of polymers containing 1,4-1,2-1,4 butadiene sequences.

Polymers having 98% cis-1,4 structure, 98% trans-1,4 structure and series of polymers containing from 11% to 75% 1,2 structure were ozonized (Table 146). The final products obtained from these polymers were succinaldehyde, 3-formyl-1,6 hexane-dial, and 4-octene-1,8-dial. Model compounds were ozonized and the products were compared with those from the polymers (Table 145).

Figures 109(a) and (b) shows chromatographic separation of the ozonolysis products from polybutadiene containing 11% and 37.2% 1,2 structure, as measured by infrared or NMR spectroscopy. Figure 109(c) shows the 1,2 content and the amount of 3-formyl-1,6 hexane-dial in the ozonolysis products.

The amounts of 1,4-1,2-1,4 sequences in polybutadienes can be estimated from the amounts of the different ozonolysis products (Table 146) if one considers the amount of 1,4 structure not detected. (Since the ozonolysis technique cleaves the centre of butadiene monomer unit, one half of a 1,4 unit remains attached to each end of a block of 1,2 units after ozonolysis; these structures do not elute from the gas chromatographic column.) Using random copolymer probability theory, the maximum amounts of these undetected 1,4 structures can then be calculated.

Hill et al.[1423] utilized ozonolysis in their investigation of butadiene methyl methacrylate copolymer. The principal products were succinic acid, succindialdehyde and dicarboxylic acids containing several

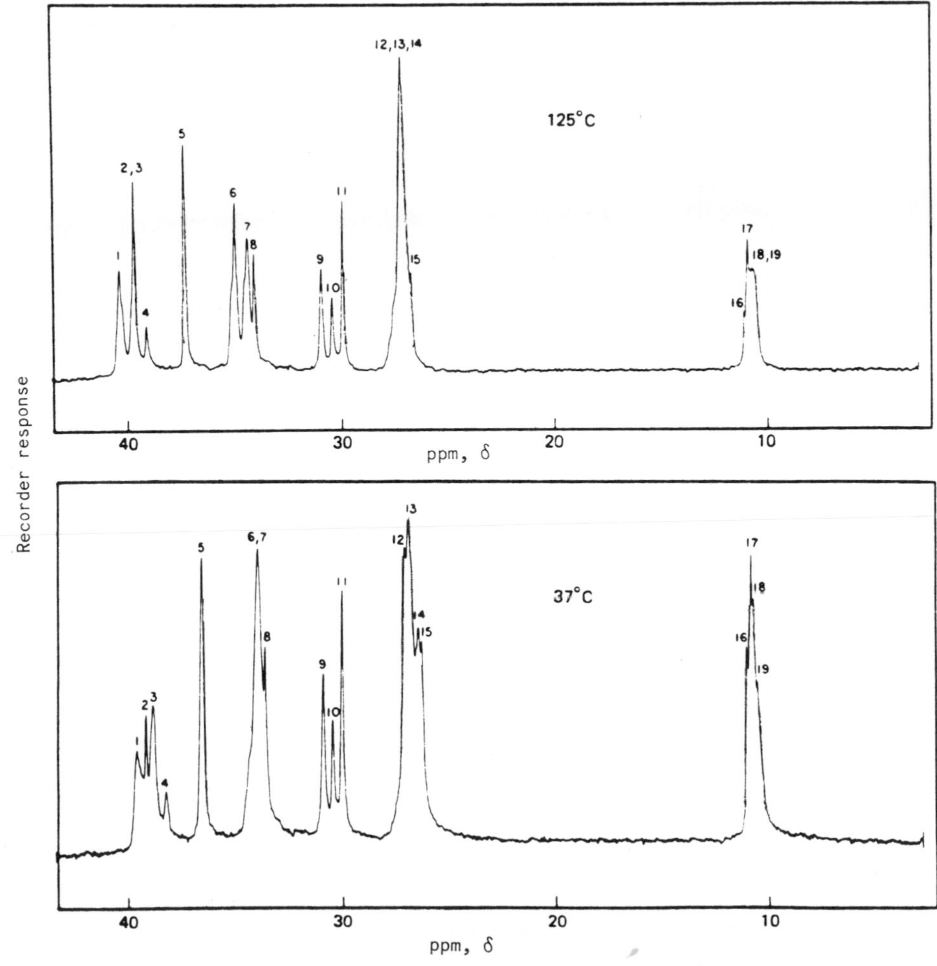

Figure 108 ^{13}C NMR spectra of hydrogenated polybutadiene (73% 1,2-additions) at 37 and 125°C shown with respect to a trimethylsilane internal standard.

methyl methacrylate residues.   The percentage of butadiene (9.2%) recovered as succinic acid and succindialdehyde provided a measure of the 1,4-butadiene-1,4-butadiene linkages in the copolymers, and the percentage of methylmethacrylate units (51%) recovered as trimethyl 2 methylbutane-1,2,4-tricarboxylate (4) n = 1, provided a measure of the methyl methacrylate units in the middle of butadiene-methacrylate-butadiene triads.

10.3.2 Polyisoprene

Isoprene has the structure

$$CH_2 = CH - \underset{3}{\overset{\overset{\displaystyle Me}{|}}{C}} = CH_2$$

$$1 \qquad 2 \qquad 3 \qquad 4$$

Table 145 - Ozonolysis of unsaturated organic compounds

1.5 hexadiene

$$CH_2 = CH - CH_2 - CH_2 - CH = CH_2 \xrightarrow[TPP]{O_3} {}^>CHO - CH_2 - CH_2 - CHO + 2CH_2O$$
succinaldehyde

4 vinyl-1-cyclohexene

$$CH_2 = CH - \underset{TPP}{\overset{O_3}{\bigcirc}} \longrightarrow CHO - CH_2 - \underset{|}{CH} - CH_2 - CH_2 - CHO + CH_2O$$
$$CHO \quad \text{3-formyl-1,6-hexane dial}$$

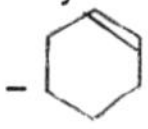

cis-cis 1,5 cyclooctadiene

$$\xrightarrow[TPP]{O_3} CHO - CH_2 - CH_2 - CH = CH - CH_2 - CH_2 CHO$$
succinaldehyde cis-4 octene-1,8-dial

cyclooctene

$$\xrightarrow[TPP]{O_3} CHO \, (CH_2)_6 \, CHO$$
1,8-octane-dial

4 methyl-1-cyclohexene

$$Me - \underset{TPP}{\overset{O_3}{\bigcirc}} \longrightarrow CHO - CH_2CH_2 - \underset{|}{CH} - CH_2 - CHO$$
$$Me$$
$$\text{3-methyl-1,6-hexane dial}$$

cyclopentene

$$\underset{TPP}{\overset{O_3}{\bigcirc}} \longrightarrow CHO - (CH_2)_3 - CHO$$
1,5-pentane dial

Polyisoprene is a completely alternating ethylene propylene copolymer;

$$- CH_2 - \underset{|}{C} = CH - CH_2 - CH_2 - \underset{|}{C} = CH - CH_2 -$$
$$Me \qquad\qquad\qquad Me$$

Its polymers can contain the following four types of unsaturation:

$$\text{trans 1,4} \qquad \left[\begin{array}{cc} - CH_2 & CH_3 \\ \diagdown & \diagup \\ C = C \\ \diagup & \diagdown \\ H & CH_2 - \end{array} \right]_n$$

$$\text{cis 1,4} \qquad \left[\begin{array}{c} -CH_2 \qquad\qquad CH_2 \\ \diagdown \qquad\qquad \diagup \\ C = C \\ \diagup \qquad\qquad \diagdown \\ H \qquad\qquad\; CH_3 \end{array} \right]_n$$

$$\text{3,4 addition} \qquad \left[\begin{array}{c} Me \\ | \\ -C - CH_2 - \\ | \\ CH \\ \| \\ CH_2 \end{array} \right]_n$$

$$\text{1,2 addition} \qquad \left[\begin{array}{c} -CH_2 - CH - \\ | \\ C - Me \\ \| \\ CH_2 \end{array} \right]_n$$

Table 146 - Microozonolysis of polybutadiene

| Sample | 1,4 vinyl (cis+(1,2), trans) (%) | | Area (from GC)% | | | |
			Succinal-dehyde	3-formyl 1,6-hex-anedial	4-octene 1,8-dial	1,2 units occurring in 1,4-1, 2-1, 4 sequences
1	98.0	2	50	1	49	0.5
2	89.1	10.9	30	10	60	5
3	89.0	11.0	43	7	50	3
4	81.0	19.0	34	14	52	6
5	76.2	23.8	36	25	39	11
6	71.8	28.2	33	27	40	11
7	69.7	30.3	48	26	26	10
8	67.7	32.3	36	26	38	9
9	64.2	35.8	38	31	31	10
10	62.8	37.2	45	27	28	8
11	50.5	49.5	26	41	33	12
12	56.0	44.0	30	39	31	11
13	26.0	74.0	33	64	3	5

Vodchnal and Kossler[1424,1425] have reported an infrared method for analysis of polyisoprenes suitable for polymers with a high content of 3,4 addition and relatively small amounts of cis-1,4 and trans-1,4 structural units.

Absorptivities of the bands which are commonly used for the determination of the amount of 1,4 structural units are about 50 times lower than absorptivity of the band at 11.26 microns which is used for the determination of the amount of 3,4-polyisoprene units. Therefore, in analyses of samples with high content of 3,4-polyisoprene units it is necessary to use two concentrations or two cuvettes with different

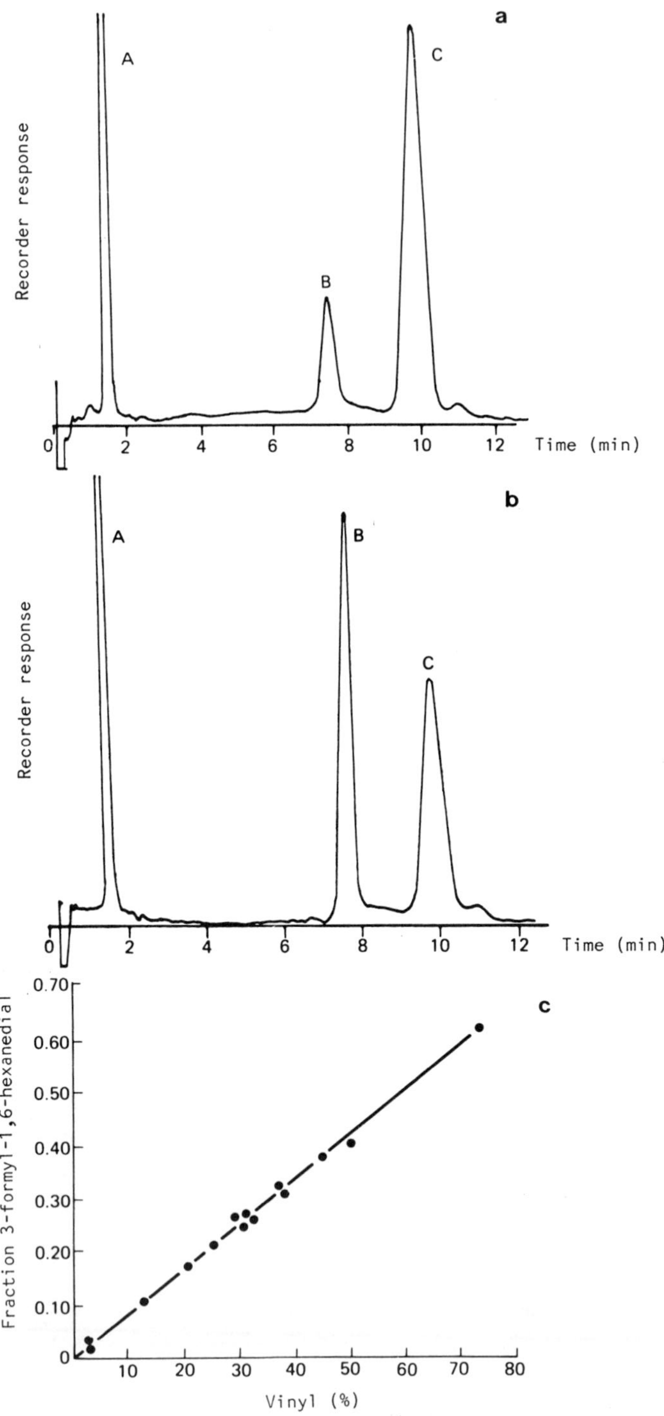

Figure 109   Ozonolysis products from polybutadiene containing (a)
11% vinyl structure; (b) 37.2% vinyl structures: (A)
succinaldehyde; (B) 3-formyl-1, 6-hexanedial; (C) 4-
octene-1, 8-dial; (c) relationship of yield of 3-formyl-
1, 6-hexanedial from ozonolysis to percentage vinyl
structures (from NMR or infrared spectra) in
polybutadienes.

Table 147 - Apparent molar absorptivities km for CS₂ solutions
(km;mol⁻¹ 1.cm⁻¹)

Sample	1,4 units			3,4 units	
	(17.48 microns) 572 cm⁻¹	(11.90 microns) 840 cm⁻¹	(11.26 microns) 888 cm⁻¹	(5.62 microns) 1780 cm⁻¹	(3.26 microns) 3070 cm⁻¹
Heavea	5.7	16.6	1.72	0	0
Balata	2.7	7.6	0.58	0	0
3,4-polyisoprene	6.5	0	110	3.46	30.6

Table 148 - Results of analyses using various absorption
bands (structure %)

3,4 units				1,4 units
	3.25 micron	5.62 micron		11.90 micron
88 cm⁻¹	3070 cm⁻¹	1780 cm⁻¹	Average	840 cm⁻¹
31.6	32.9	–	32	43
37.0	39.0	–	38	42
37.2	40.2	–	39	47
39.0	42.3	–	41	41
–	49.7	51.3	51	–
–	54.6	59.6	58	28

Table 149 - Results of analyses

cis-1,4(%)	trans-1,4(%)	3,4%		
		11.26 micron 888 cm⁻¹	3.26 micron 3070 cm⁻¹	5.62 micron 1780 cm⁻¹
0	10	–	57	62
60	9	15	11	–
45	17	–	20	–

thicknesses. Application of the 5.62 and the 3.26 micron band offers the
possibility of using only one cuvette and one concentration. The 5.62 and
3.26 micron absorption bands do not overlap with absorption bands of other
structural forms, the accuracy of analyses thus being increased. Besides
exact determination of the amount of 3,4 structural units it is possible
to estimate an approximate amount of 1,4 addition from the 11.90, 17.48
and 16.66 micron absorption bands.

Values of apparent molar absorptivities of 3,4-polyisoprene, hevea
and balata in carbon disulphide solutions for the 17.48, 11.90, 11.26,
5.62 and 3.26 micron absorption bands are summarized in Table 147.

The results of measurements of the samples in carbon disulphide solutions, obtained using absorptivities from Table 147, are summarized in Table 148. The cuvettes employed were 2 mm thick, concentration of polymer varying from 0.5 to 3 g in 100 g solution. From the value of absorption at 1190 micron the minimum amount of 1,4 structural units was estimated assuming that all 1,4 units are cis. Analysis using the 17.48 and 10.20 micron bands was inapplicable due to the presence of cyclic structure.

The results of analyses of the samples in potassium bromide pellets are presented in Table 149. In these analyses it was possible to utilize the 17.48 and 16.66 micron absorption bands only, for an approximate estimation of the relative abundance of 1,4 structural units.

Fraga and Benson[1427] have investigated a thin-film infrared method for the analysis of polyisoprene. They emphasize that clear, smooth and uniform films are necessary, and that these can be cast from a toluene solution of the polymer. Film thickness should be maintained to provide between 0.5 and 0.7 absorbance units at the peak near 7.3 micron. When good quality films are used the repeatability of the method is excellent. Thinner films will give slightly lower results. Binder[1427] found a direct correlation existed between the intensity of the 13.48 micron and the percentage net cis-1,4 for various high cis-1,4 polyisoprenes. Synthetic cis-1,4 polyisoprenes prepared with Zeigler catalysts or lithium catalysts contain a small percentage of the 3,4 structure. On the other hand, naturally occurring polyisoprenes such as natural rubber (Hevea), gutta percha, balata, and chicle consist exclusively of the 1,4 structure. The differences between the thermal and mechanical properties of the natural and synthetic polyisoprenes have been attributed to the amount of cis-1,4 units. It is reasonable to expect that the physical properties of polyisoprenes are also affected by the distribution of the isomeric structure units along the polymer chain, as well as the composition of the polymers.

Various infrared methods[1430-1433] have been reported for the analysis of polyisoprene with predominantly cis-1,4 and trans-1,4 structural units with low amounts of 3,4 addition.

Tanaka et al.[1428] determined the distribution of cis-1,4 and trans-1,4 units in 1,4 polyisoprenes by using ^{13}C NMR spectroscopy, and found that cis-1,4 and trans-1,4 units are distributed almost randomly along the polymer chain in cis-trans isomerized polyisoprenes, and the chicle is a mixture of cis-1,4 and trans-1,4 polyisoprenes. These workers[1429] have also investigated the ^{13}C NMR spectra of hydrogenated polyisoprenes and determined the distribution of 1,4 and 3,4 units along the polymer chain for n-butyllithium catalysed polymers, and have confirmed that these units are randomly distributed along the polymer chain. The polymers did not contain appreciable amounts of head-to-head or head-to-tail 1,4 linkages.

Maynard and Moobel[1434] and Ferguson[1435] have described infrared methods, respectively, for determining cis-1,4, trans-1,4,3,4 and 1,2 structures and 1,4 structures in polyisoprene.

Kossler and Vodchnal[1437] came to the conclusion that the infrared spectra of polymers containing cis-1,4, trans-1,4 and 3,4 or cyclic structural units are not additively composed of the spectra of stereoregular polymers containing only one of these structures.

It is known that stereoregular cis-1,4 polyisoprene (Hevea) and stereoregular trans-1,4 polyisoprene (balata) have absorption bands at 8.84 and 8.69 microns respectively. These workers found that a polymer having a high content of 3,4 structural units, in addition to the 1,4 structural units, has no absorption band at either 8.84 or 8.69 microns but does have a band at 8.77 microns. They attribute this band to the C-CH_3 vibration of the $-C(CH_3) = CH$ structural unit separated by other structural units.

The appearance of the absorption band at 8.77 micron in some synthetic polyisoprenes has been mentioned several times by Binder[1437-1439], with the comment that the origin of this band is not known[1439].

A similar phenomenon has been discovered by analysis of a polymer having approximately 20% trans-1,4 in addition to about 75% cis-1,4 structural units, as estimated by an analysis using the absorption bands at 17.48 and 10.20 microns. The band at 8.85 micron was shifted towards higher values. In a mixture of hevea and balata with the same content of 20% trans-1,4 structural units both the 8.85 and 8.69 micron bands are quite distinct. The behaviour of the 8.85 and 8.69 micron bands is in agreement with the finding of Golub[1440,1441] who has shown that during the cis-trans isomerization of polyisoprene the 8.80 micron absorption band appear instead of the 8.88 micron band in cis-1,4 isomers or the 8.70 micron band in trans-1,4 isomers. The statement of Maynard and Moobel[1442] that the small amount of trans-1,4 structural units may be better detected using the band pair near 7.65 micron rather than the bands at 8.84 and 8.68 micron is also in good agreement with the findings of Kossler and Vodchanal[1436].

As a consequence of these results it is possible to conclude that only polyisoprenes having long sequences of cis-1,4 or trans-1,4 units have the absorption bands at 8.85 and 8.69 micron respectively. It is also evident that the analysis of synthetic polyisoprenes using these absorption bands leads to distorted results. Kossler and Vodchnal[1436] obtained better results using the absorption bands 17.48, 10.20 and 11.26 microns for cis-1,4 and trans-1,4 and 3,4 polyisoprene structural units, respectively. The use of various combinations of different absorption bands permits one to conclude whether a polymer in question is more of the block copolymer type or a mixture of stereoregular polymers.

The diad distribution of cis-1,4 and trans-1,4 units in low molecular weight 1,4-polyisoprene has been determined from the ^{13}C NMR spectra at 350k by Morese-Seguela et al.[1443].

Gronski and co-workers[1444], Beebe[1445], Dalinskaya et al.[1446] and Duck and Grant[1447] used the chemical shift correction parameters for linear alkanes in the aliphatic region of ^{13}C NMR spectra to determine the relative amounts of 3,4 and cis-1,4 units of polyisoprene.

Microstructure studies have been carried out on the cyclic content and cis-trans isomer distribution in polyisoprene[1448].

Pyrolysis-gas chromatography has been applied to the investigation of the sequence distribution of the 1,4 and 3,4 units in poliso-prenes[1449,1450,1452]. This method is based on the structural relationship between the isoprene dimers and the diad sequences of 1,4 and 3,4 units. It has been pointed out be Tanaka et al.[1451] that it is difficult to deduce accurately the polymer structure by this technique, because of two major

problems.   First, the dimer fraction is only a minor pyrolytic product
(about 30%), the monomer being the major product and other products
constituting a small percentage.   Second, during the pyrolysis the chain-
scission reaction is often complicated by side-reactions which can alter
the structure of the products.

The ozonolysis technique has been applied[1453] to polyisoprene. This
polymer having nearly equal 1,4 and 3,4 structures, produced large amounts
of vininaldehyde, succinaldehyde and 2,5 hexanedione, indicating blocks of
1,4 structures in head-tail, tail-tail and head-head configurations.

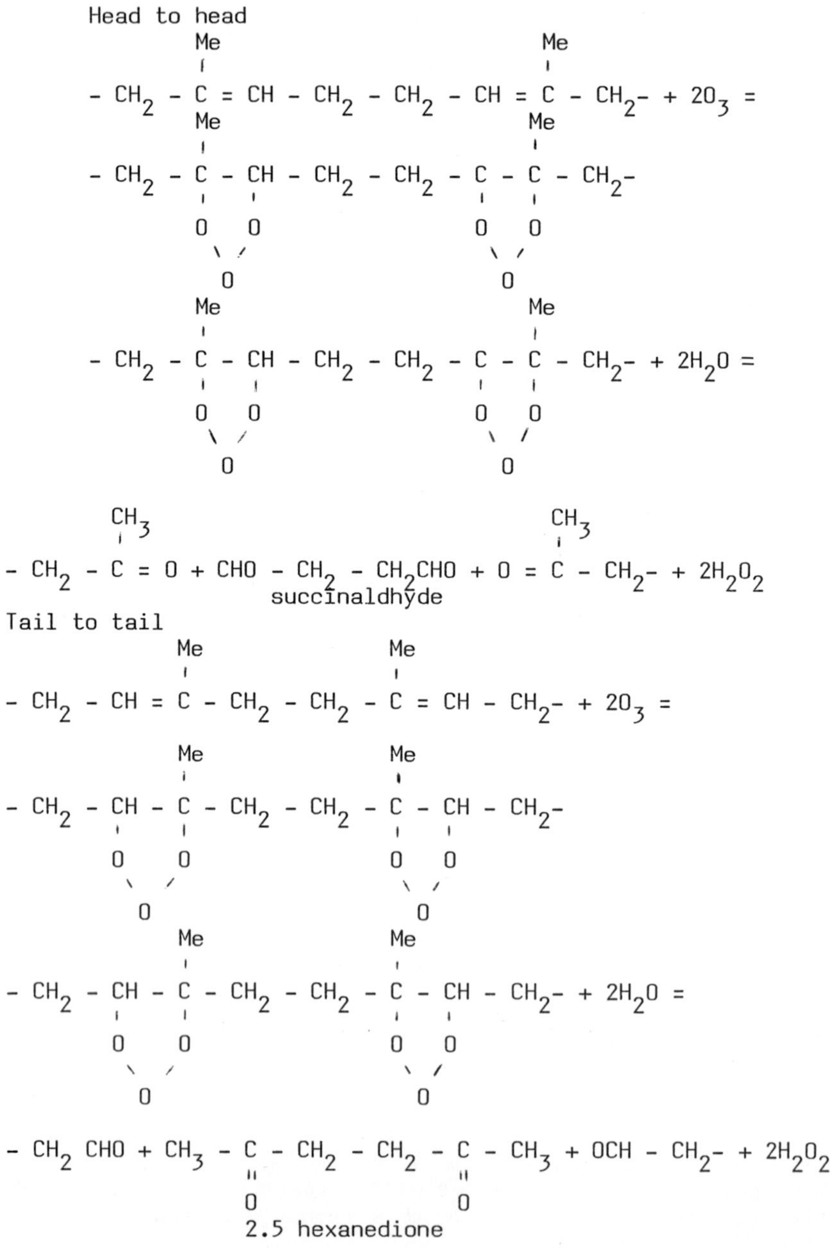

Head to head

$$- CH_2 - \overset{\overset{\text{Me}}{|}}{C} = CH - CH_2 - CH_2 - CH = \overset{\overset{\text{Me}}{|}}{C} - CH_2 - + 2O_3 =$$

$$- CH_2 - \underset{\underset{O \;\;\; O}{|\;\;\;\;|}}{\overset{\overset{\text{Me}}{|}}{C}} - CH - CH_2 - CH_2 - \underset{\underset{O \;\;\; O}{|\;\;\;\;|}}{\overset{\overset{\text{Me}}{|}}{C}} - C - CH_2 -$$

(O bridge)

$$- CH_2 - \overset{\overset{\text{Me}}{|}}{C} - CH - CH_2 - CH_2 - \overset{\overset{\text{Me}}{|}}{C} - C - CH_2 - + 2H_2O =$$

(O bridge)

$$- CH_2 - \overset{\overset{CH_3}{|}}{C} = O + CHO - CH_2 - CH_2CHO + O = \overset{\overset{CH_3}{|}}{C} - CH_2 - + 2H_2O_2$$

succinaldhyde

Tail to tail

$$- CH_2 - CH = \overset{\overset{\text{Me}}{|}}{C} - CH_2 - CH_2 - \overset{\overset{\text{Me}}{|}}{C} = CH - CH_2 - + 2O_3 =$$

$$- CH_2 - CH - \overset{\overset{\text{Me}}{|}}{C} - CH_2 - CH_2 - \overset{\overset{\text{Me}}{|}}{C} - CH - CH_2 -$$

(O bridges)

$$- CH_2 - CH - \overset{\overset{\text{Me}}{|}}{C} - CH_2 - CH_2 - \overset{\overset{\text{Me}}{|}}{C} - CH - CH_2 - + 2H_2O =$$

(O bridges)

$$- CH_2 CHO + CH_3 - \overset{\overset{}{C}}{\underset{O}{\|}} - CH_2 - CH_2 - \overset{\overset{}{C}}{\underset{O}{\|}} - CH_3 + OCH - CH_2 - + 2H_2O_2$$

2.5 hexanedione

Head to tail

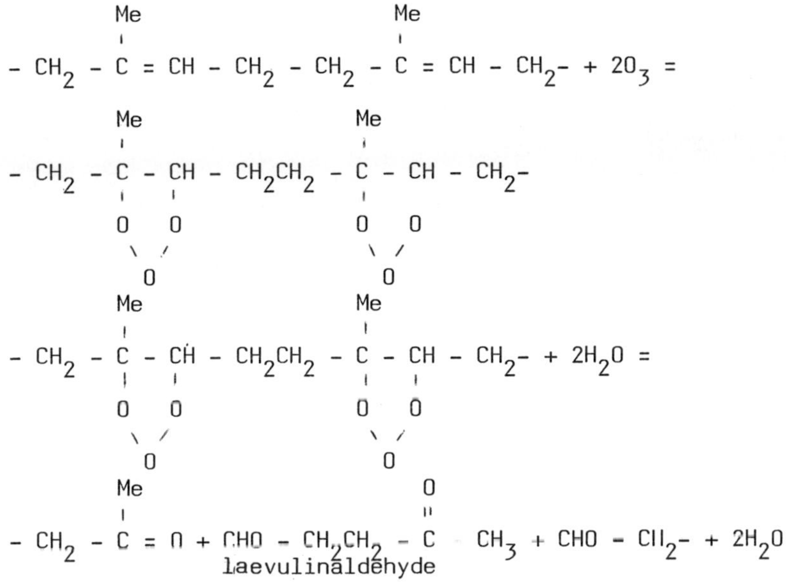

$$
\begin{array}{cc}
\text{Me} & \text{Me} \\
| & | \\
-CH_2 - C = CH - CH_2 - CH_2 - C = CH - CH_2- + 2O_3 =
\end{array}
$$

$$
\begin{array}{cc}
\text{Me} & \text{Me} \\
| & | \\
-CH_2 - C - CH - CH_2CH_2 - C - CH - CH_2- \\
\;\;|\;\;\;\;| & \;\;|\;\;\;\;| \\
\;\;O\;\;\;\;O & \;\;O\;\;\;\;O \\
\;\;\;\backslash\;/ & \;\;\;\backslash\;/ \\
\;\;\;\;O & \;\;\;\;O
\end{array}
$$

$$
\begin{array}{cc}
\text{Me} & \text{Me} \\
| & | \\
-CH_2 - C - CH - CH_2CH_2 - C - CH - CH_2- + 2H_2O = \\
\;\;|\;\;\;\;| & \;\;|\;\;\;\;| \\
\;\;O\;\;\;\;O & \;\;O\;\;\;\;O \\
\;\;\;\backslash\;/ & \;\;\;\backslash\;/ \\
\;\;\;\;O & \;\;\;\;O
\end{array}
$$

$$
\begin{array}{cc}
\text{Me} & O \\
| & \| \\
-CH_2 - C = O + CHO - CH_2CH_2 - C \;\; CH_3 + CHO - CH_2- + 2H_2O
\end{array}
$$

laevulinaldehyde

10.4 REGIOISOMERISM

10.4.1 Introductory

As well as stereoisomerism and geometrical isomerism as discussed in Chapters 10.2 and 10.3, polymers and copolymers can exhibit a third form of isomerism known as regioisomerism. Head to head, head to tail and tail to head isomerism is well known in the case of simple organic compounds thus, a dimer of styrene monomer can exist in the following three different structural forms:

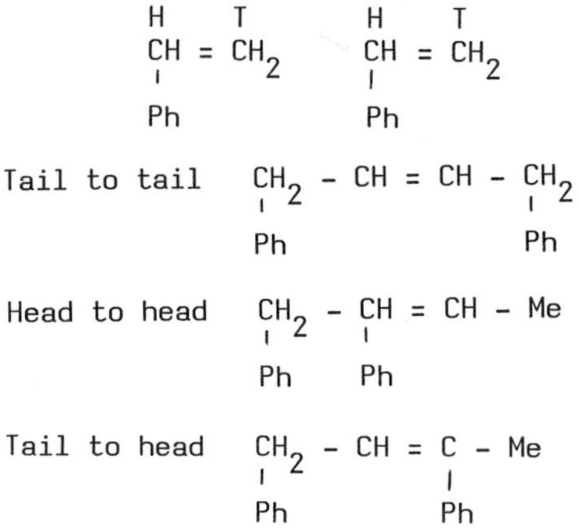

$$
\begin{array}{cccc}
\text{H} & \text{T} & \text{H} & \text{T} \\
CH & = CH_2 & CH & = CH_2 \\
| & & | & \\
Ph & & Ph &
\end{array}
$$

Tail to tail
$$
\begin{array}{c}
CH_2 - CH = CH - CH_2 \\
|\;| \\
Ph\;\;\;\;\;\;\;\;\;\;\;\;\;\;\;\;\;Ph
\end{array}
$$

Head to head
$$
\begin{array}{c}
CH_2 - CH = CH - Me \\
|\;\;\;\;\;\;\;| \\
Ph\;\;\;\;Ph
\end{array}
$$

Tail to head
$$
\begin{array}{c}
CH_2 - CH = C - Me \\
|\;\;\;\;\;\;\;\;\;\;\;\;| \\
Ph\;\;\;\;\;\;\;\;Ph
\end{array}
$$

each of which, incidentally, can exist in cis and trans geometric isomeric
forms. Similarly, regioisometric forms of polymers exist.

10.4.2 Polypropylene

Thus an isotactic polypropylene diad has the following three
possible structures:

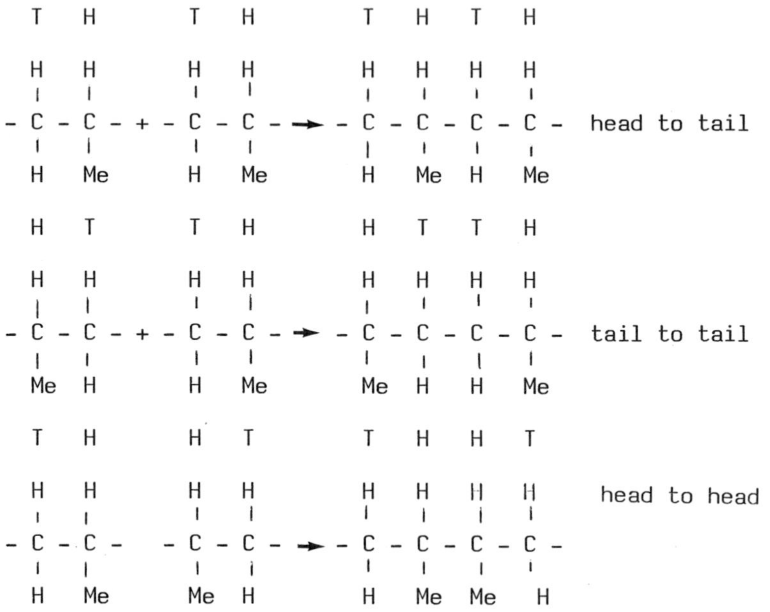

```
    T   H        T   H        T   H   T   H
    H   H        H   H        H   H   H   H
    |   |        |   |        |   |   |   |
  - C - C - + - C - C -  →  - C - C - C - C -   head to tail
    |   |        |   |        |   |   |   |
    H   Me       H   Me       H   Me  H   Me


    H   T        T   H        H   T   T   H
    H   H        H   H        H   H   H   H
    |   |        |   |        |   |   |   |
  - C - C - + - C - C -  →  - C - C - C - C -   tail to tail
    |   |        |   |        |   |   |   |
    Me  H        H   Me       Me  H   H   Me


    T   H        H   T        T   H   H   T
    H   H        H   H        H   H   H   H
    |   |        |   |        |   |   |   |
  - C - C -    - C - C -  →  - C - C - C - C -   head to head
    |   |        |   |        |   |   |   |
    H   Me       Me  H        H   Me  Me  H
```

With the entry of the third propylene monomer unit i.e., the formation of
a triad, additional possibilities for even more structures exist. Thus in
the case of isotactic polypropylene:

```
    T   H   T   H        T   H              H   H   H   H   H   H
    H   H   H   H        H   H              H   H   H   H   H   H
    |   |   |   |        |   |              |   |   |   |   |   |
  - C - C - C - C- →  - C - C -  →       - C - C - C - C - C - C -
    |   |   |   |        |   |              |   |   |   |   |   |
    H   Me  H   Me       H   Me             H   Me  H   Me  H   Me
```

                          Head to tail - head to tail

```
            H   T
            |   |
            H   H        H   H   H   H   H   H
            |   |        |   |   |   |   |   |
          - C - C  →  - C - C - C - C - C - C -
            |   |        |   |   |   |   |   |
            Me  H        H   Me  H   Me  Me  H
```

                          Head to tail - head to head

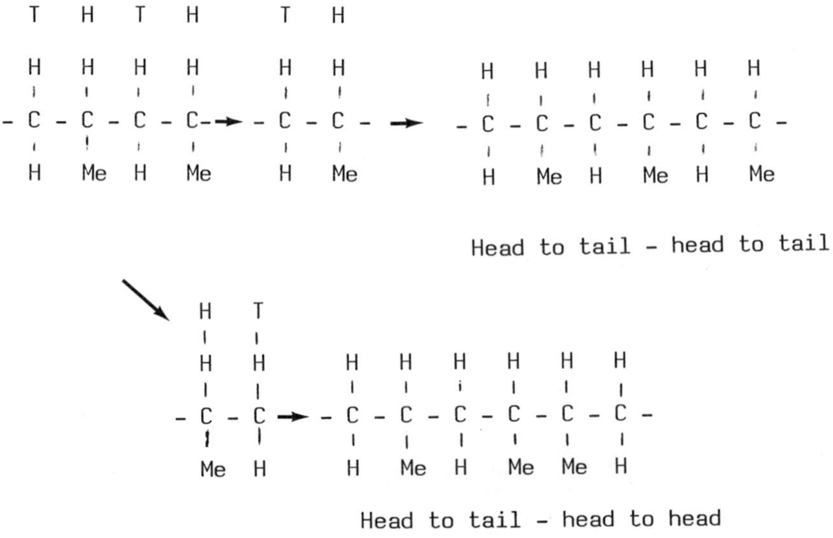

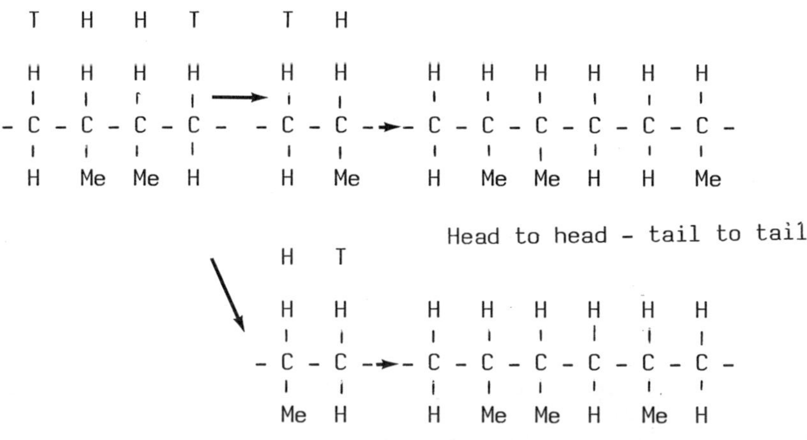

Thus with isotactic polypropylene alone six different placements are possible when considering triads ^{13}C NMR spectroscopy can be used to determine isolated head to head and tail to tail units in polypropylene[1453]. Polypropylenes produced using vanadyl catalysts possessed the normal head-to-tail structure[1454]. More detailed examination, however shows that the amorphous fractions isolated from these polypropylenes do show infrared absorption at 13.3 microns pointing to the presence of methylene sequences of two units, which can only mean tail-to-tail arrangement of propylene units does occur. A very small absorption peak at 13.3 microns was also found in the spectrum of the crystalline fractions. Polypropylenes prepared with catalysts based on VCl_3 show only the normal head-to-tail arrangement in both amorphous and crystalline fractions, as do polymers prepared from $TiCl_3$ catalyst. The amount of propylene units coupled tail-to-tail was estimated to range from 5 to 15% for the amorphous and from 1 to 5% for the crystlline fractions.

The amount of propylene units in tail-to-tail arrangement was calculated from spectra of thin films by comparing the ratio of the absorbances at 13.60 micron and 8.65 microns in the spectrum of hydrogenated natural rubber. This implies the assumption that the absorbance per CH_2 group is the same at 13.30 microns for $(CH_2)_2$ sequences as at 13.60 micron for $(CH_2)_3$ sequences.

The differences in amount of tail-to-tail coupled units between crystalline and amorphous fractions are to be expected, since every head-to-head and tail-to-tail configuration disturbs the regularity of the isotactic chain. In the polypropylenes every tail-to-tail configuration must necessarily be accompanied by a head-to-head coupling. This would be expected to show up in an absorption peak at 8.8 to 9.0 microns characteristic of the structure

```
        H   H
        |   |
      - C - C -
        |   |
        Me  Me
```

which is also found in hydrogenated poly-2,3,-dimethylbutadiene, used as a model compound and in alternating copolymers of ethylene and butene-2[1455]. In the polypropylenes examined by Van Schooten and Mostert[1456] and in ethylene-propylene copolymers, they did indeed find an absorption band near 9.0 microns, although, unlike Van Schooten et al[1456] (see above) it is much less sharp than in the model compound.

All spectra containing the 13.3 micron peak show a further small band at 10.9 micron which is also found in the spectrum of poly-2,13-dimethylbutadiene. To summarized the infrared spectrum of amorphous polypropylene prepared using vinadyl catalysts in addition to normal head to tail structures $-CH\ CH_3-CH_2-CH\ CH_3-CJ_2-$ shows the following features

Absorption	Characteristic of	also found in
13.3u * tail to tail	methylene sequences of two units, i.e. tail to tail arrange- ment of propylene units	polybutene-2 ethylene alternating polymer

```
            Me  H   H   Me                    Me  Me  H   H
            |   |   |   |                      |   |   |   |
          - C - C - C - C -                  - C - C - C - C -
            |   |   |   |                      |   |   |   |
            H   H   H   H                      H   H   H   H
```

ethylene-propylene
copolymer (tail to tail)

```
                                               Me  H   H   H
                                               |   |   |   |
                                             - C - C - C - C -
                                               |   |   |   |
                                               H   H   H   H
```

in polypropylene
every tail to tail
coupling accompanied
by head to head coupling

```
   9.0 u              H    H    H    H
 head to head         |    |    |    |
                  - C - C - C - C -
                      |    |    |    |
                      H    Me   Me   H
```

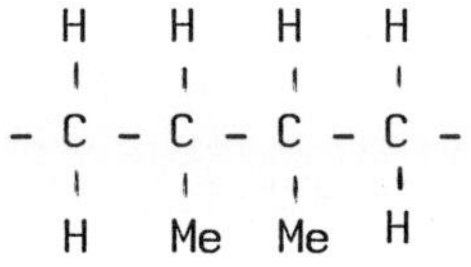

polybutene-2 ethylene
alternating copolymer
```
                  Me   Me   H    H
                   |    |    |    |
                 - C - C - C - C -
                   |    |    |    |
                   H    H    H    H
```

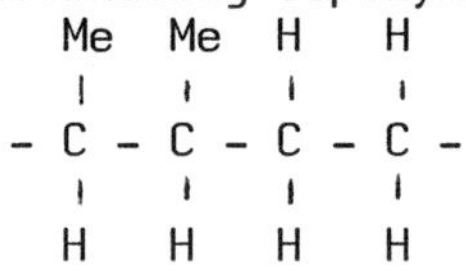

hydrogenated 2,3
dimethylbutadiene
```
                   H    Me   Me   H
                   |    |    |    |
                 - C - C = C - C -
                   |              |
                   H              H
```

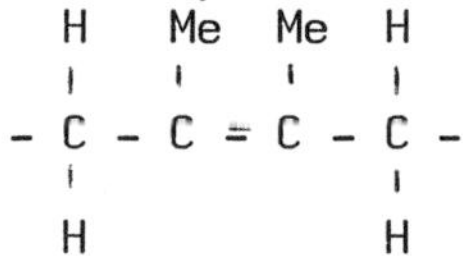

ethylene-propylene
copolymer (head to head)
```
                   H    Me   Me   H
                   |    |    |    |
                 - C - C - C - C -
                   |    |    |    |
                   H    H    H    H
```

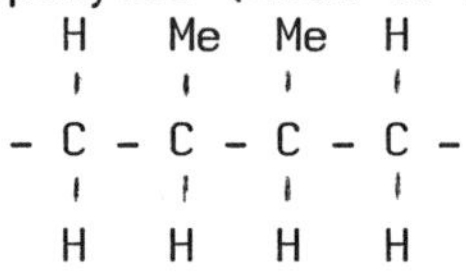

* All ethylene propylene and polypropylene polymers showing an absorbance at 13.3 u also show small absorbance at 10.9 u which is also found in poly 2,3--dimethylbutadiene.

10.4.3 Ethylene-Propylene Copolymers

Ethylene-propylene copolymers can contain up to four types of sequence distribution of monomeric units. These are propylene-propylene (head to tail and head to head), ethylene-propylene and ethylene to ethylene.

```
            H    H    H    H
            |    |    |    |
(1)      ( - C - C - C - C - )      head to tail isotactic
            |    |    |    |
            H    Me   H    Me  )n

            H    H    H    H
            |    |    |    |
(2)      ( - C - C - C - C - )      head to tail isotactic
            |    |    |    |
            H    Me   Me   H  )n
```

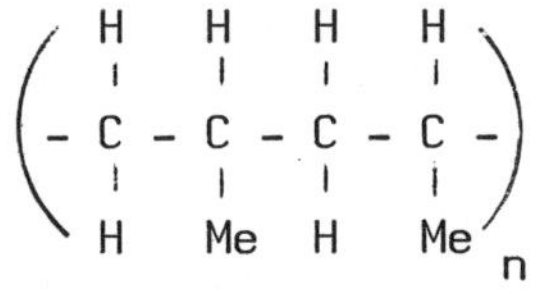

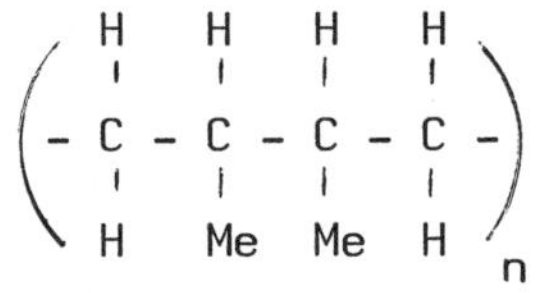

$$(3) \quad \left(\begin{array}{cccc} H & H & H & H \\ | & | & | & | \\ -C & -C & -C & -C- \\ | & | & | & | \\ Me & H & H & H \end{array} \right)_n \qquad \text{ethylene-propylene}$$

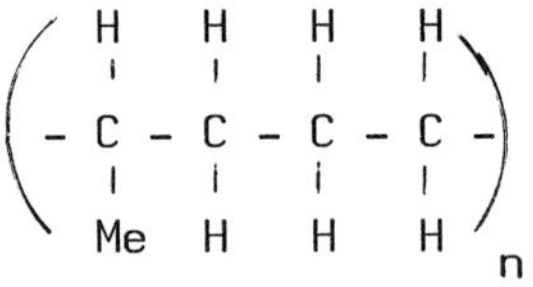

$$(4) \quad \left(\begin{array}{cccc} H & H & H & H \\ | & | & | & | \\ -C & -C & -C & -C- \\ | & | & | & | \\ H & H & H & H \end{array} \right)_n \qquad \text{ethylene-ethylene}$$

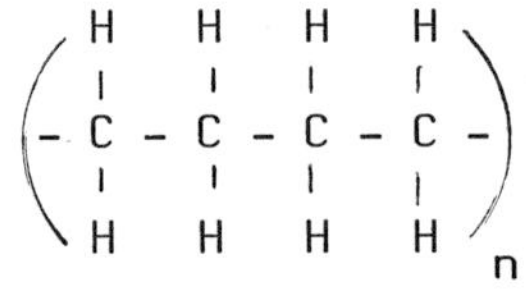

In addition, other stereochemical placements could occur, eg. syndiotactic head to tail and head to head polypropylene units;

$$(5) \quad \left(\begin{array}{cccc} H & Me & H & H \\ | & | & | & | \\ -C & -C & -C & -C- \\ | & | & | & | \\ H & H & H & Me \end{array} \right)_n \qquad \text{head to tail syndiotactic}$$

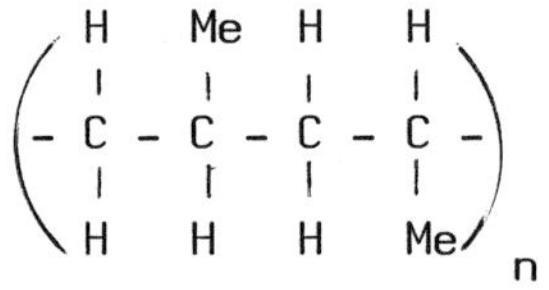

$$(6) \quad \left(\begin{array}{cccc} H & Me & H & H \\ | & | & | & | \\ -C & -C & -C & -C- \\ | & | & | & | \\ H & H & Me & H \end{array} \right)_n \qquad \text{head to head syndiotactic}$$

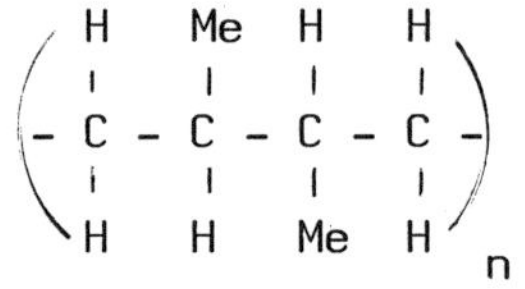

Sequences 1-4 and their average sequence lengths of both monomer units, can be measured by the Tanaka and Hatada[1457] method. Measurements were made at 15.1 MHz. Assignments of the signals were carried out by using the method of Grant and Paul[1458], and also by comparing the spectra with those of squalane, hydrogenated natural rubber, polyethylene and atactic polypropylene. The accuracy and the precision of intensity measurements, that is the deviation from the theoretical values and the scatter of the measurements, respectively, were checked for the spectra of squalane and hydrogenated natural rubber, and were shown to be at most 12% for most of the signals.

Infrared spectroscopy also provides information on regioisomerism in ethylene propylene copolymers. The fact that the polypropylenes prepared with $VOCl_{3-}$ or $VO(OR)_{3-}$ containing catalysts show tail to tail arrangement means that tail to tail coupling of propylene units may also occur in the ethylene-propylene copolymers. However, because the content of $(CH_2)_2$ sequences in the copolymers is much higher than in the polypropylenes prepared with the same catalysts, a large part of these sequences most likely stems from isolated ethylene units between two head to head oriented propylene units, their relative amount depending on the ratio of reaction rates of formation of the sequences.

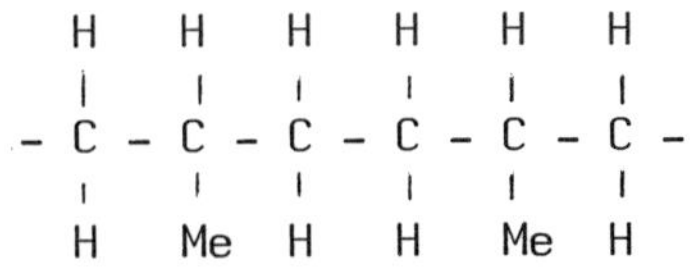

```
              H   H   H   H   H   H
              |   |   |   |   |   |
          - C - C - C - C - C - C -
              |   |   |   |   |   |
              H   Me  H   H   Me  H
   versus
              H   Me  Me  H
              |   |   |   |
          - C - C - C - C -
              |   |   |   |
              H   H   H   H
```

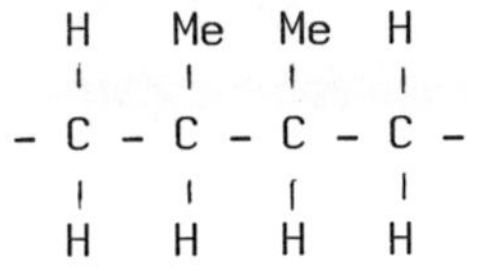

An absorption at 13.3 microns which is characteristic of methylene sequences of two units is characteristic of a tail to tail configuration of ethylene and propylene units. This absorption is also found in polybutene-2-ethylene copolymer,

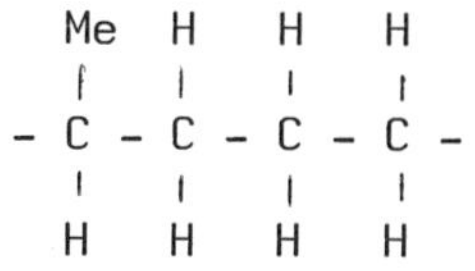

```
     Me  H   H   H                    Me  Me  H   H
     |   |   |   |                     |   |   |   |
 - C - C - C - C -                  - C - C - C - C -
     |   |   |   |                     |   |   |   |
     H   H   H   H                     H   H   H   H
```

ethylene-propylene   polybutene-ethylene
copolymer (tail to tail)   alternating copolymer

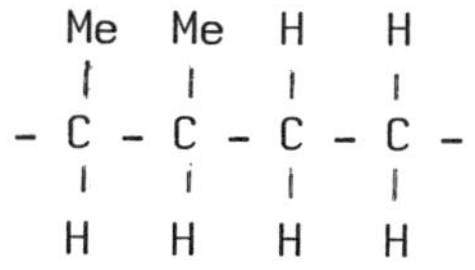

An absorption of 9.0 microns which is characteristic of a head to head configuration of two propylene units i.e.

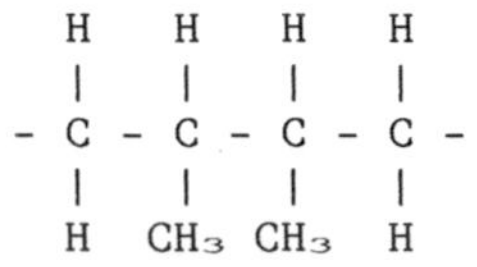

```
              H    H    H    H
              |    |    |    |
          - C -  C -  C -  C -
              |    |    |    |
              H   CH₃  CH₃   H
```

is found in amorphous ethylene-propylene copolymers and in polybutene-2-ethylene alternating copolymer and in hydrogenated poly 2,3 dimethyl butadiene;

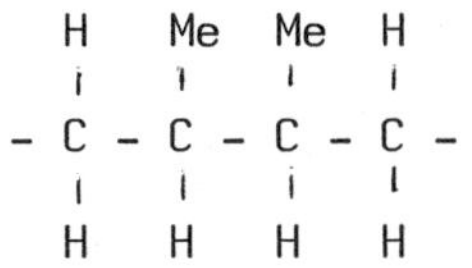

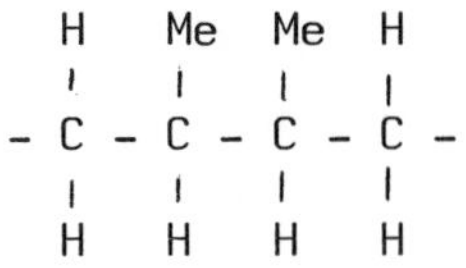

```
     H   Me  Me  H          Me  Me  H   H          H   Me  Me  H
     |   |   |   |           |   |   |   |          |   |   |   |
 - C - C - C - C -       - C - C - C - C -      - C - C - C - C -
     |   |   |   |           |   |   |   |          |   |   |   |
     H   H   H   H           H   H   H   H          H   H   H   H
```

ethylene-propylene   polybutene-1-   hydrogenated poly-2,3-
copolymer head to   ethylene alternating   dimethyl butadiene
head   copolymer

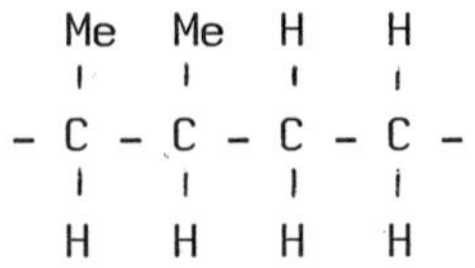

10.4.4   Polybutadiene-1-Ethylene Copolymer

The infrared spectrum of amorphous alternating polybutene-1 ethylene copolymer shows absorptions at 13.3 microns, characteristic of methylene sequences of two units and at 9 microns characteristic of the structure

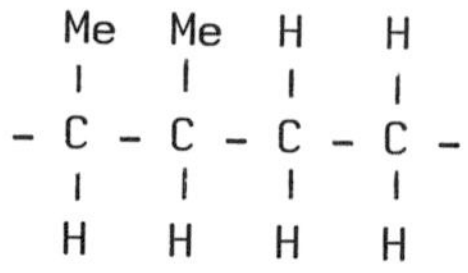

$$
\begin{array}{cccc}
Me & Me & H & H \\
| & | & | & | \\
-\,C & -\,C & -\,C & -\,C\,- \\
| & | & | & | \\
H & H & H & H
\end{array}
$$

An absorption at 10.8 microns, also found in hydrogenated poly 2.3-dimethyl-butadiene also confirms the above structure.

10.4.5   Poly 2,3 Dimethyl Butadiene

Hydrogenated poly 2,3‑dimethyl butadiene has a strong infrared absorption at 9.0 micron confirming a head to head configuration of two propylene units and the hydrogenated polymer

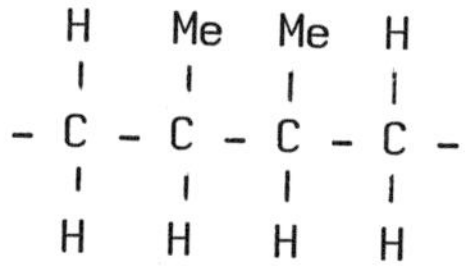

$$
\begin{array}{cccc}
H & Me & Me & H \\
| & | & | & | \\
-\,C & -\,C & -\,C & -\,C\,- \\
| & | & | & | \\
H & H & H & H
\end{array}
$$

and the following structure for the unhydrogenated polymer;

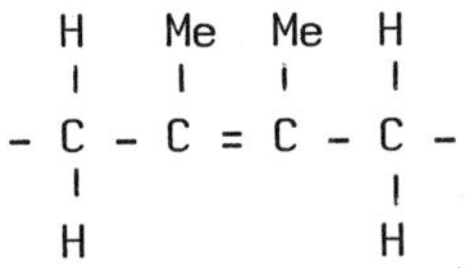

$$
\begin{array}{cccc}
H & Me & Me & H \\
| & | & | & | \\
-\,C & -\,C & =\,C & -\,C\,- \\
| & & & | \\
H & & & H
\end{array}
$$

10.4.6   Polybutadiene

In the case of dimers many regioisometric configurations are possible. Polybutadiene unsaturation occurs in three forms; trans-1,4, cis-1,4 and vinyl-1,2

$$
\text{trans-1,4} \qquad
\left[
\begin{array}{c}
-\,CH_2 \diagdown \\
\qquad\quad C = C \diagup ^{H} \\
H \diagup \qquad\quad \diagdown CH_2-
\end{array}
\right]_n
$$

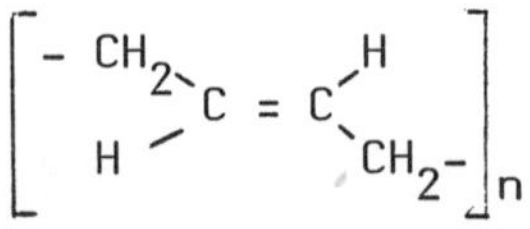

$$\text{cis-1,4} \qquad \left[\begin{array}{c} \text{CH}_2 \quad\quad \text{CH}_2 \\ \diagdown\text{C} = \text{C}\diagup \\ \text{H} \qquad\qquad \text{H} \end{array} \right]_n$$

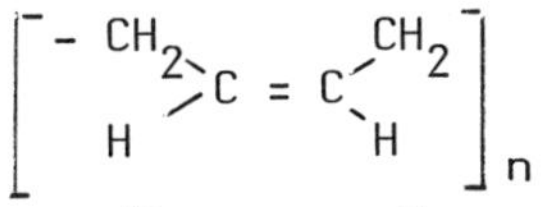

$$\text{vinyl} \qquad \left[\begin{array}{cc} \text{H} & \text{H} \\ | & | \\ -\text{C} & -\text{C}- \\ | & \\ \text{CH} & \\ \| & \\ \text{CH}_2 & \end{array} \right]_n$$

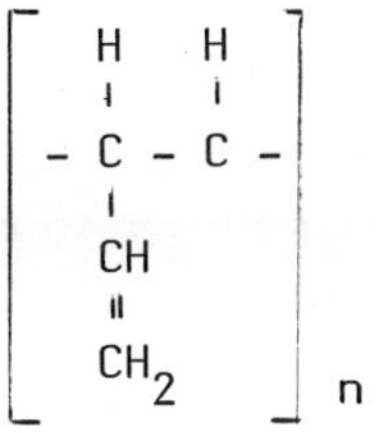

No opportunity for regioisomerism exists in the cis and trans - 1,4 configurations but it does in the case of the vinyl-1,2 unsaturation.

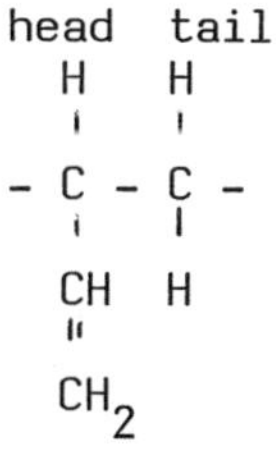

$$\begin{array}{cc} \text{head} & \text{tail} \\ \text{H} & \text{H} \\ | & | \\ -\text{C} - & \text{C}- \\ | & | \\ \text{CH} & \text{H} \\ \| & \\ \text{CH}_2 & \end{array}$$

Head to head configuration

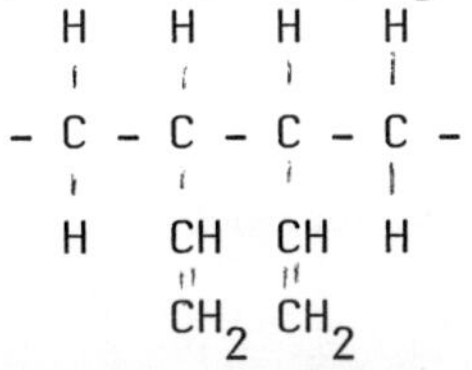

$$\begin{array}{cccc} \text{H} & \text{H} & \text{H} & \text{H} \\ | & | & | & | \\ -\text{C} - & \text{C} - & \text{C} - & \text{C}- \\ | & | & | & | \\ \text{H} & \text{CH} & \text{CH} & \text{H} \\ & \| & \| & \\ & \text{CH}_2 & \text{CH}_2 & \end{array}$$

Tail to tail configuration

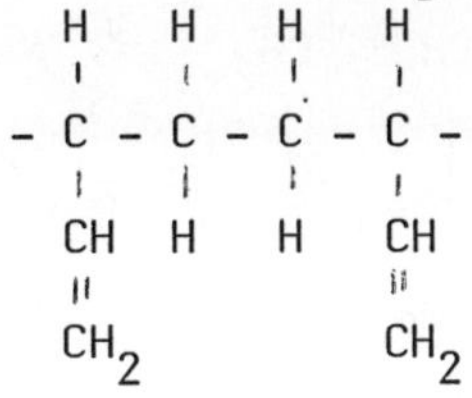

$$\begin{array}{cccc} \text{H} & \text{H} & \text{H} & \text{H} \\ | & | & | & | \\ -\text{C} - & \text{C} - & \text{C} - & \text{C}- \\ | & | & | & | \\ \text{CH} & \text{H} & \text{H} & \text{CH} \\ \| & & & \| \\ \text{CH}_2 & & & \text{CH}_2 \end{array}$$

Head to tail configuration

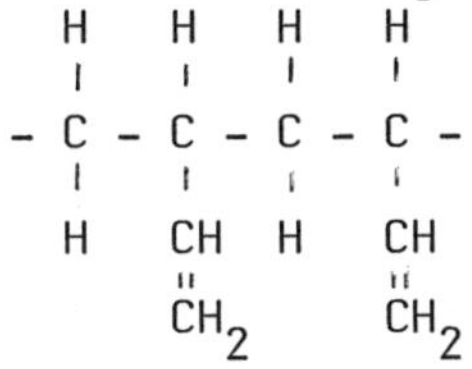

$$\begin{array}{cccc} \text{H} & \text{H} & \text{H} & \text{H} \\ | & | & | & | \\ -\text{C} - & \text{C} - & \text{C} - & \text{C}- \\ | & | & | & | \\ \text{H} & \text{CH} & \text{H} & \text{CH} \\ & \| & & \| \\ & \text{CH}_2 & & \text{CH}_2 \end{array}$$

10.4.7 Polyisoprene

Due to the methyl group, opportunities for regioisomerism exists for all four types of unsaturation present in polyisoprene.

$$
\text{trans-1,4} \qquad
\left[\begin{array}{c}
-CH_2 \diagdown \quad \diagup Me \\
\quad\quad C = C \\
H \diagup \quad \diagdown CH_2
\end{array}\right]_n
$$

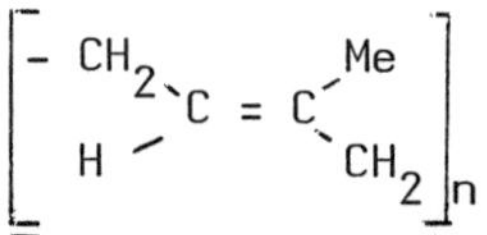

$$
\text{cis-1,4} \qquad
\left[\begin{array}{c}
-CH_2 \diagdown \quad \diagup CH_2 \\
\quad\quad C = C \\
H \diagup \quad \diagdown Me
\end{array}\right]_n
$$

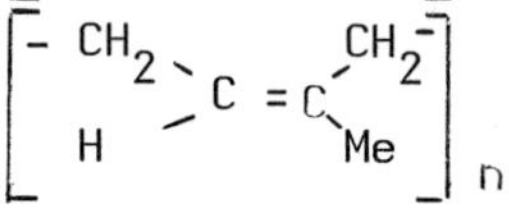

$$
\text{3,4 additions} \qquad
\left[\begin{array}{c}
Me \\
| \\
-C - CH_2 \\
| \\
CH \\
\| \\
CH_2
\end{array}\right]_n
$$

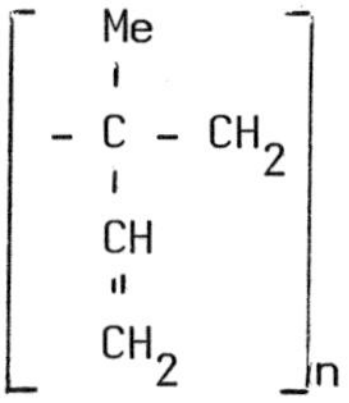

$$
\text{1,2 additions} \qquad
\left[\begin{array}{c}
H \quad\; H \\
| \quad\;\; | \\
-C - C - \\
| \quad\;\; | \\
H \quad C - Me \\
\quad\quad \| \\
\quad\quad CH_2
\end{array}\right]_n
$$

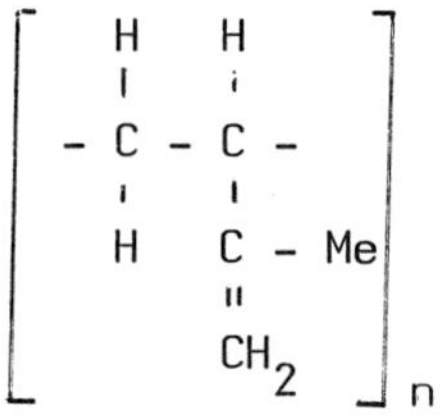

Regioisomerism in trans-1,4 polyisoprene:

Head to head

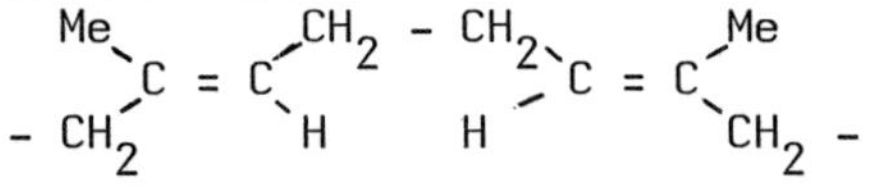

$$
\begin{array}{c}
Me \diagdown \qquad\qquad \diagup CH_2 - CH_2 \diagdown \qquad\qquad \diagup Me \\
\quad C = C \qquad\qquad\qquad\qquad C = C \\
-CH_2 \diagup \qquad \diagdown H \qquad H \diagup \qquad\qquad \diagdown CH_2 -
\end{array}
$$

Tail to tail

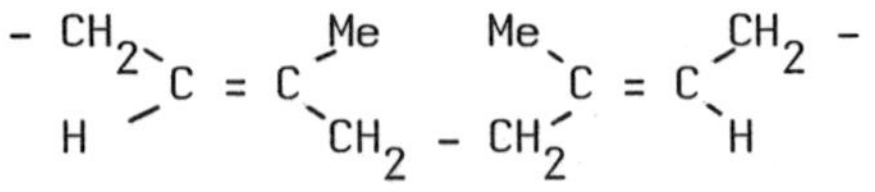

$$
\begin{array}{c}
-CH_2 \diagdown \qquad\qquad \diagup Me \qquad Me \diagdown \qquad\qquad \diagup CH_2 - \\
\quad C = C \qquad\qquad\qquad\qquad C = C \\
H \diagup \qquad \diagdown CH_2 - CH_2 \diagup \qquad\qquad \diagdown H
\end{array}
$$

Head to tail

$$
\begin{array}{c}
Me \diagdown \qquad\qquad \diagup CH_2 - \\
\quad C = C \\
\diagup \qquad \diagdown H \\
Me \diagdown \qquad \diagup CH_2 - CH_2 \\
\quad C = C \\
-CH_2 \diagup \qquad \diagdown H
\end{array}
$$

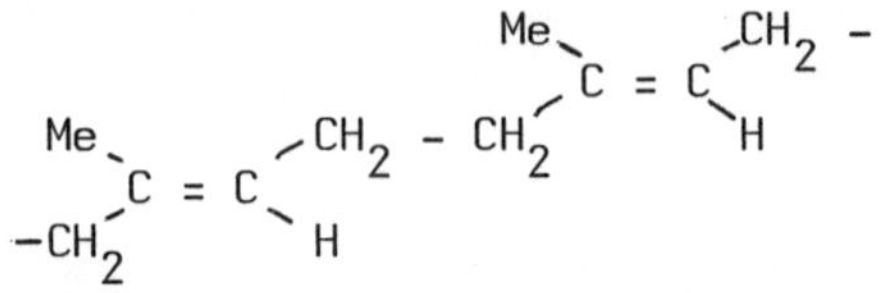

Regioisomerism in cis-1,4 polyisoprane
Head to head

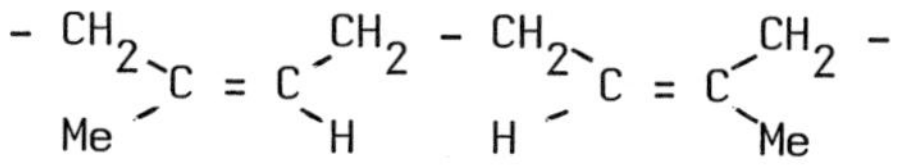

$$- CH_2-\underset{Me}{\overset{}{C}} = C\overset{CH_2}{\underset{H}{}} - CH_2-\underset{H}{\overset{}{C}} = C\overset{CH_2}{\underset{Me}{}} -$$

Tail to tail

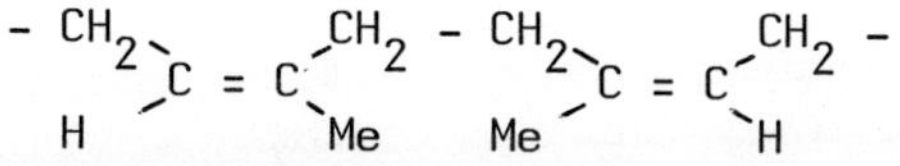

$$- CH_2-\underset{H}{\overset{}{C}} = C\overset{CH_2}{\underset{Me}{}} - CH_2-\underset{Me}{\overset{}{C}} = C\overset{CH_2}{\underset{H}{}} -$$

Head to tail

$$- CH_2-\underset{Me}{\overset{}{C}} = C\overset{CH_2}{\underset{H}{}} - CH_2-\underset{Me}{\overset{}{C}} = C\overset{CH_2}{\underset{H}{}} -$$

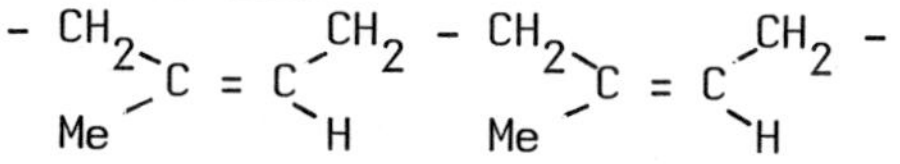

Regioisomerism in 1,2 polyisoprene:

```
                  Head to tail
                  H    H    H    H
                  |    |    |    |
              -   C -  C -  C -  C  -
                  |    |    |    |
         Me  -    C    H    H    C  - Me
                  ||             ||
                  CH2            CH2

                  Tail to tail
                  H    H         H    H
                  |    |         |    |
              -   C -  C    -    C -  C  -
                  |    |         |    |
                  H    C - Me    C -  Me
                       ||        ||
                       CH2       CH2

                  Head to tail
                  H    H              H    H
                  |    |              |    |
              -   C -  C    -         C -  C  -
                  |    |              |    |
         Me  -   CH    H    Me  -    CH    H
                  ||                  ||
                  CH2                 CH2
```

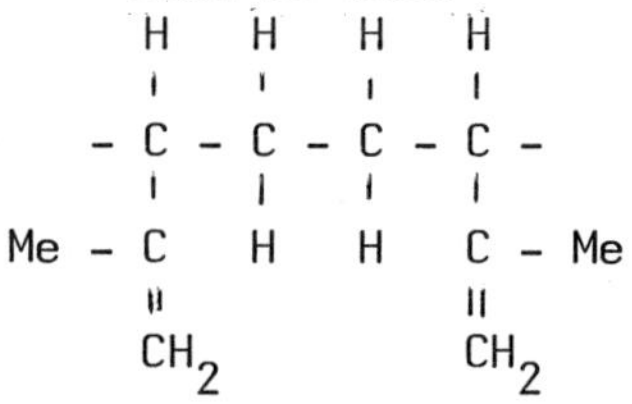

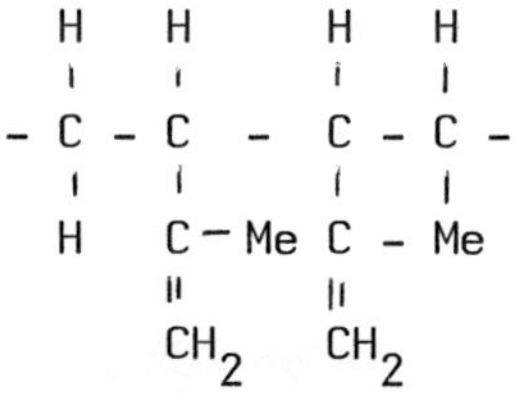

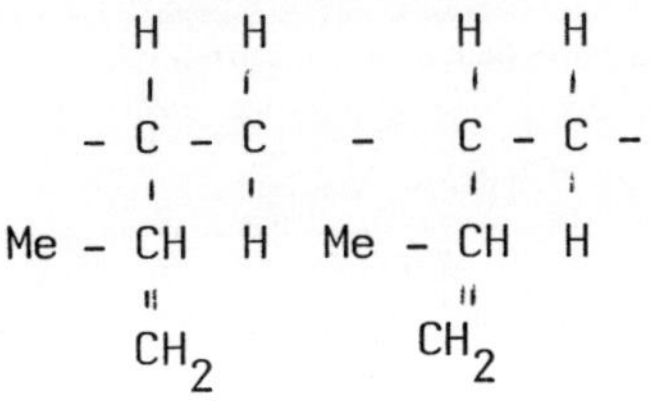

Regioisomerism in 3,4 polyisoprene:

```
                  Head to head
                  H    Me   Me   H
                  |    |    |    |
              -   C -  C -  C -  C  -
                  |    |    |    |
                  H    CH   CH   H
                       ||   ||
                       CH2  CH2
```

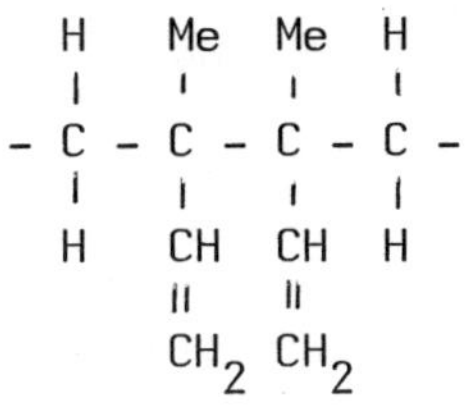

```
                    Tail to tail
                    Me   H     H   Me
                     |   |     |    |
                   - C - C  -  C - C -
                     |   |     |    |
                    CH   H     H   CH
                    ||               ||
                    CH              CH
                      2               2

                    Head to tail
                    H    Me   H    Me
                    |    |    |     |
                  - C -  C -  C -  C -
                    |    |    |     |
                    H   CH   H    CH
                         ||          ||
                        CH          CH
                          2           2
```

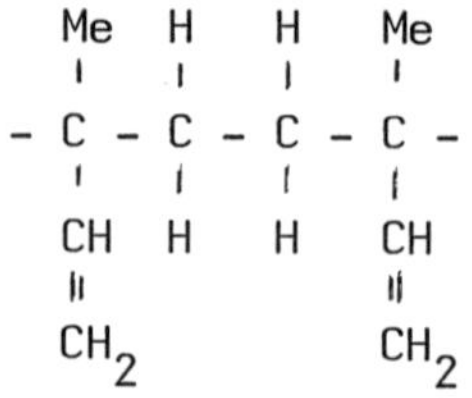

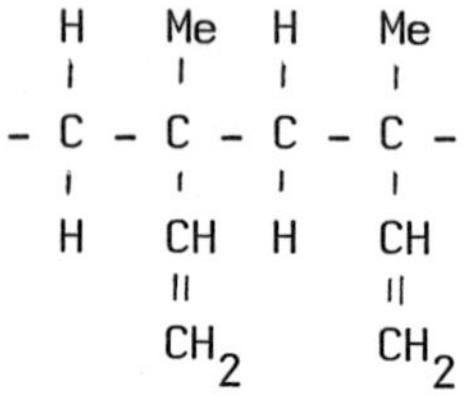

<u>10.4.8 Polypropylene Glycols</u>

Ozonization followed by lithium aluminium hydride reduction to
oxyalkylene groups has been used to study the occurrence of regioisomerism
in polypropylene glycols.

Adopting the following nomenclature for propylene glycol

```
                    Head to tail
                      H    Me
                      |    |
                  H - C -  C - H
                      |    |
                      OH   OH
```

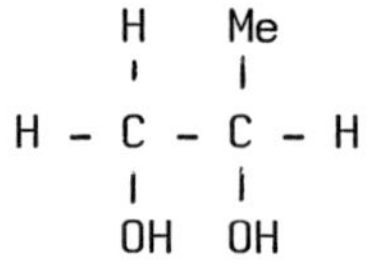

then the following three sequences are possible in polypropylene glycol:

```
                              Me   H        H    Me
                               |   |        |    |
      Head to head      - O -  C - C  - O - C  - C -
                               |   |        |    |
                              OH   H        H    OH


                              H    Me       Me   H
                              |    |        |    |
      Tail to tail     - O -  C  - C  - O - C  - C -
                              |    |        |    |
                              OH   H        H    OH


                              H    Me       H    Me
                              |    |        |    |
      Tail to head     - O -  C  - C  - O - C  - C -
                              |    |        |    |
                              OH   H        H    OH
```

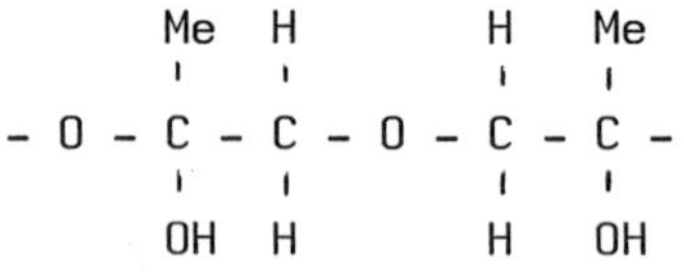

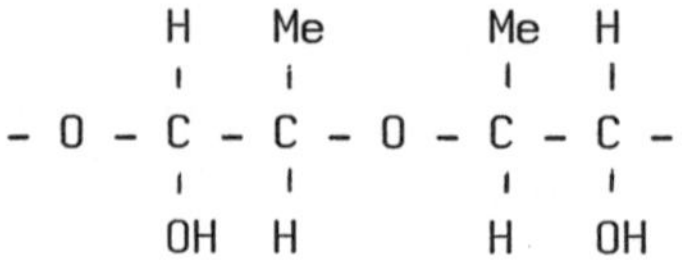

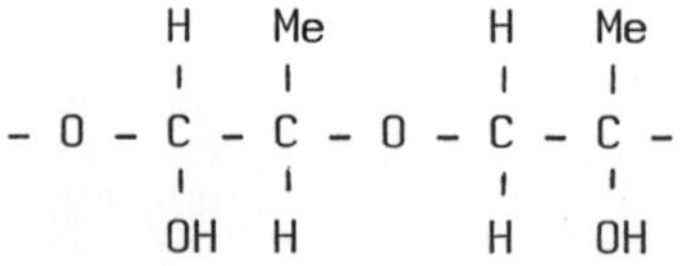

If these sequences occurred consecutively in polypropylene glycol
then it would have the structure:

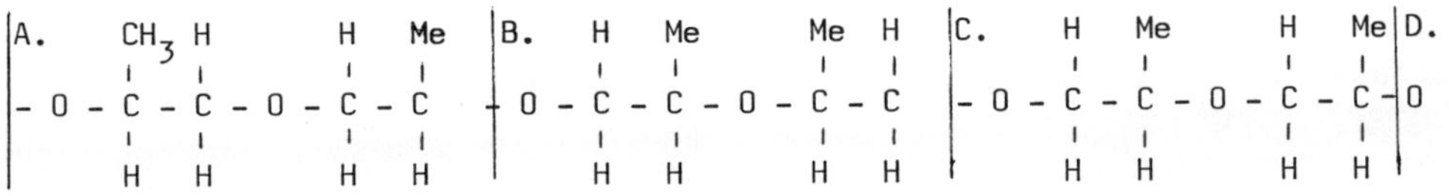

```
A.    CH3 H      H  Me |B.  H  Me      Me H |C.   H  Me      H  Me|D.
      |   |      |  |  |    |  |       |  | |     |  |       |  |
 - O - C - C - O - C - C -+ O - C - C - O - C - C | - O - C - C - O - C - C -|O
      |   |      |  |  |    |  |       |  | |     |  |       |  |
      H   H      H  H  |    H  H       H  H |     H  H       H  H |

         head to head          tail to tail           tail to head
```

Upon ozonolysis, fission occurs at points A, B, C and D to produce
aldehydic and ketonic groups.

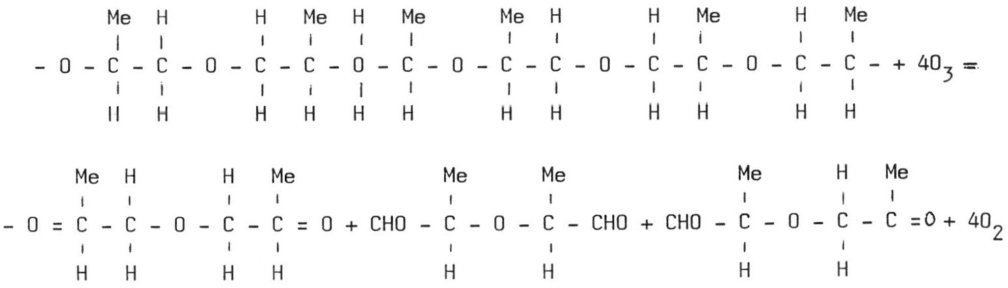

```
      Me  H      H  Me H  Me      Me H      H  Me      H  Me
      |   |      |  |  |  |       |  |      |  |       |  |
 - O - C - C - O - C - C - O - C - O - C - C - O - C - C - O - C - C - + 4O3 =
      |   |      |  |  |  |       |  |      |  |       |  |
      ||  H      H  H  H  H       H  H      H  H       H  H
```

```
      Me  H      H  Me         Me         Me         Me      H  Me
      |   |      |  |           |          |          |       |  |
 - O = C - C - O - C - C = O + CHO - C - O - C - CHO + CHO - C - O - C - C = O + 4O2
      |   |      |  |           |          |          |       |
      H   H      H  H           H          H          H       H

         head to head          tail to tail           tail to head
```

Upon reduction with lithium and aluminium hydride the following diglycols
are produced.

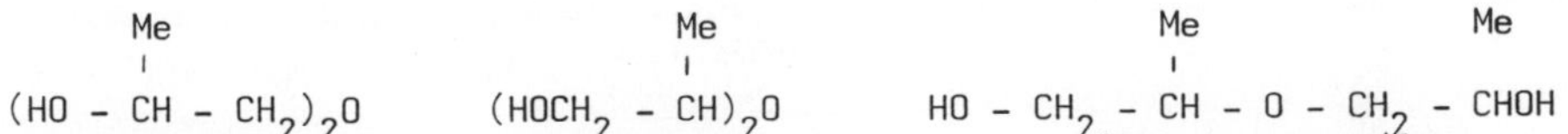

```
        Me                    Me                  Me           Me
        |                     |                   |
 (HO - CH - CH2)2O      (HOCH2 - CH)2O     HO - CH2 - CH - O - CH2 - CHOH
     secondary             primary            primary - secondary
  propylene glycol      propylene glycol      propylene glycol
   (head to head)        (tail to tail)        (tail to head)
```

From the relative amounts of these three glycols produced, determined by
gas chromatography, the amounts of head to head, tail to tail and tail to
head configurations can be deduced. Note that the di-primary propylene
glycol occurs in two optically active forms, 1(a) and (b).

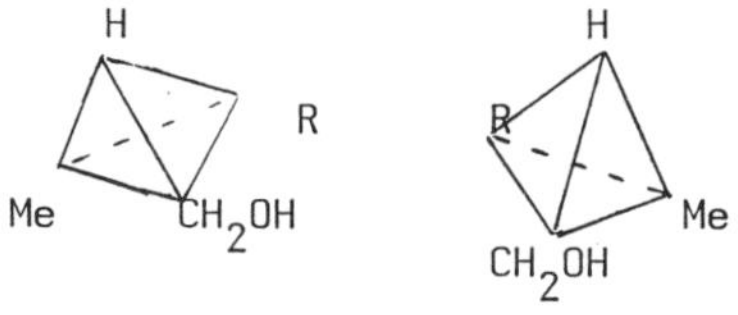

```
          H                    H
          |                    |
         /|\                  /|\                R is OCH - CH2OH
        / | \ R            R / | \                     |
   Me  /  |  \         Me   \  |  /  Me                 Me
       CH2OH                 CH2OH
```

10.4.9 Other Polymers

Regioisomerism is also exhibited by other polymers including polyvinylidene fluoride, polyvinylidene choride[1459-1462] polydienes and polyvinyl acetate[1463,1464]. Head to head and tail to tail sequences have been determined in polyvinylidene chloride[1459-1463] and polyvinyl acetate[1463,1464].

<pre>
          polyvinylidene chloride   polyvinyl acetate
                 H   T   H   T
              CH = CCl   CH = CHCOOAc
                2   2   2

                 Cl H   H   Cl   H   H   H   H
                  | | | | | | | |
head to tail - C - C - C - C -   - C - C - C - C -
                  | | | | | | | |
                 Cl H   H   Cl   COOAc H   H COOAc

                 H   Cl Cl H   H   H   H   H
                  | | | | | | | |
tail to tail - C - C - C - C -   - C - C   -   C - C -
                  | | | | | | | |
                 H   Cl Cl H   H COOAc COOAc H
</pre>

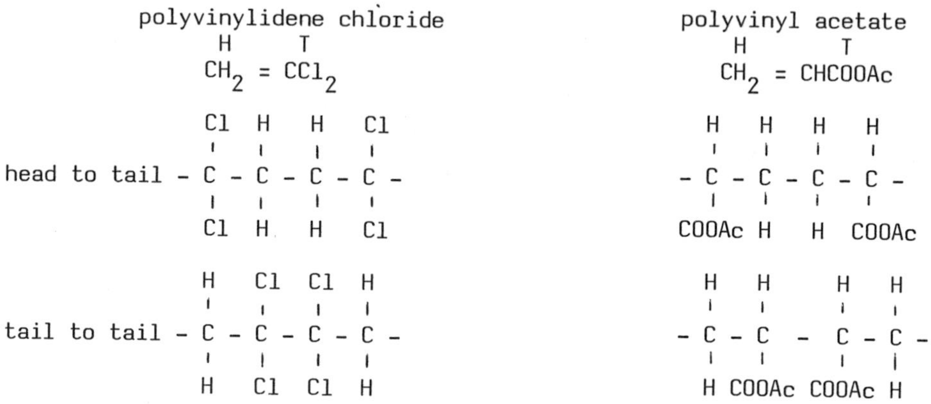

10.5 FORMS OF UNSATURATION IN POLYOLEFINS AND OTHER POLYMERS

10.5.1 Polyethylene

Much useful information regarding the types of unsaturation present in polyethylene can be gained by the application of infrared spectroscopy. Polyethylenes can contain various types of unsaturation which are of great importance from the microstructural point of view. These include, external vinylidene, terminal vinyl and internal cis and trans unsaturation.

<pre>
                                                       absorption maximum

          R R C = CH external vinylidene   8.86 cm^{-1}
           1 2   2   (11.26 u)

              eg. R(CH^2)n - CR' = CH
                                        2

          - CH = CH terminal vinyl   990 (10.10 u)
                   2

              eg. R(CH) - CH = CH   910 (10.99 u)
                     2 n   2

          H
          |
        - C ≂ C - internal trans unsaturation 964 cm^{-1} (10.37 u)

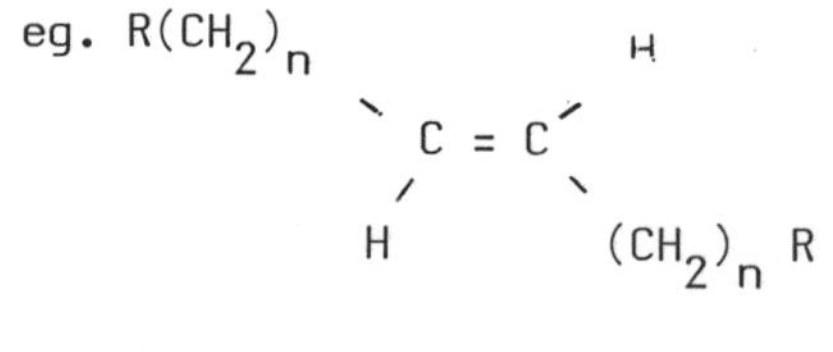

            eg. R(CH)
                    2 n   H
                              \   /
                               C = C
                              /   \
                         H   (CH) R
                                         2 n
</pre>

$$- \overset{\underset{H}{|}}{C} - \overset{\underset{H}{|}}{C} - \text{ internal cis unsaturation}$$

$$\text{eg. } R(CH_2)_n \diagdown \diagup (CH_2)_n R$$
$$C = C$$
$$\diagup \diagdown$$
$$H \qquad\qquad H$$

In branched polyethylene most of the unsaturation is of the external vinylidene type, whilst trans olefinic end groups and terminal vinyl groups are relatively low in concentration. In linear polyethylene most of the unsaturation is terminal vinyl with relatively small amounts of external vinylidene and trans olefinic end groups.

The electron irradiation of linear and branched polyethylenes causes a number of molecular rearrangements in the chemical structure of the polymer[1465]. In addition to the significant changes in the type and distribution of unsaturated groups an infrared comparison of the radiation-induced chemical changes that occur in air and in a vacuum showed that the presence of oxygen has a marked influence on the structural rearrangements that occur on irradiation.

Figure 110(a) shows the infrared spectra of the branched polyethylene before and after irradiation in vacuum and air. Strongly absorbing trans-type unsaturation (CH = CH) bands at 10.37 micron appear in both the vacuum and air-irradiated sample spectra. Vinylidene decay on irradiation is shown by the decrease in the $R_1R_2C = CH_2$ band at 11.26 microns.

Irradiation in vacuum produces a significant decrease in the methyl ($-CH_3$) content, 7.28 micron whereas in the bombardment in air there appears to be only a negligible decrease in $-CH_3$, if any. In addition a comparison of the 13.89-13.70 micron doublet shows that only the 13.89 micron component remains in the spectra of the vacuum sample, whereas there is only a slight decrease of the 13.69 micron component in the air-irradiated sample. Additional evidence of structural changes is shown in the spectra of the air-irradiated sample. Here both -OH and C=O bands appear, and there is a general depression of the spectrum background from 7.69 to 11.11 microns. In branched polyethylene before irradiation most of the unsaturation is of the external vinylidene type. After a dose of 6 MR, trans-unsaturation at 10.73 microns increases and the vinylidene at 11.26 microns decreases.

In linear polyethylene almost all the saturation is of the terminal vinyl type (CH = CH$_2$) as shown by the bands at 10.10 and 10.99 microns. Here again, after only a 6 MR dose, the trans groups form rapidly and the vinyl groups at 10.10 and 10.99 microns decrease. Because of the rapid increase of trans groups during irradiation and the simultaneous decrease of the other unsaturated groups, it might appear that the trans groups are being formed from a reaction involving the sacrifice of the other unsaturated groups in the polymer. In order to determine the validity of this observation, Luongo and Solovay[1466] exposed to similar doses of irradiation a sample of polymethylene which has no infrared detectable unsaturation or branching. They found that in polymethylene, the trans

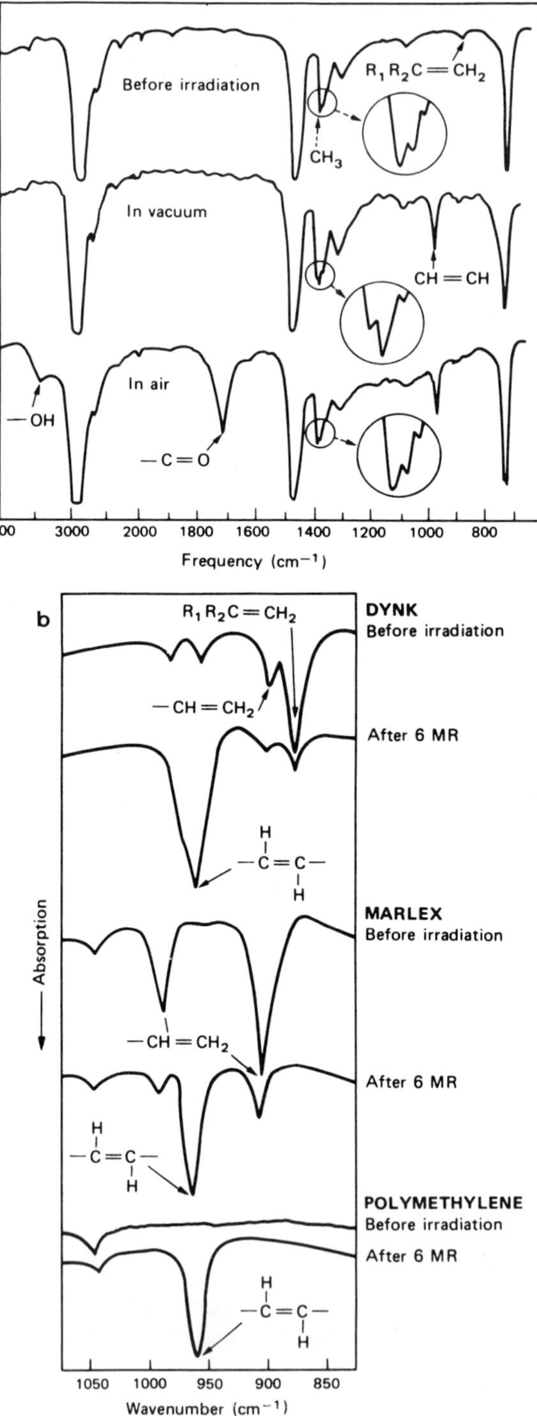

Figure 110   Irradiation of polyethylene: (a) the effect of irradiation (500 Mrad) in air and vacuum on branched polyethylene; (b) unsaturation region of three polyethylenes before and after irradiation at 6 Mrad.

unsaturation band still forms strongly after irradiation in either air or vacuum. This means that the trans groups come from a reaction that is independent of either unsaturation or branching. As for the vinyl and vinylidene decay, although there is no conclusive mechanism to explain their disappearance, they probably become saturated by atomic hydrogen in the system or become crosslinking sites.

Unsaturation in low-density polyethylene has been estimated to $\pm$ 0.003 C = C/10^3C atoms by compensating with brominated polymer of the same thickness[1467].

Rueda[1468] has used infrared spectroscopy to measure vinyl, vinylidene and internal cis or trans olefinic groups in polyethylene.

Dankovics[1469] determined the degree of unsaturation of low-density polyethylene. The total degree of unsaturation in polyethylene was determined by summing the vinyl, vinylidene and internal unsaturation derived from the differential infrared spectra using unbrominated polyethylene film as the sample and a brominated film as the reference[1468].

Faizi et al.[1470] determined unsaturation in unsaturated oligoesters and Murano[1471] used NMR to study unsaturation in polyesters of fumaric acid with ethylene glycol and sebacic or adipic acid.

10.5.2 Polypropylene

The types of unsaturation which can occur in polypropylene are listed below:

	Absorption maximum
$R_1R_2C = CH_2$ external vinylidene	1645 cm^{-1} (6.08 u)
eg. R – CH(Me) – CH$_2$ – C(Me) = CH$_2$	and 890 cm^{-1} (12.30 u)
R – CH = CH$_2$ terminal vinyl	910 cm^{-1} (10.99 u)
eg. R – CH$_2$ – CH(Me) – CH = CH$_2$	
– CH(Me) – CH = CH – C(Me) – CH$_2$ – CH(Me) – internal double bands	1665 cm^{-1} (6.00 u)

Identification of the free radicals produced when polymers are irradiated, normally by gamma-radiation from ^{60}Co or by fast electrons from an electron accelerator tube, can sometimes give vital information regarding structural features, saturated and unsaturated, of the original polymer.

Slovakhotova et al.[1472] have applied infrared spectroscopy to a study of structural changes in polypropylene in vacuum with fast electrons from an electron accelerator tube (200kV accelerating field) and with gamma-radiation from ^{60}Co. They found (see above) the infrared spectrum of irradiated polypropylene contains absorption bands in the 6.08, 11.23 and 13.60-13.51 micron regions. The first two bands correspond to $RR'C = CH_2$ vinylidene groups and the band in the 3.60-13.51 micron region to propyl branches, $R-CH_2CH_2CH_3$. When polypropylene is degraded thermally these groups are formed by disproportionation between free radicals formed by rupture of the polymer backbone. Under the action of ionizing radiations the polymer backbone ruptures, with formation of two molecules, with vinylidene and propyl end-groups, at a temperature as low as that of liquid nitrogen, because the corresponding bands are found in the infrared spectrum of polypropylene irradiated at -196°C and measured at -130°C. When polypropylene is irradiated with dosages greater than 350 Mrad a band appears at 10.99 micron, corresponding to vinyl groups, $R - CH = CH_2$, i.e. degradation of polypropylene can also involve simultaneous rupture of two C - C bonds in the main and side chains. The strength of the vinyl-group band in the spectrum of irradiated polypropylene is lower than that of the vinylidene group, although the extinction coefficients of these bands are approximately the same[1472].

In the spectrum of amorphous polypropylene irradiated with a dosage of 4000 Mrad at -196°C and measured at -130°C, in addition to the band at 6.08 microns, a weaker band appears with a maximum near 6.00 microns, possibly due to internal bonds.

$$
\begin{array}{ccccc}
Me & & Me & & Me \\
| & & | & & | \\
-\,CH\,-\,CH & = & C\,-\,CH\,-\,C\,- \\
\end{array}
$$

When the spectrum of the specimen is recorded after it is heated to +25°C this maximum disappears, leaving only a shoulder on the strong band at 6.08 microns. The extinction coefficient [1472], of the band at 6.00 microns is less than that of the 6.08 microns band by a factor of 6.7. Supplementary evidence of the formation of internal double bonds in irradiated polypropylene is provided by the presence in the spectrum of bands in the 12.27-11.69 region. In this region lie bands due to deformation vibration of CH at double bonds in

$$
\begin{array}{c}
R \\
\,\diagdown \\
\quad C = CHR \\
\,\diagup \\
Me
\end{array}
$$

groups, existing in various conformations[1473]. The appearance of bands at 12.27 and 11.69 microns in the spectrum of irradiated polypropylene can be regarded as an indication of the formation of internal double bonds in the polymer.

A study of the ESR spectra of irradiated polypropylene[1474] has shown that the alkyl radicals formed during irradiation at -196°C

$$
\begin{array}{cccc}
Me & & Me & \\
| & & | & \\
-\,CH_2\,-\,CH\,-\,CH_2\,-\,CH\,- \\
\end{array}
$$

undergo transition to alkyl radicals when the specimen is heated; i.e. on heating the radical centres migrate to internal double bonds with the formation of stable allyl radicals. Irradiation at room temperature leads immediately to the formation of allyl radicals. It is very probable that the decrease in intensity of the internal double bond valency vibration band at 6.00 microns and the broadening of its maximum after a specimen irradiated at a low temperature is heated to room temperature, is associated with the formation of allyl radicals because interaction of the π-electrons of the double bond with the unpaired electron also lowers the frequency of the double-bond vibration. Comparison of the intensities of the terminal vinyl double-bond bands at 11.23 and 10.99 microns with the band at 6.08 microns in the spectrum of irradiated isotactic polypropylene shows that the intensity of absorption in the 6.08 micron region does not correlate with the intensity of absorption in the 11.23 regions. Thus, according to the known extinction coefficients for these bands, the ratio of their optical densities should be

$$D_{11.23}/D_{6.08} = 3.7, D_{10.99}/D_{6.08} = 3.2 \text{ and } (D_{11.23} + D_{10.99})/D_{6.08} = 3.5$$

In the latter case $D_{6.08}$ is the sum of the optical densities of the vinylidene and vinyl absorption bands in this region. An optical density ratio for these bands of approximately this value was found (1.75 to 3.3) for the products of thermal degradation of polypropylene. It is seen that only in the case of amorphous polypropylene irradiated with gamma-radiation from ^{60}C is the ratio $(D_{11.23} + D_{10.99})/D_{6.08}$ close to the value calculated from the extinction coefficients of these bands. In the spectra of irradiated isotactic polypropylene, however, the intensity of the 6.08 micron band is greater than would be expected if only vibration of terminal double bonds contributes to absorption in this region. This increase in absorption in the 6.08 micron region can be related to absorption by the internal double bond in the allyl radical, the vibrational frequency of which is lowered by conjugation of the π-electrons of the double bond with the unpaired electron of the radical. In amorphous polypropylene irradiated at room temperature the alkyl radicals can combine rapidly; therefore there is obviously little formation of allyl radicals. This explains the fact that the ratio of the optical densities of the terminal double bond bands in the 11.10 and 6.08 micron regions is close to the calculated value.

The occurrence of conjugated double bonds in irradiated polypropylene is indicated by the following facts: (1) in the spectra of isotactic polypropylene irradiated with dosages of 2000-4000 Mrad at room temperature there is a band at 6.21 microns, which is the region in which polyene bands occur, whereas this band is absent from the spectrum of isotactic polypropylene irradiated with the same dosages at -196°C; (2) in the electronic spectra of these polypropylene specimens the boundary of continuous absorption is shifted to a region of longer wavelength in comparison with the spectra of polypropylene irradiated with the same dosages at -196°C.

It has been shown from ESR spectra[1475] that when specimens of isotactic polypropylene are heated above 80°C they contain polyenic free radicals

$$- CH - (C = CH)_n - CH -$$
$$\quad\ \ \overset{|}{Me} \qquad\quad\ \overset{|}{Me}$$

and this also indicates the possibility of migration of double bonds along
the polymer chain.

10.5.3 Other Polymers

ESR spectroscopy has also been applied to studies of unsaturation
and other structural features in a wide range of other polymers; including
polyethylene[1476-1487], polypropylene[1491-1497], polybutenes[1491], polystyrene[1498-1500], PVC[1501,1502], polyvinylidene chloride[1505], polymethylmethacrylate[1506-1515], polyethyleneglycol polycarbonate[1514,1518], polyacrylic
acid[1514-1517,1519,1520], polyphenylenes and polyphenylene oxides[1522],
polybutadiene[1523], conjugated dienes[1524,1525], polyester resins[1526] and
various copolymers, styrene grafted polypropylene[1527], ethylene
acrolein[1528], butadiene-isobutylene[1529], and vinyl acetate copolymers[1530],
vinyl chloride propylene copolymers[1531], polyester, polystyrene,
cellophane[1532,2122] and high density polyethylene[1533].

The elegant work of Seeger, Exner and Cantow[1534,1535] with labelled
ethylene-propylene copolymers on the relative amounts of alpha and beta
cleavage compared to statistical chain cleavage provides a foundation for
discussion of the products arising from these reactions. They concluded
that the probability of backbone cleavage at the branch sites is much
greater than statistical chain cleavage, and that alpha and beta cleavage
occur with equal frequency.

Since there are a number of different olefin products possible from
cleavage at a branch site, the determination of the type of chain branch
is best accomplished by pyrolysis-hydrogenation-gas chromatography. On-
line catalytic hydrogenation of the pyrolysis products in the injection
port of the chromatograph converts all of the olefins to saturated
hydrocarbons, therefore reducing the number of possible products.
Cleavage of the carbon-carbon bond in the polymer backbone at the branch
site (alpha cleavage) results in the formation of n-alkanes. Backbone
cleavage at the bond beta to the branch site gives methylalkanes. In the
case of methyl branch sited, beta cleavages gives 2-methylalkanes, and
chain cleavage beta to ethyl and butyl branches gives 3-methyl and 5-
methylalkanes, respectively[1536]. Statistical cleavage will produce a
mixture of branched alkanes.

10.6 SHORT CHAIN BRANCHING IN POLYMERS

10.6.1 Polyethylene

Methyl branching. A regression analysis of infrared, differential
thermal analysis and x-ray diffraction data by Laiber et al[1537] for low-
pressure polyethylene showed that, as synthesis conditions varied, the
number of methyl groups varied from 0 to 5 per 1000 carbon atoms and the
degree of crystallinity varied from 84 to 61%.

Saturated hydrocarbons evolved during irradiation of polyethylene
are characteristic of short side-chains in the polymer. Solovay and
Pascale[1538] showed that a convenient analysis is effected by programmed
temperature gas chromatography. In order to minimize the relative
concentrations of extraneous hydrocarbons, i.e. those not arising from
selective scission of complete side-chains, it is necessary to irradiate
at low temperatures and doses. Such analyses of a high-pressure

polyethylene indicated that the two to three methyls per 1000 carbon atoms detected in infrared absorption (low-pressure polyethylenes at least an order of magnitude lower) are probably equal amounts of ethyl and butyl branches. These arise by intramolecular chain transfer during polymerization. At a dose of 10 Mrad about 1-4% of the alkyl group are removed. Methane is the only hydrocarbon detected on irradiation of polypropylene, indicating little combination of methyl radicals to form ethane during irradiation.

The measurement of the methyl absorption at 7.26 microns in polyethylene can serve as a good estimation of branching. However, interference from the methylene absorption at 7.31 microns makes it difficult to measure the 7.31 micron band, especially in the case of relatively low methyl contents.

A method has been developed[1539] which utilizes the suggestion by Neilson and Holland[1540]. They associated the amorphous phase absorption of polyethylene at 7.31 and 7.69 microns with the trans-trans conformation of the polymer chain about the methylene group. Therefore the intensities of these two absorptions are proportional to one another. By placing an annealed film (ca. 10-15 mil) of high density polyethylene in the reference beam of a double beam spectrometer and a thin, quenched film of the sample in the sample beam, most of the interference at 7.31 microns can be removed. The method has the advantage that it is not necessary to have complete compensation for the 7.31 band since a correction for uncompensation at 7.25 microns can be applied based on the intensity of the 7.31 micron absorption.

A calibration for the methyl absorption based on mass spectrometric studies of such gaseous products produced during electron bombardment of polyethylene has demonstrated irradiation-induced detachment of complete alkyl units[1541]. In addition to saturated alkanes characteristic of the branches, small quantities of methane, other paraffins, and olefins were simultaneously evolved. It was suggested that 'extraneous' paraffins result from cleavage of the main chain[1541,1542] Nerheim[1543] has described a circular calibrated polymethylene wedge for the compensation of CH_2 interferences in the determination of methyl groups in polyethylene by infrared spectroscopy.

Methyl group content of low-density polyethylene has been determined with a standard deviation of 0.18% provided methylene group absorptions were compensated by polyethylene of similar structure[1544].

Nishoika et al.[1545] determined the degree of chain branching in low-density polyethylene using proton NR Fourier transform NMR at 100 MHz and ^{13}C Fourier transform NMR at 25 MHz with concentrated solutions at approximately 100°C. The methyl concentrations agreed well with those of infrared based on the absorbance at 7.25 microns.

<u>Alkyl groups higher than methyl.</u> High-density (low-pressure) polyethylenes are usually linear, although the physical and rheological properties of some high-density polyethylenes have suggested the presence of long chain branching (butyl and higher groups) at a level one to two orders of magnitude below that found for low-density polyethylenes prepared by a high-pressure process. A measurement of long chain branching in high-density polyethylenes has been elusive because of the low concentrations involved[1546] and can only be directly provided by high-field, high-sensitivity NMR spectrometers.

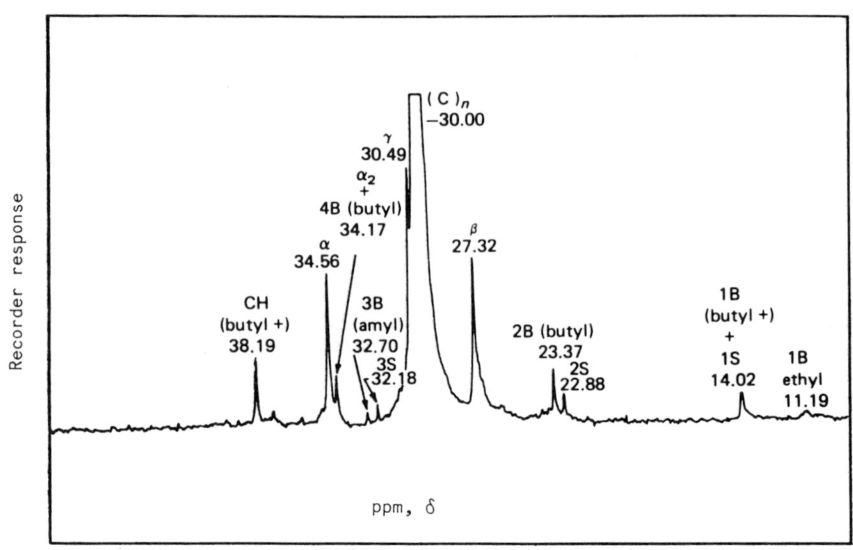

Figure 111 ^{13}C spectrum at 25.2 MHz of low density polyethylene from high pressure process.

High-density polyethylenes prepared with Ziegler type, titanium-based catalyst have predominantly n-alkyl or saturated end-groups. Those prepared with chromium-based catalysts have a propensity toward more olefinic end-groups. The ratio of olefinic to saturated end-groups for polyethylenes prepared with chromium-based catalysts is approximately unity. The end-group distribution is therefore another structural feature of interest in low-pressure polyethylenes, because it can be related to the catalyst employed, and possibly to the extent of long chain branching. It is possible not only to measure by ^{13}C NMR concentrations of saturated end-groups, but also the olefinic end-groups and, subsequently, an end-group distribution.

The capability for discerning the length of short chain branches has made ^{13}C NMR a powerful tool for characterizing low-density polyethylenes produced by free-radical, high-pressure processes. The ^{13}C NMR spectrum of such a polyethylene is shown in Figure 111. Others are also present, and Axelson, et al in a comprehensive study[1547] have concluded that no unique structure can be used to characterize low-density polyethylenes. They have found nonlinear short chain branches as well as 1,3 paired ethyl branches. Bovey et al[1548] compared the content of branches six carbons and longer in low-density polyethylenes with the long chain branching results obtained through a combination of gel permeation chromatography and intrinsic viscosity. An observed good agreement led to the conclusion that the principal short chain branches contained fewer than six carbons and the six and longer branching content could be related entirely to long chain branching. Others have reported similar observations in studies where solution methods are combined with ^{13}C NMR[1549]. However, as a result of the possible uncertainty of the branch lengths, associated with the resonances for branches six carbons and longer, ^{13}C NMR should be used in conjunction with independent methods to establish true long chain branching.

From the results we have seen thus far, it is easy to predict the ^{13}C NMR spectrum anticipated for essentially linear polyethylenes

containing a small degree of long chain branching. An examination of a
^{13}C NMR spectrum from a completely linear polyethylene, containing both
terminal olefinic and saturated end-groups, shows that only five
resonances are produced. A major resonance at 30 ppm arises from
equivalent, recurring methylene carbons, designated as 'm', which are four
or more removed from an end-group or a branch. Resonances at 14.1, 22.9
and 32.3 ppm are from carbons 1,2 and 3, respectively, from the saturated,
linear end-group. A final resonance, which is observed at 33.9 ppm,
arises from an allylic carbon, designated as 'a', from a terminal olefinic
end-group. These resonances, depicted structurally below, are fundamental
to the spectra of all polyethylenes.

$$CH_3 - CH_2 - CH_2 - CH_2 - \quad - (CH_2)_n - \quad - CH_2 - CH = CH_2$$
$$\,_{'1'}\quad\,_{'2'}\quad\,_{'3'}\qquad\qquad\,_{'n'}\qquad\qquad\,_{'a'}$$

An introduction of branching, either long or short, will create
additional resonances to those described above.

From the observed ^{13}C NMR spectrum of the ethylene-1-octene
copolymer Randall[1550] found that the alpha and beta and methine resonances
associated with branches six carbons and longer occur at 34.56, 27.32 and
38.17 ppm respectively. Thus in high-density polyethylenes, where long
chain branching is essentially the only type present, ^{13}C NMR can be used
to establish unequivocally the presence of branches six carbons long and
longer. If no comonomer has been used during polymerization, it is very
likely that the presence of such resonances will be indicative of true
long chain branching. In any event, ^{13}C NMR can be used to pinpoint the
absence of long chain branching and place an upper limit upon the long
chain branch concentration whenever branches six carbons and longer are
detected.

It can be seen from the above considerations that ^{13}C NMR is a
highly attractive method for characterizing polyethylenes. A serious
drawback is not encountered even though branches six carbons in length and
longer are measured collectively. The short branches are generally less
than six carbons in length and truly long chain branches tend to
predominate. On occasions there may be special exceptions for
'intermediate' branch lengths, so independent rheological measurements
should be sought as a matter of course. Nevertheless ^{13}C NMR is a direct
method, which possesses the required sensitivity to determine long chain
branching in high-density polyethylenes, and provide much information of
microstructural interest.

A further technique that has been used in investigations on short
chain branching in polyethylene is pyrolysis-gas chromatography.

Van Schooten and Evenhuis[1551,1552] applied their pyrolysis (at
500°C) hydrogenation-gas chromatographic technique to the measurement of
short chain branching and structural details of three commercial
polyethylene samples, a linear polyethylene, a Ziegler low-pressure
polyethylene of density 0.945 and a high-pressure low-density polyethylene
of density 0.92. Details of the pyrograms are given in Tables 150 and
151. It was observed that the sizes of the iso-alkane peaks increase
strongly with increasing amount of short chain branching, i.e. increase as
we go from linear polyethylene to high-pressure polyethylene. None of the
n-alkane peaks in the Ziegler low-pressure high density polyethylene
pyrogram is significantly greater than the corresponding peak in the
pyrogram of a linear polyethylene.

Table 150 - Relative sizes of n-alkane peaks in pyrograms
of various polyethylenes

Peak ratio	Linear polyethylene	Ziegler polyethylene	High-pressure polyethylene
$n-C_4-n-C_7$	0.71	0.74	<u>1.38</u>
$n-C_5-n-C_7$	0.59	0.57	<u>0.75</u>
$n-C_6-n-C_7$	1.52	1.25	1.28
$n-C_8-n-C_7$	0.63	0.65	0.68
$n-C_9-n-C_7$	0.67	0.64	0.72
$n-C_{10}-n-C_7$	1.02	0.91	0.88
$n-C_{11}-n-C_7$	1.00	0.71	0.75

Table 151 - Iso-alkane peaks in polyethylene pyrograms

Sample	Iso-alkane peak
Zeigler polyethylene	$i-C_5$, $3MC_5$, $(3MC_6)$, $(3MC_7$, $3MC_8$, $(3MC_9)$
High-pressure polyethylene	iC_4, $2C_5$, $2MC_5$, $3MC_5$, $2MC_6$, $3MC_6$ $2MC_7$ - $4MC_7$, $3MC_7$, $2MC_8$ - $4MC_8$, $3MC_8$, $4MC_9$ - $5MC_9$, $2MC_9$, $3MC_9$, $4MC_{10}$ - $5MC_{10}$, $2MC_{10}$ - $4MC_{10}$, $3MC_{10}$

The pattern of the increased iso-alkane peaks in the Ziegler
polyethylene pyrogram strongly suggests that these peaks are mainly due to
ethyl side-groups. This is in good agreement with the results of electron
irradiation experiments. For the high-pressure polyethylene Van Schooten
and Evenhuis found that the n-butane peak of the pyrogram showed a clear
increase in size, and the n-pentane peak a smaller, although probably
significant, increase. The increases in the n-butane and n-pentane peaks
are probably due to n-butyl and n-pentyl side-groups, respectively, pentyl
groups being much less numerous than butyl groups. However, from the
pyrograms of the ethylene-butene, ethylene-hexane-I and ethylene-octene-I
copolymers it is known that n-butyl side groups give a 2-methyl C_6 peak
which is at least equal to the 3-methyl C_7 peak. In the high-pressure
polyethylene pyrogram, however, the 3-methyl C_7 peak, therefore, is
probably due to ethyl side-groups. This is in agreement with the sizes of
the other 3-methylalkane peaks. It may be concluded that in high-pressure
polyethylene the short-chain branches are mainly ethyl and, for a smaller
part, n-butyl groups, while some n-pentyl groups may also be present.

Van Schooten and Evenhuis[1551,1552] concluded that the results
obtained by pyrolysis-hydrogenation-gas chromatography appear to be in
good agreement with those obtained by infrared and electron irradiation

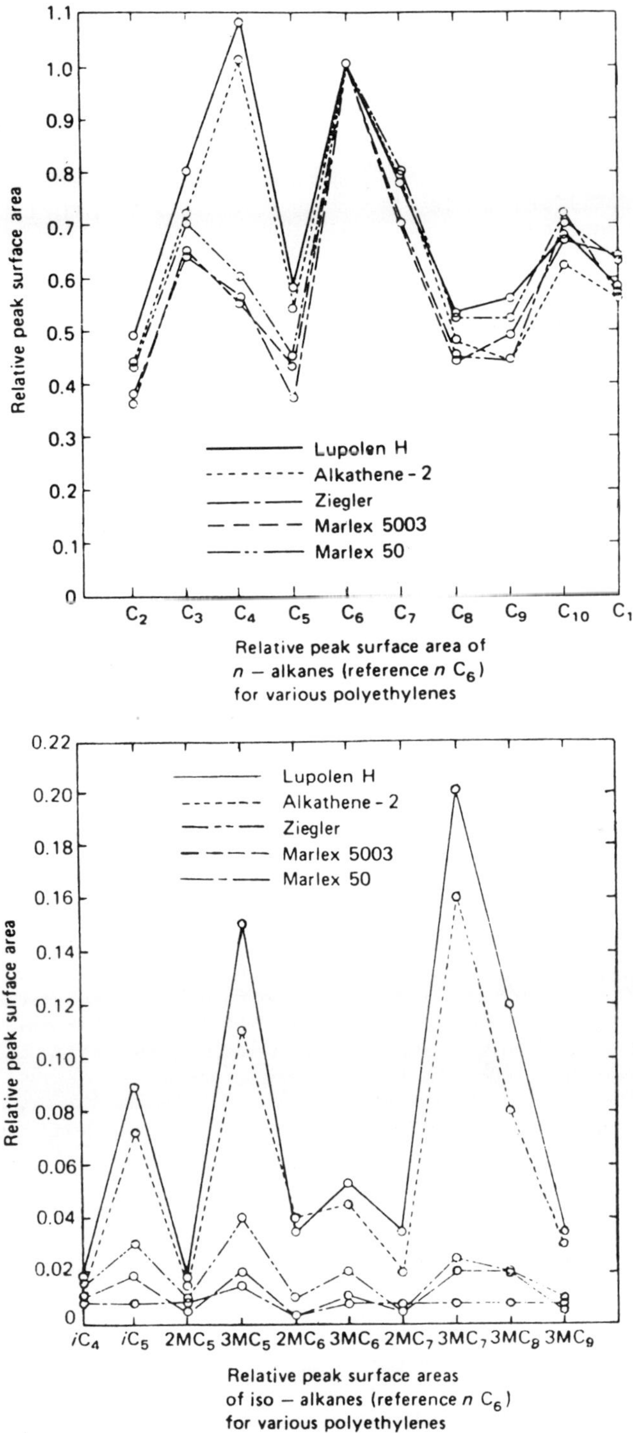

Figure 112   Pyrolysis-gas chromatography carbon number distribution (relative peak surface areas of (a) n-alkanes, reference n-C_6 of various polyethylenes: (b) iso-alkanes, references n-C_6 of various polyethylenes.

Table 152 - Branching frequency of polyethylenes estimated from pyrogram (branches/1000 carbon atoms, from peak surface ratios)

Sample	n-Butane	Iso-pentane	3-Methyl pentane	3-Methyl pentane
Marlex 50 ex. Phillips (very little branching - 1 methyl group/1000 carbon atoms	(I)	(I)	(I)	(I)
Marlex 5003 ex. Phillips (polyethylene containing a little copolymerised butane-1)	0.5	4	2	2
Ziegler low-pressure polyethylene (about 3-6 branches/1000 carbon atoms)	0.3	7	5	5
Alkathene 2 ex. ICI high-pressure poly-ethylene (20-30 branches/1000 carbon atoms)	21	19	17	19
Lupolen H ex. BASF high-pressure polyethylene (20-30 branches/1000 carbon atoms)	(24)	(24)	(24)	(24)

studies. They showed that the pyrograms of a linear polyethylene contained only very small peaks for branched and cyclic alkanes and very large n-alkane peaks. The largest peaks in the pyrogram are those for propane, n-hexane, n-heptane, n-decane and n-undecane, indicating important hydrogen exchange reactions followed by scission with the fifth (C_3 and n-C_6), ninth (n-C_7 and n-C_{10}) and thirteenth (n-C_{11} and n-C_{14}) carbon atoms. Hydrogen transfer with the sixth carbon atom would account for the rather large n-C_4 peak (n-C_4 and n-C_7), but this peak could also be due to intermolecular chain transfer reactions.

Pyrograms were also prepared for a range of polyethylene containing different amounts of short chain branching (see Table 152 for details of samples). Previous work by high-energy electron irradiation and mass spectrometry has shown that the short branches in high-pressure polyethylenes are mainly ethyl and n-butyl groups, but other short branches have also been supposed to be present.

The pyrograms obtained on these various polymers are shown in Figure 112 a) and b) (n-C_6 peak taken as reference). These pyrograms show marked differences which can be attributed to differences in short chain branching. The small amount of branching in Marlex 5003 and Ziegler polyethylenes is reflected only in the somewhat larger iso-alkane peaks, whereas the n-alkane pattern is practically the same as found for Marlex 50 (low branching). The Alkathene 2 and Lupolen H high-pressure polyethylenes show, on the other hand, larger n-butane and n-pentane peaks (Figure 112 a)). The iso-alkane peaks that show the largest increase (for

Alkathene 2 and Lupolen H) are the iso-pentane and 3-methyl alkane peaks (Figure 112 b)). The results in Figure 112 b) clearly show that the highest amount of branching is present in Lupolen H, the lowest in Marlex 50. Assuming arbitrarily that these polymers, respectively, contain 24 and 1 short side chains/1000 carbon atoms, and that the relative increase of the n-butane and the iso-alkane peaks is linearly related with the amount of branching, then the branching frequency of the other three samples can be obtained by interpolation. These values are in good agreement with those found in the literature. From the large increase in the n-butane peak and the relatively small increase in the ethane peak it is concluded that the two high-pressure polyethylenes (Alkathene 2 and Lupolen H) contain mainly n-butyl side-chains.

10.6.2 Polybutene-1

Whereas in the pyrolysis-gas chromatography of polyethylene and polypropylene the splitting up of the main chain is virtually the only result of thermal reaction, the stripping off of side-chains in polybutene-1 is clearly indicated by the nature of the pyrolysis products.

In the case of polybutene-1 a mechanism involving stripping off of side-chains is involved in the formation of ethylene:

$$- CH_2 - CH - \longrightarrow - CH_2 - CH - + CH_2 = CH_2$$
$$\mid$$
$$CH_2$$
$$\mid$$
$$CH_2$$

In actual fact a considerable yield of ethylene is obtained from polybutene-1, whereas the production of propylene is much smaller because this can only emanate from the principal chain. The complicated structure of polybutene-1, as compared with the structures of polyethylene and polypropylene, gives rise to the formation of a markedly larger number of structurally isomeric degradation products. Theoretically with a maximum molecular size of C_8, 37 different hydrocarbon decomposition products should be obtainable from polybutene-1 and, in fact, Voigt[1170] was able to identify, or indicate the probable existence of, 33 compounds for polybutene-1.

10.6.3 Ethylene-Higher Olefin Copolymers

NMR spectroscopy. Short chain branches can be introduced deliberately in a controlled manner into polyethylenes by copolymerizing ethylene with a 1-olefin. The introduction of 1-olefins allows the density to be controlled, and butene-1 and hexene-1 are used for this purpose. Once again as in the case of high-pressure process low-density polyethylene, ^{13}C NMR can be used to measure ethyl and butyl branch concentrations independently of the saturated end-groups. This result gives ^{13}C NMR a distinct advantage over corresponding infrared measurements because the latter technique can only detect methyl groups irrespective of whether the methyl group belongs to a butyl branch or chain end. ^{13}C NMR also has a disadvantage in branching measurements because only branches five carbons in length and shorter can be discriminated independently of longer chain branches[1550,1548]. Branches six carbons in length and longer give rise to the same ^{13}C NMR spectral pattern independently of the chain length. This lack of discrimination among the longer side-chain branches is not a deterring factor, however, in the usefulness of ^{13}C NMR in a determination of long chain branching.

By far the most difficult structural measurement is long chain branching. In low-density polyethylenes the concentration of long chain branches is such (<0.5 per 1000 carbons) that characterization through size exclusion chromatography in conjunction with either low-angle laser light scattering or intrinsic viscosity measurements becomes feasible[1553,1548,1549,1555,1555]. When ^{13}C NMR measurements have been compared to results from polymer solution property measurements, good agreement has been obtained between long chain branching from solution properties with the concentration of branches six carbons long and longer[1548,1549]. Unfortunately, these techniques utilizing solution properties do not possess sufficient sensitivity to detect long chain branching in a range of 1 in 10,000 carbons, the level suspected in high-density polyethylenes. The availability of superconducting magnet systems has made measurements of long chain branching by ^{13}C NMR a reality because of a greatly improved sensitivity. An enhancement by factors between 20 and 30 over conventional NMR spectrometers has been achieved through a combination of higher field strengths, 20 mm probes, and the ability to examine polymer samples in essentially a melt state. The data discussed by Randall[1556,1557] have been obtained from both conventional iron magnet spectrometers with field strengths of 23.5 kG and superconducting magnets operating at 47kG.

The ^{13}C NMR spectra from a homologous series of six linear ethylene 1-olefin copolymers beginning with 1-propene and ending with 1-octene are reproduced in Figure 113. The side-chain branches are therefore linear, and progress from one to six carbons in length. Also, the respective 1-olefin concentrations are less then 3% thus only isolated branches are produced. Unique spectral fingerprints are observed for each branch length. The chemical shifts, which can be predicted with the Grant and Paul parameters[1158,1150] are given in Table 153 for this series of model ethylene-1-olefin copolymers. The nomenclature used to designate those polymer backbone and side-chain carbons discriminated by ^{13}C NMR, is as follows:

$$- CH_2 - \overset{\gamma}{C}H_2 - \overset{\beta}{C}H_2 - \overset{\alpha}{C}H_2 - \overset{1}{C}H_2 - \overset{\alpha}{C}H_2 - \overset{\beta}{C}H_2 - \overset{\gamma}{C}H_2 - CH_2 - CH -$$

$$
\begin{array}{cc}
2 & CH_2 \\
3 & CH_2 \\
4 & CH_2 \\
5 & CH_2 \\
6 & CH_2
\end{array}
$$

The distinguishable backbone carbons are designated by Greek symbols, while the side-chain carbons are numbered consecutively starting with the methyl group and ending with the methylene carbon bonded to the polymer backbone[1458]. The identity of each resonance is indicated in Figure 113. It should be noticed in Figure 113 that the '6' carbon resonance for the hexyl branch is the same as alpha, the '5' carbon resonance is the same as beta and the '4' carbon resonance is the same as gamma. Resonances 1,2 and 3, likewise, are the same as the end-group resonances observed for linear polyethylene. Thus a six-carbon branch produces the same ^{13}C spectral pattern as any subsequent branch of greater length. Carbon-13 NMR, alone cannot therefore be used to distinguish a linear six-carbon from a branch of some intermediate length or a true long chain branch.

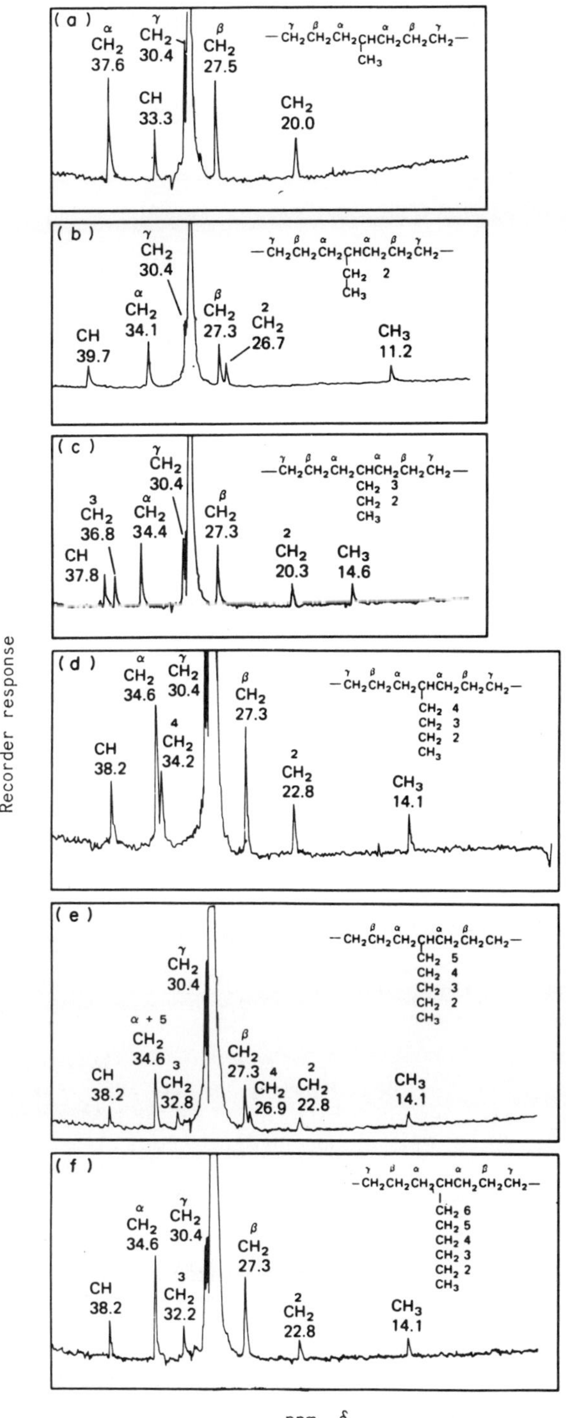

Figure 113 ^{13}C NMR at 25.2 MHz of (a) an ethylene-1-propene copolymer, (b) and ethylene-butene copolymer; (c) an ethylene-1-pentene copolymer; (d) an ethylene-1-hexene copolymer; (e) an ethylene-1-heptene copolymer and (f) an ethylene-1-octene copolymer.

Table 153 - Polyethylene backbone and side-chain ^{13}C chemical shifts
in ppm from trimethylsilane($\pm$0.1) as a function of branch length (Carbon
chemical shifts, which occur near 30.4 ppm, are not given because they
are often obscured by the major 30 ppm resonance for the 'n' equivalent
recurring methylene carbons) (solvent: 1,2,4-trichlorobenzene;
temperature: 125°C)

Branch length	Methine	6	5	4	3	2	1
1	33.3	37.6	27.5	20.0			
2	39.7	34.1	27.3	11.2	26.7		
3	37.8	34.4	27.3	14.6	20.3	36.8	
4	38.2	34.6	27.3	14.1	23.4	-	34.2	
5	38.2	34.6	27.3	14.1	22.8	32.8	26.9	34.6
6	38.2	34.6	27.3	14.1	32.2	30.4	27.3	34.6

<u>Radiolysis-gas chromatography.</u> A study of the products produced
upon vacuum radiolysis of ethylene homopolymers and copolymers is another
means of obtaining information on branching in these polymers[1558]. If a
correction is applied, to take into account the fragments arising from
scission at chain ends, the remaining products can be quantitatively
accounted for as entirely due to scission of side-branches introduced onto
the backbone chain by the alpha-olefin comonomer. The cleavage of
branches takes place, for all practical purposes, exclusively at the
branch points at which the branches are attached to the backbone chain.
The same data, together with similar radiolysis data of poly(3-methyl
pentene-1) and poly(4-methyl pentene-1), further showed that all branches
cleave with equal efficiency, regardless of their length. Radiolysis
does, therefore, provide a reliable and convenient tool for the
quantitative characterization of high-pressure polyethylene with regard to
the unique short chain branching distribution that is characteristic of
each.

The results obtained in a series of irradiations of ethylene-alpha-
olefin copolymers containing about 4 mole% comonomer are shown in Table
154.

During radiolysis of polyethylene there also takes place a certain
amount of random scission at chain ends in addition to the cleavage of
branches. The observed extraneous hydrocarbons is derived from scission
of stray branches that might have been introduced on the chains by stray
impurities during polymerization, but the fact that one also observes a
consistent decrease in the total amount of these extraneous hydrocarbons
derived from the homopolymer with an increase in its molecular weight
leads to the conclusion that the random scission at chain ends is their
main cause. Obviously, then, if one makes an appropriate allowance for
these radiolysis fragments derived from chain ends, then the only
significant paraffin left in the radiolysis products of each copolymer is
that corresponding to the branch introduced on the polyethylene backbone
by the comonomer.

Since it is generally agreed that high-pressure polyethylene
actually contains a variety of short branches, and not just one type of
branch, it is obvious that, if one wants to translate the hydrocarbon
analysis of its radiolysis products into quantitative branch-type
analysis, one will also need accurate information on the relative

Table 154 - Radiolysis products of ethylene-alpha-olefin copolymers
(^{235}U radiation source)

Copolymers	CH_4	C_2H_6	C_3H_{10}	i-C_4H_{10}	n-C_4H_{10}	n-C_5H_{12}	n-C_6H_{14}
			G value x 10^2*				
Ethylene-propylene	1.7	0.1	-	0.03	-	-	
Ethylene-butene-1	0.2	1.5	0.1	0.1	0.2	0.03	-
Ethylene-pentene-1	0.2	0.3	2.1	0.02	0.05	-	-
Ethylene-hexane-1	0.4	0.3	0.2	0.03	1.2	0.02	-
Ethylene-octene-1	0.2	0.3	0.1	-	0.1	0.1	1.2
Linear polyethylene	0.2	0.3	0.2	0.03	0.06	0.05	-

* G value defined as the number of molecules of the particular product
produced per gram of sample per 100 e.v. of incident radiation dose. This
is calculated with 10% accuracy, from the gas chromatographic analysis
data of the products, weight of the polymer sample irradiated, and
radiation dose to which the sample has been subjected.

efficiency of the scission of different branch types. To obtain this
information Kamath and Barlow[1559] irradiated some additional ethylene
alpha-olefin copolymers, differing significantly in their comonomer
content, and the hydrocarbons in their radiolysis products were analysed
as before. Table 155 lists these results. The efficiency of scission is
calculated from the known comonomer content (methyl group analysis) and
the observed G values of the principal hydrocarbons after application of
the appropriate correction against the small chain-end fragmentation. It
is clear from the data that all branches of up to six carbon atoms or more
break off with equal efficiency, and that the branch length per se exerts
little or no effect on the ease of scission.

Table 156 presents results obtained in the radiolysis of high-
pressure polyethylenes of various densities. Noteworthy is the disparity
between the total number of all branches, as determined from the
radiolysis data, and the methyl group content, as derived by the infrared
method. This may be attributed to the fact that the usual infrared method
of determining the methyl group content, based on the absorbance of 7.25
micron does not necessarily count all methyl groups. For example, if two
methyl groups are attached to one carbon atom, as in polyisobutylene, the
characteristic methyl absorption at 7.25 microns splits, giving two bands
at 7.20 and 7.40 microns respectively[1560], which consequently are lost in
the methyl group determination procedure.

In line with the observations reported by earlier workers, the two
most populous branches, according to the radiolysis method, are ethyl and
n-butyl, which occur in the ratio 2:1, as observed by others[1561].

Table 155 - Efficiency of radiation scission of
short branches (G value x 10^2)

| Copolymers | CH$_3$ per 1000 | | | | | | Average G x 10 CH$_3$ per 1000 CH$_4$ |
	CH$_2$	CH$_4$	C$_2$H$_6$	C$_3$H$_8$	n-C$_4$H$_{10}$	n-C$_6$H$_{14}$	
Ethylene-propylene	41.9	3.4					0.074
	26.4	1.7					
Ethylene-butene-1	28.7		2.0				
	21.6		1.5				0.072
	4.1		1.1				
Ethylene-pentene-1	25.6			2.1			0.072
	17.1			1.1			
Ethylene-hexene-1	23.5				1.2		0.073
	7.8				0.7		
Ethylene-octene-1	19.1					1.3	0.074
	15.1					1.2	
	11.6					0.9	

Van Schooten and Evenhuis[1562] reported on the application of their pyrolysis (at 500°C)-hydrogenation-gas chromatographic technique to the measurement of short chain branching and structural details in modified polyethylene which contain small amounts of comonomer (about 10% by weight of butene-1, hexane-1, and octene-1). A survey of the isoalkane peaks of the pyrograms of these polymers with their probable assignment is given in Table 157. The effect of the comonomer on the relative sizes of the n-alkane peaks is given in Table 158. The peaks stemming from the total side group appear to be somewhat increased in intensity (n-C$_4$ peak in the polyethylene-hexane pyrogram, n-C$_6$ peak in polyethylene-octene pyrogram). the polyethylene-octene pyrogram also shows that the peak stemming from the n-alkane with one carbon atom less is somewhat enlarged.

This suggests that chain scission may occur both at he alpha and at the beta C-C bond. Only the latter type of scission will produce methyl substituted iso-alkanes, e.g. after intramolecular hydrogen transfer. A list of iso-alkanes that may be expected to be formed in this way is compared in Table 159 with a list of those that have been found to be significantly increased in size in the copolymer pyrograms.

Seeger and Barrall[1563] have investigated the applicability of

Table 156 – Distribution of short-chain branches in high presusre polyethylene
(branches per 1000 methylene units)[1]

Resin no	Density (g/cc)	$-CH_3$	$-C_2H_5$	C_3H_7	$i-C_4H_9$	$-n-C_4H_9$	$-C_5H_{11}$	Total branches per 1000 methylene	Methyl group content per 1000 methylene[2]
A	0.934	2.9	11.8	1.7	0.3	5.9	1.6	24.2	18.8
B	0.929	2.5	15.2	2.1	0.3	8.5	2.0	30.6	24.0
C	0.924	3.8	16.7	2.2	–	9.9	1.8	34.4	30.9

[1] Calculated from B values of isolated hydrocarbons assuming scission efficiency of 0.073×10^2 per 1000 carbon atoms. As short chain branches outnumber chain ends by 30:1, no correction for the fragmentation products at chain ends was made.

[2] Determined by infrared analysis.

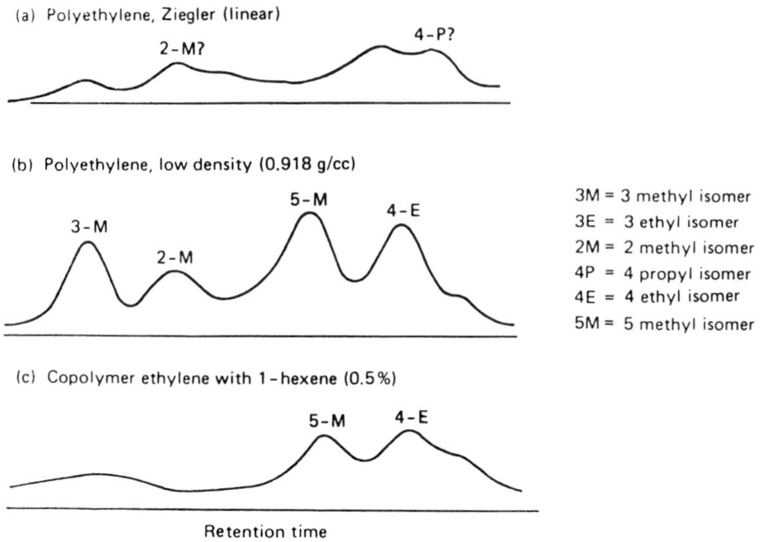

Figure 114   Single branched fragments (C_{10}).

pyrolysis-hydrogenation-gas chromatography to the elucidation of side-chain branching in ethylene-hexene-1 copolymers. Figure 114 shows the single branches fragment pattern up to ten carbon atoms obtained for this copolymer compared with those obtained for Ziegler (linear) polyethylene and low-density (0.918 g/cc) polyethylene.

Pyrolysis-hydrogenation of these model polymers shows that the areas of 3-methylakanes are increased relative to the areas of the corresponding 2-methylalkanes in the pyrograms of ethylene-1-butene copolymers and that 5-methylalkanes are indicative of butyl branching[1562,1564,1565]. Other indications of C_4 branches are given by an increase in the relative amounts of 2-methylhexane (2-MC_6) and 3-methylheptane (2-MC_7) and an increase in the relative amount of the n-C_4 peak. However, the areas of 2-MC_6 and 3-MC_7 are equal in the case of C_4 branch, while for ethyl branches the area of 3-MC_7 is enhanced relative to 2-MC_6. Thus not only must individual peaks be interpreted for the determination of the type of branch, but the overall pyrolysis pattern must be considered as well.

It is possible not only to measure the kind of branching present, but also to identify the polymer with this simple series. The distribution of isomers varies significantly with the type of branching. In the copolymer with 1-hexene (butyl branches) the 5-methyl and 4-methyl isomers are dominant. The low-density polyethylene (0.918 g/cc) shows a high 3 methyl peak as well as a high 5-methyl and 4-methyl peak yield. This indicated that both ethyl and butyl branches are present in the polyethylene material. Such evidence confirms former suggestions, mainly on the basis of infrared measurements, that low density polyethylene has branches which are the result of certain intramolecular transfer reactions during high-pressure polymerization[1566].

10.6.4 PVC

It has been recognized since the work of Cotman[1567] that be studying the polyolefin obtained from the reduction of PVC with lithium aluminium

Table 157 - Areas of the main iso-alkane peaks in the pyrograms of linear polyethylene, ethylene-butene-1, ethylene-hexene-1 and ethylene-octene-1 copolymers

Component	Peak area in arbitrary units			
	Linear PE	Ethylene-butene-1	Ethylene-hexene-1	Ethylene-octene-1
i-C$_4$	1-	20	_37_	_30_
i-C$_5$	18	_47_	_62_	_46_
2MC	9	14	18	20
3MC$_5$-CyC$_5$	31	_54_	46	76
2MC$_6$	14	13	74	22
3MC$_6$	13	_33_	24	26
2MC$_7$-4MC$_7$	16	17	25	27
3MC$_7$-3EC$_6$	16	_67_	_64_	32
2MC$_8$-4MC$_8$-ECyC$_6$	50	44	38	_119_
3MC$_8$	9	_39_	12	17
4MC$_9$-5MC$_9$-4EC$_8$	11	21	_51_	36
i-PCyC$_6$-BuCyC$_5$			94	
2MC$_9$-n-PCyC$_6$	40	35	40	65
3MC$_9$	9	_27_	6	_45_
4MC$_{10}$-5MC$_{10}$-secBuCyC$_6$	25	23	_67_	32
2MC$_{10}$-4MC$_{10}$-n-BuCyC$_6$	33	30	47	_102_
3MC$_{10}$	15	_32_	14	34

Table 158 - Relative sizes of n-alkane peaks for polyethylene and for ethylene-butene-1, ethylene-hexene-1 and ethylene-octene-1 coplymers

Peak ratio	Marlex 50	Ethylene-butene-1	Ethylene-hexene-1	Ethylene-octene-1
n-C$_4$-n-C$_7$	0.71	0.86	_1.15_	0.86
n-C$_6$-n-C$_7$	0.59	0.69	0.61	_0.91_
n-C$_6$-n-C$_7$	0.52	1.66	1.62	_1.88_
n-C$_8$-n-C$_7$	0.63	0.58	0.66	0.79
n-C$_9$-n-C$_7$	0.67	0.63	0.62	0.75
n-C$_{10}$-n-C$_7$	1.02	0.93	0.99	1.05
n-C$_{11}$-n-C$_7$	1.00	0.82	0.90	0.92

Table 159 – Expected and observed increased iso-alkane peaks in copolymer pyrograms

| Ethylene-butene-1 | | Ethylene-hexene-1 | | Ethylene-octene | | Number of backbone C atoms in fragment |
Expected	Observed	Expected	Observed	Expected	Observed	
–	–	–	$i\text{-}C_4$	–	$i\text{-}C_4$	–
–	–	–	$i\text{-}C_5$	–	$i\text{-}C_5$	–
$i\text{-}C_5$	$i\text{-}C_5$	$2MC_6$	$2MC_6$	$2MC_8$	$2MC_8$	3
$3MC_5$	$3MC_5$	$3MC_7$	$3MC_7$	$3MC_9$	$3MC_9$	4
$3MC_6$	$3MC_6$	$4MC_8$	–	$4MC_{10}$	$4MC_{10}$	5
$3MC_7$	$3MC_7$	$5MC_9$	$5MC_9$	$5MC_{11}$	–	6
$3MC_8$	$3MC_8$	$5MC_{10}$	$5MC_{10}$	$6MC_{12}$	–	7
$3MC_9$	$3MC_9$	$5MC_{11}$	–	$7MC_{13}$	–	8

hydride, valuable structural information can be gained concerning the starting molecule. This reduction reaction has been investigated and refined[1568] to the point where conditions have been established so that the chlorine may be efficiently removed from the polymer without degradation. The reduced polymer is similar to high-density polyethylene in almost all respects[1569]. Thus, studies that have been applied to polyethylene may also be applied to reduced PVC. Indeed better qualitative agreement has resulted when gamma-ray radiolysis followed by identification of the gaseous hydrocarbons by mass spectrometry used in conjunction with infrared measurements is applied to reduced PVC than when this technique is applied to conventional polyethylene. It was concluded from the large yields of methane relative to butane and ethane that the predominant side-chains along the PVC backbone are mainly one carbon long. It has been demonstrated by ^{13}C NMR that most of the short branches in PVC are pendent chloromethyl groups[1570]. This information was obtained from PVC samples reduced with lithium aluminium hydride and lithium aluminium denteride respectively.

Poly(vinylchloride) upon pyrolysis in a combustion furnace undergoes non-chain scission reactions. Non chain scission are common to short chain esters or halogenated molecules which have a hydrogen atom attached to the carbon atom beta to the substituted group. These molecules dissociate into an olefinic structure and an acid, the main pyrolysis products of polyvinyl chloride are volatile products, including hydrochloric acid and benzene. The benzene is formed as the result of a cyclization of the unsaturated chain ends.

Ohtani and Ishikawa[1572] pyrolysed polyvinylchloride at 425°C in a nitrogen atmosphere and studied the resulting degradation products by infrared and ultraviolet spectroscopy. Their findings showed that: 1) the pyrolysis products (other than hydrogen chloride) consisted of aliphatic and aromatic hydrocarbons, and 2) the types of pyrolysis products from polyvinylchloride were a function of the stereoregularity (tacticity) in the ungraded polymer. Additional qualitative or quantitative measurements were not made.

O'Mara[1573] carried out pyrolysis of polyvinylchloride (Geon 1-3, CI = 57.5%) by two general techniques. The first method involved heating the resin in the heated (325°C) inlet of a mass spectrometer in order to obtain a mass spectrum of the total pyrolysate. The second, more detailed, method consisted of degrading the resin in a pyrolysis-gas chromatograph interfaced with a mass spectrometer through a molecule enricher[1574]. Samples of polyvinylchloride and plastisols (10-20 mg) were pyrolysed at 600°C in a helium carrier gas flow. Since a stoichiometric amount of hydrogen chloride is released (58.3%) from polyvinylchloride when heated at 600°C, over half of the degradation products, by weight, is hydrogen chloride.

A typical pyrogram of a polyvinylchloride resin obtained by this method using an SE32 column is shown in Figure 115. The major components resulting from the pyrolysis of polyvinylchloride are hydrogen chloride, benzene, toluene, and naphthalene. In addition to these major products, a homologous series of aliphatic and olefinic hydrocarbons ranging from C_1 to C_4 are formed. O'Mara[1573] obtained a linear correlation between weight of polyvinylchloride pyrolysed and weight of hydrogen chloride obtained by gas chromatography.

The thermal decomposition of polyvinylchloride has usually been represented as involving the dehydrochlorination of the polymer backbone.

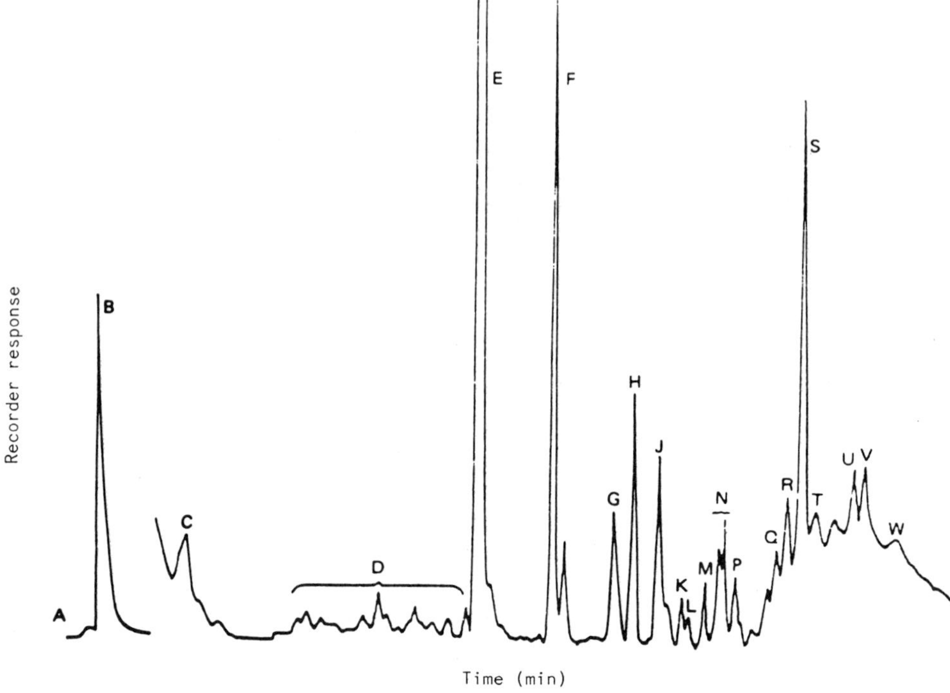

Peak identification	Components
A	CH_4 * CH, * CO_2 *, C_2H_4, * C_2H_6 *
B	HCl, C_3H_6, * C_3H_8 *;
C	Butane[a], butene[a], butadiene[a], diacetylene[a]
D	C_5 and C_6 aliphatic and olefinic hydrocarbons
E	Benzene
F	Toluene
G	Chlorobenzene
H	Xylene
J	Allylbenzene
K	C_9H_{12}
L	C_9H_{12}
M	Indane
N	Indene, ethyltoluene
P	Methylindane
R	Methylindanes
S	Naphthalene
T	Dimethylindane
U	Methylnaphthalene
V	Methylnaphthalene, acenaphthalene
W	Dimethylnaphthalene

*Separated and identified on an 8ft. Poropak QS. HCl ~ 58.3%, ash 3–4%.

Figure 115   Pyrograms of PVC resin from 20 ft x 3/16 in 10% SE32 on 80/100 CRW chromatographic column. Lettered peaks refer to identifications below.

On the other hand Stromberg et al.[1575] assumed the release of chlorine molecules in their kinetics of the decomposition of polyvinylchloride. The isolation of small amounts of hydrogen and chlorine was reported by Tsuchiva and Sumi[1576] on pyrolysis of polyvinylchloride in an inert gas. Ohta[1577] suggested that a part of the hydrogen chloride released from polyvinylchloride may recombine with the double bonds along the chain introduced by the dehydrochlorination. After pyrolysis of polyvinylchloride many kinds of hydrocarbons consisting of aliphatics and aromatics in addition to hydrogen chloride, were detected by Stromberg and other workers[1575,1578-1582].

Ahlstrom et al.[1583] applied the techniques of pyrolysis-gas chromatography and pyrolysis-hydrogenation-gas chromatography to the determination of short chain branches in PVC and reduced PVC. Their attempts to determine the short chain branches in PVC by pyrolysis-gas chromatography were complicated by an inability to separate all of the parameters affecting the degradation of the polymer. Not only does degree of branching change the pyrolysis pattern, but so do tacticity[1554] and crosslinking[1585].

In order to eliminate some of the above parameters, and improve on the then existing methods of polymer characterization, these workers examined the pyrolysis of polyethylene and studied several PVCs and lithium aluminium hydride-reduced PVCs differing in the amount of branch content, as obtained by infrared spectroscopy and ^{13}C NMR but not in tacticity.

For the pyrolysis of PVC, a ribbon probe was used. On-line hydrogenation of the pyrolysis products was accomplished by using hydrogen as the carrier gas with 1% palladium on Chromasorb-P catalyst inserted in the injection port liner. Maximum triplet formation occurred at C_{14} for low-density polyethylene and for reduced PVC, and at C_{15} for high-density polyethylene. The occurrence of the peak maxima at C_{14} for reduced PVC indicates that the total branch content is higher than that of high-density polyethylene. But aside from the C_{14}, C_{15} peak maxima difference, the pyrolysis pattern for even the most highly branched PVCs resembles high-density polyethylene more than low-density polyethylene. These data indicate that the type of short-chain branch in PVC is qualitatively more like that of high-density polyethylene, but that the sequence length between branch sites is shorter in low-density polyethylene and PVC. Since low-density polyethylene contains a large amount of ethyl and butyl branches, and PVC and high-density polyethylene contain mainly methyl and some ethyl branches, this qualitative resemblance would be expected. A relative measure of the total amount of short-chain branches for these polymers can be obtained by calculating the percentage of branched products formed (Table 160).

More information about the specific type of short-chain branch in PVC can be found from an examination of the C_1-C_{11} hydrocarbons (Figure 116). Here quantitative differences between the reduced PVCs become more apparent. The most obvious differences occur in the amounts of iso-C_7 and iso-C_8 products formed, which indicate differences in the total branch content. As the amount of short-chain branching (C_{10}%) in the reduced PVCs increases, there is a decrease in the amount of iso-alkanes formed (Table 161). The data in Table 161 show small but distinguishable differences in the short-chain branch content of the reduced PVCs.

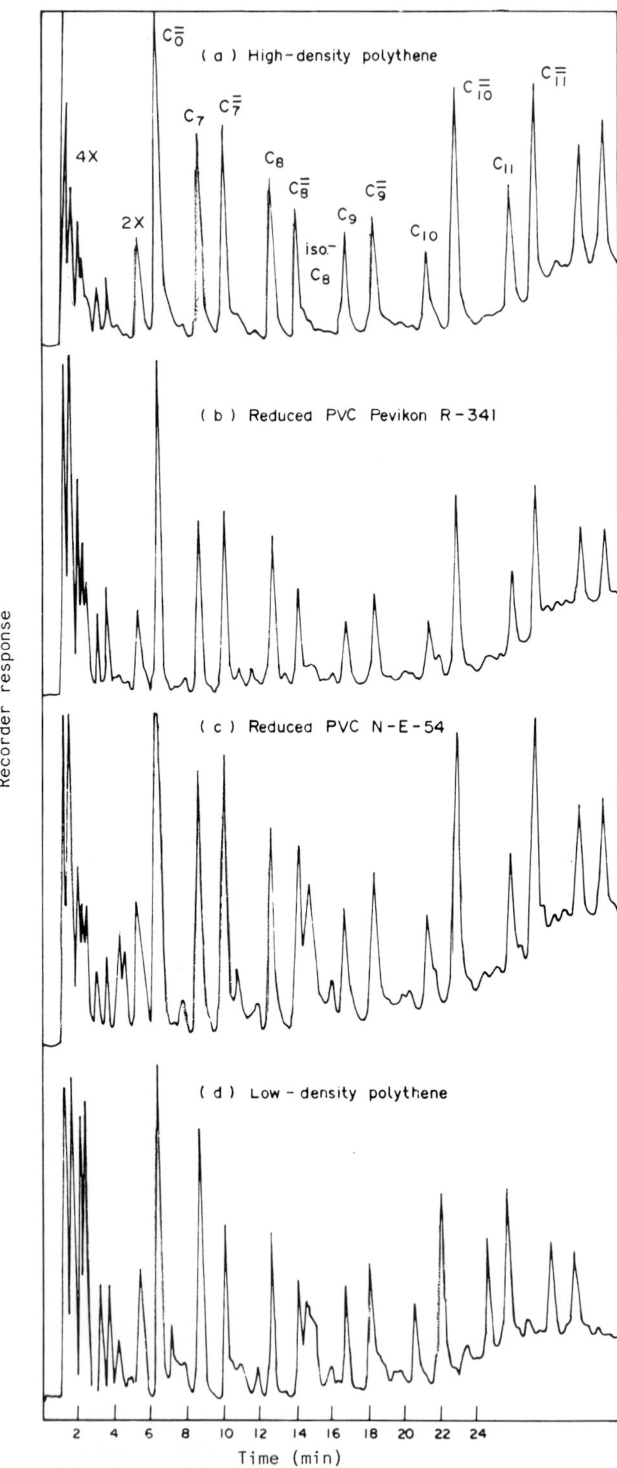

Figure 116 Pyrolysis of polyethylenes and reduced PVC; (a) high density polyethylene; (b) reduced PVC Pevikron R-341; (c) reduced PVC N-E-54; (d) low density polyethylene Durapak column fragments C_1-C_{11}.

Table 160 - Relative total branch content of high-density
polyethylene, LAH-reduced PVC and low-density polyethylene

Sample	Branched products (%)
High-density polyethylene	12.0
Reduced PVC*	19.0
Low-density polyethylene	26.0

* Average value

Table 161 - Short chain branch content of lithium aluminium
hydride reduced PVC

Sample	iso C_8^-/n-C_8^-	C_{10}^-%
high density polyethylene	0.1	8.6
Pevikon R341	0.52	7.7
Nordforsk E-80	0.71	7.2
Nordforsk S-80	1.20	7.0
Nordforsk S-54	1.59	6.8
Nordforsk E-54	1.63	6.5
Ravinil R100/650	1.63	6.5
Shinttsu TK1000	1.87	6.3
low density polyethylene	1.83*	5.2

* This ratio does not include the C_4 branch content.

THERMAL, OXIDATIVE AND PHOTOSTABILITY

11.1 THERMAL STABILITY

11.1.1   Introductory

Thermal methods of analysis predominate in the measurement of this characteristic of polymers. Thermal methods of analysis of polymers are important in that these techniques can provide information about the thermal and oxidative stability of polymers, their lifetime or shelf-life under particular conditions, phases and phase changes occurring in polymers, and information on the effect of incorporating additives in polymers. As will be seen in Chapters 12 and 13, these methods have also been applied to the measurement amongst others of melting points, percent crystallinity, crystallization temperature, glass transition temperature, melting temperature, cure rate, degree of cure, composition, thermal stability, expansion coefficient, and softening temperature.

Most of this work has involved the traditional thermal analysis technique of differential scanning calorimetry (DSC), differential thermal analysis, (DTA), thermogravimetric analysis (TGA) and thermal volatilization analysis (TVA). Recently, however, newer thermal analysis techniques have been introduced and the literature on these is growing, viz, evolved gas analysis (EGA), differential photocalorimetry (PPC), thermomechanical analysis (TMA), dynamic mechanical analysis (DMA) and, also, a pressure version of differential scanning calorimetry (PDSC).

Thermal analysis techniques for the examination of polymers can conveniently be discussed under the following eight main headings. The technique of controlled pyrolysis followed by gas chromatography and/or spectrometry is, in effect, a thermal method of analysis and is discussed elsewhere in this book (Chapters 4.4, 9.4 and 10).

Thermogravimetric analysis.
Differential thermal analysis.
Differential scanning colorimetry and pressure differential scanning colorimetry.
Thermal volatilization analysis.
Evolved gas analysis.
Differential photocalorimetry.
Thermomechanical analysis (Chapter 12).
Dynamic mechanical analysis (Chapter 12).

All of these techniques, except possibly dynamic mechanical analysis, have been used in thermal stability studies. The principles of the methods discussed below:

11.1.2 Thermal Analysis Techniques

<u>Thermogravimetric analysis.</u> Measurement of weight changes in a material as it is heated , cooled or held at a constant temperature in an inert atmosphere to determine composition and thermal stability, (weight v temperature plots).

<u>Differential thermal analysis.</u> This is a measurement of heat related phenomena which are associated with transitions in materials. In this technique the polymer sample is temperature programmed at a controlled weight and, instead of determining weight changes as in thermogravimetric analysis, the temperature of the sample is continually monitored. Just as a phase change from ice to water or vice versa is accompanied by a latent heat effect, so when a polymer undergoes a phase change from for example a crystalline to an amorphous form, heat is either evolved or absorbed.

<u>Differential scanning calorimetry.</u> Differential scanning calorimetry measures the amount of energy absorbed or released by a sample as it is heated, cooled, or held at a constant temperature. The precise measurement of sample temperature is also made with differential scanning calorimetry.

<u>Thermal volatilization analysis.</u> In this technique, in a continuously evacuated system, the volatile products are passed from a heated sample to the cold surface of a trap some distance away. A small pressure develops which varies with the rate of volatilization of the sample. If this pressure is recorded as the sample temperature is increased in a linear manner, a thermal volatilization analysis thermogram showing one or more peaks is obtained. The trace obtained is somewhat dependent on heating rate, which therefore should be standardized.

The various techniques discussed above measure heat changes, phase changes, pressure changes and weigh loss under controlled conditions. Although, in many instances these measurements are very meaningful, they are limited in the information they can provide. Thermogravimetric analysis, for example, measures the weight of a sample as a function of temperature and time as the temperature is varied. This is valuable quantitative information but tells us nothing about the chemistry of the processes occurring. To bridge this gap a field has recently opened up called evolved gas analysis (EGA) in which it is possible to monitor the by-products of reactions associated with heat.

<u>Evolved Gas Analysis.</u> In this technique the sample is heated at a controlled rate under controlled conditions and the weight changes monitored, (i.e. thermogravimetric analysis). At the same time the reaction products are lead into a suitable instrument for identification, and in some cases quantitation. Many variants of this approach have recently been developed based on three methods for thermally breaking down the samples: pyrolysis, linear programmed thermal degradation (i.e. without recording weight change) and the thermogravimetric approach (i.e. sample weight continuously recorded).

i)	Pyrolysis - gas chromatography.	
ii)	Pyrolysis - mass spectrometry.	
iii)	Linear programmed thermal degradation - computerized gas chromatography.	
iv)	Thermogravimetric analysis - Fourier transform infrared spectroscopy or infrared spectroscopy.	
v)	Linear programmed thermal degradation - mass spectrometry.	

More Recent Thermal Techniques

<u>Differential photo-calorimetry.</u> This new method measures the heat of reaction of photo initiated reactions or is used to measure the cure rate and degree of cure of photocurable polymers. The technique uses a dual sample DSC to measure the heat of reaction of one or two samples as they are exposed to light usually a high pressure mercury arc lamp with a maximum intensity in the 200-400 nm range.

<u>Pressure differential scanning calorimetry (PDSC).</u> Measures heat flow and temperature of transitions as a function of temperature, time and pressure (elevated pressure or vacuum). The ability to vary pressure from 0.01 torr to 1000 psig makes PDSC ideal for polymer oxidative studies as well as other pressure sensitive reactions.

<u>Thermomechanical analysis.</u> Measurement of dimensional changes (such as expansion, contraction) in a material by movement of a probe in contact with the sample to determine temperature related mechanical behaviour in the temperature range -180°C to 800°C as the sample is heated, cooled (temperature plot) or held at a constant temperature (time plot). This form of thermal analysis is discussed further in Chapter 12.

<u>Dynamic mechanical analysis.</u> Dynamic mechanical analysis is the measurement of the mechanical response of a material as it is deformed under periodic stress. Material properties of primary interest include modulus (E'), loss modulus (E"), tan S (E"/E'), compliance viscosity, stress relaxation, creep and damping. These properties, expressed in quantitative units of measurement, characterize the viscoelastic performance of a material and enable predictions to be made regarding the performance of a material over a wide range of conditions. Test variables include temperature, time, stress, strain and deformation frequency operating modes:

fixed frequency	stress v time
resonant frequency	oscillation amplitude v time
stress relaxation	strain v time
creep	stress v time

This form of thermal analysis is discussed further in Chapter 12.

11.1.3 Applications of Thermogravimetric Analysis

This technique involves continuous weighing of the polymer as it is subject to a temperature programme. This technique can provide quantitative information about the kinetics of the thermal decomposition of polymeric material from which the thermal stability of the polymer can be evaluated. It is used to study the influence on polymer degradation of factors such as effect of crystallinity, molecular weight, orientation, tacticity, substitution of hydrogen atoms, grafting, copolymerization and the addition of stabilizers. Figure 117 shows decomposition profiles for PTFE and a glass filled polyamide.

The lifetime or shelf-life of a polymer can be estimated from these kinetic data. Ozawa[1586] and Flynn and Wall[1587], observed that the activation energy of a thermal event could be determined from a series of thermogravimetric runs performed at different heating rates. As the heating rate increased, the thermogravimetric change occurred at higher temperatures. A linear correlation was obtained by plotting the logarithm of the heating rate or scan speed against the reciprocal of the absolute

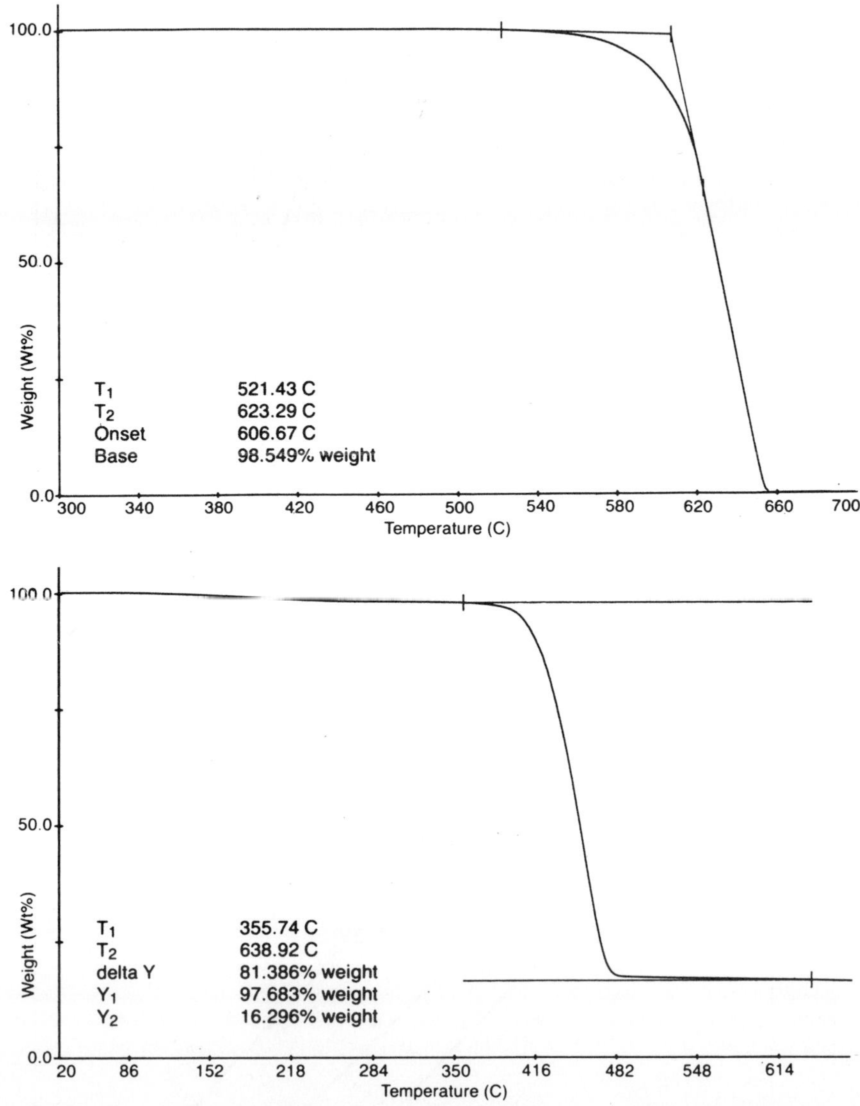

Figure 117   TGA decomposition profiles for (a) PTFE; (b) fibreglass
               reinforced nylon.

temperature at the same conversion or weight loss percentage.   The slope
was directly proportional to the activation energy and known constants.
To minimize errors in calculation, approximations were used to calculate
the exponential integral[1587-1589].   It was assumed that the initial
thermogravimetric decomposition curve (2-20% conversion) obeyed first
order kinetics.   rate constants and pre-exponential factors could then be
calculated and used to examine relationships between temperature, time and
conversion levels.   The thermogravimetric decomposition kinetics could be
used to calculate:

1.   The lifetime of the sample at selected temperatures.
2.   The temperature which will give a selected lifetime.
3.   The lifetimes at all temperatures at known percentage conversion.

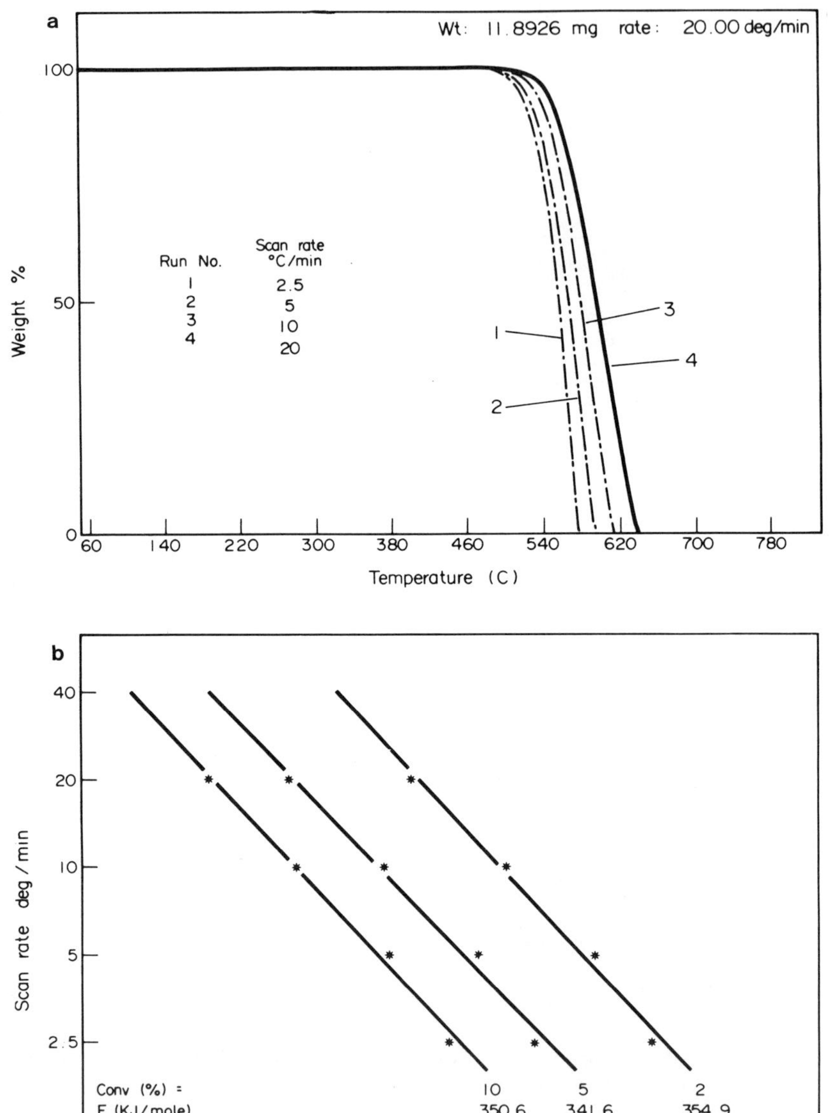

Figure 118   (a) thermogravimetric curves of PTFE; (b) Arhenius plots for thermogravimetric analysis of PTFE; (c) rate constant versus temperature curves of PTFE; (d) Half life temperature curve of PTFE; (e) conversion versus time at 500°C curves of PTFE.

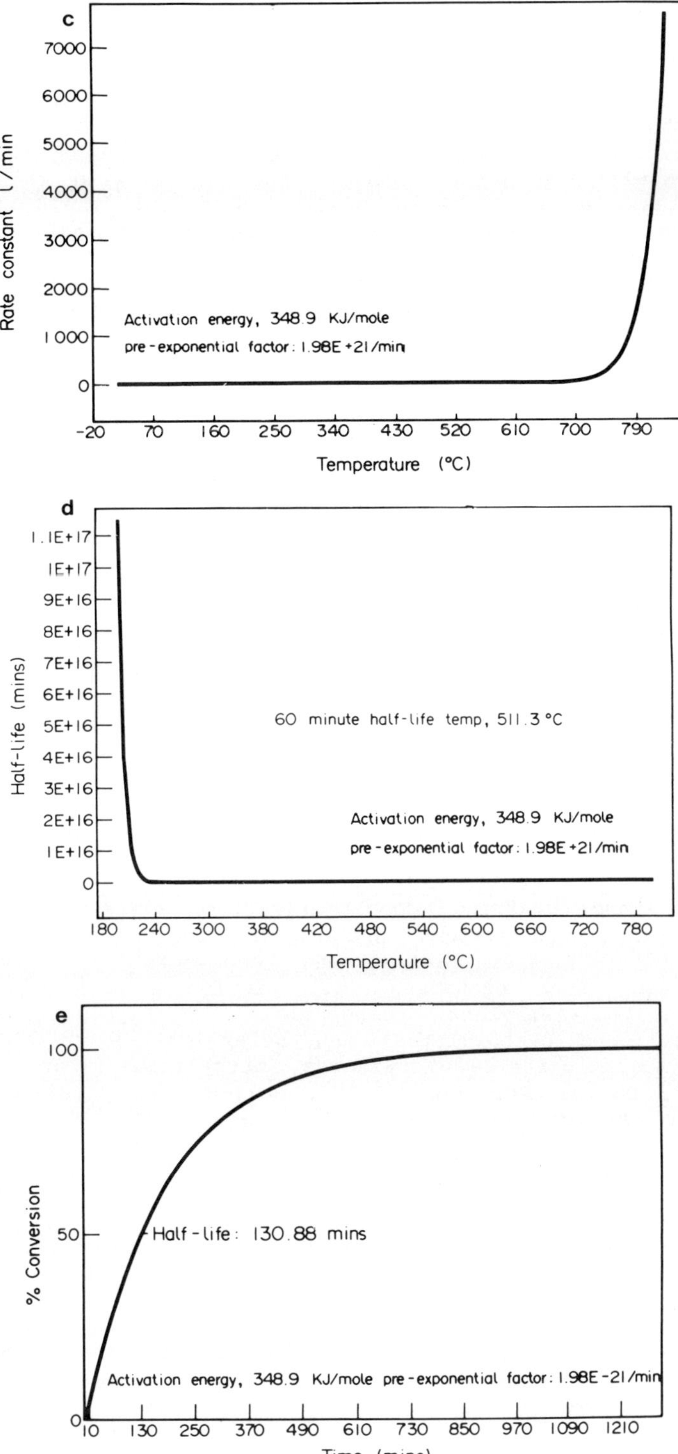
c
Rate constant l /min
7000
6000
5000
4000
3000
2000
1000
0
Activation energy, 348.9 KJ/mole
pre-exponential factor: 1.98E+21/min
-20 70 160 250 340 430 520 610 700 790
Temperature (°C)

d
1.1E+17
1E+17
9E+16
8E+16
7E+16
6E+16
5E+16
4E+16
3E+16
2E+16
1E+16
0
Half-life (mins)
60 minute half-life temp, 511.3 °C
Activation energy, 348.9 KJ/mole
pre-exponential factor: 1.98E+21/min
180 240 300 380 420 480 540 600 660 720 780
Temperature (°C)

e
100
50
0
% Conversion
Half-life: 130.88 mins
Activation energy, 348.9 KJ/mole pre-exponential factor: 1.98E-21/min
10 130 250 370 490 610 730 850 970 1090 1210
Time (mins)

Figure 118(a) shows thermograms (percentage weight versus temperature) for a 10 mg specimen of PTFE obtained at four different heating rates - 2.5, 5, 10 and 20°C per minute in a dynamic air ⁕atmosphere. From these data can be calculated the rate of decomposition of PTFE, the activation energy and the relationship between the rate constant or half-life and temperature.

The results shown in Figure 118(a) agree with the theoretical prediction that, as the rate of heating increased, the thermograms are displaced to higher temperature. Activation energies at selected percentage conversion levels were calculated using the results. The activation energy was calculated from the slope of the graph of scan time against the inverse of the absolute temperature (Figure 118(b)). After the activation energy had been determined, the rate constants, half-lives and percentage conversions could be calculated for certain temperatures. In Figure 118(c) the rate constant is plotted against temperature to provide information on the stability of the sample from ambient temperature to 800°C. The half-life can be calculated from these kinetic data and a graph of half-life versus temperature plotted (Figure 118(d)). The temperature equivalent to a half-life of 60min is also determined. These data can be printed out in tabular form (Table 162) to facilitate quantitative comparisons. The rate of percentage conversion with respect to time can be computed for any selected temperature. Figure 118(e) shows percentage conversion versus time at 500°C together with the half-life of the sample, 130.88 min.

The results calculated using the thermogravimetric decomposition kinetics can be used for comparative studies. However, if the lifetime at specific temperatures and percentage conversion are known for the sample, adjusted lifetimes can be calculated. This type of data can be obtained by elevated temperature measurements on 'bulk' samples.

Thermogravimetric analysis has been used to study degradation kinetics and various factors affecting thermal stability of polymers, such as crystallinity, molecular weight, orientation, tacticity, substitution of hydrogen atoms, grafting, copolymerization, addition of stabilizers, etc. More important systems investigated include cellulose[1590-1594], polystyrene[1591,1596], ethylene-styrene coolymers[1597], styrenedivinyl-benzene based ion exchanges[1598], vinylchloride-acrylonitrile co-polymers[1599], poly(ethylene terphthalate)[1600], polyesters such as poly(isopropylidene carboxylates) and polyglycollide[1601-1603], Nylon 6 grafted with acrylonitrile, acrylamide, methylmethacrylate, and methyl-acrylate[1604], polypyromellitimides[1605], poly-N-naphthylmaleimides[1606] and poly[benzobis(aminoiminopyrolenes)][1605], acrylics[1608,1609], PVC[1610,1611], acrylamide-acrylate copolymers and polyacrylic anhydride[1612]. One of the most common uses of thermogravimetric analysis is the quantitative determination of components in a mixture. Because of differences in thermal stability, the selection of the appropriate temperature programme and sample atmosphere results in the quantitative separation of the major components of the polymer or elastomer. Figure 119 for example shows a separation of synthetic rubber and carbon black.

In addition to polymers thermogravimetric analysis has been used for the examination of polymer additives. Al Sammerrai et al[1613] determined solvents in polymer additives. The usefulness of thermogravimetric analysis in additive degradation-volatilization studies can be illustrated in the case of polystyrene containing 1.5% ethylan MLD (lauric diethanolamide) or Catanac LSA (a quaternary ammonium compound) antistatic additives. Indications from the polymer extrusion plant were that at a polymer processing temperature of 260°C between 30 and 60% of Ethylan MLD

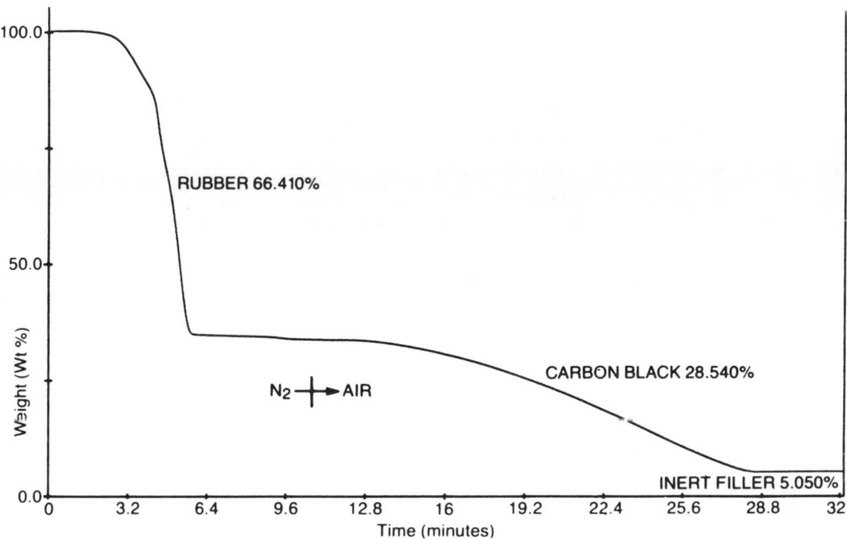

Figure 119   TGA thermogram of carbon black filled rubber.

Table 162 - Thermogravimetric Analysed PTFE Half-life Table

Temp (°C)	Temp (K)	1000/T (1/K)	K(T) (1/min)	Years	Days	Hours	Minutes
400	673.16	1.485	1.667E-06	0	288	14	26
410	683.16	1.463	4.154E-06	0	115	20	57
420	693.16	1.442	1.007E-05	0	47	18	21
430	703.16	1.422	2.383E-05	0	20	4	36
440	713.16	1.402	5.504E-05	0	8	17	51
450	723.16	1.382	1.242E-04	9	3	21	0
460	733.16	1.363	2.740E-04	0	1	18	9
470	743.16	1.345	5.920E-04	0	0	19	30
480	753.16	1.327	1.253E-03	0	0	9	13
490	763.16	1.310	2.600E-03	0	0	4	26
500	773.16	1.293	5.296E-03	0	0	2	10
510	783.16	1.276	1.059E-02	0	0	1	5
520	793.16	1.260	2.081E-02	0	0	0	33
530	803.16	1.245	4.022E-02	0	0	0	17
540	813.16	1.229	7.647E-02	0	0	0	9

Activation energy, 348.9' kJ/mole pre-exponential factor 1.98E + 21/min,
Convs 2 5 10

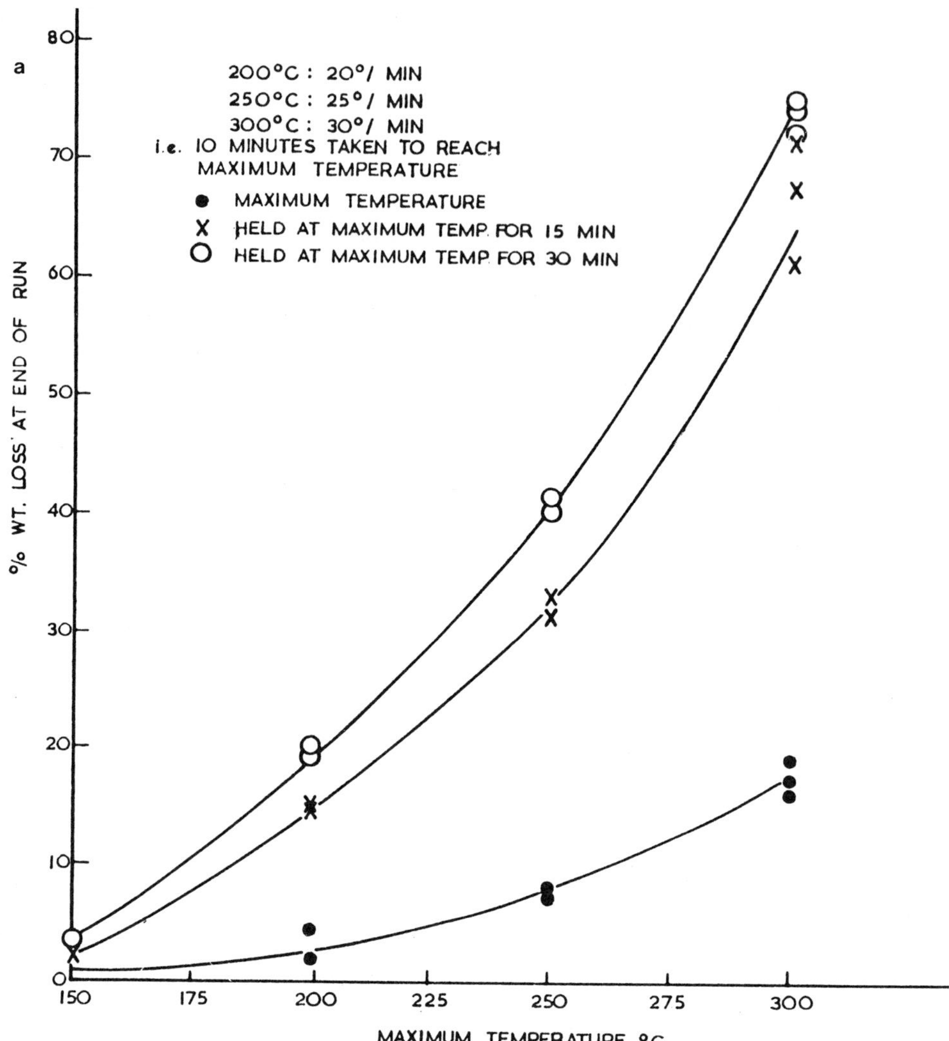

Figure 120   Temperature weight loss plots of (a) Ethylan MLD and (b)
               Catanac LSA.

degraded and/or volatilized whereas practically none of the Catanac LSA
activity was lost.   Thermogravimetric runs were performed on the two neat
additives in which the polymer sample was programmed to maximum
temperatures of 150, 200, 250 and 300°C at rates such that the sample
reached maximum temperature in 10 minutes.   The sample was then held at
300°C for a further 30 minutes.

       The weight loss of the sample was measured:

a)   When the maximum temperature is first reached i.e. 10 minutes after
       the heater is switched on; and

b)   at 15 and 30 minutes after the maximum temperature is reached, i.e.
       25 and 40 minutes after the heater is switched on.

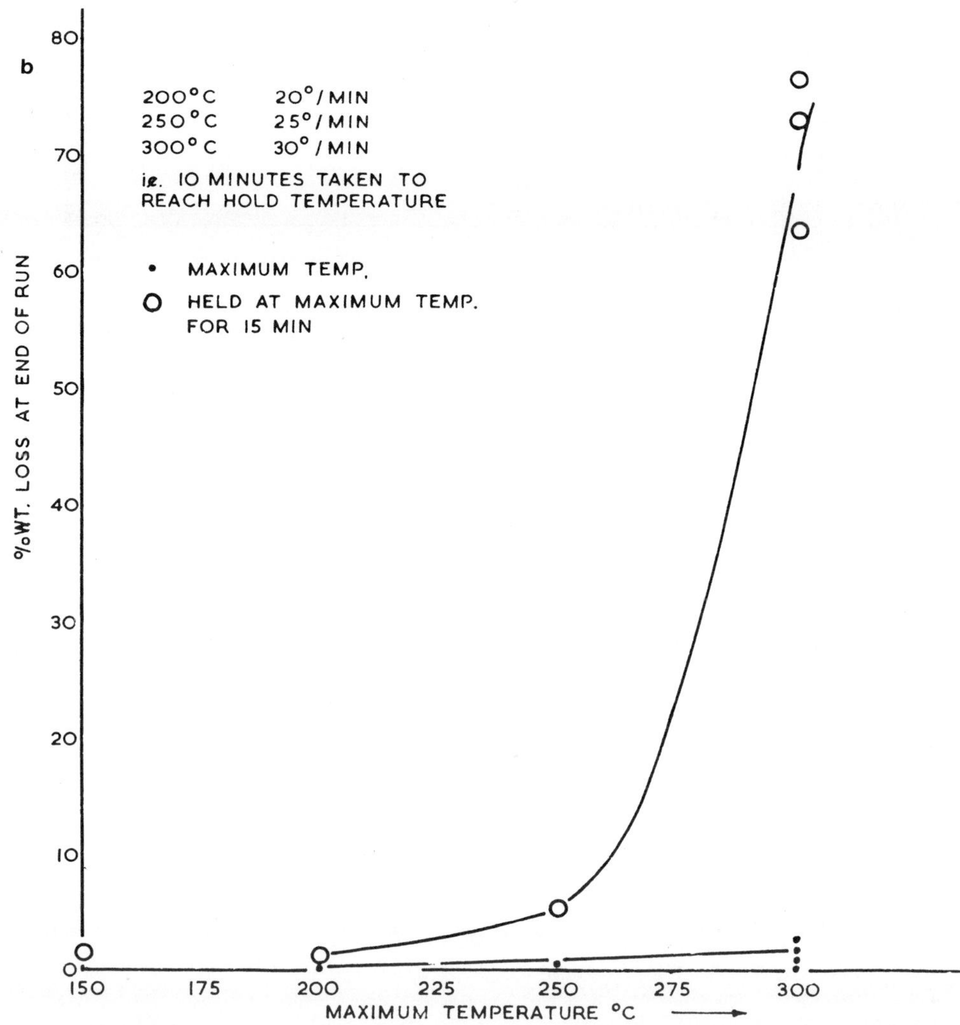

Temperature weight loss plots (Figure 120) indicates that loss of volatiles from Catanac LSA was appreciably less than from Ethylan MLD when the compounds are heated for 10 or 25 minutes in the temperature range 200°C to 250°C. However, Catanac LSA volatilises as extensively as Ethylan MLD when heated for 25 minutes at temperatures in excess of 250°C.

When Ethylan MLD was heated from room temperature to temperatures between 260 and 300°C in 10 minutes it loses between 10% and 20% of its weight as volatiles. This heating schedule was similar to that to which Ethylan MLD is exposed during processing of polystyrene, as the above volatiles loss is of a similar order to that deduced by nitrogen analyses. The T.G.A. data obtained for Catanac LSA show that under the conditions discussed above it loses very little volatile material. However, this does not mean that under these conditions Catanac LSA has not degraded to a non-volatile product.

Visual examination of the condensed volatiles and residues obtained at the end of a T.G.A. experiment showed whether degradation of the test

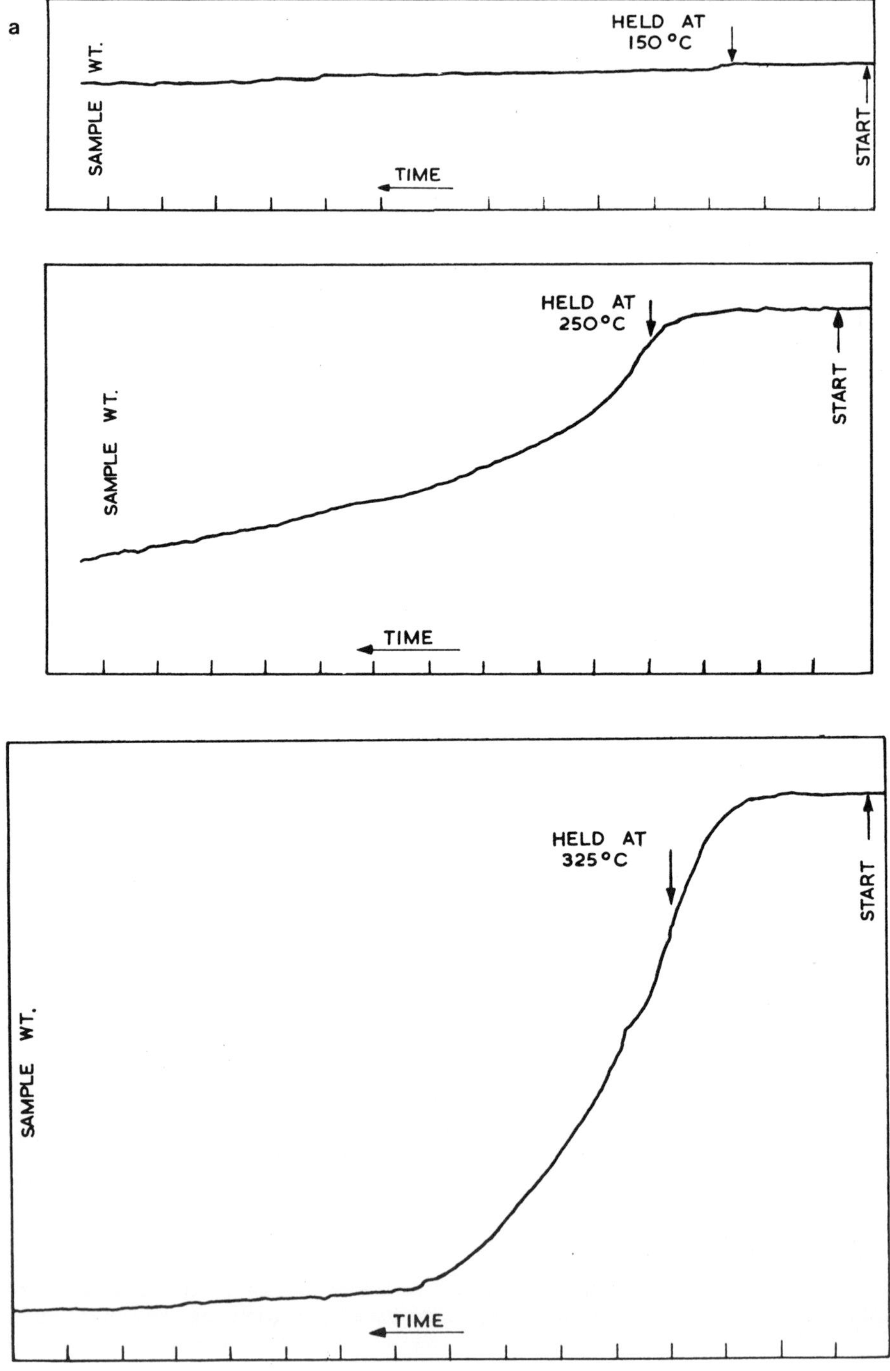

Figure 121   TGA curves for (a) Ethylan MLD and (b) Catanac LSA at
             different maximum temperatures of between 150°C and
             325°C.   Rate of heating to maximum temperature 30°C/min.
             Weight of sample 9 $\pm$ 0.5 mg.

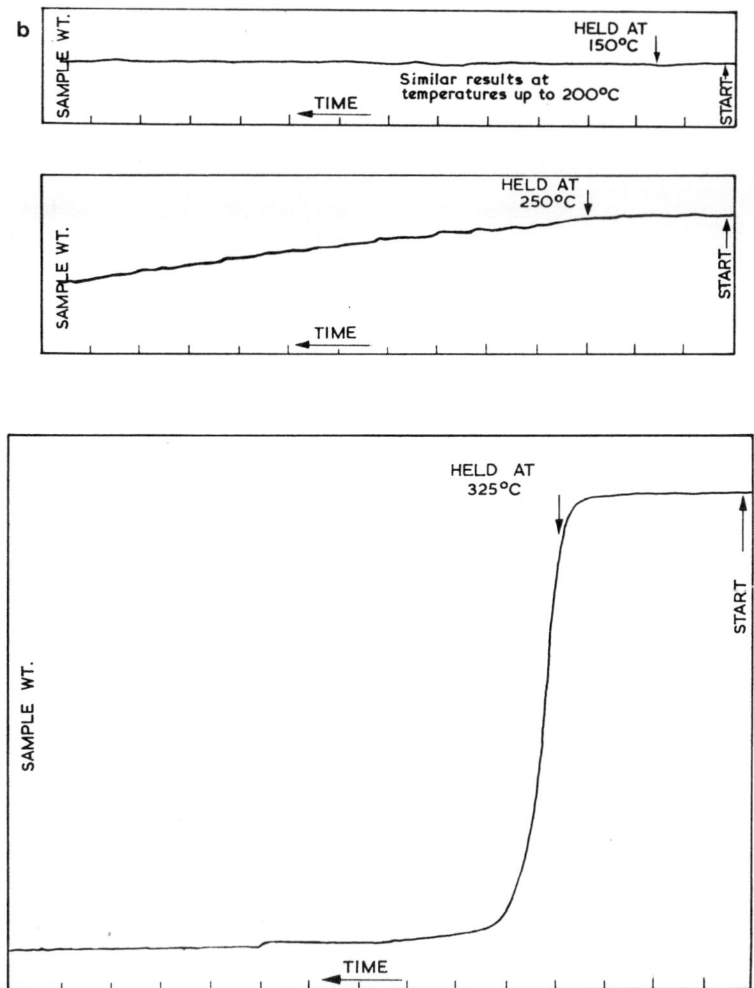

compound has occurred to any extent. Ethylan MLD and Catanac LSA gave brown-coloured condensates (indicating decomposition) when heated to 225°C and 275°C respectively, i.e., decomposition of both substances can occur at temperatures which are likely to be encountered during the processing of polystyrene (i.e. up to 200°C during Banbury mixing and up to 260°C during extrusion).

In Figure 121 are shown time weight plots for the two additives obtained when they were heated in air at 30°C min^{-1} to maximum temperatures between 150 and 325°C in separate experiments and then held for 30 minutes at the maximum temperature.

It is evident from these experiments that much more information would have been obtained had the volatile products and the residue been examined chemically, i.e. evolved gas analysis see later, undegraded ethylan MLD can be estimated by examination of its hydroxy band at 2.95 microns in the infrared without interference from its degradation product.

11.1.4 Applications of Differential Thermal Analysis

This technique has been used in polymer stabilization studies[1615,1616], polymer pyrolysis kinetics studies[1617,1618] and examination of the effects of thermal history on polyethylene[1614].

11.1.5 Applications of Differential Scanning Calorimetry

In addition to chemical reaction associated with thermal and oxidative decomposition differential scanning calorimetry can also be used to monitor the curing of thermosets. Examples are the polymerization of unsaturated polyester resins, poly addition of epoxy resins with curing agents and of isocyanates with polyols.

An isothermal differential scanning curve shows at a glance whether a reaction proceeds normally, in other words, the rate of reaction and thus the heat flow, reaches maximum upon the reaction mixture's attainment of the reaction temperature. To locate a suitable isothermal reaction temperature, a dynamic experiment is carried out at 10°C/min. The optimum isothermal temperature will lie between the start of reaction (at 20 per cent of the peak height) and the peak maximum temperature. An epoxy resin used for powder coating gives values of 180°C to 220°C for example. Conversely, an autocatalytic reaction shows an increasing reaction rate after an induction period.

To determine the extent of reaction as a function of reaction time, it is assumed that the area under the curve increases proportionally to the conversion, i.e. the conversion at a time t is equal to the partial area at the time t, divided by the total area. The graph of the extent of reaction versus reaction time is constructed by taking e.g. five calculated values.

Figure 122 shows a typical curing curve obtained from a thermoset material showing the exothermic peak produced as it is released during the curing process. A particular example is the study of curing kinetics in diallylphthalate moulding compounds[1621-1623]. Differential scanning calorimetry has also been used to monitor the cross linking[1619] rate of polyethylene - carbon black systems, and study the decomposition of polyoxy-propylene glycols[1620], the thermal stability of poly(vinylalcohol) modified at low levels by various reagents and by grafting with other vinyl monomers[1624]. The technique has been used in kinetics studies and a check on experimentally determined activation energies and Arrhenius frequency factors. A kinetic study of isothermal cure of epoxy resin has been carried out[1625,1626]. Kinetic parameters associated with crosslinking process of formaldehyde-phenol, and formaldehyde-melamine copolymers have been obtained from exotherms of a single differential scanning calorimetric temperature scan[1627].

Using this technique, kinetic studies have been made of the polymerization of styrene[1628,1629], methylmethacrylate[1629,1630], vinyl acetate[1631], bis-maleinimides[1632], and phenolformaldehyde[1633], and also curing of epoxide resins[1634-1636] and polyesters[1637].

A combination of programmed and isothermal techniques has been used for characterizing unresolved multistep reactions in polymers[1638].

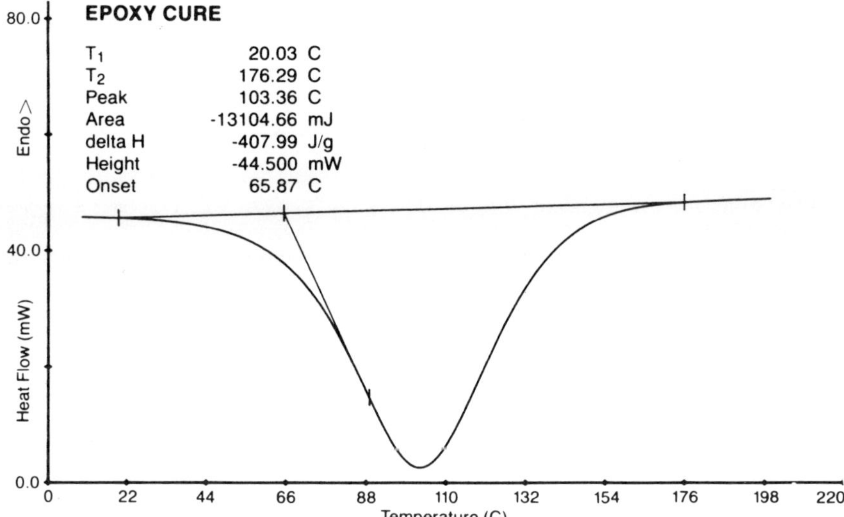

Figure 122   Use of differential scanning calorimetry in curing studies.

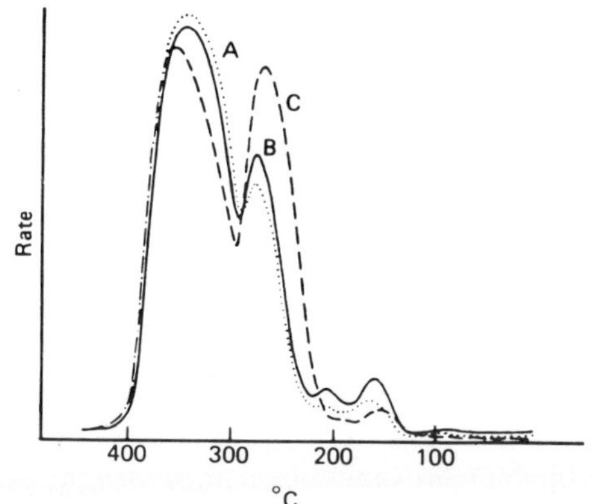

Figure 123   Thermograms (10°C/min) for samples of poly(methyl-methacrylate) of various molecular weights: (a) 820,000; (b) 250,000; (c) approx. 20,000.

11.1.6 Applications of Thermal Volatilization Analysis

Thermal volatilization analysis thermograms for various polymethylmethacrylates are illustrated in Figure 123. As in the case of thermogravimetric analysis, the trace obtained is somewhat dependent on the heating rate. With polymethyl methacrylate the two stages in the degradation are clearly distinguished (Figure 123). The first peak above 200°C represents reaction initiated at unsaturated ends formed in the termination step of the polymerization. The second, larger, peak corresponds to reaction at higher temperatures, initiated by random scission of the main chain. It is apparent that as the proportion of chain ends in the sample increases, the size of the first peak increases

also. These TVA thermograms illustrate very clearly the conclusions drawn
by Macallum[1639] in a general consideration of the mechanism of degradation
of this polymer. The peaks occurring below 200°C can be attributed to
trapped solvent, precipitant, etc. These show up very clearly, indicating
the usefulness of TVA as a method of testing polymers for freedom from
this type of impurity.

The technique has been applied to a range of polymers including
polystyrene[1640,1641],   styrene-butadiene   copolymers[1640],   PVC[1640],
polyisobutene, butyl rubber and chlorobutyl rubber and poly-alpha-methyl-
styrene[1640].

11.1.7 Applications of Evolved Gas Analysis

This is generic name for a wide variety of processes in which the
polymer is heated at controlled conditions of temperature and a chosen
atmosphere ranging from inert (e.g. nitrogen on helium) to reactive (e.g.
oxygen) and the breakdown products produced are examined by any one of a
wide   variety   of   analytical   techniques,   either   qualitatively   or
quantitatively, or both. A classical example of evolved gas analysis is
the pyrolysis-gas chromatography technique described elsewhere in this
book. Some other examples of the application of the technique include
polymer weight change-degradation product studies, the production of
polymer thermochromatograms and polymers additive degradation studies.

Evolved   gas   analysis   is   a   technique   in   thermal   analysis   for
characterizing compounds evolved during sample heating. It is, in a
sense, a reverse technique of thermogravimetric analysis, and if only one
compound is evolved during sample heating, then evolved gas analysis and
thermogravimetric analysis should principally give identical information.
Usually a number of components evolve during heating of a sample according
to complicated degradation kinetics and this valuable information is
poorly reflected on a thermogravimetric analysis curve. In view of the
last fact, it is surprising that the value of evolved gas analysis has
frequently   been   overlooked,   and   thus   used   less   often   than   thermogravimetric analysis.

To date, most significant work in the evolved gas analysis field has
been done by mass spectrometry[1642-1644] - a valuable tool in identifying
the evolved components. However, evolved gas analysis can be applied
where identification is trivial or an exact knowledge of the names of the
evolving components is not necessary. The number of components and their
evolving rate with temperature also give much information about the
thermal behaviour of a sample. Then, instead of complicated and expensive
mass spectrometry, a more simple, straightforward, and less expensive
technique - gas chromatography - can be used.

The problem with gas chromatography as an analyser of the evolved
component is that it takes much more time to separate the evolved
components than to analyse their mass in the mass spectrometer. If the
sample is heated by a linear thermal program (as is the case in thermal
analysis), then many interesting kinetic phenomena may occur during
separation of components on the gas chromatograph. This makes an on-line
gas chromatographic analysis of the evolved components difficult. The
evolved components could be trapped at moments of interest[1645], thus
overcoming the sampling rate problem However, this is not the best
solution. With computer control and some modification of the conventional
gas   chromatographic   apparatus,   increasing   radically,   the   gas
chromatographic analysis speed makes the on-line evolved gas analysis
still possible.

Modern high speed gas chromatographic systems are able to separate some light hydrocarbons within a second[1646,1647]. This approach requires special sampling valves, narrow bore columns (diam. = 0.005 cm), and detectors with a fast response. However, in evolved gas analysis these fast separations are not necessary. Taking samples on-line from the reactor and separating them within 1 minute is frequently satisfactory in order to better understand degradation kinetics. The apparatus for this approach can be constructed of commercially available parts.

Kullick et al.[1648] have described an evolved gas analyser based on thermogravimetry and gas chromatography, the sample under study is heated in the reactor. The inert gas flows through the reactor and carries the evolved products through one path of the sampling valve into the air. The pure carrier gas flows through the other path of the sampling valve into the gas chromatographic column. By a command from the computer the valve reverses for a short time and the sample flows into the column. The detector signal is recorded digitally and stored on a floppy diskette for further use.

For experimental control an Apple IIe computer was used. Apple IIe has several slots where application-dependent interface cards can be inserted. Also, it is relatively easy to program the Apple IIe to control an experiment using a 6502 uP (microprocessor) code and Applesoft Basic programming language. Two built-in interface cards are used - one for communication between Apple IIe and HP 7225 A plotter, and the other for interfacing Apple IIe and a standard CAMAC crate which contains a 24-bit analog-to-digital converter (ADC) (Institute of Cybernetics, Estonian Academy of Sciences). The ADC records the gas chromatographic detector signal. The analog signal counting frequency was 4 points/sec. Two 1-bit outputs of Apple IIe are used for controlling the solenoid valve status. The solenoid valve controls the Deans type[1649] flow switch located in the gas chromatographic oven. This sampling system was capable of performing injections of gas samples with a precision of 0.3% and duration as short as 0.1 sec.

The sample is in a quartz test tube in the flow-through reactor. The reactor temperature is programmable linearly from 50 to 500°C at a speed of up to 20°/min. For reactor temperature control a separate temperature program is used. The amount of the evolved gas depends on the sample weight which varies in the range of 1-100 mg.

The evolved products are separated in two short, open tubular columns: a 10-m x 0.125 mm column coated with OV-43, and a 45-m x 0.5-mm column coated with SE-30. The apparatus is built on the basis of a Fractovac 4200 GC (Carlo Erba Instrument 10)

Reaction temperature and chromatogram running time information are obtained from the apparatus. Therefore it is possible to produce a thermogravimetogram. There are three plotting forms possible.

(i)   Chromatograms are produced in a sequence as they appear on the chart recorder. As the chromatograms are stored on a floppy disc post processing is possible. Chromatograms can be plotted with a different attenuation and the times axes can be expanded or suppressed to obtain a better knowledge about particular feature of a chromatogram. Also, the presentation facilitates quantification and obtaining kinetic data, (see figure 124).

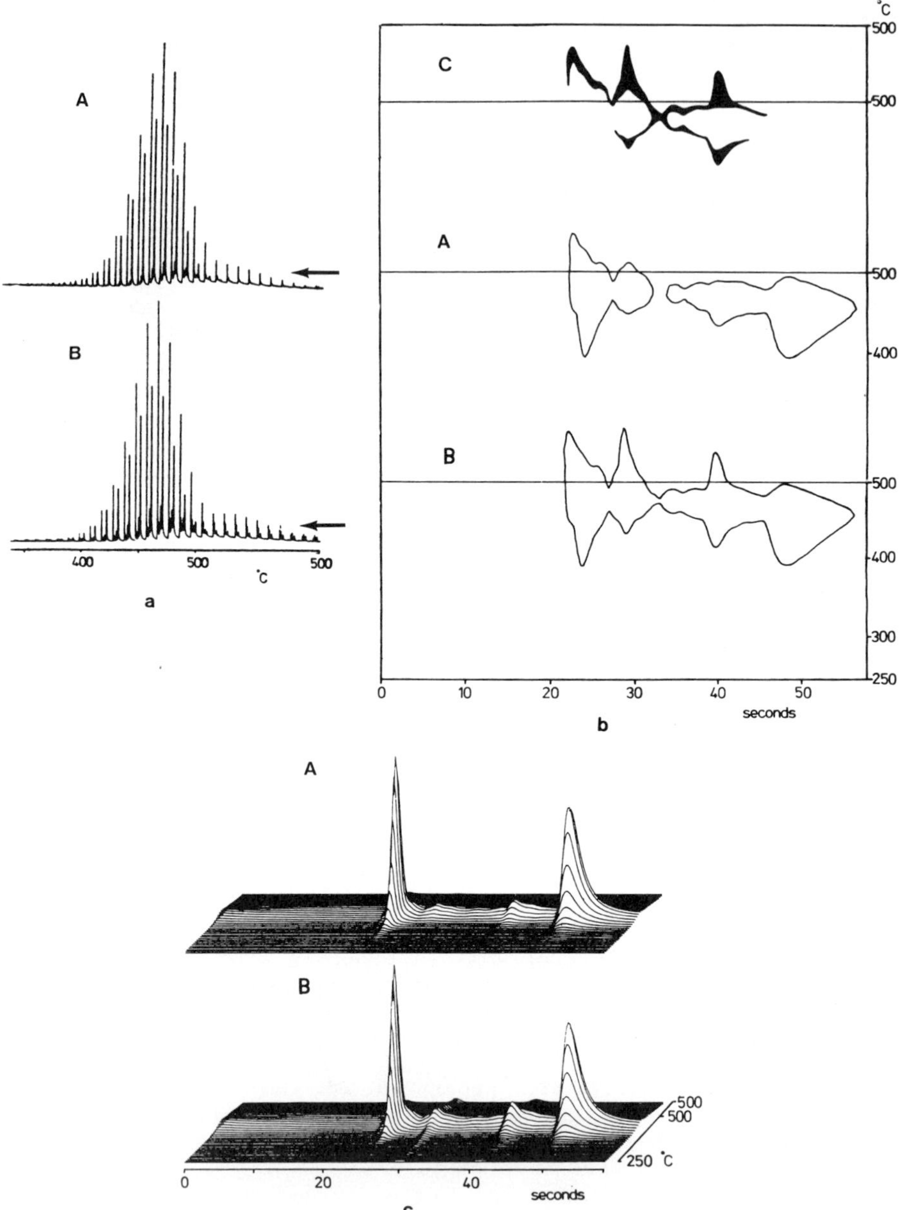

Figure 124   Plotting forms of thermocromatograms.
(a) Sequences of rubber thermochromatogram slices. The
arrows mark cutting levels for (A) cis 1,4-polyisoprene
and (B) 1,4-cis-polyisoprene with 5% 3,4-cis-
polyisoprene. Sample heating rate 10°C/min; GC column
SE-30; column temperature 150°C.
(b) Stack plots of rubber thermochromatograms. Sample
and conditions are the same as in Figure 124 (a).
(c) Contour plots of rubber thermochromatograms. Sample
and conditions are the same as in Figure 124 (b). For
cutting level see Figure 124 (a) - contour pattern
derived by subtracting B from A.

(ii)   A thermochromatogram can be considered as a surface. The isometric projection of this surface is called a stack plot. In this plot chromatograms are plotted one above the other and the upper chromatogram is shifted relative to the lower one to form a plotting angle. To control the "up" and "down" portions of the plotter pen, a reference function is formed. If a particular chromatogram point value exceeds this function value at this point, then the plotter pen gets the command "down" and the reference function gets a new value that equals the value of the chromatogram point. After that the plotter moves toward this point, drawing a line on the chart. If the chromatogram point value is lower than that of the reference function value, then the pen gets the command "up", the reference function value remains unchanged, and the line will not be drawn on the plotter chart -this simple rule enables the hidden line problem to be solved very easily. The stack plot gives a very good qualitative survey of the process, occurring during sample heating, but any kind of measurement is difficult from this plot. If the thermochromatogram is considered to be a surface, the above sequence plot of chromatograms can be regarded as a set of thermochromatogram slices made by planes parallel to the plane formed by time and evolving rate axes. The stack plots are shown in Figure 124(b).

iii)   The third plot frequently used in the 2-D data presentation is the contour plot. The contour plot is obtained by cutting the thermochromatogram surface by a plane which is parallel to the plane formed by temperature and time axes, and by plotting the cutting line on the temperature-time plane. Between two adjacent chromatograms a bilinear surface for approximation of the thermochromatogram is used. The contour plot is very useful for comparing retention times of different thermochromatograms. As an example of the contour plot is shown in Figure 124(c).

To identify the separated evolved products emerging via the gas chromatograph, Kullik et al.[648] either linked the apparatus to a mass spectrometer or collected the evolved compounds in traps for subsequent mass spectrometry.

They applied the procedure to thermoresist rubbers. Figures 124(a) are thermochromatograms and Figure 124(b) the "stack plots" for two rubber samples. Thermochromatogram A in Figure 124(a) represents synthetic, 1,4-cis-polyisoprene rubber, and B is 1,4-cis-polyisoprene rubber containing 5% 3,4-cis-isomer. Differences are evident, and at the first glance a conclusion can be drawn that rubber B is more thermostable than A, whose degradation begins at lower temperature and continues at higher temperature.

From an analytical point of view, thermochromatography should be a useful tool in differentiating samples for manufacture control. Differences between two similar thermochromatograms can be made clearly evident by using the contour plot (Figure 124c)). In this figure, A and B are contour plots for 1,4 cis polyisoprene and 1.4 cis polyisoprene containing 5% 3,4 cis isomer.   C shows a contour pattern which characterizes the deflection between the two samples, i.e. C = (B-A).

In Figure 125(a) A is a stack plot for the butadiene styrene rubber, and Figure 125a) B is a thermochromatogram for the butadiene alpha-methyl

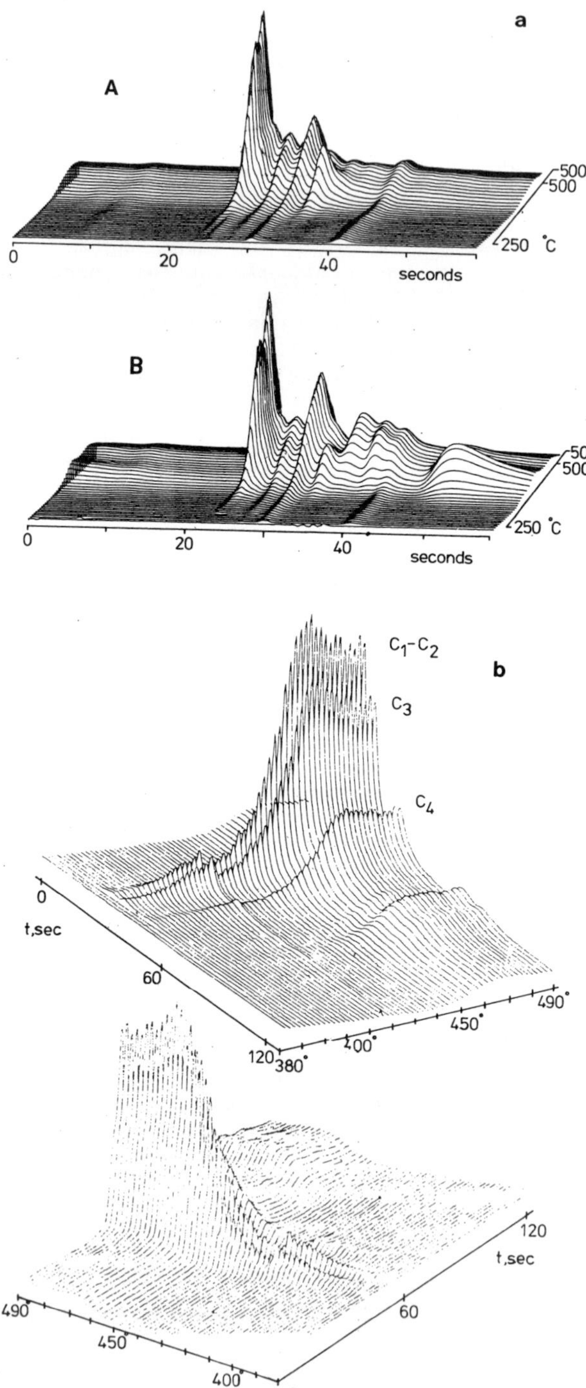
a
A
500
500
250 °C
0
20
40
seconds
B
500
500
250 °C
0
20
40
seconds
b
C$_1$–C$_2$
C$_3$
C$_4$
0
t,sec
60
120
380°
400°
450°
490°
490°
450°
400°
380°
60
120
t,sec

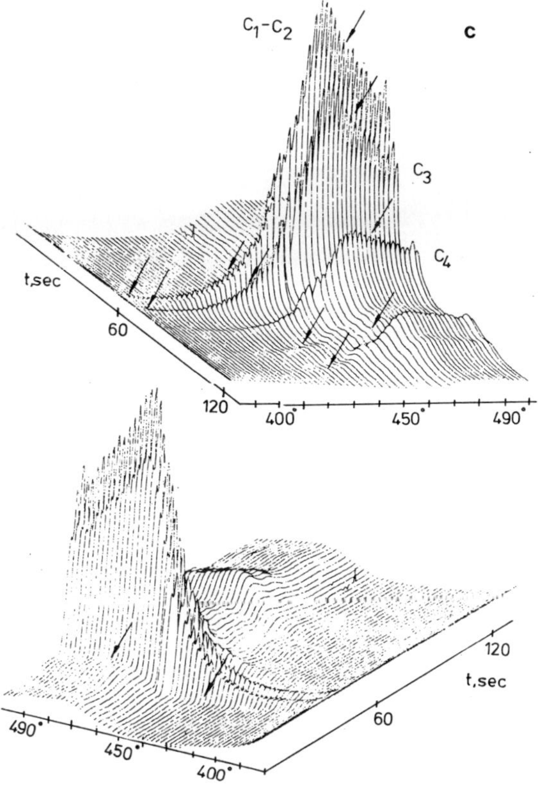

Figure 125   (a)   Stack plots of the butadiene styrene rubber
thermochromatograms. A-polybutadiene styrene copolymer;
B-poly-butadiene alpha-methyl styrene coolymer.
Conditions are the same as in Figure 124 (a).

Effects of ageing of polyethylene:
(b) stack plot of unaged cross-linked low density
polyethylene; (c) stack plot of cross linked low density
polyethylene after ageing 1 month at 100°C.
C_1, C_2, C_3, C_4 possible hydrocarbons, sample heating
rate 3°C min^{-1}, GC column SE 30, column temperature
60°C.

styrene copolymer which, as can be seen degrades in a more complicated way
than the butadiene styrene rubber.

The effect of polymer aging is reflected on thermochromatograms.
Figures 125(b) and (c) depict TGA stack plots of the same polyethylene,
while the only difference lies in the ageing of one polymer for a month at
100°C. The changes due to ageing are marked by the arrows. This points
to the conclusion that continuous heating of polyethylene makes it more
unstable. In figures 125(b) and (c) the advantage of computerized data
handling is seen. In both figures, the thermochromatograms are presented
from a different point of view. This technique allows one to discover the
features from the hidden areas and obtain more valuable information about
degradation reactions.

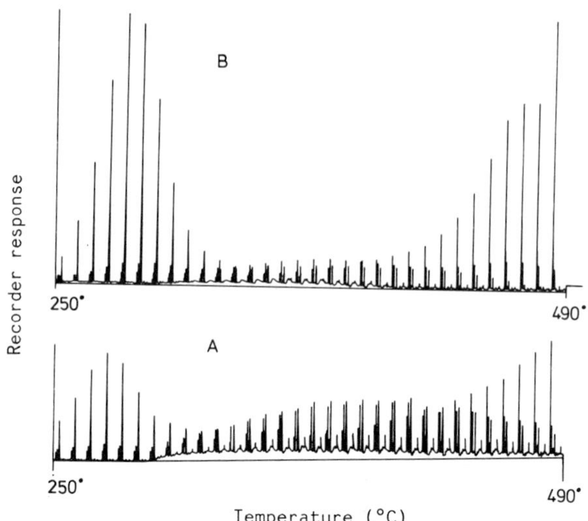

Figure 126   Sequence of slices of thermochromatograms for cellulose
pyridine phosphate; B - cellulose fiber containing
phosphoronitrilamine. Sample heating rate 7°/min; GC
column OV-43, column temperture 60°C.

Thermochromatography   is   useful   in   studying   flame-resistant
materials, e.g. cellulose fibres.   Phosphorous-containing antipyrenes
increase the amount of water present in the degradation products which, in
turn, increases flame resistance of a polymer.   The amount and composition
of   the   burning   products   should   also   be   taken   into   account   in
characterizing   the   flame   resistance   of   cellulose   fibres.   In   the
thermochromatograms shown in Figure 126; two fibres containing phosphorous
antipyrenes   are   presented.   One   sample   is   cellulose   fibre   cross-linked
with phosphonitrilamide (sample B); the second sample is a graft copolymer
of cellulose with polymethylvinylpyridine phosphate (sample A).   Both
samples   have   the   same   oxygen   index[1650]   (approx 40)   indicating   similar
flame resistance.   However as follows from Figure 126, the amount of the
evolved   light   burning   products   as   a   function   of   temperature   is   quite
different for these fibres.   Although identification of the degradation
products is complicated in this case, the components evolved are probably
light hydrocarbons and oxygen compounds because low gas chromatographic
column temperature (60°C) were used in separation.

Thermochromatography was used to compare the usefulness of different
catalysts   in   polymerization   processes.   Figure 127   shows   thermochromatograms of the degradation products of urea melamine formaldehyde resins
cured with equal amounts of $AlCl_3$, NH_4Cl, and $FeCl_3$ as catalysts.   The
thermal   destruction   of   the   above   polymers   gives   two   main   products:
formaldehyde and methanol.

A   combination   of   thermogravimetric   analysis   and   infrared
spectroscopy or Fourier transform infrared spectroscopy (FTIR) has also
been used to carry out evolved gas analysis in polymer and polymer
additive degradation studies.   The Stanton Redcroft apparatus launched in
1987 for combining a thermogravimetric apparatus with Fourier transform
infrared spectroscopy is described later under thermal analysis equipment.
This equipment combines the following features:

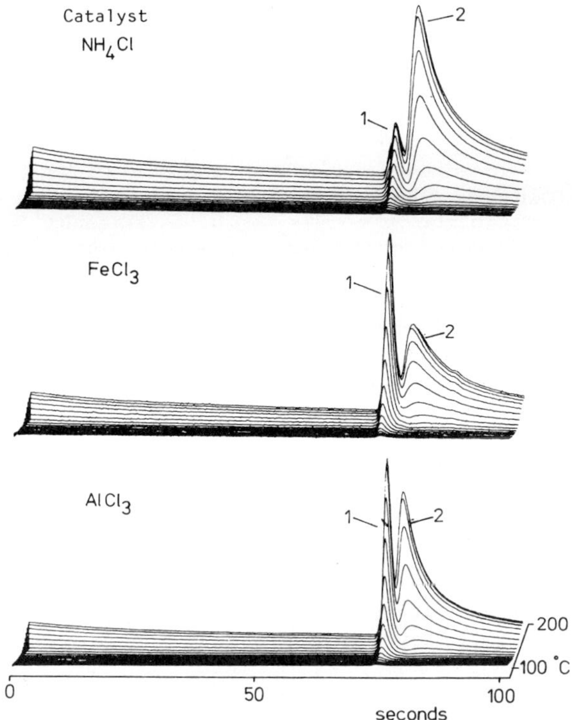

Figure 127   Thermochromatograms of the urea melamine formaldehyde
               resins. 1-formaldehyde; 2-methanol. Sample heating
               rate 5°/min; GC column SE-30, column temperature 130°C.

TG and FTIR data stored on one computer using the same time basis;

data storage sufficient for long data collections, for example over
4 hours with a 5 second resolution;

spectral resolution of at least 2 cm^{-1} to handle the gaseous
products often of low molecular weight;

high sensitivity for trace component detection and identification;

ability to plot gas evolution profiles.

This equipment has, to date, been used for thermal degradation
studies on PTFE and polyvinyl chloride and filler, monomer and plasticizer
analyses on polymers. It is worth noting that while quantitation from
FTIR results is feasible, there could be a discrepancy between the
calculated weight loss of evolved gases and that found by
thermogravimetric analysis. This discrepancy can be accounted for by the
evolution of a gas such as nitrogen which FTIR is unable to detect.

There is another approach to the use of thermal degradation. In
evolved gas analysis the production of organic volatiles is monitored by
continuously weighing the sample and identifying and determining the
volatiles produced. In the alternative approach the sample is heated

under controlled conditions in a sealed ampoule and the degradation
products in the residue are identified and determined. This approach has
been applied in studies of the thermal degradation of 0.5% of the phenolic
antioxidant Santonox R in low density polyethylene. Preliminary
thermogravimetric analysis on neat Santonox R indicated that it degrades
at temperatures as low as 160°C in air and degrades and/or volatilizes
faster at higher temperatures. In the absence of air degradation and
volatilization occur at 250°C and probably to some extent at lower
temperatures. To establish the fate of 0.5% Santonox R which had been
milled at temperatures between 160 and 250°C into low density polyethylene
the polymer was extracted with diethyl ether and examined by thin-layer
chromatography followed by infrared spectroscopy of the compounds
separated on the thin-layer plate. In this procedure 1 ml of a 1% diethyl
ether solution of the polyethylene extract was applied along the side of a
20 cm x 20 cm plate coated with a layer of GF 254 fluorescent silica gel.
The plate was developed with 9:1 petroleum ether:ethyl acetate and then
examined under a 254 nm ultra violet light source. This revealed the
presence of bands containing ultra violet absorbing substances at Rf 0.0,
0.15, 0.25, 0.30, (these four bands are also revealed by the 2:6 dibromo
phenoquinone-4-chlorimine detection reagent). The edge of the plate was
treated with 20% sulphuric acid and heated to 160°C and this revealed the
presence of two further bands at Rf 0.75 and 1.0.

The silica gel corresponding to each of the six observed bands
wasthen separately scraped off from the plate and reheated with absolute
ethyl alcohol to desorb the organic matter. The six extracts thus
obtained were then prepared into micro potassium bromide discs and
examined by infrared spectroscopy in the 2.5 to 15 micron region. The
results obtained in these tests are summarized in Table 163. These
results suggest that upon milling into polyethylene at 260°C Santonox R
degrades into at least two products, i.e. phenol (band 2) and non-phenolic
hydrocarbon, (band 4). The phenolic breakdown product (band 2) has some
resemblances to the material produced upon heating Santonox R in air for
30 minutes at 250°C.

Excess Ferric ions react with Santonox R but not with the oxidation
products to produce ferrous ions which can then be determined by the
spectrophotometric 2,2' dipyridyl spectrophotometric method, giving an
estimate of the unoxidized Santonox R content of the extract.

Santonox (reduced) + Fe^{3+} = Santonox (oxidised) + Fe^{2+}.

Obviously any oxidised Santonox in the polymer extract would not be
included in this estimation. Now it is likely that some or all of the
Santonox degradation products in polyethylene exist in the oxidised form
which would not be included in the result obtained by the dipyridyl
method. Therefore, lower analysis should be obtained for polymers in
which the Santonox is more degraded. The method was applied to these
polyethylenes in which 0.5% Santonox R had been incorporated by milling at
various temperatures. The results obtained (Table 164) show that
perceptible Santonox degradation occurred in the case of the polymer
milled at 260°C compared with polymers milled at lower temperatures.

Mass spectrometric methods. Mass spectrometry has been used quite
extensively as a means of obtaining accurate information regarding
breakdown products produced upon pyrolysis of polymers. This includes
applications to polystyrene[1651,1654], polyvinyl chloride[1653], polyethers[1653,1655,1656], PVC[1653], polycarboxypiperazine and polyureth-
enes[1657], phenolics[1658], PTFE[1658], polybenzimidazole[1659], epoxies[1660] and
ethylene-vinyl acetate copolymers[1661].

Table 163 – Development of band chromatogram of toluene extract of polyethylene milled at 250°C (0.5% Santanox R)

Band number	Rf value	254 mu ultra-violet absorption	2:6 dibromo p-benzoquin-one-4-chloro-mine detection reagent	20% sulphuric acid detect-ion reagent	Comments on infrared spectrum
1*	0.0 (start)	positive	positive	positive	Shows presence of hydrocarbon, i.e. (poly-ethylene wax) and Santonox R, (probably retained at origin due to incomplete elution by development solvent).
2	0.15	positive	positive	positive	Infrared spectrum similar but not identical to that of undegraded Santonox R, probably a Santonox degradation product, contains hydroxy group.
3	0.25	positive	positive	positive	Infrared spectrum almost identical to that of undegraded Santonox R.
4	0.30	positive	positive	positive	Shows strong hydrocarbon features, also absorption at 7.9, 8.4 and 9 microns which occur in undegraded Santonox, hydroxy groups absent.
5	0.75	negative	negative	positive	Santonox absent, shows some weak hydrocarbon features, hydroxy groups absent.
6*	1.0 (front)	negative	very slight	positive	Predominantly a hydrocarbon (polyethylene wax) (also found in toluene extract of Santonox-free polyethylene control), hydroxy groups absent

* only the Rf 0.0 and 1.0 bands show in the toluene extract of the Santonox free control polyethylene which has been milled at 260°C.

Table 164 - Influence of milling temperature on degradation of
Santonox R in polyethylene, 2 2'dipyridyl method

Milling temperature during incorporation of Santonox R °C	Santonox R %w* Nominal Santonox addition to polymer 0.5%
160	0.58, 0.55 (0.56)
200	0.58, 0.54 (0.56)
260	0.45, 0.43 (0.44)

* Zero Santonox contents obtained for Santonox R-free control polymers
which have been milled at the three temperatures.

Mass spectrometric thermal analysis, the determination of total ion
as a function of time and temperature in combination with differential
thermal analysis[1622], has proved to be very useful for distinguishing
between energy changes caused by phase transformations and those caused by
decomposition reactions. Determination of total ion current alone is not
particularly informative in most cases of polymer degradations, when
several products may be formed simultaneously or sequentially. Shulman et
al[1658] modified Gohlke's procedure by repeatedly scanning spectra as the
temperature is raised on a linear program, then plotting peak height as a
function of temperature. Mass spectrometric-thermal analysis
determinations conducted in this way proved helpful in several polymer
investigations. They present a preliminary discussion of this technique,
including suggested applications and new methods of data treatment which
permit determination of kinetic parameters from mass spectrometric-thermal
analysis.

For qualitative analysis of polymers, it is not necessary to know
the products of the reaction, since identification can be based on
temperatures and relative heights at the maximum of several of the more
prominent fragments once these have been established for known materials.
Additional information about the degradation chemistry can be secured if
one chooses peaks characteristic of specific products.

Since the Knudsen cell inlet system and the mass spectrometer are
being evacuated continuously, the pressure in the vicinity of the ionizing
filament is dependent on the rate of production of the effusing gases.
Peak height is therefore proportional to the rate of production of a given
species. For competitive first order reactions, the amount of each
product is proportional to its rate of production, so that the composition
of the product mixture at a particular temperature may be derived from a
single spectrum. Also, since peak height is proportional to the rate at
each temperature, a semilogarithmic graph of the peak height vs.
reciprocal absolute temperature (Arrhenius plot) will give a straight line
with a slope of E_a/R over a range in which the amount of polymer (or of a
functional group) does not change significantly. At heating rates of 15-
50°C/min, it is possible to obtain linear plots over a 50-200°C
temperature increment, the increment varying inversely as the activation
energy, E_a. In this way, activation energies for reactions leading to
each product can be obtained easily. A composite activation energy
comparable to that derived from thermogravimetric analysis or isothermal
kinetic studies can be obtained by taking a weighted average of the

individual activation energies. Furthermore, from areas under rate of formation vs. time graphs, the relative yield of each volatile product can be determined. Unfortunately, identification of products of polymer degradation from the mass spectrum of an unfractionated effluent cannot be considered reliable unless confirmation is secured using other analytical methods, such as gas chromatography.

Polymer samples were placed in a tungsten crucible in the Knudsen cell inlet system of the Time-of-Flight mass spectrometer. The system was evacuated, then samples were heated at a linear rate. Spectra were determined at 1 min. intervals. The thermocouple was welded to a support rod 3 mm below the crucible and had been calibrated against a thermocouple in the bottom of the crucible at each heating rate. Details are summarized in Table 165 and 166. Data is presented in Figures 128(a) to (c) for three polymers.

An Arrhenius plot for the polybenzimidazole shows that the average activation energy is temperature dependent, varying from 26-40 kcal/mole, compared to a value of 44 ± 11 kcal/mole derived from isothermal kinetics.

Table 165 - Experimental details of mass spectrometric thermal analysis

Polymer	Form	Weight, mg	Heating rate °C/min
Phenolic[b]	Powder	14.9	29
Polytetrafluorethylene[c,d]	Chips	12.7	14.5
Polybenzimidazole[e]	Powder	15.5	20

[b] Derived data is presented in ref. 1663.

[c] Du Pont Teflon.

[d] E_a = 93 kcal/mole for C_2F_4 formation.

[e] Poly-2,2'-(m-phenylene)-5,5'-bibenzimadazole (cured under 2000 psi of nitrogen to 400°C).

Similar data for the phenolic resin[1663] gives a range of 16-48 kcal/mole, compared to reported values ranging from 16-55 kcal/mole.

Risby et al.[1653] have described the design of a temperature programmable probe and computer controller and present mechanisms of thermal desorption decomposition energies and preexponential factors in the linear programmed thermal degradation mass spectrometry (time resolved pyrolysis) of polystyrene and polyvinyl chloride. Their results show a stepwise thermal degradation process. Linear programmed thermal degradation - mass spectrometry is based on a collection of sequential mass spectra during the programmed heating of the sample. The data, with temporal dependence, allow detection of subtle differences in structure which would not be apparent with data from pyrolysis - gas chromatography or pyrolysis - mass spectrometry. In addition, activation energies and preexponential factors for the decomposition processes can be obtained from linear programmed and thermal degradation - mass spectrometry by

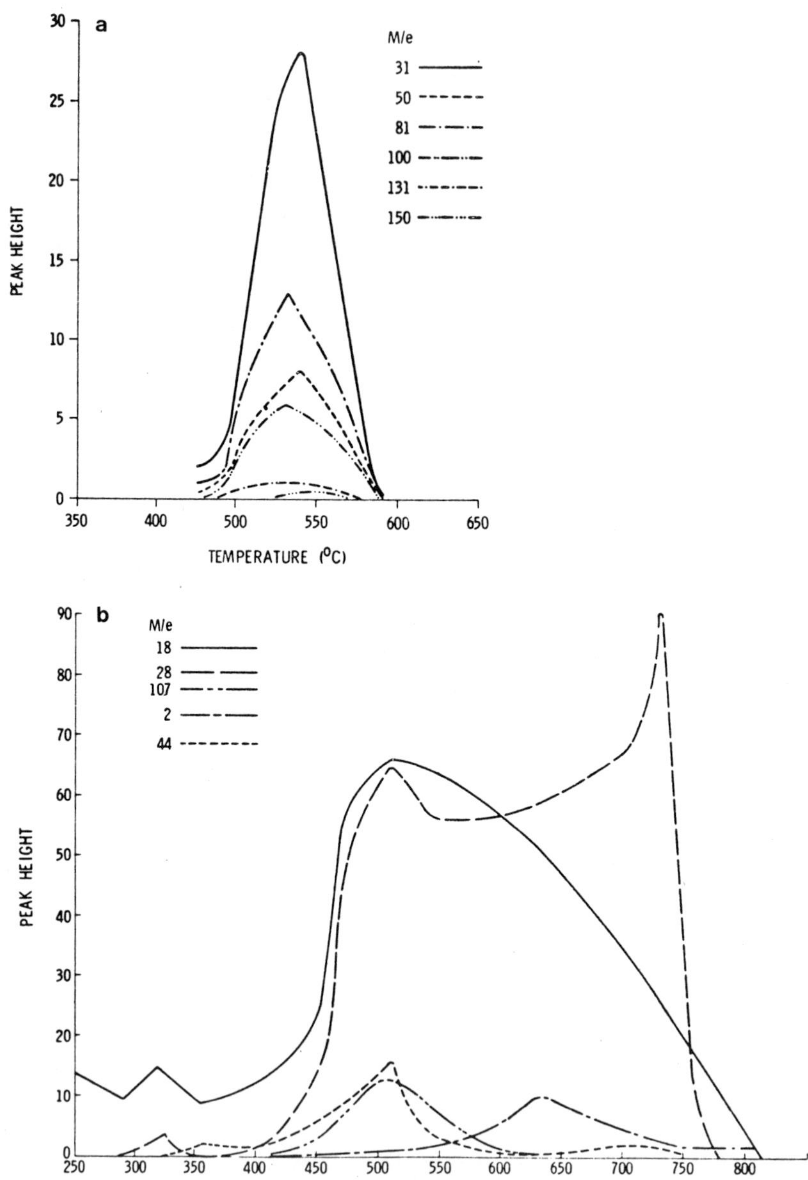

Figure 128   Mass spectrometric thermal analysis of 9a)
               polytetrafluoroethylene; (b) phenol-formaldehyde resin;
               (c) poly-2,2'(mephenylene-5,5'bibenzimidazole).

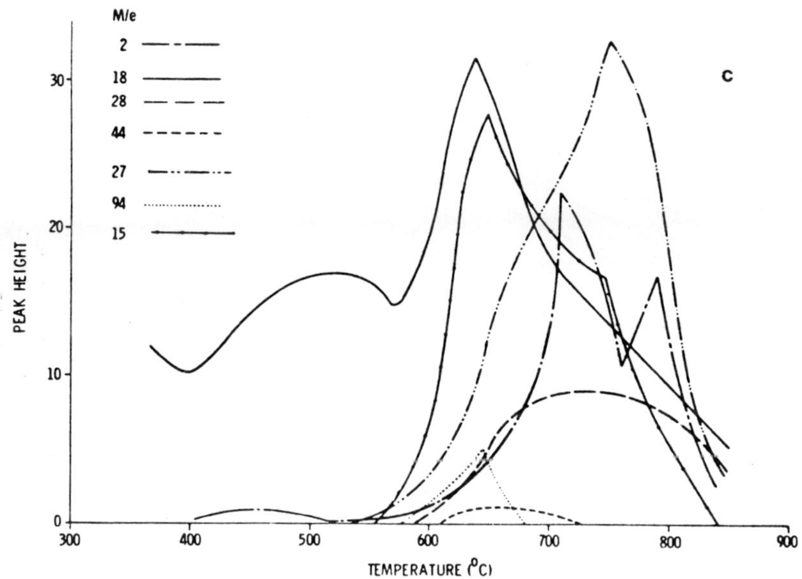

Table 166 - Products of Poly-2,2'-(m-phenylene)-5,5'
-Bibenzimidazole Pyrolysis

Product	Yield % of volatiles	Activation energy Kcal/mole
H_2	13	38
CH_4	1.5	-
NH_3	13	54
H_2O	38	22
HCN	30	48
CO	4	31
CO_2	0.3	-
C_6H_5OH	0.9	-

changing the rate of sample heating; such results cannot be obtained by pyrolysis - gas chromatography or pyrolysis - mass spectrometry.

The major advance made by Risby et al[653] is the redesign of the temperature programmable probe and controller to allow accurate measurements of decomposition temperature.

In this procedure aliquots (0.5 ul) of a solution of polystyrene (1 mg/mL) in chloroform or of poly(vinyl chloride) (1 mg/mL) in tetrahydrofuran were placed on the platinum filament by using a

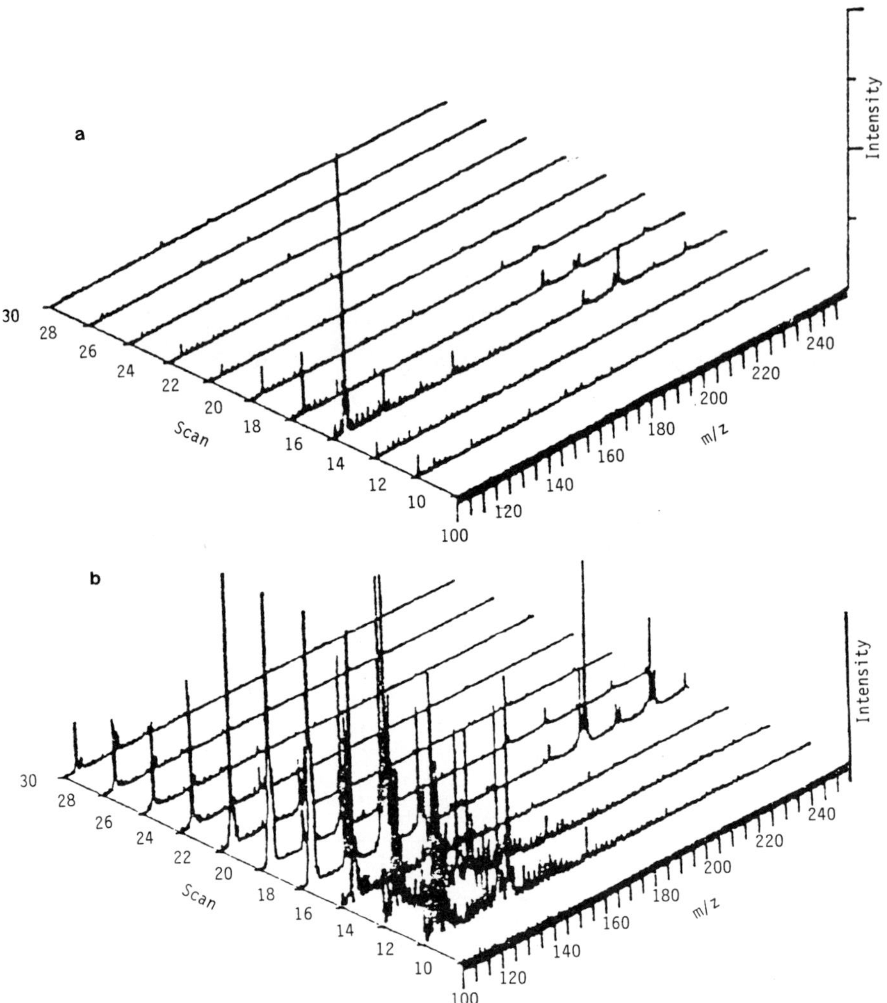

Figure 129   (a) Three-dimensional plot for thermal degradation at
16°C/s of uncross linked polystyrene (M,100,000) shown
as m/z vs. Intensity vs. scan number (10=73°C, 30 =
563°C). (b) Three-dimensional plot for thermal
degradation at 15.5°C/s of polystyrene - 12%
divinylbenzene, show as m/z v́s. Intensity vs. scan
number (10 = 97°C, 30 = 564°C). (c) Specific non
profiles for thermal degradation at 16.0°C/s of un-
cross-linked polystyrene (M,100,000) shown as intensity
vs. scan number: (A) mass 105 m/z, temperature of
evolution maximum 423°C: (B) mass 117 m/z, temperature
of evolution maximum 423°C: (C) mass 209 m/z,
temperature of evolution maximum 423°C: (D) mass 221
m/z, temperature of evolution maxima 423°C.
(d) Specific nonprofiles for thermal degradation
at 15.5°C/s of polystyrene - 12% divinilbenzene
shown as vs scan number: (A) mass 105 m/z,teperature of
evolution maximum 423°C; (B) mass 117 m/z,
temperature of evolution maximum 423°C; (C) mass
209 m/z, temperature of evolution maximum 423°C;
(D) mass 221 m/z, temperature of evolution maximum
423°C.

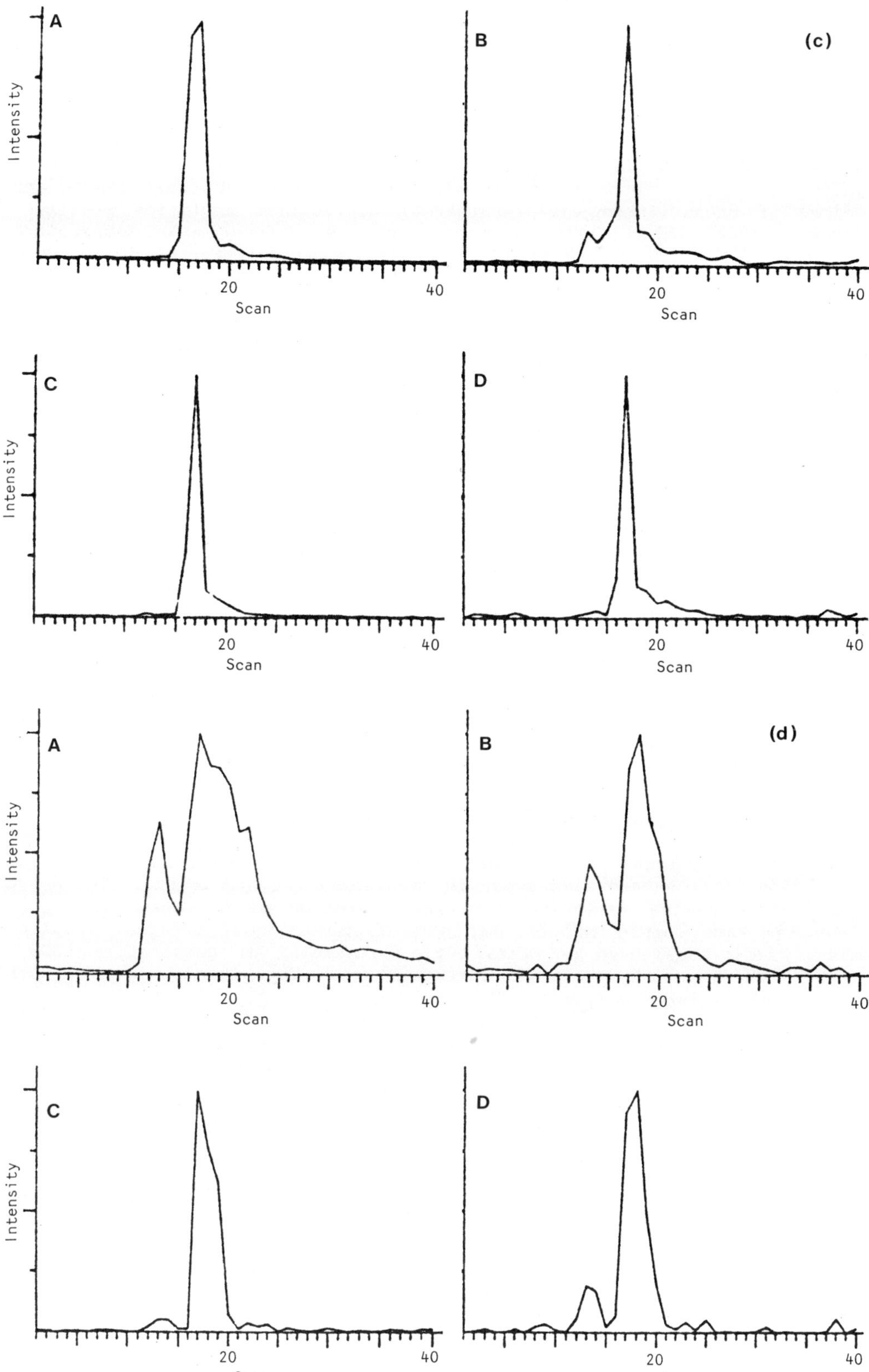

microsyringe. Most of the solvent was allowed to evaporate under ambient conditions and any residual solvent was removed by the vacuum prior to insertion into the source of the chemical ionization mass spectrometer. A reproducible coating of about 500 ng of the sample on the filament was obtained; the small sample size ensures that there is intimate contact between the sample and the platinum filament. Risby et al[1653] selected polystyrene for investigation of the effects of polymer chain length and degree of cross-linking on the thermal degradation evolution profiles. Figures 129(a) and (b) show the mass spectra as a function of temperature for the thermal degradations of un-cross-linked polystyrene, M_r100,000, and 12% divinyl benzene cross-linked polystyrene. The difference between the samples is clearly shown, as the heavily cross-linked polystyrene starts to degrade at a lower temperature and the degradation occurs over a wider temperature range. These differences appear even more dramatic in Figures 129(c) and (d) which are the profiles for specific ions (105,117,209, and 221 m/z) as a function of scan number (sample temperatures). In Figure 129(c) more than one kinetic process for the evolution of the monomer fragment (105 m/z) is indicated since the profile is clearly broader than the profiles for the specific ions at 117,209, and 221 m/z. These differences in evolution profiles may perhaps be explained by different types of microstructures within the un-cross-linked polystyrene which can lose a styrene fragment; the activation energies for the removal of the monomer from these microstructures would be expected to differ. This hypothesis of different microstructures within the polystyrene is supported by further evidence. The evolution profiles for larger fragments are sharper, suggesting less variety in kinetic processes of evolution, i.e., less variance in microstructure. In Figure 129(d) (cross-linked polystyrene) profiles for all the fragments are broad suggesting that each ion may be produced by a variety of kinetic processes. Cross-linking would be expected to produce different types of microstructures. (However, differences could be explained by the fragments having arisen from either styrene or or divinylbenzene.) Polystyrene with a lower degree of cross-linking has a more random structure and would be expected to have many different microstructures. this is indeed supported by the complex profiles shown in Figure 130.

The temperatures (°C) which correspond to the maxima of the evolution profiles are also shown in Figures 129(c) and 129(d). There are temperature maxima variations for the evolution of different fragments from the same sample and for the same fragment from different samples. Also, the cross-linked polymers begin pyrolysing at temperatures 200°C lower than the maximum evolution temperature. This is not observed with the un-cross-linked polymers.

Since it was impossible to separate and identify the kinetic processes corresponding to the loss of specific fragments from the sample of cross-linked polystyrene, and difficult to ensure that the same chemical species were involved Risby et al[1657] did not attempt to obtain values for the activation energies by varying rate of sample heating. In addition, evolution profiles for specific fragments from un-cross-linked polystyrene samples were complex, and values for the activation energies given in Table 167 are based on the 235 m/z. There is good agreement between the values for the activation energies and the preexponential factors for the two un-cross-lined polystyrenes as would be expected since they differ only in molecular weight. Also, there is reasonable agreement of the value for activation energy with a literature value of 48.04 kcal obtained with a much larger sample size.

The major fragment ions observed for the polystyrene samples arc listed in Table 168. These fragments have been observed by other workers

Table 167 - Temperature Data, Activation Energies, and
Pre-exponential Factors for Uncross-linked Polystyrenes

M_r 390 000		M_r 100 000	
Ramp rate, °C/s	Temp of maxima, °C	Ramp rate, °C/s	Temp of maxima, °C
15.9	390	16.0	388
7.79	385	7.77	386
3.92	363	3.87	365
1.97	352	1.99	349
0.99	345	0.99	343
activation energy 42 kcal		40 kcal	
Pre-exponential factor 3.6×10^{13} s^{-1}		6.1×10^{12} s^{-1}	

using pyrolysis gas chromatography and/or mass spectrometry[1664-1678] and using chemical ionization mass spectrometry[1669].

Udseth and Friedman[1679] carried out a mass spectrometric study of polystyrene with a molecular weight of 2100. The polystyrene was evaporated from a probe filament heated at 1000°C sec^{-1} under both electron impact and methane and argon chemical ionization conditions. Smooth rhenium ribbon surface direct insertion probes were used with both the electron impact and chemical ionization sources. Under electron impact conditions extensive fragmentation and depolymerization were observed but oligomers up to $(C_8H_8)_{11}$ were detected (Figure 131(a)). Under chemical ionization conditions (Figure 131(b) and (c)) oligomers up to $(C_8H_8)_{27}$ were detected and spectra that approximately reproduced the oligomers distribution were obtained. These workers also measured quantitatively temperature dependencies of rates of desorption and activation energies of evaporation rates for many of the ionic species present.

Evaporation studies were carried out in both a chemical ionization and an electron impact ion source and in each case the products were extracted and analysed by a computer-controlled quadrupole mass spectrometer. A DEC PDP 8/E computer was used to control the Extranuclear Laboratories Inc. quadrupole power supplies which operated an Extranuclear Laboratories Inc. 3/8 inch diameter quadrupole rod mass analyser assembly[1680]. Studies below 1200 amu were made with an Extranuclear Laboratories Inc. power supply operating at a frequency of approximately 1.7 MHz. A special low-frequency (292 KHz) supply capable of mass analysis up to 70,000 amu was used for the mass range of 1000-4000 amu.

Samples were evaporated from a rhenium ribbon filament that was heated at a rate of approximately 1000°C/s. The spectra were taken at

Table 168 - Polystyrene Pyrolysis Products

Basic structure of fragment	Mass of ion with max intens in the region of mol wt	% total ion current at max evolution[a]
$CH=CH_2$ | Ph	104	4.0

$CH_2=CHCH$
   |
   Ph
	117	4.7

$CH_2=CHCHCH$
   |
   Ph
	130	1.8

$CH_2=CHCH_2CH=CH$
   |
   Ph
	144	2.1

$CHCH_2CH$ | | Ph Ph	194	1.9

CH_2CHCH_2CH
   | |
   Ph   Ph
	208	9.7

$CH_2CHCH_2CHCH_2$
   | |
   Ph   Ph
	222	2.4

$CH=CHCH_2CH=CHCH_2$
   | |
   Ph   Ph
	234	6.0

[a] Uncross-linked, M_r 390,000.

scanning rates of from 2 to 11 amu/ms and scanned in steps ranging in width from 0.28 to 10 amu. The chemical ionization spectra were taken with argon and methane as reagent gases at a pressure of 0.1 torr. Temperature measurements of rates of sample evaporations were made by using the resistance of the rhenium as a thermometer. The probe filament was heated with a constant current rapid heating pulser. The voltage drop across the filament was measured and recorded in the computer during the course of the evaporation with a 5-MHz voltage to frequency converter. Accurate values for the change in resistance of the filament and correspondingly the change in temperature could thus be recorded for each 1.2-ms time period of the sample evaporation. The precision of sample probe temperature measurement is estimated at $\pm$ 3K. The sample probe surface is not uniformly heated with a considerable temperature gradient from the centre to ends supported on heavy wire leads. Consequently, the absolute temperature of sample desorption was uncertain by a value larger than the precision of measurement. This uncertainty is estimated to be less than 25 K. Measurements of slopes of plots of log relative ion intensity vs. reciprocal absolute temperature depend on differences in temperature rather than absolute measurements and are not sensitive to the larger estimated systematic error in temperature that may be encountered in experiments. Measurement of the temperature of desorption of trytophane samples from a probe equipped with a thermocouple using single heating techniques and from a filament probe rapidly heated indicated an internal consistency of temperature measurement to better than $\pm$ 5K.

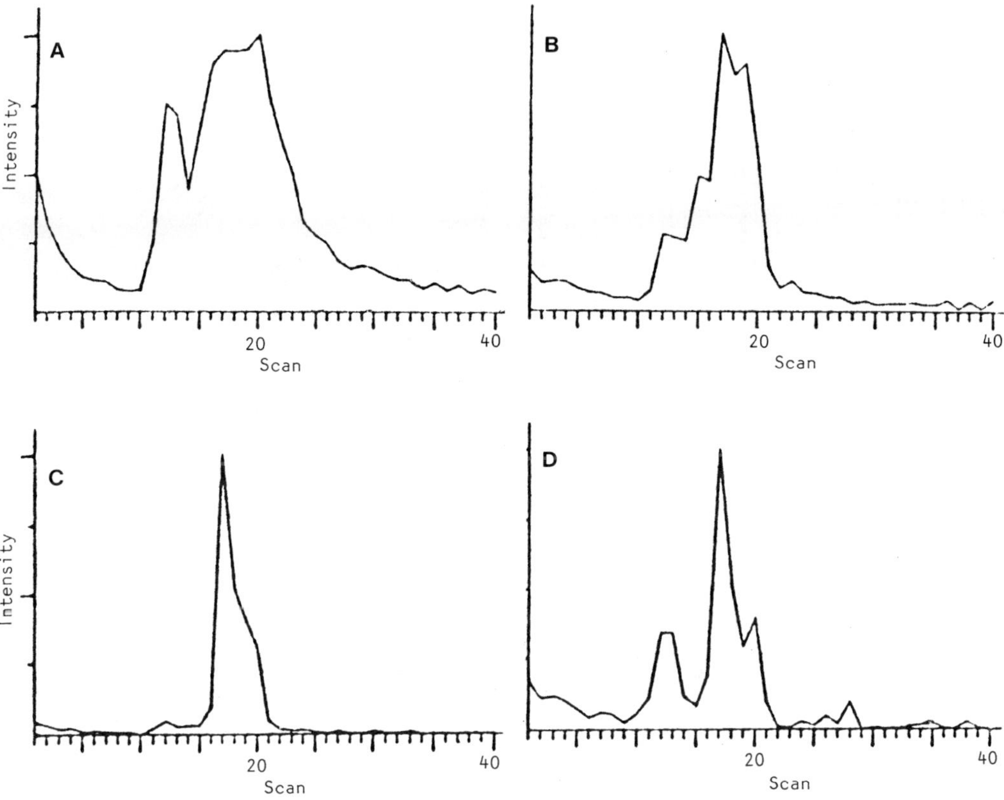

Figure 130   Specific ion profiles for thermal degradation at
              15.6°C/s of polystyrene - 1% divinylbenzene, shown as
              intensity vs. scan number: (A) mass 105 m/z,
              temperatures of evolution maxima, 129, 448, 551°C; (B)
              mass 117 m/z temperatures of evolution maxima 129, 395,
              503°C; (C) mass 209 m/z, temperature of evolution
              maximum 393°C; (D) mass 221 m/z temperatures of
              evolution maxima 175, 393°C.

Corrections were made for the transmission of the quadrupole mass
analyzer as a function of resolution by use of the equaton[1681,1682].

$$\ln (T_{r2}/T_{r1}) - a(R_1 - R_2)$$

where R_1 and R_2 are values of the resolution and T_{r1} and T_{r2} are the
respective values of ion transmission at these resolutions. The constant
a had a value of -0.0174[1683]. The resolution at each peak in the mass
spectrum was measured, and the resolution as a function of mass was used
to obtain an approximate correction of relative ion intensities.

The mass scale in the lower region of the spectra was calibrated by
use of perfluorotributylamine as an internal standard. Higher masses were
identified by extrapolation of the low mass calibration and by the
assumption in the higher mass region, that in the very clean oligomer
spectrum produced by argon chemical ionization, that the maximum and

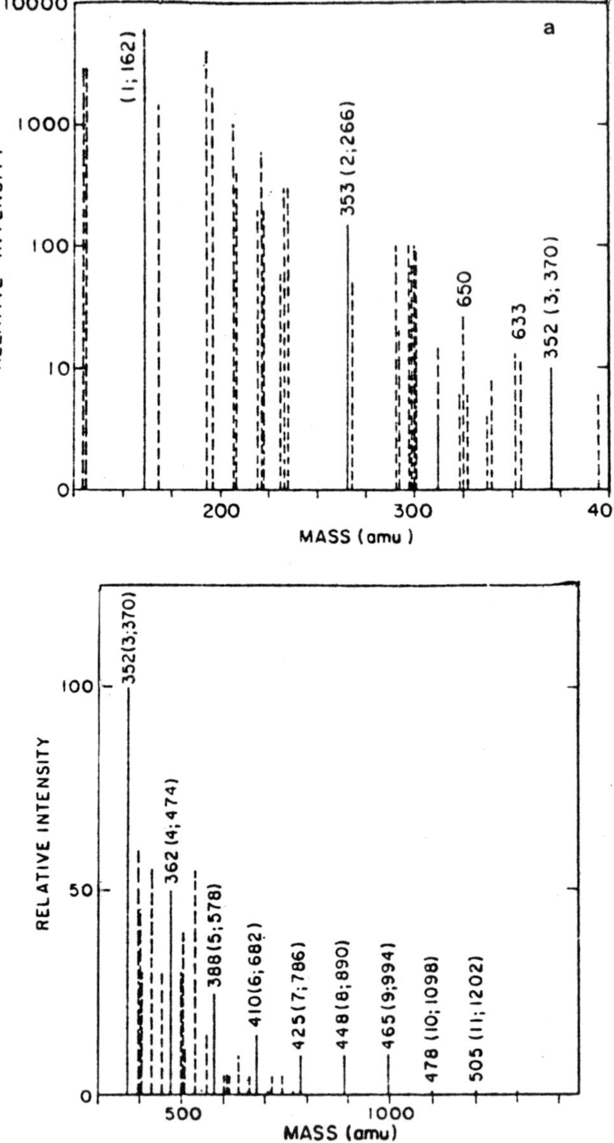

Figure 131   Electron ionization and chemical ionization spectra of polystyrene: (a) partial EI spectrum of polystyrene. The solid lines are the inner spectrum and the numbers in parentheses give first the number of monomer units present and second the assigned monoisotopic mass. The numbers not in parentheses give the evaporation tmeperature in kelvin. The dashed lines are fragment sequence peaks. (b) Partial argon CI spectrum of polystyrene. The solid lines are the monomer spectrum and the dashed lines are intermediate fragment peaks. The format for the numbers is the same as that used in Figure 131 (a). (c) Partial methane CI spectrum of polystyrene. The solid lines are the monomer sequence and the dashed lines are solvation or fragment peaks. The format for the numbers is the same as that used in Figure 131 (a).

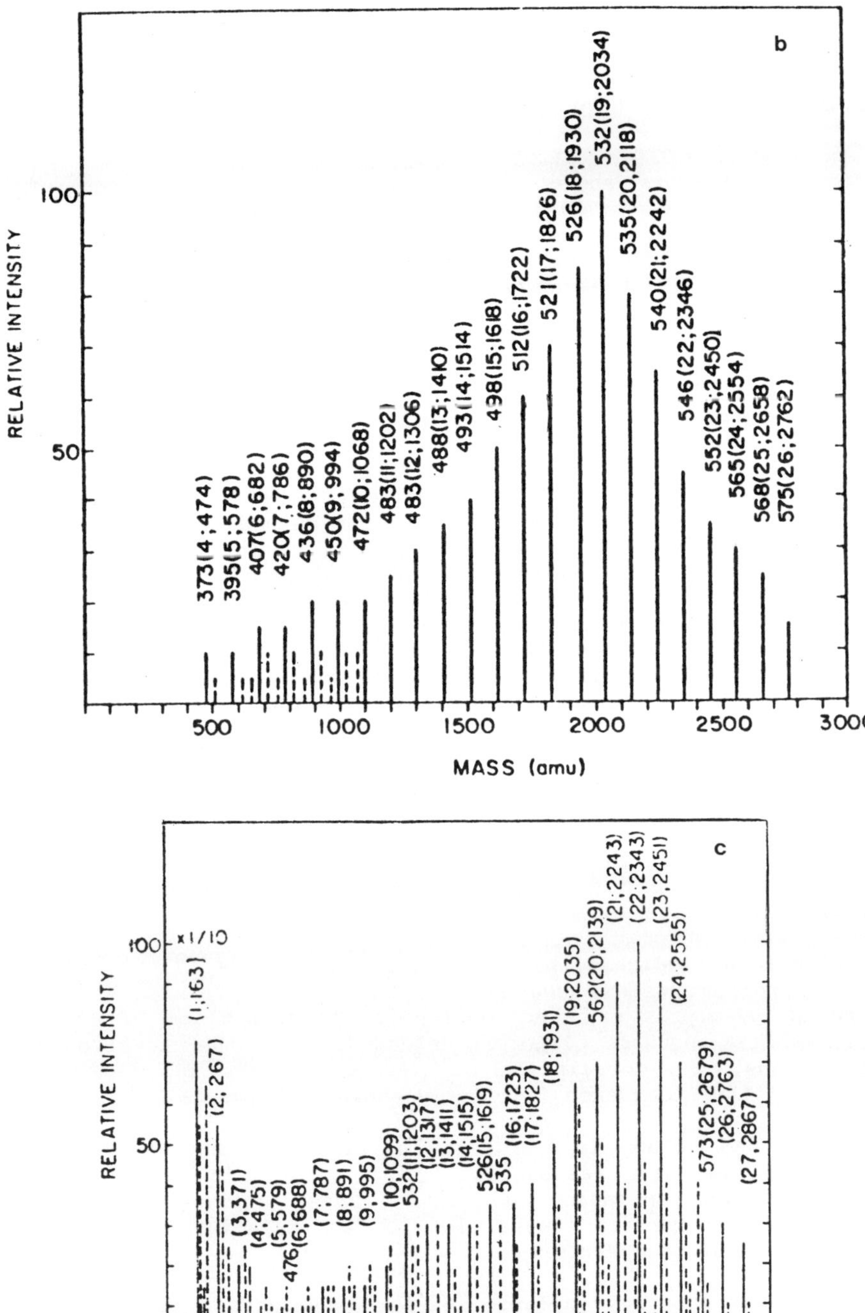
b
RELATIVE INTENSITY
100
50
373(4;474)
395(5;578)
407(6;682)
420(7;786)
436(8;890)
450(9;994)
472(10;1068)
483(11;1202)
483(12;1306)
488(13;1410)
493(14;1514)
498(15;1618)
512(16;1722)
521(17;1826)
526(18;1930)
532(19;2034)
535(20,2118)
540(21;2242)
546(22;2346)
552(23;2450)
565(24;2554)
568(25;2658)
575(26;2762)
500
1000
1500
2000
2500
3000
MASS (amu)
c
RELATIVE INTENSITY
100 ×1/10
50
(1;163)
(2;267)
(3,371)
(4;475)
(5;579)
476(6;688)
(7;787)
(8;891)
(9;995)
(10;1099)
532(11;1203)
(12;1317)
(13;1411)
(14;1515)
526(15;1619)
535 (16;1723)
(17;1827)
(18;1931)
(19,2035)
562(20,2139)
(21;2243)
(22;2343)
(23,2451)
(24,2555)
573(25;2679)
(26;2763)
(27;2867)
500
1000
1500
2000
2500
3000
MASS (amu)

minimum masses were separated by a mass difference of 104n where n - 1 was the number of equally spaced peaks between the peaks identified as min or max. Absolute masses were derived by assigning the mass 104n + 58 to a peak where n could be unambiguously determined by independent calibration. This calibration of the mass scale facilitated interpretation of the methane chemical ionization spectra with attendant solvated ion peaks.

In Figure 131(a) electron impact mass spectra are presented graphically. Numbers above selected ions indicate the temperature at which counting rates of 500 counts/ms were observed. The numbers inside the parentheses give the number of the styrene monomer units and the respective masses of the ion. Temperatures of desorption were arbitrarily defined in terms of 500 counts/ms which corresponded to approximately two-thirds of the maximum intensity of the relatively low abundance ions in Figure 131 (a) curve II. The electron-impact data in Figure 131(a) are divided into two molecular weight regions which overlap at the trimer ion at m/e = 370. The solid lines in the figure are used to indicate ions which consist of styrene monomer units and the butyl initiator, with masses given by the relation m = 104n +58. These ions are designated as the A oligomer sequence. Ions with values of n ranging from 1 to 11 are found in the electron-impact mass spectrum. The A sequence constitutes the only ions observed above m/e = 780. The dashed lines represent ions, with masses that do not fit the A sequence, produced by thermal surface reactions and/or electron-impact decompositions. Sequential mass distributions can also be found in the dashed line spectra. For example, there is a set of ions with masses, 291,395,499,603, and 707, which differ by the mass of a styrene monomer unit.

Temperatures of desorption, shown in Table 169 show a gradual increase for A oligomer ions with increasing molecular weight. The observation of lower molecular weight ions at lower temperatures and during earlier stages in the desorption is inconsistent with the hypothesis that these ions are electron-impact fragmentation products. Temperatures of desorption establish the lower molecular weight ions as primarily products of thermal depolymerization reactions. Low molecular weight fragments ions not part of the A sequence with masses at 325 and 351, respectively, are included in Table 169.

These ions, indicated by dashed lines in Figure 131(a), appear late in the desorption process and at high enough temperatures to be either products of pyrolysis or electron-impact decompositions of the highest molecular with A sequence oligomers detected. With the exception of these fragments ions, the correlation of increasing mass and temperature of desorption of the A sequence is clearly shown in Table 169.

The second column in Table 169 gives activation energies for the rate limiting processes responsible for the generation of the respective ions in the electron-impact spectrum. These activation energies are calculated from slopes of linear plots of log relative ion intensity vs. reciprocal absolute temperature.

Activation energies in Table 169 for A sequence oligomers are higher for the n = 2 and n = 3 oligomers, relatively constant for n = 4 through 8, and lowest for n = 9-11. If the respective rate limiting processes were desorption in this homologous series of oligomers, a gradual increase in activation energy with molecular weight would be expected. If oligomer ions were formed exclusively by electron-impact decomposition of the higher molecular weight polymers, then both activation energies and

Table 169 - Evaporation temperatures and activation energies for
some of the peaks in the EI Spectrum

Mass, amu	E_a, Kcal/mol[a]	temp, K
266	51	353
325	17	650
351	17	633
370	41	352
474	36	362
578	27	388
682	26	410
786	27	425
890	29	448
994	21	465
1098	18	478
1202	18	505

[a] Estimated uncertainty $\pm$ 3 kcal/mol.

temperatures of desorption of the respective ions would be expected to be
almost identical. The dimer and trimer ions have virtually the same
temperatures of desorption, but the 10 kcal/mol difference in activation
energy supports the conclusion that these ions are formed in surface
thermal depolymerization reactions at different rates. Differences in
temperature of desorption for oligomer ions with n = 3-8 support the
argument that, in spite of similar activation energies, these oligomers
are not for the most part produced from the same parent species by
electron-impact dissociation.

Sample plots of log relative intensity vs. 1/T are given in Figure
132. A complex mechanism for the production of neutral species which are
parents of the trimer ions is shown by the segmented plot of the
temperature dependence of the relative intensity of the trimer ion. There
is, at lower temperatures, the slope that gives a 41 kcal/mol activation
energy and a higher temperature process that gives an activation energy
lower than 27 kcal/mol or lower than that of the pentamer ion.

Chemical-ionization mass spectra taken with methane and argon
reagent gases are presented in Figures 131(b) and (c), respectively.
these spectra were taken to determine the effect of contact of a weakly
ionized plasma with the heated sample surface on the competitive
depolymerization and desorption processes. The possibility of chemical
interaction of neutral reagent gases with the sample is, in the case of
methane, remote. With argon, neutral molecule-surface chemical
interactions can be eliminated from consideration. The methane chemical
ionization spectrum was scanned from the mass of the monomer in the A
sequence of ions up to mass 3000. With argon the lower molecular weight
region was not investigated. The lowest oligomer shown in the argon
chemical oxinization spectrum is the tetramer, m/e = 473. The chemical
ionization spectra were taken with relative low resolution, $\pm$4 mass unit,
in the mass range between 500 and 3000 amu. Resolution limitations are
reflected by presentation of data as monoisotopic mass spectra. For ions
containing in the order of 100 or more carbon atoms with approximately
1.1% ^{13}C, ions having one or more ^{13}C atoms would be the most abundant

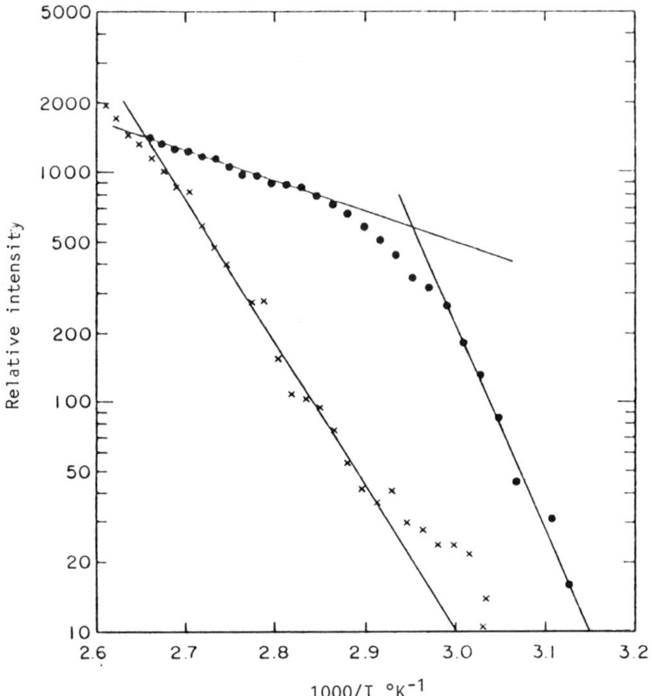

Figure 132   Evaporation plots from rapidly heated polystrene
                 obtained under E1 conditions. X are a plot of an
                 evaporation of the 5 mer and ● are a plot of an
                 evaporation of the 3 mer.

species in a polyisotopic spectrum. Higher molecular weight ions were not
sufficiently well characterized, with respect to mass, to establish the A
sequence ions in the methane chemical ionization as protonated species.
Nor were the solvated A sequence ions, indicated by the dashed lines in
the higher molecular weight region of the methane chemical ionization
spectrum identified precisely. The addition of a two or three carbon,
hydrocarbon, fragment was established. The graphical data in Figure
131(b) and (c) accurately reflect relative intensities of oligomer and
fragment ion peaks but are schematic in that they are presented as
monoisotopic masses.

     The methane and argon chemical ionization spectra both show higher
molecular weight species not seen under similar conditions of sample
heating with electron-impact source. Some small differences may be noted
but there is no conclusive evidence for volatility enhancement in which
the argon chemical ionization plasma in contact with the sample gives rise
to lower temperatures of desorption. Temperatures of desorption for
selected ions in the methane chemical ionization spectrum are slightly
higher than those found for electron impact or argon chemical ionization.
In both chemical ionization spectra the correlation of increasing
temperature of desorption with increasing ion molecular weight for the A
sequence oligomer ions is maintained. Here again the lower molecular
weight ions appear earlier in the sample process and cannot be taken as
products of exothermic gas-phase ionic decomposition reactions. Such

reactions might be expected with argon because of the relatively large difference between the recombination energy of argon ions and estimated ionization potentials of styrene polymers)approximately 4 or 5 eV).

In view of this potential exothermicity, the argon chemical ionization spectrum is unusually clean and free of fragmentation products. These results may be explained by a very efficient collisional deexcitation of oligomer ions by neutral argon in the chemical ionization source or by deexcitation of products of heterogeneous ion-molecule reactions on the surface of the sample probe. The only evidence in the mass spectrum for a somewhat more gentle ionization with methane than argon is the shift to lower molecular weight in the maximum of the argon spectrum. With argon chemical ionization, the n = 19 oligomer is the most abundant high molecular weight ion. In the methane chemical ionization spectrum the most abundant ion in this mass region has 22 styrene monomer units.

In contrast to the values found in the electron-impact mass spectrum, activation energies for the chemical ionization spectra tend to increase with increasing molecular weights. Unfortunately comparisons are limited because many of the ions detected in the chemical ionization spectra are not seen under electron-impact conditions. With argon chemical ionization one can compare the nmers with n = 4 through 8 and see a systematic 6-9 kcal increase in activation energy with argon chemical ionization. However, with higher molecular weight oligomers n = 9-11 the argon activation energies are 16-32 kcal/mol larger than those observed with the same mass ions under electron-impact ionization conditions. Temperatures of desorption and activation energies are presented for the chemical ionization experiments in Table 170 and 171.

Two sets of activation energies are given for the argon chemical ionization studies, because with higher molecular weight ions, plots of log intensity vs. 1/T gave segmented lines which were resolvable into low and high temperature processes. This phenomenon was not observed in the methane chemical ionization studies but was reproducible in the argon experiments. Comparison of the data on selected A sequence oligomers shows differences in activation energy for rate-limiting processes leading to desorption in the argon and methane chemical ionization spectra. For example, for the n = 15 ion, the respective activation energies are 62 kcal/mol with argon chemical ionization and 44 kcal/mol with methane chemical ionization. No difference was found for the activation energy of the n = 20 oligomers in the methane spectra and during the high temperature part of the argon spectra. An 18 kcal/mol difference is noted for the n = 11 oligomer ion at mass 1202 in the argon spectrum and the assigned mass of 1203 in the methane chemical ionization spectrum.

These data all indicate a significant interaction of reagent ions with molecules on the surface of the sample probe during the sample heating process. There is evidence in the lower molecular weight portion of the methane chemical ionization spectrum of production of relatively large yields of lower molecular weight depolymerization products. The major differences in spectra, temperatures of desorption, and activation energies found in the electron impact and chemical ionization studies suggest that sample exhaustion via low-temperature depolymerization reactions was in part inhibited by interaction of gaseous ions with species on the surface of the sample probe. This inhibition could have been an ionic catalysis of recombination or polymerization reactions that reduce the concentration of lower molecular weight oligomers which desorb at lower temperatures.

Table 170 - Evaporation temperatures and activation energies for
some of the peaks in the argon CI spectrum

| | E_a, kcal/mol[a] | | |
mass, amu	initial	final	temp, K
474		32	373
578		32	395
682		34	407
786		36	420
890		35	436
994		37	450
1098		45	472
1202		50	483
1306		49	483
1410		49	488
1514		62	493
1618		62	498
1722	48	49	512
1826	55	47	521
1930	41	44	526
2034	42	63	532
2138	40	52	535
2242	42	36	540
2346	56	38	546
2450	45	43	552
2554	52	38	565
2658	65	28	568
2762	64	35	575
2866	64	37	577

[a] Estimated uncertainty $\pm$ 3 kcal/mol through mass 1618 and $\pm$ 6
kcal/mol for higher masses.

Table 171 - Evaporation temperatures and activation energies for
some of the peaks in the Methane CI Spectrum

mass, amu	activation energy E_a, kcal[a]	evaporation temp, K
579	22	446
607	24	476
633	20	476
1203	31	532
1224	32	521
1253	33	532
1619	44	526
1665	42	535
2139	51	562
2659	51	573

[a] Estimated uncertainty $\pm$ 3 kcal/mol.

The lower activation energies in the electron impact spectra, particularly for nmers 9-11, may be the result of the failure of the basic assumption of zero-order kinetics for their desorption. These oligomers are detected at temperatures well above those required for desorption of 99% of the sample and are produced in the desorption of the last few layers of sample. Under these circumstances sample depletion can play an important role in giving activation energies that are too low. Correction for this type of error is difficult. Suffice it to say that activation energies for ions produced after most of the sample has been desorbed are probably lower limits of values of activation energies of desorption and are useful only to establish the presence or absence of genetic relationships between these ions and lower molecular weight species.

The result of the interaction of the gaseous ions in the chemical ionization source with the solid sample surface is to facilitate observation of spectra which can be well correlated with the structure of the solid polymer. If one assumes that the lower molecular weight oligomers of the A sequence, with values of n = 1-3, are primarily thermal depolymerization products, then the remaining mass spectrum can be used to calculate average molecular weights of the polymer on the probe surface. Number and weigh average molecular weights are calculated from the methane chemical ionization spectrum to be 1987 and 2143, respectively, on the basis of the latter assumption. Values of M_n and M_w calculated from the argon chemical ionization spectrum are 1800 and 1970, respectively. The ratios of M_w/M_n in the argon and methane chemical ionization spectra are 1.10 and 1.08, respectively in good agreement with the nominal value for this sample of polystyrene (2100).

Risby et al.[653] used polyvinyl chloride as an example of a polymer which undergoes a bimodal pyrolysis[669,1684-1689], in which initial evolution of hydrogen chloride leads to a conjugated polyene which subsequently pyrolyses to hydrocarbon fragments. The ability to observe this bimodal evolution is dependent upon the sample heating rate.

Since these evolution profiles involve fragments with different polarities, namely, hydrogen chloride and hydrocarbons, both negative and positive ion methane chemical ionization mass spectrometry were used to follow the progress of the pyrolysis. Negative ion detection was used in order to selectively detect halogenated species since there is some controversy as to whether halogenated hydrocarbons are also evolved with hydrogen chloride. Positive ion detection was used to detect the evolution of the hydrocarbon fragments.

<u>Negative ions.</u> Poly(vinyl chloride) showed a bimodal evolution profile in the negative ion detection mode, but the first evolution was due to the release of the residual solvent, tetrahydrofuran. The second evolution was due to the following major species: $[Cl]^-$ (35,37 m/z), $[HCl_2]^-$ (71,73,75 m/z), and $[PhCHCl]^-$ (125,126,127,128 m/z). Presence of the ions $[Cl]^-$ and $[HCl_2]^-$ indicates the loss of hydrogen chloride from the polymer. These differences in the widths of the evolution profiles can be rationalized: dehydrochlorination produces allylic groups that will cyclize and the subsequent losses of hydrogen chloride will have different kinetics. Examples of these profiles are shown in Figure 133. Examination of specific ion profiles shows that the onset of the evolution of hydrogen chloride (loss of Cl^-) occurs prior to the onset of the evolution of $HCl_2]^-$ or $[PhCHCl]^-$ suggesting that the latter ions maybe the products of secondary gas phase reactions. Although the evolution profiles were broad, the average activation energy could be obtained by

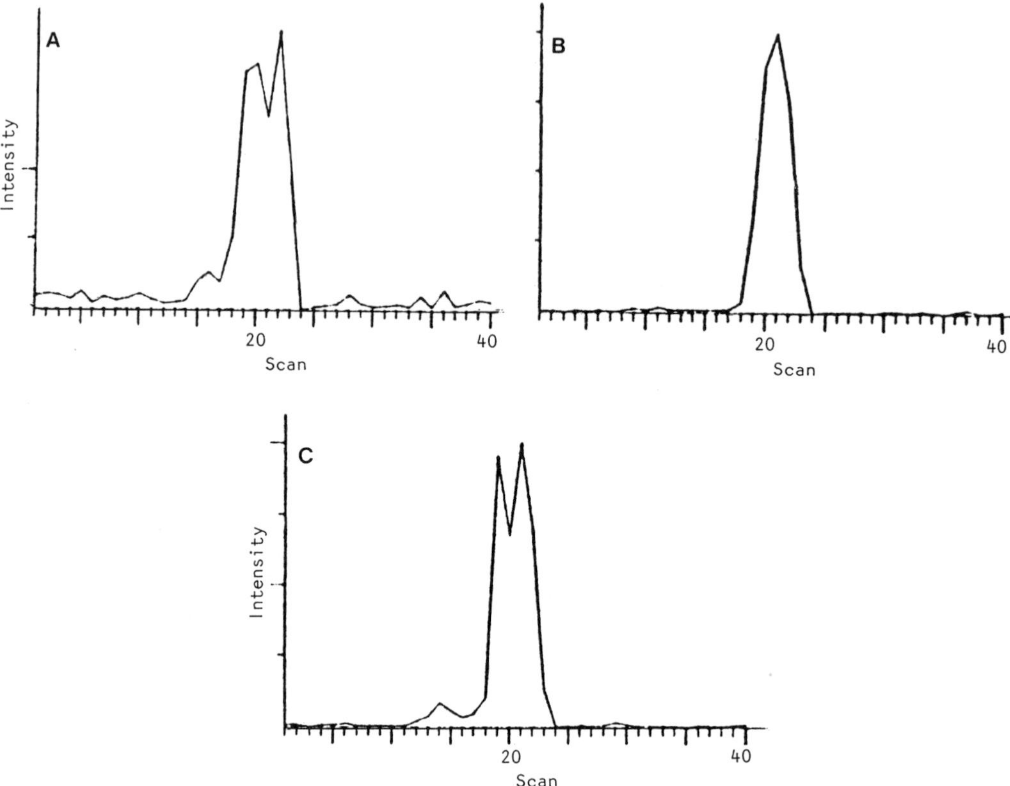

Figure 133   Specific nagative ion profiles for thermal degradation
               at 7.8°C/s of polyvinyl chloride shown as intensity vs.
               scan number; (A) mass 35 m/z, temperature of evolution
               maximum 330°C; (B) mass 71 m/z, temperature of evolution
               maximum 305°C; (C) mass 125 m/z, temperature of
               evolution maximum 305°C.

the measurement of the temperature maxima as a function of sample heating
rates. The result of these studies are given in Table 172.

      The values for activation energy obtained by Risby et al.[1653] (26-33
kcal) fall within the range of previously reported values (20-33
kcal)[1690-1692].

      Positive ions. Since the background ion current of the positive ion
methane chemical ionization mass spectrum in the region of 10-60 m/z was
intense, no attempt was made to monitor the evolution of hydrogen chloride
($^{+}H_2Cl$ m/z 36). Protonated molecular ions were observed for the major
fragments with the exception of methylnaphthalene, ethylnaphthalene,
methylanthracene, and ethylanthracene. The nominal structures for the
fragments that result from the thermal degradation of the polyene are
shown in Table 173.

      The specific ion current decreased with increasing molecular weight
as would be expected. There was some evidence of the evolution of
aliphatic hydrocarbons, but they were not characterized. The evolution
profiles for some of the aromatic hydrocarbon fragments, shown in Figure

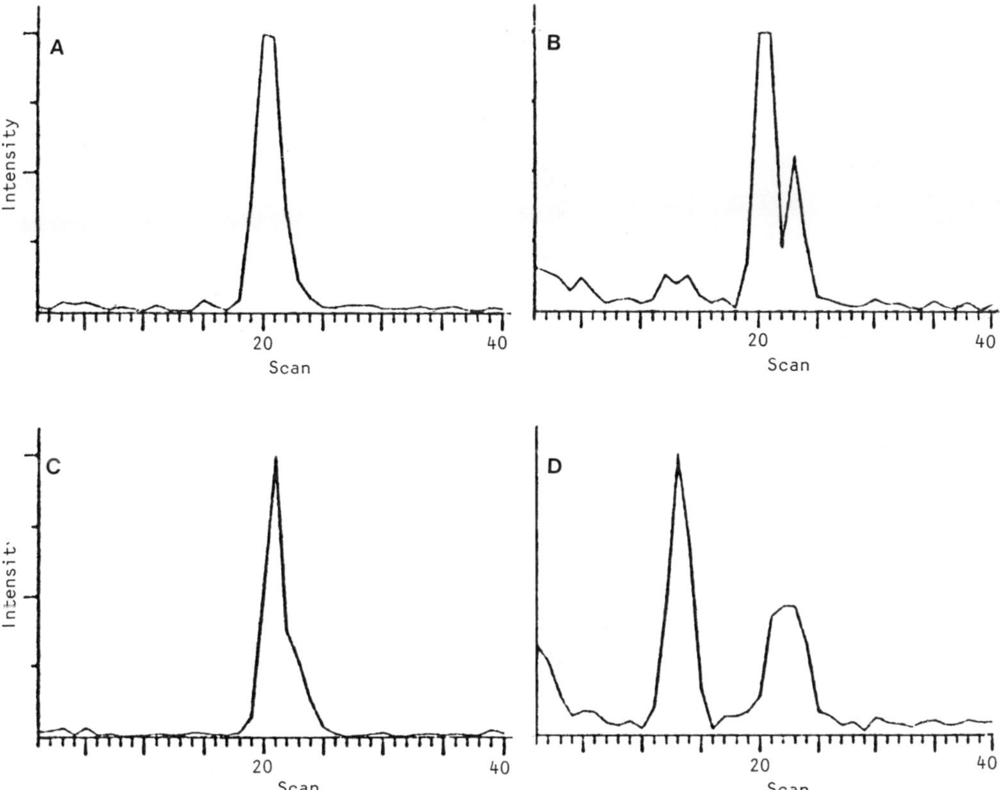

Figure 134   Specific ion profiles for thermal degradation at 8.0°C/s of polyvinyl chloride shown as intensity vs. scan number: (A) mass 79 m/z temperature range of evolution maxima 266-394°C; (B) mass 107 m/z temperature range of evolution maxima 266-394°C; (C) mass 119 m/z temperature range of evolution maxima 294°C; (D) mass 142 m/z temperature range of evolution maxima 321-348°C.

134, were bimodal and the first evolution maximum occurred concomittantly with the dehydrochlorination step. These results confirm that the ion [PhCHCl]⁻ was formed in secondary gas phase reactions since it was not observed in the positive ion spectra. It is interesting to note that not all the ions had this bimodal evolution profile.

Although some of the evolution profiles for the loss of these aromatic hydrocarbon fragments were bimodal and were clearly due to multiple kinetic processes, attempts were made to measure average activation energies. The activation energies for dehydrochlorination and polyene degradation are presented in Table 172. The activation energy of the first evolution maxima was lower than the average activation energy for the dehydrochlorination. This is difficult to rationalize since dehydrochlorination must precede or occur concomittantly with the formation of the polyene. However, this difference may not be significant and may reflect errors in the measurement of the activation energies by

Table 172 - Temperature data, activation energies and
pre-exponential factors for poly(vinyl chloride)

Dehydrochlorination		Thermal degradation of polyene	
ramp rate °C/s	temp of max °C	ramp rate °C	temp of max °C
15.5	326	15.65	319
7.80	306	7.99	294, 354
3.96	306	4.00	283, 349
1.98	286	2.01	275, 340
1.00	276	1.02	254, 334
activation energy	33 kcal		25, 72 kcal
pre-exponential factor	6.3×10^{11} s^{-1}		6.3×10^{8}, 9.1×10^{24} s^{-1}

this method. If this difference is significant, it may be because this technique measures the average activation energy for the loss of the particular fragment; the loss of hydrogen chloride has a much greater variation in activation energy than does the loss of the hydrocarbons. This may explain the wide variation in published activation energies for the same processes. The second evolution profile has a much higher activation energy similar to the value found for the thermolysis of isoprene[1693], 56-63.05 kcal, which has a similar degree of unsaturation.

In a classical example of the thermochemical analysis O'Mara[1694] studied the thermolysis of a PVC resin containing 57.4% chlorine by two techniques. The first method involved heating the resin in the heated (325°C) inlet of a mass spectrometer in order to obtain a mass spectrum of the total pyrolysate. The second, more detailed, method consisted of degrading the resin in a pyrolysis-gas chromatograph interfaced with a mass spectrometer through a molecule enricher.

Samples of PVC resin and plastisols (10-20 mg) were pyrolysed at 600°C in a helium carrier gas flow. Since a stoichiometric amount of hydrogen chloride is released (58.3%) from PVC when heated at 600°C, over half of the degradation products, by weight, is hydrogen chloride. The major components resulting from the pyrolysis of PVC are hydrogen chloride, benzene, toluene and naphthalene. In addition to these major products, an homologous series of aliphatic and olefinic hydrocarbons ranging from C_1 to C_4 are formed.

O'Mara[1694,1695] obtained a linear correlation between the weight of PVC pyrolysed and the weight of hydrogen chloride obtained by gas chromatography. Good agreement is obtained between the expected and found hydrogen chloride contents by this procedure (Table 174).

The technique was then applied to various PVC's into which different inorganic fillers had been incorporated ($CaCO_3$, CaO, $Al(OH)_3$, Na_2CO_3, Al_2O_3, LiOH[1696,1697], TiO_2, SnO_2 and ZnO_2[1696]. This work provided valuable information regarding reaction mechanisms that occur upon heating filled and unfilled PVC to high temperatures.

Table 173 - Hydrocarbon pyrolysis products from
poly(vinyl chloride) (7.80°C/s)

hydrocarbon fragment	mass of ion with max intens in region of mol.wt	% total ion current at max evolution	temp of evolution max, °C
benzene	79	5.8	266-294
toluene	93	1.0	321-348
styrene	105	2.0	294
ethylbenzene	107	3.0	266-294
propylene-benzene	119	3.4	294
propyl-benzene	121	1.4	294
naphthalcnc	129	4.2	294
methyl-naphthalene	142	0.9	321-348
ethyl-naphthalene	156	1.4	294
propylene-naphthalene	169	0.9	294
propyl-naphthalene	171	0.8	294
anthracene	179	0.7	294,321
methyl-anthracene	192	0.6	294,321
ethyl-anthracene	206	0.5	294
propylene-anthracene	219	0.2	294,321
propyl-anthracene	221	0.2	294

    Chang and Mead[1698] coupled a thermogravimetric analyser-gas
chromatograph - high resolution mass spectrometer in tandem in order to
study the thermal stability of a 75:25 by weight ethylene - polystyrene
foam. They used a stainless steel system consisting of two valves and two
traps to join a thermogravimetric analyzer to a gas chromatograph. This
arrangement has the capability of making multiple trapping at several
desired portions of a thermogram. The thermogravimetric analysis effluent
gas was collected in a cold trap and then directly injected into the gas
chromatograph for separation. The separated components were individually
introduced into the high resolution mass spectrometer for unequivocal
identification. Results indicated excellent separation between the
contents of the two traps. Gas chromatographic resolution and peak shape
were comparable to conventional injection systems. Sixteen effluent
compounds from the thermal degradation of polystyrene foam were
identified. The elemental compositions of the molecular ions of 31
components were obtained from the pyrolysis of ethylene-vinyl acetate
copolymer.

Table 174 - Pyrolytic analysis of HCl from PVC compounds,
polyethylene and chlorinated poly(vinylchloride)

	Percentage HCl	
Sample type	Found	Theoretical
PVC plastisol	30.1	29.1
PVC plastisol	37.0	37.3
PVC plastisol	44.7	44.5
PVC-vinyl acetate copolymer	53.5	52.9
PVC-vinyl acetate copolymer	46.3	46.3
PVC compound	38.5	38.6
PVC compound	47.5	47.7
Chlorinated polyethylene	34.9	34.3
Chlorinated polyethylene	40.7	41.1
Chlorinated poly(vinylchloride)*	63.9	69.2
Chlorinated poly(vinylchloride)	68.5	71.5

* Impure material, theoretical recovery not expected.

This system consisted of a thermogravimetric analyser, gas chromatograph, high resolution mass spectrometer, GC to HRMS interface unit, and TGA to GC interface unit.

The thermogravimetric analyser was a Du Pont Model 950 of the Du Pont 900 Thermal Analysis System (E.I. du Pont de Nemours & Co., Wilmington,Del.) The heating rate was 10-15°C per minute. The sample size was about 2-10 mg. helium was used as TGA carrier gas at a flow rate of 60-80 cc per minute.

The gas chromatograph was a Perkin-Elmer Model 881 (Perkin-Elmer Co., Norwalk, Conn.) with a flame ionization detector. Two GC Columns were employed namely a 6 ft x 1/8 inch stainless steel column containing 5% SE-30 on chromosorb W and a 10 ft x 3/16 inch stainless steel column containing 15% SE-30 on chromosorb W. The GC carrier gas was helium and the flow rate was approximately 60 cc per minute.

The high resolution mass spectrometer is a GEC Model 21-110B (E.I. du Pont de Nemours & Co., Wilmington Del.) Photoplates with 30-exposure capacity were used to record the mass spectra of the GC peaks. These photoplates were processed via a real time-time shared comparator computer system[1699].

The gas chromatograph to mass spectrometer interface unit includes a Biemann-Watson molecular sparator[1700], a gas chromatograph stream splitter, and two Hoke needle valves (Hoke Inc., Cresskill, N.J.). One needle valve is placed in between the stream splitter and the molecular separator to split approximately 50% of the gas chromatograph effluent into the molecular separator and the mass spectrometer. The ion source pressure was maintained at approximately 8-10 x 10^{-6} mm mercury during the operation. The temperature of the connection lines were maintained at 220-250°C.

The thermogravimetric analyser and gas chromatograph interface unit is constructed from an eight-way valve (valve A), a six-way valve, and two collecting traps. All parts are made of stainless steel except the Viton

O rings used in the valves. These multi-direction linear valves are products of Loenco Inc., Altadena, Calif. (Cat. No. L-206-6 and L-208-8). The traps are made from 8-inch x ¼ inch stainless steel tubing with a capacity of approximately 3 ml. The connection tubes are of 1/8 inch stainless steel tubing. All joints are of Swagelok fittings. The ball joint on the standard (ca. 50-ml capacity) quartz furnace tube of the Du Pont thermogravimetric analyser was removed, and to this furnace tube a ca. 1-inch length of 6-mm borosilicate glass tubing joined through a graded seal. Through a glass-to-metal seal on this borosilicate glass tube, the remaining interface is joined. The interface unit is enclosed in an oven except the traps. The oven temperature was maintained at 150°C. The oven temperature is limited to the thermal stability of the Viton O rings which start to soften at approximately 220°C. The temperature of the connection lines is maintained at 200-250°C. Traps are cooled with liquid nitrogen. A simply constructed oval shaped sleeve heater consisting of heater tape covered with asbestos is used to warm up the traps to 200-250°C. Traps are cooled with liquid nitrogen. A simply constructed oval shaped sleeve heater consisting of heater tape covered with asbestos is used to warm up the traps to 200-250°C. An external cold trap (vent trap) is connected to the exit of the TGA effluent to prevent possible backflow of moisture into the traps.

Figures 135(a) and (b) respectively show the thermogravimetric thermogram and the gas chromatogram obtained for polystyrene foam. It is seen in Figure 135 (b) that the thermally degraded products consist of a complex mixture of about 20 components. Sixteen of these components were readily identified by the high resolution mass spectrometer is listed in Table 175. Figure 136 shows a low resolution mass spectrum obtained of a component identified as styrene dimer. Both low and high resolution mass spectra are obtained from the computer printout of the photoplate data[1699]. No effort was made to identify the minor components.

It is interesting to note the presence of two oxygenated compounds, C_8H_8O (peak No. 11) and C_9H_8O (peak No. 12). Peak No. 11 may arise from oxygenation of ethylbenzene, and peak No.12 may arise from oxygenation of alpha-methylstyrene.

Ethylene-vinyl acetate (75:25 by weight) showed two distinct thermogravimetric analysis breaks (Figure 137). The thermogravimetric analysis thermogram was obtained from pyrolysis of a 9.2 mg sample in a helium atmosphere at a 15°C per minute heating rate. Two traps were employed to collect the effluents evolved which corresponded to the two breaks (Figure 137). Figures 138(a) and (b) show the gas chromatograms obtained from the contents of traps 1 and 2, respectively. Both were acquired by maintaining the column temperatures at ambient for 6 minutes, and then programming the column temperatures from 50 to 225°C at 10°C per minutes. Only negligible amounts of trap 1 contents were observed in trap 2. The exact mass of the molecular ion, its corresponding elemental composition, deviation in millimass units (observed-theoretical mass), and its probable molecular structure are tabulated in Table 176. Trap 1 contents arise from the vinyl acetate portion of the polymer, whereas trap 2 contents, with the exception of acetone, consist of aliphatic and olefinic hydrocarbons from the ethylene portion.

Peltonene[1701] applied thermogravimetric analysis followed by gas chromatography and mass spectrometry to the determination of volatile compounds arising from epoxy powder paints during the curing process carried out at 200°C. These volatile compounds may be residues from the

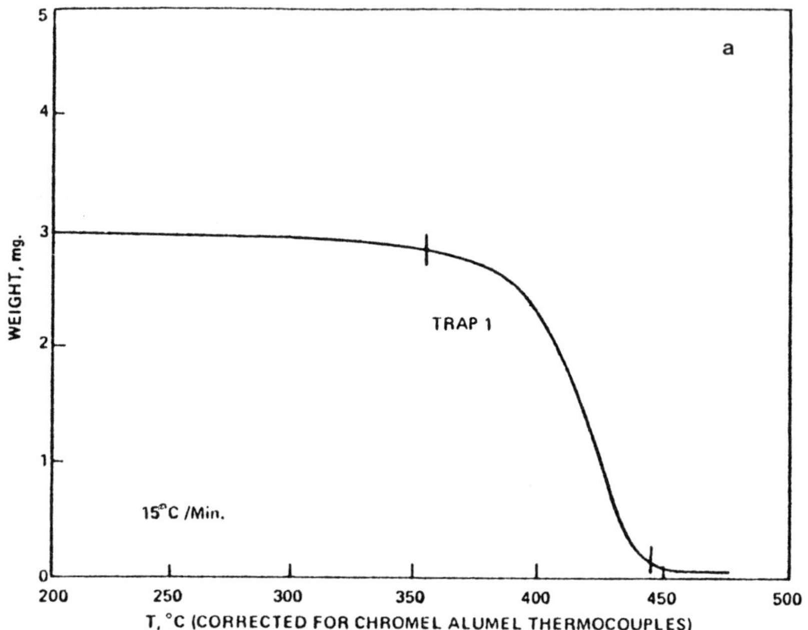

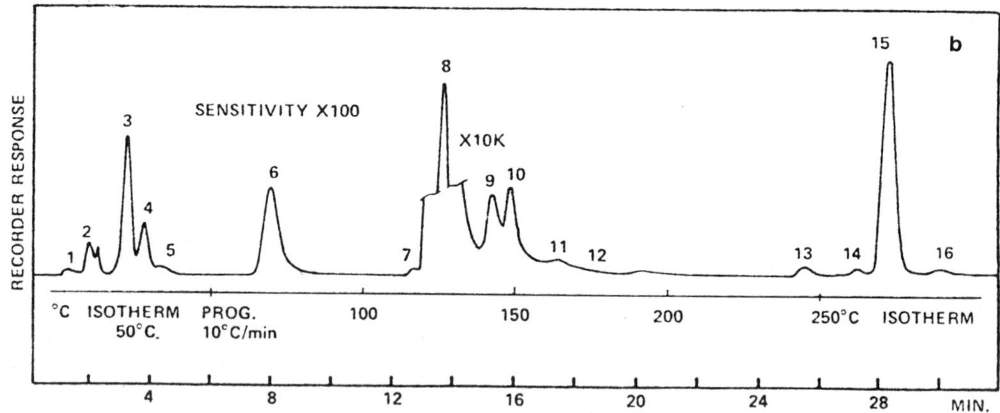

Figure 135   (a) TGA thermogram of polystyrene foam 3 mg sample, TGA heating rate 15°C min⁻¹ helium carrier gas.   (b) G.C. chromatogram of effluent components from pyrolysis of polystyrene foam.   Column temperature 50°C for 7 mins, then programmed at 24°C min⁻¹ for 250°C.

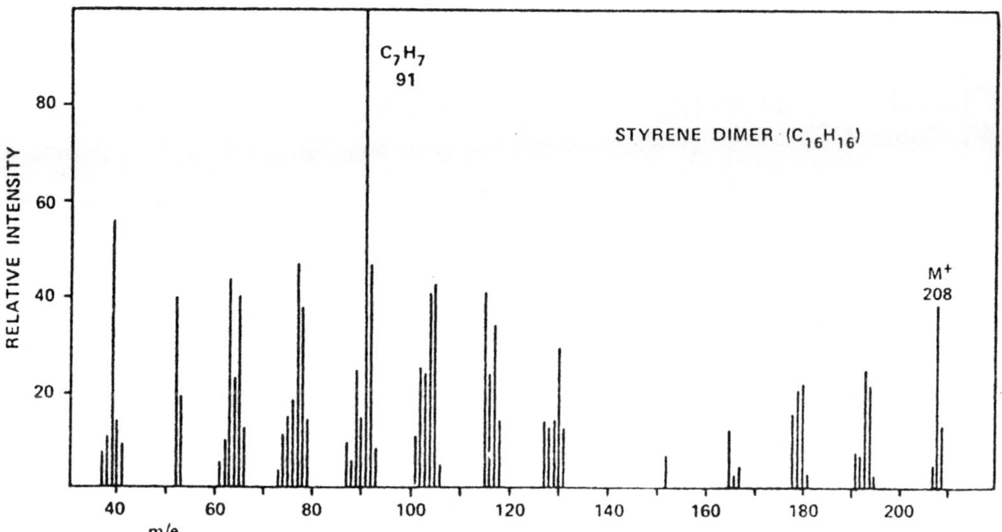

Figure 136   Low resolution mass spectroum of peak no. 15 from decomposition of polystyrene foam.

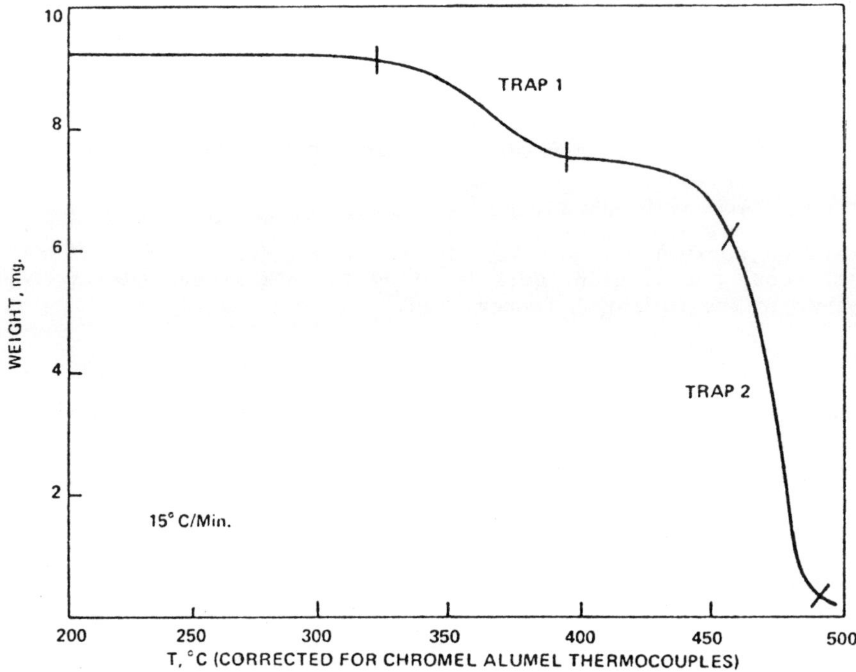

Figure 137   TGA thermogram of ethylene vinyl acetate copolymer.

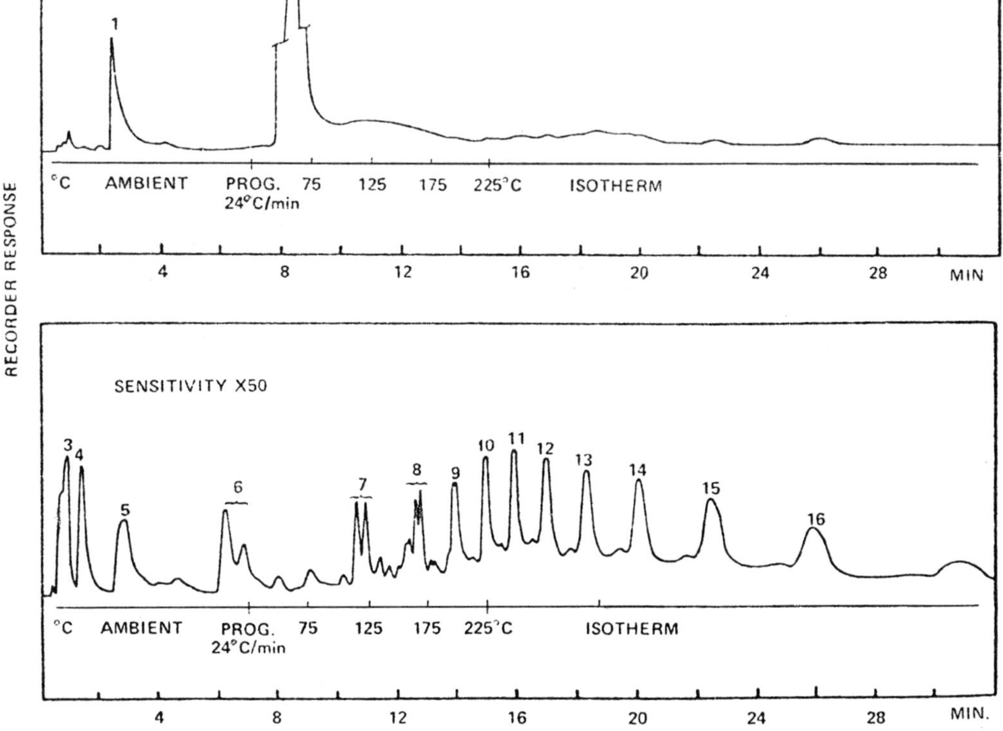

Figure 138   GC chromatogram of trap contents from pyrolysis of
ethylene-vinylacetate copolymer (a) trap 1; (b) trap 2.
Column temperature was maintained at ambient for 6
minutes, then programmed from 50°C to 225°C at 10°C/min.

synthesis of epoxy resin or they can be constituents of the paint. Two
types of epoxy powder paint were involved in this study, they differ from
each other in the form of hardener used.

Structures of (A) epoxy resin, (B) dicyandiamide hardener, (C)
pyromellitic anhydride hardener and (D) 2-methylimidazoline accelerator.

Table 175 - Components from decomposition of polystyrene foam

Peak No.	Molecular ion found	Elemental composition	Deviation mmu[a]	Probable structure
1	43.9901	CO_2	-0.30	Carbon dioxide
2	72.0950	C_5H_{12}	-1.10	Pentane
3	86.1105	C_6H_{14}	-1.36	Hexane
4	84.1105	C_6H_{12}	1.86	Hexene
5	78.0477	C_6H_6	-0.80	Benzene
6	92.0622	C_7H_8	0.38	Toluene
7	106.0785	C_8H_{10}	-0.29	Ethylbenzene
8	104.0633	C_8H_8	-0.81	Styrene
9	118.0773	C_9H_{10}	0.86	alpha-Methylstyrene
10	132.0946	$C_{10}H_{12}$	-0.77	$Ar-C_4H_7$
11	120.0582	C_8H_8O	-0.76	$Ar-C_2H_3O$
12	132.0584	C_9H_8)	-0.94	$Ar-C_3H_3O$
13	182.1088	$C_{14}H_{14}$	0.66	$Ar-C_2H_4-Ar$
14	196.1269	$C_{15}H_{16}$	1.76	$Ar-C_3H_6-Ar$
15	208.1245	$C_{16}H_{16}$	0.60	Styrene dimer
16	220.1242	$C_{17}H_{16}$	0.90	$(Ar)_2-C_5H_6$

[a] Deviation = (theoretical mass - observed mass) x 1000.

Dicyandiamide-hardened paints have a glossy surface, whereas with an acid anhydride hardener the result is semi-glossy. In the powder, the amount of hardener is about 4% and that of the accelerator 1%.

The solid resins in powder paints are produced in two different ways. In the conventional method, the resin is produced in isobutyl methyl ketone, and in the fusion method bisphenol-A is condensed catalytically without solvents to give a liquid epoxy resin.

In this study, the powders were cured in a device permitting precise control of temperature, type of atmosphere, rate of gas flow and the collection of the volatile compounds. The compounds were studied with a high-resolution gas chromatograph and a gas chromatograph-mass spectrometer. The mass loss of the powders during curing was studied by thermogravimetry.

The thermogravimetry analyses were formed with a Stanton Redcroft 770 apparatus (Stanton Redcroft, UK). In the analyses, the temperature was raised from 20 to 200°C at a rate of 20°C min^{-1} and the sample size was 5 mg.

Table 176 – Components from decomposition of ethylene-vinyl acetate copolymer

Peak no[a]	Molecular ion found	Elemental composition	Deviation mmu[b]	Probable structure	Peak no[a]	Molecular ion found	Elemental composition	Deviation mmu[b]	Probable structure
1	58.0416	C_3H_6O	0.24	Acetone	9	126.1402	C_8H_{18}	0.58	Nonene
2	60.0207	$C_2H_4O_3$	0.36	Acetic acid		128.1556	C_9H_{20}	0.83	Nonane
3	42.0470	C_3H_6	−0.15	Propene	10	140.1557	$C_{10}H_{20}$	0.79	Decene
	44.0630	C_3H_8	−0.49	Propane		142.1707	$C_{10}H_{22}$	1.38	Decane
4	56.0619	C_4H_8	0.63	Butene	11	154.1719	$C_{11}H_{22}$	0.15	Undecene
	58.0777	C_4H_{10}	0.53	Butane		156.1869	$C_{11}H_{24}$	0.82	Undecane
5	58.0407	C_3H_6O	1.08	Acetone	12	168.1863	$C_{12}H_{24}$	1.40	Dodecene
	70.0780	C_5H_{10}	0.15	Pentene		170.2034	$C_{12}H_{26}$	−0.05	Dodecane
	72.0944	C_5H_{12}	−0.59	Pentane	13	180.2109	$C_{13}H_{26}$	1.53	Tridecene
6	84.0939	C_6H_{12}	−0.01	Hexene		182.2019	$C_{13}H_{28}$	−1.37	Tridecane
	86.1098	C_6H_{14}	−0.26	Hexane	14	196.2201	$C_{14}H_{28}$	−1.09	Tetra-decene
7	98.1086	C_7H_{14}	0.88	Heptene		198.2371	$C_{14}H_{30}$	−2.41	Tetra-decane
	100.1244	C_7H_{16}	0.79	Heptane	15	210.2362	$C_{15}H_{30}$	−1.50	Penta-decene
	112.1259	C_8H_{16}	−0.75	Octene		212.2530	$C_{15}H_{32}$	−2.69	Penta-decane
	114.1417	C_8H_{18}	−0.88	Octane	16	224.2517	$C_{16}H_{32}$	−1.36	Hexa-decene
						226.2666	$C_{16}H_{34}$	−0.64	Hexa-decane

[a] Peak numbers 1 and 2 from Trap 2. [b] Deviation = (theoretical mass − observed mass) x 1000

In the curing process, temperature was adjusted to 200°C. the air flow rate was 500 ml min^{-1} (synthetic air:20% O_2, 80% N_2). The 10 g sample was spread over a distance of 800 mm along the glass tube (1500 x 17 mm). The trap (1000 x 4 mm) was cooled to -78°C with acetone and dry-ice. The volatiles were washed out from the spiral with a 1 ml of acetone.

The gas chromatograph was a Hewlett-Packard 5790-A (Hewlett-Packard, USA) equipped with flame-ionisation and nitrogen-phosphorus detection systems. The chromatograms were recorded with a Hewlett-Packard 3392-A integrator. A BP-5 fused-silica column (25 m x 0.2 mm i.d.) from SGE (Australia) was used and the carrier gas was helium at a flow-rate of 1 ml min^{-1}. The oven temperature was kept at 35°C for 1 min after injection, then raised to 240°C for 20 min. The injector temperature was 220°C, the detector temperature 280°C and the volume injected 0.1 ul.

Mass spectral data (chemical ionisation and high-resolution electron impact) were obtained with a Finnigan-MAT 8230 double-focussing system with an Incoss data system and a Varian 3700 gas chromatograph. In chemical ionisation the reactant gas was isobutane, the ionisation energy 160 eV, the emission current 0.05 mA, the resolution 1000 and the scan range 65-400 m/z at the rate of 1 scan s^{-1}. In high-resolution electron-impact mass spectrometry the ionising energy was 70ev, the emission current 1 mA, the resolution 3000 and the scan range 50-350 m/z at the rate of 0.55 scans s^{-1}. The gas chromatographic conditions were as above with the exception of the column, which was an SE-54 (25 m x 0.32 mm i.d.) (Orion Analytica, Finland). A Hewlett-Packard 5970 mass selective detector coupled to the gas chromatograph (Hewlett-Packard 5890) was used for low-resolution electron-impact mass spectrometry. The scan range was 45-350 m/z. The gas chromatographic conditions were the same as above, but here the column was a Hewlett-Packard Ultra 2 (25m x 0.2 mm i.d.)

The thermogravimetric mass losses were 0.3-0.4% and did not depend on the manufacturing process of the resin, or on paint colour (Table 177). The mass losses were similar in air and in nitrogen. Most of the mass loss may have been due to condensed moisture released during heating. No further mass loss occurred after 20 min.

The gas chromatograms of volatiles emitted from white amide-cured epoxy powder paints, cured and manufactured by the solution method (Nos. 1 and 2) are presented in Figure 139(a) and (b). The compounds identified are listed in Table 178. The most abundant compounds detected were isobutyl methyl ketone (4-methylpentan-2-one) and the various isomers of xylene, which are typical solvents used in the production of resins. The volatiles characteristic of all three amide-cured epoxy powders studied were an unidentified compound and melamine. The source of melamine is not clear but it has been suggested that dicyandiamide decomposes to cyanamide during curing. Cyanamide can react with dicyanamide to form melamine. However, cyanamide was not observed among the volatilised products.

Gas chromatograms for a black paint with a resin made by the solvent method are shown in Figure 139(c) and (b). Table 179 lists the compounds identified. Even in this instance, the most abundant compounds were isobutyl methyl ketone and xylenes. The origin of 2-phenyl-4 H-imidazoline is not certain, but it might have been added as a catalyst or it may arise from the anhydride adduct during the curing process.

Most of the identified compounds gave observable molecular ions in electron-impact mass spectrometry. Chemical ionisation mass spectrometry

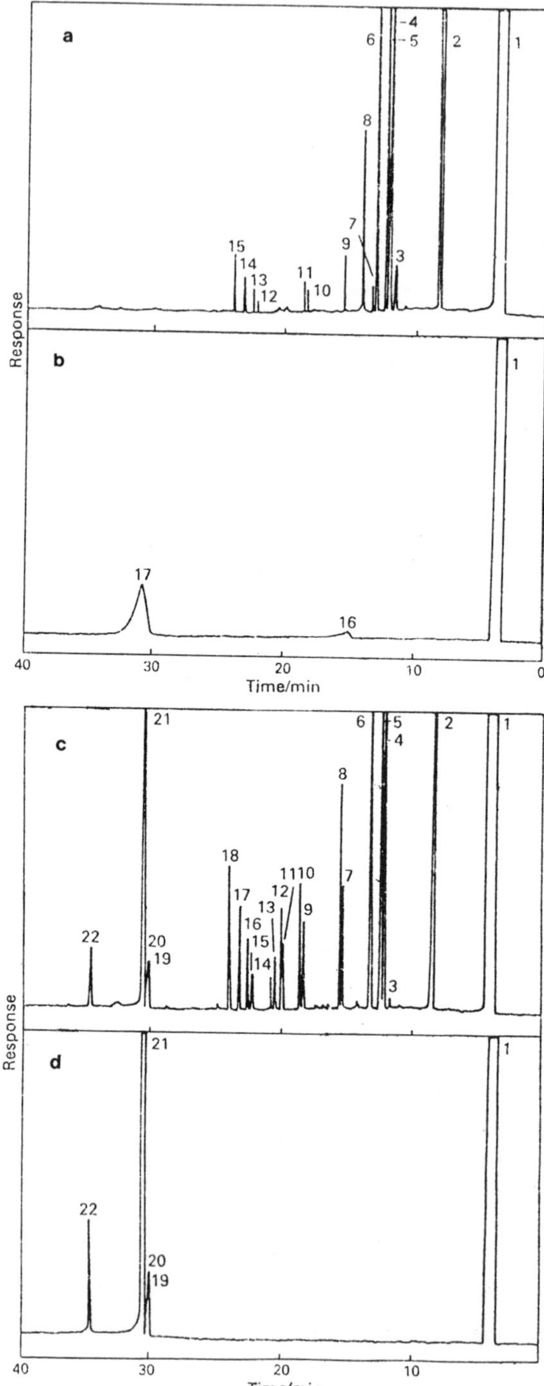

Figure 139   Gas chromatograms of compounds volatilised during curing
of amide paint. (a) flame ionization detector; (b)
nitrogen specific detector. Compounds are listed in
Table 178. Gas chromatograms of compounds volatilised
during curing of anhydride paint: (c) flame ionisation
detector: (d) nitrogen specific detector. Compounds are
listed in Table 179.

Table 177 - Mass loss of epoxy powder paints during curing

Production method and colour	Mass loss in air, %	Mass loss in nitrogen, %
Conventional:		
White 1	0.3	0.3
White 2	0.4	0.3
Black 4	0.4	0.4
Black 5	0.3	0.3
Fusion:		
White 3	0.3	0.3
Black 6	0.4	0.3

with isobutane was carried out to determine the relative molecular masses. In all instances the quasi-molecular ion was M + 1 with a relative abundance of 100%; no M + 57 was observed. Some compounds were identified on the basis of their mass spectra only because no reference compounds were obtainable.

High-resolution mass spectra were obtained for unidentified compounds a, b, c, d and f. The molecular formula of compounds a, b, c and d is $C_{12}H_{22}O$. Proposed structures are isomers of ketones; e.g., a and b might be 2-or 4-isomers of ethyl isopropylcyclohexyl ketone and c and d might be cyclohex-4-ylhexan-2-one and cyclohex-4-yl-4-methylpentan-2-one, respectively. One unidentified compound (e) had the quasi-molecular ion M + 1105 and the only fragment was M - 18 in chemical ionisation mass spectrometry. The proposed formula is $C_4H_{12}N_2O$. This might be a diamino alcohol. The molecular formula of unidentified compound f is $C_{15}H_{14}O$. The proposed structure is 2-biphenyl-2,3-epoxypropane. The eight most intense ions of the unidentified compounds and their relative abundances, together with the M + 1 ions in their chemical ionisation spectra, are given in Table 177 and 178.

Other polymers that have been examined by evolved gas analysis include PVC[1702-1704], polystyrene,[1705-1708], styrene-acrylonitrile co-polymers[1709-1711], polyethylene and polypropylene[1712-1717], polyacrylates and copolymers[1718-1721, 1727-1729], polyethylene terephthalate and poly-phenylenes, and polyphenylene oxides and sulphides[1721-1726]. Studies involving the use of chromatography include the thermal degradation of polyvinyl chloride[1730], and vinyl plastics[1731] and polysulphone[1732].

11.1.8 Thermal Analysis Equipment

The parameters that can be measured by some available commercial instrumentation are listed in Table 180.

<u>Dupont Thermal Analyst 2000/2100.</u> The 2000 is based on the new IBM Personal System/2 - Model 50 computer and offers a large software library for thermal analysis, plus numerous standard applications, including word processing, data management and storage. When combined with a wide range

Table 178 - Volatile compounds emitted from amide-hardened epoxy
powder paints during curing

No. in Figure 3	Compound	Paint 1	2	3	Method of identification
1	Solvent				
2	4-Methylpentan-2-one	x	x		GC + GC - MS
3	Chlorobenzene	x			GC + GC - MS
4	1,2-Dimethylbenzene	x	x		GC + GC - MS
5	Ethylbenzene	x	x		GC + GC - MS
6	1,4-Dimethylbenzene	x	x		GC + GC - MS
7	1,3-Dimethylbenzene	x	x		GC + GC - MS
8	Trimethylbenzene	x	x		GC + GC - MS
9	3-Isobutylbut-3-en-2-one	x	x		GC - MS
10	alpha-Methylbenzyl alcohol	x	x		GC + GC - MS
11	Acetophenone	x	x		GC + GC - MS
12	Unidentified[a]*	x			
13	Unidentified[b]*	x			
14	Unidentified[c]*	x			
15	Unidentified[d]*	x			
16	Unidentified[e]*	x	x	x	
17	Melamine	x	x	x	GC + GC - MS
-	Phenylglycidyl ether		x		GC + GC - MS
-	Unidentified[f]			x	

*[a] EI:53(8), 55(24), 57,(100), 58(120, 67(8), 69(8), 85(60). CI: M +
1183(100).

[b] EI:53(8), 55(26), 57(100), 67(10), 69(10), 85(80), 97(8), 125(6). CI: M
+ 1183(100.

[c] EI:55(32, 57(28), 68(32), 83(83), 107(25), 123(100), 167(20), 182(6).
CI:M + 1183(100).

[d] EI:55(30), 67(54), 82(16), 83(30), 107(12), 125(100), 140(10), 182,(10).
CI:M + 1183(100).

[e] EI:53(1), 55(100), 56(98), 57(24), 71(15), 73(52), 74(8) 86(6). CI:M +
1105(100).

[f] EI:85(8), 126(16), 139(10), 152(8), 165(20), 195(100), 196(16), 210(10).
CI: M + 1211.

Table 179 - Volatile compounds emitted from anhydride-hardened epoxy powder paints during curing

No. in Figure 4	Compound	Paint			Method of identification
		4	5	6	
1	Solvent				
2	4-Methylpentan-2-one	x	x		GC + GC - MS
3	Chlorobenzene	x			GC + GC - MS
4	1,2-Dimethylbenzene	x	x		GC + GC - MS
5	Ethylbenzene	x	x		GC + GC - MS
6	1,4-Dimethylbenzene	x	x		GC + GC - MS
7	3-Isobutylbut-3-en-2-one	x			GC - MS
8	Benzaldehyde	x			GC + GC - MS
9	alpha-Mehthylbenzyl alcohol	x	x		GC + GC - MS
10	Acetophenone	x	x		GC + GC - MS
11	Dec-6-en-3-one	x			GC - MS
12	Dec-6-en-5-one	x			GC - MS
13	3-Methylbenzyl alcohol	x	x		GC + GC - MS
14	4-Methylbenzyl alcohol	x	x		GC + GC - MS
15	Unidentified [a]*	x			
16	Unidentified [b]*	x			
17	Unidentified [c]*	x			
18	Unidentified [d]*	x			
19	Phenylimidazonline	x			GC - MS
20	Phenylimidazoline	x			GC - MS
21	2-Phenyl-4H-imadazoline	x		x	GC + GC - MS
22	Diphenylimidazoline	x			GC - MS
-	Phenol		x		GC + GC - MS
-	4-Methyl-2,6-di-tert-butyl phenol		x		GC + GC - MS
-	Phenyl glycidyl ether			x	GC + GC - MS

* See table 178.

Table 180 – Thermal measurements possible with commerical analysers

	Perkin Elmer Series 7	Dupont 9900 Thermal Analysis System	Stanton Redcroft Omnitherm/Digilab	Setaram (Clayton Scientific)
TGA	TGA7	951		
DTA		DTA analyser		
DSC	DSC-7	910 912 dual sample cell		Micro DSC calorimeter
EGA			Stanton Redcroft 1000°C thermo-balance Omnitherm controller digilab FTS 40 or FTS 60 Fourier transform infrared spectrometer with 3200 data system	
DPC		930		
PDSC		Dual sample PCSC cell		
TMA	TMA7	943 (discussed further in Chapter 12		
DMA		983 (discussed further in Chapter 12)		
Professional computer	Model 7700	Thermal Analyst 2000/2100		

of thermal analysis accessories the Thermal Analyst 2000 will meet the diverse needs of most laboratories.

The Thermal Analyst 2100 is based on the new IBM Personal System/2 - Model 60 computer, and offers similar software to the 2000. In addition, however, the 2100 can be easily expanded to accommodate the laboratory's future needs - only four of the 2100's eight built-in expansion slots are used in the instrument's current configuration.

Some features of these instruments are listed in Table 181. A range of attachments is available (Table 182), for the following techniques; thermogravimetric analysis, differential thermal analysis, differential scanning calorimetry, differential photocalorimetry, thermomechanical analysis and dynamic mechanical analysis (TMA and DMA discussed further in Chapter 12).

<u>Perkin Elmer Modular Series.</u> In their modular system, Perkin Elmer supply three types of thermal analyser, the TGA-7 thermogravimetric analyser, the DSC-7 differential scanning calorimeter and the TMA-7 thermomechanical analyst, (the latter is discussed further in Chapter 12). These instruments are used in conjunction with the 7700 professional computer.

The PE7700 Professional Computer incorporates 32 bit microprocessor technology, the Motorola M68000, for fast accurate computing and handling.   This operating system permits real-time multi-tasking instrument operation and data acquisition as well as true general purpose computing power in either BASIC or FORTAN. The PE7700 incorporates two double-sided, double-density micro-floppy disc drives which are used for loading new applications software and archiving data. Additionally, a Winchester hard disc drive (20 or 40 megabyte storage capacity) is used for permanent applications program storage, calibration information and data storage.

One of the main features of the PE 7700 operating system is the ability to perform multi-tasking instrument operation and data analysis. For thermal analysis, this means that multiple thermal analysis modules can be operated under independent instrument control while simultaneously taking data and storing it on the Winchester disc for later analysis. This multi-user, multi-tasking feature permits greatly increased laboratory productivity. The 7 Series thermal analysis software (Table 183) has been specifically designed for the performance of these multi-tasking operations, making the changeover from one task to another quick and easy.

An extensive library of thermal analysis and general purpose software programs is available for use with the PE 7700 Professional Computer.

Details concerning the three Perkin Elmer thermal measuring instruments mentioned above are given in Table 184.

The outstanding features and benefits of the thermogravimetric analyser TGA7 include:

Highest sensitivity microbalance offers detectability of small weight changes.

Complete computer control for unattended operation and ease of use.

Table 181 - Dupont Thermal Analyst 2000/2100

Model 2000	Model 2000/2100	Model 2100
IBM Personal System/2- model 50 Computer 1.5 MB Random Access Memory (RAM) - By being able to hold large software programs and data files, the Thermal Analyst 2000 operates quickly, and can perform complex analysis.	Intel 80286 Processor - Works at high speed (10 MHz) to complete large tasks in one-third to one-fifth the time previously required.	IBM Personal System/2-Model 60 Computer 3.0 MB Random Access Memory (RAM) - By being able to hold large software programs and data files, the Thermal Analyst 2100 operates very rapidly, and can perform complex analysis.
Du Pont Single-Tasking TA Operating System - Provides complete control of all experimental conditions for a single module, or the ability to run any Du Pont Thermal analyst data analysis program.	Intel 80287 Maths Co-Processor - Also works at 10 MHz, to speed mathematical data analysis.	Du Pont Multi-Tasking TA Operating System - This allows the operator to simultaneously run an experiment, analyse data from current or prior experiments, and set up conditions for subsequent ones. While an experiment is being run, data can be plotted on the graphics display or stored on disc.
1.44 MB Diskette Drive - The new 3½" diskette holds 4 times more information than a 5¼" floppy disk.	IBM DOS 3.30 Computer Operating System - The heart of 'IBM compatibility).	1.44 MB Diskette Drive - The new 3½" diskette holds four times more information than a 5¼" floppy disk. Its hard, durable casing improves reliability and offers protection against exposure of the magnetic medium.
20 MB Hard Drive - Provides fast, ready access to even large programs and data files	I/O Attachment Card for External Disk Drive - Permits the attachment of an external drive, so that existing IBM-compatible software or data files contained on 5¼" floppy disks can be used.	44 MB Hard Drive - The extensive capacity permits longer experiments and/or retention of more files in the permanent memory.
High-Resolution Graphics - The 12" monochrome display produces high resolution graphics and can generate up to 64 shades of grey.	Options: Additional Operating Software (2000 only) - Simultaneous Run/Analyse. Provides multi-tasking capability. Permits the operator to run an experiment, analyse data from a previous experiment - all simultaneously. Realtime Transfer and Plot - Provides automatic transfer of data	

from the analysis modules to either
the $3\frac{1}{2}$" diskette or 20 MB hard disk.
Also permits the plotting of up
to four current or prior experi-
ments on the CRT.
Colour graphics.
Dot matrix printer.
High resolution colour plotter.
Keyboard display.

Multimodul Operating System –
Allows up to four analysis modules
to be run simultaneously. Except
for the 983 DMA, which contains
built-in interface electronics, each
analysis module to be run simultaneously must have a module interface.
(NB: For multimode operation in the
2000, the Colour Graphics Display,
Simultaneous Run/Analyse and Real-
time Transfer and Plot options must
be installed.)

External $5\frac{1}{4}$" Floppy Disk Drive –
Permits the transfer of software
programs or data to and from $5\frac{1}{4}$"
floppy disks. Recommended for
laboratories with existing IBM-
compatible software.

Module Switching Accessory (MSA) –
The MSA is designed for the laboratory that has more than one analysis module but does not need
simultaneous operation. It permits
up to three analysis modules to be
connected simultaneously to one
Module Interface, with power to
each. In addition, it eliminates
the normal warm-up time.

Colour Graphics Display –
The 14" colour display produces high
resolution graphics and generates a
a palette of more than 256,000
colours. It also has an adjustable
tilt-and-swivel base.

Table 182 - Attachments for Dupont Thermal Analyst 2000/2100

Thermal technique	Features
951 thermogravi- metric analyser	Ambient to 1200°C range. High sensitivity. Horizontal design.
Differential thermal analyser	Ambient to 1600°C range. Modular plug-in cell used with 910 or 912 DSC cell base. Provides semi-quantitative heat flow over a wide temperature range.
910 Differential scanning calorimeter	-170° to 725°C range. High calorimetric sensitivity (6uW). Superior baseline performance. Pressure option to 7 MPa (1000 psi).
With 912 dual sample DSC	-170°C to 650°C range. Use with 912 DSC cell base. Pressure option to 7 MPa (1000 psi).
930 Differential photocalorimeter	UV/visible optical system. Use with 910 or 912 DSC modules. Photo feedback system for reproducible exposure.
943 Thermo- mechanical analyser (see Chapter 13)	-180° to 800°C range. Ambient to 1200°C option. High sensitivity (0.5 um). Interchangeable probes for various materials and measurements.
983 dynamic mechanical analyser (see Chapter 13)	-150° to 500°C range. Four operating modes: fixed frequency (4 decades), resonant frequency, stress relaxation and creep measurement.

Automatic furnace removal and cooling.

Compact, modular system design.

High temperature operation to 1500°C.

Wide range of sample sizes.

Simultaneous instrument operation.

Fast cool-down time.

Excellent temperature accuracy.

Excellent atmosphere control.

The features of the DSC-7 differential scanning calorimeter include:

Power-compensated DSC design.

Excellent calorimetric accuracy.

High performance at a down-to-earth price.

Ease of operation.

Complete computer control for unattended operation.

Compact, modular system design.

Simultaneous instrument operation.

Fast program cooling capability.

True isothermal operation.

Precise temperature accuracy and control.

Stanton (Redcroft) omnitherm evolved gas analyser (EGA). This apparatus introduced in 1987 bridges the gap between thermogravimetric analysis which merely determines weight changes accurately whilst a polymer is heated and Fourier transform infrared spectroscopy which provides information regarding the chemical nature of the volatiles produced when a polymer is heated. It consists of a Stanton Redcroft 1000°C thermobalance, an omnitherm controller and the Digilab FTS 40 or FTS 60 Fourier transform infrared spectrometer with a 3200 data system. Evolved gases from the 1-5 mg sample are taken from the tg oven to the ftir cell by a short, heated glass-line transfer capillary. Both cell and line are typically maintained at 230°C which, together with nitrogen purging, means that the cell and line are self-cleaning. To avoid placing the heated gas cell in the normal sample compartment, Digilab designed and built a special interface bench. This includes a 10 cm long, 6 mm internal diameter, stainless steel heated gas cell with double windows at both ends.

The advantages of dedicated tg-ftir using this system are:

Low swept volume which produces a clean supply of evolved gases.

The gas path is heated end to end.

tg and ir data can be coordinated in time.

Evolved gas profile and derivative weight losses can be examined together.

All data is available for later analysis.

High sensitivity to trace components.

Single computer.

Evolved gases can be assigned to specific weight losses.

Storage space available for long data collection runs.

Table 183 - Software available for Perkin Elmer 7 Series thermal analysers

---

TGA-7 thermogravimetric analyser	DSC-7 differential scanning calorimeter

---

SET UP

Like the DSC 7 SET UP program, the TGA7 SET UP program is used to take data from the TGA7. It includes easy to use routines for direct GA control from the computer keyboard. Other options include heat, cool and isothermal modes of operation, multi-step program, development, and operation in either temperature or time scales.

DELTA Y

The DELTA Y program permits the calculation of percent weight changes at any point on a TG curve. Options include the total percent weight change and the temperature or time limits over which the weight change has been calculated.

ONSET

The ONSET program permits the quantitative determination of the onset temperature or time of a weight loss or weight gain.

PLOT

The PLOT program is used to obtain a hard copy printout of TG data, results and calculations.

OPTIMIZE

The OPTIMIZE programs are a series of software programs which are used to optimize TG data curves. Programs are

available for rescaling the temperature or time axes and the ordinate axes.

SET UP

The SET Up program is used to set up and take data from the analyser. It includes easy to use routines for direct instrument control from the computer keyboard. Other options include heat, cool and isothermal modes of operation, multi-step program development and operation in either temperature or time scales.

PEAK

The PEAK program permits the calculation of the parameters of any peak. Options include peak area calculations, peak max/min temperature or time, peak onset temperature or time, and the calculation of the peak height.

GLASS TRANSITION

The GLASS TRANSITION program permits the calculation of the parameters of a glass transition. Options include the glass transition onset and midpoint temperature, as well as the change in the specific heat before and after the glass transition.

ONSET

The ONSET program permits the quantitative determination of the onset temperature or time of transition, such as in oxidative stability testing or decomposition testing.

PLOT

The PLOT program is used to obtain a hard copy printout of DSC7 data, results and calculations.

OPTIMIZE

The OPTIMIZE programs are a series of software programs which are used to

RECALL

The RECALL program is used to recall to the screen data, results, and run parameters which are stored on disk. Options include recalling a a single curve, recalling multiple curves and recalling curves of different techniques (i.e. TG and DSC curves).

SAVE

The SAVE program permits the storage of data, results, calculations and comments on the disk.

CALIBRATE

The CALIBRATE programs permit automatic mass and temperature calibration of the TGA7 analyser.

CALCULATIONS

The CALCULATIONS programs permit a wide variety of calculations to be performed on TG data. Programs for curve subtraction and derivative calculations are available.

optimize DSC7 data curves. Programs are available for rescaling the temperature or time axis, the energy axis, shifting curves, resloping curves and aligning curves.

RECALL

The RECALL program is used to recall to the screen data, results and run parameters which are stored on disk. Options include recalling a single curve, recalling multiple curves and recalling curves of different techniques (i.e. DSC and TG curves) for direct data comparison.

SAVE

The SAVE program permits the storage of raw data, results, calculations and comments on the disk.

BASELINE OPTIMIZATION

The BASELINE OPTIMIZATION program permits automatic baseline optimiza- of any DSC scan.

CALCULATIONS

The CALCULATIONS programs permit a wide variety of calculations to be performed on DSC data. Programs for curve subtraction, normalization and derivative calculations are available.

For certain applications, compared to a mass spectrometer combination, the ftir has the advantages that it is inherently more simple - no vacuum is involved - and it is research-grade quality, whereas the ms is likely to be a basic quadrupole device. While the ms could be more sensitive and would possibly provide more information as an evolved gas analyser, it would also have a much greater need for maintenance than the ftir.

11.2 OXIDATIVE STABILITY

11.2.1 Thermal Methods

In most cases thermal oxidation studies in polymers as discussed in the previous section would be conducted under an inert gas such as nitrogen or helium in order to avoid the formation of secondary oxidation products which would complicate the interpretation of data obtained. In some cases, however, information is required on the stability to oxidation of polymeric materials and in these instances the thermal experiment would

Table 184 - Details concerning Perkin-Elmer thermal analysers

TGA-7 thermogravimetric analyser		DSC-7 differential thermal analyser	
Balance Sensitivity	0.1 ug (10^{-4}mg).	DSC Type	Power compensated temperature null principle. Measures energy directly <u>not</u> differential temperature (T).
Balance Accuracy	Better than 0.1%.		
Weighing Precision	Up to 0.1%.	DSC Cell	Independent dual furnaces constructed of platinum-iridium alloy with independent platinum resistance heaters and temperature sensors.
Sample Capacity	Standard Furnace: Up to 50 microliters.		
	High Temperature Furnace: Up to 250 microliters.		
		Maximum Sensitivity	8 uW/cm.
Temperature Range	Standard Furnace: Ambient to 1000°C.	Dynamic Range	8 uW/CM to 28 mW/cm.
	High Temperature Furnace: 50°C to 1500°C.	Noise (RMS)	0.002 Watt.
		Calorimetric Accuracy	Better than $\pm$ 1%.
Heating & Cooling Rates	Standard Furnace: 0.1° to 200°C/min in 0.1°C increments.	Calorimetric Precision	Better than $\pm$ 0.1°C.
		Temperature Precision	$\pm$ 0.1°C.
	High Temperature Furnace: 0.1° to 100°C/min in 0.1°C increments.	Temperature Accuracy	$\pm$ 0.1°C.
		Temperature Display	0.1°C increments.
Temperature Precision	Standard Furnace: $\pm$ 2°C.	Heating & Cooling Rates	0.1 to 500°C/min in 0.1°C increments.
	High Temperature Furnace: $\pm$ 5°C.	Controlled (Program) Cooling	Ambient: 10°C/min to 50°C. 20°C/min to 65°C. 50°C/min to 100°C. 100°C/min to 170°C.
Cool Down Times	Standard Furnace: 1000°C to 50°C in less than 15 min.		

Sample Type	High Temperature Furnace: 1500°C to 100°C in less than 35 mins. Solids, liquids, powders, films and fibres.
Atmosphere	Static or dynamic including nitrogen, argon, helium, carbon dioxide, air, oxygen or other inert or active gasses. Analyses may also be made at normal or reduced pressure.
Reduced Pressure	Operation to 10^{-4} torr optional.
Temperature Sensors	Standard Furnace: chromal-alumel thermocouple. High Temperature Furnace: Platinum/10% rhodium-platinum thermocouple.
Cooling Method	Forced air cooling.

GRAPHICS PLOTTER 2

General	High resolution alphanumeric X-Y printer/plotter with graphics.
Operation	Multiple pen operation for automatic multicolour plot generation.

Cooling times	Liquid N_2	10°C to -170°C. 50°C/min to -165°C. 100°C/min to -135°C. 200°C/min to -85°C. 300°C/min to -80°C.
	Ambient:	From 725°C to 100°C in under 4 minutes.
	Liquid N_2	From +200°C to -150°C in under 2 mins.
Temperature Range		Standard unit allows operation from ambient to 725°C. With optional cooling accessories, the range may be extended to -170°C.
Temperature Sensors		Distributed platinum resistance thermometers.
Atmosphere		Static or dynamic including nitrogen, argon, helium, carbon dioxide, air, oxygen or other inert or active gasses.
Sample Type		Solids, liquids, powders, films or fibres.

Sample Containers	Type	Material	Pressure Rating
	Standard	Aluminium,gold platinum, copper graphite.	Ambient
	Sealed	Aluminium,gold	2 atmos.

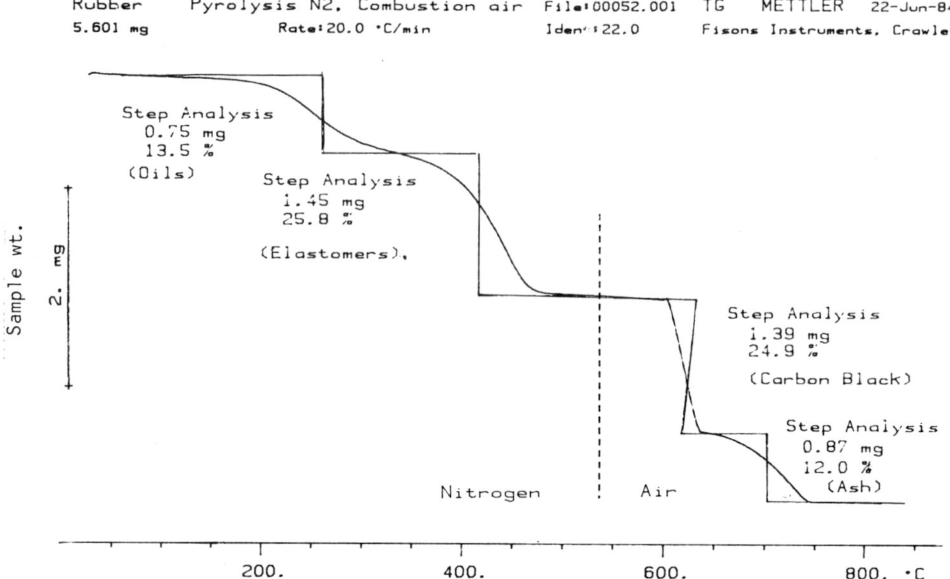

Figure 140   Derivative thermogram obtained by oxidation of synthetic
                rubber.

be carried out in an atmosphere of oxygen or air. Applications of the
various techniques listed in the previous section to the determination of
oxidative stability are discussed below.

   <u>Thermogravimetric analysis.</u> In this technique the sample is heated
in air or oxygen at a constant heating rate until complete oxidation or
combustion occurs and weight changes are continuously recorded as a weight
versus temperature plot. In Figure 140 is illustrated a derivative
thermogram obtained in the oxidation of a synthetic rubber sample. Very
often the shape of the derivative thermogram is typical for a certain
polymer and can be used for characterisation.

   Thizon et al.[1733] have studied the effect of low concentrations of
oxygen and water in nitrogen on the rate of thermal degradation of
ethylene-propylene copolymers. Using thermogravimetric analysis they are
able to obtain estimates of the rate and change of polymer weight with
time (dm/dt), hence activation energy

$$\frac{dm}{dt} = K \exp \frac{(-E)}{RT}$$

at a range of polymer temperatures (240-350°C) and oxygen contents in the
head gas (0-90 ppm) and water contents in the head gas (0-600 µpm). It
is known that oxygen reacts with hydrocarbon polymers even at moderate
temperatures[1734]. The resulting peroxides or hydroperoxides which are
thermally unstable can induce a chain degradation through reactions
involving the $RO_2°$ radical. Oxygen contamination of the head gas in a
thermogravimetric experiment is thus likely.

   The oxygen effect is severe, particularly at low temperatures. The
degradation rate can be multiplied by a factor as large as six for an
oxygen concentration of 90 ppm at 240°C; it is multiplied by 1.2 at 280°C
and 2 at 240°C for a concentration of 20 ppm. The water content of the

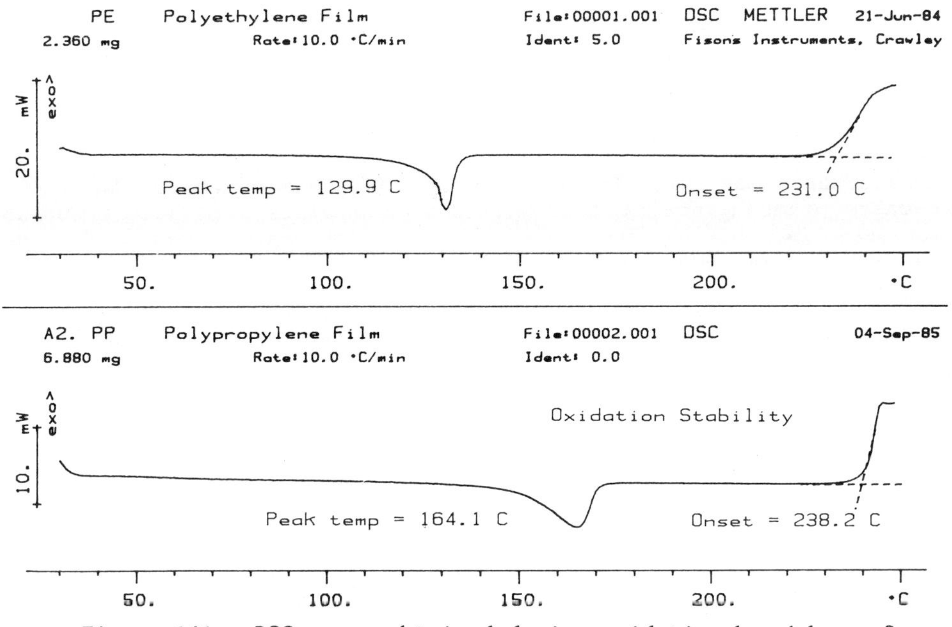

Figure 141   DSC curve obtained during oxidative breakdown of
polyethylene and polypropylene.

carrier gas is not as critical at least for the ethylene-propylene
copolymers and does not affect appreciably the thermal degradation.

   <u>Differential scanning calorimetry.</u> To examine oxidative breakdown,
a sample is tested in flowing air or oxygen. Since oxidation is highly
exothermal, onset of oxidation is clearly visible on a differential
scanning calorimeter trace. The time to the extrapolated onset of the
exothermal reaction, under isothermal conditions is called the induction
period. Dynamic determination of oxidation stability (heating the sample
at 10°C/min, for example) is much quicker than the isothermal procedure
and permits simultaneous measurement of glass transition or melting
temperatures. This type of analysis is shown in Figure 141 by the
comparative performance between polyethylene and polypropylene films. As
with the isothermal induction period, the resulting onset temperature is a
measure of polymer stability.

   This type of analysis is very useful for polyolyfins intended for
use in products that must be very durable (cable insulation, water pipes).
It is also possible to measure the performance of stabilizers, or how
metal ions, such as copper catalyse oxidation, the extent of which can be
determined by oxidizing the sample in open aluminium and copper crucibles
for comparison.

   A relatively new technique is pressure differential scanning
calorimetry; (PDSC), This technique measures heat flow and temperatures of
transitions as a function of temperature, time and pressure (elevated
pressure or vacuum). The ability to vary pressure from 0.01 torr to 1000
psig makes pressure differential scanning calorimetry ideal for the
determination of oxidative stability and for other pressure sensitive
reactions.

Dupont supply a dual sample pressure differential calorimetry cell suitable for use in their 9900 thermal analysis system (Table 180).

Antioxidants are common additives for polymers. Oven aging and thermal analysis techniques such as differential scanning calorimetry and thermogravimetric analysis have been used with varying degrees of success to measure both the concentration and effectiveness of antioxidants. In most cases, the failure of these tests to correlate with actual end-use performance is due to the volatility of the antioxidant and other components at test temperatures.

Figure 142 shows the thermogravimetric analysis of five different antioxidants at 150°C and ambient pressure. As can be seen, even at this relatively modest test temperature, the volatility of these stabilizers varies greatly. If an isothermal test were performed under these conditions, results would be greatly affected by the volatility of the antioxidant and not accurately reflect end-use performance.

Pressure differential scanning calorimetry eliminates the volatility problem in two ways. First, high pressure decreases the volatility by increasing the boiling point, and second, high pressure increases the concentration of the reacting gas, oxygen. This allows lower test temperatures or provides significantly shorter test times at equivalent temperatures.

Figure 143(a) illustrates a differential scanning calorimetry heating experiment in air on an impact-modified polypropylene sample. It shows the melt of two components followed by oxidative decomposition at 237.7°C. When run by thermogravimetric analysis in nitrogen the same material shows thermal degradation starting near 300°C (Figure 143(b)). However, an expanded view of the top 5% weight loss shows a component (0.66%) of the sample is lost between 126 and 184°C at ambient pressure. The identity of the component is unknown, but the mere fact that the weight loss occurs makes oxidative stability results suspect if they are generated at a temperature above 126°C and at ambient pressure.

Figure 143(c) shows isothermal differential scanning calorimetric and pressure differential scanning calorimetric curves obtained for the same sample. The pressure differential calorimetric curve was run at a temperature (164°C) at which the sample is still partially solid, and results show a sharp, reproducible onset for oxidation. In the solid state, additives are less likely to migrate to the surface and the sample is well below its thermal decomposition temperature. At ambient pressure, the lowest test temperature that could be used to detect the oxidation was 184°C. At this temperature, very high sensitivity was required to even see the oxidation, and results are suspect because of the 0.66% weight loss that occurs starting at 126°C at ambient pressure (thermogravimetric analysis data). Maintaining the sample in the solid phase for such stability correlations lends more credibility to correlations with real-life expectancies since the end-use failure is a solid phase decomposition mechanism.

The following advantages of pressure differential and scanning calorimetry over differential scanning colorimetry and oven aging for determination of oxidative stability have been identified as follows:[1735]

1.   At 600 psig oxygen pressure and a specimen size of 4-5 mg, the maximum oxidation rate is maintained throughout the reaction and the results are not dependent on oxygen concentration or diffusion rate.

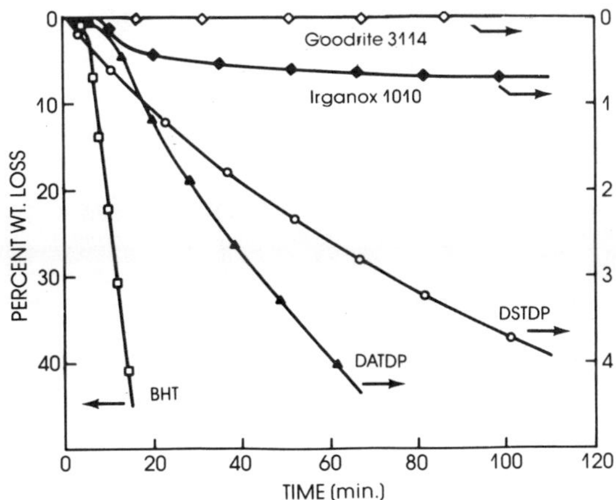

Figure 142   Relative volatility of stabilizers at 150°C.

2.   Temperature and pressure can be precisely controlled with commercial
      thermal analysis equipment.   This eliminates errors caused by
      temperature variations commonly encountered in oven testing.

3.   Pressure differential scanning calorimetric oxidation exotherms are
      sharper and better defined than the usual differential scanning
      calorimetric tests at atmospheric pressure, making extrapolation of
      the onset of autoxidation easier and more precise.

4.   The test temperature may be varied to permit analysis of samples in
      the molten, semi-molten, or non-molten state, or to permit changing
      the duration of the test; conversely, studies may be conducted at
      relatively low temperatures if a longer test period can be
      tolerated.

5.   Because high oxygen pressures accelerate auto-oxidation, the test
      can be carried out at sufficiently low enough temperatures to
      minimize volatilization of stabilizers.

      Thermal volatilization analysis.   This technique is applicable to
thermal oxidative studies on polymers.

      Evolved gas analysis.   Schole et al[1736] have reported on the
characterisation of polymers using an oxidative degradation technique.   In
this method the oxidation products of the polymers are produced in a short
pre-column maintained at 100-600°C just ahead of the separation column in
a gas chromatograph.   The oxidation products are swept on to the
separation column and detected in the normal manner.

      Figure 144 shows the chromatograms of Carbowax 20M and UCON Polar
polyglycols.   The components are the oxidation products produced in the
pre-column by the oxygen.

      An alternative technique involves infrared spectroscopic examination
of films of the polymer that have been subject to oxidative degradation at
elevated temperatures[1737-1743].   Changes in concentrations of carbonyl and
other oxygenerated functional groups can be obtained by this method.

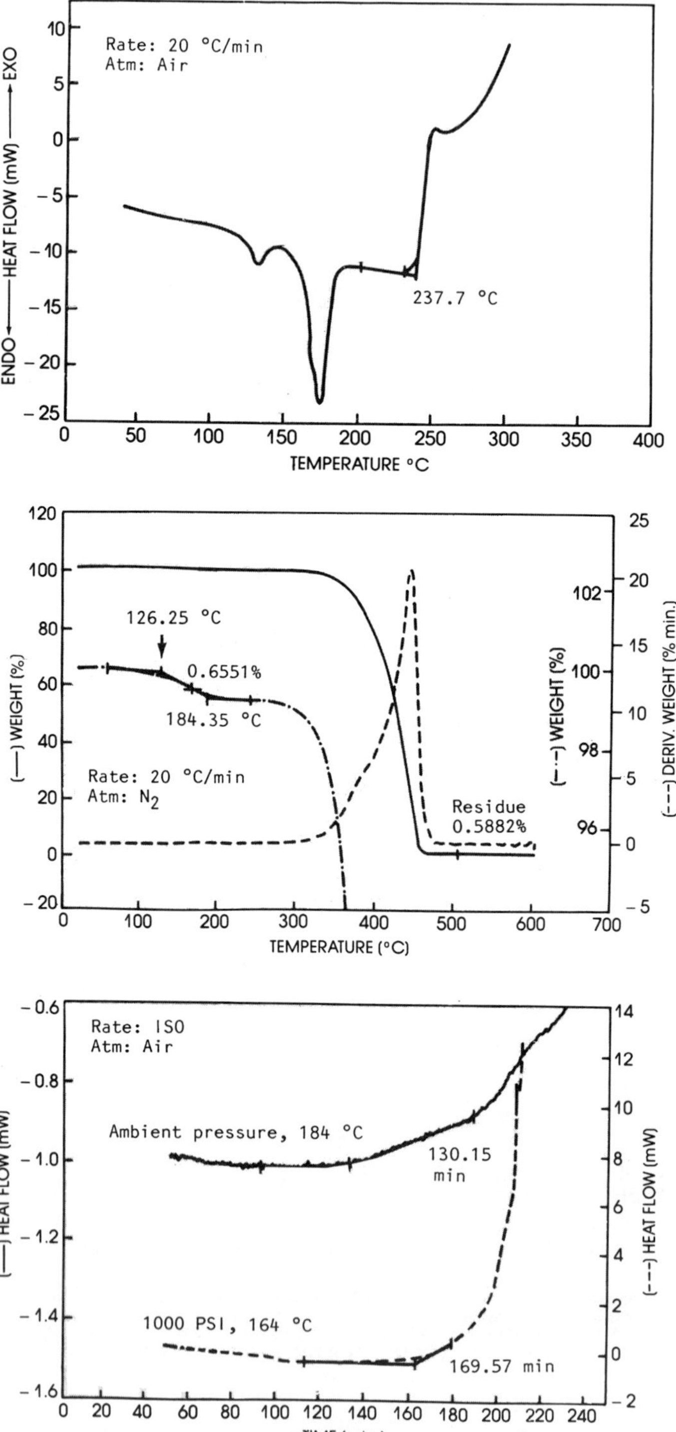

Figure 143 Thermal properities of polypropylene: (a) melt and oxidative degradation; (b) composition and thermal degradation; (c) oxidative stability.

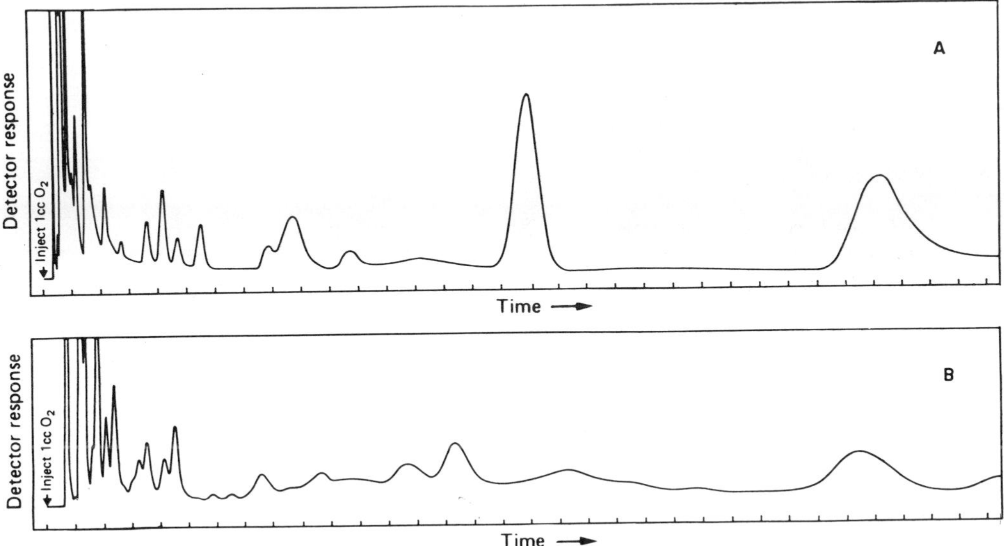

Figure 144   Gas chromatogram of oxidation products for Carbowax 20M
               and UCON polar. (A) 15% Carbowax 20 M on Chromosorb P;
               column length 1 ft; precolumn temperature, 203°C;
               attenuation, 16 (Aerograph, A-90-C with flame
               ionization). Analytical column 15% Carbowax 20 M; 5%
               stearic acid on Gas Pack S (60 to 80 mesh); column
               temrperature 98°C, helium flow rate 70cc/min; chart
               speed 3 min/in. (B) 15% UCON polar on chromasorb P;
               precolumn temperature, 202.5°C.

11.2.2 Infrared Spectroscopy of Oxidised Polymers

Extensive infrared studies have been carried out on oxidized
polyethylene[1744-1750]. Infrared absorption spectrometry has been widely
used to determine the oxidation products and the rate of formation of
these products during the thermal or photo-oxidation of polyethylene[1751,1752,1753]. Acids, ketones and aldehydes, the end-products
reported from these oxidations, have similar spectra in the 5.50-6.00
micron region. It is only in this carbonyl stretching region that the
products have suitable absorptivity to give quantitative data. The
absorption band of the acid (5.84 micron) ketone (5.81 micron) and
aldehyde (5.77 micron) groups present in oxidized polyethylene are so
overlapped as to give only a broad band on relatively low-resolution
infrared spectrometers. Interpretation of these data, based on the
increase in total carbonyl rather than on a single chemical moiety, could
lead to incorrect conclusions because of the large differences in the
absorptivity of the various oxidation products. Acid absorptivity has
been reported [1754] to be 2.4 times greater than that of ketones and 3.1
times greater than that of aldehydes. Rugg et al[1745] using a grating
spectrometer for increased resolution, have demonstrated that the carbonyl
groups formed by heat oxidation are mainly ketonic, while in highly photo-
oxidized polyethylene the amounts of aldehyde, ketone and acid are
approximately equal. This procedure is adequate for qualitative, but
suitable for accurate quantitative, data.

An infrared study of oxidative crystallization of polyethylene was made from examination of the 5.28 micron 'crystallinity' band and 7.67 micron 'amorphous' band and carbonyl absorption at 5.83 microns[1753]. Miller and co-workers[1756] used Fourier transform infrared to study the effect of irradiation on polyethylene. Aldehydic carbonyl and vinyl groups decreased and the ketonic carbonyl and trans-vinylene double bonds increased on irradiation.

Infrared reflection was used in studies of oxidation of polyethylene at a copper surface in the presence and absence of an inhibitor, N,N-diphenyl-oxamide[1757].

Cooper and Probe[1758] have used alcoholic sodium hydroxide to convert the acid groups to sodium carboxylate (6.40 micron) to analyse polyethylene oxidized with corona discharge in the presence of oxygen and oxone. This procedure requires 5 days, and has been found to extract the low molecular weight acids from the film.

Heacock[1759] has described a method for the determination of carboxyl groups in oxidized polyolefins without interference by carbonyl groups. This procedure is based upon the relative reactivities of the various carbonyl groups present, in oxidized polyethylene film, to sulphur tetrafluoride gas. The quantity of the carboxyl groups in the film is then measured as a function of the absorption at 5.45 microns.

$$R - C{\overset{O}{\diagdown}}_{OH} + SF_4 \rightarrow R - C{\overset{O}{\diagdown}}_{F} + HF + SOF_2$$

The appearance in the spectra of irradiated (gamma rays from ^{60}Co) polypropylene specimens, after storage in air, of strong band in the region of 5.85 microns, corresponding to carbonyl groups, must be explained by reaction of oxygen with the long-lived allyl radical, with formation of peroxide radicals which form carbonyl groups by decomposition.

$$\begin{matrix} & Me & & Me \\ & | & & | \\ - CH_2 - & C & - CH = & C - \\ & | & & \\ & OO & & \end{matrix}$$

The intensity of the 5.85 micron band (and consequently the degree of oxidation) increases sharply with time of storage of specimens in air. Irradiated amorphous specimens oxidize to a considerably smaller extent than isotactic polypropylene specimens. The degree of oxidation of specimens irradiated at - 196°C increases more rapidly than when the specimens are irradiated at 25°C. All these facts indicate that the lifetime of the allyl radicals is longer in crystalline polypropylene, and that the concentration of these radicals is higher in specimens irradiated at low temperature. The free radicals are destroyed only after heat treatment of the specimens in an inert atmosphere at 150°C. After this heat treatment the intensity of the 5.85 micron band ceases to increase on storage, i.e. no further oxidation occurs.

Adams[1760] has compared on a qualitative basis the non-volatile oxidation products obtained by photo-and thermal oxidation of polypropylene. He used infrared spectroscopy and chemical reactions. The major functional group obtained by a photodecomposition is ester followed by vinyl alkene, then acid. In comparison, thermally oxidised polypropylene contains relatively more aldehyde, ketone and gamma-lectone, and much less ester and vinyl alkene. Photodegraded polyethylene contains mostly vinyl alkene followed by carboxylic acid. Gel permeation chromatography determined the decrease in polypropylene molecular weights with exposure time. Adams determined that there is one functional group formed per chain scission; in thermal oxidation there are two groups formed per scission.

Adams[1760] makes the following comments regarding the infrared spectrum of oxidized polypropylene.

<u>Hydroxyl region.</u> The hydroxyl absorption in the infrared for polypropylene has a broad band centred at 2.90 microns (associated alcohols) with a definite shoulder at 2.77 microns (unassociated alcohols). At a similar extent of degradation, thermally oxidized polyolefins show hydroxyl bands of roughly half the absorbence values of the photo-oxidised polyolefins. Thus thermal oxidation produces about half as many hydroxyl groups as photooxidation in polyolefins.

A portion of the polypropylene hydroxyl absorption could be due to hydroperoxides. If so, then an exposed sheet, with the volatiles removed, heated in a nitrogen atmosphere for 2 days at 140°C, should show a decrease in the hydroxyl infrared band and an increase in the carbonyl band due to the decomposition of hydroperoxides under such treatment. The infrared spectrum of the photodegraded polypropylene sheet subjected to the thermal treatment showed at 20% decrease in the hydroxyl band. however, the broad carbonyl band at 5.75 microns did not increase but showed a 5% decrease. The small gamma-lactone (5.75 micron) and vinyl alkene (6.08 micron) bands did show a slight increase, however. Thus, these results are due not to hydroperoxide decomposition but to some carboxylic acids converting to gamma-lactones and some terminal alcohols dehydrating to vinyl alkenes at the high temperature. While hydroperoxides are undoubtedly an intermediate in the photo-oxidation process, they decompose too rapidly under ultraviolet light to build up any significant concentration.

<u>Carbonyl region.</u> The polypropylene carbonyl band after 335 h exposure is broad, with few discernible features except for the vinyl alkene band at 6.08 microns. The broadness of the carbonyl band indicates a large variety of functional groups, and makes accurate quantitative analysis difficult. The large vinyl alkene at 6.08 microns stands out clearly and distinct carboxylic acid (5.83 microns) and gamma-lactone (5.58 microns) spikes can be readily identified.

After the volatile products are remove by the vacuum oven, the carbonyl band for polypropylene decreases. Isopropanol extraction removes about 40% of the polypropylene carbonyl. The carbonyl band is then narrow and appears to centre at the ester absorption at 5.75 micron.

Treatment with base converts lactones, esters and acids to carboxylates (6.33 microns), leaving only a small band at 5.81 microns, which is due to aldehyde and ketone.

Upon reacidification of the polypropylene, some of the original esters at 5.75 microns do not re-form but become carboxylic acids and gamma-lactones. Curiously, the vinyl alkene band becomes less intense with each step and broader, shifting down to 6.10 - 6.25 microns, the vinyl groups may be isomerized into internal alkenes or become conjugated during the various treatments, although no such change occurs with either the polyethylene vinyl alkene or with the process-degraded polypropylene vinyl alkene. Wood and Statton[1761] developed a new technique to study molecular mechanics of orientated polypropylene during creep and stress relaxation based on use of the stress-sensitive 10.25 micron band and orientation-sensitive 11.12 micron band. The far infrared spectrum of isotactic polypropylene was obtained from 400 to 10 cm^{-1} and several band assignments were made[1762]. Isotacticity of polypropylene has been measured from infrared spectra and pyrolysis-gas chromatography following calibration from standard mixtures of isotactic and atactic polypropylene. The infrared spectrum of oxidised polypropylene indicated small amounts of OOH groups plus larger concentrations of stable cyclic peroxides or expoxides in the polypropylene chain[1762].

Grassie and Weir[1763] described an apparatus for the measurement of the uptake of small amounts of oxygen by polystyrene with a high degree of precision. Grassie and Weir[1764] investigated the application of ultraviolet and infrared spectroscopy to the assessment of polystyrene films after vacuum photolysis in the presence of 253.7 nm radiation using the apparatus mentioned above. During irradiation there is a general increase in absorption in the region 230-350 nm Rates of increase are relatively much greater in the 240 and 290-300 nm regions. Absorption in the 240 nm region is characteristic of compounds having a carbon-carbon double bond in conjunction with a benzene ring. Styrene, for example, has an absorption band at 244 nm.

Schole et al[1765] have applied an oxidative degradation technique to the study of polystyrene. In this technique the polystyrene sample is mixed with a support in a precolumn which is mounted at the inlet to a gas chromatographic column.

Shaw and Marshal[1766] have carried out an infrared spectroscopic examination of emulsifier free polystyrene which had been oxidized during polymerization. Evidence was found for the presence of surface carboxyl groups bound to the polymer chains, presumably formed by oxidation during polymerization. The band at 5.86 microns was assigned in part to the carbonyl stretching mode of dimeric carboxylic acid, formed by oxidation, in the polystyrene chains. Adsorption at 5.65 microns which was very weak, was tentatively attributed to the carbonyl stretching mode of the monomeric form of this acid. The structure of the acid endgroup was not established but the results obtained suggest that it was possibly a phenylacetic acid residue or a residue of standard (unoxidized) and of oxidized emulsion polymerized polystyrene in the region 12.5 to 25.0 microns.

11.2.3 Electron Spin Resonance Spectroscopy

Ohnishi et al[1767], have carried out an ESR study of the radiation oxidation of PVC. This technique has also bee applied to studies on polyethylene[1768], polymethylmethacrylate[1769], PVC[1770], polymethacrylic acid[1771] and polycarbonate[1772].

11.3 Photooxidative Stability

Differential scanning calorimetry. This technique has been used in conjunction with differential thermal analysis to investigate the kinetics of the oxidation of isotactic polypropylene[1773].

   <u>Differential photocalorimetry.</u> Differential photocalorimetry is a
new thermal analysis technique that measures the cure rate and degree of
cure of the photocurable polymers. The technique uses dual sample
differential scanning calorimetry to measure the heat of reaction of one
or two samples as they are exposed to light. The light source can be of
several types, the most common being a high-pressure mercury arc lamp with
maximum intensity in the 200-400 nm (ultraviolet) range.

   A number of users of commercial differential scanning calorimetry
instruments have built homemade versions of the differential
photocalorimetry but have only been partially successful in the analysis
of photocurable polymers. Their failure to develop a completely adequate
system has been a result of two factors. The first and most significant
is the change in the intensity of the light with time of operation - as
much as an 80% reduction in the first 1000 hr of operation. The second
reason for limited success was the lack of data analysis software to
convert raw data into easy-to-understand results that could be correlated
with actual performance.

   These problems have been solved with commercial differential
photocalorimetry systems such as the Du Pont 930 Differential
Photocalorimeter. It uses a dual sample differential scanning calorimeter
to measure heat generated during the photocure. The advantage of dual-
sample differential scanning calorimetry over standard differential
scanning calorimetry for photocure studies is that it allows use of an
internal standard. The sample and standard are subjected to identical
conditions such as light intensity and wavelength, temperature, and
atmosphere, allowing absolute comparisons even if the light intensity
changes.

   Although output of the lamp diminishes with time, the actual effect
on the sample can be minimized with a photo-feedback system that varies
voltage to maintain intensity. This allows such instruments to provide
much more consistent data over long periods of time.

   Almost all modern thermal analysis systems are computer-based.
These computers can contain a megabyte or more of memory and perform even
complex analyses in seconds. Data analysis software is usually menu
driven and contains soft-keys for easy, one-button operation, allowing
users to obtain results quickly and with minimal training.

   <u>Photocure rates.</u> The use of radiation-curable coatings and
adhesives is growing rapidly.

   UV-curable polymer systems typically contain monomer or oligomer
(unreacted polymer), a light sensitive photoinitiator, and other additives
for colour and mechanical properties. Concentrations of these components
determine viscosity, cure rate, adhesion, flexibility and abrasion
resistance. These properties are also directly affected by the degree of
cure. Therefore, to ensure consistent product, it is necessary to monitor
response of the photopolymer both before and after processing.

   Figure 145(a) compares the response of two different photopolymer.
Sample #1 cures very rapidly, while #2 takes more than 50 sec of exposure
to reach its maximum rate. Using the data analysis software, the operator
can generate a plot of percent cure versus time and quickly determine if
the material is within specification.

   The one component that significantly affects the cure rate of
photocurable polymers is the photoinitiator. Decreasing the concentration

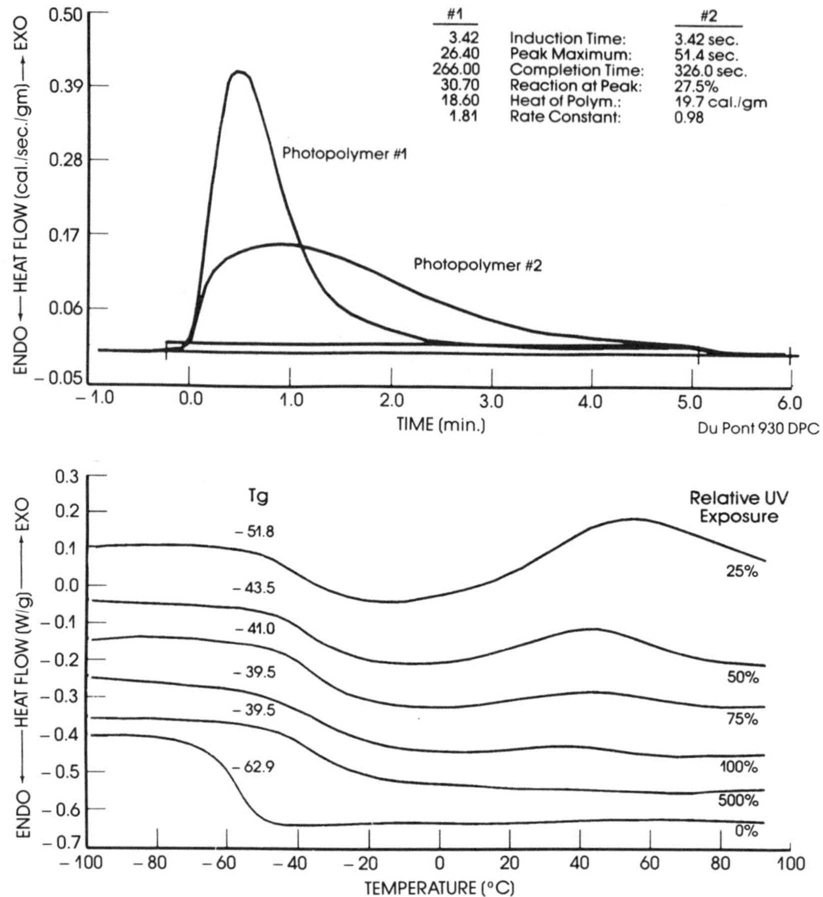

Figure 145   Differential photocalorimetry of photopolymers (a) no.1
               rapid curve polymer, no.2 slow cure polymer, (b) effect
               of exposure on glass transition of photopolymers.

of a photoinitiator from 0.5% to 0.3% can more than double the induction
time for the reaction. Since the photoinitiation is frequently the most
expensive of the photopolymer ingredients, differential photocalorimetry
offers a fast, reliable method for optimizing and verifying the
concentration of the photoinitiator.

Degree of cure. Traditionally, differential scanning calorimetry
has been used with most thermosetting polymers to measure the degree of
cure. Two techniques have been developed to make the measurement. The
first is to measure the glass transition temperature (Tg) of the material
(Chapter 12). As cure proceeds, chemical cross-linking increases and the
glass transition temperature moves toward some maximum value. The second
method for determining the degree of cure of thermosets uses the residual
heat of reaction of the sample as it is heated through the curing
temperature in the differential scanning calorimeter.   There are
advantages and disadvantages for each method.   Therefore, since both
values can be easily obtained from a single differential scanning
calorimetry experiment, it is best to measure both.

Photopolymers can be analysed in a way very similar to thermosets. The only difference is that photopolymers use light to initiate chemical cross-linking, while thermosets use heat. Figure 145(b) shows the effect of varying the exposure time on a photopolymer used for electronic applications. With no exposure (0%), the glass transition temperature is - 62.9°C. As the relative amount of UV exposure is increased from 0% to the recommended amount (100%), the glass transition temperature is seen to increase to - 39.5°C. Even at 5 times the normal exposure, no further increase in Tg is seen, indicating that the standard exposure provides nearly complete cure.

Another phenomenon seen in Figure 145(b) is the effect of heat on samples that have been only partially exposed. As these materials are heated, an exothermic peak is seen between 20 and 100°C. This peak is caused by additional curing taking place and indicates that once the sample is exposed and cure begins, cure can be continued using either light or heat. Since differential photocalorimetry can control both of these variable, either sequentially or simultaneously, it promises to be a powerful tool for quality control and research on photopolymer systems.

11.4 EXAMINATION OF POLYMER COMBUSTION PRODUCTS

When polymers are burnt or smoulder in air, the combustion products are extremely complex, often consisting of several hundred compounds. Because of the toxic or unknown nature of these products, it is important to know their composition in some detail. This information is also essential for mechanistic and modelling studies of the smoke formation process, which can lead to design of less hazardous polymers in the future.

A number of analytical methods [1774,1775] involving pyrolysis of polymers have been reported in the literature. Michal, Mitera and Tardon[1776] developed a method using direct gas chromatography - mass spectrometry for their study of the combustion of polyethylene and polypropylene. Morikawa[1777] used gas chromatography to determine polycyclic aromatic hydrocarbons in combustion of polymers. Liao and Browner[1778] described a method for the polycyclic aromatic hydrocarbons. Many other workers have studied soot and smoke formation and mechanisms in the combustion of polymers. Generally in these studies, relatively simple and specific methods were used, which were appropriate for the intended tasks. However, these methods are not suitable for complete analysis of the very complex smoke particulates resulting from combustion of many polymers. Most methods have been developed either for volatile compounds of low molecular weight or for polycyclic aromatic hydrocarbons. Joseph and Browner[1779] developed a method which can be used to analyse all classes of compounds produced by the combustion of polymers and applied this method to rigid polyurethane foams prepared by the polymerization of diphenylmethane diisocyanate and glycols:

$$R(NCO)_2 + R^1(OH)_2 - OCN \, HRN \, COOR^1O -$$

The method is directed toward the particulates produced and exclude the volatile compounds of very low molecular weight.

Smoke particulates from the smouldering of polyurethane foam in air were collected on glass fibre filters and extracted with chloroform. The concentrated extract was subjected to acid and base extractions. The acid compounds was converted to methyl esters and analysed by gas chromatography/mass spectrometry/data system. The basic and phenolic

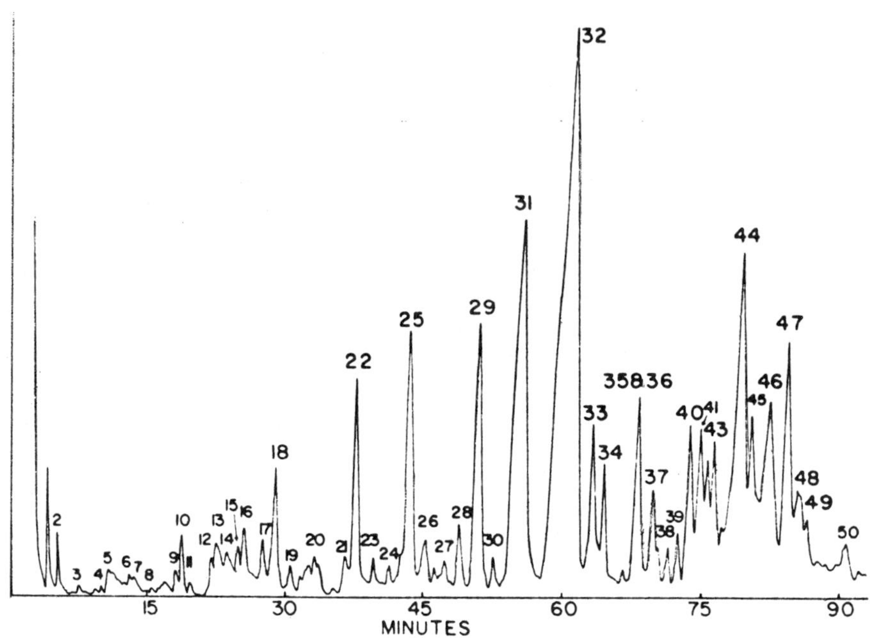

peak #	mol. wt.	name	peak #	mol. wt.	name
1	93	aniline	26	180	3,8-phenanthroline
2	107	toluidine	27	193	methyl benzoquinoline
3	135	dimethyltoluidine	28	207	dimethyl benzoquinoline
4	149	trimethyltoluidine	29	221	trimethyl benzoquinoline
5	148	dipropylene glycol methyl ether	30	207	dimethyl benzoquinoline
6	191	NI[a]	31	226	4,4'-diaminodimethyl diphenylmethane
7	143	3- or 4-methylquinoline	32	207	dimethyl benzoquinoline
8	157	2,6-dimethylquinoline	33	198	4,4'-diaminodiphenylmethane
9	157	4-ethylquinoline	34	212	4,4'-diamino methyl diphenylmethane
10	191	NI	35	226	4,4'-diamino dimethyl diphenylmethane
11	171	1,7-dimethylene-2,3-dimethylindole ?	36	240	4,4'-diamino trimethyl diphenylmethane
12	171	trimethylquinoline	37	240	isomer of above
13	206	tripropylene glycol methyl ether	38	234	NI
14	206	isomer of above	39	248	234 + 14
15	185	quinoline + 56	40	296	NI
		a propylene glycol derivative	41	248	isomer of peak 39
17	209	$C_{15}H_{15}N$	42	262	248 + 14
18	209	isomer of above	43	262	248 + 14
19	207	N-tolyl butylurethane	44	296	NI
20	207	isomer of above	45	276	248 + 28
21	249	207 + 42	46	276	262 + 14
22	249	isomer of above	47	290	276 + 14
23	183	N-phenyl p-toluidine	48	304	276 + 28
24	179	benzoquinoline	49	318	290 + 28
25	193	3-methyl benzoquinoline	50	394	318 + 76

[a] NI = Not identified.

Figure 146   Total ion chromatogram for basic fraction.

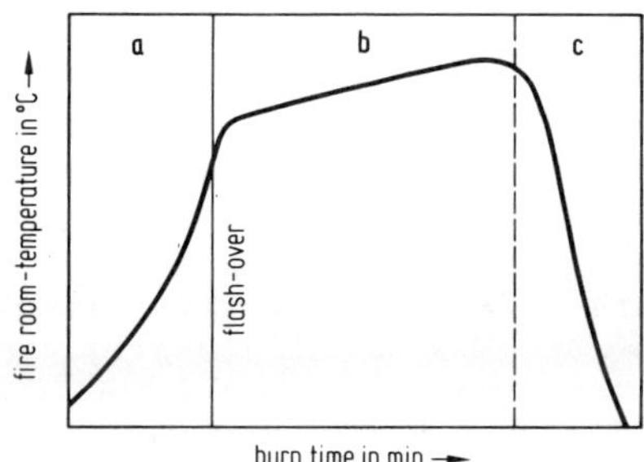

Figure 147   Temperature/time fire profile: a = fire development
phase; b = fully developed fire; c = receding fire.

compounds were analysed on the same system without derivatization. The
neutrals were separated into different classes by high performance liquid
chromatography on a bonded amine column. Different fractions were collected and each fraction is analysed on gas chromatography - mass
spectrometry with data collection.

In figure 146 is shown a total ion chromatogram with peak
identification for the basic fraction obtained from a rigid polyurethane
foam which had been heated to 340°C. Most of the peaks in the ion
chromatogram where identified from chemical ionization derived molecular
weights.

In separating the neutral compounds into various fractions, a bonded
amine column was used. The bonded amine column eliminates some of the
problems associated with silica and alumina columns, viz,
nonreproducibility of separations due to the variation of water content of
eluant and samples and loss of trace components due to the high
adsorptivity of adsorbent.

Some nitrogen-containing compounds were identified in the neutral
fractions. There were five-membered nitrogen-containing ring compounds,
including indoles, isoxazole, indazole, and carbazoles which do not show
basic properties, and consequently do not react with 1 N hydrochloric acid
in the basic extraction step. A number of phthalate esters were also
present in different fractions. It is possible for phthalates from the
original sample to be separated in fractions 2 to 4, but the
isooctylphthalate in neutral fraction 5 is probably an impurity.

11.5 COMBUSTION TESTING AND RATING OF PLASTICS

11.5.1 Introductory

The progress of a fire is often divided into three phases consisting
of 1) ignition and early development, 2) total involvement, 3) fire
recession and extinguishment. The temperature profile of such a process
is depicted schematically in Figure 147.

The fire characteristics of a material are characterized by ease of
ignition, contribution to flame spread, and heat contribution, as well as
other factors generally associated with fires, including smoke density,
toxicity, and corrosiveness of the combustion by-products. Fire
behaviour, however, cannot be considered a materials property because it

is markedly affected by both material and environmental factors. These include the distribution of material in the room, material geometry and other physical factors, temperature history, thermal conductivity, the intensity and type of ignition source, exposure time to the ignition source, integrity of the material and ventilation effects[1780,1781]. The varying nature of the fire risk situation and the influence of materials and environmental factors makes it very difficult to establish tests and ratings criteria that would be generally applicable. That is, no doubt, on the main reasons that many different tests have been developed in research, industry and by regulatory agencies. Many of these, in part material specific, procedures serve well as quality control or developmental guide-posts. The results obtained from within the regime of a test standard are, at best, of limited use in drawing conclusions concerning the performance in real fire situations[1782,1783]. Here we must insure that the environmental conditions adequately simulate the fire risk situation of concern.

The risk assessment must differentiate between the production, storage, and end-use applications. Different risks in production often are due to the amounts of combustible material. The end-use application, transportation, construction, furniture or furnishings, all have a major impact in determining risk. Duplication of natural ignition sources such as cigarettes, sparking contacts, and overheated wiring are a first step toward proper applications and risk-orientated testing.

Small laboratory tests as, for example, ASTM D-1692[1783], Oxygen Index[1784], Setchkin-Test[1785] or similar tests under certain circumstances provide a means for production quality control. Their results can only be applied toward the characterisation of fire risk assessment if it can be demonstrated that the tests adequately simulate the actual fire situation.

The small laboratory test, previously designated ASTM D-1692, was abandoned without replacement as a consequence of the 1978 FTC action[1786,1787]. As part of the UL test procedure, Subject 94, this test was used to determine classes HBF, HF-1, and HF-2[1788]. The laboratory procedure consisted of exposing a test specimen to the flame from a Bunsen burner outfitted with a wing tip flame spreader. The rate of and extent of burning were measured. The Oxygen Index Test measures the oxygen/nitrogen concentration needed to support combustion of a test specimen. The specimen is positioned vertically and ignited at the top. Results from such small laboratory tests are, by definition, not suitable for drawing conclusions in real fire situations. Enriched oxygen environments are found in the space industry. In such a case, the Oxygen Index Test could be used to study and draw conclusions about the ignition phase from ignition sources of low intensity.

The Setchkin-Test according to ASTM D-1929, is used to determine the temperature for solids at which ignitable decomposition products are formed. This is analogous to determining flashpoint temperature of liquids[1789,1790]. The term flash ignition (FI) temperature refers to ignition points determined, using a small gas pilot flame. Self-ignition (SI) temperature refers to the temperature where the gases ignite through contact with the hot oven walls.

The application of combustion testing of plastics and rubbers in different application is discussed below in further detail.

11.5.2 Mining Applications

Mining applications require that specific test criteria be met because of the extraordinary difficulties in rescue and fire extinguishment efforts underground. Highly expandable and combustible blowing agents may not be used if they increase the fire hazard underground. In case of an accidental fire, the blowing agents used should not propagate the fire. The tunnel test facility per DIN 22118[1791] is used, for example, to evaluate the contribution to fire spread of conveyor belts with textile inserts, such as would be used in anthracite mines. The test samples measure 1200 x 90 mm and are placed horizontally into the laboratory test tunnel. The test sample is exposed from underneath to the flame of a special propane burner for a period of 15 minutes. The burner is positioned 170 mm from the front edge of the sample. A maximum burn extent and afterburn are specified in order to meet the minimum test criteria. The basis for the standardization of this test procedure was a series of full-scale tests carried out by the Tremonia Mining Organization in Dortmund.

11.5.3 Electric Applications

In Germany, performance requirements for electric applications are regulated through VDE guidelines[1792-1794]. Localized smouldering or ignition may result from overheated wires, sparking contacts, or other failure of electrical equipment. Underwriter Laboratories, Incorporated, has developed universally applicable fire protection test procedures[1795]. According to UL Subject 94[1788], horizontally (or, if applicable, vertically) oriented test samples are exposed to a gas flame source. Test criteria consists of rate of burning, the burn distance, or the continued burning of test samples themselves and/or drippings from melted materials after removal of the flame source. A horizontal test procedure using a 30 second flame exposure is used for determining the HB-classification. This test is a analogous to ASTM D-635[1796] and the discontinued ASTM D-1692. Bar-like samples measuring 127 x 12.7 x 6.4 mm are used in both the vertical and the horizontal test procedures. The UL-94 vertical test procedure also requires that the test specimens be conditioned for seven days at 70°C. The sample is ignited at the lower end. The afterburn time of the sample or dripped material is used for classifying samples as V0, V1 and V2.

The ASTM E-162 test procedure[1797] normally used for construction applications is used to test larger size electrical boxes for housings. The requirements for such are outlined in UL Subject 94. In this test, the contribution of the sample to surface flame spread is measured under conditions where the sample is exposed to a specified radiant heat flux. Another applications specific procedure is the test for hot wires or contacts[1798,1799] according to VDE-0471. Potential ignition sources through failure of overheated wires or termination screws are simulated in this procedure. The performance of insulation is determined by contact with hot, glowing wires. Wire temperatures at the point of failure are determined. The test procedure specifies stepwise increasing temperatures in the range of 450 to 960°C. Test criteria include ignition time, dripping, and subsequent burning. The insulation test sample is mounted on a movable fixture and pressed against the hot wire loop with a force of 1 N. The penetration distance is restricted. The exact temperature of the wire loop is measured with a miniature thermocouple.

The failure of electrical appliances or fixtures can initiate smouldering which could lead to localized fires. In VDE-0471[1800], such conditions are simulated through, e.g., flame exposure from a Bunsen burner. Depending on the application (for example, TV cabinets, cable, etc.), other fire sources are used.

A widely utilized test procedure is the glowing rod test per VDE/DIN 53459[1801]. Here a rod shaped test sample is pressed tip-to-tip against a glowing rod heated to 960°C with a force of 1 N. Test criteria specify burning rate and extent.

11.5.4 Transportation Applications Area

The Federal Motor Safety Vehicle Standard (FMVSS) 302[1802] serves as a worldwide basis for specification in outfitting automotive interiors. This test provides for a limiting burning rate of "4 inches per minute" and falls under the jurisdiction of the Federal Highway Administration (FHA) of the United States. It has been incorporated worldwide into industrial specifications, national standards, as well as international standards[1803]. The test samples measuring 356 x 100 x d mm and are supported horizontally ("d" is the sample thickness). The free end is subjected to a specific flame exposure. Actual applications-ready materials are to be tested[1804]. These include composites, such as laminates prepared using adhesives, flame-lamination, etc. The decisive test criteria is the burning rate. To insure that the parts that might be individually exported satisfy the requirements of the American statues, the automotive industry has, in part, adopted more stringent internal requirements such as limitation of burn distance.

Combustibility-related regulations covering procurement of materials for use by the German Railroad industry is covered in Specification Books 899/35/I and II[1805]. The regulations governing tests and performance criteria with regard to combustibility, dripping characteristics, as well as smoke density, are specified by form DV 899/35[1806].

The vertically or horizontally arranged test samples are subjected to a wide specified flame for three minutes in the vertical, and two minutes in the horizontal configurations. The extent of surface destruction is used to determine the ratings ranging from "B1, leichtbrennbar" to "B4, nichtbrennbar". The dripping characteristics, as well as the visually determined smoke characteristics are respectively divided into four additional classifications. The finished compo-characteristics are respectively divided into four additional classifications. The finished composites used for padding or cushioning applications are tested. The European Organization of Railroad Authorities is currently considering a French proposal for fire safety determination and testing of seat upholstery. The origin of the procedure comes from model tests designed to simulate the actual application. The fire source is a 100 g paper cushion measuring 36 x 27 cm. The cushion is conditioned for four hours at 70°C. Performance criteria require that the test sample not be completely consumed, that the burn time be less than ten minutes and that the combination of flaming and dripping not occur.

The regulations of the Federal Aviation Administration (FAA) are applicable worldwide to the aircraft transportation industry. The test requirements are specified by FAR 25.853[1807]. Test samples are oriented vertically, horizontally, or at a 45 degree angle and exposed to a specified Bunsen burner ignition source. Test criteria consist of burn distance, occurrence of dripping behaviour, and continued burning after removal of the ignition source. These test requirements have been incorporated into national standards[1808]. Internal Industry specifications[1809] also limit, among other factors, smoke density and toxicity of combustion products. The determination of toxicity is made from an analytical perspective. The concentrations of and compositions of the combustion products are determined from gas samples obtained during smoke density measurements[1810] in the NBS chamber. Concentrations of sulphur dioxide, carbon monoxide, carbon dioxide, hydrogen cyanide, hydrogen chloride etc. are determined.

The requirements of the "Seeberufsgenossenschaft" apply to ocean liners, operating under the German flag. While many aspects are regulated through the Inter Governmental Maritime Consulting Organization (IMCO), the setting of test and performance criteria for combustible materials is left to national agencies. For ships with German registration, a limited use of combustible products is permitted. These must, however, conform to the construction class B 1 in accordance with DIN 4102[1811]. In addition, the judgement of relative toxicity must be favourable. Through animal experiments, the relative toxicity of the combustion products of the test material may be no worse than that of the combustion products of wood or cork. In order to minimize the potential for fire spread, these construction elements must additionally be covered by steel with a minimum thickness of one millimetre. Analogous safety requirements, for example, those covering combustible insulation, are specified as part of the M-Notice[1812] in the United Kingdom.

11.5.5 Furniture and Furnishing Applications

No general regulations apply to furniture or furnishings. German requirements of, for example, theatre curtains, decorations, etc. used in public meeting places or department stores[1813] are generally not well defined. Based on detailed fire statistics[1814-1816], the United Kingdom is preparing a law[1817], using test and performance criteria of BS 5852[1818]. As in the United States[1819,1820], the primary consideration is the reduction of fire risk in homes from cigarette ignition sources. In those cases where the cigarette ignition and/or the simulated match ignition conditions are not successfully passed, the furniture must be so labelled and identified. After passage of an intermediate, transition period, it is planned that all furniture would be required to pass this cigarette ignition test.

The test procedures are based on numerous, full scale model fire tests that were done worldwide. Aside from governmentally required testing[1821,1822], industry carried out extensive testing and trials[1823-1825] to insure statistically correct test and performance criteria.

The United States additionally imposes material specific requirements. Examples are the California State statutes[1827] and regulations of the New York Port Authority[1827]. Among other requirements for cushioning materials are the construction specific tests, such as ASTM E 162 and ASTM D2843[1828]. If the cushioning material fails, upholstery combinations may be substituted.

The British Property Service Agency (PSA) responsible for furniture procurement requires applications oriented tests with correspondingly staggered ignition sources ranging from cigarettes to a wooden crib[1829].

11.5.6 Construction Material Applications

The construction area is forced to comply with the most encompassing set of regulations. In Germany, the requirements for fire prevention and protection are contained in or regulated through zoning regulations, building permits and guidelines, as well as construction specific codes such as DIN 4102. Analogous regulations are promulgated through, for example, the building regulations in Great Britain, building codes in the United States, or the guidelines of the fire police in Switzerland. According to DIN-4102, combustible building materials are categorized into "B1, schwerentflammbar; B2, normal entflammbar; or B3, leichtenflammbar", using both a small burner test and a large chimney test procedure.

The small burner test consists of a vertically oriented specimen which is exposed on either edge or side to a specified ignition flame for 15 seconds. To obtain a B2 classification, the flame front may not have reached a previously marked line at 150 mm within a 20 second time interval inclusive of the 15 second flame exposure time. The test for possible B1 performance uses four vertically arranged test samples, 1000 x 190 x dmm (normally d < 80 millimetres). The samples in this chimney arrangement are exposed to a flame source at their lower edge using a gas ring burner for periods of ten minutes. Test criteria consist of undamaged sample distances and combustion gas temperatures. Dripping behaviour, if any, is noted even though it is not part of the test classification. however, such an observation may be required to obtain the necessary building permits.

Enhanced tests and performance criteria are being developed for classification of composites into B1, B2 and B3 categories.

Floor coverings are evaluated for B1 performance, using both a critical radiant flux test and a modified small burner test procedure. To obtain a B2 rating using this modified test, requires an increased time of 30 seconds. The critical radiant flux test measures the contribution to flame spread of a horizontally orientated test sample. The sample is exposed to a radiant energy flux of varying intensity. The flames may not propagate past the 0.45 W/cm^2 exposure limit. This so-called "critical limit" was determined using wood parquet flooring.

It has proven difficult to determine any correlation between the test results of the various national procedures[1830]. Because of this, the experts of TC92 of the International Standards Organization (ISO) have undertaken the development of new test procedures to independently characterize ignitability, flame spread, rates of heat release, and other fire-related parameters[1831-1834].

Worldwide efforts continue to correlate laboratory tests to real life fire behaviour through the use of applications oriented model tests and rules[1835]. Examples of such programs are the corner test program[1836] carried out by Factory Mutual and the corrugated metal tool deck trials[1837,1838], carried out by TNO. The corner test has been used to determine the fire behaviour of rigid foam materials when exposed to a severe wood crib fire.

<u>Construction elements and special construction.</u> The fire resistance of constructions that form an enclosed space is determined by the phenomenon of flashover, that is, the extension of the fire from the room of origin to adjacent spaces. Containment of the fire is through a sufficiently high fire resistance performance of the components forming

the enclosed space, such as walls, ceilings and doors. For proper classification, it is necessary to ensure that construction details, such as holes for cables, water or other piping, as well as joint details, do not result in weak spots allowing fire penetration. To properly assess fire safety of construction components, their performance is determined relative to thermal loading of the panels using a standard time-temperature profile designated (ETK) in DIN 4102. Such a profile has been adopted both nationally and as ISO Standard 834[1839]. Construction elements are assembled into the wall or ceiling test furnace in the same form as they would use. The elements are then subjected to the standard time-temperature exposure for the time interval corresponding to the rating desired. Acceptability requires that fire breakthrough does not occur, that structural integrity be preserved, and that temperature gradients remain within specified limits. In Germany, speciality constructions such as parapets have reduced fire performance requirements because of the lower fire risk associated with such applications. The reduced test conditions essentially consist of a modified (flattened) standardized curve according to DIN 4102, Section 3.

For single case evaluations of the fire performance of parapet elements, applications-like model tests can be conducted using a test device currently used for testing facades[1840]. Roofing is evaluated for fire spread caused by external fire sources and radiant heat. According to DIN 4102, Section 7[1841], the fire source consists of 600 g of wood shavings. The test is carried out at roof inclinations of 15 and 30°C. Neither burn-through nor unacceptable fire spread may occur to pass the performance criteria. The corresponding testing in the Netherlands is regulated under Standard NEN 3883[1842].

For proper risk assessment of the fire performance of metal roof constructions, model fire test studies[1843] were done. The trials carried out by TNO indicated that the fire performance classification of the insulation was the main factor influencing fire spread along the upper roof system to adjacent sections.

11.5.7 Other Fire-related Factors

In any fire, liquid, solid and gaseous combustion products are formed No prevailing official regulations for toxicity and smoke density are currently in effect. For the category of "nicht brennbare" building materials (category A2), judgments are based on "expert opinion". The smoke density measurements are made, using the XP-2 cabinet[1844] and the DIN - "pyrolysis apparatus[1845].

The test sample in the XP-2 chamber (ASTM D 2843) is supported on a wire mesh and ignited. The build-up of combustion products results in attenuation of a horizontal light beam. A strip-like test sample is decomposed in air in the combustion tube per DIN 52436. Again, a photoelectric system is used to measure the dependency on test temperature.

Apparatus similar to the XP-2 chamber is used in Switzerland to judge smoke risk of combustible construction material[1846].

The aircraft industry uses, on a worldwide basis, the smoke density apparatus developed by the National Bureau of Standards (NBS). The vertically oriented test sample is subjected to a radiant heat flux of 2.5 W/cm^2. The test is run both with and without the pilot ignition flame

The attenuation of the light beam of the vertically oriented measuring system is determined. A promising development is the ISO smoke box[1834], developed by ISO TC 92. One of the benefits of this method is the ability to also test composite materials. The smoke development is measured in the presence of a radiant energy flux in the range of 0 to 5 W/cm^2 with piloted ignition.

The combustion tube apparatus according to DIN 53436 also serves to generate the combustion products for animal experiments. Toxicologists are of the opinion that neither the determined chemical composition of the combustion products, nor the analytically determined concentrations are sufficient to determine acute toxicity[1847-1849]. Possible additive, synergistic or antagonistic effects can only be determined through animal experiments. However, such studies and assessments can only be made by using combustion products with a composition representative of that developed in an actual fire situation. The combustion tube apparatus consists of a quartz tube, heated by a movable electric ring furnace. The ring furnace is moved uniformly and at opposite direction of the air stream. The thermal load on the test specimen is determined by test temperatures in the range from 100 to 600°C. The reason for the counter-current air stream is to guarantee uniform thermal conditions for the test sample over the test interval.

The corrosiveness of combustion products has been considered in a variety of ways[1850,1851] in relation to actual fire damage as well as large-scale experiments. Electric insulation studies have been carried out[1850,1851] using the combustion tube apparatus to provide the data necessary for standards' work. VDE Draft 0472, Section 813, provides information concerning test and performance criteria for the determination of the corrosiveness of combustion products from fires of electric origin[1852].

CHAPTER 12

GLASS TRANSITION TEMPERATURE AND OTHER TRANSITIONS

12.1 INTRODUCTORY

The glass transition temperature (Tg) is defined as the temperature at which a material loses its glasslike, more rigid properties and becomes rubbery and more flexible in nature. Practical definitions of Tg differ considerably among different measurement methods; therefore, specification of Tg requires indication of the method used.

Amorphous polymers, when heated above Tg pass from the hard to the soft state. During this process, relaxation of any internal stress occurs. At Tg many physical properties change abruptly, Youngs and Shear Modulus, specific heat, coefficient of expansion and dielectric constant. For hard polymeric materials this temperature corresponds to the highest working temperature; for elastomers, it represents the lowest working temperature. Several methods exist for determining Tg, these include differential thermal analysis (Chapter 12.2), differential scanning calorimetry (Chapter 12.3), thermomechanical analysis (Chapter 12.4) dilatometry, dynamic mechanical analysis (Chapter 12.7) and spectroscopic methods (Chapter 12.6). Each method requires interpretation to determine Tg. For this reason exact agreement is frequently not obtained between results obtained by different methods.

Other types of transitions occur in polymers. In the dynamic mechanical loss spectra of polymers, the transitions which are related to different molecular motions within the polymer are called alpha, beta and gamma transitions.

The alpha transition involves long segments of the polymer chain where the movement causes other chain segments to move out of the way. These "cooperative main-chain motions" become increasingly prevalent at the glass transition temperature and can be used to define the glass transition of a material.

The beta transition involves chain segments that are shorter than those in alpha transitions. For that reason, they occur below the Tg of the material. The motion producing the gamma transition involves short-chain segments. In many polymeric systems, the gamma transition is caused by the "crankshaft rotation" of the methylene ($-CH_2-$) groups on a long polymer chain. Since the gamma transition involves short molecular segments, it occurs below the alpha and beta transitions.

12.2 Tg BY DIFFERENTIAL THERMAL ANALYSIS

Differential thermal analysis has been used to Tg measurement in polystyrene[1853]. It has also been used to investigate multiple melting peaks in isotactic polypropylene[1854-1856] and terylene terephthalate[1857],

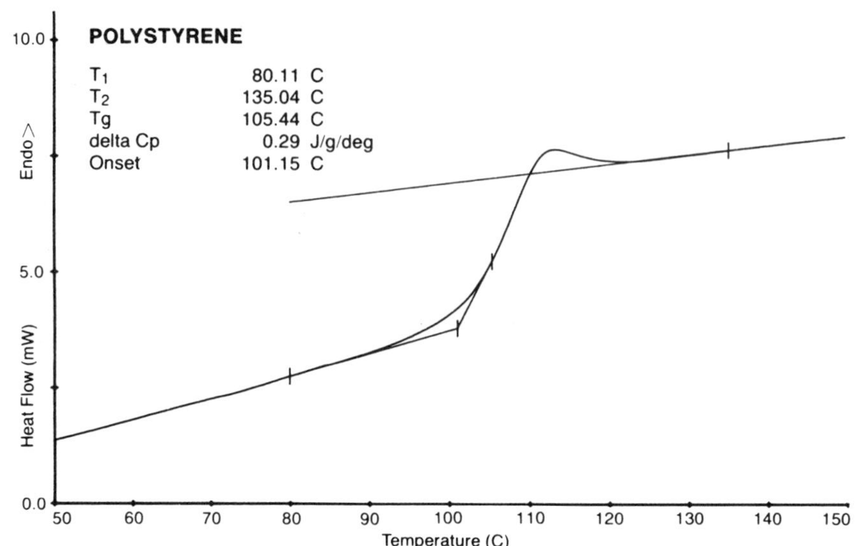

Figure 148 Glass transition termperature. Because the DSC 7
            directly measures changes in the specific heat of
            material, the parameters associated with the glass
            transition of amorphous materials are easily made. as
            shown for this sample of polystyrene, the temperatures
            where these changes occur, as well as the magnitude of
            the specific heat changes at the glass transition, are
            calculated using the TAS 7 Tg program.

first order transitions in ethylene-methacrylate copolymers[1858] and the
annealing of isotactic polybutene-1 and the three polymorphic forms of
polybutene-1[1859,1860].

12.3 Tg BY DIFFERENTIAL SCANNING CALORIMETRY

Isothermal crystallization studies by differential scanning
calorimetry provide a sensitive technique for measuring molecular weight
or structural differences between very similar materials. (Heat flow mw
versus time.) At Tg an endothermic shift occurs in the differential
scanning calorimetric curve corresponding to the increase in specific
heat, (heat flow [mw] versus temperature). At Tg an endothermic shift
occurs in the differential scanning calorimetric curve corresponding to
the increase in specific heat. Determination of the glass transition
temperature from the differential scanning calorimetric curve is based on
the following three auxiliary lines: extrapolated baseline before the
transition, inflectional tangent through the greatest slope during
transition, and the baseline extrapolated after the transition. Figure
148 shows this measurement carried out on a polystyrene sample using a
Perkin Elmer DSC-7 instrument illustrating the temperatures at which the
glass transition occurs as well as the magnitude of the specific heat
changes at the glass transition. Frequently, an endothermic peak occurs
simultaneously with the change in specific heat the first time a substance
is heated above the glass transition temperature. This event is caused by
a relaxation phenomenon and normally appears only when the transition
temperature is exceeded for the first time. To rule out this effect, a

second measurement is carried out immediately after the substance has cooled.

Johnston[1861,1862] studied the effects of sequence distribution on the glass transition temperature of alkyl methacrylate-vinyl chloride and alpha-methyl styrene-acrylonitrile copolymers by differential scanning calorimetry, differential thermal analysis and thermochemical analysis. The heat of volatilization of polymers has been determined. Values obtained for poly-(methylmethacrylate) agreed well with calculated values[1863]. Chemically crosslinked and oriented low-density polyethylenes have been investigated using differential photocalorimetry and differential scanning calorimetry[1864,1865].

Heat of fusion measurements have also been obtained by this technique. Dunlop et al.[1866] measured the specific heat of PVC compositions.

12.4 Tg AND OTHER TRANSITION TEMPERATURES BY THERMOMECHANICAL ANALYSIS

Perkin Elmer supply the TMA-7 thermomechanical analyser, details of which are given in Table 185. In this analyser a quartz probe closely monitors dimensional changes in the sample under study. The position of this probe is continuously monitored by a high sensitivity linear variable displacement transducer (LVDT). The transducer itself is temperature controlled to provide excellent stability and reproducibility. The probe mechanism is controlled through a unique new closed loop electromagnetic design circuit. This design allows precise probe control, computer controlled application of force to the sample and constant sample loading throughout the experiment. These features provide exceptional temperature control over the range from -170°C to +1000°C. Other features of the TMA7 include: multiple probe types for multiple modes of operation, computer control for unattended operation, sample load selection through the computer keyboard, automatic zero load calculation, one touch probe control and position, simultaneous independent instrument operation, precise temperature control and heating and cooling rates of 0.1 to 100°C min^{-1}.

A series of quartz probes are available that allow the TMA 7 to be used in a variety of different operating modes; expansion, compression, flexure, extension and dilatometer.

Thermomechanical analysis is the general term for measurements of deformation of changes in dimension. With thermomechanical analysis, the sudden change in the slope of the expansion curve is used to determine Tg. On the other hand, an increased load on a hemispherical probe allows the measurement of the softening point (plastic deformation), which is in the vicinity of Tg.

The glass transition temperature is used to identify polymers (see Table 186). However, additives (such as plasticizers) solvent residues, and moisture can lower Tg.

The glass transition temperature of thermosets is influenced by the curing conditions and the degree of cure. This fact is especially useful because measurements of glass transition temperature allows monitoring of degree of cure using only a small sample (approximately 20 mg).

An example of the uncured resin, followed by subsequent measurements of the cured resin, showing a glass transition around 110°C plus two

Table 185 - Perkin Elmer TMA-7 Series Thermomechanical Analyser

Software	Analyser	
SET UP The TMA 7 SET UP program is used to take data from the TMA7. It includes easy to use routines for direct TMA control from the computer keyboard. Other options include heat, cool and isothermal modes of operation, multi-step step program development and operation in either temperature or time scales	Sensitivity	0.4 um/cm
	Ordinate Linearity	$\pm$ 0.5% to 2.5 cm
	Sample Size	0.75 cm (0.3 in.) diameter (maximum) 1.91 cm (0.75 height (maximum)
	Computer Con- trolled Loading	0 to 150 grams
DELTA Y The DELTA Y program permits the calculation of the absolute change in sample dimension at any point on the curve. Options include the total dimensional change and the temperature or time limits over which the dimensional change has been calculated	Temperature Range	Standard unit allows operation from ambient to 1000°C. With optional accessories the range may be extended to -170°C
	Heating and Cooling Rates	0.1°C to 100°C/min in 0.1°C increments
	Temperature Precision	$\pm$ 2°C
	Sample Type	Solids, liquids, powders, films, fibres
ONSET The ONSET program permits the quantitiative deter-mination of the onset temp-erature or time of an event on a TMA scan	Atmosphere	Static or dynamic inclu-ding nitrogen, argon, helium, carbon dioxide, air, oxygen or other inert or active gases
EXPANSION COEFFICIENT The EXPANSION COEFFICIENT permits the automatic calculation of the expansion coefficient (alpha) over any temp-erature range	Temperature Sensors	Chromel-alumel thermo-couple
	Probe Types	Expansion Extension Penetration Flexure Compression
OPTIMIZE The OPTIMIZE programs are a series of software programs which are used to optimize TMA data curves. Programs are available for rescaling the temperature or time axes, and for resca-ling the ordinate axis	<u>TAC 7 Thermal Analysis Controller</u>	
	General	The TAC 7 is the intell-igent microprocessor con-troller which links the PE7700 to the 7 Series thermal analysis models
RECALL The RECALL program is used to recall to the screen	Microprocessor	High speed microprocessor for control of all functions

data, results, and run parameters which are stored on disk. Options include recalling a single curve, recalling multiple curves, and recalling curves of different techniques (i.e. TMA and DSC curves)

Memory — Extended memory capability is built in

Communications Ports — Two RS-232C communications ports, an analog control port and a digital control port

SAVE
The SAVE program permits the storage of data, results, calculations, and comments on the disk

CALIBRATE
The CALIBRATE programs permit automatic dimensional and temperature calibration of the TMA-7 analyzer

CALCULATIONS
The CALCULATIONS progams permit a wide variety of calculations to be performed on TMA data. Programs for curve subtraction and derivative calculations are available

---

thermomechanical curves of expansion and softening are shown in Figure 149.

In Figure 150 is shown an application of the Perkin Elmer TMA-7 thermomechanical analyser to the characterization of a composite material, viz. an epoxy printed circuit board material. From a single experiment the glass transition temperature (Figure 150(a)) and the expansion coefficient (Figure 150(b)) for the laminate are readily determined.

Wohitjen and Dessy[1867] have described a surface acoustic wave device to perform thermomechanical analysis of polymer films. Amplitude measurements of Tg for polycarbonate, polysulfone, and single and two-phase copolymers of the above agree with Rheovibron measurements. Low order transitions (at 24°C) have been measured in Teflon. Although the Tg measurements using a quartz SAW delay line[1868] were made at 30 MHz, the poor coupling with the surface made by normal disk samples resulted in the excellent agreement observed with classical low frequency methods. On the other hand, cast films, which have good surface contact with the substrate, show the shifts in Tg predicted by the time-temperature principle. They demonstrated the utility of the SAW device in photoresist investigations by monitoring the effects of solvent evaporation and photo induced cross-linking of the resist. The device affords a significant advantage in studies of this kind because it can monitor films of the same thickness used in industrial applications. The polymer clamping device is illustrated in Figure 151(a). Temperature control was over the range 0-200°C. The amplitude measurement system is illustrated in Figure151(b).

Table 186 - Tg and Tm of Polymers

Polymer	Abbr.	Tg	Tm
Polybutadiene		-86	(-20)
Polyisobutylene	PIB	-73	(44)
Poly(ethylene vinyl acetate) copolymer	EVA	-20..20[2]	40..100[2]
Polyethylene, low d.	LDPE	(-100)	120
Polybutene(P.butylene)	PB		130
Polyethylene, high d.	HDPE	(-70)	135
Poly(oxymethylene) copolymer	POM		164..168
Polypropylene	PP	(-30)	165
Poly(vinyl chloride) soft	PVC w	-40..100[1]	
Poly(vinylidene chloride)	PVDC	-17	
Poly(oxymethylene) (homopolymer)	POM		175..180
Poly(vinylidene fluoride)	PDF_2		178
Polyamide 11	PA11		186
Poly(vinyl acetate)	PVAC	30	
Poly(vinyl chloride)	PVC	85	(190)
Poly(butylene terephthalate)	PBTB	65	220
Polyamide 6	PA6	(40)	220
Polyamide 6,10	PA6,10	(46)	226
Poly(vinyl alcohol)	PVA	85	
Polystyrene	PS	90..100	
Poly(methyl methacrylate)	PMMA	105	
Epoxy resin	EP	50..150[3]	
Poly(phenylene oxide)	PPO		230
Polycarbonate	PC	155	(235)
Polyamide 6,6	PA6,6	(50)	255
Poly(ethylene terephthalate)	PETP	(69)	256

Poly(ethylene tetrafluoro-ethylene) copolymer	ETFE		270
Poly(fluoroethylene propylene)	FEP		280
Poly(phenylene sulfide)	PPS	80	280
Polyacrylonitrile	PAN	100	(320)
Polytetrafluoroethylene	PTFE	(-20)	327

A 10 ml thick film of polyethylene terephthalate produced the curve shown in Figure 152(a). As the temperature approached the glass transition, the sample began to soften and to adhere to the SAW device, thus attenuating the wave. At 75°C the wave was almost completely attenuated. This is the reported glass transition temperature of polyethylene terephthalate. Thus, points of discontinuity in the amplitude vs. temperature profiles provide a clear indication of the glass transition temperature of polymer specimens clamped to the SAW device.

Figure 152(b) presents the results of an unoriented bisphenol A polycarbonate sample. A glass transition temperature of 156°C is apparent. Another polycarbonate sample was strained by 50% and the

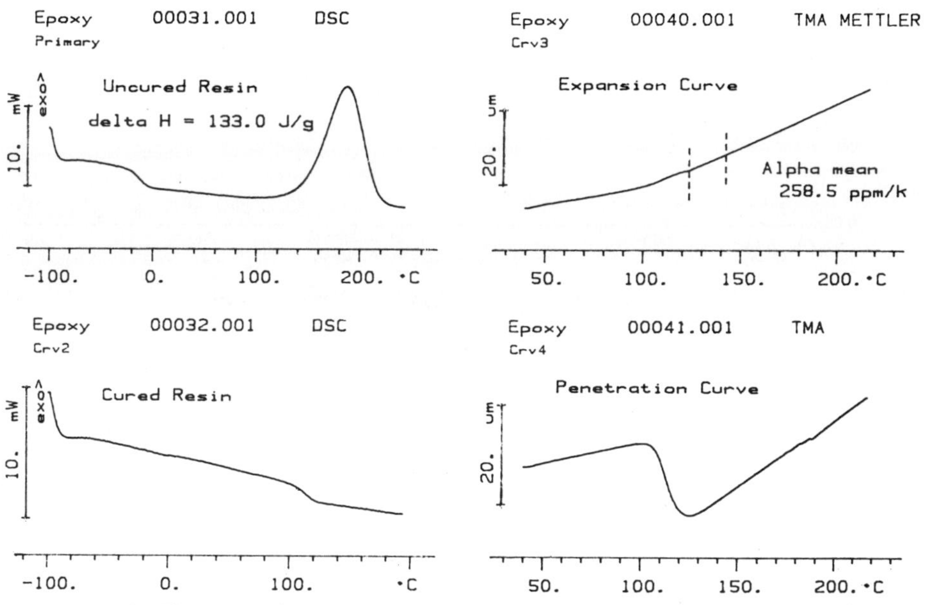

Figure 149 Thermomechanical analysis of epoxy resins. Measurements of glass transition temperature.

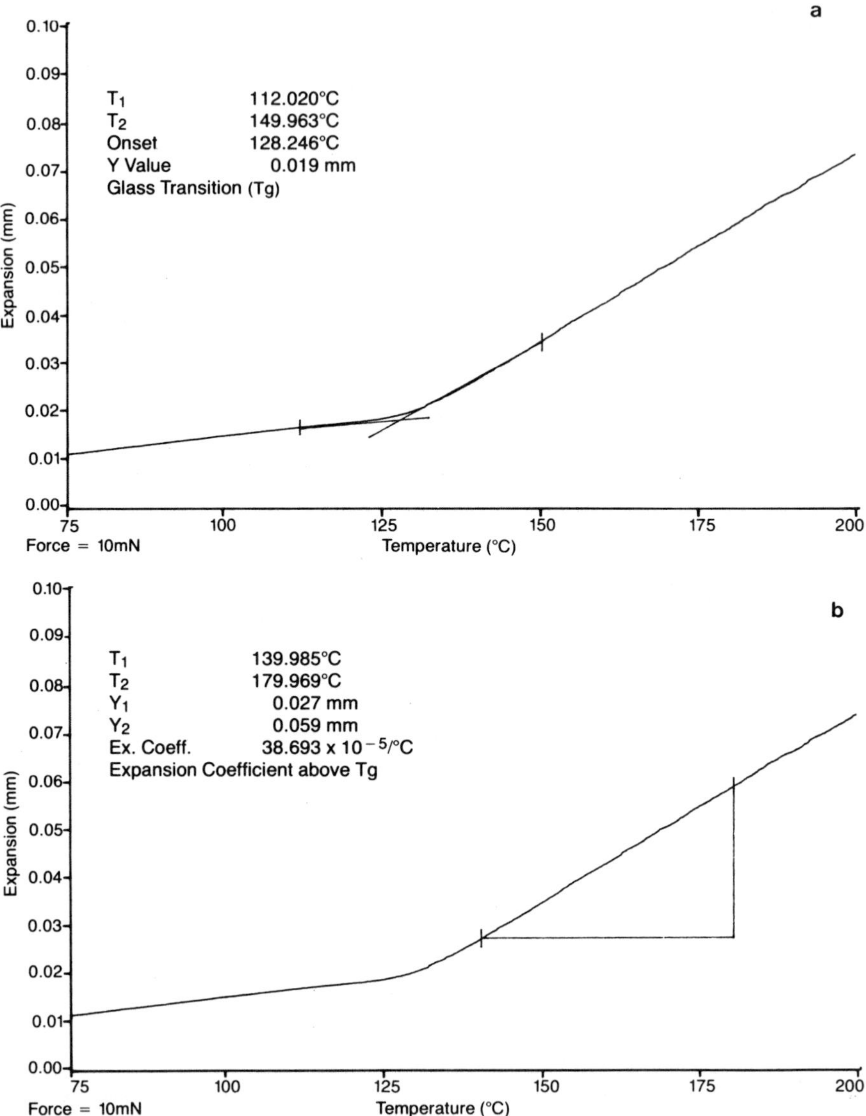

Figure 150   Thermomechanical analysis of epoxy printed board
material (a) measurement of Tg; (b) measurement of
expansion coefficient.

resulting profile appears in Figure 152(c). The results are very similar
with the strained sample, indicating a Tg of 152°C.

Figures 152(d) and (e) show the amplitude versus temperature
response of a) an unoriented polysulfone film and b) one strained by 25%.
Once again, the results are quite similar with the Tg of the unoriented
sample occurring at 186°C and that of the strained sample occurring at
189°C.

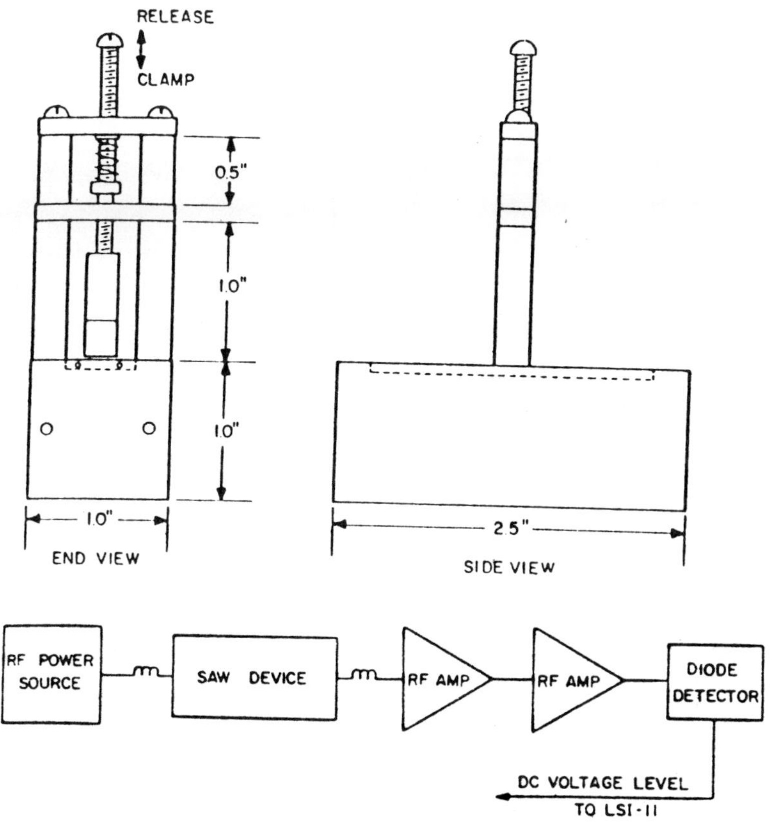

Figure 151   Thermomechanical analysis (a) polymer film clamping
system; (b) simplified amplitude measurement apparatus.

The glass transition for a single-phase polycarbonate-polysulfone
copolymer is presented in Figure 152(f).   A single Tg of 174°C is
indicated.   Figure 152(g) illustrates the results obtained from a two-
phase polycarbonate-polysulfone block copolymer.   Two discontinuities
appear in the profile at 171 and 195°C which correspond to the Tg's of the
two domains present in the film.   Figure 152(i) illustrates the glass
transition profile of a very thin (0.1 ml) unclamped file of
polymethylmethacrylate demonstrating a peak attenuation at 150°C.   The
method described by Wohitjen and Dessy[1867,1868] was capable of detecting
the subtle crystalline transition of Teflon (polytetrafluoroethylene)
occurring at 27°C (Figure 152(h).

This is a good example of the ability of the SAW device to detect
more subtle transitions and greatly increases the spectrum of potential
applications.   Most polymers experience relaxation transitions at
temperatures far below the glass transition.   These relaxations typically
involve movements of small chain segments or side groups and exhibit
modulus changes substantially less than those seen at the Tg.
Poly(tetrafluoroethylene) undergoes a first-order transition around room
temperature which is a result of order-disorder effects in the crystalline
regions of the polymer[1869].   The response obtained from the SAW device

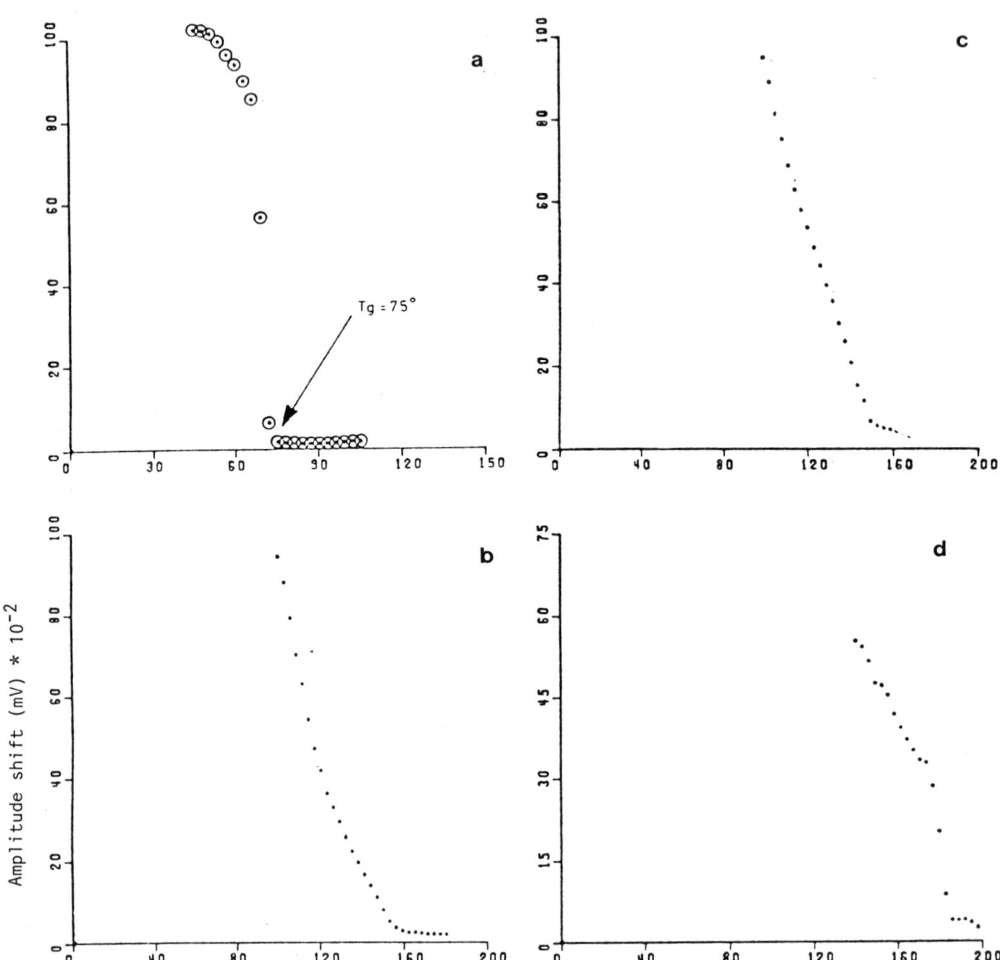

Figure 152 Glass transition profiles: (a) polyethylene
terephthalate; (b) unstrained bisphenol A polycarbonate;
(c) strained (50%) bisphenol A polycarbonate; (d)
unstrained pokysulphone; (e) strained (25%)
polysulphone; (f) single phase block polysulphone film;
(g) two phase polycarbonate polysulphone copolymer; (h)
teflon, crystalline fraction; (i) polymethylmethacrylate
cast film.

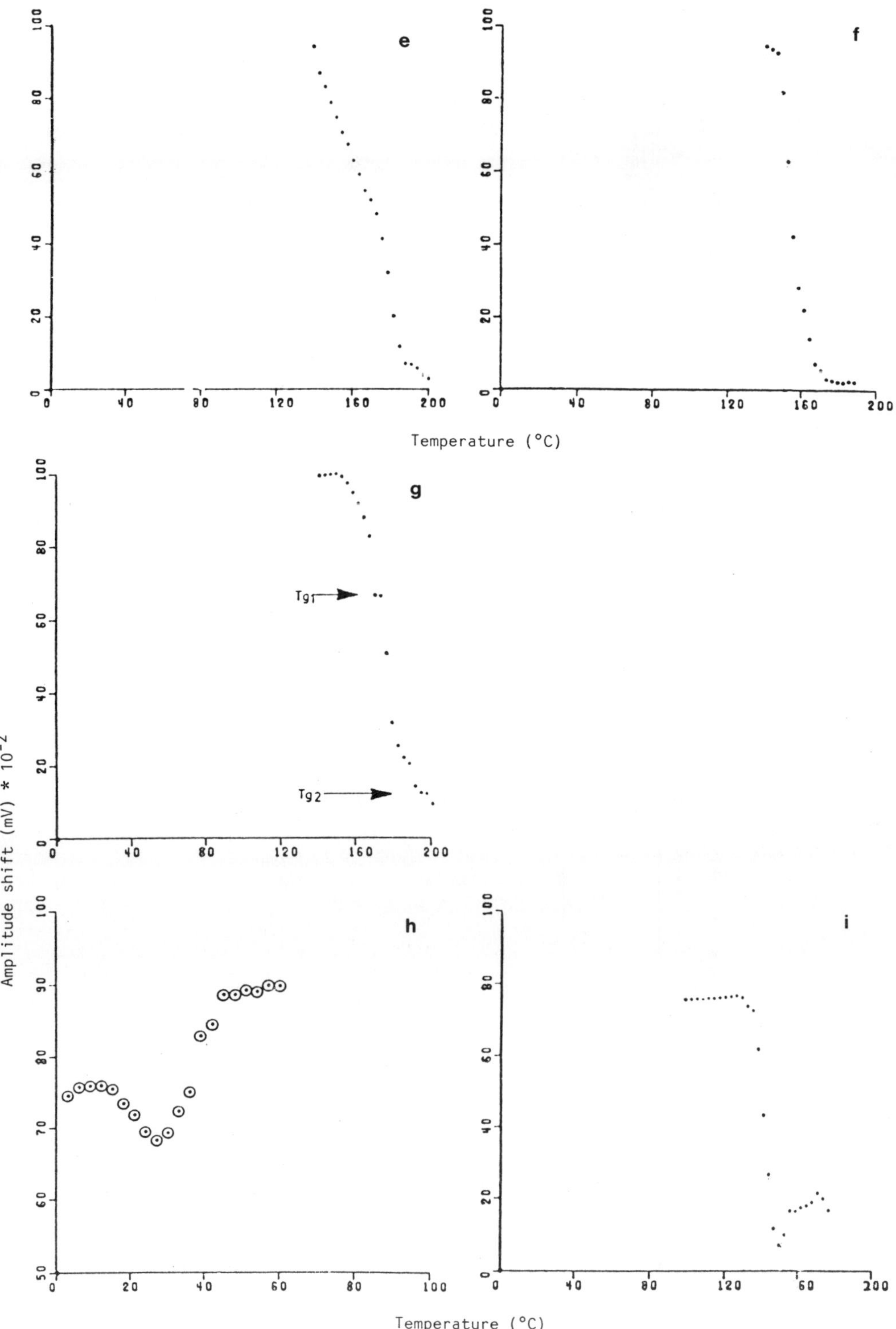
e
f
g
Tg1
Tg2
h
i
Amplitude shift (mV) * 10^-2
Temperature (°C)
Temperature (°C)

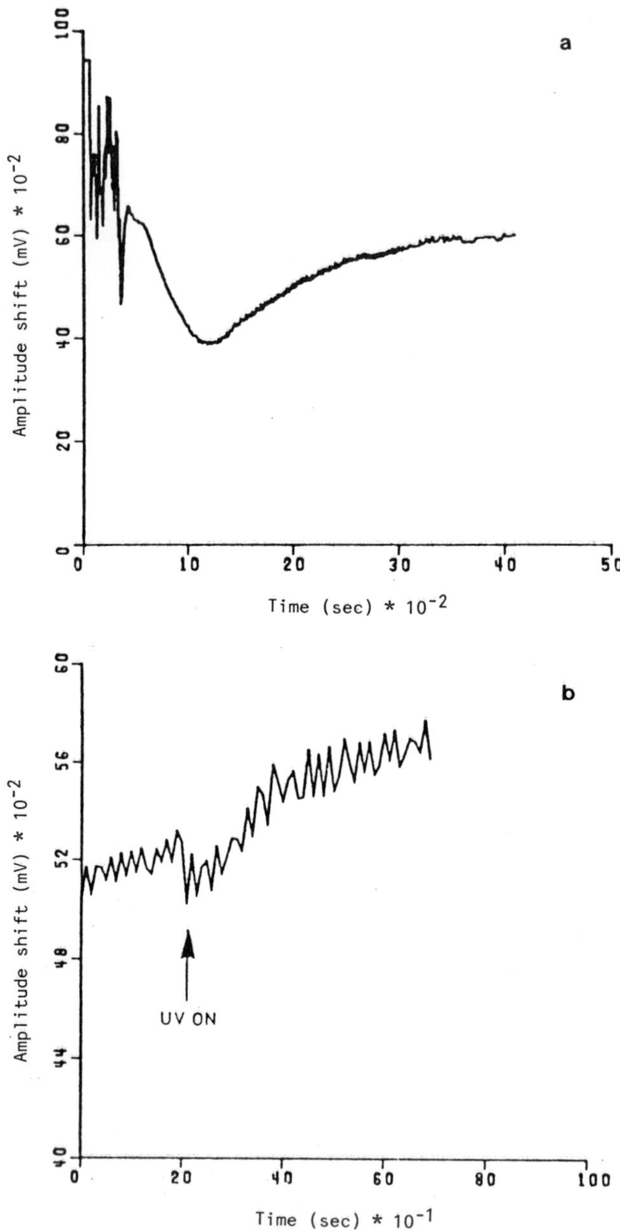

Figure 153 (a) solvent evaporation in KPR photoresist; (b) effects of UV irradiation on KPR photresist.

with a Teflon film on the surface [Figure 152(h)] agrees very closely with data describing the linear expansion coefficient of polytetrafluoro-ethylene[1870]. Between 10 and 27°C the polytetrafluoroethylene linear expansion coefficient increases from about 1.25 x 10^{-4} to 2.1 x 10^{-4} reciprocal degrees centigrade. This would cause the film to swell slightly on the surface, thus increasing the surface contact which results in greater attenuation of the surface wave. The phenomenon was reversible and reproducible as one would expect. Thus, the SAW detector seems very well suited as a sensitive monitor of expansion coefficient changes associated with subtle polymer mechanical transitions.

Wohitjen and Dessy[1868] applied the SAW device to the area of photoresist materials. Solvent evaporation rate during the baking of the photoresist can have a significant effect on the quality of the film. The use of a SAW device permits investigations of this rate to be conducted with films that are the same thickness as those used commercially. Techniques which employ microbalance monitoring of bulk photo resists do not provide direct information about film behavior[1871]. The effect of placing a drop of KPR photo-resist on the surface is shown in Figure 153(a). Wild oscillations experienced shortly after application of the liquid photoresist are caused by interference effects as the film thickness changes owing to solvent loss. A compressional wave is launched into the photoresist solution and partially reflected back into the surface from the photoresist/air boundary. This causes constructive and destructive interference effects to be observed as the boundary distance changed. The visual observation of the photoresist being dry coincides with the end of the yield oscillations. Other effects persist in the sample after the gross drying has occurred. Attenuation of the wave slowly increases until the trend reverses itself at about 1200 s. The activity observed in the film levels off after about 5000 s has elapsed. Further research will be necessary before the nature of the observed behaviour can be completely understood. The presence of this unexposed photoresist on the surface of the SAW device provided a convenient opportunity to study the effects of photo cross-linking of the resist. When polymers are cross-linked the effect is to raise the Tg (i.e. make the polymer more brittle). This is seen in Figure 153(b). After illumination with UV light, the attenuation of the wave is reduced which is precisely the response expected. The signal-to-noise ratio obtained was poor. Thus it was possible to observe the effect of UV induced cross-linking in a very small photoresist film.

There are several factors which distinguish the SAW device as a useful monitor of polymer glass transitions. The device is very sensitive and this permits very small samples to be used. Sample preparations and mounting are simple and rapid. The device is quite rugged and possesses a small thermal mass which permits fairly rapid-temperature scans to be made.

Other applications of thermomechanical analysis. Although not directly associated with the measurement of Tg these are included here for completeness.

i)   Textile fibre and thin film studies by quantitative measurement of extension, contraction and the temperature associated with these physical changes, (temperature versus extension measurements).

ii)   Rigidity studies, e.g. evaluation of thermochemical behaviour of polyurethane which showed changes in rigidity due to structural changes occurring during polymer degradation[1872].

iii)   Thermal relaxation studies on polymer films, fibres and powders[1873].

iv)   Heat deflection (distortion) studies, (temperature versus flexure plots).

v)   Tensile compliance studies[1874].

12.5 Tg BY DYNAMIC MECHANICAL ANALYSIS

This is discussed further under dynamic mechanicalanalysis, (Chapter 12.7.2).

12.6 DETERMINATION OF Tg AND OTHER TRANSITIONS BY SPECTROSCOPY

Whilst thermal methods predominated for the determination of polymer transitions, infrared spectroscopy has been used to provide information on temperature transitions.

Vorob'yev and Vettegren[1875] used infrared spectroscopy to determine temperature transitions in polycarbonate. Thermal transitions in polycarbonates were determined from the concentration variation of the residual solvent or plasticizer.

Structural transitions and relaxation phenomena of polycarbonates have been followed by plotting the absorbances at 8.13 microns (stretching vibration of C-O-C groups) and 10.64 microns against temperature[1876].

High field ^{13}C Fourier transform proton NMR spectra have indicated -40°C as the upper limit to the Tg of linear polyethylene[1877]. Pulsed NMR measurements of ultra high molecular weight linear polyethylene indicated a second transition of longer relaxation time appearing at the temperature characteristic of the gamma relaxation. The gamma relaxation meets most of the criteria for assignment of Tg[1878].

12.7 APPLICATIONS OF DYNAMIC MECHANICAL ANALYSIS

12.7.1 Characterization of Viscoelastic and Rheological Properties

Basically this new technique involves the measurement of the mechanical response of a polymer as it is deformed under periodic stress and is used to characterize the viscoelastic and rheological properties of polymers. It is used to measure Tg and because of its importance a fuller description of the applications of the technique including the prediction of polymer lifetimes is included here.

Dynamic mechanical analysis is the measurement of the mechanical response of a material as it is deformed under periodic stress. Material properties of primary interest include modulus (E'), loss modulus (E"), tan δ (E"/E'), compliance, viscosity, stress relaxation, and creep. These properties, expressed in quantitative units of measurement, characterize the viscoelastic performance of a material.

Dynamic mechanical analysis provides material scientists and engineers with the information necessary to predict the performance of a material over a wide range of conditions. Test variable include temperature, time, stress, strain, and deformation frequency. Because of the rapid growth in the use of engineering plastics and the need to monitor their performance and consistency, dynamic thermal analysis has become the fastest growing thermal analysis technique.

Recent advances in the materials sciences have been both the cause and effect of advances in materials characterization technology. The technique of dynamic mechanical analysis is one of the most important developments because it allows materials scientists and engineers to accurately predict the performance of a material under a wide range of conditions.

The theory of dynamic mechanical analysis, which measures the mechanical response of a material as it is deformed under periodic stress, has been understood for many years. But because of the complexity of measurement mechanics and the mathematics required to translate theory into application dynamic mechanical analysis did not become a practical tool until the late 1970's when Du Pont developed a device for reproducibly subjecting a sample to appropriate mechanical and environmental conditions. The addition of computer hardware and software capabilities several years later made dynamic mechanical analysis a viable tool for the industrial scientist because it greatly reduced analysis times and the labour intensity of the technique.

Dynamic mechanical analysis is a viable tool for the industrial scientist because it greatly reduce analysis times and the labour intensity of the technique.

Dynamic mechanical analysis provides punatiative information on the viscoelastic properties - modulus and damping - of materials. Viscoelasticity is the characteristic behaviour of most materials in which a combination of elastic properties (stress proportional to stain) and viscous properties (stress proportional to stain rate) are observed. The dynamic mechanical analyser simultaneously measures both elastic properties (modulus) and viscous properties (damping) of a material. Such data is particularly vital because of the growing trend toward use of polymeric materials as replacements for metals in structural applications.

Whereas elastic modulus at a single temperature (as measured by a static-mode instrument) is indicative of the properties and quality of metals, this is not necessarily true of polymeric materials in which molecular relaxation transitions dramatically influence the temperature and time (frequency^{-1}) dependence of material properties.

Dynamic mechanical analysis is the preferred tool for evaluating the effects of the following variables on the mechanical properties of materials:

<u>Temperature.</u> Temperature scanning can determine whether product performance will remain within specifications at end-use temperatures.

<u>Frequency (1/time).</u> Deformation frequency sweeps can be used to study the frequency dependence of material properties.

<u>Load.</u> Creep and stress relaxation measurements can determine the long-time mechanical properties of materials and their ability to withstand loading and deformation influences.

The Du Pont 983 Dynamic Mechanical Analyser is a versatile laboratory instrument for characterising the viscoelastic and rheological behaviour of materials. This system is designed to operate in any of four modes: fixed frequency, resonant frequency, stress relaxation, or creep.

System components are the 983 dynamic mechanical analyser module, an optional liquid nitrogen cooling accessory, a thermal analyser/temperature

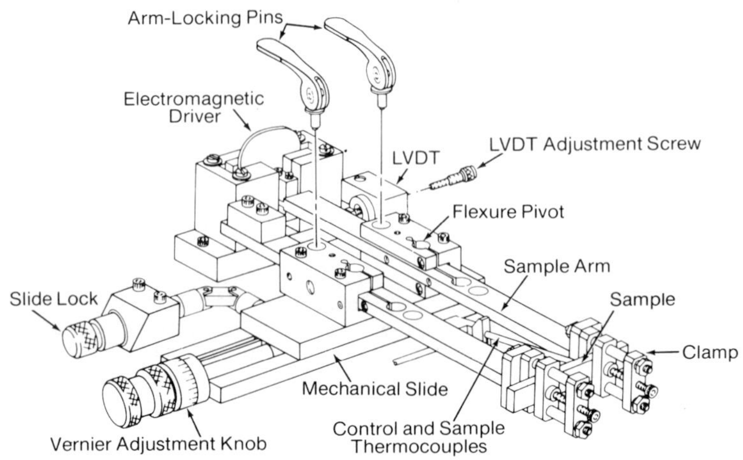

Figure 154   DMA electromechanical system.

controller, (Model 9000 computer controller or Dupont Thermal-Analyst 2100) data analysis software, and a plotter.

The 9900 controls the test temperature and collects, stores and analyses the resulting data. Data analysis software for the 983/9900 dynamic mechanical analysis system provides publication-quality hardcopy reports of all measured and calculated viscoelastic properties, including time/temperature superpositioning of the data for the creation of master curves. A separate calibration program automates the instrument calibration routine and insures highly accurate, reproducible results. In addition to measuring specific viscoelastic properties of interest, the four modes provide a more complete characterization of materials, including their structural and end-use performance properties. The system also can be used to simulate and optimize processing conditions, such as those used with thermoset resins.

The 983 dynamic mechanical analysis can evaluate a wide variety of materials, ranging from very soft, such as elastomers and supported viscous liquids, to very hard, such as reinforced composites, ceramics and metals. The clamping system will accommodate a range of sample geometries, including rectangles, rods, films and supported liquids.

The Du Pont 983 Dynamic Mechanical Analyser produces quantitative information on the viscoelastic and rheological properties of a material by measuring the mechanical response of a sample as it is deformed under periodic stress.

Figure 154 shows the mechanical components of the 983 Dynamic Mechanical Analyser. The clamping mechanism for holding samples in a vertical configuration consists of two parallel arms, each with its own flexure point, an electromagnetic driver to apply stress to the sample, a linear variable differential transformer for measuring sample strain and a thermocouple for monitoring sample temperature. A sample is clamped between the arms and the system is enclosed in a radiant heater and dewar flask to provide precise temperature control. In a fixed frequency

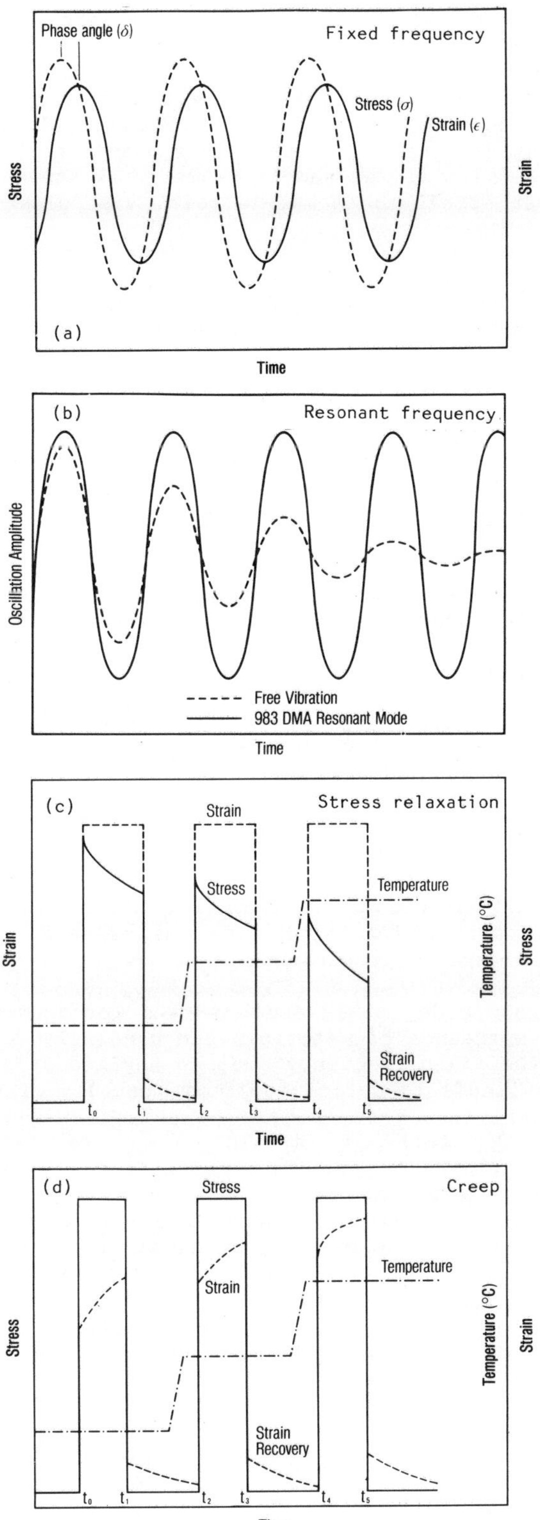

Figure 155   Modes of operation of Dupont 983 Dynamic mechanical analyser.

experiment, a sinusoidal driver signal (applied stress) forces the sample to undergo oscillatory motion at an operator selected frequency, and deformation (strain). Energy dissipation in the sample causes the resulting strain to be out of phase with the driver signal. In other words, because the sample is viscoelastic, maximum strain does not occur at the same instant as maximum stress. This phase shift or lag is defined as the phase angle (δ). If the sample geometry is known, by measuring the phase angle, driver signal and sample strain, viscoelastic properties can be calculated.

Easily interchanged clamp faces are designed to accommodate various geometries - smooth and serrated faces for rectangular shapes, and notches faces to hold cylinders or tubing. A horizontal clamping system with smooth clamp faces is available to hold soft samples and supported liquids.

The sample is clamped between the ends of two parallel arms, which are mounted on low-force flexure pivots allowing motion in only the horizontal plane. The distance between the arms is adjustable by means of a precision mechanical slide to accommodate a wide range of sample lengths. An electromagnetic motor attached to one arm drives the arm/sample system to a strain (amplitude) selected by the operator. As the arm/sample system is displaced, the sample undergoes a flexural deformation. A linear variable differential transformer (LVDT) mounted on the driven arm measures the sample's response (strain and frequency) to the applied stress, and provides feedback control to the motor.

The sample is positioned in a temperature-control chamber which contains a radiant heater and a coolant distribution system. The radiant heater provides precise and accurate control of sample temperature. The coolant distribution system uses cold nitrogen gas for smooth, controlled subambient operation and for quench cooling at the start or end of a run. Both the radiant heater and the coolant accessory are controlled automatically by the 983 system to assure reproducible temperature programming. An adjustable thermocouple, mounted close to the sample, provides precise feedback information to the temperature controller, as well as a readout of sample temperature.

<u>Fixed frequency mode.</u> This mode is used for accurate determination of the frequency dependence of materials and prediction of end-use product performance. In the fixed frequency mode an applied stress, (i.e. a force per unit area that tends to deform the body, usually expressed in Pascals (Newtons M^2) or psi) forces the sample to undergo sinusoidal oscillation at a frequency and amplitude (strain, i.e. the deformation from a specified reference state, measured as the ratio of the deformation to the total value of the dimension in which the change occurs. Strain is nondimensional, but frequently is expressed in reference values such as % strain) selected by the operator. Energy dissipation in the sample causes the sample strain to be out of phase with the applied stress (Figure 155(a)). In other words, since the sample is viscoelastic, the maximum strain does not occur at the same instant as maximum stress. This phase, shift or lag, defined as phase angle (δ), is measured and used with known sample geometry and driver energy to calculate the viscoelastic properties of the sample.

The 983 DMA can be programmed to measure a sample's viscoelastic characteristics at up to 57 frequencies during a single test. In such multiplexing experiments, an isothermal step method is used to hold the sample temperature constant while the frequencies are scanned. The sample

is allowed to reach mechanical and thermal equilibrium at each frequency before the data is collected. After all the frequencies have been scanned, the sample is automatically stepped to the next temperature and the frequency scan is repeated. Multiplexing provides more complete rheological assessment than is possible with a single frequency.

  <u>Resonant frequency mode.</u> This mode is used for detection of subtle transitions, which are essential for understanding the molecular behaviour of materials and structure/property relationships. Allowing a sample to oscillate at its natural resonance provides higher damping sensitivity than at a fixed frequency, making possible detection of subtle transitions in the material. Because of its high sensitivity the resonance mode is particularly useful in analysing polymer blends and filled polymers, such as reinforced plastics and composites.

  In the resonance mode, the 983 operates on the mechanical principle of forced resonant vibratory motion at a fixed amplitude (strain), which is selected by the operator. The arms and sample are displaced by the electromagnetic driver, subjecting the sample to a fixed deformation and setting the system into resonant oscillation.

  In free vibration, a sample will oscillate at its resonant frequency with a decreasing amplitude of oscillation (dashed line in Figure 155(b)). The resonance mode of the 983 DMA differs from the free vibration mode in that the electromagnetic driver puts energy into the system to maintain a fixed amplitude, as illustrated by the solid line in Figure 155(b). The make up energy, oscillation frequency and sample geometry information are used by the dynamic mechanical analyser software to calculate the desired viscoelastic properties.

  <u>Stress relaxation mode.</u> This mode is used for measurement of ultra-low frequency relaxation in polymers and composite structures, a primary indicator of the long-term performance of materials. This mode of operation provides definitive information for prediction of the long-term performance of materials by measuring the stress decay of a sample as a function of time and temperature at an operator selected displacement (strain).

  Figure 155(c) is a schematic representation of the process for measuring stress relaxation in a viscoelastic material. Using an isothermal step method, the sample is allowed to equilibrate at each temperature in an unstressed state. The sample is then stressed to the selected strain. The amount of driver energy (stress) required to maintain that displacement is recorded as a function of time for a period selected by the operator. After the measurements are recorded, the driver stress is removed and the sample is allowed to recover in an unstressed state. This sample recovery (strain) can be recorded as a function of time for a period selected by the operator. When the measurements at one temperature are complete, the temperature is increased and the experiment is repeated.

  <u>Creep mode.</u> This mode is used for measurement of flow at constant stress to determine load-bearing stability of materials, a key to prediction of product performance. The creep mode is used to measure sample creep (strain) as a function of time and temperature at a selected stress. Using the isothermal step method, the sample is allowed to equilibrate at each temperature in a relaxed state. After equilibration, the sample is subjected to a constant stress, as illustrated in Figure

Table 187 - Materials properties and performance characteristics determimined by dynamic mechanical analysis

Material	Composites and thermosets Prepregs	Cured laminates	Thermo plastics	Elastomers	Coatings and adhesives
Property	viscosity	glass transition temperature	polymer morphology	curing characteristics	damping and sound absorption correlations
	gelation	modulus and strength	compatability of polymer blends	damping characteristics vs. end-use frequency	compatability with substrates
	rate of cure	damping characteristics	effect of modifiers on impact properities	rate and degree of cure	rate of cure
	degree of cure	degree of cure	rate and degree of cure	effect of fillers on modulus and damping properties	degree of cure
	effects of fillers and additives	impact resist- correlations	effect of processing characteristics on crystallinity	recovery from deformation	glass transition temperature
	compatability of matrices and fillers (interface studies)	creep	long-term stability of molded parts	heat build up and dissipation	stiffness
		ply/fiber orientation effects	water and solvent effects		
		stress relaxation			
		thermal stability			
		water and solvent effects			

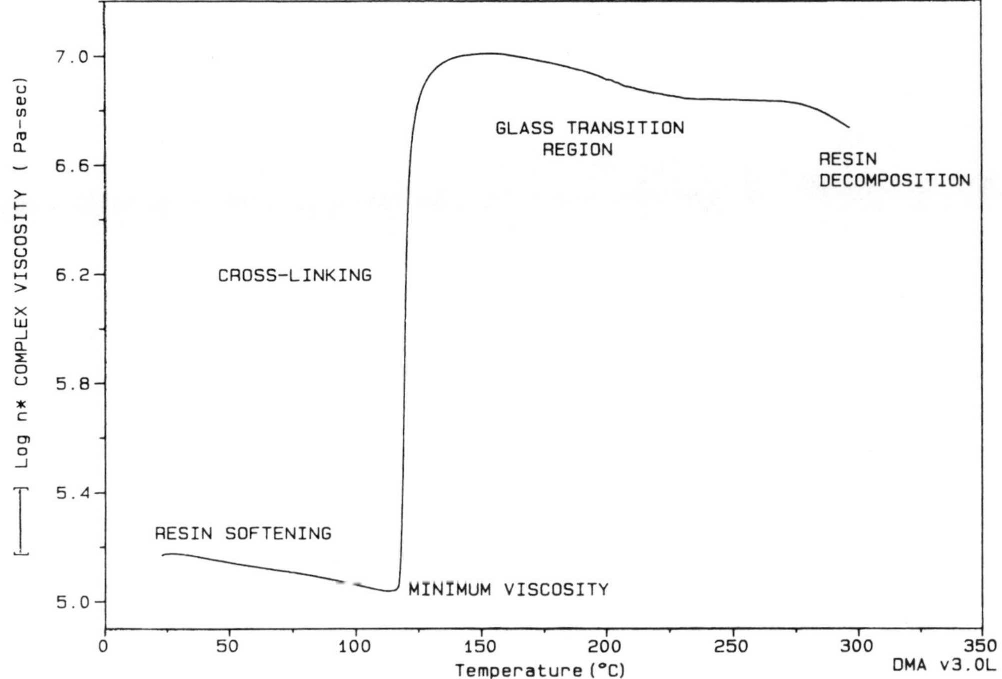

Figure 156   Temperature - complex viscosity plot for sheet moulding
compound demonstrating measurement of Tg and resin
decomposition temperature of fully cured composite.

155(d).   The resulting sample deformation (strain) is recorded as a
function of time for a period selected by the operator.   After the first
set of measurements is made, the driver stress is removed and the sample
is allowed to recover in an unstressed state.   Sample recovery (strain)
can be recorded as a function of time for any desired period.   When the
measurement at one temperature is complete, the temperature is changed and
the measurement is repeated.

Creep studies are particularly valuable for determining the long-
term flow properties of a material and its ability to withstand loading
and deformation influences.

The most frequently analysed materials properties and performance
characteristics determined by dynamic mechanical analysis are tabulated in
Table 187.   Those applications of the technique relating to Tg and other
minor transitions are now discussed.

12.7.2 Tg by Dynamic Mechanical Analysis

The thermal curve in Figure 156 represents the softening and cure of
a sheet molding compound prepreg.   By monitoring complex viscosity, the
curing process can be followed as the resin in the prepreg is transformed

from a viscous liquid to a highmodulus, cross-linked matrix. Complex viscosity (η^*) is defined as the ratio of the tangential frictional force per unit area relative to the velocity gradient perpendicular to the flow of a liquid. Complex viscosity can be calculated from dynamic mechanical data as:

$$\eta^* = \frac{\sqrt{(G')^2 + (G'')^2}}{\omega}$$

where: G' is shear storage modulus
       G" is shear loss modulus
       is angular frequency

Resin softening, minimum viscosity, crosslinking, Tg of the fully cured composite and resin decomposition temperature can be identified from a single experiment. This information can be used to maximize productivity during manufacturing by identifying optimum processing conditions and formulation variables.

   <u>Stress relaxation mode.</u> Stress relaxation is defined as a longterm property measured by deforming a sample at a constant displacement (strain) and monitoring stress decay over a period of time.

   Stress is a very significant variable that can dramatically influence the properties of plastics and composites. Induced stress is always a factor in structural applications and can result from processing conditions, thermal history, phase transitions, surface degradation, and variations in the expansion coefficient of components in a composite. Since the modulus of a material is not only temperature-dependent, but time-dependent as well stress relaxation behaviour of polymers and composites is of great importance to the structural engineer. Stress relaxation is also important to the polymer chemist developing new engineering plastics because relaxation times and modulus are affected by the polymer's structure and transition temperatures. Therefore, it is essential that polymeric materials that will be subjected to loading stress be characterized for stress relaxation and creep behaviour.

   Figure 157 shows the stress relaxation curves obtained for polycarbonate. The large drop in modulus of the material above 130°C is caused by the increase mobility of the polymer molecules as the glass transition temperature is approached.

   <u>Creep mode.</u> Creep is defined as a long term property measured by deforming a sample at a constant stress and monitoring the flow (strain) over a period of time. Viscoelastic materials flow or deform when subjected to loading (stress). In a creep experiment, a constant stress is applied and the resulting deformation is measured as a function of temperature and time. Just as stress relaxation is an important property to structural engineers and polymer scientists, so is creep behaviour.

   Shown in Figure 158 is the creep behaviour of a polyether ether ketone laminate. Below the glass transition temperature the laminate exhibits less than 2% creep, while above the glass transition temperature creep increases to greater than 10%.

   The thermal curve in Figure 159 shows the flexural storage modulus and loss properties of a rigid, pultruded, oriented-fibre-reinforced vinyl ester composite. Since the flexural modulus and glass transition temperature increase dramatically with post curing the test can be used to

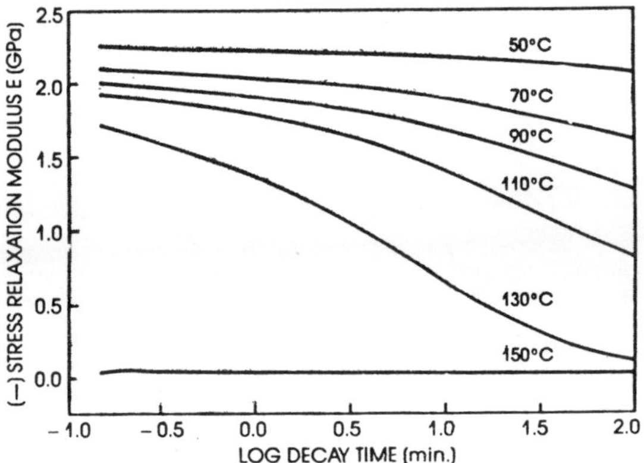

Figure 157 Stress relaxation module of polycarbonate.

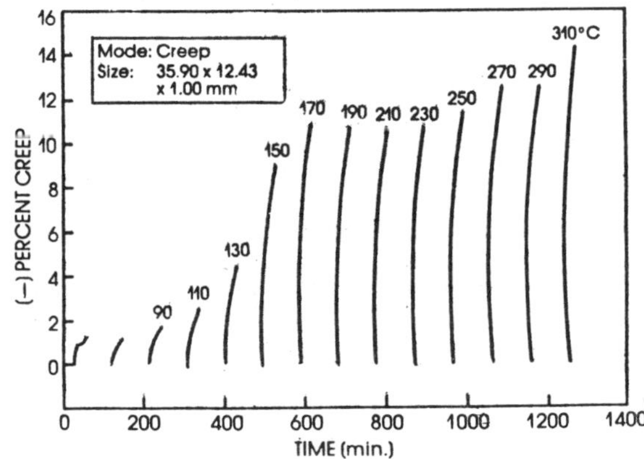

Figure 158 Creep behaviour of PEEK/carbon fibre composite.

evaluate the degree of cure, as well as to identify the high-temperature mechanical integrity of the composite.

12.7.3   Alpha, Beta and Gamma Transition Temperatures by Dynamic Mechanical Analysis

Dynamic mechanical analysis is sensitive enough to detect even weak secondary transitions in the resin matrix of a highly filled composite. In fact, all properties measured by this technique generate strong, well-defined signals that are not clouded by background noise or other interferences. In a plot of shear storage modulus (storage modulus is defined as a quantitative measure of the stiffness or rigidity of a material. For example for homogeneous isotropic elastic substances in tension, the strain ϵ is related to the applied stress σ or equation

$$E = \frac{\sigma}{\epsilon}$$ where E is defined as elastic modulus.

A similar definition of shear modulus (G) applies when the strain is

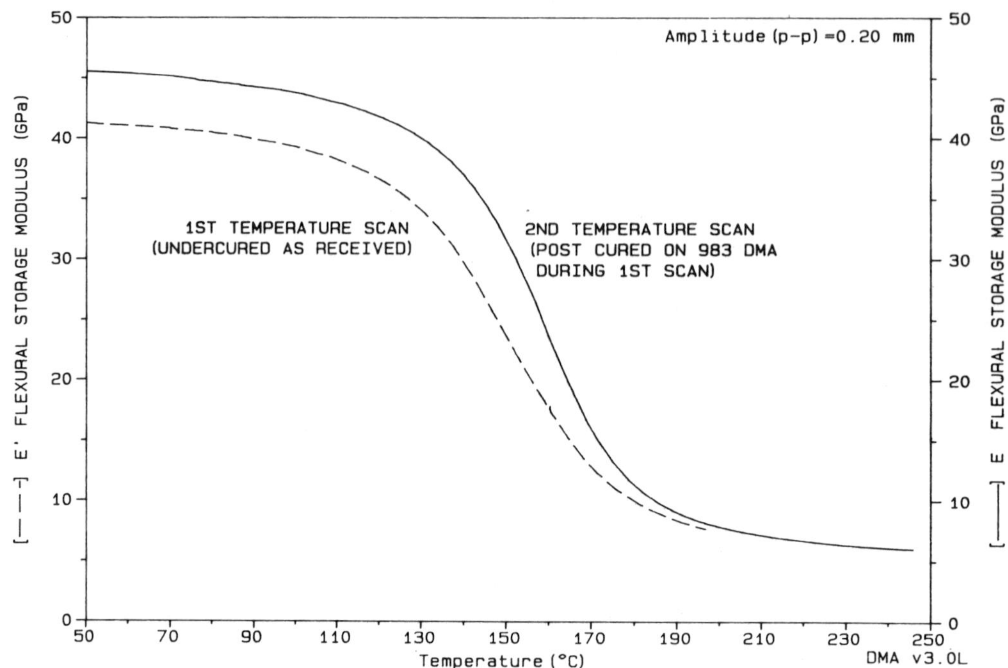

Figure 159 Flexural storage modules and loss properties of rigid
extruded oriented fibre reinforced - vinyl ester
composites.

shear.) versus polymer temperature in Figure 160 is demonstrated the three
transitions occurring in a semicrystalline 66 nylon composite. These
results were obtained with a Dupont 983 dynamic thermal analyser. The
results in Figure 160 provide information about the chemical structure of
the resin matrix. The alpha transition at 25°C is the glass transition of
the material and is associated with a significant decrease in modulus.
The beta transition is due to the local main chain motions of the amide
groups found in the polymer backbone structure. The gamma transition is
thought to be related to the molecular motion of short sequences of
methylene groups and possibly to some molecular-motion contributions from
the amide groups.

Figure 161 compares the loss moduli (damping) of linear (high
density) and branched (low density) polyethylene. Three transitions are
seen; alpha at 80°C, beta at -5°C and gamma at -110°C. Each of these
transitions corresponds to specific molecular motions which have
significance in terms of structure/property relationships:

i)   Alpha-transition is associated with crystalline relaxations
      occurring below the melting point of polyethylene.

ii)   Beta transition is due to motion of the amorphous region side chains
      or branches from the main polymer backbone. The intensity of the
      beta-transition varies with the degree of branching.

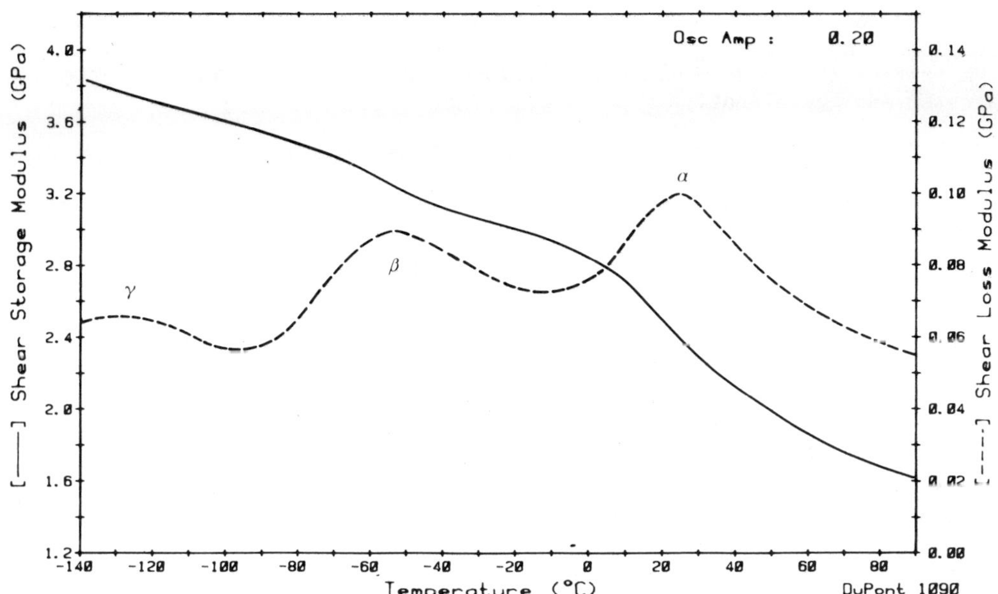

Figure 160   Shear storage modules - temperature plot for semicrystalline 66 nylon demonstrating alpha, beta and gamma
transitions.

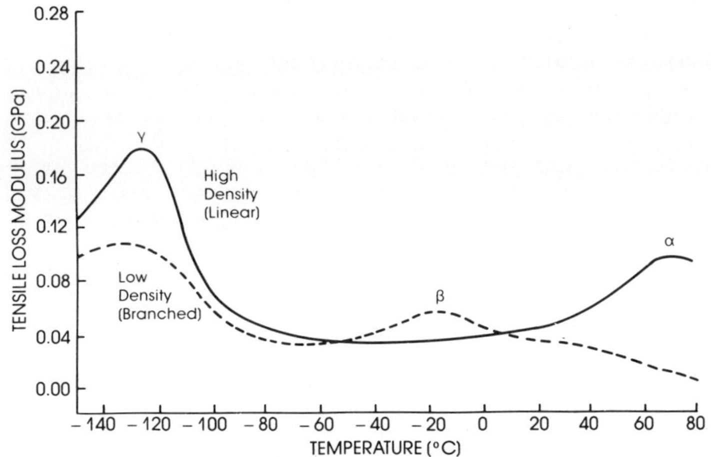

Figure 161   DMA comparison of low and high density polyethylene.

iii) Gamma-transition is a result of crankshaft rotation of short
      methylene main chain segments and can influence the low temperature
      impact stability of polyethylene.

End-use properties such as toughness, acoustical and mechanical
damping, and impact resistance are affected by the type, size and
temperature of transitions. Transitions provide the mechanism to convert
kinetic energy to heat.

12.7.4 Prediction of Polymer Lifetime by Dynamic Mechanical Analysis

A practical and accurate method for predicting the useful service
life of polymers has long been sought. The need has become increasingly
critical with the development of new materials and demanding applications,
particularly those in which engineering plastics and composites are
substituted for metals.

The proliferation of materials gives scientists and engineers new
design freedom. It also presents a considerable challenge. Before the
best material for an application can be selected, the required performance
properties (such as rigidity, strength, impact resistance and creep) and
the environment in which the product will operate must be defined. Then,
the desired life expectancy for the product must be determined. Only then
can the material selection process begin.

Traditional evaluation procedures generally are laborious, time
consuming, and expensive because they require fabrication of prototype
parts and testing under actual end-use or simulated-service conditions.
These processes are more empirical than analytical, making the results of
questionable value. The processes are generally impractical because they
require months or years to produce results.

Sichima[1879] has discussed the application of the Dupont 983 dynamic
mechanical analyser to the prediction of polymer lifetimes and longterm
performance, e.g. creep in gaskets, stress relaxation in snap-fit parts,
modulus decay in composite structural beams, and creep in bolted plates;
and heat deformation frequencies in structural parts. The ability of this
system to generate master curves makes prediction of product performance
fast and easy, and facilitates the correlation of DMA data with
evaluations performed by traditional, time- and labour-intensive methods.

Master curves can be generated in any of three modes - fixed
frequency, stress relaxation or creep - and can be used to evaluate how a
variety of critical performance properties will change over time. Using
Time/Temperature Superposition software, "master curves" can be generated
which permit the projection of performance properties to times as long as
20 years and to deformation frequencies beyond the range of practical
testing.

The underlying basis for time - temperature superpositioning is the
equivalency between time (or frequency) and temperature as they affect
polymeric materials. It has been demonstrated that viscoelastic data
collected at one temperature can be superimposed upon data obtained at a
different temperature simply by shifting one of the curves along the time
(or frequency) axis[1880].

The superposition principle is based upon the premise that the
processes involved in molecular relaxation or rearrangement occur most

rapidly at high temperatures. The time over which these processes occur can be significantly reduced by conducting the measurements at elevated temperatures, then transposing the data to lower temperatures. Thus, viscoelastic changes which occur relatively quickly at higher temperatures can be made to appear as if they occurred at lower temperatures simply by shifting the data with respect to time.

For example, consider the results of a creep (constant stress) experiment performed on polycarbonate. Figure 162 shows log creep compliance (- S, a quantitative, time dependent measure of the ability of a material to yield when a constant stress is applied:

$$S(t) = \frac{strain\ (t)}{stress}$$

- as a function of log time at two different temperatures. The creep compliance curve at 137.5°C is selected as the reference. The data obtained at 140°C can be shifted to the right (i.e., to a longer time) to superimpose them upon the chosen reference curve, as shown in Figure 163. Shifting the higher temperature data to the lower reference temperature has the effect of making the data appear to have been collected at a lower temperature, thus increasing the corresponding time scale.

Viscoelastic data can be collected by performing static measurements under isothermal conditions - creep or stress relaxation, (stress relaxation modulus, a quantitative time dependent measure of the stiffness of a material when subjected to a constant strain over a period of time:

$$E(t) = \frac{stress\ (t)}{strain}$$

- or by performing frequency multiplexing experiments in which a sample is analysed at a series of desired frequencies. By selecting a reference curve and then shifting the other data with respect to time and superimposing them upon a selected reference curve, a master curve can be generated. One benefit of the master curve is that it covers times or frequencies which often are beyond the range obtainable by practicable experiments.

The degree of horizontal (time) shifting required to superimpose a given set of data upon a reference can be described mathematically with respect to temperature. Two models are commonly used to define this relationship.

The Williams-Landel-Ferry (WLF) equation[1881,1882] is:

$$\log a_T = \frac{-C_1(T-T_0)}{C_2+T-T_0}$$

In this equation, T_0 is the reference temperature in degrees kelvin (typically defined as T_g), C_1 and C_2 are constants, T is the measurement temperature, and a_T is the shift factor. For many amorphous polymers, it has been found that $C_1 = 17.4$ and $C_2 - 51.6$.

The WLF equation typically is used to describe the time/temperature behaviour of polymers in the glass transition region. The equation is based on the assumption that, above the glass transition temperature, the fractional free volume increases linearly with respect to temperature[1880].

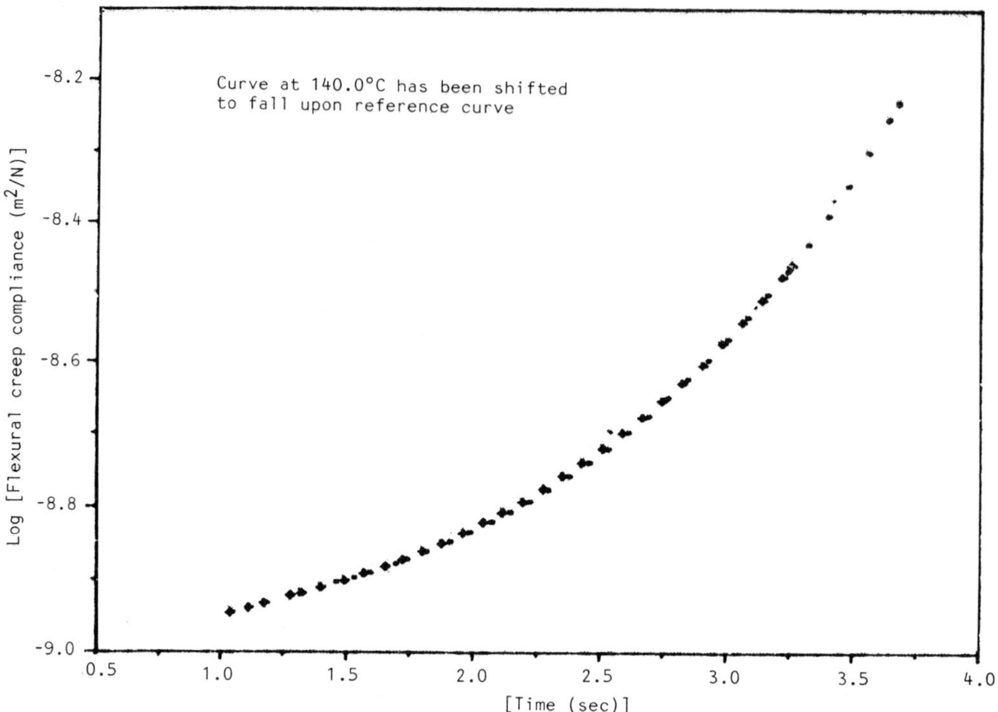

Figure 162   Log creep compliance (S) as a function of log time at
             two different temperatures.

The model also assumes that the viscosity of the material rapidly
decreases as the free volume of the material increases.

The other model commonly used to relate the shift factors to
temperature is the Arrhenius equation:

$$\log a_T = \frac{-E_a}{R(T-T_o)}$$

In the equation, E_a is the activation energy associated with the
relaxation transition, R is the gas constant (R = 8.314 J/mole°), T is the
measured temperature (generally Tg), T_o is the reference temperature, and
a_T is the time-based shift factor.   The Arrhenius equation typically is
used to describe the viscoelastic events associated with beta or gamma
relaxation transitions, or for the glass transitions associated with
semicrystalline polymers.   Frequently, it is used to obtain the activation
energy associated with the glass transition event.

When the experimentally determined shift factors are fitted to a
mathematical model, the master curve can be shifted to any desired
temperature.   Thus, data that were collected and referenced to a higher
temperature can be shifted, within reason, to a lower temperature.

To demonstrate the application of time/temperature superpositioning,
master curves were generated for polycarbonate using both frequency
multiplexing and creep techniques.   The data were generated using dynamic

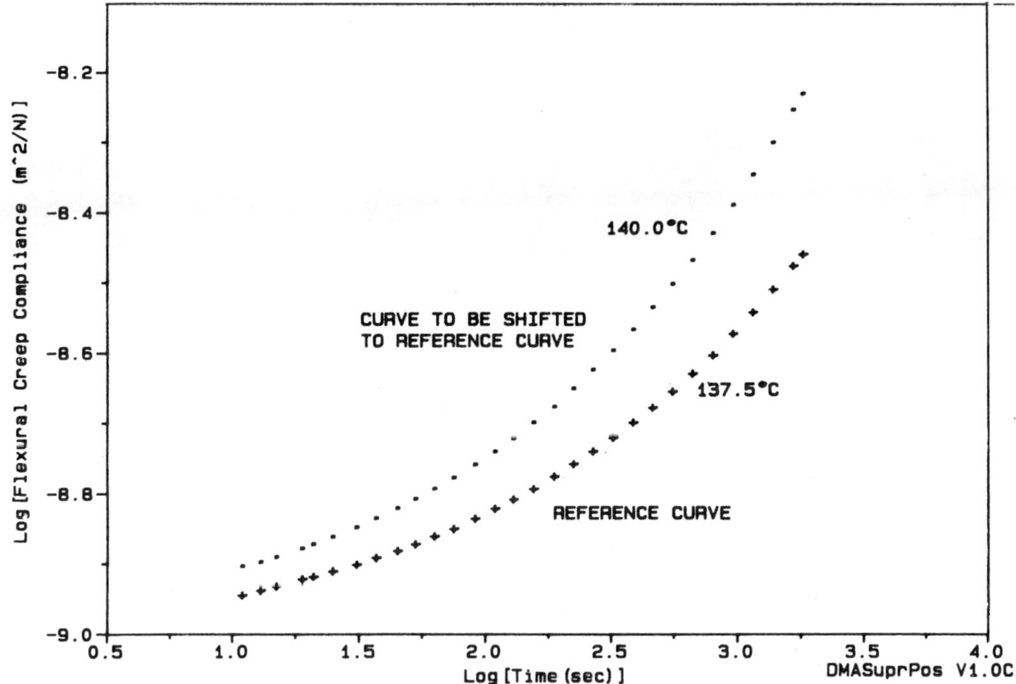

Figure 163 Curve at 140°C shifted to fall upon reference curve.

mechanical analysis and the master curves were produced using the dynamic mechanical analysis superposition software.

The polycarbonate data were obtained in the glass transition region (130°C to 160°C). When a master curve is to be generated on a given material, it is recommended that the experiment be set up to ensure that data are collected as the sample passes completely through the desired relaxation transition.

The dynamic mechanical analysis results for polycarbonate obtained using frequency multiplexing techniques are displayed in Figure 164. Analysis frequencies of 0.010, 0.020, 0.100, 0.200, 0.500, and 1.00 Hz were used. It is recommended that a minimum of 4 analysis frequencies over 2 decades be used when generating a master curve. The plot shows the flexural storage modulus (E') and loss modulus (E") as functions of temperature at the various analysis frequencies. The loss modulus peak temperatures show that the glass transition moves toward higher temperatures as the analysis frequency increases. The loss modulus data can be used to generate a master curve. Figure 165(a) shows a plot which represents all of the individual E" data points as a function of log frequency. The data column at the extreme left represents data collected at 0.010 Hz, while that at the extreme right represents the data collected at 1.00 Hz. For generation of the master curve, a reference set of data is chosen. In this experiment, 145°C was selected as shown in Figure 165(a). The remaining data sets are shifted to either higher or lower frequencies to fall upon the chosen reference. The partially complete master curve is shown in Figure 165(b). The data sets in the lower portion of the plot will be shifted to the left (to lower frequencies),

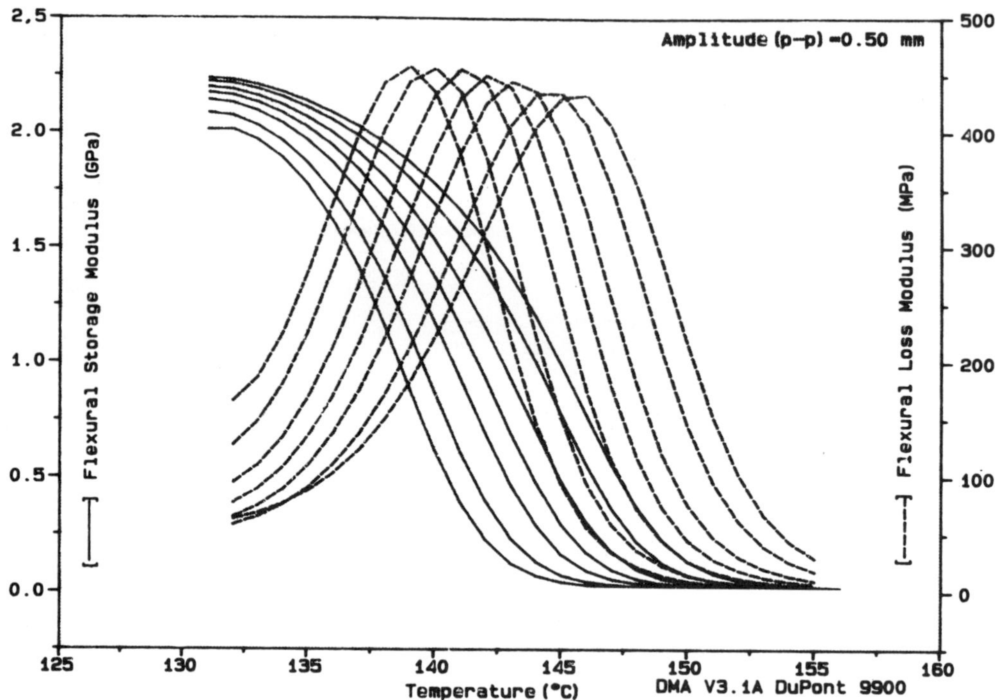

Figure 164   DMA results for polycarbonate obtained using frequency
multiplexing techniques.

while those in the upper portion will be shifted to the right (to higher
frequencies) to cause them to fall upon the intermediate (reference)
curve.   During the shifting, the computer stores the shift factors, a_T,
for later use.

The complete master curve for polycarbonate, obtained by shifting
all of the individual data points, is seen in Figure 166.   This curve
shows the change in loss properties with respect to frequency of 0.31 Hz
at 145°C, represented by the peak maximum.

The shift factors can now be fitted to one of the mathematical
models.   Figure 167 is a plot of the experimentally determined shift
factors as a function of temperature.   (The slight curvature in the shift
factor plot reflects WLF-type behaviour.)   The WLF equation was selected
to relate the shift factors (time) to temperature.   The solid line
represents the WLF model, while the boxes on the curve represent the
experimentally determined shift factors.   The close correlation
demonstrates the excellent fit obtained with the software-based
assessments of the constants: C_1 - 22.80 and C_2 - 79.06.

Once the shift factors have been described by the model, the
software can be used to transpose the master curve to any desired
temperature within the range that data is available.

A master curve for any of the other viscoelastic quantities (i.e.
flexural loss modulus [E"], flexural storage compliance [S"], and tan
delta) can be generated from the polycarbonate data, using the DMA
superposition software.

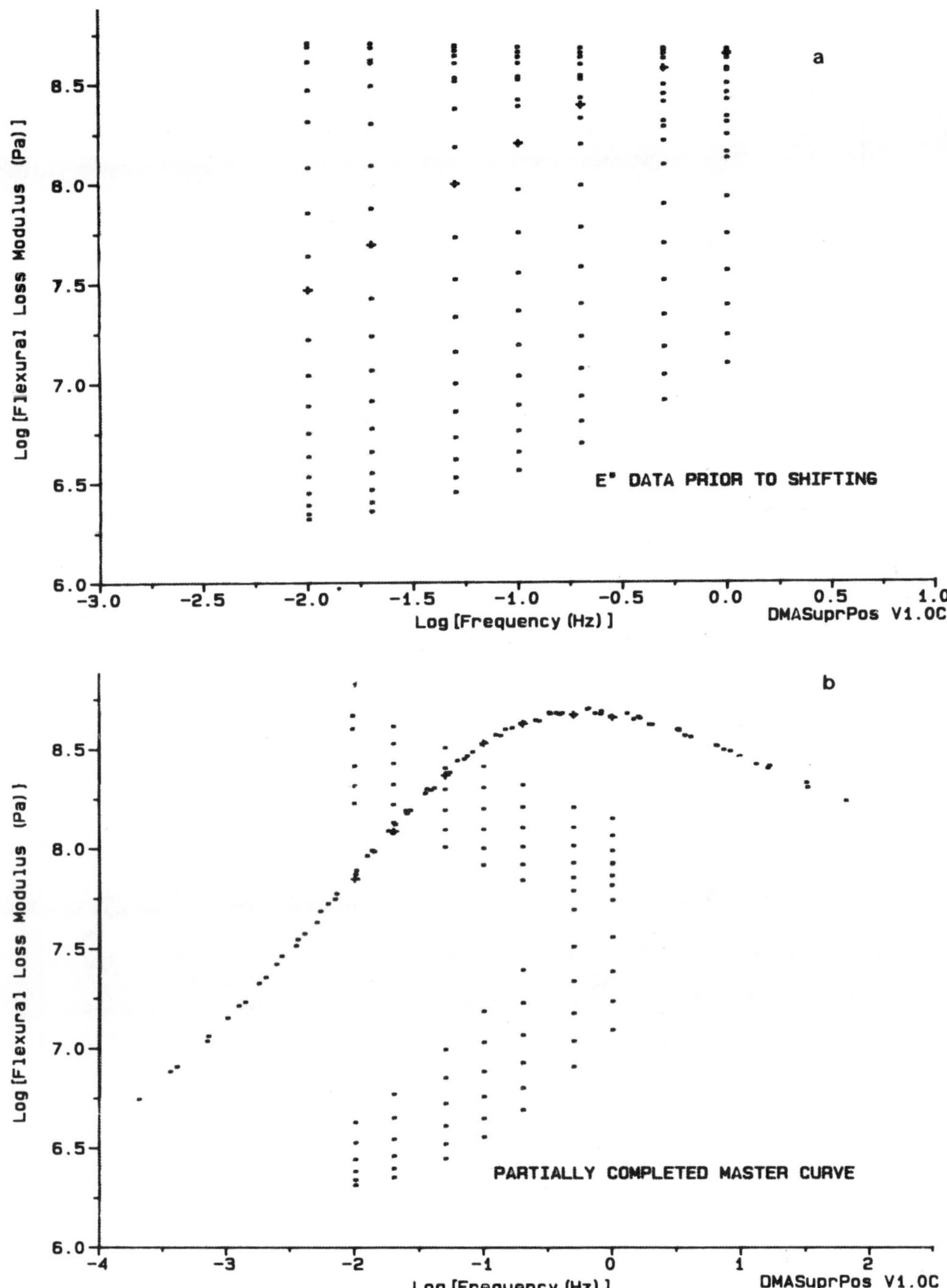

Figure 165   (a) Plot representing all of the individual "E" data
points as a function of log frequency (before shifting);
(b) Partially completed master curve based on loss
modulus data.

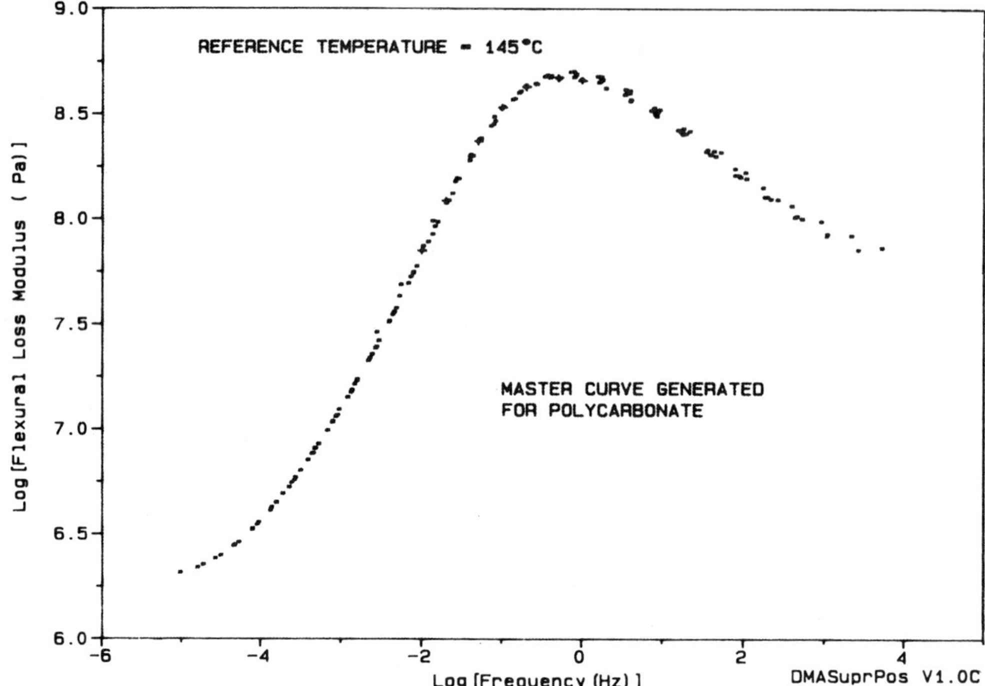

Figure 166   Complete master curve from loss modulus data for polycarbonate.

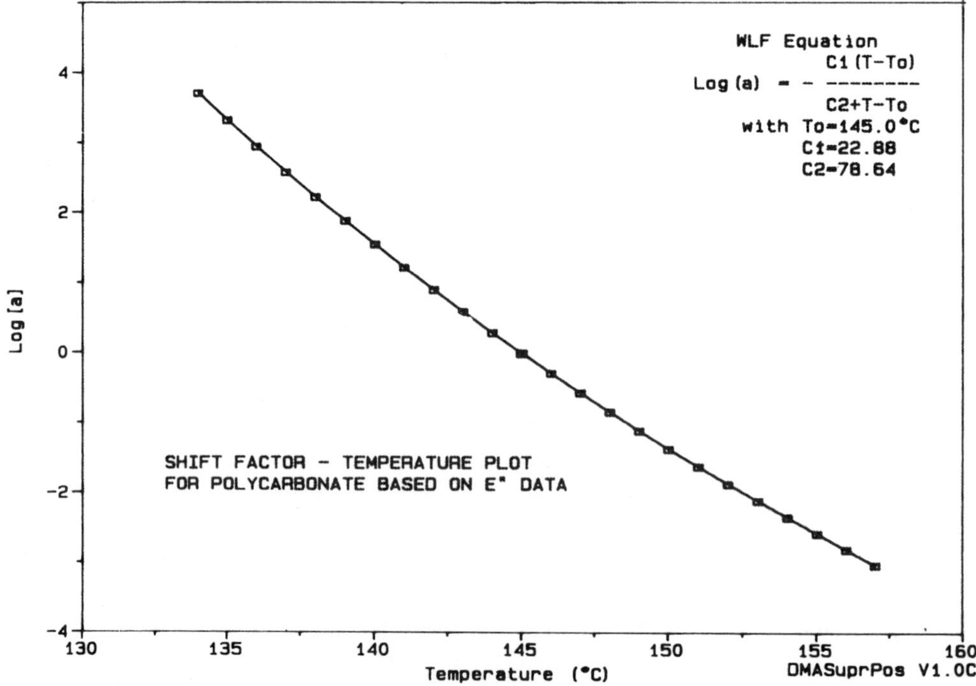

Figure 167   Plot of experimentally determined shift factors as a function of temperature.

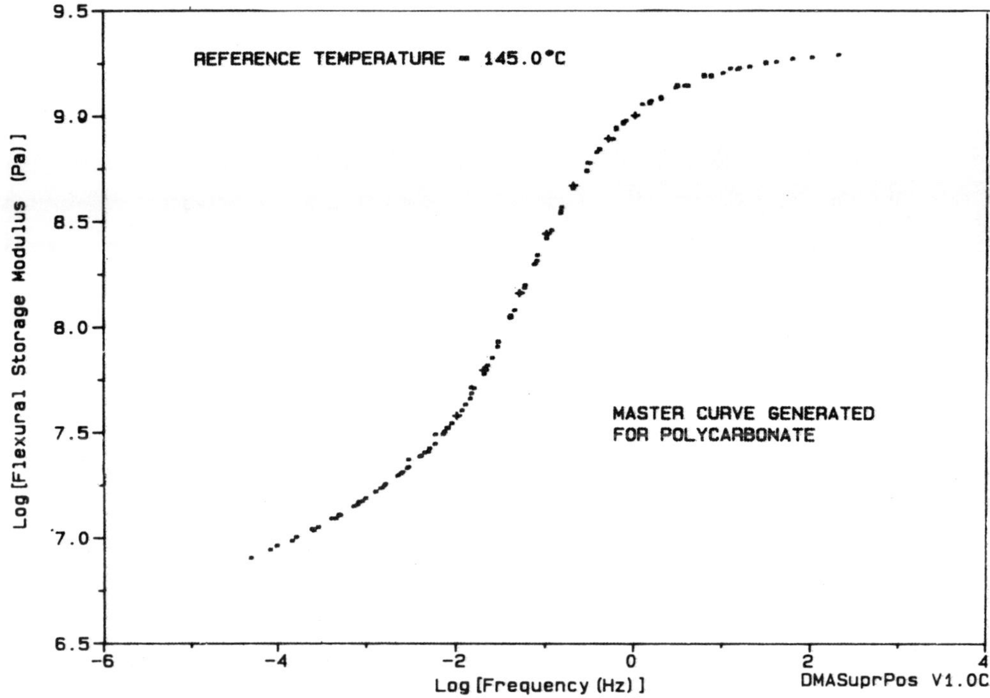

Figure 168   Master curve generated from the storage modulus data for polycarbonate.

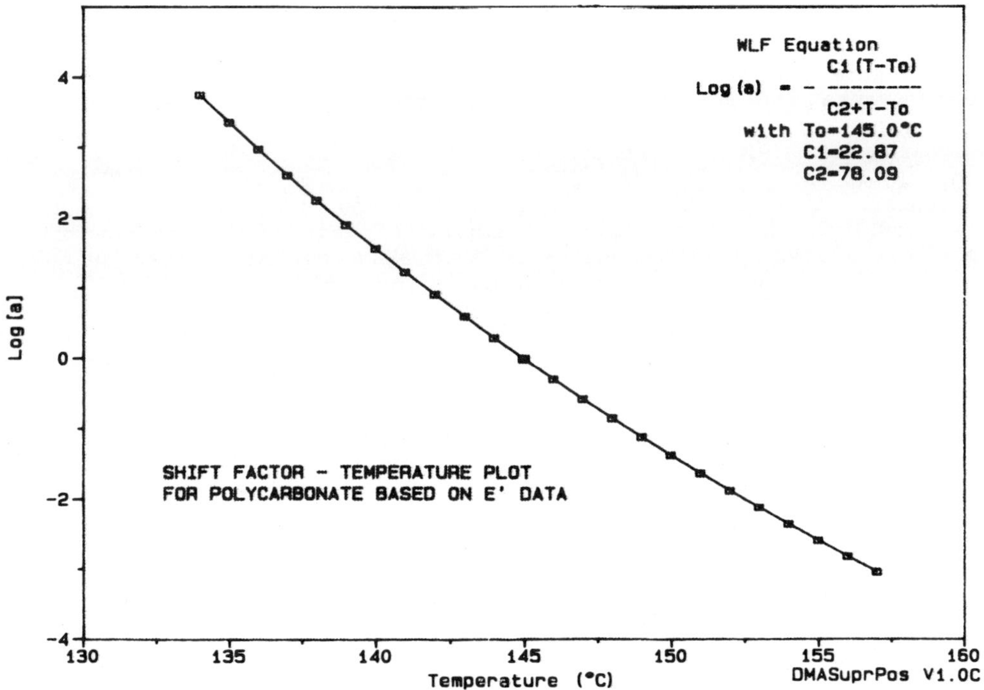

Figure 169   Shift factor plot obtained from generation of the "E" master curve.

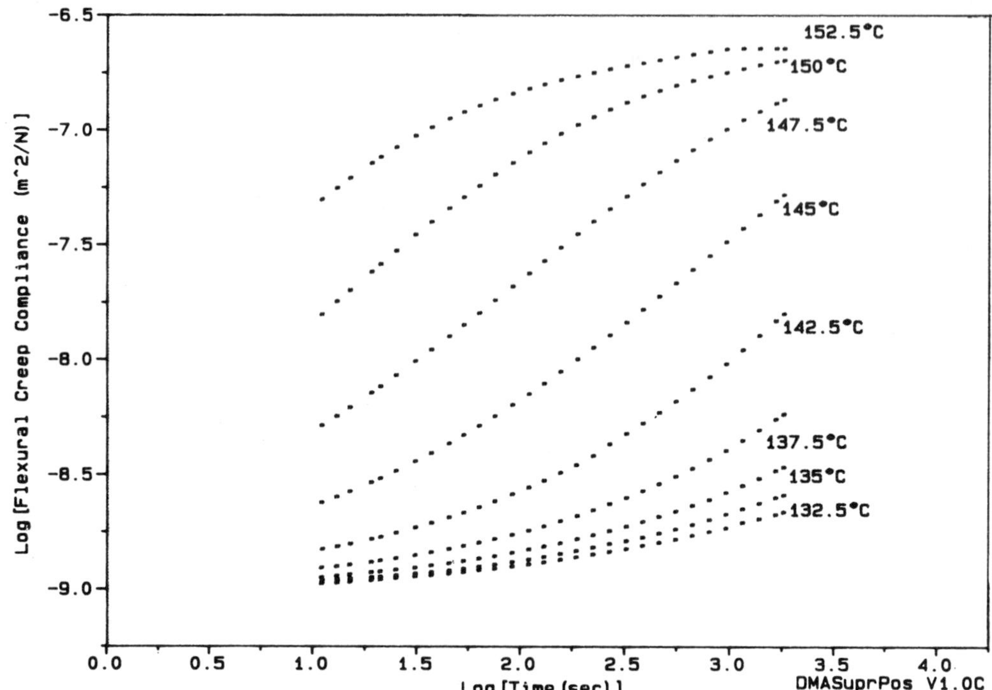

Figure 170   Creep data obtained on polycarbonate in the temperature
range of 130°C ot 155°C.

Figure 168 displays the master curve generated from the storage
modulus data for polycarbonate.   The curve has been referenced to a
temperature of 145°C.   This curve shows the effects of frequency on
polycarbonate at this temperature.   At very low frequencies (or long
times), the material exhibits a low modulus with behaviour similar to that
of rubber.   At high frequencies (or short times), the polycarbonate has a
high modulus and shows behaviour similar to that of an elastic solid.
Thus the master curve demonstrates that data collected over only two
decades of frequency can be transformed to cover nine decades.

The shift-factor plot obtained from generation of the E' master
curve is shown in Figure 169.   The WLF equation, again was selected to
model the time/temperature behaviour.   Best-fit values of the constants,
C_1 = 22.9 and C_2 = 78.1, were obtained by the software.   These constants
are in very good agreement with those obtained previously from shifting
the loss modulus data.   In addition to fixed-frequency data, creep or
stress relaxation data can be used with the software to generate master
curves.

Figure 170 shows the creep data obtained on polycarbonate in the
temperature range of 130 to 155°C.   In this curve, the log creep
compliance (S) is shown as a function of log decay time.   One of the
curves is selected as the reference (in this case, T_o = 145°C), then the
other curves are shifted along the log time axis and superimposed upon the

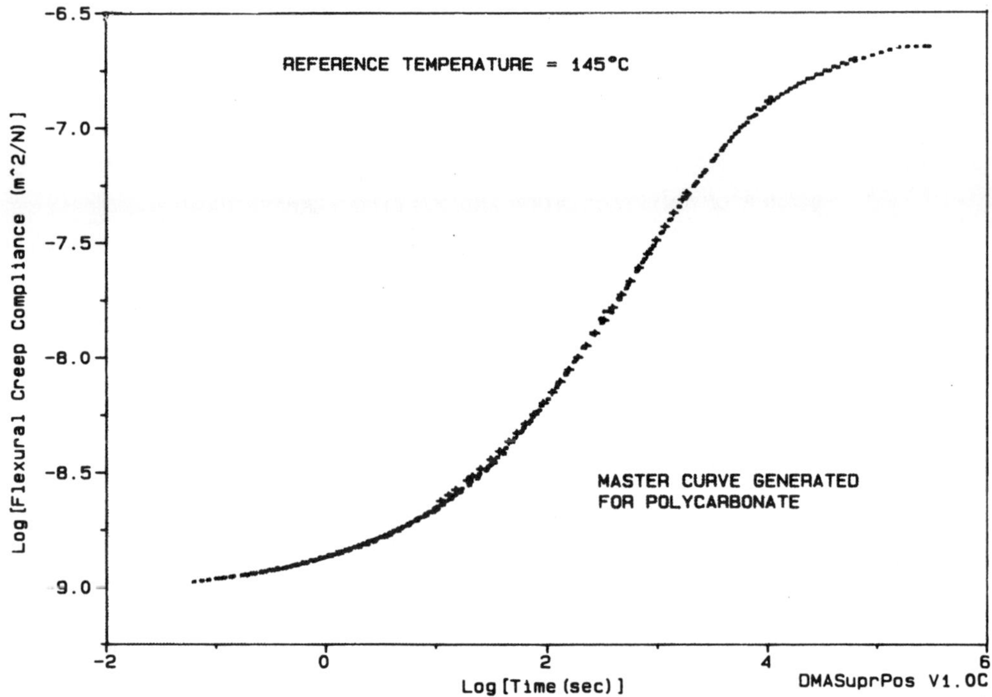

Figure 171   Final master curve based on creep data.

reference curve.   The final master curve based on creep data is shown in
Figure 171.   The curve shows that at small time intervals the material
exhibits relatively low compliance (or high modulus).   At longer times,
viscous flow occurs and the material exhibits a high compliance (or
modulus).   Thus this master curve clearly demonstrates the effects of time
on the mechanical properties of polycarbonate.

CHAPTER 13

DEGREE OF CRYSTALLINITY AND MELTING TEMPERATURE

13.1 DEGREE OF CRYSTALLINITY

13.1.1 Introductory

Crystallinity is a state of molecular structure referring to a long range periodic geometric pattern of atomic spacings. In semicrystalline polymers, such as polyethylene, the degree of crystallinity (% crystallinity) influences the degree of stiffness, hardness and heat resistance.

In semicrystalline polymers, some of the macro-molecules are arranged in crystalline regions, known as crystallites, while the matrix is amorphous. The greater the concentration of these crystallites, i.e. the greater the crystallinity, the more rigid the polymer. Morphology denotes the internal structure of a material (e.g. separate polymer phases, crystalline regions, amorphous orientation, etc). Amorphous is a term generally used to describe polymers totally lacking in long-range spatial order (crystallinity). The term is also used to denote noncrystalline regions within partially crystalline polymers.

13.1.2 Thermal Methods

Crystallinity can be calculated from a differential scanning calorimetry curve by dividing the measured heat of fusion by the heat of fusion of 100 per cent crystalline material, see figure 172. The crystalline melting point, Tm, which is a characteristic property, is used for quality control and for the identification of semicrystalline polymers.

The crystallites are destroyed upon melting and reform upon cooling. Their type and quantity depends on the sample's thermal history. Crystallinity values have been determined[1883] for poly(p-biphenyl acrylate) and poly(p-cyclohexylphenyl acrylate) from both heat of fusion and heat capacity measurements by differential scanning calorimetry. Differential scanning calorimetry has also been used to study the degree of crystallinity of Nylon 6[1884,1885] and crosslinked poly(vinylalcohol) hydrogels submitted to a dehydration and annealing process[1886].

Differential thermal analysis has been used for the measurement of crystallinity in random and block ethylene methacrylate copolymers[1887], and of polybutene[1888], polyethylene terephthalate, 1,4-cylco-hexane dimethyl terephthalate and polypropylene[1889,1890], and the examination of morphologically different structures of polyethylene ionomers[1891].

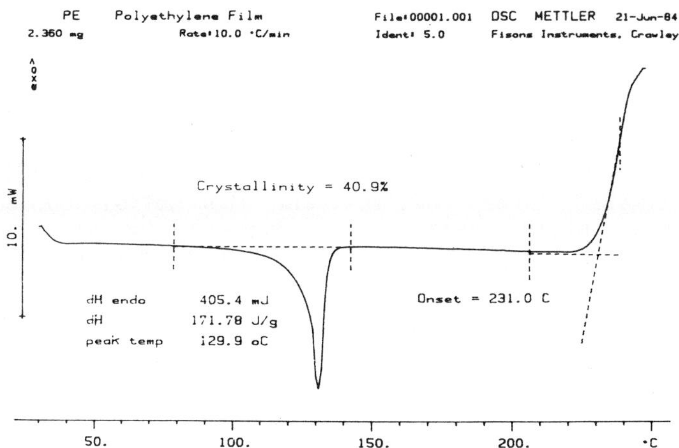

Figure 172   Determination of crystallinity of polyethylene film.

13.1.3   NMR Spectroscopy

NMR line width studies of crystalline polymers are based on the work of Wilson and Pake[1892]. This method was, however, unsuccessful due to the rather arbitrary decomposition procedures used which yielded a crystalline fraction which was not in agreement with crystallinity results obtained by the x-ray method. To overcome this difficulty Bergmann and Nawotki[1893-1895] decomposed the spectrum into three components and this resulted in an excellent agreement between NMR and x-ray crystallinities. Unfortunately, with this method it was not possible to prove the existence of the two amorphous components of the polymer examined, also the two amorphous mobilities could not be theoretically predicted[1893,1895,1896]. Bergmann[1897] succeeded eventually as discussed below in improving the separation procedure by finding more suitable line widths for the crystalline and amorphous components of the polymer. In this procedure a new method was evolved for the determination of the crystalline component and of the amorphous component based on a distribution of correlation times, instead of the two discrete correlation times as used in earlier work[1893-1895].

The line-width NMR procedure for determining crystallinity is outlined below.

Preparation of samples. One mm thick pressure moulded sheets of high density polyethylene, low density polyethylene, polypropylene and polyoxymethylene were slowly cooled from the melt, dried and filled into the spectrometer tube under a nitrogen atmosphere and the tube was was then sealed. 6.10 nylon was quenched in a carbon dioxide-heptane mixture, annealed at 205°C for one hour and quenched again in ice-water in order to produce well-ordered crystals with an alpha structure. Polyethylene terephthalate was quenched in ice-water and annealed for 16 hours at 200°C.

Apparatus. The proton NMR-measurements were performed at 60 MHz by means of the Bruker spectrometer CXP. The H_o-field inhomogeneity corresponded to the decal to 1/e of water signal in 6 ms. The H_o-field was regulated to 2.10^{-8} with respect to time constancy. The distance of the two 90° pulses of the solid echo measurements[1898] each of 1.5 microsecond length, was 6 microsecond. The probe assembly and the receiver were modified with respect to band width and dead time in such a way that the solid-echo signal was precisely identical to the Fourier transform of the broadline-spectrum of high density polyethylene at room temperature.

NMR analysis. For the analysis the solid-echo signals were digitized with a time increment of 0.2 microsecond by use of a fast analog digital converter and an averager. In order to record the crystalline component correctly, the 2000 points obtained were smoothed by computation and concentrated to 60 points, where the points in the range $t < 70$ microsecond were four times denser than the rest. The analysis of the solid-echo curves was performed according to the method of least squares by comparing them with the shape discussed below by means of iteration of the curve parameter.

Crystalline component. In cases where the amorphous phase shows a smaller relaxation time $T_{1\rho}$ than the crystalline one, the shape of the crystalline component G_c can be measured by a combination of the solid-echo and the $T_{1\rho}$ techniques. The principle of the method consists in the removal of the amorphous magnetization by the $T_{1\rho}$ experiment and the ensuing observation of the remaining crystalline magnetization. This method is explained in Figure 173 for clarity where the magnetizations of both phases are separated. At the beginning, the magnetization vector of both phases is turned by a 90° pulse from the z- to the y-axis (Figure 173(a) and (b)). Then the spin-locking pulse[1899,1900] is switched on. It causes a reduction of the magnetization with time, and for the amorphous phase this process occurs faster than for the crystalline one (Figure 173(c). Figure 173(d) shows the case in which the magnetization of the amorphous phase has vanished, but that of the crystalline phase is still maintained. Next the spin-locking pulse is switched off, and after a 90° pulse the solid-echo signal of the crystalline component is observed.

As an example figure 174 shows the progressive removal of the amorphous component of high density polyethylene at 310 K, as the duration spin locking time (d) of the spin locking pulse is varied. For d = 0 the normal solid-echo signal, i.e. the sum of the amorphous and the crystalline component, is observed. With increasing spin locking time (d) the fraction of the amorphous component progressively decreases, until at spin locking time d = 50 ms only the crystalline component is left. In order to prove that the signal for spin locking time (d) = 50 milliseconds in fact is that of the crystalline component, the solid-echo signal of a high density polyethylene sample, the amorphous phase of which had been removed by treatment with nitric acid[1901], is shown as well. Apart from the time-scaling factor s = 1.07 (which probably is a consequence of the imperfection of the spectrometer) both curves are identical.

Amorphous component. The shape of the amorphous component follows the following equation:

$$G_a(t) = \frac{\Sigma \exp\left[-(y_i/\alpha)^2\right] \cdot \exp\left\{-\omega_p^2 \tau_{ci}^2 \left[\exp(-t/\tau_{ci})-1+t/\tau_{ci}\right]\right\}}{\Sigma \exp\left[-(y_i/\alpha)^2\right]} \tag{1}$$

with

$$y_i = \ln(\tau_{ci}/\tau_{cm}) \ . \tag{2}$$

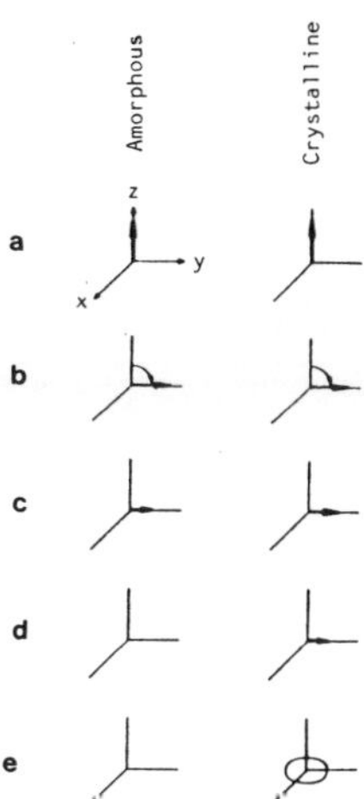

Figure 173   Method to measure the crystalline component by a
combination of solid-echo and $T_{1\rho}$ - technique.

Equation 1 is a modification with respect to a Gaussian distribution of
correlation times c of the equation:

$$\ln G(t) = -w_p^2 \tau_{ci}^2 [\exp(-t/\tau_{ci})-1+t/\tau_{ci}] \qquad (3)$$

derived for the shape of a simple motional process with an exponential
correlation function [1902]. w_p^2 is the second moment of G_a, τ_{cm} the medium
correlation time, and alpha the width of the distribution. The sum ranges
from $i = 1$ to n. The upper and lower y-limit were chosen to be such that
the probability was 10% of the maximum value. In equation 1, G_a consists
of a sum of KUBO-TOMITA curves which are supposed to be equidistant with
respect to y_1. n is chosen so high that G_a is independent of n. In this
case the distribution used cannot be distinguished from continuous
distribution. Equation 1 is able to reproduce adequately the shape of
amorphous polymers. A comparison of this shape with the "two-component"
shape introduced by Bergmann and Nawotki[1893-1895], shows that the
continuous distribution of equation 1 has been approximated by two
discrete correlation times.

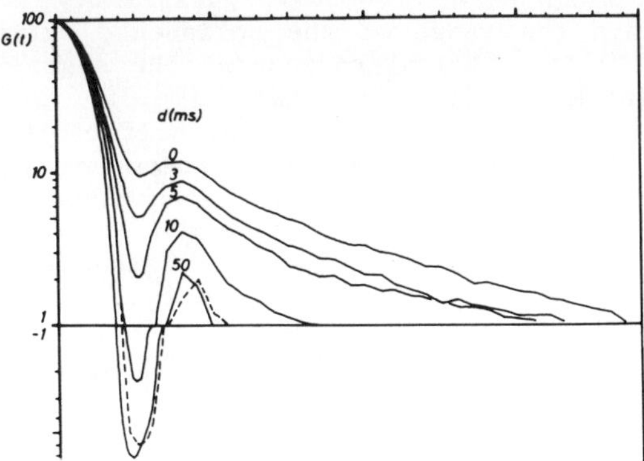

Figure 174   Elimination of the amorphous component of HDPE at 310 K
with the progressively increasing spin-locking time d.
For comparison, solid-echo signal of HNO_3 treated HDPE.

Table 188 - Comparison of cystallinities determined by NMR and
x-rays for seven polymers

Polymer	Density g/cm^3	T K	Crystallinity W_c (%)		d ms	$\omega_p\tau_{cm}$	α
			NMR	x-ray			
HDPE	0.963	310	74$\pm$1	75	100	0.7	1.1
LDPE*	0.918	300	47$\pm$1	43	30	0.4	1.9
PP	0.918	310	79$\pm$1	78	50	0.3	1.0
PP	0.899	310	60$\pm$2	57	50	0.3	1.3
POM	1.424	310	72$\pm$1	67	150	0.5	0.0
PA 6.10	1.098	370	44$\pm$3	52	30	0.3	1.8
PETP	1.410	400	51$\pm$3	60	10	3.0	3.1

* Reference crystallinity determined from density measurement[1905].

Bergmann[1897] compared crystallinities obtained for polymers by line
width NMR measurements with those obtained by x-ray crystallography (Table
188)[1903,1904]. The NMR crystallinity (Wc NMR) was determined at the
specified temperature and w_c (x-ray) at room temperature. The spin-
locking time d for the measurement of the crystalline component ranged
from 10 to 150 milliseconds. For most polymers the results in Table 188
shows an agreement of crystallinity within the limits of error, i.e. $\pm$ 2%
for w_c (x-ray). The error with regard to W_c (NMR) mainly depends on the
value of alpha. The greater error in conjunction with the broader
distribution is explained by the greater difficulty to distinguish the
immobile amorphous from the crystalline protons in these cases.

The temperature at which the NMR derived crystallinity was measured
was chosen to minimize error in the crystallinity measurements. That
means that a) within the range of the solid-echo curve not distorted by
the H$_o$ inhomogeneity (t < 300 microseconds) the signal decays to 1-3% of
its initial ordinate. At the same time a further condition b)
$\omega_p\tau_{cm} \lesssim 1$ has to be fulfilled, i.e. at least half of the distribution must
be mobile($\omega_p\tau_{ci} < 1$) to make the amorphous component separable. Despite the
difference in the lineshapes of the immobile amorphous ($\omega_p\tau_{ci} > 1$) and the
crystalline protons, the amorphous component in case $\omega_p\tau_{cm} \lesssim 1$ is still
separable, owing to the symmetrical shape of the Gaussian distribution.
As shown in Table 188 condition b is fulfilled in all cases except
polyethylene terephthalate for which an exceptionally high alpha-value is
obtained. In this case, an increase of the temperature would fulfil
condition b, but violate condition a.

13.1.4 Infrared Spectroscopy

Simak[1906] has carried out Fourier transform infrared measurements of
the specific volume of the crystalline phase (Vc cm$^{3-1}$) and the specific

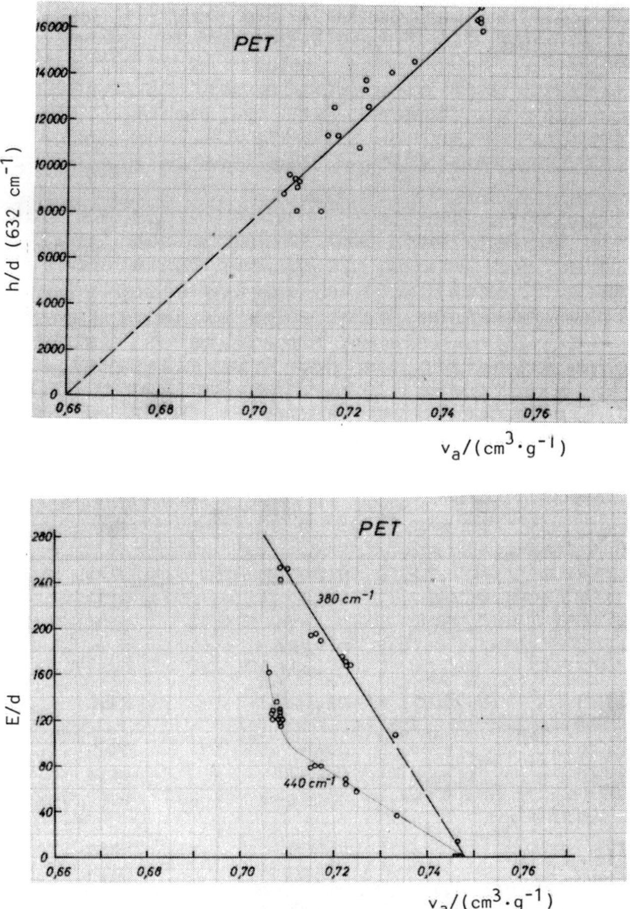

Figure 175   Calibration specific volume v infrared adsorption of
polyethylene terphthalate; (a) crystalline bands at 632
cm^{-1}, VC = 0.66, cm^3 g^{-1}; (b) amorphous bands at 440 and
380 cm^{-1}; Va = 0.7470 m^3 g^{-1}.

volume of the amorphous phase (Va cm^{3}g^{-1}) on polyethylene terephthalate.
They evaluated absorptions occurring at 632, 440 and380 cm^{-1} originating
from C-O-C groups attached to the aromatic ring found in polyethylene
terephthalate.   The 632 cm^{-1} band was used for the evaluation of Vc and it
is seen in Figure 175(a) that these results lead to a specific volume of
Vc 0.66 cm^3 g-1.   The 440 and 380 cm^{-1} bands were used for the evaluation
of Va and it is seen in figure 175(b) that the results lead to a specific
volume of Va = 0.747 cm^{3}g^{-1}.   The amount of the transconfiguration in the
amorphous phase for a well quenched sample of polyethylene terephthalate
was about 12-13%.

Numerous other workers[1907-1931] have discussed the application of
infrared spectroscopy to the determination of the crystallinity of
polyethylene terephthalate.

Infrared band assignments in the 4000-400 cm^{-1} region have been made for crystalline poly(ethylene glycol dimethyl ether)[1932]. Jansson and Yannas[1933] measured the infrared dichroism of polycarbonates at different strain levels. Below 0.6% strain, the dichroism is negligible while above this level the dichroism increases linearly. The infrared spectra of polyesters in the C-O-R stretching region has been discussed[1934]. Infrared and Raman band assignments were made by Holland Moritz and coworkers[1935-1937] on linear aliphatic polyesters including influence of the ester and CH_2 groups. The forms of crystalline poly(ethylene glycol adipate) has been explained by infrared spectroscopy[1938]. Funke and Schuh[1939] used infrared in studies of the chemical structure of radiation cured polyester resins. Elliot and Kennedy[1940] measured the infrared spectra of the three crystalline forms of cationically synthesized poly-3-methyl-1-butene. The alpha, beta and gamma (crystalline phases of cationically synthesized poly-3-methyl-1-butene been have characterized by infrared spectroscopy[1941]. McRae et al[1942] followed changes in the relative concentrations of methylene groups in crystalline regions, in gauche conformations and in tie chains in amorphous regions of cold drawn high density polyethylene by unpolarized and polarized infrared spectroscopy. Luongo[1943] studied crystalline orientation effects in the 13.69-13.88 doublet of transcrystalline polyethylene. The effects of defects in the infrared spectrum of polyethylene crystals was established by comparison with polyethylene in the extended chain form[1944]. Bell and coworkers used infrared dynamic mechanical and molecular weight measurements in studies of morphology of uni-axially oriented polyethylene terephthalate[1945] and of chain folding in annealed polymer[1946]. The Raman spectra of isotactic polypropylene has at 5-523°K showed bands characteristic of the unit cell[1947]. Relations between crystallinity and chain orientation have been observed in the Raman spectra of oriented polypropylene[1948,1949].

13.1.5 X-ray Diffraction

Turley[1950] has published a collection of x-ray diffraction patterns of polymers. X-ray diffraction methods have been applied to crystallinity and crystal structure studies in a wide range of polymers including cellulose[1951,1952], Nylon 66[1953], Nylon 6[1954,1955], Nylon 8[1954], ethylene-propylene copolymer[1956], poly-4-methylpentene[1957], polyacrylonitrile[1958] and polychloroprene[1959].

13.1.6 X-ray Diffraction/Infrared Microscopy of Synthetic Fibres

In order to establish structure-property relationships with high-performance fibres, x-ray diffraction has been the analytical technique most often chosen[1960-1964]. For most cases, x-ray diffraction was more than adequate in analysing structure in the fibres industry; however, with the discovery of the new generation of high-performance fibres with higher strength and orientation the sensitivity of x-ray diffraction to structure is being pushed to the limit[1965]. There was a definite need for a more sensitive technique capable of both orientation measurements and general fibres analysis.

High-performance fibres have historically been difficult to analyse by transmission Fourier transform infrared spectroscopy. Until a few years ago, the analysis of samples smaller than 25 micrometer was virtually impossible even with an infrared microscope because the visible-light portion did not have sufficient depth of field to focus the sample, and diffraction effects would result in sloping baselines and stray light. Young[1966] has applied infrared microscopy to Kelvar and Nomex aramid

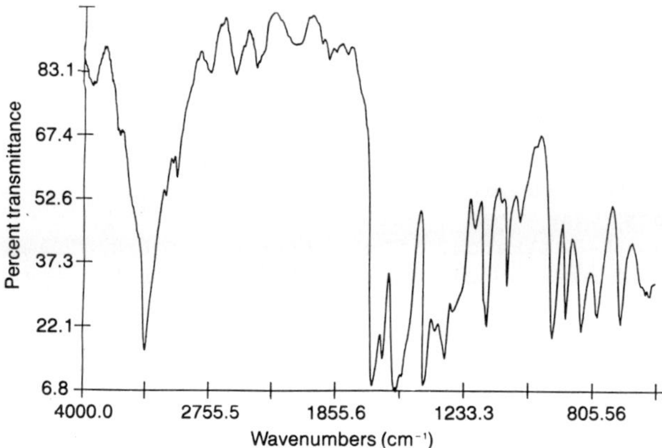

Figure 176   Infrared spectrum of single filament of Kevlar with the microscope optimized.

filaments as small as 12 micrometer. Filament physical geometry, especially for the highly oriented samples, causes optical aberrations resulting in band distortions that can lead to poor spectral quality. Some of the new infrared microscopes on the market are capable of minimizing these problems. Young[1966] used on IR-Plan infrared microscope, (obtained from Spectratech, Stamford, Conneticut), to obtain infrared spectra of the filaments and also to carry out orientation measurements. The microscope uses a 15 x IR/Vis reflecting lens with Cassegrain optics to focus the infrared beam onto the sample (Reflachromat, Spectra-Tech). A 10 x reflecting Cassegrain beam condenser is used to collect the transmitted beam after it passes through the sample. The beam condenser also has a sample compensator, which can be used to correct for optical aberrations caused by the fibre samples. The microscope uses its own narrowband liquid nitrogen-cooled MCT detector to maximize sensitivity. The FT-IR used was a Nicolet 5DXB bench model with 2 cm^{-1} maximum resolution capability (Nicolet Instrument Corporation, Madison, Wisconsin). In both the sample and the background files, 256 scans were run at 4 cm-1 resolution and spectral sweep from 2.5 to 15.3 microns. The background was acquired after the sample was scanned using the same aperture width. The microscope uses remote slits to mask above and below the sample, which speeds up sample preparation without sacrificing stray-light rejection. The dichroic ratios were measured using a AgCl wire-grid polarizer placed before the 15 x Cassegrain reflecting lens. Samples were mounted on the infrared microscope slides provided by the vendor. This will most accurately define the sampling area by optimizing the focus of the 15 x IR/vis reflecting lens. The single filament should be mounted under some tension because any slack in the filament will tend to vibrate in and out of the path of the beam. There should also be no twist in the filament, especially if orientation measurements are going to be taken.

Figure 176 is an infrared spectrum of a single 12 micrometer diameter filament of highly crystalline Kevlar. The vibrational modes that are most characteristic of Kevlar are the very intense amide vibrations[1967]. The intense band at 3.01 micron is the N-H stretching mode. The characteristic amide-I and amide-II bands are at 6.07 and 6.49 microns, respectively. The 6.07 micron band has been attributed to the C-O and C-N stretches from the carbonyl carbon of amide. The 6.49 micron band is a combination vibration due to the N-H bending, the C-N stretching from the carbonyl carbon, and an aromatic C-C stretch. The peaks that confirm the presence of an aromatic amide are the aromatic ring vibrations at 6.20 and 6.61 microns. Para substitution of the rings is confirmed by the 12.12 micron band. The intensity and sharpness of the bands are characteristic of the radial structure and the large crystallite size of Kevlar. A final feature of Kevlar is the broad base of the 3.01 micron vibration. This has been associated with moisture and also to nonhydrogen-bonded N-H stretch. The infrared spectra of laboratory Kevlar films are different from fibres in terms of band intensity and sharpness, due to crystallinity differences and the absence of radial sheets, but the same degree of orientation can be achieved in a Kevlar film cast from spinning solution.

Molecular orientation measurements of polymers are an important way of determining how a product will perform in service. The overall bulk orientation of Kevlar has been directly related to its stiffness and indirectly related to its strength. As far as structure is concerned, the orientation is a measure of the degree of order and how the macromolecular chains are placed in the crystal lattice.

Orientation as measured by x-ray diffraction is well documented as directly related to fibre stiffness (modulus). X-ray techniques yield bulk measurements that are sensitive only to the crystalline phase of the material and for most cases this is adequate. However, it has been observed that filaments of very different moduli sometimes cannot be distinguished using x-ray measurements (Table 189). Clearly, this causes a dilemma as to whether orientation controls modulus or x-ray diffraction is not sensitive enough to detect the orientation difference. Young[1966] investigated infrared microscopy's sensitivity for determining the orientation using individual bond order instead of relying on bulk orientation. To maximize sensitivity, he chose peaks that were pure vibrational transition and not a combination of modes. Deformation modes have proven not to render meaningful data because the mode will complicate the measurement by vibrating in two directions of the fibre plane. The optimum vibration to choose would be pure transition that is stretching mode because it will oscillate linearly in either the x or y direction.

A silver chloride wire-grid polarizer placed before the 15 x IR/Vis reflecting lens was used to orient the beam. The filament orientation was very sensitive to tension, so each sample was measured in a set of callipers to keep the tension constant. This was found to be most critical for the low-modulus samples because they seemed to be slightly easier to stretch, thereby changing the orientation. The standard deviation ranged from 0.3 to 0.55 and is attributed more to small fibrestructure differences than to the reproducibility of the technique. The dichroic ratios were calculated using the formula:

$$DR = \frac{A_\perp + A_{||}}{A_\perp - A_{||}}$$

where DR is the dichroic ratio, $A_\perp$ is the infrared absorbance when the electric vector of the source is perpendicular to the fibre axis, and $A_{||}$ is the infrared absorbance when the electric vector of the source is parallel to the fibre axis.

The samples were then broken on a recording stress-strain analyzer using a 1 in. break to determine the modulus. The results are shown in Figure 177 which is calculated for the 3.01 micron band of Kevlar. (Ideal

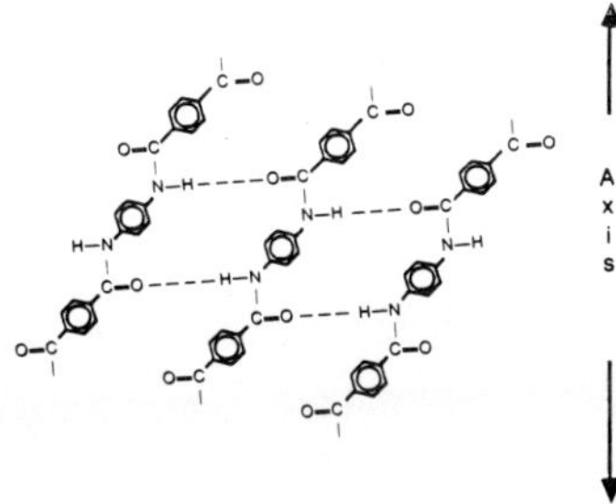

Table 189 - X-ray structure-property relationships
of the various Kevlar products

Product name	Orientation angle	Modulus (g/denier*)
Kevlar 29	16°	565
Kevlar 68	9° - 13°	780
Kevlar 49	4' - 11ᵁ	900
Kevlar 149	8° - 11°	1125

* Denier is defined as grams of polymer per 9000 meters of fibre.

orientation of Kevlar would be the polymer chains parallel to the fibre axis and the hydrogen bonding perpendicular to the fibre axis, as shown below.) What can be seen in Figure 177 is that as the modulus increases, the hydrogen bonding of Kevlar preferentially becomes more highly oriented perpendicular to the fibre axis.

13.2   MEASUREMENTS OF CRYSTALLIZATION PHENOMENA

13.2.1   Thermal Methods

Differential thermal analysis.   Figure 178 shows a typical DTA thermogram of a linear high-pressure polyethylene blend[1968]. This polymer, upon heating, undergoes three phase changes from its high-pressure form (115°C) to cocrystalline form (124°C) to a linear form (134°C).

The 115°C peak was associated with the high-pressure polyethylene, whereas the 134°C peak was shown to be proportional to the linear content of the system. Clampett[1969] also applied differential thermal analysis to a study of the 124°C peak which he describes as the co-crystal peak. His results appear to indicate that there are two classes of co-crystals in linear-high-pressure polyethylene blends with the linear component being responsible for the division of the blends into two groups. The property of the linear component which is responsible for the division is related to the crystallite size of the pure liner crystal.

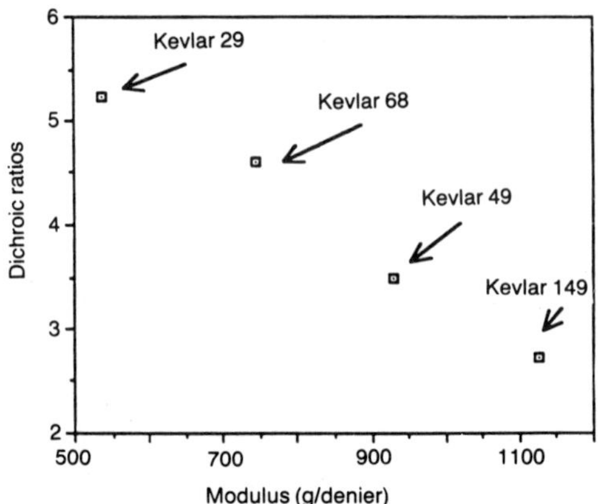

Figure 177   Plot of dichroic ratios of the 3320 cm^{-1} band measured
              by infrared microscopy of the different stiffnesses of
              Kevlar products (means of 10 measurements per sample).

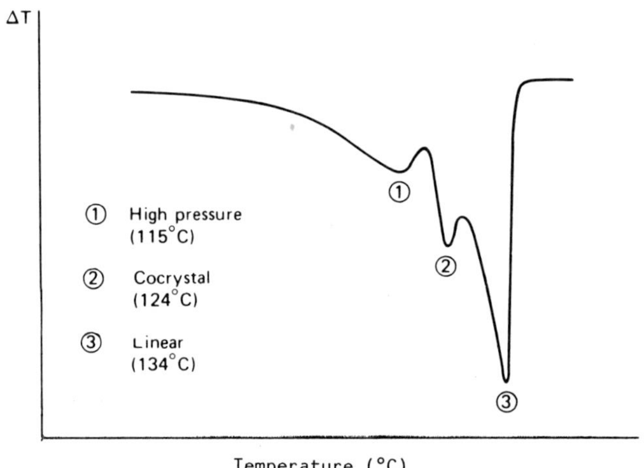

Figure 178   Typical DTA thermogram of a linear high pressure
              polyethylene blend.

<u>Differential scanning calorimetry.</u>   A method based on differential
scanning calorimetric measurement of heat of crystallization has been
reported for the determination of the molecular weight of PTFE[1977].

Differential scanning calorimetry has been used to study the crystal
structure and thermal stability poly(vinylalcohol) modified at low levels
by various reagents and by grafting with other vinyl monomers[1978].

Differential scanning calorimetry has been used to study
crystallization kinetics of many polymeric systems including amorphous

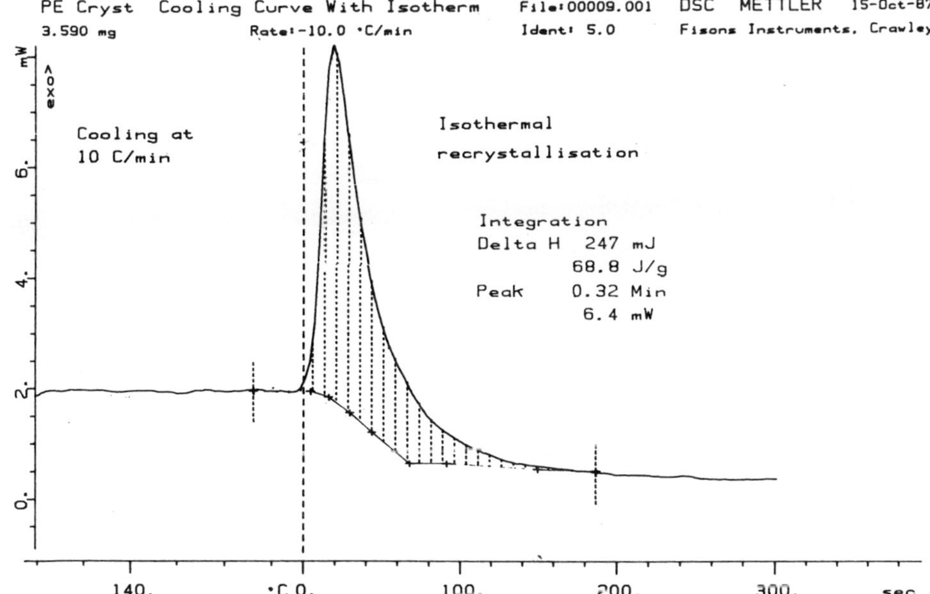

Figure 179   Differential scanning colorimetry recrystallization
curve of crystalline polyethylene.

cellulose[1970], polyethylene and chlorinated polyethylenes[1971-1973], aliphatic polyesters[1974], Nylon 8[1975], Nylon 66[1975] and 610[1976].

The development of crystal modifications of poly-1-butene has been followed using differential scanning calorimetry[1979].

Certain thermoplastics e.g. poly(ethylene terephthalate), can be frozen in the amorphous state below the glass transition temperature by quench-cooling of the melt. The occurrence of different crystal forms (polymorphism of polyamides, as an example) and melting gaps caused by tempering can be read from DSC melting curve. If recrystallization is studied, crystal growth rate and supercooling can be measured, Figure 179.

Isothermal crystallization studies by differential scanning calorimetry provide a sensitive technique for measuring molecular weight or structural differences between very similar materials. In Figure 180 is shown an isothermal crystallization curve obtained for polyethylene using the Perkin Elmer DSC-7 instrument.

Differential scanning calorimetry (heat flow versus temperature curve), is particularly well suited to the detection and study of liquid crystals. Studies of transition temperatures and enthalpies, range of stability and states of reversion from one form to another are all possible.

Differential scanning calorimetry and low-angle x-ray scattering have been used to draw phase diagrams of poly(ethylene oxide) - polystyrene block copolymers in the presence of diethyl phthalate as a preferential solvent of polystyrene[1986].

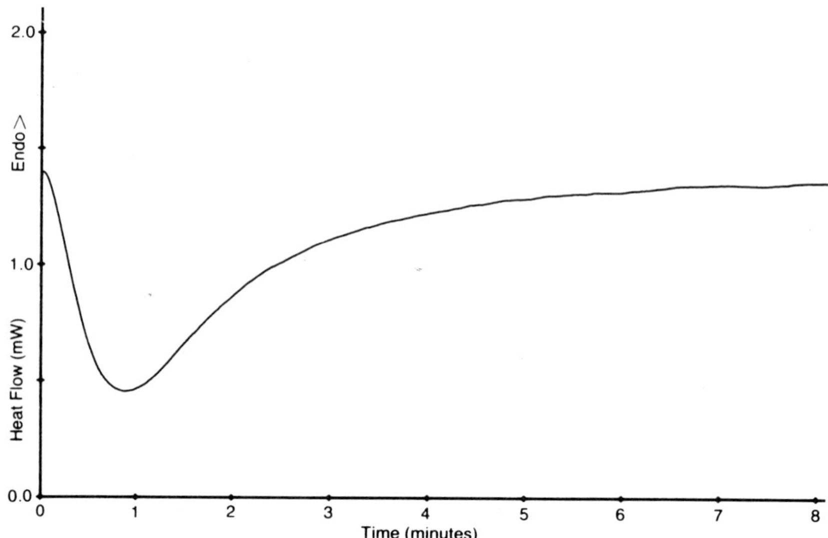

Figure 180   Differential scanning colorimetry isothermal
              crystallization curve of polyethylene.

        Illers et al.[1981] have shown that differential scanning calorimetry
can be used to prepare phase diagrams and illustrated it by work on a
binary blend of polystyrene and tetramethyl bisphenol-A-polycarbonate.
Differential scanning calorimetric curves obtained for different blends of
the same two polymers, ranging in concentration from 100% polystyrene
(lower curve) to 100% tetramethylbisphenol-A-polycarbonate are shown in
Figure 181(a). In Figure 181(b) is shown a set of curves for 1:1 mixtures
of polystyrene (Mw 320,000, Tg 109°C) and tetramethyl bisphenol-A-
polycarbonate (Mw = 41,000, Tg 200°C) which had been previously heated and
chilled up to the temperatures indicated on the curves, then chilled in
ice water. The dotted line indicates the phase diagram.

        Illers et al[1981] showed that a phase diagram of a binary polymer
blend can be derived from the glass temperatures Tg of the demixed phases
under the following conditions:

1.   The Tg's of the pure components are sufficiently different from one
        another.

2.   The Tg's of the one-phase homogeneous mixtures can be determined and
        vary monotonically with composition.

3.   The equilibrium state of the mixture is attained after sufficient
        long annealing at temperature T.

4.   The equilibrium state at the temperature T can be frozen by
        quenching.

They showed that between Tg and binodal, the one-phase mixtures exist in a
hindered thermodynamic state, because the crystallization of the
tetramethyl bisphenol-A-polycarbonate component is very slow process which

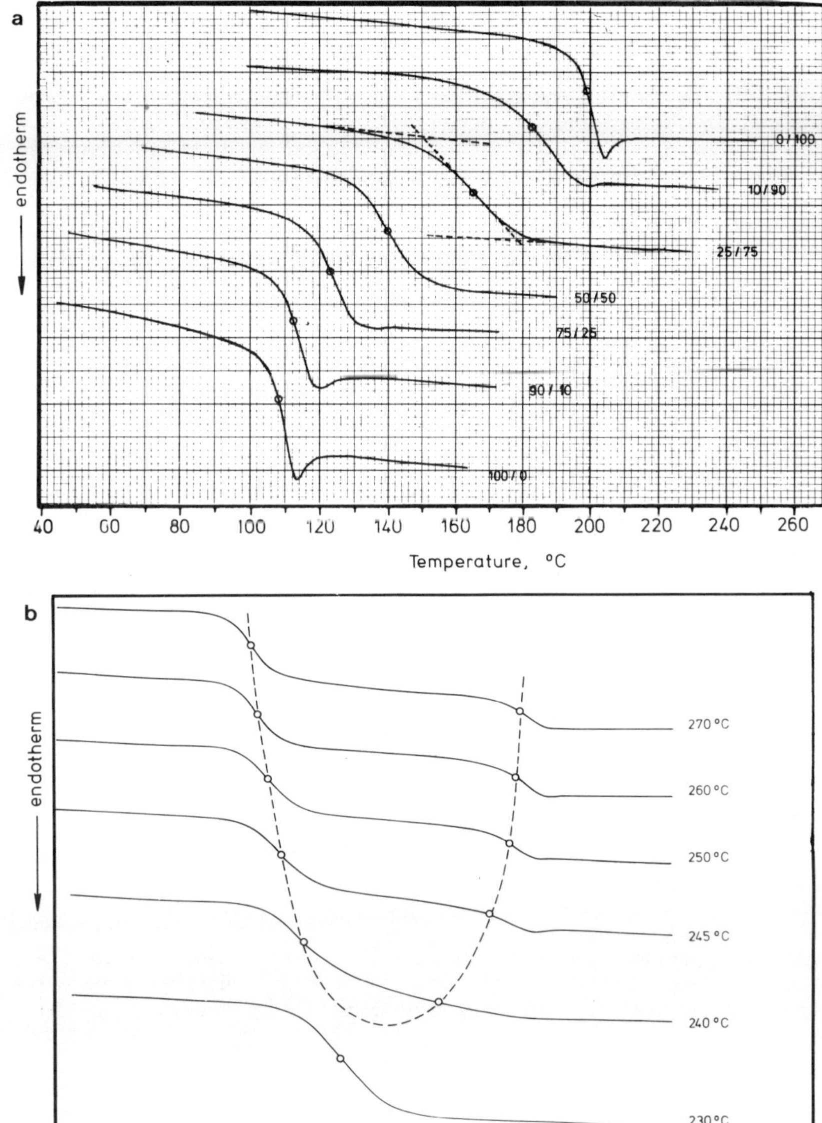

Figure 181   Differential scanning colorimetry of binary blends of
             polystyrene and tetramethyl bisphenol-A-polycarbonate:
             (a) DSC of various blands between 100% polystyrene
             (lower curve) and 100% tetramethyl bisphenol-A-
             polycarbonate (upper curve).   Heating rate 20°C min^{-1}.
             (b) 1:1 mixture of polystyrene: tetramethylbisphenol-A-
             polycarbonate, preheated to stated temperature (230-
             270°C) then cooled in ice prior to DSC run.

requires very long annealing. With increasing amounts of polystyrene in the mixture the crystallization half time of tetramethyl bisphenol-A-polycarbonate is lowered.

13.2.2 Spectroscopic Methods

Simak[1987] showed that poly (-caprolactam), (polyamide-6) crystallizes in different modifications, which show different characteristic infrared absorption bands. Based on this he prepared phase diagrams. He evaluated the absorption coefficients of the bands from the dependence of their intensity on temperature and specific volume of different polyamide-6 samples and described a method for the quantitative determination of the amounts of alpha and gamma* modifications and of the amorphous phase. On the basis of these data the densities of the samples were calculated and shown to agree with the values determined experimentally. The specific volume of the amorphous phase was $v = 0.917$ cm^3/g, and was independent of any contents of alpha, gamma* or gamma modifications the samples may contain. A typical phase diagram is illustrated in Figure 182.

Axelson [1983] showed that broad line proton NMR spectra of drawn high density polyethylene has three component lines: a broad component assigned to crystalline regions; an intermediate component assigned to high molecular weight molecules interconnecting the crystalline regions; and an isotropic narrow component assigned to mobile low molecular weight material and the ends of molecules reflected from the crystalline regions.

A laser radiation procedure has been adapted to non-destructive testing for spherulite size and deformation and crystallinity and degree of stretching[1984].

13.3 DETERMINATION OF MELTING TEMPERATURE, Tm

Tm is defined as the temperature at which crystalline regions in a polymer melt.

13.3.1 Thermal Methods

In semi-crystalline polymers, some of the macro molecules are arranged in crystalline regions, known as crystallites, while the matrix is amorphous. The greater the concentration of crystallites, i.e. the greater the crystallinity, the more rigid is the polymer, i.e. the higher is the Tm value.

In Figure 183 is illustrated a sample temperature-heat flow differential scanning calorimetric curve obtained for high density polyethylene by the Perkin Elmer DSC-7 instrument illustrating the measurement of the melting point (Tm) and the heat of melting in a single run. The true melting points of crystalline polymers can be determined by plotting the differential scanning calorimetry melting peak temperatures as a function of the square root of heating rate and linear extrapolation to zero heating rate.

Differential thermal analysis has been used to study the effect of side-chain length in polymers on melting point[1985]; and the effect of heating rate of polymers on their melting point[1986,1987]. Differential scanning calorimetry has been used to evaluate multiple melting peaks in polystyrene[1988-1990]

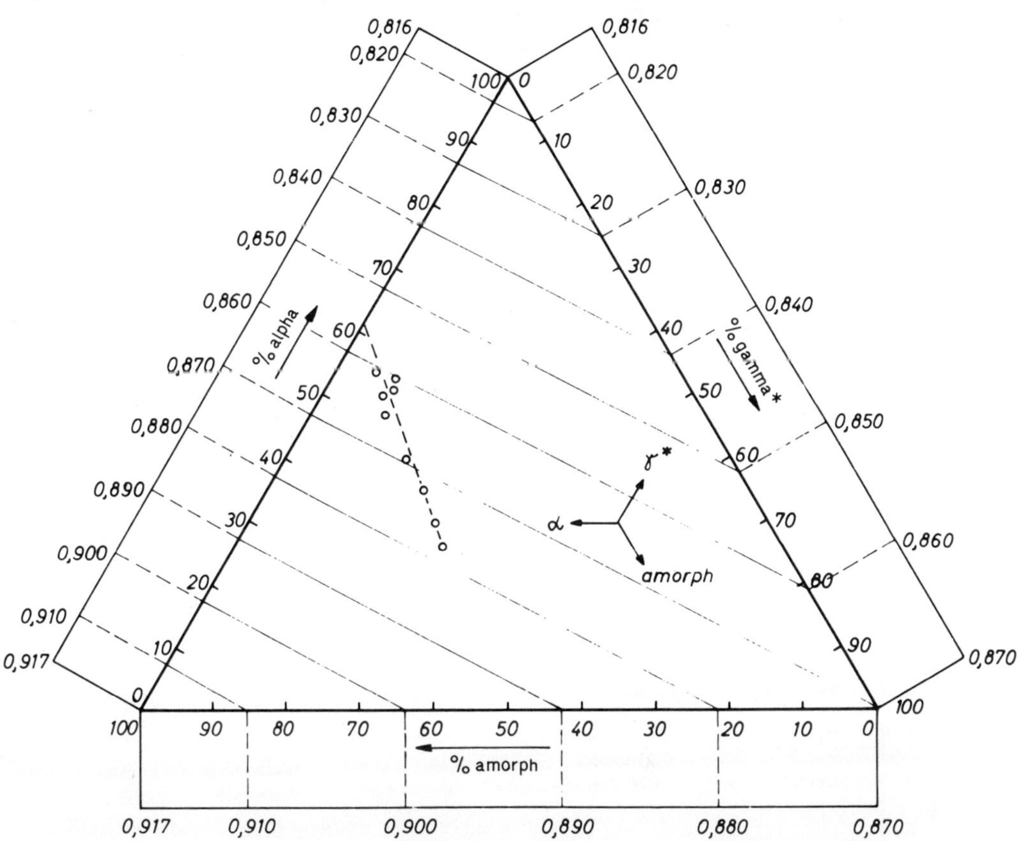

Figure 182   Relationship between specific volume of alpha, beta
crystalline modifications specific volume of amorphous
content and composition for polyamide (gamma
modification absent, all axes specific volume, cm^3 g^{-1}).

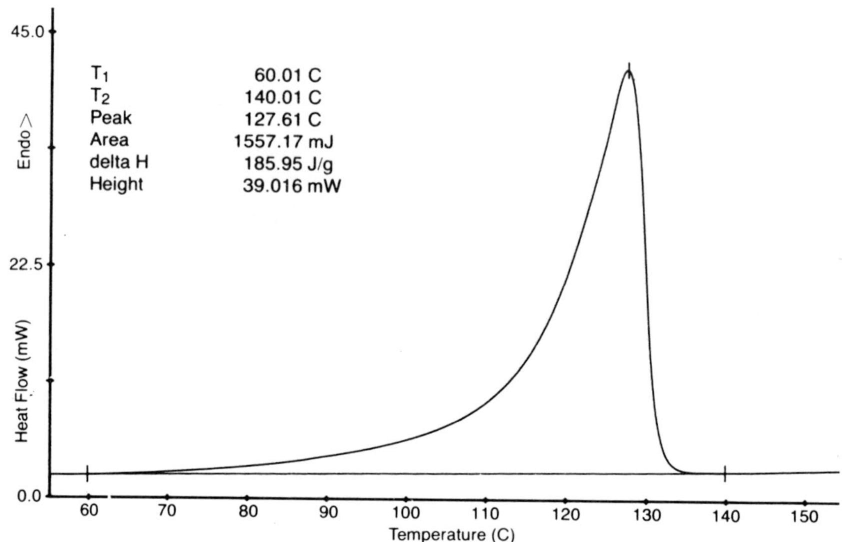

Figure 183   DSC run on high density polyethylene (Perkin Elmer DSC-7 instrument) showing measurement of Tm and heat of melting in a single run.

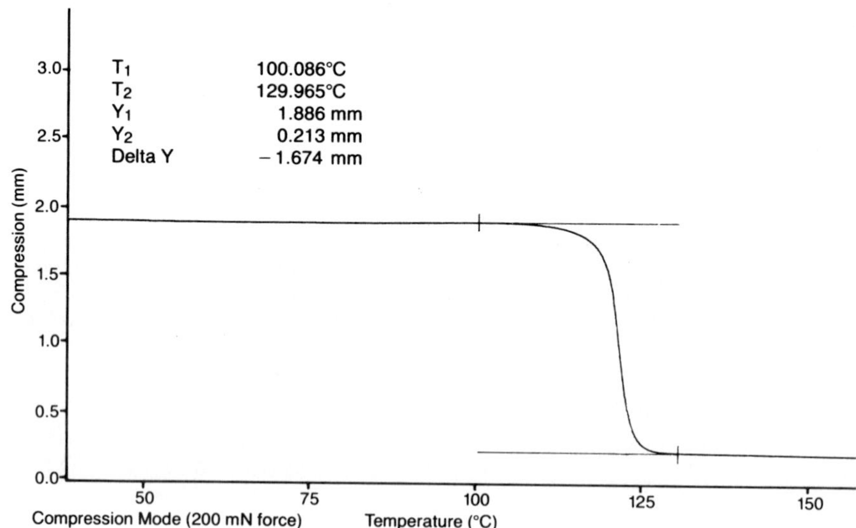

Figure 184   Use of thermomechanical analysis (Perkin Elmer TMA-7 instrument) to evaluate Tm and physical and dimensional changes in polyethylen uncrosslinked.

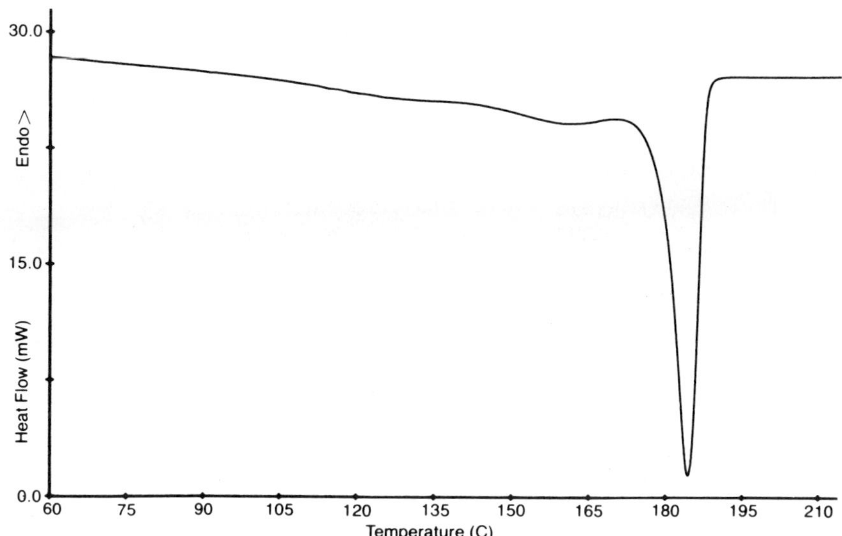

Figure 185   Controlled cooling programme on nylon 6/12 illustrating
               the softening temperature.

    Differential scanning calorimetry measurements of the energy during
melting of aqueous polymer solutions and gels yield heats of mixing and
sorption[1991].   The technique has also been used to study the heat changes
occurring in a polymer as it is cooled, (Figure 185).

    Thermomechanical analysis has been used for softening measurements
of polymers and the measurement of the amount of probe penetration into
the polymer at particular applied forces as a function of temperature.
This technique allows evaluations of Tm (Figure 184) and of the evaluation
and dimensional properties over the temperature range of use or under
actual accelerated conditioning cycles.

APPENDIX 1

TYPICAL PYROLYSIS - GAS CHROMATOGRAPHY AND IR AND NMR
SPECTRAL CURVES FOR POLYMERS

a)   Pyrolysis - gas chromatograms.

b)   Photolysis - gas chromatography.

c)   Infrared Spectroscopy.

d)   Nuclear magnetic resonance spectroscopy.

PYROLYSIS GAS CHROMATOGRAPHY

In figure 186 is shown a collection of pyrograms obtained by filament and furnace methods[1992-1995] (Figure 186.1 to 186.43), by Curie point pyrolysis[1996] (Figures 186.44 - 186.45) and by laser pyrolysis[1997,1998] (Figures 186.46 to 186.52).

INFRARED SPECTRA

Infrared spectra of thin films of polymers and copolymers in the region up to 2.5 microns are characteristic of the polymer. Some infrared spectra are shown in Figures 187.1 to 187.62. Other information on infrared spectra of polymers and copolymers are tabulated in Table 190.

NUCLEAR MAGNETIC RESONANCE SPECTRA

NMR spectra of polymers are shown in Figures 188.1 to 188.9 and of copolymers in Figures 188.10 to 188.23. PMR spectra of some polymers and copolymers are reproduced in Figures 188.4, 188.5, 188.9 and 188.22 - 188.23.

Various other sources of information on ^{13}C NMR and PMR spectra of polymers are quoted in Table 191.

Table 190 - Infrared spectra of polymers

Polymers	Reference
Polyvinyl phenyl ether	1999
Polychloroprenes	2000
PVC	2001-2006, 2021
Chlorinated PVC	2007

Polyisoprenes	2008,2018
Isotactic polypropylene	2009
Polymethacrylic acid	2010
Polyethylenes	2011, 2012
Polyethylene oxide	2013
Polyepichlorohydrin	2014
Polycarbonate	2015
Polymethyl methacrylate	2016
Polymethacrylonitrile	2017
Polyisobutylene terephthalate	2017
Polyisoprene	2018
Polybutadienes	2019, 2026
Cellulose	2020
PVC	2021
Polyacrylonitrile	2022, 2024
Polychlorprenes	2023
Fluoro polymers	2025
Cyclopolyisoprenes and cyclized polyisoprenes	2027
Polyvinylformal	2028
Polybenzyls	2029
General Reviews	2030-2063

<u>Copolymers</u>

Vinyl chloride sulphur dioxide	2064
Vinylcyclohexane-styrene	2065
4-methyl-1-pentene-vinyl cyclohexane	2066
Vinyl bromide-vinylidene chloride	2067
Polyethylene oxybenzoate	2068
Styrene-acrylonitrile	2069

Table 191 - ^{13}C NMR and PMR Spectra of Polymers

Polymer	Comments	Reference
^{13}C NMR POLYMERS		
Chlorinated polyethylene	High resolution PMR and ^{13}C NMR	2070,2071,2081 2085
Polyphenyl ethers	PMR and ^{13}C NMR	2072
Acrylics	-	2073
Partially hydrogenated natural rubber, gutta percha and Cis 1,4 polybutadiene	-	2074,2075
Polyethylene glycol	-	2076
1,4 polybutadiene	-	2076
Epoxidized 1,4 polyisoprene and 1,4 polybutadiene	-	2077
Polybutadienes	-	2078,2079
Rubbers	-	2080
Cross linked chloromethylated polystyrene	-	2082
Polymethylmethacrylate	-	2083,2084
Chorinated PVC	-	2085,2087
Polystyrene	-	2086
Polycarbonate	-	2088
Chlorinated PVC	-	2087
^{13}C NMR COPOLYMERS		
Ethylenepropylene coplymers	computerization of ^{13}C NMR	2089
Methylmethacrylate-styrene alternating polymers	-	2090
Styrene-butadiene	-	2091
Trichloroethylene-vinylchloride	-	2092,2097
Ethylene-methacrylate	-	2093
Vinyl chloride-sulphur dioxide	-	2094
Vinyl copolymers	-	2095

Ethylene-vinyl chloride & Ethylene-vinylidene chloride	-	2096
Trichloroethylene-vinyl chloride	-	2097

PMR POLYMERS

Polymethylmethacrylate	-	2098
Polystyrene	-	2099,2100
PVC	-	2101
Vinyl ethers	-	2102
PVC tetrahydrofuran &	-	2103
PVC-n-butyraldehyde complexes	-	2104
Polyacrylic acid	-	2104
Poly(methyl-alpha-chloro acrylate)	300 MHz spectra	2105
Polymethylmethacrylate	-	2106-2108
Carboxy terminated poly- butadienes	-	2109
Poly alpha-methylstyrene	-	2110
Natural rubber	-	2111
Ion-exchange resins	-	2112
Chorinated poly-isobutenes	-	2113
Sulphonated polystrene resins	-	2114,2115
Polyvinyl phenyl ether	60 and 100 MHz	2116
PVC	-	2117
Lactone polyester	-	2118
Polymethylmethacrylate	-	2119
Chlorinated PVC	-	2120
Poly-alpha-methyl styrene	-	2121
Polycarbonate	-	2122
Poly 1,3 pentadiene	60 MHz and 300 MHz spectra	2123
Poly 2-allylphenylacrylate	-	2124

Table 191 - Continued

Poly-(4-methyl-pentene-1)	-	2125
Polymethylmethacrylate	-	2126,2127
Polymethylmethacrylate & Polyethylmethacrylate	-	2128
Polymethacrylic acid	-	2129
Polypropylene	-	2130
Natural rubber	300 MHz spectra	2131
Polystyrene	-	2132
Cyclic ethers	-	2133
Polymethacrylonitrile	-	2134
Alpha-methyl styrene tetramer Poly-p-isopropyl alpa-methyl styrene	-	2135
Poly(ethylene glycols)	-	2136
Polyethylene	-	2137
Polyacrylamide & Polymethacrylamide	-	2138
Polypyrrolidone	-	2139
Polychloroprene	-	2140
PVC	-	2141,2142
Phenol formaldehyde resins		2143,2145
Nylon 66	-	2144
Polyvinylidene fluoride	-	2146
Polyvinyl formate	-	2147
Polyacrylonitrile	-	2148
Polymethylmethacrylate	-	2149
Polyvinyl-trifluoroacetate	^{19}F NMR	2150
Diene copolymers	-	2151
Carboxyl terminated poly-butadienes	-	2152
Poly-vinyl-formal	-	2153
Polypropene sulphone	-	2154

Table 191 - Continued

Carboxy terminated adipic acid- ethylene glycol Adipic acid-neopentyl glycol	–	2176
Chloroprene-methylmethacrylate	–	2177
Vinylidene chloride-vinyl chloride	–	2178,2179
Propylene-styrene	–	2180
Alpha-methyl styrene-p-methyl alpha-methyl styrene	–	2181
Methyl methacrylate-styrene	–	2182

Figure 186   Pyrolysis - gas chromatograms;   F = filament pyrolysis;
(on pp. 655-677)   Fu = furnace pyrolysis; C = curve point pyrolysis; L =
                    lazer pyrolysis.

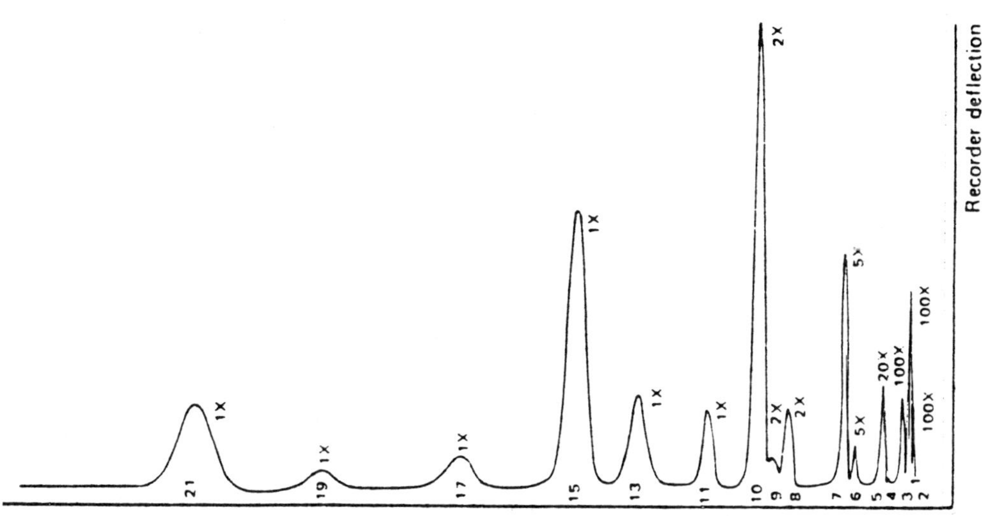

Figure 186.1 Polyethylene (Fu), (1) methane; (2) ethane; (3) propane;
              (4) isobutane; (5) n-butane; (6) isopentane; (7) n-
              pentane; (8) 2-methyl pentane and/or methyl
              cyclopentane; (9)3-methyl pentane; (10) n-hexane; (11)
              2,4 dimethyl-pentane and/or methyl cyclopentane; (13) 3-
              methyl hexane and/or cyclohexane; (15) n-heptane; (17)
              2,4-dimethylhexane and/or toluene; (19) 3-methyl
              heptane; (21) n-octane.

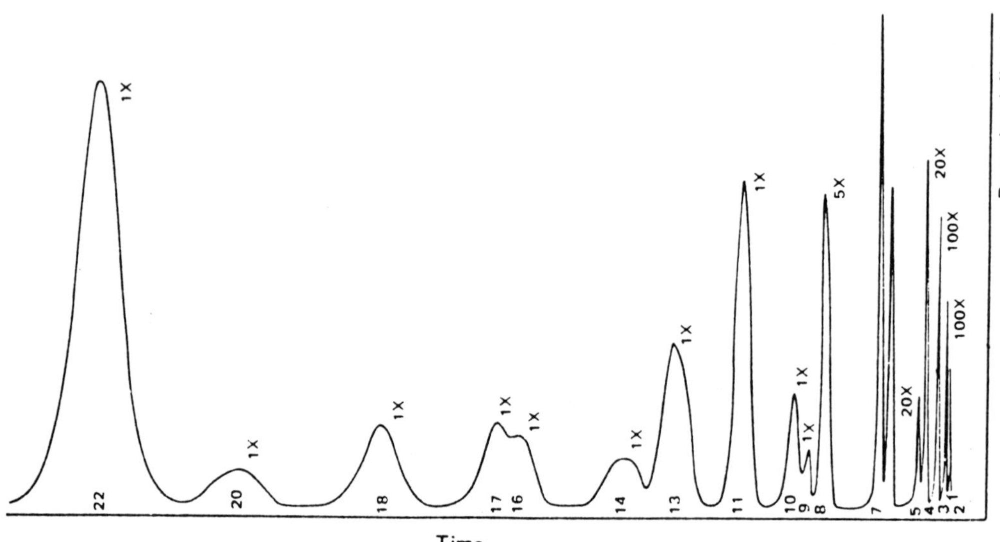

Figure 186.2 Polypropylene (FU); (1) methane; (2) ethane; (3) propane; (4) isobutane (5) n-butane; (7) n-pentane; (8) 2-methyl pentane; (9) 3-methyl pentane; (10) n-heptane; (11)2,4-dimethyl pentane and/or methyl cyclopentane; (13) 3-methyl pentane and/or methyl cyclopentane; (13) 3-methyl pentane and/or cyclohexane; (14) 1,3 dimethyl cyclopentane cis/or trans; (16) 2,5-dimethyl hexane; (1) 2,4-dimethyl hexane and/or toluene; (18) 2-methylheptane, 4-methyl heptane and/or methyl cyclohexane; (20) 1,3-dimethylcyclohexane and/or 1,4-dimethylcyclohexane trans.

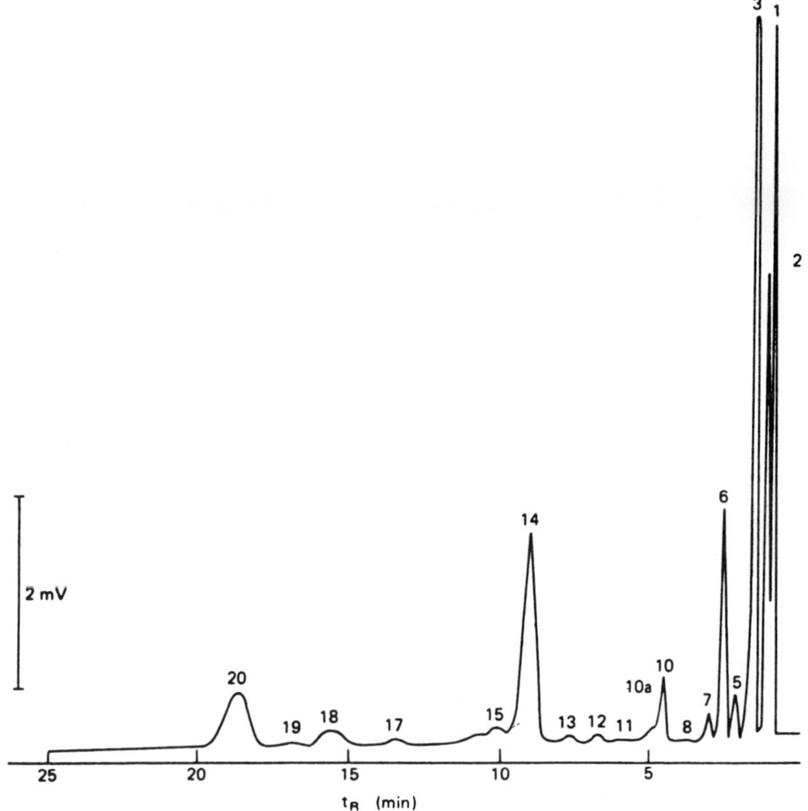

Figure 186.3   Polybutene-1: (1) ethylene (acetylene, ethane); (2) propylene (propane); (3) butene-1 (butane); (4) butene-2; (5) isopentane; (6) 2 methylbutene-1-1, (pentane pentene); (7) 3 methyl pentene-1; (8)3-methyl pentane; (9) hexene-1 (hexane); (10) hexene-3 (2-ethyl butene-1); (10a) hexene-2, hexadiene (1.4); 2 ethylbutadiene 1,3; (11) hexadiene 1,3; (12) cyclic products; (13) 3-methyl hexane, 4-methyl hexene-1; (14) heptene-3, heptane, 2-ethyl pentene-1; (15) cyclic products; 916) heptene-2; (17) heptadiene 2,4; (18) 3-methyl (p-heptane); (19) 5 methyl heptene-3, 5-methyl heptene-2; (20) 2-ethyl hexene-1.

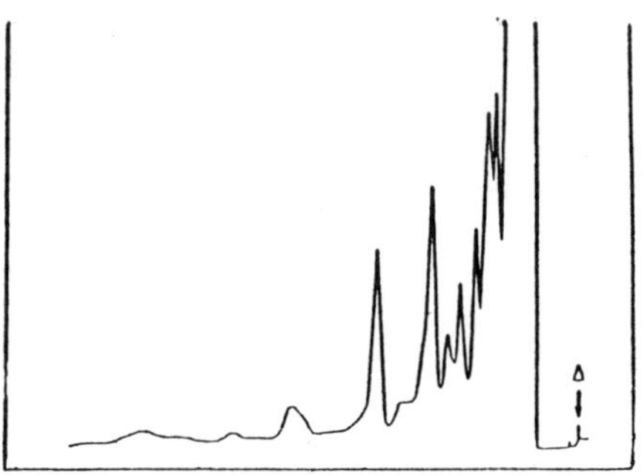

Figure 186.4 Poly-3 methyl butene-1 (F).

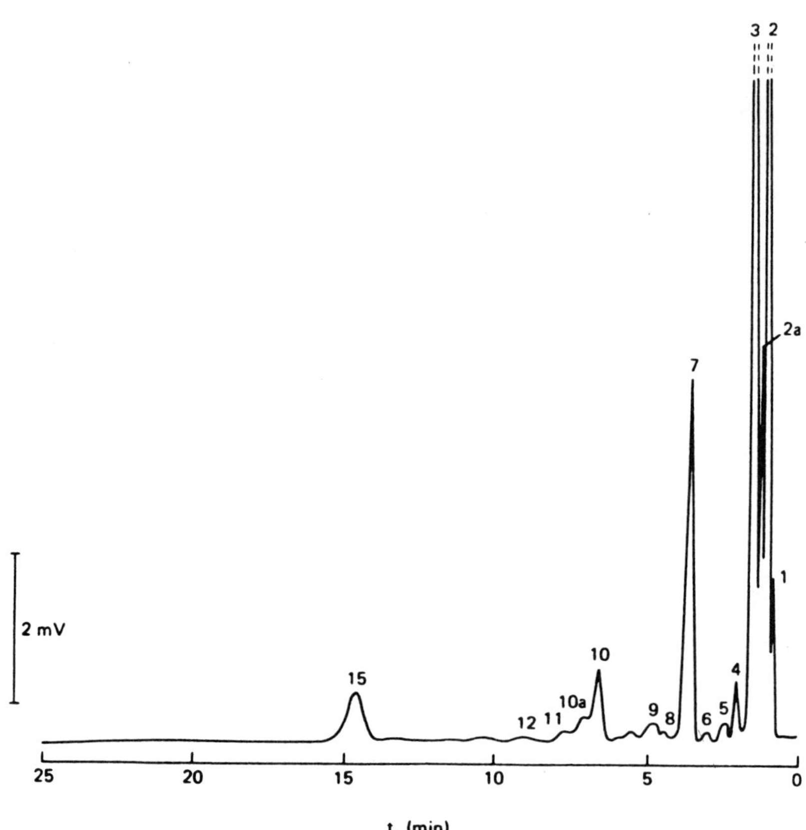

Figure 186.5 Poly-4-methyl pentene-1 (F): (1) ethylene (acetylene, ethane); (2) propylene (propane); (3) iso butane; (4) 2-butane; (4) isopentane, 3-methyl butene-1; (5) 2-methyl butene-1, pentane, pentene, pentadiene; (6) 2-methyl pentane; (7) 4-methyl pentene-1, 4 methyl pentene-2; (8) 2-methyl pentene-1, 2-methyl pentadiene 1,4; (9) 2,4 dimethyl pentane; (10) 2,4 dimethyl pentene-1; (10a) 5-methylhexene-1; (11) 5-methyl hexene-3, 2-methyl hexane, 2,4-dimethyl pentadiene 1,4; (12) 5-methyl hexene-2; (13) cyclic secondary products; (14) 2,4 dimethyl hexane, 5-methyl hexadiene 1,3; (15) 2-methyl heptane, 6-methyl heptene-1.

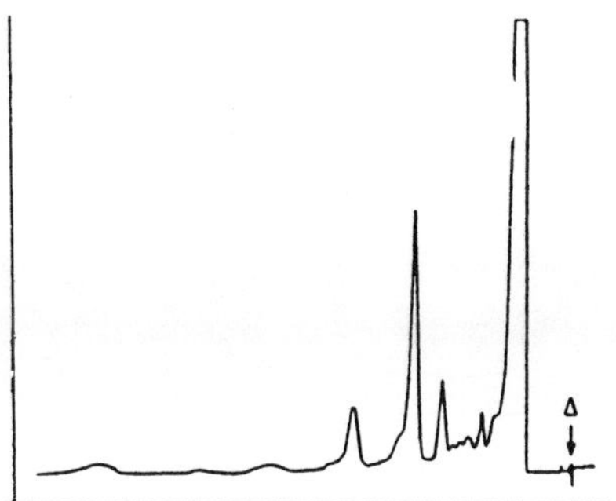

Figure 186.6 Polyvinyl acetate (F)

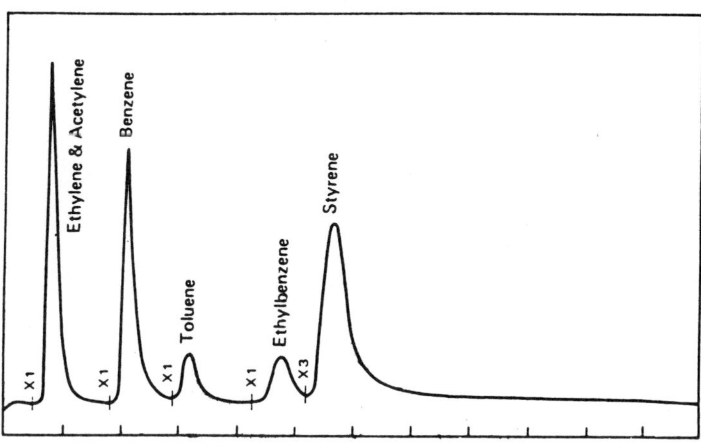

Figure 186.7 Polystyrene.

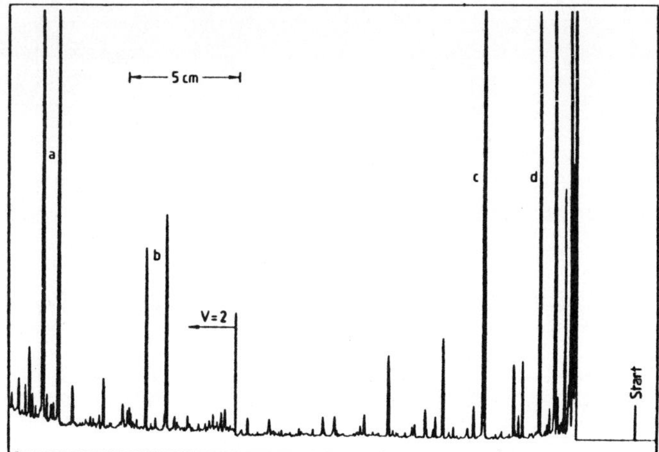

Figure 186.8 Polyethylacrylate polyisobutyl acrylate (physical
mixture) (Fu); (a) isobutyl acrylate; (b) ethyl acrylate
dimer; (c) isobutyl acrylate; (d) ethyl acrylate.

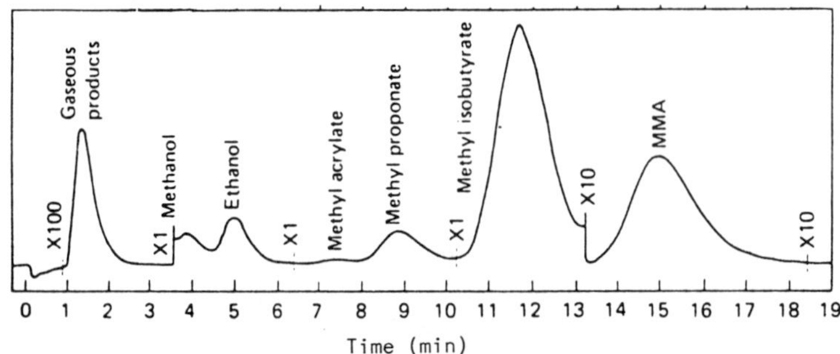

Fiugre 186.9 Polymethyl methacrylate (Fu).

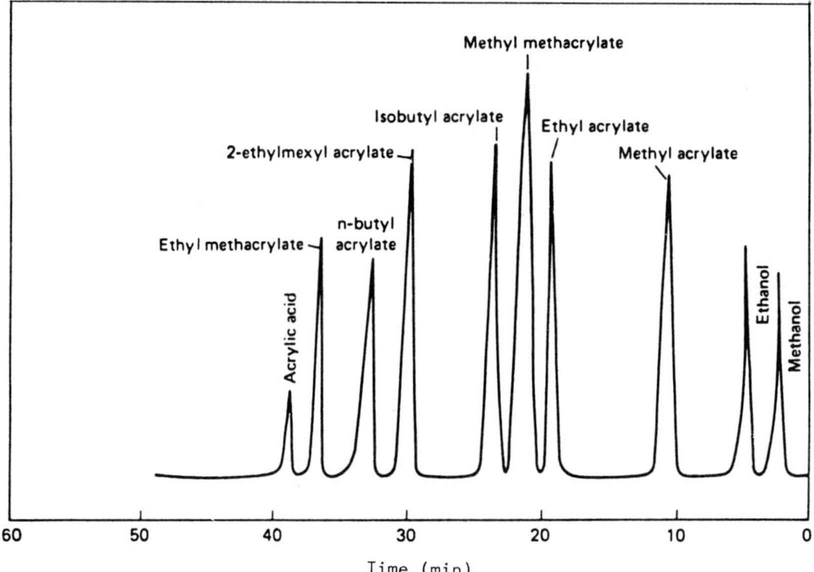

Figure 186.10 Polyacrylates and polymethacrylates, (Fu).

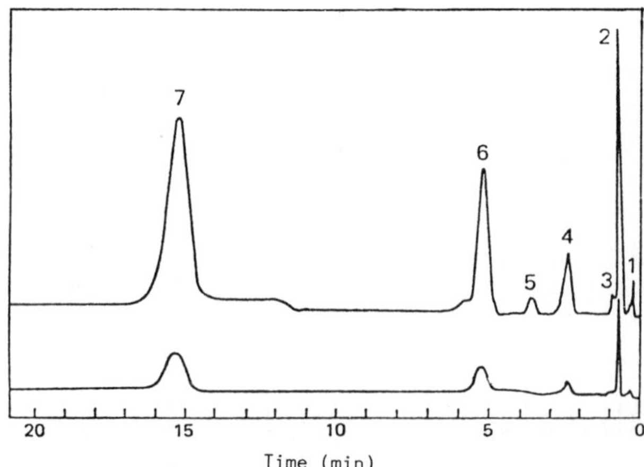

Figure 186.11 Acrylic emulsion (Fu); (1) methane; (2) ethylene; (3) ethane; (4) methanol; (5) acetaldehyde; (6) ethanol; (7) methyl methacrylate; 7 ug pyrolysed.

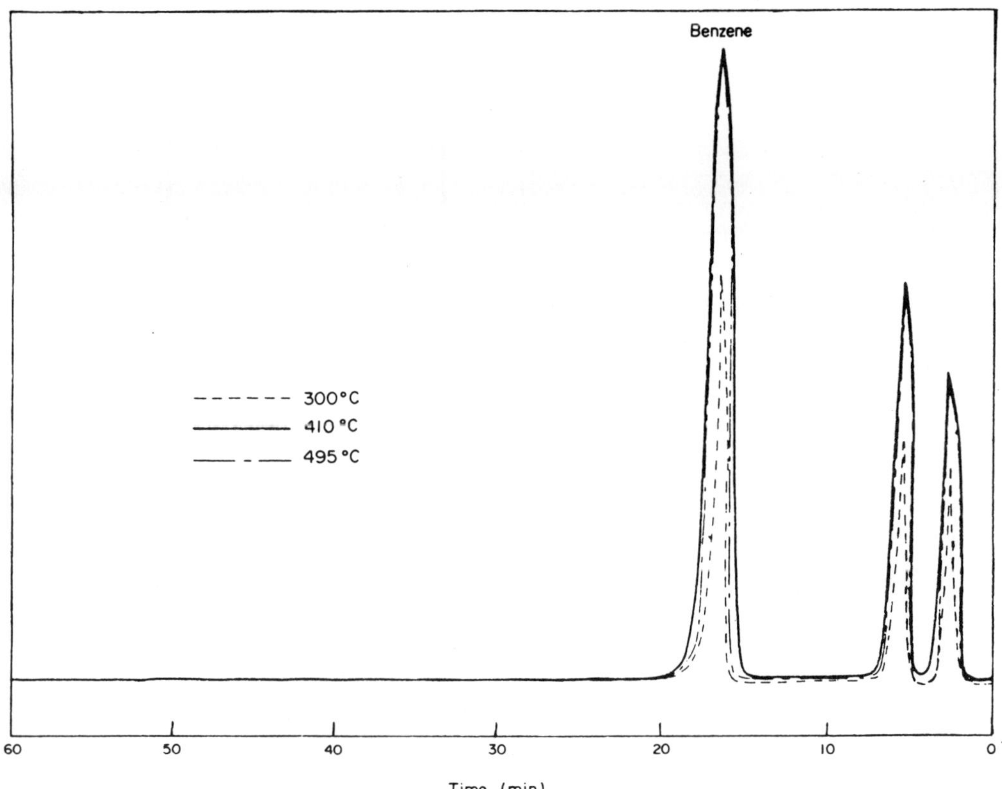

Figure 186.12 Polyvinyl chloride, effect of pyrolysis temperature
(Fu).

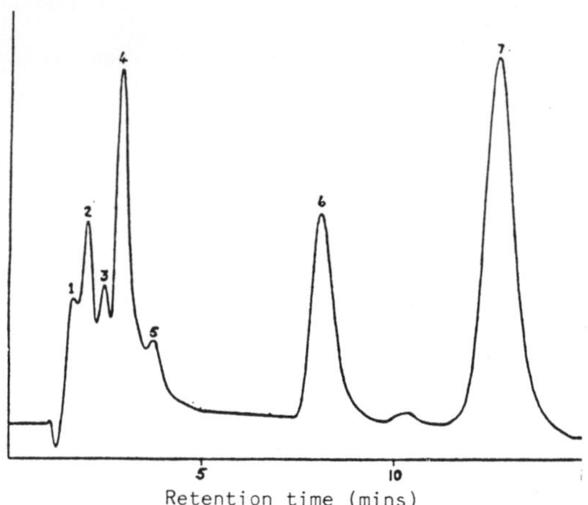

Figure 186.13 Polyvinyl chloride, dioctyl phthalate (50:50), (Fu).

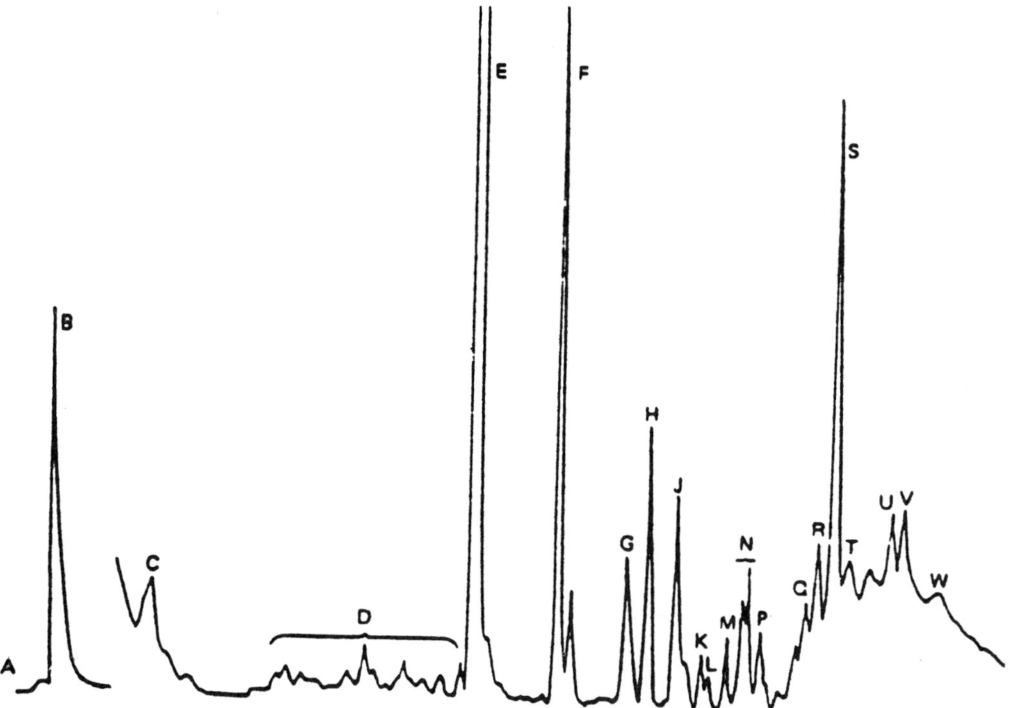

Figure 186.14 Polyvinyl chloride; (A) methane, carbon dioxide, ehtylene, ethane; (B) HCl, C_3H_6, C_3H_8; (C) butane, butene, butadiene, diacetylene; (D) C_5/C_6 aliphatic/aromatic hydrocarbons; (E) benzene; (F) toluene; (G) chlorobenzene; (H) xylene; (J) alkylbenzene; (K) C_9H_{12}; (L) C_9H_{12}; (M) indane; (N) indene; ethyl toluene; (P) methyl indane; (R) methyl indanes; (S) naphthalene; (T) dimethyl indane; (U) methylnaphthalene; (V) methylnaphthalenes, acetonaphthalene; (W) dimethylnaphthalene.

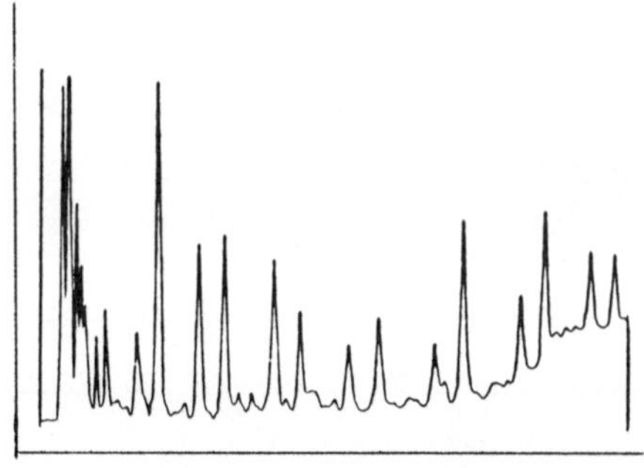

Figure 186.15 Reduced polyvinyl chloride (Pevikon R-34).

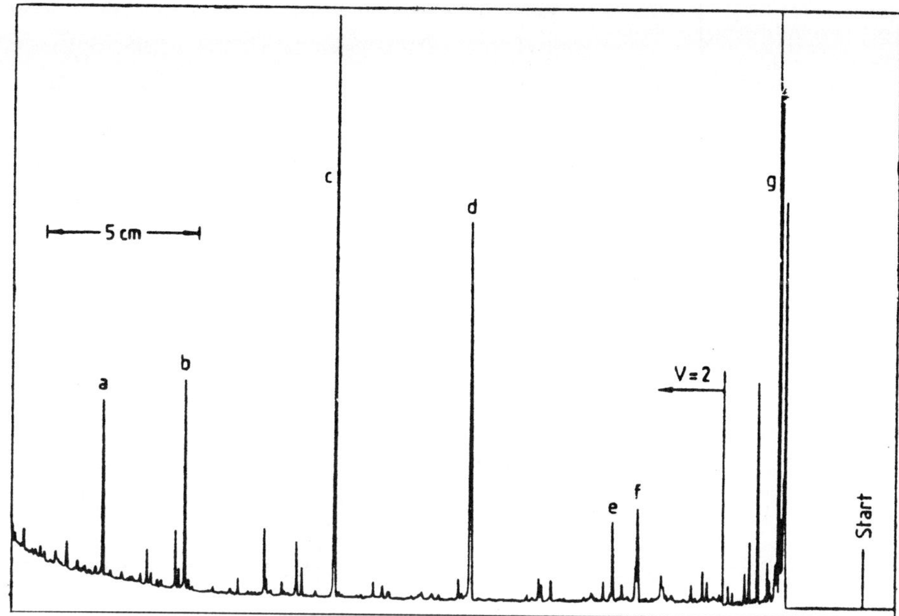

Figure 186.16   Polyvinylidene chloride: (a) tetrachlorostyrene; (b) trichlorstyrene; (c) 1,3,5-trichlorobenzene; (d) m-dichlorobenzene; (e) trichlorobutadiene; (f) chlorobenzene; (g) vinylidene chloride.

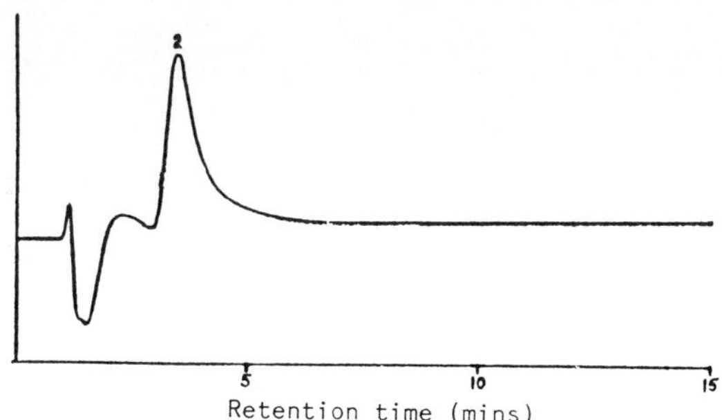

Figure 186.17 Polytetrafluoroethylene (Fu).

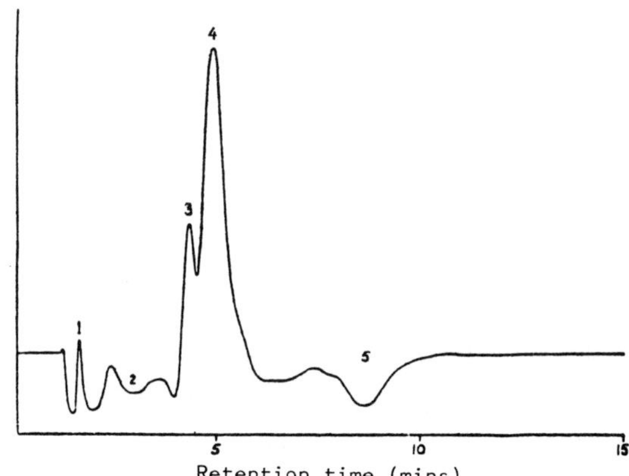

Figure 186.18 Polychlorofluorethylene (Fu).

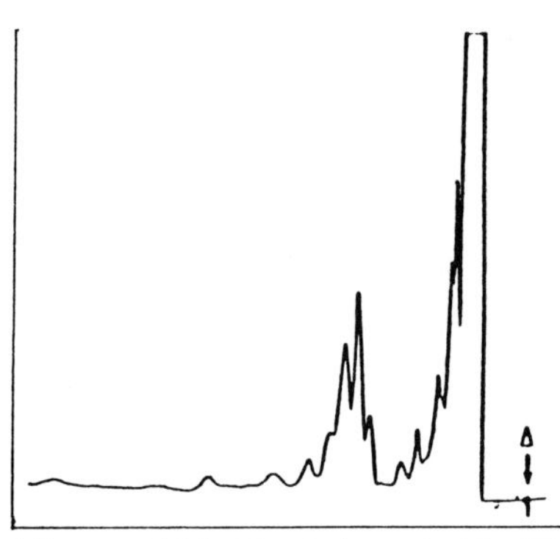

Figure 186.19 Nylon 6.10 (F)

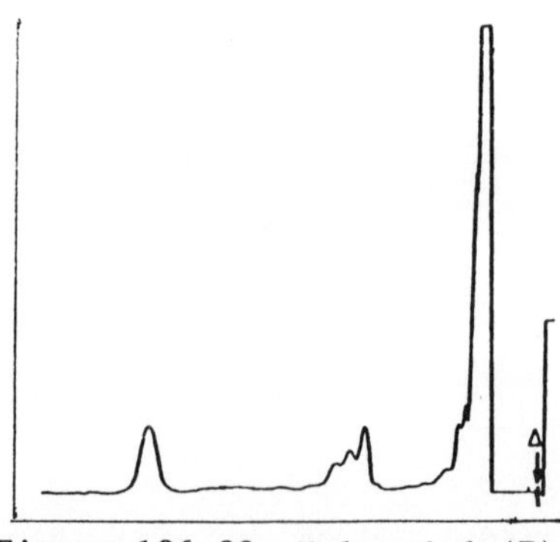

Figure 186.20 Nylon 6.6 (F).

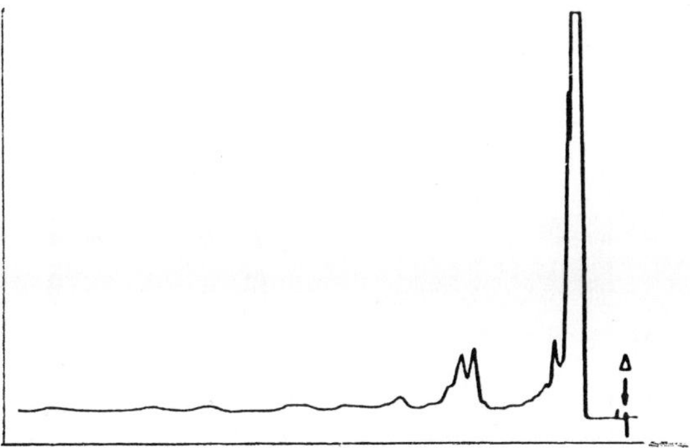

Figure 186.21 Nylon (F).

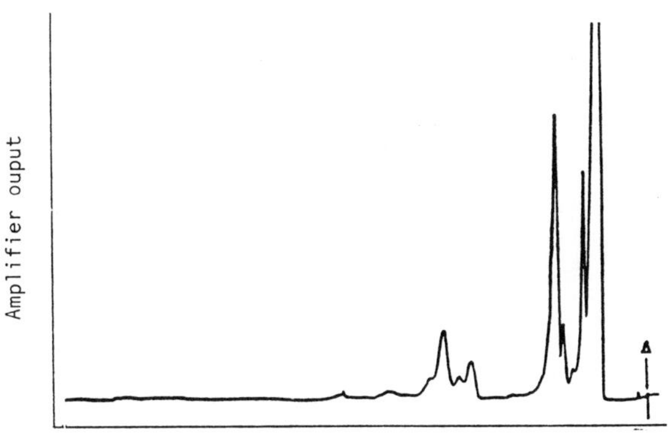

Figure 186.22 Polyurethane (F).

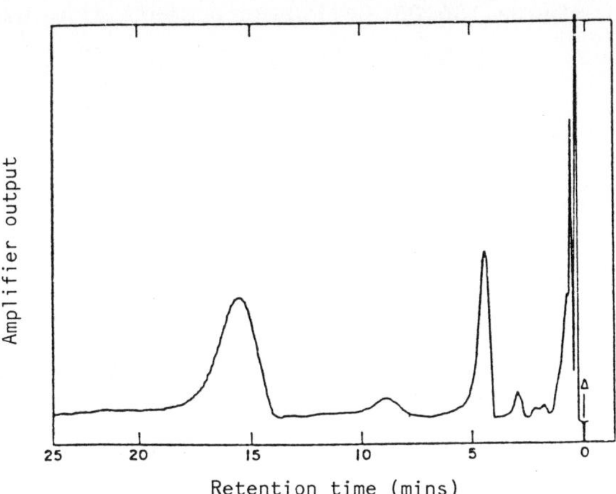

Figure 186.23 Cellulose butyrate (F).

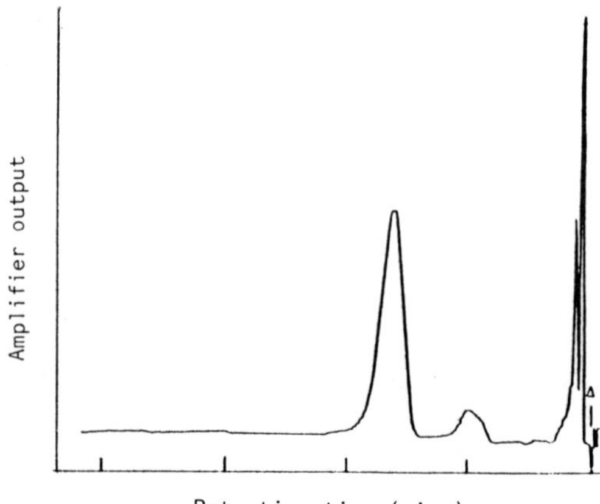

Figure 186.24 Cellulose propionate (F).

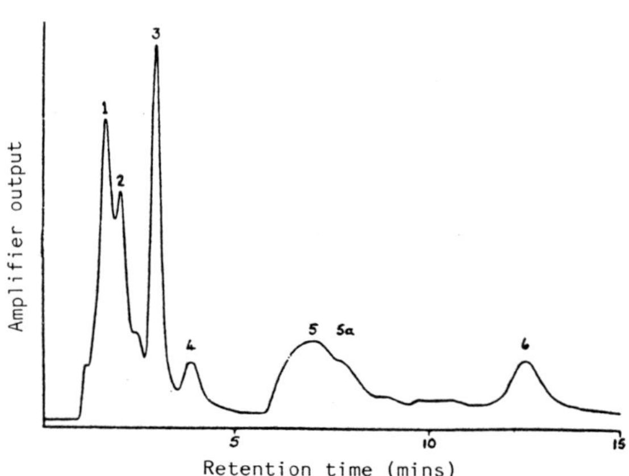

Figure 186.25 Cellulose acetate (F).

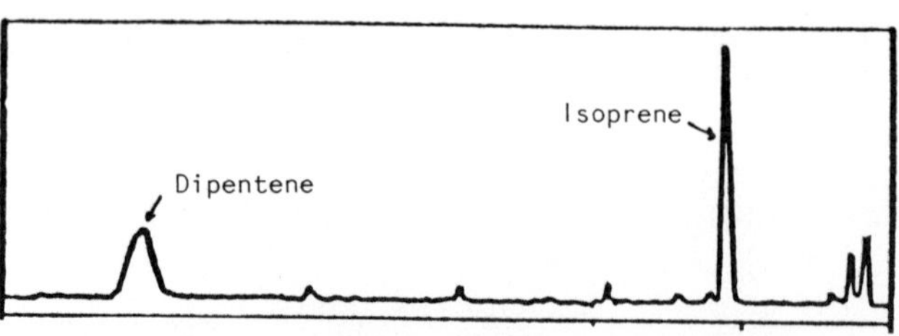

Figure 186.26 Natural rubber (polyisoprene).

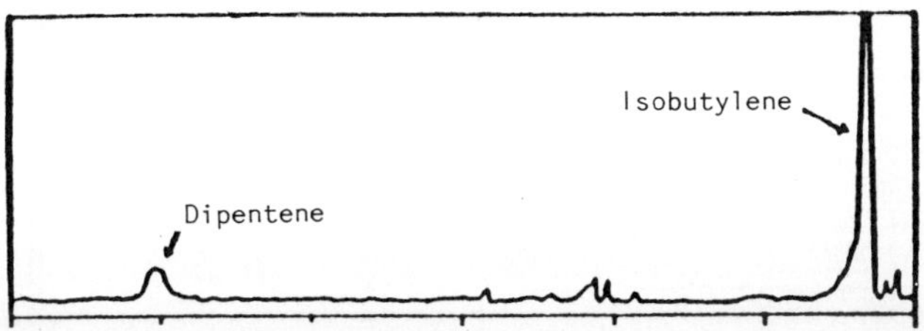

Figure 186.27 Isobutylene - isoprene rubber (Fu).

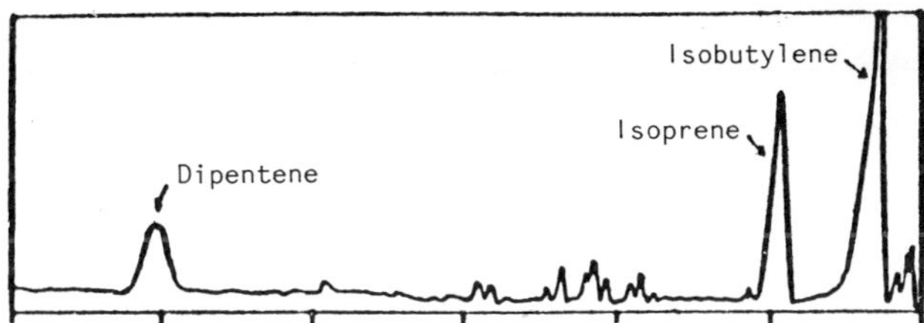

Figure 186.28 Natural rubber, isobutylene-isoprene (1:1) (Fu).

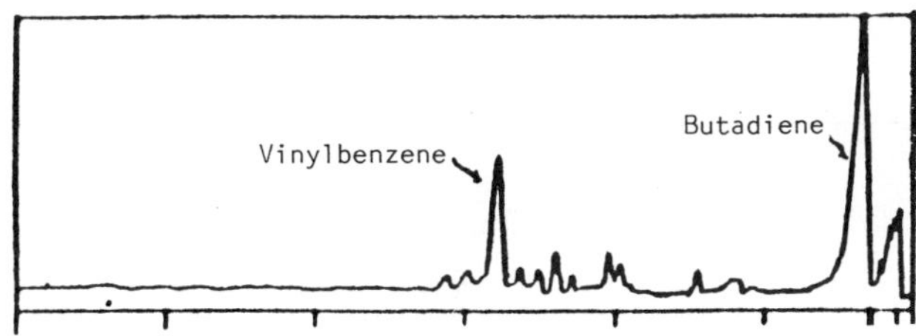

Figure 186.29 Cis polybutadiene (Fu).

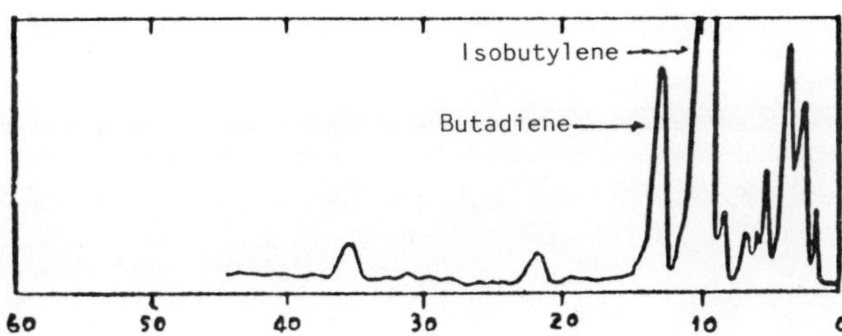

Figure 186.30 Cis polybutadiene - isobutylene isoprene (1:1) (Fu).

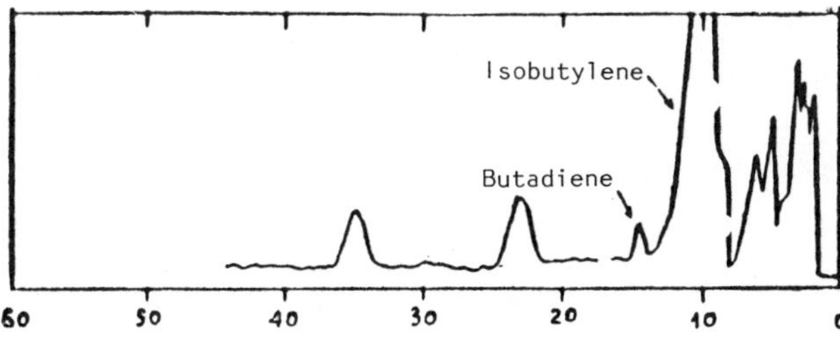

Figure 186.31 Cis polybutadiene: isobutylene isopreen (1:9) (Fu).

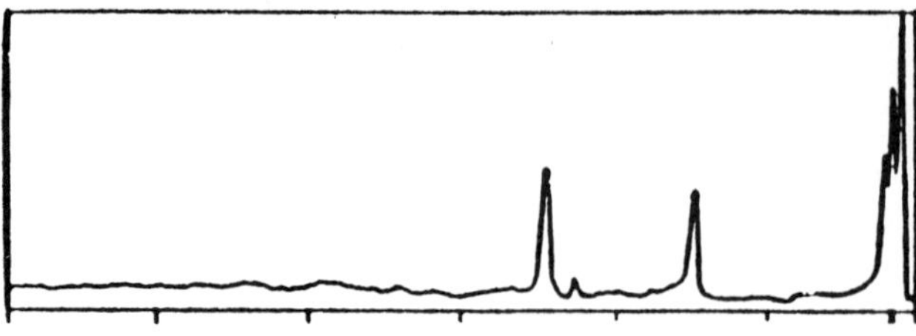

Figure 186.32 Polychloroprene (Fu).

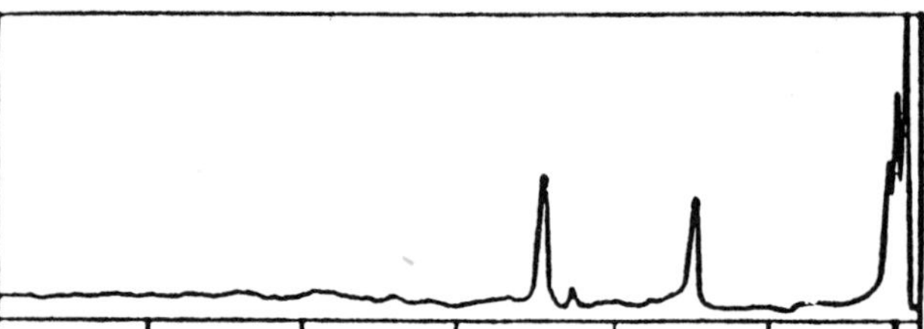

Figure 186.33 Butadiene - acrylonitrile (Fu).

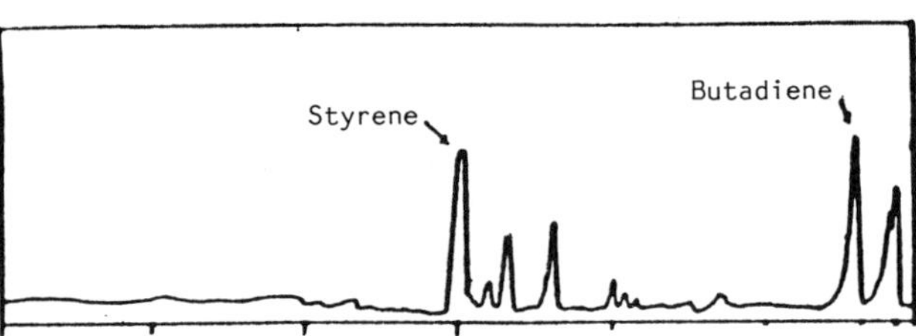

Figure 186.34 Polybutadiene, effect of pyrolysis temperature.

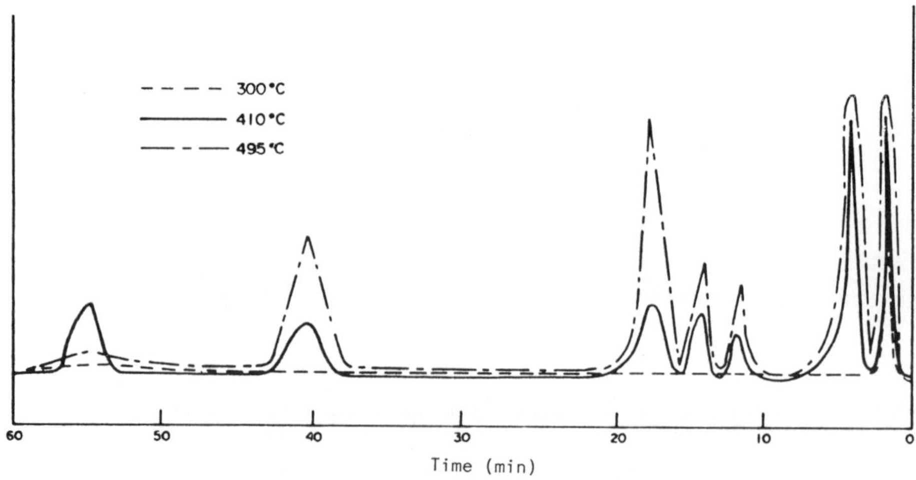

Figure 186.35 Styrene-butadiene (Fu).

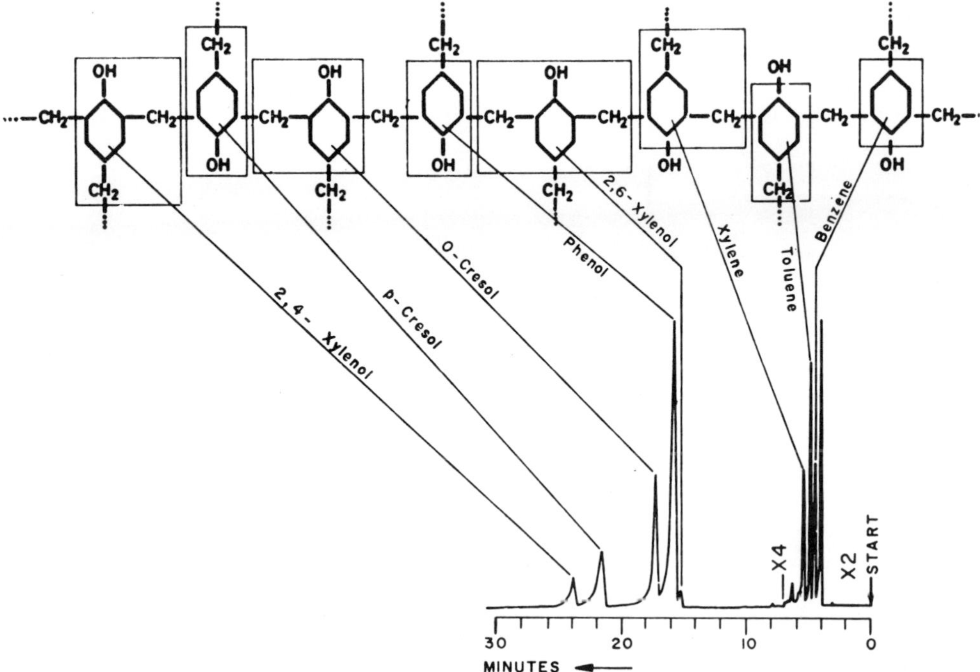

Figure 186.36 Phenol formaldehyde resin (F); pyrolysed at 900°C, PP9 stationary phase, column 170°C, 100 m capillary column, FID detector.

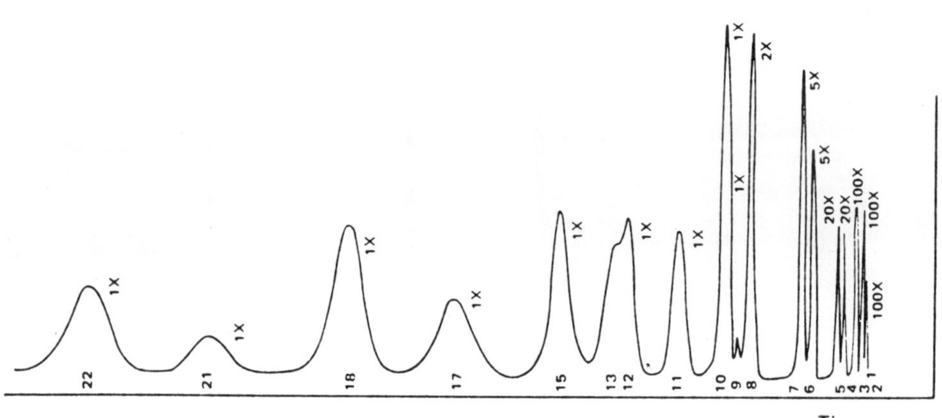

Figure 186.37 Ethylene-propylene copolymer (Fu); (1) methane; (2) ethane; (3) propane; (4) isobutane; (5) n-butane; (6) isopentane; (7) n-pentane; (8) 2-methyl pentane and/or cyclopentane; (9) 3-methyl pentane; (10) n-hexane; (11) 2,4-dimethyl-pentane and/or methyl cyclopentane; (12) 2-methyl hexane; (13) 3-methyl hexane and/or cyclohexane; (15) n-heptane; (17) 2,4 dimethylhexane and;or toluene; (18) 2-methyl heptane, 4-methyl heptane and/or methylcyclohexane; (21) n-octene; (22) 2,4 dimethylheptane.

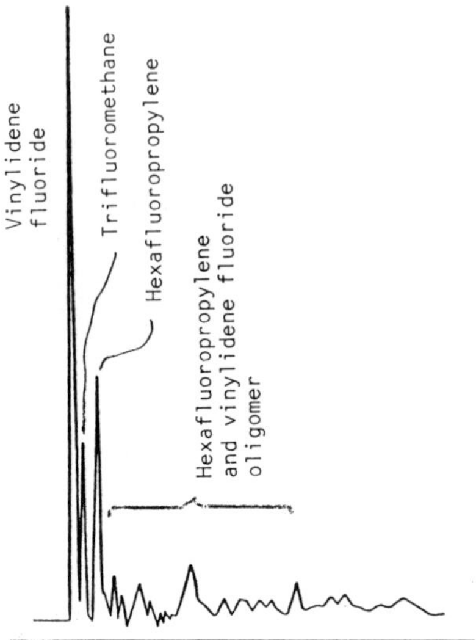

Figure 186.38 Hexafluoropropylene vinylidene fluoride copolymer (60:40 w/v).

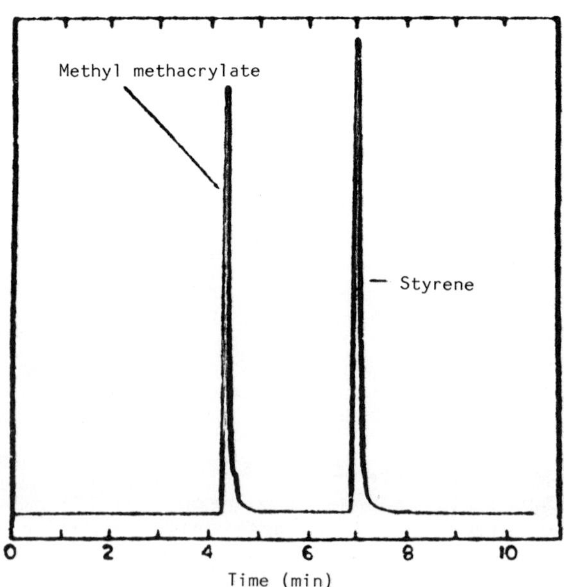

Figure 186.39 Styrene-methylmethacrylate copolymer.

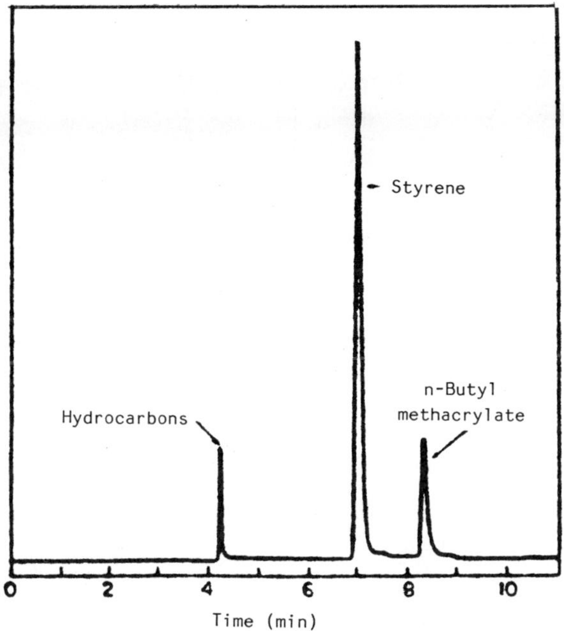

Figure 186.40 Styrene-n-butyl acrylate copolymer.

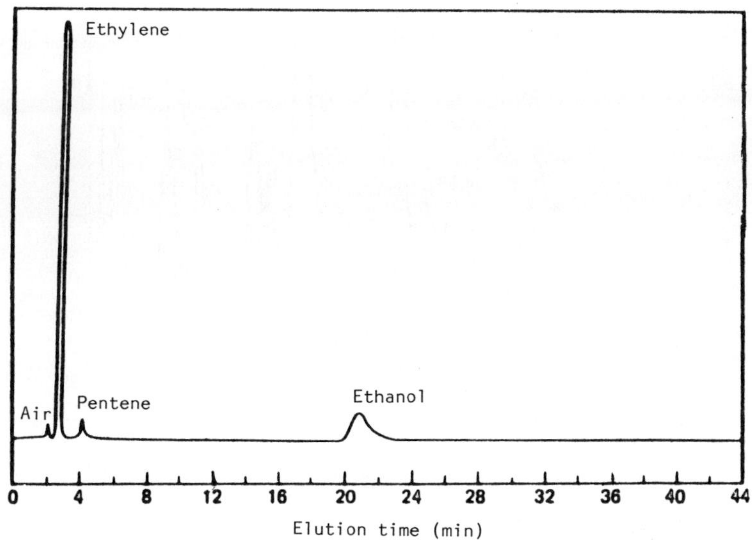

Figure 186.41 Ethylene-ethyl acrylate copolymer, pyrolysed at 475°C.

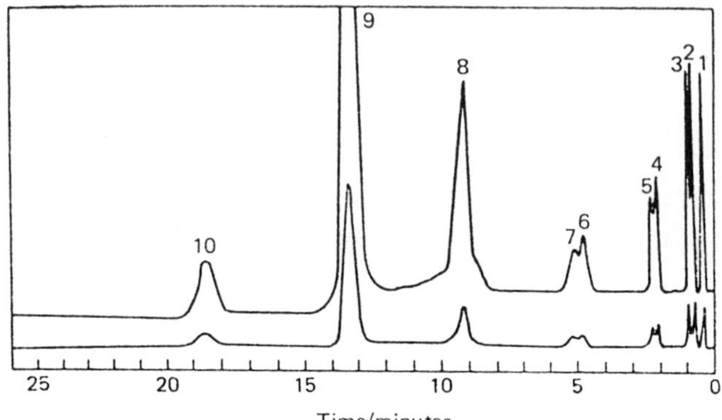

Figure 186.42 Polyvinyl chloride-polyvinyl acetate copolymer (15:5 w/w) 91) methane; (2) ethylene; (3) ethane; (4) propylene; (5) propane; (6) n-butene; (7) n-butane; (8) acetic acid; (9) benzene; (10) toluene (44 ug pyrolysed).

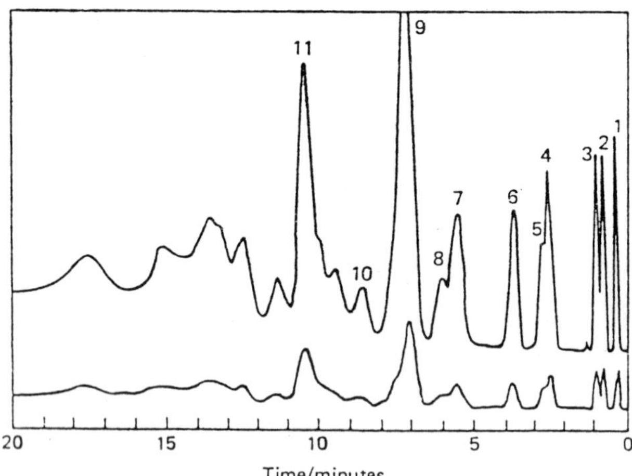

Figure 186.43 Linseed pentaerythritol-o-phthalate alkyl resin: (1) methane; (2) ethylene; (3) ethane; (4) propylene; (5) propane; (6) acetaldehyde; (7) n-butene; (8) n-butane; (9) acrolein; (10 alkyl alcohol; (11) methacrolein; (15 ug pyrolysed).

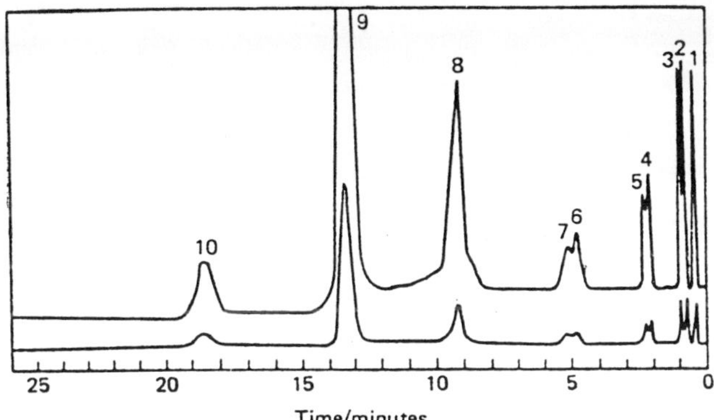

Figure 186.44   Vinyl chloride vinyl acetate copolymer (95.5 w/w): (1)
                methane; (2) ethylene; (3) ethane; (4) propylene; (5)
                propane; (6) n-butene; (7) n-butane; (8) acetic acid;
                (9) benzene; (10) toluene (44 ug pyrolysed).

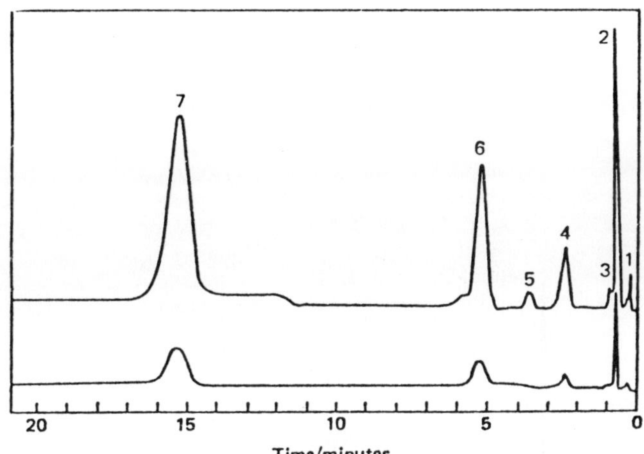

Figure 186.45 Acrylic emulsion (c): (1) methane;(2) ethylene; (3)
               ethane; (4) methanol; (5) acetaldehyde; (6) methanol.
               (7) methyl methacrylate (7 ug pyrolysed).

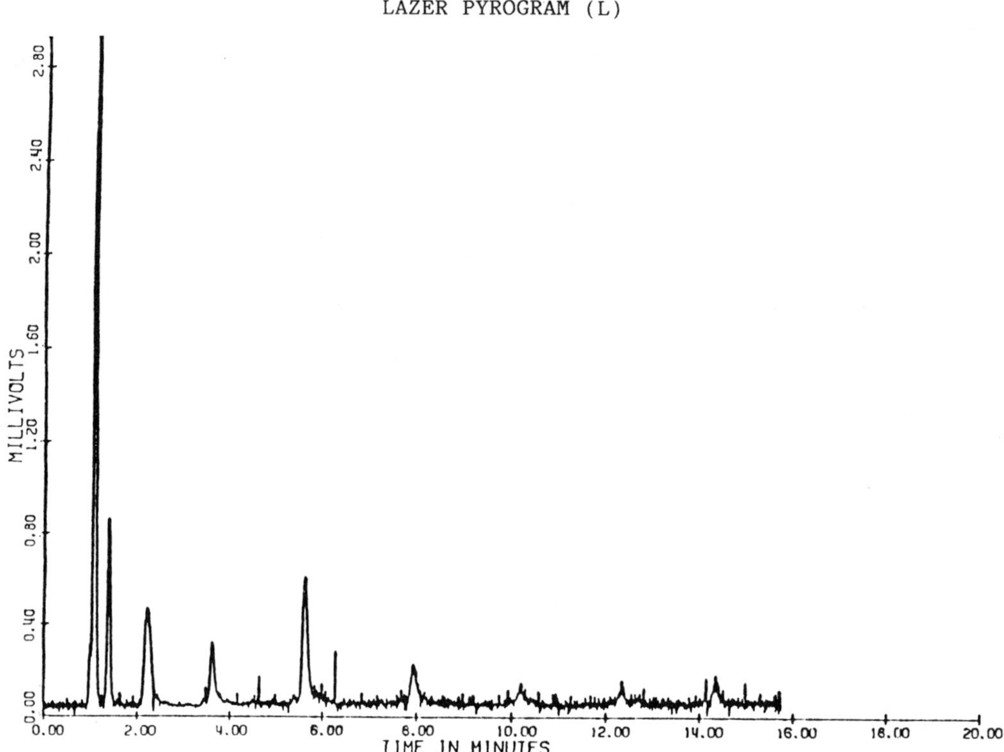

Figure 186.46 Green Eastman Imperial polyethylene.

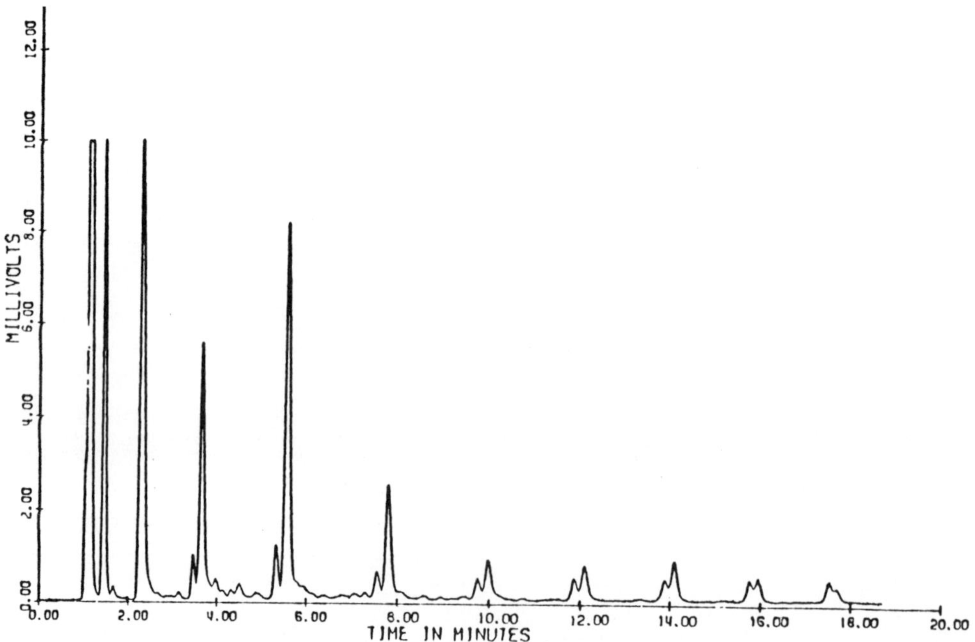

Figure 186.47 Black Eastman Imperial polyethylene.

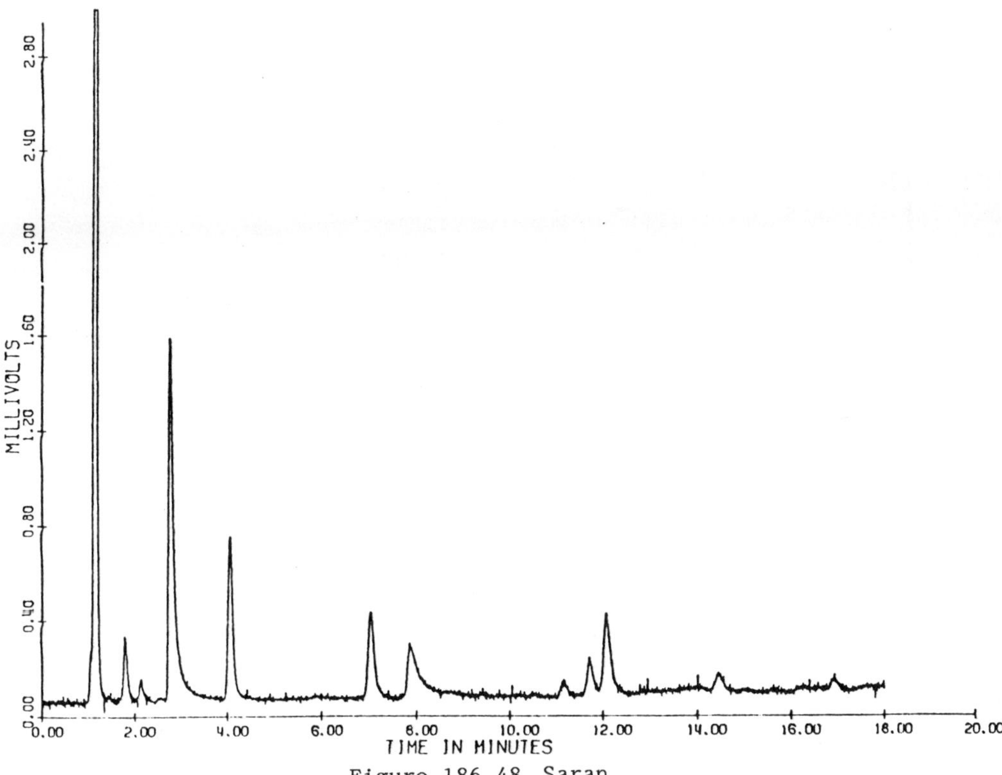

Figure 186.48 Saran.

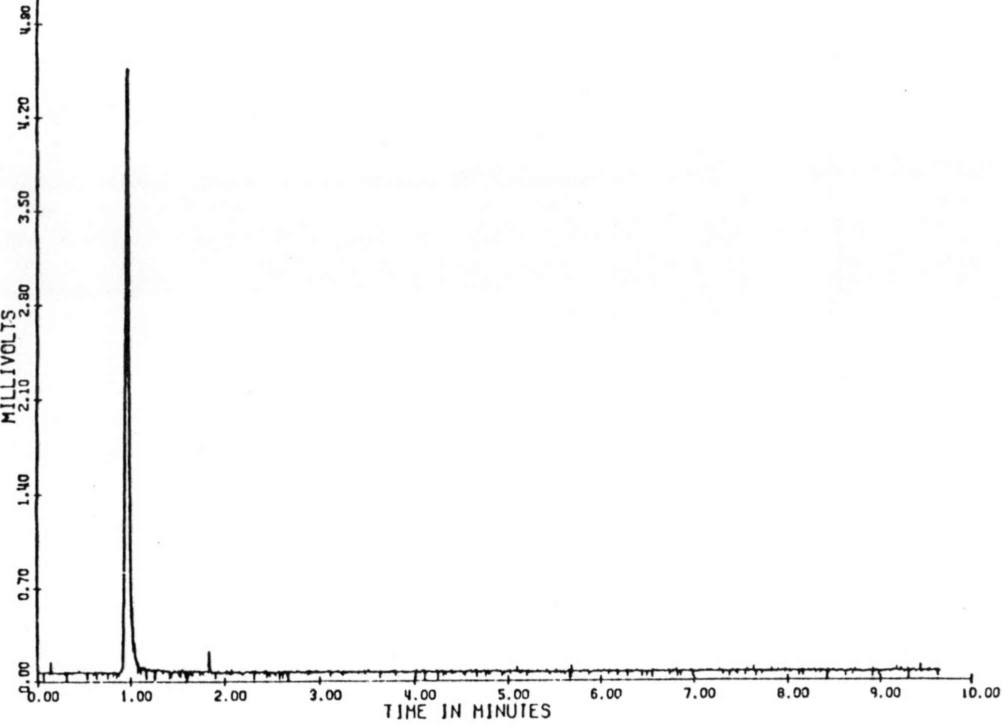

Figure 186.49 Teflon.

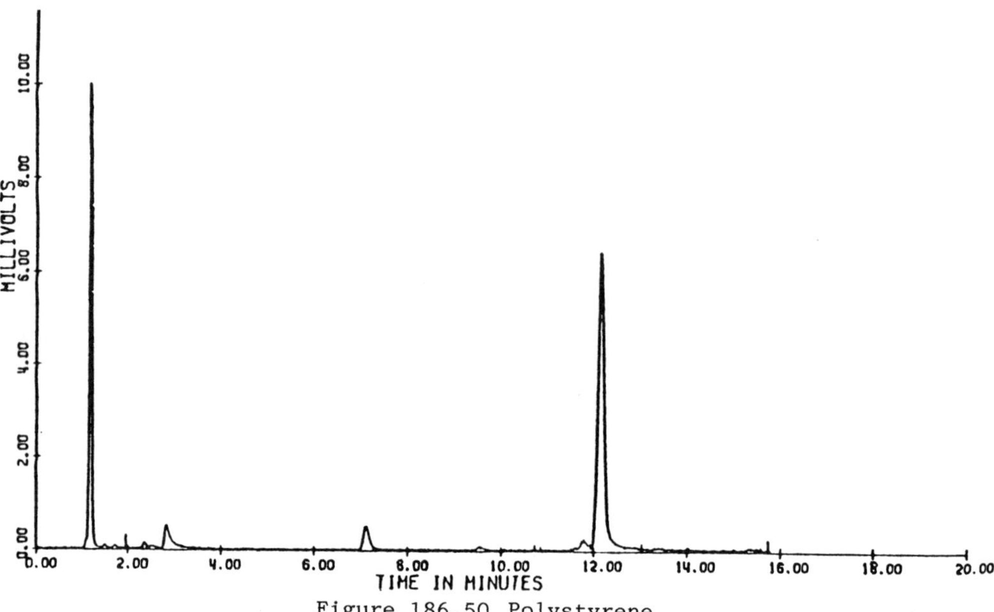

Figure 186.50 Polystyrene.

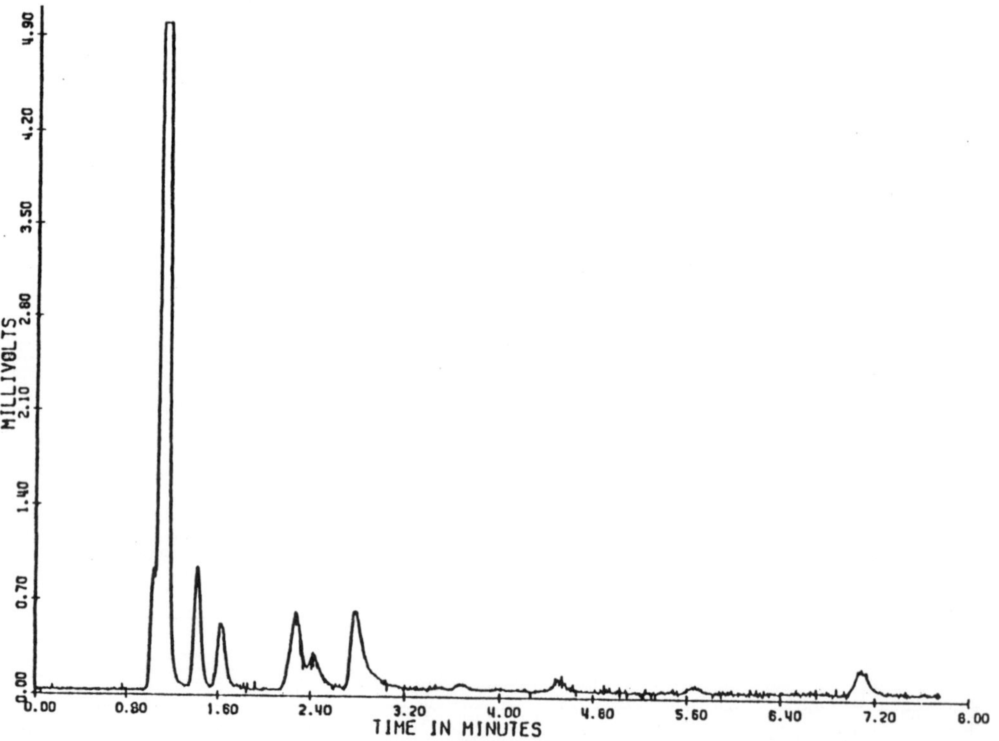

Figure 186.51 Tennesse Eastman Epolene (96%) coke (4%) mixture.

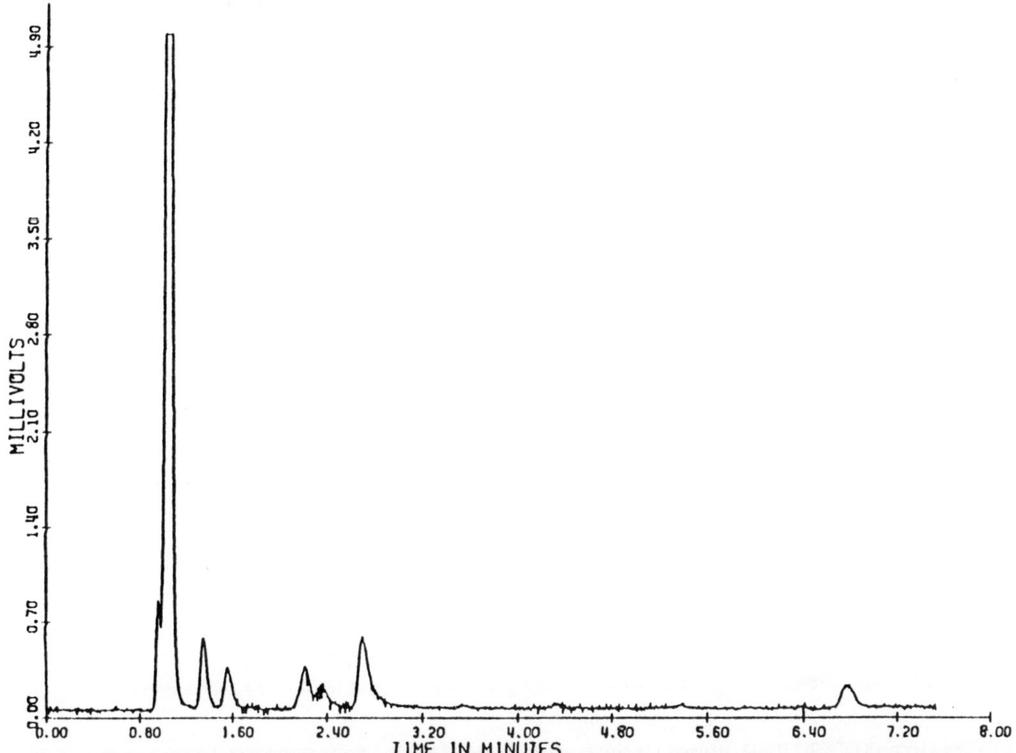

Figure 186.52 Union Carbide low density linear DYNH polyethylene (94.5%) (coke 5.5%).

Figure 187   Infrared Spectra.
(on pp. 678-701)

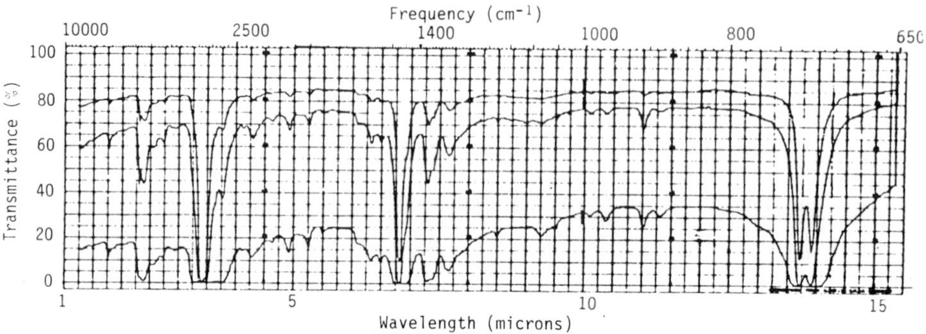

Figure 187.1   High density polyethylene (Kuva 2), 0.03, 0.1, 0.5 mm films.

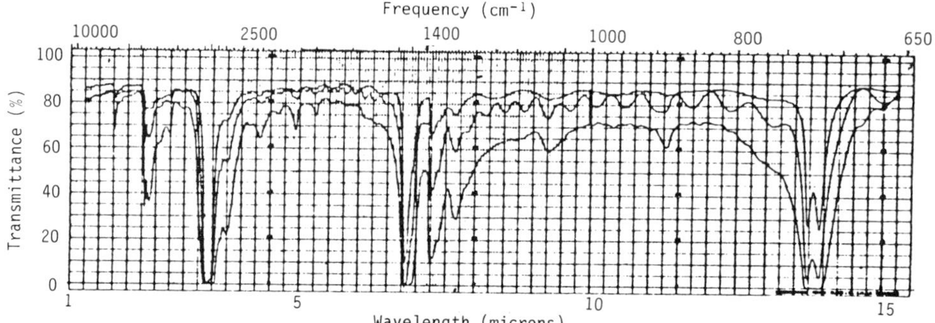

Figure 187.2   Low density polyethylene (Kuva 1), 0.03, 0.075, 0.2 mm films.

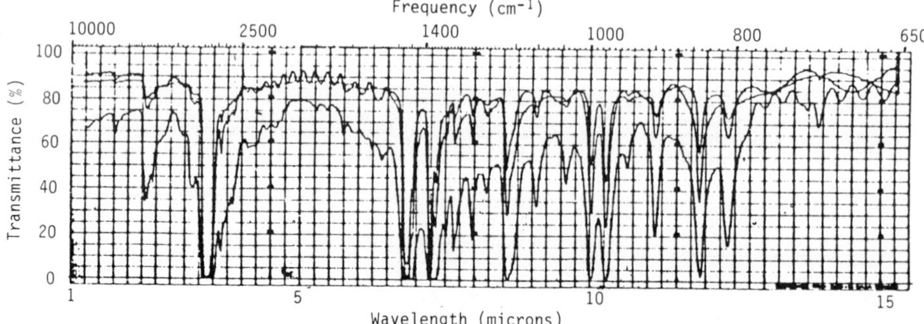

Figure 187.3 Polypropylene (Kuva 4), 0.02, 0.05, 0.20 mm films.

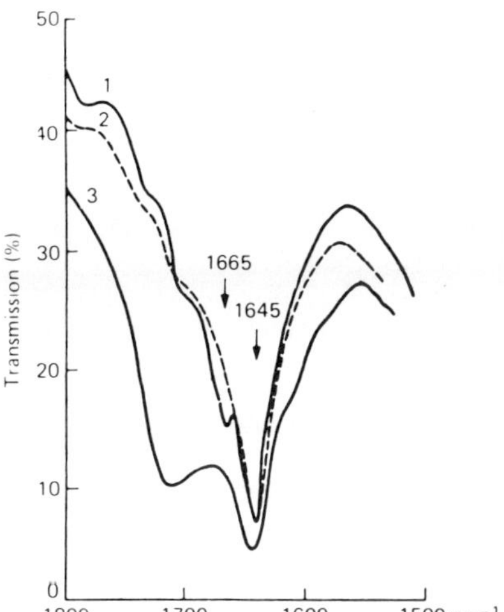

Figure 187.4 Amorphous polypropylene (d = 300 u); (1) irradiated by
fast electrons with a dosage of 4000 Mrad at 196°C
(spectrum recorded at 130°C); (2) the same specimen
after heating (spectrum recorded at 25°C); (3) the same
specimen 2 weeks after irradiation (spectrum recorded at
25°C).

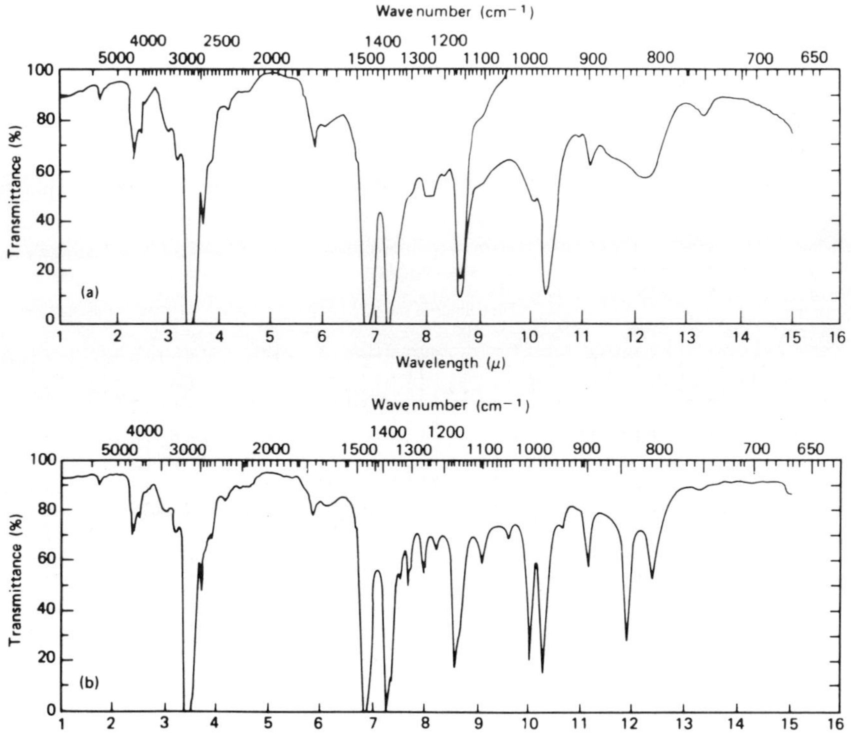

Figure 187.5 Polypropylene prepared with vanadyl-base catalyst; (a)
amorphous part; (b) crystalline part.

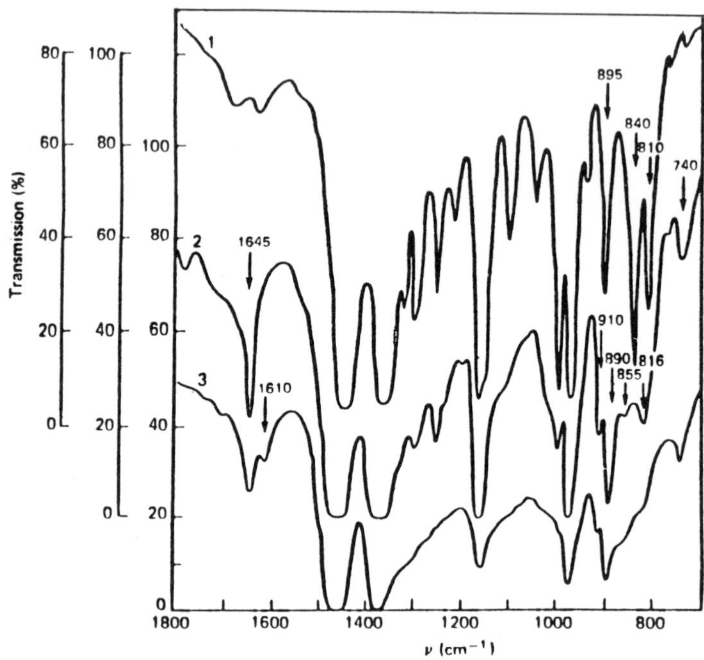

Figure 187.6 Isotactic polypropylene: (1) original (d = 200 u); (2) irradiated by fast electrons at 25°C, dosage ·500 Mrad (d = 300 u); (3) dosage 400 Mrad (d = 200 u).

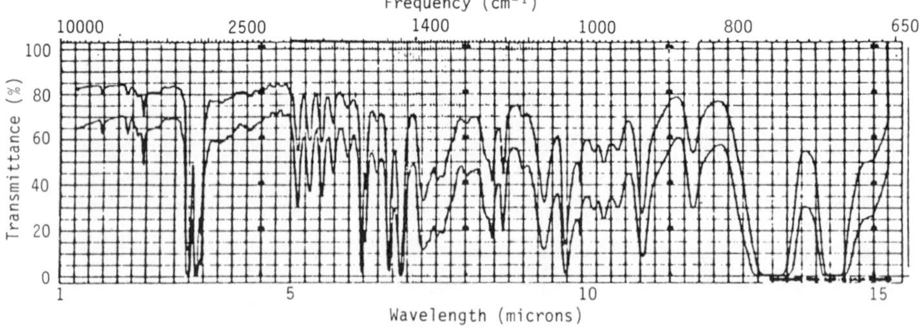

Figure 187.7   Polystyrene, 0.06 mm film.

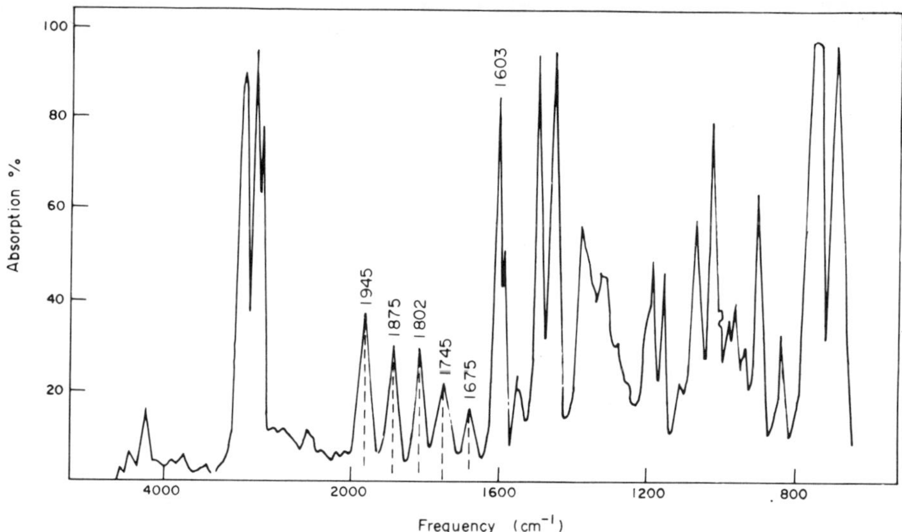

Figure 187.8 Polystyrene.

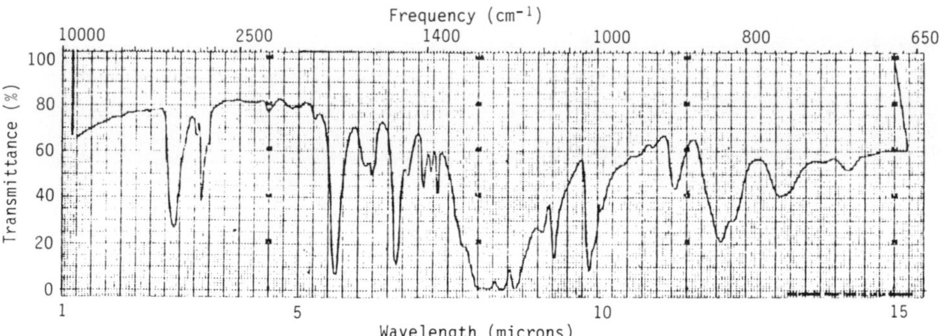

Figure 187.9 Polycarbonate, KBR pellet (2 mg in 400 mg KBR).

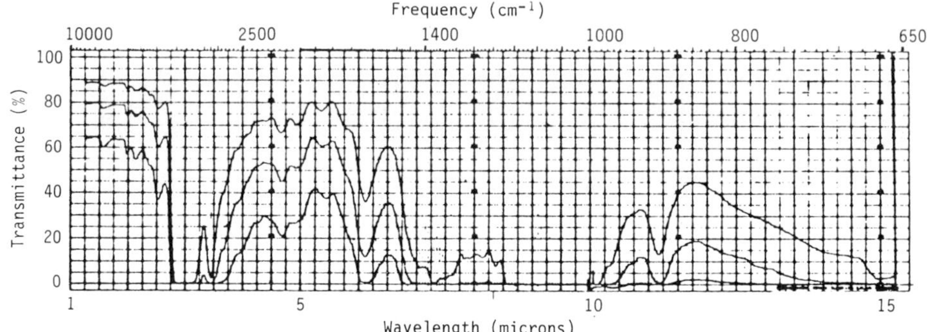

Figure 187.10 Cellophane (Kuva 1), 0.022, 0.044, 0.088 mm.

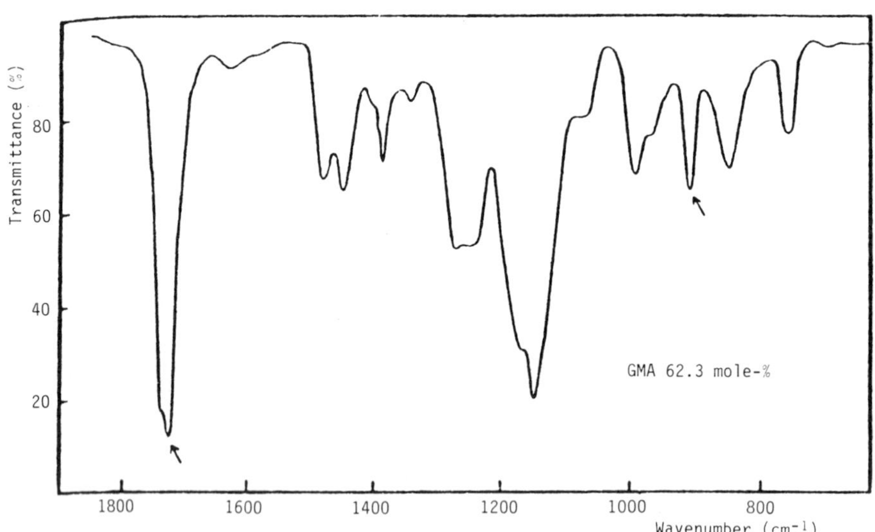

Figure 187.11   Methyl methacrylate (33.7% mole %) - glycidyl
methacrylate (62.3% mole %) copolymer.

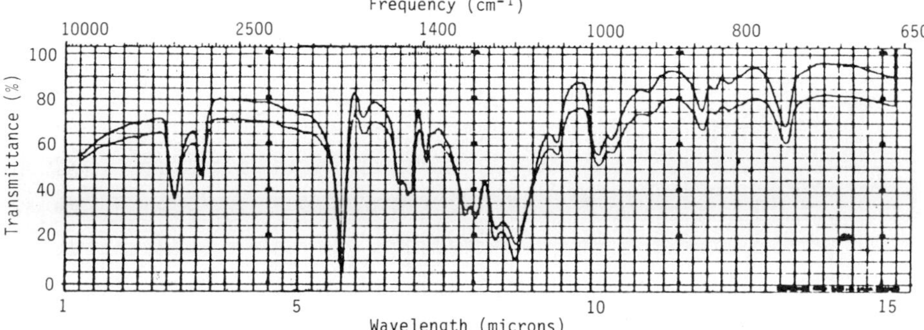

Figure 187.12 Polymethyl methacrylate (Perspex), KBR disc.

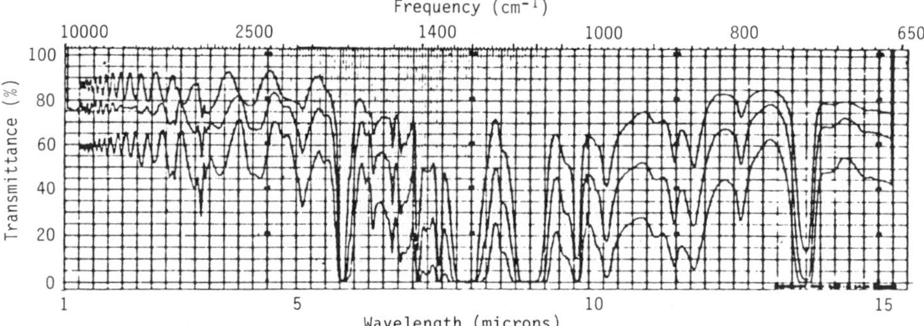

Figure 187.13   Polyethylene terphthalate (Mylar, terylene, trevira), KBR disc.

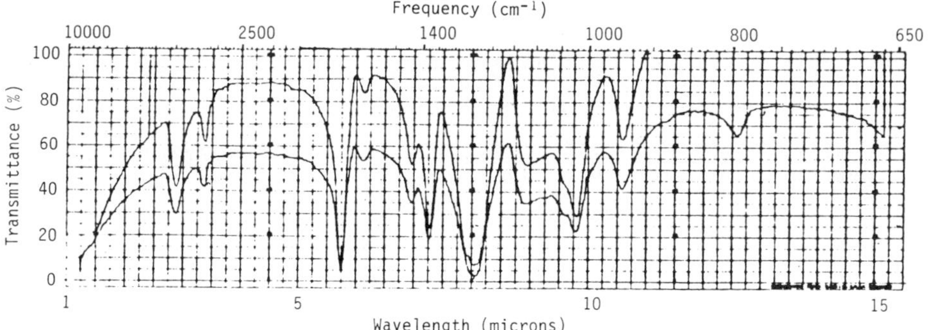

Figure 187.14 Polyvinyl acetate, KBR disc.

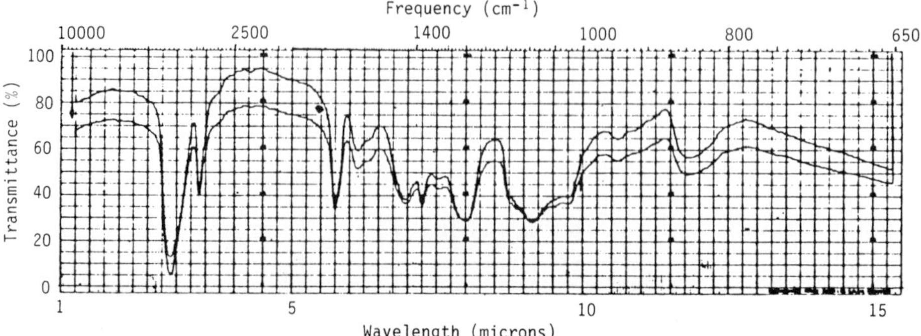

Figure 187.15 Polyvinyl alcohol, KBR disc.

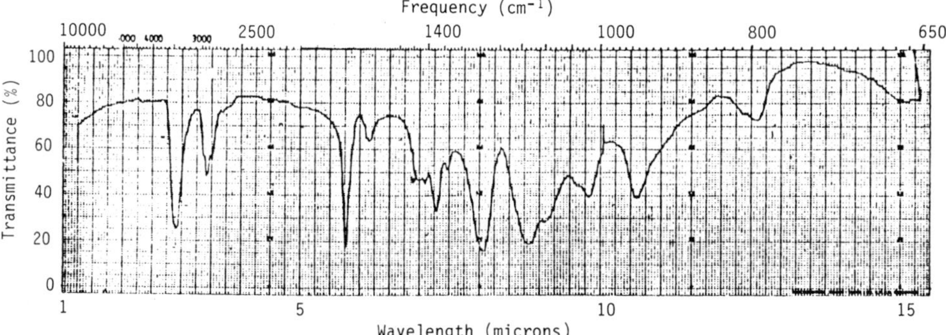

Figure 187.16 Polyvinyl acetate, (Alvar Shauingin Chemicals). KBR pellett, 1.5 mg in 400 mg KBR scale expanded.

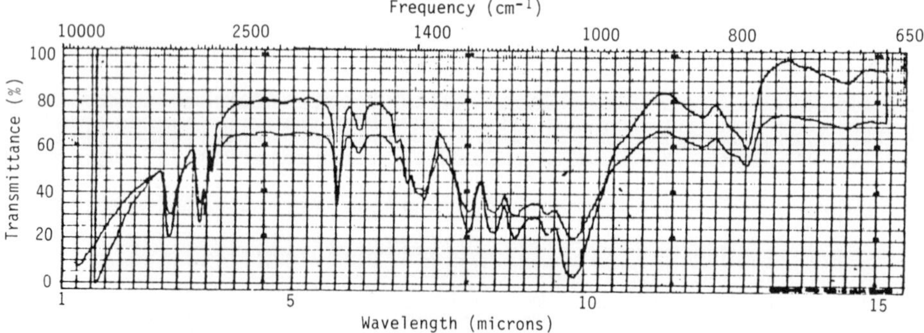

Figure 187.17 Polyvinyl formal, KBR disc.

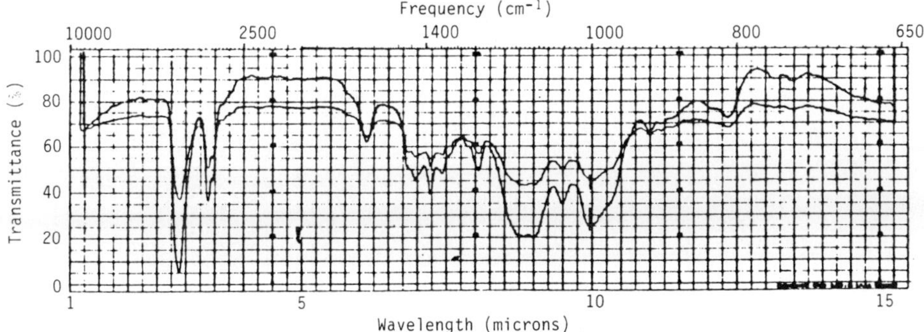

Figure 187.18 Polyvinyl butyrate, KBR disc.

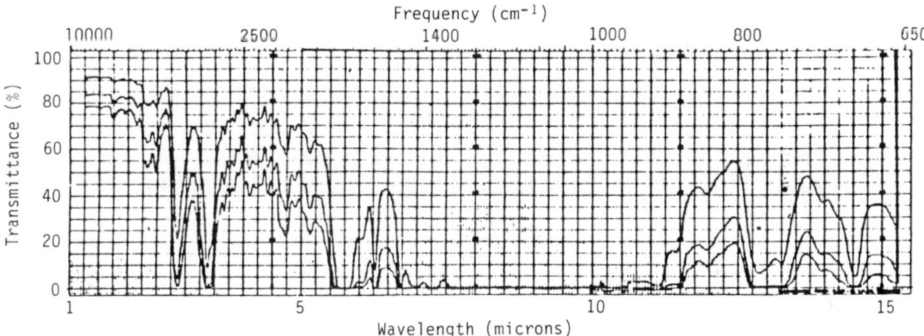

Figure 187.19 Cellulose actetate, 0.06, 0.12, 0.18 cm film.

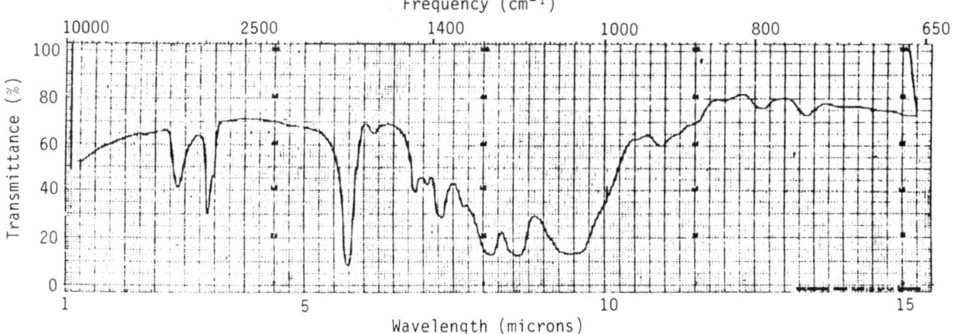

Figure 187.20 Cellulose acetate butyrate (Tenite butyrate, Eastman).
KBR pellet (1.5 mg in 400 mg KBR).

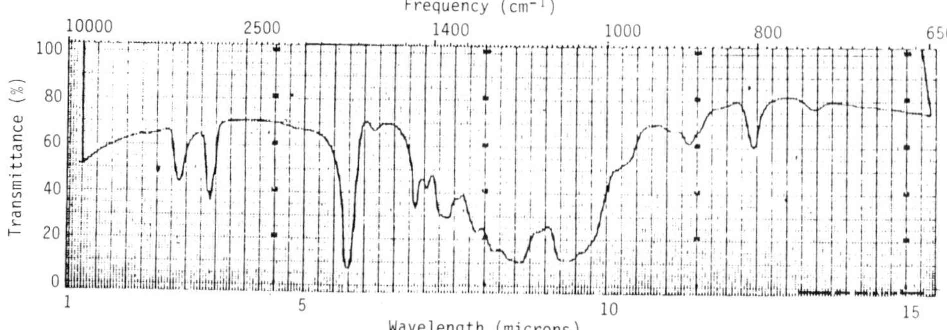

Figure 187.21   Cellulose propionate (Forticel W-60099A, Amcel, Europe)
KBR pellet (1.5 mg in 400 mg KBR).

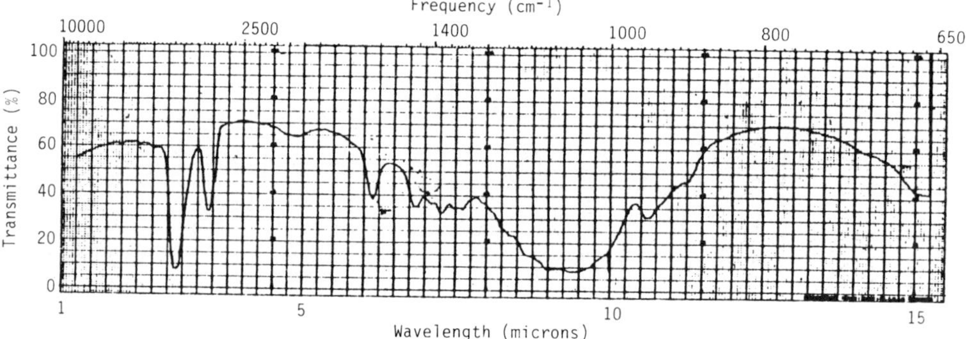

Figure 187.22 Methyl cellulose (L light) KBR pellett.

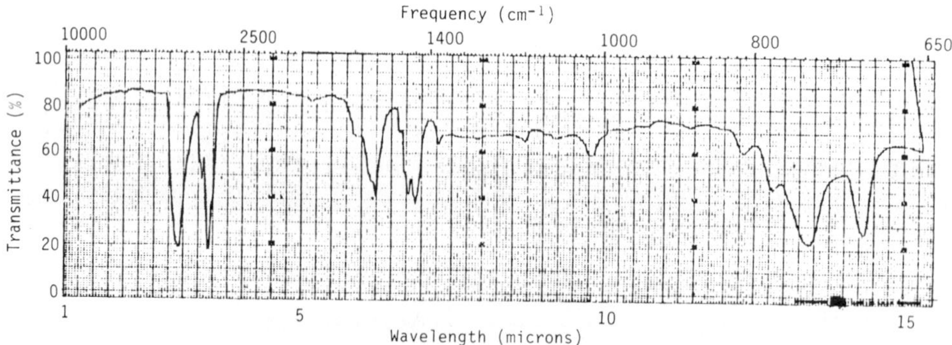

Figure 187.23   Coumarone-indene resin (Paradene-2, Neville Cindu,
Chemicals, N.V., Holland).   KBR pellet (2 mg in 400 mg
KBR.

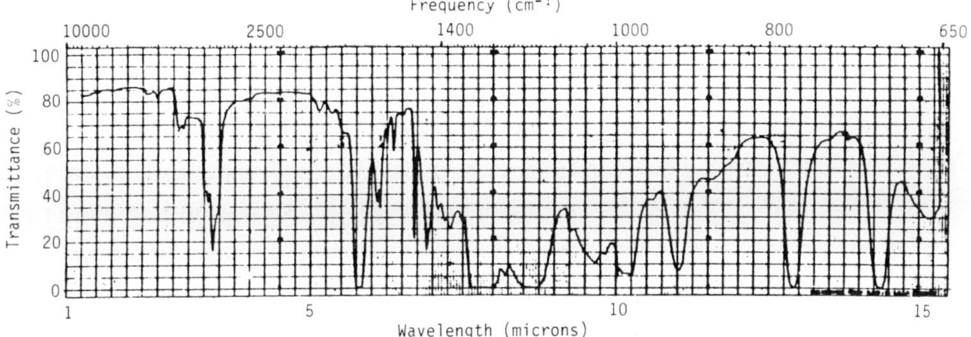

Figure 187.24 Polyester resin (Vestopal P) capillary film between sodium chloride plates.

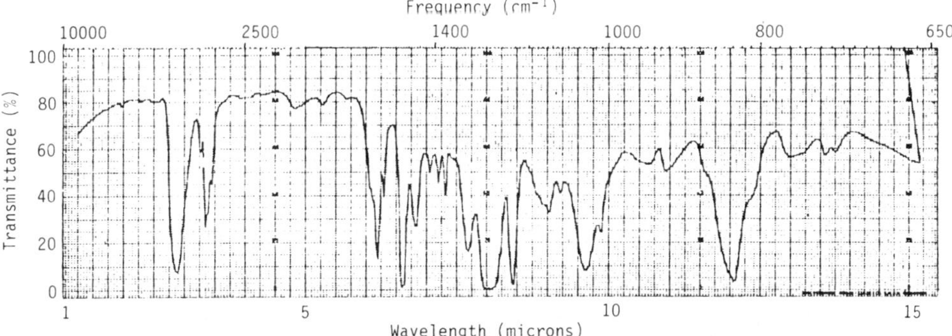

Figure 187.25 Epoxy resin (Araldite 6071, Ciba) KBR pellet (2 mg in 400 mg KBR).

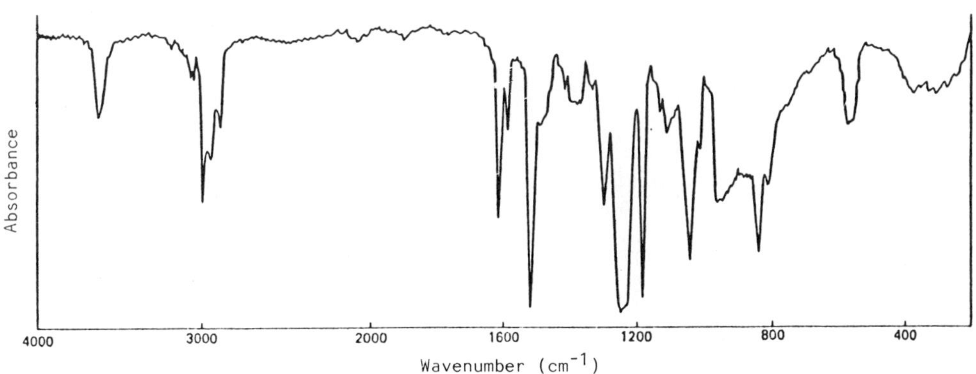

Figure 187.26 Epoxy resin (Epicote 1004) chloroform solution.

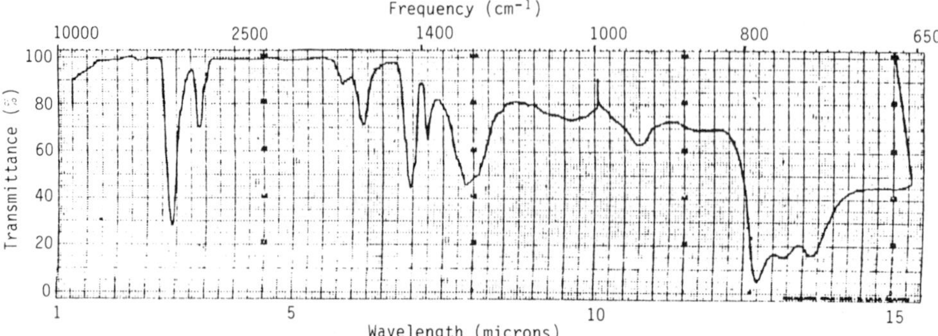

Figure 187.27 Chorinated polypropylene (Parlon P, Hercules Co.) KBR pellet (1.5 mg in 400 mg KBR, scale expanded.

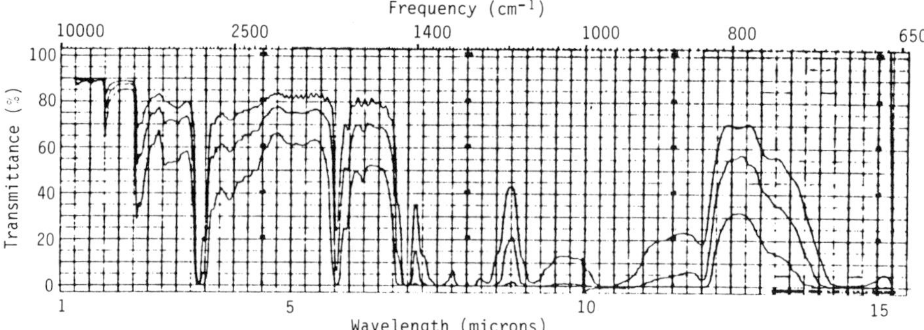

Figure 187.28 Rigid PVC, 0.4 mm film.

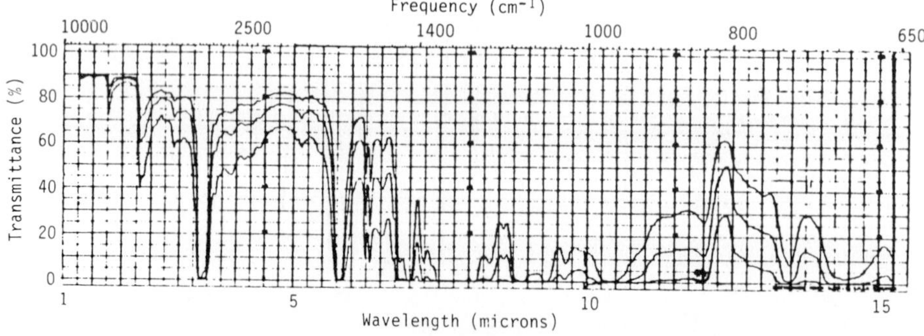

Figure 187.29 Plasticized PVC, 0.3 mm film.

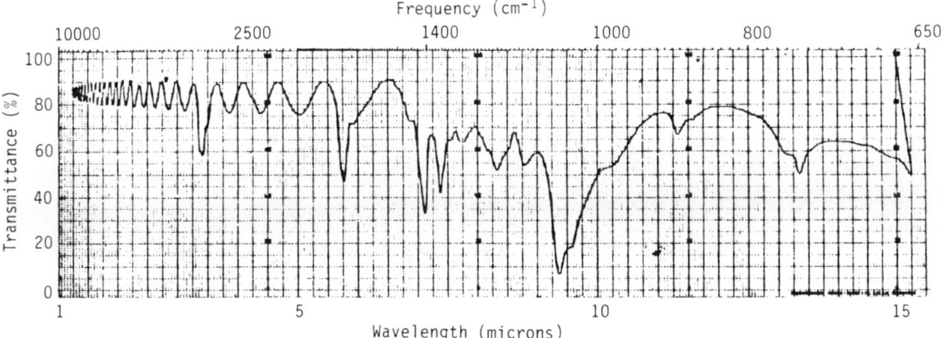

Figure 187.30   Polyvinylidene chloride (Saran, Dow Chemical Co.), 12 u thick film.

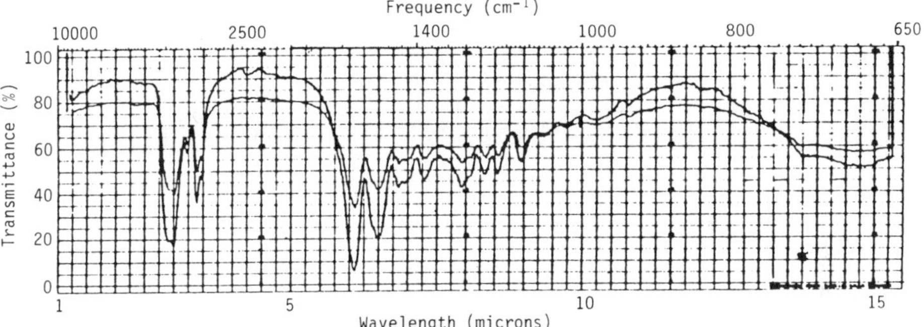

Figure 187.31 Nylon, KBR disc.

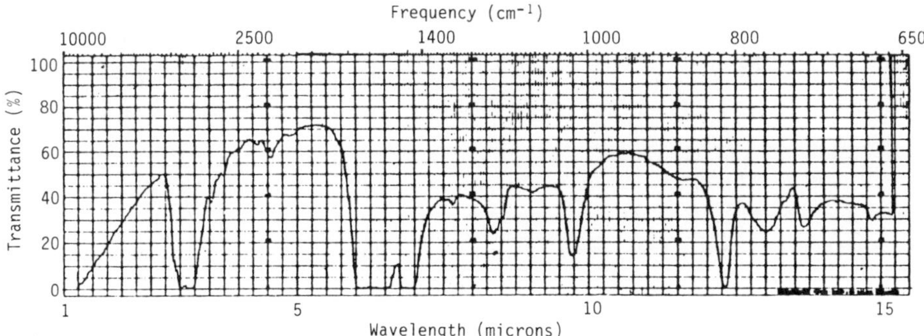

Figure 187.32 Melamine resin, KBR disc.

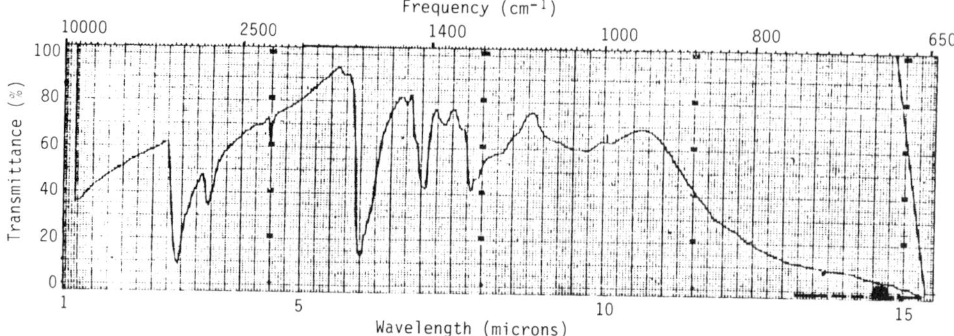

Figure 187.33 Polyacrylamide, KBR pellet, 1.5 mg in 400mg KBR, scale expanded.

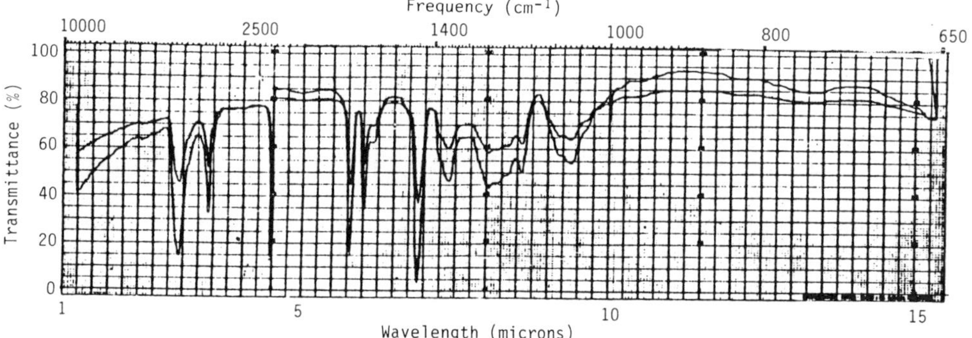

Figure 187.34 Polyacrylonitrile (Dralong) KBR pellet, scale expanded.

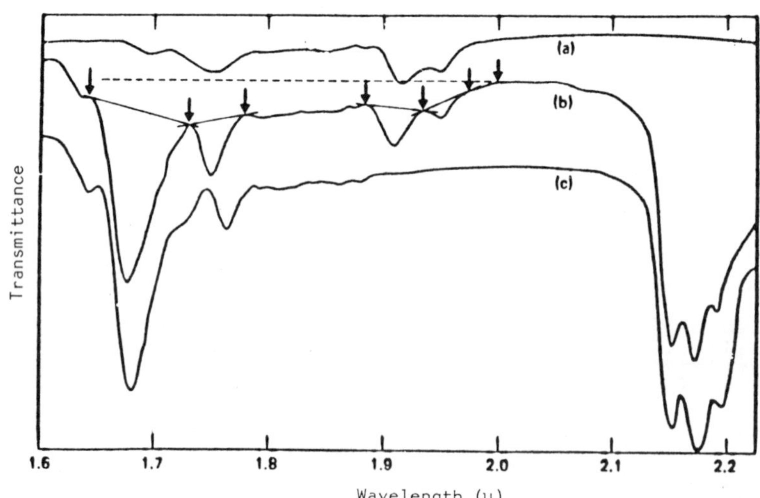

Figure 187.35 Near infrared spectra (a) polyacrylonitrile; (b) styrene acrylonitrile copolymer (acrylonitrile content 25.7 wt%); (c) polystyrene.

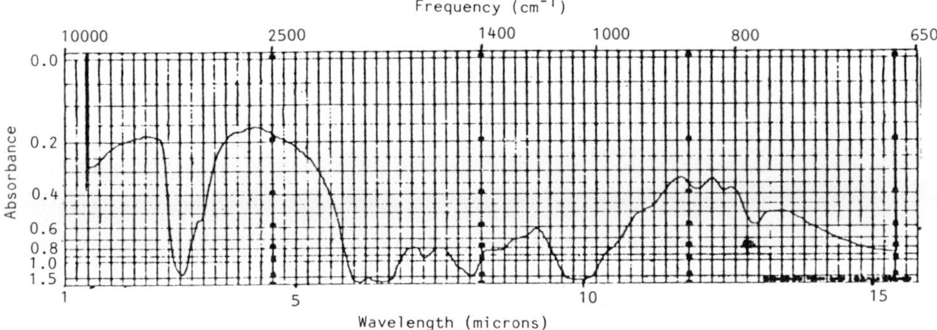

Figure 187.36 Urea formaldehyde resin, KBR disc.

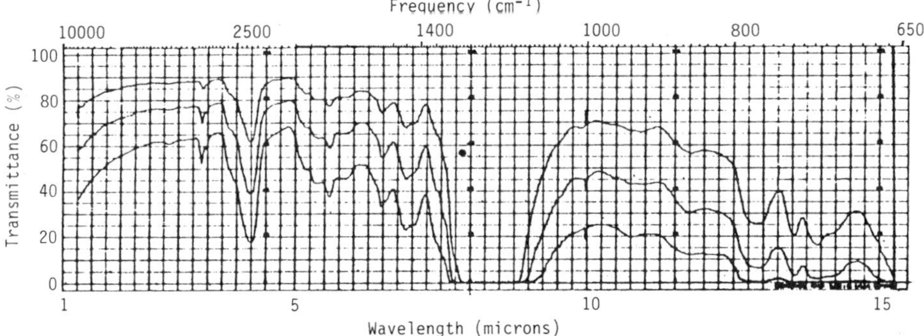

Figure 187.37 Polytetra fluoroethylene (Teflon) 0.055, 0.11, 0.22 mm
film.

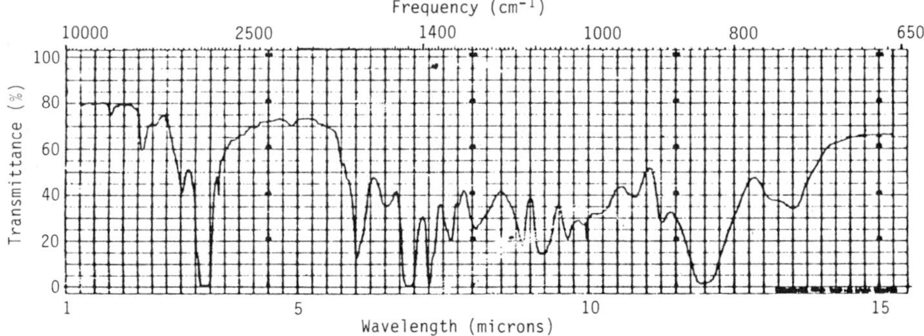

Figure 187.38 Natural rubber, KBR disc.

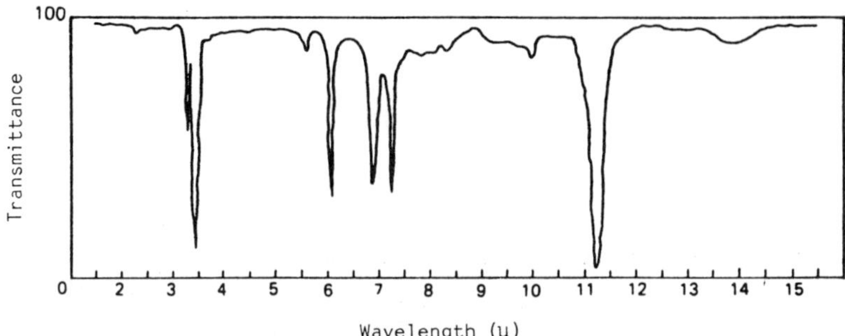

Figure 187.39 Poly 2,4-isoprene, KBR disc.

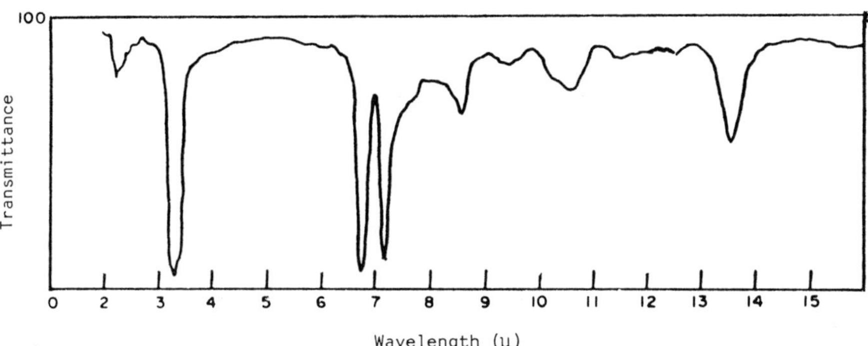

Figure 187.40 Hydrogenated polyisoprene, KBR disc.

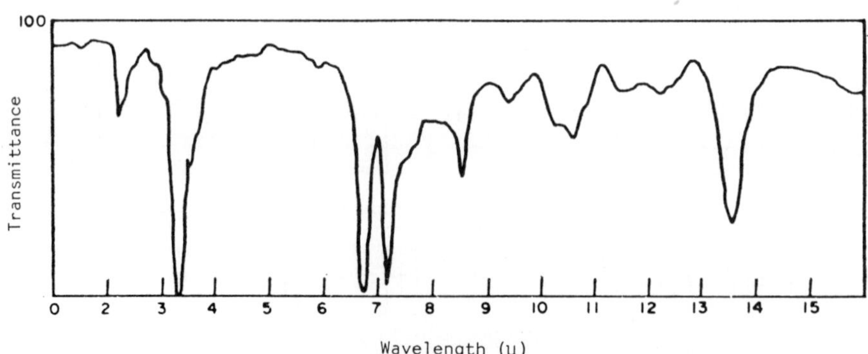

Figure 187.41 Hydrogenated natural rubber, KBR disc.

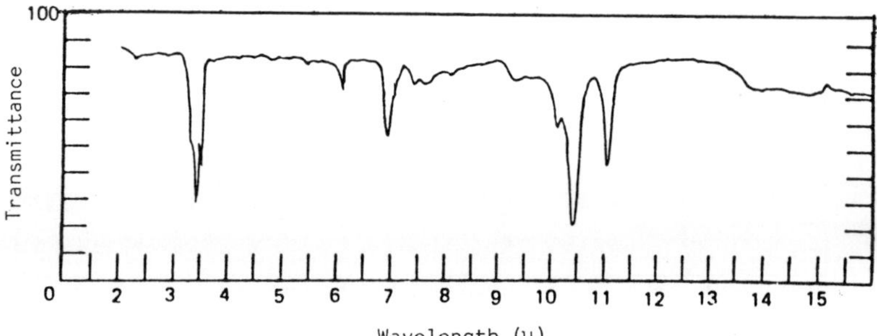

Figure 187.42 Emulsion polybutadiene, KBR disc.

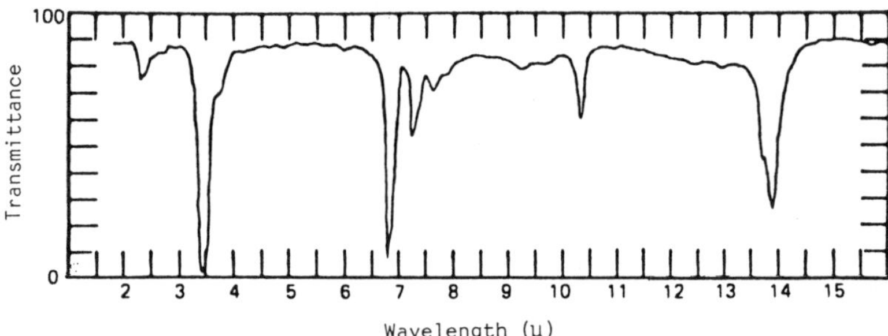

Figure 187.43 Hydrogenated emulsion polybutadiene, KBR disc.

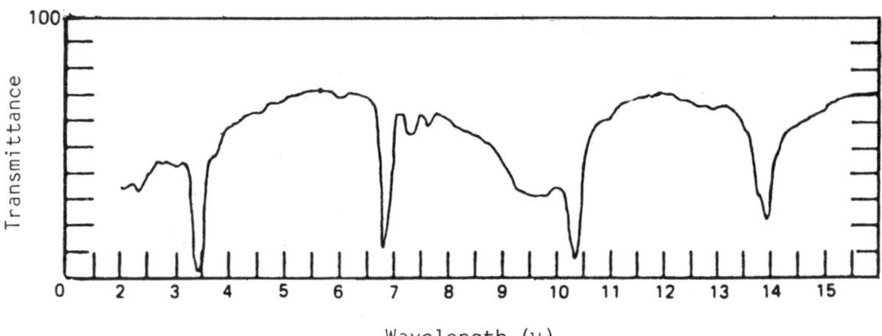

Figure 187.44 Hydrogenated lithium polybutadiene, KBR disc.

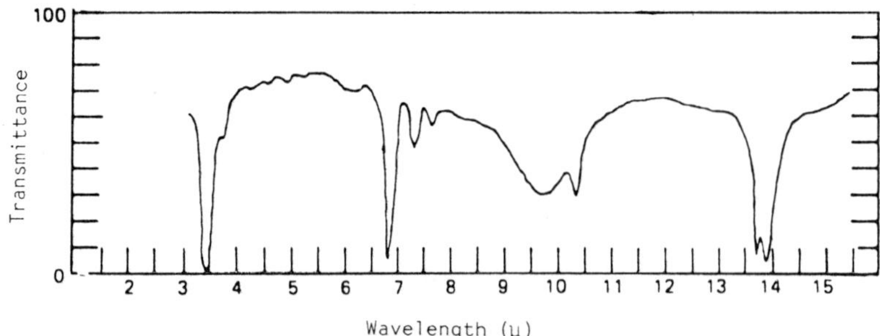

Figure 187.45 Hydrogenated Ziegler polybutadiene, KBR disc.

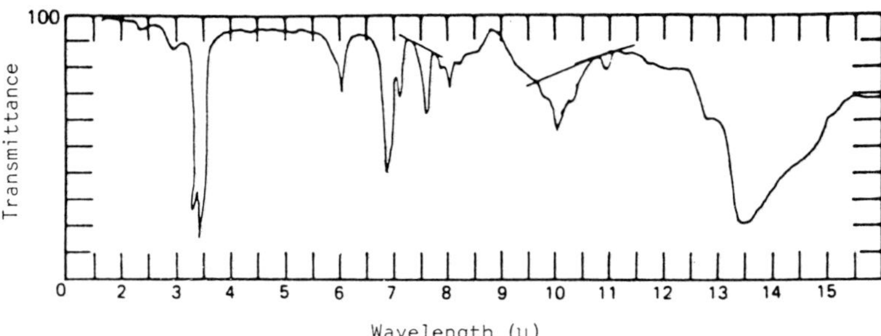

Figure 187.46 Cis-1,4-polybutadiene, KBR disc.

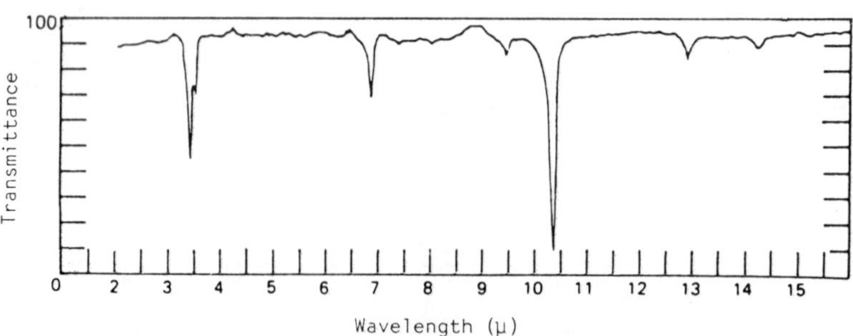

Figure 187.47 Trans 1,4 polybutadiene, KBR disc.

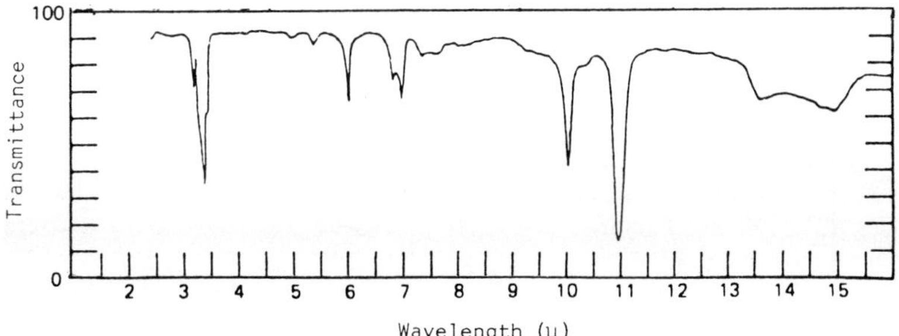

Figure 187.48 1,2 polybutadiene, KBR disc.

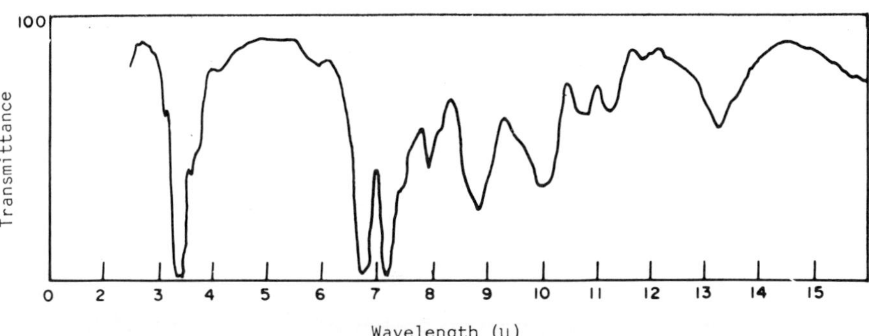

Figure 187.49 Hydrogenated polydimethyl butadiene, KBR disc.

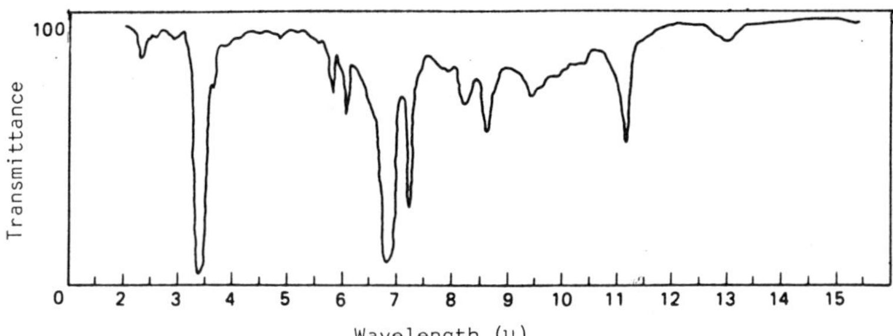

Figure 187.50 Polydimethyl butadiene, KBR disc.

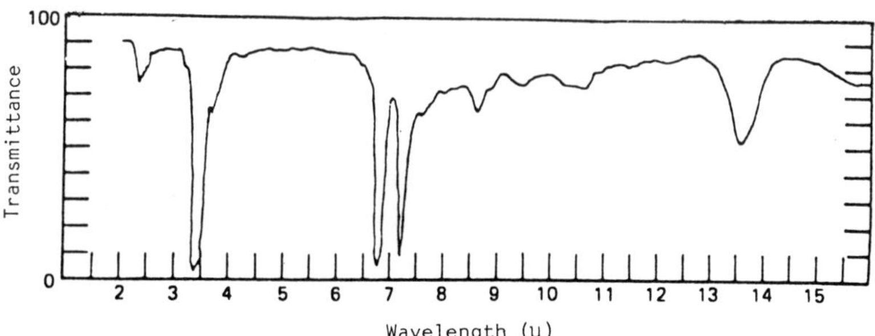

Figure 187.51 Hydrogenated polypiperylene, KBR disc.

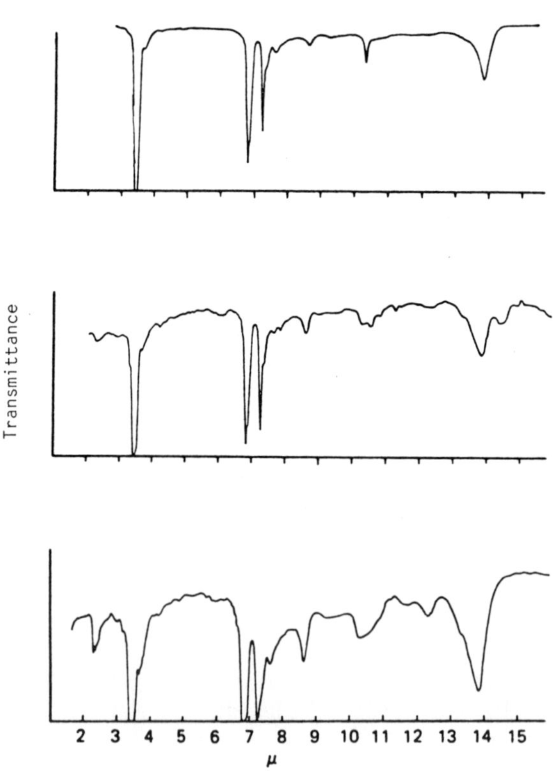

Figure 187.52 Ethylene propylene-diene elastomers, KBR disc.

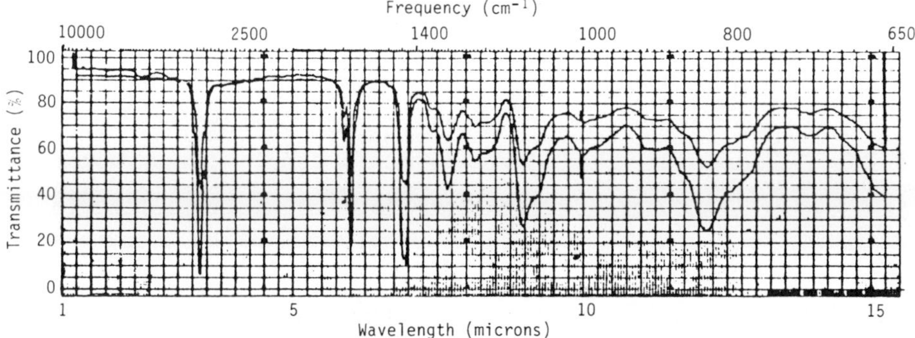

Figure 187.53   Neoprene (chloroprene), KBR disc.

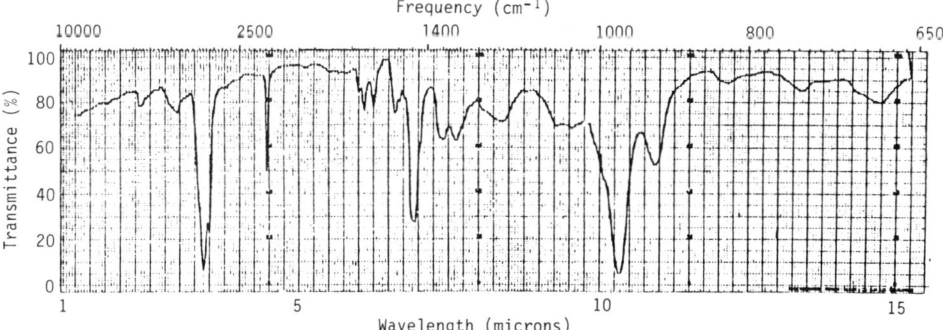

Figure 187.54   Nitrile rubber (Hycar 1042, Gooderich Chemical Co.), benzene solution evaporated on sodium chloride window, scale expanded.

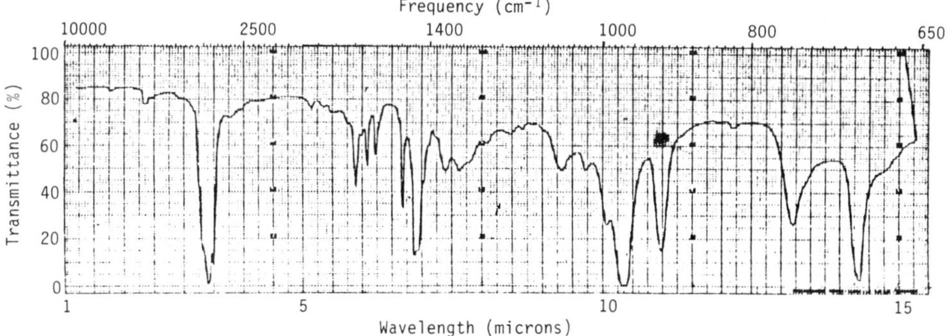

Figure 187.55 Styrene butadiene (SBR 15 M Phillips Chemical Co.), benzene solution evaporated on sodium chloride window.

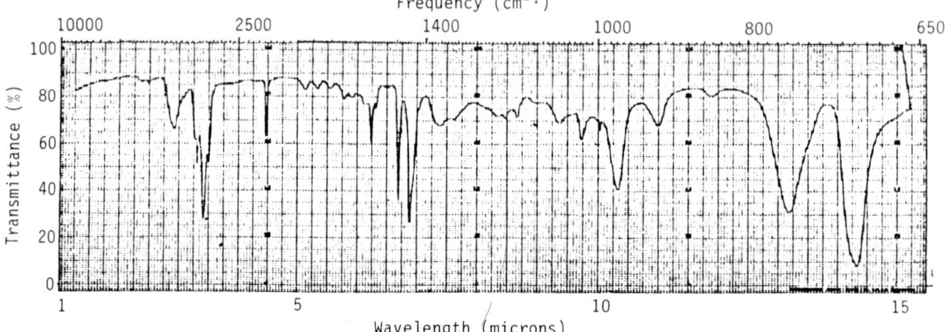

Figure 187.56   Acrylonitrile-butadiene-styrene terpolymer (Nowadur W
                Bayer Chemicals), benzene solution evaporated on sodium
                chloride windows.

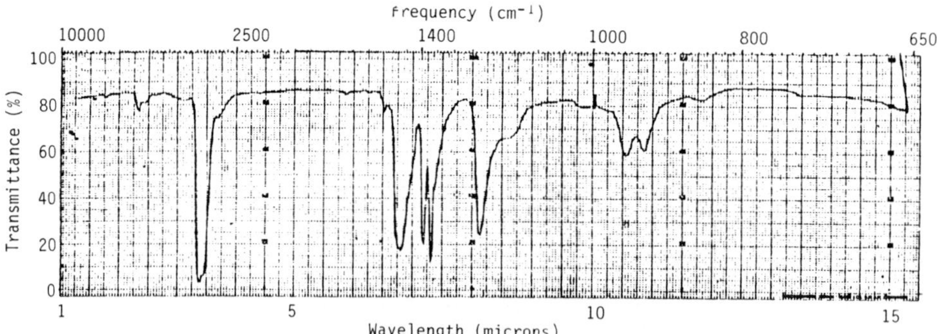

Figure 187.57   Butyl rubber (Polysar butyl 301, Polysar Co.), benzene
                solution evaporated on sodium chloride window, scale
                expanded.

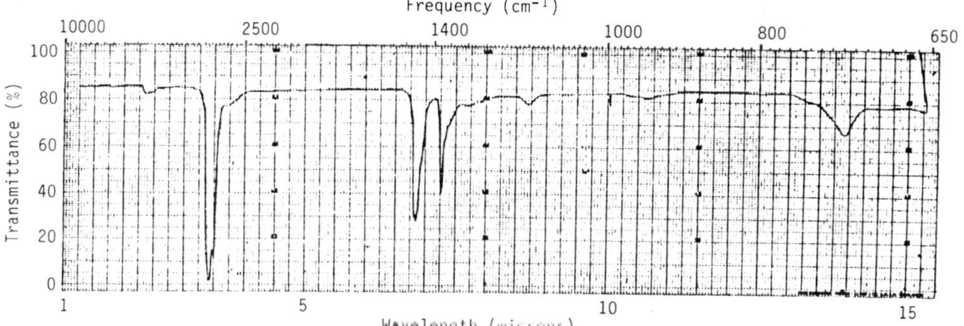

Figure 187.58   Ethylene propylene-diene terpolymer (APUK, Bunewerke Huls) (GmbH), benzene solution evaporated on sodium chloride window.

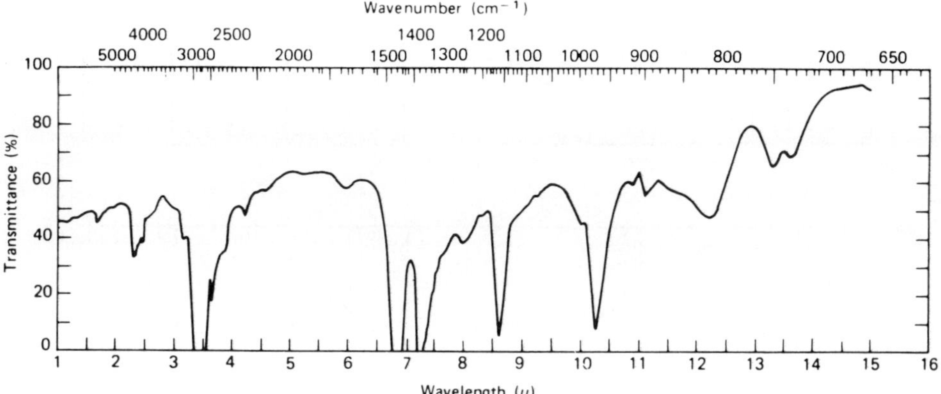

Figure 187.59   Infrared spectrum of ethylene propylene copolymer containing 85 mole % propylene.

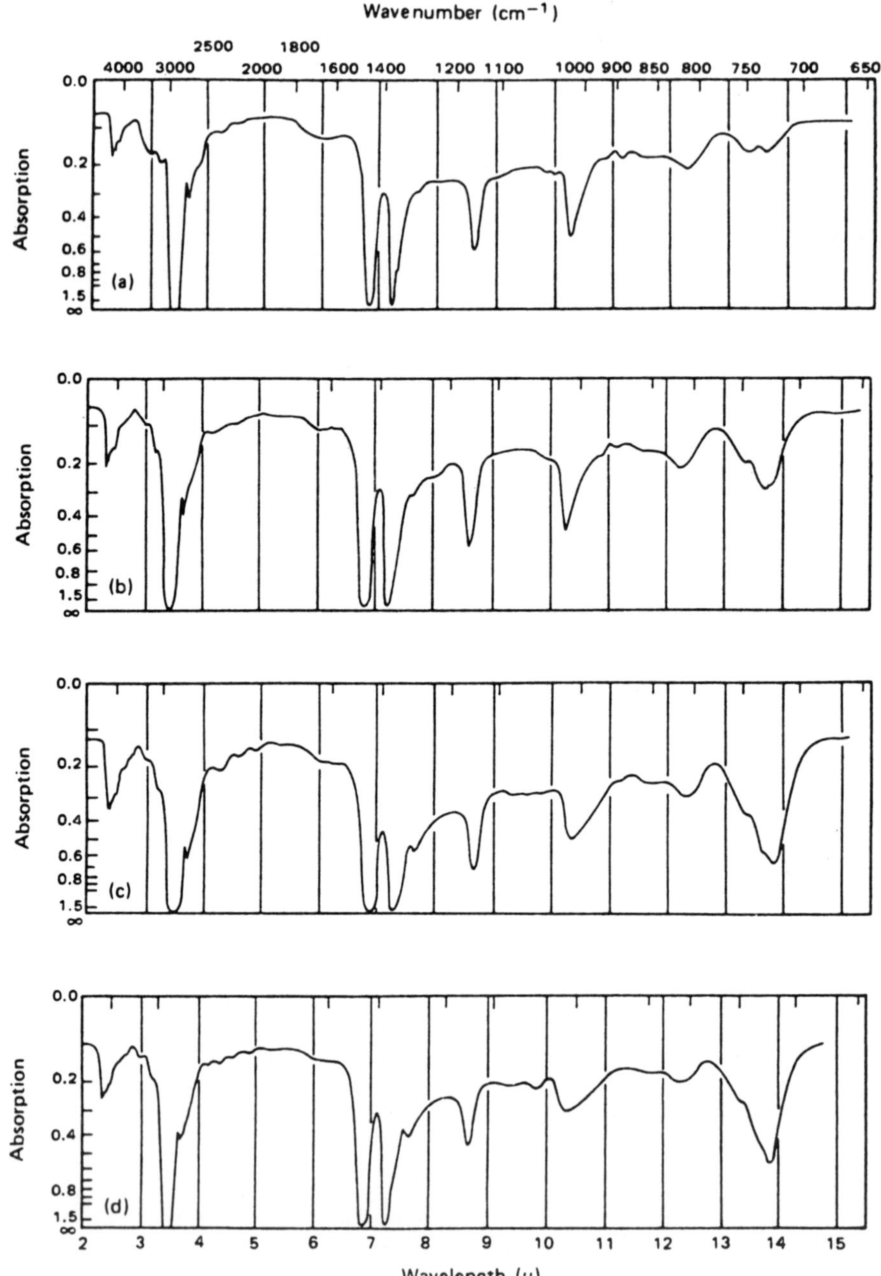

Figure 187.60   Infrared spectra of ethylene propylene copolymers of various compositions: (a) 85.5% polyethylene; (b) 74.0% polyethylene; (c) 65.9% polyethylene; (d) 55.5% polyethylene.

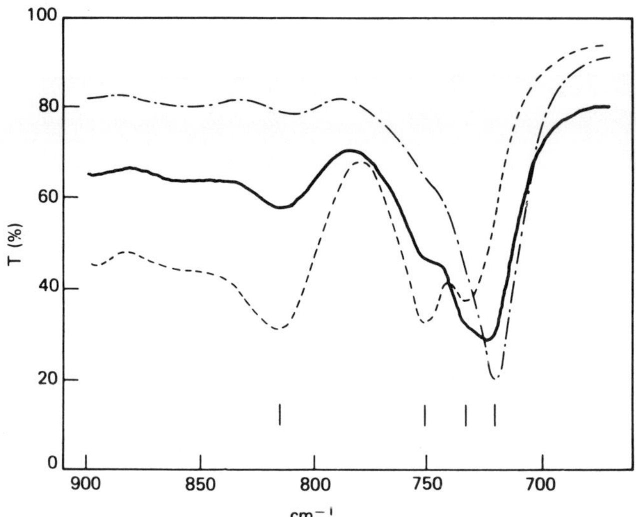

Figure 187.61   Ethylene propylene copolymers at various compositions --
(25% (3 mol));_ (50% (3 mol));... (75% (3 mol)).

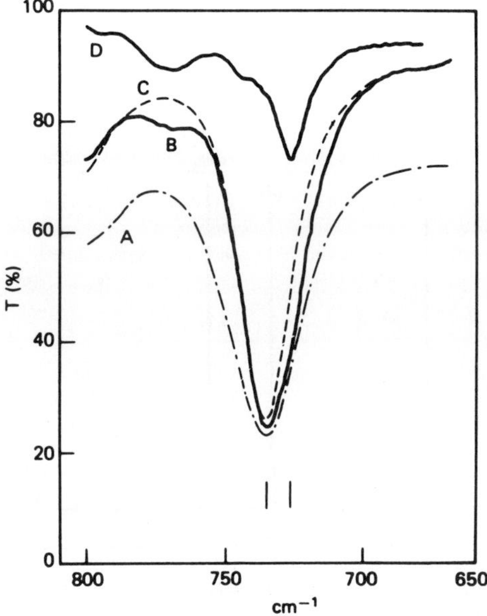

Figure 187.62   Ethylene propylene copolymer in cast film in the C-H
bending region of the spectrum. Base line points shown
by arrows. Grating to prism changeover at 1416 cm^{-1}
(7.06 u).

Figure 188   NMR Spectra.
(on pp. 702-712)

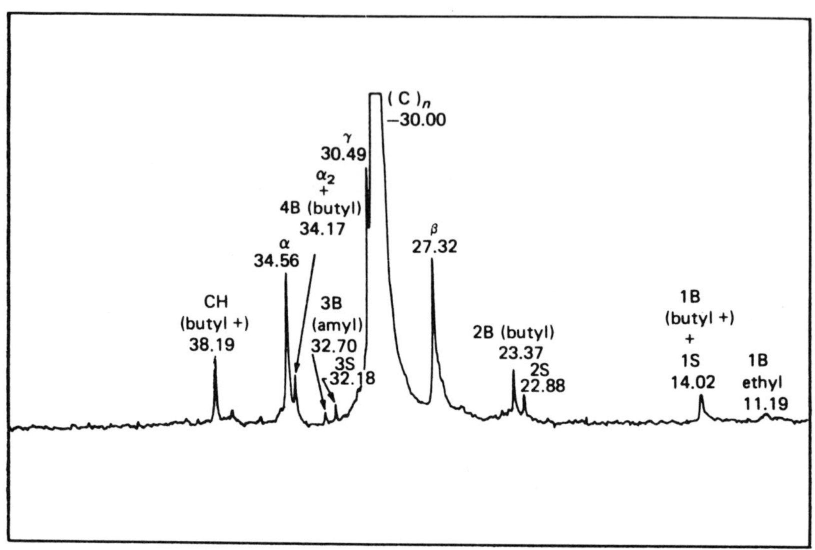

Figure 188.1 ^{13}C NMR spectrum at 25.2 MHz of a low density polyethylene from a high pressure process.

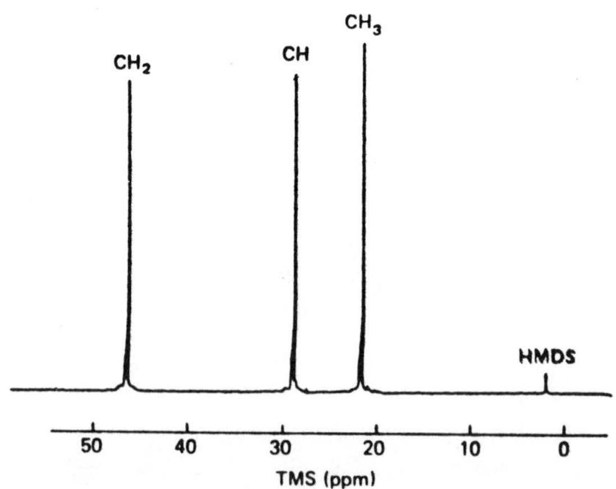

Figure 188.2 ^{13}C NMR spectrum at 25.2 MHz of predominantly isotactic crylstalline polypropylene.

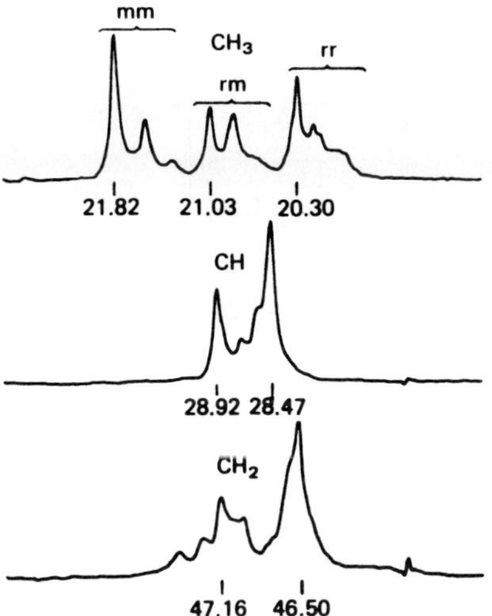

Figure 188.3   ¹³C NMR spectrum of non-crystalline polypropylene in 1,2,4-trichlorobenzene showing methylene (46 ppm) methine (28 ppm) and methyl (20 ppm) regions.

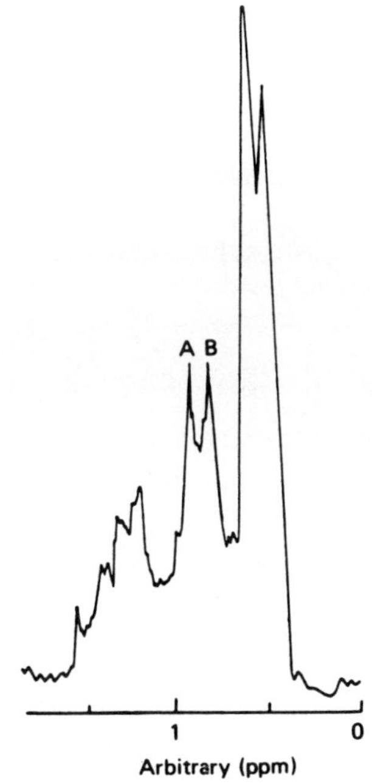

Figure 188.4   PMR spectrum of atactic polypropylene in hot heptane containing up to 40% syndiotactic placement.

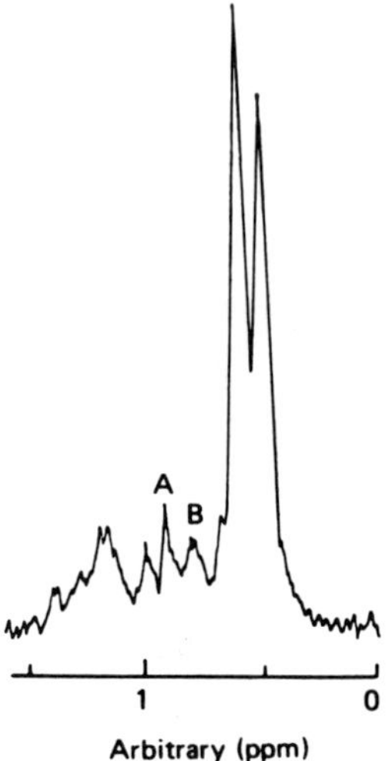

Figure 188.5   PMR spectrum of (>95%) isotactic polypropylene in hot
heptane.

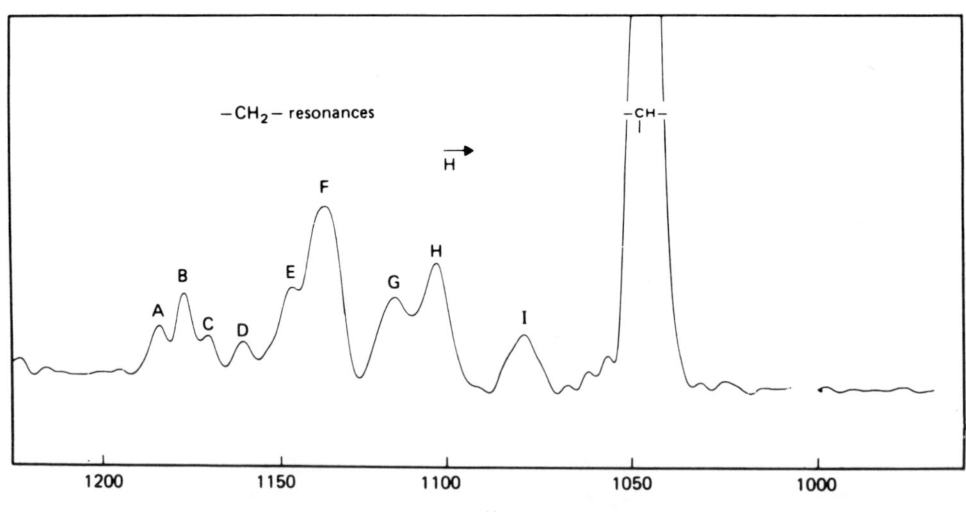

Figure 188.6 ^{13}C NMR spectrum of an amorphous polystyrene at 25.2 MHz
and 120°C.   The MHz values are relative to an internal
tetramethylsilane (TMS) standard.   A-I methylene
resonances.

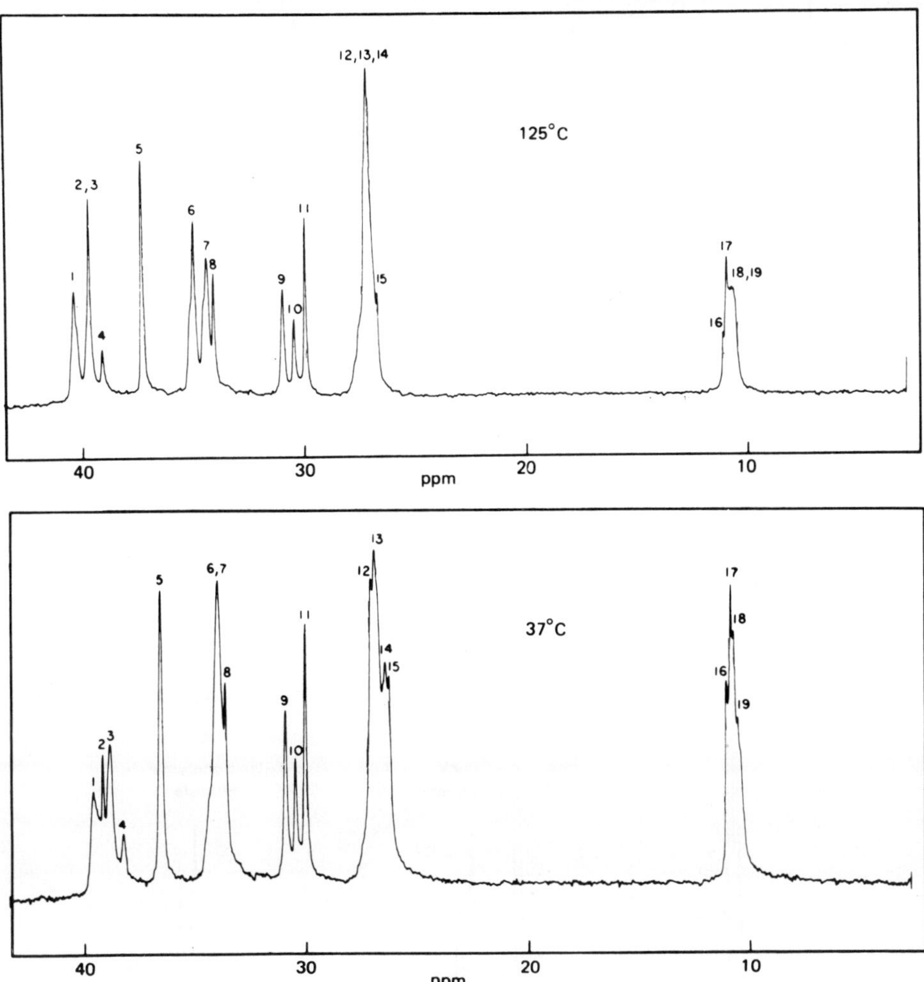

Figure 188.7 ^{13}C NMR spectra of a hydrogenated polybutadiene (73% 1,2-additions) at 37 and 125°C shown with respect to a tremethylsilane internal standard.

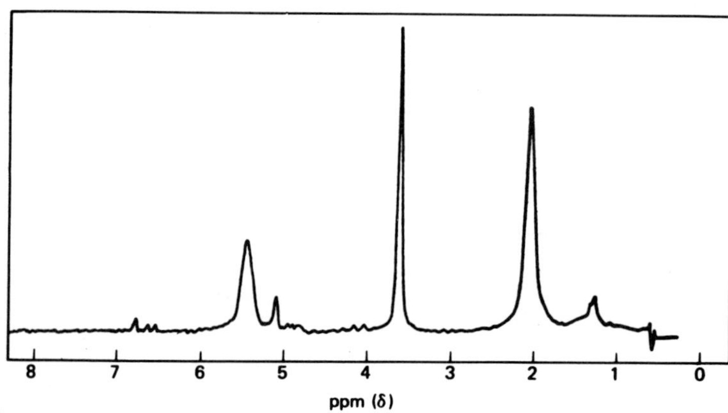

Figure 188.8   Typical NMR spectrum of extracted ungrafted rubber.
Signal at 5.4 ppm - olefinic protons of 1,4-
polybutadiene.
Signal at 5.1 ppm - olefinic protons of 1,2-
polybutadiene.
Signal at 3.5 ppm - dioxane (internal standard).
Signal at 2.0 ppm - aliphatic protons of 1,4-
polybutadiene.
Signal at 1.2 ppm - methylene protons of the soap.
Signals between 6 and 7 ppm impurities in
hexachlorobutadiene.

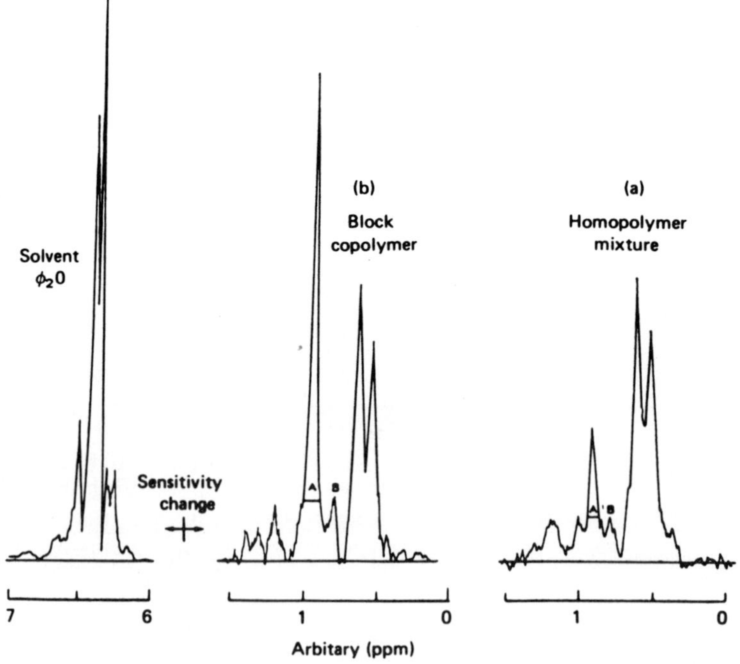

Figure 188.9   PMR spectra of (a) ethylene and propylene copolymers and
(b) mixture of polyethylene and polypropylene
copolymers.

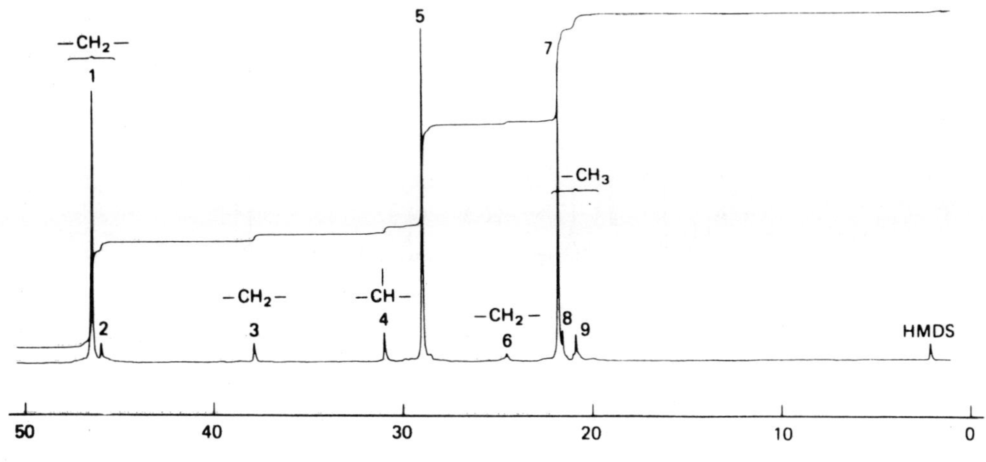

Figure 188.10   252 MHz ¹³C proton noise decoupled NMR spectrum at 125°C of 97:3 w/v ethylene-propylene copolymer in 1,2,4 trichlorobenzene and perdeutrobenzene.   Internal standard hexamethyldisoloxane.

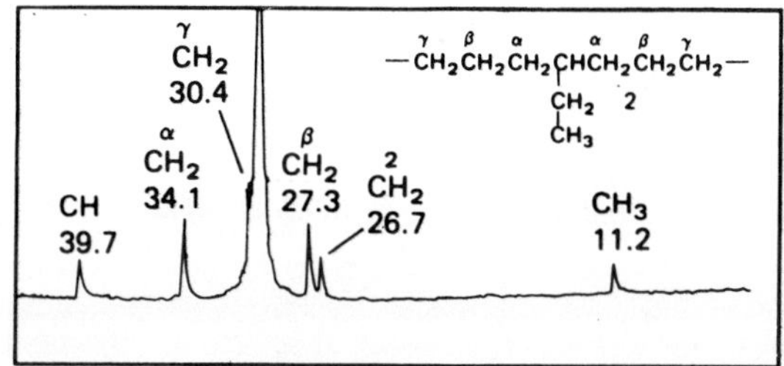

Figure 188.11   25.2 MHz ¹³C NMR spectrum of ethylene butene copolymer.

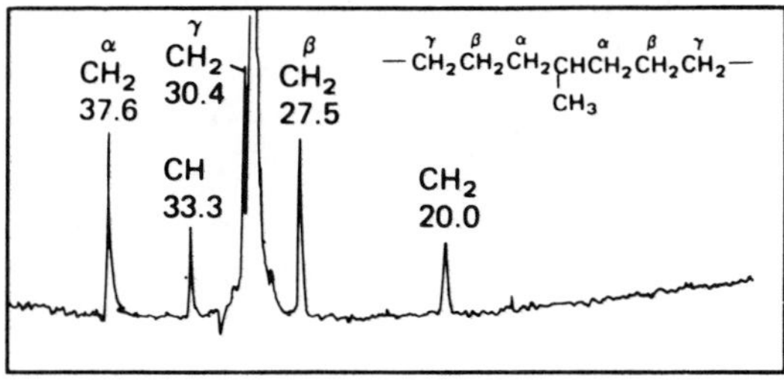

Figure 188.12   25.2 MHz ¹³C NMR spectrum of ethylene-propylene. copolymer.

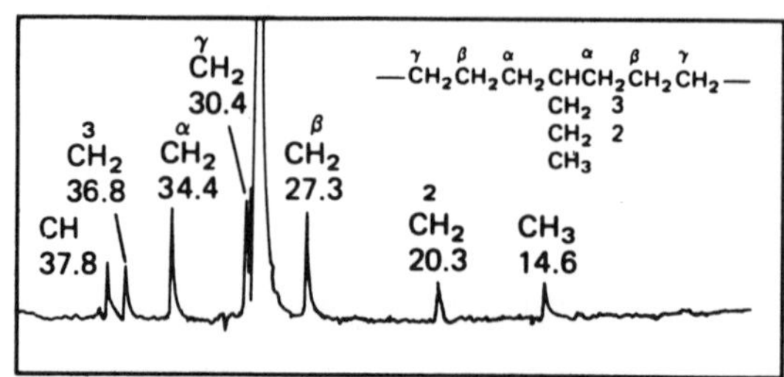

Figure 188.13   25.2 MHz ¹³C NMR spectrum of ethylene pentene-1 copolymer.

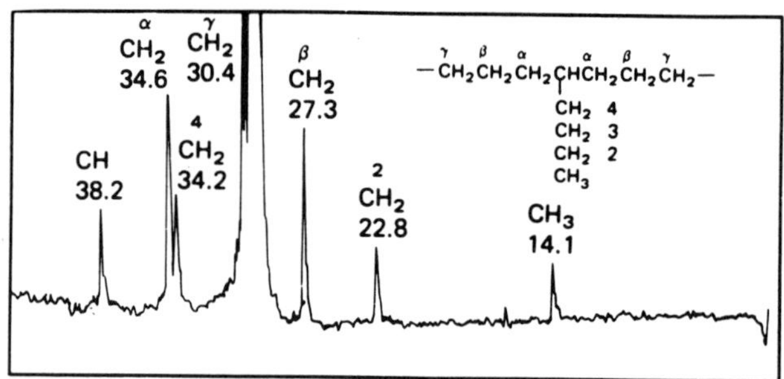

Figure 188.14   25.2 MHz ¹³C NMR spectrum of ethylene-1-hexene copolymer.

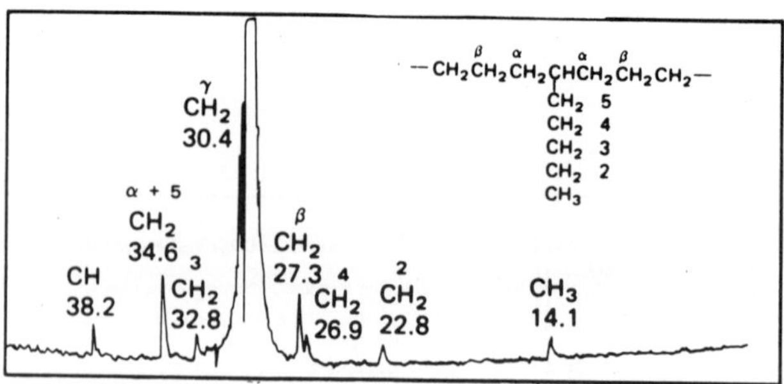

Figure 188.15   25.2 MHz ¹³C NMR spectrum of ethylene-heptane copolymer.

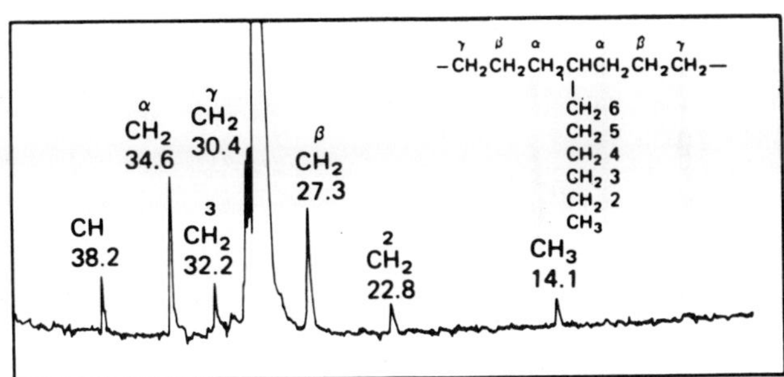

Figure 188.16   25.2 MHz ^{13}C NMR spectrum of ethylene-1-octene copolymer.

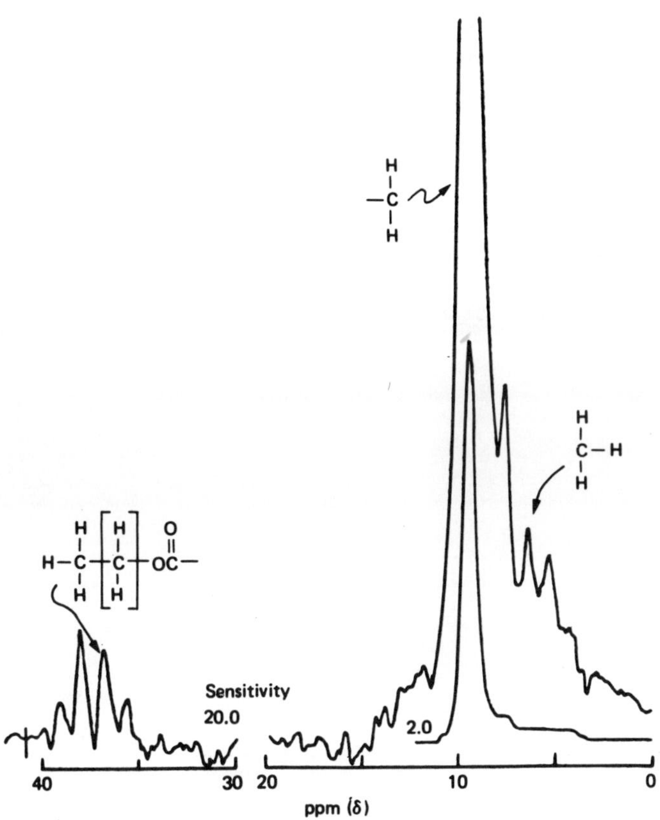

Figure 188.17 NMR spectra.   Polyethylene-ethyl acrylate.

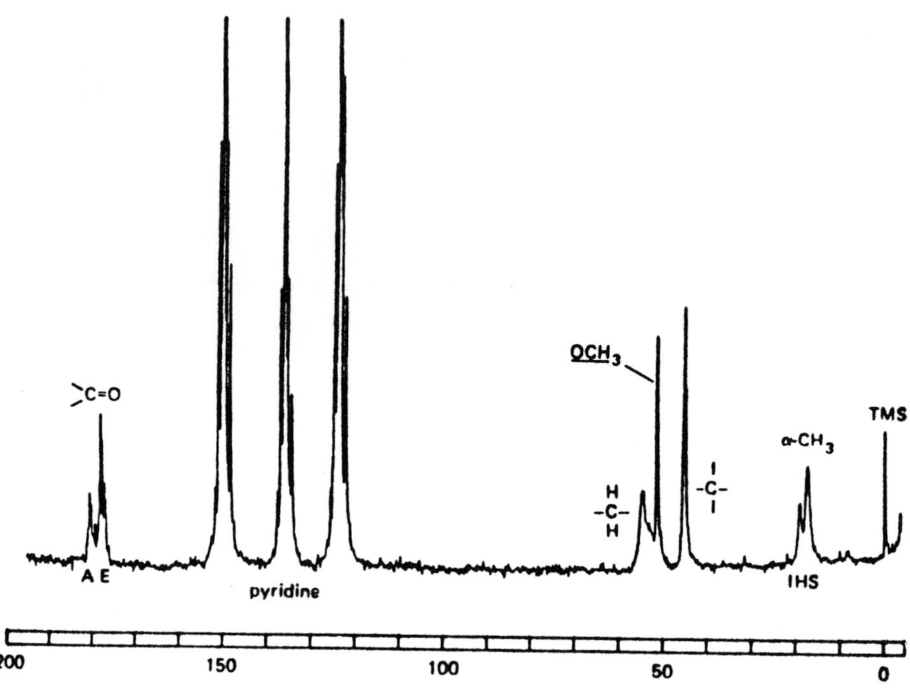

Figure 188.18   The 20 MHz fully proton decoupled ^{13}C NMR spectrum of
                polymethylmethacrylate (67%) methacrylic acid (33%)
                dissolved in 50:50 v/v pyridine/pyridine d$_5$.   The I, H,
                S rotation refers to isotactic, heterotactic and
                syndiotactic triads.   A and E refer to acid and ester.

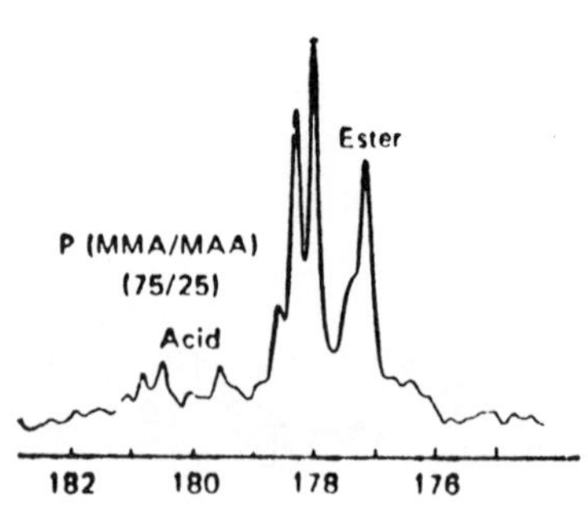

Figure 188.19   The 20 MHz ^{13}C NMR spectrum of polyethyl methacrylate
                (75%) - methacrylic acid (25%) in pyridine.

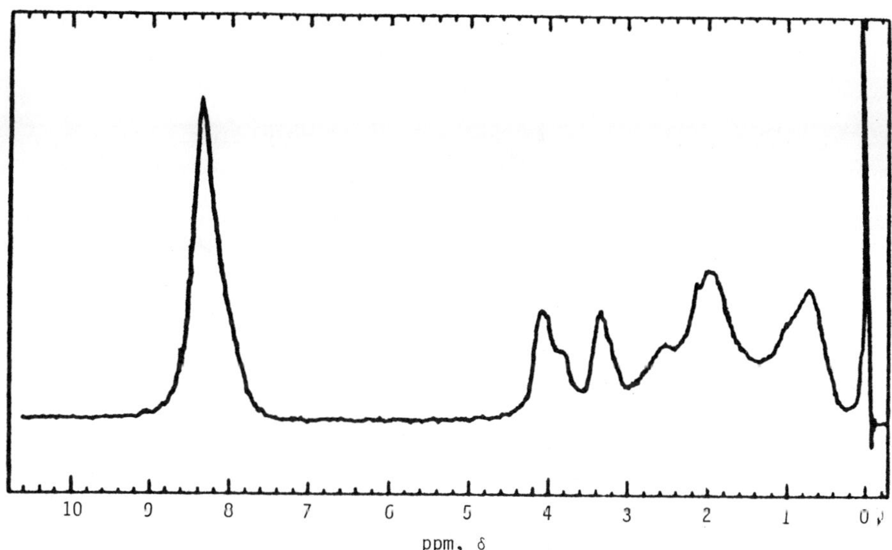

Figure 188.20 50 H₃, proton NMR spectrum of styrene methyl
methacrylate copolymer (10% w/v) in deutrochloroform).

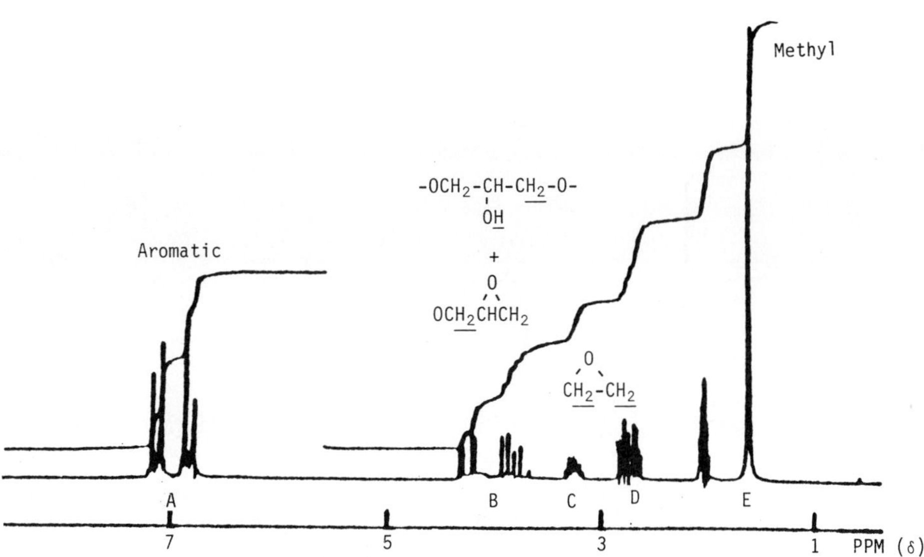

Figure 188.21 PMR spectrum of Bakelite epoxy resin ERL-2774 - biphenol
A- epichlorohydrin (DGEBA).

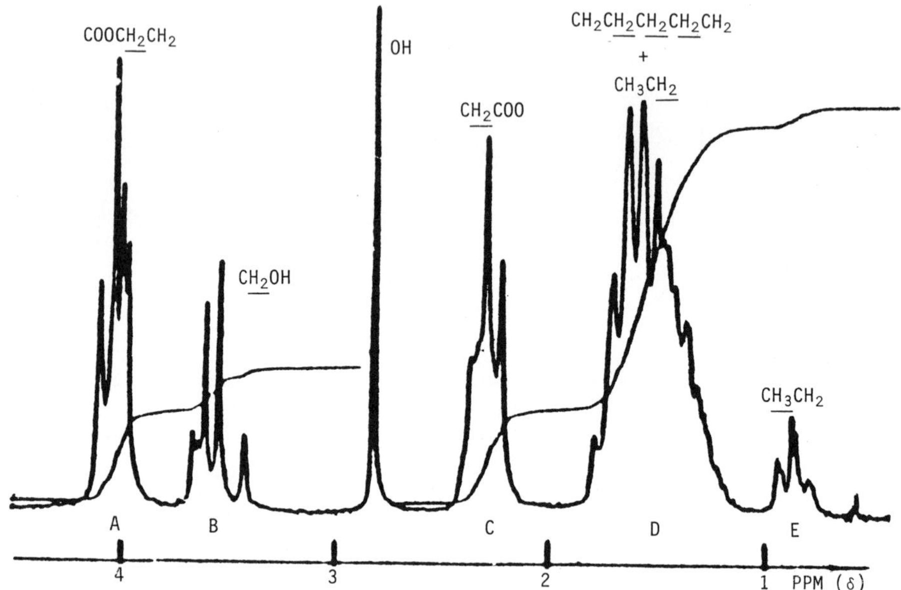

Figure 188.22   PMR spectrum of NIAX polyol PCP-0310 (TMP PCP) trimethylol propane started caprolactam ester triols.

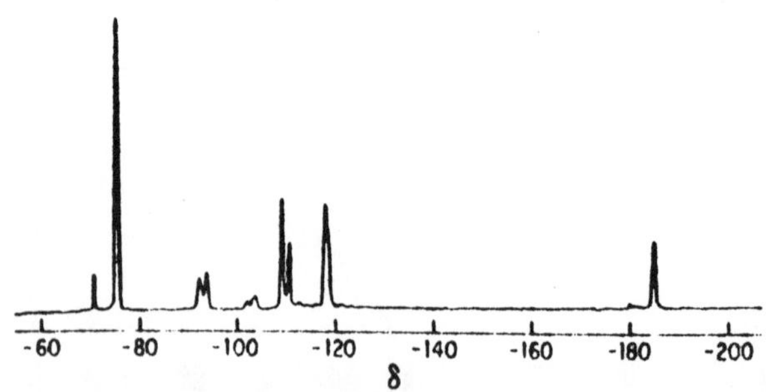

Figure 188.23   94.1 MHz 19F NMR spectrum of polyhexafluoropropylene (60% v/w) vinylidene fluoride (40%) copolymer.

APPENDIX 2

Chemical and physical properties of polymers, useful data relating to various polymer is tabulated in Table 192.

Table 192 (a) – Physical & Chemical Properties of Polymers
Thermal Properties

	T_g °C	T_m °C	Thermal expansion ASTM D696 (mm mm^{-1} °C^{-1} x 10^{-5})	Specific heat (kJkg^{-1})	Thermal conductivity ASTM C 177 (Wm^{-1} °C x 10^{-3})	Softening point (vicat) (°C)
POLYCONDENSATION TYPES						
Phenol formaldhyde						
(a) Unfilled	–	–	2.5–6	1.6–1.76	1.26–2.52	–
(b) Woodflour/Cotton flock filled	–	–	3–4.5	1.45–1.68	1.68–2.94	–
Urea formaldehyde						
Cellulose filled	–	–	2.7	1.68	2.94–4.2	–
Melamine formaldehyde						
(a) Unfilled	–	–	–	–	–	–
(b) alpha-cellulose filled	–	–	4	1.68	2.94–4.2	–
Polyesters						
(a) Resin	65–69	220–256	5.5–10	–	3.4–3.6	–
(b) Dough moulding compound	–	–	1.1–5	1.04	9.4–10/6	–
(c) Sheet moulding compound	–	–	2	1.04	–	–
Epoxides						
(a) Rigid	50–150	–	5–9	–	1.68–2.1	–
(b) Flexible	–	–	5–9	–	1.68–2.1	–
Nylons						
(a) Type 6	40	220	8–13	1.68	–	M.P 215

(b) Type 6/6	50	255	10–15	1.68	2.18–2.43	M.P.264
(c) Type 11	–	186	15	2.4	4.2	185
(d) Type 12	–	–	10.4	2.1	2.9	175
POLYMERIZATION TYPES						
Polyethylenes						
(a) Low density	–100	120	16–18	2.3	3.37	85–87
(b) High density	–70	135	11–13	2.22–2.3	4.63–5.22	120–130
Polypropylene	–30	165	11	1.93	1.38	150
Polystyrenes						
(a) Conventional	90–100	–	6–8	1.34–1.45	1.1–1.38	82–103
(b) Toughened	–	–	3.4–21	1.34–1.45	1.26	78–100
Styrene acrylonitrile	–	–	6–8	1.34–1.45	1.21	85–103
Acrylonitrle/butadiene/ styrene vinyl polymers	–	–	6–13	1.38–1.68	0.62–3.62	85
(a) Rigid Polyvinyl chloride	85	190	5	0.84–2.1	1.47	82
(b) Rigid vinyl chloride/ vinyl acetate	–	–	7	0.84–2.1	1.55	–
(c) Rubber modified PVC	–	–	5	0.84–2.1	1.89	78
Polyacetals	–	–	8.1	1.45	2.3	175
Polycarbonates	155	235	7	1.26	1.93	165
Ioromers	–	–	12	2.3	2.4	68–76
Polyphenylene oxide	–	230	5.2–6	1.33	5.2–6	125–148
Polyphenylene sulphide	80	280	–	–	–	–

Table 192(a) (continued)

Acrylics	–	–	5–9	1.47	1.68–2.52	80–98
Ethylene vinylacetate fluor- inated polymers	–20 to 20	40/100	16–20	2.8	0.3	64
(a) Fluorinated ethylene propylene	–	280	8.3–10.5	1.18	2.1	MP 285–295
(b) Polytetrafluorethylene	–20	327	9–22^a	1.05	2.52	Trans pt 372
(c) Polytrifluorochloroethylene	–	–	0.7	0.93	–	Dec pt > 400
Cellulosic plastics						
(a) Cellulose acetate	–	–	8–16	1.26–1.76	1.68–3.37	70
(b) Cellulose acetate butyrate	–	–	11–17	1.26–1.68	1.68–3.37	70
(c) Cellulose propionate	–	–	11–17	1.26–1.68	1.68–3.37	–
Metals and alloys						
(a) Brass (Cu–Zn, 70/30)	–	–	1.99	0.379	1210	*
(b) Mild Steel (0.6% Carbon)	–	–	1.26	0.463	631	*
(c) Carbon Steel (0.4%) Carbon)	–	–	1.12	0.463	505	*
(d) Aluminium (99% Pure)	–	–	2.4	0.965	2180	*
(e) Duralumin (Al–Si–Cu–Mg)	–	–	2.25	0.965	1470	*
(f) Phosphor bronze (Cu–Sn–P)	–	–	1.78	0.379	840	*

^{a} varies with temperature.

Table 192 (b) – Physical & Chemical Properties of Polymers
Miscellaneous Properties

	Specific gravity ASTM D792	Dielectric constant ASTM D 150 (10^6 Hz)	Burning rate ASTM D635 (in/min)*	Specific volume ASTM D542 (D)	Refractive ASTM D542 (1/8" thickness % 24 hrs)
POLYCONDENSATION TYPES					
Phenol formaldhyde					
(a) Unfilled	1.25-1.3	4.5-5.0	–	800-771	1.5-1.7
(b) Woodflour/Cotton flock filled	1.32-1.45	4.0.7.0	–	765-645	–
Urea formaldehyde					
Cellulose filled	1.47-1.52	6.4-6.9	–	682-659	1.54-1.56
Melamine formaldehyde					
(a) Unfilled	1.48	–	–	678	–
(b) alpha-cellulose filled	1.47-1.52	7.2-8.2	–	682-659	–
Polyesters					
(a) Resin	1.1-1.4	2.8-4.1	–	910-715	1.52-1.58
(b) Dough moulding compound	1.7-2.3	5.2-6.4	–	690-435	–
(c) Sheet moulding compound	1.7-2.6	4.2-5.8	–	590-400	–
Epoxides					
(a) Rigid	1.0-3.2	3.0-4.0	–	832	1.6
(b) Flexible	1.2	3.0-4.0	–	832	–
Nylons					
(a) Type 6	1.13	3.0-7.0	–	886-875	–
(b) Type 6/6	1.14	3.6-6.0	–	921-875	1.53

Table 192(b) (continued)

(c) Type 11	1.04–1.05	3.2	–	960–950	1.52
(d) Type 12	1.01–1.02	3.1	–	990–980	–
POLYMERIZATION TYPES					
Polyethylenes					
(a) Low density	0.91–0.93	2.25–2.35	1.02–1.06	1091–1061	1.51
(b) High density	0.941–0.965	2.25–2.35	1.02–1.06	1061–1037	1.54
Polypropylene	0.9–0.91	2.25–2.3	0.75–0.83	1119–1048	1.49
Polystyrenes					
(a) Conventional	1.04–1.11	2.4–3.1	–	953–896	1.59–1.60
(b) Toughened	0.98–1.1	2.4–3.8	–	1030–911	–
Styrene acrylonitrile	1.075–1.1	2.75	3.1	932–911	1.57
Acrylonitrle/butadiene/ styrene vinyl polymers	0.99–1.1	2.7–4.75	0.6–1.0	1012–911	–
(a) Rigid Polyvinyl chloride	1.38–1.4	3.0	0.04–0.4	728	1.52
(b) Rigid vinyl chloride/ vinyl acetate	1.37–1.45	3.9–3.5	0.04–0.4	939–691	1.55
Modified PVC	1.35	–	–	742	–
Polyacetals	1.425	3.7	1.0–1.1	706	1.48
Polycarbonates	1.2	2.6	–	832	1.58
Ionomers	0.93–0.96	2.4–2.5	0.9–1.1	1060	1.51
Polyphenylene oxide	1.06–1.1	2.64	–	945–910	–
Polyphenylene sulphide	1.6–1.9	3.8–6.6	–	–	–
Acrylics	1.17–1.2	2.2–3.2	0.6–1.3	857–835	1.49

Ethylene vinylacetate fluor-inated polymers	0.925–0.95	2.6–3.2	1.02–1.06	1070	–
(a) Fluorinated ethylene propylene	2.14–2.17	2.1	–	466–459	1.34
(b) Polytetrafluorethylene	2.1–2.2	2.1	–	475–473	1.35
(c) Polytrifluorochloroethylene	2.1–2.15	2.5	–	475–463	1.43
Cellulosic plastics					
(a) Cellulose acetate	1.23–1.34	3.2–7.0	–	815–746	1.46–1.50
(b) Cellulose acetate butyrate	1.15–1.22	3.2–6.2	–	867–821	1.46–1.49
(c) Cellulose propionate	1.18–1.24	3.4–3.6	1.0–1.3	846–815	1.46–1.49
Metals and alloys					
(a) Brass (Cu–Zn, 70/30)	8.5	–	*	126	
(b) Mild Steel (0.6% Carbon)	7.87	–	*	127	
(c) Carbon Steel (0.4%) Carbon)	7.85	–	*	134	
(d) Aluminium (99% Pure)	2.82	–	*	395	
(e) Duralumin (Al–Si–Cu–Mg)	2.8	–	*	359	
(f) Phosphor bronze (Cu–Sn–P)	8.98	–	*	113	

* Numerical flame spread ratings cannot directly reflect behaviour under actual fire conditions and can be no more than a very general indicator. Performance of test pieces may be radically different from that of plastic products in fire situations and all polymers will burn with varying degrees of intensity under suitable conditions.

Table 192 (c) – Physical & Chemical Properties of Polymers
Chemical resistance properties

	Water absorption ASTM D 570 (1/8" thickness % 24 hours)	Resistance (chemical) ASTM D 543 (See Key at foot of table)	Resistance (solvent) ASTM D 543 (See Key at foot of table)
POLYCONDENSATION TYPES			
Phenol formaldhyde			
(a) Unfilled	0.1–0.2	A/B, d/c, a/b, a/c	unaffected
(b) Woodflour/Cotton flock filled	0.3–1	A/B/, A/C, a/b, A/c	unaffected
Urea formaldehyde			
Cellulose filled	0.4–0.8	d/B, D/C, a/b, a/c	unaffected
Melamine formaldehyde			
(a) Unfilled	0.3–0.5	A/B, A/C, U/b, U/c	unaffected
(b) alpha-cellulose filled	0.1–0.6	A/B, A/C, U/b, U/c	unaffected
Polyesters			
(a) Resin	0.15–0.6	a/B, A/C, U/b, A/c	attacked by most solvents
(b) Dough moulding compound	0.06–0.28	U/B, A/C, U/b, a/c	A/t, A/K, a/T
(c) Sheet moulding compound	0.1–0.15	U/B, A/C, U/b, a/c	a/t, a/K, a/T
Epoxides			
(a) Rigid	–	U/B (mineral), U/b, U/C, U/c	U/H, U/H, a/t, a,k
(b) Flexible	–	Increased attach com-pared with rigid	U/p
Nylons			
(a) Type 6	0.9–3.3	A/B, U/C, U/b, U/C	unaffected

(b) Type 6/6	0.4-1.5	A/B, U/C, U/b, U/c	unaffected
(c) Type 11	0.4	A/B, U/C, U/b, U/c	a/t
(d) Type 12	0.25	A/B, U/C, U/b, U/c	a/t

POLYMERIZATION TYPES

Polyethylenes			
(a) Low density	<0.015	A/B, U/C, U/b, U/c	S/T above 60°C
(b) High density	<0.01	A/B, U/C, U/b, U/c	unaffected below 80°C
Polypropylene	<0.01	A/B, U/C, U/b, U/c	unaffected below 80°C
Polystyrenes			
(a) Conventional	0.03-0.4	A/B, U/C, U/b, U/c	S/e, h, T, t, K
(b) Toughened	0.1-0.3	A/B, U/C, U/b, U/c	S/e, h, T, t, K
Styrene acrylonitrile	0.2-0.3	A/B, U/C, A/b, U/c	S/e, t, K
Acrylonitrle/butadiene/ styrene vinyl polymers	0.1-0.3	A/B, U/C, A/b, U/c	S/e, s/t, s/K
(a) Rigid Polyvinyl chloride	0.05	U/B, U/C, U/b, U/c	S/e, S/K, A/T
(b) Rigid vinyl chloride/ vinyl acetate	0.08	U/B, U/C, U/b, U/c	S/e, S/K, A/T
(c) Rubber modified PVC	0.1	U/B, U/C, U/b, U/c	S/e, S/K, A/T
Polyacetals	0.12	A/B, A/C variable bc	unaffected
Polycarbonates	0.3	a/b, a/c, U/b, U/c	unaffected
Ionomers	0.1-0.4	A/B, a/b, a/C, a/c	slightly attached
Polyphenylene oxide	0.07-0.08	U/B, U/C, U/b, U/c	a/c, a/t, u/p, U/t

Table 192(c) (continued)

Polyphenylene sulphide	0.05-0.03	-	-
Acrylics	0.3-0.4	A/B, A/C, a/b, a/c	S/K, S/e, S/t
Ethylene vinylacetate fluor-inated polymers	0.05-0.13	A/B, U/b, U/C, U/c	A/t, A/T
(a) Fluorinated ethylene propylene	0	U/B, U/C, U/b, U/c	unaffectd
(b) Polytetrafluorethylene	0	U/B, U/C, U/b, U/c	unaffected
(c) Polytrifluorochloroethylene	0	U/B, U/C, U/b, U/c	a/t in some cases
Cellulosic plastics			
(a) Cellulose acetate	1.9-6.5	d/B, d/C, a/b, a/c	S/K, S/e, S/H
(b) Cellulose acetate butyrate	0.9-2.2	d/B, d/C, a/b, a/c	S/K, S/e, S/H
(c) Cellulose propionate	1.2-2.8	d/B, d/C, a/b, a/c	S/K, S/e, S/H

KEY: Interpretation of this section, which acts as a guide to chemical and solvent resistance, note that considerations of temperature are all-important.

s = slightly soluble	d = decomposed	p = paraffins
S = soluble	h = higher alcohols	b = dilute acids
i = insoluble	H = alcohols	B = strong acids
U = unaffected	T = aromatics	c = dilute alkalies
a = slightly affected	C = strong alkalies	t = chlorinated hydro-carbons
A = Attacked	K = ketones	e = esters

APPENDIX 3

SUMMARY OF METHODS

ANTIOXIDANTS

PHENOLIC TYPE

Compounds	Polymer	Method	Reference
Halogenated biphenyls p-tert butyl phenol Ionox 330 Butylated hydroxy anisole Butylated hydroxy toluene 2,6 di tert butyl p-cresol 2,6 di tert butyl 4-methyl phenol	Polyethylene	Solvent extraction glc	2183-2196
Bis-(3,5-di-t-butyl-4-hydroxy phenol) methane	Polyethylene	Solvent extraction oxidation of alkaline solution and spectrophotometric evaluation	2197
Ionol	Polyethylene	Cyclohexane extraction, visible spectrophotometric finish (low levels)	2198
Phenolic antioxidants including di-cresylol propane and Santonox R	Polyethylene	Toluene extraction coupling with sulphanilic acid spectrophotometric finish (low levels)	2199

Phenolic antioxidants including Agerite Alba (hydroquinone mono benzyl ether), Agerite Spar (styrenated phenol), Agerite Superlite (polyalkyl polyphenol) Antioxidant 425, Antioxidant 2246, Deenax, Ionol, 1-Naphthol, 2-Naphthol, Naugawhite, (alkylated Phenol), Nevastain A, (not disclosed), Nevastain B, (not disclosed), Nonyl phenol, p-Phenyl phenol, Polygard, Santovar A, Santovar O, Santowhite crystals, Santowhite MK (reaction product of 6-tert-butyl-m-cresol and SCl_2), Santowhite powder, Solux, Stabilite white powder (not disclosed), Styphen 1 (Styrenated phenol), Wingstay S (Styrenated phenol), Wingstay T (a hindered phenol)	Polyethylene	Methanol or ethanol extract-coupling with diazotized p-nitroaniline-spectrophotometic finish 2200
Butylated hydroxytoluene	Polyethylene	Solvent extraction, spectrophotometric finish using 2,6 di-chloro-p-benzoquinone-4-chloromine 2201
Miscellaneous	Polyethylene	Solvent extraction, spectrophotometric finish using alpha alpha' beta pycryl-hydrazyl 2202,2203
Topanol OC Borox M	Polyethylene	Solvent extraction, thin-layer chromatography, then ultra violet spectroscopy 2204

Ionox 330	Polyethylene	Solvent extraction oxidation in PbO$_2$ or nickel peroxide - ultra violet spectroscopy in neutral and alkaline solution.	2204-2205
p-Methoxyphenol 4,4' methylene bis-(2,6-di-tert-butyl-phenol) and Santonox R	Polyethylene	Solvent extraction, ultraviolet spectroscopy (using bathochromic shift in sodium hydroxide solution)	2206
Ionol, Santonox R, Irganox 1010	Polyethylene	Chloroform extraction, ultraviolet spectroscopy	2207,2210
Ionol, Santonox R	Polyethylene	Ultraviolet spectroscopy solvent extraction - ultra violet spectroscopy solvent extraction	2211
Santowhite powder, Butylated hydroxy toluene Santonox R	Polyethylene	Chloroform extraction column chromatography, successive elution with chloroform (removes BHT) 1:10 aqueous methanol (removes Santowhite and Santonox R, eluates analysed by ultraviolet spectroscopy	2212
Butylated hydroxytoluene, Santonox R, Irganox 1076, CAO 14	Polyethylene	Freeze grinding, diethyl ether extraction, column chromatography, with refractive index and ultra-violet detectors at column outlet	2213

	Polyethylene	Solvent extraction, column chromatography	2214
-	Polyethylene	Solvent extraction, column chromatogrphy on silica gel using a range of solvents and various detectors.	2215-2217
Ionol, Santonox R	Polyethylene	Isooctane extraction – infrared spectroscopy	2218
-	Polyethylene	Solvent extraction, various finishes including spectrophotometry, NMR analysis and mass spectrometry	2219-2225
-	Polyethylene	Extraction with cloroform or hexane or diethyl ether or toluene or carbon disulphide-iso octane mixtures or water followed by various instrumental finishes	2226-2235
Agerite D	Polyethylene	Solvent extractions fluorescence method	2236
Ionol, Santonox R	Polyethylene	Hexane extraction, ethanol solution of residue examined by thin-layer chromatography	2237,2238
Ionol	Polyethylene	Solvent extraction, thin-layer chromatography	2239,2240

–	Polyethylene	Solvent extraction, thin-layer chromatography	2241–2245
–	Polyethylene	Solvent extraction, paper chromotography using chromogenic reagents	2246,2251
Irganox 1076 4,4' thio bis (6-t-butyl-m-cresol), Pentaerythritol tetrabis 3,5,di-t-butyl-4-hydroxyhydrocinn-amate, 2,2' methylene bis (4-methyl-6-t butyl phenol), Octadecyl (3,5-di-t-butyl-4-hydroxyphenyl) acetate, Ionol	Polyethylene	n-Heptane extraction, thin-layer chromotograpny	2252
2,6 di-tert-butyl-4-methylphenol	Polyethylene	Thermal analysis-gas chromatography	2253
2-t-butyl-4-methylphenol 2,6-di-t-butyl-4-methylphenol 2,2'-bis(4-methyl-6-butylphenyl) methane	Polyethylene	Solvent extraction, bromometric estimation	2254
Santonox R	Polyethylene	Solvent extraction, bromometric estimation	2255
Ionox 330	Polypropylene	Solvent extraction glc	2256–2257
–	Polypropylene	Solvent extraction glc	2258
2,6,di-t-butyl-p-cresol Laurylthiodipropionate 4,4' methylene bis (2,6 di-t-butyl-	Polypropylene	p-Xylene extraction glc	2259

phenol)
$2,2^1$ methylene bis(6-C^1-methylcylco-
hexyl)-p-cresol)
$4,4^1$ thio bis(6-t-butyl-m-cresol)
$4,4^1$ butylidene bis(6-t-butyl-m-cresol)
Ionox 330
Octadecyl 3,5 di-t-butyl 4-hydroxy-
hydrocinnamate
2-hydroxy 4(octyloxy)benzophenone
2-hydroxy-4-methoxybenzophenone
4-dodecyloxy-2-hydroxybenophenone
2-(2,4 benzotriazol-2 yl)-p-cresol
2,4 di-t-butyl-6-(5-chloro-2,4-
benzotriazol-2yl) phenol

Alkyl phenol and bis phenol anti-oxidants	Polypropylene	Volatile preconcentration mass spectrometry	2260
$4,4^1$butylidene antioxidants (2-t-butyl-5-methyl) phenol $4,4^1$thio bis(6-t-butyl-m-cresol) Pentoacryltritol tetra bis 3,5,di-t-butyl 4-hydroxyhydrocinnamate $2,2^1$ methylene bis(4 methyl-6-t-butyl phenol) Octadecyl (3,5 di-t-butyl-4-hydroxy-phenyl) acelate 2,6 di-t-butyl-p-cresol	Polypropylene	n-Heptane extraction, thin-layer chromatography for quantitative analysis	2261
p-tert-butyl phenol	Polycarbonate	solvent extraction, diazotization with p-nitroaniline	2262
Phenyl salicylate resorcinol benzoate	Rubbers and polyisoprenes	diethylether extraction, spectro-scopy	2263,2264

	Rubbers and polyisoprenes	acetone extraction, spectroscopy	2265

AMINE ANTIOXIDANTS

–	Polyethylene	Solvent extraction, thin-layer chromatography	2266–2269
2,6 di-t-butyl phenol, diphenyl-amine	Polyethylene	Solvent extraction glc	2194
Phenyl-2 napthylamine	Polyethylene	Solvent extraction glc	2270
Amine antioxidants e.g. Nonox CI	Polyethylene	Solvent extraction spectro-photometric finish using dipyridyl	2271
Succononx 19 Nonox CI	Polyethylene	Toluene extraction, methanol precipitation of polymer, spectrophotometric determin-ation after reaction with hydrogen peroxide	2272,2273
Phenyl-napthylamines	Polyethylene	Solvent extractions fluores-cence method	2274
N-phenyl-beta napthylamine	Polyethylene	Volatile preconcentration-mass spectrometry	2275
(N-phenyl-2 napthylamene) Subst'd p-phenylene Subst'd diamine type Subst'd Santoflex 13 sec heptyl phenyl-p-phenylene-diamine	Polyethylene	Ethanol extraction, glc	2276

Antioxidants, phenylnapthyl amines	Rubbers	Solvent extraction, fluorescence spectroscopy	2277
Amine, types of antioxidants and antiozonants	Rubbers	Solvent extraction, gel permeation chromatogrpahy using tetrahydrofuran dilution solvent	2278-2282
Antioxidant, plasticizers, accelerators	Rubbers	Carbon tetrachloride solution of sample put through column chromatography, ultraviolet monitoring of effluent successive elution with various solvents	2283
N-N'diethyl aniline, N-ethylaniline, diphenylamine, N-phenyl-2-napthylamine	Rubbers	Freeze grinding, diethyl ether extraction, column chromatography with refractive index and ultraviolet detectors	2284
Antioxidants	Rubbers	Solvent extraction glc	2285-2287
1,2 dihydroxy-2,2,4 trimethyl-6-ethoxy quinoline N-isopropyl-N'-phenyl-p-phenylenediamine N,phenyl-2-napthylamine N,N'di-2-octyl-p-phenylene diamine N,N' diphenyl-p-phenylene diamine	Rubbers	Acetone extraction glc	2285
Di-2-mapthyl-p-phenylene diamine Poly(trimethyl-dihydro-quinoline)	Rubbers	Solvent extraction glc	2285

AMINE AND PHENOLIC ANTIOXIDANTS

Topanol C C, Topanol CA, Polygard, Irganox 1010, Irganox 1076, Santonox R Annulex PBA15, Nonox DCP, Nonox-WSP, Ionox 330	Polyethylene	Toluene extraction glc	2288
Phenyl-2-napthylamine Santonox R, phenyl-2-naphthylamine	Polyethylene	Solvent extractions phosphorescence method	2236
Amine acid, phenolic types	Polyethylene	Solvent extraction-thin-layer chrcmatography for quantitative analysis	2289
Phenolic and amine antioxidants	Polyethylene	Solvent extraction, high performance liquid chromatography	2290
Phenolic and amine	Polyethylene	Solvent extraction - TLC	2240
Antioxidants, phenolic acid amine type	Polystyrene	Solvent, extraction, thin-layer chromatography for quantitiative analysis	2289
Amine and phenolic antioxidants	Rubbers and polyisoprene	Ethanolic hydrochloric acid extraction spectroscopy	2291
Amine and phenolic type	Vulcanizates	Solvent extraction, acetylation, glc	2292
Amine and phenolic type	Vulcanizates	Solvent extraction, TLC	2293,2294
Amine and phenolic types, N-phenyl-napthylamine, acylated diphenylamines 4,4' diemthyoxydiphenylamine	Vulcanizates	Solvent extraction, TLC for identification	2293

Amine and phenolic	Vulcanizates	Solvent extraction, paper paper chromatography	2295,2296
Phenolic and amine antioxidants	Vulcanizates		2297
Agerite range of antioxidants, Stabilite Alba Santoflex 66 Santoflex DD Santoflex 13 N-phenyl-beta-napthylamine Wingstay S 2,6,di-tertbutyl-p-cresol N,N' diphenyl-p-phenylene diamine	Vulcanizates	acetonitrile extraction – then mass spectrometry, infrared spectroscopy and NMR	2298

DIALKYLTHIODIPROPIONATE SECONDARY ANTOXIDANTS

Dilaurylthiodipropionate	Polyethylene	GPC	2237
Dilaurylthiodipropionate	Polyethylene	hydrolysis to lauryl alcohol extraction with chloroform-ethanol-hexane and thin-layer chromatography, then glc	2188,2299
Dilaurylthiodipropionate	Polyethylene	IR spectroscopy of film	2300
Dilaurylthiodipropionate	Polyethylene	Solvent extraction, polarography	2301
Dilaurylthiodipropionate	Polypropylene	Hydrolysis to lauryl alcohol chloroform-ethanol-alcohol-hexane extraction. Thin-layer chromatography then glc.	2302

Dilaurylthiodipropionate	Ethylene-vinyl acetate and acrylonitrile buta-diene-styrene	Hydrolysis to lauryl alcohol hydrolysis with chloroform-ethanol-hexane then TLC or GLC 2302

TRISNONYLATED PHENYL PHOSPHITE ANTIOXIDANTS

Trisnonylated phenyl phosphite	Polyethylene	Methanol or ethanol extrac-tion, coupling with diazotized p-nitroaniline spectrophoto-metry 2303
Trisnonylated phenyl phosphite	Rubbers, polyisoprene	Solvent extraction, hydrolysis to nonyl phenol, diazotization, spectrophotometry 2304
Trisnonylated phenyl phosphite	Styrene butadiene	Methanol or ethanol extraction coupling with diazotized-p-nitro-aniline-spectrophotometric finish 2305
Trisnonylated phenyl phosphite	Styrene butadiene	Solvent extraction, hydrolysis to nonyl phenol-diazotization and spectrophotometric evaluation 2306
Trisnonylated phenyl phosphite	Styrene butadiene	Solvent extraction addition of sodium hydroxide and measurement of bathochromic shift spectro-photometrically 2307

MULTIANTIOXIDANT METHODS

Antioxidants; uv absorbers, lubricants antistatic agents, optical brightners	Polyethylene	Solvent extraction glc 2308,2309

6 dodecyl-2,2,4-tri methyl-1,2 di-hydro-quinaline trisnonyl phenyl phosphate dilaurylthio dipropionate	Polyethylene	Hydrolysis to lauryl alcohol chloroform-ethanol-hexane extraction and thin-layer chromatography then glc	2302
Butylated hydroxy toluene "2246" Santoflex range, (hydroquinone type) Superlite, Polygard, Ionox 1076, 4'4 butylidene (2-tert butyl 5-methyl) phenol	Polyethylene	n-Heptane extrction - TLC	2310
Diorgano-sulphide, tert phosphite types, distearyl thiodipropionate dilauryl thiodipropionate, 4,4' thio bis (6-tert butyl-m-cresol), 1'1 thobis (2-napthol), triphenyl phosphite triethylphosphite tri-p-tolyl phophite tris(dinonyl phenyl) phosphite 2,2' thio bis (-6-tert-butyl-p-cresol) triisopropylphosphite	Polyethylene	Oxidative with m-chloro-peroxybenzoic acid, unreacted oxidant estimated iodimetrically	2311
Diorgano sulphides and tert phosphites e.g., distearylthio dipropionate, dilaurlythiodipropionate, triphenylphosphite, didecylphosphite	Polypropylene	Heptane extraction, oxidation with m-chloro-peroxy-benzoic acid, unreacted oxidant estimated iodimetrically	2312
Ionol, (2,6 di-t-butyl-p-cresol)	Polystyrene	Methylene dichloride solution glc	2313

| Optical brightner Tinuvin P (hydro-xybenzylbenzotriazole) | Polystyrene | Methylene dichloride solution glc | 2313 |
| Ionol, Tinuvin P | Polystyrene | Methylene dichloride solution glc | 2314 |

Key:

Ionox 30	1,3,3-trimethyl-2,4,6 tris (3,5 di-tert-butyl-4 hydroxy benzyl) benzene
Toponol OC	2,6 di-tert butyl 4-methyl phenol
Borox M	bis-(3,5 di-tert-butyl-4-hydroxy phenyl methane)
Antioxdant 425 & Antioxidant 2246	2,2' methylene-bis(6-tert-butyl-4-methyl phenol)
Deenox & Ionol	2,6 di-tert butyl-p-cresol
Santovar A	2,5 di-tert-butyl hydroquinone
Polygard	tris (nonylated phenol) phosphite
Santovar 0	2.5 di-tert butyl hydroquinone
Santowhite crystals	4.4'-thio-bis (6-tert-butyl-2-methyl phenol)
Santowhite powder	4,4' butylidene bis (-3-methyl-6-tert-butyl m-cresol)
Agerite D	Polymeric dihydroxy quinone
Irganox 1076	4,4' butylidene (2-tert butyl-5-methyl) phenol

Nonox CI	N N' di-2-naphthyl-p-phenylene diamine
Succononox 18	N-stearoyl-p-amino phenol
Topanol CA	1,1,3-tris(2-methyl-4-hydroxy 5-tert butyl(phenyl) butane
Organox 1010	Pentaerythriteol tetra-3(3,5 di-tert butyl-4-hydroxy phenyl propionate)
Annulex PBA15	1,1 di(3-tert butyl-4-hydroxy-6-methylphenol) butane
Nonox DCP	2,2 bis(3-methyl-4-hydroxy phenyl)-propane
Nonox WSP	bis(2-hydroxy-5-methyl-3-(methylcyclohexane)phenyl) methane
Stabilite Alba	di-o-tolyl ethylene diamine
Santoflex 66	n-phenyl-N'cyclohexyl-p-phenylene diamine
Santoflex DD	6-dodecyl-1,2 dihydro-2,2,4 trimethyl quinoline
Santoflex 13	N-Phenyl-beta-naphthalene
Wingstay S	Styrenated phenol

APPENDIX 4

SUMMARY OF METHODS

MONOMERS

POLYMERS

Monomer	Polymer	Method	Reference
Styrene, ethyl benzene	Polystyrene	Solution in o-dichlorobenzene or methylene dichloride, GLC	2315
Styrene, alpha-methyl styrene benzene, toluene, ethyl benzene, xylene, cumene, propyl benzenes, ethyl toluenes, butyl benzenes	Polystyrene	Solution in propylene oxide, GLC	2316,2317
Styrene	Polystyrene	Solution in THF, GLC	2318,2319
Styrene	Polystyrene	Solution in DMF, GLC	2319,2320
Styrene, acrylonitrile butadiene	Polystyrene	Solution in DMP, GLC	2320
Styrene	Polystyrene	Solution in benzene GLC	2321
alpha-Methylstyrene	Polystyrene	Solvent solution polarography	2322,2323
Styrene	Polystyrene	Solvent solution polarography	2322-2325

Styrene	Polystyrene	Chloroform extraction, methanol precipitation, ultraviolet spectroscopy	2326
Methyl methacrylate	Polymethacrylates	Polarography	2327,2328
Methyl methacrylate	Polymethacrylates	IR spectroscopy	2329
Methyl methacrylate	Polymethacrylates	glc	2330
Ethylacrylate styrene	Polyacrylates	Distilled in presence of toluene solution	2331
Methyl acrylate, ethyl acrylate	Polyacrylates	Solution in methylene chloride, glc	2331
2-ethyl hexyl acrylate, butyl acrylate, ethyl acrylate, vinyl propionate	Polyacrylates	Solution in solvent, glc	2333
2-ethyl hexyl acrylate, vinyl acetate	Polyacrylates	Solution in propyl acetate cylcohexanol, glc	2334
Butyl acrylate, methyl acrylate ethyl acrylate	Polyacrylates	Solution in isopropanol, glc	2335
Methyl methacrylate, ethyl acrylate styrene	Polyacrylates	Solution in solvent, glc	2336,2331
Vinyl acetate, 2-ethyl hexyl acrylate	Polyacrylates	Solution in propyl acetate, glc	2336

Ethyl acrylate styrene	Polyacrylates	Distilled in presence of toluene, glc	2331,2332
Vinyl chloride	PVC	Head space, glc	2337,2338
Vinyl chloride	PVC	glc-mass spectrometry	2339,2340
COPOLYMERS			
Divinylbenzene	Styrene-divinylbenzene	Pyrolysis-mass spectrometry	2341
Styrene	Polyester laminates, styrene butadiene acrylics	Polarorographic	2342
Styrene & butadiene	Styrene butadiene	Head space analysis and gas chromatography of IMF solution	2343,2344
Benzene & toluene	Rubber adhesives	Gas chromatography	2345
Styrene	Latex	Gas chromatography	2346,2347
Styrene, dialkyl phthalate, methyl	Polyester resin mouldings	Gas chromatography	2348,2349
Styrene ethyl acrylate	Styrene ethyl acrylate	Radioactive tracer methods	2350,2351
Styrene, methacrylic acid	Styrene methacrylic acid	Mercurimetric	2352,2353
Ethylene glycol, dimethyl terephthalate, dimethyl isophthalate, dimethyl adipate	Isophthalic acid, terephthalic acid, ethylene glycol	Gas chromotography	2354

Ethyl acrylate, styrene, vinyl acetate	Styrene-ethyl acrylate	Distill in presence of toluene, glc	2355
Methyl acrylate, ethyl acrylate, 2-ethylhexyl acrylate, butyl acrylate, ethyl acrylate, vinyl propionate	Mixed polyacrylates	Solution, glc	2356,2357
2-ethylhexyl acrylate, vinyl acetate, butyl acrylate, methyl acrylate, methacrylic acid	Mixed polyacrylates	Solution in propyl acetate cyclohexanol, glc, solution in isopropanol, glc	2358,2359
Methyl methacrylate, ethyl acrylate styrene	Styrene acrylate and styrene methacrylate	Solution, glc	2360
Vinyl acetate, 2-ethyl hexyl acrylate	Vinylacetate-2-ethyl-hexyl acrylate	Solution in propyl acetate, glc	2356
Styrene butadiene, alpha methyl-styrene acrylonitrile	Styrene butadiene & acrylonitrile-butadiene-styrene	Head space glc	2361
Acrylonitrile	Styrene-acrylonitrile	Polarography	2362
Styrene acrylonitrile	Styrene-acrylonitrile	Solution in dimethyl-formamide polarography	2363,2365
Styrene acrylonitrile butadiene	Acrylonitrile-butadiene-styrene & styrene acrylo-nitrile	Solution in dimethyl-formamide, glc	2354,2366
Styrene acrylonitrile	Styrene acrylonitrile	Brommetric titration	2367

REFERENCES

1   Squirrel, D.C.M, Analyst (London), 106, 1042 (1981).
2   Cambell, R.H, Wize, R.W., J. Chromatography, 12, 178 (1963).
3   Slonaker, D.F., Sievers, D.C., Anal. Chem., 36, 1130 (1964).
4   Van der Heide, R.F., Wouters, O., Lebensm Untersuch Forsch, 117
     129 (1982).
5   Schroder, E., Rudolph, G., Plaste Kautschuk, 10, 22 (1963).
6   Metcalf, K., Tomlinson, R., Plastics (London), 25, 319 (1960).
7   Yushkevichyute, S.S., Shlyapnikov, Y.U.A., Plastichskie Massy, 1,
     54 (1967)
8   Stafford, C., Plasticheskie Massy, 34, 794 (1962).
9   Zilio-Grande, F., Libralesso, G., Sassu, G., Svegidao, G., Mater
     Plast Elast, 30, 643 (1964).
10   Squirrel, D.C.M., Analyst (London), 106, 1042 (1981).
11   Haslam, J., Soppet, W.W., J. Chem. Soc., 67, 33 (1948).
12   Doering, H., Kunststoffe, 28, 330 (1938).
13   Thinius, K., Analystisch Chemie der Plaster, Springer, Berlin
     (1952).
14   Haslam, J., Squirrel, D.C.M., Analyst (London), 80, 871 (1955).
15   Wake, W.C., The Analysis of Rubber and Rubber like Polymers,
     McLaren, London, (1958).
16   Clark, A.D., Bazill, G., British Plastics, 31, 16 (1958).
17   Korn, D., Waggon, H., Plaste Kautechuk, 11, 278 (1964).
18   Myers, LW., Lord, E.B., Shell Chemical Co., Carrington, U.K.,
     private communication.
19   Eggerston, F.T., Dimba, M., Stross, F.H., Shell Chemical Co.,
     Emeryville, California, private communication.
20   How-Kissin, Y.V., Gol'dfarb, Y.Y., Krentsel, B.A., Vysokmol. Soedin.
     Ser A, 14, 2229 (1972).
21   Bakuyutov, N.K., Nissin, Y.V., Yu, V., Vasilova, I., Arkhipova,
     Z.V., Vusokimol. Soedin. Ser A, 17, 2163 (1975).
22   Kornet, S., Plasty, A., Kamcuk, 13, 176 (1976).
23   Urbanski, J., Handbook Analytical Synthesis Polymer Plastics 372-
     87, Ellis Horwood Ltd., Chichester, U.K., (1977).
24   Bird, V., Plastics Technology, 40, September (1963).
25   Mano, E.B., Anal. Chem., 32, 291 (1960).
26   Winterscheidt, H. & Seifen, O., Fettes Wachse, 80, 239 71 (1954);
     81, 16 (1965).
27   Hinteraier, A., Fette, U., Seifen, 51, 367 (1944).
28   Hummel, D., Kunstostoff, Rundschau, 5, 85 (1958).
29   Samsel, E.P, Anal. Chem., 29, 574 (1957).
30   Liebermann, C., Chem. Berichte, 17, 1884 (1984).
31   Morawski, T., Chemiker Zeitung, 12, 1321 (1988).
32   Storch, L., Ber Osterruch Ges. 3 Forderung D. Chem. Ind., 9, 93
     (1887).
33   Kappelmeier, C.P.A., Farber. Zeitung, 40, 1141 (1935); 41, 161
     (1936); 42, 561 (1937).

34   Moir, S., Iu Mateillo, J.J., Protect. and Decorat. Coatings Bd. v.
      105, J Wiley & Sons, New York (1956).
35   Gibbs, H.D. & Matiello, J.J., Protect. and Decorat. Coatings, $\underline{v}$,
      105, J. Wiley & Sons, New York (1956).
36   Swann, M.H., and Weil, D.C., Anal. Chem., $\underline{28}$ 1463 (1956).
37   Berl, E.U.K., hefter. Cellulosechemie, $\underline{14}$, 67 (1933).
38   Foucry, M.J., Peintures, Pigments, Verniz, $\underline{30}$, 925 (1954).
39   Rudd, H.W. & Zonsveld, J.J., J. Oil Col. Chem. Assoc., $\underline{39}$, 314
      (1956).
40   Swann, M.H. & Esposito, G.G., Anal. Chem., $\underline{28}$, 1006 (1956).
41   Gunter, F.A., Mitarb, U., Anal. Chem., $\underline{23}$, 1835 (1951).
42   Lohmann, H., J. Prakt. Chem., $\underline{135}$, 57 (1939).
43   Liebermann, C., Chem. Berichte., $\underline{17}$, 1884 (1884).
44   Morawski, T., Chamber. Zeitung, $\underline{12}$, 1321 (1888).
45   Mano, E.b., Anal. Chem., $\underline{32}$, 291 (1960).
46   Heermann, P., in Agster, A. Barberei u textil-chem.,
      Untersuchungen Berlin - Gottinger-Heidelbuerg., Springer, S-139
      (1956).
47   Freytag, H., Diese, S., $\underline{145}$, 24 (1955).
48   Hummel, D., Kunstostoff Rundshau, $\underline{5}$, 85 (1958).
49   Weschsler, H., J. Polymer Sci., $\underline{11}$, 233 (1953).
50   Cook, W.S., Jones, C.O. & Altenau, M.G., Canadian Spectroscopy,
      $\underline{13}$, 64 71 (1968).
51   Houk, W.W. & Solverman, L., Anal. Chem., $\underline{31}$, 1069 (1959).
52   Mitchell, B.J. & O'Hear, H.J., Anal. Chem., $\underline{34}$, 1621 (1962).
53   Bergman, J.S., Ehrhart, C.H., Granatelli, L. & Janck, J.L., 153rd
      National American Chemical Society Mtg., Miami Beach, Florida,
      April (1967).
54   Haslam, J., Squirrel, D.C.M., Analyst (London), $\underline{86}$, 239 (1961).
55   Falcon, J.Z., Lowe, J.L., Gacta, L.J. & Altenan, A.G.G., Anal.
      Chem., $\underline{47}$, 171 (1975).
56   Colson, A.F., Analyst (London), $\underline{88}$, 26 (1963).
57   Colson, A.F., Analyst (London), $\underline{88}$, 791 (1965).
58   Colson, A.F., Analyst (London), $\underline{90}$, 35 (1965).
60   Johnson, C.A. & Leonard, H.A., Analyst (London), $\underline{86}$, 101 (1961).
61   Yoshizaki, T., Anal. Chem., $\underline{35}$ 2177 (1963).
62   Colson, A.F., Analyst (London), $\underline{35}$ January (1969).
63   Mittenberger, W.D. & Gross, H., Kunstsofftecknick, $\underline{12}$ (7), 176;
      (8) 219; (9) 281; (10) 277; (11) 303; (1973).
64   Tanaka, Y. & Morikawa, T., Kagaku to Kogyo (Osaka), $\underline{48}$, (10) 387
      (1974); Chem. Abstr. $\underline{82}$, 98702w.
65   Johnson, C.A., Leonard, H.A., Analyst (London), $\underline{86}$, 101 (1961).
66   Colson, A.S., Analyst (London), Oct 791 (1963).
67   Bosch, K., Beitr Gericht. Med., $\underline{33}$, 280 (1975).
68   Jones, C.E.R. & Moyles, A.F., Nature (London), $\underline{191}$, 663 (1961).
69   Barlow, A.R., Lehrle, S. & Robb, J.C., Polymer, $\underline{2}$, 27 (1961).
70   Cogliane, J.a., Rev. Sci. Instr., $\underline{34}$, 439 (1963).
71   Dalnogare, S., Paper presented at 147th Meeting American Chemical
      Society, Philadelphia. Abstract of papers p. 14B (1964).
72   Giacobbo, H. & Simon, W., Pharm. Acta. Helv., $\underline{39}$, 162 (1964).
73   May, R.W., Pearson, E.F., Proter, J. & Scothern, M.D., Analyst
      (London), $\underline{98}$, 364 (1973).
74   May, R.W., Pearson, E.F., Scothern, M.D., Pyrolysis Gas
      Chromatography, Analytical Sciences Monography, SAC, London.
75   Voigt, J., Kunststoffe, $\underline{54}$, 2 (1964).
76   Voigt, J., Kunststoffe, $\underline{51}$, 18 (1961).
77   Voigt, J., Kunststoffe, $\underline{51}$, 314 (1961).
78   Barlow, A.K., Lehrler & Robb, J.C., Polymer, $\underline{2}$, 27 (1961).
79   Voigt, J., Kunststoffe, $\underline{54}$, 2 (1964).
80   Strassburger, J., Brauer, G.M., Tryon, M. & Farziati, A.F., Anal.
      Chem., $\underline{32}$, 454 (1960).

81   Bombaugh, K.J., Cook, C.E. & Clampitt, B.H., Anal. Chem., 35, 1834 (1963).
82   Nelson, D.F., Yee, J.L. & Kirk, P.L., Microchem. J., 6 225 (1962).
83   Groten, B., Anal. Chem., 36, 1206 (1964).
84   Lehman, F.F. and Branch, G.M., Anal. Chem., 33, 673 (1961).
85   Wampler, T.P., Levy, E.J., Analyst (London), 111, 1065 (1986).
86   Cox, B.C. & Ellis, B., Anal. Chem., 36, 90 (1964).
87   Kolb, A., Kaiser, K.H., J. Gas Chromatography, 2, 233 (1964).
88   Cieplinski, E.W., Ettre, L.S., Kolb, B. and Keinmer, G., Anal. Chem., 705 357 (1964).
89   Wampler, T.P. & Levy, E.J., Analyst (London), 111 1065 (1986).
90   Folmer, O.F., Anal. Chem., 43, 1057 (1971).
91   Folmer, O.F. & Azarranga, L.V., J. Chromatogr. Sci., 7, 665 (1969).
92   Juvet, R.S., Smith, J.L.S. and Pang Li, K., Anal. Chem., 44, 49 (1972).
93   Hughes, J.C., Wheals, B.B. & Whitehouse, M.J., Analyst (London), 102 143 (1977).
94   Menzelaar, H.L.C., Posthumus, M.A., Kistemaker, P.G. & Kiskinaker, J., Anal. Chem., 45, 1546 (1973).
95   Jackson, M.T. & Walker, J.Q., Anal. Chem., 43, 74 (1971).
96   Mattern, D.E., Tyan Lin, F., Hercules, D.M., Anal. Chem., 56, 2762 (1984).
97   Chin-An Hu, J., Anal. Chem., 53, 942 (1981).
98   Razinger, M., Penca, M., Zapan, J., Anal. Chem., 53, 1107 (1981).
99   Visappa, A., Tekn, Kemian Aikakauslechti, 19, (1962), 121-137; The State Institute for Technical Research, Report Series IV - Chemistry 42, (1962); Perkin Elmer Application & New Bulletin 1963 no. 1, 1-8.
100   Deschant, J., "Ultraospektroskopische Untermethungen on Polymerern", Akadamie Verlag, Berlin, 516 pp (1972).

101   Hippe, Z. & Kerste, A., Bull. Akad. Pol. Sci. Ser. Sci. Chim., 21, 395 (1973).
102   Leukoth, G., Gummi. Asbest. Kumst., 27, 794 (1974).
103   Alexander, R., The Analytical Report, Perkin-Elmer Ltd., Beaconsfield, Bucks, U.K. No 7 February (1985).
104   Osland, R., Laboratory Practice, 37, 73 (1971).
105   Siewierski, S., and Koziowski, W., Polimery, 18, 221 (1973).
106   Gardella, J.A. & Grobe, G.L., Anal. Chem., 56, 1169 (1984).
107   Gabbay, S.M., Stivala, S.S., Polymer, 17, 121 (1976).
108   Quernum, B.M., Berticat, P., Vallet, G., Polymer Journal, 7, 277 (1975).
109   Stanescu, G., Revista de Chemie, Bucharest, 64, 42 (1963).
110   Sloane, H.J, Johns, T., Ulrich, W.F, Cadman, W.J., Applied Spectroscopy, 19, 130 (1965).
111   Yoshini, T., Shimoniya, M.J., Polymer Science A3, 2811 (1965).
112   Aliev, S.M., Bairomov, M.R., Azizov, A.G., Aliev, S.A., Akhmedov, S.T., Azerb Khim Zl., 3,70 (1976).
113   Forette, J.E., Pozek, H.L., J. Appl. Polymer Science, 18, 2973 (1974).
114   Aslanova, R., Yul'chivaev, A.A., Usmanov, K.U., Uzb Khim Zl., 17, 31 (1973).
115   Chia, K.S., Chen, G.M.S., Chin, J., Chem. Soc. Taipei, 20, 241 (1973).
116   Mirabella, F.M., Barrall, E.M., Johnson, J.F., Polymer, 17 17 (1976).
117   Dawkins, J.V., Hemming, M.J., Appl. Polymer Science, 19, 3107 (1975).
118   Littke, W.H., Flebec, W., Schmote, R., Kimmer, B.W., Faserforsh Textiltech, 26, 503 (1975).

119   Brako, F.D., Wexler, A.S., Anal. Chem., 35 1944 (1963).
120   Yoshino, T., Shinomiya, M., J. Polymer Science A, 3, 2811 (1965).
121   Sloane, H.J., Bramstone-Cooke. R., Applied Spectroscopy 27, 217
       (1973).
122   Schrader, B., Angew. Chem. Internat. Ed., 12, No 11, 884 (1973).
123   Boerlo, F.J., Keonig, J.L., Polym. Character., Interdisciplinary
       Appraches, Proc. Symp. 1970 (Pub 1971), 1-13 Ed. Craver, C.
       Plenum, New York, N.Y.
124   Boerio, F.J., Koenig, J.L., J. Polym. Sci., Symp. No. 43, 205
       (1973).
125   Keonig, J.L., Chem. Technol., 2, 411 (1972).
126   Slovokhotova, N.A., Metody Ispyt., Kontr. Issled. Madhinostroit.
       Mater. 3, 41 (1973).
127   Schutte, C.J.H., Fortschr. Chem. Forsch., 36, 57 (1972).
128   Zerbi, G., Chim. Ind. (Milan), 55, 334 (1973).
129   Kumpanenko, I.V., Kazanskil, K.S., Usp. Khim. Fiz. Polim., 1973,
       64.
130   Andrews, R.D., Hart, T.R., Charact. Met. Polm. Surf., (Symp.)
       (1976) (Pub.1977), 2,207-40; Lee, L.H., Ed.: Academic: New York.
131   Chauser, M.G., Ermakova, V.D., Mishenko, O.B., Gunova, V.S.,
       Cherkashin, M.I., Arm. Khim. Zh., 26, 608 (1973).
132   Keonig, J.L. Proc. Polym. Charact. Conf. 1974 (Pub. 1975), 73-93;
       Cleveland State Univ: Cleveland Ohio.
133   Chalmers, J.M., Polymer, 18, 681 (1977).
134   Crompton, T.R., European Polymer Journal, 4, 473 (1968).
135   Lorenz, O., Scherle, W., Dummer, W., Kautschuk Gummi, 7, 273
       (1954).
136   Cornish, P.J., J. Appl. Polymer Sci., 7, 727 (1963).
137   Fiorenza, A., Bonomi, G., Saredi, A., Mater. Plast. Elast., 31,
       1045 (1965).
138   Auler, H., Asbest Kunststoffe, 14, 1024 (1961).
139   Kawaguchi, T., Ueda, K., Koga, A., J. Soc. Rubber Ord. (Japan),
       28, 525 (1955).
140   Ruddle, L.H., Wilson, J.R., Analyst (London), 94, 105 (1969).
141   Hilton, C.L., Rubber Age, 84, 263 (1958).
142   Burchfield, H.P., July, J.N., Anal. Chem., 19, 383 (1960).
143   Cielesky, C., Nagy, F.Z., Lebensm Unter-Forsch, 114 13 (1961).
144   Spell, H.L. & Eddy, R.D., Anal. Chem., 32, 1811 (1960).
145   Miller, R.G.J. & Willis, H.A., Spectochim. Acta., 14, 119 (1959).
146   Guichon, G. & Henniker, J., British Plastics, 37, 74 (1964).
147   Luongo, J.P., Appl. Spectry., 4, 117 (1965).
148   Drushel, H.V. and Sommers, A.L., Anal. Chem., 36, 836 (1964).
149   Van der Heide, R.F., Wouters, O., Lebensum Untersuch Forsch, 117,
       129 (1962).
150   Slonaker, D.F., Sievers, D.C., Anal. Chem., 36 1130 (1964).
151   Waggon, H., Jehle, D., Die Nahrung, 4 495 (1965).
152   Schroder, E., Hagen, E., Plaste Kautshuk, 15, 625 (1968).
153   Dobies, R.S., J. Chromatography, 40 110 (1969).
154   Roberts, C.B., Swank, J.D., Anal. Chem., 36, 271 (1964).
155   Nosikov, Yu.D., Vetchinkina, U.N. Neftekhimya, 5, 284 (1965).
156   Helf, C., Bockwan, D., Plaste Kautschuk, 11, 624 (1964).
157   Pocaro, P.J., Anal. Chem., 36 1664 (1964).
158   Freedman, R.W., Charlier, G.O., Anal. Chem., 36, 1880 (1964).
159   Duval., A.H., Tully, W.F., J. Chromatog., 11, 38 (1963).
160   Brooks, V.T., Chem. Ind., 1317 (1959).
161   Knight, H.S., Siegel, H., Anal. Chem., 38, 1221 (1966).
162   Grant, D.W., Vaughan, G.H., Gas Chromatography, P.305.
       Butterworths, London (1962).
163   Long, R.E., Guvernator, G.C., Anal. Chem., 39, 1493 (1967).

164   Roberts, C.B., Swank, J.D., Anal. Chem., _36_, 271 (1964).

165   Schroder, E., Rudolph, G., Plaste Kautschuk, _1_, 22 (1963).

166   Buttery, R.G., Stuckey, G.N., J. Agr. Food Chem., _9_, 283 (1961).

167   Styskin, E.L., Gurvich, Ya.A., Kumack, S.T., Khim Prom., _5_, 359
       (1973); Zavod Lab.; _39_, 27 (1973).

168   Schabon, J.F. & Fenska, L.E., Anal. Chem., _52_, 1411 (1980).

169   Majors, R.E., J. Chromotographic Science, _8_, 338 (1970).

170   Luongo, J.P, Appl. Spectrosc., _19_, 117 (1965).

171   Miller, R.G.J., Willis, H.A., Spectrochim Acta, _14_ 119(1959).

172   Guichon, G., Henniker, J. Brit. Plast., _37_ 74 (1964).

173   Drushel, H.V., Sommers, A.L., Anal. Chem., _36_, 836 (1964).

174   Albarino, R.V., Applied Spectroscopy, _27_, 47 (1973).

175   Wexler, A.S., Anal. Chem., _35_, 1926 (1963).

176   Soncek, J. & Jelinkova, E., Analyst (London), _107_, 623 (1982).

177   Ruddle, L.H. and Wilson, J.R., Analyst (London), _94_ 105 (1969).

178   Knight, H.S., Siegel, H., Anal. Chem., _38_ 1221 (1966).

179   Budyina, V.V., Marinin, V.G. & Vodzinsku, Yu V., Zavod Lab., _36_,
       1051 (1970).

180   Hilton, C.L., Anal. Chem., _32_, 383 (1960).

181   British Standard 2782, Part 4, Method 405B (1965).

182   Crompton, T.R., Appl. Polymer Science, _6_ 538 (1962).

183   Majors, R.E., J. Chromatographic Science, _8_, 338 (1970).

184   Gaeta, L.J., Schleuter, E.M., Altenau, H.G., Rubber Age, 101, 47
       (1968).

185   Wize, R.W., Sullivan, A.B., Rubber Chemistry and Technology, _335_
       No. 3, July-Sept (1962).

186   Wize, R.W. Sullivan, A.B., Rubber Chemistry Technology, _35_, 684
       July/Sept (1962).

187   Wize, R.W., Sullivan, A.B., Rubber Age, _91_, 773 (1962).

188   Sawada, M., yamagi, I and Yamashina, T., J. Soc. Rubber Ind.
       (Japan), _35_, 284 (1962).

189   Glavind, J., Acta. Chem. Scand., _17_ 1635 (1963).

190   Kabota, T., Kuribayashi, S. and Furuhama, T.S., Soc. Rubber Ind.
       (Japan), _35_ 662 (1962).

191   Mayer, H., Deut, Lebensm Rundshau, _57_, 170 (1961).

192   Glavind, J. Acta Chem. Scand, _17_ 1635 (1963).

193   Blois, M.S., Nature (London), _181_, 1199 (1958).

194   Hilton, C.L., Rubber Age, _84_ 263 (1958).

195   Hilton, C.L., Rubber Chem. Technol., _32_ 844 (1959).

196   Metcalf, K., Tomlinson, R.F., Plastics, _25_ 319 (1960).

197   Schroder, E. & Rudolph, G., Plaste Kautschuk, _10_ 22 (1963).

198   British Standard 2782, Part 4, Method 405D (1965).

199   Hilton, C.L., Anal. Chem., _32_, 383 (1960).

200   Parker, C.A., Barnes, W.J., Analyst (London), _82_, 606 (1957).

201   Parker, C.A., Anal. Chem., _37_ 140 (1961).

202   Drushell, H.V., Sommers, A.L., Anal. Chem., _36_, 836 (1964).

203   Clark, A.D., Bazill, G., British Plastics, _31_ 16 (1958).

204   Simpson, D. & Currell, B.R., Analyst (London), _96_ 515 (1971).

205   Slonaker, D.F. & Sievers, D.C., Anal. Chem., _36_ 1130 (1964).

206   Neubert, G., Z. Anal. Chem., _203_ 265 (1964).

207   Van der Heide, R.F. & Wouters, O., Lebensun Untersuch Forsch, _117_,
       129 (1962).

208   Van der Heide, R.F., Z. Lebensum Untersuch Forsch., _124_, 198
       (1964).

209   Davidek, J., & Pokorney, J., Rev. Univ. Ind. Santander., _4_ 11
       (1962).

210   Wandel, M. and Tengler, H., Felte. Seifen. Anstrichmittel., _66_ 815
       (1964).

211   Ishikaway, S. and Katsun, G, Bitamin, _30_, 203 (1964).

212   Davidek, J., Pokorney, J., Z. Lebensum Untersuch Forsch., 115 113
       (1961).
213   Waggon, H., Korn, O. and Jehle, D., Nagrung, 9 495 (1965).
214   Cook, C.D., Nash, N.G. & Flanagan, H.R., J. Amer. Chem. Soci., 75
       6242 (1953).
215   Yuasa, T. and Kamiya, K., Japan Analyst, 13 966 (1964).
216   Van der Heide, R.F., Maagdenberg, A.C. and Van der Neut, J.H.,
       Chem Weekblad, 61, 440 (1965).
217   Copius-Peereboom, J.W., Nature (London), 204 748 (1964).
218   Mocker, F., Kautschuk Gummi, 11, 1161 (1964).
219   Mocker, F., Kautschuk Gummi, 12, 155 (1959).
220   Mocker, F., Kautschuk Gummi, 13, 91 (1960).
221   Mocker, F., Old, J., Kautschuk Gummi, 12, 190 (1959).
222   Adams, R.N., Rev. Polarog., 11, 71 (1963).
223   Zweig, A., Lancaster, E., Neglia, M.T., Jura, W.H., Amer. Chem.
       Soc., 86, 413 (1964).
224   Vodzinskii, Yu, Semchikova, G.S., T. Po Khim i Khim Teknol., 272,
       (1963).
225   Vermillion, F.J., Pearl, T.A., J Electrochem. Soc., 111 1392
       (1964).
226   Barendrecht, E., Anal. Chim. Acta, 24, 498 (1961).
227   Gaylor, V.F., Elving, P.J., Contrad, A.L., Anal. Chem., 25, 1078
       (1953).
228   Gaylor, V.F., Conrad, A.L., Landerl, J.H., Anal. Chem., 29, 228
       (1957).
229   Gaylor, V.F., Conrad, A.F., Landerl, J.H., Anal. Chem., 29, 224
       (1957).
230   Hedenberg, J.F., Freiser, H., Anal. Chem., 25, 1355 (1953).
231   Hamilton J.W., Tappel, A.L., J. Amer. Oil Chem. Soc., 40, 52
       (1963).
232   Analytical Applications Report No 637D, Southern Analytical,
       Camberley, Surrey, U.K.
233   Analytical Applications Report No 651/2, Southern Analytical Ltd.,
       Camberely, Surrey, U.K.
234   Yamaji, O., Yamashina, T., J. Soc. Rubber Ind. (Japan), 35, 284
       (1962).
235   Sawada, M., Yamagi, I., Yamashina, T., J Soc. Rubber Ind. (Japan),
       35, 284 (1962).
236   Vasil'eva, A.A., Vodzinskii, Yu.V., Korshunov, I.A., Zavod. Lab.,
       34 1304 (1968).
237   Budke, C., Bannerjee, D.K., Miller, G.D., Anal. Chem., 36, 523
       (1964).
238   Ward, G.A., Talanta, 10, 261 (1963).
239   Elving, P.J., Krivis, A.F., Anal. Chem., 30, 1654 (1958).
240   Elving, P.J., Krivis, A.F., Anal. Chem., 30 1648 (1958).
241   Lintner, C.J., Schleif, R.H., Higuchi, T., Anal. Chem., 22 534
       (1950).
242   Simpson, D. & Currell, B.R., Analyst, 96 515 (1971).
243   Millingen, M.B., Anal. Chem., 46 746 (1974).
244   Gaeta, L.J. & Scheluter, E.W. & Altenau, A.G., Rubber Age., 47 101
       (1969).
245   Budyina, V.V., Marinin, V.G., Vodzinskii, Yu.V., Kalinina, A.I.,
       Korschunov, I.A., Zavod Lab., 36 1051 (1970).
246   Sedlar, F., Feniskova, E., Pac, J., Analyst (London), 99, 50
       (1974).
247   Kellum, G.E., Anal. Chem., 43, 1843 (1971).
248   Nawakowski, A.C., Anal. Chem., 30 1868 (1958).
249   Hilton, C.L., Anal. Chem., 32, 383 (1960).
250   Brandt, B.J., Anal. Chem., 33, 1390 (1961).

251   Robertson, M.M., Rowley, R.H., British Plastics, 26 January
       (1960).
252   Haslam, J., Soppet, W.W., J. Chem. Soc., 67, 33 (1948).
253   Doebring, H., Kunststoffe, 28, 230 (1938).
254   Thinius, K., Analytische Chemie der Plaste, Springer, Berlin
       (1952).
255   Haslam, J., Squirrel, D.C.M., Analyst (London), 80, 871 (1955).
256   Wake, W.C., The Analysis of Rubber and Rubber like Polymers,
       McLaren, London, 1958.
257   Clark, A.D., Bazill, G., British Plastics, 31 16/1958.
258   Krisher, A., Anal. Chem., 43 1130 (1971).
259   Majors, R.E., J. Chromatographic Science, 8 338 (1970).
260   Mansfield, P.B., Chemy. Ind., 28 792 (1971).
261   Haslam, J. and Soppet, W.W., J. Chem. Soc., 67 33 (1948).
262   Dahring, H., Kunststoffe, 28 230 (1938).
263   Robertson, M.W. & Rowley, R.M., British Plastics, January 26
       (1960).
264   Haslam, J. & Squirrell, D.C.M., Analyst, 8 8/1 (1955).
265   Wake, W.C., The Analysis of Rubber and Rubber-like Polymers,
       McLaren, London (1958).
266   Hasse, H., Kautl. Gummi Kunstoffe, 20 501 (1967).
267   Schroeder, E. & Hagen, E., Plaste Kautsch., 15 625 (1968).
268   Zulake, J. & Cuichon, G., Revista de Plastices Modernos, 13 13
       (1963).
269   Zulake, J., Landault, C. & Guichon, G., Bull. Soc. Chim. France,
       1294, (1962).
270   Cook, C.D., Elgood, E.J. and Solomon, D.H., Anal. Chem., 34 1177
       (1962).
271   Veres, L., Kunststoffe, 59 13 (1969).
272   Veres, L., Kunststoffe, 59 241 (1969).
273   Turnstall, F.I.H., Anal. Chem., 42 542 (1970).
274   Zulaica, J. & Guichon, G., Anal. Chem., 35 1724 (1963).
275   Yoshita, T., Toyoda Gosei Giho, 17 64 (1975); Chem. Abstr., 88
       170952w (1978).
276   Barla, F., Muayag Gumi, 14, 330 (1977). Chem. Abstr., 88 1705955z
       (1978).
277   Takada, T., Fukui-ken Kogyo Shikenjo Nempo., 49 66 (1974); Chem.
       Abstr., 87 118636m (1977).
278   Takada, T., Fukui-ken Kogyo Shikenjo Nempo, 50 77 (1975); Chem.
       Abstra., 87 118638m (1977).
279   Krishen, A., Anal. Chem., 43 1130 (1971).
280   Guicon, G. and Henniker, J., British Plastics, 37 74 (1964).
281   Stuerurle, H. & Pfab, W., Deutsch Lebensmittel Rdach., 65 113
       (1964).
282   Peereboom, J.W.C., J. Chromatography, 3 323 (1960).
283   Von Nagy, F.Z., Lebensmittel Unterschung and Forschung, Part 4,
       126 (1965).
284   Ligotti, I., Piazantini, R. & Bonomi, G., Hass. Chim. No. 1. 16
       (1964).
285   Von Braun, D., Chimia, 19 February (1965).
286   Jaminel, F., I.C. Formaco Edizons Practica, Anno XVIII No. 12
       December (1963).
287   Burns, W., J. Appl. Chem., 5 599 (1955).
288   Schevechenko, Z.A. and Favorskaya, I.A., Vestu. Lenigr. Univ., 19
       2 (1964).
289   Ruddle, L.H., Swift, S.D., Udris, J. & Arnold, P.E. in P.W.
       Shallis (Editor) "Proceedings of the SAC Conference, Nottingham
       1955", W. Heffer & Sons Ltd., Cambridge Press.
290   Walker, W.H. & Gaenshirt, K.H., Z. Analyt. Chem., 267 127 (1973).

291   Veres, L., Kunststoffe, _59_ 13 (1969).
292   Veres, L., Kunstostoffe, _59_ 241 (1969).
293   Cambell, G.E., Foxton, A.A., & Wordsall, R.L., Lab. Pract., _19E_ 369 (1970).
294   Stuele, W. & Pfab, W., Deutsch Lebensmittel Rundschau, Sch., _65_ 113 (1969).
295   Burns, W., Polymer J. Singapore, _3_ 58 (1971).
296   Von Braun, D. Kunststoffe, _52_ 2 (1962).
297   Gomaryova, A., Elekriviz. Kahl. Tech., _23_ 207 (1970).
298   Sokolowska, R., Roczn. Panst. Zakl. Hig., _595_ 20 (1969).
299   Kreiner, J.G., J. Chromatography, _75_ 271 (1973).
300   Mazur, H., Rocz. Panst. Zakl. Hig., _23_ 263 (1972).
301   Kula, H., Rocz. Panst. Zakl. Hig., _20_ 307 (1969).
302   Guichon, G. & Henniker, J., British Plastics, February 14 (1974).
303   Udris, J., Analyst, _96_ 130 (1971).
304   Robertson, M.W. & Rowley, R.M., British Plastics, January 26 (1960).
305   Mansfeld, P.B., Chemy. Ind., _28_ 792 (1971).
306   Hutibese, R.J. & Latz, H.W., J. Agric. Food Chem., _18_ 377 (1970).
307   Zilio-Grandi, F., Libraesso, J., Sassu, G. & Svegidao, G., Mater Plast. Elast., _30_ 643 (1964).
308   Wexler, T., Jaks, I. & Jucan., Mater. Plat. (Bucharest), _9_ 6. 268 (1972).
309   Mol'kova, L.N., Kalinin, A.I. and Perepletchikova, E.M., Zh. Anal. Kim., _27_ 1924 (1972).
310   Denning, J.A. & Marshall, J.A., Analyst (London), _97_ 710 (1972).
311   Lappin, G.R. & Zannucci, J., Anal. Chem., _41_ 2076 (1969).
312   Bundergesund Heitschlatte, _16_ 155 (1973).
313   Kirchner, J.G., Thin-layer Chromatography, Interscience, New York, pp 168-169, 679-684 (1966).
314   Simpson, D & Currell, B.R., Analyst, _96_ 515 (1971).
315   Stahl, E., Thin-layer Chromatography, Academic Press, New York, pp 349-352, 487 (1965).
316   Randerath, K., Thin-layer Chromatography, Academic Press, New York, pp 179-181 (1966).
317   Robbitt, J.M., Thin-layer Chromatography, Reinhold, New York, p 180 (1963).
318   Zijp, J.W.H., Rec. Trav. Chim., _76_ 313 (1957).
319   Zijp, J.W.H., Rec. Trav. Chim., _77_ 129 (1958).
320   Wandel, M. & Tengler, H., Fette. Seifen Anstrichmittell, _66_ 815 (1964).
321   Dobies, R.S., J Chromatography, _35_ 370 (1968).
322   Cumpelik, B.A., Drug Cosmet. Ind., _113_ 44 (1973).
323   Schroder, E. & Hagen, E., Plaste Kautsch, _15_ 625 (1968).
324   Van der Heide, R.F., Ernahrungforschung, _24_ 239 (1966).
325   Jentzsch, D., Kruger, H., Lebricht, G. & Deucks, G., Gut. I.Z. Anal. Chem., _236_ 92 (1968).
326   Denning, J.A., & Marshall, J.A., Analyst, _97_ 710 (1972).
327   Romano, S.J., Renner, J.A. & Leitner, P.H., Anal. Chem., _45_ 2327 (1973).
328   Dengreville, M., Analysis, _5_ 195 (1977).
329   Udris, J., Analyst (London), _96_ 130 (1971).
330   Groagova, A. & Pribyl, M., Z. Analyst Chem., _234_ 423 (1968).
331   Belpaire, F., Revie Belg., Mat. Plast., _6_ 201 (1965).
332   Simpson, D. & Currell, D.B., Analyst (London), _96_ 515 (1971).
333   Udris, J., Analyst (London), _96_ 130 (1971).
334   Fassy, H., Lalet, P, Chim. Analyst, _52_ 1281 (1970).
335   Havranek, E., Bumbarlova, A., Kapisinka, V., Chemicky Prumsyl, _20_ 536 (1970).

336   Udris, J., Analyst (London), $\underline{96}$ 130 (1971).
337   Simpson, D. & Currell, B.R., Analyst, $\underline{96}$ 515 (1971).
338   Stafford, C., Anal. Chem., $\underline{34}$ 794 (1962).
339   Thinus, K., Stabilizierung and Alterung von Plastwerkstoffen
       Akademie Verlag Berlin (1960).
340   Schmidt, W., Beckman Report, $\underline{4}$ 6 (1961); 3, 13 (1962) & 3, 13
       (1962).
341   Neubert, G., Beckman Report, $\underline{15}$ 203, 265 (1964).
342   Burger, K., Beckman Report, $\underline{15}$ 192, 280 (1963).
343   Franzen, V. & Neubert, G., Chemiber, Zeitung, Chem. Apparatur., $\underline{89}$
       801 (1965).
344   Berger, K.G., Rudt, U. & Mack, D., Deutsch. Lebensmittell Rdsch.,
       $\underline{6}$ 180 (1967).
345   Mazur, H., Rocnz. panst. Zakl. Hig., $\underline{24}$ 551 (1973).
346   Udris, J., Analyst, $\underline{96}$ 130 (1971).
347   Groagova, A. 7 Pribyl, M.Z., Anal. Chem., $\underline{234}$ 423 (1968).
348   Stapfer, C.H. & Dvorkin, R.D., Anal. Chem., $\underline{40}$ 1891 (1968).
349   Schroeder, E., Pure Appl. Chem., $\underline{36}$ 233 (1973).
350   Mal'kova, L.M., Kalanin, A.I. & Derepletchikova, E.M., Zhur.
       Analit. Khim., $\underline{27}$ (1972).
351   Verfungenegen & Mittalungen des Ministeriums for Gesundheitwesen
       No. 2 (1965).
352   Schroeder, E. & Thinius, K., Deutsch Farben Z., $\underline{14}$ 146 (1960).
353   Schroeder, E. & Malz, S. Deutsch Farben Z., $\underline{5}$ 417 (1958).
354   Schroeder, E. & Hagen, E., Plaste. Kautsch., $\underline{15}$ 625 (1968).
355   Verfunger and Mitteilungen des Ministeriums fur Geunleitesau, No.
       2 (1965).
356   Schroeder, E. & Thinius, K, Deutsch, Farben Z., $\underline{14}$ 146 (1960).
357   Schroeder, E. & Malz, S., Deutsch, Farben, $\underline{5}$ 417 (1958).
358   Malik, W.U., Hague, R. & Palverma, S., Bull. Chem. Soc. Japan, $\underline{36}$
       746 (1963).
359   Schmidt, W., Beckman Report, $\underline{4}$ 6 (1961); $\underline{3}$ 13 (1962).
360   Mal'kova, L.N., Kalanin, A.I. & Perepletchikova, E.M., Zhur.
       Analit. Khim., $\underline{27}$ 1924 (1972).
361   Schroeder, E., Hagen, E. & Zysik, M., Beckman Report, $\underline{13}$ 720
       (1966).
362   Mazur, H., Rocz. panst. Zakl. Hig., $\underline{23}$ 263 (1972).
363   Hagen, E., Deutsch. Lebensmittel. Rdsch., $\underline{14}$ 158 (1967).
364   Schroeder, E., Hagen, E. & Frimel, S., Deutsch Lebensmittel
       Rdsch., $\underline{3}$ 814 (1967).
365   Korn, O. & Waggon,H., Nahrung, $\underline{8}$ 351 (1964).
366   Waggon, H., Kohler, U. & Korn, O., Ernahrungforschung., $\underline{11}$ 548
       (1966).
367   Simpson, D. & Currell, B.R., Analyst, $\underline{96}$ 515 (1971).
368   Veres, L., Kunststoffe, $\underline{59}$ 241 (1969).
369   Weatherhead, R.G., Analyst, $\underline{91}$ 445 (1966).
370   Kreiner, J.G., J. Chromatography, $\underline{75}$ 271 (1973).
371   Schroeder, E. & Hagen, E., Plaste Kautsch, $\underline{15}$ 625 (1968).
372   Kucera, UJ., Collect. Czechoslavakia Chem. Commun., $\underline{28}$ 1344
       (1963).
373   Wandel, M., Tengler, H. & Ostromov, H., Die Analyse von
       Weichmachern, Springer Verlag, Berlin, Heidelbeg, New York (1967).
374   Zilio-Grandi, R., Libraleso, G., Sassu, G. & Svegidao, G., Mater.
       Plast. Elast., $\underline{30}$ 643 (1964).
375   Korn, O. & Waggon, H., Plaste Kautschuk, $\underline{11}$ 278 (1964).
376   Tittareli, P., Zerlia, T., Colli, A. & Ferrari, G., Anal. Chem.,
       $\underline{55}$ 220 (1983).
377   Smith, T.L. & Whelihan, B.N., Text. Chem. Color., $\underline{10}$ 35 (1978).
378   Cope, J.R., Anal. Chem., $\underline{45}$ 562 (1973).
379   Brammer, J.A., Frost, S. & Reid, V.W., Analyst, $\underline{92}$ 91 (1967).

380   Protivova, J & Pospisil, S., J. Chromatography, 88 99 (1974).
381   Yoshikova, T. & Kimura, K., J. Appl. Polymer Sci., 15 2513 (1971).
382   Colburn, A.P., & Pigford, R.L., J.H. Perry (Editor) Chemical
      Engineers Handbook, McGraw Hill, New York, 34d Ed. p. 538 (1950).
383   Edwards, G.D., J. Polymer Sci., Part 21, 105 (1967).
384   Henrickson, J.G & Moore, J.C., J. Polymer Sci., Part A-I, 4 167
      (1966).
385   Lattimer, R.P., Harris, R.E., Rhoe, C.K. & Schulten, H.R., Anal.
      Chem., 58 3188 (1986).
386   Kriner, J.G., Warner, W.C., J. Chromatography, 44 315 (1969).
387   Mirksch, R., Prole, O., Gummi Asbest. 13 250 (1960).
388   Zijp, J.W.H., Ric. Trav. Chim., 75 1155 (1956).
389   Gaczynski, R. & Stephen, M., Przemylek Chem., 38 9 (1959).
390   Burger, V.L., Rubber Chem. Technol., 32 1452 (1959).
391   Kreiner, J.G., Rubber Chem. Technol., 44 381 (1971).
392   Kreiner, J.G. & Warner, W.C., J. Chromatography, 44 315 (1969).
393   Auler, j., Identification of antioxidants, antiozonants and
      accelerators by means of thin-layer chromatography ETDC - Aachen
      Materials and Research Dept. (1967).
394   Delves, R.B., J. Chromatography, 26 296 (1967).
395   Gororyova, A., Elekt. Kabl. Tech., 23 207 (1970).
396   Fiorenza, M., Bonomi, G. & Saradi, A., Rubber Chem. Techol., 41
      630 (1968).
397   Keriner, G. & Warner, W.C., J. Chromatography, 44 315 (1969).
398   Millingen, M.B., Anal. Chem., 46 746 (1974).
399   Hutton, A.S & Altenau, A.G., Dublin Chemistry & Technology, 46
      1035 (1973).
400   Hayes, M.W. & Altenau, A.G., Rubber Age, 102 59 (1970).
401   Bellamy, L.J., Laurie, J.H. & Pren, E.W.S., Trans. Inst. Rubber
      Industry, 15 (1947).
402   Mann, J., Trans. Inst. Rubber Industry, 27 232 (1951).
403   Parker, C.A. & Berriman, J.M., Trans. Rubber Industry, 28 279
      (1952).
404   Freiner, J.B., Rubber Chem. Techol., 44 381 (1971).
405   Yasa, T. & Kamiya, K., Benseki Kagsku, 13 966 (1964).
406   Kreiner, J.G. & Warner, W.C., J. Chromatogr., 44 315 (1969).
407   Kreiner, J.G. & Warner, W.C., J. Chromatogr., 44 315 (1969).
408   Miksch, R., Prole, O., Gummi Asbest., 13 250 (1960).
409   Zijp, J.W.H., Rec. Trav. Chim., 75 1155 (1956).
410   Hilton, A.S & Altenau, A.G., Rubber Chemistry & Technology, 46
      1035 (1973).
411   Mocker, F, Kautchuk Gummi., 11 281 (1958).
412   Mocker, F, Kautchuk Gummi., 12 155 (1959).
413   Mocker, F, Kautchuk Gummi., 13 92 (1960).
414   Adams, R.N., Rev. Polarog, 11 71 (1963).
415   Zweig, A., Lancaster, E., Neglis, M.T. & Jura, W.H., J. Am. Chem.
      Soc., 86 413 (1964).
416   Vodzinskii, Yu. & Semchikova, G.S., T. Po Khim Teknol., 272
      (1963).
417   Vermillion, F.J. & Pearl, T.A., J. Electrochem. Soc., 11 1392
      (1964).
418   Varma, J.P., Suryanaraya, N.P. & Sircar, A.K., J. Sci. Ind. Res.,
      India, 21 49 (1962).
419   Cook, W.S., Jones, C.O., Altenau, G. Can. Sepctrosc., 13 64, 71
      (1968).
420   Hank, W.W., Silverman, L., Anal. Chem., 31 1069 (1959).
421   Mitchell, B.J., O'Hear, H.J., Anal. Chem., 34 1621 (1962).
422   Bergmann, J.S., Ehrhart, C.H., Granatelli, L., Janik, J.L, 153rd
      National American Chemical Society Meeting, Miami Beach, Florida,
      April (1967).

423   Haslam, J., Soppet, W. & Willis, H.A., Appl. Chem., $\underline{1}$ 112 (1951).
424   Liberti, A., Costa, G. & Pauluzzi, E., Chem. & Ind., $\underline{38}$ 674
       (1956).
425   Kendall, D.N., Appl. Spectroscopy, $\underline{7}$ 179 (1953).
426   Cachia, M., Southwart, D.W. & Davison, E., J. Appl. Chem., $\underline{8}$ 291
       (1958).
427   Bellamy, L.J. & Williams, R.L., Paper presented at International
       Symposium on Microchemistry, Birmingham, August (1958).
428   Martin, A.J.P., 1st International Gas Chromatographic Symposium,
       Instrument Society of America, Michigan State University (1957).
429   Kuelemans, A.I.M. & Kwantes, A., 4th World Petroleum Congress,
       Rome (1955).
430   Anderson, D.M.W., Analyst (London), $\underline{84}$ 50 (1959).
431   Haslam, J., Jeffs, A.R. & Willis, H.A., Analyst (London), $\underline{86}$ 44
       (1961).
432   Rudewicz, P., Munson, B., Anal. Chem., $\underline{58}$ 358 (1986).
433   Lattimer, R.P., Harris, R.E., Rhee, K., Anal. Chem., $\underline{58}$ 3188
       (1986).
434   Vargo, J.D., Olson, K.L., Anal. Chem., $\underline{57}$ 672 (1985).
435   Likens, S.T. & Nickerson, G.B., Proc. Amer. Soc. Brow-Chem., 5
       (1984).
436   Sharp, J.L. & Paterson, G., Analyst (London), $\underline{105}$ 517 (1980).
437   Smuts, T.W., VanNiekerk, F.A., Pretorius, V.J., Gas Chromatog., $\underline{5}$
       190 (1967).
438   Giddings, J.C. Anal. Chem., $\underline{35}$ 2215 (1963).
439   Giddings, J.C. Anal. Chem., $\underline{37}$ 61 (1965).
440   Locke, D.C. Anal. Chem., $\underline{39}$ 921 (1967).
441   Snyder, L.R., Anal. Chem., $\underline{39}$ 698, 705 (1967).
442   Schroder, E, Pure Anal. Chem., $\underline{36}$ 233 (1973).
443   Kirkland, J.J., Anal. Chem., $\underline{41}$ 218 (1969).
444   Kirkland, J.J., J. Chromatogr. Sci., $\underline{7}$ 7 (1969).
445   Halasz, I., Walking, P., J. Chromatogr. Sci., $\underline{7}$ 129 (1969).
446   Waters Associated "New Developments in Chromatography" No. 1
       (1970).
447   McDonell, H.L., Anal. Chem., $\underline{40}$ 221 (1968).
448   Haulein, A., Corning Glass Works, personal communication (1970).
449   Waters Associates, New Development in Chromatography, No. 1
       (1970).
450   Kirkland, J.J., J. Chromatogr. Sci., $\underline{7}$ 361 (1969).
451   Majors, R.E., J, Chromatogr. Sci., $\underline{8}$ 338 (1970).
452   Griddle, W.J., British Plastics, $\underline{242}$ May (1963).
453   Campbell, R.H., Wize, R.W., J. Chromatography, $\underline{12}$ 178 (1963).
454   Schabaum, J.F., Fenska, L.E., Anal. Chem., $\underline{52}$ 1411 (1980).
455   Rudewicz, P., & Munson, B., Anal. Chem., $\underline{58}$ 358 (1986).
456   Vargo, J.D. & Olsen, K.L., Anal. Chem., $\underline{57}$ 672 (1985).
457   Dohmann, K., Laboratory Practice, July (1965).
458   Slonaker, D.F & Sievers, D.C., Anal. Chem., $\underline{36}$ 1130 (1964).
459   Van der Neut, J.H., Maagdenberg, A.C., Plastics, $\underline{31}$ 66 January
       (1966).
460   Van der Heide, R.F., Wouters, O.Z., Fur Lebenmittelforschung, $\underline{17}$
       129 (1962).
461   Waggon, H., Jehle, D., Die Nahrung, $\underline{4}$ 495 (1965).
462   Braun, D., Chimical Ind., $\underline{19}$ 77 (1965).
463   Wandel, M., Tengler, H., Kunststoffe, $\underline{55}$ 11 (1965).
464   Haalpaap, H., Chem. Org. Tech., $\underline{35}$ 488 (1963).
465   Braun, D., Geenen, I.T., J. Chromatog., $\underline{7}$ 56 (1962).
466   Knappe, E., Peteri, D.Z., Anal. Chem., $\underline{188}$ 184 (1961).
467   Lynes, A.J., Chromatography, $\underline{15}$ 108 (1964).
468   Lane, E.S., J. Chromatog., $\underline{18}$ 426 (1965).

469   Crump, G.B., Anal. Chem., $\underline{36}$ 2447 (1964).
470   Dallas, H.S.H., J. Chromatog., $\underline{17}$ 267 (1965).
471   McCoy, R.N., Fiebig, E.C., Anal. Chem., $\underline{37}$ 593 (1965).
472   Juvet, R.S., Smith, L.S. & Li, K.P., Anal. Chem., $\underline{44}$ 49 (1972).
473   Crompton, T.R., Myers, L.W., Blair, D., Brit. Plast., December (1965).
474   Crompton, T.R., Myers, L.W., Europ. Polymer J., $\underline{4}$ 355 (1968).
475   Rohrshneider, L.Z., Anal. Chem., $\underline{255}$ 345 (1971).
476   Shanks, R.A., Pye Unicorn Newsletter, (1975).
477   Clover, G.C., Murphey, M.E., Anal. Chem., $\underline{31}$ 1682 (1959).
478   Crompton, T.R., Buckley, D., Analyst (London), $\underline{90}$ 76 (1965).
479   Steichen, R.J., Anal. Chem., $\underline{48}$ 1398 (1976).
480   Pfab, W. & Noffz, D., Z. Anal. Chem., $\underline{37}$ 195 (1963).
481   Betso, S.R. & McLean, J.D., Anal. Chem., $\underline{48}$ 766 (1976).
482   Brown, L., Analyst (London), $\underline{104}$ 1165 (1979).
483   Skelly, N.E. & Husser, E.R., Anal. Chem., $\underline{50}$ 1959 (1978).
484   Husser, E.R. Stehl, R.H., Price, D.R. & Delap, R.A., Anal. Chem., $\underline{49}$ 154 (1977).
485   Shiono, S.J., Polymer Sci., A-1, $\underline{17}$ 4120 (1979).
486   Zaborsky, L.M., Anal. Chem., $\underline{49}$ 1166 (1977).
487   Gankina, E.S., Val'chikhina, M.D., Belen'kii, G., Vysokomol. Soedin. Ser., A, $\underline{18}$ (5), 1170 (1976).
488   Van der Maeden, FPB., Biemond, M.E.F., Janssen, P.C.G.M., J. Chromatog., $\underline{149}$ 539 (1978).
489   Zaborsky, L.M., Anal. Chem., $\underline{49}$ 1166 (1977).
490   Ludwig, J.F., Bailie, A.G., Anal. Chem., $\underline{56}$ 2081 (1984).
491   Mouney, T.H., Smith, G.A., Anal Chem., $\underline{56}$ 1773 (1984).
492   Mourey, T.H., Anal. Chem., $\underline{56}$ 1777 (1984).
493   Utterback, D.F., Millington, D.S., Gold, A., Anal. Chem., $\underline{56}$ 470 (1984).
494   Crompton, T.R., Myers, L.W., Blair, D., British Plastics, December (1965).
495   Crompton, T.R., Myers, L.W., European Polymer Journal, $\underline{4}$ 355 (1968).
496   Crompton, T.R., Myers, L.W., Plastics and Polymers, Page 205, June (1968).
497   Roper, J.N., Anal. Chem., $\underline{42}$ 688 (1970).
498   Davies, J.T., & Bishop, J.R., Analyst (London), $\underline{96}$ 55 (1971).
499   Guthrie, J.L., & McKinney, R.W., Anal. Chem., $\underline{49}$ 1676 (1977).
500   Di Pasquale, C., Di Iorio, G. & Capaccioli, T., J. Chromatogr., $\underline{152}$ 538 (1978).
501   Shanks, R.A., Pye Unican Newsletter (1975).
502   Rohrschneider, L.Z., Analyt. Chem., $\underline{255}$ 345 (1971).
503   Shapras, P. & Clover, G.C., Anal. Chem., $\underline{34}$ 433 (1962).
504   Tweet, O., Miller, W.K., Anal. Chem., $\underline{35}$ 852 (1963).
505   Wilkinson, L.B., Norman, C.W., & Brettner, N.P., Anal. Chem., $\underline{36}$ 1759 (1964).
506   Nowak, P. & Klemmett, O., Kunststoffe, $\underline{52}$ 604 (1962).
507   Adcock, L.H., Patra, $\underline{3}$ 5 (1962).
508   Ragelis, E.P. & Gajan, R.J.F., Assoc. Offic. Agric. Chem., $\underline{45}$ 918 (1962).
509   Pfab, W. & Noffz, D., Z. Anal. Chem., $\underline{37}$ 195 (1963).
510   Shapras, P. & Claver, G.C., Anal. Chem., $\underline{36}$ 2282 (1964).
511   Crompton, T.R., Myers, L.W., & Blair, D., British Plastics, December (1965).
512   Crompton, T.R., Myers, L.W., European Polymer Journal, $\underline{4}$ 355 (1968).
513   Schwoetzer, G., Z. Anal. Chem., $\underline{260}$ 10 (1972).
514   Podzeeva, R.M., Lukhovitskii, U.I. & Forpov, V.L., Zavod. Lab., $\underline{37}$ 168 (1971).

515   Crompton, T.R., Myers, L.W., Plastics and Polymers, 205 (June 1968).
516   Shapras, P., Claver, G.C., Anal. Chem., $\underline{36}$ 2282 (1964).
517   Shanks, R.A., Scan, $\underline{6}$ 20 (1975).
518   Berens, A.R., Crider, L.B., Tomamek, C.M., Whitney, J.M., B.P. Goodrich Tyne Centre. Brecksville Ohio. Manuscript circulated to members of the Vinyl Chloride Safety Association. 14 November 1974, entitled "Analysis of Vinyl Chloride in PVC Powders by Head Space Chromatography".
519   Berens, A.R., paper delivered to the 168th American Chemical Society Meeting, Atlantic City, New Jersey, September 1974, entitled "The Solubility of Vinyl Chloride in Polyvinyl Chloride".
520   Steichen, R.J., Anal. Chem., $\underline{48}$ 1398 (1976).
521   Hartley, A.J., Lueng, Y.K., McMahon, J., Booth, C., Shepherd, I.W., Polymer, $\underline{18}$ 336.
522   Crompton, T.R., Myers, L.W., Plastics and Polymers, 205 June (1968).
523   Jeffs, A.r. Analyst (London), $\underline{94}$ 249 (1969).
524   Maltese, P., Clementini, L., Panizza, S., Mater. Plaste. Elastomerie, $\underline{35}$ 1669 (1969).
525   Takashimia, S., Okada,F., himeji Kogyo Daigaku Kenkyu Hokoku, $\underline{12}$ 34 (1960).
526   Haslam, J., Jeffs, A.R., Willis, H.A., J Oil Col. Chem. Asoc., 45 325 (1962).
527   Haslam, J., Chem. Age., $\underline{82}$ 169 (1959).
528   Haslam, J., Jeffs, A.R., J. Appl. Chem., $\underline{7}$ 24 (1957).
529   Haslam, J., Jeffs, A.R., Analyst (London) $\underline{83}$ 455 (1958).
530   Ernes, D.A., Hashumaker, Anal. Chem., $\underline{55}$ 408 (1983).
531   Welsch, T., Engewald, W. & Kowash, E., Plaste. u Kaut., $\underline{23}$ 584 (1976).
532   Urbanski, J., Anal. Chem., (Warsaw), $\underline{22}$ 749 (1977).
533   Tengler, H. & Von Falkai, B., Kunstoffe, $\underline{62}$ 759 (1972).
534   Maltese, P., Clementini, L., Panizza, S., Mater. Plaste. Elastomerie, $\underline{35}$ 1669 (1969).
535   Jeffs, A.R., Analyst (London), $\underline{94}$ 249 (1969).
536   Hoffman, E.R., Anal. Chem., $\underline{48}$ 445 (1976).
537   Squirrel, D.C.M., Analyst (London), $\underline{106}$ 1042 (1981).
538   Jeffs, A.R., Analyst (London), $\underline{94}$ 249 (1969).
539   Maltese, P., Clementini, L., Panizza, S., Mater. Plaste. Elastomerie, $\underline{35}$ 1669 (1969).
540   Armieau, V. & Costain, D., Mater Plast. (Bucharest), $\underline{9}$ 606 (1972).
541   Harrington, R.C. & Keister, D.D., Paper presented at the Plastics Paper Conference of the Technica Association of the Pulp and Paper Manufacturers Association.
542   Rice, D.D. & Trowell, J.M., Anal. Chem., $\underline{39}$ 157 (1967).
543   Schamalz, E.O., Faserforsch Tex. Tech., $\underline{21}$ 209 (1970).
544   Maltese, P., Clementini, L., Panizza, S., Mater. Plaste. Elastomerie, $\underline{35}$ 1669 (1969).
545   Schmalz. E.O., Anal. Abstracts, $\underline{17}$ 2449 (1969).
546   Narasaki, H. & Umezawa, K., Kobunshi Kagaku, $\underline{29}$ 438 (1972).
547   Smith, A.J., Anal. Chem., $\underline{36}$ 944 (1964).
548   Henn, L., Anal. Chim. Acta., $\underline{73}$ 273 (1974).
549   Gorsuch, T., Analyst, $\underline{87}$ 112 (1962).
550   Gorsuch, T., Analyst, $\underline{84}$ 135 (1959).
551   Korenaga, T., Analyst (London), $\underline{106}$ 40 (1981).
552   Ogure, H., Bunseki Kagaku, $\underline{24}$ 197 (1975).
553   Mitterberger, H., Gross, H., Kunststofftecknic $\underline{12}$ 76 (1973).
554   Bakroni, M.L., Chakavarty, N.K., Chopra, S., Indian J., Technology, $\underline{13}$ 576 (1975).
555   Tanaka, K., Morikawat, T., Kagaku., To Kogyo (Osaka), $\underline{48}$ 387 (1974).

556   Falcon, J.Z., Love, J.L., Yacta, L.T., Altenau, A.G., Anal. Chem.,
      47 171 (1975).
557   Narasaki, H., Hijaji, V., Unno A., Bunseki Kagaku, 22 541 (1973).
558   Kalinina, L.S., Nikitina, N.I., Matorina, M.A. & Sedova, I.V.,
      Plast. Massy, 5 66 (1976).
559   Hernadez, H.A., International Laboratory, 84 September (1981).
560   Caselta, B., Di Pasquale, G., Soltientini, A., Atomic
      Spectroscopy, 6 62 (1985).
561   Di Pasquale, G., Caselta, B., Atom C. Spectroscopy, 5 209 (1984).
562   Radovici, A., Radovici, R & Giornei, T., Mater. Plast.
      (Bucharest), 11 360 (1974).
563   Sokolov, D.M.U., Nesterenko, G.N. & Golubeva, L.K., Zavod. Lab.,
      39 939 (1973).
564   Bahrani, M.L., Chakravarty, N.K & Chopra, S.C., Indian J.
      Technol., 13 576 (1975).
565   Monett, S. & Horowicz, S., Anal. Chem., 52 1529 (1980).
566   Cook, W.S., Jones, C.O., & Altenau, A.G., Canadian Spectroscopy,
      13 64 (1968).
567   Bergmann, J.S., Ehrhart, C.H., Grantelli, L. & Janik, L.J., 153rd
      National ACS Meeting, Miami Beach, Florida (April 1967).
568   Rowe, W.A. & Yates, K.P., Anal. Chem., 35 368 (1963).
569   Shott, J.E., Jr., Garland, T.J. & Clark, R.O., Anal. Chem., 33 506
      (1961).
570   Gorsuch, T.T., Analyst, 84 135 (1959).
571   Gamble, L.W. & Jones, W.H., Anal. Chem., 27 1456 (1955).
572   Leyden, D.E., Lennox, J.C. & Pittman, C.U., Anal. Chim. Acta., 64
      143 (1973).
573   Kabayashi, O., J. Polymer Sci, A-1, 17 293 (1979).
574   Hull, D., Gilmore, J., Division of Fuel Chemistry, 141st Meeting
      ACS, Washington DC, March 1962.
575   Bergmann, J.S., Ekhart, C.H., Grantelli, L., Janik, L.J., 153rd
      National ACS Meeting, Miami Beach, Florida, April 1967.
576   Rowe, W.A., Yates, K.P., Anal. Chem., 35 368 (1963).
577   Shott, J.E., Garland, T.J., Clark, R.O., Anal. Chem., 33 506
      (1961).
578   Battiste, D.R., Butler, J.P., Cross, J.B. & McDaniel, M.P., Anal.
      Chem., 53 2232 (1981).
579   Kuta, E.J., Quackenbush, F.W., Anal. Chem., 32 1069 (1960).
580   Bukata, S.W., Zabrocki, L.L. & McLaughlin, M.F., Anal. Chem., 35
      886 (1963).
581   Hyden, S., Anal. Chem., 35 133 (1963).
582   Citovicky, P., Simek, I., Mikulasova, D & Chrassova, V., Chem.
      Zvesti, 30 (3), 342 (1976)., Chem. Abstr., 88 105960h (1978).
583   Bradley, A. & Heagnay, J.R., Anal. Chem., 42 894 (1970).
584   Kunyyatnikov, T.C., Visgert, R.V., Berlin, A.A., Vijn. L'viv
      Politkh Inst. No. 57 36 (1971).
585   McNeill, I.E., Polymer, 4 15 (1963).
586   McGuchan, R., McNeill, I.W., J. Polymer Sci., Al. 4 2051 (1966).
587   Pepper, D.C., Reilly, P.H., Proc. Chem. Soc., 1 460 (1961).
588   Gallo, S.g., Weiss, H.K., Nelson, J.F., Ind. Eng. Chem., 40 1277
      (1948).
589   Lee, T.S., Kolthoff, I.M. Johnson, E., Anal. Chem., 22 995 (1950).
590   Dinsmore, H.C., Smith, D.C., Rubber Chem. Technol., 22 572 (1949).
591   Harms D.C., Anal. Chem., 25 1140 (1953).
592   Hummel, D., Rubber Chem. Technol., 32 854 (1959).
593   Lermer, M., Gilbert, R.C., Anal. Chem., 36 1382 (1964).
594   Tyron, M., Horowicz, E., Mandel, J.J., Res. Nat'l Bur. Stds., 55
      219 (1955).
595   McKillop, D.A., Anal. Chem., 40 607 (1968).

596   Shibata, C., Yamazaki, M. & Tabenuchi, T., Bull. Chim. Soc.
       (Japan), 50 311 (1977).
597   Brako, F.D., Wexler, A.S., Anal. Chem., 35 1944 (1963).
598   Majewski, J., Polimery, 18 142 (1973).
599   Stelzer, R.S., Smullin, C.F., Anal. Chem., 34 194 (1962).
600   Hazzezyc, B., Walczyk, W., Chem. Anal. (Warsaw), 20 885 (1975).
601   Fritz, J.S & Schlenk, G.H., Anal. Chem., 31 1808 (1959).
602   Floria, J.A., Dobratz, J.W. & McClure, J.H., Anal. Chem., 36 2053
       (1964).
603   Horia, J.M., Dobratz, J.W., McClure, J.H., Anal. Chem., 36 2053
       (1964).
604   Siggia, S., Hanna, J.G. & Culmo, R., Anal. Chem., 33 900 (1961).
605   Spririn, Yu L. & Yatsimirskaya, T.A., Vysokomol. Soedin Ser. A.15
       2595 (1973).
606   Stenmark, G.A. & Weiss, F.T., Anal. Chem., 28 1784 (1956).
607   Motorina, M.A., Kalinina, L.S., & Metalkina, E.I., Plast. Massy.,
       6 74 (1973).
608   Law, R.D., J. Polymer Science, A-1, 9 589 (1971).
609   Fritz, D.E, Sahil, A., Keller, H.P. & Kovat, E., Anal. Chem., 51 7
       (1979).
610   Kaduji, I.I. & Rees, J.H., Analyst (London), 99 435 (1974).
611   Smith, R.M. & Dauson, M., Analyst (London), 105 85 (1980).
612   L, Ho, F.F., Anal. Chem., 45 603 (1973).
613   Kim, C.S.Y., Dodge, A.I., Lau, S., Kawsaki, A., Anal. Chem., 54
       232 (1982).
614   Cohen, J.L., Fong, G.P., Anal. Chem., 47 313 (1975).
615   Conners, K.A., Pandit, K., Anal. Chem., 50 1542 (1978).
616   Dee, I.A., Biggers, B.L., Fyke, M.E., Anal. Chem., 52 572 (1980).
617   Kreshkov, L.N., Shvelsovr, L.N., Emelin, E.A., Soc. Plast. (10),
       53 (1968), Chemical Abstracts, 70 21345q (1969).
618   Mathieson, A.R., McLaren, J.V., J. Polymer Sci. A-3 25555 (1965).
619   Kathalsky, A., Shavit, N. & Eisenberg, H.J., Polymer Science, 13
       69 (1954).
620   Law, R.D., J Polymer Science, 9 589 (1971).
621   Connor, A.Z. & Egler, F.W., Anal. Chem., 22 1129 (1950).
622   Francis, C.V., Anal. Chem., 25 941 (1953).
623   Gemerenko, E.n., Perepletchikova, E.M. Zh. Anal Khim., 29, (4),
       830 (1974).
624   Kato, K., J. Appl. Polymer Sci., 17 105 (1973).
625   Leukrath, G., Gummi Asbest. Kunstst., 29 (9) 585 (1976); Chem.
       Abstr., 86 44183S (1977).
626   Miller, D.L., Samsel, E.P., Cobler, J.G., Anal Chem., 33 677
       (1961).
627   Samsel, E.P., McHard, J.A., Ind. End. Chem. Anal. Ed., 14 750
       (1942).
628   Haslam, J., Hamilton, J.B., Jeffs, A.R., Analyst (London), 83 983,
       66 (1958).
629   Haslam, J., Jeffs, J., Anal. Chem., 7 24 (1957).
630   Fritz, R., Fresenius, Z., Anal. Chem., 176 421 (1960).
631   Steyermark, A., J. Assoc. Off. Angric. Chem., 38 367 (1955).
632   Ehrlich, Rogozinsky, S. & Patchornick, A., Anal. Chem., 36 849
       (1964).
633   Viebock, F. & Brechner, C., Ber. Dtsch. Chem., 63 3207 (1930).
634   Gyenes, I., "Titration in Non-Aqueous Media", J Iliffe Books,
       London, P. 261 (1967).
635   Urbanski, J., Plaste Kantsch, 15 260 (1978).
636   Cross, C.K., Mackay, A.C.J., Amer. Oil Chem. Soc., 50 249 (1973).
637   Cohen, M., Ind. Eng. Chem., 47 2096 (1955).
638   Durbetaki, A.J., Anal. Chem., 28 2000 (1956).

639   Stenmark, G.A., Anal. Chem., 29 1367 (1957).
640   Kappelmeier, C.P.A., Farbon Ztg., 40 1141 (1935).
641   Kappelmeier, C.P.A., Verfkroniek, 27 291 (1954).
642   Ferckow, F.W., Farben Ztg., 44 33 (1939).
643   Bandel G., Angew Chem., 51 570 (1938).
644   Malm, C.J., Gening, L.B. & Williams, R.F., Ind. Eng. Chim. Anal.
      Ed., 14 935 (1942).
645   Malin, C.J., Tonghe, L.J., Laird, B.C. & Smith, G.D., Anal. Chem.,
      26 188 (1954).
646   Schleuter D.D., Siggia, S., Anal. Chem., 49 2343 (1977).
647   Schleuter D.D., Siggia, S., Anal. Chem., 49 2349 (1977).
648   Parker, R.D., Dow Corning Corporation, Barry, Wales, unpublished
      procedure.
649   Gritz, G. & Hurcht, H., Z. Anorg. Allg. Chem., 317 35 (1962);
      Chem. Abstr., 57 1443g (1962).
650   Krasikova, V.M. & Kaganova, A.N. & Lobtov, V.D., J. Anal. Chem.
      USSR. (Engl. Trans.), 28 1458-1461 (1971).
651   Voronkov, M.G. & Shemyatenkova, V.T., Bull. Acad. Sci. USSR. Div.
      Chem. Sci., 178 (1961), Chem. Abstr., 55 16285b (1961).
652   Franc, J. & Placek, K., Collect. Scech. Chem. Commun., 38 513-515
      (1973).
653   Franc, J., Chem. Abstr., 82 67923q (1975).
654   Kreshkov, A.P., Shemyatenkova, V.T., Syavtsillo, S.V &
      Palamarchuck, N.A., J. Anal. Chem. USSR, (Engl. Transl.) 15 727
      (1960).
655   Heymun, G.W., Bujalski, R.L. & Bradley, H.B., J. Gas Chromatogr.,
      2 300 (1964).
656   Krasikova, V.M. & Kaganova, A.N., J. Anal. Chem. USSR (Engl
      Transl.) 25 1212 (1970).
657   Bissell, E.R. & Fields, D.B., J. Chromatog. Sci., 10 164 (1972).
658   Franc, J & Placek, K., Mikrochim. Acta 11., 1 31 (1975).
659   Hanson, C.L. & Smith, R.C. Anal. Chem., 44 1571 (1972).
660   Dubiel, S.V., Griffith, G.W., Long, C.L., Baker, G.K. & Smith,
      R.E., Anal. Chem., 55 1533 (1983).
661   Eckstein, Y., Dreyfuss, Anal. Chem., 52 537 (1980).
662   Gerschenko, Z.V., Blinov, V.F. & Zimin, Yu B., Plast. Massy. 12 12
      (1975).
663   Poharelsky, L & Heran, J., Chem. Prumsyl., 27 630 (1977).
664   Berl, E. & Hefter, K., Cellulose Chemie., 14 67 (1933).
665   Kanjilal, C., Mitra, B & Palit, S.R., Makromol. Chem., 178 1707
      (1977).
666   Purdon, J.R., Mate, R.D., J. Polymer Sci., B, 1, 451 (1963).
667   Purdon, J.R., Mate, R.D., J. Polymer Sci., A, 1, 8 1306 (1970).
668   Rogozinski, M., Kramer, M., J. Polymer Sci., A-1, 10 3110 (1972).
669   Turner, R.R., Carlson, P.W., Altneau, A.G., J. Elastomers Plast.,
      9 94 (1974).
670   Nakajima, A., Chem. High Polymers Japan, Kobunski Kagaku, 7 64
      (1950).
671   Nakajima, A., Fujiwara, H., High Polymers Japan, Kobunski Kagaku,
      37 909 (1964).
672   Wijga, P.W.O., Van Schooten, J., Beerma, J., Makromol Chemie, 36
      115 (1960).
673   Kenyon, A.C., Salyer, C.O., Kirz, J.E., Brown, O.R., J. Polymer
      Sci., C-8, 205 (1965).
674   Lovric, L.J., Grubisic-Gallot, Z., Kunst, B., Europ. Polymer J.,
      12 (3),189 (1976).
675   Kolke, V., Billmeyer, F.W., J. Polymer Sci., C-8, 217 (1965).
676   Akutin, M.S., Goldberg, V.H., Lavruchin, F.G., Vysokmore. Scr., A,
      19 (5), 1113 (1977).

677   Ogawa, T., Hoshino, S., J. Appl. Polymer Sci., 17 2235 (1973).
678   Ogawa, T., Tanaka, S., Inaba, T., J. Appl. Sci., 18 1351 (1974).
679   Loconti, J.D., Cahill, J.W., J. Polymer Sci., 49 152 (1961).
680   Loconti, J.D., Cahill, J.W., J. Polymer Sci., A-1, 3163 (1963).
681   Ruskin, A.M., Parravano, G., J. Appl. Polymer Sci., 8 565 (1964).
682   Bryson, A.P., Hawke, J.G., Parts, A.G., J. Polymer Sci., 12, 1323 (1974).
683   Kalal, J., Marousek, V., Svec, F. Sb V ys, Sk Chem. Technol. Praze Org. Chem. Technol., C22 57 (1975); Chem. Abstr., 83 132443v (1975).
684   Gal'perin, V.M., Zak, A.G., Kuznetsov, N.A., Roganova, Z.A., Smolyanskii, A.L., Vysokomol. Soedin. Ser. A., 17 (3), 575 (1975).
685   Kossler, L.D., Vodchnal, J., J. Polymer Sci., A-3, 2511 (1965).
686   Baker, C.A., Williams, R.J.P., J. Chem. Soc. (London), 2, 852 (1965).
687   Sutherland, J.E., Research Laboratories Polymer Prep., R 17, (2) (1976).
688   Teramachi, S., Fukao, T., Polymer J., 6 (6), 532 (1974).
689   Hori, F., Ikada, Y., Sakurada, J.J., J. Polymer Sci. Polymer Chem. Ed., 13 755 (1975).
690   Peebles, L.H., Huffman, M.W., Ablett, C.T., J. Polymer Sci., A-1, 479 (1969).
691   Repina, L.P., Khalatur, P.G., Zavod Lab., 41 (3), 287 (1975); Chem. Abstr., 83 60334n (1975).
692   Katada, T., White, J.L., Macromolecules, 7 (1), 106 (1974).
693   Gleoeckner, G., Kahle, D., Plast. Kautsch, 23 (8), 577 (1976).
694   Gloeckner, G., Kahle, D., Plast Kautsch, 23 (5), 338 (1976).
695   Vakhtina, I.A., Khrenova, R.F., Tarnkov, O.G. Zh-Anal. Khim., 28 (8), 1625 (1973).
696   Hori, F., Ikada, Y., Sakurada, I.J., J. Polymer Sci. Polymer Chem. Ed., 13 755 (1955).
697   Buter, R., Tan, Y.Y., Challa, G., Polymer, 14 171 (1973).
698   Belen'skii, B.g., Gankina, E.S., Nefedov, P.P., Lazareva, M.A., Savitskaya, T.S., Voichikhina, M.D., J. Chromatogr., 108 (1), 61 (1975).
699   Inagaki, H., Kamiama, F., Macromolecules, 6 (1), 107 (1973).
700   Kotaka, T., Uda, T., Tanaka, T., Inagaku, H., Makromol. Chem., 176 (5), 1273 (1975).
701   Geymer, D.O., Shell Chemical Co. Ltd., Emeryville, California, private communication.
702   Inagaki, H., Miyamoto, T., Kamiyama, F.J., Polymer Sci., B-7, 329 (1969).
703   Mikkelsen, L., Characterization of High Molecular Weight Substances, Pittsburgh Conference, Anal Chem., 5-9 March (1962).
704   Puschmann, O., Fette Seife Anstrichmittel, 65 (1963).
705   Celedes Pacquot, C.Rev. France Corps. Gras., 9 145 (1962).
706   Sweeley, C.C., Bentley, R., Makita, M., Wells, W.W., J. Amer. Chem. Soc., 85 2497 (1973).
707   Fletcher, J.P., Persinger, H.E., J. Polymer Sci., A-1, 6 1025 (1968).
708   Myers, L.W., Shell Research Ltd., Carrington, Cheshire, U.K., private communication.
709   Utterbach, D.F., Millington, D.S., Gold, A., Anal. Chem., 56 470 (1984).
710   Kuzaev, A.I., Susiova, E.N., Entelis, S.G., Dokl. Akad. Nauk. S.S.R., 208 (1), 142 (1973).
711   Chang, M.S., French, D.M., Rogers, P.L., J. Macromol. Sci. Chem., 7 (8), 1727 (1973).
712   Law, R.D., J. Polymer Sci. Polymer Chem. Ed., 11 175 (1973).

713   Law, R.D., J. Polymer Sci., A-1, $\underline{9}$ 589 (1971).

714   Law, R.D., J. Polymer Sci., A-1, $\underline{7}$ 2097 (1971).

715   Teramachi, S., Hasegawa, A., Shima, Y., Akatsuka, M., Nakajima, M., Macromolecules, $\underline{12}$ 992 (1979).

716   Glockner, G., Van den Berg, J.H.M., Meijerink, N.L.J., Scholte, T.G., Koningsveld, R., Macromolecules, $\underline{17}$ 962 (1984).

717   Sato, H., Takeuchi, H.; Suzuki, S., Tanaka, Y., Makroroml. Chem. Rapit Commun., $\underline{5}$ 719 (1984).

718   Danielewicz, M., Kubin, M., J. Appl. Polm. Sci., $\underline{26}$ 951 (1981).

719   Tanaka, T., Omoto, M., Donkai, N., Inagaki, H., J. Macromol. Sci. Phys., B17, 211 (1980).

720   Balke, S.T., Sep. Purif. Methods, $\underline{11}$ (1982).

721   Mori, S., Uno, Y., Anal. Chem., $\underline{59}$ 90 (1987).

722   Mori, S., Uno, Y., Anal. Chem., $\underline{58}$ 303 (1986).

723   Mourey, T.H., Smith, Snyder, L.R., Anal. Chem., $\underline{56}$ 1773 (1984).

724   Mourey, T.H., Anal. Chem., $\underline{56}$ 1777 (1984).

725   Ludwig, F.J., Anal. Chem., $\underline{56}$ 2081 (1984).

726   Zaborsky, L.M., Anal. Chem., $\underline{49}$ 1166 (1977).

727   Glockner, G., Polymercharakterisierung durch Flussigkeitschromatographi; VEB Deutscher Verlag der Wissenschaften: Berlin, 1981; pp 91-92.

728   Englehardt, H., Elgass, H., In High-Performance Liquid Chromatography, Advances and Perspective, Horvath, C., Ed., Academic Press, New York, 1980, Vol. 2, pp 57-111.

729   Zaborski, L.M., Anal. Chem., $\underline{49}$ 1166 (1977).

730   Atkinson, E.R., Calouche, S.I., Anal. Chem., $\underline{43}$ 460 (1971).

731   Yamanis, J., Vilenchich, R., Adelman, M., J Chromatogr., $\underline{108}$ 79 (1975).

732   Mourney, T.H., Smith, G.A., Snyder, L.R., Anal. Chem., $\underline{56}$ 1774 (1984).

733   Lattimer, R.P., Harmon, D.J., Welch, K.R., Anal. Chem., $\underline{51}$ 1293 (1979).

734   Snyder, L.R., Glajch, J.L., Kirkland, J.J., J. Chromatography, $\underline{3}$ 157 (1983).

735   Mourney, T.H., Anal. Chem., $\underline{56}$ 1777 (1984).

736   Snyder, L.R., High-Perform. Liq. Chromatogr. $\underline{3}$ 157 (1983).

737   Snyder, L.R., Principles of Adsorption Chromatography, Marcel Dekker, New York, (1968).

738   Snyder, L.R., J. Chromatogr., $\underline{184}$ 363 (1980).

739   Moore, J.C., J. Polymer Science Part $\underline{2}$ 835 (1964).

740   Benningfield, L.V., "A dielectric constant detector for liquid chromatography and its applications." Paper presented at the 30th Pittsburgh Conference on Analytical Chemistry and Applied Spectroscopy, Cleveland, Ohio, 5-9 March, 1979, p 123 (1979).

741   Benningfield, L.V., Mowry, R.A., J Chromatogr. Sci., $\underline{19}$ 115 (1981).

742   Fuller, E.N., Porter, G.T., Roof, L.B., J. Chromatogr. Sci, $\underline{17}$ 661 (1979).

743   Roof, L.B., Porter, G.T., Fuller, E.N., Mowrey, R.A., "On-line Polymer Analysis by Liquid Chromatography". Instrumentation and Automation in the Paper, Rubber, Plastics and Polymerization Industries. 4th TFAC Conference, Ghent, Belgium 3-5 June 1980. (A. Van Canwenberge, ed.) Pergamon Press, New York, pp 47-53 (1980).

744   Ross, J.H., Shank, R.C., Adv. Chem. Ser., $\underline{125}$ 108 (1971).

745   Bode, R.K., Benningfield, L.V., Mowrey, R.A., Fuller, E.N., International Laboratory, $\underline{40}$ Nov./Dec. (1981).

746   Grubisic, Z., Rempp, P., Benolt, H., J. Polymer Sci. B, $\underline{5}$ 753 (1967).

747   Moore, J.C., Hendrickson, J.G., J. Polymer Sci. C, $\underline{8}$ 233 (1985).
748   Mori, S., J. Chromatogr. $\underline{157}$ 75 (1978).
749   Mori, S., J. Appl. Polymer Sci., $\underline{18}$ 2391 (1974).
750   Balke, S.T., Hamielec, A.E., Leclair, B.P., Pearce, S.L., Ind. Eng. Chem. Produ. Res. Dev., $\underline{8}$ 54 (1969).
751   Loy, B.R., J. Polymer Sci. Polymer Chem. Ed., $\underline{14}$ 2321 (1976).
752   Vrijbergen, R.R., Soeteman, A.A., Smit, J.A.M., J. Appl. Polymer Sci., $\underline{22}$ 1267 (1978).
753   McCrackin, F.L., J. Appl. Polymer Sci., $\underline{21}$ 191 (1977).
754   Mahabadi, H.K., O'Driscoll, K.F., J. Appl. Polymer Sci., $\underline{21}$ 1283 (1977).
755   Balke, S.T., Hamielec, A.E., Leclair, B.P., Pearce, S.L., Ind. Eng. Chem. Prod. Res. Dev., $\underline{8}$ 54 (1969).
756   Loy, B.R., J. Polyer Sci. Polymer Chemistry Edition $\underline{14}$ 2321 (1976).
757   Vrijbergen, R.R., Soeteman, A.A., Smit, J.A.M., J. Applied Polymer Sci., $\underline{22}$ 1267 (1978).
758   Mori, S., Susuki, T., J. Liquid Chromatography, $\underline{3}$ 343 (1980).
759   McCracken, F.L., J. Appl. Polymer Sci., $\underline{21}$ 191 (1977).
760   Mahabadi, H.K., O'Driscoll, K.F., J. Appl. Polymer Sci., $\underline{21}$ 1283 (1977).
761   Weiss, A.R., Cohn-Ginsberg, E., J. Polymer Sci. B, $\underline{7}$ 379 (1969).
762   Belinskii, B.G., Nefedov, P.F., Vysokomol Soedin, A-14, 1568 (1972).
763   Kalinsky, M., Janca, J., J. Polymer Sci., A-1, $\underline{12}$ 1181 (1974).
764   Morris, M.C., J. Chromatogr., $\underline{55}$ 203 (1971).
765   Hamilec, A.E., Omoridion, S.N.E., ACS Symposium Series No. 138, 183 (1980).
766   Mori, S., Anal. Chem., $\underline{53}$ 1813 (1981).
767   Mori, S., Yamakawa, A., Liquid Chromatography, $\underline{3}$ 329 (1980).
768   Hamielec, A.E., Ray, W.H., J. Applied Polymer Sci., $\underline{13}$ 1319 (1969).
769   Mori, S., Yamakawa, A., J. Liquid Chromatogr., $\underline{3}$ 329 (1980).
770   Provcler, T., Rosen, E.M., Separation Science, $\underline{5}$ 437 (1970).
771   Quano, A.C., Dawson, B.L., Johnson, D.E., "Liquid Chromatography of Polymers and Related Materials", Cazes, J., Ed., Marcel Dekker, New York, 1977, pp 1-9.
772   Goedhardt, D., Opshoor, A., J. Polymer Sci., Polymer Phys. Ed., $\underline{8}$ 1227 (1970).
773   Ogawa, T., Suzuki, Y., Inaba, T., J. Polymer Sci., A-1, $\underline{10}$ 737 (1972).
774   Crouzet, P., Fine, O., Mangin, P., Paper presented at the 5th International Seminar, London (1968). Preprint 10, J. Appl. Polymer Sci., $\underline{13}$ 1205 (1969).
775   Dubin, P.L., Koontz, S., Wright, K.L., J. Polymer Sci., $\underline{15}$ 2047 (1977).
776   Kranz, D., Pohl, H.U., Bnmann, H., Angew. Makromol. Chem., $\underline{26}$ 67 (1972).
777   Gaubsic, Z., Rempp, P., Benoit, H., J. Polymer Sci., Part B, $\underline{5}$ 753 (1967).
778   Mori, S., Anal. Chem., $\underline{55}$ 2414 (1983).
779   Tally, C.P., Bowman, L.M., Anal. Chem., $\underline{51}$ 2239 (1979).
780   Belenkii, B.G., Gankina, E.S., J Chromatogr., $\underline{141}$ 13 (1977).
781   Teramachi, S., Hasegawa, A., Yoshida, S., Macromolecules, $\underline{16}$ 542 (1983).
782   Glockner, G., Van den Berg, J.H.M., Meljerink, N.L.J., Scholte, T.G., Koningsveld, R., Macromolecules, $\underline{17}$ 962 (1984).
783   Glockner, G., Van den Berg, J.H.M., Meljerink, N.L.J., Scholte, T.G., Koningsveld, R., J. Chromatogr., $\underline{317}$ 615 (1984).

784   Balke, S.T., Patel, R.D., J. Polymer Sci., Polymer, Lett. Ed., $\underline{18}$
       453 (1980).
785   Balke, S.T., Patel, R.D, Adv. Chem. Ser. No 203, 281 (1983).
786   Tanaka, T., Omoto, M., Donkal, N., Inagaki, H., J. Macromol. Sci.
       Phys. B17, 211 (1980).
787   Danielewicz, M., Kubin, M.,j. Appl. Polym. Sci, $\underline{26}$ 951 (1981).
788   Mori, S., Uno, Y., Suzuki, M., Anal. Chem., $\underline{58}$ 303 (1986).
789   Williamson, G.R., Cervenka, A., European Polymer Journal, $\underline{8}$ 1009
       (1972).
790   Williamson, G.R., Cervenka, A., European Polymer Journal, $\underline{10}$ 295
       (1974).
791   Cervenka, A., Williamson, G.R., European Polymer Journal, $\underline{10}$ 305
       (1974).
792   Cervenka, A., Die Makromolekulane Chemid, $\underline{170}$ 239 (1973).
793   Grubisio, Z., Rempp. R., Benoit, H., J. Polymer Sci., A2 $\underline{8}$ 1803
       (1970).
794   Cervenka, A., Bates, T.W., J. Chromatography, $\underline{53}$ 85 (1970).
795   Drott, E.E., Mendelson, R.A., J. Polymer Sci., A2 $\underline{8}$ 1361 (1970).
796   Williamson, G.R. Cervanka, A., European Polymer Journal, $\underline{8}$ (8),
       1009 (1972).
797   Nakajima, N., Adv. Chem. Ser., 125, $\underline{98}$ (1971) (unpublished 1973).
798   Eldarov, E.G., Goldberg, V.M., Parkratova, G.V., Akutin, M.S.,
       Toptygin, D. Ya. Zavod Lab., $\underline{40}$ (3) 269 (1974).
799   Lovric, L., Nafta (Zagreb), $\underline{23}$ (12), 606 (1972).
800   Lovric, L., Nafta (Zagreb), $\underline{24}$ (1) 47 (1973).
801   Ross, J.H., Shank, R.L., Adv. Chem. Ser., $\underline{125}$ 108 (1971).
802   Maley, J. Polymer Sci., C-8, 253 (1965).
803   Starck, P., Kantola, P. Kem-Kemi, $\underline{3}$ (2) 100 (1976).
804   Gianotti, G., Gaita, A., Romanini, D., Polymer, $\underline{21}$ 1087 (1980).
805   Maley, I.E., J. Polymer Sci., C-8, 253 (1965).
806   Cooper, W., Vaughan, G., Eaves, D.E., Madden, R.W., J. Polymer
       Sci., $\underline{50}$ 159 (1961).
807   Ambler, M.r., Mate, R.D., Durden, J.R., J. Polymer Sci., $\underline{12}$ 1771
       (1974).
808   Ogawa, T., Inaba, t., J. Polymer Sci., $\underline{21}$ (11), 2979 (1977).
809   Vaughan, M.F., Ind. Polym. Charact. Mol. Weight Proc. Meeting, $\underline{111}$
       (1973).
810   Winns, A.M., Swarin, S.J., J. Appl. Polymer Sci., $\underline{19}$ (5), 1243
       (1975).
811   Mori, S., Anal. Chem., $\underline{53}$ 1813 (1981).
812   Otaka, E.P. J. Chromatogr., $\underline{76}$ (1), 149 (1973).
813   Uglea, C.V., Makromol. Chem., $\underline{166}$ 275 (1973).
814   Beden'kii, B.G., Gankina, E.S., Nefedov, P.D., Kuznetsova, M.A.,
       Valchikhina, M.D., J. Chromatogr., $\underline{77}$ 209 (1973).
815   Kranz, D., Rahl, Hu, Baumann, H.,Angew Makromol. Chem., $\underline{26}$ 67
       (1972).
816   Tung, L.H., Runyon, J.R., J. Appl. Polymer Sci., $\underline{17}$ 1589 (1973).
817   Moore, J.C., Hendrickson, J.G., J. Polymer Sci., C-8, 233 (1965).
818   Zhdanov, S.P., Belenkii, B.G., Nekdov, P.P., Kormal'di, E.V., J.
       Chromatogr., $\underline{77}$ (1), 149 (1973).
819   Moore, J.G., J. Polymer Sci., A-2, 835 (1964).
820   Belen'kii, B.G., Gankina, E.S., Nefedov, P.P., Kuznetsova, M.a.,
       Valchikhina, M.D., J. Chromatogr., $\underline{77}$ (1), 209 (1973).
821   Guenet, J.M., Gallot, Z., Picot, C., Bennett, D., J. Appl. Polymer
       Sci., $\underline{21}$ (8), 2181 (1977).
822   Chang, F.S.C., Adv. Chem. Ser., $\underline{125}$ 154 (1971) (published (1973).
823   Hoffmann, M., Urban, H., Makromol. Chem., $\underline{178}$ (9), 268 (1977).
824   Amber, M.R., Mate, R.D., Durdon, J.R., J. Polymer Sci., $\underline{12}$ 1771
       (1974).

825   Stojanov, K, Shirazi, Z.H., Audu, T.O.k., Chromatographia, $\underline{11}$ (5), 274 (1978).

826   Stojanov, K., Shirazi, Z.H., Audu, T.O.K., Ber. Bunsenges Phys. Chem., $\underline{81}$ (8), 767 (1976).

827   Regnier, F.E., Noel, R., J. Chromatogr. Sci., $\underline{14}$ 316 (1976).

828   Janca, j., Kolinsky, M., Ustav Makromol. Chem. Lesk. Akad. Ved. (Prague, Czeckoslovakia), Plsty Kauc., $\underline{13}$ (5), 138 (1976).

829   Daley, L.E., J. Polymer Sci., C-8, 253 (1965).

830   Lin, F.T., Tung, L.X., Liu, F.Y., Hsu, W.W., J. Chim., Inst. Chem. Eng., $\underline{4}$ 43 (1973).

831   Nakao, L, Kuramoto, K., Nippon Secchaku Kyokal Shi., $\underline{8}$ (4), 186 (1972).

832   Mori, S., Anal. Chem., $\underline{53}$ 1817 (1981).

833   Janca, J., Kolinsky, M., J. Appl. Polymer Sci., $\underline{21}$ (1), 83 (1977).

834   Revillon, A., Dumont, R., Guyot, A., J. Polymer Sci., Polymer Chem. Ed., $\underline{14}$ 2263 (1976).

835   Fritzsche, P., Klug, P., Grobe, V., Nuova Chem., $\underline{49}$ 39 (1973).

836   Krans, D., Pohl, Hu., Baumann, H., Angew Makromol. Chem., $\underline{26}$ 67 (1972).

837   Mori, S., Anal. Chem., $\underline{55}$ 2414 (1983).

838   Cooper, D.R., Semylen, J.A., Polymer, $\underline{14}$ 185 (1973).

839   Paschke, E.E., Bidingmeyer, B.A., Bergmann, J.C., J. Polymer Sci. Polymer Chem. Ed., $\underline{15}$ (4), 983 (1977).

840   Miarzk, M., Sir, Z., Coupek, J., Angew Makromol. Chem., $\underline{64}$ (1), 147 (1977).

841   Shiono, S., Anal. Chem., $\underline{51}$ 2398 (1979).

842   Meyerhoff, G., J. Polymer Sci., $\underline{9}$ 596 (1971).

843   Miyamoto, Y., Tomoshige, S., Inagak, H., Polymer J., $\underline{6}$ 564 (1974).

844   Grubisic, Z., Rempp, P., Benoit, H.J., Polymer Sci., B, $\underline{5}$ 753 (1967).

845   Weiss, A.R., Cohn-Ginsberg, E., J. Polymer Sci., A-2, $\underline{8}$ 148 (1970).

846   Belinski, B.G., Nefedov, P.P., Vysokomol. Soedin., A-14, 1568 (1972).

847   Kolinsky, M., Janca, J., J. Polymer Sci., A-1, $\underline{12}$ 1181 (1974).

848   Janca, J., Vlcek, P., Trekoval, J., kolinsky, M., J. Polymer Sci. Polymer Chem. Ed., $\underline{13}$ 1471 (1975).

849   Robertson, A.P., Cook, J.A., Gregory, J.T., Kinetics Symposium Boston, Massachusetts, pp. 258-273 (1972).

850   Moore, J., Hillman, D.E., Brit. Polymer J., $\underline{3}$ 259 (1971).

851   Kato, Y., Sasaki, H., Aiuru, M., Hashimato, T., J. Chromatogr., $\underline{153}$ 546 (1978).

852   Berck, D., Nova, L., Chem. Prumsyl, $\underline{23}$ 91 (1973).

853   Vakhtina, I.T., Tarakanov, O.G., Plast. Kaut, $\underline{21}$ 28 (1974).

854   Klesper, E., Hartmann, W., J. Polymer Sci., Polym. Lett. Ed. $\underline{15}$ 707 (1977).

855   Klesper, E., Hartmann, W., J. Polymer Sci., Polym. Lett. Ed., $\underline{15}$ 9 (1977).

856   Bartle, K.D., Davies, I.L., Raynor, H.M., Hunt, B., Holding, S., (editors), In "Size Exclusion Chromatography, Blackie, Glasgow (1988).

857   Bartle, K.D., in Smith, R.M. (Ed.), Supercritical Fluid Chromatography, Royal Society of Chemistry, London, pp. 1-26 (1988).

858   Chester, T.L., Burkes, L.J., Delaney, T.E., Innis, D.P., Owens, G.D. and Pinksont, J.D., in Charpentier, B.A. and Sevenants M.R. (Eds.), Supercritical Extraction and Chromatography, ACS Symposium Series No. 366, Amercan Chemical Society, Washington DC, pp. 144-160 (1988).

859   Raynor, M.W., Bartle, K.D., Davies, I.L., Williams, A., Clifford,
      A.A., Chalmers, J.M., Cook, B.W., Anal. Chem., 60 427 (1988).
860   Raynor, M.W., Davies, I.L., Bartle, K.D., Williams, A., Chalmers,
      J.D., Cook, B.W., Euro. Chrom. News, 1 18 (1987).
861   Frew, N.M., Johnson, C.G., Bromund, R.M., in Charpentier, B.A. &
      Sevenants M.R. (Eds.), Supercritical Fluid Extraction and
      Chromatography, ACS Symposium Series No. 366, American Chemical
      Society, Washington DC, pp. 208-228 (1988).
862   Taylor W.C., Tung, L.H., Paper presented at 140th Meeting of
      American Chemical Society, Chicago, Illinois, September 1961 (also
      SPE transactions, P. 119. April 1962).
863   Morey, D.R., Tamblyn, J.W., J. Appl. Phys., 16 419 (1945).
864   Claesson, S., J. Polymer Sci., 16 193 (1955).
865   Gamble, L.W., Nipke, W.t., Lane, T.L., J. Appl. Polymer Sci., 9
      1503 (1965).
866   Wesslau, H., Makromol. Chem., 20 111 (1956).
867   Beattie, W.H., private communication.
868   Gooberman, G., J. Polymer Sci., 40 469 (1959).
869   Tanaka, S., Nakamura, A., Morikawa, H., Die Makromol. Chem., 85
      164 (1965).
870   Morey, D.R., Tamblyn, J.W., J. Applied Physics., 16 419 (1945).
871   Gruber, U., Elias, H.G., Die Makromol. Chem., 78 58 (1964).
872   Taylor, W.C., Graham, J.P., Polymer Letters, 2 169 (1964).
873   Taylor, W.c., Tung, L.H., SPE Society Plastics Engineers Trans., 2
      119 (1963).
874   Gamble, L.W., Wipke, W.t., Lane, T., J. Amer. Chem. Soc. Polymer
      Chem. Preprints, 4 No.2, 162 (1963).
875   Wesslau, H., Makromol. Chem., 20 111 (1956).
876   Klenin, V.I., Schegolev, S. Yu., J. Polymer Sci. Polymer Symp., 42
      part 2, 965 (1973).
877   Hall, O., Techniques of Polymer Characterization (P.M. Allen, ed.)
      Academic Press, New York, chapter II (1959).
878   Beattie, W.H., J. Polymer Sci., A, 3 527 (1965).
879   Klenin, V.I., Padol'skii, A.F., Shchegolev, S. Yu., Shvortsburd,
      B.I., Petrova, N.E., Vysokomol. Soedin. Ser., A, 16 (5), 974
      (1974).
880   Cantow, H.H., Kowalski, m., Krozer, C., Angew Chem. Int. Ed.
      Engl., 11 (4), 336 (1972).
881   Tashumuklamedov, S.A., Khasankhanova, M.n., Tillaev, R.S., Uzb.
      Kim Zh., 17 (2), 35 (1973).
882   Grechanovskii, V.A., Vysokomol. Soedin., 17 2721 (1975).
883   Lanikova, J., Hlensek, H., Chem. Prumsyl., 23 (6), 10 (1973).
884   Banicka, E., Ciganekova, V Sh Prednasek., Makrotest 1973, 2 64
      (1973).
885   Vasile, C., Onu, A., Popa, O., Matel, T., Mater. Plast.
      (Bucharest), 10 (12), 631 (1973).
886   Takashima, K., Nakae, K., Shibata, M., Macromolecules, 7 641
      (1974).
887   Kallistov, O.V., Zavod Lab., 38 (6), 711 (1972).
888   Kamata, T., Nakahara, T., J. Colloid Interface Sci., 43 89 (1973).
889   Dautzenberg, H., J. Polymer Sci., C., 39 123 (1972).
890   Pogorel'skii, K.V., Asanov, A., Akhmedov, K.S., Doklady Akad.
      Nauk. Uzh. S.S.R., 27 (3)., 28 (1970).
891   Bradley, J.H., J. Polymer Sci., C-8, 305 (1965).
892   Machtle, W., Klodwig, U., Makromol. Chem., 180 2507 (1979).
893   Blair, W.E., J. Polymer Sci., C-8, 287 (1965).
894   Suzuki, H., Leonis, C.G., Brit. Polymer J., 5 (6), 485 (1973).
895   Mattern, D.E., Hercules, D.M., Anal. Chem., 57 2941 (1985).
896   Williamson, G.R., Cervenka, A., European Polymer Journal, 10 295
      (1974).

897   Ambler, M.R., Mate, R.D., J. Polymer Sci., A-1, 10 (9), 2677 (1972).

898   Sakurada, I., Ikada, Y., Kawahara, T.J., Polymer Sci., 11 2329 (1973).

899   Berens, A.R., Paper delivered to the 168th American Chemical Society Meeting, Atlantic City, New Jersey, entitled, 'The solubility of vinyl chloride in polyvinylchloride'. September 1974.

900   Gilbert, M., Hylart, F.J., J. Polymer Sci., A-1, 9 227 (1971).

901   Kalinina, L.S., Motorina, H.A., U.S.S.R. Patent No. 340, 962 (6/5/72).

902   Fritz, D.F., Sahil, A., Keller, H.P., Koval, E., Anal. Chem., 51 7 (1979).

903   Bezdadea, E.C., J. Polymer Science Polymer Physics Edition, 15 611 (1977).

904   Motorina, M.A., Kalinina, L.S., Metalkina, E.I., Plast. Massy 6 74 (1973).

905   Smironova, O.V., Kirshak, V.V., Salonim, I.Y., Urman, Y.G., Alekseeva, S.G., Bayamov, V.A., Vysolomal Soedin Ser. A., 17 2415 (1975).

906   Law, R.D., J. Polymer Science, A-1, 9 589 (1971).

907   Tsefanov, K.L.B., Panalotov, I.M., Erusalimskii, B.L., European Polymers Journal, 10 557 (1974).

908   Manaff, D.L., Ingham, J.D., Miller, J.A., Orgamic Magnetic Resonance, 10 198 (1977).

909   Valuev, V.I., Shiykhter, R.A., Dmitrieva, I.S., Isvetovshii, I.B., Zhur Anal. Khim., 30 (6), 1235 (1975); Chem. Abstr., 83 131156d (1975).

910   Nissen, D., Rossback, V., Zahn, H., J. Appl. Polymer Sci., 18 1953 (1974).

911   Ghosh, P., Chadha, S.C., Mukoyee, A.R., Palit, R., J. Polymer Sci., A-2, 4443 (1964).

912   Palit, S.R., Ghosh, P., Microchem. J. Symp. Ser., 2 663 (1961).

913   Bartlett, P.D., Nozaki, K., J. Polymer Sci., 3 216 (1948).

914   Ghosh, P., Mukherjee, A.R., Palit, S.R., J. Polymer Sci., A-2, 2807 (1964).

915   Maiti, S., Saha, M.K., J. Polymer Sci., A-1, 5, 151 (1967).

916   Palit, S.R., Makromol. Chem., 36 89 (1959).

917   Palit, S.R., Makromol. Chem., 38 96 (1960).

918   Maiti, S., Ghosh, A., Saha, M.K., Nature (London), 210 513 (1966).

919   Ghosh, P., Sengupta, P.K., Pramanick, A., J. Polymer Sci., A, 3 1725 (1965).

920   Ghosh, P., Chadha, S.C., Mukhejee, A.R., Pait, S.R., J. Polymer Sci., A, 2 4433 (1964).

921   Saha, M.K., Ghosh, P., Palit, S.R., J. Polymer Sci., A, 2 1365 (1964).

922   Palit, S.R., Makromol. Chem., 36 89 (1959).

923   Palit, S.R., Makromol. Chem., 38 96 (1960).

924   Stetnagel, W.J., Palit, S.R., J. Polymer Sci., A-1, 15 945 (1977).

925   Ghosh, P., Chadha, S.C., Mukherjee, A.R., Palit, S.R., J. Polymer Sci., A-2, 4443 (1964).

926   Banthia, a.K., Mandal, B.M., Palit, S.R., J. Polymer Sci. Polymer Chem. Ed., 15 (4), 945 (1977).

927   Kanjilal, C., Mitra, B.C., Palit, S.R., Makromol. Chem., 178 1707 (1977).

928   Mukhopadhyay, S., Mitra, B.C., Palit, S.R., J. Polymer Sci., A-1, 7 2442 (1969).

929   Palit, S.R., Ghosh, P., Palit, S.R., J. Polymer Sci., A-2, 1365 (1964).

930   Saha, M.K., Ghosh, P., Palit, S.R., J. Polymer Sci., A-2 1365
        (1964).
931   Platonov, J.P., Belyaev, V.M., Grigor'eva, F.P., Plast. Massy, 4
        77 (1975).
932   Stark, P., Kantola, P., Kem-Kemi, 3 (2), 100 (1976).
933   Lanikova, J., Hlousek, M., Chem. Prumsyl, 27 (12), 628 (1977).
934   Ross, J.H., Shank, R.L., Ad. Chem. Ser., 125 108 (1971).
935   Chiang, R., J. Polymer Sci., A-3, 3679 (1965).
936   Cuesta de la, Billmeyer, F.W., J. Polymer Sci., A-1, 1721 (1963).
937   Das, N., Palit, S.R., J. Polymer Sci., A-1, 11 1025 (1973).
938   Nakajuima, A., Chem. High Polymers, Japan, Kobunski Kagaku, 7 64
        (1950).
939   Nakajuma, A., Fujiwara, H., High Polymers, Japan, Kobunski Kagaku,
        37 909 (1964).
940   Welsh, T., Engewald, W., Kowash, E., Plaste u kaut, 23 584 (1976).
941   Wozniak, T., Pezegl Wlok, 30 (1), 18 (1976).
942   Feldman, P., Ughea, C.W., Nuta, V., Popa, M., Rev. Roum Chim
        (Bucharest), 17 1033 (1972).
943   Simonova, M.I., Alzenshtein, E.M., Zavod Lab.,40 (4), 435 (1974).
944   Shemalz, E.O., Fas Forsch Tex. Tech., 21 209 (1970).
945   Longhorst, M.A., Stanley, F.W., Cutie, S.S., Sugarmon, J.H.,
        Wilson C.R., Hoagland, D.A., Prud'homme, R.K., Anal. Chem., 58
        2242 (1986).
946   Machtle, W., Makromol. Chem., 183 2215 (1982).
947   Williamson, G.R. Cervenka, A., European Polymer Journal, 8 1009
        (1972).
948   Williamson, C.R., Cervenka, A., European Polymer Journal, 10 295
        (1974).
949   Moore, W.R.A.D., Milln, S.W., British Polymer Journal, 1 181
        (1969).
950   Hama, T., Yamasuchi, K., Susuki, T., Makromol. Chem., 155 283
        (1972).
951   Chiang, R., J. Phys. Chem., 70 2348 (1966).
952   Beckey, H.D., "Principles of Field Ionization and Field Dedoprtion
        mass Spectrometry", Pergamon Press, New York (1977).
953   Schulten, H.R., Adv. Mass Spectrom. 7 83 (1977).
954   Schulten, H.R., Methods Biochem. Anal., 24 313 (1977).
955   Mead, W.L., Anal. Chem., 40 743 (1968).
956   Wiley, R.H., Cook, J.C., Jr., J Machromol. Sci., Chem., A10, 811
        (1976).
957   Wiley, R.H., Macromol. Rev., 14 379 (1979).
958   Lattimer, R.P., Welch, K.R., Pausch, J.B., Rapp, U., Varian MAT
        311A Application Note No 27, Varian MAT Mass Spectrometry, Florham
        Park, NJ, (1978).
959   Lattimer, R.P., Harmon, D.J., Welch, K.R., Anal. Chem., 51 1293
        (1979).
960   Matsuo, T., Matsuda, H., Katakuse, I., Anal. Chem., 51 1329
        (1979).
961   Lattimer, R.P., Welch, K.R., Rubber Chem. Technol., 51 925 (1978).
962   Lattimer, R.P., Welch, K.R., Rubber Chem. Technol., 53 151 (1980).
963   Lattimer, R.P., Harmon, D.J., Anal. Chem., 52 1808 (1980).
964   Florey, P.J., in "Principles of Polymer Chemistry", Cornell
        University Press, Ithica, N.Y., Chapter 7.
965   Altares, T., Wymon, D.P., Allen, V.R., J. Polymer Sci., Part A 2
        4533 (1964).
966   Yeager, F.W., Becker, J.W., Anal. Chem., 49 722 (1977).
967   Sionim, I.Y., Urman, Y.G., Klyuchnikov, U.N., Plaste Massy, 4 68
        (1973).
968   Crompton, T.R, Reid, V.W., J. Polymer Sci., Part A, 1 347 (1963).

969   Albert, N.K., Woodbury Research Laboratory, Shell Chemical Co.,
       Woodbury, U.S.A., private communication.
970   Basch, A., Lewin, M., J. Polymer Sci. Poly., 11 1707 (1973).
971   Dollimore, D. Holt, B., J. Polymer Sci. Polym. Phys. Ed., 11 1703
       (1973).
972   Varma, S.D., Narashimhan, V., J. Appl. Polym. Sci., 16 3325
       (1972).
973   Funt, J.M., Magill, J.H., J. Poly. Csi. Polm. Phys. Ed., 12 217
       (1974).
974   Kokta, B.V., Valade, J.L., Martin, W.N., J. Appl. Polm. Sci., 17
       (1973).
975   Hirozawa, S.T., In Kolthoff, I.M. & Elving, P.J. Editors "Treatise
       on Analytical Chemistry" part II volume 14. John Wiley, New York,
       p.23 (1971).
976   Van Houwelingen, G.D.B., Analyst (London), 106 1057 (1981).
977   Fraga, D.W., Emeryville Research Laboratory Shell Chemical Co.
       Ltd., Emeryville, California, U.S.A., private communication.
978   Chujo, S., Sathoh, S., Ozeka, T., Nagai, E., J. Polymer Sci., 61
       512 (1962).
979   Chujo, R., Satoh, S., Nagai, E., J. Polymer Sci., A-2, 895 (1964).
980   Turner, R.R., Carlson, S.W. Altenau, A.G., unpublished work.
981   Shibota, C., Yamazaki, M., Tabenchi, T., Bull. Chem. Soc. Japan,
       50 311 (1977).
982   Sloane, H.J., Branstone-Cooke, R., Appl. Spectroscopy, 27 217
       (1973).
983   Sewell, R.P., Skidmore, D.W., J. Polymer Sci., A-I, 2425 (1968).
984   Chujo, S., Satoh, S., Ozeka, T., Nagai, E., J. Polymer Sci., 61
       512 (1962).
985   Chujo, S., Satoh, S., Nagai, E., J. Polymer Sci., A-2, 895 (1964).
986   Svob, N., Flajsman, F., Croat. Chem. Acta., 42 417 (1970).
987   Hank, r., Rubber Chem. Technol., 40 936 (1967).
988   Van Schooten, J., Evenhuis, J.K., Polymer (London), 6 561 Nov.
       (1965).
989   Van Schooten, J., Evenhuis, J.K., Polymer (London), 6 343 (1965).
990   Hoer, H., Kooyman Anal. Chim. Acta., 5 550 (1951).
991   Sewell, R.P., Skidmore, D.W., J. Polymer Sci., A-I, 6 2425 (1968).
992   Cooper, W., Eaves, D.e., Tunnicliffe, M.E., Vaughan, G., European
       Polymer J., 1 121 (1965).
993   Tunnicliffe, M.E., Mackillop, D.A., Hank, R., European Polymer J.,
       1 259 (1965).
994   Altenau, H.G., Headley, L.M., Jones, S.O., Rensaw, S.C., Anal.
       Chem., 42 1280 (1970).
995   Lee, T.C., Kolthoff, I.M., Johnson, E., Anal. Chem., 22 995
       (1950).
996   Kemp, A.R., Peters, H., Ind. Eng. Chem. Anal. Ed., 15 52 (1943).
997   Hank, R., Rubber Chem. Techno., 40 936 (1967).
998   Majer, J., Sodomka, J., Chem. Prumsyl., 25 (11) 601 (1975), Chem.
       Abstr., 84 45101 (1976).
999   Munteanu, D., Savu, N., Rev. Chim. (Bucharest), 27 (10) 902
       (1976). Chem. Abstr. 86 121973d (1977).
1000   Aydin, V., Kaczmar, B.U., Schulz, R.C., Angew Makromol. Chem., 24
       171 (1972).
1001   Lenkrath, G., Gummi Asbest. Kunststoffe., 29 585 (1976).
1002   Bevington, J.C., Eaves, D.E., Vale, R.L., J. Polymer Sci., 32 317
       (1958).
1003   Bevington, J.C., Trans. Faraday Soc., 56 1762 (1960).
1004   Miller, D.L., Samsel, E.P., Cobler, J.G., Anal. Chem., 33 677
       (1961).
1005   Samsel, E.P., McHard, J.A., Ind. Eng. Chem. Anal. Ed., 14 750
       (1942).

1006   Haslam, J., Hamilton, J.B., Jeffs, A.R., Analyst (London), 83 66
        (1958).
1007   Haslam, J., Jeffs, A.R., J. Anal. Chem., 7 24 (1957).
1008   Anderson, D.G., Isakson, K.E., Snow, D.L., Tessari, D.J.,
        Vandeberg, J.T., Anal. Chem., 43 894 (1971).
1009   Barrall, E.M., Porter, R.S., Johnson, D.E., Anal. Chem., 35 73
        (1963).
1010   Porter, R.S., Hoffman, A.S., Johnson, J.F., Anal. Chem., 34 1179
        (1962).
1011   Squirrel, D.C.M., "Automatic methods in volumetric analysis"
        Hilger and Watts, London, p. 94-99 (1964).
1012   Anderson, D.G., Isakson, K.E., Snow, D.L., Tessari, D.J.,
        Vandeberg, J.T., Anal. Chem., 43 894 (1971).
1013   Puchalsky, C.B., Anal. Chem., 51 1343 (1979).
1014   Haslam, J., Wills, M.A., Squirrel, D.C.N., Identification and
        analysis of plastics, 2nd edition, Ilcliffe, London, pp 57-63
        (1972).
1015   Porter, R.S., Nicksic, S.W., Johnson, J.F., Anal. Chem., 35 1948
        (1963).
1016   Aydin, V., Kaczmar, B.U., Schulz, R.C., Angew Makromol. Chem., 24
        171 (1972).
1017   Leukroth, G., Gummi Asbest. Kunststoffe, 29 585 (1976).
1018   Majer, J., Sodomka, J., Chem. Prumsy., 25 (11), 601 (1975), Chem.
        Abstr., 84 45101 (1976).
1019   Munteanu, D., Savu, N., Rev. Chim. (Bucharest), 27 (10) 902
        (1976), Chem. Abstr., 86 121973d (1977).
1020   Campbell, D.R., Anal. Chem., 47 1477 (1975).
1021   Phillips, M.A., J. Chem. Soc. (London), 2393 (1928).
1022   Helmuth, J., Poly. Vehromarium Plast., 3 7 (1973).
1023   Czaja, K., Nowadowska, Y., Zubec, I., J. Polymery (Warsaw), 21 158
        (1976).
1024   Majer, J., Sodemnka, J., Chem. Pruimsyl., 25 601 (1975).
1025   Sirynk, A.G., Bulgakova, R.A., Vysokomol Soedin, Ser. B., 19 152
        (1977).
1026   Aydin, V., Kaczmar, B.U., Schulz, R.C., Angew Makromol Chem., 24
        171 (1972).
1027   Esposito, C.G., Swann, M.H., Anal. Chem., 34 1048 (1962).
1028   Percival, D.F., Anal. Chem., 35 236 (1963).
1029   Van Lingen, R.L.M., Fresenius, Z., Anal. Chem., 169 410 (1969).
1030   Van Houwelingen, G.D.B., Albaro, J.G.M., Do Hoog, A.J., Fresenius,
        Z., Anal. Chem., 300 112 (1980).
1031   Nissen, D., Rossbach, V., Zahn, H., J. Appl. Polymer Sci., 18 1953
        (1974).
1032   Crisp, S., Lewis, B.G., Wilson, A.D.J., Dental. Res., 54 (6), 1238
        (1975).
1033   Tan, J.S., Gasper, S.P., Macromolecules, 6 (5), 741 (1973).
1034   Garrett, S.R., Guile, R.L., J. Am. Chem. Soc., 73 4533 (1951).
1035   Douglas, J., Timnick, A., Guile, R.L., J. Polymer Sci., A-I 1609
        (1963).
1036   Mathieson, A.R., McLaren, J.V., J. Polymer Sci., A, 3 2555 (1965).
1037   Katelhalsky, A., Shavit, N., Eisenberg, H., J. Polymer Sci., 13 69
        (1954).
1038   Mathieson, A.R., McLaren, J.V., J. Chem. Soc., 3581 (1960).
1039   Leyte, J.C., Mandel, M., J. Polymer Sci., A, 2 1879 (1974).
1040   Mandel, M., Leyte, J.C., J. Polymer Sci., 56 823 (1962).
1041   Johnson, D.E., Lyeria, J.R., Horikawa, T.T., Pederson, L.A., Anal.
        Chem., 77 49 (1977).
1042   Sharp, J.L., Paterson, G., Analyst (London), 105 517 (1980).
1043   Germanenko, E.N., Perepletchikova, E.M., Zhur. Anal. Khim., 29 830
        (1974).

1044   Van Houwelingen, G.D.B., Analyst (London), 106 1057 (1981).
1045   Van Houwelingen, G.D.B., Peters, M.W.M.G., Huysmans, W.G.B.,
        Fresenius, Z., Anal. Chem., 293 396 (1978).
1046   Tankaka, M., Kojima, I., Anal. Chem. Acta., 41 75 (1968).
1047   Groom, T., Babicc, J.S., Van Leunwen, B.G., J. Cell Plast., 10 53
        (1974).
1048   Dickie, R.A., Hammond, J.S., de Vries, J.E., Holubka, J.W., Anal.
        Chem., 54 2045 (1982).
1049   Crompton, T.R., unpublished work.
1050   Anderson, D.G., Isakon, K.W., Vandeberg, J.T., Jao, M.Y.T.,
        Tessari, D.J., Afremow, L.C., Anal. Chem., 47 1008 (1975).
1051   Haslam, J., Hamilton, J.B., Jeffs, A.R., Analyst (London), 83 66
        (1958).
1052   Miller, D.F., Samsel, E.P., Cobler, J.G., Anal. Chem., 33 677
        (1961).
1053   Neumann, E.W., Nadeau, H.G., Anal. Chem., 10 1454 (1963).
1054   Swann, W.B., Dux, J.P., Anal. Chem., 33 654 (1961).
1055   Zeman, I., Novak, L., Mitter, L., Stekla, J., Holendova, O., J.
        Chromatogr., 119 581 (1976).
1056   Newmann, E.W., Nadeau, H.G., Anal. Chem., 10 1454 (1963).
1057   Swan, W.B., Dux, J.P., Anal. Chem., 33 654 (1961).
1058   Zeman, T., Novak, L., Mitter, L., Stekla, J., Holendova, O., J.
        Chromatography. 119 581 (1976).
1059   Mathias, A., Mellor, N., Anal. Chem., 38 472 (1966).
1060   Stead, J.B., Hindley, A.H., J. Chromat., 42 470 (1969).
1061   Konishi, K., Kanoh, Y., Japan Analyst, 15 1110 (1966).
1062   Karger, M.H., Mazur, Y., J. Amer. Chem. Soc., 90 3878 (1968).
1063   Tsuji, K., Kounishi, K., Analyst (London), 99 54 (1974).
1064   Cervenka, A., Merrall, C.H., J. Chem. Phys., 62 120 (1975).
1065   Swaraj, S., Ranby, B., Anal. Chem., 47 1428 (1975).
1066   Peltonen, K., Pfaffi, P., Itkonen, A., Analyst (London), 110 1173
        (1985).
1067   Hammerich, A.D., Willeboordse, F.G., Anal. Chem., 45 1096 (1973).
1068   Dobinson, B., Hofman, W., Stark, B.P., In "The Determination of
        Epoxide Groups", Pergamon Press, New York, (1969).
1069   Van, Houwelingen, G.D.B., Analyst (London), 106 1057 (1981).
1070   Byenes, I., "Titration in Non-Aqueous Media", Iliffe Books,
        London, p.261 (1967).
1071   Schleuter, D.D., Siggia, S., Anal. Chem., 49 2349 (1977).
1072   Schleuter, D.D., Ph. D. Dissertation University of Massachussets,
        Amherst Mass (1976).
1073   Schleuter, D.D., Siggia, S., Anal. Chem., 49 2343 (1977).
1074   Frankoski, S.P., Siggia, S., Anal. Chem., 44 507 (1972).
1075   Kreshkov, A.n., Shvelsova, L.N., Emelin, E.A., Sov. Plast., (Engl.
        Transl.) (10) 53 (1968), Chem. Abstra., 70 20345q (1969).
1076   Emelin, E.a., Savinov, V.M., Sokolov, L.B., J. Anal. Chem.,
        U.S.S.R., (Engl. Transl.)., 28 1188 (1973); Chem. Abstr., 79
        146956m (1973).
1077   Curnuck, P.A., Jones, M.E.B., Br. Polym. J., 5 21 (1973).
1078   McGowan, R.J., Anal. Chem., 41 2074 (1969)., 42 942 (1970).
1079   Kulikova, N.P., Shablygin, M.V., Utervskii, L.E., Khim. Volokna, 3
        24 (1973), Chem. Abstr., 79 92637v (1973).
1080   Teleshova, A.S., Teleshov, E.N., Pravednikov, A.N., Vysokomol.
        Soedin., Ser. A., 13 2309 (1971), Chem. Abstr., 76 46644k (1972).
1081   Gomoryova, A., Chem. Abstr., 83 59692u (1975).
1082   Stahl, E., Dey, L.S., Kunststoffe, 64 657 (1974), Chem. Abstra.,
        82 112514v (1975).
1083   Mlejnek, O., Cveckova, L., J. Chromatogr., 94 135 (1974).
1084   Kalinina, L.S., Doroshina, L.I., Chem. Abstr., 79 54100g (1973).

1085   Scheddel, R.T., Anal. Chem., 30 1303 (1958).

1086   Tukuchi, T., Tauge, S., Sigimura, Y., J. Polymer Sci., A-1, 6 3415
        (1968).

1087   Kranz, D., Dinges, K., Wending, P., Angew Makromol. Chem., 51 25
        (1976).

1088   Krishen, A., Anal. Chem., 44 494 (1972).

1089   Natta, G., Mazzanti, G., Valvassori, A., Pajaro, A., Chim. Ind.
        (Milan), 29 733 (1957).

1090   Drushell, H.V., Iddings, F.A., Anal. Chem., 35 28 (1963).

1091   Drushell, H.V., Iddings, F.A., Division of Polymer Chemistry 142nd
        Meeting American Chemical Society, Atlantic City, September
        (1962).

1092   Wei, P.W., Anal. Chem., 33 215 (1961).

1093   Gossl, T., Makromol Chem., 42 1 (1961).

1094   Corish, P.J., Anal. Chem., 33 1798 (1961).

1095   Tosi, G., Simonazzi, T., Die Angewandte Makromolecular Chemie., 32
        153 (1973).

1096   Ciampelli, F., Bucci, G., Simonazzi, A., Santambrogio, A., La
        Chimica L'Industria, 44 489 (1962).

1097   Bucci, G., Simonazzi, T., La Chemica et L'Industria, 44 262
        (1962).

1098   Lomonte, J.N., Tirpak, G.A., J. Polymer Sci., A2 705 (1964).

1099   Drushel, H.V., Crit. Rev. Anal. Chem., 1 161 (1970).

1100   Tosi, C., Ciampelli, F., Adv. Polym. Sci., 12 88 (1973).

1101   Brown, J.E., Tyron, M., Mandel, J., Anal. Chem., 35 2173 (1963).

1102   Tyron, N., Horowicz, Z., Mondel, E., J. Research National Bureau
        of Standards 55 219 (1955).

1103   Fraser, G.V., Hendra, P.J., Walker, J.H., Cudby, M.E.A., Willis,
        H.A., Makromol. Chem., 173 205 (1973).

1104   Tosi, C., Makromol. Chem., 170 231 (1973).

1105   Popov, U.P., Duvonov, A.P., Zhur. Prikl. Spektrosk., 18 1077
        (1973).

1106   Porter, N., Nickless, O., Johnson, P., Anal. Chem., 35 1948
        (1963).

1107   Wilkes, C.E., Carmen, C.J., Harrington, R.A., J. Polymer Sci., 43
        237 (1973).

1108   Carmen, C.J., Harrington, R.A., Wilkes, C.E., Macromolecules, 10
        536 (1977).

1109   Ray, G.J., Johnson, P.E., Knox, J.R., Macromolecules, 10 773
        (1977).

1110   Randall, J.C., Macromolecules, 11 33 (1978).

1111   Crain, Jnr., W.O., Zambelli, A., Roberts, J.D., Macromolecules, 14
        330 (1971).

1112   Carman, C.J., Wilkes, C.E., Rubber Chem. Technol., 44 781 (1971).

1113   Schaefer, J., Makromolecules, 4 107 (1971).

1114   Grant, D.M., Paul, E.G., J. Amer. Chem. Soc., 86 2984 (1964).

1115   Paxton, J.R., Randall, J.C., Anal. Chem., 50 1777 (1978).

1116   Yur'eva, F.A., Wasanova, F.O., Pant-Yanskii, A.E., Seidov, N.M.,
        Mamedova, V.M., Abasov, A.T., Malova, H.G., Mysakomol Soedin Seo
        A., 19 2401 (1977).

1117   Popov, V.P., Duvanova, A.P., Zh. Prikl. Spectrosk, 22 1115 (1975).

1118   Saunders, J.M., Komoroski, K., Makromolecules, 10 1214 (1977).

1119   Van Schooten, J., Berkenbosch, R., Polymer (London), 2 357 (1961).

1120   Van Schooten, J., Mostert, S., Polymer (London), 4 135 (1963).

1121   Verdier, J.C., Guyot, A., Macromol. Chem., 175 1543 (1974).

1122   Neumann, E.W., Nadeau, H.G., Anal. Chem., 35 1454 (1963).

1123   Barrall, E.M., Porter, R.S., Johnson, J.F., J. Appl. Polym. Sci.,
        9 3061 (1965).

1124   Barrall, E.M., Porter, R.S., Johnson, J.F., Appl. Polm. Sci., 9
        3061 (1965).

1125   Turner Jons, A., J. Polym. Sci. Polym. Lett. Ed., $\underline{3}$ 591 (1965).
1126   Slonaker, D.F., Combs, R.L., Coover, Jr., H.W., J. Macromol. Soi. Chem., $\underline{1}$ 539 (1967).
1127   Huff, T., Bushman, C.J., Cavender, J.W., J. Appl. Polym. Sci., $\underline{8}$ 825 (1964).
1128   Tosi, C., Lachi, M.P., Pinto, A., Macromol. Chem., $\underline{120}$ 225 (1968).
1129   Wegemer, N.J;., J. Appl. Polym. Sci., $\underline{14}$ 573 (1970).
1130   Lomonte, J., J. Polymer Sci. Polym. Lett. Ed., $\underline{1}$ 645 (1963).
1131   Brown, J.E., Tryon, M., Mandel, J., Anal. Chem., $\underline{35}$ 2172 (1973).
1132   Myers, L.W., Lord, E.B., Shell Chemical Co., Carrington, U.K., Private Communication.
1133   Willbourne, J., J. Polymer Sci., $\underline{34}$ 569 (1959).
1134   Neumann, E.W., Nadeau, H.G., Anal. Chem., $\underline{10}$ 1454 (1963).
1135   Madorsky, S.L., Straus, S., J. Research National Bur. Std., $\underline{53}$ 361 (1954).
1136   Anderson, D.G., Isakson, K.E., Snow, D.L., Tessari, D.J., Candeberg, J.T., Anal. Chem., $\underline{43}$ 894 (1971).
1137   Evans, D.L., Weaver, J.L., Mukherji, A.K., Beatty, C.L., Anal. Chem., $\underline{50}$ 857 (1978).
1138   Nishoika, A., Katoy Ashikari, N., H. Polymer Sci., $\underline{62}$ 510 (1962).
1139   Nishoika, A., Katoy Mitsuoka, H., J. Polymer Sci., $\underline{62}$ 59 (1962).
1140   Bovey, F.A., J. Polymer Sci., $\underline{62}$ 197 (1962).
1141   Ito, K., Yashimata, Y., J. Polymer Sci., $\underline{83}$ 625 (1965).
1142   Harwood, H.J., Ritchey, W.M., J. Polymer Sci., $\underline{B3}$ 419 (1965).
1143   Kranz, D., Dinges, K., Wendling, P., Anal. Makromol. Chem., $\underline{51}$ (1) 25 (1976).
1144   Paul, S., Ranby, B., Anal. Chem., $\underline{47}$ 1428 (1975).
1145   Johnson, D.E., Lyerla, J.R., Horikawa, T.T., Pederson, L.A., Anal. Chem., $\underline{49}$ 77 (1977).
1146   Nishoika, A., Kato, Y., Ashikari, N., J. Polymer Sci., $\underline{62}$ 510 (1962).
1147   Nishoika, A., Kato, Y., Mitsuoka, H., J. Polymer Sci., $\underline{62}$ 59 (1962).
1148   Bovey, F.A., J. Polymer Sci., $\underline{62}$ 197 (1962).
1149   Ito, K., Yashimata, Y., J. Polymer Sci., B-3, 625 (1965).
1150   Harwood, H.J., Ritchey, W.M., J. Polymer Sci., B-3, $\underline{419}$ (1965).
1151   Allen, B.J., Elsea, G.M., Keller, K.P., Kinder, H.D., Anal. Chem., $\underline{49}$ 471 (1977).
1152   Brame, E.G., Yeager, F.W., Anal. Chem., $\underline{48}$ 709 (1976).
1153   Blackwell, J.T., Anal. Chem., $\underline{48}$ 1883 (1976).
1154   Fisch, M.H., Dannenberg, J.J., Anal. Chem., $\underline{49}$ 1408 (1977).
1155   Barrall, E.M., Porter, R.J., Johnson, J.F., J. Chromatogr., $\underline{11}$ 177 (1963).
1156   Barrell, E.M., Porter, R.S., Johnson, J.F., Anal. Chem., $\underline{35}$ 73 (1963).
1157   Porter, R.S., Hoffman, A.S., Johnson, J.F., Anal. Chem., $\underline{34}$ 1179 (1962).
1158   Yur'eva, F.A., Guseinova, F.O., Portyanskii, A.E., Seidov, N.M., Mamedova, V.M., Abasov, A.T., Malova, H.G., Vysokomol Soedin. Ser. A., $\underline{19}$ (10) 2410 (1977); Chem Abstr., $\underline{87}$ 202315d (1977).
1159   Popov, V.P., Duvanova, A.P., Zh. Prikl Sepctrosk., $\underline{22}$ 1115 (1975); Chem. Abstr., $\underline{83}$ 115221d (1976).
1160   Gol'benberg, A.L., Zh. Prikl. Spektrosk., $\underline{19}$ 510 (1973).
1161   Manlus, G.G., J. Polymer Sci., C8, 137 (1965).
1162   Grasley, M.H., Barnum, E.R., Martinez Reasearch Laboratory, Shell Chemical Co. Ltd., private communication.
1163   Wasilewska, W., Grzywa, E., Rajkiewicz, M., Chem. Anal. (Warsaw), $\underline{19}$ 89 (1974).

1164   Kubic, I., Singbar, M., Navratil, M., Petrochemia, _17_ 10 (1977); Chem Abstr., _88_ 51308f (1978).

1165   Kubic, I., Singbar, M., Navratil, M., Petrochemia, _17_ 15 (1977).

1166   Munteanu, D., Toader, M., Mater Plast. (Bucharest), _13_ 97 (1976).

1167   Manlus, G.G., J. Polymer Science, _62_ 263 (1962).

1168   Chiang, T.C., Pham, Q.T., Guyat, A., J. Polymer Science, A-1, _15_ 2173 (1977).

1169   Nishoika, A., Kato, Y., Mitsuoka, H., J. Polymer Science, _62_ 510 (1962).

1170   Voigt, J., Kunststoffe, _54_ 2 (1964).

1171   Voigt, J., Kunststoffe, _51_ 18 (1961).

1172   Voigt, J., Kunststoffe, _51_ 314 (1961).

1173   Van Schooten, J., D.E.W. Berkenbosch, R., Polymer (London), _2_ 357 (1961).

1174   Van Schooten, J., Mostert, S., Polymer (London), _4_ 135 (1963).

1175   Barlow, A., Lehrie, R.S., Robb, J.C., Polymer (London), _2_ 27 (1961); S.C.I. Monogr. No. 17, "Techniques of Polymer Science", p. 267, London (1963).

1176   Van Schooten, J., Mostert, S., Polymer (London), _4_ 135 (1963).

1177   Strassburger, J., Bramer, C.M., Tyron, M., Ferziati, F., Anal. Chem., _32_ 454 (1960).

1178   Van Schooten, J., Evenhuis, J.K., Polymer (London), _6_ 561 (Nov. 1965).

1179   Van Schooten, J., Evenhuis, J.K., Polymer (London), _561_, Nov. 1965 and Polymer (London), _6_ 343 (1965).

1180   Lehmann, F.A., Brauer, G.M., Anal. Chem., _33_ 676 (1961).

1181   Brauer, G.M., J. Polymer Science, C_8_ 3 (1965).

1182   Strassburger, J., Brauer, C.M., Tyron, M., Ferziati, A.F., Anal. Chem., _32_ 454 (1960).

1183   Van Schooten, J., Evenhuis, J.K., private communication.

1184   Van Schooten, J., D.E.W. Berkenbosch, R., Polymer (London), _2_ 357 (1961).

1185   McMurray, H.L., Thornton, V., Anal. Chem., _24_ 318 (1952).

1186   Van Schooten, J., Duck, E.W., Berkenbosch, R., Polymer, _2_ 357 (1961).

1187   Bucci, G., Simonazzi, T., J. Polymer Sci., C-7, 203 (1964).

1188   Bucci, G., Simonazzi, T., Chimica Industria, _44_ 262 (1962).

1189   Natta, G., Mazzanti, G., Valvassori, A., Sartori, G., Merero, D., Chem. Ind. (Milan), _42_ 125 (1960).

1190   Veerkamp, Th. A., Veermans, A., Makromol. Chem., _50_ 147 (1961).

1191   Van Schooten, J., Mostert, S., Polymer, _4_ 135 (1963).

1192   Bucci, G., Simonazzi, T., J. Polymer Sci., C-7, 203 (1964).

1193   McMurray, H.L., Thornton, V., Anal. Chem., _24_ 138 (1952).

1194   Natta, G., Dall'Asta, G., Mazzanti, G., Ciampelli, F., Kolloid, Z., _182_ 50 (1962).

1195   Van Schooten, J. Duck, E.W., Berkenbosch, R., Polymer (London), _2_ 357 (1961).

1196   Van Schooten, J., Mostert, S., Polymer (London), _4_ 135 (1963).

1197   McMurray, H.L., Thornton, V., Anal. Chem., _24_ 138 (1952).

1198   Natta, G., Dall'Asta, G., Mazzanti, G., Ciampelli, F., Kolloid Z., _182_ 50 (1962).

1199   Sheppard, N., Sutherland, G.B., Nature (London), _159_ 739 (1947).

1200   Bucci, G., Simonazzi, T., J. Polymer Science, _C-7_ 203 (1964).

1201   Rugg, F.M., Smith, J.J., Martman, L.H., Ann. N.Y. Acad. Sci., _57_ 398 (1953); J. Polymer Sci., _11_ 1 (1953).

1202   Sutherland, G.B., Disc. Faraday Soc., _9_ 279 (1950).

1203   Natta, G., Mazzanti, G., Valvanori, A., Sartori, G., Merero, D., Chim. Ind. (Milan) _42_ 125 (1960).

1204   Liang, C.Y., Lytton, M.R., Bodre, C.J., J. Polymer Science, _54_ 523 (1961).

1205   Liang, C.Y., Watt, W.R., J. Polymer Science, 51 514 (1961).
1206   Tanaka, T., Hatada, K., J. Polymer Science, 11 2057 (1973).
1207   Grant, D.M., Paul, E.C., J. Amer. Chem. Soc., 86 2984 (1964).
1208   Crain, W.O., Zambelli, A., Roberts, J.P., Makromolecules, 4 330 (1971).
1209   Cannon, C.J., Wilkes, C.E., Rubber Chem. Technol, 44 781 (1971).
1210   Ray, G.J., Johnson, R.E., Knox, J.R., Makromolecules, 10 773 (1977).
1211   Porter, R.S., J. Polymer Science, A-1, 4 189 (1966).
1212   Schaeter, J., Natusch, D.F.S., Macromolecules 5 416 (1972).
1213   Inoue, Y., Nishoika, A., Chujo, R., J. Polymer Science Polymer Physics Edition, 11 2234 (1973).
1214   Wilkes, C.E., Carmen, C.J., Harrington, R.A., J. Polymer Science, 43 237 (1973).
1215   Carmen, C.J., Harrington, R.A., Wilkes, C.E., Macromolecules 10 536 (1973).
1216   Sanders, J.M., Kamoroski, R.A., Macromolecules, 10 1214 (1977).
1217   Paxton, J.R., Randall, J.C., Anal. Chem., 50 1777 (1978).
1218   Randall, J.C., Macromolecules, 11 33 (1978).
1219   Carmen, C.J., Wilkes, C.E., Rubber Age Technology, 44 781 (1971).
1220   Schaefer, J., Macromolecules, 4 107 (1971).
1221   Yur'eva, F.A., Gusonova, F.O., Pant-Yanskii, A.E., Seidov, N.M., Mamedova, V.M., Abasov, A.T., Malova, H.G., Vysolomol Soedin Sect. A., 19 2401 (1977).
1222   Popov, V.P., Dubanova, A.P., Zhur Prikl Spectroscop, 22 1115 (1975).
1223   Wall, L.A., Madorsky, S.L., Brown, D.W., Straus, S., Simka, R. J. Amer. Chem. Soc., 76 3430 (1954).
1224   Tsuchiya, J., Sumi, K., J. Polymer Sci., B, 6 356 (1968).
1225   Tsuchiya, Y., Sumi, K., J. Polymer Sci., A-1, 6 415 (1968).
1226   Wall, L.A., Strauss, S., J. Polymer Science, 44 313 (1960).
1227   Van Schooten, J., Evenhuis, J.K., Polymer (London), 6 343 (1965).
1228   Hachathorn, M.J., Brock, M.J., J. Polymer Sci. Polymer Chem. Ed., 13 (4), 945 (1975).
1229   Kawasaki, A. Paper presented at the 27th Autumn meeting, Japan Chemical Society 1672. Pre-prints, 2 20 (1972).
1230   Hackathorn, M.J., Brock, M.K., Rubber Chem. Technol., 45 1295 (1972).
1231   Sugimura, Y., Takenchi, T., Anal. Chem., 50 1173 (1978).
1232   Harwood, H.J., Angew Chem. International Edition 4 1051 (1965).
1233   Tosi, C., Makromol. Chem., 170 231 (1973).
1234   Popov, V.P., Duvanova, A.P., Zh. Prikl. Spektrosk., 18 1077 (1973).
1235   Seeger, M., Exner, J., Cantow, H.J., Quad. Ric. Sci., 84 102 (1973); Chem. Abstr., 81 50225 (1973).
1236   Seno, J., Tsuge, S., Takeuchi, T., Makromol. Chem., 161 195 (1972).
1237   Bakuyutov, N.G., Kissin, Yu V., Vavilova, I., Arkhipova, Z.V., Vysokmol. Soedin. Ser., A, 13 2163 (1975).
1238   Modric, I., Holland-Moritz, K., Hummel, D.O., Colloid Polymer Sci., 254 342 (1976).
1239   Kamida, K., Yamaguchi, K. Makromol. Chem., 162 205 (1972).
1240   Simak, P., Ropte, E., Makromol. Chem. Suppl., 1 507 (1975).
1241   Oi, N., Moriguchi, K., Bunseki Kagaki, 23 798 (1974); Chem. Abstr., 82 17308x (1974).
1242   Oi, n., Myazaki, K., Moriguchi, K., Shimada, H., Kabunski Kagaki, 29 388 (1972).
1243   Ebdon, J.J., Kandil, S.H., Morgan, K.J., J. Polymer Sci., A-1, 17 2783 (1979).

1244   Shirakawa, H., Yamagaki, N.,Kambara, S. Personall communication
       for Dr. N. Yamazaki, Tokyo Institute of Technology.
1245   Kalal, J., Honska, M., Seycek, O., Adamek, P., Makromol. Chem. 164
       249 (1973).
1246   Arg, T.L., Harwood, H.J., J. Amer. Chem. Soc., Polymer Preprints,
       5 306 (1964).
1247   Yaragisawa, K., Chem. Hisg. Polymers (Tokyo), 21 312 (1964).
1248   Germar, M., Makromolecular Chem., 84 36 (1965).
1249   Enomoto, S., J. Polymer Sci., 55 95 (1961).
1250   Oswald, H.J., Kubu, E.T., Soc. Plastics Eng. Trans. 3 168 (1963).
1251   Bruck, D., Hummel, D., Makromolecular Chem., 163 245 (1973).
1252   Kunpanenko, I.I., Kazananskii, K.S., J. Polymer Sci., Polymer
       Symp. no. 42. PT. 2, 973-80 (1973).
1253   Carmen, C., Macromolecules, 7 789 (1974).
1254   Ritter, W., Elgert, K.F., Cantow, H.J. Makromol. Chem., 178 557
       (1977).
1255   Chen, H.Y., J. Polymer Sci. Polymer Letters Ed., 12 427 (1973).
1256   Elgert, K.F., Cantow, H.J., Stutzel, B. Frenzel Angew Chem. Int.
       Ed., 12 427 (1973).
1257   Brame, E.G., Khan, A.A., Rubber Chem. Technol., 50 272 (1977).
1258   Lindsay, G.A., Santee, E.R., Harwood, H.J., Polymer Preprints,
       American Chemical Society Divn. Polm. Chem., 14 646 (1973).
1259   Araki, K., Fukui-ken Kogyo Shinenjo Nempo, 50 63 (1975); Chem.
       Abstr. 87 118514 (1977).
1260   Wilkes, C.E., J. Polymer Sci. Polymer Symp., 60 161 (1977); Chem.
       Abstr., 89 24952a (1978).
1261   Keller, F., Opitz, H., Hoesselbarth, B., Beckert, D., Reigherdt,
       W., Faserforsch Textiltech., 26 329 (1975).
1262   Keller F., Hoesselbarth, B., Faserforsch Textiltech, 29 152
       (1978).
1263   Keller, F., Zepnik, S., Hoesselbarth, B., Faserforsch Textiltech.,
       28 287 (1977).
1264   Shipman, J.j., Golub, M.A., J. Polymer Sci., A-1, 832 (1963).
1265   Isa, I.A., j. Polymer Sci., A-1, 10 881 (1972).
1266   Fsa, I.A., Meyers, M.E., J. Polymer Sci., A-1, 11 2125 (1973).
1267   Brame, E.G., J. Polymer Sci., A-1, 9 2051 (1971).
1268   Collins, G.C.S., Lowe, A.S., Nicholas, D., European Polymer J., 9
       1173 (1973).
1269   Buchak, B.E., Ramey, K.C., J. Polymer Sci. Polymer Lett. Ed., 14
       401 (1976).
1270   Ibrahim, R., Katritzky, A.R., Smith, A., Weiss, D.E., J. Chem.
       Soc. Perkin Trans., 2 1537 (1974).
1271   Keller, F., Plaste Kautsch, 22 8 (1975).
1272   Delfini, M., Segre, A.C., Conti, F., Macromolecules, 6 645 (1973).
1273   Wu, T.K., Ovenall, D.W., Reddy, G.S., J. Polymer Sci. Polymer
       Phys. Ed., 12 901 (1974).
1274   Mori, Y., Ueda, A., Tamzawa, H., Matsuzaki, K., Kobayoshi, H.,
       Makromol. Chem., 176 699 (1975).
1275   Wang, A., Suzuki, T., Harwood, H.J., J. Polymer Prep. Amer. Chem.
       Soc. Div. Polymer Chem., 16 644 (1975).
1276   Pogorel'skii, K.Y., Asanov, A., Akhmedov, K.S., Dokl. Akad. Nauk.
       Uzh. S.S.R., 27 28 (1970).
1277   Keller, F., Fukrmann, C., Roth, H., Findetsen, G., Raezsch, M.
       Plaste Kautsch, 24 626 (1977); Chem. Abstr., 88 38232g (1978).
1278   Roussel, R., Galin, J.C., J. Macromol. Sci. Chem., A, 11 (2), 347
       (1977).
1279   Okada, T., Otsurn, M., J. Polymer Sci. Polymer Lett. Ed., 14 (10),
       595 (1976).
1280   Wu, T.K., Ovenall, D.W., J. Polymer Prep. Amer. Chem. Soc. Div.
       Polymer Chem., 17 (2), 693 (1976).

1281   Natta, G., Dannusso, F., J. Polymer Sci., $\underline{34}$ 3 (1959).
1282   Bovey, F.A., Polymer Conformation and Configuration, Academic
        Pres, New York, N.Y., p.8 (1969).
1283   Randall, J.C., J. Polymer Sci. Polymer Phys. Ed., $\underline{12}$ 703 (1974).
1284   Zambelli, A., Locatelli, P., Bajo, G., Bovey, F.A., Makro-
        molecules, $\underline{8}$ 687 (1975).
1285   Stehling, F.C., Knox, J.R., Macromolecules, $\underline{8}$ 595 (1975).
1286   Schaefer, J., Natusch, D.F.S., Macromolecules, $\underline{5}$ 416 (1972).
1287   Axelson, D.E., Mandelkern, L., Levy, G.C., Macromolecules, $\underline{10}$ 557
        (1977).
1288   Provasoli, A., Ferre, D.R., Macromolecules, $\underline{10}$ 874 (1977).
1289   Randall, J.C., ACS Symposium Series No. 103, Carbon 13 NMR Polymer
        Science, Wallace, M., Pasika (ed.), American Chemical Society
        (1979).
1290   Randall, J.C., J. Polymer Sci. Polymer Phys. Ed., $\underline{14}$ 2083 (1976).
1291   Stehling, F.C., J. Polymer Sci., A-1, $\underline{4}$ 189 (1966).
1292   Barrall, E.M., Porter, R.S., Johnson, J.F., Paper presented to
        Division of Polymer Chemistry 148th National Meeting, American
        Chemical Society, Chicago, September 1964. Preprints $\underline{5}$, No. 2, 816
        (1964); J. Appl. Polymer Sci., $\underline{9}$ 3061 (1965).
1293   Satoh, S.R., Chujo, T., Nagai, E., J. Polymer Sci., $\underline{62}$ 510 (1962).
1294   Reilly, G.A., Shell Chemical Co., Emeryville, California, private
        communication (1964).
1295   Mitani, K., J. Makromol. Sci. Chem., A, $\underline{8}$ (6), 1033 (1974).
1296   Inoue, Y., Mishoika, O., Chuka, R., Makromelecular Chem., $\underline{152}$ 15
        (1972).
1297   Zambelli, A., Dorman, D.E., Brewster, A.I.R., Bovey, F.A.,
        Makromolecules, $\underline{6}$ 925 (1973).
1298   Randall, J.C., J. Polymer Sci., Polymer Phys. Ed., $\underline{12}$ 703 (1974).
1299   Randall, J.C., J. Polymer Sci., Polymer Phys. Ed., $\underline{14}$ (11), 283
        (1976).
1300   Randall, J.C., J. Polymer Sci., Polymer Phys. Ed., $\underline{14}$ 208 (1976).
1301   Randall, J.C., J. Polymer Sci., $\underline{14}$ 1693 (1976).
1302   Cavelli, L, Relaz Corso Teor-Prat Risonenza Magh. Nuci., 351
        (1973).
1303   Brosio, E., Delfini, M., Conti, F., Nuova Chim., $\underline{48}$ 35 (1972).
1304   Stehling, F.C., Knox, J.R., Makromolecules, $\underline{8}$ 595 (1975).
1305   Stehling, F.C., J. Polymer Sci., A-2, 1815 (1964).
1306   Peraldo, M., Gazz. Chem. Ital., $\underline{89}$ 798 (1959).
1307   McDonald, M.P., Ward, I.M., Polymer, $\underline{2}$ 341 (1961).
1308   Brada, J.J., J. Polymer Sci., $\underline{3}$ 370 (1960).
1309   Luongo, J.P., J. Appl. Polymer Sci., $\underline{3}$ 302 (1960).
1310   Liang, C.V., Pearson, F.G., J. Mol. Spect., $\underline{5}$ 290 (1960).
1311   Sibilia, J.P., Wincklhofer, R.C., J. Appl. Polymer Sci., $\underline{6}$ 557
        (1962).
1312   Schneider, B., Storr, J., Daskoulova, D., Kolinsky, M., Sykora,
        S., Lim, D., Paper presented at International Symposium on
        Macromolecular Chemistry, Prague, (1965).
1313   Schneider, J.S.B., Kalinsky, M., Ryska, M., Lim, D., J. Polymer
        Science, A-1, $\underline{5}$ 2013 (1967).
1314   Abe, Y., Tasumi, M., Shimanouchi, T., Satoh, S., Chujo, R., J.
        Polymer Sci., A-1, $\underline{4}$ 1413 (1966).
1315   Ando, E., Nishoika, A., Wanatabe, S., Polymer Journal, $\underline{3}$ 403
        (1972).
1316   Randall, J.C., J. Polymer Sci., $\underline{13}$ 889 (1975).
1317   Inone, Y., Nishioka, A., Chujo, R., Macromolecular Chem., $\underline{156}$ 207
        (1972).
1318   Matsuzaki, K., Urya, T., Osada, K., Kawamura, T., Macromolecules,
        $\underline{5}$ 816 (1972).

1319   Kissin, Yu B., Gol'dfarb, Yu Ya, Novoderzhkin, Yu V, Krentsel,
        B.A., Vysokomol Soedin. Ser. B, 18 167 (1976).
1320   Strassila, D., Klewsper, E., J. Polymer Science, Polymer Letters
        edition, 15 199 (1977).
1321   Richards, D.H., Williams, R.L., J. Polymer Science, Polymer
        Chemistry Edition, 11 89 (1973).
1322   Leonard J., Malhotra, S.L., J. Polymer Sci., Polymer Chemistry
        Edition, 12 2391 (1974).
1323   Tiers, G.V.D., Boven, F.A., J. Polymer Sci., A, 1 833 (1963).
1324   Smets, G., Van Humbeeck, W., J. Polymer Sci., A, 1 1227 (1963).
1325   Brownstein, S., Wiles, D.M., J. Polymer Sci., A, 2 1901 (1964).
1326   Murano, M., Yamadera, R., J. Polymer Sci., A-1, 5 1855 (1967).
1327   Matsuzaki, K., Okada, M., Uryu, T., J. Polymer Sci., A-1, 6 1701
        (1971).
1328   Fujii, K., Brownstein, S., Eastham, A.M., J. Polymer Sci., A-1, 6
        2387 (1968).
1329   Cooper, W., Johnson, F.R., Vaughan, G., J. Polymer Sci., A, 1 1509
        (1963).
1330   Fujii, K., Machizuki, T., Imoto, S., Ukida, J., Matsumoto, J., J.
        Polymer Sci., A, 2 2327 (1964).
1331   Solve, H., Urya, T., Matsuzake, K., Tabata, Y., J. Polymer Sci.,
        A, 2 3333 (1964).
1332   Matsuzaki, K., Uryu, T., Ishida, A., Ohki, T., Takenchi, M., J.
        Polymer Sci., A-1, 5 2875 (1967).
1333   Svegliado, G., Talamini, G., Vidotto, G., J. Polymer Sci., A-1, 5
        2875 (1967).
1334   Matsuzaki, K., Uryu, T., Okada, M., Hiroyuki, S., J. Polymer Sci.,
        A-1, 6 1475 (1968).
1335   Aylward, N.N., J. Polymer Sci., A-1, 8 319 (1970).
1336   Chiellini, E., Salvadori, P., Osgan, M., Pin, O., J. Polymer Sci.,
        A-1, 8 1589 (1970).
1337   Ito, K., Yamashita, Y., J. Polymer Sci., A-1, 4 631 (1966).
1338   Bacsekai, R., Lindeman, P.L., Robeinstein, D.L., J. Polymer Sci.,
        A-1, 10 1297 (1972).
1339   Randall, J.C. Makromolecules, 11 (3), 592 (1978).
1340   Carmen, C., Macromolecules, 7 789 (1974).
1341   Mauzac, M., Vairon, J.P., Simalt, P., Polymer, 18 1193 (1977).
1342   Okada, T., Ikushige, T., J. Polymer Sci., Polymer Chem. Ed., 14
        (8), 2059 (1976).
1343   Abe, A., Nishoika, A., Kobunski Kagaku, 29 402, 448 (1972).
1344   Carman C.J., Tarpley, A.R., Goldstein, J.H., Macromolecules, 4 445
        (1971).
1345   Carman, C.J., Macromolecules, 6 725 (1973).
1346   Bovey, F.A., Anderson, E.W., Douglas, D.C., Mason, J.A., J. Chem.
        Phys., 39 1199 (1963).
1347   Ramey, K.C., J. Phys. Chem., 70 2525 (1966).
1348   Bovey, F.A., Hood, F.P., Anderson, E.W., Kornegay, R.L., J.
        Polymer Chem., 71 312 (1967).
1349   Heatley, F., Bovey, F.A., Macromolecules, 2 241 (1969).
1350   Cavelli, L., Borsini, G.C., Carraro, G., Confalonieri, G., J.
        Polymer Sci., A-1, 8 801 (1970).
1351   Baker, C., Maddams, W.F., Park, G.S., Robertsson, H., Makromol.
        Chem., 165 231 (1973).
1352   Millan, J.L., De la Pena, J.L., Rev. Plast. Mod., 26 232 (1973).
1353   Kelen, T., Galambos, G., Tudos, F., Balint, G., European Polymer,
        J., 5 617 (1969).
1354   Bezadea, E., Braun, D., Burinana, E., Caraculacu, A., Istrate-
        Robila, G., Angew Makromol Chem., 37 35 (1974).
1355   Schroeder, E., Byrdy, M., Plaste Kautsch, 24 757 (1977).

1356   Wu, T.K., Macromolecules $\underline{6}$ 737 (1973).
1357   Isa, I.A., J. Polymer Sci., A-1, $\underline{10}$ 881 (1972).
1358   Fsa, I.A., Meyers, M.E., J. Polymer Sci., A-1, $\underline{11}$ 2125 (1973).
1359   Elgart, K.F., Cantow, H.J., Stutzel, B., Frenzel Angew Shem. Int.
        Ed., $\underline{12}$ 427 (1973).
1360   Inoeu, Y., Nishoika, A., Chujo, R., Makromol. Chem., $\underline{156}$ (1972).
1361   Jasse, B., Laupretre, F., Monnerie, L., Makromol. Chem. $\underline{178}$ 1987
        (1977).
1362   Borsa, F., Lanzi, G., J. Polymer Sci. Polymer Phys. Ed., $\underline{14}$ 283
        (1976).
1363   Randall, J.C., J. Polymer Sci.,, Polymer Phys. Ed., A-1, $\underline{14}$ 283
        (1976).
1364   Yakata, K., Hirabayashi, T., J. Polymer Sci., A-1, $\underline{14}$ 57 (1976).
1365   Elgert, K.F., Stuetzel, B., Polymer, $\underline{16}$ (10), 758 (1975).
1366   Sandner, B., Keller, F., Roth, H., Faserforsch Textiltech., $\underline{26}$ 278
        (1975).
1367   Yamashita, Y., Yoshida, M., Kawasc, J., Ito, K., Aschi Garasu
        Kogyu Gijutsu, Shoreeikal Kenkvo Hakoku, $\underline{25}$ 87 (1974); Chem.
        Abstr., $\underline{84}$ 60070s (1974).
1368   Regel, W., Westfeld, L., Cantow, H.J., Angew Chem. Int. Ed. Engl.,
        $\underline{12}$ 434 (1973).
1369   Abe, A., Nishoika, N., Kobunshi Kagaku, $\underline{29}$ 402 448 (1972).
1370   Ibrahim, R., Katritzky, A.R., Smith, A., Weiss, D.E., J. Chem.
        Soc. Perkin Trans., $\underline{2}$ 1537 (1974).
1371   Keller, F., Plaste Kautsch, $\underline{22}$ 8 (1975).
1372   Delfini, M., Segre, A.C., Conti, F., Macromolecules, $\underline{6}$ 645 (1973).
1373   Wu, T.K., Ovenall, D.W., Reddy, G.S., J. Polymer Sci. Polymer
        Phys. Ed., $\underline{12}$ 901 (1974).
1374   Kusakov, M.M., Koshevnick, A., Yu, Mekenitskaya, L.I., Shulipine,
        L.M., Amerik, Yu., B., Golova, L.K., Vysokomol. Soedin. Ser. B.,
        $\underline{15}$, (3), 150 (1973).
1375   Suzuki, T., Horwood, H.J., Polymer Prec. Amer. Chem. Soc. Div.
        Polymer Chem., $\underline{16}$ 638 (1975).
1376   Cavelli, L., Relaz, O., Corso Teor. Prat. Rizonenza Magn. Nucl.,
        351 (1973).
1377   Spevacek, J., Schneider, B., Makromol. Chem., $\underline{176}$ 729 (1975).
1378   Matsuzaki, K., Kanai, T., Kawamura, T., Matsumoto, S., Uryu, T.,
        J. Polymer Sci. Polymer Chem. Ed., $\underline{11}$ 961 (1973).
1379   Hateda, K., Ohata, K., Okamoto, Y., Kitayama, T., Umemeuru, Y.,
        yuki, H., J. Polymer Sci. Polymer Lett. Ed., $\underline{14}$ 531 (1976).
1380   Heublein, G., Boerner, R., Schnetz, H., Faserforsch Textiltech.,
        $\underline{29}$ 317 (1978); Chem. Abstr., $\underline{88}$ 191695c (1978).
1381   Johnson, A., Klesper, E., Wirthlin, T., Makromol Chem., $\underline{177}$ 2398
        (1976).
1382   Klesper, E., Johnson, A., Gronski, W., Wehrilr, F.W., Makromol.
        Chem., $\underline{176}$ 1071 (1975).
1383   Suzuki, T., Mitani, K., Takegami, Y., Furukawa, J., Kobayashi, E.,
        Arai, Y., Polymer J., $\underline{6}$ 496 (1974).
1384   Ebdon, J.R., J. Macromol. Sci. Chem., 417 (1974).
1385   Katritzky, A.R., Smith, A., Weiss, D.E., J. Chem. Soc. Perkin
        Trans., $\underline{2}$ 1547 (1974).
1386   Heublein, G., Freitag, W., Schuetz, H., Faserjorsch Textil., $\underline{26}$
        (10), 498 (1975).
1387   Matsuzaki, K., Ito, H., Kawamura, T., Uryu, T., J. Polymer Sci.
        Polymer Chem. Ed., $\underline{11}$ 971 (1973).
1388   Whipple, E.B., Green, P.J., Macromolecules, $\underline{6}$ 38 (1973).
1389   Yamadira, R., Murano, M., J. Polymer Sci, A-1, $\underline{5}$ 2259 (1967).
1390   Lafeyre, W., Cheradame, H., Spaasky, N., Sigwait, P.J., Chem.
        Phys. Physio Chem. Biol., $\underline{70}$ 838 (1973).

1391   Uryu, T., Shimazu, H., Matsuzak, K., J. Polymer Sci. Polymer Lett. Ed., $\underline{11}$ 275 (1973.)
1392   Schaefer, J., Macromolecules, $\underline{4}$ 590 (1972).
1393   Kawamura, I.H., Uryu, I., J. Polymer Sci., A-1, $\underline{11}$ 971 (1973).
1394   Matsuzaki, k., Okazono, S., Kanai, T., J. Polymer Sci., A-1, $\underline{17}$ 3447 (1979).
1395   Keller, F., Plaste Kautsch, $\underline{22}$ 8 (1975).
1396   Keller, F., Findeisen, G., Raetzsch, M., Roth, H., Plaste Kautsch, $\underline{22}$ 722 (1975).
1397   Elgert, R.F., Ritter, W., Makromol. Chem., $\underline{177}$ 2021 (1976).
1398   Binder, J.L., J. Polymer Sci., A-1, $\underline{42}$ (1963).
1399   Natta, G., 15th Annual Technical Conference of the Society of Plastic Engineers, New York, January 1959.

1400   Fraga, D.W. Emeryville Research Laboratories, Shell Chemical Co. Ltd., Emeryville, U.S.A., private communication.

1401   Carlson, D.W., Altenau, A.G., Anal. Chem., $\underline{41}$ 969 (1969).
1402   Mochel, U.D., Rubber Chem. Technol. $\underline{40}$, 1200 (1967).
1403   Carlson, D.W., Ransaw, H.C., Altenau, A.G., Anal. Chem., $\underline{42}$ 1278 (1970).
1404   Binder, J.L., Anal. Chem., $\underline{26}$ 1877 (1954).
1405   Binder, J.L., J. Polymer Sci., A-1, 47 (1963).
1406   Braun, D., Canji, E., Angew Makromol Chem., $\underline{33}$ 143 (1973).
1407   Braun, D., Canji, E., Angew Macromol Chem., $\underline{35}$ 27 (1974).
1408   Hast, M., Deur Siftar, D., Chromatographia, $\underline{5}$ 502 (1972).
1409   Silas, R.D., Yates, J., Thornton, V., Anal. Chem., $\underline{31}$ 529 (1959).
1410   Cornell, S.W., Kownig, J.L., Makromolecules, $\underline{2}$ 540 (1969).
1411   Neto, N., diLauro, C., European Polymer J., $\underline{3}$ 645 (1967).
1412   Binder, J.L., Anal. Chem., $\underline{26}$ 1877 (1954).
1413   Clark, J.K., Chen, H.Y., J. Polymer Sci., $\underline{12}$ 925 (1974).
1414   Harwood, H.J., Ritchey, W.M., J. Polymer Sci., $\underline{B2}$ 601 (1964).
1415   , Hatada, K., Terewaki, Y., Okuda, H., Tanaka, Y., Sato, H., J. Polymer Sci. Polymer Lett. Ed., $\underline{12}$ (6), 305 (1974).
1416   Toy, M.S., Stringham, R.S., Polymer Prep'n American Chemical Society Division of Polymer Chemistry, $\underline{17}$ 672 (1976).
1417   Hackathorne, M.J., Brock, M.J., J. Polymer Sci. Polymer Chem. Ed., $\underline{13}$ (4), 945 (1975).
1418   Furukawa, J., Haga, K., Kobayashi, W., Iseda, Y., Yoshimoto, T., Sakamato, K., Polymer J., $\underline{2}$ 371 (1971).
1419   Hill, R., Lewis, J.R., Simonsen, J.L., Trans. Faraday Soc., $\underline{35}$ 1067 (1937).
1420   Alekseeva, E.N., J. Gen. Chem. (USSR), $\underline{11}$ 353 (1941).
1421   Rabjohn, N., Bryan, C.E., Inskip, G.E., Johnstone, H.W., Lawson, J.K., J. Amer. Chem. Soc., $\underline{69}$ 314 (1947).
1422   Yakubehik, A.I., Spaaskova, A.J., Zak, A.G., Shotatskaya, I.D., Zhur. Obshsh Khim., $\underline{28}$ 3080 (1958).
1423   Hill, R., Lewis, J.R., Simenson, L.L., Trans Faraday Society (London), $\underline{35}$ 1073 (1939).
1424   Vodchnal, J., Kossler, I., Coll. Czech. Communs., $\underline{29}$ 2428 (1964).
1425   Vodchnal, J., Kossler, I., Coll. of Czech Communs, $\underline{29}$ 2859 (1964).
1426   Fraga, D.W., Benson, L.H., Shell Chemical Co. Ltd., Torrance, California, private communication.
1427   Binder, J.L., J. Polymer Sci., A., $\underline{37}$ (1963).
1428   Tanaka, Y., Sato, H., Sumiya, T., Polymer J., $\underline{7}$ 264 (1975).
1429   Tanaka, Y., Sato, H., Ogura, A., Nagoya, I., J. Polymer, Sci. Polymer Chem. Ed., $\underline{14}$ 73 (1976).
1430   Richardson, W.S., Sacker, A., J. Polymer sci., $\underline{10}$ 353 (1953).
1431   Plkovskiu, E.I., Volkenstejn, M., Dokl. Acad. Nauk, S.S.S.R., $\underline{95}$ 310 (1954).

1432   Binder, J.L., Ransaw, H.C., Anal. Chem., $\underline{29}$ 503 (1957).
1433   Corish, P.J., Spectrochim. Acta., $\underline{15}$ 585 (1959).
1434   Maynard, J.T., Moobel, W.E., J. Polymer Sci., $\underline{13}$ 251 (1954).
1435   Ferguson, R.C., Anal. Chem., $\underline{36}$ 2204 (1964).
1436   Kosler, I., Vodchnal, J. Polymer Lett., $\underline{4}$ 415 (1963).
1437   Binder, J.L., Ransaw, H.C., Anal. Chem., $\underline{29}$ 503 (1957).
1438   Binder, J.L., Rubber Chem. Technol., $\underline{35}$ 57 (1962).
1439   Binder, J.H., J. Polymer Sci., A, $\underline{1}$ 37 (1963).
1440   Golub, M.A., J. Polymer Sci., $\underline{36}$ 523 (1959).
1441   Golub, M.A., J. Polymer Sci., $\underline{36}$ 10 (1959).
1442   Maynard, J.T., Moobel, W.E., J. Polymer Sci., $\underline{13}$ 251 (1954).
1443   Morese-Seguela, B., St. Jacques, M., Renaud, J.M., Prud'Lomme, J.,
        Makromolecules, $\underline{10}$ 431 (1977).
1444   Gronski, W., Murayamo, N., Carlow, H.J., Miyamaot, T., Polymer, $\underline{17}$
        358 (1976).
1445   Beebe, D.H., Polymer, $\underline{19}$ (2), 231 (1978).
1446   Dalinskaya, E.R., Khachaturov, A.S., Poletaeva, I.A., Kormer,
        V.A., Makromol Chem., $\underline{179}$ 409 (1978).
1447   Duck, E.W., Grant, D.M., Macromolecules, $\underline{3}$ 165 (1970).
1448   Van, Stratum Dvorak J., J. Chromatogr., $\underline{71}$ 9 (1972).
1449   Galin, M.J., Macromol. Sci-chem., A, $\underline{7}$ 783 (1973).
1450   Hackathorn, M.J., Brock, M.J., J. Polymer Sci. Polymer Lett. Ed.,
        $\underline{8}$ 617 (1970).
1451   Tanaka, Y., Sato, H., Ogura, A., Nagoya, I., J. Polymer Sci.,
        Polymer Chem. Ed., $\underline{14}$ 73 (1976).
1452   Hackathorn, M.J., Brock, H.J., J. Polymer Science Polymer
        Chemistry Edition, $\underline{13}$ 945 (1975).
1453   Asakurin, T., Ando, I., Nishoika, A., Doi, Y., Keii, T., Makromol.
        Chem., $\underline{178}$ 791 (1977).
1454   Van Schooten, J., Duck, E.W., Berkenbosch, R., Polymer (London), $\underline{2}$
        357 (1961).
1455   Natta, G., Dall'Asta, G., Mazzanti, G., Ciampelli, F., Kolloid,
        Z., $\underline{182}$ 50 (1962).
1456   Van Schooten, J., Mostert, S., Polymer (London), $\underline{4}$ 135 (1963).
1457   Tanaka, T., Hatada, K., J. Polymer Sci., $\underline{11}$ 2057 (1973).
1458   Grant, D.M., Paul, E.C., J. Amer. Chem. Soc., $\underline{86}$ 2984 (1964).
1459   Bovey, F.A., Schilling, F.C., Kwei, T.K., Frisch, H.L., Macro-
        molecules, $\underline{10}$ 559 (1977).
1460   Okuda, K.J., Polymer Sci., A-2, 1749 (1964).
1461   Chujo, R., Satoh, S., Nagai, E., J. Polymer Sci., A-2, 895 (1964).
1462   McClanalan, J.L., Privitera, S.A., J. Polymer Sci., A-3, 3919
        (1965).
1463   Abe, A., Nishoika, N., Kobunski. Kagaku., $\underline{29}$ 402 448 (1972).
1464   Ibrahim, R., Katritzy, A.R., Smith, A., Weiss, D.E., J. Chemical
        Society Perkin Trans., $\underline{2}$ 1357 (1974).
1465   Luengo, J.P., Solovay, R., J. Applied Polymer Science, $\underline{7}$ 2307
        (1963).
1466   Luongo, J.P., Solovay, R.J., Appl. Polymer Sci., $\underline{7}$ 2307 (1963).
1467   Barral, M.J., Polymer Sci., $\underline{13}$ 1515 (1975).
1468   Rueda, D.R., Balta-Calleja, F.J., Hidalgo, A., Spectrochimica.
        Acta, $\underline{30A}$ 1545 (1974).
1469   Dankovics, A., Muanay Gumi, $\underline{11}$ 380 (1974); Chem. Abstr., $\underline{82}$ 86773g
        (1975).
1470   Faizi, N.K., Chikin, Y.A., Zh. Prikl. Specrosk, $\underline{28}$ 167 (1978).
1471   Murano M., J. Polymer Sci., A-1, $\underline{9}$ 567 (1971).
1472   Slovakhotova, N.A., Il'netheva, Z.F., Vasiliev, L.A., Kangin,
        V.A., Karpov Physio-chemical Scientific Research Institute,
        unidentified publication, (1963).
1473   Binder, Z.Z., Ransow, H.C., Anal. Chem., $\underline{29}$ 503 (1957).

1474   Fisher, N., Hellwege, R.H., J. Polymer Sci., $\underline{56}$ 33 (1962).
1475   Fisher, H., Hellwege, R.H., J. Polymer Sci., $\underline{56}$ 33 (1962).
1476   Browning, H.L., Ackerman, H.D., Patton, H.W., J. Polymer Sci., A4, 1433 (1966).
1477   Tsuji, K., Polymer Sci. Polymer Chem. Ed., $\underline{11}$ 1407 (1973).
1478   Tsuji, K., J. Polymer Sci. Polymer Chem. Ed., $\underline{11}$ (2), 467 (1973).
1479   Cambell, D., Araki, K., Turner,. D.T., J. Polymer Sci., A-1, $\underline{4}$ 2597 (1966).
1480   Cambell, D., Turner, D.T., J. Polymer Sci., A-1, $\underline{5}$ 2199 (1967).
1481   Hama, Y., Hosano, Y., Shinohara, K., J. Polymer Sci., , A9, 1411 (1971).
1482   Seiki, T., Takeshita, T., J. Polymer Sci., A-1, $\underline{10}$ 3119 (1972).
1483   Tsuji, K., J. Polymer Sci., A-1, $\underline{11}$ 467 (1973).
1484   Tsuji, K., J. Polymer Sci., A-1, $\underline{11}$ 1407 (1973).
1485   Hori, Y., Shimaka, S., Kashiwabara, H., Polymer, $\underline{18}$ (6), 567 (1977).
1486   Hori, Y., Shimada, S., Kashiwabara, H., Polymer, $\underline{18}$ (6), 1143 (1977).
1487   Olnishi, S.I., Sugimato, S.I., Nitta, I., J. Polymer Sci., Part A., $\underline{1}$ 605 (1963).
1488   Ono, Y., Feii, T., J. Polymer Science, A-1, $\underline{4}$ 2429 (1966).
1489   Segushi, T., Tamura, N., J. Polymer Sci., A-1, $\underline{12}$ 1671 (1974).
1490   Seguchi, T., Tamura, N. J. Polymer Sci., A-1, $\underline{12}$ 1953 (1974).
1491   Loy, B.R., J. Polymer Sci., A-1, 225 (1963).
1492   Wall, I.A., J. Polymer Sci., $\underline{17}$ 141 (1955).
1493   Buck, T., Ph D Thesis, North Western University Microfilm MIC 60-4761 Ann Arbor, Michigan (1960).
1494   Forrestal, L.J., Hodgson, W.G., J. Polymer Sci., A-2, 1275 (1964).
1495   Ohnishi, S., Sugimoto, S., Nitta, I., J. Polymer Sci., A-1, 625 (1963).
1496   Kusumoto, N., Matsumoto, K., Tabayagni, M., J. Polymer Sci., A-1, 1773, (1969).
1497   Ooi, T., Schiotsula, M., Hama, Y., Shinokarie, K., Polymer, $\underline{16}$ 510 (1975).
1498   Florin, R.E., Wall, L.A., Brown, D.W., J. Polymer Sci., A-1, 1521 (1963).
1499   Bullock, A.T., Cameron, G.g., Smith, P.M., Polymer, $\underline{14}$ 525 (1973).
1500   Florin, R.E., Wall, L.A., J. Chem. Phys., $\underline{57}$ (4), 1791 (1972).
1501   Ouchi, I., J. Polymer Sci., A-3, 2685 (1965).
1502   Liebman, S.A., Renwer, J.F., Gollatz, K.A., Nauman, C.D., J. Polymer Sci., A-1, $\underline{9}$ 1823 (1971).
1503   Sakaguchi, M., Yamakawa, H., Shomo, J., J. Polymer Sci. Polymer Lett. Ed., $\underline{12}$ 193 (1974).
1504
1505   Hay, J.N., J. Polymer Sci., A-1 $\underline{8}$ 1201 (1970).
1506   Bowden, M.J., O'Donnell, J.H., J. Polymer Sci., A-1, $\underline{7}$, 1665 (1969).
1507   Sakai, Y., Iwasaki, M., J. Polymer Sci., A-1, $\underline{7}$ 1749 (1969).
1508   Harris, J.A., Horojosa, O., Author, J.C., J. Polymer Sci., A-1, $\underline{11}$ 3215 (1973).
1509   Gueskens, G., David, C., Makromol. Chem., $\underline{165}$ 273 (1973).
1510   Yashoika, H., Matsumotoh, H., Uno, S., Higashide, F., J. Polymer Sci., A-1, $\underline{14}$ 1331 (1976).
1511   Sakaguchi, M., Kodama, S., Ediund, O., Sohma, J., J. Polymer Sci. Polymer Lett. Ed., $\underline{12}$ 609 (1974).

1512   Bullock, A.T., Cameron, G.G., Elson, J.M., Polymer, $\underline{15}$ 74 (1974).
1513   Michel, R.E., Chapman, F.W., Mao, T.J., J. Polymer Sci., A-1, $\underline{5}$ 1077 (1967).

1514   Hama, Y., Shinohara, K., J. Polymer Sci., A-1, $\underline{8}$ 651 (1970).
1515   Clay, M.R., Charlesby, A., European Polymer J., $\underline{11}$ 187 (1975).
1516   Placek, j., Szocs, F., Borsig, E.J., J. Polymer Sci., A-1, $\underline{14}$ 1549 (1976).
1517   Shioji, J., Ohnishi, S., Nitta, I., J. Polymer Sci., A-1, 3373 (1963).
1518   Hajimoto, Y., Tamura, N., Okamoto, S., J. Polymer Sci., A-3, 255 (1965).
1519   Iwasaki, M., Sakai, Y., J. Polymer Sci., A-1, $\underline{7}$ 1537 (1969).
1520   Thyron, F.C., Baijal, M.D., J. Polymer Sci., A-1 $\underline{6}$ 505 (1968).
1522   Lerner, L.R., J. Polymer Sci., A-1, $\underline{12}$ 2477 (1974).
1523   Hiraki, K., Inoue, T., Hirai, H., J. Polymer Sci., A-1, $\underline{8}$ 2545 (1970).
1524   Hirai, H., Kiraki, K., Hoguchi, I., Inoui, T., Makishima, S., J. Polymer Sci., A-1 $\underline{8}$ 2393 (1970).
1525   Takeda, K., Yoshida, H., Hayashi, K., Okamura, S., J. Polymer Sci., A-1, $\underline{4}$ 2710 (1966).
1526   Wiechec, L, Anal. Chem. (Warsaw), $\underline{18}$ 853 (1973).
1527   Eda, B., Numone, K., Iwasaki, M., J. Polymer Sci., A-1, $\underline{8}$ 1831 (1970).
1528   Tanaka, O., J. Polymer Sci., A-1, $\underline{11}$ 2069 (1973).
1529   Pilar, J., Toman, L, Marek, M., J. Polymer Sci., A-1, $\underline{14}$ 2399 (1976).
1530   Fleischer, G., Hellebrand, J., Win, Z., Karl Marx Univ., Leipzig, Math., Naurweiss, Reiche, $\underline{21}$ 653 (1972).
1531   Alebarino, V., Otacka, E.P., Luongo, J.P., J. Polymer Sci., A-1, $\underline{15}$ 1517 (1977).
1532   Vigneron, J.M., Deschreider, A.R., Lebensin-Wiss Technol, $\underline{5}$ 198 (1972).
1533   Kawshima, T., Shimada, S., Kashiwabara, H., Sohma, J., Polymer Journal, $\underline{5}$ 135 (1973).
1534   Seeger, M., Exner, J., Cantow, H.J., Paper presented at 23rd International Congress of Pure and Applied Chemistry, Boston, 1971, Macromolecular Preprints, $\underline{11}$ 739 (1971).
1535   Exner, j., Seeger, M., Cantow, J.J., Angew Chem. (International Ed.), $\underline{10}$ 346 (1971).
1536   Stehliag, F.C., Knox, J.R., Macromolecules, $\underline{8}$ 595 (1975).
1537   Laiber, M., Opera, H., Toader, M., Laiber, V., Rev. Chem. (Bucharest), $\underline{28}$ (9) 881 (1977); Chem. Abstr., $\underline{88}$ 23539p (1978).
1538   Solovay, R., Pascale, J.W., J. Polymer Sci., $\underline{2}$ 2041 (1964).
1539   D-ASTM-2283-64T. Tentative Methods of Test for the Absorbance of Polyethylene due to Methyl Groups at 1378 cm-1. Method A.
1540   Neilson, O., Holland, O., J. Mol. Spect., $\underline{4}$ 448 (1960).
1541   Boyle, D.A., Simpson, W., Waldron, J.D., Polymer, $\underline{2}$ 323 (1961).
1542   Willbourne, A.H., J. Polymer Sci., $\underline{34}$ 509 (1959).
1543   Nerheim, A.G., Anal. Chem., $\underline{47}$ 1128 (1975).
1544   Barral, M., J. Polymer Sci., $\underline{13}$ 1515 (1975).
1545   Nishoika, A., Ando, I., Matsumoto, J., Bunseki Kayaku, $\underline{26}$ (5), 308 (1977); Chem. Abstr., $\underline{87}$ 102729h (1977).
1546   Kraus, G., Stacy, C.J., J. Polymer Sci., Symposium No. 43, 329 (1973).
1547   Axelson Levy, G.C., Mandekern, L., Macromolecules, $\underline{12}$ 41 (1979).
1548   Bovey, F.A., Schilling, F.G., McCrackin, F.I., Wagner, H.L., Macromolecules, $\underline{9}$ 76 (1976).
1549   Foster, G.N., Polymer Preprints, $\underline{20}$ 463 (1970).
1550   Randall, C., J. Polymer Sci., Polymer Physics Edition, $\underline{11}$ 275 (1973).
1551   Van Schooten, J., Evenhuis, J.K., Polymer (London), $\underline{6}$ 561 November (1965).

1552 Van Schooten, J., Evenhuis, J.K., Polymer (London), <u>6</u> 353 (1965).

1553 Kraus, G., Stacey, C.J., J. Polymer Science Symposium No. 43, 329 (1973).

1554 Barlow, A., Wild, A., Ranganath, R., J. Appl. Polymer Sci., <u>21</u> 3319 (1977).

1555 Wild, L, Ranganath, R., Barlow, A., J. Appl. Polymer Sci., <u>21</u> 3331 (1977).

1556 Randall, J.C., J. Polymer Sci., Polymer Phys. Ed., <u>11</u> 275 (1973).

1557 Randall, J.C., J. Appl. Polymer Sci., <u>22</u> 585 (1978).

1558 Kamath, P.M., Barlow, A., J. Polymer Sci., A-1 <u>5</u> 2023 (1967).

1559 Kamath, P.M., Barlow, A., J. Polymer Sci., A-1, <u>5</u> 2023 (1967).

1560 Bellamy, L.J., Infrared Spectra of Complex Molecules. Wiley, New York, <u>27</u> 13 (1958).

1561 Willbourn, A.J., J. Polymer Sci., <u>34</u> 569 (1959).

1562 Van Schooten, J., Evenhuis, J.K., Polymer (London), <u>6</u> 561 (1965).

1563 Seger, M., Barrall, E., J. Polymer Sci., A-1, <u>13</u> 1515 (1975).

1564 Van Schooten, J., Evenhuis, J.K., Polymer (London), <u>5</u> 61 Nov. (1965).

1565 Van Schooten, J., Evenhuis, J.K., Polymer (London), <u>6</u> 343 (1965).

1566 Willbourn, A.H., J. Polymer Sci., <u>34</u> 569 (1959).

1567 Catman, J.D., J. American Chemical Society, <u>77</u> 2790 (1955).

1568 Carrega, M., Bonnebat, C., Zednik, G., Anal. Chem., <u>42</u> 1807 (1970).

1569 De Vries, Bonnebat, C., Carrega, M., Pure Applied Chemistry, <u>26</u> 209 (1971).

1570 Bovey, F.A., Abbas, K.B., Schilling, F.C., Starnes, W.H., Macromolecules, <u>8</u> 437 (1975).

1571 Cobler, J.G., Samsel, E.P., SPE Transactions 145 April (1962).

1572 Ohtani, S., Ishikawa, T., Kogyo Kagaku Zasshi, <u>65</u> 617 (1962).

1573 O'Mara, M.M., J. Appl. Polymer Sci., <u>13</u> 1887 (1970).

1574 Watson, J.T., Biemann, K., Anal. Chem., <u>36</u> 1135 (1964).

1575 Stromberg, R.R., Staus, S., Ackhammer, B.G., J. Polymer Sci., <u>35</u> 355 (1959).

1576 Tsuchiya, F., Sumi, K., J., Appl. Chem., <u>17</u> 364 (1967).

1577 Ohta, M., Kogyo Kagaku Zasshi, <u>55</u> 31 (1952).

1578 Boettner, E.A., Ball, G., Weiss, B., J. Appl. Polymer Sci., <u>13</u> 377 (1969).

1579 Noffz, D., Benz, W., Pfab, W., Z. Anal. Chem., <u>235</u> 121 (1968).

1580 Liebman, S.A., Ahlstrom, D.H., Quinn, E.J., Geighley, A.G., Meluskey, J.T., J. Polymer Sci., A-1, <u>9</u> 1921 (1971).

1581 Boettner, E.A., Weiss, B., Amer. Ind. Hyg. Assoc. J., <u>28</u> 535 (1967).

1582 Carrega, M., Bonnebat, C., Zedovik, G., Anal. Chem., <u>42</u> 1807 (1970).

1583 Ahlstrom, D.H., Lubman, S.A., Abbas, K.P., J. Polymer Sci. Polymer Chem. Ed., <u>14</u> 2479 (1976).

1584 Chang, E.P., Salovey, R., J. Polymer Sci. Polymer Chem. Ed., <u>12</u> 2927 (1974).

1585 Suzuki, M., Tsage, S., Takeuchi, T., J. Polymer Sci., A-1, <u>10</u> 1051 (1972).

1586 Ozawa, T., Bull. Chem. Soc. Japan, <u>38</u> 1881 (1965).

1587 Flynn, J.H., Wall, L.A., Polymer Letters, <u>4</u> 323 (1966).

1588 Doyle, C.D., J. Appl. Polymer Sci., <u>5</u> 285 (1961).

1589 Zsako, C.D., J. Appl. Polymer Sci., <u>19</u> 333 (1980).

1590 Basch, A., Lwein, M., J. Polymer Sci. Polymer Chem. Ed., <u>11</u> 3071 (1973).

1591 Basch, A., Lwein, M., J. Polymer Sci. Polymer Chem. Ed., <u>11</u> 3095 (1973).

1592 Basch, A., Lwein, M., J. Polymer Sci. Polym., <u>11</u> 1707 (1973).

1593   Dollimore, D., Holt, B., J. Polymer Sci. Polymer Phys. Ed., $\underline{11}$
        1703 (1973).

1594   Varma, D.S., Narasimhan, V., J. Appl. Polymer Scil., $\underline{16}$ 3325
        (1972).

1595   Funt, J.M., Magill, J.H., J. Polymer Sci. Polymer Phys Ed., $\underline{12}$ 217
        (1974).

1596   Kikta, B.V., Valade, J.L., Martin, W.N., J. Appl. Polymer Sci.,
        $\underline{17}$, 1 (1973).

1597   Judd, M.D., Norris, A.C., J. Therm. Anal., $\underline{5}$ (2), 179 (1973).

1598   Tulupov, P.E., Karpov, O.N., Zh. Fiz. Khim., $\underline{47}$ (6), 1420 (1973).

1599   Joesten, B.L., Johnston, N.W., J. Macromol. Sci. Chem., $\underline{8}$, 83
        (1974).

1600   Halip, V., Stan, V., Biro, A., Radovici, R., Mater. Plast.
        (Bucharest), $\underline{10}$ (11), 601 (1973).

1601   Cooper, D.R., Sutton, G.J., Tighe, B.J., J. Polymer Sci., Polymer
        Chem. Ed., $\underline{11}$ 2045 (1973).

1602   Patterson, A., Sutton, G.J., Tigh, B.J., J. Polymer Sci. Polymer
        Chem. Ed., $\underline{11}$ 2343 (1973).

1603   Sutton, G.J., Tighe, B.J., J. Polymer Sci. Polymer Chem. Ed., $\underline{11}$
        1069 (1973).

1604   Varma, D.S., Ravisankar, S., Angew, Makromol. Chem., $\underline{28}$ 191
        (1973).

1605   Kromalte, R., Malegina, N.D., Kotov, B.V., Oksent'evich, L.A.,
        Pravednikov, A.N., Vysokomol. Soedin, Ser. A., $\underline{14}$ (10), 2148
        (1972).

1606   Barrales-Rienda, J.M., Ramos, J.G., J. Polymer Sci. Symp., No. 42,
        1249 (1973).

1607   Hodd, K.A., Holmes-Walker, W.A., J. Polymer Sci. Symp. No. 42,
        1435 (1973).

1608   Gedemer, T.J., J. Macromol. Sci. Chem., $\underline{8}$ (1), 95 (1974).

1609   Kiran, E., Gillham, J.K., Gipstein, E.J., Makromol. Sci-Phys. B,
        9, (2) 341 (1974).

1610   Ahlstrom, D.H., Foltz, C.R., J. Polymer Sci., A-1, $\underline{16}$ 2703 (1978).

1611   Gedemer, T.J., J. Macromol. Sci. Chem., $\underline{8}$ 95 (1974).

1612   Dassanayake, N.L., Philips, R.W., Anal. Chem., $\underline{56}$ 1753 (1984).

1613   Al-Sammerrai, D., Al-Najjar, H., Selam, W., Analyst (London), $\underline{110}$
        1267 (1985).

1614   Holden, H.W., J. Polymer Sci., C-6, 53 (1964).

1615   Dick, N., Westerberg, C.J., Macromol Sci. Chem., A, $\underline{12}$ 455 (1978).

1616   Poshet, G., Kunstst-Plast. (Solothurn, Switz.), $\underline{25}$ (1), 24 (1978).

1617   Reich, L., Thermochimica Acta, $\underline{5}$ 433 (1973).

1618   Kotoyori, T., Thermochimica Acta, $\underline{5}$ 51 (1972).

1619   Caillot, C., Fournie, R., Alidoux, C., Briver-Rev Therm. $\underline{11}$ 461
        (1972).

1620   Van, K.V., Malhotra, S.L., Blanchard, L.P. J. Macromol. Sci.
        Chem., $\underline{8}$ 843 (1974).

1621   Slysh, R., Hettinger, A.C., Guyler, K.E., Polymer Eng. Sci., $\underline{14}$
        264 (1974).

1622   Willard, P.E., J. Macromol. Co. Chem., $\underline{8}$ 33 (1974).

1623   Willard, P.E., SPEJ, $\underline{29}$ 38 (1973).

1624   Radhakrishnan, N.G., Padbye, M.R., Angew Makromol. Chem., $\underline{43}$ 177
        (1975).

1625   Sickfield, J., Heinze, B., J. Therm. Anal., $\underline{6}$ 689 (1974).

1626   Sourouur, S., Kamal, M.R., Thermochim. Acta, $\underline{14}$ (1-2), 41 (1976).

1627   Kay, R., Westwood, A.R., Europ. Polymer J., $\underline{11}$ (1), 25 (1975).

1628   Ebdon, J.R., Hunt, B.J., Anal. Chem., $\underline{45}$ 1410 (1973).

1629   Godard, P., Mercier, J.P., J. Appl. Polymer Sci., $\underline{18}$ 1493 (1974).

1630   Malavasic, T., Vizovisek, I., Lananje, S., Moze, A., Makromol.
        Chem., $\underline{175}$ 873 (1974).

1631   Moze, A., Vizovisek, I., Malavasic, T., Cernec, F., Lapanje, S.,
       Makromol. Chem., 175 1507 (1974).
1632   Heinen, K.U., Hummel, D.O., Kolloid-Z. Z. Polym., 251 901 (1973).
1633   Sebenik, A., Vizovisek, I., Lapanje, S., Europ. Polymer J., 10 273
       (1974).
1634   Barton, J.M., J. Macromol Sci. Chem., 8 25 (1974).
1635   Barton, J.M., Makromol Chem., 171 247 (1973).
1636   Crane, L.W., Dynes, P.J., Kaelble, D.H., J. Polymer Sci. Polymer
       Lett. Ed., 11 533 (1973).
1637   Sacher, E., Polymer, 14 91 (1973).
1638   Duwalt, A.A., Thermochimica Aita, 57 8 (1974).
1639   Macallum, J.R., Makromol. Chem., 83 137 (1965).
1640   McNeill, I.C., J. Polymer Sci., A-1, 4 2479 (1966).
1641   Mehmet, Y., Roche, R.S., J. Appl. Polymer Sci., 20 (7), 1955
       (1976).
1642   Hmelnitsii, R.A., Lukashenko, I.M., Brodskii, E.S., Pyrolysis Mass
       Spectrometry of Macromolecules (Khimija, Moscow, 1980).
1643   Meuzelaar, H.L.C., Haverkamp, I, Hileman, F.D., Pyrolyisis Mass
       Spectrometry of Recent and Fossil Biomaterials, Elsevier,
       Amsterdam (1982).
1644   Risby, T.H., Yergey, I.A., Socca, I.I., Anal. Chem., 54 2228
       (1982).
1645   Yamada, K., Oura, T., Haruki, T., "Proceedings of IV Int. Conf. of
       Thermal Analysis", (Budapest, 1974), part 3, p.1029.
1646   Gaspar, G., et al., Anal. Chem., 50 1572 (1978).
1647. Schutjes, C.P.M., et al., J. Chromatogr., 279 269 (1983).
1648   Kullik, E., Kalyurand, M., Lambery, M., Laboratory Practice, p.73,
       Jan-Feb (1978).
1649   Deans, J.R., J. Chromatography, 289 43 (1984).
1650   Heinsoo, E., et al., Anal. Appl. Pyrol., 2 131 (1980).
1651   Coloff, S.G., Vanderborgh, N.E., Anal. Chem., 45 1507 (1973).
1652   Schmitt, C.R., J. Fire Flammability, 3 303 (1972).
1653   Risby, T.H., Yegey, J.A., Anal. Chem., 54 2228 (1982).
1654   Foti, S., Liguori, A., Moravigra, Pl, Montando, G., Anal. Chem.,
       54 647 (1982).
1655   Moe, G.J., Thermochimica, Acta, 10 259 (1974).
1656   Hughes, J.C., Wheals, B.B., Whitehouse, M.J., Analyst (London),
       102 143 (1977).
1657   Joseph, K.T., Browner, R.F., Anal. Chem., 52 1083 (1980).
1658   Schulman, G.P., Polymer Letters, 3 911 (1965).
1659   Risby, T.H., Yergey, J.A., Scocca, J.J., Anal. Chem., 54 2228
       (1982).
1660   Peltonen, K., Analyst (London), 111 819 (1986).
1661   Chang, T.L., Mead, T.E., Anal. Chem., 431 534 (1971).
1662   Langer, H.G., Gohlke, R.S., Smith, D.H., Anal. Chem., 37 433
       (1965).
1663   Shulman, G.P., Lochte, H.W., Polymer Preprints, 6 36 (1965).
1664   Madorsky, S.L., Strauss, S.J., Res. Natl. Bur. Stand., (U.S.) 40
       417 (1948).
1665   Bradt, Pl, Dibeler, V., Mohler, F.L., J. Res. Natl. Bur. Stand.,
       (U.S.), 50 201 (1953).
1666   Wiley, R.H., Smithson, L.H. Jr., J. Macromol. Sci. Chem., 2 589
       (1968).
1667   Zeman, A., In "Thermal Analysis", Wiedeman, H.G., Ed., Birhauser
       Verlag, Basel, (1972).
1668   Stenhagen, E., Abrahamson, S., McLafferty, F.W., "Atlas of Mass
       Spectral Data", Wiley-Interscience, New York, 4 645 (1979).
1669   Shimizu, Y., Munson, M.S. B., J. Polym. Sci., Polym. Chem. Ed., 17
       1991 (1979).

1670   Cascaval, C.n., Mater. Plast. (Bucharest), 16 120 (1979).

1671   Blazso, M., Varhegyi, G., Jakab, E., J. Anal. Appl. Pyrolysis, 2 177 (1980).

1672   Oehme, G., Baudisch, H., Mix, H., Makromol. Chem., 177 2657 (1976).

1673   Kiran, E., Gilliham, J.K., J. Appl. Polym. Sci., 21 1159 (1977).

1674   Cameron, G., C., MacCallum, J.R., J. Macromol. Sci., Rev. Macromol. Chem., Cl, 327 (1967).

1675   Rudin, A., Samanta, M.C., Reilly, P.M., J. Appl. Polym. Sci., 24 171 (1979).

1676   Cameron, G.g., Meyer, J.M., McWalter, I.T., Macromolecules, 11 696 (1978).

1677   Cascaval, C.N., Straus, S., Brown, D.W., Florin, R.E., J. Polym. Sci., 57 81 (1976).

1678   Rudin, A., Samanta, M.C., Van der Hoff, B.M.E., J. Polym. Sci., Polym. Chem. Ed., 17 493 (1979).

1679   Udseth H.R., Friedman, N., Anal. Chem., 53 29 (1981).

1680   Radus, T.P., Udseth, H.R., Friedman, L., J. Phys. Chem. 83 2969 (1979).

1681   Vasile, M.J., Jones, G.R., Falconer, W.E., Int. J. Mass Spectrom. Ion. Phys. 10 457 (1972).

1682   Brubaker, W.M., Adv. Mass Spectrom., 4 293 (1968).

1683   Gaffney, J.S., Pierce, R.C., Friedman, L., International J. of Mass Spectrometry Ion Physics, 25 439 (1977).

1684   Burille, P., Bert, M., Michel, A., Guyot, A., J. Polym. Sci., Polym. Lett. Ed., 16 181 (1978).

1685   O'Mara, M.M., J. Polym. Sci., Poly. Chem. Ed., 8 1887 (1970).

1686   Chang, E.P., Salovey, R., J. Polym. Sci., Polym. Sci., Polym. Chem. Ed., 12 2927 (1976).

1687   Ahlstrom, D.H., Liebman, S.A., Abbas, K.B., J. Polym. Sci., Polym. Chem. Ed., 14 2479 (1976).

1688   Alajberg, A., Arpino, P., Deursiftar, D., Ginochow, G., J. Anal. Appl. Pyrol., 1 203 (1980).

1689   Ballisteri, A., Foti, S., Montaudo, G., Scamporrino, E.J., Polym. Sci. Polym. Chem. Ed., 18 1147 (1980).

1690   Guyot, A., Benevise, J.P., Trambouze, Y., J. Appl. Polym. Sci., 6 103 (1962).

1691   Talamini, G., Pezin, G., Makromol. Chem. 39 26 (1960).

1692   Stromber, R.R., Straus, S., Achhammer, B.G., J. Poly. Sci., 35 355 (1959).

1693   Straus, S., Madorsky, S.L., Ind. Eng. Chem. 48 1212 (1956).

1694   O'Mara, M.M., J. Polymer Sci., A-1, 8 1887 (1970).

1695   O'Mara, M.M., J. Polymer Sci., A-1, 9 1387 (1971).

1696   Watson, S.T., Biemann, K., Anal. Chem., 36 1135 (1964).

1697   Iida, T., Nakanishi, M., Goto, K., J. Polymer Sci., 12 737 (1974).

1698   Chang, T.L., Mead, T.E., Anal. Chem., 43 534 (1971).

1699   Desiderio, D.M. Jr., Mead, T.E., Anal. Chem., 40 2090 (1968).

1700   Watson, J.T., Biemann, K., Anal. Chem., 37 844 (1965).

1701   Peltonen, K., Analyst (London), 111 819 (1986).

1702   Abney, G., Head, B.C., Poller, R.C., J. Polymer Sci. Macromol. Rev., 8 1 (1974).

1703   Danforth, J.D., Takeuchi, T.J., J. Polymer Sci., A-1, 11 2091 (1973).

1704   Guyot, A., Bert, A., Spitz, R., J. Polymer Sci., A-1, 8 1596 (1970).

1705   Shibazaki, Y., Kamebe, H., Kabish Kayaku, 21 65 (1964).

1706   Beckewitz, F., Housinger, H., Angew Makromol. Chem., 46 143 (1975).

1707   Wall, L.A., J. Elastoplast, 5 36 (1973).

1708   Mitera, J., Kubelka, V., Novak, J., Mosteckey, J. Plasty Kavc, 14
       (1), 18 (1977); Chem. Abstr., 86 172232u (1977).
1709   Chaigneau, M., Analysis 5 (5), 223 (1977).
1710   Grassie, N., Bain, D.R., J. Polymer Sci., A-1, 8 2683 (1970).
1711   Grassie, N., Bain, D.R., J. Polymer Sci., A-1, 8 2669 (1970).
1712   Schmitt, C.R., J. Fire Flammability, 3 303 (1972).
1713   Beachel, C., Beck, D.L., J. Polymer Sci., A-3, 457 (1965).
1714   Seeger, M., Gritter, R.J., J. Polymer Sci., A-1, 15 1393 (1977).
1715   Tsuchiya, A., Sumi, K., J. Polymer Sci., A-1, 7 1599 (1969).
1716   Moiseeva, U.D., Nieman, M.H., Vu Soed, 3 1383 (1961).
1717   Wall, L.A., Soc. Petrol Engrs. J., 16 1 (1960).
1718   McGaugh, M.C., Kottle, S., J. Polymer Sci., A-1, 6 1243 (1968).
1719   Grassie, N., Torrance, B.D.J., J. Polymer Sci., A-1, 6 3303
       (1968).
1720   Grassie, N., Torrance, B.D.J., J. Polymer Sci., A-1, 6 3315
       (1968).
1721   Nagasawa, M., Holtzer, A., j. Amer. Chem. Soc., 86 538 (1964).
1722   Troitskii, B.B., Varyukhin, V.A., Khokhlova, L.V., Trudy Khim,
       Khim Teknol, 2 115 (1974).
1723   Gilland, J.C., Lewis, J.S., Angew Makromol. Chem., 54 49 (1976).
1724   Ehlers, G.F.C., Fisch, K.R., Powell, W.R., J. Polymer Sci., A-1, 7
       2969 (1969).
1725   Ehlers, G.F.C., Fisch, K.R., Powell, W.R., J. Polymer Sci., A-1, 7
       2955 (1969).
1726   Ehlers, G.F.C., Fisch, K.R., Powell, W.R., J. Polymer Sci., A-1, 7
       2931 (1969).
1727   Clark, J.E., Jellinek, H.H.G., J. Polymer Sci., A, 3 1171 (1965).
1728   Ny, T.H., Williams, H.L., Makromol. Chem., 182 3323 (1981).
1729   Bambough, K.J., Clampitt, B.H., J. Polymer Sci., A, 3 805 (1965).
1730   Boetnner, E.A., Weiss, B., Amer. Ind. Hyg. Assoc. J., 28 535
       (1967).
1731   Boettner, E.A., Ball, G., Weiss, B., J. Appl. Polym. Sci., 13 377
       (1969).
1732   Hale, W.F., Farnham, A.G., Johnson, R.N., Clendinning, R.A., J.
       Polymer Sci., A-1, 5 2399 (1967).
1733   Thizon, M., Fon, C., Valentin, P., Guichon, G., Anal. Chem., 48
       1861 (1976).
1734   Norling, P.M., Tabocsky, A.V., "Thermal Stability of Polymers",
       R.T. Conley editor, Dekker, New York, N.Y. Vol.1, Chap. 5, (1970).
1735   Dugan, G., McCarty, J.D., NATAS Conference Proceedings 525 (1984).
1736   Schole, R.G., Bednarczyk, J., Yamanchi, T., Anal. Chem., 38 331
       (1966).
1737   Neiman, M.P., Ageing & Stabilization of Polymers (translated from
       Russian), Consultants Bureau, New York, Chapter 4 (1965).
1738   Paulik, J., Paulik, G., G.I.T., (Glas-Ontrum-Tech), Fachz Lab., 16
       (9), 1043 (1972).
1739   Meyer, C.S., Ind. Eng. Chem., 44 1095 (1952).
1740   Carlsson, P.j., Kato, Y., Wiles, D.N., Macromolecules, 1 459
       (1968).
1741   Scheehan, W.C., Cole, T.B., J. Appl. Polymer Sci., 8 2359 (1964).
1742   Ross, S.E., J. Appl. Polymer Sci., 9 2729 (1965).
1743   Adams, J.H., Goodrich, J.E., J. Polymer Sci., A-1, 8 1269 (1970).
1744   Cross, L.H., Richards, R.B., Willis, H.A., Disc. Faraday Soc.
       (London), 9 235 (1950).
1745   Rugg, R.M., Smith, J.J., Bacon, R.C., J. Polymer Sci., 13 535
       (1954).
1746   Pross, A.W., Black, R.M., J. Soc. Chem. Ind. (London), 69 115
       (1950).
1747   Beachell, A.C., Nemphos, S.P., J. Polymer Sci., 21 113 (1956).
1748   Beachell, A.C., Tarbet, G.W., J. Polymer Sci., 45, 451 (1960).

1749   Luongo, J.P., J. Polymer Sci., 42, 139 (1960).
1750   Cooper, G.D., Prober, X., J. Polymer Sci., 44 397 (1960).
1751   Thompson, H.W., Tarkington, P., Proc. Royal Soc. (London), A3 184
        (1945).
1752   Pross, A.W., Black, R.M., J. Soc. Chem. Ind. (London), 69 113
        (1950).
1753   Kveder, H., Ungar, G., Nafta (Zagreb), 24 (2), 85 (1973).
1754   Pentin, Yu A., Tarasevich, B.N., El'tsefon, B.S., Vestn. Mosk
        Univ. Khim., 14 (1), 13 (1973).
1755   Miller, P.J., Jackson, J.F., Portor, R.S., J. Polymer Sci. Polymer
        Phys. Ed., 11 2001 (1973).
1756   Tabb, D.L., Sevcik, J.J., Koenig, S.L., J. Polymer Sci. Polymer
        Phys. Ed., 13, (4) 815 (1975).
1757   Chan, M.G., Allara, D.I., Polymer Eng. Sci., 14 (1), 12 (1974).
1758   Cooper, G.D., Prober, M., J. Polymer Sci., 44 397 (1960).
1759   Heacock, J.F., J. Appl. Polymer Sci., 7 2319 (1963).
1760   Adams, J.H., J. Polymer Sci., A-1, 8 1279 (1970).
1761   Wood, R.P., Statton, W.O., J. Polymer Sci. Polymer Phys. Ed., 12
        1575 (1974).
1762   Goldstein, M., Seeley, M.E., Willis, H.A., Zichy, V.J.I., Polymer,
        14 (11), 530 (1973).
1763   Grassie, N., Weir, N.A., J. Appl. Polymer Sci., 9 963 (1965).
1764   Grassie, N., Weir, N.A., J. Appl. Polymer Sci., 9 975 (1965).
1765   Schole, R.G., Bednorszyk, J., Tamanei, T., Anal. Chem., 38 331
        (1966).
1766   Shaw, J.N., Marshall, M.C., J. Polymer Sci., A-1, 6 449 (1968).
1767   Ohnishi, S.I., Sugimoto, S.I., Nitta, I., J. Polymer Sci., A-1,
        625 (1963).
1768   Soliovey, R., Yager, A., J. Polymer Sci., A, 2 219 (1964).
1769   Hajimoto, Y., Tamura, N., Okamoto, S., J. Polymer Sci., A, 3 255
        (1965).
1770   Buchi, I., J. Polymer Sci., A, 3 2685 (1965).
1771   Sakar, Y., Iwasaki, H., J. Polymer Sci., A-1, 7 1749 (1969).
1772   Hama, Y., Shinohara, K., J. Polymer Sci., A-2, 8 651 (1970).
1773   Duswalt, A.A., Thermochimica Acta, 8 57 (1974).
1774   Hileman, F.D., Vorhees, K.J., Wojeik, L.H., Birky, M.M., Ryan,
        P.W., Einhorn, I.N., J. Polymer Sci., 13 571 (1975).
1775   Boettner, E.A., Ball, G., Weiss, B., J. Appl. Polymer Sci., 13 377
        (1969).
1776   Michal, J., Mitera, J., Tardon, S., Fire & Materials, 1, 160
        (1976).
1777   Morikawa, J., Combust. Toxicol., 3 349 (1978).
1778   Liao, J. C., Browner, R.F., Anal. Chem., 50 1683 (1978).
1779   Joseph, K.T., Browner, R.F., Anal. Chem., 52 1083 (1980).
1780   Becker, W., in J. Troitzsch: Brandverhalten von Kunststoffen.
        Munchen, Wien, Carl hanser Verlag, 1982, G. Schreyer:
        Konstruieren mit Kunststoffen. Munchen, Wien, Carl Hanser Verlag,
        1972.
1781   Prager, F.H., Kunstst. Bau 13 (1978), No.2, 45.
1782   DIN Standard 75200, Bestimmung des Brennverhaltens von Werkstoffen
        der Kraftfahrzeuginnenausstattung.
1783   ASTM Standard D 1692, Standard Method of Test for Flammability of
        Plastic Sheeting and Cellular Plastics.
1784   ASTM Standard D 2863-76, Standard Method for Test of Flammability
        using the Oxygen Index Method.
1785   ASTM Standard D 1926, Ignition Properties of Plastics.
1786   Federal Trade Commission (FTC), File No.7323040, Agreement
        Containing Consent Order to Cease and Desist.
1787   FTC Proposed Trade Regulation Rule Concerning Disclosure of

Combustion Characteristics in the Marketing and Certification of
          Cellular Plastics Products, 39 Fe. Reg. 28 292, Aug. 6, 1974, 40
          Fed. Reg. 30 842, July 23, 1975.
1788   UL Subject 94, Test for Flammability of Plastic Materials.
1789   DIN Standard 51755, Bestimmung des Flammpunktes im geschlossenen
          Tiegel nach Abel-Pensky.
1790   DIN Standard 53213, Flammpunktbestimmung im geschlossenen Tigel.
1791   DIN Standard E22118, Fordergurte mit Textileinlagen fur den
          Steinkohlebergbau unter Tage.
1792   VDE 0472, Leitsatze fur die Durchfuhrung von Prufungen an
          isolierten Leitungen: Sect. 814, Funktionserhalt bei
          Fammeinwirkung.
1793   VDE 0860, Bestimmung fur netzbetriebene Flammwidreigkeit.
1794   VDE 0304, Bestimmungen fur Prufverfahren zur Beurteilung des
          thermischen Verhaltens fester isolierstoffe: Brandverhalten.
1795   UL 1416, Overcurrent and Overtemperature Protection for Radio and
          Television Type Appliances.
1796   ASTM D 635, Standard Method of Test for Flammability of
          Selfsupporting Plastics.
1797   ASTM E 162, Standard Method of Test for Surface Flammability of
          Materials Using a Radiant Head Energy Source.
1798   VDE 0471/DIN 57471, Sect. 2, Gluhdrahtprufung.
1799   VDE 0471/DIN 57471, Sect. 3, Entwurf Gluhkontaktprufung.
1800   VDE 0471/DIN 57471, Sect. 5, Prugung mit Flammen.
1801   DIN Standard 53459, Beurteilung des VEhaltens wahrend und nach
          Beruhrung mit einem Glustab.
1802   Federal Motor Vehicle Safety Standard (FMVSS) 302 - Test procedure
          and Specimen Preparation, Docket 3-3, Notice 6: Fed. Reg. 38, No.
          95, May 17, 1973.
1803   ISO R 3795, Road Vehicles - Determination of Burning Behaviour of
          Interior Materials for Motor Vehicles.
1804   FMVSS 302, Notice 7: Proposed Covered Component, Fed. Reg. 38,
          No. 133, July 12 1973.
1805   DV 899/23/I, Merkbuch A fur Bauwerke, und DV 899/35/II, Merkbuch B
          fur Fahrzeuge, Behalter und Container.
1806   DV 899/35. Merkblatt der Deutschen Bundesbahn fur die Prufung des
          Brandverhaltens fester Stoffe.
1807   Federal Air Regulation FAR, Part 25: Fed. Reg. 34 No. 135,
          12.8.1968, Fed. Reg. 37, No. 37, 24.2.1972.
1808   Luftfahrt-Tauglichkeitsvorshriften: LTV 001500 Brandverhalten von
          Luftfahrtwerkstoffen.
1809   Airbus Technical Specification, ATS 1000.01.
1810   ASTM-STP 422.
1811   DIN Standard 4102, Brandverhalten von Baustoffen und Bauteilen.
1812   Merchant Shipping Notice M. 592, Board of Trade, Marine Div.,
          1970.
1813   VFDB Vorbeugender Brandshutz. Wiesbaden, Verlag Kultur und Wissen,
          GmbH., 1982.
1814   Chandler, S.E., The Incidence of Residential Fires in London - The
          Effect of Housing and Other Social Factors. BRE Information Paper
          IP 20/79, Aug 1979.
1815   Chandler, S.E., Some Trends in Furniture Fires in Domestic
          Premises - BRE-CP 66/76/
1816   Chandler, S.E., Baldwin, R., Fire and Materials 1 (1976) 76.
1817   The Upholstered Furniture (Safety) Regulation 1980 - Consumer
          Protection.
1818   BS 5852, Part 1, 1979, Fire Tests for Furniture.
1819   PPF-6-1974, Proposed Standard for the Flammability of Upholstered
          Furniture.

1820   Berl, W.G., Halpan, B.M., Fire J. 105 (1979).

1821   Palmer, K.N., Taylor, W., Paul, K.T., Fire Hazards of Plastics in Furniture and Furnishings - Characteristics of the Burning, BRE-CP 3/75, Jan 1975.

1822   Palmer, K.N., Taylor, W., Paul, K.T., Fire Hazards of Plastics in Furniture and Furnishings: Fully Furnished Rooms. BRE-CP 21/76, Feb. 1976.

1823   Wilson, W.J., Large Scale Fire Tests. SPI - 4th Ann. Combustion Symp., 1975.

1824   Prager, F.H., Untersuchungen zum Brandverhalten von Polstermobeln, 5. Int. Brandshutseminar Karlsruhe, 1976.

1825   Marchant, R.P., The Ignitibility of Upholstery by Smokers Materials, FIRA, UK, 1977.

1826   State of California, Techn. Inf. Bulletins No. 116 and 117.

1827   The New York Port Authority, Specifications Governing the Flammability of Upholstery Materials and Plastics Furniture.

1828   ASTM D-2843, Measuring the Density of Smoke from the Burning or Decompostion.

1829   DOE/PSA, Fire Retardant Specifications.

1830   Emmons, H.W., Fire Res. Abstr. and Rev., 10 133 (1968).

1831   ISO TR 3814, The Development of Tests for Measuring "Reaction to Fire" of Building Materials.

1832   ISO DP 5657, Ignitibility of Building Materials.

1833   ISO DP 5658, Spread of Flame Test.

1834   ISO DP 5659, Smoke Generated by Solid Materials.

1835   ISO TR 6585, Fire Hazards and the Design and Use of Fire Tests.

1836   Factory Mutual Research Corp. A Fire Study of Rigid Cellular Plastic Materials for Insulated Wall and Roof/Ceiling Constructions, Project 22000.

1837   Zorgman, H., Behaviour of Insulated Steel Roofs in Fully Developed Fires, 5. Int. Brandshutzseminar. Karlsruhe, 1976.

1838   Prager, F.H., H. Zorgmann, Kunstst. Bau 14 (1979). Vol. 2, 4.

1839   ISO Standard 834, Fire Resistance Tests - Elements of Building Construction.

1840   Kloker, W., H. Niesel, F.H., Prager, H.W., Schiffer, O. Bokenkamp, H.E., Klingelhofer, Kunststoffe 67 438 (1977).

1841   DIN Standard 4102, Brandverhalten von Baustoffen und Bauteilen, Sect. 7.

1842   NEN 3883, Assessment of the Hazard to roofs of Flying Brands.

1843   Kunstst. Bau 14 Vol. 2, 51 (1979).

1844   ASTM D 2843, Measuring the Density of Smoke from the Burning or Decomposition of Plastics.

1845   DIN Standard 53436/37, Verschwelungsapparatur und Rauchichtemessung.

1846   Brandverjutungsdienst Zurich/Schweiz, Wegleitung fur Polzeivorshriften.

1847   ISO TR 6543, The Development of Tests for Measuring Toxic hazards in Fire.

1848   Kimmerle, G., J.F.F. Com. Tox. 1 (1974).

1849   Oettel, H., H. Hoffmann, Z VFDB 3 (1968).

1850   Locher, F.W., S. Sprung, Beton Vol. 2, 63, Vol.3, 99 (1970).

1851   Hammer, Chr., Fischer, K., Beton, Vol 9 (1971).

1852   VDE Draft 0472, Sect. 813, Prufung von Kabeln und isolierten Leitungen - Korrosivitat von Brandgasen.

1853   Wunderlich, B., Bodily, D.M., J. Appl. Polymer Science, C-6, 137 (1964).

1854   Kamida, K., Yamaguchi, K., Makromol. Chem., 162 205 (1972).

1855   Duswalt, A.A., Cox, W.W., Polymer Characterization, Inter-disciplinary Approaches.

1856   Marchetti, A., Martuscelli, E., J. Polymer Sci. Polymer Phys. Ed.,
        $\underline{12}$ 1649 (1974).
1857   Miller, G.W., Themrochimica Acta, $\underline{8}$ 129 (1974).
1858   Smith, O.F., Anal. Chem., $\underline{35}$ 1835 (1963).
1859   Holden, H.W., J. Polymer Sci., C-6, 209 (1964).
1860   Clampitt, B.H., Hughes, R.Hl., J. Polymer Sci., C-6, 43 (1964).
1861   Johnston, N.W., J. Macromol. Sci. Chem., $\underline{7}$ 531 (1973).
1862   Johnston, N.W., Macromolecules, $\underline{6}$ 453 (1973).
1863   Frederick, W.J. Jr., Mentzer, C.C., J. Appl. Polymer Sci., $\underline{19}$ (7),
        1799 (1975).
1864   Haberfeld, J.L., Reffner, J.A., Soc. Plast. Eng. Tech. Pap., $\underline{21}$
        858 (1975).
1865   Pope, D.P., J. Polymer Sci. Polymer Phys. Ed., $\underline{14}$ (5), 811 (1976).
1866   Dunlap, L.H., Foltz, C.R., Mitchell, A.G., J. Polymer Sci. Physics
        Edition, $\underline{10}$ 2223 (1972).
1867   Wohitjen, H., Dessy, R., Anal. Chem., $\underline{51}$ 1470 (1979).
1868   Wohitjen, H., Dessy, R.E., Anal. Chem., $\underline{51}$ 1463 (1979).
1869   Haldon, R.A., Schell, W.J., Simha, R., "Transitions in Glasses at
        Low Temperatures," "Cryogenic Properties of Polymers", Koenig and
        Serafini, Ed., Marcel Dekker, New York, 152 (1968).
1870   Haldon, R.A., Schell, W.J., Simha, R., "Transitions in Glasses at
        low Temperatures. Cyrogenic properties of Polymers", Koenig and
        Sekafina editions, Marcel Dekker, New York, p.148 (1968).
1871   Hatzakis, M., Thomas, J., Watson Research Laboratories, P.O. Box
        218, Yorktown Heights, N.Y. 10598, private communications, April
        25, 1978.
1872   Yano, S., J. Appl. Polymer Science, $\underline{21}$ 2645 (1977).
1873   Maruta, M., Ito, K., Kunimatsu, Y., Yamada, K., Therm. Anal.
        (Proc. Int. Conf.), 483 (1977).
1874   Gillen, K.T., J. Appl. Polymer Sci., $\underline{22}$ (5), 1291 (1978).
1875   Varob'yev, V.M., Vetteghen, V.T., Polymer Science, U.S.S.R., $\underline{17}$
        520 (1975).
1876   Andronikashivili, G.G., Samsoniya, S.A., Zhamierashvili, M.G.,
        Sooleshch. M., Akad Nauk. Gruz. SSSR., $\underline{85}$ 73 (1977), Chem. Abstr.,
        $\underline{86}$ 190545C (1977).
1877   Beatty, C.L., Froix, M.F., Polymer Prep'r American Chemical
        Society, Division Polymer Chemistry, $\underline{16}$ 628 (1975).
1878   Smith, J.B., Manuel, A.J., Ward, I.M., Polymer, $\underline{16}$ 57 (1975).
1879   Sichina, W.J., International Laboratory, p.36, May (1988).
1880   Ferry, J.D, In "Viscoelastic Properties of Materials" 3rd edition,
        J. Wiley, New York, (1960).
1881   Tobolsky, A.V., Properties and Structures of Polymers, Wiley, New
        York, 1960.
1882   Williams, M.L., Landel, R.F., Ferry, J.D., J. Am. Chem. Soc., $\underline{77}$
        3701 (1955).
1883   Frosini, V., Magagnini, P.L., Newman, B.A., J. Polymer Sci.
        Polymer Physics Ed., $\underline{12}$ 23 (1974).
1884   Coppola, G., Filippini, R., Pallesi, B., Polymer, $\underline{16}$ (7), 546
        (1975).
1885   Kamide, K., Imanaka, A., Kobunshi Ronbunshu, $\underline{32}$ (9), 537 (1975).
1886   Peppas, N.A., Merrill, E.W., J. Appl. Polymer Sci., $\underline{20}$ (6), 1457
        (1976).
1887   Bamnbaugh, K.J., Clampett, B.H., J. Polymer Sci., A-3, 805 (1965).
1888   Geacinfov, C., Scholtland, R.S., Miles, R.B., J. Polymer Sci., C-
        6, 197 (1964).
1889   Schewnken, F.R.F., Zuccarello, R.K., J. Polymer Sci., C-61 (1964).
1890   Donald, H.J., Humes, E.S., White, L.W., J. Polymer Sci., C-6, 93
        (1964).

1891   Marx, C.L., Cooper, S.L., Makromol Chem., 168 339 (1973).
1892   Wilson, C.W., Pake, G.E., J. Polymer Sci., 10 503 (1953).
1893   Bergmann, K., Nawotki, K., Kolloid, Z.Z., Polymer, 219 132 (1967).
1894   Bergmann, K., Kolloid, Z.Z., Polymer, 251 962 (1973).
1895   Bergmann, K., J. Polymer Sci. Polymer Physics Ed., 16 1611 (1978).
1896   Raudelkerm, L, J. Polymer Sci. Symposium, No. 50, 457 (1975).
1897   Bergmann, K., Polymer Bulletin 5 355 (1981).
1898   Powles, J.G., Strange, J.H., Proceedings Physical Society
        (London), 82 6 (1963).
1899   Solomen, I., CR Acad Sci. Paris, 248 92 (1959).
1900   Farrer, T.C., Becker, E.D., "Pulse and Fourier Transform NMR"
        (1971).
1901   Palmer, R.P., Cobbold, A.J., Makromol Chemie, 24 174 (1966).
1902   Kubo, R., Tomita, K., J. Physical Society, Japan, 9 888 (1954).
1903   Haberkorn, H., Illers, K.H., Simak, P., Coll. Polymer Sci., 257
        820 (1979).
1904   Hermans, P.H., Weidmiger, A., Makromol Chemie, 44-46 24 (1961).
1905   Hendus, H., Illers, K.H., Kunststoffe, 57 193 (1967).
1906   Simak, P., Makromol Chemie, Makromol Symposium, 5 61 (1986).
1907   Cobbs, W.H., Burton, R.L., J. Polymer Sci., 10 275 (1953).
1908   Thomson, A., Woods, D.W., Nature (London), 176 78 (1955).
1909   Miller, R., Willis, H., J. Polymer Sci., 19 485 (1956).
1910   Novak, J., Suskov, V., Zooin, L., Beitr. Faserforsch. Textiltech.,
        6 513 (1969).
1911   Edelmann, K., Wyden, H., Kautsch. Gummi, Kunstst., 25 353 (1972).
1912   Statton, W., Koenig, J., Hannon, M., J. Appl. Phys., 41 4290
        (1970).
1913   Illers, K.H., Colloid Polym. Sci., 258 11 (1980).
1914   Pitha, J., Jones, R.N., N.R.C.C. Bull. No. 12, 1968.
1915   Danz, R., Dechant, I., Ruscher, Ch., Faserforsch. Textiltech., 21
        503 (1970).
1916   Miyake, A., J. Polymer Sci., 38 497 (1959).
1917   Manley, T.R., Williams, D., J. Polymer Sci., Part C., 22 1009
        (1969).
1918   Boerio, F., Bahl, S.K., McGraw, G., J. Polymer Sci., 14 1029
        (1976).
1919   Mehta, R., Bell, I.P., J. Polymer Sci. Polymer Phys. Ed., 11 1793
        (1973).
1920   Baumgarter, A., Blasenbrey, S., Pechhold, W., Kolloid Z & Z.
        Polm., 250 1026 (1972).
1921   Schonherr, F., Faserforsch. Tectiltech., 21 246 (1970).
1922   Schmidt, G., J. Polymer Sci., Part A., 1 1271 (1963).
1923   Astbury, W., Brown, C.I., Nature (London), 158 871 (1946).
1924   Daubeny, R., Bunn, C.W., Brown, C.I., Proc. R. Soc. London, A, 226
        531 (1954).
1925   Zahn, H., Krzikalla, R., Makromol. Chem., 23 1 (1957).
1926   Kinoshita, Y., Nakamura, R., Kitomo, Y., Ashida, T., Polym. Prepr.
        (Am. Chem. Soc., Div. Polym. Chem.), 20 454 (1979).
1927   Kilian, H.G., Halboth, H., Jenkel, E., Kolloid Z. & Z. Polym., 172
        166 (1960).
1928   Fakirov, S., Fischer, E.W., Schmidt, G.W., Makromol Chem., 176
        2459 (1975).
1929   Aharoni, S.M., Sharma, R.K., Szobota, J.S., Vernik, D., J. Appl.
        Polym. Sci., 28 2177 (1983).
1930   Lin, S.B., Koenig, J.L., J. Polymer Sci. Polym. Phys. Ed., 20 2277
        (1982).
1931   Lin, S.B., Koenig, J.L., J. Polymer Sci. Polymer Symposium, 71 121
        (1984).
1932   Matsura, H., Miyazawa, T., Machida, K., Spectrochim. Acta., A. 29
        771 (1973).

1933   Jansson, J.F., Yannas, I.V., Int. Long Rheology, 7th Ed., Chalmers
        E., University of Technology Geoteborg, Sweden, 274 (1976).
1934   Smolyanskii, A.L., Gusakova, G.V., Trudy Vologod Moloch. Inst. No.
        60 105 (1970).
1935   Holland-Moritz, K., Kolloid, Z.Z., Polymer 251 906 (1973).
1936   Holland-Moritz, K., Hummel. Do. J., Mol Struct., 19 289 (1973).
1937   Holland-Moritz, K., Hummel. Do., Quad Ric. Sci., 84 158 (1973).
1938   Kadyrmstova, F.N., Malkakov, L.I., Zhur Prikl. Spektrosk., 18 928
        (1973).
1939   Funke, W., Schuh, H., J. Polymer Sci. Symp. No. 42, 379 (1973).
1940   Elliot, J.J., Kennedy, J.P., J. Polymer Sci., A-1, 11 2993 (1973).
1941   Elliot, J.J., Kennedy, J.P., J. Polymer Sci., Chemistry Edition,
        11 299 (1973).
1942   McRae, M.A., Maddams, W.F., Preedy, J.E., J. Mater Sci., 11 2036
        (1976).
1943   Luongo, J.P., J. Polymer Sci., Polymer Chemistry Edition, 12 1203
        (1974).
1944   Zerbi, G., Phenous. Proc. Int. Conference, 248 (1971).
1945   Mocheria, K.K., Bell, J.P., J. Polymer Sci. Polymer Physics Ed.,
        11 1779 (1973).
1946   Mehta, R.E., Bell, J.P., J. Polymer Sci. Polymer Physics Ed., 11
        1793 (1973).
1947   Fraser, G.V., Hendra, P.J., Watson, D.S., Gall, M.J., Willis,
        H.A., Cudby, M.E.A., Spectrochim. Acta., A, 29 1525 (1973).
1948   Bailey, R.T., Hyde, A.J., Kim, J.J., Advances in Raman
        Spectroscopy, 1 296 (1972).
1949   Bailey, R.T., Hyde, A.J., Kim, J.J., Spectrochimia Acta., A, 30 91
        (1974).
1950   Turley, J.W., J. Polymer Sci. A, 3 2400 (1965).
1951   Heritage, K.J., Mann, J., Roldan Gonzalez, L., J. Polymer Sci. A,
        1 671 (1963).
1952   Ellis, K.C., Warwicker, J.O., J. Polymer Sci. A, 1 1185 (1963).
1953   Johnson, R., J. Polymer Sci. A, 1 715 (1963).
1954   Vogelsang, D.C., J. Polymer Sci. A, 1 1055 (1963).
1955   Hirami, M., J. Polymer Sci., A-1, 4 967 (1966).
1956   Jackson, J.F., J. Polymer Sci. A, 1 2119 (1963).
1957   Litt, M., J. Polymer Sci. A, 1 2219 (1963).
1958   Chiang, R., J. Polymer Sci. A, 1 2765 (1963).
1959   Krigbaum, W.R., Roe, F.J., J. Polymer Sci. A, 2 4391 (1964).
1960   Kaiya, N., Takahar, A., Kajiyama, T., Takayanagi, M., Rep. Prog.
        Polym. Phys. Jpn. 27 329 (1984).
1961   Northolt, M., Van Aartsen, J., Polym. Lett., 11 333 (1973).
1962   Northolt, M., Eur. Polym. J., 10 799, (1974).
1963   Shen, D., Molis, S., Hsu, S., Polm. Eng. Sci., 23 543 (1983).
1964   Haraguchi, K., Kajiyama, T., Kayanagi, M., J. Appl. Polym. Sci.,
        23, 915 (1979).
1965   Riewald, P., Dhingra, A., Chern, T., "Recent Advances in Aramid
        Fiber and Composite Technology", paper given at the International
        Conference on Composite Materials, London, July 1987.
1966   Young, P.H., Spectroscopy International, 1 50 (1989).
1967   Kim, P., Chang, C., Hsu, S., Polymer, 27 34 (1986).
1968   Clampett, B.H., Anal. Chem., 35 1834 (1963).
1969   Clampett, B.H., J. Polymer Sci., A-3, 671 (1965).
1970   Kimuru, M., Hatakeyama, T., Nakano, J., J. Appl. Polymer Sci., 18
        3069 (1974).
1971   Era, V., Venalainen, H., J. Polymer Sci. Sump., No. 42, 879
        (1973).
1972   Heyns, H., Heyer, S., in Thermal Analysis, vol. 3 (Wiedermann
        H.G., ed.) Birkhaeuser, Basel Switzerland p.341 (1972).

1973   Johnsen, U., Spilgies, G. Kolloid-Z Z. Polym., 250 1174 (1972).
1974   Gilbert, M., Hybart, F., J. Polymer, 15 407 (1974).
1975   Ceccorulli, G., Manescaichi, F., Makromol Chem., 168 303 (1973).
1976   Savolainen, A., J. Polymer Sci. Symp., No. 42 885 (1973).
1977   Suwa, T., Takehisa, M., Machi, S., J. Anal. Polym. Sci., 17 3253
        (1973).
1978   Radhakrishnan, N.G., Padhye, M.R., Angew Makromol. Chem., 43 177
        (1975).
1979   Era, V.A., Jauhiainen, T., Angew Makromol. Chem., 43 157 (1975).
1980   Gervais, M., Gallot, B., Makromol. Chem., 171 157 (1973).
1981   Illers, K.H., Heckmann, W., Hambreckt, J., Colloid and Polymer
        Science, 262 557 (1984).
1982   Simak, P., Makromol. Chem., 178 2927 (1977).
1983   Axelsen, D.E., Mandekkern, L.J., Polymer Science Polymer Physics
        Ed., 16 1135 (1978).
1984   Diab, J., Schokk, L, Siemens Forsch Entwickhingsber, 1 375 (1972).
1985   Guseinov, T.I., Seldov, N.M., Abasov, A., Kasumov, K., Mater.
        Vses. Soveshch relaksatsionnym yaulenlyam Polim. 2nd, 1971, 2 218-
        24 (1974).
1986   Illers, K.H., Europ. Polymer J., 10 911 (1974).
1987   Peppas, N.A., J. Appl. Polymer Sci., 20 1715 (1976).
1988   Lemstra, P.J., Schouten, A.J., Challa, G., J. Polymer Phys. Ed.,
        10 2301 (1972).
1989   Lemstra, P.J., Schouten, A.J., Challa, G., J. Polymer Phys. Ed.,
        12 1565 (1974).
1990   Lety, A., Noel, C., J. Chim. Phys. Physiochim. Biol., 69 875
        (1972).
1991   Ahad, E., J. Appl. Polymer Sci., 18 1587 (1974).
1992   Denrig, R., Kunststoffe., 75 8 (1985).
1993   Groten, B., Anal. Chem., 36 1206 (1964).
1994   Cox, B.C., Ellis, B., Anal. Chem., 36 90 (1964).
1995   Fiorenza, A., Benomi, G., Rassanga Chimica, 5 197 (1963).
1996   May, R.W., Pearson, E.F., Porter, F., Scothern, M.D., Analyst
        (London), 98 364 (1973).
1997   Folmen, O.F., Anal. Chem., 43 1059 (1971).
1998   Folmen, O.F., Azarraga, L.V., J. Chromatogr. Sci., 7 665 (1969).
1999   Kruglova, V.A., Ratousku, G.V., Borisenko, A.A., Kababina, A.V.,
        Vysokmol Soedin Ser. A., 14 1967 (1973).
2000   Ferguson, R.C., J. Polymer Sci. A, 2 4735 (1964).
2001   Krimm, S., Folt, V.L., Shipman, J.J., Berens, A.R., J. Polymer
        Sci. A, 1 2621 (1963).
2002   Krimm, S., Enomoto, S., J. Polymer Sci. A, 2 669 (1964).
2003   Enomoto, S., Kogura, M., Asakina, M., J. Polymer Sci. A, 2 5355
        (1964).
2004   Enomoto, S., Asakina, M., Satah, S., J. Polymer Sci. A, 4 1373
        (1966).
2005   Miller, D.L., Samsel, E.P., Cobler, J.G., Anal. Chem., 33 677
        (1961).
2006   Marks, G.C., Benten, J.L., Thomas, C.M., Chem. Soc. (London) Ind.,
        Monograph, 26 204 (1967).
2007   Germar, M., Makromolekulare Chem., 86 89 (1965).
2008   Saunders, R.A., Smith, D.C., J. Appl. Physics, 20 953 (1949).
2009   Painter, P.C., Watzel, M., Koenig, J.L., Polymer (London), 18 1169
        (1977).
2010   Belopol'skaya, T.V., Vestn. Leningr. Univ. Fiz. Khim., 22 44
        (1976).
2011   Painter, P.C., Havens, J., Hart, W.W., Koenig, J.L., J. Polymer
        Sci. Polymer Physics Ed., 15 1237 (1977).
2012   Painter, P.C., Havens, J., Hart, W.W., Koenig, J.L., J. Polymer
        Sci. Polymer Physics Ed., 15 1223 (1977).

2013   Yakoyama, M., Ochi., H., Tadakoro, H., Price, C.C.,
       Makromolecules, $\underline{5}$ 690 (1972).
2014   Gasan-Zade, V.G., Sutovskii, S.M., Kaplan, M.Y., Lakukras, Mater
       Ikh., Primen., $\underline{5}$ 11 (1976).
2015   Iyer, P.B., Iyer, K.R.K., Patil, N.B., J. Appl. Polymer Science,
       $\underline{20}$ 591 (1976).
2016   Fujimoto, T., Kababata, N., Furukawa, J., J. Polymer Sci., A-1, $\underline{6}$
       1200 (1968).
2017   Alter, U., Bonart, R., Colloid Polymer Science, $\underline{254}$ 348 (1976).
2018   Binder, J., J. Polymer Sci., A, $\underline{1}$ 37 (1963).
2019   Binder, J.L., J. Polymer Sci., A, $\underline{1}$ 47 (1963).
2020   Bassett, K.H., Liang, C.Y., Marcheisault, J. Polymer Sci., A, $\underline{1}$
       1687 (1963).
2021   Krimm, S., Folt, V.L., Shipman, J.J., Berens, A.R., J. Polymer
       Sci., A, $\underline{1}$ 2621 (1963).
2022   Tadokoro, H., Muralashi, S., Yamattera, R., Kamei, T., J. Polymer
       Sci., A, $\underline{1}$ 3029 (1963).
2023   Ferguson, R.C., J. Polymer Sci. A, $\underline{2}$ 4735 (1964).
2024   Enomoto, S., Koguro, M., Aschina, J. Polymer Sci., A, $\underline{2}$ 5355
       (1964).
2025   White, H.F., J. Polymer Sci., A, $\underline{3}$ 309 (1965).
2026   Binder, J.L., J. Polymer Sci. A, $\underline{3}$ 1587 (1965).
2027   Kossler, I., Vodehnal, J., Storka, M., J. Polymer Sci. A, $\underline{3}$ 2081
       (1965).
2028   Shibatani, K., J. Polymer Sci., A-2, $\underline{8}$ 1693 (1970).
2029   Young, P.R., Fernandez, J.E., J. Polymer Sci., A-1, $\underline{9}$ 1771 (1971).
2030   Emission, Molecular and Mass Spectroscopy; Chromatography;
       Resinography; Microscopy; Computerized Laboratory Systems. ASTM
       Annual Standards, American Society for Testing Materials, Vol. 42
       (1978).
2031   Komissarov, S.A., Bulenkov, T.I., Krasnov, B. Ya., Orlova, S.P.,
       Kozh-Obuv. Prom., $\underline{14}$ 41 (1972).
2032   Pivcova, H., Kirackova, L., Popisil, J., J. Polym. Sci., Symp. No.
       40, 283 (1973).
2033   Ivanchev, S.S., Zherebin, Yu. L., Zh. Prikl. Spectrosk., $\underline{19}$ (2),
       269 (1973).
2034   Ivin, K., J., "Structural Studies of Macromolecules by
       Spectroscopic Methods", Wiley, Chichester, Engl. (1976).
2035   Tadokoro, H., Kobayashi, M., Monogr. Mod. Chem., 1974, 6, 3-1,
       (1974).
2036   Hummel, D.O., Monogr. Mod. Chem., $\underline{6}$ 112 (1974).
2037   Sastre, R., Acosta, J.L., Rev. Plast. Mod., $\underline{31}$ 203 (1976).
2038   Deraouault, J., Hendra, P.J., J. Chim. Phys. Phys.-Chim. Biol.,
       $\underline{71}$ 1395 (1974).
2039   Rabolt, J.F., Diss. Abstr. Int. B $\underline{35}$ 3507, (1975).
2040   Yu, Huai-Gen, Hua Hsueh Tung Pao, (5), 304, 296 (1975).
2041   Hendra, D., Monogr. Mod. Chem., $\underline{6}$ 151 (1974).
2042   Holland-Moritz, K., Siesler, H.W., Appl. Spectrosc. Rev., $\underline{11}$ 1
       (1976).
2043   Read, B.E., Struct. Prop. Oriented Polym., 150-86, Ed., Ward,
       A.M., Wiley, New York, (1975).
2044   Bower, D.I., "Struct. Prop. Oriented Polym", 187-218, Ed., Ward,
       I.M., Wiley, New York, (1975).
2045   Coleman, M.M., Polym. Prepr. Am. Chem. Soc. Div. Polym. Chem., $\underline{17}$
       732 (1976).
2046   Wool, R.P., Statton, W.O., Polym. Prepr. Am. Chem. Soc. Div.
       Polym. Chem. $\underline{17}$ 749 (1976).
2047   Koberstein, J.T., Cooper, S.L., Shen, M.C., Rev. Sci. Instrum., $\underline{46}$
       1639 (1975).

2048   Oberbeck, W.F. Jr., Mayhan, K.G., Anal. Chem., 47 1216 (1975).
2049   Janicka, K., Handb. Anal. Synth. Polm. Plast., 64-100, Ellis
        Horwood Ltd., Chichester, England, (1977).
2050   Brown, S.C., Harvey, A.B., Pract. Spectrosc., 1 pt. C., 873
        (1977).
2051   D'Eposito, L., Koenig, J.L., Fourier Transform Infrared
        Sepctrosc., 61 (1978).
2052   Nagasawa, K., Toyota Chuo Kenkyusho R & D Rebyu, 13 61 (1977).
2053   Willis, H.A., Mol. Spectrosc., Proc. Conf., 6th (pub 1977), 555-
        69, West, A.R., Ed., Heyden, London (1976).
2054   Truett, W.L., Am. Lab., 9 33 (1977).
2055   Wiles, D.M., Tech. Charact. Polym. Mater., ADA 036082, 35-41
        (1976).
2056   Makarova, S.B., Shabanova, N.V., Bezeuvskaya, S.I., Zhadanov, B.,
        V. Vysokomol. Soedin., Ser. B 18 832 (1977).
2057   Haering, M., Chem. Rundsch., 29 15 (1976).
2058   Haering, M., Chem. Rundsch., 29 13, (1976).
2059   Haering, M. Chem. Rundsch., 29 11 (1976).
2060   Noskov, A.M., Novikov, N.I., Izy. Vyssh. Uchebn. Zaved., Khim.
        Teknol., 20 1804 (1977).
2061   Koenig, J.L., Polym. Charact. Conf., (Pub 1975), 51-72, Cleveland
        State univ., Cleveland, Ohio (1974).
2062   Koenig, J.L., Antoon, M.K., Appl. Opt., 17 1374, (1978).
2063   Mironov, D.P., Sokolova, L.N., Zavod. Lab., 43 451 (1977).
2064   Cais, R.E., O'Donnell, J.H., Makromolekulare-Chemie, 176 3517
        (1975).
2065   Kharas, G.B., Kissin, Y.V., Kiezner, V.I., Krentsel, B.A.,
        Stotsbaya, L.L., Zakharyan, R.Z., European Polymer Journal, 9 315
        (1973).
2066   Ho, W., Kissin, Y.V., Gol'dfarb, Y.Y., Krentsel, B.A., Ysokomal.
        Soedin. Ser. Y.A., 14 2229 (1972).
2067   Enomoto, S., J. Polymer Sci., 55 95 (1961).
2068   Ishibashi, M., J. Polymer Sci., A-2, 3657 (1964).
2069   Mikhailov, M., Dirlikov, S.K., Georgieva, Z., Peeva, N.,
        Panalotova, M., Dokl. Bolg. Akad. Nauk., 27 1525 (1974).
2070   Keller, F., Muegge, C., Plaste Kautsh, 24 88 (1977).
2071   Keller, F., Muegge, C., Plaste Kautsh, 24 239 (1977).
2072   Hiyashimura, T., Hoshiuo, M., Kirokawa, Y., Matsuzuki, K., Uryu,
        T., J. Polymer Sci., A-1, 15 2691 (1977).
2073   Nagata, S., Moritani, T., J. Polymer Sci., A-1, 12 1799 (1974).
2074   Shahabo, Y.A., Basheer, R.A., J. Polymer Sci., A-1, 16 2667
        (1978).
2075   Shahabo, Y.A., Basheer, R.A., J. Polymer Sci., A-1, 17 919 (1979).
2076   Holmes, B.S., Moniz, W.P., Polymer Preprints, 20 389 (1979).
2077   Gemmer, R.V., Golub, M.A., J. Polymer Sci., A-1, 16 2985 (1978).
2078   Suman, P.T., Werstler, D.D., J. Polymer Sci., A-1, 13 1963 (1975).
2079   Mochel, V.D., J. Polymer Sci., A-1, 10 1009 (1972).
2080   Katritzky, A.R., Smith, A., Rubber Journal, 154 30 (1972).
2081   Keller, F., Muegge, C., Faserforsch Textiltech., 27 347 (1976).
2082   Manalt, S.L., Horowicz Anal. Chem., 52 1529 (1980).
2083   Schaefer, J., Stejskal, E.O., Buchdahl, R., Macromolecules, 8 291
        (1975).
2084   Brosio, E., Delfino, M., Conti, F., Nuova Chim., 48 35 (1972).
2085   Kemp, A.R., Peters, H., Ind. Eng. Chem., Anal. Ed., 15 52 (1943).
2086   Ebdon, J.R., Huckerby, T.N., Polymer (London), 17 170 (1976).
2087   Pozdeeva, R.M., Lukhovitskii, U.I., Korpov, V.L., Zaved Lab., 37
        160 (1971).
2088   Jones, A.A., Bisceglia, M., Polymer Preprints, 20 227 (1979).
2089   Cheng, H.N., Anal. Chem., 54 1828 (1982).

2090   Hirai, H., Koinuma, H., Tanave, T., Takeuchi, K., J. Polymer Sci.,
        A-1, 17 1339 (1979).
2091   Conti, F., Delfini, M., Segre, A.L., Polymer, 15 539 (1974).
2092   Kleinpeter, E., Keller, F., Z. Chem., 18 222 (1978).
2093   Roth, H., Roth, H.K., Keller, F., Plaste Kautsch, 22 256 (1975).
2094   Caise, R.E., O'Donnell, J.H., Makromol. Chem., 176 3517 (1975).
2095   Cheng, H.N., Bennett, M.A., Anal. Chem., 56 2320 (1984).
2096   Keller, F., Muegge, C., Faserforsch. Textiltech., 27 347 (1976).
2097   Kleinpeter, E., Keller, F., Z. Chem., 18 222 (1978).
2098   Nagai, M., Nishoika, A.J., J. Polymer Sci., A-1, 6 1655 (1968).
2099   Morita, S., Shen, M., Ieda, G.S.M., J. Polymer Sci. Polymer
        Physics Ed., 14 1917 (1976).
2100   Morita, S., Shen, S., Polymer Prepr. American Chemical Soc.,
        Division Polymer Chemistry, 17 545 (1976).
2101   Satoh, S., J. Polymer Sci., A-2, 5221 (1964).
2102   Yuki, N., Hatada, K., Takeshita, M., J. Polymer Sci., A-1, 7 667
        (1969).
2103   Tho., P.Q., Taieb, M., J. Polymer Sci., A-1, 10 2925 (1972).
2104   Forobioni, A., Chachaty, C., J. Polymer Sci., A-1, 10 1923 (1972).
2105   Dever, G.R., Karasz, F.E., MacKnight, W.J., Teuz, R.W., J. Polymer
        Sci. Polymer Chemistry Ed., 13 1803 (1975).
2106   Peat, R., Reynolds, W.F., Tetrahedron Letters, No. 14, 1359
        (1972).
2107   Girad, H., Monjol, P., Cr. Hebd. Sciences Acad. Sci., Ser. C., 279
        553 (1974).
2108   Matsuzaki, H., Kausi, T., Kawamura, T., Matsumoto, S., Ilryu, T.,
        J. Polymer Sci. Polymer Chemistry Ed., 11 961 (1973).
2109   Tompa, A.S., Barefoot, R.D., Price, E., J. Polymer Sci., A-1, 6
        2785 (1968).
2110   Elgert, K.E., Wicke, E., Stuzzel, B., Ritter, W., Polymer, 16 465
        (1975).
2111   Kusomato, H., Gutowsky, H.S., J. Polymer Sci., A-1, 2905 (1963).
2112   De Villiers, J.P., Parrish, J.R., J. Polymer Sci., A-2, 1331
        (1964).
2113   McGuchan, R., McNeill, I.C., J. Polymer Sci., A-1, 6 205 (1968).
2114   Gordon, J.E., J. Phys. Chem., 66 1150 (1962).
2115   Gordon, J.E., Chemistry and Industry (London), 267 (1962).
2116   Kruglova, V.A., Ratovskii, G.V., Borisenko, A.A., Kalabina, A.V.,
        Vysokomal. Soedin. Ser. A., 15 1967 (1973).
2117   Cavalti, L., Relaz Corso-Toer-Prat Rizonenza Mana Nucl., 351
        (1973).
2118   Cellarelli, G., Andruzzi, F., Paci, M., Polymer (London), 20 605
        (1979).
2119   Miyamoto, T., Cantow, H.J., Makromol. Chem., 162 43 (1972).
2120   Doskcilova, D., Schneider, B., Draoradova, E., Stokr, J., J.
        Polymer Sci., A-1, 9 2753 (1971).
2121   Woodward, A.E., J. Polymer Sci., 8 137 (1965).
2122   Smirnova, O.V., Kirkak, V.V., Shonim, T.Ya., Urman, Ya.G.,
        Alekseeva, S.G., Bairomov, V.A., Vysokmol Soedin. Ser. A., 17 2415
        (1975).
2123   Beebe, D.H., Gordon, C.E., Thudum, R.N., Throckmorton, M.C.,
        Haulon, T.L., J. Polymer Sci., A-1, 16 2285 (1978).
2124   Jasse, B., Lanpretre, F., Monnerie, L., Makromol Chem., 178 1987
        (1977).
2125   Crouzet, P.1, Mangin, P., J. Appl. polymer Sci., 13 205 (1969).
2126   Amiya, S., Ando, I., Chujo, R., Polymer (London), 4 385 (1973).
2127   Amiya, S., Ando, I., Wanatebe, S., Chujo, R., Polymer (London), 6
        194 (1974).
2128   Woodward, A.E., J. Polymer Sci. C., 8 137 (1965).

2129   Bovey, F.A., J. Polymer Sci. A., $\underline{1}$ 843 (1963).
2130   Woodward, A.E., J. Polymer Sci. C., $\underline{8}$ 137 (1965).
2131   Campos Lopez, E., Palacios, J., J. Polymer Sci., A-1, $\underline{14}$ 1561 (1976).
2132   Woodward, A.E., J. Polymer Sci., $\underline{8}$ 137 (1965).
2133   Druckmay, R.G., Wu, T.K., Makromolecules, $\underline{6}$ 33 (1973).
2134   Koma, Y., Timura, K., Kondo, S., Takeda, M., J. Polymer Sci. A-1, $\underline{15}$ 1697 (1977).
2135   Richards, D.h., Williams, R.L., J. Polymer Sci., Polymer Chemistry Ed., $\underline{11}$ 89 (1973).
2136   Okada, T.J., J. Polymer Sci. A-1, $\underline{17}$ 155 (1979).
2137   Woodward, A.E., J. Polymer Sci. C, $\underline{8}$ 137 (1965).
2138   Bovey, F.A., Tiers, G.V.D., J. Polymer Sci. A, $\underline{1}$ 848 (1963).
2139   Filopovich, G., J. Polymer Sci. A, $\underline{1}$ 2279 (1963).
2140   Ferguson, R.C., J. Polymer Sci. A, $\underline{2}$ 4735 (1964).
2141   Satoh, S., J. Polymer Sci. A, $\underline{2}$ 5221 (1964).
2142   Abe, Y., Tasumi, M., Shimanonchi, T., Satoh, S., Chujo, R., J. Polymer Sci. A-1, 4 1413 (1966).
2143   Woodbrey, J.C., Higginbottom, H.P., Culbertson, H.M., J. Polymer Sci. A, $\underline{3}$ 1079 (1965).
2144   Glick, R.E., Phillips, R.C., J. Polymer Sci. A, $\underline{3}$ 1885 (1965).
2145   Hirst, R.C., Grant, D.M., Holt, R.E., J. Polymer Sci. A, $\underline{3}$ 2091 (1965).
2146   Lando, J.B., Olf, H.G., Peterlin, A., J. Polymer Sci. A-1, $\underline{4}$ 941 (1966).
2147   Ramey, K.C., Lini, D.C., Statton, G.L., J. Polymer Sci. A-1, $\underline{5}$ 257 (1967).
2148   Yamadena, R., Murano, M., J. Polymer Sci. A-1, $\underline{5}$ 1059 (1967).
2149   Nagai, M., Nishoika, A., J. Polymer Sci. A-1, $\underline{6}$ 1655 (1968).
2150   Fujii, K., Brownstein, S., Eastham, A.A., J. Polymer Sci. A-1, $\underline{6}$ 2387 (1968).
2151   Sewell, P.R., Skidmore, D.W., J. Polymer Sci. A-1, $\underline{6}$ 2425 (1968).
2152   Tompa, A.C., Barefoot, R.D., Price, E., J. Polymer Sci. A-1, $\underline{6}$ 2785 (1968).
2153   Shibatani, K., J. Polymer Sci. A-2, $\underline{8}$ 1693 (1970).
2154   Ivin, K.J., Navratil, M., J. Polymer Sci. A-2, $\underline{8}$ 3373 (1970).
2155   Ivin, K.J., Navratil, M., J. Polymer Sci. A-2, $\underline{9}$ 1 (1971).
2156   Goodlett, V.W., Dougherty, J.T., Patten, H.W., J. Polymer Sci., A-2, $\underline{9}$ 155 (1971).
2157   Murano, M., J. Polymer Sci. A-2, $\underline{9}$ 567 (1971).
2158   Kobayashi, S., Kato, Y., Wanatabe, H., Nishoika, A., J. Polymer Sci. A-1, $\underline{4}$ 245 (1966).
2159   Baras, E.M., Juveland, O.O., J. Polymer Sci. A-1, $\underline{5}$ 397 (1967).
2160   Furukawa, J., Kobayashi, E., Arai, Y., Suzuki, T., Takegami, Y., J. Polymer Sci. A-1, $\underline{14}$ 2553 (1976).
2161   McGrath, J.E., Robeson, L.M., Polymer Prep'n American Chemical Society Division of Polymer Chemistry, $\underline{17}$ 706 (1976).
2162   Yabumoto, S., Ishi, K., Arila, K., J. Polymer Sci. A, $\underline{8}$ 295 (1970).
2163   Elgert, K.F., Seiler, G., Puschendorf, G., Cantow, H.J., Makromal. Chem., $\underline{165}$ 245 (1973).
2164   Elgert, K.F., Seiler, G., Puschendorf, G., Cantow, H.J., Makromal. Chem., $\underline{165}$ 261 (1973).
2165   Froix, M.F., Goedde, A.O., Pochau, M.J., Makromolecules, $\underline{10}$ 778 (1977).
2166   Udipi, K., Harewood, H.J., Fribolin, H., Cantow, H.J., Makromol. Chem., $\underline{164}$ 283 (1973).
2167   Sabottka, J., Keller, G., Wienderlich, K., Faserforsch. Textiltech., $\underline{25}$ 352 (1974).

2168   Keller, F., Muegge, C., Faserforsch. Textiltech., 27 347 (1976).
2169   Furukawa, J., Kabayashi, E., J. Polymer Sci. Polymer Chem. Ed., 14
        2553 (1976).
2170   Stefan, D., Williams, H.L., J. Appl. Polymer Sci., 18 1279 (1974).
2171   Stefan, D., Williams, H.L., J. Appl. Polymer Sci., 18 1451 (1974).
2172   Cabasso, T., Jugur-Grodzinski, J., Vofsi, D., J. Appl. Polymer
        Sci., 18 1969 (1974).
2173   McClanahan, O., Previtera, S.A., J. Polymer Sci. A, 3 3919 (1965).
2174   Okuda, K., J. Polymer Sci. A, 2 171 (1964).
2175   Kimmer, W., Schmalke, R., Plaste Kaut., 20 274 (1973).
2176   Barshtein, R.S., Urman, Y.A., Gorbunova, V.G., Khramova, T.S.,
        Bulai, A.K., Sionin, I.Y., Dokl. Acad. Nauk., SSSR., 206 1140
        (1972).
2177   Ebdon, J.R., Polymer (London), 15 782 (1974).
2178   Chujo, R., Satoh, S., Nagai, E., J. Polymer Sci. A, 2 895 (1964).
2179   McClanahan, J.L., Previtera, S.A., J. Polymer Sci. A, 3 3919
        (1965).
2180   Kabayashi, S., Kato, Y., Wanatabe, H., Nishoika, A., J. Polymer
        Sci. A-1, 4 245 (1966).
2181   Baras, E.M., Juveland, O.O., J. Polymer Sci. A-1, 5 397 (1967).
2182   Yabumoto, S., Ishi, K., Arita, K., J. Polymer Sci. A-1, 8 295
        (1970).
2183   Kardos, E., Kosljar, V., Hudec, J., Ilka, P., Chemicke, Zvesti, 22
        768 (1968).
2184   Buttery, R.G., Stuckey, B.N., J. Agr. Food Chem., 9 283 (1961).
2185   Schroeder, E., Randolph, G., Plat. Kautschuk, 1 22 (1963).
2186   Duvall, A.H., Tully, W.F., J. Chromatograph, 11 38 (1963).
2187   Knight, H.S., Siegel, H., Anal. Chem., 38 1221 (1966).
2188   Long, R.E., Guvernator, G.C., Anal. Chem., 39 1493 (1967).
2189   Denning, J.A., Marshall, J.A., Analyst (London), 97 710 (1972).
2190   Lappin, G.R., Zannucci, J., Anal. Chem., 41 2076 (1969).
2191   Denning, J.A., Marchall, J.A., Analyst (London), 97 710 (1972).
2192   Wheeler, D.A., Talenta, 15 1315 (1968).
2193   Roberts, C.B., Swank, J.D., Anal. Chem., 36 271 (1964).
2194   Nosikov, Yu.D., Vetchinkina, V.N., Neftckhimiya, 5 284 (1965).
2195   Helf, C., Bockwan, D., Plate Kauteschuk, 11 624 (1964).
2196   Gaeta, L.J., Schleuter, E.W., Altenau, A.G., Rubber Age, 101 47
        (1969).
2197   Kharasch, M.S., Joshi, B.S., J. Org. Chem., 22 1439 (1957).
2198   Stafford, C., Anal. Chem., 34 794 (1962).
2199   British Standard 2782, Part 4, Method 405 D, (1965).
2200   Hilton, C.L., Anal. Chem., 32 383 (1960).
2201   Mayer, H., Deut. Lebunsun Rundshau, 57 170 (1961).
2202   Glavind, J., Acta. Chem. Scand., 17 1635 (1963).
2203   Blois, M.S., Nature (London), 181 1199 (1958).
2204   Ruddle, L.H., Wilson, J.R., Analyst, 94 105 (1969).
2205   Haslam, J., Willis, H.A., Identification and Analysis of Plastics,
        The Iliffe Group, London, p.307 (1965).
2206   Wexler, A.A., Anal. Chem., 35 1926 (1963).
2207   Ruddle, L.H., Wilson, J.R., Analyst, 94 105 (1969).
2208   Hilton, C.L., Rubber Age, 84 263 (1958).
2209   Burchfield, H.P., July, J.N., Anal. Chem., 19 383 (1960).
2210   Cieleszky, C., Nagy, F., Lebensm Unters-Forsch, 114 13 (1961).
2211   Spell, H.L., Eddy, R.D., Anal. Chem., 32 1811 (1960).
2212   Campbell, R.H., Wize, R.W., J. Chromatography, 12 178 (1963).
2213   Majors, R.E., J. Chromato. Sci., 8 338 (1970).
2214   Dengreville, M., Analysis, 5 195 (1977).
2215   Berger, K.G., Sylvester, N.D., Haines, D.M., Analyst, 85 341
        (1960).

2216   British Standard 2782, Part 4, Method 405 D, (1965).
2217   Van der Neut, J.H., Maagdenberg, A.C., Plastics, $\underline{31}$ 66 (1966).
2218   Spell, H.L., Eddy, R.D., Anal. Chem., $\underline{32}$ 1811 (1960).
2219   Szalkowski, C., Garber, J., Agr. Food Chem., $\underline{10}$ 110 (1962).
2220   Hilton, C.L., Rubber Age, $\underline{84}$ 263 (1958).
2221   Burchfield, H.P. July J.N., Anal. Chem., $\underline{9}$ 383 (1960).
2222   Drushel, H.V., Sommers, A.L., Anal. Chem., $\underline{36}$ 836 (1964).
2223   Webster, p.V., Franks, M.C., J. Inst. Petrol, London, $\underline{56}$ 50
         (1970).
2224   Wake, W.C., The Analysis of Rubber and Rubber-like Polymers,
         McLaren, London (1969).
2225   Carlson, D.W., Hayses, M.W., Bansaw, H.G., McFadden, A.S.,
         Altenau, A.G., Anal. Chem., $\underline{43}$ 1874 (1971).
2226   Campbell, R.H., Wise, R.W., J. Chromat., $\underline{12}$ 178 (1963).
2227   Slonaker, D.F., Sievers, D.C., Anal. Chem., $\underline{36}$ 1130 (1964).
2228   Van der Heide, R.F., Wouters, O., Lebensm. Unteruch, Forsch, $\underline{117}$
         129 (1962).
2229   Schroder, E., Rudolph, G., Plaste Kautschuk, $\underline{10}$ 22 (1963).
2230   Metcalf, K., Tomlinson, R., Plastics (London), $\underline{25}$ 319 (1960).
2231   Waggon, H., Korn, O., Jehle, D., Nahrung, $\underline{9}$ 495 (1965).
2232   British Standard 2782, Part 4, Method 405 D (1965).
2233   British Standard 2782, Part 4, Method 405 B (1965).
2234   Vusherichyute, S.S., Shlyapnikov, Yu. A., Plsaticheskie Massy. No.
         1, 54 (1967).
2235   Stafford, C., Anal. Chem., $\underline{34}$ 794 (1962).
2236   Drushel, H.V., Sommers, A.L., Anal. Chem., $\underline{36}$ 836 (1964).
2237   Waggon, H., Korn, O., Jehle, D., Die Nahrung, $\underline{4}$ 495 (1965).
2238   Sokolowska, R., Rocza Panst. Zakl, Hig., $\underline{20}$ 661 (1969).
2239   Novitskaya, L., Kararinova, N., Zhur. Analit. Khim., $\underline{28}$ 1233
         (1973).
2240   Simpson, D., Currell, B.R., Analyst, $\underline{96}$ 515 (1971).
2241   Van der Heide, R.F., Wouters, O., Lebenam. Untersuch. Forsch., $\underline{117}$
         129 (1962).
2242   Waggon, H., Korn, O., Jehle, D., Nahurung, $\underline{9}$ 495 (1965).
2243   Simpson, D., Curnell, B.R., Analyst, $\underline{96}$ 515 (1971).
2244   Kirchner, J.G., Thin-layer chromatography, Interscience, New York,
         pp 168-169, 679-684 (1966).
2245   Cumpelik, B.M., Drug Cosmet. Ind., $\underline{113}$ 44 (1973).
2246   Williamson, F.B., Rubber, S., Intern, Plastics, 148 (No.2). 24
         (1966).
2247   Zijp, J.W.H., Rec. Trav. Chim., $\underline{75}$ 1155 (1956).
2248   Zijp, J.W.H., Dissertation, Technical University, Delft (1955).
2249   Zijp, J.W.H., Rec. Trav. Chim., $\underline{75}$ 1129 (1956).
2250   Auler, H., Rubber Chem. Technol., $\underline{37}$ 950 (1964).
2251   Zijp, J.W.H., Kautschuk, U., Gummi, $\underline{10}$ 14 (1957).
2252   Dobies, R.S., J. Chromatography, $\underline{40}$ 110 (1969).
2253   Reed, P.R., Warren, P.L., Tech. Paper Reg. Tech. Conf. for Plast.
         Wnf., $\underline{11}$ 139 (1975).
2254   Schroder, E., Rudolph, G., Plate Kautschuk, $\underline{10}$ 22 (1963).
2255   Morgenthalen, L.P., unpublished work.]
2256   Knight, H.S., Siegel, H., Anal. Chem., $\underline{38}$ 1221 (1966).
2257   Styskin, E.L., Gurvic, Ya.A., Kumak, S.I., Khim. Prom., $\underline{5}$ 359
         (1973).
2258   Denning, J.A., Marshall, J.A., Analyst, $\underline{97}$ 710 (1972).
2259   Lappin, G.R., Zannucci, J., Anal. Chem., $\underline{41}$ 2076 (1969).
2260   Yashikawa, T., Ushimi, K., Kimura, K., Tamuro, N.T., J. Appl.
         Polym. Sci., $\underline{15}$ 2065 (1971).
2261   Dobies, R.S., J. Chromat., $\underline{40}$ 110 (1969).

2262 Kalinina, L.S., Doroshina, L.I., Metody Ispyt. Kontr. Issled.
      Madhinostroit, Mater., $\underline{3}$ 5 (1973); Chem. Abstr., $\underline{80}$ 54100 (1974).
2263 Wandel, M., Tengler, H., Fette, Seifen, Anstrichmittel, $\underline{66}$ 815
      (1964).
2264 Varmier, J.P., Suryanaraya, N.P., Sircar, A.K., J. Sci. Ind. Res.
      India, $\underline{20}$ 79 (1961).
2265 Yuasa, T., Kamiya, K., Japan Analyst, $\underline{13}$ 966 (1964).
2266 Kreiner, J.G., Rubber Chem. Technol., $\underline{44}$ 381 (1971).
2267 Kreiner, J.G., Warner, W.C., J. Chromatography, $\underline{44}$ 315 (1969).
2268 Auler, J., Identification of Antioxidants, Antiozonants and
      Accelerators by Means of Thin-layer Chromatography. ETDC Aachen,
      Materials and Research Dept., (1967).
2269 Kreiner, G., Warner, W.C., J. Chromatography, $\underline{44}$ 315 (1969).
2270 Pocaro, P.J., Anal. Chem., $\underline{36}$ 1664 (1964).
2271 Metcalf, K., Tomlinson, R.F., Plastics, $\underline{25}$ 319 (1960).
2272 British Standard 2782, Part 4, Method 405 B (1965).
2273 Crompton, T.R., J. Appl. Polymer Sci., $\underline{6}$ 538 (1962).
2274 Carlson, D.W., Hayes, M.W., Bansaw, H.C., McFadden, A.S., Altenau,
      A.G., Anal. Chem., $\underline{43}$ 1874 (1971).
2275 Hayes, M.M., Altenau, A.G., Rubber Age., $\underline{102}$ 59 (1970).
2276 Gaeta, L.J., Schleuter, E.W., Altenau, A.G., Rubber Age, $\underline{101}$ 47
      (1969).
2277 Parker, C.A., Barnes, W.J., Analyst, $\underline{82}$ 606 (1957).
2278 Protivova, J., Pospisil, J., J. Chromatography, $\underline{88}$ 99 (1974).
2279 Coupek, J., Pekorny, S., Pospisil, J., IUPAC Microsymposium on
      Makromolecules 11th Prague, Sept. (1972).
2280 Coupek, J., Pokorny, S., Jirachova, L, Pospisil, J., J. Chromat.,
      $\underline{75}$ 87 (1973).
2281 Coupek, J., Kalovec, M., Krivakova, M., Pospisil, J., Angew.
      Makromol. Chem., $\underline{15}$ 137 (1971).
2282 Coupek, J., Pokorny, S., protivova, J., Holcik, J., Karvas, M.,
      Pospisil, J., J. Chromatogr., $\underline{65}$ 279 (1972).
2283 Firoenza, A., Bonomi, G., Seradi, A., Materie. Plastiche ed.
      Elassteromerie, $\underline{31}$ 1045 (1965).
2284 Majers, R.E., J. Chromatogr. Sci., $\underline{8}$ 338 (1970).
2285 Wize, R.W., Sullivan, A.B., Rubber Age, $\underline{91}$ 773 (1962).
2286 Tswrugi, J., Murakaus, S., Goda, K., Rubber Chem. Technol., $\underline{44}$
      857 (1971).
2287 Wize, R.W., Sullivan, A.B., Rubber Chem. Technol., No. 3, 35 July-
      Sept. (1962).
2288 Denning, J.A., Marshall, J.A., Analyst, $\underline{97}$ 710 (1961).
2289 Simpson, D., Currell, B.R., Analyst (London), $\underline{96}$ 515 (1971).
2290 Majors, R.E., J. Chromatographic Science, $\underline{8}$ 338 (1970).
2291 Brock, M.J., Louth, G.D., Anal. Chem., $\underline{35}$ 1575 (1955).
2292 Ligottil, L, Materie Plast. Elastomerie, $\underline{39}$ 889 (1973).
2293 Kriner, J.G., Warner, W.C., J. Chromatography, $\underline{44}$ 315 (1969).
2294 Kueda, I., Fernandez, G., Revta. Plast., $\underline{24}$ 82 (1973).
2295 Zijp, J.W.H., Rec. Trav. Chim., $\underline{75}$ 1155 (1956).
2296 Burger, V.L., Rubber Chem. Technol., $\underline{32}$ 1452 (1959).
2297 Miksch, R., Prolsi, O., Gummi. Asbest., $\underline{13}$ 250 (1960).
2298 Carlson, D.W., Hayes, M.W., Ransaw, H.C., McFadden, R.S., Altenau,
      A.G., Anal. Chem., $\underline{43}$ 1874 (1971).
2299 Seldar, J., Finiokova, E., Pac, C., Analyst, $\underline{99}$ 50 (1974).
2300 Major, J., Kocmanova, V., Chem. Prum., $\underline{17}$ 372 (1967).
2301 Neureiter, N.P., Bown, D.E., Ind. Eng. Chem. Prod. Res. Dev., $\underline{1}$
      236 (1962).
2302 Sedlar, J., Feniokova, E., Pac, J., Analyst, $\underline{99}$ 50 (1974).
2303 Hilton, C.L., Anal. Chem., $\underline{32}$ 383 (1960).
2304 Nawakoswki, A.C., Anal. Chem., $\underline{30}$ 1868 (1958).

2305	Hilton, C.L., Anal. Chem., $\underline{32}$ 383 (1960).
2306	Nawakowski, A.C., Anal. Chem., $\underline{30}$ 1868 (1958).
2307	Brandt, H.J., Anal. Chem., $\underline{33}$ 1390 (1961).
2308	Schroder, E., Hagen, E., Plaste Kautsch, $\underline{15}$ 625 (1968).
2309	Wandel, M., Tengler, H., Ostromov, H., Die Analyse von Weischmachern, Springer Verlag, Berlin, Heidelberg, New York (1967).
2310	Dobies, R.S., J. Chromatography, $\underline{40}$ 110 (1969).
2311	Kellum, G.E., Anal. Chem., $\underline{43}$ 1843 (1971).
2312	Kellum, G.E., Anal. Chem., $\underline{43}$ 1843 (1971).
2313	Roberts, C.B., Swank, J.D., Anal. Chem., $\underline{36}$ 271 (1964).
2314	Roberts, C.B., Swank, J.D., Anal. Chem., $\underline{36}$ 271 (1964).
2315	Carsen, B.B., Haintzelman, W.J., Moe, H., Roussseau, C.R., J. Organic Chem., $\underline{27}$ 1636 (1962).
2316	Markelov, M.A., Semenko, E., Plast. Massy., (8), 65 (1973).
2317	De Forero, I.B., De Rascouvsky, E.G., De Ruiz, M., Del Carmen, S., Plasticos, $\underline{19}$ 122 (1971).
2318	Nowak, P., Klemm, H., Kunststoffe, $\underline{52}$ 604 (1962).
2319	Shanks, R.A., Pye Unicam Newsletter, (1975).
2320	Klesper, E., Hartmann, W., European Polymer Journal, $\underline{14}$ 77 (1978).
2321	Rohrachneider, L., Z. Anal. Chem., $\underline{255}$ 345 (1971).
2322	Novak, V., Chem. Prumsyl, $\underline{22}$ 298 (1972).
2323	Pozdeeva, R.M., Lukhovitskii, U.I., Korpov, V.L., Zavod Lab., $\underline{37}$ 160 (1971).
2324	Crompton, T.R., Myers, L.W., Blair, D., British Plastics, December (1965).
2325	Claver, G.C., Murphy, M.E., Anal. Chem., $\underline{31}$ 1682 (1959).
2326	Miller, D.F., Samuel, E.P., Cobler, J.G., Anal. Chem., $\underline{33}$ 677 (1961).
2327	Katelin, A.I., Komleva, V.N., Mal'kova, L.N., Metody. Anal. Kontrolya, Proizvod. Khim. Prom. - st., $\underline{11}$ 62 (1977).
2328	Crompton, T.R., Myers, L.W., Plastics and Polymers, 205, June (1968).
2329	Waters, D.N., Proc. Conf. Raman Spectros. 5th, 500 (1976).
2330	Cobler, J.G., Samsel, D.D., SPE Transactions, April 1962.
2331	Shapras, P., Claver, G.C., Anal. Chem., $\underline{34}$ 433 (1962).
2332	Tweet, O., Miller, W.K., Anal. Chem., $\underline{35}$ 852 (1963).
2333	Wilkinson, L.R., Norman, C.W., Brettner, N.P., Anal. Chem., $\underline{36}$ 1759 (1964).
2334	Nowak, P., Klemett, O., Kunststoffe, $\underline{52}$ 604 (1962).
2335	Adcock, L.H., Patra, $\underline{3}$ 5 (1962).
2336	Ragelis, E.P., Gajan, R.J.F., Assoc. Offic. Agric. Chem., $\underline{45}$ 918 (1962).
2337	Berens, A.R., Brider, L.B., Tomanek, C.M., Whitney, J.M., BP Goodrich Tyre Centre, Brecksville, Ohio. M/S circulated to members of the Vinyl Chloride Safety Association. Nov. 14th 1974 entitled "Analysis of Vinyl Chloride in PVC powders by head-space chromatography".
2338	Berens, A.R., Paper delivered to the 168th ACS Meeting, Atlantic City, New Jersey, Sept. 1974 entitled " The Solubility of Vinyl Chloride in Polyvinylchloride".
2339	Baba, T., Shokuhin Eisegaku Aasshi., $\underline{18}$ 6, 500 (1977); Chem. Abstr., $\underline{88}$ 153342e (1978).
2340	Gilbert, S.G., Giacin, J.R., Morano, J.R., Rosen, J.D., Package Dev. Syst., $\underline{5}$ 20 (1975).
2341	Ymasa, T., Kamiya, K., Japan Analyst., $\underline{13}$ 966 (1964).
2342	Novak, V., Siedl, J., J. Chem. Prumsyl., $\underline{28}$ 186 (1978).
2343	Shapras, P., Claver, G.C., Anal. chem., $\underline{36}$ 2282 (1964).
2344	Shanks R.A., Scan., $\underline{6}$ 20 (1975).

2345   Takashima, S., Okada, F., Nimeji, Kogyo Daigaku Kenkyu Hokoku, _12_
        34 (1960).
2346   Brodsky, J., Kunststoffe, _51_ 20 (1961).
2347   Neldon, F.M., Eggertsen, F.T., Holst, J.J., Anal. Chem., _33_ 1150
        (1961).
2348   Holtmann, R., Souren, J.R., Kunststoffe, _67_ 776 (1977).
2349   Yamaoka, A., Matsui, T., Himeji Kogyo Daigaku Kenkyu Hokoku, _30_
        101 (1977); Chem. Abstr., _89_ 7000w (1978).
2350   Acosta, J.L., Sastre, R., Rev., _29_ (224) 212 (1975).
2351   Tweet, O., Miller, W.K., Anal. Chem., _35_ 852 (1963).
2352   Kreshkov, A.P., Balyatinskaya, L.N., Chesnokova, S.M., Tr. Mosk.
        Khim Teknol. Inst., _70_ 146 (1972)., Chem. Abstr., _80_ 18144 (1974).
2353   Kreshkov, A.P., Balyatinnskaya, L.N., Chesnokova, S.M., Zh. Anal.
        Khim., _28_ (8), 1571 (1973).
2354   Oprea, N., Pogorevic, A., Rev. Chim. Bucharest, _25_ 244 (1974).
2355   Tweet, O., Miller, W.K., Anal. Chem., _35_ 852 (1963).
2356   Schwoetzer, G., Anal. Chem., _10_ 260 (1972).
2357   Streichen, R.J., Anal. Chem., _48_ 1398 (1976).
2358   Shiryaev, B.V., Kozmenkhova, E.B., Zavod. Lab., _38_ 1303 (1972).
2359   Mel'nikova, S.L., Tishehenko, U.T., Sazonenko, U.V., Lakokras
        Mater. Ikh. Primen., _4_ 56 (1977).
2360   Rosenthal, R.W., Schwartzman, L.H., Greco, N.P., Proper, P.J., J.
        Org. Chem., _28_ 2835 (1963).
2361   Shanks, R.A., Scan., _6_ 20 (1975).
2362   Schwoetzer, Z., Anal. Chem., _260_ 10 (1972).
2363   Claver, G.C., Murphy, M.E., Anal. Chem., _31_ 1682 (1959).
2364   Shapras, P., Claver, G.C., Anal. Chem., _36_ 2282 (1964).
2365   Mayo, F.R., Lewis, F.M., Walling C., J. Amer. Chem. Soc., _70_ 1529
        (1948).
2366   Doak, K.W., J. Amer. Chem. Soc., _70_ 1525 (1948).
2367   Roy, S.S., Analyst (London), _102_ 302 (1977).